COLLEGE BOTANY

Volume II

INCLUDING PTERIDOPHYTA, GYMNOSPERMS, PALAEOBOTANY AND ANGIOSPERMS

For Undergraduate and Postgraduate Students

Dr. B.P. PANDEY
MSc, PhD, FPSI
Former Head
Department of Botany
J.V. College, Baraut, Uttar Pradesh

S. CHAND
PUBLISHING

S Chand And Company Limited
(ISO 9001 Certified Company)

S Chand And Company Limited

(ISO 9001 Certified Company)

Head Office: D-92, Sector–2, Noida – 201301, U.P. (India), Ph. 91-120-4682700

Registered Office: A-27, 2nd Floor, Mohan Co-operative Industrial Estate, New Delhi – 110 044, Phone: 011-49731800

www.**schandpublishing.com;** e-mail: **info@schandpublishing.com**

Marketing Offices:

Chennai : Ph: 23632120; chennai@schandpublishing.com
Guwahati : Ph: 2738811, 2735640; guwahati@schandpublishing.com
Hyderabad : Ph: 40186018; hyderabad@schandpublishing.com
Jalandhar : Ph: 4645630; jalandhar@schandpublishing.com
Kolkata : Ph: 23357458, 23353914; kolkata@schandpublishing.com
Lucknow : Ph: 4003633; lucknow@schandpublishing.com
Mumbai : Ph: 25000297; mumbai@schandpublishing.com
Patna : Ph: 4011400; patna@schandpublishing.com

First Edition 1979
Subsequent Editions and Reprints 1980, 81, 82, 86, 89, 90, 92, 93, 94, 96, 97, 98, 99, 2001, 2003, 2004, 2006, 2007, 2009, 2010
Revised Edition 2012; Reprints 2013, 2014, 2015, 2016, 2017, 2018, 2020 (Twice), 2021

Low Priced Students' Paperback Edition (LPSPE), 2022

Reprint 2023, 2024
Reprint 2026

ISBN: 978-93-550-1061-2 **Product Code:** H6BTY68BOTN10ENAA21L

PRINTED IN INDIA

By Shri Shyamlal Printing Press Private Limited, Plot 20/4, Site-IV, Industrial Area Sahibabad, Ghaziabad – 201 010 and Published by S Chand And Company Limited, A-27, 2nd Floor, Mohan Co-operative Industrial Estate, New Delhi – 110 044.

PREFACE TO THE REVISED AND ENLARGED EIGHTH EDITION

I am glad to present, the revised and enlarged eighth edition of College Botany, volume two, in hands of students of Botany of graduate level. This edition of book contains Pteridophyta, Gymnosperms, Palaeobotany and Angiosperms. The newly added matter of Angiosperms, consists of thirty chapters, dealing with Systematics of Flowering Plants (Chapter 1–5), How to describe an Angiospermic Plant (Chapters 6–8), Diversity of Flowering Plants (Chapters 9–25), Diversity and Morphology, of Angiosperms (Chapters 26–29) and Reproduction in Flowering Plants (Chapter 30).

Actually this edition is intended to fulfil the long felt need of the students of Botany of graduate level of all Indian Universities. The syllabi of all the universities have been kept in view, during revision of the text of this edition.

The text has been written in simple and graspable language and profusely illustrated with self explanatory diagrams. The figures are provided with detailed legends. Some of the diagrams, of the prepared slides and specimens have been drawn by author himself, and most of the diagrams and figures have been quoted from authentic work of various authors.

My sincere thanks are due to Dr. S. Kumar, Dr. J. Mohan, Dr. Rajeshwari Sharma and Dr. Ashok Kumar of J.V. College, Baraut, and Dr. S.S. Sharma and Dr. Shashi, of H.P. University, Shimla (H.P.). Sanjeeva Pandey, I.F.S., Rahul Pandey, I.F.S. and Shanti Priya, I.F.S. deserve special thanks for their valuable suggestions and co-operation. I also express my deep sense of gratitude to my friends, botanists, and educationists who have accorded their full co-operation and well wishes during preparation of manuscript of this edition.

My sincerest thanks are forwarded to management, DTP and Editorial team of S. Chand & Company Ltd., for their encouragement and neat execution to the revised & enlarged edition of this book in a suitable form.

The healthy suggestions and criticisms from all corners of country are invited for the improvement of subsequent editions.

Dr. B.P. Pandey
Former Reader and Head,
Department of Botany,
J.V. College, Baraut, 250 611

'Neelkanth Kuteer'
Subhasnagar, Baraut,
(Baghpat Dt.) U.P.
January, 1, 2011

PREFACE TO THE REVISED AND ENLARGED EIGHTH EDITION

[illegible] edition of [illegible] [illegible] [illegible] [illegible] Bacteria [illegible] [illegible] Flowering Plants [illegible] [illegible] of Angiosperms [illegible] [illegible] in Flowering [illegible]

Actually this edition is intended to fulfil the long felt need of the students of [illegible] of all Indian [illegible] [illegible] have been kept in view [illegible] revision of the [illegible]

The text [illegible] simple and [illegible] language and [illegible] illustrations [illegible] Some of the diagrams [illegible] [illegible] and [illegible] [illegible]

[illegible] to Dr. [illegible] [illegible] [illegible] [illegible] [illegible] [illegible] [illegible] [illegible] [illegible] [illegible] [illegible] [illegible] [illegible] [illegible] [illegible] this edition.

My [illegible] thanks [illegible] [illegible] [illegible] [illegible] [illegible]

[illegible] suggestions [illegible] for the improvement of subsequent editions.

[illegible] P. [illegible]

[illegible] Senior [illegible]

[illegible] [illegible]

CONTENTS

PTERIDOPHYTA

GYMNOSPERMS

ANGIOSPERMS

PTERIDOPHYTA

1 Introduction

The pteridophytes occupy the intermediate position in between the bryophytes on one hand and the phanerogams on the other. They possess certain combinations of major characters which are not found in bryophytes or in phanerogams. The most characteristic feature of the pteridophytes is the presence of independent gametophytes and sporophytes at maturity. The pteridophytes are also known as 'vascular cryptogams'. The sporophytes of these plants possess a well developed system of conducting tissue which consists of xylem, phloem and other mechanical elements.

Origin of the pteridophytes. As regards their origin, the pteridophytes are known from as far back as the Silurian age of Palaeozoic era, *i.e.,* they are somewhat 380,000,000 years old.

According to one school of thought, the pteridophytes like bryophytes have been evolved from algae. According to this view both the groups are parallel to each other in their origin and they are not connected phylogenetically. Scott (1900), Eames (1936), however donot trace their origin from any particular groups of algae. The workers (Church, 1919; Arnold, 1947) trace the origin of pteridophytes from a particular group of marine algae, *i.e.,* Phaeophyceae. According to Bohlin (1901), Fritsch (1916) and Lotsy (1909) they are thought to be originated from a filamentous Chaetophorales of Chlorophyceae. Later on in 1945, Fritsch suggested that the pteridophytes originated from an erect and parenchymatous green alga with an isomorphic life-cycle.

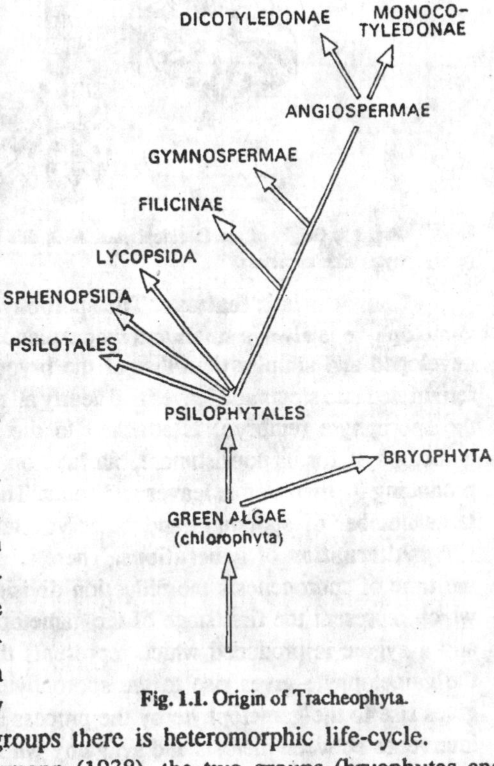

Some of the workers believe that the pteridophytes have been originated from bryophytes. The pteridophytes resemble the bryophytes in many respects. These resemblances are as follows: the general similarity in vegetative structure of gametophytes of thallose bryophytes and pteridophytes similarity in the ontogeny of antheridia and archegonia; the mature sporophyte is partially parasitic upon the gametophyte; in both the groups there is heteromorphic life-cycle.

Fig. 1.1. Origin of Tracheophyta.

According to Bower (1935) and Zimmermann (1938), the two groups (bryophytes and pteridophytes) are phylogenetically connected but they are divergent evolutionary lines evolved from hypothetical land plants of primitive archegonia type. According to Campbell (1899) they have arisen from an anthocerotean type of bryophyte.

This way, several hypotheses have been propounded by various workers from time to time.

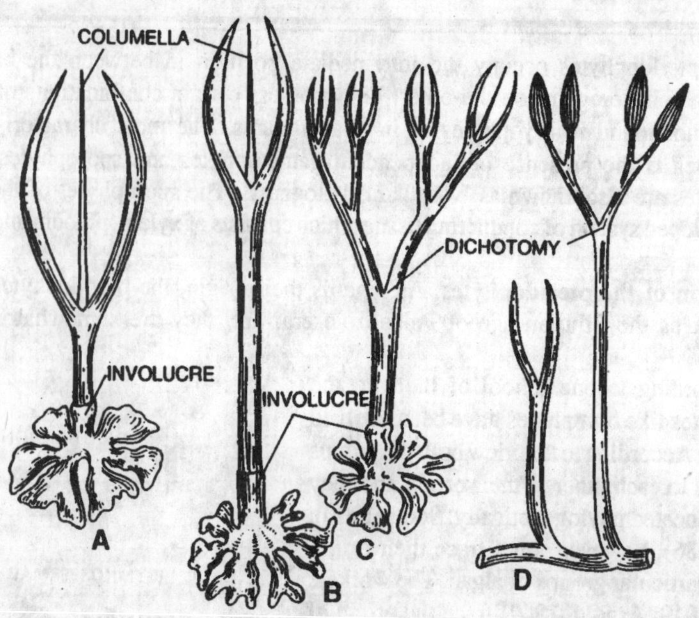

Fig. 1.2. Origin of the Pteridophytes. A-D, this scheme shows the origin of early land plants from *Anthoceros* (a bryophyte) like ancestors.

Characteristic features. The sporophyte is the conspicuous and familiar plant, while the gametophyte is always small and inconspicuous. In pteridophytes, the gametophyte is much less developed and simpler than that of the bryophytes. The sporophyte is a herbaceous plant differentiated into stems and leaves and nearly all cases have roots. In earlier stages of the development the sporophyte (embryo) is attached to the gametophyte by a foot and is dependent on the gametophyte for its nourishment, but later on, it becomes quite independent of the gametophyte, producing its own stems, leaves and roots. The sporophytes are long-lived plants which produce large number of sporangia and other vegetative organs.

Alternation of generations. There is a regular feature of alternation of generations. At the time of sporogenesis the reduction division takes place and the haploid spores are formed which represent the first stage of the gametophyte generation. Later on fertilization takes place and a zygote is produced which represents the first stage of sporophyte generation. This way the gametophyte gives rise to the sporophyte by the process of syngamy and the sporophyte gives rise to the gametophyte by the process of meiosis. The gametophyte of sexual generation intervenes between meiosis and syngamy while the sporophyte or asexual generation intervenes between syngamy and meiosis. The haploid phase (x) is the gametophyte and the diploid phase ($2x$) is the sporophyte. These two phases follow one another in the life-cycle, in regular succession.

Under normal circumstances there is a regular alternation between a gametophyte (sexual) phase and a sporophyte (asexual) phase. The male gametes,, produced in numbers from antheridia, are known as antherozoids, since they are flagellated and are able to swim in water, while the female gametes (oospheres) are non-motile and are borne singly in flask shaped archegonia. Fusion between an egg cell and an antherozoid results in the formation of a zygote, which contains the combined nuclear material of the two gametes. Its nucleus contains twiee as many chromosomes as either of the gamete nuclei and it is therefore described as diploid. The zygote develops directly by mitotic divisions into the sporophyte which is likewise diploid. The spore mother cells within sporangia divide meiotically so that the non-motile haploid spores are formed, the life-cycle is then completed when these spores germinate and grow, by mitotic division, into haploid gametophytes.

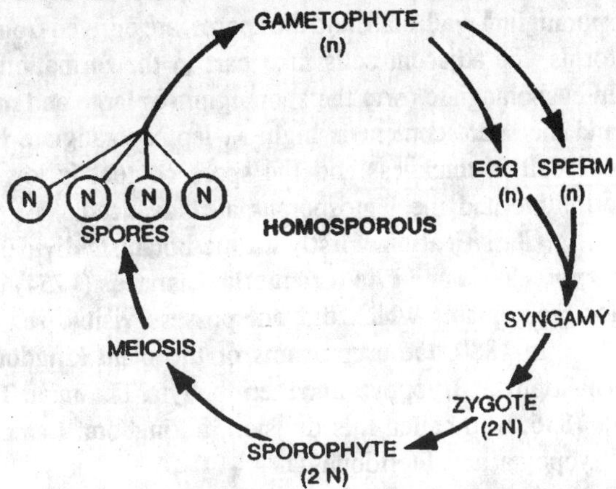

Fig. 1.3. Alternation of generations in Pteridophytes. Life-cycle pattern in homosporous pteridophytes.

In some members of pteridophytes all spores produced by the sporangia are of one type, such a plant is known as **homosporous** (*e.g., Lycopodium, Dryopteris* etc.) On the other hand, in **heterosporous** pteridophytes (*e.g., Selaginella, Marsilea* etc.). The spores are of two kinds-the smaller spores are known as microspores or male spores and are developed in microsporangia, while the larger spores are called megaspores or female spores and are formed in megasporangia. The microspores on germination produce the male gametophytes on which, the sperms are produced. The megaspores produce the female gametophytes on which, eggs are produced.

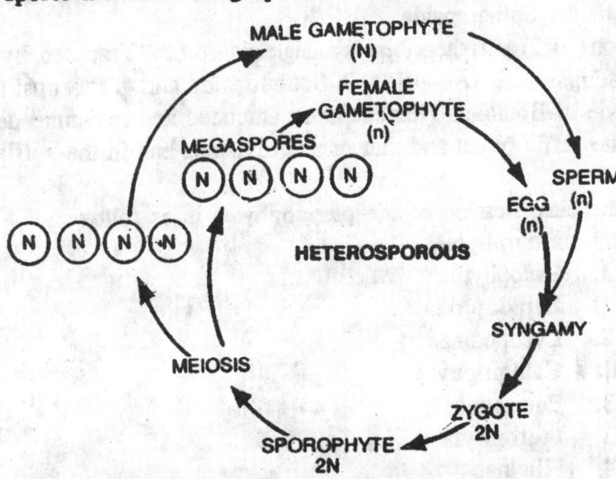

Fig. 1.4. Alternation of generations in Pteridophytes. Life cycle pattern in heterosporous pteridophytes.

In Bryophytes, the dominant phase in the life-cycle is the gametophyte, for the sporophyte is retained upon it throughout its life and is either partially or completely dependent on it for nutrition. Contrary to it in pteridophytes the sporophyte is the dominant generation, for it very soon becomes independent of the gametophyte (prothallus) and grows to a much greater size. The sporophytes are being differentiated into stems, leaves and roots. Only the sporophyte shows well developed conducting tissues (*i.e.*, xylem and phloem).

The prothalli. The gametophytes, in some pteridophytes are subterranean and in others they are retained within the resistant wall of the spore and are thus able to survive in a much wider range of habitats. It is notable that wherever the gametophyte is retained within the spore the spores are of different sizes (heterosporous), the larger megaspores give rise to female prothalli which bear only archegonia, and the smaller microspores giving rise to male prothalli which bear only antheridia. The division between heterospory and homospory is one of the criteria used in the classification of pteridophytes.

The sporoganium. The sporangium, in all pteridophytes, is initiated by the laying down of cells. Since this wall is periclinal each initial cell is divided into an outer and inner daughter cell. If the sporogenous tissue is derived from the inner daughter cell, the sporangium is described as **'eusporangiate'** and if from the outer, as **'leptosporangiate'**. In leptosporangiate forms, the sporangium wall, stalk and the spores, are derived from the outer daughter cell, but in eusporangiate forms, the adjacent cells take part in the formation of part of sporangium wall and the stalk. In eusporangiate form the sporangium is large and massive, the wall is several cells in thickness and the spore content is high. In leptosporangiate forms, the sporangium is small, the wall is one cell in thickness and the spore content is low. Of these two types, the eusporangiate is primitive and the leptosporangiate advanced.

Classification. Firstly all the botanists divided the plant kingdom into two large groups-**Cryptogamia** and **Phanerogamia,** Linnaeus (1754) used the term **Cryptogams** for the first time for those plants which did not possess visible sex organs.

In 1880, the cryptogams of the plant kingdom were divided into three large divisions: Thallophyta, Bryophyta and Pteridophyta. The name-Thallophyta was first introduced by Endlicher in 1836, who called this division, a kingdom. Later on in 1866 Haeckel introduced the names Bryophyta and Pteridophyta.

For a long time the division Pteridophyta included-the ferns, lycopods and horsetails. On the basis of their anatomy Jeffrey in 1902 divided the vascular plants into two types or 'stocks'. He included ferns, gymnosperms and angiosperms in a 'stock' known as **Pteropsida,** and lycopods and horsetails in another 'stock' known as **Lycopsida.** Jeffrey, however, could not give the names divisions or sub-divisions to these stocks. Scott in 1923 ranked them as divisions. He seggregated the horsetails as a separate division, the **Sphenopsida.**

According to Eames (1936), Tippo (1942) and others there is a single division, the **Trachaeophyta** consisting of four subdivisions-**1. Psilopsida, 2. Lycopsida, 3. Sphenopsida** and **4. Pteropsida.**

In 1950, the International Code of Botanical Nomenclature amended and recommended that all names of divisions end in the suffix-phyta and that of sub divisions end in the suffix-opsida.

According to Smith (1955) the classification of the pteridophytes is as follows:

Division I **Psilophyta**
Class 1. Psilophytinae
Division II **Lepidophyta**
Class 2. Lycopodinae
Division III **Calamophyta**
Class 3. Equisetinae
Division IV **Pterophyta**
Class 4. Filicinae

The classification followed in the present text is as follows:

Division I **Psilophyta**
Class 1. Psilotopsida
Division II **Lycophyta**
Class 1. Ligulopsida
Class 2. Eligulopsida
Division III **Arthrophyta**
Class 1. Sphenophyllopsida
Class 2. Calamopsida
Division IV **Filicophyta**
Class 1. Primofilicopsida
Class 2. Eusporangiopsida

Class 3. Protoleptosporangiopsida
Class 4. Leptosporangiopsida

The characteristic features of these divisions are as follows:

Psilophyta. The sporophyte is differentiated into a rhizoid bearing subterranean rhizome and an aerial portion. The aerial portion is branched. The vascular system is of protostelic type. Leaf gaps are absent from vascular cylinder. The terminal sporangia are borne singly at the tips of short or long branches. The gametophyte is subterranean and colourless. They are homosporous. Antherozoids are multiciliate.

Lycophyta. The sporophyte is differentiated into stem, roots, and leaves. The leaves are microphyllous, and with a single vein. The vascular strands or steles may be protostelic, siphonostelic, or polystelic. The leaf-gaps are always absent, sporophylls produce a single sporangium on the adaxial side near its base. The sporophylls are borne in strobili. They are homosporous or heterosporous. The antherozoids are biflagellate or multiciliate.

Arthrophyta. The sporophyte is differentiated into stem. roots and leaves. The stem possesses distinct ridges and furrows. The foliage leaves are borne in transverse whorls upon stems and their branches. The vascular cylinder is protostelic or siphonostelic. The leaf-gaps are absent. The sporangia are produced upon a specialized structure, the sporangiophores present at the apex of the stem. The antherozoids are multiciliate. They are homosporous.

Filicophyta. The sporophyte is differentiated into stem, leaves and roots. With the exception of protostelic forms, the other siphonostelic forms possess leaf-gaps in their vascular cylinders. The leaves are macrophyllous. The leaf bears many sporangia on either the margin or the abaxial face of the leaf lamina. They are homosporous. The antherozoids are multiciliate. The sex organs are found on the ventral surface of the heart-shaped prothallus (gametophyte).

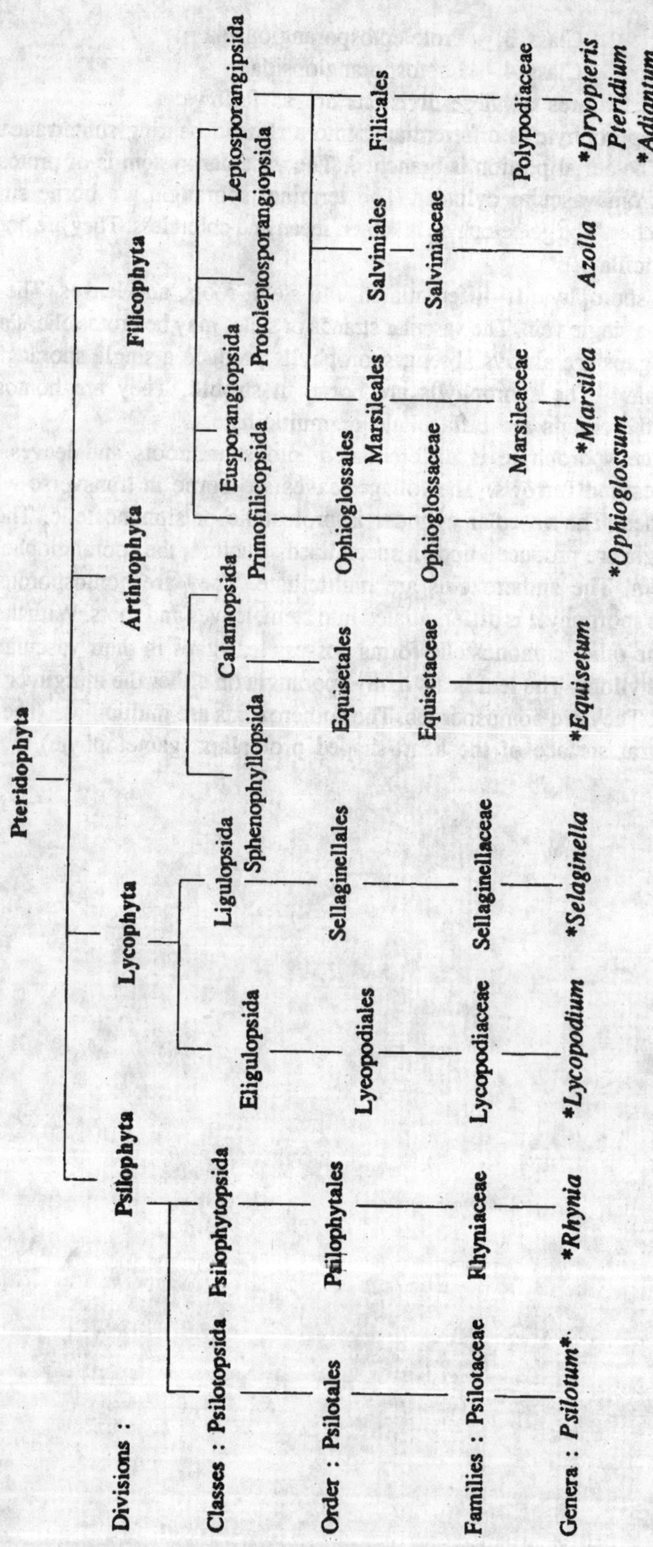
Divisions :
Pteridophyta
Psilophyta
Lycophyta
Arthrophyta
Filicophyta
Classes : Psilotopsida
Psilophytopsida
Eligulopsida
Ligulopsida
Sphenophyllopsida
Calamopsida
Primofilicopsida
Eusporangiopsida
Protoleptosporangiopsida
Leptosporangiopsida
Order : Psilotales
Psilophytales
Lycopodiales
Sellaginellales
Equisetales
Ophioglossales
Marsileales
Salviniales
Filicales
Families : Psilotaceae
Rhyniaceae
Lycopodiaceae
Sellaginellaceae
Equisetaceae
Ophioglossaceae
Marsileaceae
Salviniaceae
Polypodiaceae
Genera : Psilotum*.
*Rhynia
*Lycopodium
*Selaginella
*Equisetum
*Ophioglossum
*Marsilea
Azolla
*Dryopteris
Pteridium
*Adiantum
*Genera described in the text

2

Division-Psilophyta
Class-Psilotopsida

Order-Psilophytales

There are about 25 known genera in this order. All genera are fossil. The fossil record indicates that the Psilophytales became extinct during the Upper Devonian. They however, lived from the Middle Silurian to the Upper Devonian. The fossil remains of the Psilophytales have been found in widely scattered regions, such as Canada, Scotland, France, United States of America, Australia etc.

The characteristic features of the order are as follows:

The plant body of the sporophyte was rootless, dichotomously branched and was differentiated into a prostrate subterranean and an erect aerial portion. The aerial portion was leafless or with small simple leaves.

In majority of the genera, the sporangia were borne at the tips of the branches. However, a few genera had the sporangia in clusters, at the tips of their branches.

The stele was simple and protostelic type.

Classification. The order Psilophytales has been divided into nine families. Some of the

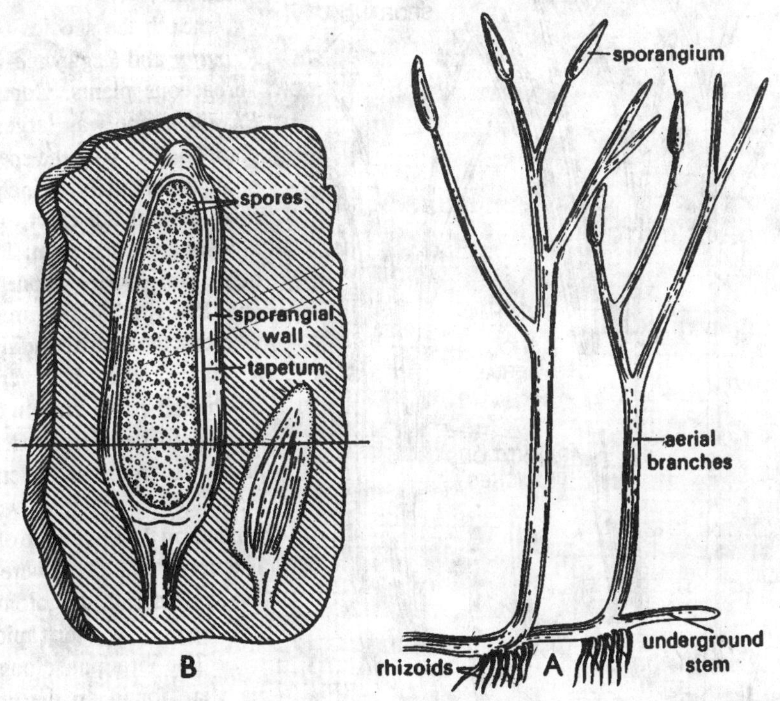

Fig. 2.1. ***Rhynia major*****. A, reconstruction of a plant; B, sporangia (a fossil).**

families consist of very imperfectly known genera. The important families are-Rhyniaceae, Psilophytaceae, Pseudoporochnaceae, Zosterophyllaceae, Asteroxylaceae etc. Here the genus *Rhynia* of Rhyniaceae has been described.

Genus RHYNIA

Systematic Position- Pteridophyta
Division-Psilophyta
Class-Psilophytopsida
Order-Psilophytales
Family-Rhyniaceae
Genus-*Rhynia*

Till now only two species of *Rhynia* named as *Rhynia major* and *R. gwynne-vaughani* have been described by Kidston and Lang (1917, 1921) from the Old Red Sandstone Beds of Middle Devonian. As the genus has been derived from the place Rhynie situated in Scotland and, therefore, it is named after the name of the place of its origin and known as *Rhynia.* A number of specimens of these two species have been found at Rhynie. The specimens are so well preserved that they give detailed information about the form and structure of this very primitive vasular plant.

Kidston and Lang (1921), thought and supported by evidences that in those times the plants grew in swampy marshes near the volcanoes.

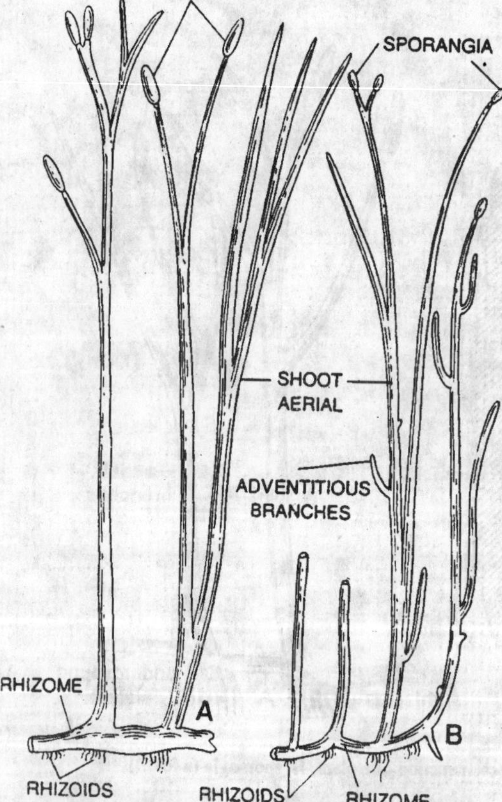

Fig. 2.2. *Rhynia.* A, reconstruction of *R. major.* B, reconstruction of *R. gwynne-vaughani* (After Kidston and Lang)

Both the species of *Rhynia, i.e., R. major* and *R. gwynne-vaughani* were herbaceous plants. Comparatively the *Rhynia major* was larger in size. The plant body of both the species consisted of dichotomously branched horizontal rhizome and the erect, aerial dichotomously branched stem. The aerial stem of *R. major* were about 50 cms. high and 1.5 mm. to 6 mm. in diameter, whereas, the corresponding structures of *R. gwynne-vaughani* were 20 cms. in height and 1 to 3 mm. in diameter. The aerial branches were leafless shoots. The plants were rootless. Instead of the roots, the unicellular rhizoids were present in patches on the underside of the rhizome. The aerial branches were naked, leafless, cylindrical, dichotomously forked and tapering at their apices.

The terminal, elongated sporangia were found on the tapering vegetative apices.

The stems of *R. gwynne-vaughani* bore Adventitious branches at the bases of aerial shoots and rhizome. It is thought that these branches got detached readily from the parent plant and helped in vegetative propagation.

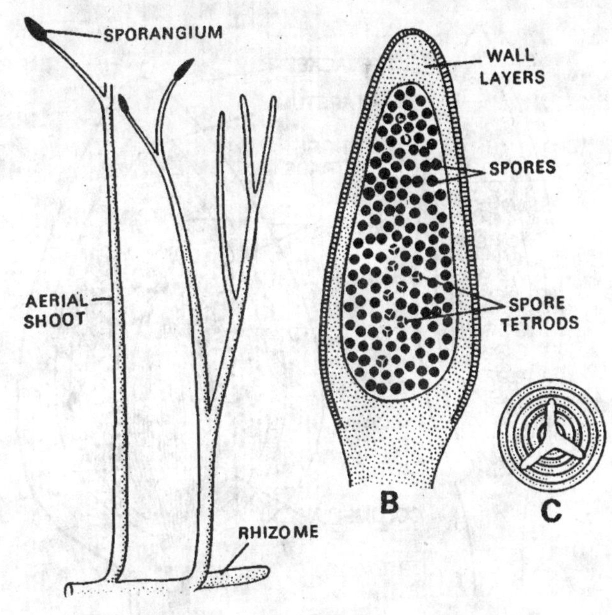

Fig. 2.3. *Rhynia gwynne-vaughani.* A, reconstruction of plant; B, L.S. of sporangium containing spores and spore tetrads; C, a spore tetrad.

Anatomy. The internal structure of the rhizome and aerial stem was quite similar to each other. This was differentiated into epidermis, cortex and stele. The details of the structures were as follows:

1. Epidermis. The epidermis was one cell in thickness with well-defined thick outer wall and conspicuous cuticle on the outer side. The cells of the epidermis were broadly fusiform from the surface view. The aerial stem bore the stomata on the epidermis here and there with pairs of guard cells. However, the rhizome did not possess any stomata. The rhizome possessed unicellular rhizoids on its surface.

2. Cortex. The cortex consists of two zones (a) the outer cortex (hypodermis) and the (b) inner cortex.

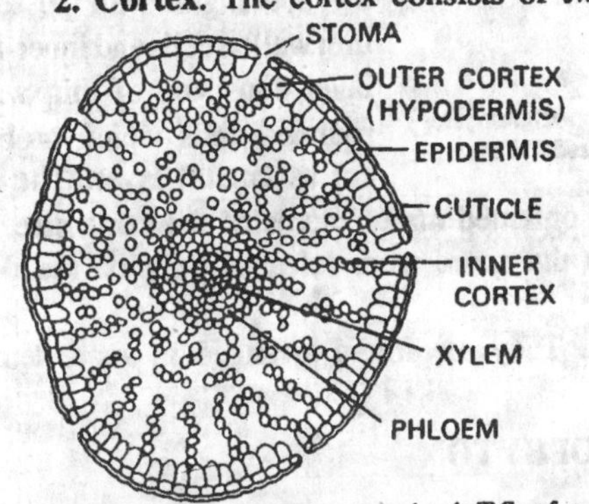

Fig. 2.4. *Rhynia gwynne-vaughani.* T.S. of an aerial branch.

(a) Outer cortex or hypodermis. Immediately beneath the epidermis there was a zone one to four cells in thickness consisted of angular parenchymatous cells having no intercellular spaces known as hypodermis. This zone was also known as the outer cortex.

(b) Inner cortex. Inner to the hypodermis there was broad zone of inner cortex. In comparison of the central stele the cortex was ten fold. The inner cortex consists of rounded cells separated from each other by large intercellular spaces which communicate to the stomata through the gaps in hypodermis. It is also believed that this zone of the inner cortex acted as the chief photosynthetic tissue.

The endodermal and pericyclic layers were altogether absent.

3. Stele. There was a protostelic type (haplostele) of stele. The central xylem core was completely surrounded by phloem layer four or five cells in thickness. The xylem consists of tracheids only. In majority of the specimens the xylem had imperfectly preserved. According to Kidston and Lang (1917), the tracheids had annular thickenings at their walls. Kidston and Lang (1920) have reported that in some of the specimens the tracheids were isodiametric, whereas in some other specimens the central tracheids were smaller in diameter and the peripheral ones larger in diameter.

The phloem consists of elongated thin walled cells with oblique end walls.

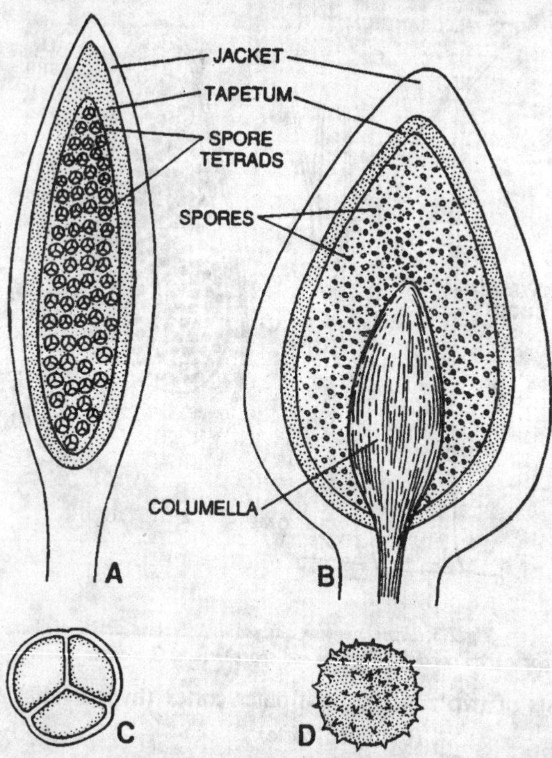

Fig. 2.5. *Rhynia gwynne-vaughani.* A, longitudinal section of sporangium; B, L.S. of sporangium of *Hornea(Horneophyton lignieri)*; C, spore tetrad of *Rhynia*;

The reproductive structures (the sporangia). The elongated or oval sporangia were terminally developed at the apices of the branches of the aerial shoot. It is not clear whether the sporangia were borne on all the apices of the branches or on certain branches of the plant. Each apex bore a single sporangium.

As regards the shape the sporangia were cylindrical and they possessed the greater diameter than those tips of the branches on which they were found. The jacket layer of the sporangium consists of several cells in thickness. The outermost cells of this layer were thick-walled and cutinized at their outer face. The middle layer or the sporangial wall possessed thin walled cells and inner-most layer consisted of thin walled rounded cells which probably act as tapetal layer. The sporangium had no columella with in it. The cavity contained many spores of the same size. The sporangia were indehiscent and it is thought that the spore were released only on the decay and disintegration of the thick sporangial wall.

The spores are spherical and cuticultarized. They are found arranged in spore tetrads.

Genus HORNEOPHYTON

Systematic position. Pteridophyta
Division-Psilophyta
Class-Psilophytopsida
Order-Psilophytales
Family-Rhyniaceae
Genus-*Horneophyton*

Habit

Horneophyton formerly known as *Hornea* was similar to *Rhynia*, but was a smaller more slender plant with somewhat more freely branched stems, which were without appendages of any sort. The rhizomes were shorter and jointed, with thick tuberous segments.

Anatomy

Rhizome. The stele of the rhizome was apparently without vascular tissue consisting wholly of parenchymatous cells.

Stem. The stele of the stem was similar to that of *Rhynia*. The steles of the aerial stems passed into the fleshy mass, expanded bell-like and faded out.

Reproduction.

Sporangia. The small sporangia were terminal on the stem tips and sessile, being continuous with the aerial stem. They had no uniform shape, being merely fertile stem tips. In many cases the sporangia were forked dichotomously like the stem. The wall of sporangium was thick and without evidence of a means of dehiscence. The spores lay in a dome-shaped cavity around and above a central columella which seems to represent a continuation of the phloem of stele.

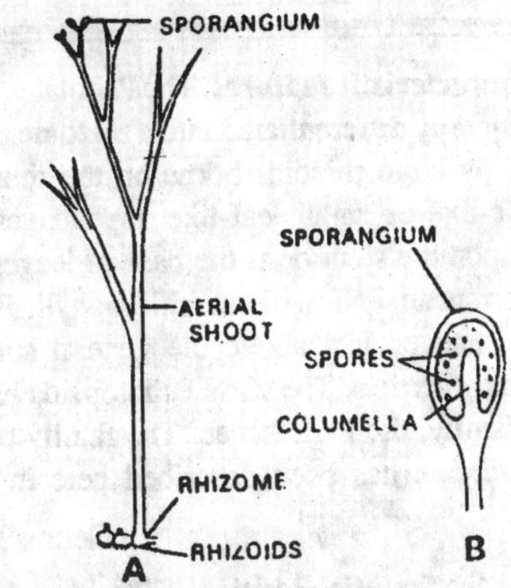

Fig. 2.6. *Horneophyton.* A, external structure; B, L.S. of sporangium.

3

Division-Psilophyta
Class-Psilotopsida

Order-Psilotales

Characteristic features. The Psilotales have a rootless, dichotomously branched sporophyte that is generally differentiated into a rhizome and an aerial shoot. The function of water absorption is served by large rhizoids borne on the rhizome. The aerial shoots may be leafless or it may bear scale-like or small leaf-like appendages. The sporangia are borne in diads or triads and directly upon the stem or at the base of leaves. The radially organized cylindrical gametophytes are subterranean and without chlorophyll. The antheridia remain partially embedded in the prothallus and project above the general surface of the prothallus.

Classification. The class Psilotopsida has a single order, the Psilotales. The order includes a single family, the Psilotaceae. The family comprises of two genera-*Psilotum* and *Tmesipteris.* Genus *Psilotum* has been described here in detail.

Genus PSILOTUM

Systematic Position. Pteridophyta
Division. Psilophyta
Class. Psilotopsida
Order. Psilotales
Family. Psilotaceae
Genus. *Psilotum.*

The genus has two well defined but polymorphic species-*Psilotum nudum* (Syn. *P. triquetrum)* and *P. flaccidum* (Syn.*P. complanatam).*

Distribution and habitat. This genus is frequently found in tropical and sub-tropical regions of both eastern and western hemispheres. *Psilotum nudum* is the commoner species of the two. *P. flaccidum* is a rare species and is found in the tropical islands.

Both the species are usually epiphytic in habit and grow upon tree ferns. These species may also be terrestrial and grow in humus or in the crevices of the rocks.

The Sporophyte

External features. The plant is slender and shrubby, rarely as much as a metre in height. It possesses green, ridged and dichotomously branched stems. The stems are perennial and somewhat xerophytic in structure.

The stems merge below the surface into slender dichotomous rhizomes which branch freely and may be one metre or more long. There are no roots. The rhizome generally contains a mycorrhizal fungus. Any dichotomy of rhizome may develop into a green aerial shoot. The rhizomes remain covered with hair-like absorbing structures or rhizoids.

The green aerial shoots of epiphytic plants are commonly pendant and those of terrestrial plants are usually erect. The aerial shoot regularly has several dichotomous branchings. The basal portion of the shoot is cylindrical. The aerial green portion of *Psilotum nudum* is radially symmetrical and longitudinally ribbed. In *P.flaccidum* the aerial shoot is markedly flattened.

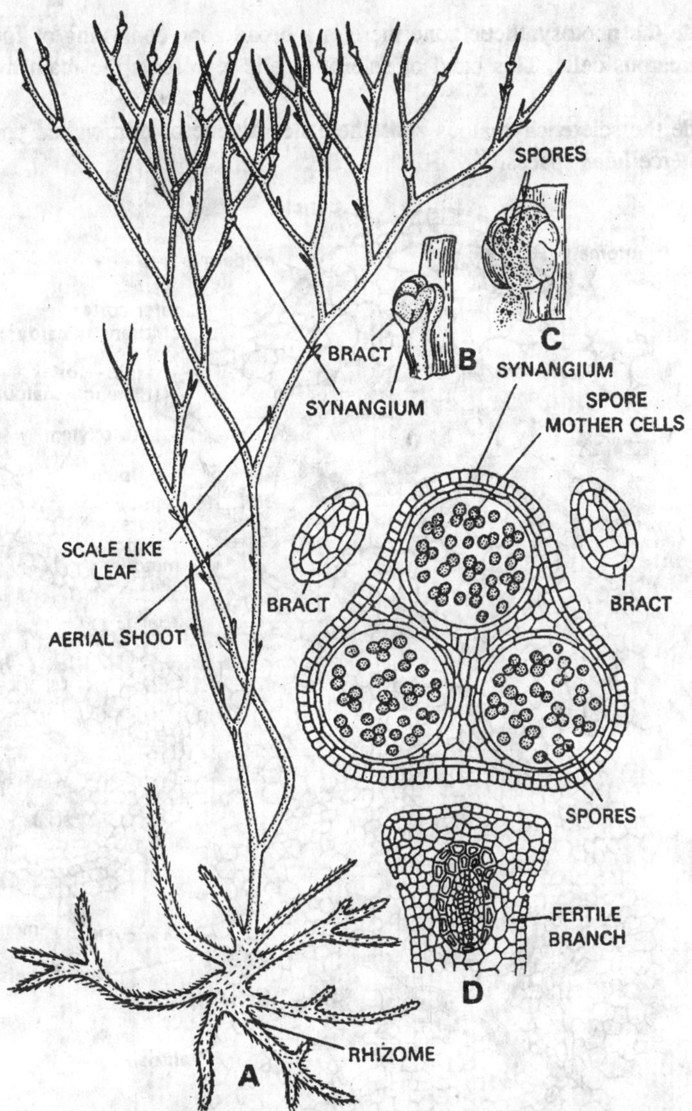

Fig. 3.1. ***Psilotum nudum.*** A, the sporophyte showing rhizome and aerial shoot; B, a closed synangium; C, an opened synangium liberating spores; D, transverse section of synangium with bracts and axis.

The aerial shoots bear small, scale-like leaves which remain more or less irregular in distribution. These leaves are without veins. These minute leaves are not found on the rhizomes and the extreme base of the aerial shoot.

Anatomy of aerial stem. In section the aerial stem shows the well-marked epidermis, one cell in thickness. This layer consists of elongated cells possessing greatly thickened external walls. The stomata are found in the grooves of the surface.

The cortex consists of three zones. The outermost zone consists of cells which are thin-walled, vertically elongated and having well-defined intercellular spaces among them. This zone is an assimilatory tissue and contains abundant chloroplasts. This zone is composed of two to five cells in breadth.

Inside the photosynthetic zone there is a broad zone consisting of four to five layers of sclerenchymatous cells. This band of sclerotic cells represents the main mechanical tissue of the plant.

Inside the sclerenchymatous zone there lies a zone of thin walled parenchymatous cells without intercellular spaces.

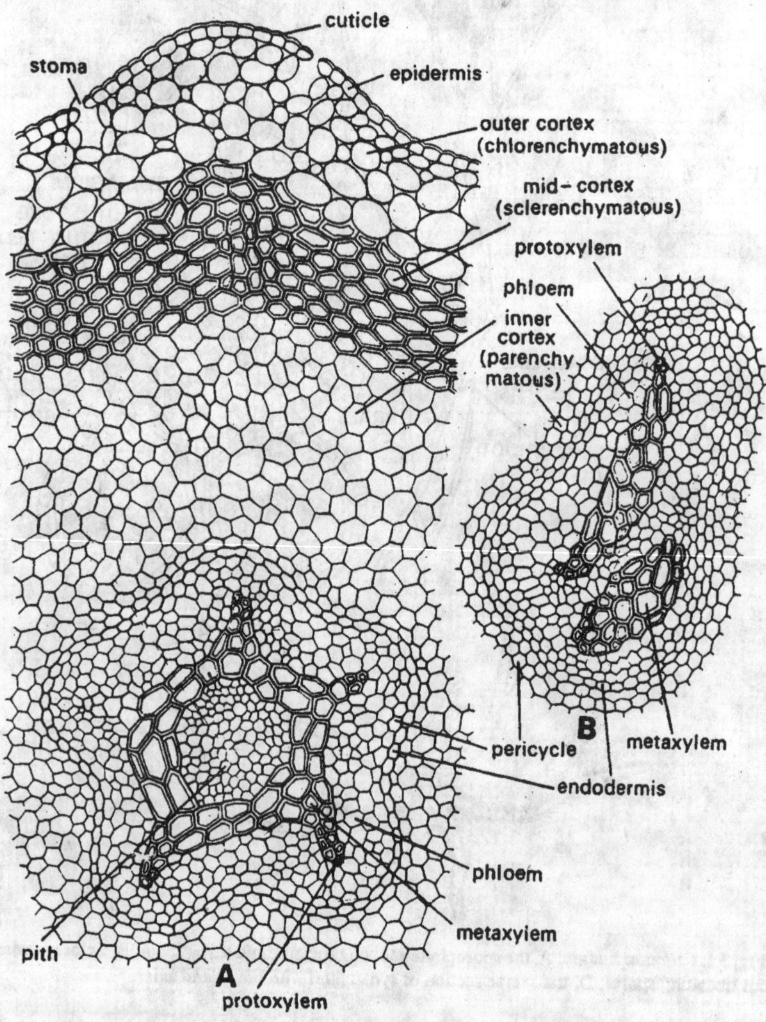

Fig. 3.2. A, T.S. of aerial stem; B, T.S. of stele of aerial stem near apex.

The well marked endodermis surrounds the stele. There lies a core(pith) of sclerenchymatous cells in the centre of the stele. This central sclerenchymatous core (pith) remains surrounded by a narrow band of xylem elements. The xylem cylinder that surrounds the sclerotic pith is stellate or star-shaped when seen in transverse section. The xylem is exarch and the protoxylem lies at the tip of each ray. The metaxylem tracheids are composed of scalariform xylary elements while the narrow protoxylem elements show spiral thickenings. In between the endodermis and the xylem there lies a mass of thin-walled cells that represent the phloem. The phloem is of very indefinite character. The cells of the phloem are tubular but they donot possess the structure of sieve tubes. The pericycle is ill-defined. This is a siphonostelic type of structure, though the pith is sclerenchymatous and not parenchymatous as found in ferns.

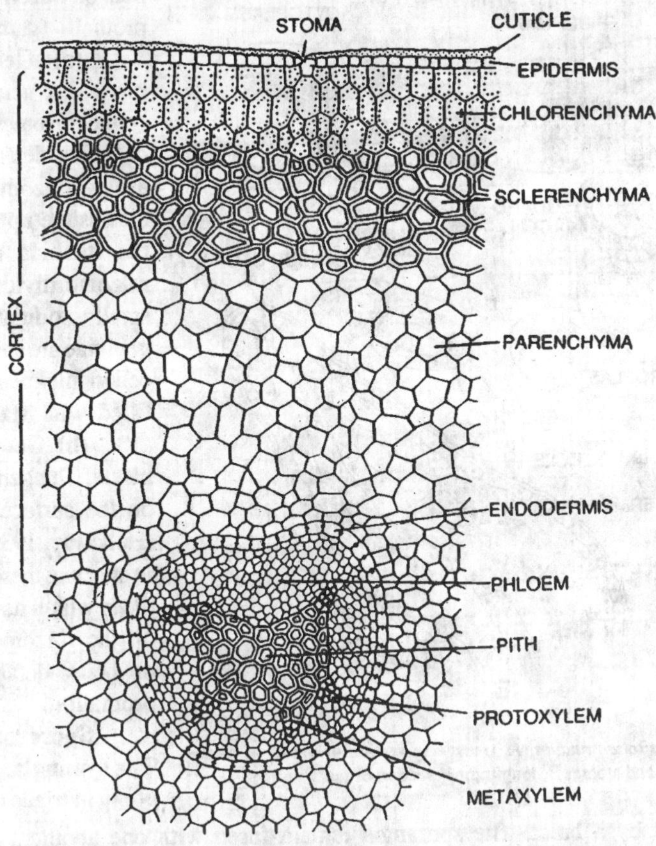

Fig. 3.3. ***Psilotum nudum.*** **Transverse section of aerial stem with actinostele.**

In the upper part of the stem the central group of sclerenchyma is not formed and there are not xylem rays, so that the structure here is typically protostelic.

Anatomy of leaf. The leaves receive no vascular tissue. They possess a very simple structure. It remains surrounded by a typical cuticularized epidermis. The stomata are not present. Inside the epidermis there lies the parenchymatous tissue which possesses well developed (*e.g.*, in *P. nudum*) or little developed (*e.g.*, in *P. flaccidum*) intercellular spaces.

Anatomy of rhizome. The epidermis is inconspicuous. All the cells of the outermost cortical layer extend into two celled absorptive rhizoids.

The major portion of the cortex is thin-walled and mycorrhizal. However, outside the endodermis there are three or four layers of thick, brown-walled cells.

The stele is of protostelic type. It has no pith and the xylem mass is usually circular in outline and donot bear any protoxylem rays. There is very little phloem and the endodermis is conspicuous.

Apical growth. It takes place by the activity of an apical cell both in the case of rhizome and aerial shoot.

Vegetative propagation. The vegetative reproduction takes place by means of **gemmae.** These vegetative structures develop both on rhizomes (sporophyte) and prothalli (gametophyte).

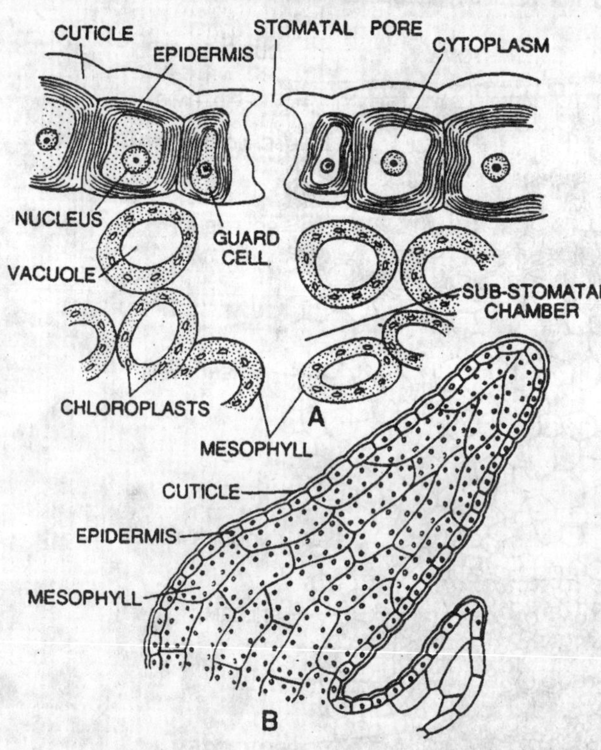

Fig. 3.4. *Psilotum nudum.* A, transverse section of a part of stem showing epidermis and stoma; B, longitudinal section of leaf.

(a) Gemmae on sporophyte. The gemmae develop freely on the rhizomes. Each such gemma on being detached develops into a new rhizome. In starved conditions the upper cell of many of the rhizoids developed on the rhizome divides and produces a small nodular gemma. These gemmae are small oval bodies one cell in thickness. The cells remain filled with starch.

(b) Gemmae on gametophyte. The gemmae also develop on the surface of the prothallus (Holloway, 1939, Bierhorst, 1953). In structure these gemmae resemble those which have been developed on the rhizome. These gemmae on being detached develop into prothalli.

Spore producing organs. The synangia. The sporangia are borne in triads on minute appendages subtended by a bract. The sporangia remain fused with one another, and therefore, the group is called a **synangium.** The so-called synangium is generally interpreted (Bower, 1935, Eames, 1936) as a sporangiophore fused with a subtending bract. The sporangia seem to be borne on the adaxial side of the appendage at the point of dichotomy and are slightly raised on broad short stalks. The slender branches of the fertile appendage are more or less erect and embrace the synangium. The fertile appendages are mostly grouped on the upper part of the stem, but both these regions are not strictly limited and sterile appendages may occur here and there among the fertile ones.

Structure of mature synangium. The synangium is three lobed when examined from the outside internally it remains divided into three chambers. The wall is composed of several layers of cells. The sporangial wall, cells have thickened considerably, except along one vertical line running from the distal end to the base of each sporangium. These are the lines of dehiscence, along which the synangium opens into three segments releasing spores. There is no true tapetum. The spores within the sporangium are bean-shaped with finely reticulated walls.

Development of synangium. The development is apparently of the eusporangiate type. The earliest stages of development have shown, that each loculus arises separately from a single epidermal cell of the sporangiophore. It shows, that this structure is a union of three sporangia, the synangium. The primary initial divides periclinally separating a **jacket initial** and an **archesporial**

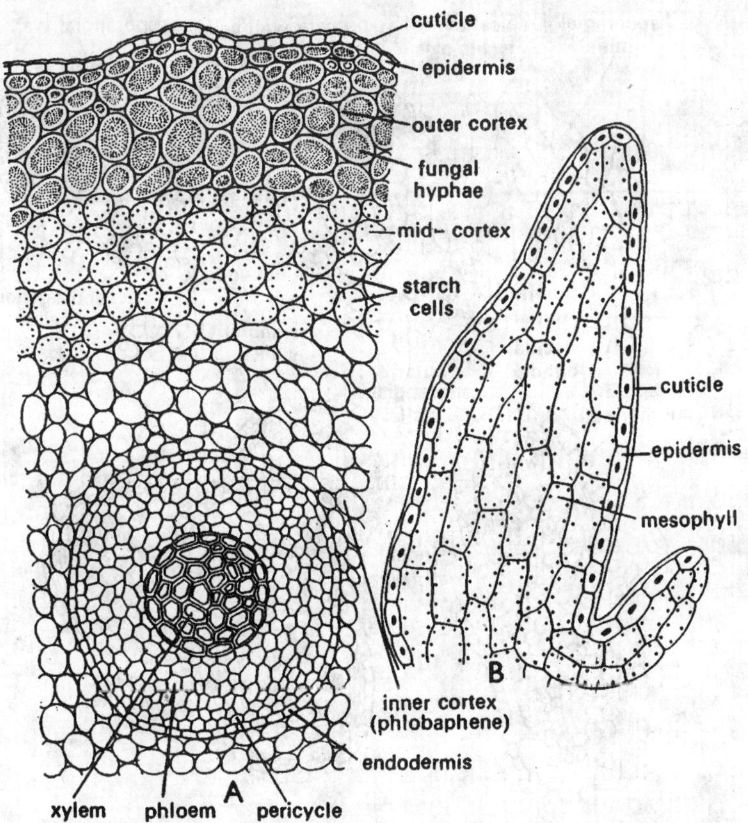

Fig. 3.5. ***Psilotum nudum.*** **A, portion of a transverse section of a rhizome showing mycorrhizal fungus; B, L.S. of leaf.**

initial. The jacket initial gives rise to a wall three to five cells in thickness, while the archesporial initial produces a central mass of sporogenous cells. Unlike most other pteridophytes, neither the outermost sporogenous cells nor the innermost jacket cells develop into a tapetum. The three sporangia are grouped around a central sterile tissue, the sporangiophore axis.

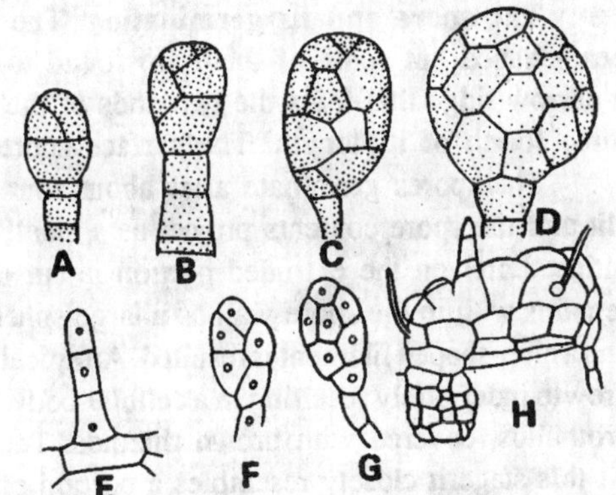

Fig. 3.6. ***Psilotum nudum.*** **A-D, the sporophytic gemmae; E-G, young gametophytic gemmae; H, germinating gemma.**

Division and redivision of the archesporial cell produce a large number of sporogenous cells. As the sporogenous tissue matures, some of the sporocytes are disorganized forming a fluid that nourishes the surviving mother cells and spores develop. The spores are bean shaped with finely reticulated walls. Simultaneously the cells of the wall of sporangium become thickened, except along one vertical line from the apex to the base of the sporangium; these act as the lines of dehiscence.

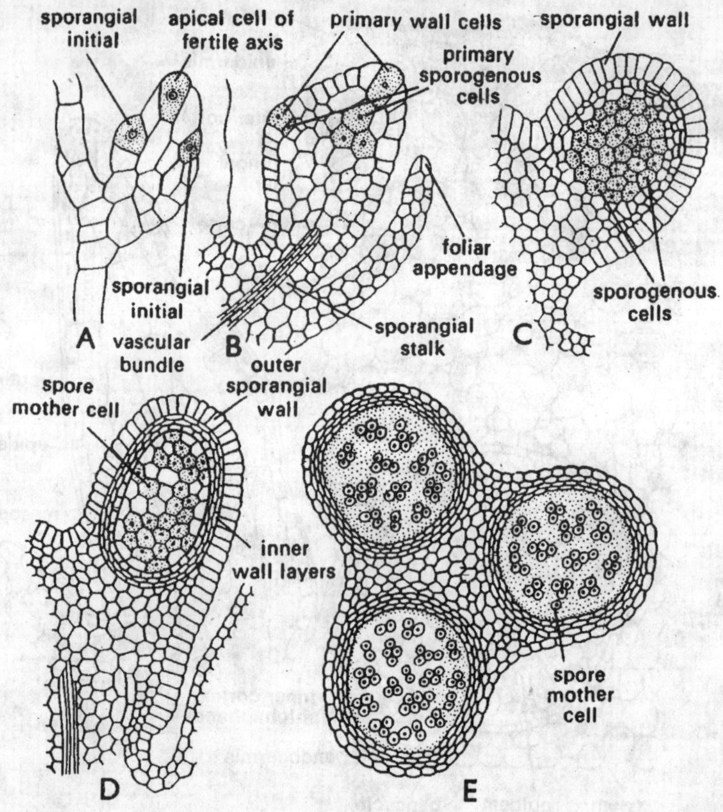

Fig. 3.7. ***Psilotum nudum.*** A-D, successive stages in the development of sporangium; E, trilocular synangium.

The Gametophyte

The spore and its germination. The spores are of equal size. Usually the spores are bean-shaped but some of them are found to be of the tetrahedral type. Each spore possesses a narrow ridge that joins the two ends of the curve. There is a median slit extending for about three fourth of its length. The surface spore is delicate and finely reticulate.

The spores germinate after about four months by rupture of its wall along the median slit and the spore contents project as a small protuberance covered with a thin membrane, the intine. Later on the extruded portion is cut off by a transverse wall so that the basal portion remains with in the spore wall as a large spherical cell. The apical cell divides by oblique walls and a club-shaped filament is resulted. An apical meristem is established which apparently continues growth indefinitely resulting in a cellular body. This elongates into a cylindrical, slightly branched prothallus, covered with brown rhizoids. The prothallus grows by means of apical meristem. At this stage it closely resembles a portion of the rhizome. The prothallial tissue is colourless, saprophytic and mycorrhizal as in the sporophyte but it lacks vascular tissue, though cases are on record (Holloway, 1939) in which tracheids were found in large prothalli.

The gametophyte is monoecious that it bears both antheridia and archegonia. The sex organs are irregularly distributed on the prothallus. The number of archegonia is lesser than that of antheridia.

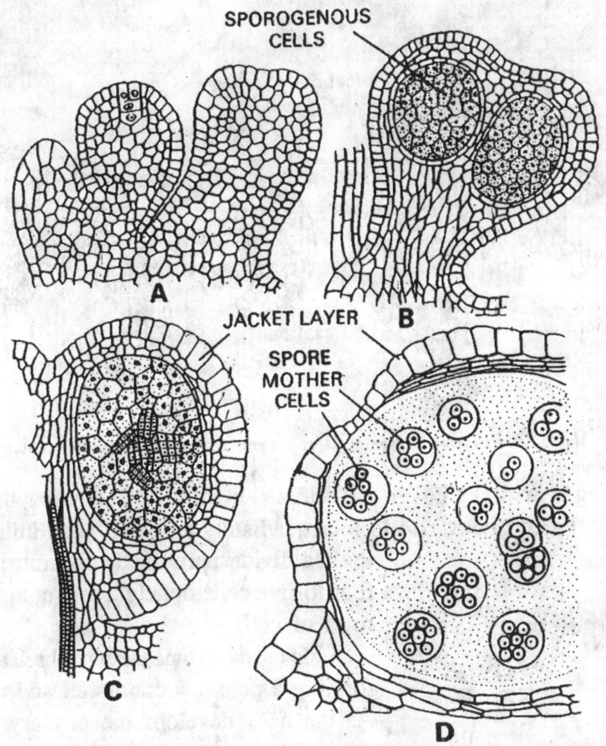

Fig. 3.8. ***Psilotum*** **spp. A, a vertical section of stem apex bearing a young leaf and a sporangiophore; B-C, successive stages in the development of sporangia; D, transverse section of a portion of an older sporangiophore and through the time of dehiscence of a sporangium.**

Development of antheridia. The antheridia appear first. They are prominently projecting, spherical bodies, with a wall of single layer of cells.

The antheridium arises from a superficial cell of the gametophyte. This cell acts as the antheridial initial. The antheridial initial divides periclinally into an outer cell, the **jacket initial** and an inner cell, the **primary androgonial.** The primary androgonial cell divides and redivides forming a considerable number of androgonial cells, the last generation of which are **androcytes.** The androcytes metamorphose into spirally coiled multiflagellate antherozoids. In the mean time jacket initial also divides anticlinally forming a **jacket layer** one cell in thickness which lies outside the outer face of androgonial tissue. Their lies a triangular opercular cell in the centre of the antheridium which by its disintegration provides an opening for the escape of antherozoids.

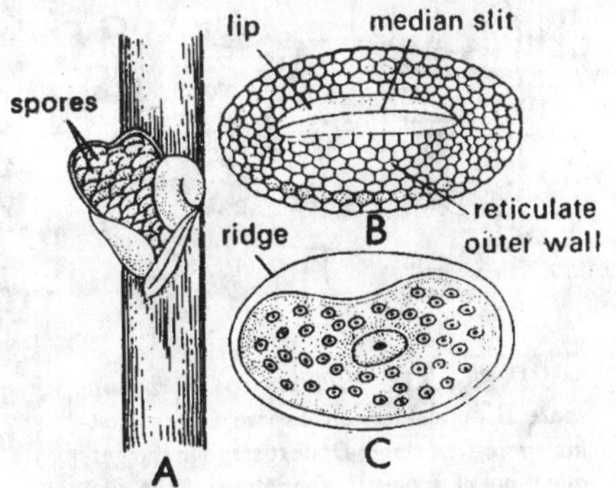

Fig. 3.9. ***Psilotum nudum.*** **A, dehisced sporangium; B, single spore; C spore in optical section.**

The archegonium and its development. The archegonium develops from a superficial cell of the gametophyte. This cell is known as archegonial initial. First of all this initial divides

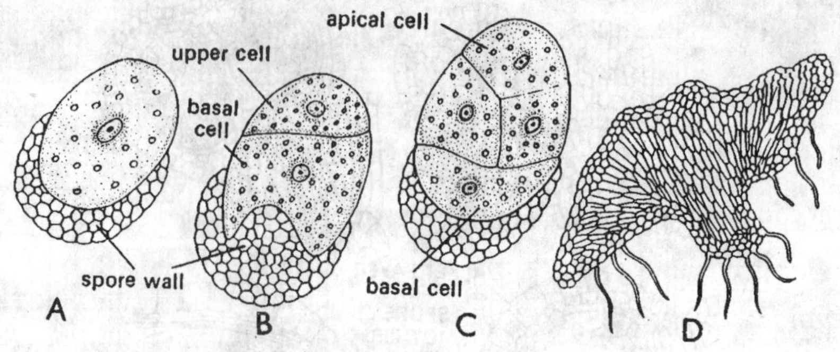

Fig. 3.10. *Psilotum nudum.* A-D, spore germination showing earlier stages.

periclinally forming a **primary cover cell** and a **central cell.** Thereafter the primary cover cell divides anticlinatly twice giving rise to four quadrately arranged neck initials. Now these neck initials divide periclinally forming an archegonial neck four to five cells in height and composed of four vertical rows of cells.

In the mean time the central cell also divides periclinally forming a **primary canal cell** and **primary venter cell.** As the neck develops the primary canal cell elongates vertically. According to Holloway (1939) and Lawson (1917) the division of neck canal cell is uncertain but it divides and gives rise to two neck canal nuclei which disintegrate later on. The behaviour of the primary venter cell is also doubtful. It is reported that it functions directly as the egg and does not undergo the usual division forming venter canal, cell and egg.

Fig. 3.11. *Psilotum* sp. A, mature spore; B, optical section of spore; C, germinating spore-first stage; D, next stage in the germination of spore; E, formation of the first cells of the prothallus; F-J, successive stages of the development of the prothallus.

As the archegonia approach maturity, the cell walls of the lowermost tier of neck cells become cutinized and thickened. With the result the sloughing off of the upper portion of the neck takes place. Simultaneously the neck canal cell or cells also disintegrate bearing a passage way for the entrance of antherozoids in the venter of the archegonium. One of the antherozoids penetrates the egg and the fusion of male and female nuclei takes place forming the oospore.

Development of embryo. The development of embryo is very simple. The fertilized egg enlarges downward. It divides periclinally and the upper cell develops into the axis, the lower cell into a foot, which sends out finger-like projections into the prothallial tissues. No root or cotyledon is formed. There is a marked line of division between axis and foot. When the developing axis breaks through the surface of the prothallus the foot and axis separate, the foot remains attached to the prothallus.

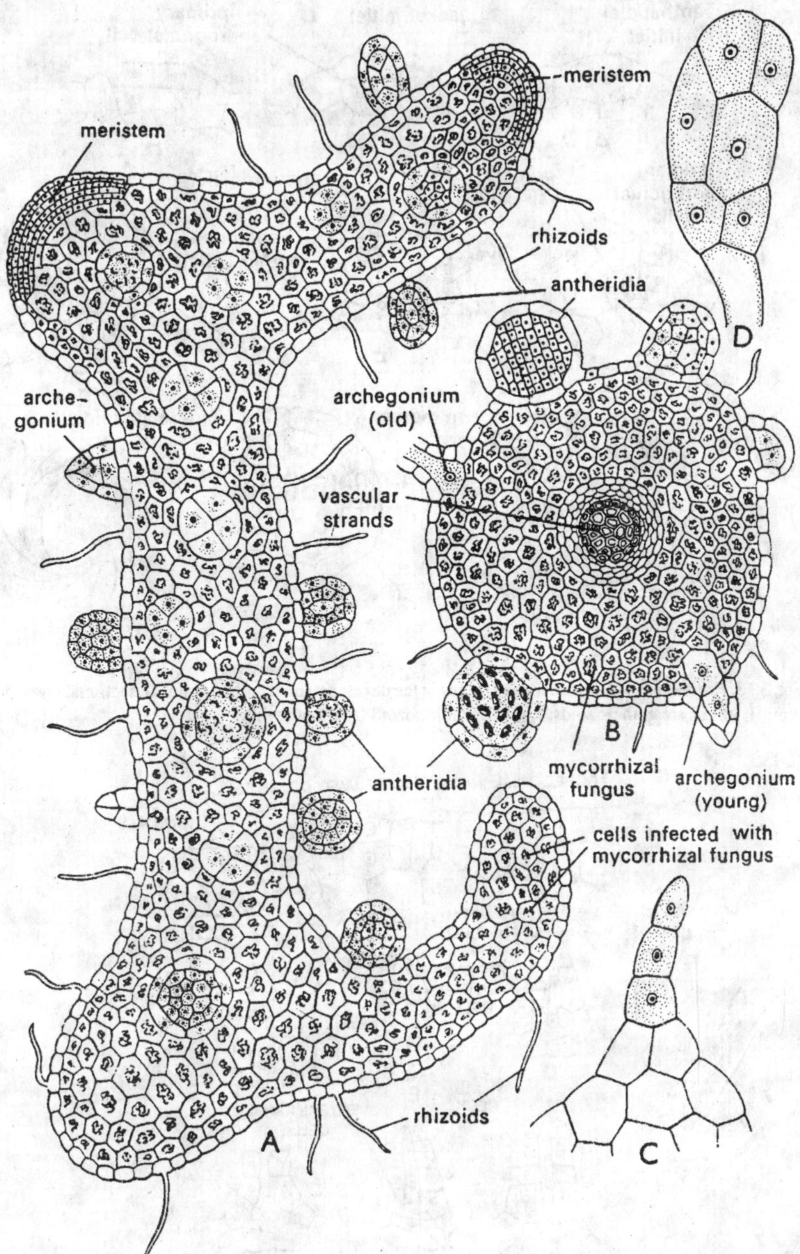

Fig. 3.12. *Psilotum nudum.* A, the entire prothallus showing apical meristem and distribution of archegonia; B, transverse section of prothallus showing antheridia, archegonia and mycorrhizal fungus; well-formed conducting strand present in the centre; C and D, gemmae on gametophyte.

The first division in development of the epibasal cell into a foot is vertical. Thereafter the divisions take place in an irregular sequence in the two halves of the foot. As development of the foot continues it becomes a cylindrical structure of equal length and breadth. Superficial cells of the foot elongate into projections that enter the prothallus like haustoria.

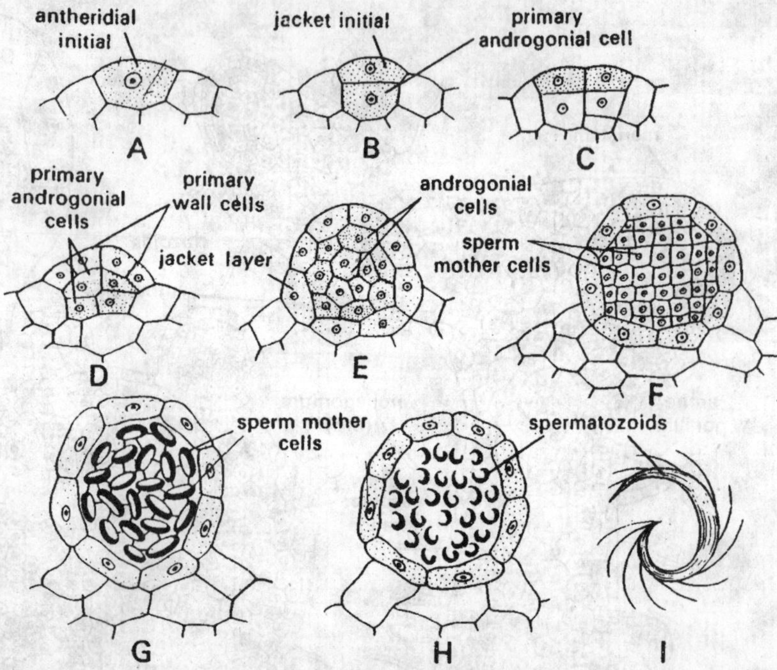

Fig. 3.13. *Psilotum* spp. A-F, successive stages in the development of antheridium; G, sectional view of a nearly mature antheridium; H, immature antherozoids; I, mature antherozoid (spermatozoid).

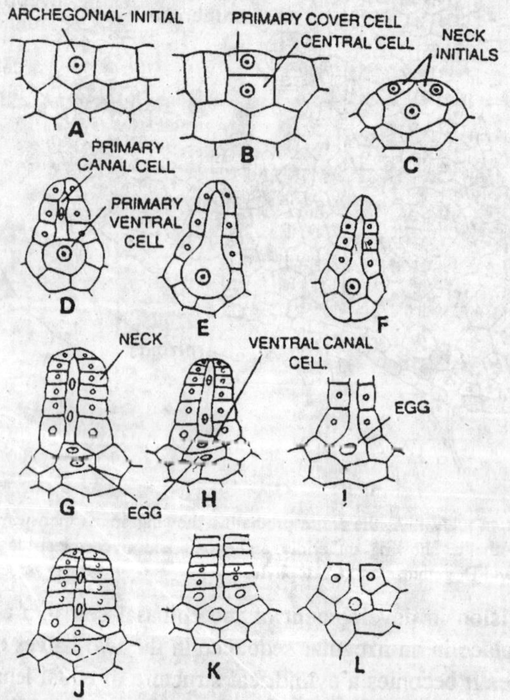

Fig. 3.14. *Psilotum* spp. A-L, successive stages in the development of archegonium

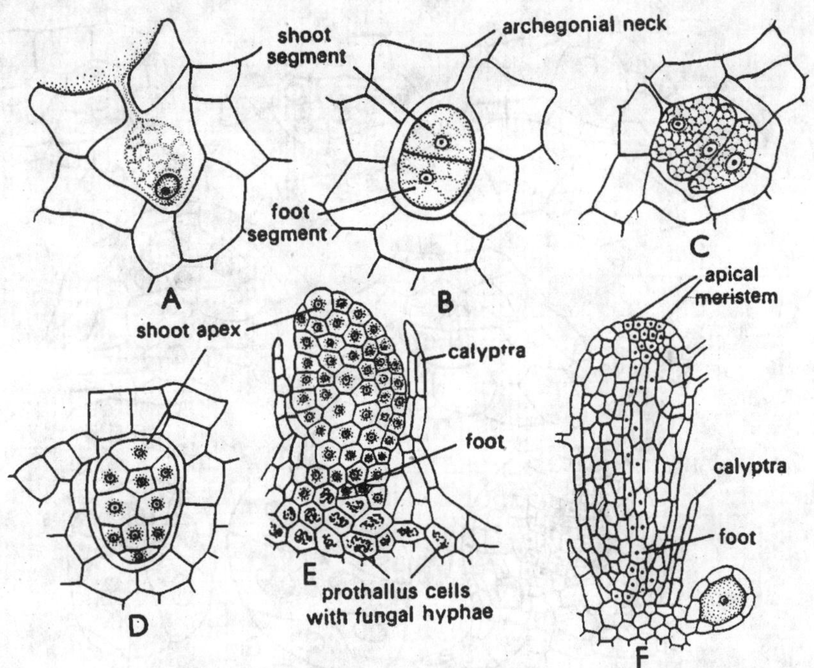

Fig. 3.15. ***Psilotum nudum.*** **A-F, successive stages in the development of embryo; A-E, after Bierhorst and F, after Holloway.**

The development of shoot takes place as follows. The epibasal daughter cell of the zygote divides vertically. The two daughter cells thus formed divide transversely. Subsequent divisions take place in irregular way. The short portion of a young embryo is hemispherical. It is differentiated vertically into two halves, derived, respectively from the first two cells of the shoot. The axis elongates vertically before it becomes free and develops an apical cell in one or in both halves of the young shoot. The apical cell always lie in the middle between base and apex of the half in which it is formed. Further development of the shoot takes place because of the activity of the apical cell or cells. When there are two apical cells the axis is dichotomous from the beginning, but in any case the first dichotomy soon takes place. In the early stages of the development of embryo the prothallus tissue also grows up like a calyptra round the young axis, but the growth activity of the embryo soon breaks through it. The protruding portion of the sporophyte soon develops rhizoids on its surface and becomes infected with a mycorrhizal fungus. Now this becomes a new independent plant. The rhizome continues to branch either dichotomously or by the formation of adventitious side branches. The original branch or some of the adventitious branches later on grow above the soil and develop into aerial shoots.

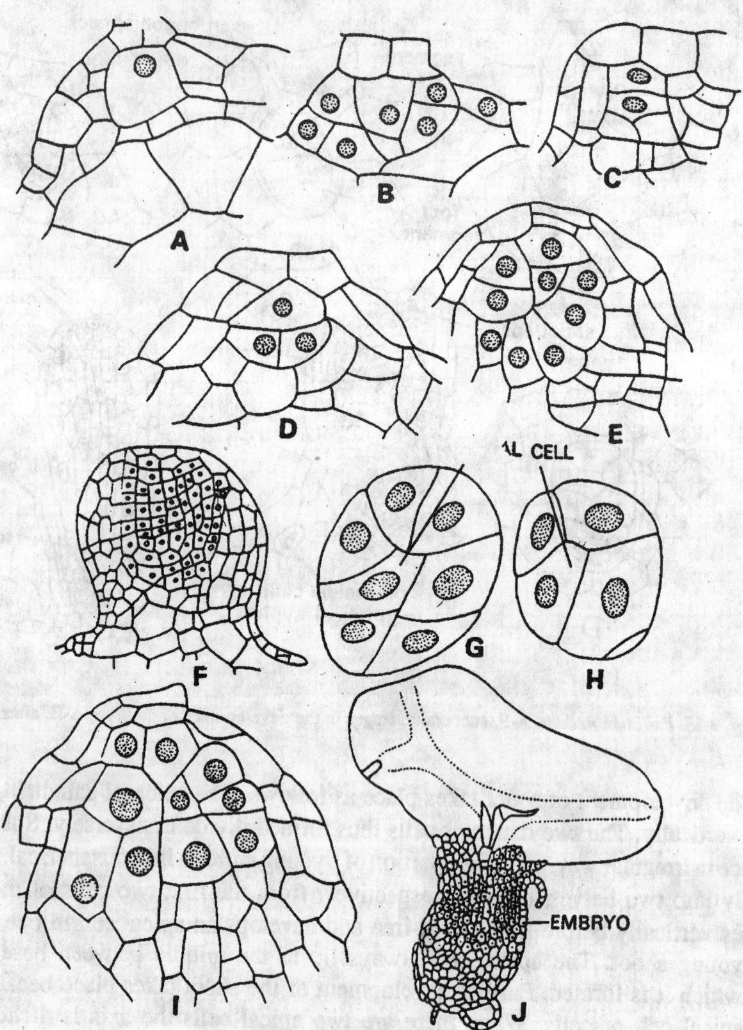

Fig. 3.16. ***Psilotum* spp. A-I, successive stages in the development of embryo.**

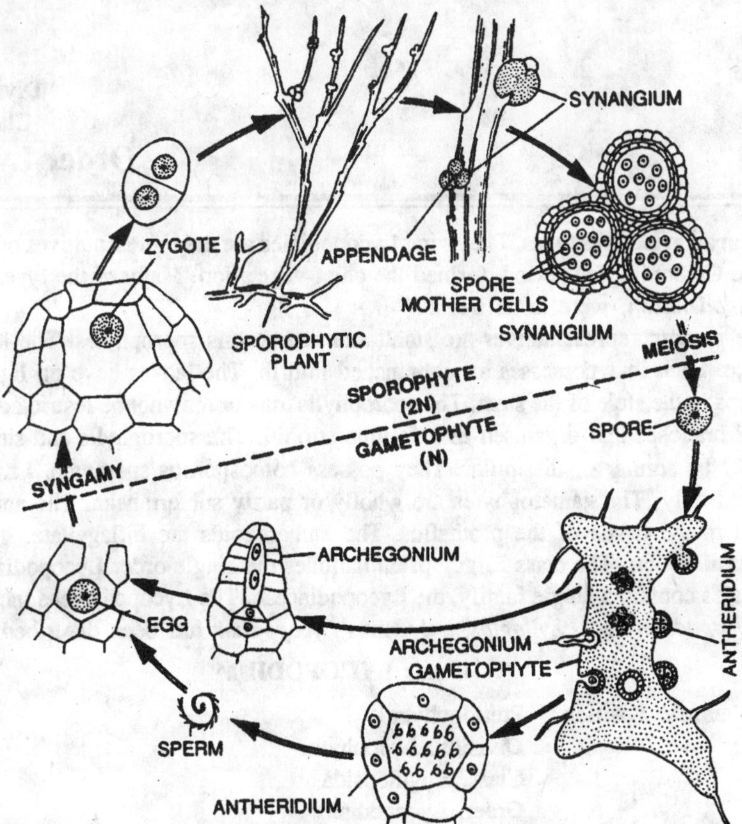

Fig. 3.17. *Psilotum.* Diagrammatic life-cycle (A-K).

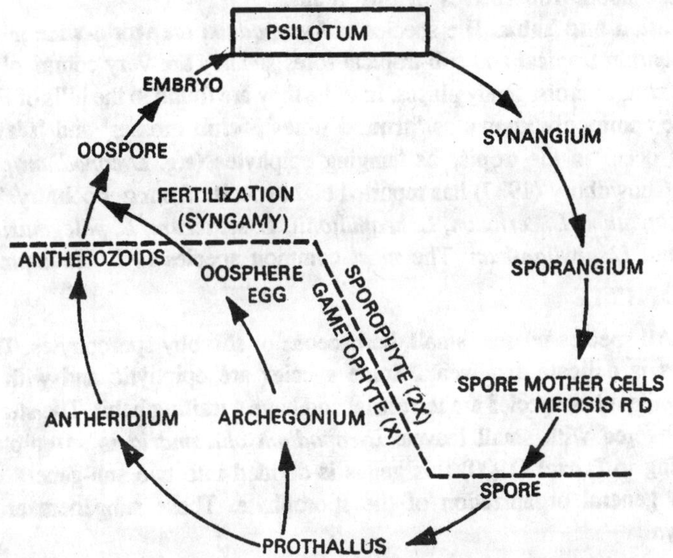

Fig. 3.18. *Psilotum* sp. Graphic life-cycle.

4

Division-Lycophyta
Class-Eligulopsida

Order-Lycopodiales

Characteristic features. The living Lycopodiales are the representatives of a group which, during the Carboniferous period, formed the chief vegetation. Many of the types then growing, *e.g., Lepidodendron*, were large trees.

The modern representatives are small and herbaceous sporophytes. The leaves are small and simple. Each leaf possesses an unbranched midrib. The leaves have no ligules. There are no leaf gaps in the stele of the stem. The sporophylls may or may not be restricted to the terminal portion of branches and organised into definite strobili. The sporophylls and simple vegetative leaves may be similar or dissimilar. They possess homosporous sporangia, *i.e.*, all the spores of one kind only. The gametophytes are wholly or partly subterranean. The antheridia remain embedded in the tissue of the prothallus. The antherozoids are biflagellate.

Classification. The class Eligulopsida includes the single order Lycopodiales. The living Lycopodiales contain a single family, the Lycopodiaceae. The Lycopodiaceae include two living genera *Lycopodium* and *Phylloglossum*. Genus *Lycopodium* has been described here in detail.

Genus LYCOPODIUM

Systematic Position. Pteridophyta
Division. Lycophyta
Class. Eligulopsida
Order. Lycopodiales
Family. Lycopodiaceae
Geneus. *Lycopodium*

There are about 180 species in this genus.

Distribution and habit. The species of *Lycopodium* are world-wide in distribution. They are mainly found in tropical and sub-tropical forests. They are very commonly found on heaths and on humus soils in moist shady places. In India they are found in the hills of Eastern Himalayas. The plants are commonly known as **'ground pines'**, **'club mosses'** and **'trailing evergreens'**. Many species occur in the tropics as hanging epiphytes (*e.g. Lycopodium phlegmaria* and *L. squarrosum*). Chowdhury (1937) has reported eight species from our country. These species are: *Lycopodium clavatum; L. cernuum, L. heamiltonii; L. setaceum; L. phlegmaria; L. wightianum; L. serratum* and *L. phyllanthum*. The most common species is *L. clavatum*.

THE SPOROPHYTE

Habit. All species possess small, herbaceous or shrubby sporophytes. The stem in almost all the species is delicate and weak. Some species are epiphytic and with erect or pendant sporophytes while other species are terrestrial and have a trailing habit. The stem and its branches are densely covered with small leaves. *Lycopodium phlegmaria* is an epiphytic species.

According to Pritzel (1900) this genus is divided into two sub-genera which differ from each other in general organization of the sporophyte. These subgenera are-*Urostachya* and *Rhopalostachya*.

Stems. Species referred to the sub-genus *Urostachya* possess branched or unbranched stems that are erect or pendant but never creeping. This subgenus includes the species, *e.g., L. selago, L. lucidulum, L. phlegmaria* and others. If the stem is branched, the branching is always dichotomous. Usually the successive dichotomies are found at right angles to one another. The species belonging to this subgenus donot bear the adventitious roots along the stem.

The subgenus *Rhopalostachya* includes the species, *e.g., L. clavatum, L. inundulatum, L. complanatum, L. cernuum* and others. The species referred to this sub genus possess prostrate stems bearing upright branches. The branching of the stem in the first-formed portion may be dichotomous, but in later-developed portion it is always monopodial.

Leaves. The leaves are small, simple, sessile, numerous and cover the axis closely. Typically the leaves are 2 to 10 mm. long. Usually the leaves are arranged in closed spirals (*e.g.*, in *L. clavatum* and *L. annotinum*) while in other cases they are arranged in whorls (*e.g.*, in *L. verticillatum* and *L. cernuum*). In some species the leaves are found to be arranged in opposite

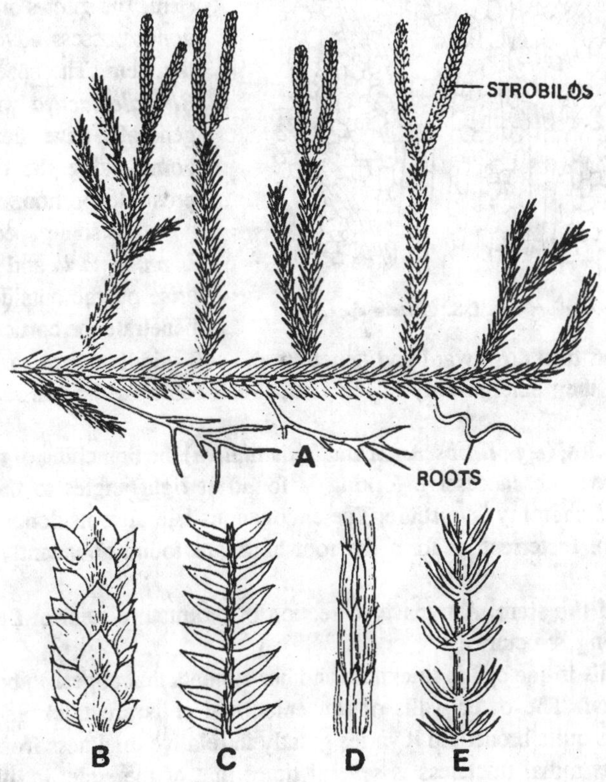

Fig. 4.1. *Lycopodium.* A, part of a plant of *L. clavatum* showing strobili; B-E, leaf form and arrangement in *Lycopodium;* B, *L. refescens;* C, *L. volubile;* D, *L. complanatum;* E, *L. cernuum.*

pairs (*e.g., L. alpimum*); in others they are irregularly arranged. Usually the leaves are lanceolate in outline possessing the broad bases. Usually the first-formed leaves on the stem donot bear the mid-rib, the later formed leaves usually possess a single unbranched vein that starts from the base but doesnot reach the apex. The leaves are all alike in size and shape, but in few species, such as *L. complanatum, L. volubile* and *L. chamaecyparissus* the leaves are **dimorphic** like that of *Selaginella* and found to be arranged in four vertical rows on the stem two being smaller and two larger in size.

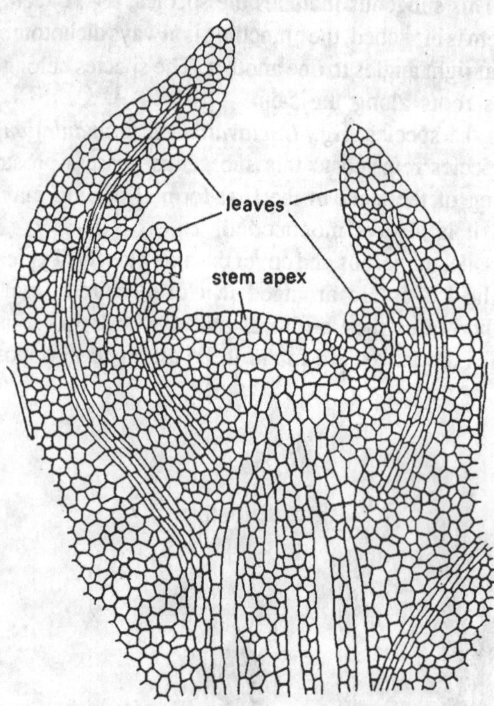

Fig. 4.2. *Lycopodium* spp. L.S. of stem apex.

Apical growth. The apical growth of the shoot takes place by means of an apical meristem which consists of a group of apical cells.

Roots. The first root is short and atleast in some species doesnot live long. The older plant bears the adventitious roots that arise singly or in groups acropetally along the lower side of the stem. The species of subgenus *Urostachya* donot possess adventitious roots along the stem. The species of the subgenus *Rhopalostachya* with a creeping stem generally have their adventitious roots borne along the entire length of the prostrate portion.

In some species, *e.g., L. selago, L. phlegmaria* and others the roots that arise on the outside of the stele donot penetrate the cortical region of the stem atonce. These roots turn downward and penetrate the soft middle cortex making canals through it, and ultimately they emerge only at the stem. Such roots are known as **'cortical roots'** or **'inner roots'.**

In some species, (*e.g., L. obscurum* and *L. lucidulum*) the branching of the roots is strikingly dichotomous. Here each successive forking is found at right angles to the preceding one. In many species the dichotomy is obscure. The endogenous lateral roots donot arise from the roots (as in seed plants). In terrestrial form the root hairs are found abundantly which persist over a long period.

Anatomy of the stem. A transverse section of the mature stem of *Lycopodium clavatum* shows the following structures:

The epidermis is one cell in thickness and has stomata that appear to be similar in structure to those of the leaf. The outer walls of the epidermal cells are thick and cutinized.

The cortex is quite broad and it varies greatly in relative thickness from species to species. In some species its radial thickness is several times that of the stele; in other species the two are approximately equal. The cortex has three zones. The peripheral and central zones consists of thickened sclerotic cells while the middle zone consists of larger and thin-walled cells containing a few chloroplasts.

The single stele occupies about half the area of the section. It consists of a number of irregularly shaped, parallel strands of xylem alternating with phloem bands. The xylem consists mainly of scalariform tracheids, with no vessels and no parenchymatous cells. Each Xylem plate remains surrounded by a single layer of parenchyma, with bordered pits on the inner walls, which lies between the xylem and the phloem. The phloem is composed of sieve tubes and

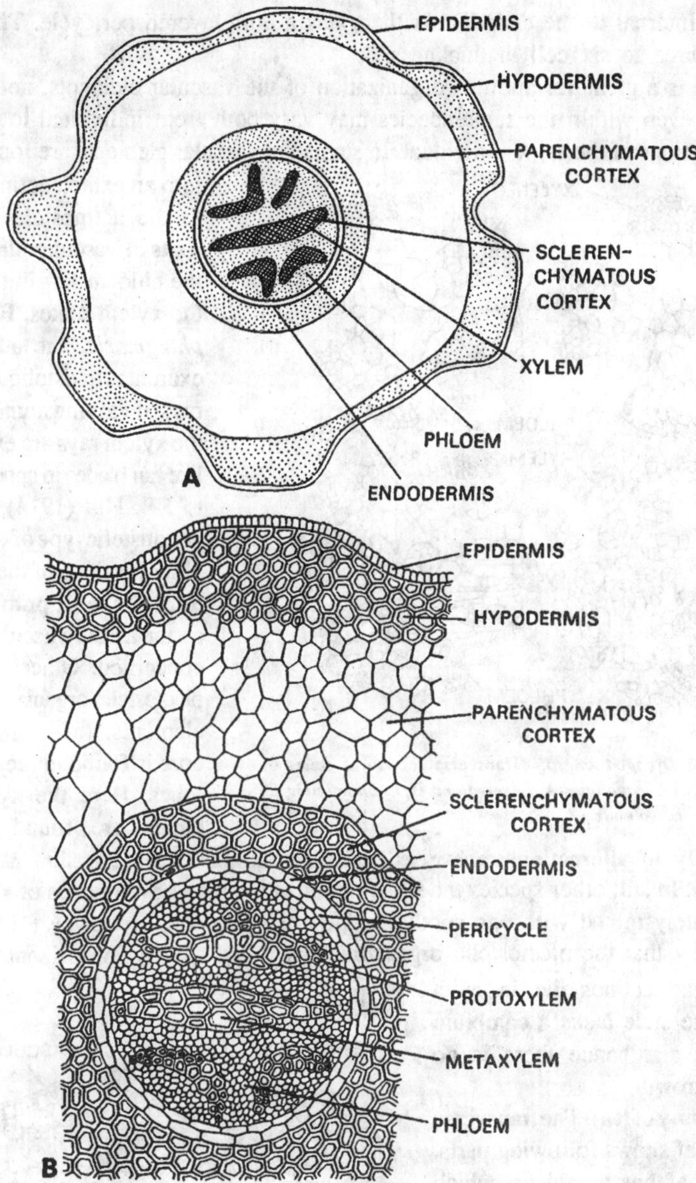

Fig. 4.3. ***Lycopodium*** **sp. A, transverse section of stem (diagrammatic); B, transverse section of stem (detailed).**

parenchyma only. The development of both xylem and phloem strands is centripetal, that is, from the periphery towards the cente. The protoxylem consists of spiral and annular tracheids. The protoxylem is found to be situated at the outer edges of each xylem mass, that is to say, in the exarch position.

At the inner face of the cortex there may be an endodermal layer which, at least in younger portions of the stem has characteristic thickened radial walls.

The endodermis of *Lycopodium* is thought to be originated from the stele and not from the cortex. Internal to the endodermis there lies a multilayered pericycle. The pericycle layer is usually three to six cell in thickness.

There is a great variation of organization of the vascular elements, not only in different species but even within the same species may vary both from individual to individual and in different portions of the same individual. In stems the vascular elements are found to be arranged into an exarch actinostele resembling to the actinosteles characteristic of roots of vascular plants. In such steles the phloem lies in the space between the xylem plates. For example, in *L. phlegmaria*, the adult stem has the exarch actinostelic organization with rays of varying number. In *L. serratum* the xylem rays are expanded outwards like fan blades in appearance. According to J.B. Hill (1914), the species with actionstelic type of organization in the older portions of the stem are thought to be the most primitive of the genus as far as the vascular organization is concerned. Other species may have plectostelic organization of the vascular elements. In such stems the xylem core is found in the form of plate like lobes. Here the xylem and phloem bands are found to be arranged symmetrically in alternating transverse bands across the stele (*e.g.*, in *L. volubile* and *L. wightianum*). In still other species the stelic organization may have masses of xylem and phloem indiscriminately mixed with one another, *e.g.*, in *L. cernuum*. Holloway (1910), Jones (1905) hold the view that the plectostelic organization is found chiefly in horizontal portions of the plant and the actinostelic in erect portions. The stele lacks a cambium in all cases and hence there is no secondary growth.

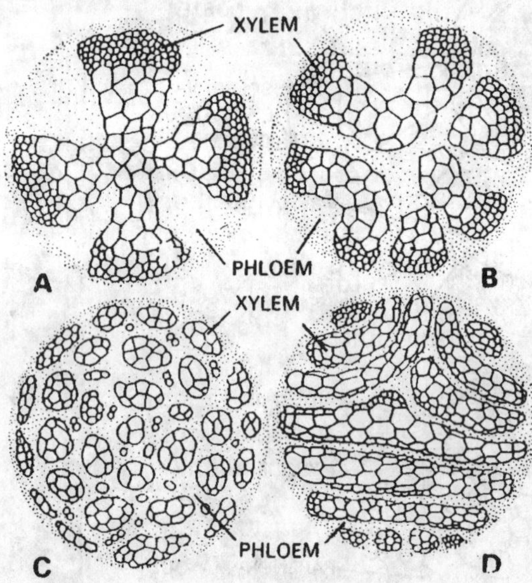

Fig. 4.4. *Lycopodium* sp. Transverse section of steles of various species of *Lycopodium*-A, *L. serratum*; B, *L. annotinum*; C, *L. cernuum*; D, *L. volubile*.

Anatomy of leaf. The transverse section of leaf shows following parts.

The leaf has a mid-rib which consists of a single concentric amphicribral bundle. The central xylem core is quite small. It is not always differentiated into meta-and protoxylem. The xylem is composed of thick walled annular and spiral tracheids. The phloem is composed of phloem parenchyma and narrow sieve-tubes. The endodermis is inconspicuous and may hardly be recognized. The vascular bundle remains surrounded by sclerenchymatous pericycle.

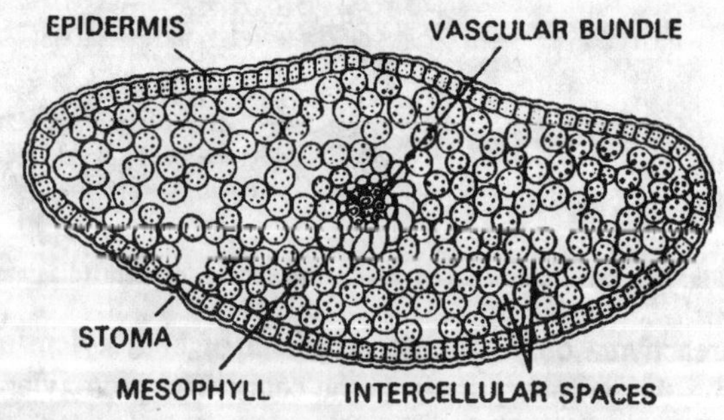

Fig. 4.5. ***Lycopodium clavatum.*** **Transverse section of leaf.**

The mesophyll tissue lies between the vascular bundle and the epidermis. The epidermis of most species has stomata on both faces of a leaf (*e.g.*, in *L. clavatum, L. selago* and others). However, in species with dimorphic leaves the stomata are frequently found on one face of the leaf (*e.g.*, in *L. complanatum* and *L. volubile*).

Anatomy of root. The transverse section of the root shows the following parts:

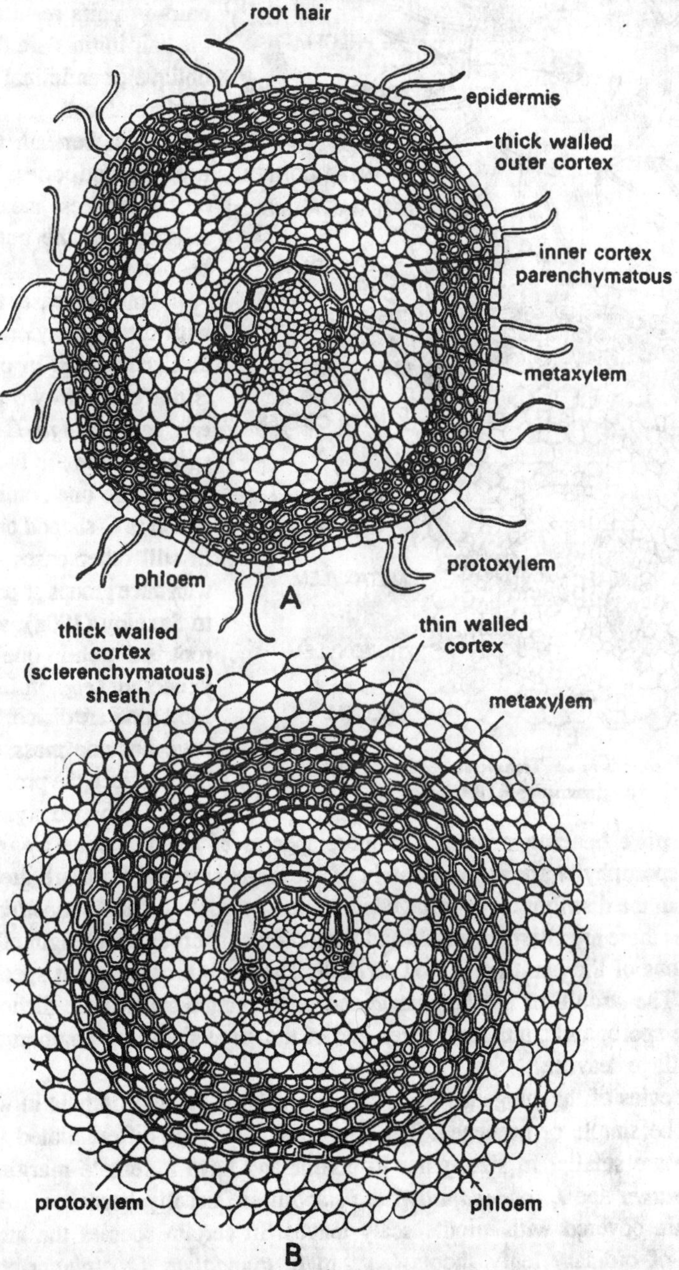

Fig. 4.6. *Lycopodium selago*. A, T.S. of aerial root; B, T.S. of cortical root.

The root is differentiated into epidermis, cortex and stele like other vascular plants. The single layered epidermis gives rise to numerous root hairs. The root hairs are formed to be arranged in pairs. According to Stokey (1907), the occurrence of root hairs in pairs results from the fact that the hair initials are found in twos by the oblique or anticlinal division of a young epidermal cell.

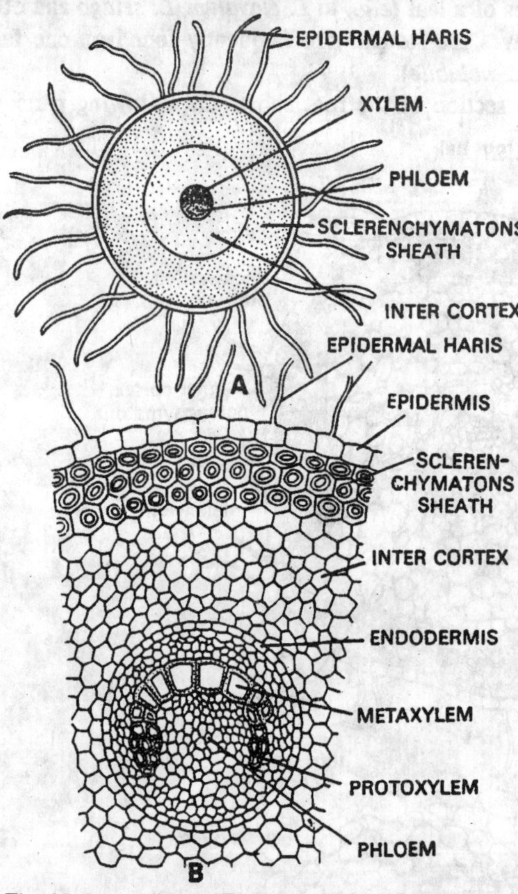

Fig. 4.7. *Lycopodium* sp. Transverse section of root. A, diagrammatic; B, detailed.

Just beneath the epidermis there lies the broad cortex. The cortex is several cells in thickness and often becomes greatly sclerified in the outer half as the root becomes older.

In some cases the stele is **monarch** with the protoxylem in one mass, *e.g.*, in *L. clavatum*. In other cases the stele is **diarch** with two protoxylem masses, *e.g.*, in *L. selago*, *L. reflexum* and other species. Here the two xylem masses are united into one continuous xylem group which is C-shaped or horse shoe shaped. In still other cases the stele is **triarch** with three groups of protoxylem. According to Saxelby (1908), sometimes the same root is diarch in one portion and tetrach in another (*e.g.*, in *L. selago* and others). Most roots are diarch. These roots generally have but one mass of phloem and this lies between the protoxylem points of the C or U shaped xylem.

The spore bearing organs. In most species of the subgenus *Urostachya*, the foliage leaves and sporophylls are approximately of the same size and both are green. However, there is variation in the distribution of foliage leaves and sporophylls. In *Lycopodium selago* and many other species there are alternate sterile and fertile regions throughout length of stem and branches. These portions of the plant are called strobili. The strobili of different species are varied from each other. The strobili of *L. phlegmaria* and other related species are dichotomously branched and here the sporophylls, although green and of the same shape are comparatively much smaller than the foliage leaves.

All species of the subgenus *Rhopalostachya* have the sporophylls in well defined strobili which may be simple or branched. The sporophylls may be differentiated from foliage leaves in that they are smaller in size, paler in colour and have a dentate margin. In some species, *e.g.*, *L. clavatum* and *L. complanatum* the strobili are usually borne on erect elongated stalks which remain covered with minute scaly leaves. In certain species the strobili are sessile at the apices of ordinary leafy shoots (*e.g.*, in *L. inundatum*, *L. alpinum* etc.)

Vegetative propagation of the sporophyte. The vegetative propagation is frequently found in the species of *Lycopodium*. Several methods of vegetative propagation are found in different

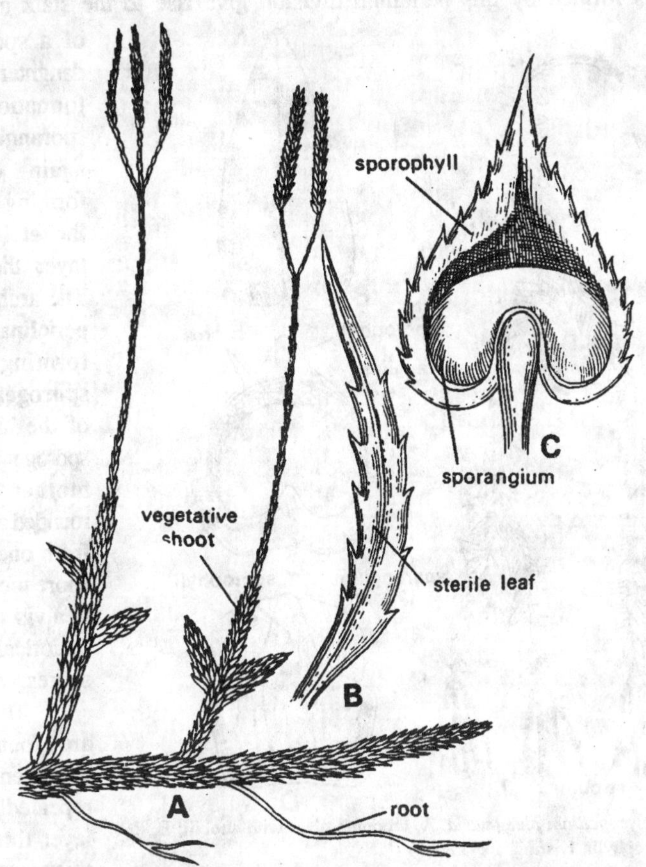

Fig. 4.8. *Lycopodium* **spp. A, shoot bearing strobili; B, sterile leaf; C, sporophyll with sporangium.**

species of *Lycopodium*. The species possessing creeping stems have the apical growth branching and the progressive death of the older portions. According to Holloway (1917) the new plants may be developed from (a) vegetative propagation of the gametophyte, (b) vegetative propagation of the juvenile stage of the sporophyte, (c) gemmae produced from the cortical cells of the root, (d) tubers developed at the apices of roots and (e) bulbils. All above mentioned vegetative bodies are capable to develop into new plants.

The bulbils which are found only in species belonging to subgenus *Urostachya* resemble superficially to the protocormous stage (juvenile stage) of young sporophytes. The bulbils may develop any where on the stems. According to R.W. Smith (1920), these bulbils are interpreted as modified leaves while other workers interpret them as modified lateral branches or modified sporangia.

A sporophyll produces but one sporangium. The sporangium remains adaxial in its position and found near the base of the sporophyll. A mature sporangium is always kidney-shaped. Each sporangium possesses a long and slender or short massive stalk, according to species.

Development of sporangium. The sporangia begin to develop at a time when the sporophyll is composed of embryonic cells. First of all there is a periclinal division of a small group of epidermal cells on the adaxial side of a sporophyll and near its base (Bower, 1894). The inner daughter cells formed by this periclinal division give rise to the stalk and the basal portion of a sporangium. The outer daughter cells contribute to the formation to the bulk of the sporangium. The outer cells again divide periclinally forming an outer layer, the **jacket initials;** and an inner layer, the **archesporial cells.** The archesporial cells divide periclinally and anticlinally forming a massive **sporogenous tissue.** The cells of the last generation of the sporogenous tissue act as **spore mother cells.** They become rounded and are being separated from one another. Now these spore mother cells float about in a viscous liquid and divide meiotically into tetrads of spores.

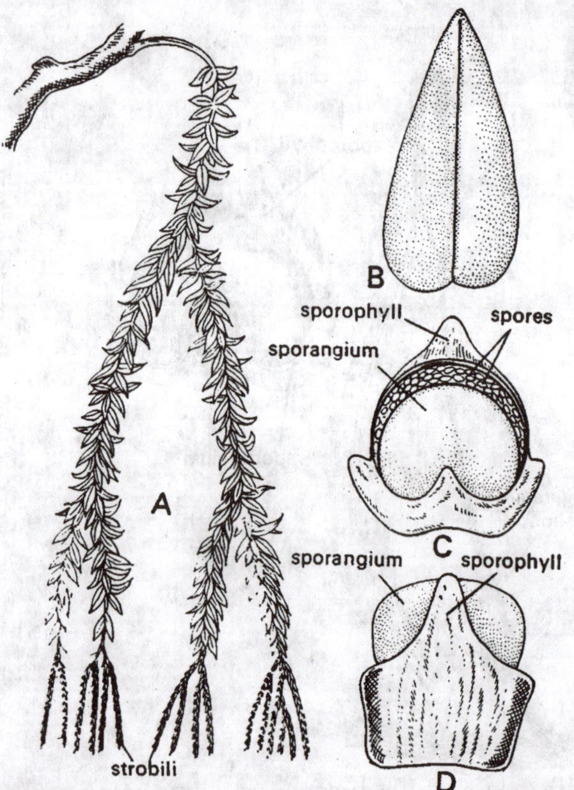

Fig. 4.9. *Lycopodium phlegmaria.* A, drooping plant with strobili; B, a leaf; C-D. sporophylla bearing sporangia.

The jacket initials, which are found external to the sporogenous tissue divide repeatedly forming a jacket layer three of more cells, in thickness. Shortly before the development of the spore mother cells a nutritive **tapetal layer** is formed around the sporogenous tissue. This layer is partly formed from the inner-most layer of jacket cells and partly from sporangial cells found just beneath the sporogenous tissue. As found in most other pteridophytes, in *Lycopodium* there is no disintegration of the tapetum during spore formation.

Dehiscence of sporangium. The mature sporangium is about 2 mm. across and kidney-shaped. On the maturity of the sporangium narrow transverse strip of cells, the **stomium** is formed across the apical portion of the outermost jacket layer. The cell walls of

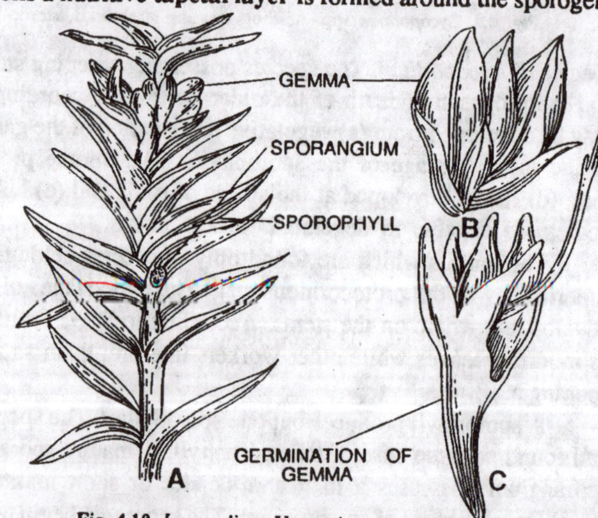

Fig. 4.10. *Lycopodium.* Vegetative reproduction. A, part of stem of *Lycopodium lucidulum* bearing gemmae; B, single gemma; C, germinating gemma of *L. selago.*

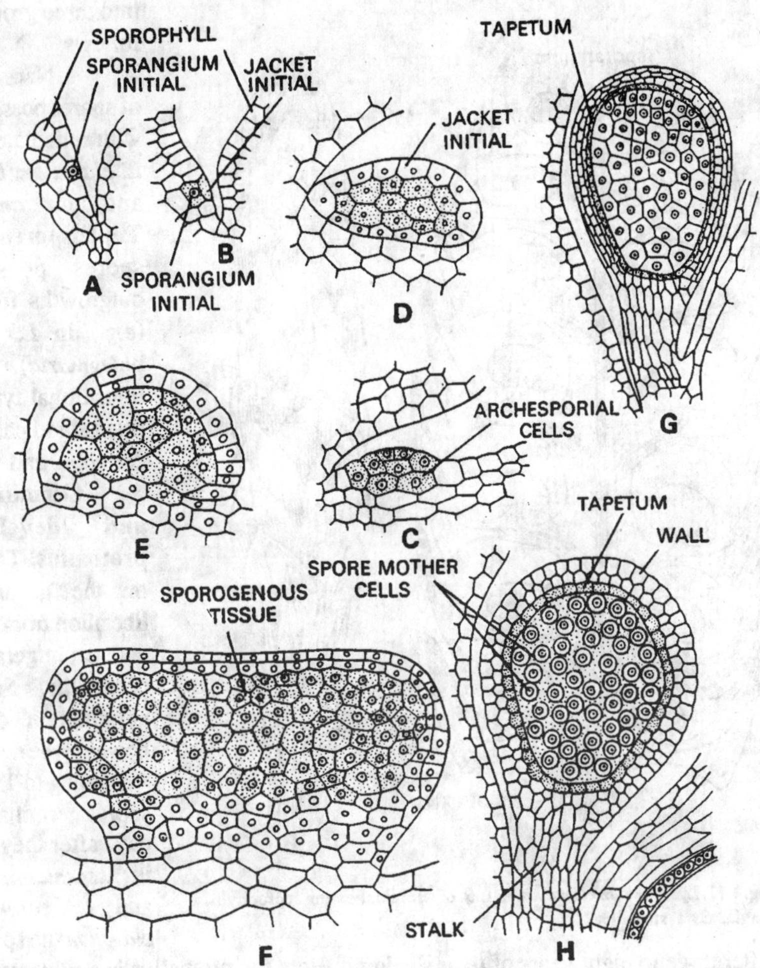

Fig. 4.11. *Lycopodium* **sp. Development of sporangium-A, radial section through base of young sporophyll showing sporangial initial; B, slightly later stage; C-E, successive stages in the development of sporangium; F, sectional view of young sporangium showing sporogenous tissue; G, radial section of a later stage, showing development of tapetum; H, stalked sporangium about to mature containing spore mother cells.**

the stomial portion become thickened and may easily be differentiated from those of other cells present in the jacket layer of the sporangium. The mature sporangium ruptures by a transverse slit at its apex along the line of the stomium. The sporangium divides into two valves which remain united at the base and dehiscing the yellow spores.

THE GAMETOPHYTE

The spore. The small spores (about .03 to .05 mm. diameter) are uniform in size and shape, *i., e.,* **homosporous.** This way *Lycopodium* is similar to most of the ferns and *Equisetum.* The spores are round or tetrahedral in shape. They are very light possessing thin exine formed of concentric lamellae, and a granular intine. Lustner (1898) classified the spores of *Lycopodium* into three groups. They are as follows:

1. Netzsporen. This type of spores possess an outer wall with reticulate ridges (*e.g.*, in *L. clavatum, L. complanatum* and *L. annotinum*). **2. Tupfelsporen.** This type of spores posses knob-like outgrowths from the surface (*e.g.*, in *L. selago* and *L. phlegmaria*). **3.** This is a transitional type possessing a fine ridged pattern (*e.g.*, in *L. cernuum* and *L. inundatum*).

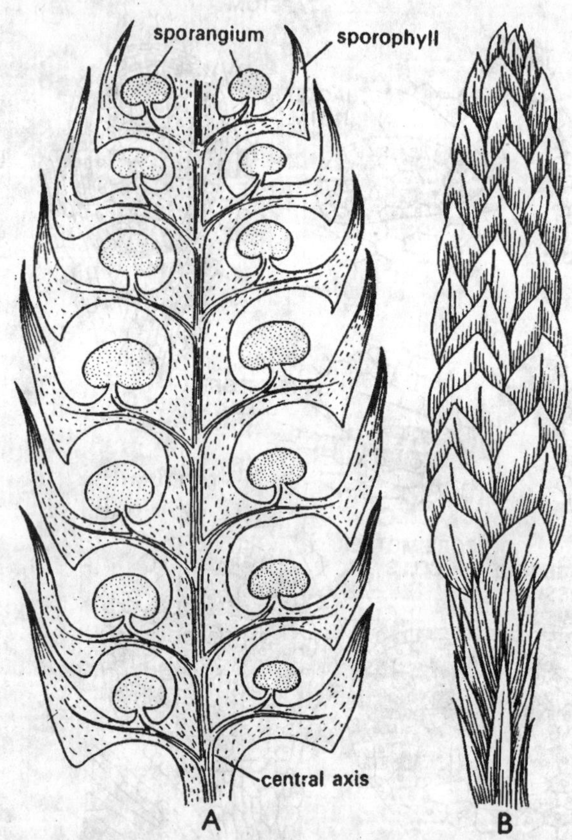

Fig. 4.12. *Lycopodium clavatum*. A, L.S. of strobilus bearing sporophylls and sporangia; B, a strobilus.

Germination of spore and development of prothallus. The spores settle on the ground after their liberation from the sporangium and each germinates into a prothallus. Sometimes the germination of spore may, however, be delayed for years. According to Treub (1888), the spores germinate within a few days after they are shed (*e.g.*, in *L.cernuum, L. inundatum* and *L. salakense*). In *L. clavatum* the spore germination begins after three to eight years of their shedding. Here the prothallus is a subterranean structure, solid and tuberous which takes from six to fifteen years to mature. The prothallus of *L. clavatum* is colourless and saprophytic and contains a mycorrhizal fungus.

According to Bruchmann (1910), the first cell division of a germinating spore takes place before the outer spore wall (exine) is ruptured and results in two hemispherical daughter cells of approximately equal size. The lower or inner daughter cell also known as **basal cell** cuts off a small lens shaped rudimentary **rhizoidal cell** while the outer daughter cell divides twice successively producing an **apical cell** with two cutting faces. In the species with subterranean colourless prothalli a long period of rest, about one year is required between the five-celled stage and the mature prothallus. Upto four to five celled stage the early growth of the prothallus takes its nourishment from the reserve material contained in the mature spores. Thereafter 4-6 celled gametophytes may have an entrance of a symbiotic phycomycetous fungus into the basal cell. If there is no entrance of the fungus there is not further development of the gametophyte. After the establishment of the endophytic mycorrhiza within the cells of the young prothallus, the apical cell cuts off some half dozen segments and then becoming replaced by a group of meristematic cells. The segments thus cut off by the apical cell divide periclinally, and the outer cells thus formed become infected with the symbiotic fungus in the same way as does the basal cell. The apical group of meristematic cells, which are formed early in the development of a gametophyte produces the major portion of the adult prothallus.

There is great diversity in form and structure of mature gametophytes in the different species of *Lycopodium*. There are three main types of mature prothalli. They are as follows:

The first type of prothallus is mainly found in *L. cernuum, L. obscurum* and *L. inundatum*. Here the prothallus is inconspicuous and reaches to the maximum height of 2 to 3 mm. and usually 1 to 2 mm. in diameter. The prothallus grows on the surface of the ground. It is an upright fleshy structure whose lower portion remains embedded in the soil. On its upper exposed part the prothallus bears green lobes with meristematic plates. The sex organs are found between the green expanded lobes. The prothallus can manufacture its own food, though an endophytic mycorrhiza is always present. Rhizoids are also produced from the prothallus.

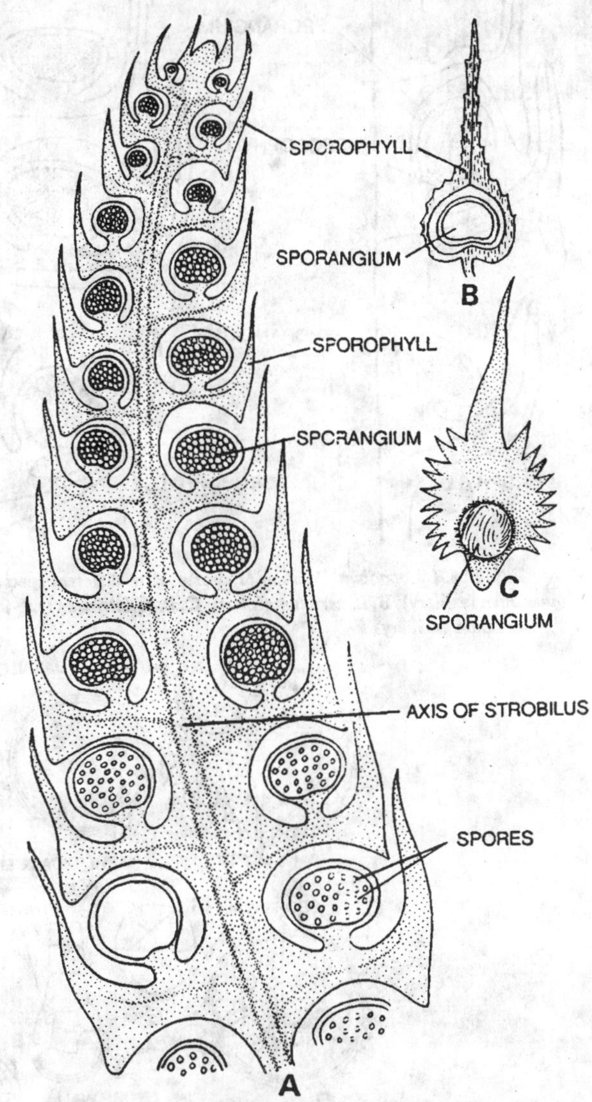

Fig. 4.13. *Lycopodium* spp. A, longitudinal section of the terminal portion of strobilus; B, sporophyll with sporangium of *L. clavatum*; C, sporophyll with sporangium of *L. cernuum*.

The second type of prothallus is found in the species, *e.g., L. clavatum, L. complanatum, L. obscurum* and *L. annotinum*. The prothalli are of various shades. They are brownish yellowish or colourless. They are saprophytic and found underground at various depths from one to eight centimetres. Much food material remains deposited in these tuberous prothalli. In *L. clavatum* the prothallus is much convoluted structure 1 to 2 cm. long with a pointed basal part. The surface is covered with an epidermis. Inside the epidermal layer there is a cortex of several layers. Just beneath the cortex there lies a palisade tissue composed of a single layer of elongated cells. These outer layers contain the endophytic fungus. This region is followed by a central tissue of hexagonal cells which store up food material. Long unicellular **rhizoids** are formed on the under surface of the prothallus. The sex organs are borne on the top surface of the prothallus. It is called the generative tissue. The cells of this tissue donot contain any food reserves and they remain meristematic for some time. The antheridia develop in the central portion and the archegonia towards the rim. The antheridia develop first and the archegonia appear later. The development is centripetal.

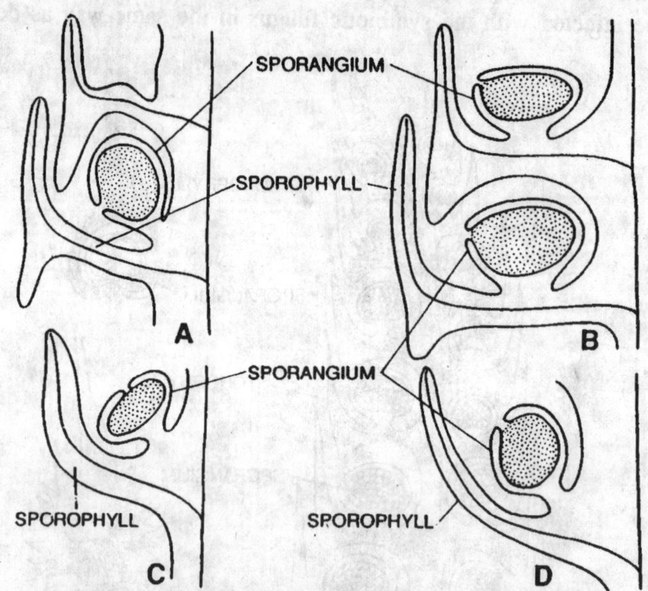

Fig. 4.14. *Lycopodium*. Position of sporangia in different species; A, *L. inundatum* (axillary); B, *L. cernuum* (foliar); C, *L. squarrosum* (sub-foliar); D, *L. lucidulum* (axillary).

The third type of prothallus is found in *L. phlegmaria* and other epiphytic

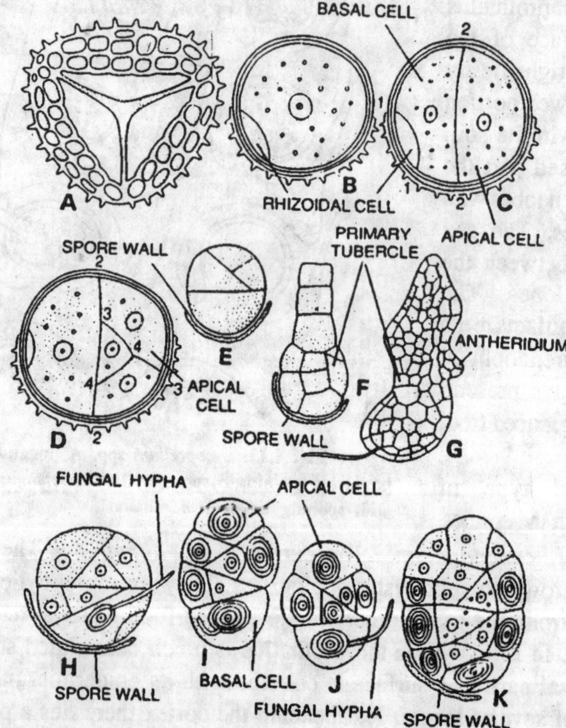

Fig. 4.15. *Lycopodium* spp. Spore germination and development of gametophyte. A, spore of *L. clavatum*; B-D, early stages in the development of gametophyte of *L. clavatum*; E-G, early stages in the development of gametophyte of *L. cernuum*; H-K, gametophytes of *Lycopodium* showing origin of mycorrhizal association.

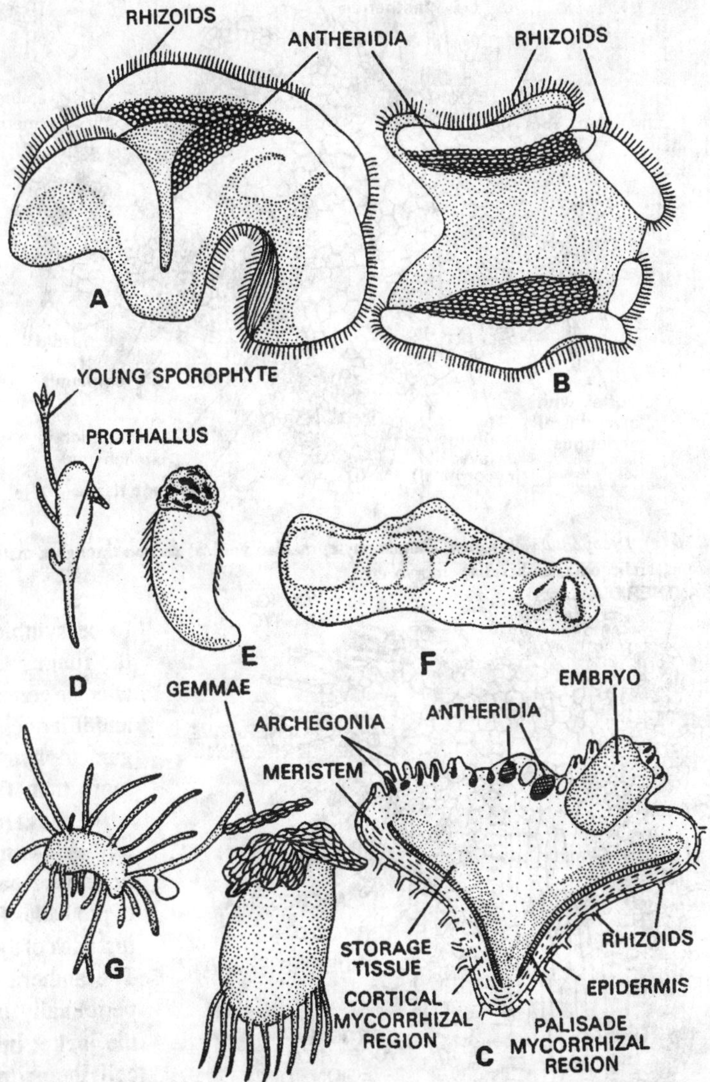

Fig. 4.16. *Lycopodium*. Different shapes of prothalli in various species. A-B, side and top aspects of mature prothallus of *Lycopodium clavatum* showing numerous antheridia and rhizoids; C, vertical section through the prothallus of *L. clavatum* showing mycorrhizal zone and reproductive organs; D, prothallus of *L. selago* with young sporophyte; E, *L.complanatum*; F, *L. annotinum*; G, *L. phlegmaria*; H, *cernuum*.

species. These saprophytic colourless prothalli develop beneath the surface of the humus found on the tree trunks. The prothallus consists of an irregularly shaped tuberous body from which colourless, cylindrical branches are given out. The sex organs are borne on the upper surface of the enlarged branches. These branches of the prothallus are interspersed with slender paraphyses.

Nutrition of prothallus. The mode of nutrition of the prothallus is remarkable. The cells of the cortical region and the palisade tissue remain filled with endophytic fungus, and these form a mycorrhizal association with the cells of the prothallus. This association is supposed

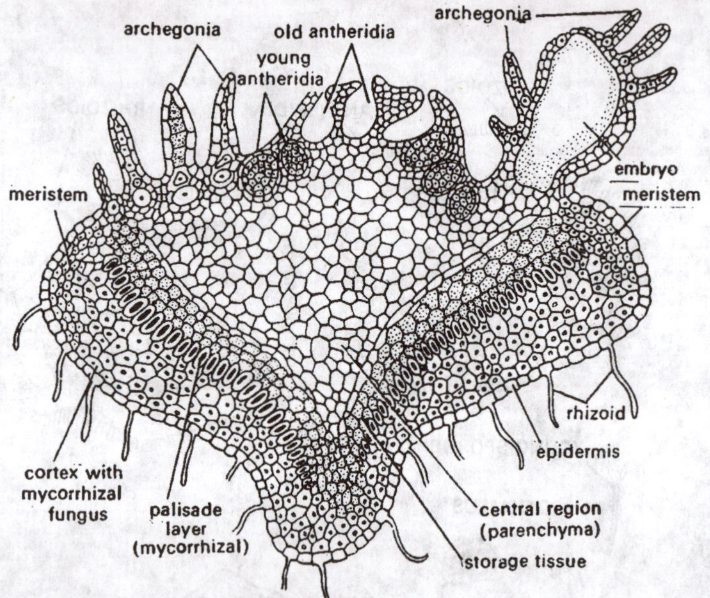

Fig. 4.16 (a). *Lycopodium clavatum.* Gametophyte; median vertical section through a mature prothallus showing mycorrhizal zone, archegonia, antheridia, embryo etc.

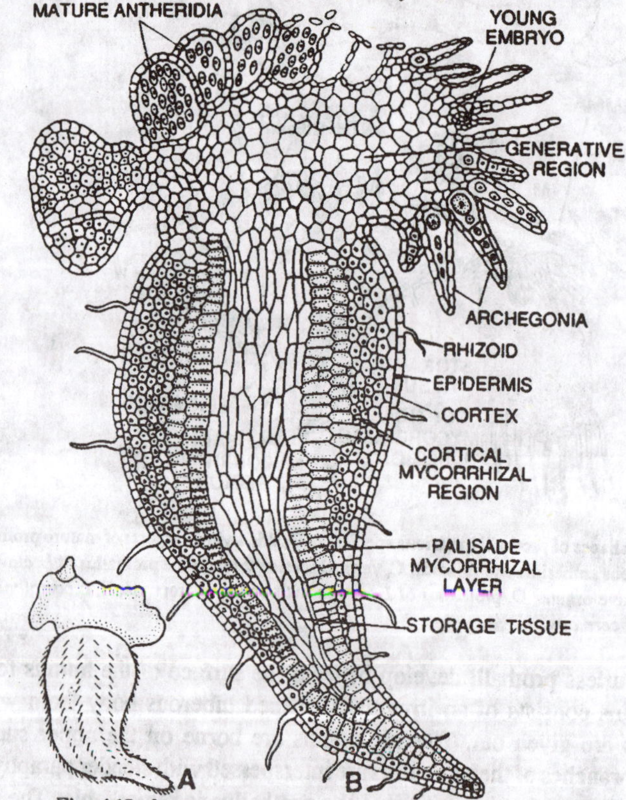

Fig. 4.17. *Lycopodium complanatum.* L.S. of a mature prothallus showing details of internal structure and the position of male (antheridia) and female (archegonia) sex organs.

to be symbiotic. The entry of the fungus takes place shortly after the germination of the spore, and if it does not take place the gametophyte never develops more than five cells.

Development of antheridium. The antheridium develops from a superficial cell of prothallus found to be situated just back of the apical meristem. The antheridial initial divides periclinally into an outer cell, the **jacket initial** and an inner cell, the **primary androgonial initial.** The primary androgonial cell divides repeatedly forming a large mass of androgonial cells that lie embedded within the prothallus. Simultaneously the jacket initial divides and redivides anticlinally forming a jacket layer around the mass of androgonial cells, one cell in thickness. The jacket layer in *L. phlegmaria* is two layered towards the periphery. The

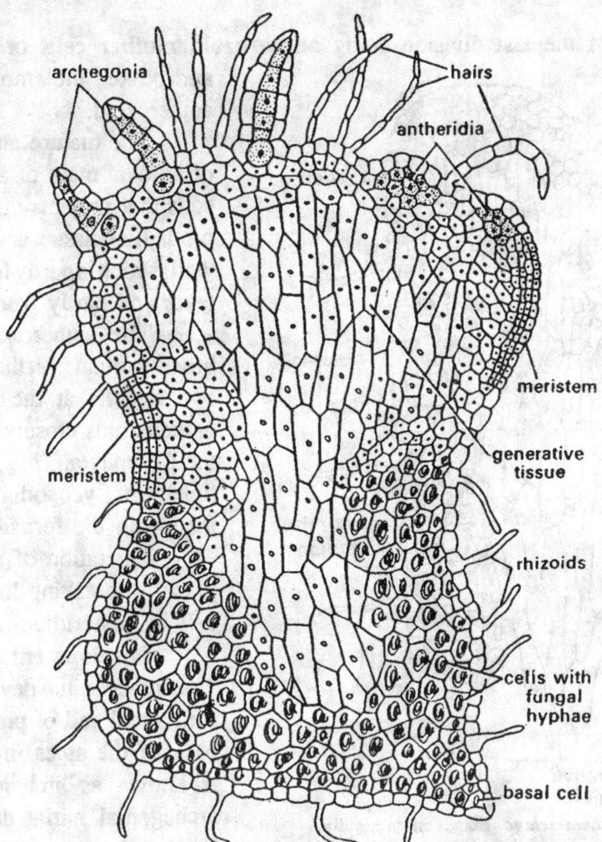

Fig. 4.18. *Lycopodium selago*. A mature prothallus (gametophyte) in longitudinal section showing details of internal structure and reproductive organs (*i.e.*, antheridia and archegonia).

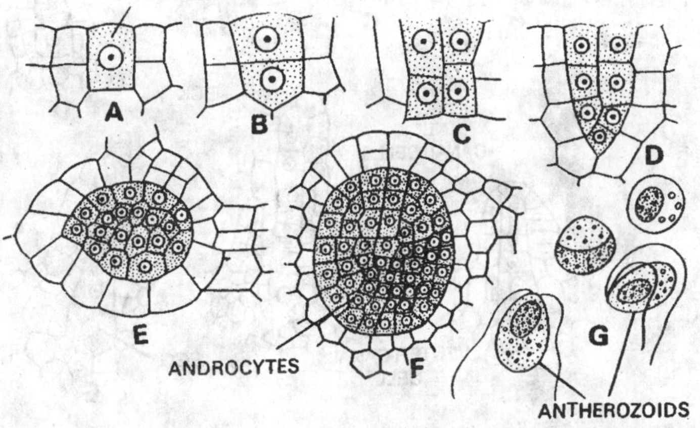

Fig. 4.19. *Lycopodium* sp. Development of antheridium. A-E, successive stages in the early development of antheridium; F, young antheridium with androcytes; G, androcytes and mature biciliate antherozoids.

androgonial cells of the last division being **antherozoid mother cells** or **androcytes.** Each androcyte metamorphoses into an antherozoid.

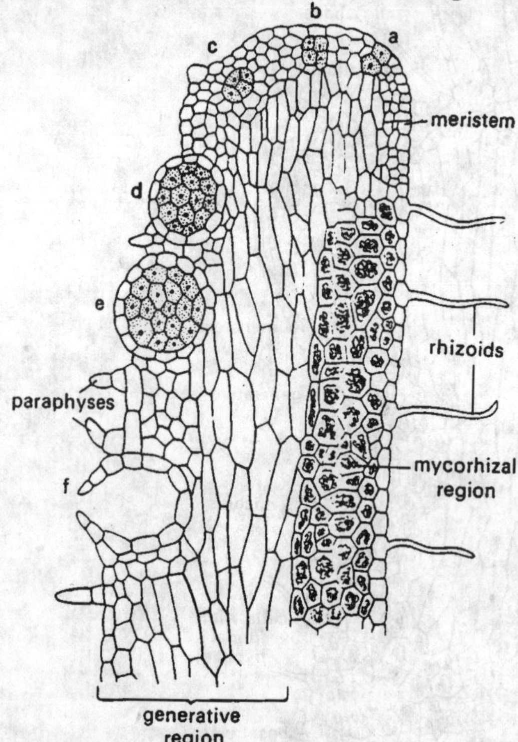

Fig. 4.20. *Lycopodium selago.* Stages in antheridial development on prothallus.

The mature antheridium consists of an oval mass of antherozoid mother cells that projects out slightly from the prothallial tissue and surrounded by a wall, which is partly formed by the jacket layer and partly from prothallial cells.

The antherozoids are fusiform, broadly rounded at the posterior end with two flagella at the anterior end. The antherozoids closely resemble those of the Bryophyta. This suggests the closer linkage of Lycopodiales with the mosses than with the ferns.

Liberation of antherozoids takes place by a breaking down of the opercular cell in antheridium.

Development of archegonium. The archegonium develops from a single superficial cell of prothallus found just back of the apical meristem. This cell is known as archegonial initial. The archegonial initial divides periclinally into a **primary cover cell** and a **central**

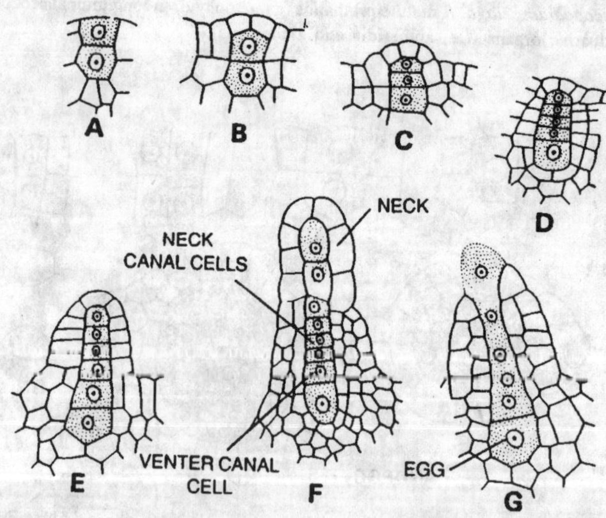

Fig. 4.21. *Lycopodium.* Development of archegonium. A, division of archegonial initial into primary neck cell and inner cell; B, vertical division of primary neck cell and transverse division of inner cell forming basal and central cells; C, central cell divides to form primary canal cell, and primary ventral cell; D-E, later stages showing basal cell, primary ventral cell and four neck canal cells; F, nearly mature archegonium with egg, ventral canal cell and neck canal cells; G, mature archegonium with egg and open neck canal.

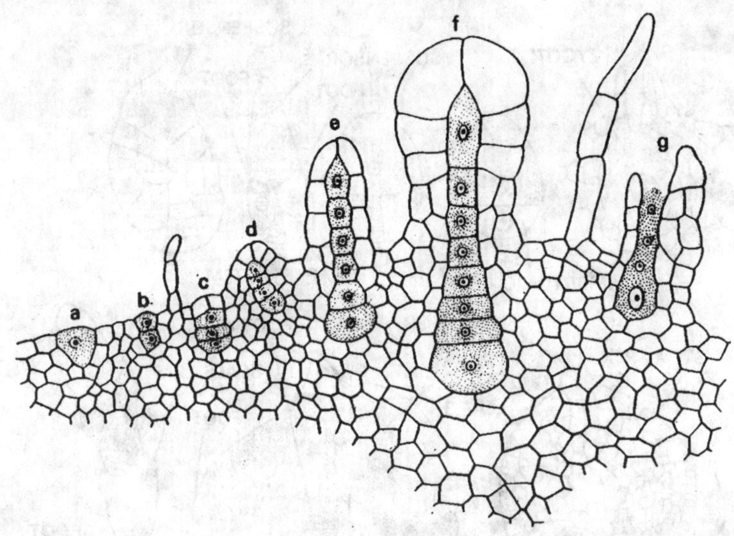

Fig. 4.22. *Lycopodium selago.* **Developmental stages (a-g) of archegonia on prothallus.**

cell (Lang, 1899; Bruchmann, 1910; Spessard, 1922). The primary cover cell divides anticlinally forming four **neck initials**, which undergo transverse division to form a long and straight neck, 3 to 4 cells in height and composed of four vertical rows of cells. The central cells divides periclinally into a **primary canal cell** and a **primary venter cell**. The former cell gives rise to about six **neck canal cells**, while the latter functions directly as egg or it divides to form a venter canal cell and an egg.

The portion of the mature archegonium derived from the primary cover cell, together with the neck canal cells it ensheaths, projects vertically above the prothallus. The remaining canal cells and the oosphere (egg) lie embedded within the prothallus. The cells lateral to the egg and lower portion of the row of neck canal cells are not portions of the archegonium as they are formed by division and redivision of prothallial cells immediately adjacent to the archegonial initial. As found in the archegonia of bryophytes, all cells of the axial row except the egg disintegrate prior to the maturation of the archegonium, the apex of the neck of the archegonium opens by the spreading apart and partial disintegration of the uppermost neck cells.

The antherozoids make their way down the neck of the archegonium and one fuses with the oosphere thus effecting the fertilization. The oospore (2x) is formed.

Development of embryo. According to Bruchmann (1910) the first division of a zygote is always transverse and separates an upper cell, the **suspensor** which doesnot divide again. The lower cell also known as **embryonic cell** divides thrice. The first division of embryonic cell is vertical and this is followed by a vertical division of two daughter cells thus formed. Transverse division of each of these cells results in an eight-celled embryo. The cells of this octant are arranged in two superimposed tiers of four cells each. The tier of cells adjacent the suspensor develops into the **foot** of the young sporophyte. The foot remains as an intraprothallial haustorium which derives food from the prothallus until the embryo at length becomes independent. The distal tier of cells develops into the remaining portions of the sporophyte, *viz.*, stem, leaf and root. According to Bruchmann (1910), the two cells facing the apical meristem of the prothallus develop into the **stem** and the two cells facing away from the meristem develop into the **cotyledon**

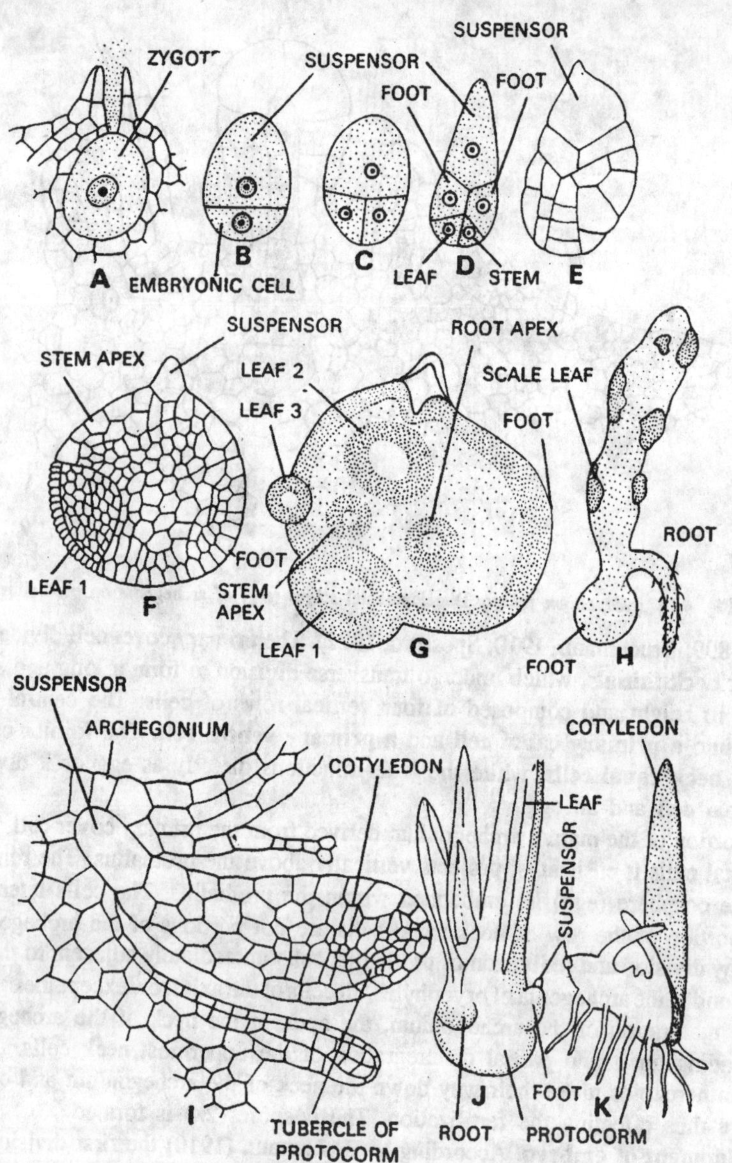

Fig. 4.23. *Lycopodium* sp. Development of embryo. A, zygote; B, first division of zygote; C, division of embryonic cell into two cells; D-F, further development in the embryo; G, embryo from above, showing origin of apex; H, young plant with foot and stem with reduced leaves; I, young embryo of *L. cernuum* emerging from the prothallus; J-K, embryo of *L. cernuum* showing protocorm.

and the primary root. The young sporophyte becomes many celled structure prior to the differentiation of the cotyledon and the embryonic stem. The primary root becomes evident even later and at one side of the region where cotyledon and foot adjoin each other. Until this time the embryo remains embedded within the prothallus and grows downward and forward toward the meristem of the prothallus. Thereafter the growing stem and cotyledon burst through the apical region of the gametophyte. Now the cotyledon and the stem start to grow upward into the air.

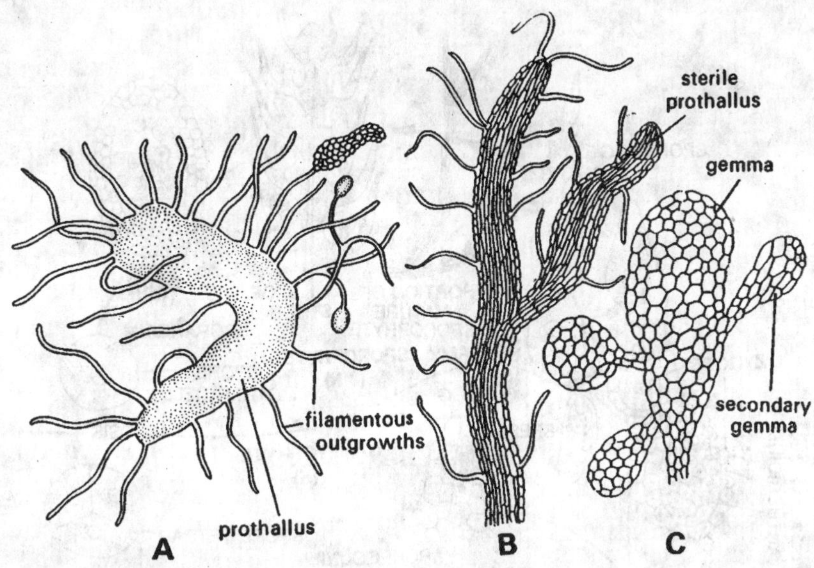

Fig. 4.24. *Lycopodium phlegmaria.* A, mature prothallus covered with filamentous outgrowths and gemmae; B, a sterile prothallus; C, a gemma with secondary gemmae.

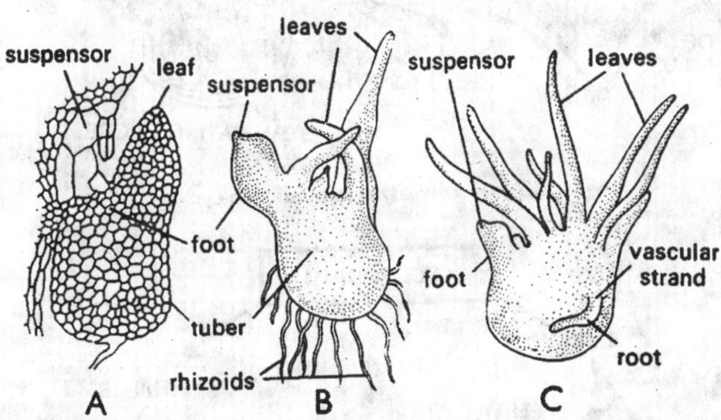

Fig. 4.25. *Lycopodium cernuum.* A, young embryo developing from prothallus; B, embryo differentiated into leaf, foot and tuber (protocorm); C, embryo with three leaves.

Morphology of protocorm. In certain cases the distal four cells of the octant-stage form a massive spherical structure, the **protocorm.** The protocorm becomes green and develops through the prothallus. The rhizoids develop on the lower surface of the protocorm, the upper surface of the protocorm gives rise a few to many erect, conical outgrowths known as **protophylls.** These structures are leaf-like in function and bear stomata in their epidermal layer. After the formation of many protophylls, the protocorm is then differentiated a meristematic region which develops into the stem of the adult plant.

According to Treub (1890), the protocorm is a structure of great antiquity and the phylogenetic precursor of the stem. However, Bower (1908), Holloway (1910) hold the opinion that is a late phylogenetic specialization and represents a 'gouty interlude' in the early ontogeny of the sporophyte.

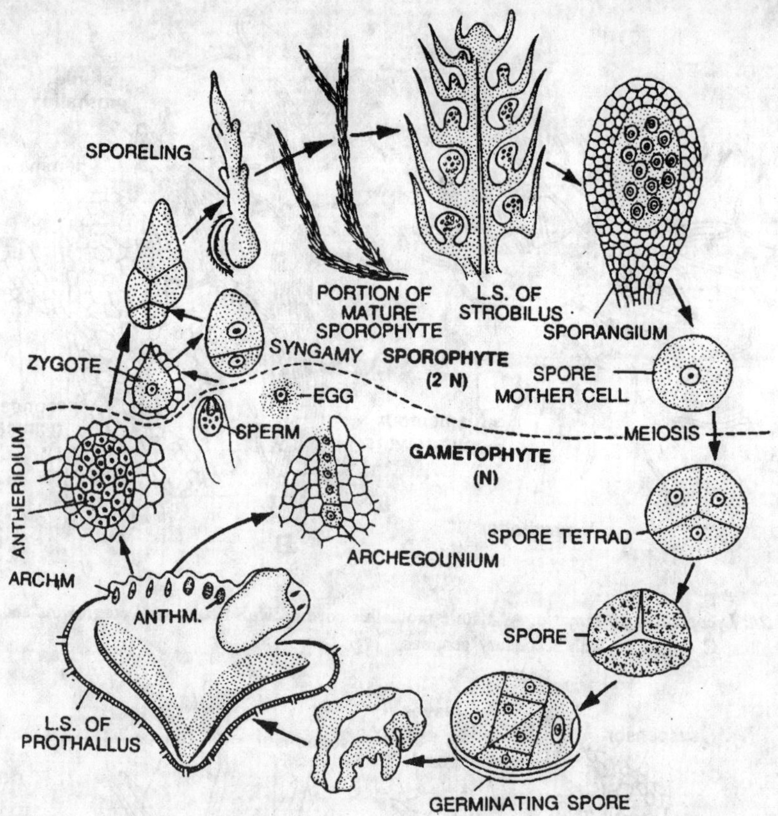

Fig. 4.26. *Lycopodium*. Diagrammatic life-cycle.

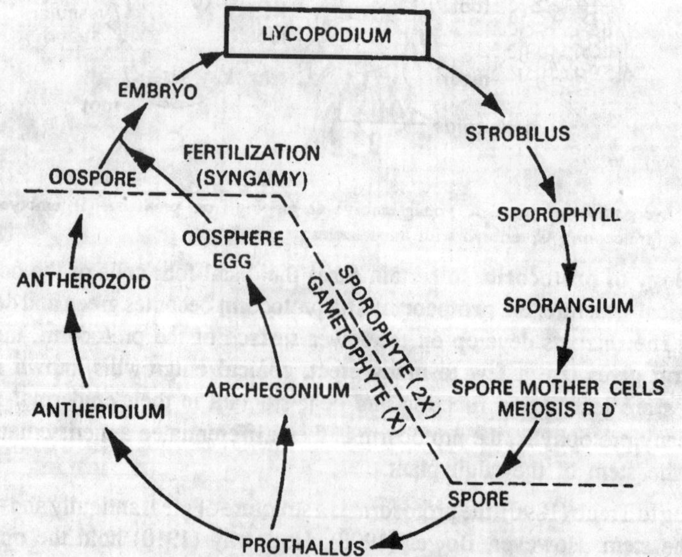

Fig. 4.27. *Lycopodium* sp. Graphic life-cycle.

5

Division-Lycophyta
Class-Ligulopsida

Order-Selaginellales

Characteristic features. The sporophytes possess herbaceous stems which may be creeping or erect. The foliage leaves are small (microphyllous). The leaves may be all alike or dimorphic. A ligule is present at the base of each leaf. The roots are borne at the distal ends of the specialized branches, the rhizophores of the stem. They are heterosporous. The micro and megaspores develop within micro and megasporangia respectively. The sporophylls are aggregated in strobili at the apices of the branches. The mega and microgametophytes are conspicuously different in size. They are dioecious. The antherozoids are biflagellate.

Genus **SELAGINELLA**

Systematic position. Pteridophyta
Division. Lycophyta
Class. Ligulopsida
Order. Selaginellales
Family. Selaginellaceae
Genus. *Selaginella.*

Selaginella is world-wide in distribution and comprises of about 700 species. Majority of the species are tropical and grow in rain forests. Some of the species are found in temperate regions. Some species are xerophytic which grow upon barren rocks and dry soils. The xerophytic species may survive for months together in the conditions of desiccation. Some of the species are moist loving and found in the shady places on the hills. In India, majority of the species of *Selaginella* are common in the Himalayas and the hills of South India. Some of the xerophytic species, like *S. lepidophylla* and *S. rupestris* remain folded or ball-like in dry conditions, but as soon as they come in contact of water they become unfolded and normal, such plants are called 'resurrection plants'. Some of the species are epiphytic, and found growing upon other plants. *S. oregana* grows upon the trunks and branches of the trees covered with moss plants. About 58 species of *Selaginella* have been reported from different parts of India. Some of the most important Indian species are-*S. kraussiana, S. chrysocaulos, S. chrysrorhizos, S. sanguinolenta, S. pallidissima* etc., and many others.

External Morphology

Habit. The plant itself is a sporophyte. The size and the shape of the sporophyte varies from species to species. Some of the species are moss-like, *e.g., S. spinulosa* and very small in size, whereas, on the other hand they attain the length of twenty metres or so, *e.g., S. willdenovii.* Some of the species possess prostrate creeping system horizontally on the surface of the soil, *e.g., S. kraussiana.* Some species possess sub-erect, system, *e.g., S. trachyphylla,* some are scandent, *e.g., S. willdenovii* and some are even erect, *e.g., S. erythropus.* Some species are climbing, and climb with the help of rhizophores, *e.g., S. alligans.* A very few species of *Selaginella* are annual, *e.g., S. pygmaea* and *S. gracillima.* Most of the species are perennials.

On the basis of the general structure of the stem and arrangement of leaves upon it, the species have been divided into two groups. Hieronymus (1900) has placed these two groups of the species under the sub-genera, *Homoeophyllum* and *Heterophyllum*.

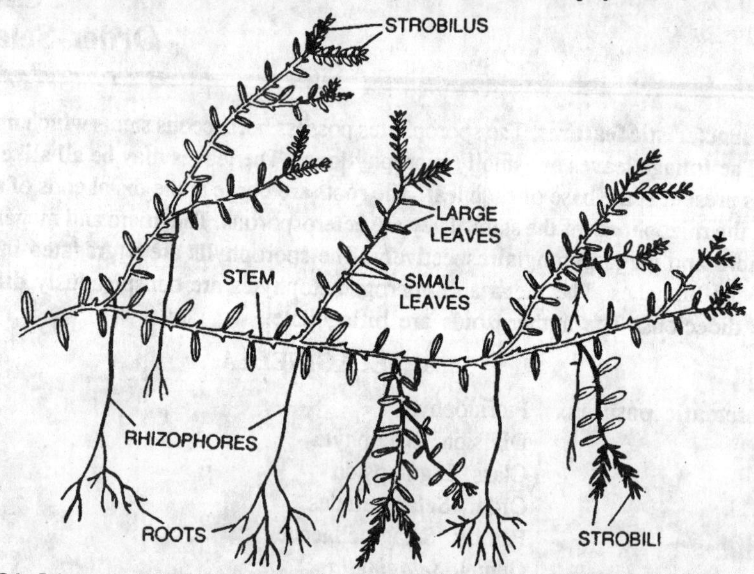

Fig. 5.1. *Selaginella kraussiana*. A, sporophyte with strobili, rhizophores roots,ligules, leaves etc.

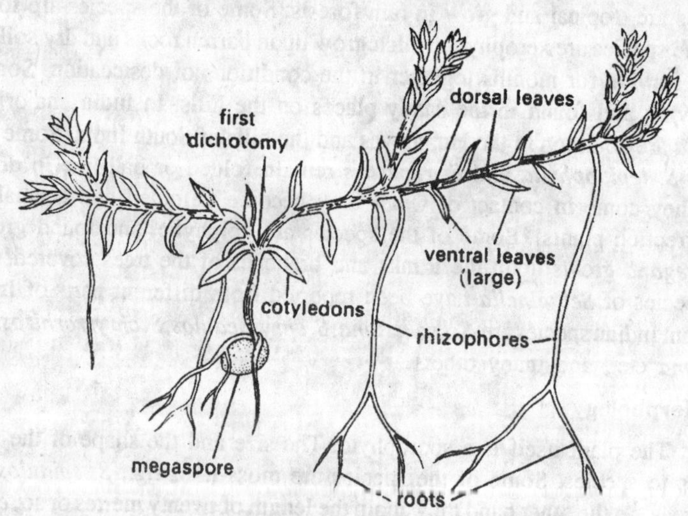

Fig. 5.2. *Selaginella kraussiana*. A, sporophyte developing from a megaspore with strobili, rhizophores, roots, leaves etc.

The sub-genus *Homoeophyllum* comprises of about 50 or more species, *e.g., S. pygmaea, S. spinulosa* and *S. rupestris* etc. These species possess erect stems and the spirally arranged leaves of the same size covering the stems.

The sub-genus *Heterophyllum* comprises of majority of the species, *e.g., S. kraussiana, S. lepidophylla, S. martensii* and many others. These species possess prostrate stems, short erect branches and the dimorphic leaves.

Stem and leaves. In each case the stem is covered over with four rows of leaves. The leaves may be of same size or of different size as the case may be. As mentioned above the species of sub-genus *Homoeophyllum* possess, the erect stems and the spirally arranged leaves of the same size covering the stems. The species belonging to the sub-genus *Heterophyllum* possess prostrate stems, short erect branches and the different sized (dimorphic) leaves. The prostrate stem is dorsiventral. The leaves are arranged in pairs. Each pair consists of two different sized leaves. The larger leaf of each pair is inserted on the ventral side of the axis. The pairs of the leaves are arranged in such a way, that small leaf alternates the small leaf and the large

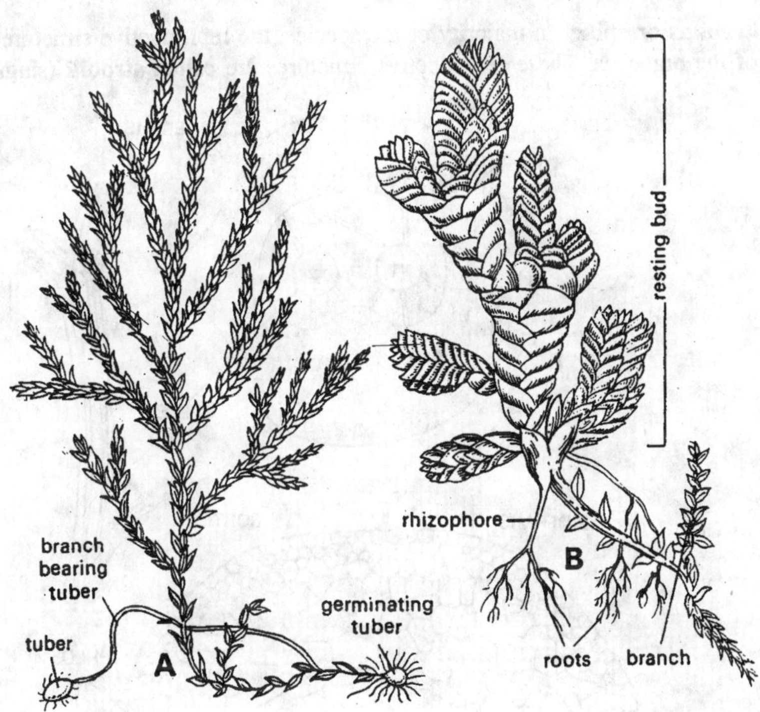

Fig. 5.3. ***Selaginella* spp. Vegetative propagation. A, plant developed from a tuber in *S. chrysorrhizos*; B, portion of *S. chrysocaulos* showing a resting bud.**

leaf alternates the large leaf, *Selaginella kraussiana* belongs to sub-genus *Heterophyllum* ano possesses all the typical characters of the sub-genus. The leaves are quite small in size, somewhat triangular or lanceolate. They may be thin in moist loving species and thick in xerophytic species.

In every case the leaves are ligulate. In all the species each leaf bears a thin, membranous small ligule at its base. The ligule is found on the upper or ventral surface. The ligules vary in their shape and size from species to species.

The structure and the function of the ligule is as follows:

The mature ligule is thin, membranous, tongue like to the fanshaped. There is a cup-shaped sheath at its base. Just adjacent to the sheath, there is a hemispherical mass of thin-walled cells, known as **glossopodium.** The function of the ligule is controversial. It has been thought by many workers, that they are waterabsorbing or water secreting organs and they however, prevent the undue drying of the stem apex and the leaves by keeping them moist.

Rhizophores and roots. In many species, the prostrate axis bears an elongate, colourless, leafless, cylindrical, downwardly growing structure known as **rhizophore.** The rhizophores arise from the axis of the ramifications. Usually a single rhizophore comes out downward from the ramification of the stem, *e.g., S.kraussiana.* But in some cases two rhizophores are given out from each branching of the stem, *e.g., S. martensii.* In such cases the rhizophore coming from ventral side develops fully whereas from dorsal side develops only in the form of a small outgrowth. At the terminal end of the rhizophore, a tuft of adventitious roots develops, *e.g., S. kraussiana.* In *S. lepidophylla,* a knot like structure develops at the end of the rhizophore and from this knot like structure the adventitious roots are given out. The adventitious roots are thin, slender and delicate. They are dichotomously branched. The primary root perishes after a very short duration.

Strobilus or spike. In majority of the species, the reproductive structures develop at the apices of the branches. These reproductive structures are called **strobili** (singular = strobilus)

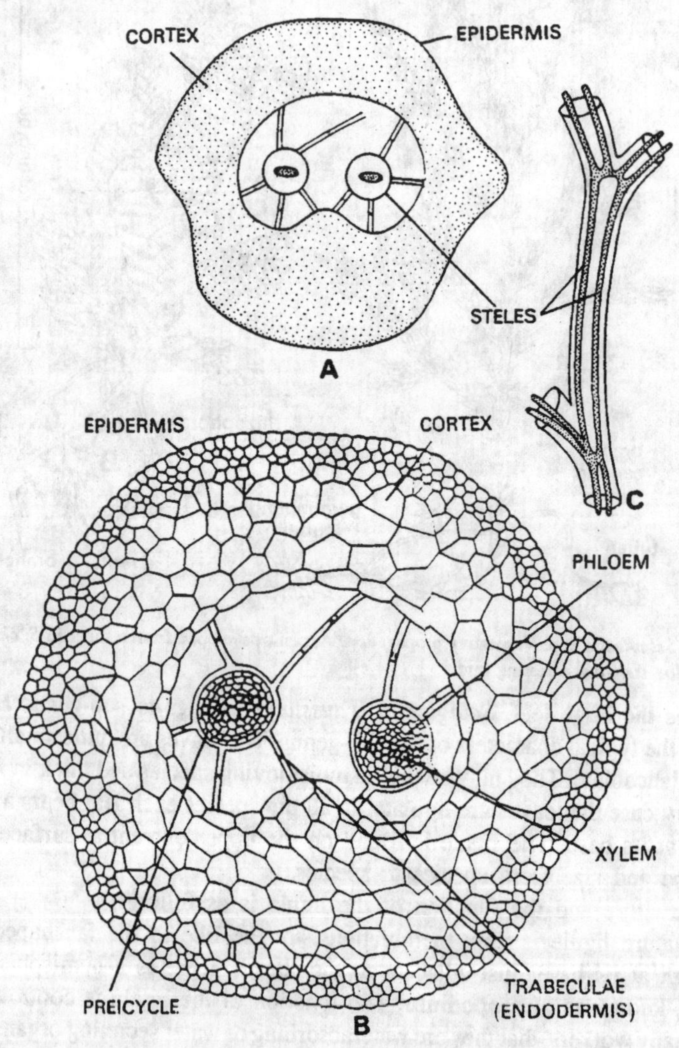

Fig. 5.4. *Sellaginella kraussiana.* Anatomy of stem. A, transverse section of stem (diagrammatic); B, transverse section of stem (detailed); C, the stem in longitudinal sectional view to show the position of steles.

or spikes. The strobilus is a compact structure of sporophylls. The sporophylls are quite similar in structure to that of ordinary vegetative leaves. Each sporophyll bears a ligule at its base. Each sporophyll bears a single stalked sporangium. The sporangia are of two types, *i.e.*, **microsporangia** and **megasporangia.** The sporophyll which bears microsporangium is microsporophyll, and which bears megasporangium is megasporophyll. Both micro and megasporophylls may be found in same strobilus or in different strobili. The sporophylls are always spirally arranged on the axis of the strobilus but in a very much condensed form. In majority of the species the micro and megasporophylls borne in the same spike, *e.g., S. kraussiana.* According to Mitchell (1910), in certain species, *e.g., S. gracilis* and *S. atroviridis* the strobili bear only one type of sporophylls. When the strobilus bears both kinds of sporophylls, the megasporophylls may be found in the lower portion of the strobilus and the microsporophylls in the upper portion of the strobilar axis, *e.g., S. spinosa* and *S. rupestris.* In certain other species the mega and microsporophylls may be arranged in different vertical rows in the strobilus.

Internal Morphology

Stem. The anatomy of the stem may be studied by cutting it in transverse sections. Leaving aside the stems of a few species, the rest of the species possess the radially elongaed endodermal cells, having well developed spaces among them known as **trabeculae.** The 'Casparian strips', characteristic of the endodermis are found on these stretched endodermal cells so-called trabeculae.

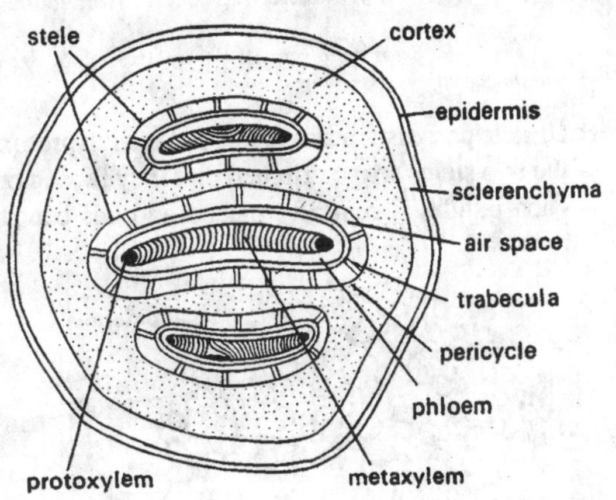

Fig. 5.5. T.S. of stem of *Selaginella willdenovii* showing three protosteles (diagrammatic).

The epidermis is the outermost superficial layer of the stem which consists of prosenchymatous cutinized cells.

Just beneath the epidermis, the cortical region is found. In some species the cortex is composed of thin walled parenchymatous cells having no intercellular species, *e.g., S. kraussiana.* Majority of the species which possess rigid stems have compact sclerenchyma in the hypodermal region beneath the epidermis. The most of the part of the cortex is found to be sclerenchymatous in xerophytic species such as *S. lepidophylla* and *S. rupestris.*

Leaving aside the stems of a few species such as *S. lepidophylla* and *S. rupestris*, the stems of the rest of the species possess stretched endodermal cells (trabeculae) having large intercellular spaces among them. This large intercellular space or lacuna is formed in the stem because of the rapid enlargement of the innermost cortical cells in comparison to the cells of vascular tissue. This way the endodermal cells become stretched and a lacuna or cavity is resulted.

The organization of the stele varies from species to species and sometimes even in the same species. This may range from a simple protostele to a complex polycyclic siphonostele. According to Wardlaw (1925), the stem may possess 2, 3, 4 or sometimes even upto 16 steles running through the intervening region between two successive branchings of the stem. Each stele is surrounded by its trabecular endodermis. In the stem of *S. kraussiana*, two steles are found running through the region between two successive branchings. Both the steles are protostelic.

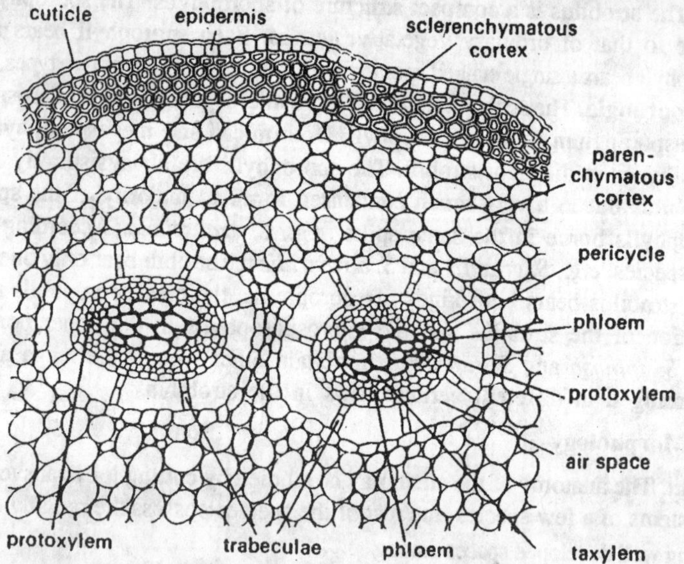

Fig. 5.6. *Selaginella* spp. T.S. of stem of *S. kraussiana* showing distelic condition. Note the trabeculae (endodermis).

Each stele possesses a single mass of exarch protoxylem in it. The masses of the exarch protoxylem of the two steles are so oriented that they face in opposite directions to each other. The metaxylem in each bundle is centrally placed and lies centripetal to the protoxylem. The centrally located

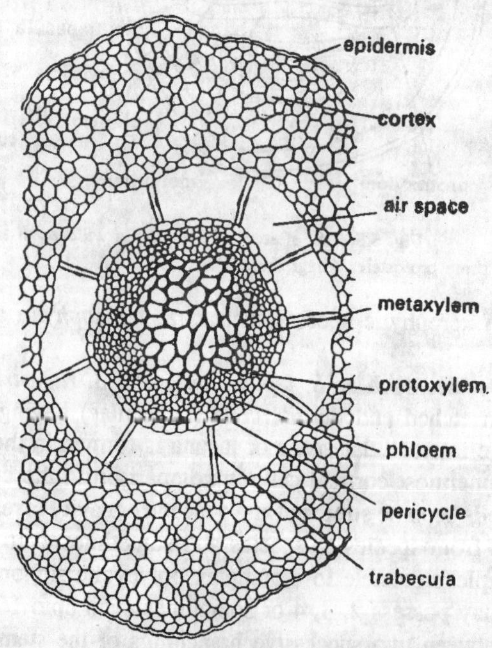

Fig. 5.7. *Selaginella kraussiana*. T.S. of stem showing monostelic condition.

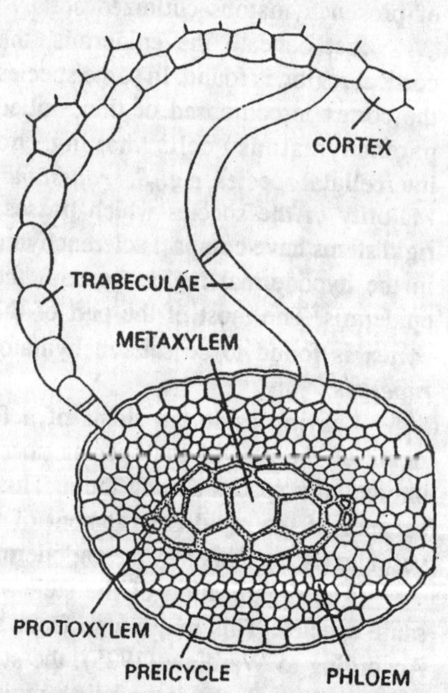

Fig. 5.8. *Selaginella kraussiana*. Part of a transverse section of stem to show inner cortex trabeculae and single stele. (After Harvey-Gibson).

xylem of the stele is ensheathed by a few layered phloem. The phloem is composed of sieve tubes. However, the companion cells are absent. According to Harvey-Gibson (1894) the xylem is surrounded by two or three layers of parenchyma and external to this parenchyma there is single layer of sieve tubes. The parenchyma and the sieve tubes of each stele are surrounded as a single layered parenchymatous pericycle. The protoxylem is composed of annular and spiral tracheids and the metaxylem of scalariform tracheids. The outermost layer of the stele is single layered endodermis. Here, the endodermis is laterally stretched and consisted of trabeculae. Much has been said about the trabeculated endodermis in preceding pages.

Apical growth. In *Selaginella* apical growth occurs from a single three or four-sided apical cell or from a group of initials, and the two situations may be found in the same plant.

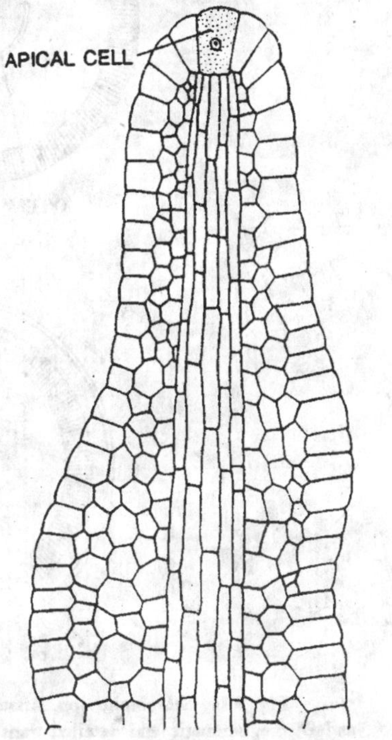

Root. The internal structure of the root may be studied by cutting its transverse sections. The structure is quite simple.

The outermost superficial layer of the root is single-layered and bear many unicellular root hairs on its surface and, therefore it is also known as piliferous layer. This layer consists of thinwalled parenchymatous cells.

Inner to the piliferous layer, there is wide cortical region. In certain species, *e.g., S. willdenovii* there is three to five layered hypodermis constituted of sclerotic cells, and thereafter the thin-walled parenchymatous cells. The last layer of the cortex is endodermis. In majority of the species the endodermis is not very clear but in certain species, *e.g., S. willdenovii* and *S. rubella*, it is well-defined and quite conspicuous.

Fig. 5.9. *Selaginella*. Stem apex showing a single apical cell in median longitudinal section.

Just beneath the endodermis, there is one to three layered pericycle composed of parenchymatous cells.

The stele is of single protostelic nature. It is monarch. The protoxylem is exarch and situated towards the periphery. The xylem is surrounded by phloem. The protoxylem is constituted of spiral and annular tracheids whereas the metaxylem of scalariform tracheids.

Rhizophore. The general anatomy of the rhizophore is like that of the root. However it lacks the piliferous layer and the root hairs.

Morphology of rhizophore. The morphological nature of the rhizophore is still a matter of controversy. Many workers held as it a root, some held it as the stem and some held that it is neither root nor stem and *sui generis* in nature. The points in the support of its root-like nature are as follows: According to Harvey-Gibson (1902) and Uphof (1920) they are roots because of their positive geotrophic nature and the root-like anatomical characters. One of the strongest points in support of its root-like nature, as regards anatomy, that even the stems of polystelic nature bear the monostelic rhizophores.

According to Wordell (1910) the rhizophore is the portion of the stem. This argument seems to be more strong and conclusive. The points in support of its stem-like nature are as follows-(a) Its origin is exogenous, (b) the cap is not found, (c) its branching or dichotomy

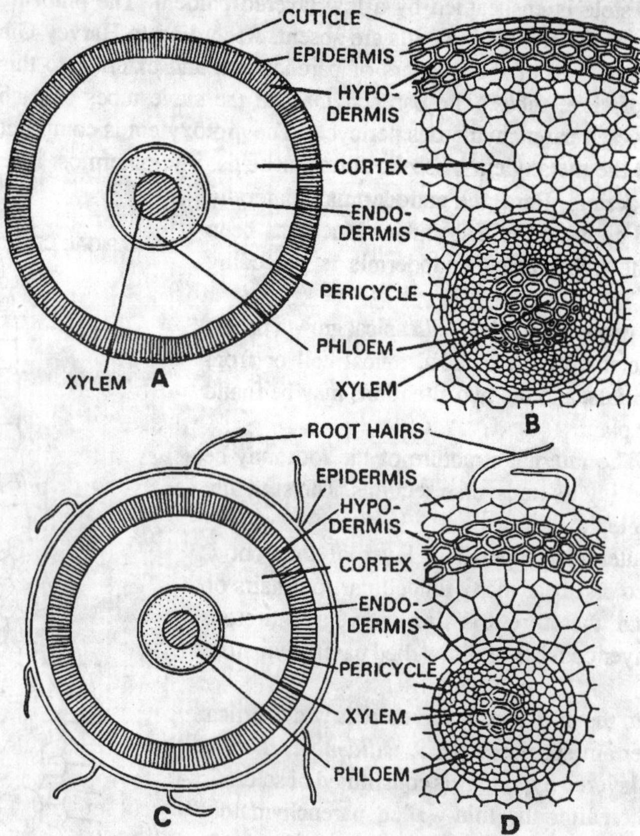

Fig. 5.10. ***Selaginella*** **spp. Anatomy. A and B diagrammatic and detailed transverse sections of rhizophore; C and D, diagrammatic and detailed transverse sections of the root.**

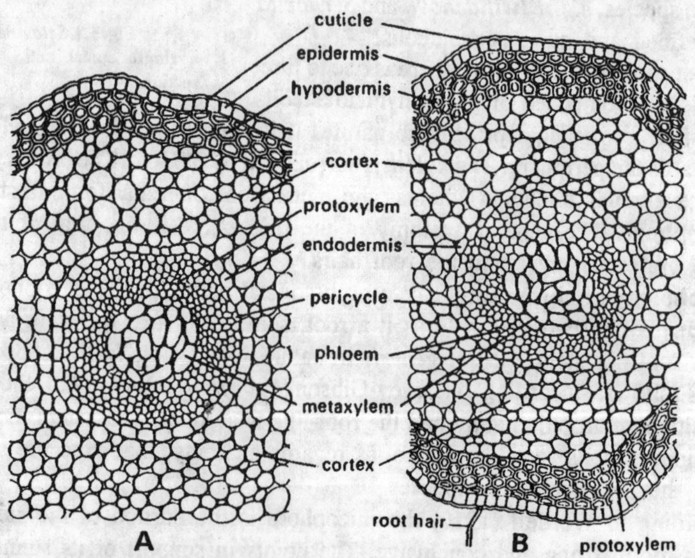

Fig. 5.11. ***Selaginella*.** **A, T.S. of rhizophore; B, T.S. of root.**

is like that of the stem and, (d) experimentally this has been proveu uy Wordell (1910) that the decapitated apices of the rhizophores develop into the leafy shoots.

According to Bower (1908, 1935) and Goebel (1905) it is neither a root nor a stem, it is an organ *sui generis*. But Schoute (1938) has said it is wrong to say them *sui generis* organs. He pointed out that they are specialized stems and behave like roots.

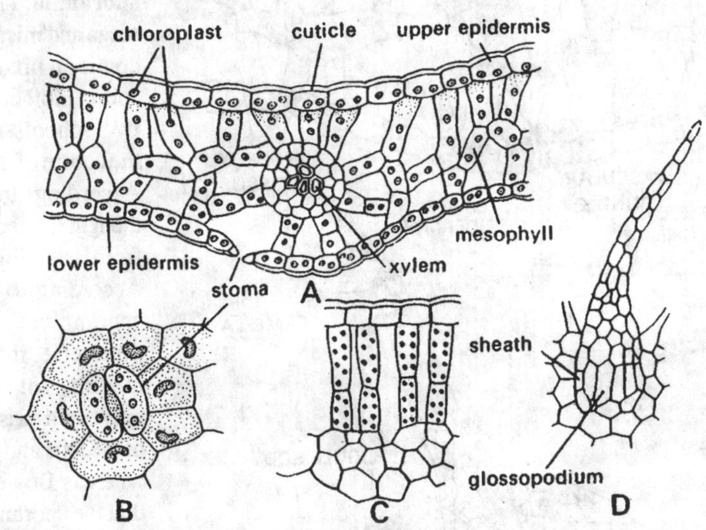

Fig. 5.12. *Selaginella kraussiana*. Anatomy of leaf. A, V.S. leaf; B, stoma; C, mesophyll tissue; D, a ligule.

Leaf. The anatomy of leaf is quite simple. To study the anatomy one has to cut the vertical microtome sections of the leaf. There is definite upper and lower epidermis. The epidermis is single-layered and composed of thin walled cells. The epidermal layers of both the surfaces contain discoid chloroplasts in them. The stomata are confined to the lower epidermis and sometimes also found on the upper epidermis. All the mesophyll cells are quite alike and contain chloroplasts in them. The epidermal layers and the mesophyll tissue are photosynthetic. The vascular bundle of the leaf is quite simple. It is concentric. The leaf traces connect the stele of the stem. In *S. kraussiana*, the xylem consists of four to five tracheids, one of which is annular and the rest spiral. The xylem is surrounded by phloem, composed of parenchyma cells and one or two sieve tubes.

Ligule. The internal structure of ligule is as follows: The ligule is a small, papillate, pointed tongue-like outgrowth. At the base of the ligule there is a hemispherical **glossopodium.** The glossopodium consists of large, thin walled, vacuolate protoplasts. This structure is covered over by a covering sheath, known as **glossopodial sheath** consisted of tubular cells in continuation of that of the epidermis of the leaf. In *S. kraussiana* the Casparian strips are present in the sheath cells. Just above the glossopodium, the body of the ligule lies. It is composed of large, polygonal, rich in granular protoplasm and uninucleate cells. The terminal pointed end of the ligule is very thin and sometimes single-celled in thickness, is known as the **terminal lamella.**

The sporophylls and sporangia. The sporophylls are found arranged in the strobili or spikes at the terminal ends of the branches. Each sporophyll bears a single stalked sporangium situated in the axil of it, on the adaxial side between ligule and its base. There are two kinds of sporangia. The sporangia bearing megaspores are called megasporangia and the sporangia with microspores are microsporangia. In the same way the sporophylls bearing megasporangia are megasporophylls and bearing microsporangia are microsporophylls.

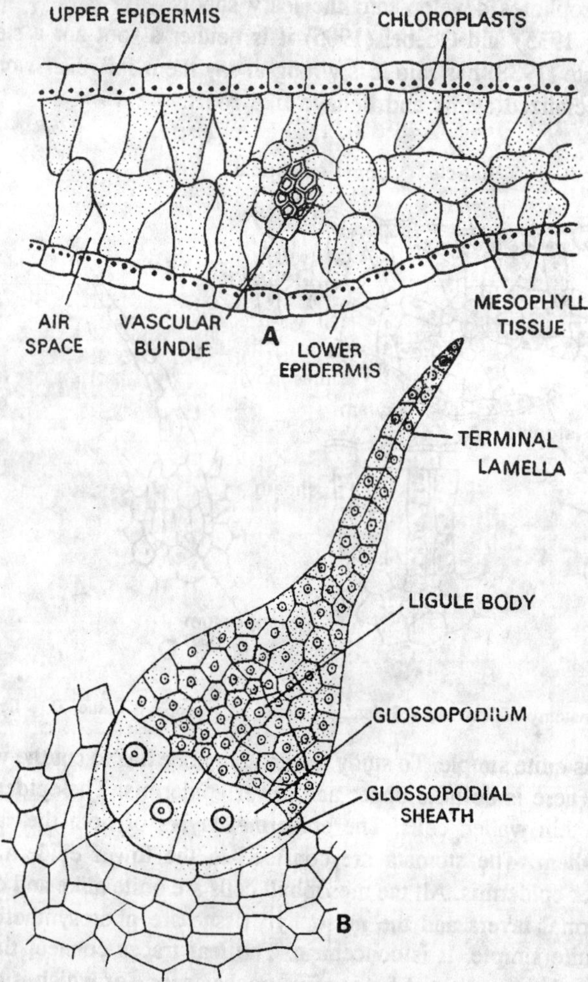

Fig. 5.13. *Selaginella* sp. A, vertical section of leaf showing chloroplasts in the epidermis; B, sectional view of a mature ligule.

Development of the sporangia. The development of mega and microsporangia is quite common upto the formation of spore mother cells. There are two schools of thought about the nature of their development. According to one school of thought the sporangium develops from a single initial cell. According to other school, the sporangium arises from a transverse row of initials. In majority of the species the sporangia arise from the cells of the axis and said to be cauline whereas Bower holds the view that the sporangium arises from the base of the sporophyll and foilar in origin. Any how, the mature sporangium is a distinctly axillary structure at the base of the sporophyll.

Irrespective of whether there is single initial cell or a group of initial cells the further development proceeds on the same line.

The initial divides first periclinally giving rise to two cells. The outer cell (or cells) is **jacket initial** (or initials) and the inner cell (or cells) is **archesporial initial** (or initials). The jacket initial divides and redivides developing a two layered jacket. The archesporial initial divides vertically and transversely again and again and this way a mass of sporogenous tissue is resulted. The jacket is two layered, the outer layer is composed of thick-walled cells and the inner layer of thin-walled cells. The jacket completely covers the sporogenous tissue. The outermost cells of the sporogenous tissue develop into a nutritional layer the **tapetum.** The cells of the tapetum remain in close contact of the developing spores to nourish them. The tapetal cells degenerate after the maturity of the spores.

The sporangia which are to develop into microsporangia contain all the sporogenous cells. The sporogenous cells of the last generation are the sporocytes. During this process 10 to 20 percent of the sporogenous cells are disintegrated. This way the sporocytes become loosened within the sporangial chamber. The functional sporocytes become rounded and separated from each other. Each sporocyte (spore mother cell) divides meiotically giving rise to a tetrad of four spores. This way numerous haploid microspores are developed within the sporangium.

The sporangia which are to develop into megasporangia contain many sporocytes in them in the beginning. Later on all sporocytes but one are disintegrated. The functional megaspore mother cell undergoes the usual reduction division and a tetrad of megaspores is developed. According to Duerden (1929) in some species, 8, 12, 24 or even more megaspores have been recorded in one sporangium. In such cases more than one megaspore mother cells are functional. According to Duerden (1929) and Mitchell (1910) there are still other species in which all the four megaspores of the tetrad donot develop as usual. In these species the megasporangium contains one, two or three functional megaspores, rest of the spores are degenerated. As the megaspores develop less in number in each megasporangium they get much of nourishment from the degenerating cells and increase a lot in size and more or less fill up the sporangium. The megasporangium of *S. willdenovii* contains the highest recorded number of the megaspores, *i.e.*, 42 whereas *S. monospora* and *S. rupestris* possess monosporic megasporangia.

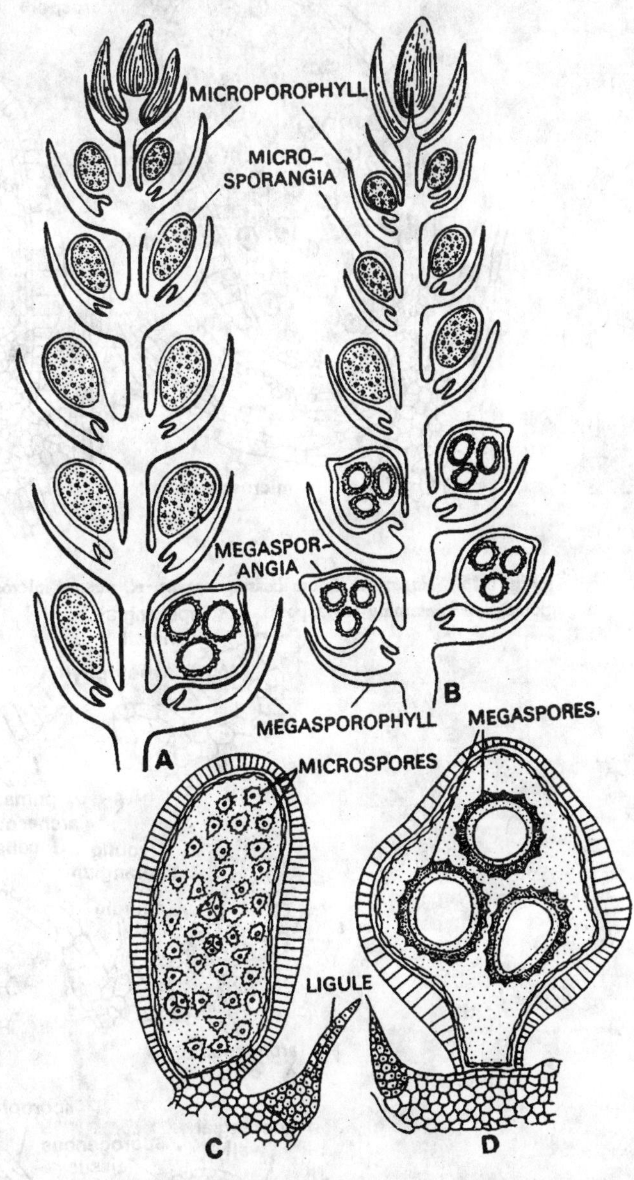

Fig. 5.14. *Selaginella* sp. Spore bearing parts. A and B, vertical sections of the spikes of two different species *S. helvetica* and *S. Kraussiana* showing the arrangement of micro and megasporangia; C, longitudinal section of microsporangium; D, longitudinal section of megasporangium.

Structure of sporangia. Both mega and microsporangia are stalked. In both the cases the wall of the mature sporangium is two layered. The outer layer of the wall is composed of thick walled columnar cells having chloroplasts in them. The inner layer of the wall is composed of thin walled cells. Just beneath the two layered sporangial wall there is a nutritional layer, the tapetum. This layer nourishes the developing spores and persists until the maturity of the spores and thereafter degenerates.

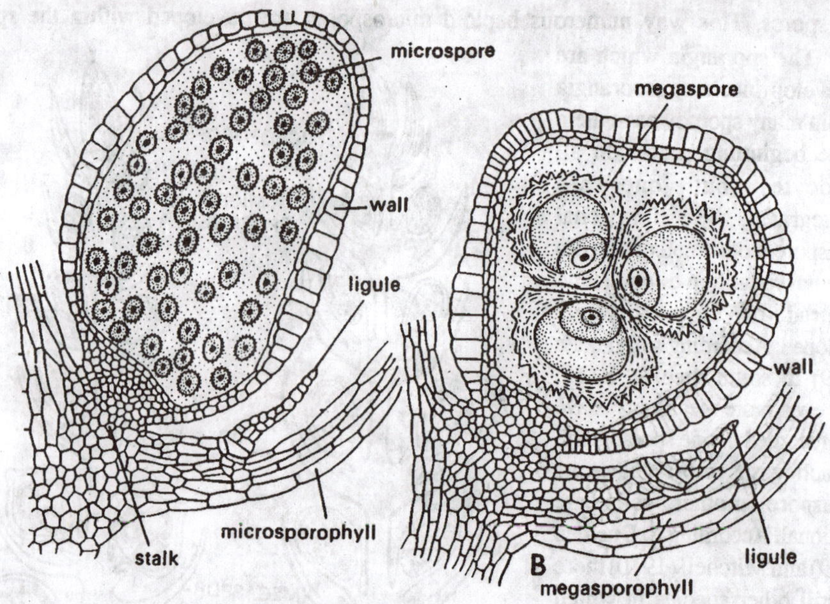

Fig. 5.15. ***Selaginella*****. Spore bearing organs. A, mature microsporangium containing microspores; B, mature megasporangium containing megaspores.**

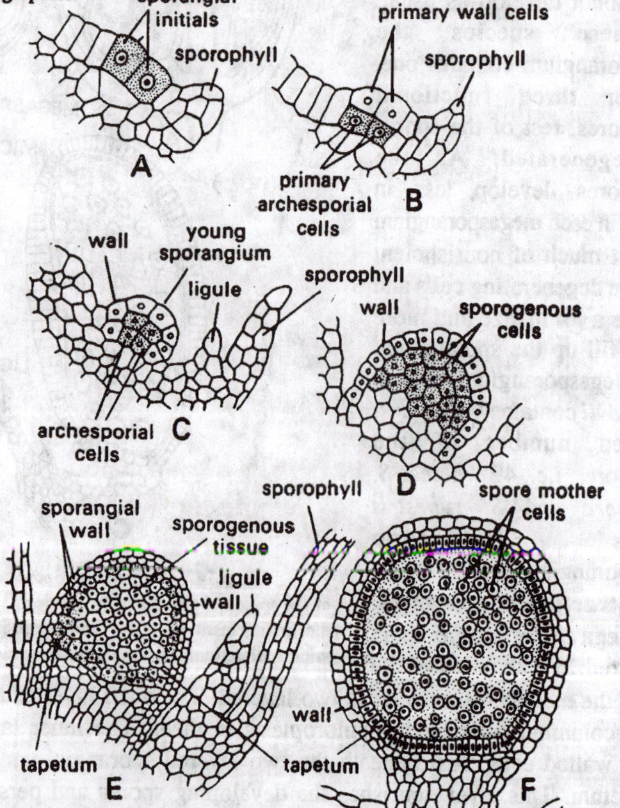

Fig. 5.16. ***Selaginella*** **sp. Development of sporangium. A-D, different successive stages in the development of sporangium and ligule; E, young sporangium and ligule; F, mature sporangium.**

The ripe microsporangia are red yellow or brown in colour. This colour of the microsporangial wall is because of the coloured miscrospores within it. Comparatively the microsporangia are smaller in size. In *S. kraussiana*, each microsporangium contains about 600 microspores in it.

The megasporangia are pale or whitish in colour. The megaspore found within the megasporangium are either pale, whitish or yellow in colour. The ripe megasporangium is fourlobed as seen in the top view. Each lobe bears a single megaspore. Comparatively, the megasporangium is larger in size and contains four megaspores within it in usual cases.

Dehiscene of sporangia. On the maturation of the sporangia the spores are liberated. Unlike most other pteridophytes, the microspore begins to develop into the microgametophyte even before its liberation from the microsporangium. The sporangial jacket splits vertically into two valves. This splitting is caused by the different hygroscopic changes in apical and lateral parts of the sporangial wall. The sporangial wall splits with a jerk and eject out the developing microgametophytes (spores) to a distance of few centimetres. This way, megaspores may be thrown out 6 to 10 centimetres and the microspores 1 to 1.5 centimetres. The microspores eject out from the microsporangium gradually by subsequent openings and closings of the split of the jacket. According to Duthie (1923, 1924) and Mitchell (1910) the spores are dispersed from the splitted sporangium by wind currents.

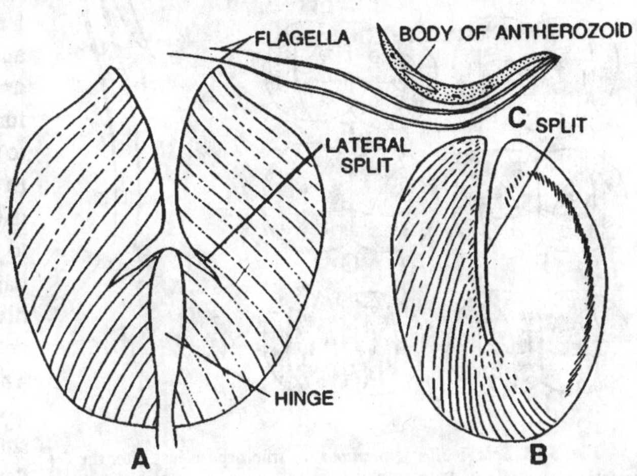

Fig. 5.17. *Selanginella erythropus.* A, dehisced megasporangium; B, dehisced microsporangium; C, antherozoid. (A and B, After Goebel).

The spores. There are two kinds of spores found in the separate sporangia. The microspores are found in the microsporangium and the megaspores in the megasporangium. Hundreds of microspores are found in each microsporangium whereas, on the other hand, the megasporangium usually contains only four but big megaspores. The presence of two different sized spores in the same species is known as **heterospory.** The phenomenon of heterospory has its own biological advantages which has been discussed in the end of the description of this genus.

The microspores. The microspores are quite small in size. The diameter of the microspores ranges from 0.015 to 0.06 millimetres. The shape of each microspore is spherico-tetrahedral. The distal face of the spore is somewhat spherical. Each microspore is covered by two spore coats. The outer coat, which is thick and ornamental is known as exine or exospore, whereas, the inner coat is thin and known as intine or endospore. The microspores contain oil droplets. They are devoid of any trace of chlorophyll.

The megaspores. Comparatively they are quite bigger in size. The diameter of each megaspore ranges from 1.5 to 5 millimeters. The shape of each spore is tetrahedral. A triradiate ridge is found on each spore. Each megaspore is surrounded by two spore coats. The outer spore coat is known as exospore which is quite thick and sculptured, whereas, the inner coat is endospore which is quite thin. The ornamentation of the megaspores varies in different species. This may be tuberculate, spinose or reticulate. On germination, the megaspore develops into a female gametophyte (megagametophyte).

Germination of microspore and formation of male or microgametophyte. The microspore germinates *in situ* and the microgametophyte is developed within it. Here, the development of the microgametophyte of *S. kraussiana* has been discussed in detail. As already mentioned the development of the microgametophyte begins even before the dispersal of the spore from the microsporangium. First of all an asymmetrical division takes place in the microspore, giving rise to two unequal cells. The small lens shaped cell is **prothallial cell** and the remaining large cell is **antheridial cell.** The antheridial cell further divides by a wall perpendicular to the prothallial cell, thus giving rise to two daughter cells. Both of these daughter cells divide at right angles to the first plane of division. One daughter cell of this pair does not divide further and becomes the first cell of the jacket of the antheridium. The other daughter cell of the pair divides by a plane parallel to the small prothallial cell. This way, two daughter cells are formed. The cell nearer to the prothallial cell divides periclinally giving rise to a **primary androgonial cell** and a **jacket cell.** The other cell of the two cells divides anticlinally and again one of the daughter cells divides periclinally resulting in the formation of two jacket cells and a primary androgonial cell. This way, a thirteen-celled microgametophyte is developed. It consists of one prothallial cell, four primary androgonial cells and eight jacket cells. In *S. kraussiana* only at this stage, the microgametophyte is shed from the microsporangium. After shedding of the thirteen-celled microgametophyte, further divisions take place. Each of the four androgonial cells divides successively six times thus producing the 256 androcytes. Each such androcyte metamorphoses into a spindle-like biflagellate antherozoid. Later on the jacket cells and the prothallial cell are disintegrated. This way the mature gametophyte possesses a mass of antherozoids swimming within the limits of the spore wall. **Duthie** (1923) supported in his statement that these later stages in the development of the microgametophyte take place in the soil. According to miss Lyon (1901), in certain other species this later development of the microgametophyte takes place on the partially opened megasporangium.

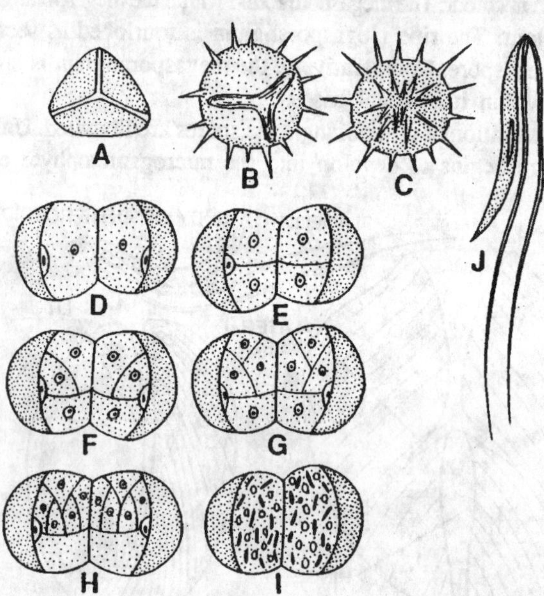

Fig. 5.18. *Selaginella kraussiana*. A, microspore just after the dissociation of a tetrad; B, completed microspore in apical view; C, completed microspore in basal view; D-I, schematic representation of the stages in the development of the microgametophyte; J, mature biflagellate antherozoid.

It is a notable point that the entire microgametophyte of *Selaginella* is very much reduced in size and developed within the limits of the sporangial wall. On account of its limited sphere and reduced size, there is difference of opinion regarding its homologies with the gametophytes of other homosporous pteridophytes such as *Equisetum* and *Lycopodium.* According to most of botanists, the prothallial cell is vegetative and homologous with the entire vegetative prothallus of the homosporous pteridophytes such as *Equisetum* and *Lycopodium.* The nature of the antheridial cell is also a matter of controversy. According to Campbell (1918) and Slagg (1932) it is the initial of a single antheridium. Bruchmann (1912) holds the opinion that antheridial cell divides and forms two or four antheridial initials according to the species.

On the maturation of the antherozoids the microspore bursts along its triradiate ridge, and the antherozoids are liberated in the film of water. They swim in the film of water with the help of their flagella. The water film is usually present on the surface of the soil on which the microspores rest.

Germination of megaspore and formation of female or megagametophyte. The megaspore also germinates *in situ.* The development of megagametophyte of *S. kraussiana* has been studied in detail. It has been discussed here in the following lines. The process of the development of the megagametophyte begins even before the maturity of the megaspore. The megaspore enlarges in size and a two-layered wall is secreted round it. When the megaspore is still enlarging, its nucleus divides, but there is no cell division. Their is a continuous nuclear division and ultimately a multinuclear protoplast with a large central vacuole is resulted. Just beneath the cell wall, the cytoplasm is found in a thin layer. After some time the enlargement of the multinucleate gametophyte is ceased, simultaneously the cytoplasmic layer underneath the cell wall becomes thick and thicker and this way the central vacuole is reduced in size. Ultimately the central vacuole disappears. Long before all this, the free cell formation takes place at the pyramid like end of the megagametophyte and a two to three layered thick cellular tissue is resulted. This cellular tissue remains separated from rest of the gametophyte by a diaphragm. The portion of the megagametophyrte below this

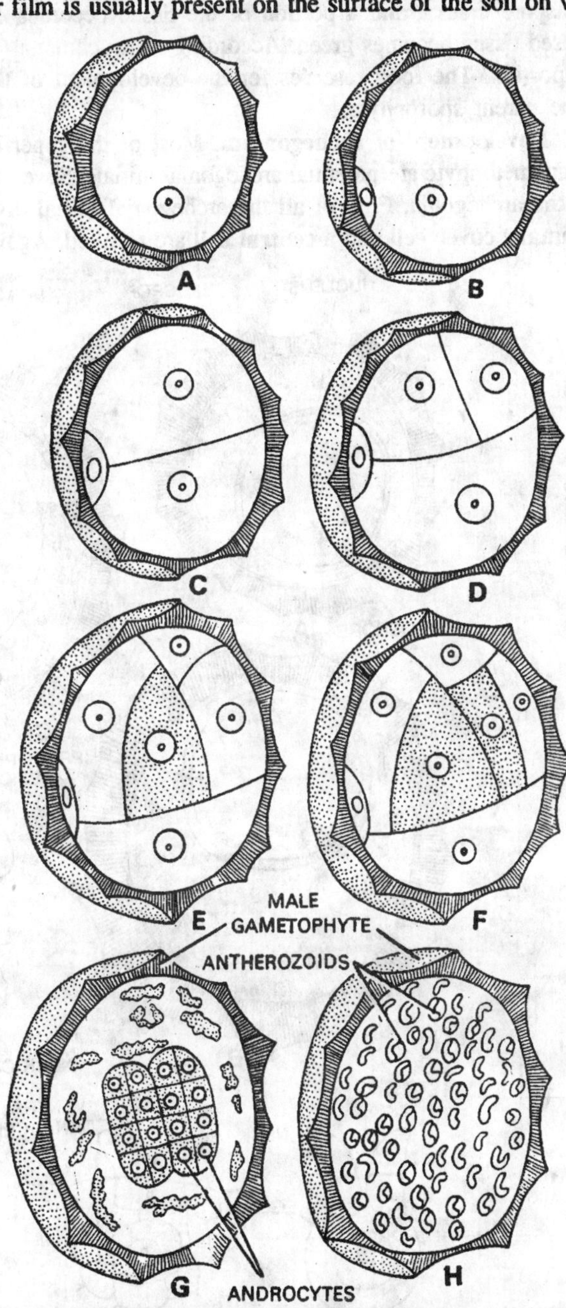

Fig. 5.19. *Selaginella kraussiana.* Development of the male gametophyte, A-G, successive stages in development of microgametophyte; H, mature male gametophyte with antherozoids.

diaphragm remains multinucleate unless and until the development of the embryo has begun. Later on this region also becomes multicellular by free cell formation. All this region beneath the diaphragm is packed with starch. As soon as the apical tissue is formed, the wall of the megaspore cracks, and a portion of the tissue becomes exposed to the outside. Probably the exposed tissue becomes green. According to Bruchmann (1912) the rhizoids also develop from this portion. The food reserves for the development of the embryo have already been stored in the parent sporophyte.

Development of archegonium. Most of the superficial cells of the apical tissue of the megagametophyte are potential archegonial initials. Several of these archegonial initials develop into the archegonia. First of all the archegonial initial divides by a periclinal division so that a **primary cover cell** and a **central cell** are resulted. Again the central cell divides periclinally

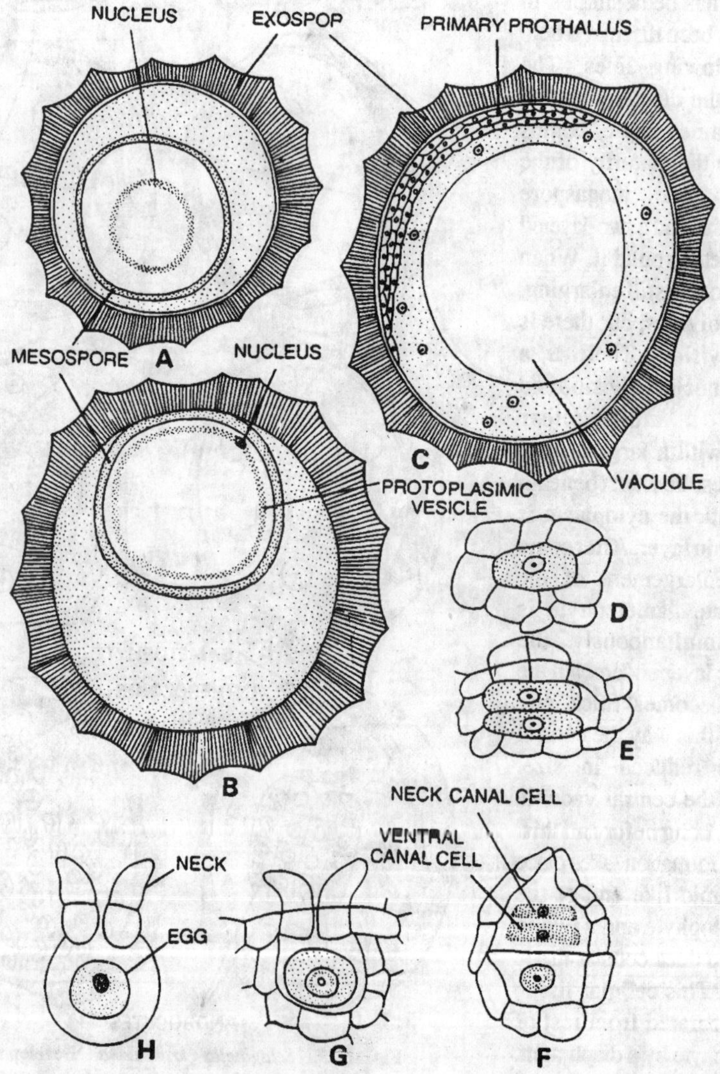

Fig. 5.20. *Selaginella kraussiana.* A, T.S. of megaspore; B, L.S. of old megaspore; C, L.S. of old megaspore before the appearance of the diaphragm; D-G, successive stages in the development of archegonium; H, mature archegonium.

resulting into a **primary canal cell** and a **primary venter cell**. The primary venter cell again divides periclinally producing a **venter canal cell** and an egg. This primary canal cell does not divide further and directly acts as a **canal cell** of the archegonium. The primary cover cell formed in the beginning divides twice anticlinaly at right angles to each other producing four **neck initials**. Each of the neck initials divides transversely, giving rise to a neck two cells in height. However, the outermost tier of four cells of the neck protrudes beyond the surface of the megagametophyte. In the meanwhile the neck canal cell and the venter canal cell are disintegrated; the terminal cells of the neck become separated, leaving a free passage for the entrance of antherozoids.

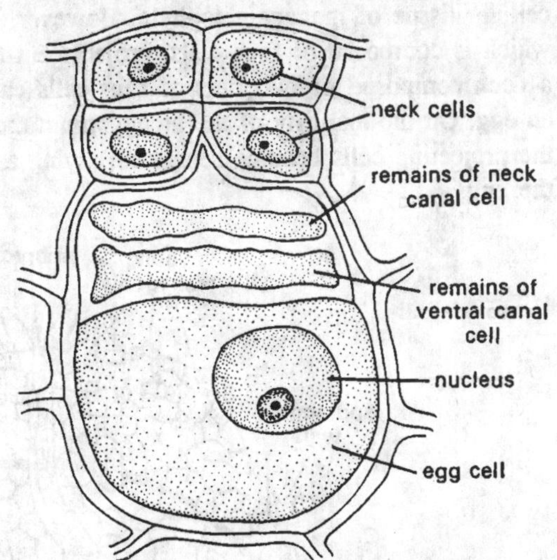

Fig. 5.21. *Selaginella*. Structure of archegonium.

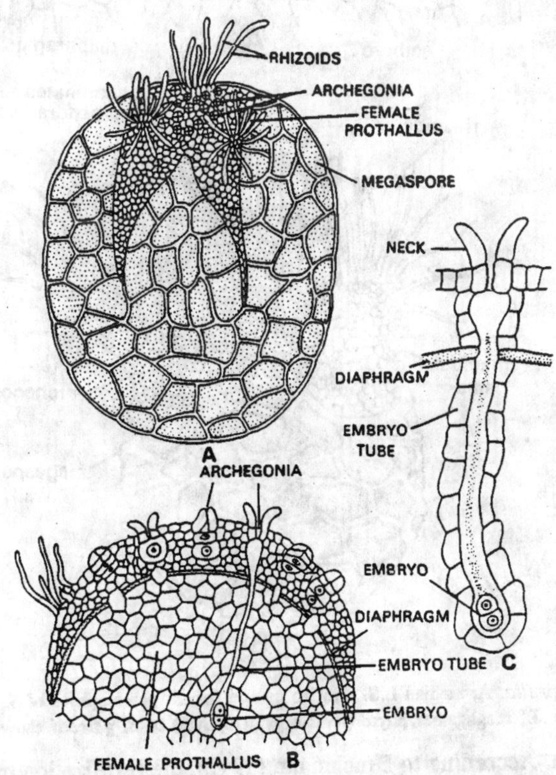

Fig. 5.22. *Selaginella* sp. Female gametophyte. A, splitted megaspore showing rhizoids, prothallus and archegonia; B, a part of the female gametophyte with diaphragm, rhizoid and young embryo; C, detailed structure of archegonium with young embryo.

Structure of archegonia. The archegonia are quite small and embedded in the apical cellular tissue of megagametophyte. However, the outer tier of the neck of the archegonium, which is composed of four cells, projects out the gametophyte. The archegonium consists of a neck composed of two tiers of four cells each, a neck canal cell, a venter canal cell and an egg. On the maturity of the archegonium the neck and venter canal cells are disintegrated; the projecting cells become separated, laying a free passage for swimming the antherozoid to the egg.

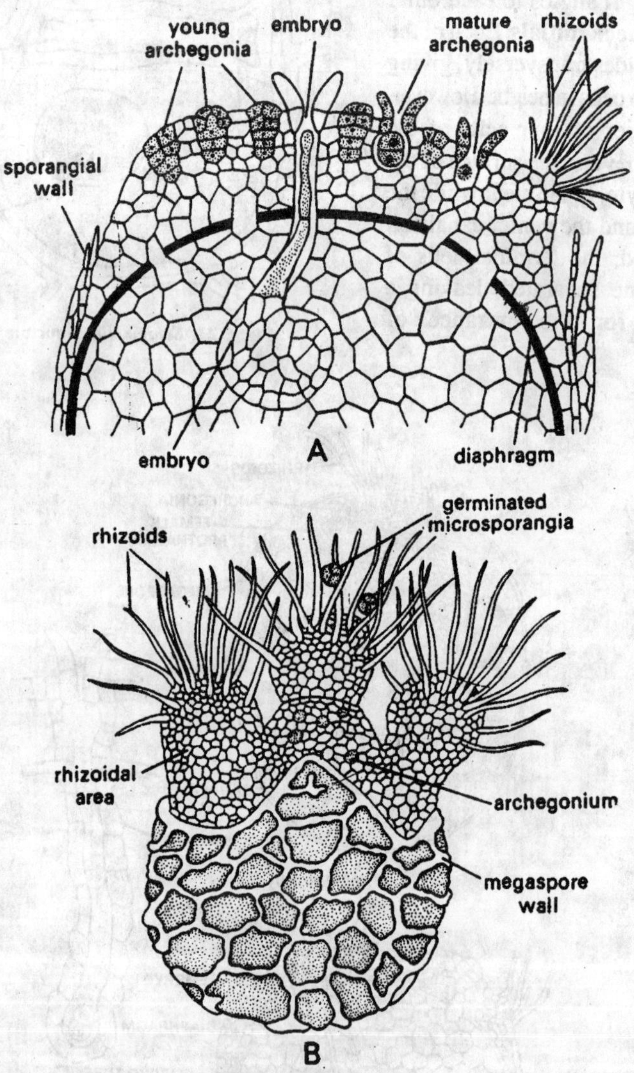

Fig. 5.23. *Selaginella.* A, median L.S. through mature female gemetophyte of *S. kraussiana* showing rhizoids, archegonia and embryo; B, megagametophyte (female gametophyte) of *S. galeotti* showing rhizoidal mounds.

Fertilization. According to Bruchmann (1912), the fertilization may take place while the megagametophytes are still in the megaspores situated in the megasporangia which may have fallen on the ground. The water is essentially required for the fertilization. After shedding from the microspore the antherozoids come in contact of the female gametophyte and approach the archegonia. The antherozoids swim down in the neck of the archegonium through the medium of water and approach the egg. Only one antherozoid penetrates the egg and the fertilization is effected. With the result the oospore is formed after fusion.

Apogamy. In some of the species the embryos develop apogamously. According to Miss Lyon (1904), the apogamy may take place due to the development of an archegonial initial into the embryo.

Parthenogenesis. This phenomenon has also been reported in several species of *Selaginella*. Bruchmann (1912, 1919) and Gieger (1934) have reported that in some species, *e.g., S. helvetica, S. anoçordia* and *S. rubricaulis* the archegonia donot open at all, and the eggs develop into the embryo parthenogenetically. According to Goebel (1915), there are still other species which produce only megaspores, *e.g., S. intermedia* , and the eggs develop into the embryos parthenogenetically.

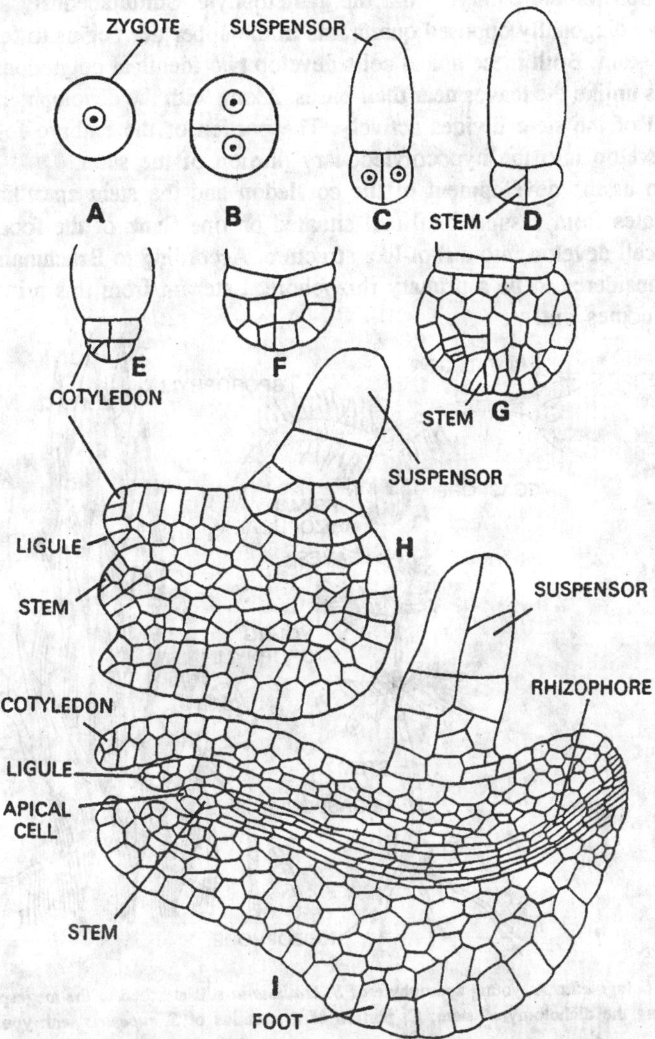

Fig. 5.24. *Selaginella* **sp. Development of embryo. A, zygote; B, transverse division of zygote; C-I, different successive stages in the development of the embryo.**

Development of embryo. As soon as the fertilization is over the zygote secretes a wall around it and develops into an embryo. First of all, generally, the zygote divides by a transverse wall giving rise to two cells. They are called the **epibasal and hypobasal** cells. Here, the further development of the embryo has been discussed of *Selaginella martensii*: The upper epibasal cell develops into a **suspensor,** whereas the hypobasal cell forms all the other parts of the embryo. The hypobasal cell divides vertically and this way two cells are formed. These two cells again divide by a vertical division and four cells are resulted. One of these four cells, divides diagonally by a vertical division forming the apical cell of the stem to be developed. Rest of three cells and the sister cell of the apical cell together divide by a transverse division, this way two upper and lower tiers of four cells each are formed. Now the cells of both these tiers divide again and again vertically and transversely. There is more active cell division in one of the lower quardrants, and this leads to the formation of the **foot** of the embryo. The foot acts as haustorium and absorbs food for the embryo from the gametophyte. Simultaneously a superficial cell, in each of the two diagonally opposed quardrants in the upper tier begins to function as the apical cell of a **cotyledon.** Both these apical cells develop two identical cotyledons. These cotyledons possess ligules unlike the leaves near their bases. Along with the development of the cotyledons, the apical cell of the stem divides actively. The portion of the embryo found just behind the cotyledons develop into the hypocotyledonary portion of the stem.

As soon as the development of the cotyledon and the stem apex has begun, an apical cell differentiates from a superficial cell situated on one flank of the foot. The segments cut off from this cell develop into a root-like structure. According to Bruchmann (1909, 1912) this structure is considered to be a primary rhizophore. Later on from this primary rhizophore, the primary root comes out.

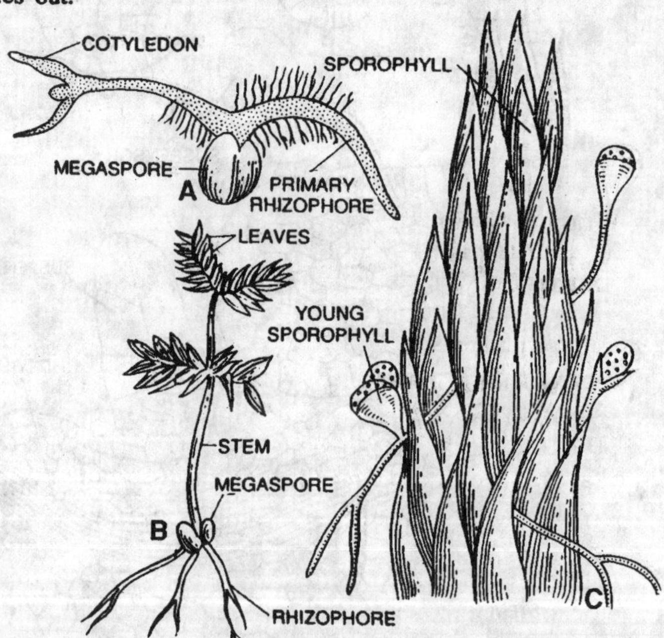

Fig. 5.25. *Selaginella*. A, young sporophyte of *S. kraussiana* still attached to the megaspore; B, somewhat older sporophyte showing the dichotomy of stem; C, portion of a strobilus of *S. rupestris* with young sporophytes.

The young sporophyte remains attached to the megagametophyte for sometime still situated within the megaspore wall. The young sporophyte grows further by sending its stem and cotyledons upward and the rhizophore and the roots downward in the soil. The most peculiar features of the juvenile sporophyte of *Selaginella* are as follows :

(a) The cotyledons are directly borne upon the stem (not found in other pteridophytes).

(b) Just posterior to the cotyledons a conspicuous hypocotyledonary part of the stem develops (not found in other pteridophytes). The apical cell of the stem is situated in between the cotyledons.

Later on, the gametophyte is finished, and the young sporophyte becomes independent by establishing its root system in the soil.

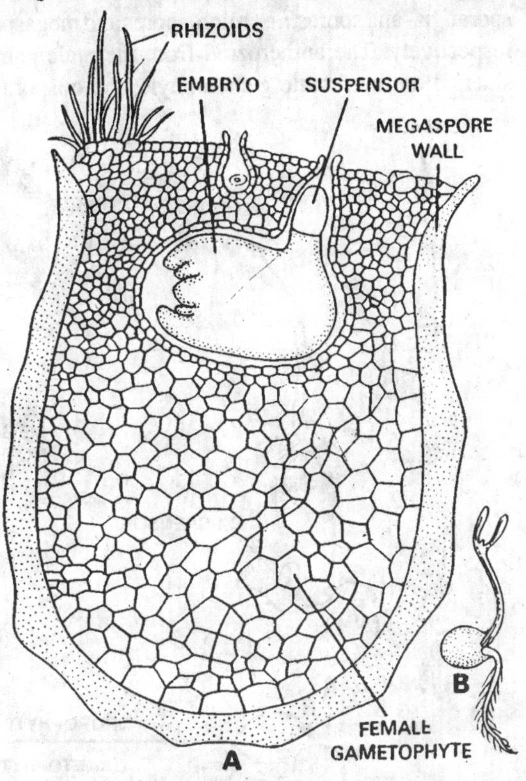

Fig. 5.26. *Sellaginella* sp. A, female gametophyte with embryo; B, young sporophyte with germinating megaspore.

Specialization of spores and heterospory. The sporangia show greater specialization than those of ferns. They are differentiated into micro and megasporangia. The microsporangia contain microspores whereas megasporangia contain megaspores in them. Each microsporangium contains a large number of microspores whereas each megasporangium contains usually only four megaspores. The condition of the production of two kinds of dissimilar spores, differing in shape and size in the same species, is termed **heterospory.** When one kind of spores are present, *e.g.. Equisetum* and ferns, the condition is said to be **homospory.** According to modern pteridologists the heterospory is found in seven genera: *Selaginella, Isoetes, Marsilea, Salvinia, Azolla, Regnellidium* and *Pilularia.* Here one thing may be noticed that the microspores develop male gametophytes whereas the megaspores germinate into the female gametophytes which may also be called as microgametophytes and megagametophytes. In the case of homospory the sex may be differentiated only at the gametophytic stage, but here the sex may be differentiated even at the sporophytic stage by the presence of small and large sized microspores and megaspores which produce antherozoids and eggs respectively.

The differentiation of microspores and megaspores and their dependence upon the sporophyte has certain advantages. The gametophytes of the ferns are, however, dependent for their nutrition upon soil and environmental conditions, whereas in the case of *Selaginella,* as for as the nutrition of gametophytes is concerned they derive it from the sporophyte, and therefore they are more independent to the external conditions than those of ferns.

Seed habit. The condition of heterospory in *Selaginella* constitutes one of the most important phases in the evolution of the plants. In certain species of *Selaginella* only one megaspore develops within the megasporangium. The highest evolved species, *i.e., S. apoda* has advanced almost upto the rank of a seed plant. Here, the megaspore remains within the sporangium; it itself fertilizes and this way for the first time the gametophyte of *Selaginella* shows complete dependence of the gametophyte upon sporophyte as in angiosperms.

Alternation of generations. Like other pteridophytes, there is an alternation of sporophytic and gametophytic generations. In *Selaginella* the sporophyte bears two kinds of sporophylls-sporangia- and spores,*i.e.*, microspores and megaspores giving rise to male and female gametophytes respectively. The antherozoid from the male gametophyte fertilizes an egg in the archegonium situated on the female gametophyte; the oospore is resulted which develops into a sporophyte.

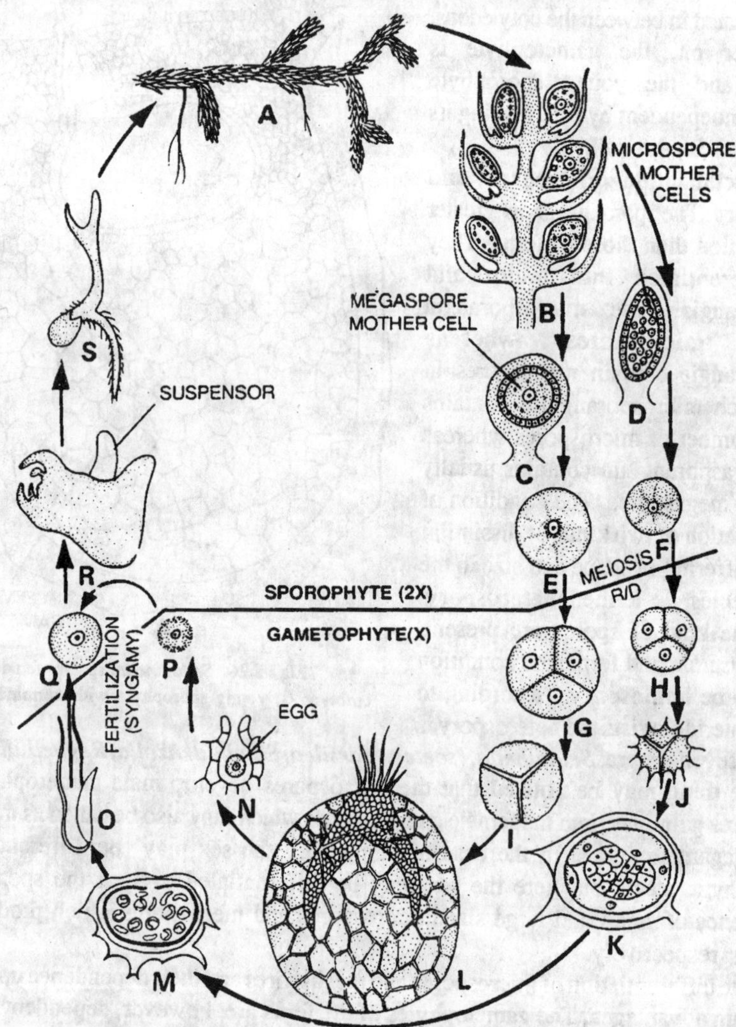

Fig. 5.27. *Selagingella* sp. Diagrammatic life-cycle. A, sporophyte with spikes and rhizophore; B, L.S. of spike; C, megasporangium with megaspore mother cell; D, microsporangium with microspore mother cells; E, megaspore mother cell; F, microspore mother cell; G, megaspore tetrad; H, microspore tetrad; I, megaspore; J, microspore; K, developing microgametophyte; L, megagametophyte; M, microgametophyte; N, archegonium with egg; O, antherozoid; P, egg; Q, oospore; R, embryo; S, young sporophyte.

Here, in comparison to the sporophyte, the gametophyte is very much reduced. On the other hand the gametophytes are never independent for their nourishment. They take their nourishment from the sporophyte. Here in *Selaginella*, the gametophyte is greatly reduced in size, especially in the male, where it is represented by a single prothallial cell and a single antheridium. The sporophytic stage begins with the formation of oospore (2x) and ends with the formation of spore mother cells (2x), whereas the gametophytic stage starts with the formation of spores (micro and mega) (x) and ends in the formation of gametes, *i.e.*, antherozoid and egg (x).

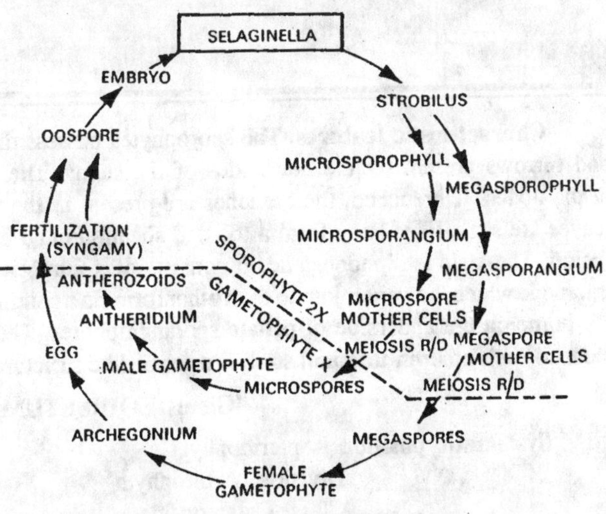

Fig. 5.28. *Selaginella* sp. Graphic life-cycle.

Economic importance. Many species of *Selaginella* are grown as potted plants for their beautiful foliage. The foliage of *S. willdenovii* and *S. caesia* are coloured and they are grown in the pots for their ornamental value. The species of *S. lepidophylla* and *S. pilifera* are xerophytic and sold as resurrection plants. The foliage of *S. serpens* changes its colour periodically. In the morning the leaves are bright green, they change to the pale green in the day time and again in the night the leaves become dark green.

6

Division-Arthrophyta
Class-Calamopsida
Order-Equisetales

Characteristic features. The sporophytes possess distinctly articulated stems. The ridges and furrows present on the internodes of the stems. The stems may or may not by branched at the nodes. If branched, the branches are present in the whorls alternating to the leaves. The leaves are scale-like. They form a toothed sheath around each node of the stem by their lateral fusion. The stele is siphonostelic. It consists of a number of collateral vascular bundles in the internode which, however, join to each other forming a continuous ring at each node. The sporangia are borne on the underside of peltate sporangiophores. The sporangiophores are arranged quite close to each other in terminal strobili or cone-like structures. The antherozoids are multiciliate.

Genus EQUISETUM

Systematic position - pteridophyta
Division. Arthrophyta.
Class. Calamopsida
Order. Equisetales
Family. Equisetaceae
Genus. *Equisetum.*

The genus comprises of about 25 species. They are world wide in distribution, however no species have been recorded from Australia and NewZealand. They grow in a variety of habitats. Majority of the species are found in north temperate zone. On the other hand many species are tropical and found in South America and West Indies. The common Indian species are: *E. debile, E. arvense* and *E. ramosissimum.* Certain species grow in ponds and marshy places, *e.g., E. pálustre;* some species are shade loving and grow in damp shady places, *e.g., E. pratense* whereas some other species grow in exposed conditions. *E. arvense* is very widely distributed; *E. debile* is found along the banks of the rivers, canals and pools in the Indian plains. They are commonly known as horse tails.

External Characters

Habit. About all the species of *Equisetum* are perennial. They bear much branched, horizontal, creeping and subterranean rhizomes which sometimes penetrate 3 or 4 feet deep in the soil. The rhizomes bear the erect and aerial branches. Some of the aerial branches are sterile and some are fertile. The plants vary in their height from species to species. *E. scirpoides* is only few inches high, whereas, on the other extreme a species of American tropics, *i.e., E. giganteum* reaches upto 12 metres or 40 feet in height. However, its diameter is less than an inch. The plant of *E. debile* grows along the banks of the river, may attain the height of 10 to 15 feet when grows in dampy and shady places. Its diameter is about one and half centimetres.

Stem. The rhizome of *Equisetum* is horizontal, perennial, much-branched, subterranean and penetrates in the soil upto 3 to 4 feet and in many species it is spread over in the area of 10 to 15 feet. It consists of well differentiated nodes and internodes. At each node there is a whorl of small, scale-like leaves united laterally to each other forming a brown sheath.

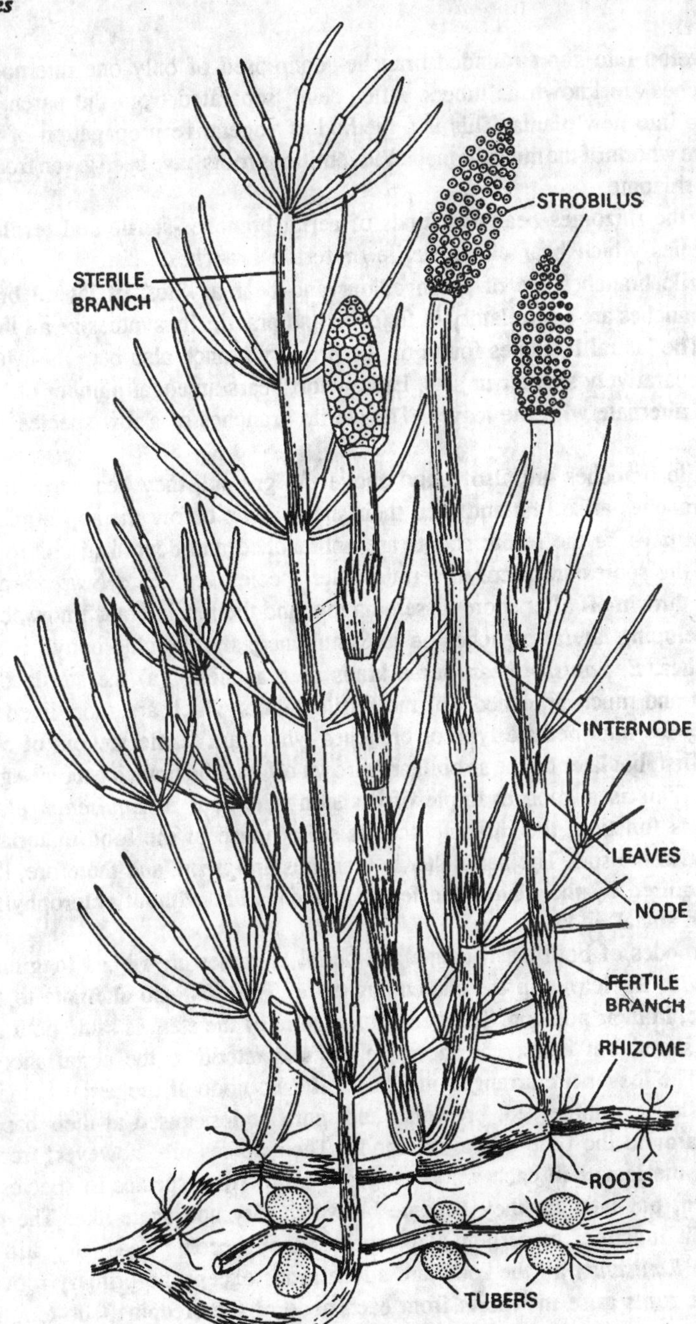

Fig. 6.1. ***Equisetum arvense*****. External features, the sporophyte showing rhizome, roots, tubers sterile and fertile branches.**

Alternating with each leaf of sheath at each node of the rhizome, there is a branch primordium or bud. These primordia may develop into aerial on subterranean branches immediately, or they may remain dormant for a considerable long period and develop into new branches on the approach of favourable conditions. Sometimes in certain species, *e.g.*, *E. telmateia* and *E. arvense* the

primordia develop into short rounded branches composed of only one internode. These short rounded branches are known as tubers. After being separated from the parent rhizome, these tubers develop into new plants. This is a method of vegetative propagation of the sporophytic generation. The whorls of the much branched adventitious roots have been given from the successive nodes of the rhizome.

Usually the rhizomes bear two kinds of aerial branches-sterile and fertile, but there are still other species which bear either sterile or fertile branches.

The sterile branches are of green colour and bear a whorl of lateral branches at each node. These branches are assimilatory in function and practically synthesize all the food required by the plant. The lateral branches found on the primary branch also bear the whorl of branches which are comparatively smaller in size. Each whorl bears an equal number of branches to that of leaves, and alternate with the leaves. The sterile branches of a few species are unbranched, *e.g., E. hyemale*.

The fertile branches are also found above the ground, they bear strobili at their apical ends. These branches arise first and shed their spores even before the appearance of the sterile branches. In *E. arvense* the fertile shoots are unbranched and devoid of chlorophyll, and they die as soon as the spores are shed. In certain other species, *e.g., E. cylvaticum* and *E. pratense* the strobili are thrown off after spore dissemination, and the fertile branch now acts as the sterile branch by developing lateral branches on it. Simultaneously, the chlorophyll is also developed in these branches. *E. palustre* bears three kinds of branches - (a) the sterile branches which are deep green and much branched, (b) the fertile shoots which are short lived and devoid of chlorophyll, (c) the intermediate type of branches which are fertile, devoid of chlorophyll and unbranched at first, but later on the strobili are thrown off and the branch becomes green, branched and persistent. This is a good example of division of labour in *Equisetum* plant.

As regards function, the rhizome acts as store house of the food material and helps in anchoring plant to the soil. The aerial sterile branches are green, and therefore, they synthesize all the food required by the plant. The fertile branches are without chlorophyll and bear the strobili at their apical ends.

The internodes of both the rhizome and aerial branches are ribbed longitudinally. These ribs alternate with the leaves in the subtending node. The ribs also alternate to the ribs of the successive node, in their position. The general structure of the stem is same both in the rhizome and the aerial shoot, but it may be more clearly understood in the aerial shoots.

Leaves. The leaves are arranged in whorls at each node of the aerial branches. They are simple, scaly, slender, uninerved, brownish and more or less fused at their bases laterally to form a sheath around the base of the internode. Their apices are, however, free and pointed. The number of the leaves in each whorl ranges 3 to 40 from species to species. At first, the leaves are green, but later on they become brownish, dry and scale-like. The mature leaves are protective in function. In certain other species they become dead, dry and scaly.

Roots. In *Equisetum* all the roots are adventitious except the primary root of the young sporophyte. The roots arise in whorls from each node of the creeping, sub-terranean rhizome. The roots may be many years together, but donot increase in their length. The roots donot arise directly, but from the lateral branch primordia situated on the nodes of the rhizomes and main aerial shoots.

Internal Structure

Stem. To study the internal structure of the stem, the transverse and longitudinal sections, passing through the nodes and internodes are required. This way the internodal and nodal anatomy of the aerial stem may be studied separately. It is as follows :

Internal anatomy. To study the anatomy of the node, one has to cut the aerial stem through internode in transverse sections. The outer surface of the stem looks wavy, because of the presence of ridges and furrows.

The stem is covered over by a single layered epidermis interrupted by stomata situated in the grooves of furrows. The epidermis is always impregnated with a thick layer of silica. The deposition of the silica on the epidermal layer gives the rough appearance to the stem, and therefore, the *Equisetum* plants are also known as 'scouring rushes'. The stomata are found in grooves of the aerial shoots. The development of the stomata is somewhat peculiar in the sense that the initial divides twice by successive longitudinal divisions and this way the two innermost cells develop into the gaurd cells, whereas, the two outermost cells develop into accessory cells on being mature of the stomata, the two accessory cells completely overarch the guard cells and the stoma. In majority of the species, *e.g., E. hymeale, E. ramosissimum* etc., the stomata are sunken, whereas in some other species, *e.g., E. palustre, E. pratense* etc., they are situated on the surface of the epidermis. The silica is deposited in the wall of guard cells in transverse radial bands.

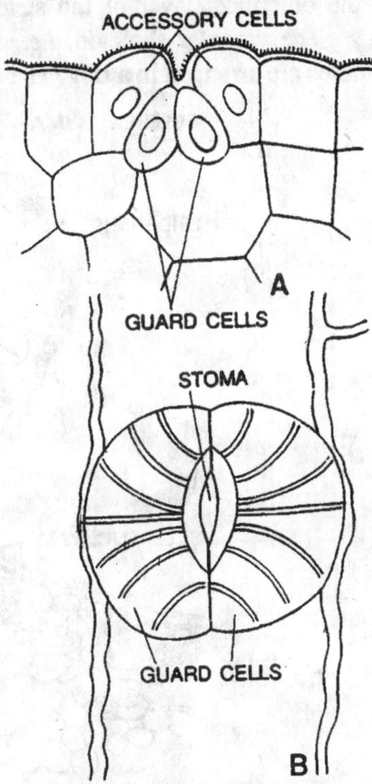

Fig. 6.2. *Equisetum* sp. A, vertical section of an older stoma; B, inner surface of an older stoma showing silicious bars on the inner walls of guard cells (A, after Strasburger; B, after Campbell).

Just underneath the epidermis, there is broad cortex. The cortex consists of mechanical and assimilatory tissue. In the outer cortex, just beneath each of the ridges, there is a strand of sclerenchyma. Usually the sclerenchyma is restricted to the periphery but in *E. giganteum* it extends inward to the endodermis. These columns of sclerenchyma are chief mechanical elements of the shoot. In addition to the above mentioned sclerenchymatous columns, the sclerenchyma strands are also found in the furrows. Here, each strand is situated in between the curved strands of chlorenchyma. The chlorenchyma possess well developed intercellular spaces. These spaces are very much distinct below the stomata. These chlorenchymatous bands are found beneath the sclerenchymatous strands situated underneath the ridges. The ends of the chlorenchymatous bands touch the epidermis in the grooves. Since the leaves are small, scaly and have less number of chloroplasts the chloroenchyma of the cortex of the shoot is the major photosynthetic tissue. The inner cortex is composed of thin walled parenchyma with well developed intercellular spaces. In this region of the cortex the air spaces known as **vallecular canals** are also present. Usually vallecular canals are found opposite the furrows in the deeper tissue of the cortex. They are alternate to the vascular bundles. These canals are filled up with air.

The endodermis is the last layer of the cortex. In *E. arvense, E. telmateia, E. palustre* and certain other species, the endodermis surrounds the entire stele. In *E. giganteum, E. limosum* and some other species, each vascular bundle is surrounded by separate endodermal layer. In still other cases, *e.g., E. hyemale, E. ramosissimum* etc., there is a common internal endodermis inside the ring of bundles delimiting the pith and common outer endodermis found outside the ring of bundles. Just beneath the endodermis a single layered pericycle is found. The pericycle is the outermost layer of the stele.

The vascular skeleton, *i.e.*, stele of *Equisetum* is siphonostelic. The separated vascular bundles are arranged in a ring. The vascular bundles are situated opposite the ridges and alternate

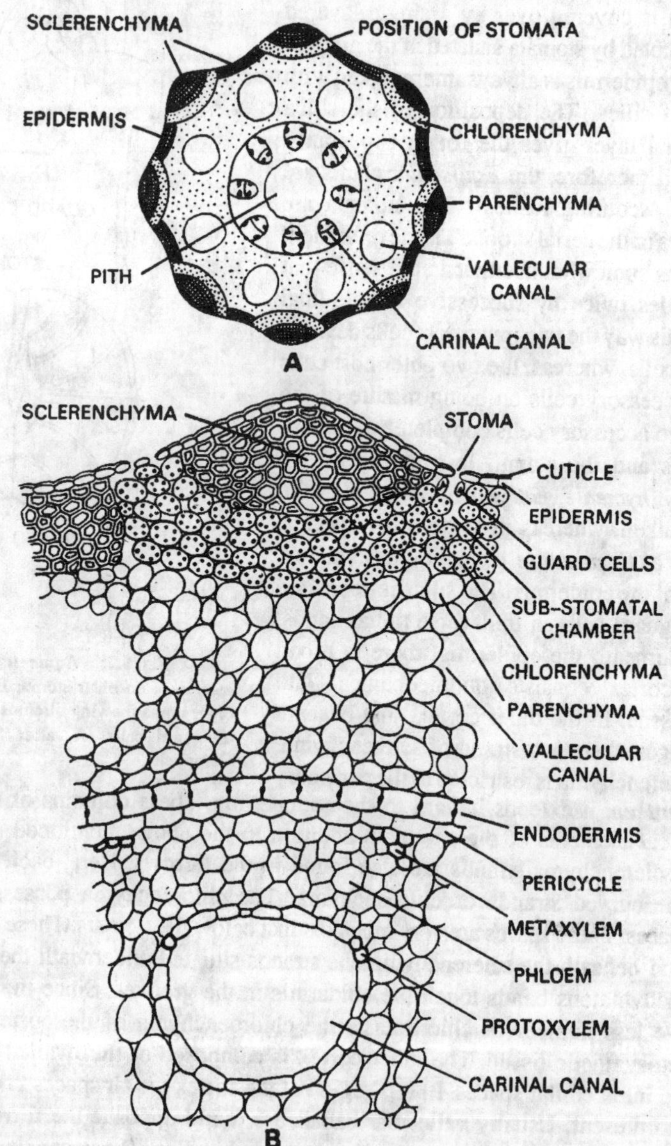

Fig. 6.3. ***Equisetum arvense*. Anatomy of stem. A.T.S. of stem (diagrammatic structure); B, T.S. of stem (detailed structure).**

to the vallecular canals. The number of bundles varies from species to species. In each internode, as many vascular bundles are found as there are leaves at the node. The vascular bundles alternate to each other in their position in the successive nodes. The internodal portions are longitudinally perforate, and each perforation runs the length of an internode. The perforations in the internode are situated above the traces. These perforations are not branch gaps. The vascular bundles are collateral type and resemble to some extent to that of monocotyledons. The vascular bundles contain both metaxylem and protoxylem. A carinal canal is developed in each bundle, because of the disintegration of the early-formed tracheids of the protoxylem during elongation of the surrounding cells of the internode. The remaining protoxylem elements are composed of few tracheids. These protoxylic elements are found arranged to the margin of carinal canal. The metaxylem elements are found in two groups. The two groups are found arranged on the margin of the carinal canal towards outside. The protoxylem lies in between the two groups of metaxylem. The metaxylem elements are composed of reticulate, scalariform or pitted tracheids. Sometimes spiral and annular tracheids are also found. The protoxylem is endarch and as supported by Eames (1909) and Miss Barrat (1920) the development of xylem is centrifugal. The phloem is composed of sieve tubes and phloem parenchyma. The sieve plates may also be seen. The companion cells are not found. The secondary growth is altogether absent. The central region of the stem is occupied by a hollow pith. The carinal canals are filled up with water.

Nodal anatomy. The alternate vascular bundles of the successive internodes are connected to each other by short branches and this way a continuous ring of the vascular cylinder is found in the node. Eames (1909) reported that the bundles at the nodes do not have carinal canals. Here, the protoxylem elements are intact and completely occupy the lacuna or carinal canal. At the node, the pith is not hollow and it forms a diaphragm separating the two successive internodes.

Rhizome. The anatomy of the rhizome is quite identical to that of the aerial shoot. However, the assimilatory tissue and the stomata are not found in the rhizome. The mechanical tissue, *i.e.*, sclerenchyma is poorly developed as compared to that of the aerial shoot. In *E. arvense*, the pith is solid, whereas in certain other species, the pith and the vallecular canals are very much reduced.

Leaf. The anatomy of the leaf is quite simple. The leaves are uninerved, this means that each leaf contains a single vascular bundle. The vascular bundles of the leaf sheath are simple and collateral. The carinal canals are not found. Individual bundles are surrounded by separate

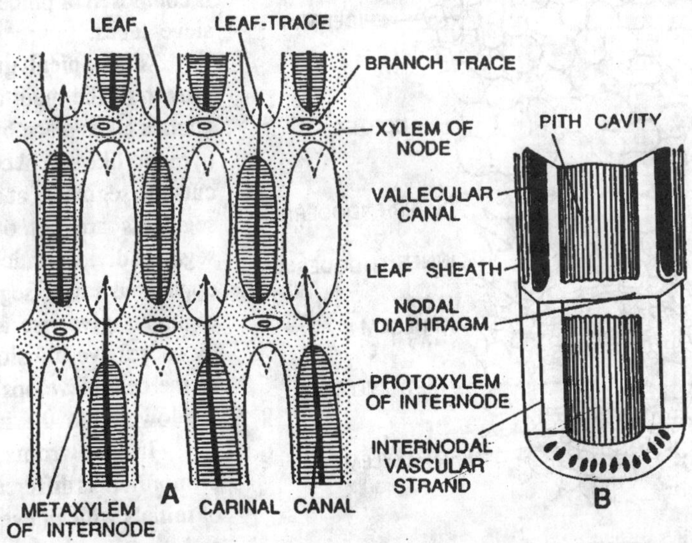

Fig. 6.4. ***Equisetum*** **sp. A, vascular structure of the stem; B, rhizome split longitudinally through nodal region.**

endodermal layers (endodermis). The outer tissues of the leaf sheath are composed of narrow sclerenchymatous bands. These bands of sclerenchyma pass up the leaf ridges and alternate with the strips of chlorophyllous tissue associated with stomata.

Root. The adventitious roots are borne at the nodes of the rhizome and aerial shoots. The anatomy of adventitious roots is quite simple. To study the anatomy the cross sections of the root are required. The outermost layer is known as piliferous layer which bears unicellular hairs upon it. Just beneath the piliferous layer there is a multilayered cortex. In small roots the cortex is divided into two zones. The outer zone is composed of three to four layered lignified exodermis. The inner zone of the cortex consists of thin walled parenchyma with well developed intercellular spaces.

The endodermis is two celled in thickness. The inner layer of the endodermis acts as pericycle. However, the pericycle is absent. The lateral roots originate from the inner layer of the endodermis. In between the two layers of the endodermis, intercellular spaces are found. The Casparian strips are strictly found in the outer layer of the endodermis.

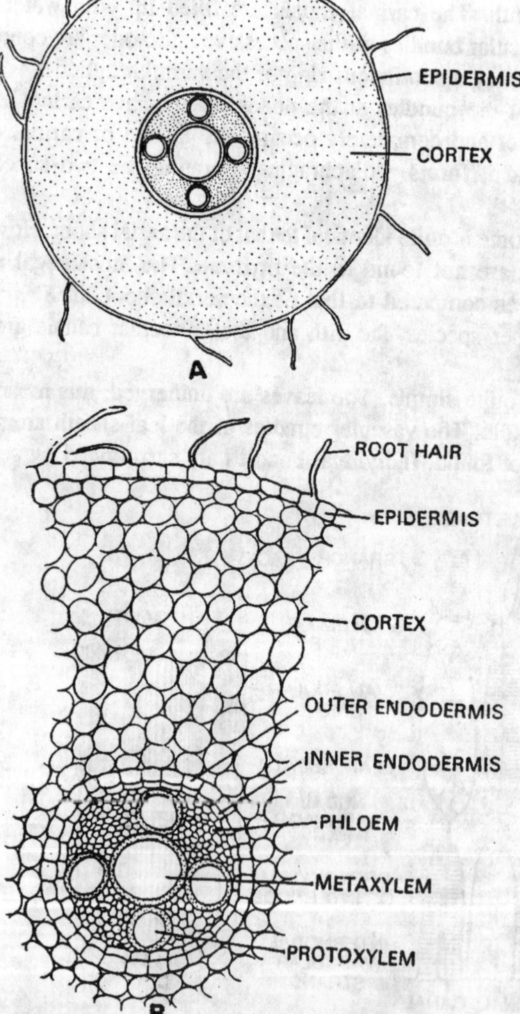

Fig. 5.5. *Equisetum arvense*. Anatomy of root. A, T.S. of root (diagrammatic structure); B, T.S. root (detailed structure).

The stele of the root varies in its nature from species to species. It is triarch to hexarch. There is a big central axial metaxylem element surrounded by three to six narrow points of protoxylem. Each protoxylem point is represented by a single tracheid. The tracheids of xylem elements are spirally thickened. The angles between the protoxylem points are completely filled with phloem. The phloem is composed of phloem parenchyma and sieve tubes.

The apical growth. The apical growth of the main aerial shoot and its laterals takes place by means of a single pyramid like apical cell. This apical cell cuts the segments at its three faces. The segments are cut off regularly. Each segment divides anticlinally forming two upper and lower segments. Both these segments give rise to two tier of cells by successive divisions. The upper tier of the cells develops into the node and the lower into the internode.

In the roots, a young root primordium is differentiated into a group of initials which develop into a root cap and a pyramid-like apical cell, which develops into the root proper. Here, the behaviour of apical cell differs from that of the stem that it cuts off segments from all its four faces, whereas, in the stem only three faces of the tetrahedral apical cell cut off the segments. The lateral segments form the tissues of the root proper, and the terminal segment forms the root-cap.

The strobilus and sporangia. In *Equisetum*, each strobilus consists of a thick axis, having several whorls of densely crowded sporangia bearing peltate appendages situated on the sporangiophores. The sporangiophores are arranged in whorls. Usually, each such whorl is composed of twenty sporangiophores or so. In many cases, immediately below the sporangiophores, the central axis of the strobilus bears a small ring-like outgrowth, known as the **annulus.**

Each sporangiophore is composed of a single slender, stalk of which the terminal end is expanded into a flattened peltate disc situated at right angles to the stalk. The peltate disc is usually hexagonal in outline in its surface view. This hexagonal appearance of the discs is attained due to the mutual pressure of the discs of the crowded sporangiophores situated on the strobilus. Each peltate disc bears a ring of 5 to 10 isolated sporangia on its under side. Each sporangium is elongated and sac-like. On their maturity, the sporangia open by longitudinal slits for dispersal of the spores.

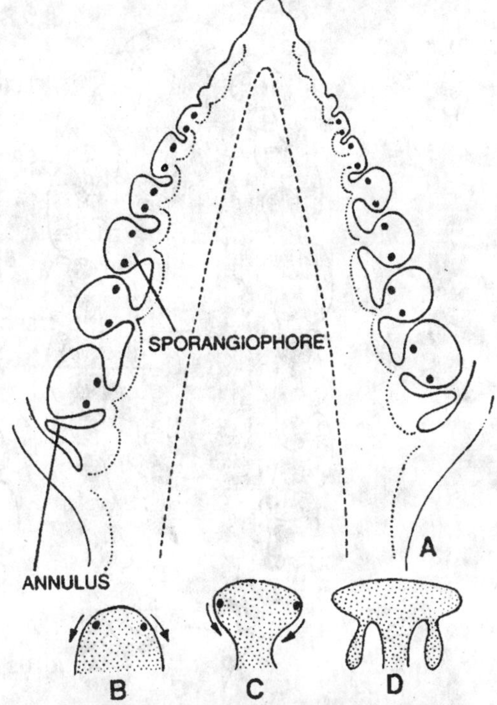

Fig. 6.6. *Equisetum arvense*. A, longitudinal section of immature strobilus; B-D, development of sporangiophore and inversion of orientation of developing sporangium.

Development of sporangium. The development of the sporangia of *Equisetum* is eusporangiate type. This means, that they are not entirely derived from a single sporangial initial. According to Bower (1894) all the sporogenous tissue in a sporangium are developed from a single superficial cell of the young sporangiophroe.

The development of the sporangia begins as the strobilus ceases to grow in length. The young sporangiophores are found in the form of convex outgrowths in acropetal succession on the conical axis of the strobilus. In the beginning of the development the sporangiophore is hemispherial, but later on this soon becomes constricted at its base developing a short stalk and sporangia bearing peltate disc at its apical end.

The development of the sporangium begins from a single superficial cell known as sporangial initial situated in the peripheral region of the young hemispherical sporangiophore. The centre of the tip of young sporangiophore increases in size actively and the sporangial initials are shifted to its lateral sides. Sooner or later, because of further development, the outer edge of the sporangiophore becomes inverted and faces the axis, this way the sporangial initials have also changed their position, and now facing the axis.

At first, the sporangial initial divides periclinally forming the two daughter cells. The inner daughter cell gives rise to the sporogenous tissue by repeated division, whereas the outer cell gives rise to both the sporogenous tissue and the portion of the jacket of the sporangium radial to the sporogenous tissue. The remaining portion of the jacket layer is developed from

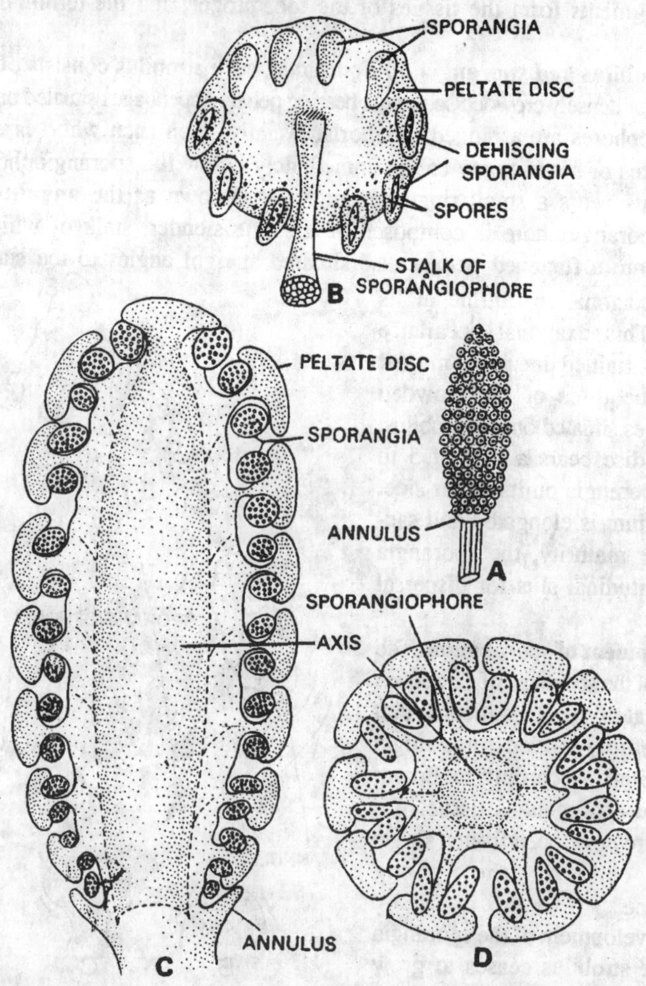

Fig. 6.7. ***Equisetum arvense*****. Structure of strobilus; A, a strobilus; B, single sporangiophore with ten sporangia on the underside of the peltate disc; C, L.S. of strobilus showing sporangia; D, T.S. of strobilus showing sporangia.**

the cells laterally placed to the original initial. Ultimately the jacket layer becomes mutilayered. The innermost layer is the tapetum and supplies nourishment to the developing spores. During the development of the spores, the tapetum and all but two layers of the jacket become disintegrated. This way, the mature sporangium is surrounded by a two layered jacket. The cells of the outermost layer of the jacket are spirally thickened.

During the development the cells of the sporogenous tissue become separated and rounded and act as spore-methor cells. About one third of the mother cells disintegrate. The disintegration of the mother cells, tapetum and the other jacket layers, together form a plasmodial fluid in which the remaining two third spore mother cells float about. Ultimately the spore mother cells undergo usual meiotic division developing the spore tetrads. Each tetrad is composed of four haploid spores.

Structure of sporangium. Each mature sporangium is elongated, sac-like and rounded at its apex. On maturity, it consists of single layered intact layer. By now, rest of the jacket layers are aborted. The sporangium contains within it the mature spores all alike. The *Equisetum* is homosporous. The sporangia are attached on the underside of the hexagonal peltate disc in a ring. Each sporangiophore bears 5-10 sporangia.

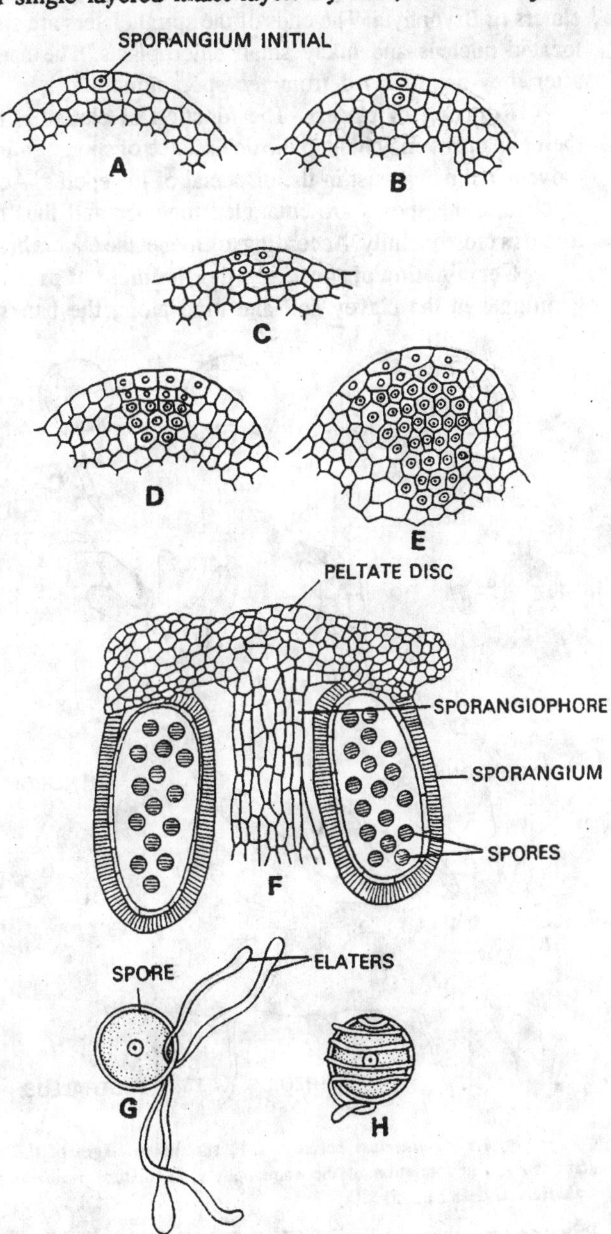

Fig. 6.8. *Equisetum arvense*. A-E, successive stages in the development of sporangium; F, sporangiophore with peltate disc and sporangia in longitudinal section; G, spore with uncoiled elaters; H, spore with coiled elaters.

Dehiscence of sporangium. On the maturation of the spores within the sporangium, it bursts by a longitudinal slit. Prior to the opening of the sporangia, the internodes of the strobilus elongate separating the sporangiophores quite apart from each other. The walls of the cells of the jacket layer are spirally thickened they shrink and lose water and the longitudinal slit develops on the sporangium for the release of the spores.

The spores. According to Beer (1909) the spore secretes a wall around it, composed of four concentric layers. The outermost layer is known as exosporium and the innermost endosporium. Outside the exosporium there is delicate cuticle called the middle layer. On the extreme outside of the spore there lies epispore which later on develops into four strips which get separated from each other from rest of the wall but for one common point of attachment. These strips are **elaters.** The strips or elaters remain coiled around the spore unless and until it is not disseminated from the sporangium. The elaters are hygroscopic. They remain coiled and uncoiled around the spore due to the changed conditions of the moisture of the air. When the air is moist they remain coiled around the spore and as soon as the air is dry the elaters are uncoiled. The elaters of *Equisetum* do not resemble in any respect to the elaters of Bryophyta. The ends of the spiral elaters are spoon-like. Each spore contains a centrally located nucleus and many small chloroplasts. The spores remain viable only for a few days after they are released from the sporangium.

Function of elaters. The function of elaters is not very certain, but it is assumed that their expansion may help in the dehiscence of sporangium. It is also thought that their hygroscopic movements may assist in the dispersal of the spores. According to Buchtien due to the presence of elaters, the spores are entangled together and thus dispersed in groups so the prothalli are found in close vicinity. According to Goebel the elaters help in fastening the spore to the substratum.

Germination of spore and development of prothallus (gametophyte). Usually the spores germinate in the clayey soil and mud along the banks of the rivers and streams. As already mentioned the spores contain chloroplasts and therefore, they can survive only for few days after their liberation from the sporangium. Prior to the germination the spore increases to some extent in its size. Within 10 to 12 hours, the first cell division in the development of the gametophyte takes place, after the spore is sown. This way, the two daughter cells are resulted one small and the other large in size. The small daughter cell develops into the first rhizoid and the large one gives rise to the remainder of the thallus (Campbell, 1918; Kashyap, 1914 and Walker, 1931). The large cell divides irregularly several times giving rise to one to several celled thick cushion-like tissue. This cushion-like structure bears numerous rhizoids on its lower surface. This cushioned structure is known as gametophyte or prothallus. Upon this cushion-like massive structure, the green lobes or branches are found. The prothallus is elongated and lobed. The cells of the lower portion of the cushion are colourless whereas those of portion bear chloroplasts and are green. The entire margin of the prothallus is meristematic. When the growth of the entire margin is uniform the prothallus becomes quite smooth and circular, whereas on the other hand when growth is more active on one side or the other, the irregularly lobed prothallus is resulted. The upright green lobes arising from the upper side of the prothallus are densely crowded and almost cover the whole upper surface of the gametophyte. In certain species the gametophytes are quite big

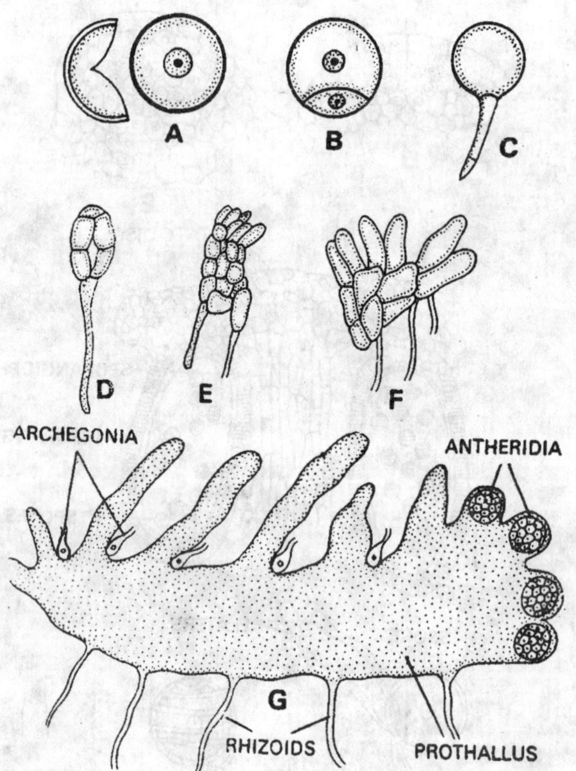

Fig. 6.9. *Equisetum arvense*. A-F, successive stages of the spore germination and formation of the gametophyte; G, mature prothallus with antheridia and archegonia.

in size, *e.g.*, in *E. debile* Rox, the gametophyte is of 3 centimetres in diameter (Kashyap, 1914), and two to three millimetres in thickness. In majority of the cases the gametophytes are less than a centimetre in size. The gametophytes are long lived. According to Walker (1921), it has been reported on the basis of culture experiments that they can survive for more than two years. The large gametophytes survive for a long period, whereas small prothalli die soon.

Nature of prothalli. There are two types of prothalli-the small male prothalli bear only-antheridia whereas the large hermaphrodite prothalli bear both the antheridia and archegonia. Whether *Equisetum* is monoecious or dioccious is a matter of controversy. According to Kashyap (1917), in *E. debile* when spores are sown thickly, the prothalli, remain small in size and bear only one type of sex organs. On the other hand, if the spores are shown quite separate from each other the prothalli become large in size and bear both types of sex organs, *i.e.*, antheridia and archegonia. In such cases, however, the archegonia develop first and thereafter the antheridia. According to Walker (1921), *E. laevigatum* is a monoecious species. In *E. arvense*, under unfavourable conditions the prothalli remain small in size and only antheridia are developed, upon them. In favourable conditions, half of the prothalli are developed as males and the remaining half as females. According to Walker (1931) the normal prothalli are monoecious, which develop archegonia first and the antheridia later on. He also stated that the prothalli developing in crowded conditions remain small and dioecious. In *E. arvense*, the normal gametophytes produce archegonia at first and thereafter antheridia. In conclusion, the *Equisetum* is monoecious. This view has been supported by several eminent bryologists.

According to Joyet Lavergne (1931), morphologically the spores are alike, but physiologically they are different, and in the same lot some of the spores are male and some are female, so according to him, there is internal heterospory.

Development of archegonium. If the favourable conditions prevail, the first sex organs appear when the normal separated gametophytes are 30 to 40 days

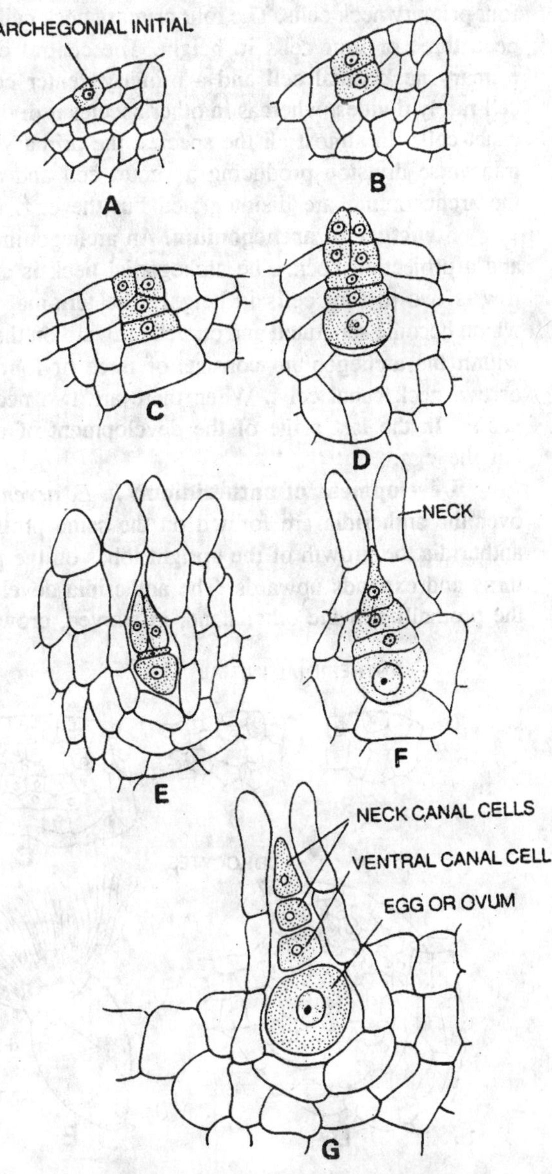

Fig. 6.10. *Equisetum sp.* A-F, successive stages in the development of archegonium; G, mature archegonium.

old. The sex organs appear only when the prothallus is differentiated into two regions-the upper green lobed region and the lower massive colourless tissue. Normally the archegonia develop first on the prothallus. The archegonia develop in the meristematic margin of the prothallus. They are situated among the up right green lobes.

The archegonial initial is a superficial cell situated in the meristematic margin of the prothallus. The archegonial initial divides first by a periclinal wall giving rise to two daughter cells. The outer daughter cell is known as the primary cover cell and the inner daughter cell is known as the mother cell of the central cell. The primary cover cell divides further producing four primary neck cells. The four primary neck cells divide transversely producing an archegonial neck three or four cells in height. The central cell divides by a transverse wall producing a primary neck canal cell and a primary venter cell. In some species, the primary neck canal cell rarely divides, whereas in other species it divides regularly producing two boot-shaped neck canal cells. In almost all the species, the primary venter canal cell divides by an asymmetrical transverse division producing a venter cell and an egg. Lastly, all the axial cells situated in the archegonium are disintegrated but the egg.

Structure of archegonium. An archegonium consists of a sunken base in the prothallus, and a projecting neck. The archegonial neck is composed of four vertical rows of cells. Each row is two to four cells in height. The terminal cells of the archegonial neck are quite long, which become separated and curved outward on their maturity. The axial row of the cells situated within the archegonium consists of three or four cells-the egg, the venter canal cell and one or two neck canal cells. When there are two neck canal cells, they are boot shaped, *e.g., E. debile*. In the last stage of the development of archegonium the axial cells are disintegrated but the egg.

Development of antheridium. In *E. arvense* as soon as the formation of archegonia is over the antheridia are formed on the same prothalli. At the time of the development of the antheridia the growth of the upright lobes on the prothallus is stopped, and the prothallus itself turns and expands upwards. The antheridia develop in these expanded and turned portions of the prothalli. On the other hand, in starved, crowded prothalli, only antheridia are developed. These prothalli may also be called the male prothalli. Usually, the antheridia are of two types-(a) the embedded type and (b) the projecting type. The former type of antheridia usually develop on the massive tissue of normal gametophytes and the later type of antheridia are developed in the meristematic margin of the starved crowded and dwarfed male prothalli. The development of the antheridium is as follows :

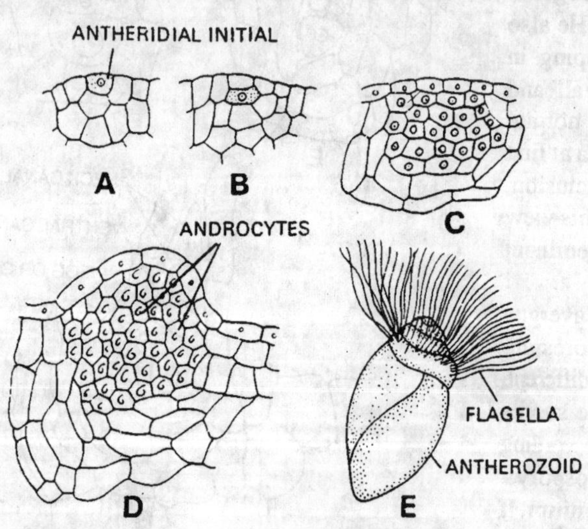

Fig. 6.11. *Equisetum* sp. A-D, successive stages in the development of antheridium; E, antherozoid.

A superficial cell of the meristematic margin of the prothallus acts as an antheridial initial. The first division of the initial is periclinal. This way, two daughter cells are formed. The outer daughter cell is known as the jacket initial and the inner daughter cell is known as the primary androgonial cell. The jacket initial divides repeatedly giving rise to single layered jacket. Sometimes there lies a triangular opercular cell at the apex of the jacket layer. The primary androgonial cell divides repeatedly giving rise to a group of androgonial cells. Ultimately, the androcyte mother cells are formed. Near the nucleus of each androcyte mother cell two blepharoplasts appear. Now, each androcyte mother cell divides and gives rise to two androcytes each possessing a blepharoplast.

Each androcyte metamorophoses into a spirally coiled multi-flagellate antherozoid. In the beginning, the rounded nucleus and the blepharoplast of the androcyte become elongated. The blepharoplast is composed of many granules. Later on all the granules become united. The nucleus

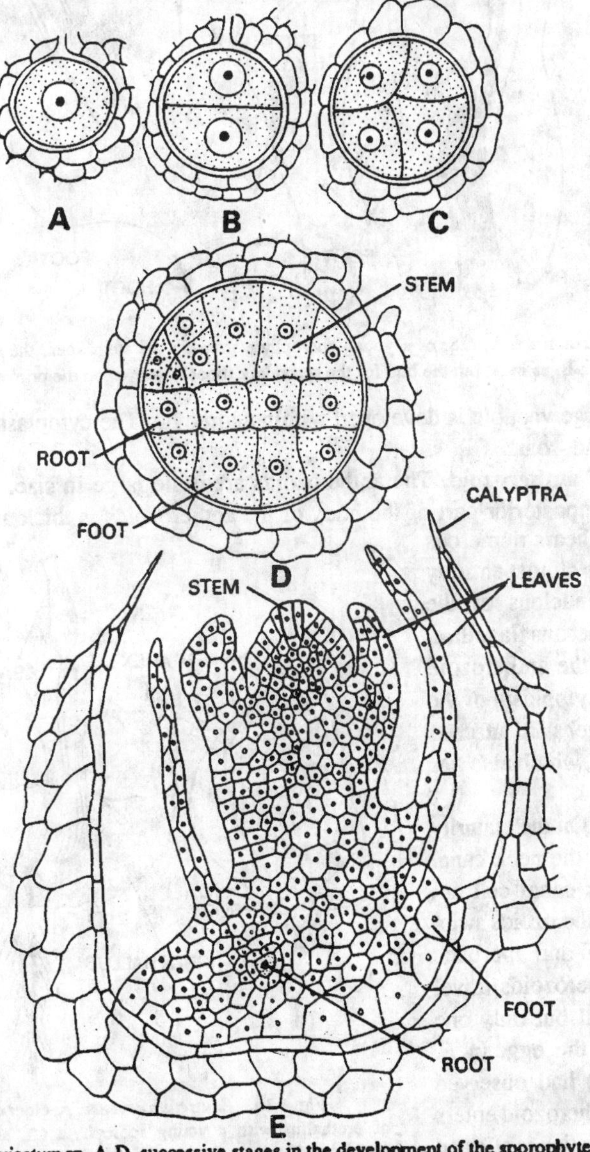

Fig. 6.12. ***Equisetum*** **sp. A-D, successive stages in the development of the sporophyte (embryo); E, young embryo**

elongates continuously and ultimately becomes crescent shaped. In the same way, the blepharoplast also becomes spirally elongated. Later on, the posterior end of the nucleus becomes expanded

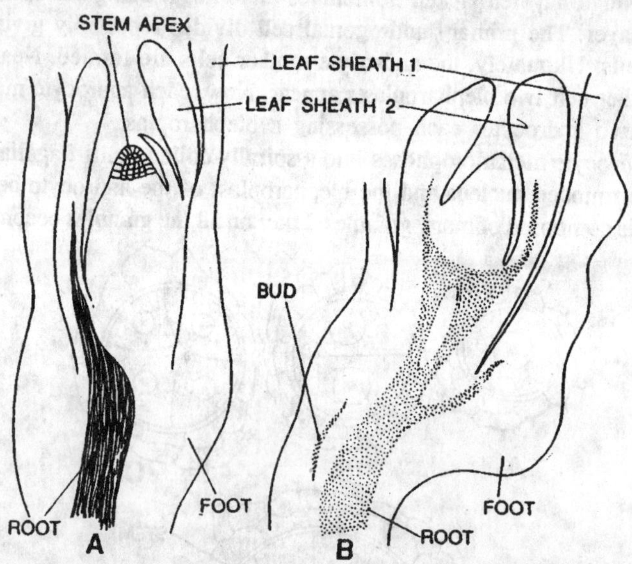

Fig. 6.13. *Equisetum* sp. A, L.S. of a young sporophyte showing the stem apex, the two first leaf sheaths, the foot and the base of root; B, as in A, but the bud for the secondary shoot is seen upon the primary root. (After Campbell).

and flattened. A large vacuole is developed within androcyte. The cytoplasm of the antherozoid becomes foamy and coarse.

Structure of antherozoid. The antherozoids are quite large in size. Each antherozoid is spirally coiled. The posterior part of the body of the antherozoids is thick and flat. The anterior part is conical and bears numerous flagella. The body is almost entirely derived from the nucleus of the androcyte. The numerous flagella at the anterior end of the antherozoid originate from the cytoplasm of the mother cell. The flagella are attached to the blepharoplast, which also lies at the antherior end.

Fertilization. On the maturity of the archegonium the neck canal cells and the venter canal cell are disorganised. The antherozoids swim down the neck canal and approach the egg. Several antherozoids travel down the neck canal but only one of them penetrates the egg. In *E. debile*, Sethi (1928) had observed that the complete antherozoid enters the egg and the fertilization is effected

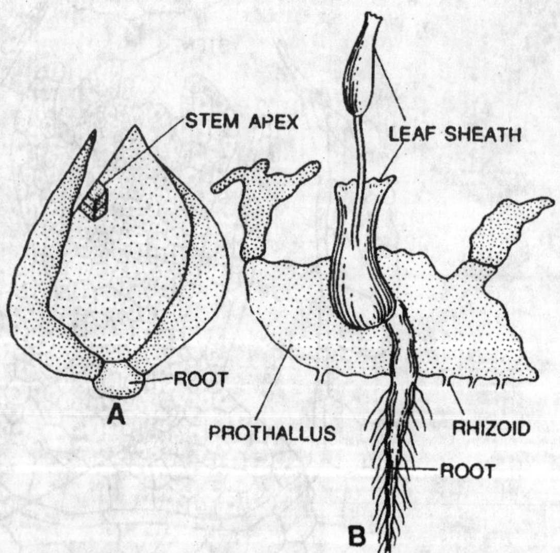

Fig. 6.14. *Equisetum arvense*. A, older embryo; B, V.S. of a lobe of prothallus with a young sporophyte (A, after Sadebeck; B, after Hofmeister).

forming a diploid oospore (2x). Unlike other pteridophytes usually more than one archegonia are fertilized and several sporophytes are developed on the same prothallus. Usually 8 to 10 normal sporophytes develop on a single prothallus, *e.g., E. debile*. Kashyap (1914) has recorded as many as 15 sporophytes on a single gametophyte in *E. debile*.

Development of embryo (young sporophyte). As supported by Campbell (1918, 1928); Jeffery (1899) and Walker (1921), usually the first division of the oospore is transverse, giving rise to two cells-the upper one is epibasal and the lower one the hypobasal. Here in the case of *Equisetum* both the halves of the oospore take part in the development of the embryo proper and no suspensor is developed like that of *Selaginella*. The next division takes place at right

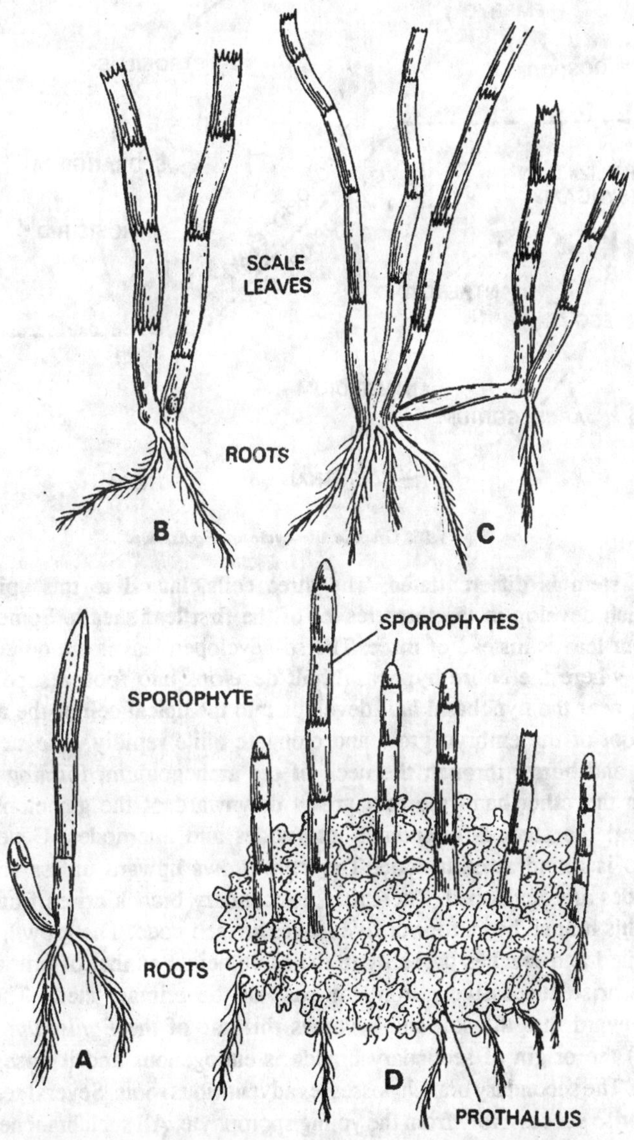

Fig. 6.15. *Equisetum* **sp. Young sporophytes. A-C, young sporophytes of** *Equisetum hiemale;* **D, a large prothallus with six developing young sporophytes.**

angles to the first and the quadrant stage of the embryo is attained. According to Sadebeck (1878) in *Equisetum* the first division of the oospore is followed by two successive divisions at right angles to each other giving rise to eight cells. According to Campbell (1921), in *E. debile*, the entire hypobasal half develops into a foot, whereas in *E. arvense* (Sadebeck, 1878) the hypobasal half of the embryo gives rise to both foot and root. Early in the development, in *E. arvense* the epibasal cell divides by three intersecting walls in such a way that a tetrahedral

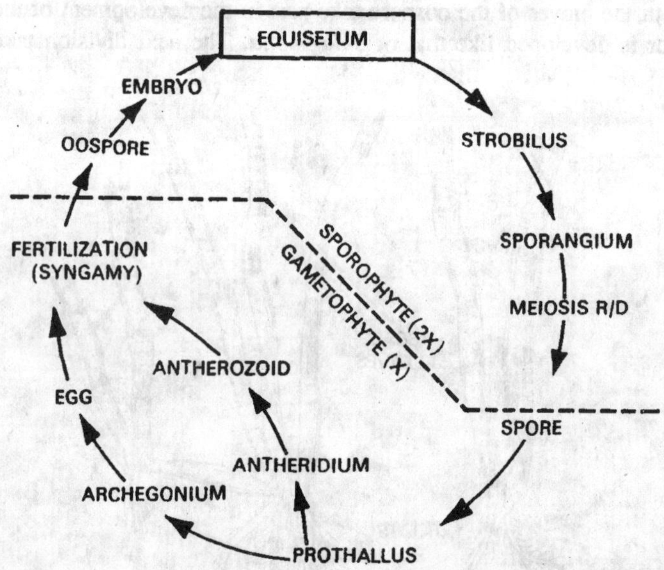

Fig. 6.16. Graphic life-cycle of *Equisetum*.

apical cell of the stem is differentiated. The three cells, lateral to this apical cell are also differentiated, which develop in the three leaves of the first leaf sheath. Sometimes, there may develop two or four leaves instead of three. The so-developed leaves are quite small and scale-like. In *E. debile*, where the entire hypobasal half develops into foot, a superficial cell of the epibasal half lying near the hypobasal half develops into the apical cell of the root. The portions of the stem and root of the embryo grow and elongate quite rapidly. The stem portion of the embryo elongates and bursts through the neck of the archegonium, forming the calyptra and grows upright. On the other hand the root grows downward of the gametophyte and reaches the soil. Soon after, the stem differentiates into nodes and internodes. Generally a whorl of three leaves, or so is found at each node. The stem grows upward unless and until 10 to 15 nodes and internodes are developed. Thereafter a secondary branch arises from the base of the primary branch. This branch bears four or five leaves at each node. The growth of the secondary branch is also limited and reaches probably to the same height as that of primary branch. More vigorous branches arise successively from the base of the primary stem. The third or fourth branch turns downward into the soil and becomes rhizome of the *Equisetum* plant. According to Cambell (1928) the origin of secondary branch is endogenous and it arises from the tissue of the primary root. The secondary branch possesses adventitious roots. Several secondary branches of limited growth arise successively from the young sporophyte. All such branches are endogenous in their origin. From the point of view of its anatomy, the primary stem is solid and without carinal canals which develop later on in the adult stem. However, the xylem is better developed in the primary stem. The base of the seedling bears protostele, but later on the stem bears siphonostele at the level of the first branch.

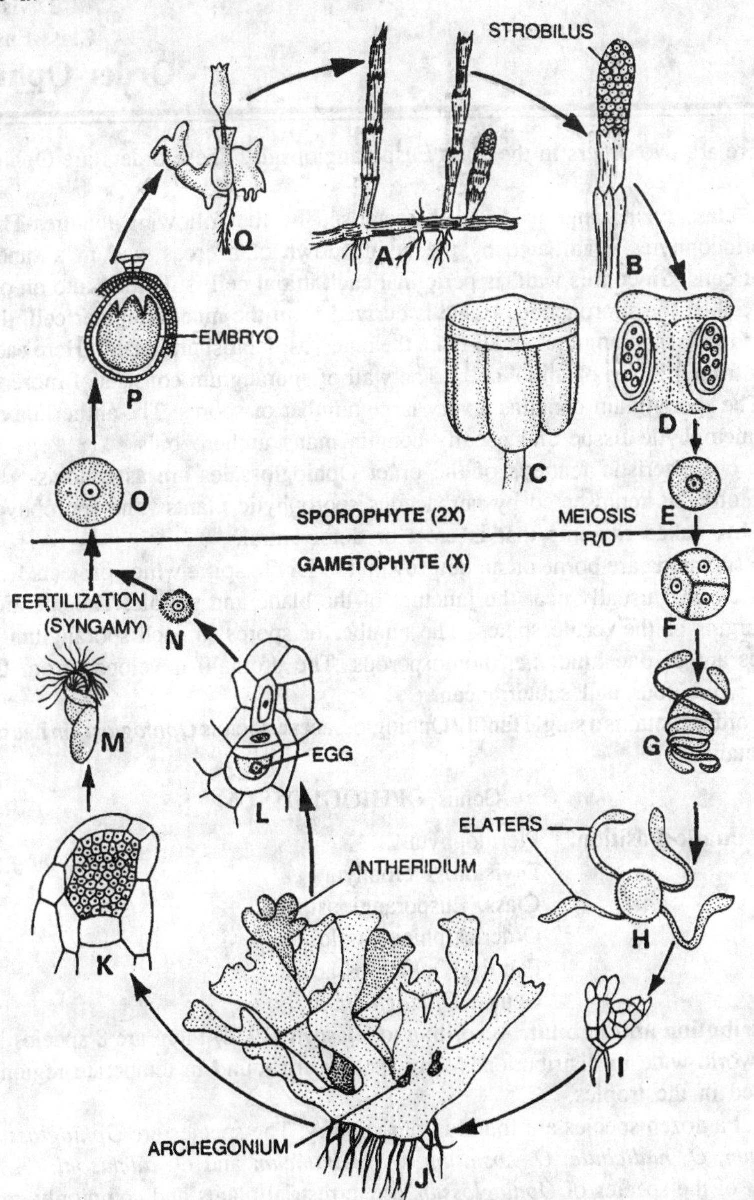

Fig. 6.17. ***Equisetum*** **sp. Diagrammatic life-cycle-A, sporophyte; B, strobilus; C, sporangiophore with sporangia; D, L.S. of sporangiphore showing sporangia and peltate disc; E, spore mother cell; F, spore tetrad; G, spore with coiled elaters; H, spore with uncoiled elaters; I, young prothallus; J, mature prothallus with archegonia and antheridia; K, antheridium; L, archegonium; M, antherozoid; N, egg; O, oospore; P, young embryo; Q, young sporophyte.**

Economic importance. Economically, they are of less value. Some of the species are used in the preparation of the indigenous medicines. *E. arvense* is used as a diuretic whereas *E. debile* is used as a cooling medicine.

7

Division-Pterophyta
Class-Eusporangiopsida

Order-Ophioglossales

There are two orders in the class Eusporangiopsida. These orders are-Ophioglossales and Marattiales.

The class Eusporangiopsida is characterised by the following features-The sporangium in all pteriodophytes is initiated by the laying down of a cross wall in a superficial cell or a group of cells. Since this wall is periclinal each initial cell is divided into an outer and inner daughter cell. If the sporogenous tissue is derived from the inner daughter cell, the sporangium is described as 'eusporangiate' and if from the outer, as 'leptosporangiate'. Here each sporangium originates from a group of initial cells. The wall of sporangium consists of more than one layer of cells. The sporangium contains a very large number of spores. The antheridia remain sunken in the gametophytic tissue and usually contain many antherozoids.

The characteristic features of the order Ophioglossales are as follows-

The order is represented by herbaceous sporophytic plants. The sporophyte possesses a short, fleshy, naked rhizome that is erect or dorsiventral.

The sporangia are borne on an outgrowth, the **fertile spike** which projects from the adaxial surface of the leaf usually near the juncture of the blade and petiole. The sporangia are found on the margins of the fertile spikes. The number of spores in each sporangium is very large. The spores are of one kind, *i.e.*, homosporous. The prothalli developed from the spores are saprophytic, tuberous and subterranean.

The order contains a single family Ophioglossaceae, Genus *Ophioglossum* has been discussed here in detail.

Genus OPHIOGLOSSUM

Systematic Position. Pteridophyta
Division. Pterophyta
Class. Eusporangiopsida
Order. Ophioglossales
Family. Ophioglossaceae
Genus. *Ophioglossum*

Distribution and habitat. According to Clasuen (1938) there are 8 species in this genus. They are world-wide in distribution. Many species are found in temperate regions, but this is least thrived in the tropics.

About a dozen species are found in our country. The species are-*Ophioglossum vulgatum, O. pendulum, O. nudicaule, O. costatum, O. lusitanicum* and *O. aitchisoni.*

Most of the species of *Ophioglossum* are terrestrial plants and commonly grow in humus soil. Two species are epiphytic, *e.g.*, *O. pendulum* and *O. palmatum.* These species grow on trunks and branches of trees.

This genus has been divided into three sub-genera-1. *Euophioglossum;* 2. *Ophioderma* and 3. *Cheiroglossa.*

THE SPOROPHYTE

The sporophyte of most species of *Ophioglossum* possesses a short, erect, subterranean rhizome. A few species (*e.g., O. pendulum; O. intermedium*) have a markedly dorsiventral stem. The branching of the sporophyte is generally due to a formation of axillary buds in the axils of leaves. Petry (1915) has also described the branching of rhizome as dichotomous.

Root. The rhizome bears numerous Adventitious roots. These roots are commonly thick, smooth and devoid of hairs. As a rule the roots are borne singly below the leaf or leaves of the current season and the leaf scars of previous seasons. The roots are mycorrhizal. They arise endogenously.

Leaves. Most of the species of *Ophioglossum* produce but one leaf each year. Some tropical species may produce four or five leaves in a season. The leaves are usually borne in an irregular

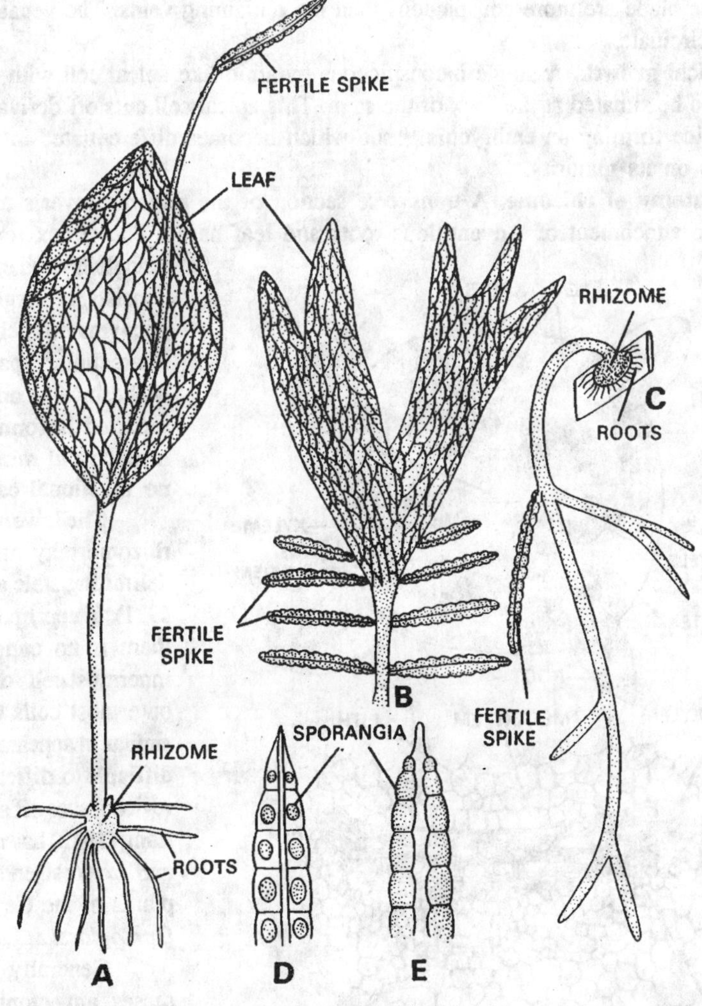

Fig. 7.1. A, sporophyte of *Ophioglossum vulgatum;* B, sporophyte of *O. palmatum;* C, sporophyte of *O. pendulum;* D-E, small portion of spike.

spiral on the stem. Where growth is seasonal the development of the leaf is very slow. In such cases the embryonic leaf at a stem apex develops slowly for three years and then grows above ground and unfolds the fourth year.

A fully expanded leaf is sharply differentiated into blade and petiole. The blade of the leaf in all but one species (*O. palmatum*) is entire and narrowly linear to broadly oval in outline. At the juncture of the blade and petiole, in all but one species (*O. palmatum*) there is a single fertile spike. In *O. palmatum* there are several small fertile spikes in the region where blade and petiole adjoin. In *O. palmatum* the blade is palmately incised into a number of narrow segments. In *O. pendulum* the leaf may attain a length of 1.5 metres or more and its petiole merges gradually into a strap-shaped lamina. The venation of the blade is always reticulate. There is no distinct midrib, but according to the species the small veinlets may or may not present. These veinlets end blindly within the large meshes. However, the veins near the central axis of the blade are more conspicuous than the remaining veins. The venation of leaf is erect and not circinate.

Apical growth. A single inconspicuous pyramid-like apical cell with three cutting faces is found to be situated at the apex of the stem. This apical cell cuts off derivatives which divide and redivide forming an embryonic tissue which becomes differentiated into stele, cortex and epidermis on its maturity.

Anatomy of rhizome. A transverse section of the rhizome reveals an irregular outline due to the attachment of Adventitious roots and leaf bases. The cortex remains bounded by a single-layered epidermis. The cortex consists of large thin-walled parenchymatous cells having abundant intercellular spaces among them. Maheshwari and Singh (1934) reported the formation of a periderm layer in old rhizomes; but there is no functional cambium.

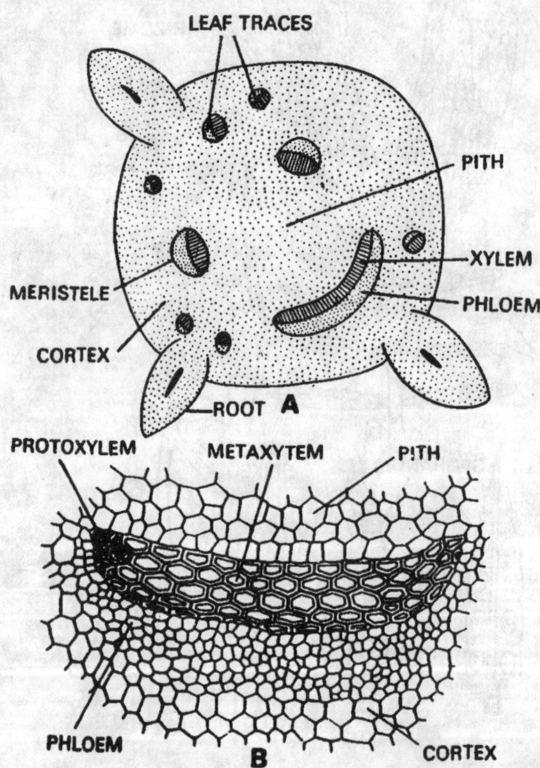

Fig. 7.2. ***Ophioglossum*** **sp. A, transverse section of rhizome; B transverse section of meristele.**

The lowermost portion of the rhizome may have an endodermis delimiting stele and cortex (Bower, 1911). In upper portion of the rhizome there is no endodermis. Here the innermost cells of the cortex and the outermost cells of the stels are so similar in appearance that it becomes difficult to differentiate where one region leaves off and the other begins. Lang (1912) has reported an internal and external endodermis in young plants of the Ceylonese species of *Ophioglossum*.

Generally the rhizome possesses an ectophloic siphonostele throughout its entire length. This siphonostele may have but one leaf gap when cut at any particular level.

More often two or more leaf gaps may overlap forming a dictyostelic condition. Most of the species have a single leaf trace departing at each gap and traversing the cortex without branching. A few species, *e.g., O. palmatum* and *O. pendulum* have two traces departing at each gap. There is a division of each trace into several strands before it runs into the leaf base.

The xylem of each meristele is endarch in arrangement. The protoxylem remains situated on the inner face of the xylem mass. The reticulate tracheids of irregular shape form the metaxylem. Boodle (1899) has reported a small amount of secondary thickening in the stem of *O. vulgatum*. The protoxylem is composed of spiral and scalariform elements. In *O. palmatum*, the protoxylem is ill defined. The relation of the xylem to the phloem is collateral, that is the phloem lies external to the xylem. Commonly the phloem consists of a layer four or five cells in thickness. However, in *O. pendulum* (Petry, 1914) the phloem consists of a single layer of cells separated from the xylem by a layer of parenchyma three to five cells in thickness. The pith internal to the xylem is composed of thin-walled large parenchymatous cells. According to Petry (1914), sometimes (*e.g.,* in *O. pendulum*) the tracheids occur in the pith region.

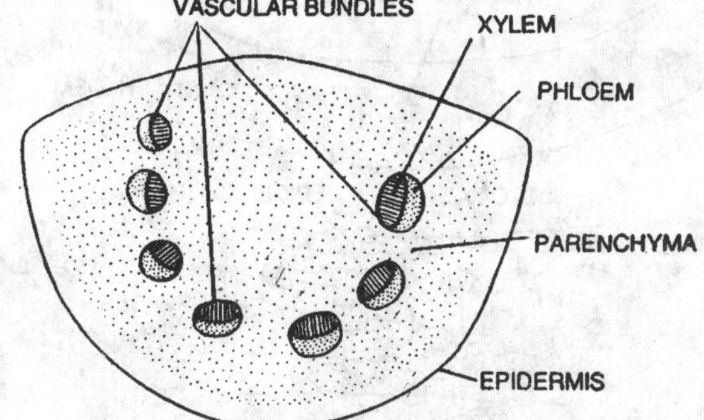

Fig. 7.3. *Ophioglossum vulgatum.* Transverse section of petiole (diagrammatic)

Anatomy of leaf. The epidermis of the petiole is single layered. The interior of a petiole is composed of a spongy parenchyma that may or may not possess conspicuous cavities in the

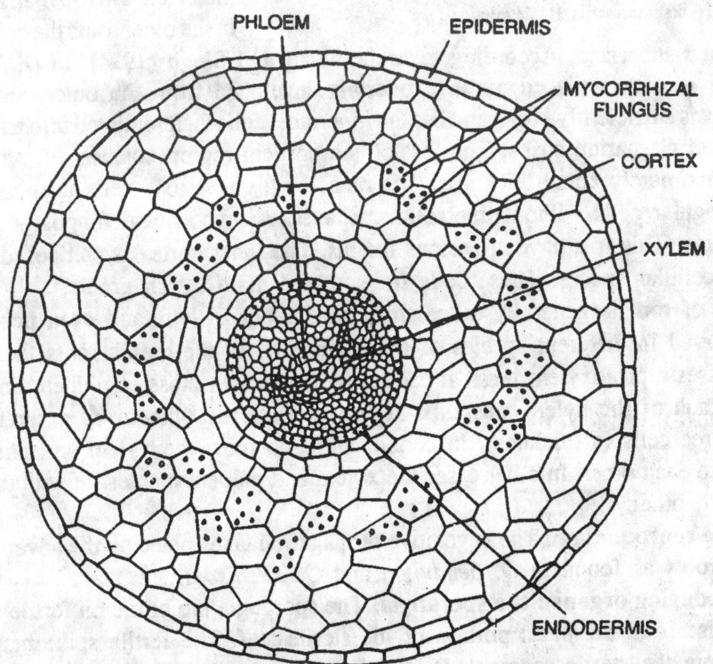

Fig. 7.4. *Ophioglossum vulgatum.* **Transverse section of petiole (detailed).**

central region. Just beneath epidermis, the parenchymatous cells contain numerous disoid chloroplasts. The number of vascular bundles is one or two at the base. The trace, or the pair of traces, departing to a leaf become divided into strands which lie parallel to one another throughout the length of the petiole. The bundles may be arranged in a crescent or C-shaped outline. The bundles are collateral. The xylem is endarch in development. The phloem is found toward outer surface.

The internal structure of the sterile leaf lamina resembles with that of petiole. In the species with erect leaves, stomata are present in equal numbers on both faces of the blade. In *O. reticulatum* where the leaf blade is more or less horizontal, the upper epidermis contains more stomata. There is no distinct hypodermis. The bulk of lamina is composed of thinwalled green cells with frequent intercellular spaces. The structure of bundle is similar to that of petiole. Several veinlets are found to be present in the leaf lamina. All cells of mesophyll tissue contain chloroplasts.

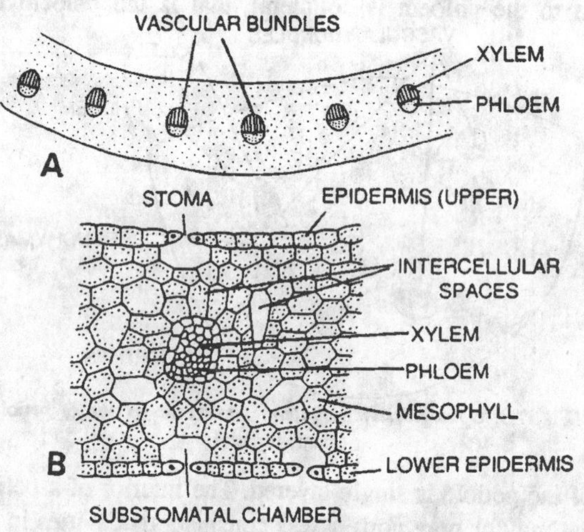

Fig. 7.5. *Ophioglossum vulgatum.* Transverse section of leaf. A, diagrammatic; B, detailed.

Anatomy of root. A transverse section of the root reveals a single-layered epidermis, a broad cortical region and a central stele. In young root the epidermis remains intact and the root hairs are lacking. In the older roots the epidermal cells are thickened and suberized. According to Maheshwari and Singh (1934), in *O. fibrosum* the epidermal layer is soon replaced by an exodermis developed from the outer cortex:

The cortex is sufficiently broad and has the innermost cells differentiated into an endodermis with well-defined Casparian strips. The rest of the cortical region consists of two concentric regions of approximately equal thickness. The outer cortex is 4 to 7 cells in thickness and is composed of angular cells without intercellular spaces. A mycorrhizal fungus is also present in the outer cortex. This is a phycomycetous fungus. The inner cortex is composed of rounded cells with intercellular spaces. These cells frequently contain starch grains.

The stele of root is generally monarch or diarch but tri-tetra-and even pentarch steles have been observed in different species of *Ophioglossum*. In the diarch roots the usual plate of xylem possesses a protoxylem mass at each pole. There is a mass of phloem on either side of the xylem. Each of the xylem masses is separated from the phloem by a parenchymatous tissue two to three cells in thickness. In certain monarch roots the xylem and phloem are to be found radial to each other. In other cases the xylem and phloem masses lie with their broader sides facing each other.

Vegetative reproduction. The vegetative propagation takes place by the adventitious buds formed on the roots as found in *O. pendulum* and *O. aitchisoni*.

Spore producing organs : the sporangia. The sporangia are borne on fertile spikes. The fertile spikes arise from the basal portion of the sterile leaf. The fertile spike may easily be differentiated from the sterile segment. It is usually erect and stalked.

The sporangia are found within the tissue of the fertile spike. The sporangia are found to be arranged on the margins of the spike in two longitudinal rows. The number of sporangia varies from 6-20 on a fertile spike.

Development of the spike. A pyramid-like apical cell with four cutting faces gives rise to a conical structure of meristematic tissue that represents a very young fertile spike. This conical outgrowth turns upwards. According to Bower (1896), the four quadrants cut just back of the apex of the spike corresponding to the four cutting faces of the apical cell. These quadrants are oriented in such a way that two lie in a plane perpendicular to the sterile lamina, and two in a plane parallel to it.

Development of sporangium. As development of spike continues and it becomes older, a vertical strip of cells in the epidermal layer of the two quadrants in the parallel to the leaf blade, becomes differentiated. Each such strip two to three cells broad and several cells tall, is a **sporangiogenic band.** The cells of the sporangiogenic band divide periclinally and anticlinally forming the two bands that become three or four cells in thickness. Now the hypodermal region of the sporangiogenic band differentiates into alternate blocks of sterile cells and archesporial cells. Each group of archesporial cells divides repeatedly forming a large number of sporogenous cells. The portion of the band external to each archesporium divides periclinally to form a multilayered jacket layer of the sporangium.

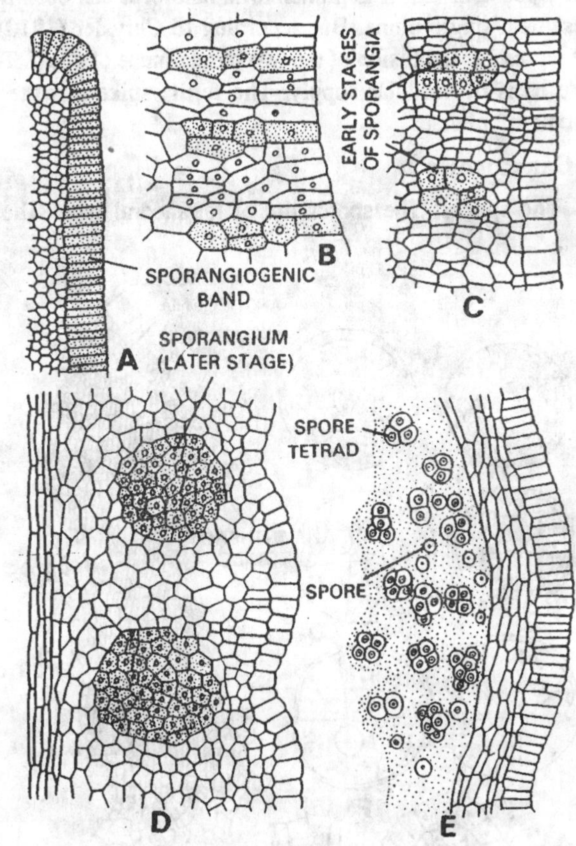

Fig. 7.6. *Ophioglossum.* Development of sporangium. A, part of a longitudinal section of fertile spike having sporangiogenic band; B, L.S. of an older spike; C, L.S. of a fertile spike showing early stages of sporangia; D, L.S. of fertile spike showing later stages of sporangia; E, tangential section of spike showing spore tetrads and spores.

The sterile cells also divide vertically again and separating the sporangia one from another.

As the sporogenous tissue continues development, it becomes surrounded by a poorly developed tapetal layer. According to Bower (1908) the tapetum has been developed from outermost sporogenous cells. But it seems more probable that the tapetum arises from cells found immediately external to the sporogenous mass. The tapetum consists of several poorly developed layers of cells that ultimately, break down into a plasmodial mass with persistent free nuclei. This plasmodial mass surrounds the isolated spore mother cells and supplies nourishment to them.

The spore mother cells divide meiotically forming the tetrads of haploid spores. The mature sporangia remain embedded within the tissues of the spike. Each sporangium contains a very large number of spores, sometimes this number reaches upto 10,000.

Dehiscence of the sporogonium. Each sporangium of the fertile spike splits by a transverse slit on the maturity of the spores. However, there is no special mechanism by means of annulus causing the opening of the sporangial wall. Dehiscence takes place by drying out and shrinking of the sterile tissues within the spike.

Morphological nature of fertile spike. The morphological nature of the fertile spike is controversial. According to Bower (1896, 1908) it is a single septate sporangium but this hypothesis has been rejected later on by bower (1911, 1926) himself. The pteriodologists of modern days are in agreement that it is pinnalike in nature. It has been held by Goebel (1915) that the spike represents a single pinna. But according to Chrysler (1910) it is more probable that the fertile spike represents the fusion of the two basal pinnae of a leaf. This interpretation has been supported on the basis of vascular supply. The fertile spike has the same vascular supply as would go to a pair of pinnae.

The Gametophyte

The spores. The spores are of tetrahedral type. The spore wall consists of two layers, the exospore and the endospore. The exospore is thick, colourless and sculptured with pits. The endospore is thin and delicate. The spore contains granular cytoplasm, a centrally placed large nucleus and oil droplets.

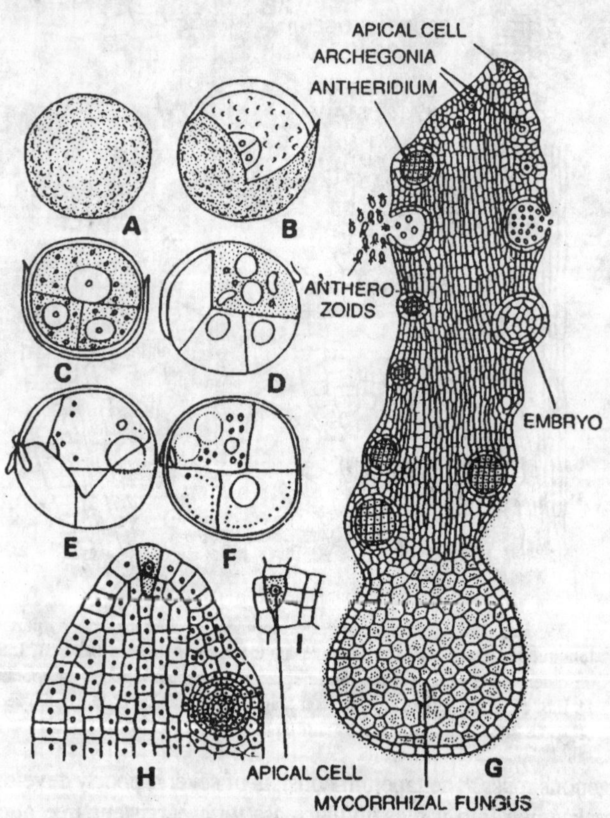

Fig. 7.7. *Ophioglossum* sp. A, spore; B, germinating spore; C, optical section of germinating spore; D-F, different views of young gametophyte; G, L.S. of prothallus; H, L.S. of prothallus apex; I, apical cell.

Germination of spore and development of prothallus (gametophyte). Germination of spores to form gametophytes takes place immediately after shedding. Sometimes the germination of spore is greatly delayed (Campbell, 1911). On absorption of water the exospore ruptures along the triradiade ridge and the inner contents of the spore surrounded by the endospore come out in the form of a germinal tube. This divides transversely into two cells. The next division is vertical in the lower cell and divides it into two equal cells. At this stage the young gametophyte consists of three cells. The young gametophyte does not develop beyond the three or four-celled stage unless it becomes infected with the phycomycetous fungus. The fungus always enters through the lower cell and establishes as mycorrhizal endophyte.

Mature prothallus. The mature gametophytes (Bruchmann, 1904; Campbell, 1907; Lang, 1902) are irregularly cylindrical to conical. According to the species the gametophytes may be unbranched (*e.g., O. vulgatum*) or profusely branched (*e.g., O. pendulum*) the gametophyte is subterranean in habit and obtain all its food through the agency of a mycorrhizal fungus. Sometimes it develops small green lobes above ground. The size of the prothallus ranges from 5 to 10 cms. In most species the gametophyte is perennial, but in *O. moluccanum* the prothalli seem to be annual (Campbell, 1911). The growing apex of a prothallus has a single apical cell with three or four cutting faces. The prothalli are usually pale, grey, yellowish or brownish in colour. However, the region immediately posterior to the apical cell is colourless. According to species the prothalli may or may not bear rhizoids. Internally there is little differentiation of tissue within a gametophyte, but for the cells towards interior may be somewhat elongated. Interior cells in the older parts of a prothallus are infected with a mycorrhizal fungus. However, this endophytic fungus is not found in the apical portion of the thallus and the outer cells of the older portion of the thallus.

The sex organs. The prothalli are monoecious and bear sex organs in acropetal succession. The antheridia and archegonia are found to be developed in irregular sequence. These sex organs remain scattered all over the prothallus.

Development of antheridium. The antheridium develops from a single superficial cell of the prothallus. This antheridial initial divides periclinally to form an outer cell, the jacket initial and an inner cell, the primary androgonial cell. According to Campbell the jacket initial always divides anticlinally to form two daughter cells of equal size. One of these daughter cells again divides anticlinally forming two daughter cells of which one is approximately triangular and the other approximately rectangular. The triangular cell functions in the same manner as an apical cell and cuts off a series of cells from each of its three sides. After it has stopped cutting off cells the triangular cell itself is known as opercular cell.

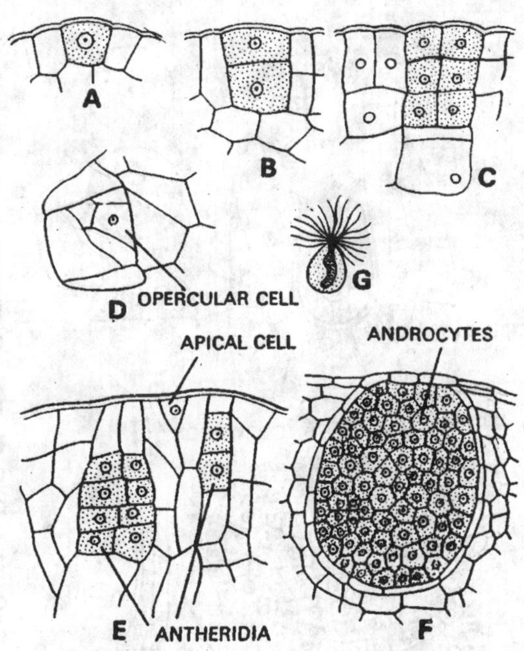

Fig. 7.8. ***Ophioglossum pendulum.*** **Development of antheridium-A, antheridial initial; B-C, median section of young antheridia; D, surface view of a mature antheridium showing opercular cell; E, L.S. of gametophyte apex showing two young antheridia and an apical cell; F, mature antheridium; G, an antherozoid.**

The primary androgonial cell divides anticlinally or periclinally to form two androgonial cells. The next division takes place in the opposite plane and this way the four androgonial cells are found to be arranged in two tiers. These four androgonial cells divide and redivide continuously to form numerous (several thousand) androgonial cells. Members of the last cell generation of androgonial cells are known as androcyte mother cells. The androcytes metamorphose into antherozoids. Each androcyte has two blepharoplasts at one side of the nucleus. Campbell (1911) has studied the metamorphosis. The first noticeable change in the metamorphosis is an elongation

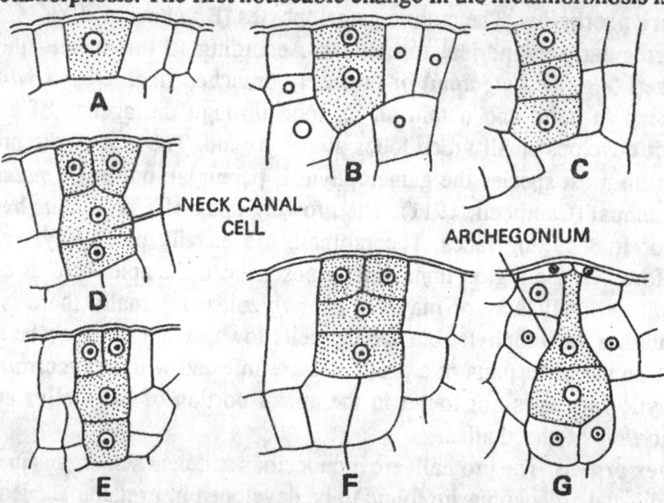

Fig. 7.9. ***Ophioglossum pendulum.*** **Development of archegonium. A, archegonial intial; B-F, successive stages in the development of archegonium; G, nearly mature archegonium.**

of the blepharoplast. There after the nucleus elongates slightly. As metamorphosis proceeds the blepharoplast becomes coiled and the nucleus also becomes coiled. Shortly after this several flagella develop at the anterior portion of the blepharoplast.

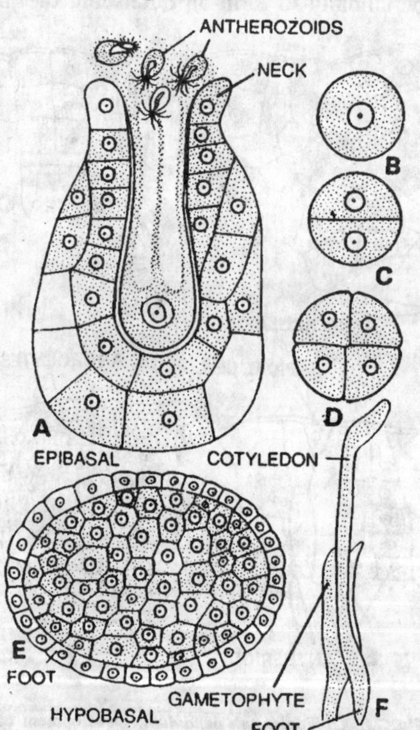

Fig. 7.10. ***Ophioglossum*** **sp. Development of sporophyte- A, fertilization; B-D, successive stages in the development of embryo; E, more advanced embryo; F, gametophyte and young sporophyte in median section.**

The mature antheridia are usually large and sunken within the tissue of the prothallus. Sometimes they are slightly projecting. The jacket layer of antheridium is single layered. On being mature of the antheridium the opercular cell becomes brownish in colour and disintegrates. Eventually the antherozoids liberate through this triangular pore.

The mature antherozoids are coiled and multiflagellate. The narrowly coiled anterior portion of the antherozoid consists mainly of cytoplasm and blepharoplast. The broad, loosely coiled posterior portion is mainly nucleus.

Development of archegonium. The archegonium develops from a single superficial cell found to be situated near the apex of the prothallus. This cell is archegonial initial. The initial divides periclinally to form a primary cover cell and an inner cell. Now the inner cell divides periclinally to form a central cell and a basal cell. The central cell lies next to the primary cover cell. Thereafter the primary cover cell divides anticlinally to form two daughter cells that divide anticlinally forming the four quadrately arranged neck initials. Now each of the neck initials divides diagonally by a transverse division resulting into the formation of neck composed of two tiers of four cells each. Later more transverse divisions take place in each tier and a mature neck three or four cells in height is formed.

As the neck develops the central cell divides periclinally to form a primary canal cell and a primary venter cell. The primary canal cell elongates vertically and pushes out into the neck. The large and conspicuous nucleus of the primary canal cell divides to form two canal cells. Shortly before the maturation of the archegonium the primary venter canal cell undergoes a division forming a venter canal cell and an egg. The venter canal cell immediately disintegrates after its formation. The venter canal cell is a transitory structure and it is rarely demonstrable (Campbell, 1907).

A nearly mature archegonium consists of a neck composed of four vertical rows of cells and it is three to four cells in height. The archegonium remains embedded within the tissue of the prothallus or it projects out very slightly. On the maturation of the archegonium the cells disintegrate and a canal is formed. One of the multiflagellate antherozoids enters through the neck canal and fuses with the egg thus forming the diploid oospore.

Development of sporogonium. The zygote divides transversely to form an outer epibasal and inner hypobasal cell. According to Campbell (1907) the hypobasal cell develops only into the foot while according to Bruchmann (1904) it gives rise both the foot and primary root. However, an apical cell becomes differentiated in the root portion of an embryo. According to Bruchmann (1904) in *O. vulgatum* development of root and foot so overmasks that of cotyledon and stem that the young sporophyte appears to be all root for a long time after it comes out from the prothallus. According to Campbell, in *O. moluccanum*, however, the root and the cotyledon grow at an equal rate. The root grows downwards while the cotyledon upwards. The method of formation of the cotyledon from the epibasal cell is not known. However, the cotyledon develops by means of an apical cell. After the earlier development of the primary root and cotyledon, there appears an apical cell of the stem. This appears in the region where cotyledon and root adjoin. This apical cell seems to be superficial cell of the embryo. The stem from this apical develops very slowly.

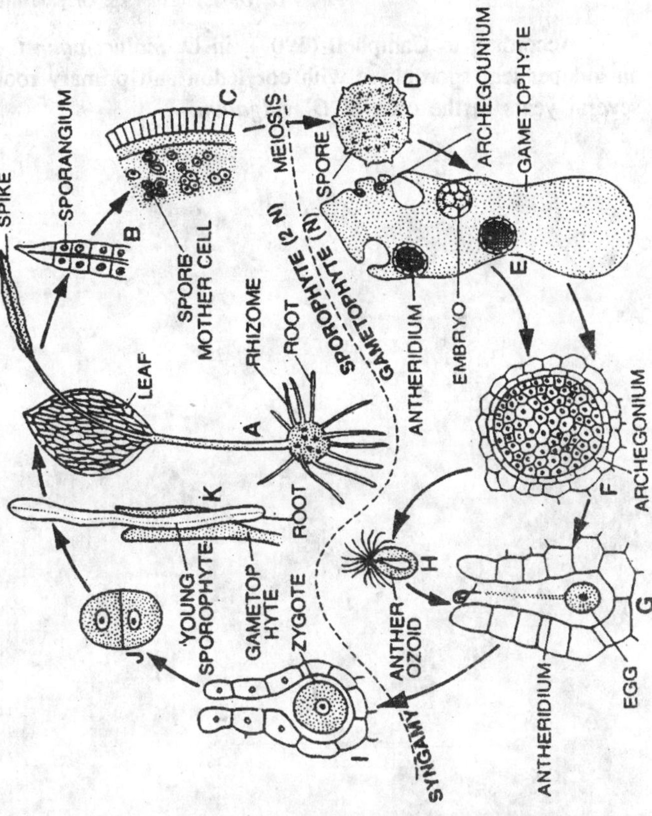

Fig. 7.11. *Ophioglossum* sp. A-K, diagrammatic life-cycle.

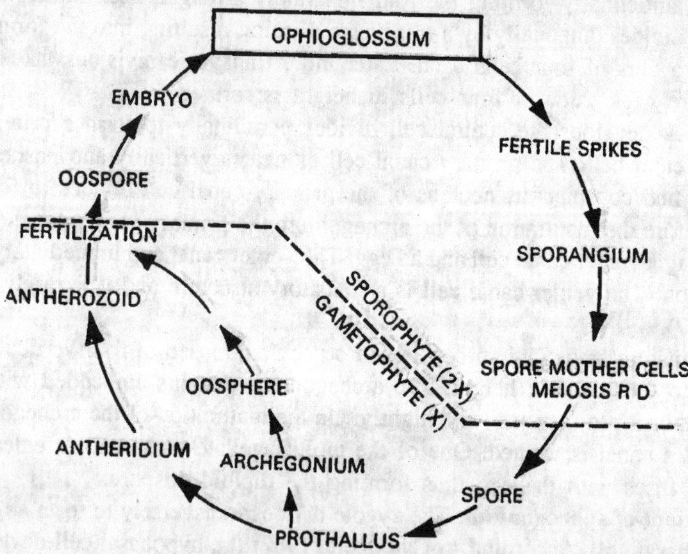

Fig. 7.12. *Ophioglossum* sp. Graphic life-cycle.

According to Campbell (1907), in *O. moluccanum* the development of an embryo into an independent sporophyte with cotyledon and primary root may take one year. It may take several years in the case of *O. vulgatum.*

8

Division-Filicophyta
Class-Leptosporangiopsida

Order-Marsileales

Characteristic features. This order includes a single family, Marsileaceae. The family includes the living genera-*Marsilea*, *Pilularia* and *Regnellidium*. They are heterosporous aquatic ferns. The sporangia of these are produced within special structures known as **sporocarps.** Each sporocarp possesses many sori which bear microsporangia and megasporangia. The sori are **gradate,** *i.e.*, the oldest sori are found at their terminal end of the receptacle and youngest at the base.

We shall consider the life-history of *Marsilea* in detail.

Genus MARSILEA

Systematic position. Pteridophyta
Division. Filicophyta
Class. Leptosporangiopsida
Order. Marsileales
Family. Marsileaceae
Genus. *Marsilea.*

There are about 65 species of *Marsilea* distributed all over the world. They are more commonly found in tropical regions, such as, Africa and Australia. Gupta and Bhardwaja (1957) have recorded about ten species of *Marsilea* from our country. About six species of *Marsilea* are known to occur in the United States. They are hydrophytic or amphibious plants. They grow rooted in the mud of marshes and shallow pools. *Marsilea vestita* and some other species grow in shallow ponds. A single plant of *M. vestita* may grow in all directions until it covers an area twenty-five metres or more in diameter. In *M. vestita* and most of other species the sporocarps develop to maturity only on the plants that are not submerged. *M. minuta* and some other species produce sporocarps in the aquatic habitat, on the other hand *M. aegyptiaca* never produces the sporocarps in the aquatic habitat. An Australian species *M. hirsuta* can survive in dry conditions throughout the year.

The Indian species recorded by Gupta and Bhardwaja (1957) are as follows-

1. M. quadrifolia; 2. M, minuta; 3. M. aegyptiaca; 4. M. brachypus; 5. M. gracilenta; 6. M. poonensis; 7. M. rajasthanensis; 8. M. brachycarpa; 9. M. condensata and *10. M. coromandelica.*

External Morphology

Stem. The species of *Marsilea* possess a rhizome which creeps on or just beneath the surface of the soil. The rhizome is slender, branched and possesses nodes and internodes. The leaves are borne alternately along the upper side of the rhizome usually at the nodes. One or more adventitious roots come out from each node on the underside of the rhizome. The rhizome is dichotomously branched and is capable of indefinite growth in all directions and covers an area more than twenty-five metres in diameter.

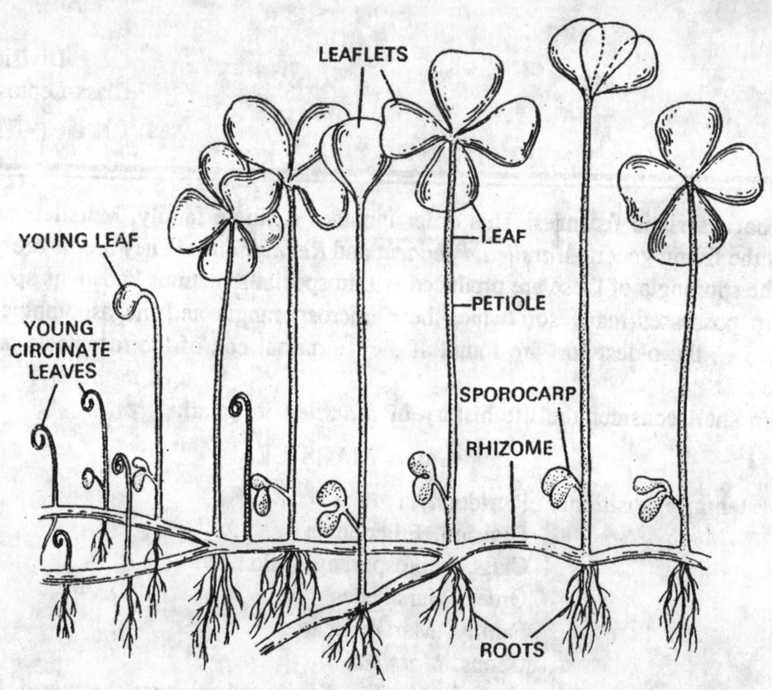

Fig. 8.1. *Marsilea quadrifolia*. A sporophyte with sporocarps, rhizome, leaves and roots.

Leaves. The leaves are borne alternately along the upper side of the rhizome at the nodes. The leaves possess circinate vernation typical of most Filicales (ferns). The young parts of the leaves of *M. minuta* and others are covered with numerous multicellular hairs. The leaves of submerged plants possess long flexible petioles and leaf lamina that float on the water surface.

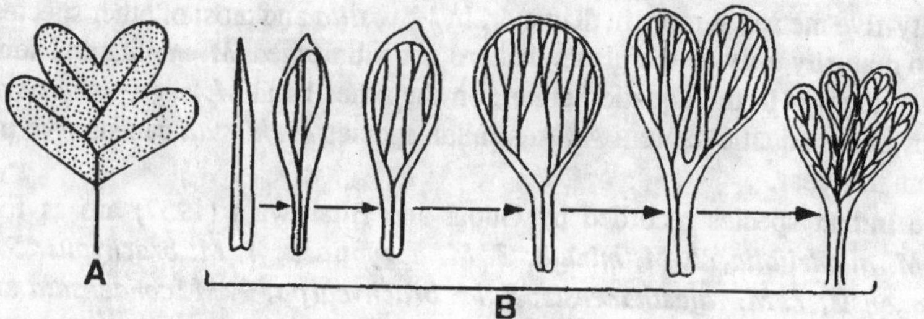

Fig. 8.2. *Marsilea* sp. A, leaf showing arrangement of leaflets as a result of three dichotomies; B, successive types of juvenile leaves of *Marsilea*. (A, after Bower; B, after Braun).

When the plants grow on mud or marshy places the leaves have got shorter and the petioles erect and stouter which spread the leaves in the air. The leaves are compound. The lamina of each leaf is divided into four leaflets pinnae arising from the apex of the petiole. Division of a leaf-blade into four leaflets results from two dichotomies arising in close succession to each other. According to Puri and Garg (1953) the leaf of *Marsilea* consists of a single leaflet bearing four pinnules and not of four leaflets. Usually these pinnules have been referred as leaflets by many authors in their text-books. The veins of each pinna are dichotomously branched and have numerous cross connections resulting in a close reticulum of veins. The shape of pinnae varies from obovate to obcuneate. The margin also varies from entire to crenate, *e.g., M. minuta* or from crenate to lobed as in *M. aegyptiaca* a xerophytic form. The leaflets may be once or twice deeply lobed dichotomously, *e.g.*, in *M. macrocarpa* and *M. biloba.* In *M. minuta,* the pinnae become coarsely toothed. At night the pinnae of aerial leaves become folded upwards and exhibit the sleeping movement.

The **sporocarps** are borne on short peduncles near the base of the petiole. In majority of cases the peduncle or stalk of the sporocarp is unbranched and bears a single sporocarp at its apex. *Marsilea quadrifolia* possesses dichotomously branched peduncle, which bears two to five sporocarps. In *M. polycarpa* single dichotomously branched peduncle bears about 6 to 26 sporocarps.

Roots. One or more adventitious roots are borne at each node on the underside of the rhizome. The adventitious roots may arise even from internodes, *e.g.*, in *M. aegyptiaca.* In *M. minuta* the roots may develop laterally. Dr. Pande (1923) pointed out that the roots of *M. minuta* secrete carbonic acid and because of this secretion the roots could penetrate the shells of pond snails.

Internal Morphology (Anatomy)

Rhizome. The rhizome of *Marsilea* possesses an amphiphloic siphonostelic vascular cylinder. The outermost layer is the single-layered epidermis without any stomata. The cortex is differentiated into outer and inner cortical regions. The outer cortical region consists of compact parenchymatous tissue. This region may be one to several celled in thickness. Just beneath this the cortical region contains large lacunae or air spaces (aerenchyma). The lacunae are separated from each other by one layered parenchymatous septa. This region may be considered the middle cortex. Internal to this region the inner cortex is found. This consists of a sclerotic zone. The cells consisting this region are thick-walled sclerenchymatous cells (fibres). Beneath this region the cortex again consists of compact parenchymatous tissue acting as storage tissue containing starch in them. Some of the tannin cells may be found in this innermost region.

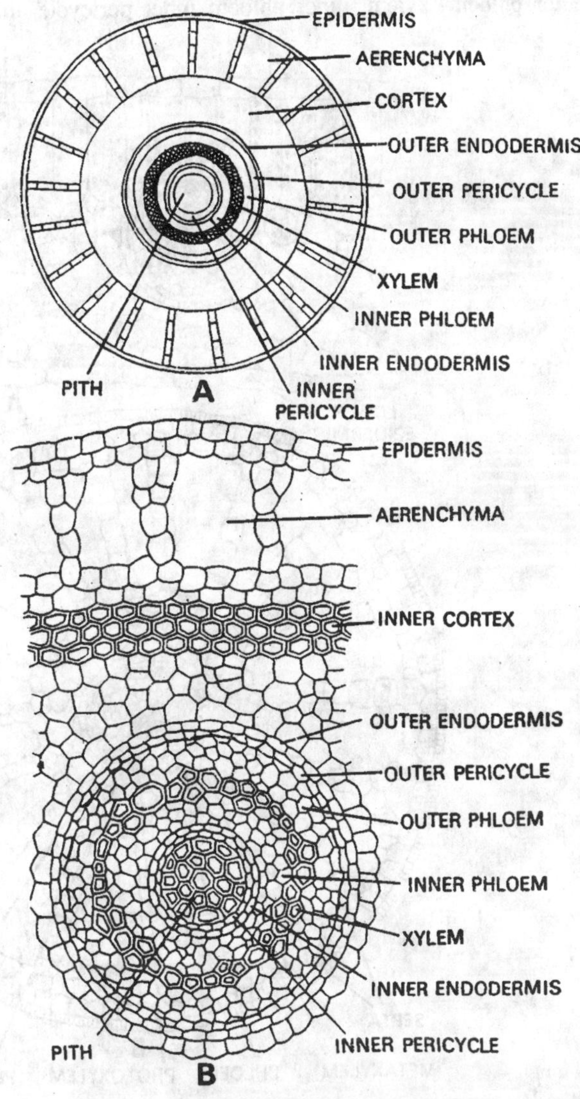

Fig. 8.3. *Marsilea quadrifolia.* **Anatomy of rhizome. A, T.S of rhizome (diagrammatic); B, T.S. of rhizome (detailed).**

The stele is amphiphloic siphonostele. All species possess the siphonosteles externally and internally limited by a single layered endodermis. Just beneath the outer endodermis there is a single layered pericycle. In the centre is the pith whose structure depends upon environmental conditions. In submerged species, the rhizomes possess usually a parenchymatous pith, whereas the plants growing on mud have a more or less sclerenchymatous pith. Outside the pith there is a sindle-layered inner endodermis, inner pericycle and then inner phloem in a continuous succession. There after, there is a ring of xylem which is surrounded by the successive rings of outer phloem, outer pericycle and outer endodermis. In *M. quadrifolia* there is no definite protoxylem but in *M. vestita* there are well-defined protoxylem groups which are exarch. Gupta and Bhardwaja (1956) reported the mesarch protoxylem in *M. aegyptiaca*. This way, the continuation of different tissue in the amphiphloic siphonostele is as follows-outer endodermis, outer pericycle, outer phloem, xylem, inner phloem inner pericycle and then inner endodermis.

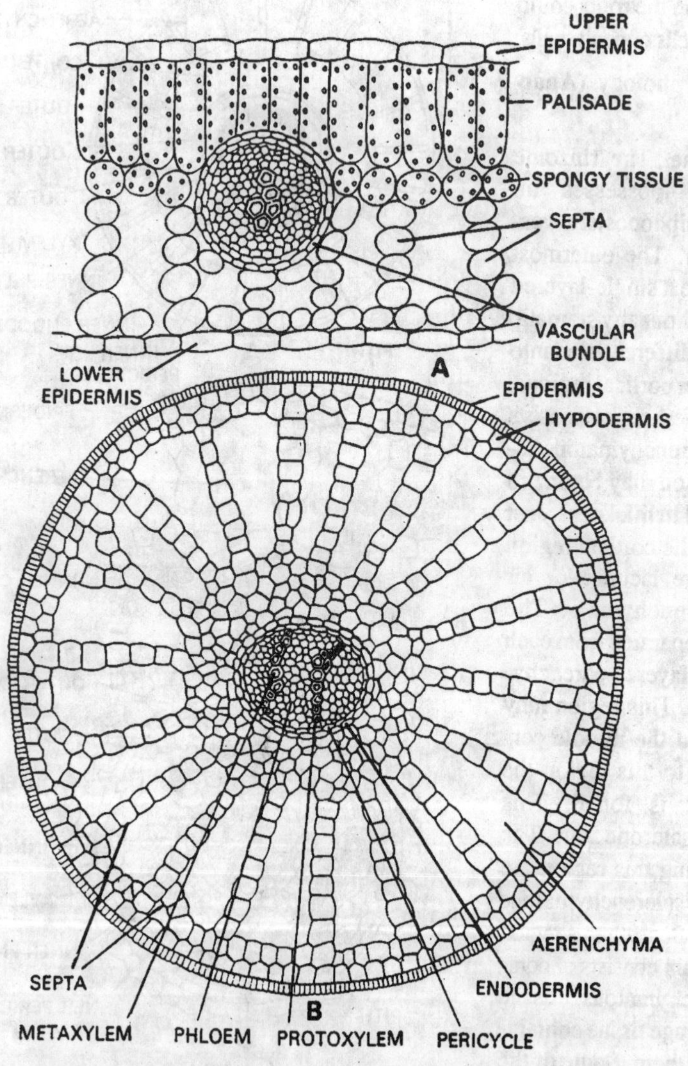

Fig. 8.4. *Marsilea quadrifolia*. A, T.S. of a portion of leaflet; B, T.S. of petiole.

Petiole. The internal structure of a petiole as seen in transverse section is as follows- The outermost layer is single-layered epidermis. Just beneath the epidermis there is hypodermal layer. Then there is other cortex constituting of large air spaces or lacunae (aerenchyma) separated from each other by septa. The inner cortex is composed of compact parenchymatous cells. In this region some of the tannin cells may be present here and there. In the central region of the petiole there is stele which is more or less triangular in outline. The stele is surrounded by a single-layered endodermis. It possesses V-shaped mass of xylem with the exarch protoxylem. The two arms of V are quite separate and some what curved from each other. The V-shaped mass of xylem is surrounded by phloem and single-layered pericycle successively. The ends of each arm of V-shaped xylem are protoxylem groups whereas the middle regions of the arms of V consisted of large tracheids representing the metaxylem groups. The opening of the V is always towards the axis, *i.e.*, adaxial side.

Leaflet. As shown in transverse section of the leaflet, it is bounded by two upper and lower epidermal layers. In the case of floating leaves the stomata are confined to the upper surface but in the case of the plants growing in mud and moist soil, the leaves are aerial, and the stomata are found on both the upper and lower surfaces. Just beneath the upper epidermis there is mesophyll differentiated into palisade and spongy tissues. The vascular bundles are concentric, *i.e.*, xylem elements are surrounded by phloem. On the outside of the vascular bundle there is singlelayered endodermis. Towards lower side there are big air spaces separated by septa.

Root. In the transverse section of the root, the epidermis, cortex and stele may be seen. The epidermis is single layered and composed of thin-walled parenchymatous cells. The cortex of the root is differentiated into two regions. The outer region consists of large air spaces or lacunae separated by septa from each

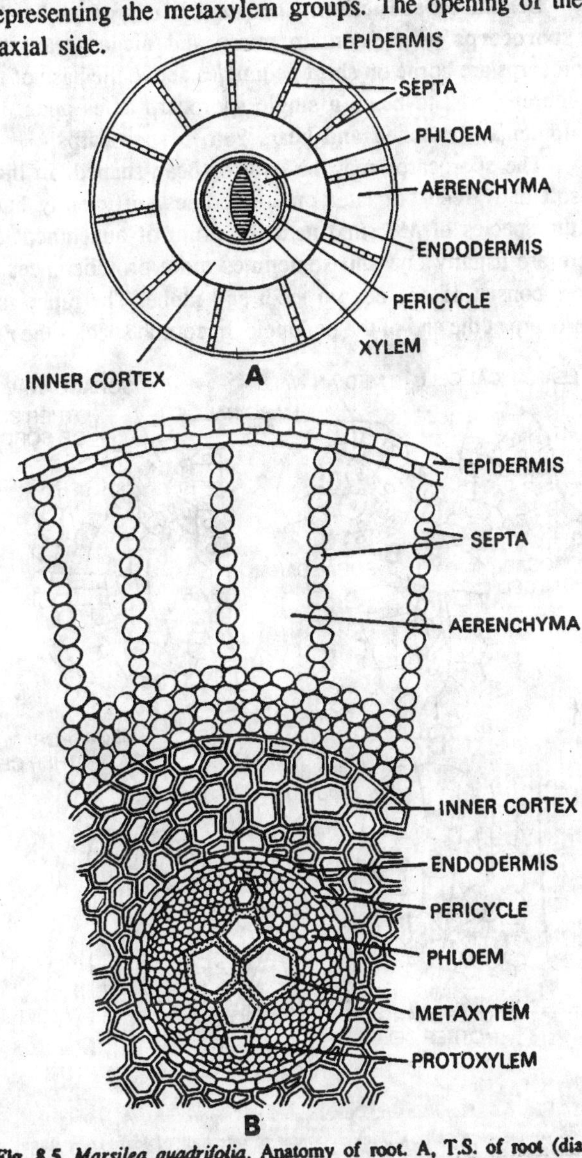

Fig. 8.5. *Marsilea quadrifolia*. Anatomy of root. A, T.S. of root (diagrammatic); B, T.S. of a portion of root (detailed structure).

other, the inner zone of the cortex consists of a few layers of sclerotic cells. Inner to the cortex there is single layered endodermis. Just beneath the endodermis there is single-layered pericycle. In the centre of the stele there is typical diarch and exarch structure with a diametric plate like strand of xylem. The phloem bands are found on either side of the xylem plate.

Reproduction. The reproduction in *Marsilea* takes place by vegetative and sexual methods.

Vegetative propagation. The vegetative reproduction takes place by means of specially developed structures known as tubers. In the dry conditions certain branches develop from the rhizome, which swell because of the storage of food material, and are termed as tubers. These tubers survive in the unfavourable and drought conditions, and on the return of favourable conditions they germinate into new plants, *e.g., M. hirsuta.*

Sexual reproduction. The plant is a sporophyte. It bears the special structures known as **sporocarps** which contain micro and megasporangia in them. It is **heterosporous.** The sporocarps are borne on short peduncles above the base of the petiole. In most cases the peduncle is unbranched and bears a single sporocarp at its apex. In *M. quadrifolia* the peduncle is dichotomously branched and bears 2 to 5 sporocarps.

The sporocarps may be oval or bean-shaped. In the earlier stages of its development it is soft and green but later on it becomes sufficiently hard and brown in colour. In majority of the species of *Marsilea* near the point of attachment of the stalk or peduncle at the base, there are usually one and sometimes more protuberances in the median plane. Such protuberances consist of one or two teeth and raphe. The raphe is found laterally fused to base of the sporocarp at the end of the peduncle. In certain species the raphe is not found, where the sporocarp is directly attached to the end of the peduncle, *e.g., M. polycarpa, M. caribaea* and others. Beyond the raphe two teeth or projections are found. These teeth are unequal in size. The lower tooth is stouter than the upper one. In some species the teeth are very prominent, *e.g., M. vestita* while in other species they are altogether absent.

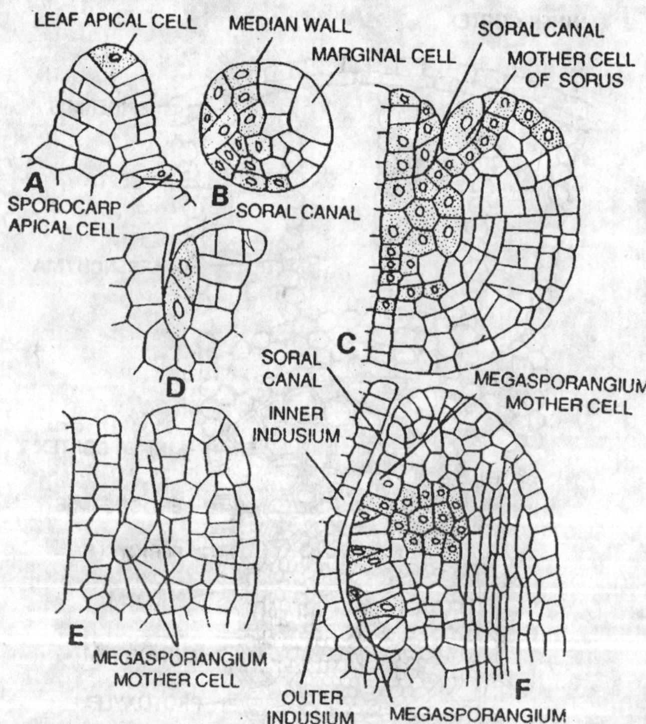

Fig. 8.6. *Marsilea* sp. Development of sporocarp. A, young leaf; B, T.S. of young sporocarp; C, development of one half of leaf; D-F, stages of development of sorus and soral canal (After Johnson).

The vascular supply of the sporocarp consists of a main vein along the narrow side facing the peduncle from which numerous lateral veins perpendicular to the main vein (dorsal bundle) are given off. These lateral veins are given off alternately right and left. Each lateral vein divides dichotomously about middle of its base and the apex. A receptacle develops at the region where each lateral vein forks dichotomously. The receptacles of one valve alternate with those on the opposite valve of the sporocarp. A flap-like outgrowth two-celled thick extends like a hood over the entire receptacle. Each receptacle projects inward from the wall of the sporocarp.

Development of sporocarp. The sporocarp of *Marsilea* develops from a marginal cell of a very young leaf. This marginal cell is situated on the inner face of the leaf and is being differentiated only after the apical cell of the leaf has cut off 6-8 segments. The marginal cells of the sporocarp initials divide and make a mass of undifferentiated cells which become curved so that the distal end gets directed horizontally. As the distal part enlarges to form the sporocarp, two rows of soral mother cells appear on the ventral side, arising directly from the marginal cells of the earlier stage. Later the marginal cell and their derivatives are pushed downwards to the lower surface and slightly towards the median line due to unequal growth of the cells. As increase in diameter continues the cells lateral to the soral mother cells grow more rapidly than the developing sorus with the result that the sporangial mother cells become burried. Simultaneously each sorus towards the median line a small cresent-shaped depression appears on the side facing downwards which deepens quickly with the growth of tissue around the sorus (Fig. 8.6 C-F). These depressions become spaces above and around the sori opening to the outside by the small pores, which are called the soral canals (Figs. 8.6 D-E). The soral canals appear linear in the longitudinal section and crescent-shaped in cross section. Each soral canal is lined on its central face by an **indusium** of which two layers may be recognized, *i.e.*, outer indusium and inner indusium (Fig. 8.6 F). In each soral canal the receptacle of the sorus faces the indusium, and sporangia arise on the receptacle in acropetal succession. The first sporangial initials appearing at the tip of the receptacle develop into megasporangia whereas the other initials appearing along the sides produce microsporangia. Lateral to the sori the massive tissue of the sporocarp develops and forms the massive outer wall which ultimately closes the soral canals, and the sori are completely enclosed within the sporocarp.

Internal Structure of the Sporocarp

To study the internal structure of the sporocarp the sections cut in longitudinal and transverse planes at right angles to the flat surface are required. The internal structure of the sporocarp in transverse, longitudinal and dorsiventral sections is as follows:-

Transverse section. The following tissues are seen in the transverse section. Outside the sporocarp there is a very thick wall consisting of three layers. The outermost layer is known as epidermis which consists of broad and columnar cells. The epidermis is interrupted by a number of stomata. Below the epidermis there are two hypodermal layers. The outer hypodermal layer consists of elongated thick-walled cells, while the lower hypodermal layer consists of elongated and palisade like cells. In the inner region of the sporocarp only two chambers of sori are seen surrounded by separate indusia. The dorsal bundles are also seen. Each sorus possesses a receptacle which bears either micro or megasporangia.

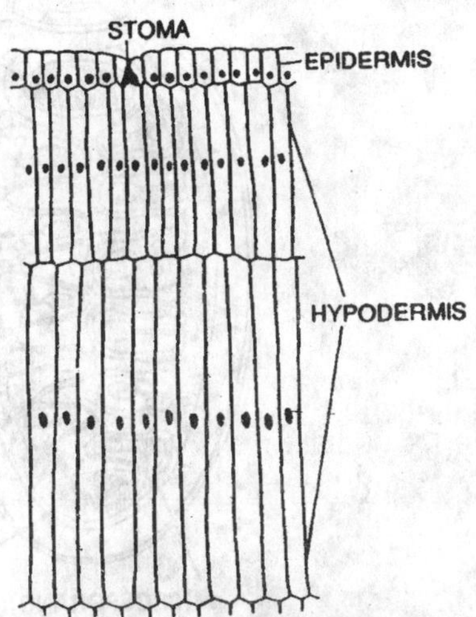

Fig. 8.7. *Marsilea*, portion of a transverse section of wall of sporocarp.

Longitudinal section. The structure of the wall of sporocarp and the gelatinous ring is the same as in transverse section. In the interior of the sporocarp there are two alternating rows of sori each surrounded by a two layered indusium. A placental branch goes to each sorus from the lateral bundles. The receptacle bears a single megasporangium at the tip and the microsporangia on the lateral sides. Each megasporangium contains a large single megaspore while the microsporangium contains many small rounded microspores.

Dorsiventral section. The structure of the outer wall of the sporocarp and gelatinous ring is the same as seen in transverse and longitudinal sections. Below the outer wall a continuous gelatinous ring is found. Within the sporocarp there is a group of sori. Each sorus appears to be formed in a cavity and is surrounded by a delicate, membranous indusium. The sorus contains micro and megasporangia. The megasporangia may easily be seen if the sporocarp is cut dorsiventrally somewhat to one side of the median line of the sporocarp, whereas, the microsporangia may

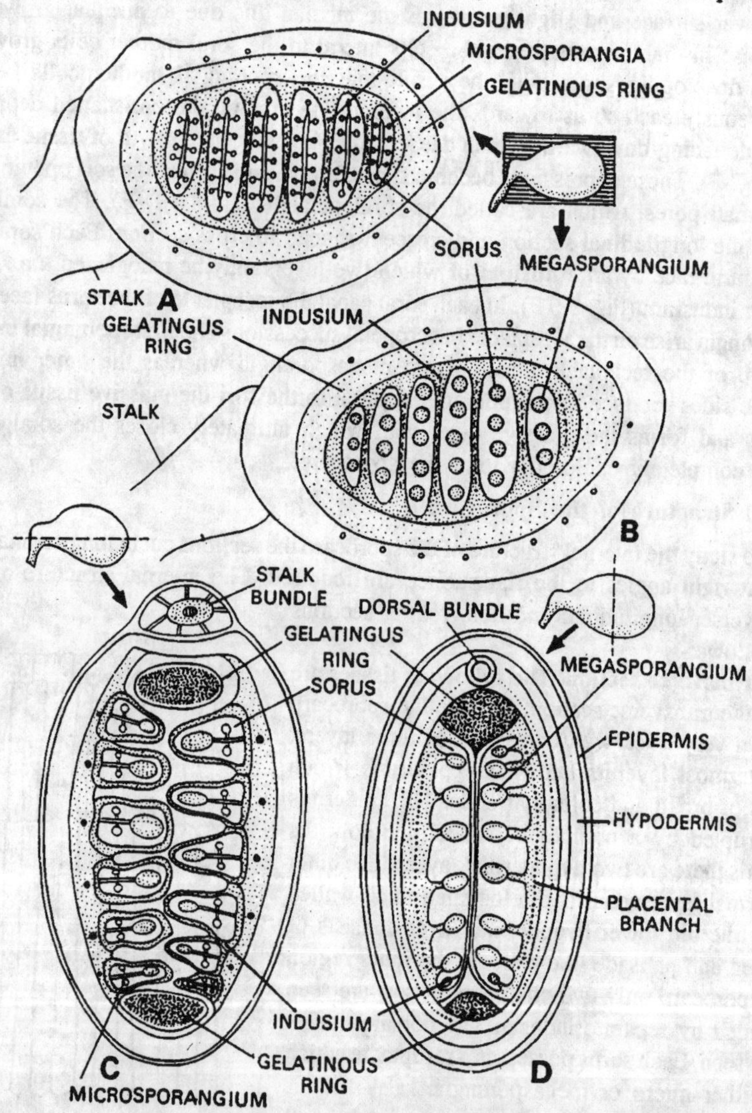

Fig. 8.8. ***Marsilea*** **sp.** Structure of sporocarp. A, section cut in dorsiventral plane somewhat to one side of the median line showing microsporangia; B, section cut in dorsiventral plane showing magasporangia; C, section cut in longitudinal section showing sori containing micro-and megasporangia; D, transverse section of sporocarp.

be easily seen if the section of the sporocarp is cut in the same plane but slightly farther away than the previous one.

Morphological nature of the sporocarp. As regards the morphological nature of the sporocarp, the interpretations have been given by various workers from time to time. There are two main hypotheses-

1. Leaf-segment or laminar hypothesis and 2. petiolar hypothesis.

The sporocarp of *Marsilea* has been interpreted as a lateral modified segment of the leaf. According to Bower (1926), Busgen (1890), Campbell (1905), **leaf segment or laminar hypothesis**-the sporocarp has been interpreted homologous with a modified fertile segment from the lower part of a leaf. **petiolar or whole leaf hypotheseis**-Johnson (1898, 1933) has interpreted the sporocarp to be homologous with an entire leaf. The argument for the latter interpretation by Johnson is that the apical growth of a sporocarp resembles to that of an entire leaf rather than to of a leaflet or leaf-segment. The leaf-segment or laminar hypothesis is supported by several workers and seems to be more correct. This hypothesis has been discussed in detail in the following lines:

Leaf segment or laminar hypothesis. However, the vascular supply to the peduncle of the sporocarp and the vascular supply to the interior of the sporocarp prove that the sporocarp is a modification of leaf-segment (pinna) rather than that of an entire leaf. The advocates of

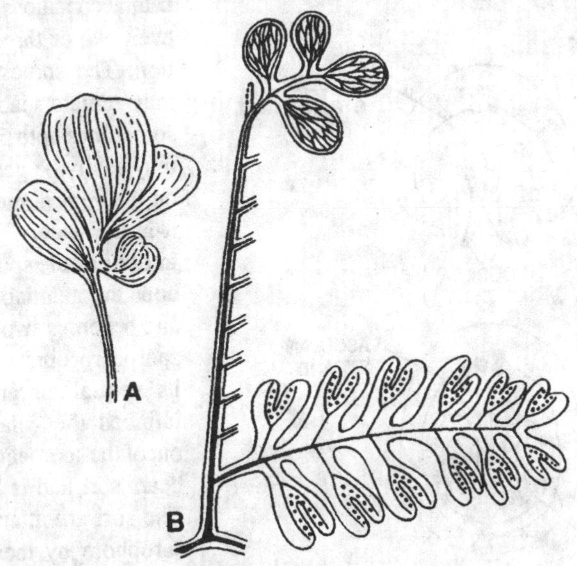

Fig. 8.9. ***Marsilea.*** A, abnormal leaf of *M. hirsuta* showing condition transitional to sporocarp; B, simple case with a single fertile leaflet at the base to develop into a sporocarp, Puri and Garg's view.

the leaf segment theory interpret it in various ways. Bugen (1890) interpreted that the sporocarp has been resulted from the opposition of two leaflets (pinnae), whereas Bower (1926) and Campbell (1905) interpreted that the sporocarp has been resulted from a single pinna by its enfolding. However, Puri and Garg (1953) interpreted that the sporocarp has been resulted from the enfolding of a pinna with several pinnules each with a marginal sorus. The presence of single dorsal main vein in the sporocarp and the single bundle in the peduncle indicate clearly that the sporocarp has been derived from a single pinna. It has been argued by Bower (1908, 1926) the ancestors of Marsileales were probably ferns with a gradate sorus surrounded by an involucroid indusium and not Schizaeaceae. In Schizaeaceae the sporangia are borne singly and develop simultaneously whereas in Marsileales the sporangia are borne in sori of a gradate type in which the sporangia at the apex of a receptacle develop earlier than those at the base of the receptacle. This way the relationships of the Marsileales appear to be more with the homosporous leptosporangiate ferns. There is no evidence that the heterospory of Marsileales arose earlier of the evolution of sporocarp. In all other ferns there is homospory and this point indicates that the origin of heterospory is somehow or other connected with evolution of the sporocarp.

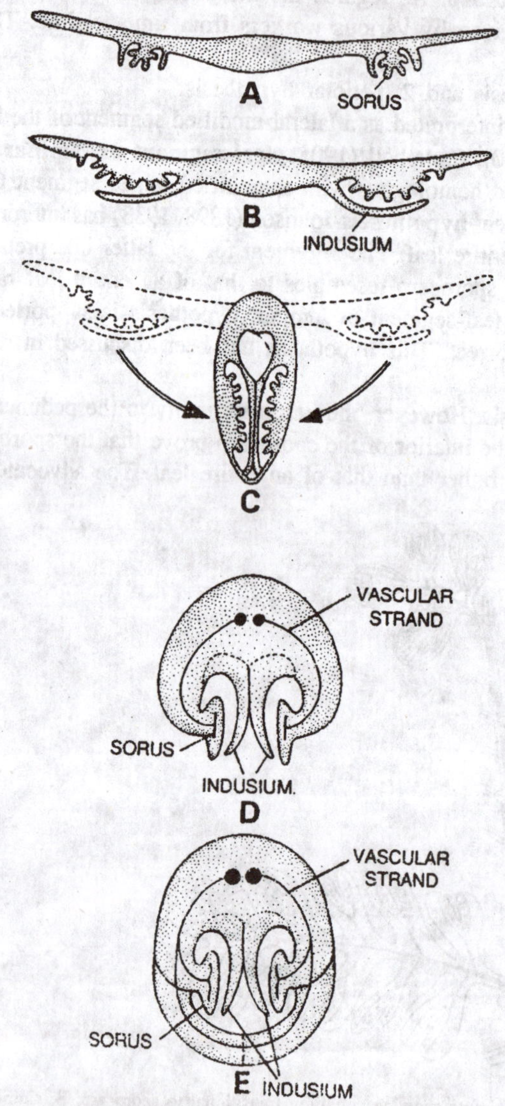

Fig. 8.10. *Marsilea* sp. Morphological nature of sporocarp. A-C, Smith's concept of the morphological nature of sporocarp; D-E, Eames concept of the nature of sporocarp.

Liberation of sori from the sporocarp. The outer wall of the sporocarp is extremely hard and strongly resistant to mechanical injury and drying out. In natural conditions the sporocarp may burst even two or three years later its formation. The spores within the sporocarp may remain viable 20 to 30 years and sometimes upto 50 years. The sporocarp may easily be germinated by injuring it on the ventral median line and keeping it in the water. The water enters to the interior of the sporocarp. Within half an hour the gelatinous ring imbibes water and becomes swollen with the result the sporocarp bursts into two valves from its ventral margin. As more water is imbibed the gelatinous ring protrudes out of the sporocarp. This gelatinous ring bears sori and is known as **sorophore.** The sori are attached to the gelatinous sorophore by their ends. As the sorophore expands and pushes out, it pulls out the sori, which are attached to it by their ends. Usually the attachment of the gelatinous ring breaks from the ventral side of the sporocarp and the dorsal part of the gelatinous ring remains attached by one end to the sporocarp. The gelatinous sorophore bears two alternating rows of soral sacs one on either side of it. A fully elongated sorophore is a long gelatinous cylinder 15 to 20 times longer than the length of the sporocarp. At the tip of the sorophore there are certain small gelatinous projections, also alternate with one another. They are the rudimentary sori which could not develop. The gelatinous ring elongates and straightens and becomes a worm-like structure. At the time of their separation from the gelatinous

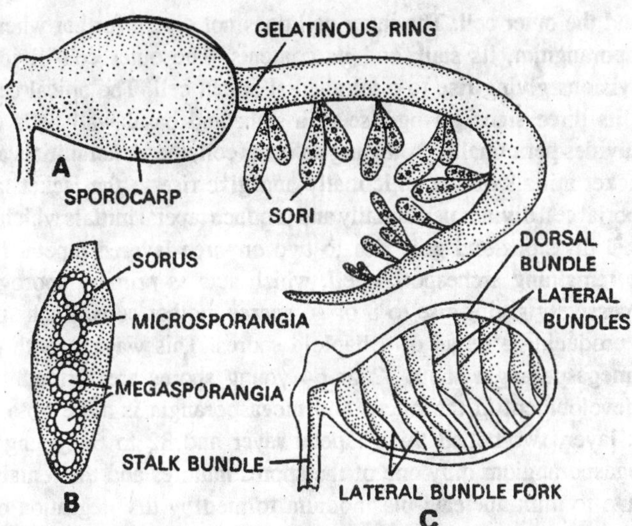

Fig. 8.11. *Marsilea quadrifolia*. A, extrusion of the gelatinous ring and sori; B, a sorus surrounded by indusium; C, vascular supply of the sporocarp in lateral view.

ring the ventral ends of the sori are torned off and sporangia with spores escape from the ventral ends of the sori. However, the indusia and the jackets of sporangia become gelatinized and the spores are liberated. The germinating spores and the developing gametophytes remain embedded in the gelatinous matrix upto their maturity.

Development of sporangia. The development of micro and megasporangia is of **leptosporangiate type**. The leptosporangiate differ from other ferns in having a sporangium with a jacket layer one cell in thickness. There is a definite number of spores within a sporangium. The sporangium develops from a single initial cell and the tapetum differentiates from a single internal cell of a developing sporangium.

One of the superficial cells acts as a sporangial initial. The sporangial initial divides periclinally producing two cells, the

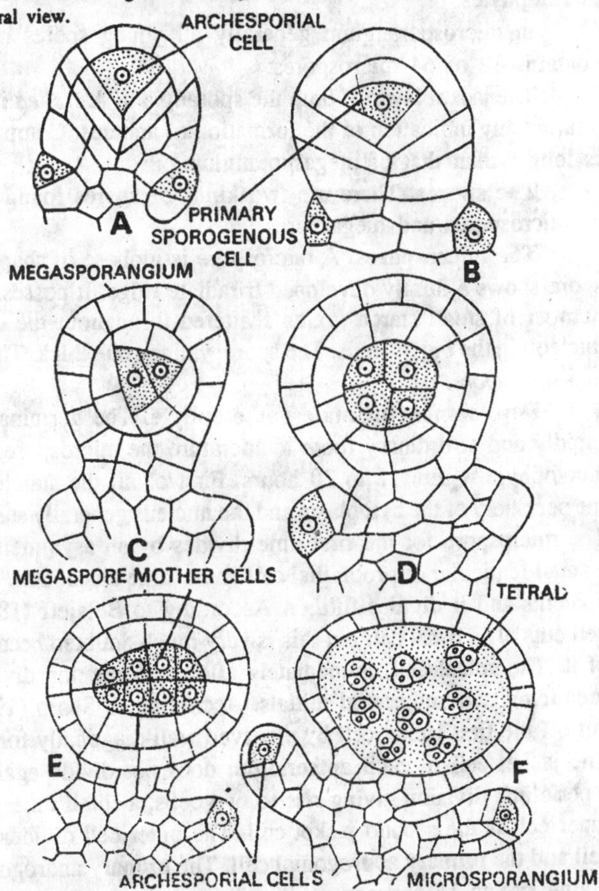

Fig. 8.12. *Marsilea quadrifolia*. Successive stages in the development of sporangia. (A-F)-micro and megasporangia.

inner cell and the outer cell. The inner cell does not divide further whereas the outer cell forms the entire sporangium, its stalk and its contents. The outer cell divides thrice by successive diagonal divisions giving rise to a tetrahedral apical cell. The apical cell cuts of two segments on each of its three sides giving rise to a stalk and basal wall cells of the sporangium. The apical cell divides periclinally producing an outer cell, the jacket initials and an inner archesporial cell. The jacket initials divide anticlinally and give rise to the jacket layer of the sporangium. The archesporial cell divides periclinally and produce tapetal initials which divide both by anticlinal and periclinal divisions and give rise to two or three layered tapetal layer. After forming the tapetum the remaining archesporial cell, which acts as primary sporogenous cell divides 3 or 4 times successively giving rise to 8 or 16 spores mother cells. Each spore mother cell divides meiotically producing a tetrad of 4 haploid spores. This way, in both types of sporangia, *i.e.*, micro and megasporangia about 32 or 64 young spores are formed.

The development in the micro and megasporangia is alike upto the formation of a one celled thick layer, two-celled thick tapetal layer and 32 to 64 young spores.

In megasporangium only one of the spores matures and the remaining spores disintegrate and contribute to multinucleate plasmodium formed by disintegration of the tapetum. As only one megaspore survives it becomes quite large in size and thick walled as in other heterosporous pteridophytes.

In microsporangium generally all young spores mature so that each microsporangium contains 32 or 64 microspores.

The jacket layer of both the sporangia of *Marsilea* is a thin walled homogeneous structure without any indication of the formation of annulus. Comparatively the stalk of microsporangium is longer than that of megasporangium.

The spores. There are two kinds of spores found in micro and megasporangia, known as microspores and megaspores.

The microspores. A microspore is globose in shape with small pyramid-like apex. Each spore shows a faintly developed triradiate ridge. It possesses a large central nucleus and a large number of small starch grains scattered throughout the cytoplasm. There may be one or two nucleoli in the big nucleus. The spore wall is quite thick. The endospore is surrounded by epispore and endospore.

Development of male gametophyte. The germination of microspore takes place quite rapidly and at ordinary room temperature the microspore develops into a fully mature microgametophyte within 12 to 20 hours. First of all the starch grains of the microspore migrate to the periphery of the cytoplasm and the nucleus generally shifts to the broad side of the microspore. The microspore for the first time divides by an asymmetrical wall forming two unequal cells- a small lenticular or prothallial cell and a large apical cell. The large apical cell contains granular contents and it divides further. According to Belajeff (1898) and Campbell (1892), the apical cell cuts off a basal cell but this is very much doubted because of the lack of evidence in support of it. The apical cell, immediately after its formation divides diagonally to the prothallial cell producing two antheridial initials. According to Sharp (1914) both the antheridial initials then cut off the first jacket cell by a curved wall diagonally formed to the wall of previous division. The jacket cell in each antheridium does not divide again. The curved sister cell divides by a periclinal division giving rise to two cells, a small inner cell and a large outer cell. The small inner cell is the second jacket cell. The outer cell divides periclinally forming the third jacket cell and the primary androgonial cell. The primary androgonial cell divides four times in a series giving rise to 16 androcytes. Simultaneously the prothallial cell distintegrates completely and the jacket cell disintegrate partially. The two androcyte masses lie within the old spore wall.

In *Marsilea* two groups of androcytes have been produced in the two hemispheres.

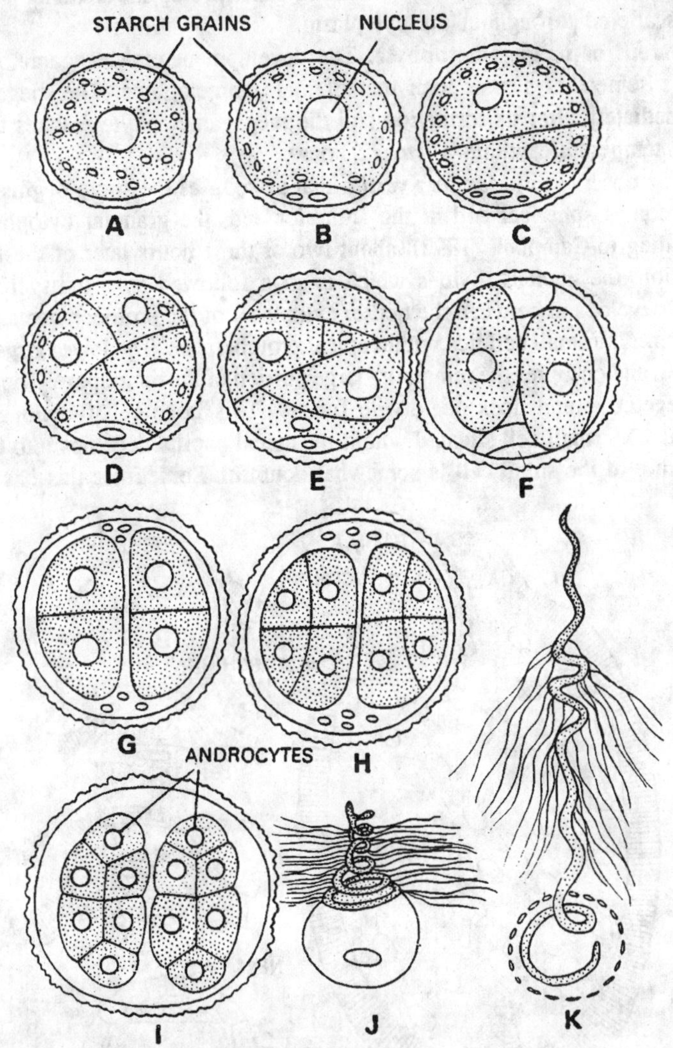

Fig. 8.13. *Marsilea quadrifolia*. A-I, successive stages in the development of microgametophyte (male gametophyte); J, antherozoid shortly after escape from microspore; K, mature elongated antherozoid.

The androcytes are metamorphosed into antherozoids. Shortly before the antherozoids are mature, the spore wall bursts and the antherozoids are liberated in the water. The antherozoids are liberated by the disintegration of the jacket cell of each antheridium

Structure of antherozoid. The antherozoids of *Marsilea* have many coils, sometimes a dozen or more. They are cork-screw like in shape. Usually the upper three or four coils consist of blepharoplast and donot bear any flagella. The broad posterior coils bear 50 to 60 flagella. At the posterior end a globular vesicle is found which contains the cytoplasm.

The megaspore. The megaspore is ellipsoidal and possesses a small projection at the anterior end. The mucilaginous spore wall is quite thick around the megaspore but thin at its protuberant anterior end. In the remaining part of the megaspore the middle spore wall layer is much thicker and provided with radial fibrils. The anterior protuberant portion contains densely granular cytoplasm and a nucleus. The remaining basal protoplast contains a large number of starch grains scattered throughout the cytoplasm.

Development of megagametophyte. The development of Megagametophyte or female gametophyte is somewhat slower than that of male gametophyte. The megaspore begins to germinate immediately after its formation and the megagametophyte reaches to its maturity at ordinary room temperature within 14 to 22 hours.

It has been observed in *Marsilea vestita* that as soon as the spore begins its germination, the nucleus becomes spherical and at the antherior end, the granular cytoplasm increases in amount. According to Campbell (1892), about two or three hours later of the beginning of the spore germination, the nucleus divides which is soon followed by a transverse division. This transverse wall develops in between the denser cytoplasm of the protuberance and the remaining watery cytoplasm. The small cell lies within apical protuberance. The lower large cell is nutritive or prothallial cell and occupies the remaining large portion of the megaspore. The large prothallial cell is quite vegetative and does not divide further. This large cell functions as store house of food material. The small cell situated within the apical papilla develops into the gametophyte proper. The nature of the small cell is somewhat doubtful. Sometimes this has been interpeted

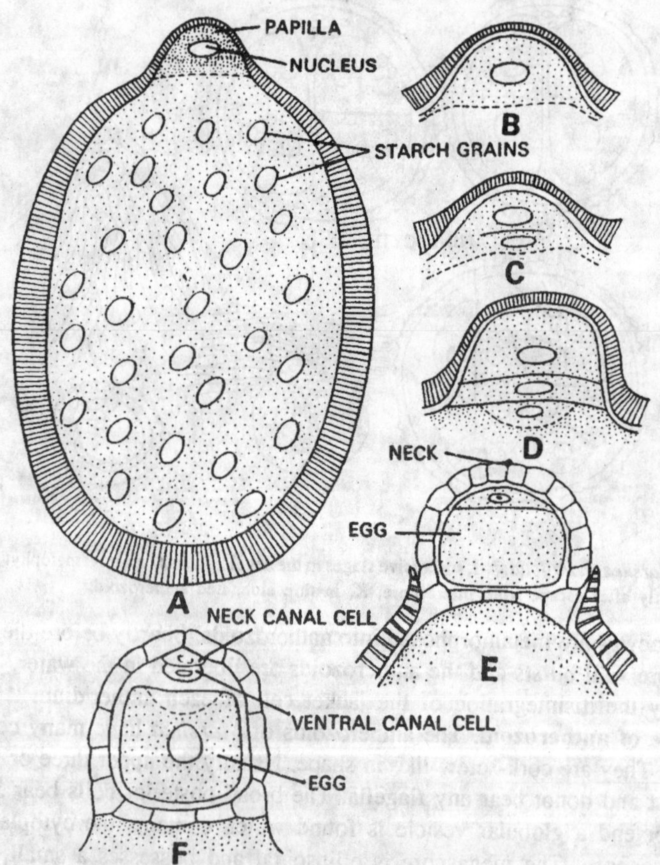

Fig. 8.14. *Marsilea* sp. A, L.S. of mature megaspore; B-F, successive stages in the development of megagametophyte (female gametophyte) and archegonium; F, mature archegonium.

as an archegonial initial but it is much more probable that the apical cell bears four cutting faces, three lateral and one basal. Each of these cuts a single segment at each face before it acts as an archegonial initial. All the four segments divide by anticlinal divisions several times giving rise to a one-celled thick vegetative tissue. According to Campbell (1892) and Strasburger (1907) the archegonial initial divides periclinally forming a primary cover cell and central cell.

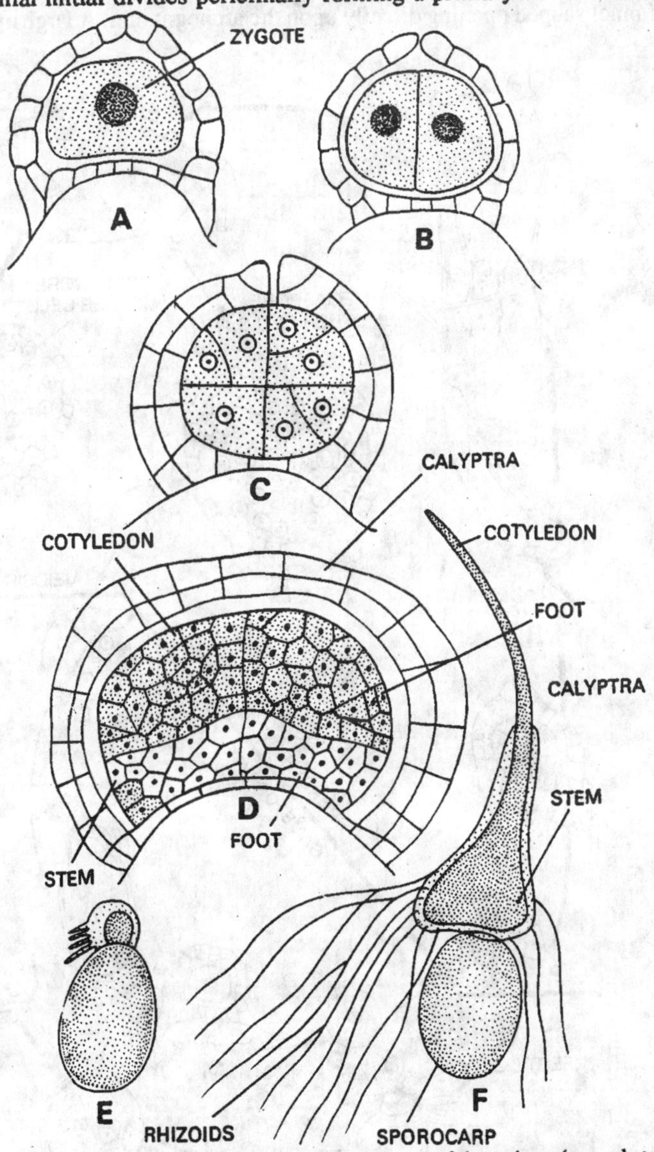

Fig. 8.15. *Marsilea* sp. A-D, successive stages in the development of the embryo (sporophyte); E, sporocarp with young embryo; F, more advanced stage of the embryo in median longitudinal section.

The primary cover cell further divides by two successive anticlinal divisions forming four quadrately arranged neck initials. Each neck initial divides by an oblique wall forming a short neck comprised of two tiers of four cells each. Simultaneously, the central cell divides into two cells, a small primary canal cell and a large primary venter cell. Straburger (1907) reported that in *M. drummondii* and some other species the primary canal divides producing two small canal cells. In all species the **primary venter** cell divides into two cells, a small venter canal cell and a large spherical oosphere or egg. On the maturation of the archegonium the neck canal cell or cells and the venter canal cell disintegrate and the archegonial neck remains open.

A mature megagametophyte is surrounded by a broad, ovoid gelatinous zone which there in a watery funnel shaped opening directly upon the archegonium. A large number of antherozoids

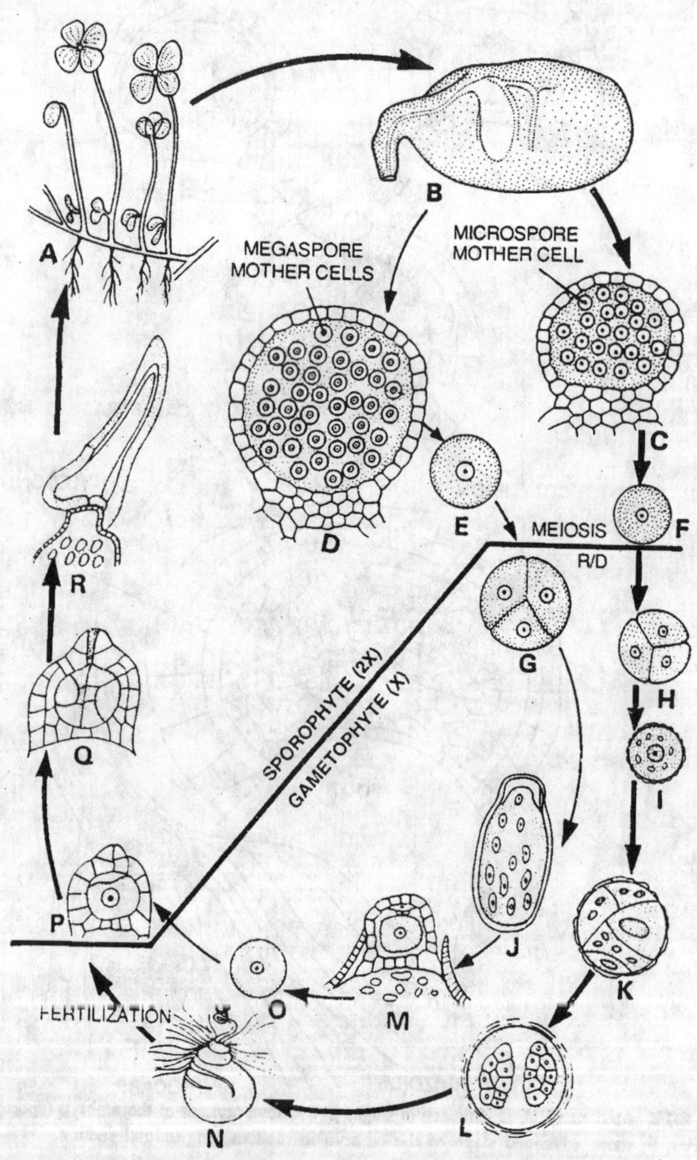

Fig. 8.16. *Marsilea*, diagrammatic life-cycle. A, plant with sporocarps (sporophyte); B, a sporocarp; C, microspore mother cells within microsporangium; D, megaspore mother cells within megasporangium; E, megaspore mother cell; F, microspore mother cell; G, megaspore tetrad; H, microspore tetrad; I, microspore; J, megagametophyte; K-L, formation of microgametophyte; M, archegonium with egg; N, antherozoid; O, the egg; P, oospore; Q, early stage of embryo formation; R. young sporophyte.

swim about the funnel shaped opening. Some antherozoids enter the gelatinous envelope, some reach the neck of the archegonium and others fail to do so. Many of the antherozoids die and their uncoiled elongated, none-motile bodies may be seen in the gelatinous envelope whereas the living actively swarming antherozoids may be seen in the gelatinous envelope coiled and moving. This way, the active and living antherozoids may be easily differentiated from the dead antherozoids which have become non-motile and inactive.

Fertilization. Many antherozoids are attracted towards the funnel shaped structure of megagametophyte. The antherozoids enter gelatinous envelope and reach to the archegonium Only one antherozoid enters the egg's cytoplasm anteriorly, but later on its orientation becomes reversed (Atkinson, 1943) and it penetrates the egg nucleus posterior end forward.

Marsilea drummondii shows parthenogenesis. Strasburger (1907) stated that the megaspore on germination produced a diploid (2x) megagametophyte which without fertilization developed into a sporophyte.

Development of embryo. In *Marsilea* the development of the embryo proceeds with great rapidity and only two to four days old sporophyte possesses the first leaf several millimetres long. In *Marsilea vestita* the zygote divides for the first time after fertilization only two or three hours later. The zygote divides first by a vertical division. Thereafter the two daughter cells divide vertically in a plane perpendicular to that of first division (Campbell, 1892). This

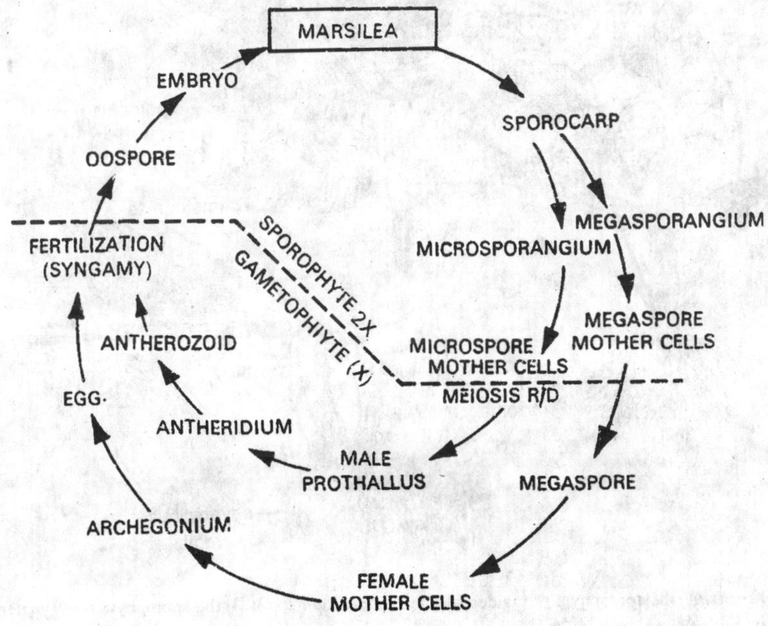

Fig. 8.17. *Marsilea.* Graphic life-cycle.

way a four-celled embryo is formed. Two sister cells of outer side develop into cotyledon and stem and the other two of inner side develop into the foot and root of the embryo. The adjoining cells of the megagametophyte divide periclinally forming a two to three-celled thick calyptra. Simultaneously, many superficial cells on the side of the calyptra facing downwards send out the rhizoids. For the first four or five days, the growth of the calyptra keeps pace with that of the embryo but later on the cotyledon and the primary root penetrate through the calyptra are going upward and the other downward respectively.

9

Division-Filicophyta
Class-Leptosporngiopsida

Order-Filicales

Characteristic features. They are homosporous ferns. They produce usually heart-shaped prothalli (gametophytes) which bear sex organs (antheridia and archegonia) on their ventral surface. The prothalli are monoecious. The sporangia are produced in sori either on the margin or on

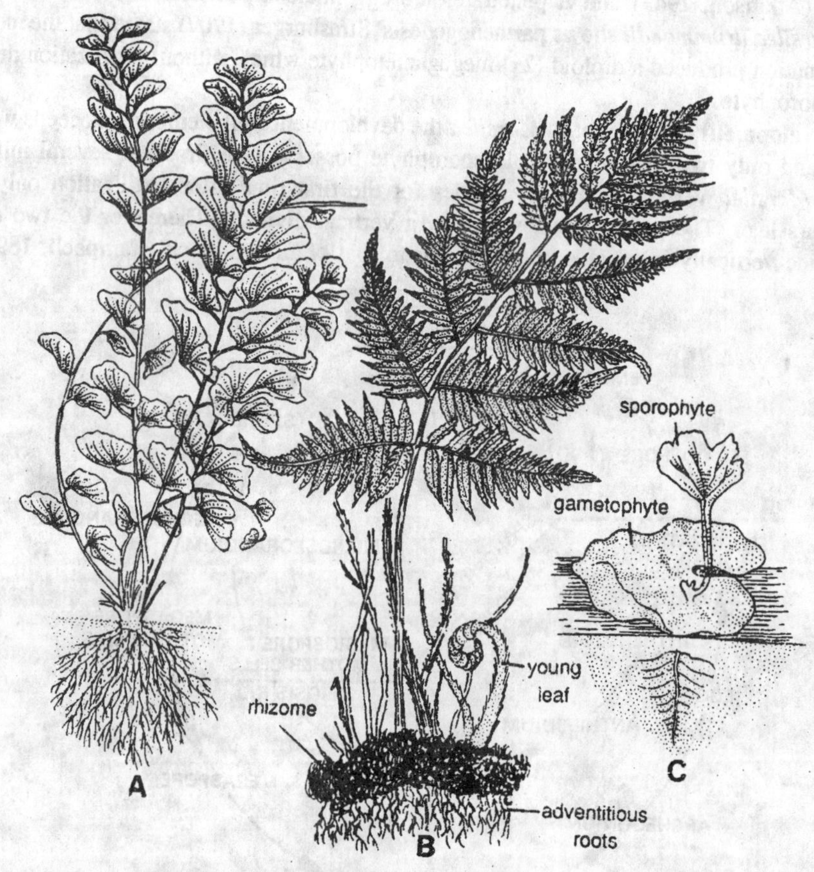

Fig. 9.1. Ferns. A, the sporophyte of maiden hair fern (*Adiantum* spp.); B, the sporophyte of *Dryopteris;* C, heart-shaped prothallus with young sporophyte.

the surface of the sporophyll. This order includes seven families, which are as follows :

1. Schizaeaceae
2. Gleicheniaceae
3. Matoniaceae

4. Hymenophyllaceae
5. Dicksoniaceae
6. Cyatheaceae
7. Polypodiaceae

Genus DRYOPTERIS
(The Male Fern)

Systematic position. Pteridophyta
Division. Filicophyta
Class. Leptosporangiopsida
Order. Filicales
Family. Polypodiaceae
Genus. *Dryopteris*

Here, we will discuss the life-history of *Dryopteris filix-mas* (male fern) of Polypodiaceae in detail.

The ferns are the best known of the living pteridophytes. The male fern-*Dryopteris filix-mas* is found in the shady and moist places both in the hills and in the plains. It commonly occurs in the thick forests. This fern is very common and widely distributed. In India, this is found throughout the country. It is very commonly found in the hills from 2,000 to 8,000 feet elevation.

External Morphology

Habit and stem. This fern possesses a short, blunt, stout and unbranched rhizome which grows obliquely upwards through the soil and appears but little above the surface of the ground. The rhizome is seldom more than a foot in length. It is four or five inches in diameter and completely covered by the persistent leaf bases of the old leaves. The top of the rhizome is broader than the lower end and it becomes conical in appearance. Sometimes, the adventitious buds also appear on the rhizome. These buds develop laterally just above the leaf bases

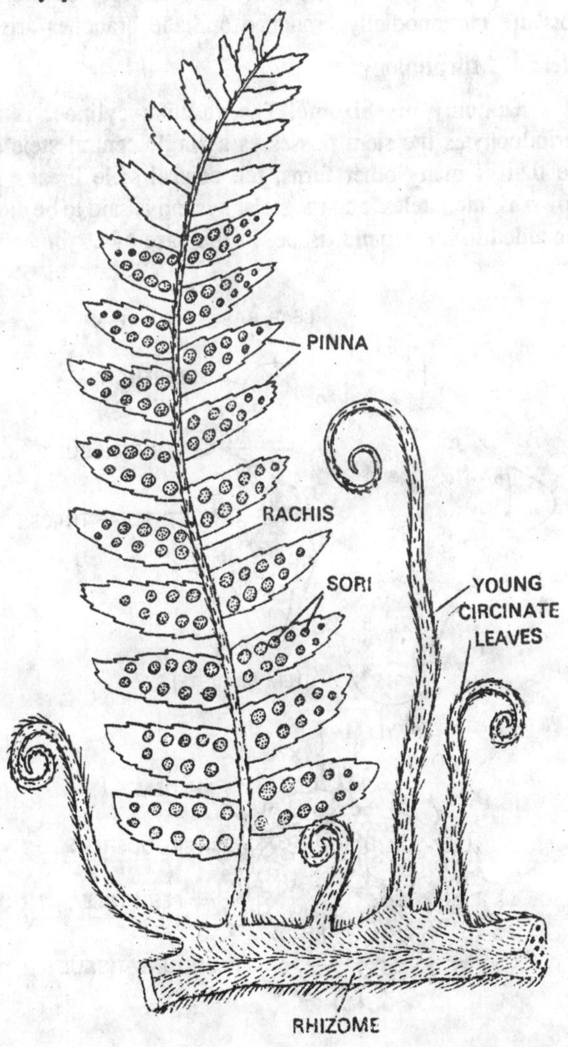

Fig. 9.2. *Dryopteris filix-mas* (Fern). Habit of fern showing rhizome, young circinate leaves and sporophyll.

The leaves. The leaves are also known as **frond.** The leaf is compound and one to three feet in length. Each leaf is composed of numerous leaflets or **pinnae.** The pinnae are again sub-divided into small **pinnules.** The leaf is petiolate. The petiole ends in a long rachis which bears pinnae. In young leaves, both the rachis and the petiole are covered with small, brown, hairy scales known as **ramenta.** The young leaves are **circinnate.** The leaf matures in two years. In the first year, the petiole, rachis and the leaflets remain coiled. Such young leaves are called circinnate leaves. The mature leaves last only for one season and again ten to twenty leaves come out, which remain spirally arranged on the rhizome.

The root. The roots are of two kinds-primary and secondary. The primary root, produced by the embryo dies away earlier. The secondary roots remain, which are also called the adventitious roots. The adventitious roots are produced secondarily from the tissues of the stem. Usually three adventitious roots develop at the base of each petiole. The roots are slender and profusely branched, so that an old stem becomes densely covered by a mass of adventitious roots. The roots are monopodially branched and the branches arise acropetally.

Internal Morphology

Anatomy of rhizome. The vascular cylinder is a typical dictyostele. In many simple pteriodophytes the stem possesses a single central stele surrounded by the cortex. In this fern, like that of many other ferns, the central stele breaks into a ring of steles. These steles are known as 'meristeles' and the stelar system is said to be dictyostelic. The meristeles are irregularly embedded in the ground tissue. In the case of *Dryopteris*, the stem is completely covered with leaf bases and there are no internodes. Any transverse section of the stem is surrounded by the leaf bases, showing the obliquely cut-leaf traces. The leaf-traces traverse the cortex and enter the petioles.

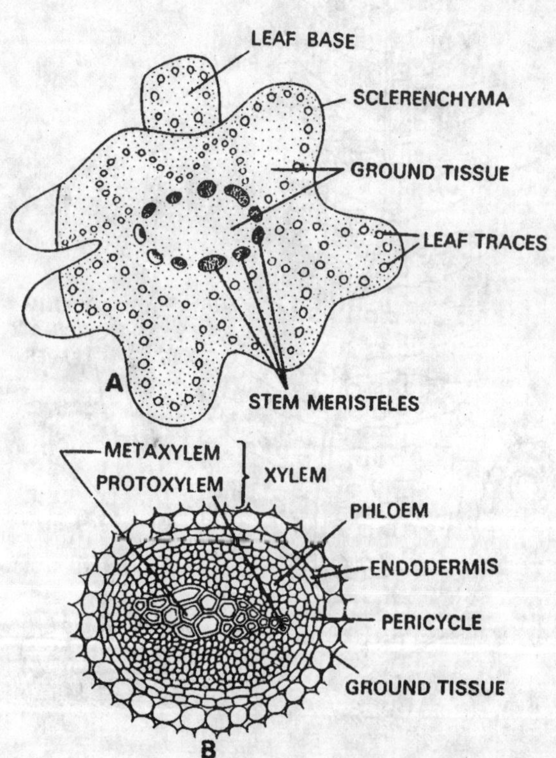

Fig. 9.3. *Dryopteris filix-mas.* A, T.S. of rhizome showing meristeles ; B, detailed structure of a meristele.

Each meristele is composed of a central mass of xylem surrounded by phloem. The scalariform tracheids and xylem parenchyma form the xylem. In the transection of the stem the tracheids appear polygonal and sufficiently wide. The vessels are altogether absent. In the centre of the meristele the tracheids bear spiral and reticulate thickenings. These tracheids represent the protoxylem. Rest of the wide tracheids of the xylem represent metaxylem. The phloem is composed of a single ring of large sieve tubes. The sieve tubes are quite long and tapering at their overlapping ends. At the junction of the two

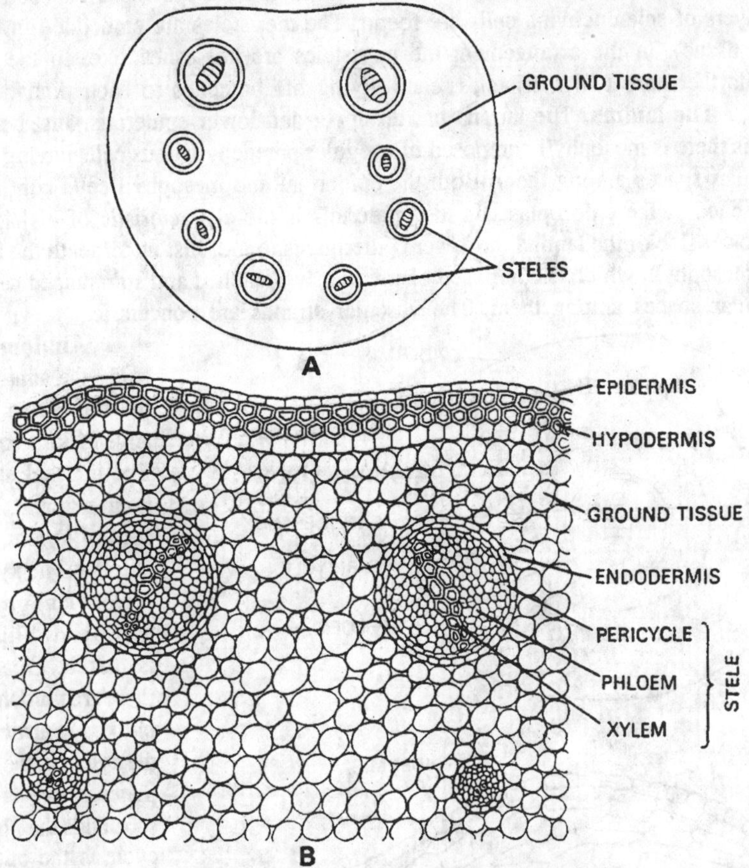

Fig. 9.4. Fern. A, T.S. of petiole (diagrammatic) ; B, T.S. of petiole showing magnified structure.

sieve tubes, the sieve plates are developed. The companion cells are not found. Just below the ring of sieve tubes, there is good deal of phloem parenchyma. The phloem is surrounded by a single-layered pericycle constituted of thin walled, squarish cells. The pericyele is the outermost layer of the cells of meristele. Each meristele is surrounded by an endodermis. The endodermis is the last layer of the cortex. The Casparian bands are found on the radial walls of the endodermal cells. Outside the endodermis the cortical parenchyma is found which extends upto the epidermis.

The meristele is concentric and amphicribral. In *Dryopteris* the meristeles are oval and independent. There is no common endodermis surrounding the complete dictyostele.

The cortex is composed of a mass of parenchyma (ground tissue) in which the meristeles are embedded. The outermost few layers of the cells of the cortex are sclerenchymatous. The outermost layer of the stem is epidermis, which is composed of the cells with thick outer walls. The epidermis bears one celled thick ramenta on it. These scale-like rementa are considered to be modified hairs.

Anatomy of leaf. From the point of view of its anatomy, the leaf may be divided into two parts (a) the petiole and (b) the lamina.

(a) The petiole. To study the anatomy of the petiole, one has to cut the transverse sections of it. The petiole is approximately semicircular in outline. The upper (adaxial) side of the petiole is flat and remains towards the stem. The outermost layer is the epidermis. Just below the epidermis a few layers of sclerenchyma cells are found. The meristeles are embedded in the parenchyma (ground tissue). In the arrangement the meristeles are horseshoe like. In the adaxial side of the rachis there are two meristeles, each giving off branches to each pinna.

(b) **The lamina.** The lamina has an upper and lower epidermis. Just beneath the upper epidermis there is mesophyll composed of irregular parenchymatous cells having well-developed intercellular spaces among them. Both the epidermal and mesophyll cells contain chloroplasts. The presence of the chloroplasts in the epidermis is the characteristic of a shade loving plant. On the lower side of the lamina the lower epidermis is found. Just underneath the lower epidermis there is mesophyll, which is composed of irregularly branched and star-shaped cells having large intercellular spaces among them. The vascular strands are concentric.

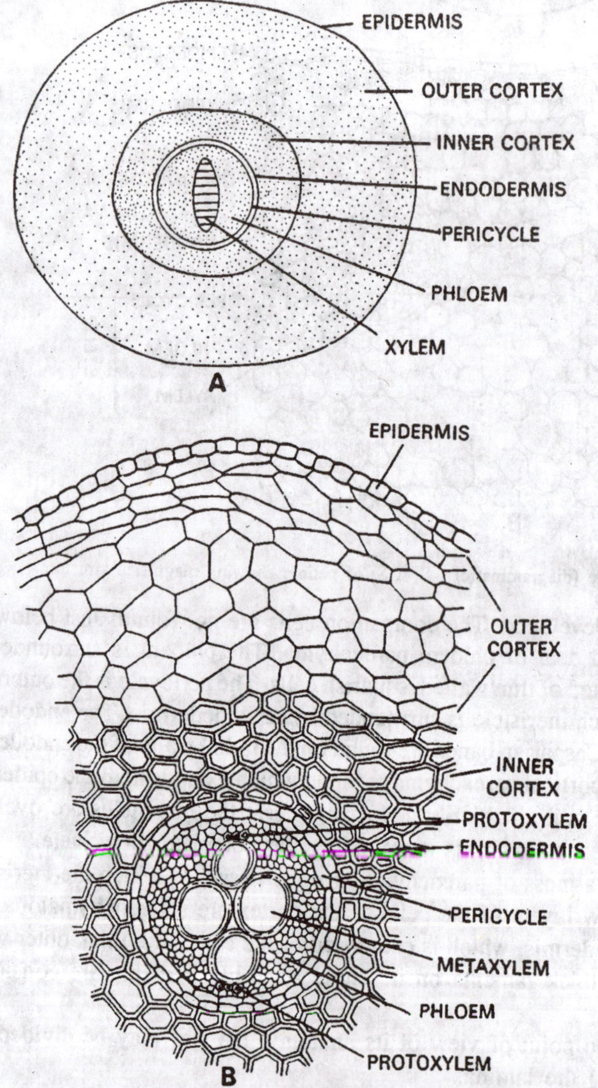

Fig. 9.5. Fern. A, T.S. of the root (diagrammatic structure) ; B, T.S. of a portion of the root (magnified structure).

Anatomy of root. To study the anatomy of the root, the transverse sections of the adventitious root are required. The internal structure of root is quite simple. There is a central core of xylem, surrounded by phloem. The xylem is with two protoxylem groups situated at its two ends. This way, the xylem is diarch. Outside the xylem there are phloem elements. The phloem is arranged laterally to the xylem in the form of phloem plates. A two-layered pericycle surrounds the phloem. The pericycle is the outermost part of the stele. Thereafter a single-layered endodermis is found around the stele. The Casparian strips are present on the radial walls of the cells of the endodermis. The outermost layer of the transverse section of the root is the piliferous layer. The piliferous layer is single-layered and bears root hairs here and there upon it. The cortical region is differentiated into two regions. The outer region consists of thin-walled cells whereas the inner region of thick-walled cells.

The apical growth. In about all the ferns the stems grow from a three-sided pyramid like single apical cell. The apical cell divides again and again, parallel to each of its three sides developing the cells of the mature tissue.

The roots also develop from a similar-three-sided pyramid like apical cell. From these three sides of the cell, all the organs of the root develop by their repeated divisions. However, the cells of the root cap are cut from the base of the apical cell.

Spore producing organs. The sporangia are produced in sori situated on the leaves, known as sporophylls. The sori are found on the lower surface of pinnae of the leaves. In *Dryopteris* the vegetative leaves and the sporophylls are quite alike, but the sporophylls occur in the old plants. The small kidney-shaped sori of sporangia are arranged in two rows on the underside of the pinnules of the leaf. The sporangia of a sorus are attached to a swollen placenta by their stalks. A delicate membranous structure, known as indusium also arises from the placenta and covers the sorus of sporangia like a hood. The mature sporangia are brown in colour.

Development of sporangium. The development of sporangium is of leptosporangiate type as in all other genera of Polypodiaceae. The sporangium develops from a single superficial cell of the placenta. This cell divides by a transverse wall forming two cells. The upper cell divides again by an oblique wall to the first wall. Another oblique division takes place, giving rise to a tetrahedral cell. This cell again divides by a transverse wall. The inner cell is known as archesporial cell. The lower cell of the pair formed by the first division divides several times and develops into the multicellular stalk. The cells surrounding the tetrahedral archesporial cell divide further giving rise to the sporangial wall. Thereafter, the archesporial cell divides to form a tapetal layer of the cells. At the same time it also produces about a dozen spore mother cells. The spore mother cells give rise to spores. Each spore mother cell divides twice developing four spores. During division, there is meiosis and the spores are haploid. Usually each sporangium contains 48 haploid spores.

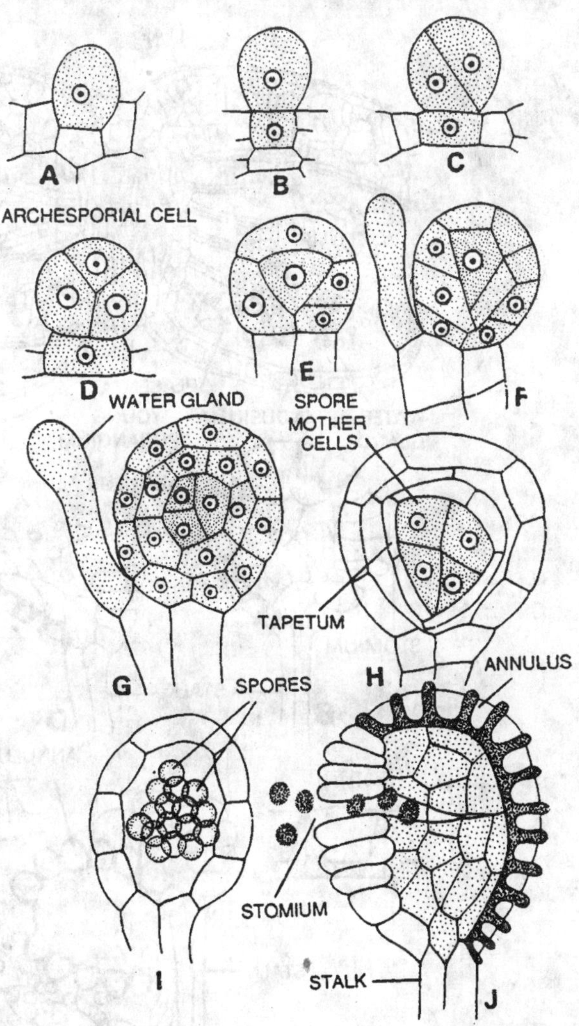

Fig. 9.6. *Dryopteris filix-mas* (Fern). A—I, successive stages in the development of sporangium ; J, mature sporangium.

Structure of sporangium. Each sporangium is composed of a multicellular stalk and a capsule. The capsule of the sporangium is of the shape of a biconvex lens. The wall of the capsule is composed of a single layer of thin walled cells. A row of specially thickened cells, known as annulus partially surrounds the capsule. At one side of the annulus, there is a stomium composed of thin walled cells. On maturity, the sporangium contains about 32 to 64 dark, ragged walled spores in it.

Dehiscence of sporangium. On the maturity of the sporangia, the indusium shrivels up and the sporangia are exposed to the outer atmosphere. The sporangium bursts by a special mechanism. The cells of the annulus are thickened in a peculiar way. The inner tangential and radial walls of the annulus are thick while the outer tangential walls are thin. The bursting of the sporangium is caused by the behaviour of the water in relation to the cells of the annulus. In young sporangia the water glands are present and they remain moist. On the maturity of

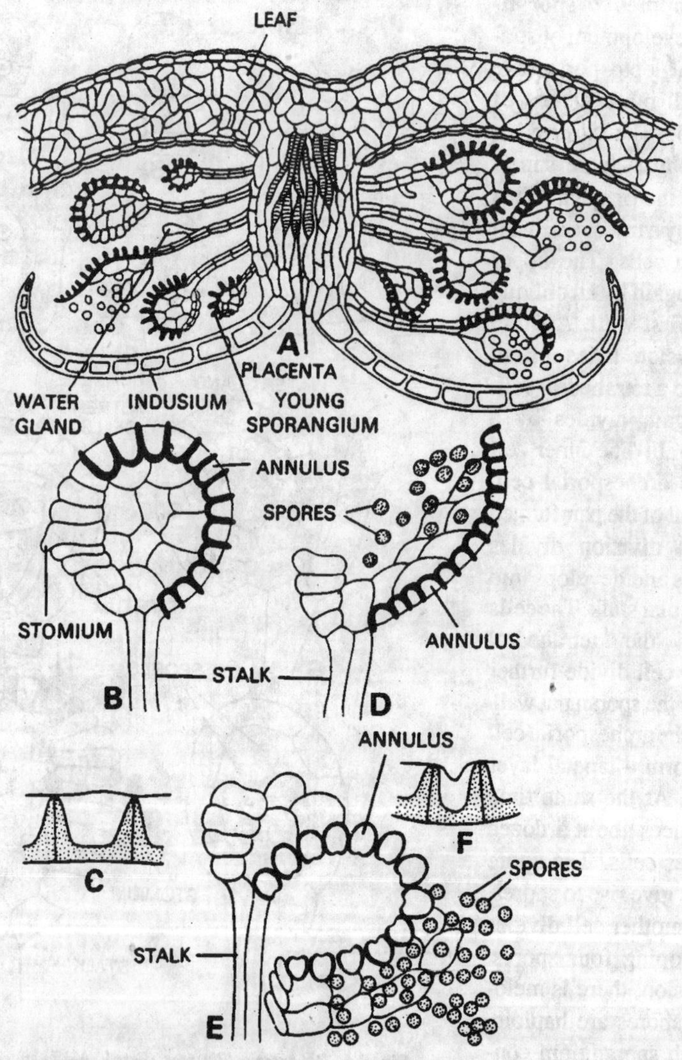

Fig. 9.7. *Fern.* A, T.S. of sporophyll showing sporangia ; B, mature sporangium ; C, cells of the annulus before eversion; D, sporangium liberating spores; E, sporangium releasing spores ; F, cells of the annulus after eversion.

the sporangia, the sporangia are exposed to dry air and the water is lost from the cells of the sporangia. This way, the thin outer walls of the annulus begin to contract, the stomium opens and the annulus along with some portion of the sporangium turns backward with a jerk, throwing the spores out of the sporangium.

The spore. The spores are double-walled. The outer wall is thick, ragged and brown in colour. It is known as exine. The inner wall is smooth and thin, and known as intine. Each spore contains a big central nucleus. The gametophytic stage of the fern begins from this spore.

Germination of the spore and development of the prothallus. After their dispersal the spores lie dormant for sometime and under the favourable conditions of moisture and temperature, they germinate. First of all the outer spore wall (exine) breaks and thereafter the intine protrudes out in the form of green germ tube. Very shortly this tube is cut off by a transverse wall and

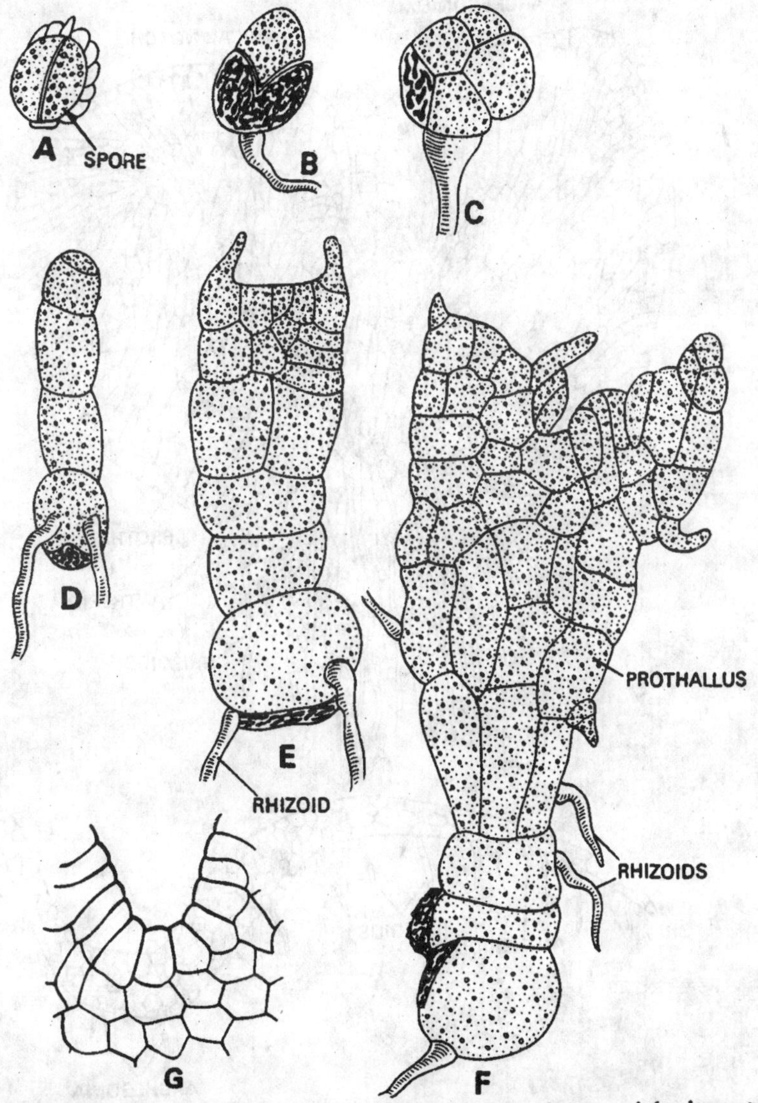

Fig. 9.8. *Dryopteris filix-mas*. A—F, successive stages in the germination of spore and development of gametophyte (prothallus).

two unequal cells-are resulted. The small colourless cell is known as rhizoidal cell. From this cell a colourless branch grows into the soil and develops a rhizoid. The large green cell elongates and divides transversely several times developing into a filament 3 to 6 cells in length. The terminal cell of the filament known as an apical cell cuts off cells behind it and gives rise to this filamentous structure. The apical cell also divides laterally and flat prothallus is formed. The segments cut off from the apical cell divide again and again developing into the young prothallus. Usually the prothallus is heart-shaped. The prothallus is green and one celled thick at its margins. In well developed prothalli the central part of the prothallus is more than one cell in thickness. This region is called cushion. Many rhizoids develop on the underside of the cushion anchoring the prothallus to the soil. The rhizoids also serve as absorbing organs.

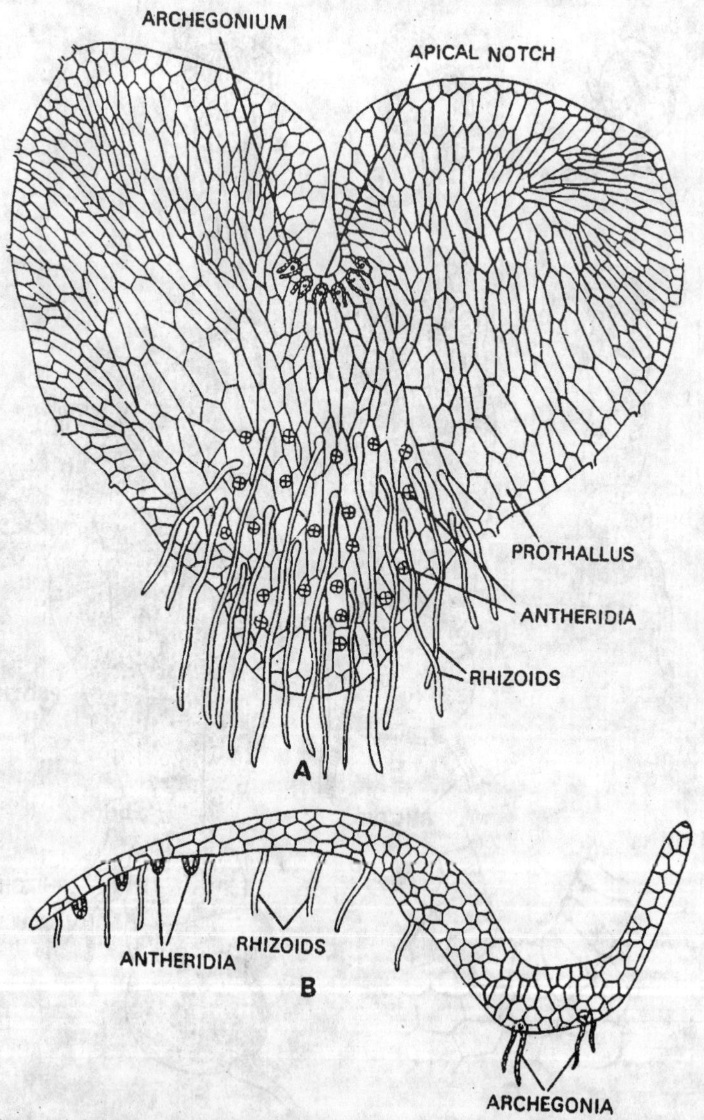

Fig. 9.9. *Dryopteris filix-mas*. A, mature prothallus bearing archegonia, antheridia and rhizoids; B, vertical longitudinal section of the prothallus showing antheridia and archegonia.

During the development of the gametophyte the original apical cell is replaced by a group of marginal initials. The Further growth of the prothallus takes place by these initials and ultimately the heart-shaped prothallus is formed.

Structure of prothallus. The prothallus is flat, thin, heart-shaped and green mass of tissue. There is a notch at its posterior end. It is one cell in thickness at its margins, but in the central region, near about the notch, it is many celled in thickness. This region is called the cushion. Several colourless rhizoids are developed from the underside of the cushion. The cells of the prothallus are thin walled. Each cell bears discoid chloroplasts and a nucleus in it. The sex organs (archegonia and antheridia) are confined in the central region of the prothallus. Each prothallus is probably 1/8 to 1/3 of an inch in diameter. The prothallus is self supporting. It absorbs water and nutrients from the soil and photosynthesize the food material as the numerous chloroplasts are found in its cells. The prothallus represents the gametophytic generation of the fern as it bears the sex organs upon it.

The sex organs. The sex organs develop on the underside of the prothallus (gametophyte). Usually the prothalli are monoecious and bear both male and female sex organs, *i.e.*, antheridia and archegonia. In the normal prothallus the archegonia develop around the notch upon the cushion and the antheridia in the basal central region of the prothallus among the rhizoids. The antheridia develop first and the archegonia later on.

Development of antheridium. One of the superficial vegetative cells of the prothallus protrudes out and divides by a transverse wall giving rise to two cells. The terminal cell acts

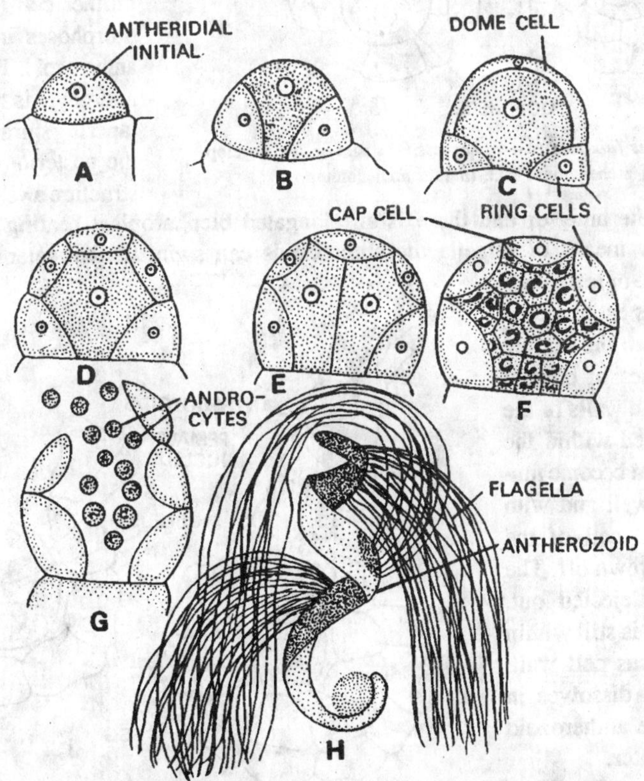

Fig. 9.10. *Dryopteris filix-mas*. A—F, successive stages in the development of antheridium; H, multiciliate antherozoid.

as an antheridial initial. First of all the antheridial initial divides by an oblique funnel-shaped wall with the wide end upwards. By successive such oblique division a central cell is formed. This central cell is surrounded by a number of peripheral cells containing the chloroplasts in them of the original initial. The first division is followed by a dome shaped wall across the top of the funnel and thus a central cell is enclosed. Thereafter, the upper cell divides by a ring-shaped wall. This way, the semispherical antheridium is composed of two ring cells, a cap cell and a central cell. The central cell divides further irregularly, and about thirty-two cells known as antherozoid mother cells (androcytes) are produced. Each antherozoid mother cell (androcyte) metamorphoses into a multiciliate antherozoid. The nucleus of each androcyte is transformed into a tapering spiral of three turns. At the posterior end of this spiral structure a vesicle of cytoplasm is attached. At the anterior end there is an elongated blepharoplast bearing a group of very long flagella. By means of flagella the antheozoids can swim in the water film.

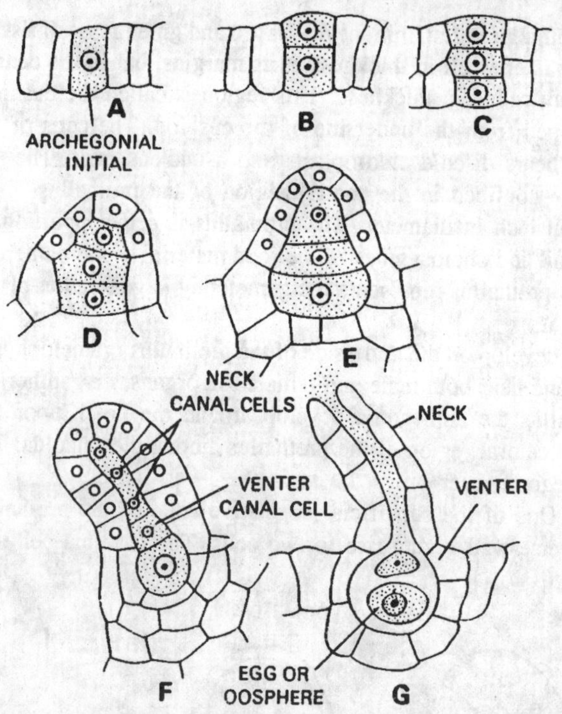

Fig. 9.11. *Dryopteris filix-mas*. A—F, successive stages in the development of archegonium ; G, mature archegonium.

Dehiscence of antheridium. The water is essential for this process. When the mature antheridium comes in contact with the water, the walls of the androcytes situated within the mature antheridium become mucilaginous and swell and with the result the lid cell of the antheridium is thrown off. The antherozoids are ejected out. Each antherozoid is still within a thin membranous cell wall, which very soon dissolves in the water and the antherozoid is free in the water.

Development of archegonium. The archegonium devel-

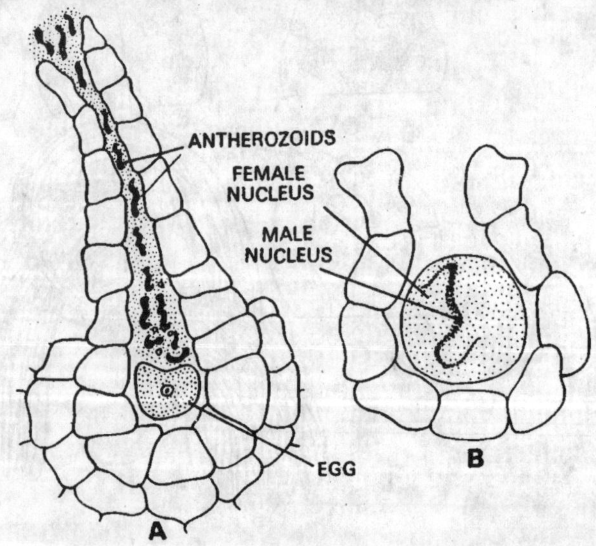

Fig. 9.12. *Dryopteris filix-mas*. A and B, fertilization.

ops from a single superficial cell of the prothallus, known as the archegonial initial. The initial divides into three cells by two transverse walls. Of these three cells the lower one is the basal cell which forms the wall around the venter cell. The remaining two cells are the central cell and the cover cell. The cover cell divides vertically twice forming four neck initials. Each neck initial divides transversely developing four vertical rows of the cells of the neck. The neck is 5 to 7 cells in height. The central cell grows up between the neck cells. This divides several

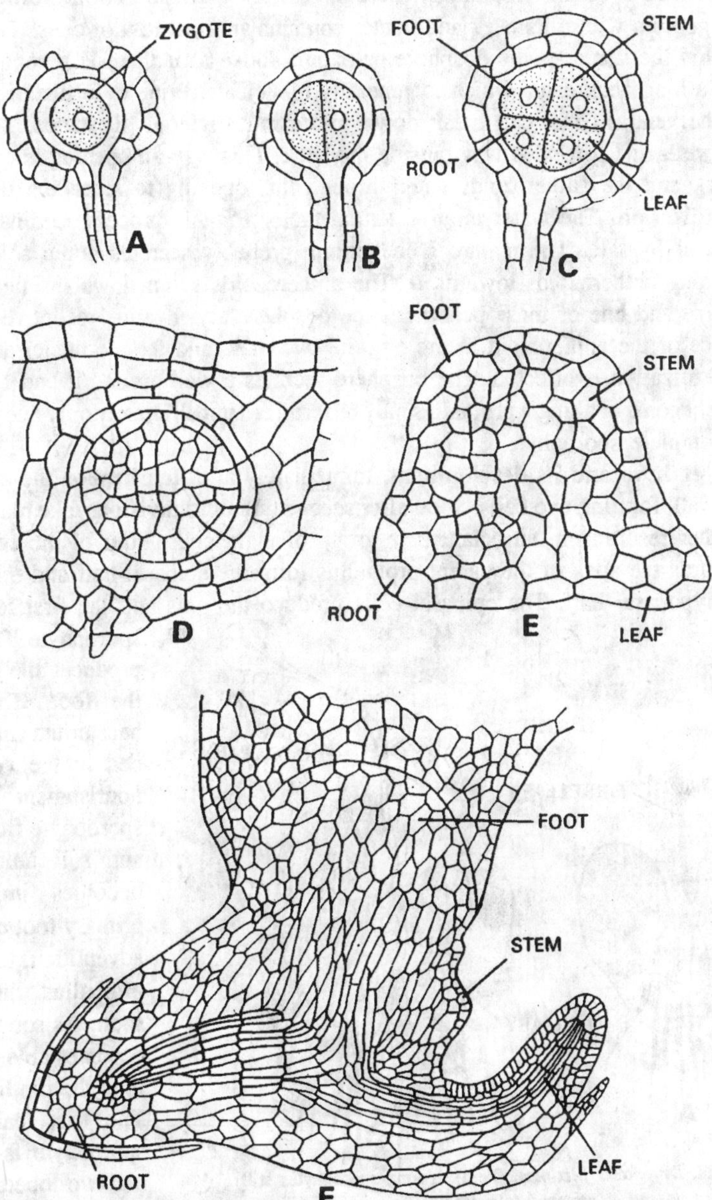

Fig. 9.13. ***Dryopteris filix-mas.*** A—F, successive stages in the development of the embryo (sporophyte) ; F, young sporophyte coming out of the prothallus.

times transversely producing a narrow neck canal cells a ventral canal cell and an egg or oosphere. The neck canal cell may or may not divide further into two cells. The neck is sharply curved. The two rows of the neck of the convex side contain 5 to 7 cells, whereas the rows of concave side have only four cells and thus the neck is strongly curved.

Structure of archegonium. The archegonium is flask-like, elongated structure consisted of a venter embedded in the tissues of the prothallus and a sharply curved neck, which projects outside the tissues of the prothallus. The curved neck consists of four vertical rows of cells. The archegonium within its neck and venter contains an axial row of cells. The lowermost cell found within the venter is the oosphere or ovum; above to it there is a ventral canal cell and thereafter a long neck canal which is generally binucleate. Prior to fertilization the neck canal cell and the ventral canal cell break down into a mucilaginous substance. This mucilaginous material swells and forces its way outside the neck. This way an opening is caused at the apex of the neck and the antherozoids enter through this opening to reach the oosphere.

Fertilization. The water is an essential factor for this process. Ordinarily the prothalli receive water by rain. The mature archegonium probably secretes malic acid, which attracts the swimming antherozoids towards it. The antherozoids swim down the neck canal through the mucilage and one of those penetrates the oosphere at a hyaline receptive spot. Soon after the entrance of the antherozoid in the egg the two male and female nuclei are fused together and the fertilization is effected. The oosphere secretes a wall around it and becomes oospore. Several archegonia of a single prothallus may be fertilized but ultimately only one oospore develops into the complete sporophyte.

The embryo and its development. Immediately after fertilization the zygote divides by a vertical wall, forming two cells. Soon after second and third divisions take place at right angles to each other resulting in an octant or a group of eight cells. Four of the cells of the octant found towards the apex of the parent prothallus form the epibasal half and the remaining four cells the hypobasal half. The epibasal cells produce the axis and the first leaf of the young sporophyte. The hypobasal half produces the primary root and the foot. The foot acts as haustorium and remains embedded in the venter. It absorbs nourishment for the young sporophyte from the prothallus and soil unless and until it becomes independent. The primary root dies soon and the adventitious roots develop. The prothallus disintegrates very soon. As soon as the first leaf becomes able to synthesize the food the prothallus disappears. The first leaf of the young sporophyte is generally small and two-lobed. Soon after other larger leaves develop which are dichotomously forked. Only after the formation of the fifth and sixth leaves, the pinnately compound leaves are developed.

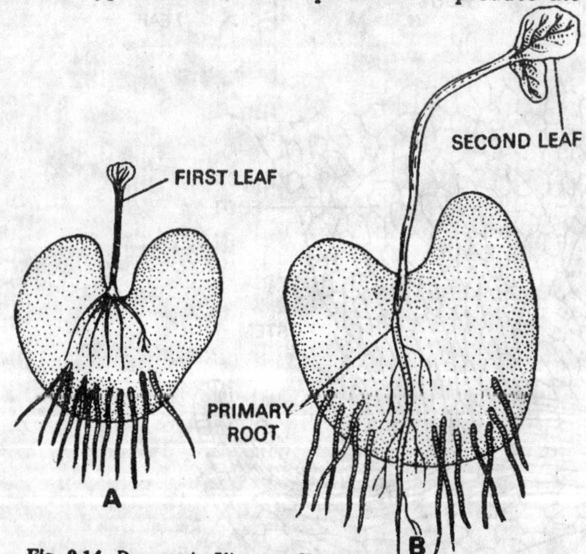

Fig. 9.14. *Dryopteris filix-mas*. (Fern). Young sporophytes still attached to prothalli.

Alternation of generations. The number of chromosomes becomes doubled, *i.e.*, diploid (2x) after fertilization is over. This diploid stage prevails upto the formation of spore mother cells. Meiotic division takes place at this stage and the haploid (x) spores are formed. The spores

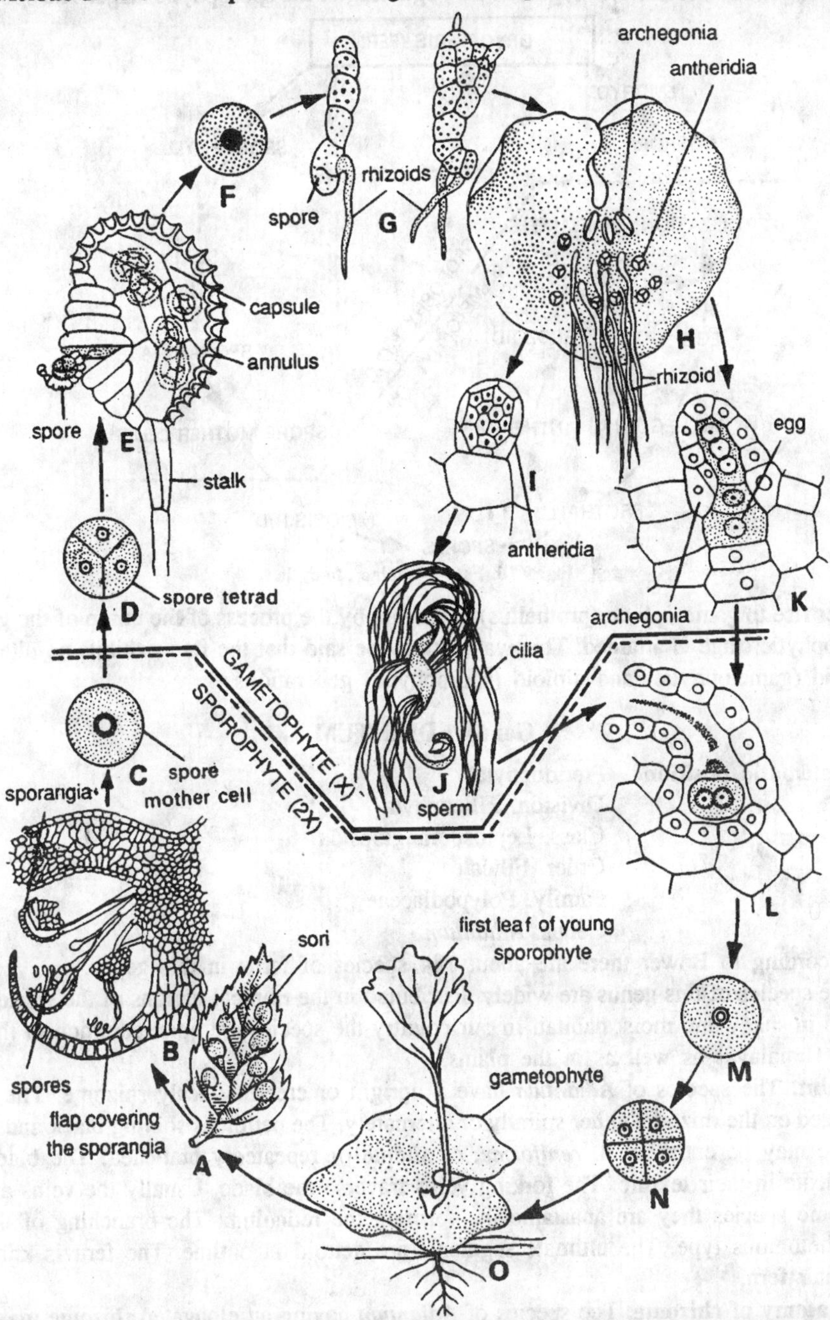

Fig. 9.15. Fern. Diagrammatic life-cycle. A, sporophyte-sporophyll ; B, T.S. of sporophyll with sporangia ; C, spore mother cell ; D, spore tetrad ; E, sporangium containing spores ; F, spore ; G, germinating spore ; H, prothallus; I, antheridium; J, antherozoid ; K, archegonium with egg ; M, oospore ; N—O, development of embryo and young sporophyte with prothallus.

are the beginners of the gametophytic stage. The spore germinates and prothallus is formed. The antheridia and archegonia develop on the prothallus. The antherozoid and ovum are fused and fertilization is effected. The zygote is the beginning of the sporophytic stage. The sporophyte

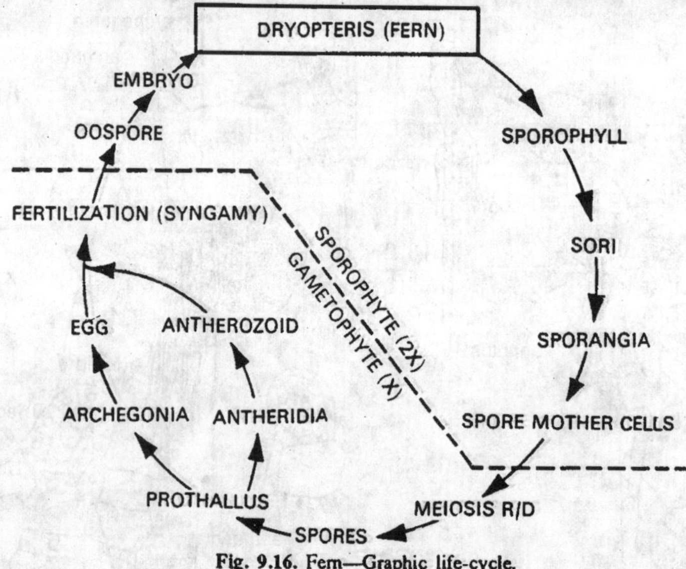

Fig. 9.16. Fern—Graphic life-cycle.

(2x) gives rise to gametophyte (prothallus) and again by the process of the union of the gametes the sporophytic stage is attained. This way, it may be said that the fern exhibits an alternation of haploid (gametophytic) and diploid (sporophytic) generations.

Genus ADIANTUM

Systematic position. Pteridophyta
Division. Filicophyta
Class. Leptosporangiopsida
Order. Filicales
Family. Polypodiaceae
Genus. *Adiantum*

According to Bower there are about 184 species of ferns in this genus.

The species of this genus are widely distributed in the tropical regions of the world. They are found in shady and moist habitat. In our country the species are quite common in the hilly tracts of Himalayas as well as in the plains.

Habit. The species of *Adiantum* have a upright or creeping scaly rhizome. The leaves are arranged on the rhizome either spirally or alternately. The petiole is shining black and brittle. The blade may be entire in *A. reniforme*, or simple or repeatedly branched. The blades are quite delicate in their texture. The forking veins traverse the blade. Usually the veins are free but in some species they are anastamosing forming the reticulum. The branching of the leaf is of dichotomous type. The ultimate segments are deltoid in outline. The fern is known as maiden hair fern.

Anatomy of rhizome. The species of *Adiantum* having an elongated rhizome may show actual solenostely, *e.g.*, in *A. pedatum*, but commonly the stele is gutter-shaped due to the great elongation of the leaf gaps. In these rhizomes which possess short internodes and where the rhizome is upright, the leaf gaps overlap and several meristeles are found arranged in a ring as seen in the transverse section of the stele.

Anatomy of leaf. The leaf base receives two closely related strands, which unite upwards to form a single four-angled meristele. The 'spicullar cells' have been found in the epidermal cells of the blade of *Adiantum.*

Anatomy of root. To study the anatomy of the root, the transverse sections of the adventitious root are required. The internal structure of root is quite simple. There is a central core of xylem, surrounded by phloem. The xylem is with two protoxylem groups situated at its two ends. This way, the xylem is diarch. Outside the xylem there are pholem elements. The pholem is arranged laterally to the xylem in the form of phloem plates. A two-layered pericycle surrounds the phloem. The pericycle is the outermost part of the stele. Thereafter a single-layered endodermis is found around the stele. The Casparain strips are present on the radial walls of the cells of the endodermis. The outermost layer of the transverse section of the root is the piliferous layer. The piliferous layer is single-layered and bears root hairs here and there upon it. The cortical region is differentiated into two regions. The outer region consists of thin-walled cells whereas the inner region of thick-walled cells.

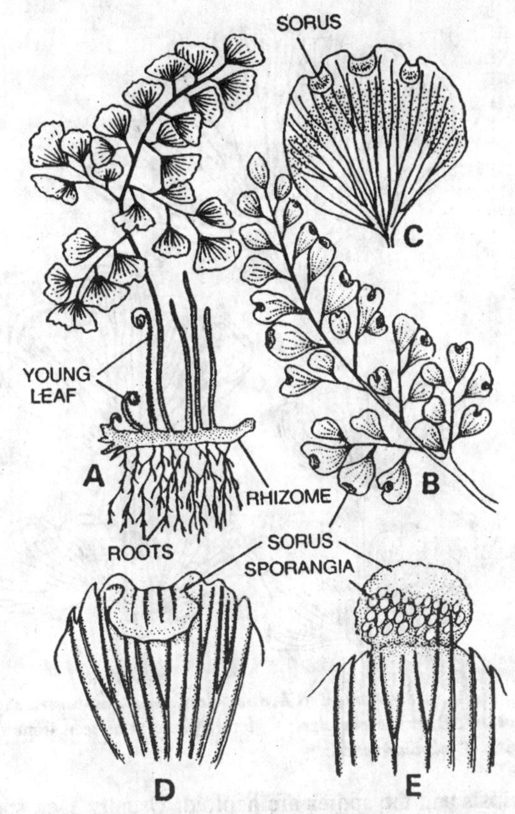

Fig. 9.17. *Adiantum.* A, habit ; B, underside of primary pinna of *Adiantum venustum* ; C, underside of leaf of *Adiantum parishii* Hook; D, parts of a fertile pinna.

The apical growth. In about all the ferns the stems grow from a three-sided pyramid like single spical cell. The apical cell divides again and again, parallel to each of its three sides developing the cells of the mature tissues.

The roots also develop from a similar three-sided pyramid like apical cell. From these three sides of the cell, all the organs of root develop by the repeated divisions. However, the cells of the root cap are cut from the base of the apical cell.

Spore producing organs. The characteristic of *Adiantum* is the presence of apparently marginal sori which are superficial in origin, and covered, by a sharply reflexed leaf margin which looks like an indusium because of its membranous nature and brown colour on its maturity. Actually there is no true indusium. The sporangia are inserted upon the distal region of the veins traversing the fertile lobe. The fertile region of the blade itself becomes sharply reflexed and serves as indusium. Each such fertile lobe bears a group of sori situated upon parallel veins. In certain cases the sporangia are found spread on to the surface of the blade in between the veins.

Development of sporangium. The development of sporangium is of leptosporangiate type as in all other genera of Polypodiaceae. The sporangium develops from a single superficial cell of the placenta. This cell divides by a transverse wall forming two cells. The upper cell divides again by an oblique wall to the first wall. Another oblique division takes place, giving rise to a tetrahedral cell. The cell again divides by a transverse wall. The inner cell is known as archesporial cell. The lower cell of the pair formed by the first division divides several times and develop into the multicellular stalk. The cells surrounding the tetrahedral archesporial cell divide further giving rise to the sporangial wall. Thereafter, the archesporial cell divides to form a tapetal layer of the cells. At the same time it also produces about a dozen spore mother cells. The spore mother cells give rise to spores. Each spore mother cell divides twice developing four spores. During division there is meiosis and the spores are haploid. Usually each sporangium contains 48 haploid spores. (Also see fig. 9.6 on page 121).

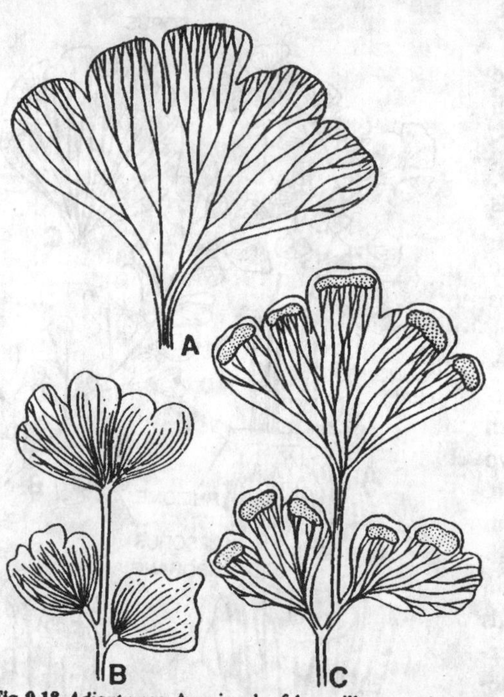

Fig. 9.18. *Adiantum* sp. A, a pinnule of *A. capillus-veneris* showing venation ; B, sterile pinnules ; C, fertile pinnules as seen from ventral side in *A. capillus-veneris*.

Structure of sporangium. Each sporangium is composed of a multicellular stalk and a capsule. The capsule of this sporangium is of the shape of biconvex lens. The wall of the capsule is composed of a single layer of thin walled cells. A row of specially thickened cells, known as annulus partially surrounds the capsule. At one side of the annulus, there is a stomium composed of thin walled cells. On maturity, the sporangium contains about 32 to 64 dark, ragged walled spores in it.

Dehiscence of sporangium. On the maturity of the sporangia, indusium shrivels up and the sporangia are exposed to the outer atmosphere. The sporangium bursts by a special mechanism. The cells of the annulus are thickened in a peculiar way. The inner tangential and radial walls of the annulus are thick while the outer tangential walls are thin. The bursting of the sporangium is caused by the behaviour of the water in relation to the cells of the annulus. In young sporangia the water glands are present and they remain moist. On the maturity of the sporangia, the sporangia, are exposed to dry air and the water is lost from the cells of the sporangia. This way, the thin outer walls of the annulus begin to contract, the stomium opens and the annulus along with some portion of the sporangium turns backward with a jerk throwing the spores out of the sporangium.

The spore. The spores are double walled. The outer wall is thick, ragged and brown in colour. It is known as exine. The inner wall is smooth and thin, and known as intine. Each spore contains a big central nucleus. The gametophytic stage of the fern begins from this spore.

Germination of the spore and development of the prothallus. After their dispersal the spores lie dormant for sometime and under favourable conditions of moisture and temperature they germinate. First of all outer spore wall (exine) breaks and thereafter the intine protrudes out in the form of a green germ tube. Very shortly this tube is cut off by a transverse wall and two unequal cells are resulted. The small colourless cell is known as rhizoidal cell. From this cell a colourless branch grows into the soil and develops a rhizoid. The large green cell elongates and divides transversely several times developing into a filament 3 to 6 cell in length. The terminal cell of the filament known as an apical cell cuts of cells behind it and give rise to this filamentous structure. The apical cell also divides laterally and a flat prothallus is formed. The segments cut off from the apical cell divide again and again developing into the young prothallus. Usually the prothallus is heart-shaped. The prothallus is green and one-called thick at its margins. In well developed prothalli the central part of the prothallus is more than one cell in thickness. This region is called cushion. Many rhizoids develop on the underside of the cushion anchoring the prothallus to the soil. The rhizoids also serve as absorbing organs.

Fig. 9.19. *Adiantum* (Fern). T.S. of stem—A, diagrammatic ; B, detailed diagram showing a siphonostele.

During the development of the gametophyte the original apical cell is replaced by a group of marginal initials. The further growth of the prothallus takes place by these initials and ultimately the heart-shaped prothallus is formed.

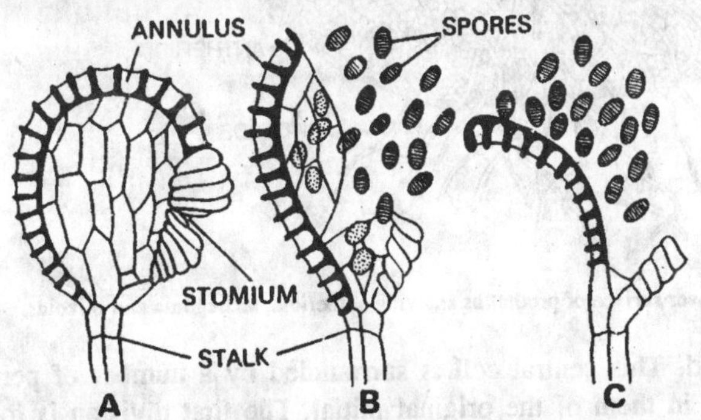

Fig. 9.20. *Adiantum* (Fern). A, sporangium ; B—C, escaping spores from sporangium.

Structure of prothallus. The prothallus is flat, thin, heart-shaped and green mass of tissue. There is a notch at its posterior end. It is one cell in thickness at its margins, but in the central region, near about the notch, it is many celled in thickness. This region is called the cushion. Several colourless rhizoids are developed from the underside of the cushion. The cells of the prothallus are thin walled. Each cell bears discoid chloroplasts and a nucleus in it. The sex organs (archegonia and antheridia) are confined in the central region of the prothallus. The prothallus is probably 1/8 to 1/3 of an inch diameter. Each prothallus is self supporting. It absorbs water and nutrients from the soil and photosynthesize the food material as the numerous chloroplasts are found in its cells. The prothallus represents the gametophytic generation of the fern as it bears the sex organs upon it.

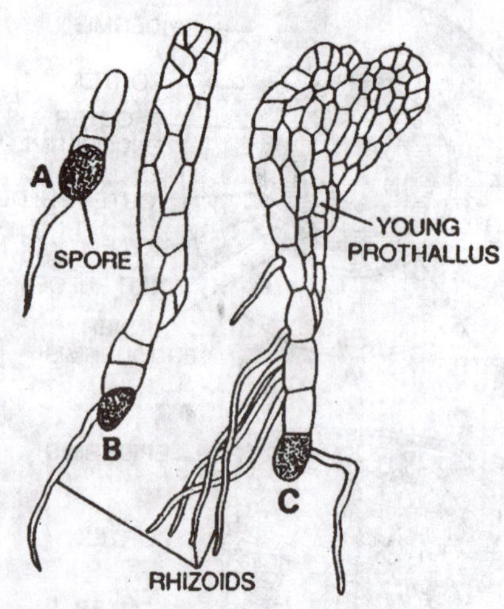

Fig. 9.21. *Adiantum* (Fern). A, spore germination and development of young prothallus.

The sex organs. The sex organs develop on the underside of the prothallus (gametophyte). Usually the prothalli are monoecious and bear both male and female sex organs, *i.e.*, antheridia and archegonia. In the normal prothallus the archegonia develop around the notch upon the cushion and the antheridia in the basal central region of the prothallus among the rhizoids. The antheridia develop first and the archegonia later on.

Development of antheridium. One of the superficial vegetative cells of the prothallus protrudes out and divides by a transverse wall giving rise to two cells. The terminal cell acts as an antheridial initial. First of all the antheridial initial divides by an oblique funnel-shaped wall with the wide end upwards. By successive such oblique

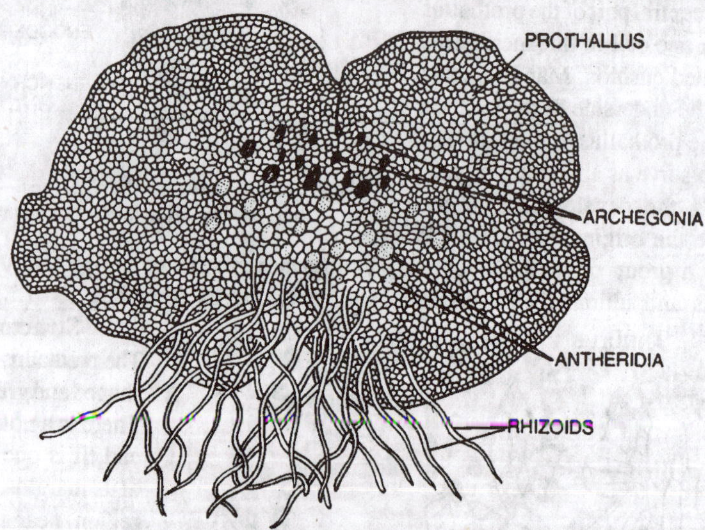

Fig. 9.22. *Adiantum* (Fern). Lower surface of prothallus showing antheridia, archegonia and rhizoids.

divisions a central cell is formed. This central cell is surrounded by a number of peripheral cells containing the chloroplasts in them of the original initial. The first division is followed by a dome shaped wall across the top of the funnel and thus a central cell is enclosed. Thereafter,

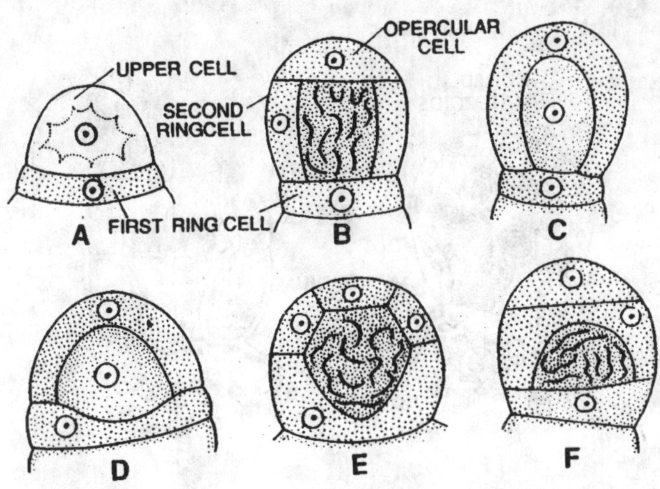

Fig. 9.23. *Adiantum sp*. A—F, stages in the development of antheridium.

the upper cell divides by a ring-shaped wall. This way, the semispherical antheridium is composed of two ring cells, a cap cell and a central cell. The central cell divides further irregularly, and about thirty two cells known as antherozoid mother cells (androcytes) are produced. Each antherozoid mother cell (androcyte) metamorphoses into a multiciliate antherozoid. The nucleus of each androcyte is transformed into a tapering spiral of three turns. At the posterior end of this spiral structure a vesicle of cytoplasm is attached. At the anterior end there is an elongated blepharoplast bearing a group of very long flagella. By means of flagella the antherozoids can swim in the water film.

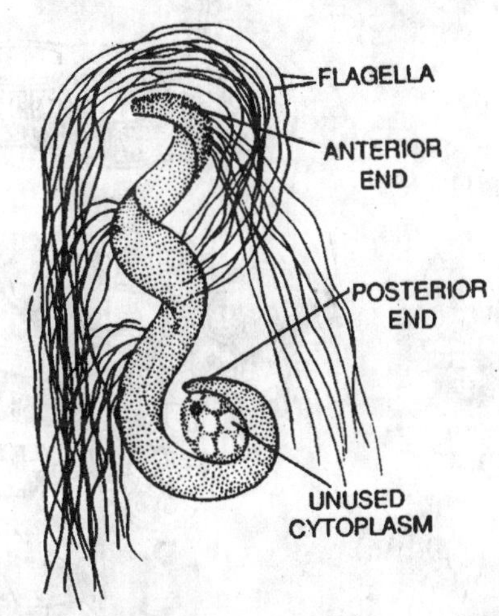

Fig. 9.24. *Adiantum* (fern). A single spermatozoid with flagella and unused cytoplasm.

Dehiscence of antheridium. The water is essential for this process. When the mature antheridium comes in contact with the water, the walls of the androcytes situated within the mature antheridium become mucilaginous and swell and with, the result the lid cell of the antheridium is thrown off. The antherozoids are ejected out. Each antherozoid is still within a thin membranous cell wall, which very soon dissolves in the water and the antherozoid is free in the water.

Development of archegonium. The archegonium develops from a single superficial cell of the prothallus, known as the archegonial initial. The initial divides into three cells by two transverse walls. Of these three cells the lower one is the basal cell which forms the wall around the venter. The remaining two cells are the central cell and the cover cell. The cover cell divides vertically twice forming four neck initials. Each neck initial divides transversely developing four vertical rows of the cells of the neck. The neck is 5 to 7 cells in height. The central cell grows up between the neck cells. This divides several times transversely producing a narrow neck canal cell, a ventral canal cell and an egg or oosphere. The neck canal cell may or may not divide further into two cells. The neck is sharply curved. The two rows of the neck of

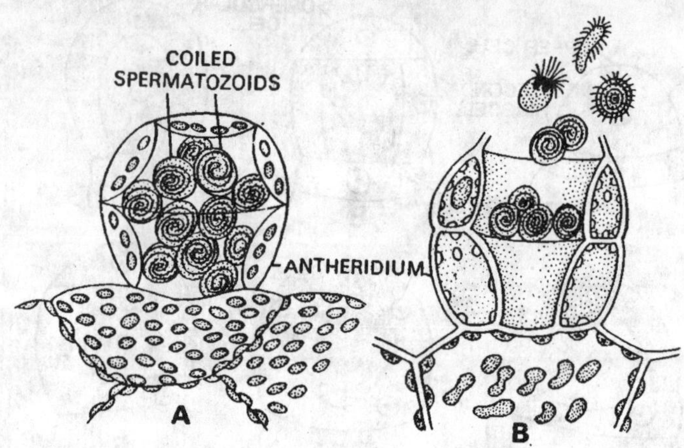

Fig. 9.25. *Adiantum* (fern). A, an antheridium on the side of a prothallus of a fern ; B, the spermatozoids are escaping from the antheridium.

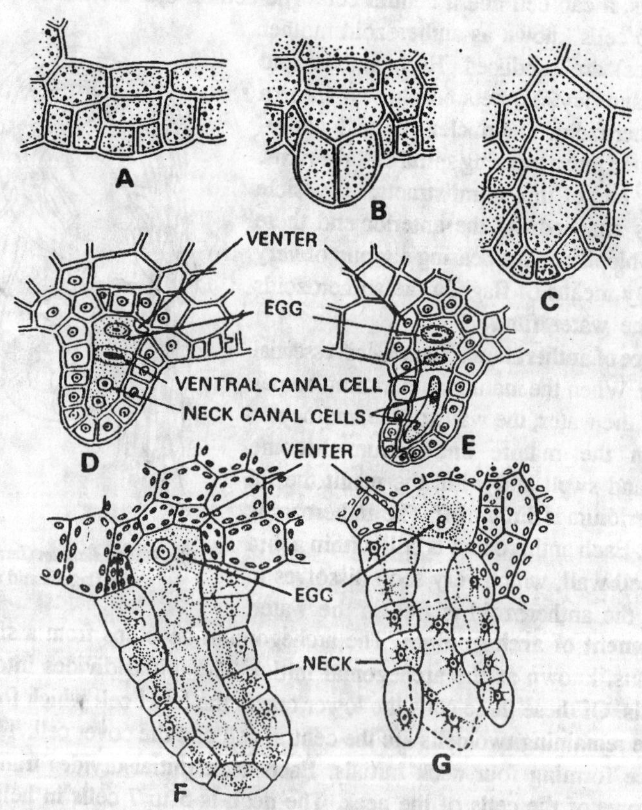

Fig. 9.26. *Adiantum* (fern). Development of archegonium—A-C, early stages of development ; D, the venter contains an-egg, and exterior to it is the ventral canal cell, and in the neck two neck canal cells ; E, the neck and venter canal cells have begun to disintegrate ; F, the cells have been disorganized ; G, archegonium open.

the convex side contain 5 to 7 cells, whereas the rows of concave side have only four cells and thus the neck is strongly curved.

Structure of archegonium. The archegonium is flask-like, elongated structure consisted of a venter embedded in the tissues of the prothallus and a sharply curved neck which projects outside the tissues of the prothallus. The curved neck consists of four vertical rows of cells. The archegonium with its neck and venter contains an axial row of cells. The lowermost cell found within the venter is the oosphere or ovum, above to it there is a ventral canal cell and thereafter a long neck canal cell which is generally binucleate. Prior to fertilization the neck canal cell and the ventral canal cell break down into a mucilaginous substance. This mucilaginous material swells and forces its way outside the neck. This way an opening is caused at the apex of the neck and the antherozoids enter through this opening to reach the oosphere.

Fertilization. The water is an essential factor for this process. Ordinarily the prothalli receive water by rain. The mature archegonium probably secretes malic acid, which attracts

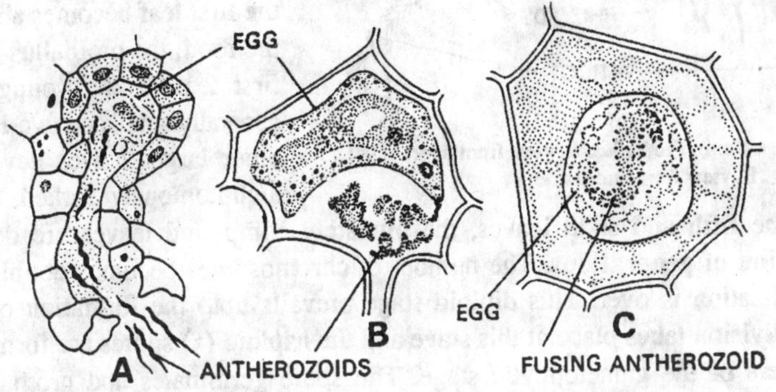

Fig. 9.27. ***Adiantum*** **(fern). Fertilization—A, spermatozoids entering the neck of archegonium ; B, single spermatozoid entering the egg nucleus ; C, union of spermatozoid with egg nucleus.**

the swimming antherozoids towards it. The antherozoids swim down the neck canal through the mucilage and one of those penetrates the oosphere at a hyaline receptive spot. Soon after the entrance of the antherozoid in the egg the two male and female nuclei are fused together and the fertilizations is effected. The oosphere secretes a wall around it and becomes oospore. Several archegonia of a single prothallus may be fertilized but ultimately only one oospore develops into the complete sporophyte.

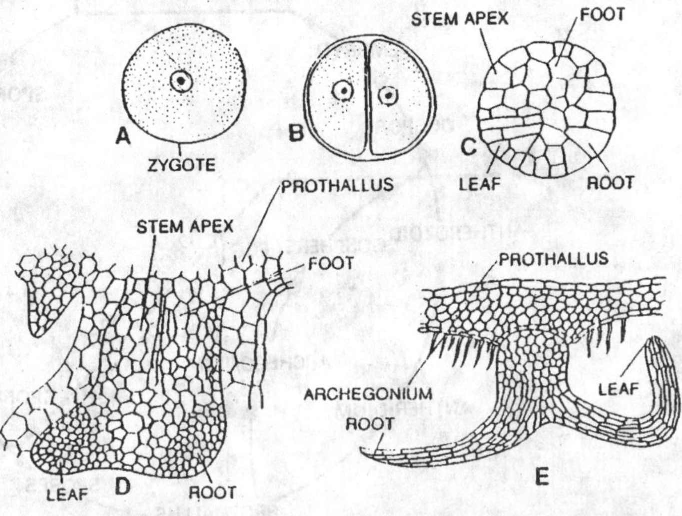

Fig. 9.28. ***Adiantum*** **sp. A—E, stages in the development of embryo.**

The embryo and its development. Immediately after fertilization the zygote divides by a vertical wall, forming two cells. Soon after second and third divisions take place at right angles to each other resulting in an octant or a group of eight cells. Four of the cells of the octant found towards the apex of the prothallus form the epibasal half and the remaining four cells the hypobasal half. The epibasal cells produce the axis and the first leaf of the young sporophyte. The hypobasal half produces the primary root and the foot. The foot acts as haustorium and remains embedded in the venter. It absorbs nourishment for the young sporophyte from the prothallus and soil unless and until it becomes independent. The primary root dies soon and the adventitious roots develop. The prothallus disintegrates very soon. As soon as the first leaf becomes able to synthesize the food, the prothallus disappears. The first leaf of the young sporophyte is generally small and two-lobed. Soon after other larger leaves develop which are dichotomously forked. Only after the formation of the fifth and sixth leaves, the pinnately compound leaves are developed.

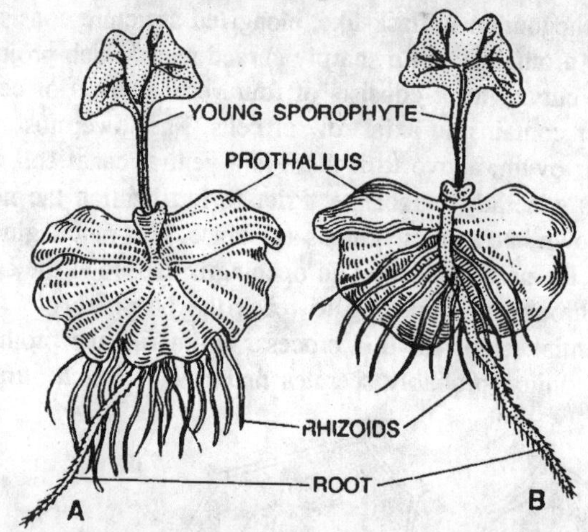

Fig. 9.29. *Adiantum* (fern), Prothallus. A, view from upper side ; B, view from under side.

Alternation of generations. The number of chromosomes becomes doubled, *i.e.*, diploid (2x) after fertilization is over. This diploid stage prevails upto the formation of spore mother cells. Meiotic division takes place at this stage and the haploid (x) spores are formed. The spores are the beginners of the gametophytic stage. The spore germinates and prothallus is formed. The antheridia and archegonia develop on the prothallus. The antherozoid and ovum fuse and fertilization is effected. The zygote is the beginning of the sporophytic stage. The sporophyte

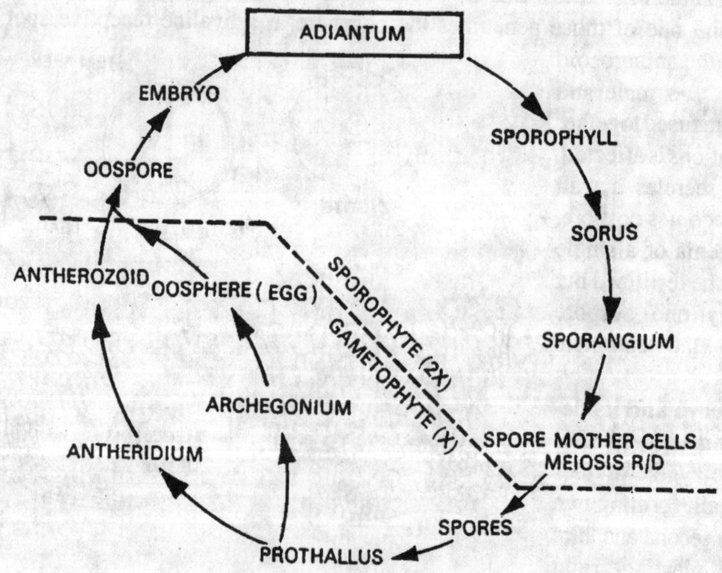

Fig. 9.30. *Adiantum* sp. Graphic life cycle.

(2x) gives rise to gametophyte (prothallus) and again by the process of the union of the gametes the sporophytic stage is attained. This way, it may be said that the fern exhibits an alternation of haploid (gametophytic) and diploid (sporophytic) generations.

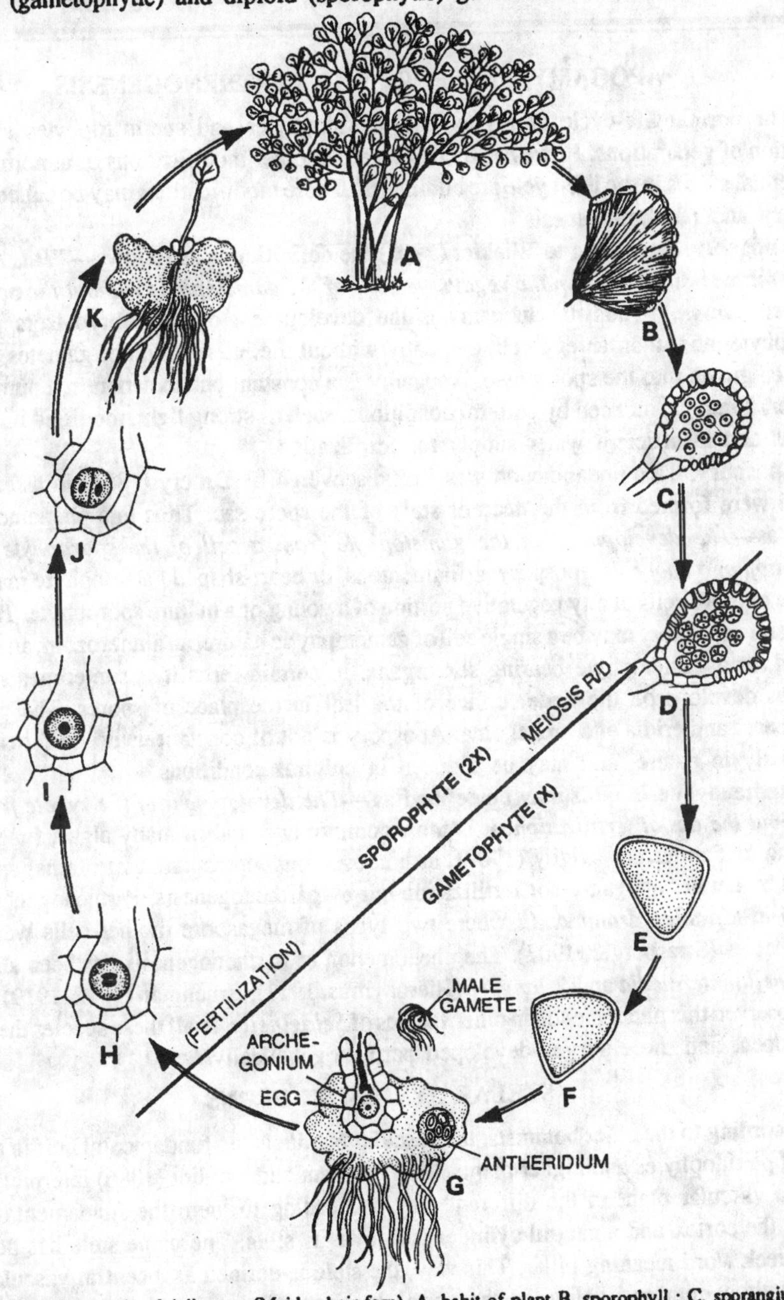

Fig. 9.31. Life cycle of *Adiantum* (Maiden hair fern). A, habit of plant B, sporophyll ; C, sporangium with spore mother cells ; D, sporangium with spore tetrads ; E, spore ; F, germinating spore ; G, prothallus with antheridia and archegonia; H, fusing male and female nuclei ; I, oospore ; J, embryo and K, prothallus with young sporophyte.

10
General Topics

APOGAMY, APOSPORY AND PARTHENOGENESIS

The normal life-cycle of bryophytes, pteridophytes and spermatophytes shows distinct alternation of generations. However, there are certain other modifications or abnormalities where the essential stages in the life-cycle are eliminated. These modifications may be called—**apogamy, apsopory** and **parthenogenesis.**

Apogamy. According to Winkler (1908) the definition is as follows— "*It is the formation of a sporophyte directly, from the vegetative cells of the gametophyte without the act of syngamy or gametic union.*" Thus the apogamy is the development of an embryo from a cell of the gametophyte other than an egg-cell, evidently without the intervention of gametes. The embryo so formed grows into the sporophyte. Apogamy is a constant phenomenon in a number of ferns. However, it may be induced by cultural conditions, such as strong light, modified nutrient culture solutions and insufficient water supply for fertilization.

Apospory. This phenomenon was first discovered by Druery (1884) in a fern where the prothalli were formed from the head or stalk of the spore sac. Thus this phenomenon may be defined as—*The development of the gametophyte from a cell of the sporophyte without the intervention of a spore.* In apospory, a filamentous, or heart-shaped gametophyte may be formed from one or more cells of any vegetative portion of a young or a mature sporophyte. The structures taking part in apospory may be a single cell of gametophytic nature, an antherozoid, an antheridium, a rhizoid, or a gametophyte bearing sex organs. In certain ferns it is sometimes seen that the prothallus develops on the undersurface of the leaf in the place of spores. The prothallus so formed bears antheridia and archegonia. Apospory is not of constant nature in ferns, and occurs occasionally in nature, and may be induced in cultural conditions.

Parthenogenesis. This may be defined as —*The development of the zygote from the egg-cell without the act of fertilization.* It is most commonly found in many algae, fungi and ferns. According to Farmer and Digby (1907), in homosporous leptosporangiate ferns, apospory was followed by neither apogamy nor fertilization but by parthenogenesis. Parthenogenesis has been observed in *Marsilea drummondii* where two types of megaspore mother cells were produced by this species (Strasburger, 1907). The phenomenon of parthenogenesis has been also observed in *Selaginella intermedia* and *S. langere* (Hieronymus, 1911). Bruchman (1912, 1919) and Goebel (1915) observed this phenomenon in other species of *Selaginelia.* In all these species the archegonia did not open, and the embryos developed parthenogenetically.

STELAR SYSTEM IN FERNS

According to the older botanists, the vascular bundle is the fundamental unit in the vascular system of pteridophytes and higher plants. Van Tieghem and Douliot (1886) interpreted the plant body of a vascular plant in the different way. According to them the fundamental parts of a shoot are the cortex and a central cylinder is known as **stele.** The name stele has been derived from a Greek word meaning pillar. This way, the stele is defined as a central vascular cylinder, with or without pith and delimited the cortex by the endodermis. Van Tieghem and Douliot

(1886) recognized only three types of steles. They also thought that the monostelic shoots were rare in comparison of polystelic shoots. It is an established fact that all shoots are monostelic and polystelic condition rarely occurs.

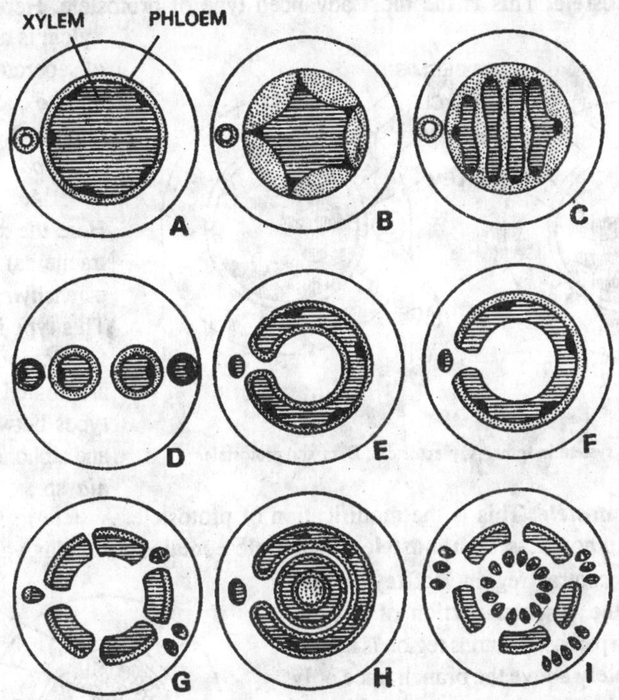

Fig 10.1. Different types of the steles found in various ferns. A, most primitive type-haplostele; B, actinostele; C, plectostele; D polystele; E, amphiphloic solenostele; F, ectophloic solenostele; G, dictyostele; H, polycyclic solenostele; I, polycyclic dictyostele.

The stele of the stem remains connected with that of leaf by a vascular connection known as the **leaf supply.**

The steles may be of the following types :

A. Protostele. Jeffrey (1898), for the first time pointed out the stelar theory from the point of view of the phylogeny. According to him the primitive type of stele is **protostele**. In protostele, the vascular tissue is a solid mass and the central core of the xylem is completely surrounded by a layer of phloem. This is the most primitive and simplest of stele.

There are several forms of the protostele which are as follows :

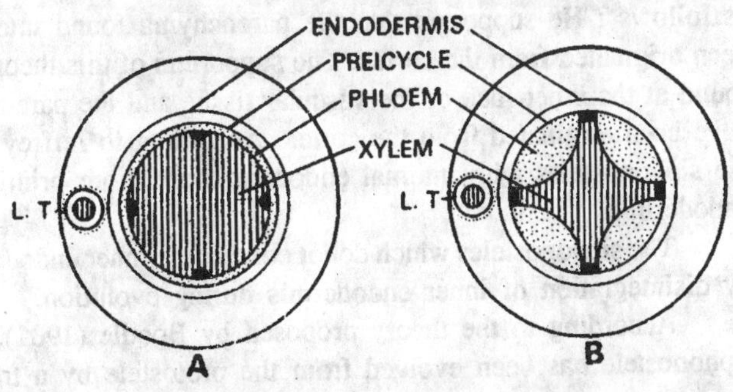

Fig. 10.2. Stelar system in ferns. A, haplostele; B, actinostele.

1. Haplostele. This is

the most primitive type of protostele. Here the central solid smooth core of xylem is surrounded by a layer of phloem, *e.g., Salaginella* sp.

2. Actinostele. This is the modification of the haplostele and somewhat more advanced in having the central xylem core with radiating ribs, *e.g., Psilotum* sp.

3. Plectostele. This is the most advanced type of protostele. Here the central core of xylem is divided into a number of separate plates arranged parallel to each other. The phloem alternates the xylem, *e.g., Lycopodium* sp.

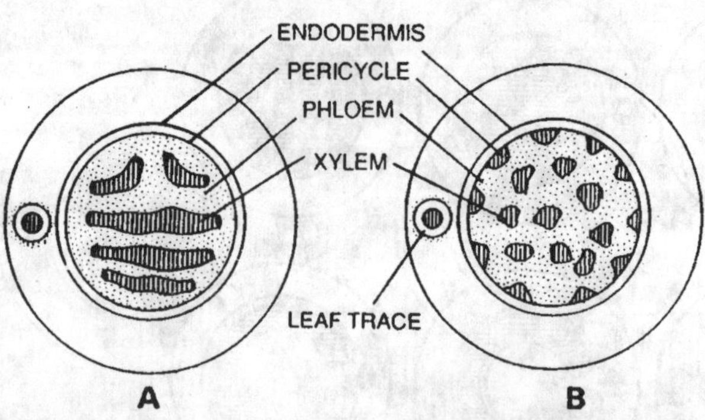

Fig 10.3. Stelar system in ferns. A, plectostele; B, mixed protostele.

4. Mixed-pith stele. Here the xylem elements (*i.e.*, tracheids) are mixed with the parenchymatous cells of the pith. This type is found in primitive fossils and living ferns. They are treated to be the transitional types between true protosteles and siphonosteles, *e.g., Gleichenia* sp., *Osmunda* sp.

B. Siphonostele. This is the modification of protostele. A stele in which the protostele is medullated is known as **siphonostele.** Such stele contains a tubular vascular region and a parenchymatous central region. Jeffrey (1898) interpreted that the vascular portion of siphonostele possessing a parenchymatous region is known as a **gap** immediately above the branch trace only or immediately above leaf and brânch trace. On the basis of these branch and leaf gaps, Jeffrey (1910), distinguished two types of siphonosteles. In one type however, the leaf gaps are not found and they are known as **cladosiphonic** siphonosteles. In the other type both leaf and branch gaps are present and they are known as **phyllosiphonic** siphonosteles.

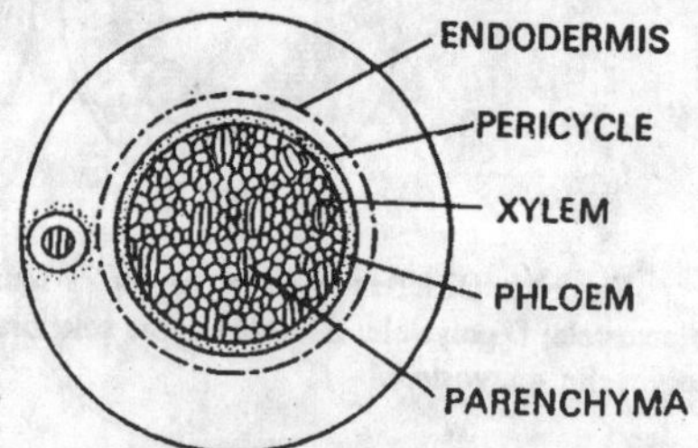

Fig 10.4. Stelar system in ferns. Protostele with mixed pith.

Jeffrey (1902, 1910, 1917) interpreted the evolution of the siphonostele from the protostele as follows : He supported that the parenchyma found internal to the phloem and xylem has been originated from the cortex. The supporters of this theory believe that the inner endodermis found at the inner face of the vascular tissue and the parenchyma encircled by this endodemis have been originated from the cortex. According to Jeffrey and other supporters of this theory the siphonosteles with internal endodermis are more primitive than those without an internal endodermis.

The siphonosteles which donot possess the inner endodermis are believed to have originated by disintegration of inner endodermis during evolution.

According to the theory proposed by Boodle (1901), and Gwynne-vaughan (1903), the siphonostele has been evolved from the protostele by a transformation of the inner vascular tissue into parenchyma.

A siphonostele may be of the following types:

1. **Ectophloic.** In this type of siphonostele, the pith is surrounded by concentric xylem cylinder and next to xylem the concentric phloem cylinder.

2. **Amphiphloic.** In this type of siphonostele the pith is surrounded by the vascular tissue. The concentric inner phloem cylinder surrounds the central pith. Next to the inner phloem is the outer phloem cylinder, *e.g.*, *Marsilea.*

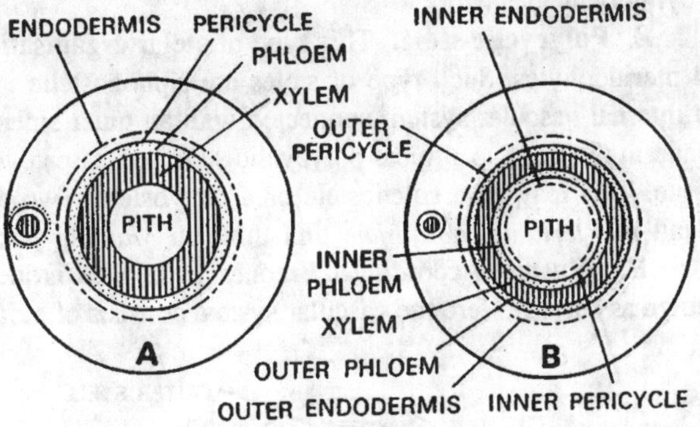

Fig 10.5. Stelar system in ferns. A, ectophloic siphonostele; B, amphiphloic siphonostele.

C. **Solenostele.** The vascular plants have been divided into two groups on the basis of the presence or absence of the leaf gaps. These groups are 1. Pteropsida and 2. Lycopsida. The ferns, gymnosperms and angiosperms are included in Pteropsida, whereas the lycopods, horse tails etc., are included in Lycopsida. The simplest form of siphonostele has no leaf gaps, such as, some species of *Selaginella*. However, among the simplest siphonostelic Pterospsida and siphonostelic Lycopsida, the successive leaf gaps in the stele donot overlap each other and are considerably apart from each other.

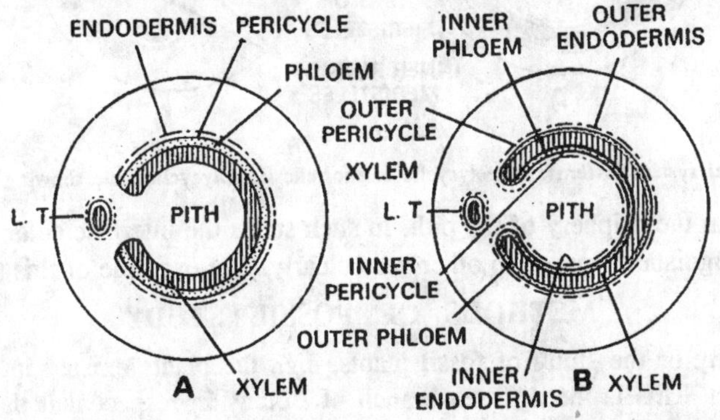

Fig 10.6. Stelar system in ferns. A, ectophloic solenostele; B, amphiphloic solenostele.

According to Brebner (1902), Gwynne - Vaughan (1901) such siphonosteles which lack overlapping of gaps are known as **solenosteles.** They may be ectophloic or amphiphloic in nature. Some authors (Bower, 1947; Wardlaw, 1952; Esau, 1953), however, interpreted the solenostele as an amphiphloic siphonostele.

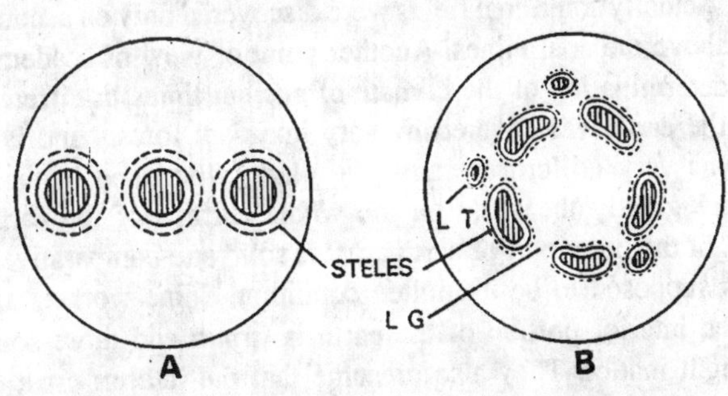

Fig 10.7. Stelar system in ferns. A, polystelic stem; B, dictyostele.

D. **Dictyostele.** In the more advanced siphonosteles of Pteropsida, the successive gaps may overlap each other. Brebner (1902) called the siphonosteles with overlapping gaps as **distyosteles.** In such cases the intervening portion of the vascular tissue meristele is of protostelic type. The dictyostele with may meristeles look like a cylindrical meshwork.

E. Polycyclic stele. This type of stelar organisation is the most complex one amongst all pteridophytes. Such type of steles are siphonostelic in structure. Each such stele possesses an internal vascular system connected with an outer siphonostele. Such connections are always found at the node. A typical polycyclic stele possesses two or more concentric rings of vascular tissue. This may be a solenostele or a dictyostele. Two concentric rings of vascular tissue are found in *Pteridium aquilinum* and three in *Matonia pectinata*.

F. Eustele. According to Brebner (1902) there is one more modification of the siphonostele, known as eustele. Here the vascular system consists of a ring of collateral or bicollateral vascular

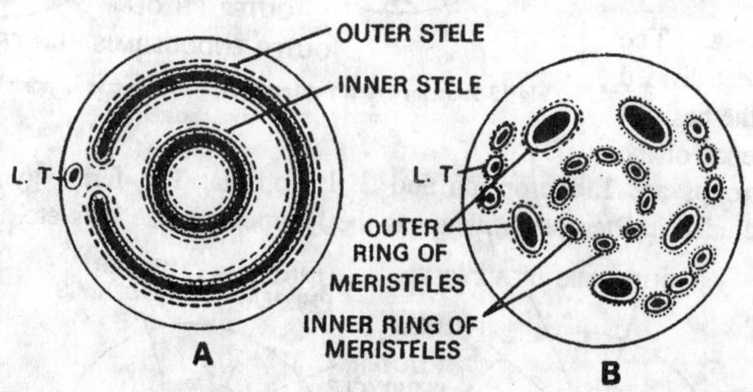

Fig 10.8. Stelar system in ferns. A, polycyclic solenostele; B, polycyclic stele showing the meristeles.

bundles situated on the priphery of the pith. In such steles the interfascicular areas and the leaf gaps are not distinguished from each other very clearly. The example of this type is *Equisetum*.

METHODS OF FOSSIL STUDY

Palaeo-botany or the study of fossil plants, *e.g.*, the plants existed in the past and now are entirely extinct. This is the difficult branch of Botany in respect that the fossil plants are difficult to obtain and they are rather scarce. Whenever the fossil plants are found, they are in parts which are to be co-ordinated. This is a tough process of the study. The fossils are cut in sections with a great difficulty and thereafter the preparations are made which require great labour, time and technique. The study of fossils is useful academically as well as economically. The academic interest lies in that their study clears up to a great extent the inter-relationships and evolution of the ancient groups of the plants. The economic interest lies in that some fossils are confined to definite strata of earth crust and they are associated with petroleum, coal and similar other things of economic value. Actually some coal fields were discovered only on account of the presence of certain fossils just above the coal mines. Another point of view of academic interest is that the fossils help in the determination of the climate of ancient times in different regions. During carboniferous period the earth was covered by very luxuriant forests and it is assumed that climate in those days was very uniform because the plants flourished in those times (*e.g.*, Carboniferous period) were greatly the same on the whole surface of the earth.

According to one school of thought the outer part of earth crust is solid and comparatively thinner than interior of earth which is supposed to be in molten condition. Some workers are of opposite view. They believe that the interior portion of the earth is firmer and more solid while the outer crust of the earth is light matter. They also presume that the interior crust of the earth consists of many heavy metals like iron and lead. Whatsoever might have been the position but it is very clear that the first crust which formed the first surface must have been very uniform.

Later on the water and air appeared and with the appearance of these two destructive factors the surface of the earth began to change. Rain and air affected the surface of the earth to great extent and at certain places the surface rose up in high mountains by bursting of interior of matter and at other places it sank down and formed seas and oceans. The rocks withered off, the volcanic eruptions and other changes took place and the general topography of the earth was completely changed. Rivers coming down from mountains bringdown pieces of rocks and large quantity of sand when they flow in the plains. This sand settles down at the bottom of the water. Along with this sand the parts of the plants and animals which get into that sand would have a chance of being preserved. This preservation takes place in different forms.

The fossils are mainly found in the sand which has been brought down by the rivers from the mountains. This sand has become compressed and sedimentary rocks have been formed in the bottom of water. They contain the pieces of plants or entire bodies or bodies of animals which have been surrounded by mud and salt and they got the chance of being preserved.

This preservation takes place in two forms :

1. **Impressions:** These impressions may be of the surface markings of the plant material or be the markings of the internal cavities in the plant material or gets into the cavity and their impressions are left. In some fine fossils even details of the venation, structure of stomata and epidermal hairs have been studied. In such impressions the plant material itself has decayed or turned into coal. However, the surface impressions cannot be relied upon as we know that the plants, *e.g.*, *Equisetum* (a pteridophyte), *Ephedra* (a gymnosperm) and *Casuarina* (an angiosperm) gave same type of impressions, but all of them the widely differentiable and have no affinities and relationships with each other.

2. **Petrification.** Another method of preservation of great importance is that the plant material when lying in water has become infiltrated with mineral matter such as lime, silica, magnesium salts and other similar substances. These salts penetrate the minute cells and the organic matter is being replaced by mineral salts. The original organic matter may have completely or partially changed but its outline and general structure is left in the form of these infiltrated substances.

In coal mines generally the coal balls are found which are formed by the infiltration of calcium bicarbonate.

In another method of preservation in which rare and small parts of plant material such as seeds, pollen grains, spores etc., have been preserved. In this method the plant material happens to fall in resin or amber extruded by some plants, such as pine trees and others.

Technique of Fossil Study

The different methods of the study of fossil plants are as follows :

It is laborious process and requires sufficiently great time. Usually the petrified specimens are cut in serial sections which give an idea of the actual structure of the fossil plant. These petrified pieces are cut into very fine slices by different methods. In one method each such piece is attached to glass plate and grounded to sufficient thinness and thereafter studied under the microscope.

Another improved method of the study of these petrified specimens is to prepare the films of the material by special techniques. The methods of preparing thin films is as follows: First of all the surface of the section of the petrified material is made smooth. If the material consists of calcium carbonate, then on the smooth surface of the slice a film of 5 percent hydrochloric acid is allowed to act for five minutes. If the slice of the petrified material is silica then the film of 10 percent bydroflouric acid (HF) is allowed to act on the smooth surface for ten minutes so that the silica is dissolved. The surface of the petrified section by the action of these acids becomes rough on account of the dissolution of the mineral matter. If any organic matter remains on the surface, now put hot gelatin on the surface. As soon as the things dry up, they are removed and studied under the microscope. This process may be successful only in the case when organic matter is left in petrified specimens. In case where organic matters are already decayed, such preparations are never good.

The fossils naturally would be pieces of plants. It is very rare that entire plant could have preserved. This way, only pieces can be studied. In such type of study the individual pieces are given botanical names, just as in living plants, the botanical names of the fossil plants are not so significant as those of living ones. As they are represented by the pieces of the plants and, therefore, their generic names would be according to stem, leaf and root or any reproductive structure. The stems are usually given the generic names which end with **'dendron'** (tree) or **'xylon'** such as, *Lygenodendron* or *Cladoxylon*. The leaves end with **'pteris'** or **'phyllum'** and reproductive parts end with strobilus. This way, the palaeobotany is the study of the parts of fossil plants and in certain case marvellous results have been obtained.

In the case of *Lygenopteris*, one of the Cycadofilicales of Carboniferous period which was found in pieces and later on the palaeobotanists supported that all the pieces belonged to the particular plant. Later on after few years the complete intact plant was found.

Important Strata

The Palaeozoic and Mesozoic strata are very important from the study point of view of the fossil plants. It is in the upper middle Palaeozoic, *i.e.*, the Devonian strata, we come across with the first land plants such as lycopods belonging to Lycopodiales, Equisetales, the seeded ferns, the primitive gymnosperms, the pteridophytes etc. In the late Palaeozoic, *i.e.*, both in the upper and lower, Carboniferous strata the earth was covered up by the very luxuriant forests. These forests were formed by lycopods, horsetails, seeded ferns and later on with primitive gymnosperms. The Carboniferous strata is most important. The coal mines are situated in this strata. The coal mines are the result of dense forests having got submerged in those times. The Mesozoic is also very important from the point of view that the first angiosperms made their appearance, otherwise the higher gymnosperms formed luxuriant forest in these times. In the later Mesozoic some of the gymnosperms disappeared. Majority of the Cycads disappeared and only a few forms have been left upto the present day.

In India the most important strata is described technically the **'Godwana system'**named after Godwana Kingdom. For the first time the rocks of this period were discovered near Narmada river.

Work on Fossils in India

The first most work on Indian fossils was done by Fiestmental, the director of geological survey of India. He published a monograph of Indian fossil forms. Thereafter Seward and Bencroft made a number of additions. From 1895 to 1909 most of the work on Indian fossils was carried on in Europe. It has been since 1919 till the death of Dr. Birbal Sahni, *i.e.*, 1949, that work of Indian fossils had been added to a great deal. He first revised the known Indian fossils and then published a number of monographs describing numerous fossil forms. The most important Indian fossil of Cycads is *Williamsonia sewardiana*. In India it is only in few places that there is a good collection of fossil plants, *e.g.*, Museum of Calcutta, the Lucknow University Museum; Presidency College, Madras, Prince of Wales College, Jammu. The collections are made from

Raniganj coal field, in Bengal, places near about Nagpur, Reeva state, Kotah state, in some parts of Kashmir near about the Satlaj river, salt range in Punjab, and Rajmahal hills, Bihar.

THE TELOME THEORY

The interpretation of the plant body as a branched axis with its divisions specialized for various functions, but all like in fundamental structure, is called the **telome theory.** The telome

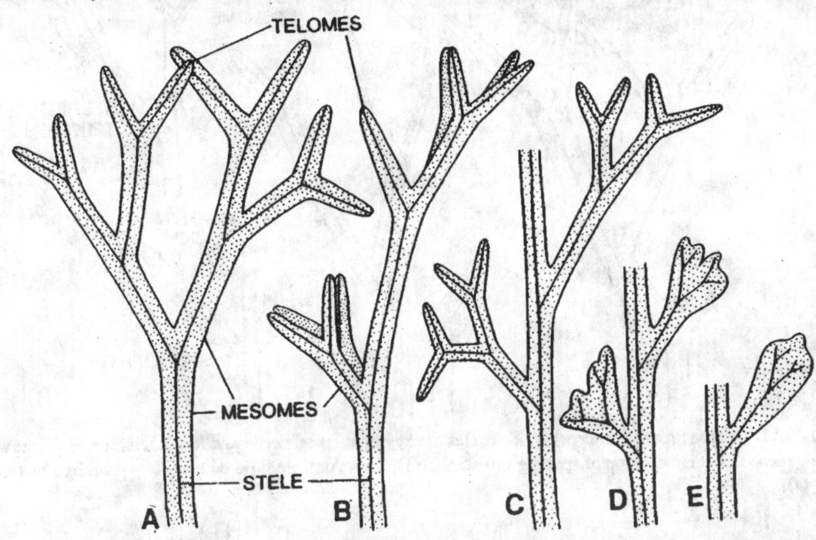

Fig 10.9. Zimmermann's Telome concept. A, illustrates hypothesis of telomes and mesomes; B, overtopping-two sister branches of a dichotomy become unequal, the stronger one forms the axis, the weaker one is pushed aside and forms the leaf; C, *planation*-branching in more than one plane is being replaced by a dichotomy in a single plane, thus arranging the telomes and mesomes in a plane; D-E, *webbing* and *syngenesis*-the telomes and mesomes, become connected by the formation of parenchymatous tissue between them (parenchymatous webbing); this is accompanied by the fusion of their steles also.

theory is due to the work of Zimmermann. He goes back to the earliest and simplest type of vascular plant known, that in the Psilophytales in which the body consists of branching leafless axes, which are all substantially alike. In some cases these may be arranged dichotomously in others monopodially, and the difference may have arisen by 'overtopping', but in either case these naked axes as the primitive units of construction of the cormophytes, which is thus regarded as primarily a system of axes. Each of the terminal branches in such a system is called a **telome.** Each telome is an ultimate branch on an older axis or **mesome.** Some telomes are sterile, others bear terminal sporangia and are therefore fertile. Telomes also tend to unite into groups, called, **syntelomes,** which may be all sterile, all fertile or mixed. According to telome theory the earliest levees are flattened telomes or **phylloids,** and the sporangium in all cases, together with its stalk or sporangiophore, is a fertile telome. Compound sporangiophores, like those of *Equisetum,* are syntelomes. The phylloids are the prototypes of the microphyllous foliage leaves. In the most primitive condition the sterile

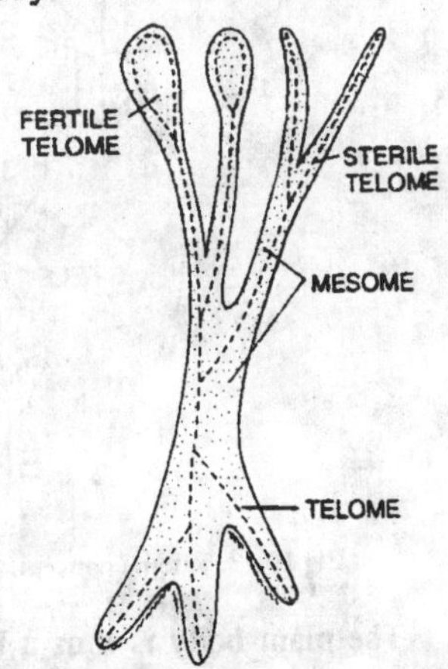

Fig 10.10. Diagram of the organization in a theoretically primitive vascular plant showing a body built of teleme and mesome limbs, the telomes being both fertile and sterile. (After Zimmermann).

and fertile telomes were apparently quite independent of each other. To axillary position which the sporangial telomes occupy in Lycopsida has been arrived at by secondary shifting of the telomes and shows three ways in which the change may have taken place.

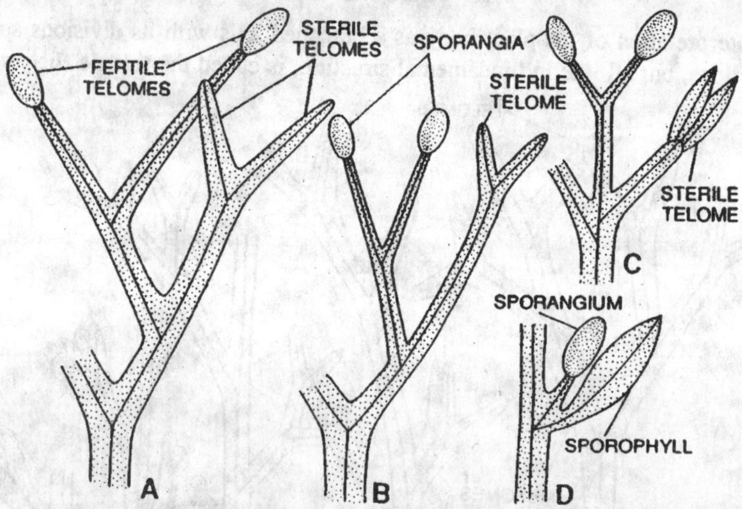

Fig 10.11. Telome theory (concept). A, illustrates the aggregation of sterile and fertile telomes; B-C, shows reduction of supporting mesomes and reduction of sporangium number; D, shows one phylloid with axillary sporangium, (seen in origin of Lycophyta).

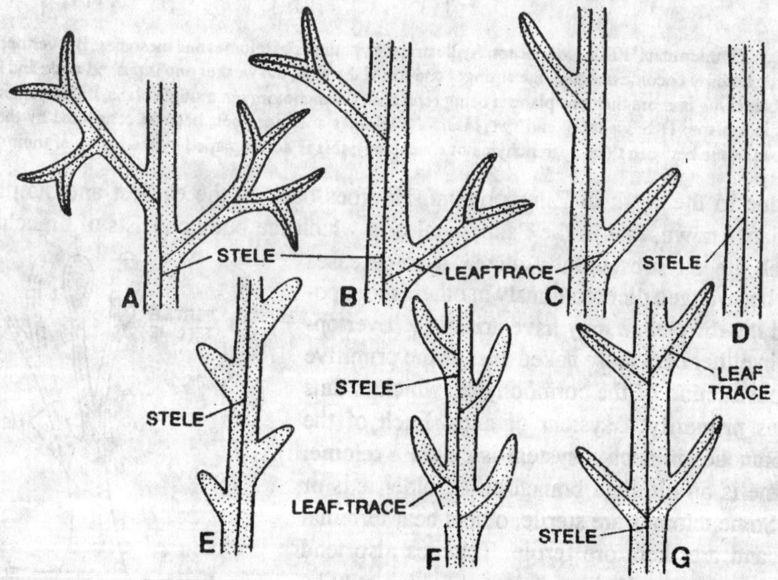

Fig 10.12. Telome concept. A-D, reduction of syntelomes to single needle like lateral appendage.

The plant body is thus a branching axis, its specialized underground segments are roots, its aerial tips, if fertile, sporangia, if sterile, phylloids. "Leaves" consist under this interpretation, of one or more sterile telomes associated in may cases with one or more fertile telomes.

Leaf morphology thus supports the division of vascular plants into four major groups- the Psilopsida are leafless or show leaves only partly differentiated; the leaf of the Lycopsida is an enation enlarged and elaborated; that of the Sphenopsida is a minor branch system; that of Pteropsida a major one. The leaf is thus of different nature in vascular plans; the evolutionary development of a photosynthetic organ has come about in more than one way.

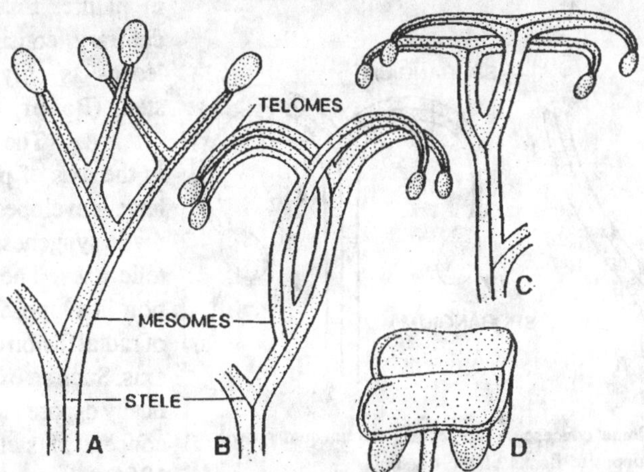

Fig 10.13. Telome Concept. A-B, *incurvation*-downward bending of the sporangia toward a common axis; C-D, fusion and expansion of telomes and mesomes into a peltate structure (seen in Sphenopsida).

Merits. This is a simple concept and explains most of the morphological problems about different organs of a plant. According to Bierhorst (1971) this theory is too simple and too easily applicable but unfortunately its excessive use has greatly diminished its value.

Demerits. 1. Telome has been considered as a readymade unit. This difficulty was realized by Zimmermann (1949, 52) and subsequently he recognized several other elementary processes as (a) interconnection of cells ; (b) rotation of cell axis; (c) differentiation of apical cell etc., which have led to the formation of an apical cell with three cutting faces. however, these elementary processes donot satisfy the plant morphologists (Puri, 1956).

2. Many other plants of much greater complexity, than *Rhynia* fossils have been discovered in beds of the same age or even earlier, *e.g.*, *Zosterophyllum, Baragnnathia* (Leclercq, 1954). Lyon Hueber (1964), Hueber and Banks (1967) and

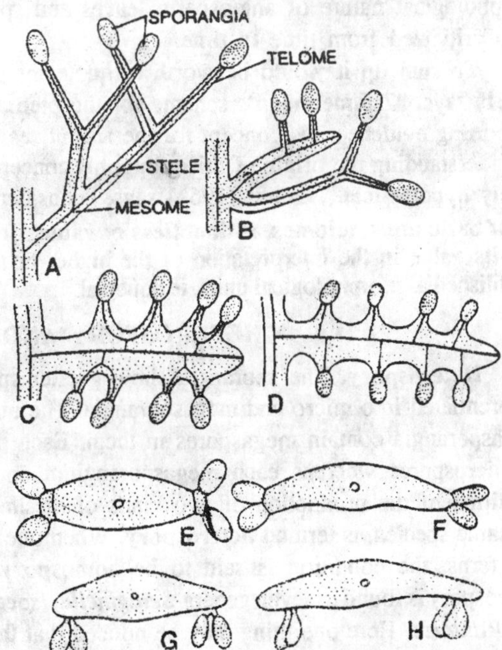

Fig 10.14. Telome concept. Origin of Pteropsida. A-B, illustrates that overtopping leads to pinnate sporophyll, C, shows the fusion of mesomes laterally with each other ; D, shows pinnately veined sporophyll with marginal sporangia; E-H, shows un-equal growth on the two opposite flanks of an organ (*i.e.*, incurvation).

Lyon (1964) observed lateral sporangia on short vascularized stalks in *Psilophyton* and *Asteroxylon* respectively instead of usual terminal sporangia.

3. According to the supporters of the telom theory all the leaves in plants are telomic in nature. Enation theory on the contrary considers microphyllous leaves as only outgrowths of the stem (Bower, 1935).

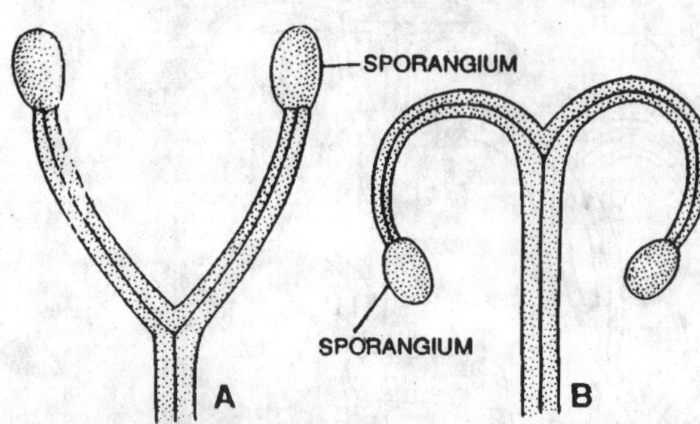

Fig. 10.15. Telome concept. A—B, shows the unequal growth of the tissue on two opposite flanks of the organ.

4. The polystelic condition in the axis of plants is supposed to have developed due to parenchymatic syngenesis of several monostelic axes. The actinostelic condition is supposed to be the product of radial fusion of steles in polystelic axis. Such an explanation is diametrically opposed to the widely accepted concept of stelar theory (Stewart, 1964).

5. Andrews (1963) has given a series of diagrams of Palaeozoic seeds to explain the origin of cupule. Pettitt (1970) found them to be more or less of the same age.

6. The theory has received little attention by angiosperm centred morphologists. Its application to stamens (Puri, 1947, 1951, 1955), venation pattern of levees (Foster, 1950), morphological nature of angiosperm leaves and sporophylls, and carpels (Eames 1961) have been criticized from time to time.

To sum up it would be worth to quote the reaction of Andrews and Eames. Andrews (1961) says, "Zimmermann's scheme for the pteropsids, or atleast some pteropsids, has much supporting evidence, his concept for the articulates may be valid, but we are only in the verge of understanding the origin of this group, his concept for the lycopsids is, so far as I am aware purely hypothetical." Eames (1961) says, "consideration of the primitive plant body as made up of basic units, telomes, is doubtless of value for the understanding of more primitive taxa, but its value in the interpretation of the higher taxa, where axis and appendages have become established as morphological units, is doubtful."

HETEROSPORY AND SEED HABIT

Heterospory. The sporangia show greater specialization than those of ferns. They are differentiated into micro and megasporangia. The microsporangia contain microspores whereas megasporangia contain megaspores in them. Each microsporangium contains a large number of microspores whereas each megasporangium contains usually only four megaspores. The condition of the production of two kinds of dissimilar spores, differing in shape and size in the same species, is termed **heterospory**. When one kind of spores are present, *e.g., Equisetum* and ferns, the condition is said to be **homospory**. According to modern Pteridologists the heterospory is found in seven genera-*Selaginella, Isoetes, Marsilea, Salvinia, Azolla, Regnellidium* and *Pilularia.* Here one thing may be noticed that the microspores develop male gametophytes whereas the megaspores germinate into the female gametophytes which may also be called as microgametophytes and megagametophytes. In the case of homospory the sex may be differentiated only at the gametophytic stage, but here the sex may be differentiated even at the sporophytic stage by the presence of small and large sized microspores which produce antherozoids and eggs respectively.

The differentiation of microspores and megaspores and their dependence upon the sporophyte has certain advantages. The gametophytes of the ferns are, however dependent for their nutrition upon soil and environmental condition, whereas in the case of *Selaginella*, as far as the nutrition of gametophytes is concerned they derive it from the sporophyte, and therefore they are more independent to the external condition than those of ferns.

Seed habit. The condition of heterospory in *Selaginella* constitute one of the most important phases in the evolution of the plants. In certain species of *Selaginella* only one megaspore develops within the megasporangium. The highest evolved species, *i.e.*, *S. apoda* has advanced almost upto the rank of a seed plant. Here, the megaspore remains within the sporangium ; it itself fertilizes and this way for the first time the gametophyte of *Selaginella* shows complete dependence of the gametophyte upon sporophyte as in angiosperms.

Biological importance of heterospory. The phenomenon of heterospory is of great biological advantage because of the fact that a large megaspore which contains female gametophyte derives its food from the sporophyte, and is independent of the external conditions as might interfere with the growth of a free living gametophyte. It thus forms better starting point for the new embryo, than an independent green prothallus which has to manufacture its own food.

In brief, the origin of seed habit is associated with the following important prerequisites:

1. The production of two kinds of spores (*i.e.*, heterospory).
2. The retention and germination of the fertilization of the egg and embryo formation, which is still within germinating megaspore.
3. Development of only one megaspore per megasporangium.

In *Selaginella*, there is remarkable approach to the seed habit, on account of the following important features:

1. *Selaginella* shows heterospory.
2. The megaspore usually germinates within the megasporangium and is not shed for a time which may very from species to species.
3. There is reduction to one megaspore in some species, *e.g.*, *S. rupestris* and *S. monospora* and a confirmed tendency to reduction in others.

It becomes quite evident that *Selaginella* has considerably advanced towards the seed habit in a few species, but its approach to the true seed is not complete due to the following features:

1. The megasporangium lacks an integument or covering.
2. The permanent retention of the megaspore within the megasporangium has not become established.
3. After the development of the embryo, the resting period is not there.

GENERAL CHARACTERS OF PTERIDOPHYTES

Occurrence

1. Most pteridophytes are terrestrial and grow in moist and shady places while some flourish well in open, dry places especially in xeric conditions. Some pteridophytes are aquatic and some are epiphytes.

The plant body (The adult sporophyte).

2. The sporophyte is the conspicuous and familiar plant body. It develops from the zygote, a diploid cell which results from the fertilization of the egg and antherozoid.
3. The sporophytic plant body remains differentiated into true roots, stem and leaves. Some primitive members lack true roots and well developed leaves (*e.g.*, in Psilophytales and Psilotales).

4. The branching of the stem may be of monopodial or dichotomous type.

5. There are two main categories of form and structure, one category comprises of megaphyllous types, in which the leaves are large in relation to the stem (*e.g.*, ferns); the second comprises of microphyllous types in which the leaves are quite small in relation to the stem (*e.g.*, Lycopods and Horse-tails).

6. All the vegetative parts of the sporophyte possess vascular supply.

Reproduction.

7. The sporophytes reproduce by spores which are produced within spoangia.

8. In some pteridophytes the sporangia develop on stems (*i.e.*, cauline in origin) while in other they are borne either on the leaves (foliar) or in the axils of the leaves. The leaves that bear sporangia are known as sporophylls.

9. The sporophylls may be widely scattered on a plant (*e.g.*, ferns) or may be clustered in definite areas and structures called cones or strobili (*Selaginella, Equisetum*).

10. In certain pteridophytes the sporangia are produced within specialized structure, the sporocarps (*e.g.*, *Marsilea, Salvinia, Azolla*).

11. The sporangium in all pteridophytes, is initiated by the laying down of a cross-wall in a superficial cell or a group of cells. Since this wall is periclinal each initial cell is divided into an outer and inner daughter cell. If the sporogenous tissue is derived from the inner daughter cell, the sporangium is described as "**eusporangiate**" (*e.g.*, in most of pteridophytes) and if from the outer as "**leptosporangiate**" (*e.g.*, in advanced pteridophytes).

12. The sporophyte plant may be homosporous (*e.g.*, *Lycopodium, Dryopteris*) or heterosporous (*e.g.*, *Selaginella, Isoetes, Marsilea*).

The gametophyte.

13. The spores on germination give rise to the haploid gametophytes or prothalli which are small and inconspicuous. The gametophytes in some pteridophytes are subterranean and in others they are retained within the resistant wall of the spore. It is notable that wherever the gametophyte is retained within the spore the spores are of diffrent sizes (heterosporous). The larger megaspores give rise to female prothalli which bear only archegonia, and the smaller microspores giving rise to male prothalli which bear only antheridia.

14. The gametophyte or prothallus bears the sex organs, antheridia and archegonia. Normally, the gametophytes formed from the homospores are monoecious, that is both antheridia and archegonia are borne on the same gametophyte or prothallus. The gametophytes formed from the heterospores are dioecious, *e.g.*, the antheridia and archegonia develop in separate male and female gametophytes,

The antheridia.

15. The antheridia may be embedded in the gametophyte or they may project from it. The embedded antheridia are commonly found in eusporangiate pteridophytes while the projecting ones are usually found in the leptosporangiate ferns.

The mature antheridium is globular and consists of an outer sterile wall inside which are found a large number of androcytes. Each androcyte metamorphoses into a single motile antherozoid.

The archegonia.

16. The archegonia are flask-shaped. Each archegonium consists of a basal swollen, embedded portion the venter and a short neck. The venter encloses the egg and ventral canal cell. The neck contains the neck canal cells.

At maturity the apical cells of arciiegunium separate, the neck canal cells disintegrate forming a passage for antherozoids to reach the egg.

Fertilization.

17. In all cases the fertilization is accomplished by the agency of water. With the result of the fusion of male gamete and female egg a diploid zygote (2x) is formed.

The embryo (The young sporophyte)

18. The zygote undergoes repeated divisions to form a new sporophyte. The young sporophyte remains attached to the gametophyte by means of a foot and draws nourishment from the prothallus until it develops its own stem, roots and leaves. The sporophyte is dependent on the gemetophyte only during its early stages.

EUSPORANGIATE AND LEPTOSPORANGIATE SPORANGIA

The sporangium. The sporangium, in all pteridophytes is initiated by the laying down of a cross-wall in a superficial cell or group of cells. Since this wall is periclinal (*i.e.*, parallel to surface) each initial cell is divided into an outer and inner daughter cell. If the sporogenous tissue is derived from the inner daughter cell the sporangium is described as **'eusporangiate'** and if from the outer, as **'leptosporangiate'**. In leptosporangiate forms, the sporangium wall, stalk and the spores, are derived from the outer daughter cell, but in eusporangiate forms, the adjacent cells take part in the formation of part of sporangium wall and the stalk. In eusporangiate form the sporangium is large and massive, the wall is several cells in thickness and the spore content is high. In leptosporangiate forms, the sporangium is small, the wall is one cell in thickness and the spore content is low. Of these two types, the cusporangiate is primitive and the leptosporangiate advanced.

Eusporangiate sporangium	Leptosporangiate sporangium
A superficial cell divides by a periclinal wall forming an outer and inner daughter cell. The sporogenous tissue is derived from the inner daughter cell.	A superficial cell divides by a periclinal wall forming an outer and inner daughter cell. The sporogenous tissue is derived from the outer daughter cell.
The adjacent cells are involved in the formation of part of the sporangium wall and the stalk.	The sporangium wall and the stalk, as well as th spores are derived from the ouuter cell.
The sporangium is large and massive; the wall is several cells thick and the spore content is high.	The sporangium is small, the wall is only one cell thick and the spore content is low.
Primitive.	Advanced.

WORK ON PTERIDOPHYTES IN INDIA

Lycopodiaceae

Chaudhary (1937) reported 33 species of *Lycopodium* from, India, he also produced a key for identification of these species. Ghosh (1942) reported twin sporangiate sporophylls in *L. phlegmaria*. Mahabale (1948) did the exhaustive study on the prothallus of *L. hamiltonii*, an epiphytic species. Mehra and Verma (1957) and Ninan (1958) did some cytological work and reported the haploid chromosome number in *Lycopodium*. Bhambhie and Puri (1963) and

Bhambhie (1965) studied the shoot apex organization, initiation and arrangement of leaves in a number of species of *Lycopodium*. Bhambhie (1965) reported that a group of cells is responsible for the formation of a leaf primordium instead of a single cell.

Selaginellaceae

Majumdar (1942) studied the origin of medullation or siphonostele in three species of *Selaginella* and thought it to be the polystelic condition. Chaudhary (1943) studied the anatomy of a xerophytic species, *S. bryopteris*, possessing the horse shoe-shaped stele or dorsiventrally flattened, ribbon-shaped protostele with two protoxylem groups. Alston (1945) reported 46 species of *Selaginella* in India. He studied their distribution and nomenclature. Panigrahi and Chaudhary (1962) reported eight additional specis of *Selaginella* from eastern regions of our country. S.C. Sharma (1966) made investigations on *Selaginella* with special reference to the epidermal structures of some Indian species. Abraham and others (1962) did much work on the cytology of *Selaginella*. The haploid chromosome number of *Selaginella* is quite low, *i.e.*, 9, 10 or 11. It is thought that probably *Lycopodium*, (n = 165 — 170) and *Ophioglossum* (n = 630) have been evolved from *Selaginella* (n = 9, 10 or 11). It also shows that polyploidy has been a very important factor in fern evolution as pteridophytes are quite rich in polyploids (Abraham *et al.* 1962). The work on trabeculae in *Selaginella* proved that if trabeculae are made up of more than one cell only the cell adjacent to the stele represents the real endodermal cell other cells are derived from cortical region (Bhambhie, 1966).

Equisetaceae

Pant and Mehra (1964) studied the development of stomata in *Equisetum* and some allied genera. The haploid chromosome number of *Equisetum* is 108 (Ninan, 1955). Y. Mohan Ram and Chatterji (1969) collected prothalli of *E. ramasissimum* from the banks of Yamuna river and reported that the spores are potentially bisexual and the environment determines the sex.

Polypodiaceae

Nayer and Kachroo (1952) studied the development of antheridium in advanced leptosporangiate ferns. Wardlaw (1958) and Wardlaw and Sharma (1961) have studied the sorus placenta and thought it to be a special kind of meristem which orignates as a growth centre in the leaf margin. Nayer (1964) has emphasized on the types of spores found in pteridophytes. Nayer and Mrs. Kaur (1972) reported various types of development in the gametophytes of different ferns. Systematics of ferns and allied species have been worked out by many workers from time to time. The names of Chowdhary (1937), Alston (1945), Gupta and Bhardwaj (1957), Panigrahi and Chowdhary (1961) and Abraham *et al*, (1962) are specially mentioned. Morphology and Anatomy of ferns is worked out by Puri and Garg (1953), Bhambhie and Puri (1963), Bhambhie (1965, 66); Palynology by Nayer (1964); development of gametophyte and their use in systematics by Nayer and Mrs Kaur (1972), morphogenesis by Sharma (1960 - 72), and Ecology by B. Gopal (1968-72). Moreover, the contribution of Mahabale is of varsatile nature and covers the various aspects of pteridophytes.

Marsileaceae

Kolhatkhar (1937, 40) worked out the anatomy and life history of *Marsilea poonensis*. The haploid chromosome number is 16. Majumdar (1944) reported upto five adventitious roots borne at regular intervals. Mahabale and Gorji (1948) recorded different types of juvenile leaves in *Marsilea*. Mahabale also reported latex containing cells in the phloem and inner cortex of stem of *Regnellidium*. Gupta (1953, 62) established one new species of *Marsilea* from Rajasthan, *M. rajasthanensis*, Gupta (1962), published his famous monograph dealing with systematics and life-history of *Marsilea*. Puri and Garg (1953) investigated much about the morphological nature of the sporocarp of *Marsilea* and concluded that originally the leaf of *Marsilea* had several pinnae all arising from one side. In the case of fertile leaf all the pinnae disappeared with the exception of basal one which persisted. It has 15 - 20 lateral pinnules forked at the tips and having sori on one margin only. Lateral fusion of the pinnules of one side and thereafter fusion with that of the other side resulted in the formation of a closed structure known as sporocarp. Foster and Gifford (1969), and Sporne (1963) have accepted the interpretation propounded by Puri and Garg.

B. Gopal (1968) recorded a new species of *Marsilea, i.e., M. maheswarii* from Pondicherry and described the formation of propagules is unfavourable conditions. B. Gopal (1968) also worked out the ecology of *Marsilea* and has shown that the spores germinate in dark at 25°C to 30°C.

Comparative Morphogenesis of Pteridophytes

A peridophyte normally exists in two diverse morphological forms during its life history. The small, simple haploid gametophyte and the large morphologically complex, diploid sporophyte. In the typical life-cycle the two generations are isolated from each by meiosis and syngamy. Each has its origin in a single cell, the spore and zygote respectively. Two of the important related morphogenetic problems that have baffled botanists since long are (i) how can the same genetic constitution produce plants of such morphological dissimilarity by very different course of development, and (ii) what are the factors responsible for initiating these two different courses of develelopment ?

The first possibility is that the germ cells, spores and zygote are intrinsically different and they necessarily give rise to unlike individuals. The second possibility is that the germ cells are equivalent and the different environments in which the spores and zygote develop determine these courses.

Bell (1959) has suggested that the initiation and maintenance of the two different forms of growth in the heteromorphic life cycle in pteridophytes may be due to the different states and potentialities of the zygote and the spores. He has further (1963) shown that the egg in ferns is cytochemically and structurally quite distinct from the spore. The experimental work with gametophytes and inventions with electron microscope led Bell (1970) to suggest that the ability of the egg to develop as a sporophyte is a consequence of intracellular changes during oogenesis.

Ward and Wetmore (1954), Jayasekera and Bell (1959), De Maggio and Wetmore (1961), have attempted to test experimentally the hypothesis that a proper physio chemical environment is necessary for the development of the sporophyte organization. It has been indicated that formative effects are normally induced in the developing embryo by the prothallus and the archegonium.

On the basis of results of other experiments conducted on *Thelypteris palustris* and *Pteris longifolia* by Jayasekera and Bell (1959) and Reviews (1959) respectively, it has been concluded that auxins diffusing from the apex of the gametophyte promote the normal differentiation of the embryo in general and the formation of the first root in particular.

De Maggio and Wetmore (1961), after conducting surgical experiments on the embryos of *Todea barbata* have concluded that the zygote normally develops into a plant with sporophytic organization mainly because of the factors, physical pressure, gradient of growth hormones, etc. According to Wittiers (1971) the manifestation of sporophytic organization is attributable to morphogenetic processes and environmental factors interacting from the time of initiation of oogenesis to the emergence of the sporophyte from the archegonium.

In the present stage of our knowlege, we are justifed in considering alternation of generations as a morphogenetic problem which still requires more investigations and information with a newer techniques for its complete elucidation.

Rhynia is a fossil member with simple plant body having prostrate rootless rhizome and erect dichotomous branches that were leafless *R. major* is larger than *R. gwynne-vaughani*. Reproductive structures were sporangia produced at the tips of aerial branches. Sporangia of *R. major* are larger.

Selaginella is herbaceous with plant body erect, creeping or climbing. Leaves may be isophyllous or heterophyllous. They are microphyllous and ligulate and arranged spirally or in four rows. Rhizophore may or may not be present. Roots are produced at the terminal ends of the rhizophores. Sometimes they arise from knot-like perennating structures. Stems are mostly protostelic and exarch. One species shows secondary growth and sometimes xylem may have vessels.

Reproductive structures are sporangia produced on sporophylls which are aggregated to form strobili. It is heterosporous. Usually the megasporophylls are at the bases and microsporophylls above but sometimes they may be mixed up. Microsporangia contain many spores whereas megasporangia have 1-4 megaspores that enlarge enormously.

Equisetum is also herbaceous commonly known as "horse tail". Plant body is provided with prostrate rhizome and erect branches that are branched at three levels. Different species, can be arranged in a reduction series. Stem that is hollow is distinguishable into nodes and internodes. Leaves are microphyllous, simple, isophyllous, extremely reduced and form sheath around the stem. Tubers may also be formed on the nodes of stem.

Sporophylls (sporangiophores) are arranged in compact strobili or cones at the ends of erect branches. Four to six sporangia are borne transversely to the cone axis on peltate sporangiophores.

Adiantum (maiden hair fern) is also herbaceous having upright or creeping scaly rhizome. Leaves are arranged spirally or alternately on the rhizome. Leaves are circinately coiled in young condition. Petioles are shining black and long blade of leaflets may be entire or repeatedly branched, delicate in texture. The forking veins traverse the blades. Usually veins are free but in some species they are anastamosing forming reticulum, *e.g.*, maiden hair fern. Sori appear to be marginal but actually the sporangia are borne on the lower (abaxial) surface of leaves near the margin and are covered by sharphly reflexed leaf lobes.

Marsilea possesses a rhizome which creeps just beneath the surface of the soil. The rhizome is slender, dichotomously branched and possesses nodes nd internodes. It is capable of indefinite growth in all directions and may cover an area more than twenty five metres in rhizome usually at the nodes, and have long petioles. One or more adventitious roots come out from each node on the underside of the rhizome. The reproductive structures are mega-and microsporangia produced in a complex structure called sporocarp and borne on short peducles near the base of the petiole. Peduncle may be branched rarely. The number of sporocarps varies considerably. In *M. polycarpa* there may be from 6 - 26 sporocarps.

The mechanism of dispersal of spores is of special interest. In *Rhynia* these were perhaps dispersed after decay of sporangial wall. In *Selaginella* and *Equisetum* they are liberated by splitting of sporangia-vertical apical splitting (*Selaginella*) or along a longitudinal line in the sporangia. In *Marsilea* the sporocarp liberates spores and sporangia by dissolution of the mucilage wall of parietal cells. In *Adiantum* the sporangia open at the stomium by hygroscopic nature of the annulus.

The mature spores are small and tetrahahedral in *Rhynia*. The triradiate spores (arising from tetrahederal tetrads) are characteristic of the group. The microspores are smaller (ranging from .005 - .006 mm in *Selaginella*, .075 mm and pyramidate or globose with pyramidate apex in *Marsilea* .03 mm in *Adiantum*). The megaspores are large (usually ranging between .5 - 5 mm in different members). The spores of *Equisetum* are somewhat spherical globular or sub-spherical with no trace of tri-radiate ridge. These spores have around them a compound spiral of 4 elaters that have a hygroscopic nature and help in dispersal of spores.

COMPARATIVE STUDY -I

	Rhynia	Psilotum	Lycopodium	Selaginella
Systematic position	Division-Psilophyta; Class-Psilophytopsida; Order-Psilophytates ; Family-Rhyniaceae; *Genus-Rhynia.*	Div.-Psilophyta; Cl.-Eligulopsida; Or.-Psilotales; Fa.-Psilootaceae; Gen.-*Psilotum.*	Div.-Lycophyta; Cl.-Eligulopsida; Or.-Lycopodiales; Fa.-Lycopodiceae; Gen.-*Lycopodium.*	Div.-Lycophyta; Cl.-Ligulopsida; Or.-Selaginellales; Far.-Selaginellaceae; Gen.-*Selaginella.*
No. of species	2	2	180	600
Distribution and Habitat	Old red Sandstone beds of middle Devonian in Rhynie, Scotland; peaty habitat near volcanoes.	Tropical and subtropical regions in eastern and western hemispheres; epiphytic or terrestrial in humus rich places.	Mainly distributed in tropical and subtropical forests; usually moist loving.	Throughout world, mainly in rain forests of tropical regions; humid, shady habitats.
SPOROPHYTE External Features Habit & Stem	Herbaceous, rootless, leafless plants; axis divided into rhizome and on erect aerial part; tufted rhizoids on rhizome.	Herbaceous, rootless, prennials with subterranean rhizome and on erect aerial part.	Herbaceous, usually perennial trailing, erect or semi-erect; epiphytes; stems branched.	Herbaceous, perennial or annual, prostrate, sub-erect, scandent, erect, climbing; stem branched.
Leaves and their arrangement	leafless	scale-like leaves without veins; arranged irregularly on the aerial stem.	simple small, lanceolate; sessile, spiral, whorled or opposite (microphyllous).	small simple ligulate, lanceolate to ovate; spiral or in four rows, all alike in size or of two size or of two sizes (microphyllous).
Root and its arrangement	Rootless	Rootless	adventitious cortical roots in some sp; no endogenous lateral roots.	adventitious; borne on the stem or at the tips of rhizophores.
Stem anatomy	Protostele (haplostele); sometimes the tracheids in the centre of xylem strand were smaller in diameter than the peripheral tracheids; no leaf trace; no leaf gaps.	aerial stem -siphonostele; rhizome protostele; xylem exarch; polyarch; leaf traces absent; leaf gaps absent.	protostele-actino stele, plectostele mixed haplostele; xylem exarch; polyarch; leaf-traces small and simple; leaf gaps absent.	protostele; siphonostele to polycyclic siphonostele; polystelic ; xylem exarch; mono or diarch; vessels sometimes present; leaf traces small and simple; leafgaps absent.

	Rhynia	Psilotum	Lycopodium	Selaginella
Root anatomy			exarch, hexarch to decarch, or diarch.	monarch and exarch.
Leaf Anatomy		Vein lacking; mesophyll undifferentiated.	Simple, concentric vascular bundle, mesophyll undifferentiated.	Single, concentric vascular bundle, mesophyll usually not differentiated.
Strobilus			Present in some species.	Usually present
Sporophylls			Resemble leaves in some species, or differ from them in size, form and colour.	Resemble leaves or differ from them slightly.
Sorus				
Sporangium	Cauline, terminal; homosporous; oval or slightly cylindrical; wall is massive; three layered; annulus absent; tapetum is probably onelayered; formed from the innermost layer of the wall; no special device for dehiscence.	Eusporangiate; cauline, terminal; homosporous; trilobed and trilocular; wall is massive; four to five cells in thickness; annulus absent; tapetum absent dehiscence along longitudinal dehiscence lines.	Eusporangiate; foliar or cauline; homosporous; large, kidney shaped with a short massive stalk; wall is three or more layered; annulus absent; tapetum onelayered; formed from the innermost layer of the wall; dehiscence by a transverse slit.	Eusporangiate foliar, axillary, or cauline; heterosporous; microsporangium-stalked, small with a smooth outline; megasporangium-stalked, large, four-lobed, wall is two layered in both type of sporangia; annulus absent; tapetum is onelayered, formed from the outermost layer of sporogenous tissue, persists until the spores are nearly mature and then degenerates; dehiscence by vertical apical splitting of the upper part of sporagial wall.
Average spore output from each sporangium.	Many	Many	Many	Megasporangium-4 megaspores; microsporangium many microspores.
GAMETOPHYTE	One kind spores; numerous; small tetrahedral in shape.	One kind spores, numerous, usually bilateral, occasion-	One kind spores, numerous, small in size, tetrahedral in	Two kinds-microspores and megaspores, microspores

	Rhynia	Psilotum	Lycopodium	Selaginella
		ally tetrahedral also 0.065 x 0.32 mm.	shape, (0.03 to 0.05 mm).	numerous, megaspores-four; microspores-small (0.015) to 0.06 mm), tetrahedral, megaspores-large (1.5 to 5 mm), tetrahedral.
Prothallus		Monoecious, subterranean, 0.5 to 2 mm. in diameter and from 1 to 18 mm, in length, more or less cylindrical, elongated, irregularly brached, radially symmetrical, mature portion varies from pale yellow to dark brown, growing apices white. Monoecious; surfae	-living or subterranean, or partially subterranean, short, 2 mm., usually radial or dorsiventral, of various shapes, green, brownish, yellowish or nearly colourless. Dioecious, microg-	ametephyte wholly within micropore wall, megagametophyte also within megaspore wall with only the apical cushion protruding; both much reduced in sizes; non-green.
Sex organs : Antheridium		Irregularly distributed or the cylindrical prothallus, many in number, project conspicuously from the surface of the prothallus, antherozoids spirally coiled, multiciliate.	On crown or on flattened or hollowed upper surface, many in number, completely embedded or slightly projecting, antherozoids usually fusiform, biciliate.	Inside the microspore wall, single antheridium, spirally coiled, biciliate antherozoids.
Archegonium		Irregularly distributed on the prothallus, many in number, sunken with the neck slightly projecting, neck 4 to 7 cells in height and composed of 4 vertical rows of cells, neck deciduous, 2 neck canal cells or merely two neck canal nuclei which disintegrate	On crown or on flattened or hollowed upper surface, many in number, sunken with neck projecting, neck is short 3 to 4 cells in height, neck canal cells 1, 2, 3, 4, 7, 14, to 16.	On the apical cushion, many in number, sunken with outer tier of neck slightly projecting, neck is short, 2 cells in height, neck canal cell -1.
Embryo		First division of zygote is transverse, suspensor present.	First division of zygote is transverse, suspensor present, weakly developed.	First division of zygote is transverse, suspensor usually well-developed.

COMPARATIVE STUDY—II

	Equisetum	Ophioglossum	Marsilea	Fern
Systematic Position	Div-Arthrophyta; Cl-Calamopsida; Or.-Equisetales; Fa.-Equisetace; Gen. *Equisetum.*	Div.-Filicophyta; Cl-Eusporangiopsida; Or.-Ophioglossales; Fa.-Ophioglossaceae; Gen.-*Ophioglossum.*	Div.-Filicophyta; Cl.-Leptosporangiopsida; Or.-Marsileales; Fa.-Marsileaceae; Gen-*Marsilea*	Div.-Filicophyta; Cl.-Leptosporangiopsida; Or. Filicales; Fa.-Polypodiaceae. Gen-*Dryopteris; Adiantum etc.*
No of Species	25	28	65	
Distribution and Habitat	Throughout world except Australia and NewZealand; grow in exposed places, marshes meadows and along river and canal banks.	Throughout world chiefly in temperate and tropical regions; terrestrial or epiphytic, prefers humus rich soil.	Cosmopolitan, mainly in warmer parts ; aquatic or amphibious.	Cosmopolitan; mainly in humus rich places.
SPOROPHYTE				
External Features Habit & Stem	Herbaceous, perennial with creeping, horizontal, jointed, ridged rhizomes.	Short erect subterranean, branched rhizome.	Herbaceous, rhizomatous, creeping.	Underground creeping rhizome.
Leaves and their arrangement	Simple, small, scalelike uninerved fused laterally to form a sheath (microphyllous).	Simple, petiolate, entire, narrowly linear to broadly oval; venation reticulate; vernation erect.	Compound, lamina divided into four leaflets, arranged alternately in two rows, dichotomous venation, circinate vernation (megaphyllous).	Pinnately compound; alternate, circinate vernation (megaphyllous).
Root and its arrangement	Adventitious, developing from the base of lateral branch primordia.	Adventitious, numerous, thick, smooth, devoid of root hairs, mycorrhizic, branched or unbranched.	Adventitious, borne at the nodes.	Adventitious, slender, little branched.
Stem anatomy	Perforated ectophloic siphonostele with nodal rings. In the internode separate endarch collateral bundles. Xylem endarch; leaftraces small and simple; leaf gaps absent	Siphonostele, or solenostele or dictyostele. Xylem endarch, leaf-traces large, one or two leaf-traces given off from the rhizome at each leaf-gap, leaf-gaps present.	Amphiphloic solenostele. Xylem exarch, leaf-traces large, leaf-gaps present, scattered.	Polycyclic dictyostele. Xylem mesarch, vessels present, leaf-traces large; inner and other series of vascular strands contribute to leaf traces; leaf-gaps and perforations present.

	Equisetum	Ophioglossum	Marsilea	Fern
Root anatomy	Tri or tetrarch, exarch; two layers of endodermis.	Monarch, di-, tri-, tetra-or pentarch, exarch.	Diarch, exarch.	Diarch, exarch.
Leaf anatomy	Single vascular bundle; collateral.	Many vascular bundles, vascular bundles endarch and collateral, mesophyll not differentiated.	Many vascular bundles, concentric mesophyll differentiated.	Many vascular bundles, concentric or collateral, mesophyll differentiaied.
Strobilus	Present.			
Sporophylls	Sporangiophores.	"Fertile spike".	Sporocaarp contains a large number of sori.	Resemble foliage leaves.
Sorus			Many marginal, number of indusia equal soral sacs.	Superficial, coenosours, surrounded by false and true in dusium.
Sporangium	Eusporangiate; borne on peltate sporangiophores; homosporous; elongated sac like with arounded apex; wall is singlelayerd or sometimes two - layered in mature condition, annulus absent, tapetum one-layered, irregular, ill defined, formed from the innermost layer of jacket, disorganises, dehisce along a longitudinal line facing inwards to wards the stalk.	Eusporangiate, terminal and marginal on the leafdivisions, homosporous, subspherical, sporangia fused not only with the leaf segment but at least in their outer part with one another along the margin, wall is massive, several layered thick, annulus absent, tapetum consists of several ill-defined layers of cells, origin controversial, either from the outer cells of sporogenous tissue or the wall, dehiscence by transvere clefts perpendicular to the wide surface of fertile spike.	Leptosporangiate, borne in sporocarps, heterosporous, both micro and megasporangia stalked, wall is singlelayered, annulus absent, tapetum two to three-layered, formed from archesporial cell, disorganises, dehiscence by dissolution of the mucilaginous wall of the parietal cells of wall on coming in contact with water.	Leptosporangiate. on abaxial surface of sporophylls, homosporous, long-stalked, capsule biconvex lens-shaped, slightly flattened, wall is single layered, annulus vertical, incomplete, tapetum two layered, formed from the archesporial cell, disorganises, transverse dehiscence by means of annulus and stomium which form the opening mechanism.
Average spore output from sporangium	Many	Many	Megasporangium one megaspore, microsporangium many microspores.	32 to 64

	Equisetum	Ophioglossum	Marsilea	Fern
GAMETOPHYTE	One kind spores; numerous; small in size; globular or subspherical with no trace of triradiate ridge.	One kind spores; numerous; small in size, tetrahedral in shape.	Two kinds-megaspores and microspores 64, megaspore-1, microspores-small 0.075 mm., globose with a pyramidate apex, megaspore ellipsoidal to oval, tetrahedral.	One kind spores, usually 64 in number, small in size, tetrahedral in shape, in shape 0.03 mm. in diameter.
Prothallus	Usually monoecious or partially dioecious; surface living, usually less than 1 cm in diameter; usually dorsiventral consisting of massive cushion like base and upright green branched; green (autotrophic)	Monoecious, subterranean, 0.5 to 10 cm. in length, linear to stellate but usually irregularly cylindrical to conical, non-green.	Dioecious, both megagametophyte and microgametophyte within spore wall, both much reduced, non-green.	Usually monoecious, surface living, about 6 mm. in diameter, dorsiventral, cordate, green (autotrophic).
Sex organs : antheridium	On upturned extensions in the margin of apex of delicate lobes; many in number; projecting, spirally coiled and flattened, multiciliate antherozoids.	Scattered and intermingled over the surface or gametophyte, many in number, mainly sunken or slightly projecting, antherozoids spirally coiled, multiciliate.	Inside the microspore wall, two in number, cork screw like, spirally coiled, multiciliate, antherozoids.	On the basal or lateral regions of the ventral surface of the prothallus, many in number, slightly projecting, spirally coiled, multiciliate antherozoids.
Archegonium	In the meristematic rim where upright green lobes are being formed; many in number; base sunken in the prothallus with neck projecting, neck is short 3-4 cells in height, neck canal cells 1 or 2, in many species boot shaped.	Scattered and intermingled over the surface of prothallus, many in number, usually sink in the prothallus tissue, but neck may project slightly, neck is 2 to 4 cells in height, single binucleate neck canal cell or 2 neck canal cells.	Embedded in a cap or tissue fitted over the apex of the spore cavity, one in number, neck is short 2 cells in height, neck canal cell-I	On the massive cushion just posterior to the apical notch on the ventral surface of the prothallus, many in number, sunken with neck projecting, neck -5 to 7 cells in height, neck recurved, neck canal cell 1, usually binucleate.
Embryo	First division of zygote is transverse, suspensor absent.	First division of zygote is usually transverse, *i.e.*, perpendicular to the long axis of archegonium, suspensor absent.	First division of zygote is vertical, merely coincident with the long axis of archegonium, suspensor absent.	First division of zygote is vertical, nearly coincident with the long axis of archegonium, suspensor absent.

Selected Readings

Arnold. C.A. 1947. *An introduction to Palaeobotany.* New York.

Bower, F.O. 1908. *The origin of a Land Flora.* London.

Bower, F.O. 1930. *Size and Form in Plants.* London.

Bower, F.O. 1935. *Primitive Land Plants.* London.

Bower, F.O. 1947. *Botany of the living plant 4th ed.* London.

Campbell, D.H. 1918. *Mosses and Ferns 3rd ed.* New York.

Campbell, D.H. 1940. *The Evolution of the Land Plants (Embryophyta).* California.

Eames, A.J. 1936. *Morphology of Vascular Plants.* New York.

Eames, A.J. and L.H. Mac Daniels 1947. *An introduction to Plant Anatomy.* New York.

Esau, K. 1953. Plant Anatomy. New York.

Foster, A.S. 1949. *Practical Plant Anatomy.* 2nd ed. New York.

Gupta, K.M. 1960. *Botanical Monograph No. 2. Marsilea.* New Delhi.

Haupt, A.W. 1953. Plant Morphology. New York.

Jeffery, E.C. 1917. *The anatomy of woody plants.* Chicago.

Manton, I. 1950. *Problems of Cytology and Evolution in the Pteridophytes.* London.

Parihar. N.S. 1955. *An introduction to Embryophyta, vol. II- Pteridophytes.* Allahabad.

Scott, D.H. 1923. *Studies in Fossil Botany,* 3rd ed. London.

Seward, A.C. 1910. *Fossil Plants.* II - Cambridge.

Seward, A.C. 1933. *Plant life through the ages.* Cambridge.

Sinnott, E.W. 1946. *Botany, Principles and Problems* 4th ed. New York.

Smith, G.M. 1955. *Cryptoganic Botany,* vol II. 2nd ed. New York.

Sporne, K.P. 1963. *The morphology of Pteridophytes.* London. Verdoorn, F. 1938. *Manual of Pteridology.* The Hague.

Wardlaw, C.W. 1955. *Embryogenesis in Plants.* London.

Wardlow, C.W. 1952. *Phylogeny and Morphogenesis.* London.

Wardlow, C.W. 1965. *Organization and Evolution in plants.* Manchester.

Willis, J.C. 1931. A dictionary of the flowering plants and ferns. Cambridge.

Abaxial—The surface of any structure which is turned away from the axis.

Acicular—Needle like in shape, sharp pointed.

Acotyledon—A plant without a cotyledon.

Acrogenous—Increasing in growth at apex.

Acropetal—Ascending, *e.g.*, leaves, flowers or roots, developing successively from an axis so that youngest arise at apex.

Actinostele—Stele with xylem radiating outwards and forming ridges, *e.g.*, ferns.

Adaxial—Turned towards the axis.

Adherent—Attached to some substratum.

Adventitious—Found in an unusual place, *e.g.*, Adventitious roots.

Aerenchyma—Aerating cortical tissue in floating plants.

Alternation of generations—The occurrence in one life history of two or more different forms differently produced, usually as alternation of a sexual with an asexual form.

Alveolus—A small pit or depression.

Amphicribral—Amphiphloic.

Amphiphloic—With phloem both peripheral and central to xylem.

Anacrogynous—Certain liverworts in which female reproductive bodies do not arise at or near apex of shoot.

Anatomy—The science which treats of the internal structure of plants.

Androcyte—A cell arising by growth from an antheridium and giving rise to antherozoid.

Annulus—Any ring like structure; special ring in fern sporangium, by action of which sporangium bursts.

Anterior—Nearer head end.

Antheridia—Plural of antheridium.

Antheridial cell—The cell representing an antheridium, as in Gymnospermae.

Antheridiophore—A gametophore bearing antheridia

Antheridium—Male gametangium in which male sexual cells are produced in many cryptogams.

Antherozoids—Male sexual cells in antheridia.

Anticlinal—Line of division of cells at right angles to surface of apex of a growing point.

Apical—Cell at tip of growth point.

Apogamy—Reproduction without intervention of sexual organs.

Apophysis—Small protuberance at apex or ovuliferous scale in pine.

Apospory—Production of a diploid gametophyte from a sporophyte without intervention of spore formation.

Aquatic—Living in water.

Archegoniophore—Gametophore bearing archegonia, *e.g., Marchantia.*

Archegonium—A female gametangium in which oospheres are formed, and in which the young plant begins development.

Archesporium—A cell or mass of cells dividing to form spore mother cells and elater forming cells.

Asexual—Having no sexual organs.

Asymmetrical—Having two sides disproportionate.

Autoecous - Bearing antheridia and archegonia on the same plant but different branches, *e.g., Funaria.*

Axial—Pertaning axis or stem.

Axillary—Growing in axil, as buds.

Basal—Near the base.

Basipetal—An order of development of organs in which the youngest structures are at the base and the oldest at the apex.

Biennial—Lasting for two years.

Biflagellate—Having two flagella.

Bifoliar—Having two leaves.

Blepharoplast—A basal granule which gives rise to the cilia of antherozoid.

Body cell—An antheridial cell.

Bordered pit—A form of pit developed on wall of tracheids and wood vessels with overarching border of secondary cell wall.

Botany—The branch of biology dealing with plants.

Bract scales - Small scales developed directly on axis of cones, *e.g.,* ovuliferous scales.

Bulbous—Like a bulb.

Calyptra—Tissue enclosing developing sporogonium in liverworts.

Cambium—The tissue from which secondary growth arises in stem.

Capsule—A sac containing spores; sporogonium in Bryophyta.

Carinal—Canals in protoxylem beneath ridges of stem in Equisetales.

Casparian band—Casparian strip; a strip encircling radial walls of endodermis cells.

Cauline—Belonging to a stem.

Centrifugal—Turning away from axis.

Chloroplast—A plastid containing chlorophylls a and b.

Columella—A prolongation of stalk into sporogonium.

Cone—The female flower of Coniferae, with woods axis and spirally arranged carpels.

Coralloid—Branching like a coral, *e.g., Cycas* roots.

Cortex—The extrastelar fundamental tissue of the sporophyte.

Cosmopolitan—World wide in distribution.

Cotyledon—The seed-leaf, primary or first leaf of an embryonic sporophyte.

Cryptogam—A plant without apparent reproductive organs.

Cuneate—Wedge shaped.

Cupule—The gemmae bearing cup of *Marchantia.*

Cushion—The central thick region in prothallus of fern.

Decumbent—Lying on the ground but rising at apex.

Dehiscence—The spontaneous opening of an organ releasing spores etc.

Development—The changes undergone by an organism from its beginning to maturity.

Diarch—With two xylem and two phloem bundles.

Dichotomy—Repeated forking.

Dictyostele—A net work formed by meristeles.

Dioecious—Having male and female sexes on different individuals.

Diploid—Having a double set of chromosomes (2x).

Dorsal—Upper surface of thallus or prothallus.

Dorsiventral—With upper and lower surfaces distinct.

Ectophloic—With phloem surrounding xylem.

Egg—Oosphere, mature female germ cell.

Elater—One of the cells with a spiral thickening which assists in dispersing spores from capsule in liverworts, *e.g., Marchantia.*

Elaterophore—Tissue bearing the elaters, in some liverworts, *e.g., Pellia.*

Embryo—A young organism in early stages of development.

Endodermis—Innermost layer of cortex in plants.

Endosperm—The nutritive tissue of certain seeds.

Endospore—Inner coat of spore wall.

Endothecium—The central region of an epibasal octant of oospore of liverworts and mosses.

Entire—With continuous margin.

Epidermis—The outermost protective layer of stems.

Epithelium—Any cellular tissue lining a cavity, *e.g.*, resin canal.

Erect—Directing upwards.

Evolution—The gradual development of organisms from pre-existing organisms since the dawn of life.

Exarch—With protoxylem strands outside metaxylem towards periphery.

Exine—Outer layer of spore wall ; exospore, exosporium.

Fertilization—The union of male and female nuclei.

Foot—An embryonic structure in vascular cryptogams through which nourishment is obtained from prothallus or thallus ; basal portion of sporophyte in mosses.

Fossil—Petrified plant portion found in rocks.

Frond—A leaf especially of fern.

Gametangium—A structure producing sexual cell.

Gametes—Sexual cells.

Gametogenesis—Gamete formation.

Gametophyte—The gamete forming phase in alternation of plant generations : sexual generation of plants.

Gemma—A bud or outgrowth of a plant which develops into a new organism, *e.g., Marchantia, Funaria* etc.

Gemma-cup—Cupule containing gemmae, *e.g., Marchantia.*

Genera—Plural of genus.

Genus—A group of closely related species, in classification of plants.

Germ tube—Short filamentous tube put forth by a germination spore.

Germination—Process of initial development.

Glossopodium—The sheathing leaf base of the ligule of *Selaginella.*

Groove—Furrow or depression.

Ground tissue—Conjunctive parenchyma.

Growing point—A part of plant body at which cell division is localized, generally terminal and composed of meristematic cells.

Guard cells—Two cells surrounding stomata of aerial epidermis of plant tissue.

Gymnosperms—The plants having seeds not enclosed in true ovary, *e.g., Pinus.*

Haploid—Having *x* number of chromosomes.

Haplont—An organism having haploid somatic nuclei.

Hepatic—Liverworts.

Heterosporous—Producing two kinds of spores, *e.g.*, *Selaginella*, *Marsilea* etc.

Homosporous—Bearing spores of one kind, *e.g.*, Fern.

Hypobasal—The lower segment of a developing zygote.

Indusium—An outgrowth of plant epidermis covering and protecting a sorus, *e.g.*, Ferns.

Integument—Coat of ovule.

Internode—The part between two successive nodes.

Intine—The inner covering membrane of a pollen grain or of a spore.

Karyogamy—Union of nuclei after cytoplasmic fusion (plasmogamy).

Lacuna—A space between cells.

Lanceolate—Lance-shaped.

Leaf gap—Gap in vascular cylinder of stem, a parenchymatous region associated with leaf traces.

Leaflet—Individual unit of a compound leaf ; pinna.

Leptosporangiate—With sporogenous tissue developing from outer cell of periclinal division.

Life-cycle—The various phases through which an individual species passes to maturity.

Ligule—Small scale on upper surface of leaf-base in Selaginellaceae and *Isoetes*.

Macrospore—Macrospore ; a larger spore of heterosporous plants.

Medullary rays—A number of strands of connective tissue extending between pith and pericycle.

Megagametophyte—The female gametophyte developed from a megaspore, *e.g.*, *Selaginella, Marsilea* etc.

Megaspore—Larger spore of heterosporous plants, regarded as female, *e.g.*, *Selaginella*, *Marsilea*.

Megasporophyll—A spore bearing leaf developing megasporangia, *e.g.*, *Selaginella*.

Mesarch—Xylem having metaxylem developing in all directions from the protoxylem, characteristic of ferns.

Mesophyll—The internal parenchyma of a leaf; chlorophyllous cells.

Metamorphosis—Change of form and structure, *e.g.*, the formation of antherozoid from an androcyte.

Microgametophyte—The male gametophyte developed from a microspore, *e.g.*, *Selaginella, Marsilea*.

Micropyle—Aperture for admission of pollen tube at ovule apex.

Microsporangium—A sporangium bearing a number of microspores, *e.g.*, *Selaginella*, *Marsilea*.

Microspore—The spore developed in a microsporangium of heterosporous plants.

Mid-rib—The large central vein of a leaf.

Monoecious—With sex organs on one gametophyte; hermaphrodite.

Morphology—The science of form and structure of plants.

Mycorrhiza—Association of fungal mycelium with roots of a higher plant

Neck canal—The hollow within the elongated part of an archegonium.

Node—The joint of a stem at which leaves arise.

Nucellus—Parenchymatous tissue between ovule or megaspore and its inner integument, and extending from chalaza to micropyle.

Nurse cells—Cells providing nourishment to the developing spores in the sporogonium.

Octant—One of the units in eight-celled stage in segmentation of ovum.

Oosphere—An egg before fertilization.

Oospore—The zygote or fertilized egg cell.

Operculum—A lid or covering flap at apex of capsules of mosses.

Order—In classification, group of organisms closely allied, ranking between family and class.

Ovule—The megasporangium of seed plant.

Ovuliferous—Containing ovule; scales, each bearing one or more ovules, developed on bract scales, as in Coniferae.

Palaeobotany—Botany of fossil plants and plant impressions.

Palisade—Arrangement of opposed elongated cellular structures.

Paraphysis—A slender filamentous epidermal outgrowth occurring among sex organs, *e.g.*, *Funaria*

Parthenogenesis—Reproduction without fertilization by a male element.

Peduncle—Stalk.

Perichaetium—One of membranes enveloping archegonia of bryophytes, *e.g.*, *Marchantia.*

Periclinal—System of cells parallel to surface of apex of a growing point.

Pericycle—The external layer of stele, the layer between endodermis and conducting tissues.

Perigynium—Membranous envelope of archegonium in liverworts, *e.g.*, *Marchantia.*

Peristome—The region surrounding mouth; term used in connection with moss capsule.

Pinna—A leaflet of a pinnate leaf.

Pinnate—A compound leaf, having leaflets on each side of an axis or mid-rib.

Pit—A depression formed in course of cell-wall thickening in plant tissue.

Pollen—The powder produced by anthers, consisting of pollen grains.

Pollen chamber—Pit formed at apex of nucleus below micropyle.

Pollen grain—The haploid microspore of seed plants.

Pollen-tube—A tubular process developed from pollen grains, represents male gametophyte.

Pollination—Transference from anther to ovule.

Pollination drop—Mucilaginous drop exuded from micropyle and which detain pollen grains, as in gymnosperms.

Polyembryony—Formation of several embryos in the ovule, *e.g.*, *Pinus.*

Polystelic—With several steles.

Posterior—Situated behind.

Primitive—Of earliest origin.

Prothallus—A small thin structure, the gametophyte or haploid sexual generation of pteridophytes.

Protonema—The filamentous thallus of mosses from which the moss plant buds.

Pseudo elater—One of the chains of cells of sporogonium of liver worts.

Pteridology—The branch of botany dealing with ferns.

Pteridophyte—A vascular cryptogam, any of the ferns, club mosses or horse tails.

Quadri foliate—Four leaved, compound palmate leaf, with four leaflets arising at a common point, *e.g.*, *Marsilea.*

Rachis—The axis bearing leaflets

Ramenta—Brown scale-like structures found on fern leaves.

Receptacle—A sporophore, terminal disc of mosses.

Resin—An acidic excretion product of certain plants.

Resin canals—Ducts in bark, wood, mesophyll, etc., particularly of conifers, lined with glandular epithelium excreting essential oils.

Rhizoid—A root-like outgrowth of thallus, *e.g.*, liverworts, mosses, ferns, unicellular hairs on lower side of prothallus,

Rhizome—A thick horizontal stem, partly along and partly underground, sending out shoots above and roots below, *e.g.*, Ferns.

Rhizophore—A naked branch which grows down into soil and develops roots from apex, *e.g.*, *Selaginella.*

Root—Descending portion of plant, fixing it in soil, and absorbing moisture and nutrients,

Secondary growth—Development of secondary meristem or cambium producing new tissue on both sides.

Seed—A mature fruit containing an embryo ready for germination under suitable conditions.

Segment—A division formed by cleavage of an ovum.

Septa—Plural of septum.

Septum—A partition separating two cavities of tissue.

Seta—Sporophore of liverworts and mosses.

Sexual cell—Ovum or sperm.

Shoot—Stem of a vascular plant derived from the plumule.

Sieve tubes—Phloem vessels, long slender structure consisting of elongated cells placed end to end.

Siphonostele—The hollow vascular cylinder of a stem, which may contain pith.

Solenostele—A stage after the siphonstele in fern stem development.

Sori—Plural of sorus.

Sorophore—Stalk bearing sori, *e.g.*, *Selaginella.*

Sporangia—Plural of sporangium.

Spore—A highly specialized reproductive cell of plants.

Spore-mother cells—Sixteen cells produced by repeated division of an archesporium, each in turn dividing into four spores.

Sporocarp—Special spore bearing structure of Marsileales.

Sporogenesis— Spore formation.

Sporogenous—Spore producing.

Sporophyll—A sporangium bearing leaf.

Sporophyte—The diploid (2x) spore-producing phase in alternation of plant generations.

Spur—Dwarf shoot, *e.g.*, *Pinus.*

Stalk-cell—The barren cell of two into which the antheridial cell of gymnosperms divides.

Stele—A bulky strand or cylinder of vascular tissue contained in stem and root of plants.

Stomium—Group of thin walled cells in fern sporangium where it ruptures on maturity.

Suspensor—A modified structure developing in the embryo development of *Selaginella* and *Pinus.*

Syngamy—Sexual reproduction; fertilization.

Tapetum—Special nutritive layer investing sporogenous tissue of sporangium.

Terminal—Situated at the end.

Tetrad—A group of four spores formed by meiotic division of spore mother cells.

Thallus—A combination of cells presenting no differentiation of leaf and stem.

Theca—A spore-case.

Trabeculae—Plates of sterile cells extend across cavities of pteridophytes.

Tracheid—One of the cells with spiral thickening or bordered pits, conducting water and solutes, and forming woody tissue.

Tracheidal cells—Pericycle cells resembling tracheids.

Vallecular canal—One of canals in cortical tissue of stem of horse-tails (*Equisetum*).
Vascular bundle—A group of special cells consisting of two parts, xylem and phloem.
Vertical—Standing upright; lengthwise in direction of axis.
Villose—Pubescent.
Wing—Broad membranous expansion of pollen grain of *Pinus*.
Xylem—Lignified portion of vascular bundle.
Zone—An area characterized by similar flora.
Zygote—Cell formation by union of two gametes or reproductive cells.

Question Bank

Long Answer Type

1. Draw the life-cycles of a homosporous and a heterosporous pteridophyte. How do they differ ? Which is more evolved and why ?

2. Define the term 'stele'. By means of labelled diagrammatic sketches only, depict the origin of leaf trace and structure of the various types of stele in pteridophytes. Arrange your illustrations in a evolutionary sequence.

3. Give in a tabular form, a comparative account of the spore producing organs of *Lycopodium, Selaginella* and *Equisetum*. Comment briefly upon features of special interest in them.

4. Give an illustrated account of the development of the male and the female gametophyte of *Selaginella* and state the relation between heterospory and the origin of seed habit in plants.

5. Describe briefly the various types of steles met within pteridophytes studied by you; add a note on their evolution.

6. With the help of a labelled diagram, describe the strobilus of *Equisetum* and how it differs from the cone of *Lycopodium*.

7. Draw a vertical section of the strobilus of *Selaginella* and name the parts. What are the main differences between the strobili of *Lycopodium* and *Selaginella*.

8. What is heterospory ? Illustrate your answer with reference to life-history of *Selaginella*. Discuss the significance. Kanpur, 1981.

9. Where will you search for the plant of *Equisetum* is nature? Give a detailed account of the structure of its cone. Kanpur, 1981.

10. Describe the structure of sporocarp of *Marsilea* and write a short note on its morphological nature. Kanpur, 1981.

11. Describe the various types of steles found in Pteridophytes you have studied. Kanpur, 1982, 1984, 1985, 1987.

12. Describe the male and female gametophytes of *Lycopodium*. Kanpur, 1982.

13. Write an essay on heterospory and seed habit in pteridophytes. Kanpur, 1983.

14. Describe the internal structure of *Lycopodium* stem giving different types of steles found in it. Kanpur, 1983.

15. Give a brief account of the life-history of *Selaginella*. Kanpur, 1983, 1986.

16. Describe the male and female gametophytes of *Marsilea*. Kanpur, 1984.

17. Give an illustrated account of arrangement of various types of spores in sporocarp of *Marsilea*. Kanpur, 1984.

18. Describe the significant features in the life-history of *Lycopodium*. Kanpur, 1985.

19. What is alternation of generations ? Explain it in the life-history of *Equisetum*. Kanpur, 1986.

20. Describe the significant features in the life-history of *Aspidium* (fern). Kanpur, 1987.

21. Define the stele and describe the different types of steles in *Lycopodium* species with the help of neat and labelled diagrams only. Kanpur, 1988.

22. Describe the development of male gametophyte of *Selaginella* inside the sporangium. Kanpur, 1988.

23. Discuss the morphological nature of sporocarp of *Marsilea*. Kanpur, 1989.

24. Describe *Rhynia* on the basis of plant body and vascular structure. From where this genus discovered? How does it differ from *Horneophyton*. Kumaon, 1981.

25. What do you think were the essential features of first land plants ? How far does *Rhynia* fit in your conception? Kumaon, 1982, 1985.

26. Give an account of the structure and position of sex-organs of *Lycopodium, Equisetum*, and *Marsilea*. In which genus do you find the most reduced type of antheridium ? Give its significance, if any. Kumaon, 1983.

27. What is stele? Give an illustrated account of the various types of stelar structures seen in pteridophytes. Kumaon, 1984, 1986.

28. Describe the stelar system in pteridophytes. Rohilkhand, 1979.

29. Compare the gametophytes of *Selaginella* and *Marsilea*. Rohilkhand, 1979.

30. Give a comparative account of the strobili of *Lycopodium, Selaginella* and *Equisetum*. Rohilkhand, 1979.

31. With the help of well-labelled diagrams, discuss the anatomy of stem and root of *Equisetum*. Rohilkhand, 1981.

32. What is stele ? Give an account of the stelar organization in the pteridophytes studied by you. Rohilkhand, 1982, 1984, 1986.
33. Describe the male and female gametophytes of *Selaginella*. Rohilkhand, 1984.
34. Compare the spore producing organs of *Selaginella* and *Equisetum*. Rohilkhand, 1985.
35. Describe the morphology of *Marsilea* sporocarp. How does it differ from a true seed. Rohilkhand, 1986.
36. Compare the spore producing organs of *Selaginella* and *Lycopodium*. Rohilkhand, 1987.
37. Give a brief account of the salient features of Psilophytales and discuss the systematic position of *Rhynia*.

OR

Describe the stem structure and reproduction in *Rhynia*. Awadh , 1981.
38. Compare structures of stem of *Equisetum, Selaginella* and *Marsilea* lying emphasis on their stelar system. Awadh, 1981.
39. Describe the different types of steles present in various members of pteridophytes studied by you. Awadh, 1983.
40. How many types of gametophytes found in *Lycopodium ?* Illustrate your answer with suitable diagrams. Awadh, 1984, 1987.

Diagrammatic Type

1. Give labelled diagrams only to show the main structural features of the following : (i) T.S. *Aspidium* rhizome, (ii) T.S. *Marsilea* rhizome, (iii) T.S. *Equisetum* rhizome. Kanpur, 1981.
2. Draw well labelled diagrams of the following (description not needed) : (i) T.S. *Marsilea* petiole, (ii) V.S. of fern leaf having sori, Kanpur, 1982.
3. Draw well labelled diagrams of the following : (i) T.S. *Marsilea* rhizome, (ii) T.S. *Selaginella* rhizophore. Kanpur, 1983.
4. Draw well-labelled diagrams of the following : (i) L.S. of *Equisetum* cone, (ii) V.S. of fern leaf with sori, Kanpur, 1984.
5. By means of labelled diagrams only illustrate the life-history of *Aspidium* (fern). Kanpur, 1985.
6. By means of labelled diagrams only illustrate the life-history of *Marsilea*. Kanpur, 1986.
7. By means of labelled diagrams only illustrate the life - history of *Equisetum*. Kanpur, 1987.
8. With the help of neat and labelled diagrams only, show the internal structure of the *Marsilea* sporocarp. Kanpur, 1988.
9. Draw neat and labelled diagrams of the following : (i) T.S. of *Equisetum* internode, (ii) L.S. of *Selaginella*, cone, (iii) V.S. of fern sporophyll. Kanpur, 1988.
10. With the help of neat and labelled diagrams only, show the different types of steles in pteridophytes. Kanpur, 1989.
11. Draw well labelled diagrams of the following : (i) T.S. of *Selaginella* stem, (ii) T.S. of *Marsilea* rhizome. Kanpur, 1989.
12. Differentiate the following with the help of neat and labelled diagrams only : (i) T.S. of stem of *Lycopodium clavatum* and *Lycopodium cernuum*. (ii) T.S. of microsporophyll of *Selaginella* and megasporophyll of *Selaginella*, (iii) T.S. of internode of *Equisetum* and node of *Equisetum*. Kanpur, 1989.
13. Draw well-labelled diagrams of longitudinal sections of strobili of *Equisetum, Lycopodium* and *Selaginella*. (Description not needed). Kumaon, 1986.
14. Give well-labelled diagrams of any two of the follwoing : (i) cone of *Equisetum*, (ii) rhizome of *Marsilea*, (iii) stem of *Lycopodium*. Rohilkhand, 1980.
15. Draw suitable diagrams of the following: (i) antherozoid of *Selaginella*, (ii) archegonia of fern, (iii) sporophyll of *Lycopodium*. Rohilkhand, 1982.
16. Illustrate the life-history of *Lycopodium* with the help of lebelled diagrams only. Rohilkhand, 1983, 1985.
17. Draw well-labelled diagrams of the following : (i) T.S. *Selaginella* stem, (ii) T.S. *Marsilea* rhizome, (iii) T.S. fern petiole. Rohilkhand, 1983.
18. Draw labelled diagrams of the following : (i) T.S. *Equisetum* petiole, (ii) T.S. *Marsilea* petiole, (iii) V.S. leaf *Aspidium* passing through sori. Rohilkhand, 1984.
19. Draw labelled diagrams of T.S. stem of *Selaginella* and *Lycopodium*. Rohilkhand, 1985.
20. Draw only labelled diagrams of the following : (i) T.S. petiole of *Marsilea*, (ii) T.S. stem of *Lycopodium*. Rohilkhand, 1986.
21. Illustrate the life-cycle of *Equisetum* with the help of necessary diagrams only. Rohilkhand, 1987.
22. Draw neat and labelled diagrams only of the following : (i) T.S. of *Equisetum* rhizome, (ii) T.S. of *Marsilea* petiole, (iii) T.S. of *Lycopodium* stem, (iv) T.S. of *Selaginella* stem. Awadh, 1980.
23. Give labelled sketches of any three of the following : (i) L.S. gametophyte of *Lycopodium*, (ii) T.S. internode of *Equisetum* stem, (iii) L.S. strobilus of *Selaginella*. (iv) T.S. petiole of *Marsilea*. Awadh, 1981.

24. With labelled diagrams only, illustrate the structure of spore bearing parts of *Lycopodium* and *Equisetum*. Awadh, 1982.

25. With the help of labelled diagrams only, give the details of the following : (i) *Rhynia*, (ii) HLS and VLS of sporocarp of *Marsilea*, (iii) T.S. stem of *Selaginella*. Awadh, 1983.

Short Answer Type

1. Give the classification, geological distribution and characteristic features of *Rhynia*.
2. Explain how the complex type of stele evolved from the simplest type. Give an example for each.
3. What is the essence of the Telome concept.
4. Enumerate the views on the morphology of the sporocarp of *Marsilea*.
5. What is the significance of heterospory ?
6. Write note on morphological nature of the rhizophore in *Selaginella*.
7. Write short note on *Rhynia*.
8. What are the external features by which you will identify the sporophytes of *Selaginella, Equisetum, Ophioglossum* and *Marsilea* ?
9. Discuss in brief the morphology of the fertile spike of *Ophioglossum*.
10. Describe how elaters are produced in *Equisetum*.
11. What are the xerophytic characters exhibited morphologically and anatomically by *Equisetum* ?
12. Discuss in brief the morphology of the sporangiophore of *Equisetum*.
13. Write detailed note on stelar structure in *Lycopodium*.
14. Distinguish the rhizophores from that of roots in *Selaginella*.
15. Enumerate the root like characters of rhizophores of *selaginella*.
16. How can the strobilus of *Lycopodium* be distinguished from that of *Selaginella* ?
17. Explain the leaf arrangement in *Selaginella*.
18. Explain the stages of development of megasporangium which makes it different from microsporangium in *Selaginella*.
19. Mention any two root characteristics and any two stem characteristics of rhizophore of *Selaginella*.
20. Comment briefly on the morphological nature of the protocorm of *Lycopodium*.
21. Write notes on the gametophytes of *Lycopodium*.
22. Write short notes on : (i) amphiphloic siphonostele, (ii) false indusium, (iii) trabeculae, Kanpur, 1981.
23. Write short notes on : (i) heterospory, (ii) cone of *Equisetum*. Kanpur, 1982.
24. Write notes on : (i) *Equisetum* cone, (ii) *Marsilea* sporocarp. Kanpur, 1983.
25. Write short notes on : (i) cone of *Lycopodium*, (ii) rhizophore of *Selaginella*. Kanpur, 1984.
26. Write short notes on any three : (i) sorus, (ii) siphonostele, (iii) seed habit in *Selaginella*, (iv) trabeculae. Kanpur, 1986.
27. How would you distinguish between the following : (i) rhizophore and root, (ii) true indusium and false indusium (iii) *Equisetum* rhizome and *Marsilea* rhizome. Kanpur, 1987.
28. Write detailed notes on any two of the following : (i) prothallus of *Lycopodium*, (ii) protocorm of *Lycopodium*, (iii) strobilus of *Lycopodium*. Kanpur, 1988.
29. Write short notes on : (i) seed habit in *Selaginella*, (ii) *Equisetum* cone.
30. Write short notes on any four : (i) abaxial, (ii) basal, (iii) axillary, (iv) adaxial.
31. Write notes on: (i) telome, (ii) protostele, (iii) heterospory, (iv) *Rhynia*. Rohilkhand, 1981.
32. Write notes on any four : (i) trabeculae, (ii) sorus, (iii) spike of *Selaginella*, (iv) annulus, (v) fossil. Rohilkhand, 1982.
33. Write notes on : (i) meristele, (ii) heterospory, (ii) ligule, (iv) spore of *Equisetum*. Rohilkhand, 1983.
34. Write notes on : (i) *Rhynia*, (ii) cone of *Equisetum*, (iii) sporocarp. Rohilkhand, 1984.
35. Write short notes on : (i) *Rhynia*, (ii) rhizophore, (iii) rhizoids. Rohilkhand, 1987.
36. Distinguish between : (i) homospory and heterospory, (ii) apogamy and apospory, (ii) leaf traces and leaf gaps. Kumaon, 1981.
37. Distinguish between the following : (i) homospory and heterospory, (ii) sporophyll and sporangium, (iii) spore mother cell and spore as far as their chromosome number is concerned, (iv) strobilus and sori. Kumaon, 1983.
38. Explain the following : (i) carinal canal, (ii) vallecular canal, (iii) sporocarp, (iv) ligule, (v) annulus. Kumaon, 1984.
39. Distinguish between the following : (i) homospory and heterospory, (ii) apogamy and aposproy, (iii) leaf trace and leaf gap, (iv) protophylls and protocorm, (v) ectopholoic and amphiphloic siphonostele. Kumaon, 1985.
40. Write short notes an any three : (a) protostele, (b) dictyostele, (c) siphonostele, (d) heterospory. Awadh, 1980.
41. Write brief and illustrated notes on : (i) protocorm of *Lycopodium*, (b) gametophyte of *Equisetum*, (c) ligule of *Selaginella* and its morphological nature, (d) sporocarp of *Marsilea*. Awadh, 1981

42. Write short notes on any three : (i) fossil types, (ii) rhizophore in *Selaginella*, (iii) nature of stele in *Marsilea* stem, (iv) spore of *Equisetum*, (v) leptosporangiate development. Awadh, 1982.

43. Point out the differences in between : (i) T.S. of rhizome and petiole of *Marsilea*, (ii) T.S. of rhizome and stem of *Equisetum*, (iii) give the botanical names of atleast three species of each of the four representatives of pteridophytes studied by you. Awadh, 1983.

44. Write short notes on any three : (a) heterospory, (b) sporangium of *Rhynia*, (c) rhizophore, (d) dehiscence of sporocarp of *Marsilea*. Awadh, 1984.

45. Point out the differences in between : (a) T.S. of rhizome and petiole of *Marsilea*, (b) T.S. of rhizome and stem of *Equisetum*, (c) T.S. of root and rhizophore of *Selaginella*. Awadh, 1985.

46. Write short notes on any three : (a) heterospory ; (b) ligule of *Selaginella* ; (c) morphological nature of *Marsilea* sporocarp; (d) dehiscence of sporangium in *Equisetum* ; (e) protocorm of *Lycopodium*. Awadh, 1986.

47. Discuss any two of the following : (i) *Equisetum* has both hydrophytic and xerophytic characters; (ii) protocorm; (iii) morphological and anatomical differences occurring between the plants of *Marsilea* growing in water and land surface. Awadh, 1988.

Multiple Choice Type

1. The living example of near telome is : (i) *Marsilea*, (ii) *Adiantum*, (iii) *Osmunda*, (iv) *Psilotum*.
2. Mixed protostele is found in : (i) *Selaginella*, (ii) *Lycopodium*, (iii) *Equisetum*, (iv) *Adiantum*.
3. Simple collateral vascular bundles are found in : (i) *Equisetum*, (ii) *Lycopodium*, (iii) *Pleopeltis*, (iv) *Selaginella*.
4. Common element of xylem in pteridophytes is : (i) tracheae, (ii) tracheids, (iii) sclereids, (iv) vessels.
5. Spores of pteridophytes are : (i) haploid, (ii) diploid, (iii) triploid, (iv) tetraploid.
6. Ligule and cone are present in : (i) *Selaginella*, (ii) *Lycopodium*, (iii) *Isoetes*, (iv) *Psilotum*.
7. The sporangia are developed in *Equisetum* on : (i) upper surface of sporophylls, (ii) sporangiophores, (iii) vegetative leaves, (iv) lower surface of sporophylls.
8. In the root of *Equisetum* the endodermis is : (i) absent, (ii) one layered, (iii) two layered, (iv) three layered.
9. Vallecular canal in *Equisetum* is situated : (i) below the ridges, (ii) below the furrows, (iii) between ridges and furrows, (iv) between pith and endodermis.
10. A cavity formed by the disintegration of protoxylem tracheids in *Equisetum* stem is called : (i) vallecular canal, (ii) carinal canal, (iii) central cavity, (iv) air chamber.
11. In *Equisetum* sporangia are born in : (i) microsporophylls, (ii) megasporophylls, (iii) sporangiophore, (iv) fertile leaves.
12. Presence of air cavities in the stem of *Equisetm* indicate : (i) xerophytic feature, (ii) hydrophytic feature, (iii) mesophytic feature, (iv) halophytic feature.
13. *Equisetum* is commonly called :(i) club moss, (ii) horse tail, (iii) orill wort, (iv) stone wort.
14. Intercalary meristem is seen in : (i) *Ophioglossum*, (ii) *Lycopodium*, (iii) *Psilotum*, (iv) *Equisetum*.
15. *Equisetum* is : (i) incipiently heterosporous, (ii) homosporous, (iii) heterosporous, (iv) aposporous.
16. The stale of *Ophioglossum* is : dictyostele, (ii) polystele, (iii) protostele, (iv) solenostele.
17. The sorus of *Pteris* is : (i) simple type, (ii) mixed type, (iii) gradate type, (iv) none of above.
18. The sporangia of *Ophioglossum* are seen in the form of a : (i) strobilus, (ii) sporophyll, (iii) spike, (iv) tassel.
19. Leaf gaps and leaf traces are seen in the rhizome of : (i) *Psilotum*, (ii) *Ophioglossum*, (iii) *Pteris*, (iv) *Marsilea*.
20. Amphiphloic siphonostele is found in: (i) *Psilotum*, (ii) *Ophioglossum*, (iii) *Pteris*, (iv) *Marsilea*.
21. Stele in *Pteris* rhizome is :(i) dictyostele, (ii) polystelic, (iii) polycyclic, (iv) meristelic.
22. Gradate sorus is found in : (i) *Hymenophyllum*, (ii) *Marsilea*, *(iii) Pteris*, (iv) *Glychenia*.
23. Stele in the stem of *Salaginella* is : (i) endarch, (ii) exarch, (iii) mesarch.
24. Stele in *Lycopodium* is : (i) dictyostelic, (ii) protostelic, (iii) solenostelic, (iv) siphonostelic.
25. *Rhynia* was discovered by : (i) Arnold, (ii) Kidston and Lang, (iii) Birbal Sahni, (iv) Campbell.
26. *Rhynia* belongs to a geological period caled : (i) Carboniferous, (ii) Permean, (iii) Devonian, (iv) Cambrian.
27. In *Lycopodium* and *Selaginella* though the sporangia are developed on the adaxial surface of the sporophylls, they can be distinguished by : (i) different types of spores, (ii)nature of the sporophyll, (iii) pattern of vasculature, (iv) arrangement of sporangia.
28. The stele in the stem of *Selaginella* is : (i) protostele, (ii) solenostele, (iii) polystele, (iv) dictyostele.
29. The male gametes of *Selaginella* are : (i) biflagellate, (ii) multiflagellate, (iii) monoflagellate, (iv) aflagellate.
30. On which side is the sporangium seen on the sporophyll in *Lycopodium* : (i) abaxial, (ii) basal, (iii) axillary, (iv) adaxial.
31. Wall of sporangium in *Lycopodium* usually consists of : (i) single layer, (ii) two layers, (iii) three layers, (iv) many layers.
32. An adaxial outgrowth from the base of the leaves in *Selaginells* is called : (i) stipule, (ii) ligule, (iii) trabeculae, (iv) velum.

33. *Lycopodium* is commonly called as : (i) horse tail, (ii) club moss, (iii) quill wort, (iv) stone wort.

34. Trabeculae in *Selaginella* is modified : (i) pericycle cells, (ii) endodermal cells, (iii) pericycle and endodermal cells, (iv) cortical cells.

35. Trabeculated endodermis is seen in :(i) *Psilotum*, (ii) *Gleichenia*, (iii) *Ophioglossum*, (iv) *Selaginella*.

36. Double layer of endodermis is found in the root of : (i) *Pteris*, (ii) *Equisetum*, (iii) *Selaginella*, (iv) *Hymenophyllum*.

37. The rhizophore of *Selaginella* has : (i) triarch stele, (ii) monarch stele, (iii) diarch stele, (iv) tetrarch stele.

38. Rhizophore of *Selaginella* is like a : (i) root, (ii) organ *suigeneris*, (iii) stem, (iv) part of leaf.

39. A fern differs from a moss in having :(i) an independent gametophyte, (ii) an independent sporophyte, (iii) sperms, (iv) archegonia. C.P.M.T.1977.

40. Fern gametophyte is : (i) homothallic, (ii) heterothallic, (iii) heteroecious, (iv) autoecious. C.P.M.T.1977.

41. Formation of a sporophyte from a vegetative portion of prothallus is called: (i) apocarpy, (ii) apogamy, (iii) apomixis, (iv) apospory. C.P.M.T. 1977.

42. Young sporophyte of fern draws nourishment from the prothallus through : (i) haustoria, (ii) foot, (iii) root, (iv) rhizoids. C.P.M.T. 1978.

43. *Cyas* and *Pteris* resemble each other in the presence of : (i) cambium, (ii) ciliated sperm, (iii) vessels, (iv) seeds. C.P.M.T. 1979.

44. Pteridophytes differ from bryophytes in having : (i) motile sperm, (ii) alternation of generations, (iii) archegonia, (iv) vascular tissue. C.P.M.T. 1980.

45. *Cyas* differ from *Pteris* in having : (i) tracheids, (ii)motile sperms, (iii) pollen tubes, (iv) archegonia. C.P.M.T. 1980.

46. In *Dryopteris* the antherozoids are : (i) biciliate and coiled, (ii) biciliate and sickle shaped, (iii) multiciliate and coiled, (iv)multiciliate and sickle shaped. C.P.M.T. 1980.

47. A fern differs from a moss in having : (i) alternation of generations, (ii) swimming sperms (iii) independent gametophyte, (iv) independent sporophyte. C.P.M.T.1980.

48. Prof. Birbal Sahni is famous for his researches in the field of : (i) palaeobotany, (ii) physiology, (iii) cytology, (iv) ecology. C.P.M.T. 1983.

49. In ferns the xylem is : (i) exarch, (ii) mesarch, (iii) endarch, (iv) polyarch. C.P.M.T. 1983.

50. Which of the following conditions found in *Cyas* is also found in *Pteris* : (i) naked ovules, (ii) circinate vernation of leaves, (iii) presence of coralloid roots, (iv) presence of secondary growth. C.P.M.T. 1983.

51. The vascular bundle in which the xylem is surrounded by phloem is known as : (i) radial, (ii) collateral, (iii) amphicribral concentric, (iv) conjoint. C.P.M.T. 1985.

52. The spores of the fern plant germinate to form : (i) thick-walled zygospore, (ii) an embryo, (iii) the new fern plant, (iv) a prothallus. C.P.M.T. 1985.

53. The stele in the rachis of *Dryopteris* is : (i) protostelic, (ii) siphonostelic, (iii) dictyostelic, (iv) haplostelic. C.P.M.T. 1985.

54. The sori of *Pteris* are : (i)discontinous and circular, (ii) continuous and linear, (iii) discontinuous and vermiform, (iv) discontinuous and reniform. C.P.M.T. 1988.

Answers

1 (iv) , 2 (ii), 3 (i), 4 (ii), 5 (i), 6 (i), 7 (ii), 8 (iii), 9 (ii), 10 (ii), 11 (iii), 12 (ii), 13 (ii), 14 (iv), 15 (ii), 16 (i), 17 (i), 18 (iii), 19 (iii), 20 (iv), 21 (i), 22 (ii), 23 (iii), 24 (ii), 25 (ii), 26 (iii), 27 (i), 28 (i), 29(i), 30 (iv), 31 (ii), 32 (ii), 33 (ii), 34 (ii), 35 (iv), 36 (ii), 37 (ii),38 (i), 39 (ii), 40 (i), 41 (ii), 42 (ii), 43 (ii), 44 (iv), 45 (iii), 46 (iii), 47 (iv), 48 (i), 49 (ii), 50 (ii), 51 (iii), 52 (iv), 53 (iii), 54 (i).

INDEX

GYMNOSPERMS

1 Introduction

The gymnosperms have long been considered as a group of seed-bearing plants. Together with the angiosperms they are known as phanerogams or spermatophytes. They are distinguished from the angiosperms by having naked ovules and seeds. The ovules are not enclosed in an ovary and the seeds are not enclosed in a pericarp.

The gymnospermae comprises of trees and shrubs which are characterized by xeromorphic acicular or fern-like leaves and unisexual inflorescence.

The ovules lie exposed on the magasporophylls which are open and therefore, devoid of a style and stigma. The nakedness of the ovules was clearly recorded by Robert Brown in 1928.

The pollen grains are spherical or oval, often with a bladder-like extention of the exine. The pollination is direct in the sense that the pollen grains or at least the pollentubes reach the nucleus of the ovule without the intervention of the stigma and styles through which the pollen tubes have to penetrate in angiosperms.

The pollen grains possess a prothallus of two or more cells, one of which produces two non-motile or rarely motile male gametes or spermatozoids (*e.g., Cycas, Ginkgo*).

The pollen grains are disseminated by wind. They directly settle upon the micropyle. The embryo sac (female gametophyte) contains a tissue before fertilization which is known as rudimentary prothallium, and two or more archegonia are formed in it, each consisting of a large egg cell and a short neck. Only one embryo usually develops to maturity. The embryo consists of an axis bearing two to several cotyledons and ending in a radicle. The whole embryo remains enclosed in a mass of endosperm which represents a continuation of the gametophytic tissue. The sporophylls are usually arranged in a cone-like structure.

The following key may be used in the indentification of the gymnosperms:

Trees or shrubs usually resinous; ovules naked, not enclosed in an ovary; flowers unisexual, rarely bisexual; leaves needle-shaped scale-like, linear, pinnate, rarely fan-shaped or oblong elliptic, mostly evergreen.

From the point of view of their botany the interest in the group is several fold. They are all wood plants. They range from the tallest known trees, *e.g., Sequoia* sp. to dwarf shrubs, *e.g., Ephedra.* The diversity of their vegetative parts is very great. They possess greater diversities in their reproductive organs. They are, in many points of structure and details of behaviour, intermediate between vascular cryptogams and angiosperms. Several major groups of gymnosperms are extinct and are known only from a study of their fossils in Palaeozoic or Mesozoic strata.

Comparison of Ferns and Gymnosperms

1. In both the groups there is alternation of generations but there is reduction in the gametophytic generation of the gymnosperms. In gymnosperms the gametophyte is reduced and is unlike that of ferns as it is completely dependent upon the sporophyte.

2. In gymnosperms there is gradual increase in the complexity of the sporophyte as it differentiates into root, stem and leaf. The tap root system becomes very well developed in gymnosperms and its continuous growth gives better anchorage and absorption of nutrients unlike the ferns in which the roots are always adventitious.

3. The members of Cycadales resemble the ferns further in the compound nature of their leaves and in circinate vernation.

4. The companion cells remain absent from the phloem of both gymnosperms and ferns. The vessels are also not found in the xylem of all gymnosperms except the Gnetales where the vessels are present. Similarly phloem is composed of sieve cells and phloem parenchyma; companion cells are absent.

5. The gymnosperms resemble *Selaginella, Isoetes, Marsilea* and some fossil ferns in having heterospory. But in gymnosperms the spores germinate *in situ* the megaspore is retained within the megasporangium.

6. Multiflagellate and motile antherozoids are found in ferns as well as in Cycadales, but in other higher gymnosperms, such as the Coniferales and the Gnetales they are non-flagellate.

7. There is excessive secondary growth in gymnosperms.

8. The seed-habit predominates in the gymnosperms and the angiosperms. However, the seeds were present in the fossil pteridophytes but not in the living ones.

9. In gymnosperms the pollen grains are light, and develop partly before their release from the pollen sac.

10. In gymnosperms the fertilization is effected by pollen tube (*i.e.*, siphonogamy) in contrast to zooidogamy in pteridophytes where the motile sperms reach the eggs through neck canals of archegonia.

11. In gymnosperms the female gametophyte is permanently retained inside the megaspore. The archegonia of gymnosperms lack the neck canal cells and frequently the ventral canal cell in eliminated.

Comparison of Angiosperms and Gymnosperms

1. All gymnosperms are woody, although herbs are quite common in angiosperms.

2. Generally two types of branches are found in gymnosperms, *i.e.*, long shoots and dwarf shoots.

3. The leaves of gymnosperms are evergreen and with thick cuticle.

4. Gymnosperms lack true vessels. Xylem vessels are absent and xylem parenchyma is scanty. Sieve plates are present on lateral walls. There are no sieve-tubes and companion cells. Only sieve cells and phloem parenchyma are present.

5. In gymnosperms the xylem is either mesarch or endarch. In angiosperms it is never mesarch.

6. Resin and gum canals are very common in gymnosperms.

7. In gymnosperms the leaves generally possess transfusion tissue and sunken stomata.

8. The aggregation of sporophylls in cones or strobili, in more advanced form, develop flower-like structure, but coloured bracts, perieanth lobes and nectaries are absent.

9. Development of sporangium in gymnosperms is of eusporangiate type as also found in angiosperms.

10. The male gametophyte has two prothallial cells in Coniferales and Ginkgoales while one in Cycadales. No prothallial cells are found in angiosperms. One stalk cell and two male gametes which are flagellated in Cycadales and Ginkgoales while non-flagellate in rest of gymnosperms like that of angiosperms.

11. All gymnosperms are anemophilous and the pollen grains come in direct contact with the ovules (direct pollination). Here the pollen grains fall directly on the pollen drop which is sucked into the nucleus. In angiosperms the pollen grains are deposited on the receptive surface of the stigma (indirect pollination).

12. In gymnosperms, generally there is long interval between pollination and fertilization; in Amentiferae (angiosperms) long interval between pollination and fertilization exists. In *Ginkgo* and *Cycas* the pollen tube has haustorial function while in others it has sperm carrying function. The pollens possess wings in *Pinus*.

13. In gymnosperms, the ovule remains covered by one integument while in angiosperms there are two integuments. In majority of gymnosperms the integument has a vasular supply while in angiosperms the vascular strand terminates at the base of the integuments.

14. In gymnosperms, the ovules are always naked and never enclosed in ovary as found in angiosperms.

15. In gymnosperms, germination of megaspore is accompanied by nuclear free division, ultimately resulting in the formation of female gametophyte.

MAIN EVENTS OF EARTH'S HISTORY

GEOLOGICAL ERA AND TIME IN MILLIONS OF YEARS BEFORE THE PRESENT	PERIODS	DURATION IN MILLIONS YEARS	MAIN GVMNOSPERM GROUPS
0	QUATERNARY RECENT		
CENOZOIC 70	TERTIARY	62	
MESOZOIC	CRETACEOUS	72	
200	JURASSIC	45	
	TRIASSIC	50	
PALAEOZOIC	PERMIAN	50	
	CARBONI FEROUS	65	
	DEVONIAN	60	
	SILURIAN	20	
	ORDOVICIAN	75	
600	CAMBRIAN	100	
PROTEROZOIC 2000	PRE-CAMBRIAN		
ARCHAEOZOIC 3000			
5000			ORIGIN OF EARTH

CAYTONIALES
CYCADALES
GINKGOALES
CONIFERALES
GNETALES
TAXALES
PTERIDOSPERMALES
CYCADEOIDALES
CORDAITALES

Fig. 1.1. Geological history of the Gymnosperms.

Classification of gymnosperms. Until recently it was generally accepted, and expressed in schemes of classification, that the Gymnospermae formed a group co-ordinated with the Angiospermae. From time to time suggestions have been made for splitting the Gymnospermae into a number of major groups taxonomically equivalent one to another and to the angiosperms.

Several systems of classification of gymnosperms have been proposed from time to time by various workers. A few systems which have been supposed to by modern are given in the following paragraphs.

Coulter and Chamberlain (1910) divided the gymnosperms into seven orders. Some of the orders are quite extinct and not found present day. These orders are as follows:

1. Cycadofilicales–extinct
2. Bennettitales–extinct
3. Cycadales–Mesozoic to present day
4. Cordaitales–extinct
5. Ginkgoales–Palaeozoic to present day
6. Conferales–Palaeozoic to present day
7. Gnetales–Recent.

Chamberlain (1934) however, divided the gymnosperms into two large groups A. **Cycadophyta** and B. **Coniferophyta**.

In the **Cycadophyta**, sporophylls are in cones. They include three orders-

Cycadofilicales(Pteridospermae)–extinct

Bennettitales (Cycadeoideales)–extinct

Cycadales–Mesozoic to present day.

The **Coniferophyta** is divided into four orders

Ginkgoales–Palaeozoic to present day

Cordaitales–extinct

Coniferales–Palaeozoic to present day

Gnetales–Recent.

Prof. B. Sahni (1920) divided the gymnosperms into two large groups and gave them the terms:

1. Phyllospermae and 2. Stachyospermae.

1. **Phyllospermae**. They comprise of the Pteridosperms and the Cycadophyta (Cycadales and Bennettitales). The seeds are being inserted on the modified leaves.

2. **Stachyospermae**. This group includes the orders of Coniferophyta of Chamberlain. They are more or less microphyllous plants with seeds inserted on the stems.

Raizada and Sahni (1938) have summarized the clasification of gymnosperms as follows:-

Cycadophytes.

Pteridospermae (Cycadofilicales)–Carboniferous. Extinct.

Cycadeoideales (Bennettitales)–Mesozoic. Extinct.

Cycadales–Mesozoic to present day.

Pentoxylales. Jurassic

Coniferophytes.

Cordaitales–Palaeozoic. Exinct

Ginkgoales–Palaeozoic to present day.

Coniferales–Palaeozoic-Mesozoic to present day

Gnetales–Recent.

The characteristic features of these orders are as follows:

1. **Pteridospermales** (Cycadofilicales). Trees or small plants with fern-like leaves.

Sporangia borne on frond-like leaves, more or less modified, but never formed a cone.

2. Cycadeoideales (Bennettitales). Trees with stems covered with an armour of persistent leaf bases. Male sporophylls frond-like forming a loose crown. Female sporophylls found in a cone-like structure, no longer leaf-like.

3. Cycadales. Plants with palm-like habit, male and female sporophylls grouped into cones, except in the genus *Cycas* where the female sporophylls are collected into a loose crown. Male cells (gametes) motile.

4. Pentoxylales. Low shrubs. Female inflorescence consisting of numerous cones spirally attached on a peduncle, each bearing numerous ovules on a central axis. No megasporophylls,

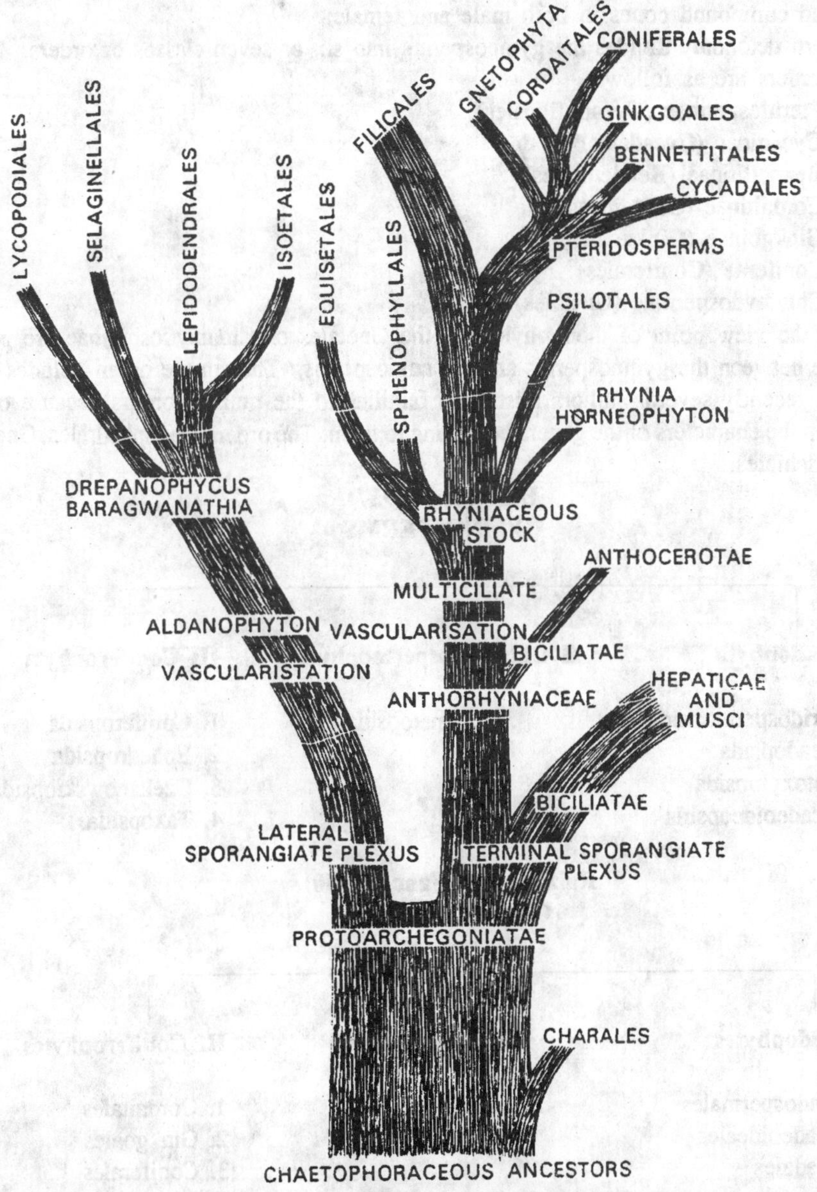

Fig. 1.2. Evolution in plants.

ovuliferous scales or interseminal scales. Male flowers terminally borne on dwarf shoots. Unilocular sporangia terminate short branchlets of the appendages or microsprophylls.

5. Cordaitales. Large trees with flat strap-shaped leaves. Fructifications in cones.

6. Coniferales. Large trees or shrubs. Leaves needle or scale-like, sometimes flattened, male cells never motile. The male flowers are arranged in more or less compact clusters or catkins and female flowers are usually clustered to form a cone.

7. Ginkgoales. Medium sized trees. Leaves flattened and lobed in various ways with dichotomous venation. Male and female flowers collected into strobili.

8. Gnetales. Small trees or shrubs or climbing shrubs.Opposite leaves, dicotyledonous embryos and compound cones in both male and female.

Modern taxonomy divides the gymnosperms into six or seven classes or orders. These classes or orders are as follows:

1. Pteridospermae (Cycadofilicales)
2. Cycadinae (Cycadales)
3. Bennettitinae (Bennettitales)
4. Cordaitinae (Cordaitales)
5. Ginkgoinae (Ginkgoales)
6. Coniferae (Coniferales)
7. Chlamydospermae (Gnetales) ???

From the view point of their phylogeny the Gnetales or Chlamydospermae are placed intermediate between the gymnosperms and the angiosperms. Though the order includes three families yet recently several authors raised the families to the rank of orders because of the divergence in the characters of the genera belonging to them. The orders are-Ephedrales, Gnetales and Welwitschiales.

D.D. Pant (1957)

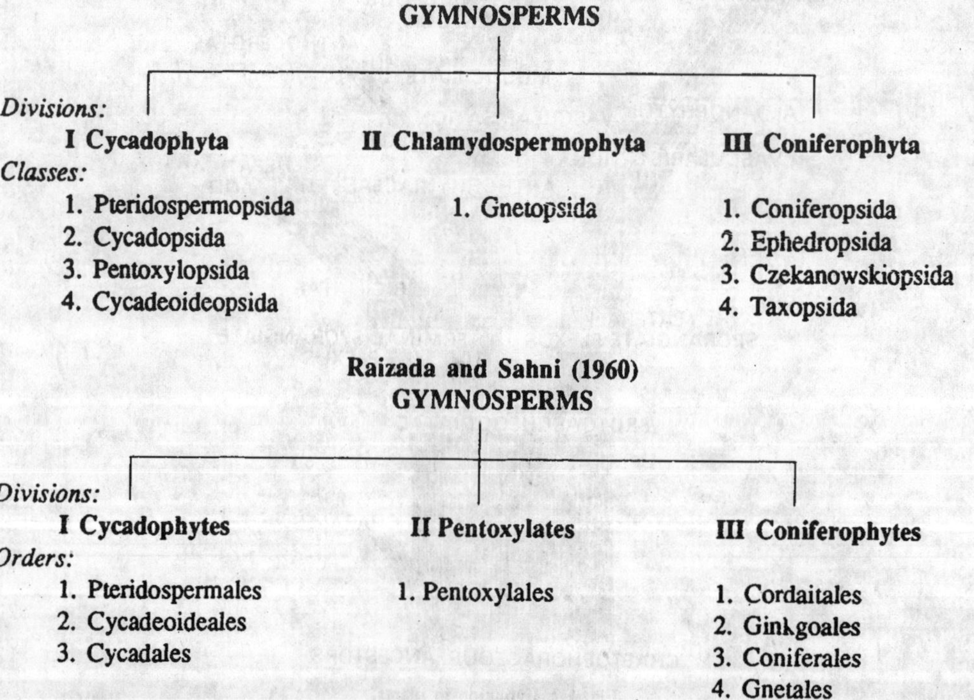

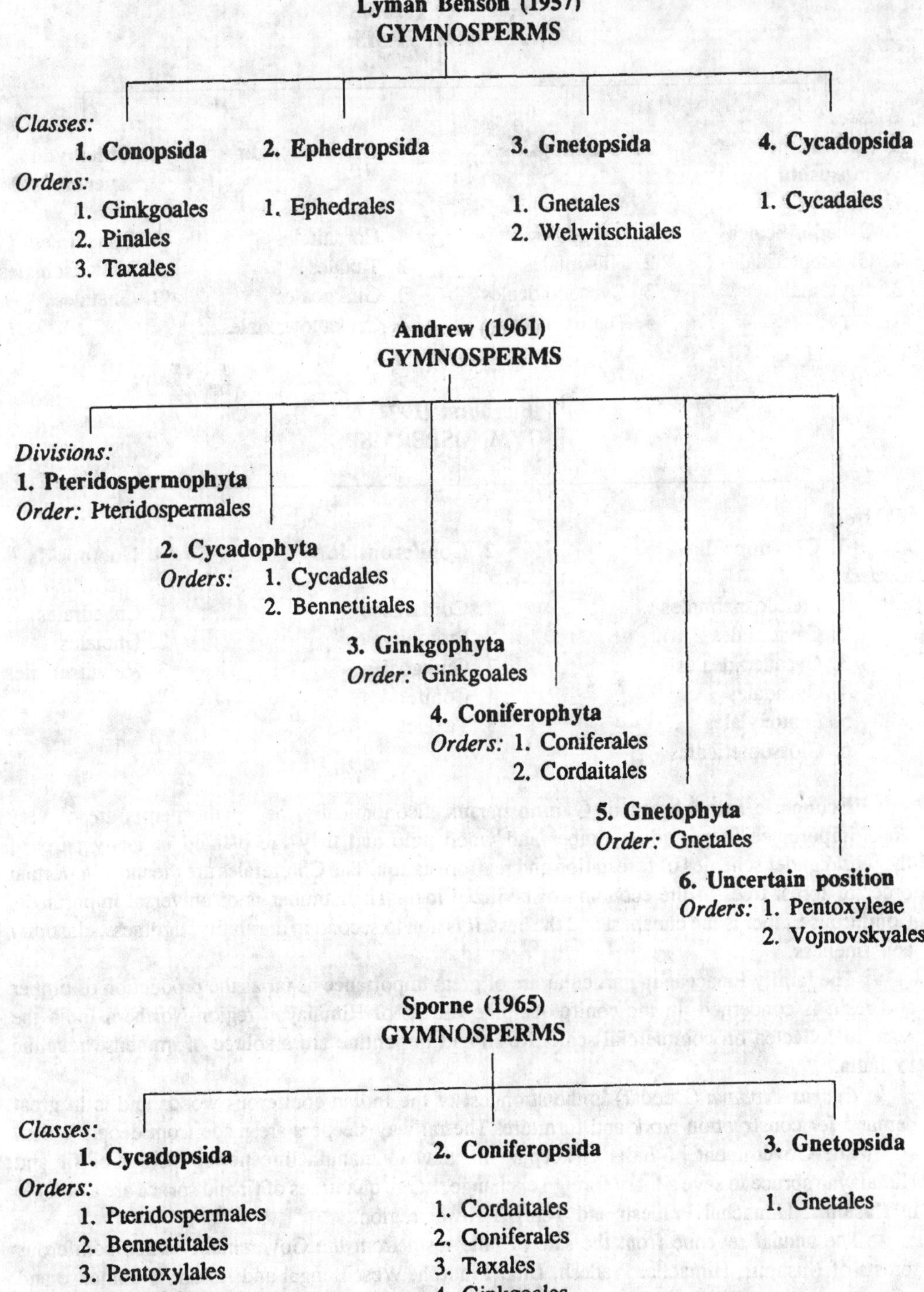
Lyman Benson (1957)
GYMNOSPERMS
Classes:
1. Conopsida
2. Ephedropsida
3. Gnetopsida
4. Cycadopsida
Orders:
1. Ginkgoales
2. Pinales
3. Taxales
1. Ephedrales
1. Gnetales
2. Welwitschiales
1. Cycadales
Andrew (1961)
GYMNOSPERMS
Divisions:
1. Pteridospermophyta
Order: Pteridospermales
2. Cycadophyta
Orders: 1. Cycadales
2. Bennettitales
3. Ginkgophyta
Order: Ginkgoales
4. Coniferophyta
Orders: 1. Coniferales
2. Cordaitales
5. Gnetophyta
Order: Gnetales
6. Uncertain position
Orders: 1. Pentoxyleae
2. Vojnovskyales
Sporne (1965)
GYMNOSPERMS
Classes:
1. Cycadopsida
2. Coniferopsida
3. Gnetopsida
Orders:
1. Pteridospermales
2. Bennettitales
3. Pentoxylales
4. Cycadales
1. Cordaitales
2. Coniferales
3. Taxales
4. Ginkgoales
1. Gnetales

Gangulee (1968)

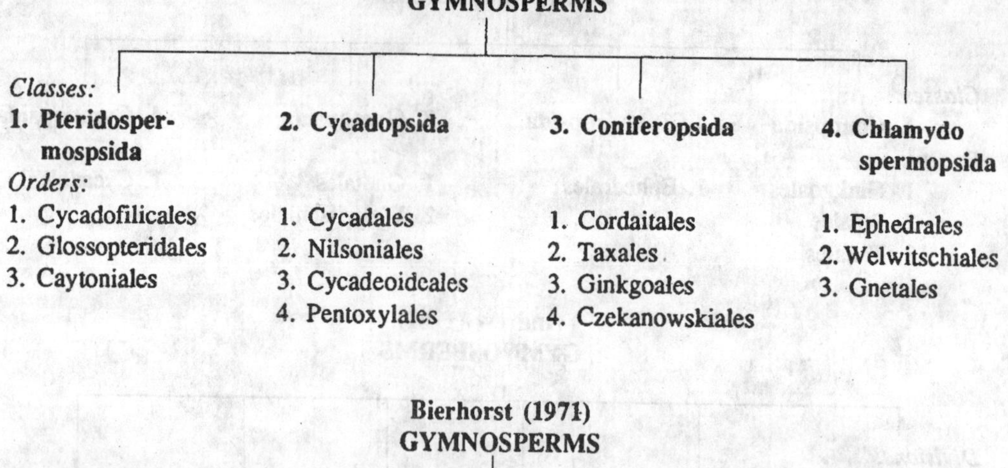

Bierhorst (1971)
GYMNOSPERMS

Classes:

1. Cycadopsida

Orders:

1. Pteridospermales
2. Caytoniales
3. Cycadeoideales
4. Cycadales
5. Pentoxylales
6. Glossopteridales

2. Coniferopsida

1. Cordaitales
2. Protopityales
3. Ginkgoales
4. Coniferales
5. Taxales

3. Gnetopsida

1. Ephedrales
2. Gnetales
3. Welwitschiales

Economic Importance of Gymnosperms. Economically the gymnosperms are of very great importance in providing timber and wood pulp and they are planted in many parts of the world under schemes of forestation and reafforestation. The Coniferales are the most important order of forest trees in the economy of civilized man. Their timber is of universal importance. Coniferous timber is the cheapest and the best. It is non to second in durability, lightness, elasticity and fineness.

The family Pinaceae in particular are of great importance as far as the production of timber and resin is concerned. In the coniferous-pine forests of Himalayan region-Northern India the resin is collected on commercial scale. Resin and turpentine are a source of immense revenue to India.

Cedrus deodara (Deodar) is the strongest of the Indian coniferous woods and is in great demand for construction work and furniture. The railway sleepers are made from deodar wood.

The Government of India is keeping in view to manufacture newsprint out of fir and Himalayan spruce to save a lot of foreign exchange. Large quantities of fir and spruce are available in Kashmir, Himachal Pradesh and Tehri-Garhwal region.

The annual revenue from the sale of pine resin from the Government owned coniferous forests of Kashmir, Himachal Pradesh, Uttar Pradesh, West Bengal and Assam is immense and millions of rupees. The revenue derived from the sale of coniferous timber of Government owned forests of above mentioned states adds much to the economy of country.

Besides, the coniferous forests provide a cool and soothing climate suitable for health. Several health resorts and sanatoria run in the hills under the shadow of pine trees where hundreds of people and patients go every year for recreation and treatment.

The pith of *Cycas revoluta* yields sago, and thus, the plant is called the sago palm. The seeds of *Pinus gerardiana* are edible and known as chilgoza. They are generally found in the dry arid zones of Himalayas. The famous drug known as ephedrine is obtained from different species of genus *Ephedra*. The young leaves and strobili of *Gnetum* are eaten as vegetable.

2

Class-Cycadopsida

Order-Cycadales

The Cycadales comprise a group of the plants with a number of primitive characters than are possessed by any other group of living gymnosperms, and because of the retention of these primitive characters sometimes they are supposed to be 'the living fossils'.

They are of great importance because they occupy the position intermediate between higher cryptogams on one hand and the angiosperms on the other.

The Pteridosperms and the Cycadeoideales are now extinct. It is thought that the Cycadales, were probably derived from the Pteridospermae. The Cycadales is an isolated group among the gymnosperms and in the Cretaceous this group was represented by larger number of species than today.

Fig. 2.1. Cycadales. ***Dioon spinulosum.*** **A tree showing habit.**

Fig. 2.2. Cycadales. ***Bowenia spectabilis.*** **Stem and bipinnate leaves.**

Geographical distribution : The Cycadales comprise a single family Cycadaceae which includes nine genera and seventy species. The more interesting point as regards genera is that with the exception of two genera, *i.e., Zamia* and *Cyas*, the rest of the genera are very much restricted in their distribution. The genera are found in tropical and sub-tropical regions of the world. The four genera, *i.e., Zamia, Microcyas, Dioon* and *Ceratozamia* are found exclusively in Western Hemisphere, whereas the remaining five, *i.e., Cycas, Macrozamia, Bowenia, Encephalartos* and *Stangeria* in Eastern Hemisphere.

Zamia is found in both northern and southern parts of Western Hemisphere, and the remaining three genera, *i.e., Microcyas, Dioon* and *Ceratozamia* are confined in other parts of Western Hemisphere. This way, *Zamia* is found in both North and South America whereas the remaining three genera grow in Iocalized places of Mexico and West-Indies.

Zamia includes 28 species; *Microcycas* one species; *Dioon* three species and *Ceratozamia* two species.

Cycas, Macrozamia, Bowenia, Encephalartos and *Stangeria* are confined in Eastern Hemisphere. The genus *Cycas* is widely distributed and found both in Northern and Southern Hemispheres. They are extended in east from Madagascar to Japan including Australia. The

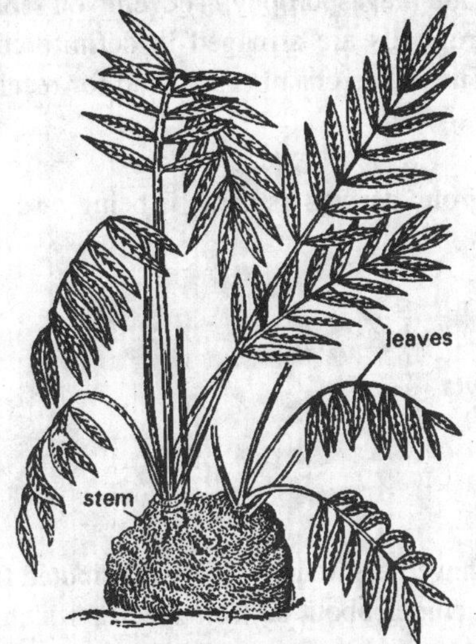

Fig. 2.3. Cycadales. ***Stangeria paradoxa.*** **Showing habit.**

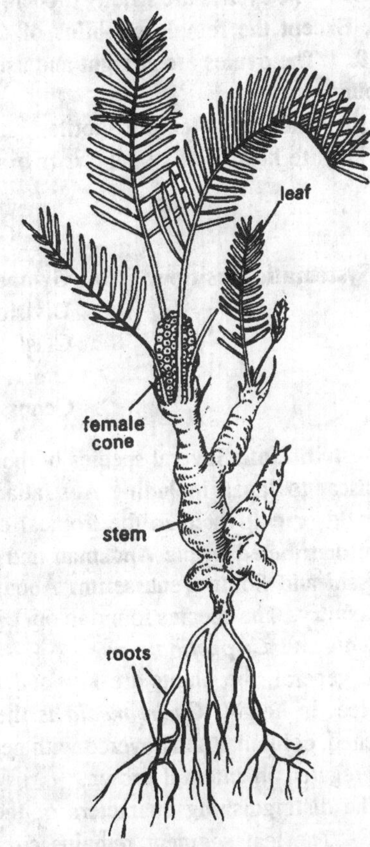

Fig. 2.4. Cycadales. ***Zamia floridiana.*** **A plant showing habit and bearing a female cone.**

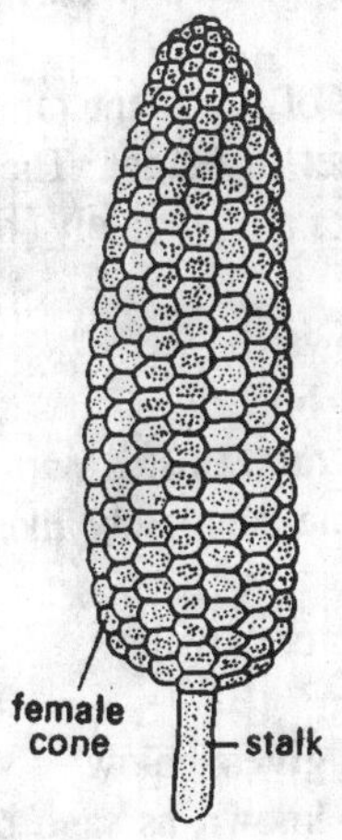

Fig. 2.5. Cycadales. ***Zamia floridiana.*** **A male cone.**

species of *Macrozamia* are found only in Australia. *Encephalartos* and *Stangeria* are found only in South Africa. *Cycas* is widely distributed whereas the remaining genera are localized.

Cycas includes 16 species; *Macrozamia* 14 species; *Bowenia* one species; *Encephalartos* 12 species and *Stangeria* one species.

The most characteristic features of the group Cycadales are as follows:

1. All cycads are typical xerophytes.
2. The plants are low and palm-like in habit.
3. The stem is short, unbranched, columnar and covered with dense persistent leaf bases.
4. The leaves are pinnately compound and arranged in a terminal crown.
5. The plants grow very slowly but they live for ages.
6. Comparatively the pith is large and cortex is broad.
7. There is a narrow zone of conducting tissue whereas in conifers the case is reverse. The conducting strand is represented by conjoint, collateral, endarch and open vascular bundles around the pith separated from each other by medullary rays.
8. The cycads are strictly dioecious, *i.e.*, micro and megasporophylls develop on separate plants. Except the female strobilus of *Cycas* the sporophylls are arranged in definite cones.
9. The ovules are straight and usually sessile. The pollen chamber is found for receiving the pollens.
10. Male gametes are motile.

The life history of genus *Cycas* from which the group derives its name is being discussed here.

Genus CYCAS

Systematic position. Gymnospermae
Division. Cycadophyta
Class. Cycadopsida
Order. Cycadales
Genus. *Cycas*

Distribution. Several species of the genus *Cycas* have been found widely distributed from Madagascar to Japan including Australia. The genus includes about sixteen species which are found wild or cultivated in the tropical and sub-tropical regions of the world. In our country they are distributed in the Andaman and Nikobar Islands, Tamil Nadu, Nepal, Sikkim,Bengal and Assam and South Tenasserim. About five species have been reported from various parts of our country. The species found in our country are: *Cycas revoluta, C. beddomei, C.circinnalis, C. rumphii* and *C. pectinata.*

In general, the plants are low and palm-like. The normal size of the plant ranges from 4 to 8 feet in height. *Cycas media* is the tallest species upto 20 feet in height. The stem is unbranched, columnar and covered with persistent leaf bases. The leaves are pinnately compound and aggregated in internal crown.

The distinguishing characters of the genus *Cycas* are as follows:

1. The leaf segment remains circinately involute within the bud.
2. The pinnae are provided with central mid-rib but no lateral veins.
3. The megasporophylls are not aggregated in cones but borne separately like foliage leaves pectinate in their upper part.
4. The megasporophyll bears on its lower margins two or more ovules.
5. The ovules are ascending.

A key for the identification of the common Indian species is given below:

1. *Cycas revoluta* Thunb.- This is native of China and Japan known as sago palm. In India, it is cultivated in the gardens. A plam-like dioecious small tree. Trunk about 1.8 m.to 3 m. high. Suckers present; the stem clothed with old persistent leaf bases. Leaves pinnate, .6 to 1.8 m. long; petiole thick quadrangular; leaflets numerous, sub-opposite, curved downwards,

narrow, stiff, dark green, margin revolute, terminating in a spine-like tip. Carpophylls 10 to 23 cm. long, blade ovate, lacinate into villous segments nearly to the centre, wooly, stalk longer than the blade with 4 to 6 ovules. It flowers in rainy season.

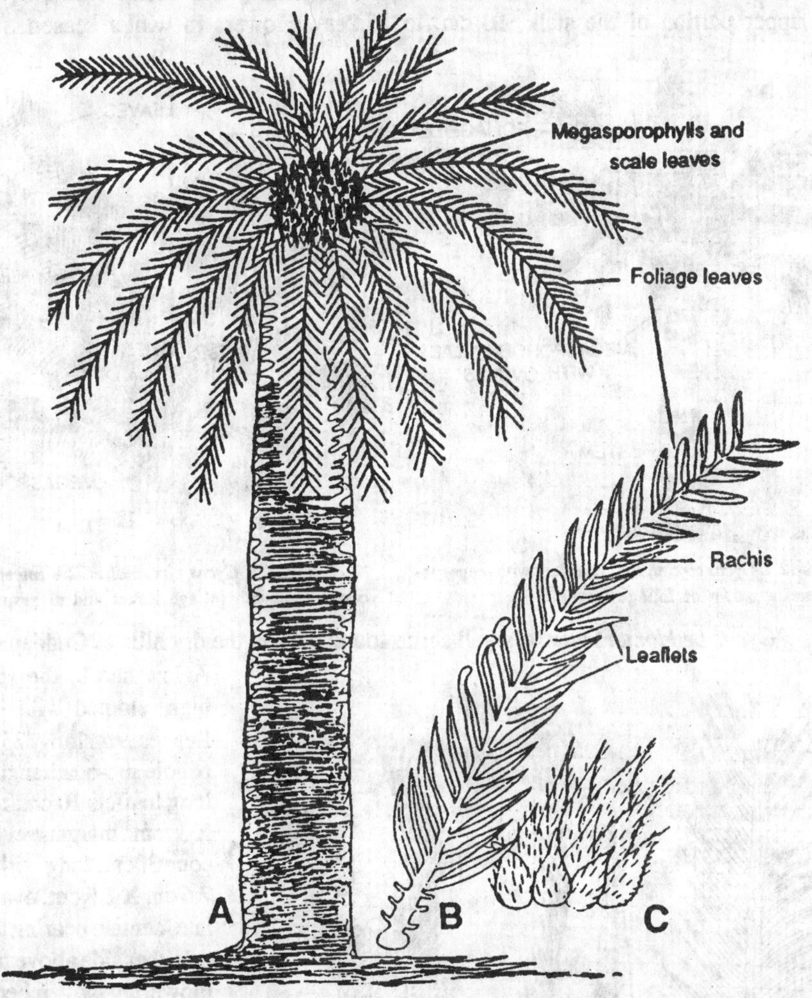

Fig. 2.6. *Cycas*. External structure. A, a female plant of *cycas revoluta;* B, leaf of *C. revoluta;* C, scale leaves of *C. revoluta*.

2. *Cycas circinnalis* Linn. This species is commonly found in South India, also cultivated in the gardens. Evergreen palm-like tree upto 8 m. high, cylindrical trunk with a crown of fern-like curved pinnate leaves clothed with prominent annular leaf scars. Leaves pinnate 1.5 m.to 2.5 m. long; petiole .4 m to.6 m long;leaflets 80 to 100 pairs, alternate, 15 cm to 30 cm. long, margins revolute; carpophylls 18 to 33 cm, X 2.5 to. 4 cm.; blade margins triangular with sharp narrow teeth; stalk long bearing in its upper portion 5-18 ovules; male cone about 33 cm long It flowers from February to March.

3. *Cycas rumphii* Miq. It is found in Andaman and Nicobar Islands, also found in Burma, Malaya, and N. Australia. Evergreen palm -like tree, 3m, to 7.6m. high trunk very tough, cften

branched. Leaves darkgreen and glossy, .6m to .8m leaflets 50-60 pairs, 20cm. to 38cm. long, 1.2cm. to 1.7cm.broad, petiole at base about 4cm broad; carpophylls about 23 cm. long X 1.2cm. broad, blade densely villous, ovate male cone 30.4 cm. long and 10cm. broad; ovules 6to 10 on the upper portion of the stalk, 10 cm. to 15.2cm. Flowers in winter season.

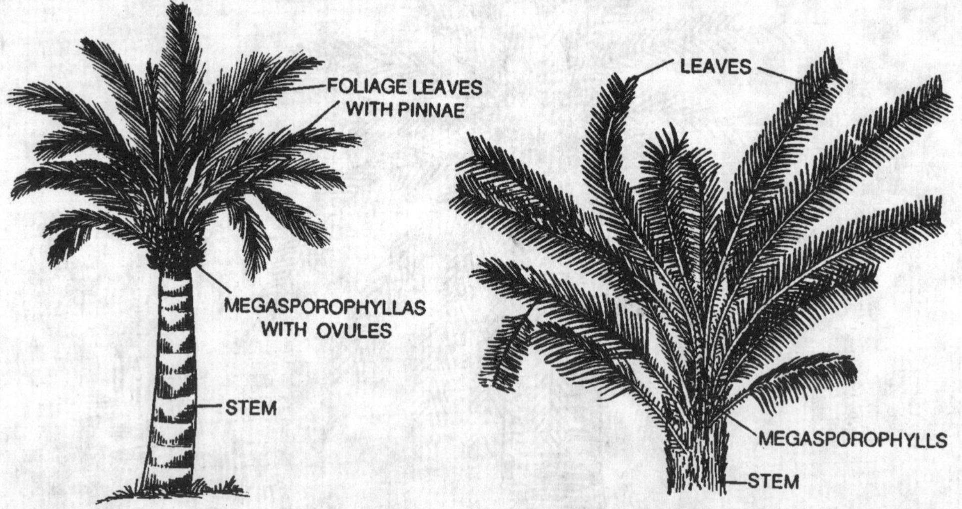

Fig. 2.7. *Cycas circinnalis*. The plant with megasporophylls and a crown of foliage leaves.

Fig. 2.8. *Cycas circinnalis*. The apical portion of a young plant with foliage leaves and megasporophylls.

4. *Cycas beddomei* Dyer. A small shrub found only in the dry hills of Cuddapah District. A low shrub, the stem 15cm. high, clothed with leaf bases. Leaves .9m. long, 23 cm. wide, petiole sub-quadrangular, 15cm. long leaflets 10 cm. to 17.7 cm. X 3 mm., margins revolute; male cone 30cm. long with the blade 7.6 cm X 2.5 cm. ovate-lanceolate, dentate bearing two ovules on either side above the middle. Flowering November - December.

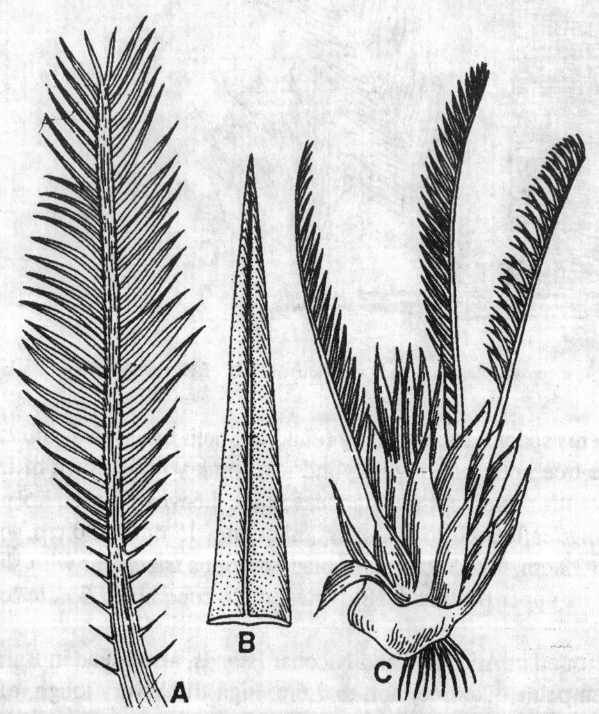

Fig. 2.9. *Cycas revoluta*. A, single leaf; B, part of leaflet; C, a bulbil.

5. *Cycas pectinata* Griff. The species is found in the Someshwar hills of Bihar. It is also found in the hills of Assam. It is distributed in Sikkim, Nepal and Pakistan. It is evergreen palm like, 2.4 m. to 3 m. high tree. In Assam,however it attains a height of 6.7m. Sometimes branched. Leaves 1.5 m. to 2.1 m. long, petiole about 45cm. long with spines; leaflets

17.5 cm. to 25.5 cm. long. Male cone about 45.5 cm. X 15.2 cm. Carpophylls 15.2 cm. long tawny silky, 2 or 3 pairs of ovules above the midrib. Flowers in May.

External Morphology

The *Cycas* is an Eastern genus. *Cycas revoluta* is found in South India and some parts of Gujarat in natural conditions. This genus differs from other members of Cycadaceae that

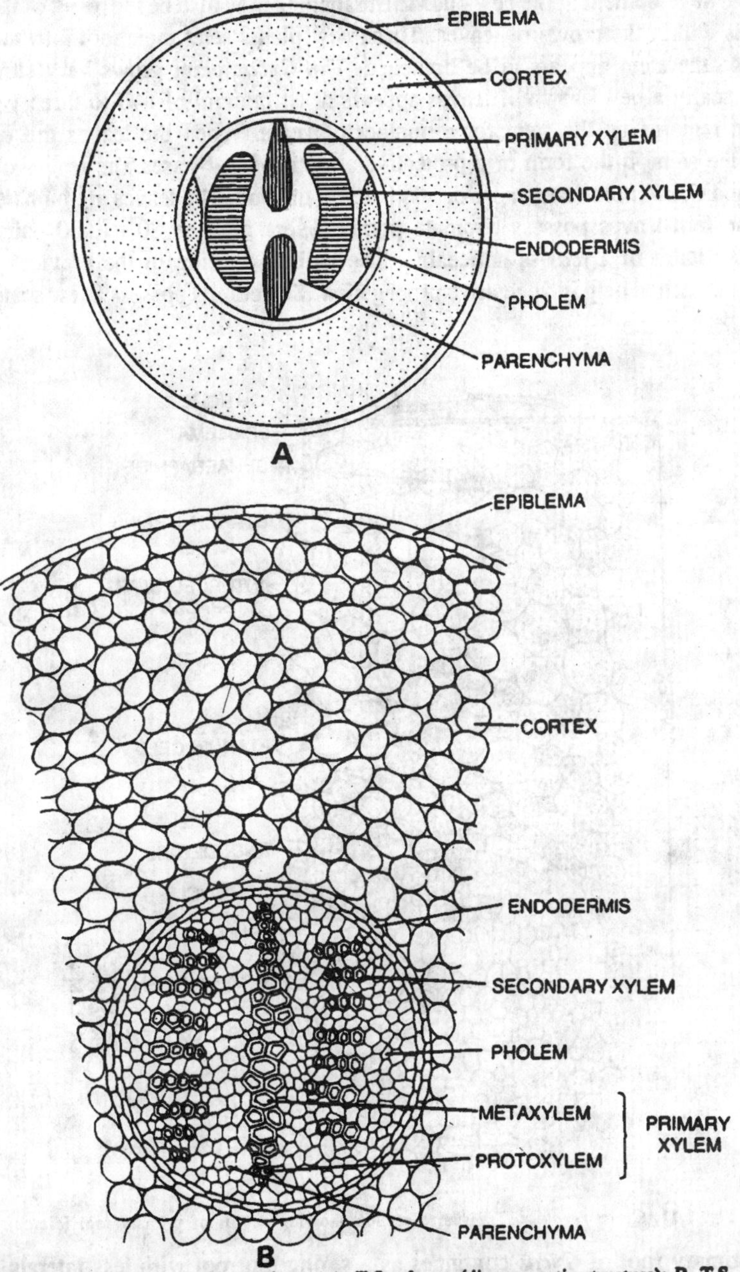

Fig. 2.10. *Cycas* sp. Anatomy of normal root. A, T.S. of root (diagrammatic structure); B, T.S. of root (detailed structure).

here female organs do not assemble in a cone while in all other genera the female structure is represented by a simple cone. In general the *Cycas* plant is unbranched and bearing crown of large fern- like leaves. The stem of all the species is tuberous when young, but in some species the tuberous body passes into the columnar stem. The stem is usually unbranched but sometimes the branches are seen. The branches are either the result of injury to the stem or it may be the development of the new bud. The branching may also be the result of the germination of the seeds within the crown of leaves. These new plants send their root into the parent plant body and so the stem appears to be branched. The *Cycas* plant grows very slowly, but lives for ages. Usually a new crown of leaves appears at an interval of two to three years, and each such crown remains on the stem for a number of years. When the leaves fall off their bases remain on the stem in the form of a protective armour of leaf-bases.The leaves of *C. revoluta, C. media* and *C. circinnalis* are more or less hairy in the bud. The leaves are pinnately compound and like the fern-leaves possess circinate ptyxis. There may be 80 to 100 pairs of leaflets. The fern-like habit of *Cycas* is noticeable. The bulbils develop in the crevices of the scales which when detached help in vegetative propagation. The bulbils possess a few scales and foliage leaves.

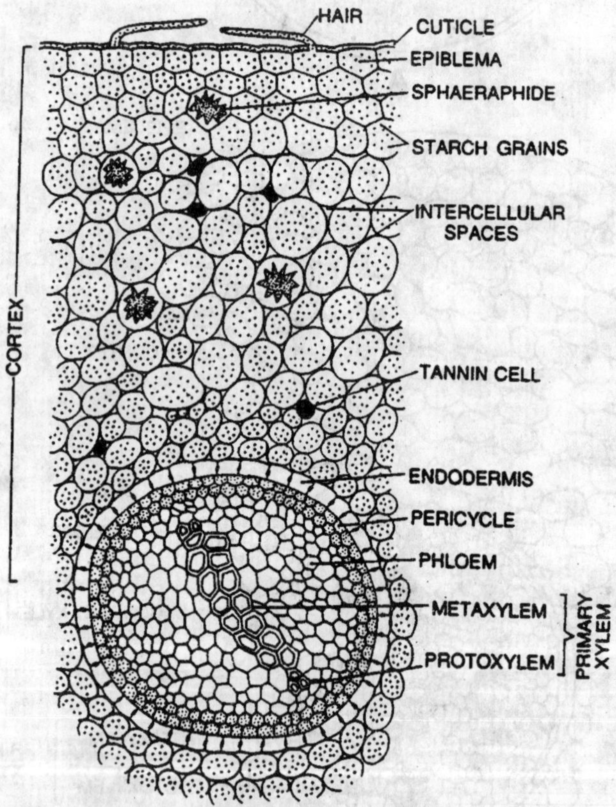

Fig. 2.11. *Cycas revoluta*. Transverse section showing details of primary root (diarch).

The primary root of *Cycas* continues as a strong tap root with less laterals. Besides the numerous small secondary soft negatively geotropic ***coralloid roots*** develop. These roots are coral-like in appearance and, therefore known as coralloid roots. These roots occur abundantly below the surface of the soil and protruding above it. Reinke described them as special organs for aeration. Schneider suggested a symbiotic association.

Schaede(1914) investigated the coralloid roots of various cycads and their symbiosis with blue-green algae. It was found that in the species investigated, the coralloid roots whether inhabited by Cyanophyceae or not, contain no bacteria as formerly suggested by Bottomley and Life and other workers. Schaede (1944) suggested that the bacteria observed by previous workers were probably introduced from the outside by an imperfect technique.

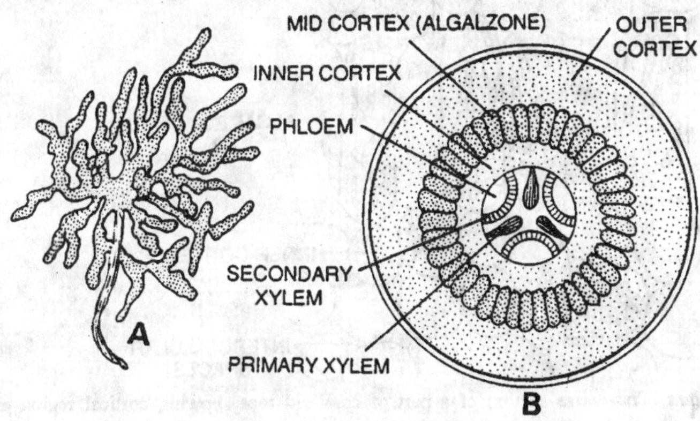

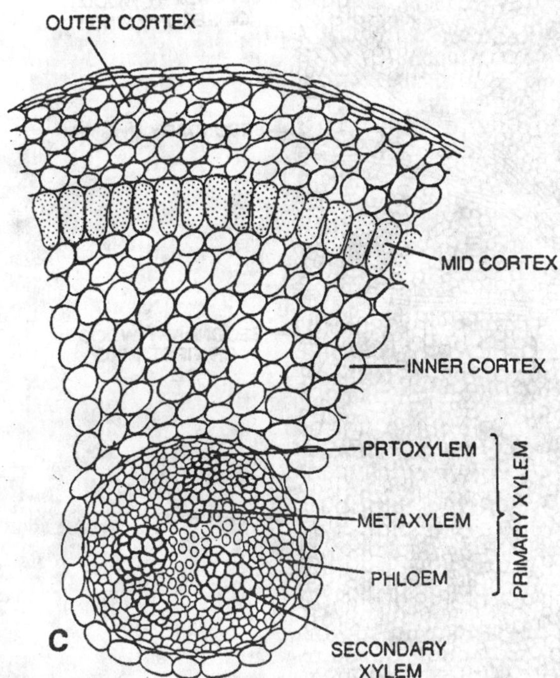

Fig. 2.12. *Cycas* sp. Anatomy of coralloid root. A, coralloid roots; B, T.S.of coralloid root (diagrammatic structure); C, T.S. of coralloid root (detailed structure).

Internal Morphology (Anatomy)

Normal root. The anatomy of normal root is very simple, the root resembles to that of

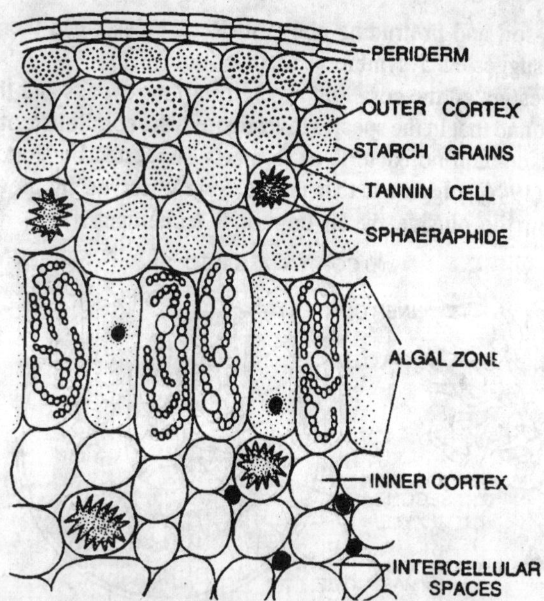

Fig. 2.13. *Cycas revoluta*. Transverse section of a part of coralloid root showing cortical region and algal zone.

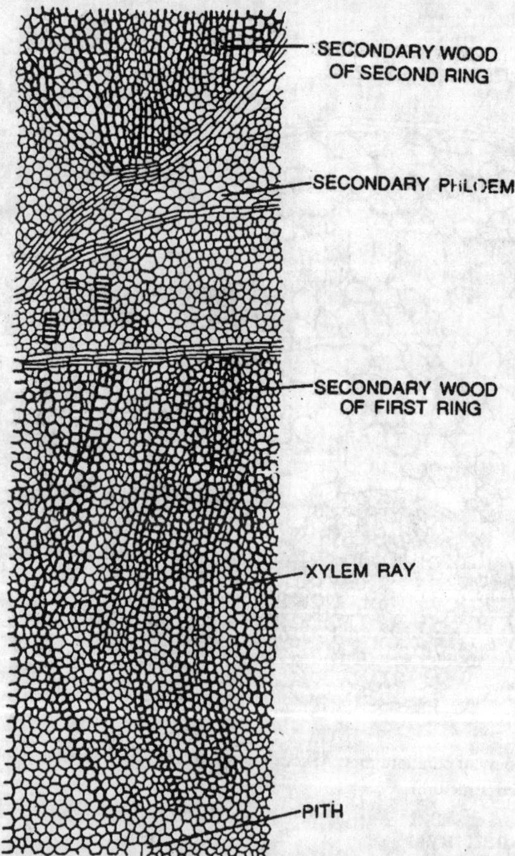

Fig. 2.14. *Cycas* sp. Transverse section of an old stem showing two rings of secondary wood.

a dicotyledonous plants in its internal structure. The outermost layer of the roots is known as **piliferous layer** possessing numerous unicellular **root hairs**. The root hairs perish soon. The roots are diarch, triarch or tetrarch. The bundles are radially arranged and the xylem is exarch. In young roots the pith is present but it soon disappears or becomes smaller as soon as the secondary xylem develops due to secondary growth. The secondary xylem is traversed by many parenchymatous rays. Outside the roots the cork is developed. The cork cambium develops from the pericyclic region.

Coralloid root. The anatomy of the coralloid root is similar to that of normal root. The differences are as follows: In coralloid root the cortex is divided into three zones, *i.e.*, the outer cortex, middle cortex and the inner cortex. In the outermost region of the root the cork is developed by cork cambium. The outer and inner cortical regions are comprised of parenchyma. The middle cortical zone harbours the blue-green alga *Anabaena cycadaeae* and is termed algal zone. The stele is diarch, triarch or tetrarch and surrounded by an endodermis which is followed by pericycle.

Schneider (1894) suggested a symbiotic association of alga and bacteria in the coralloid root. Life (1901) isolated from the roots an alga and some bacteria and believed them to be concerned with the assimilation of nitrogen as well as aeration. Bottomley could isolate nitrogen fixing bacteria from the nodules of *Cycas* roots.

Stem. The anatomy of the stem of *Cycas* resembles to the stem of dicotyledonous angiosperms. The most characteristic feature which distinguishes, the stem of *Cycas* from other gymnosperms is the presence of a large pith, broad cortex and a narrow zone of conducting

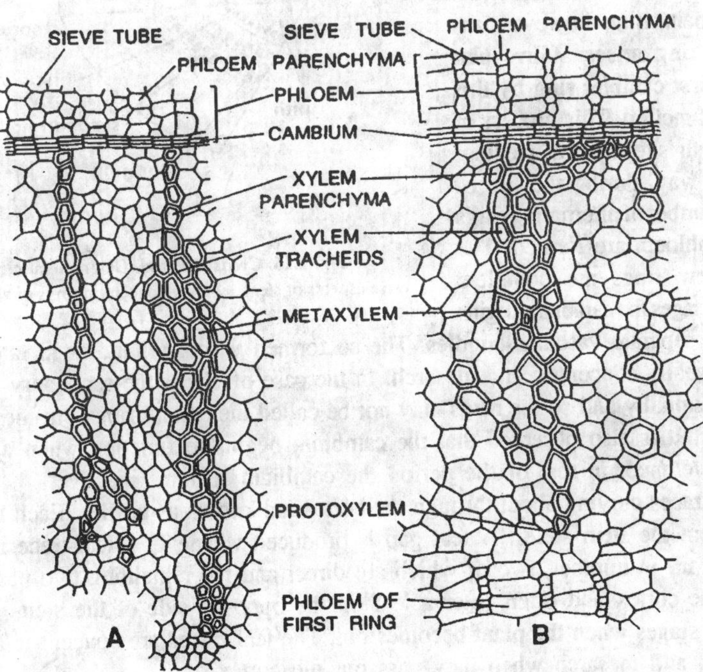

Fig. 2.15. *Cycas revoluta*. Antomy of stem. A, transverse section of a part of vascular bundle of first ring; B, T.S. of a part of vascular bundle of second ring.

tissue. The cortex is thickly clothed on the outside by the numerous leaf bases and woody scales. The systems of mucilage canals and leaf traces are found in the cortex. The leaf traces are known as the 'girdles'. In *Cycas* the leaf traces are concentric and mearch in structure. The mucilaginous canals are also found in the pith region. The canals of pith are joined to those of cortex by canals running through the medullary rays of the stem.

In central pith region there are many open, endarch vascular bundles separated from each other by medullary rays. The amount of xylem is quite less in the primary stem. The collateral and cauline medullary bundles are found distributed from the pith portion of the stem. At intervals these medullary bundles fuse with each other and form a dense network of anastamosing strands running in pith portion.

Secondary thickening in Cycas. In *Cycas* the first cambial ring remains active for a few years and cuts on xylem and phloem in usual way, *i.e.*, xylem towards inside and phloem towards outside. After sometime outside the secondary phloem ring, however, another cambial ring arises at irregular interval. The first cambial ring by this time ceases to function. This process of secondary growth is repeated at irregular intervals. This way, during the life of *Cycas* plant a number of alternating rings of xylem and phloem are formed. The first formed few rings are complete whereas in later stages the cambium begins to cut off only separate vascular bundles. The so formed vascular bundles in later stages by cambium are usually concentric and mesarch. In the case of *Cycas* the secondary growth does not take place annually and so the rings may not be called the annual rings but they are known as growth rings. It is also observed that the cambium begins to function when a new crown of leaves is developed, in rest of the period the cambium remains inactive.

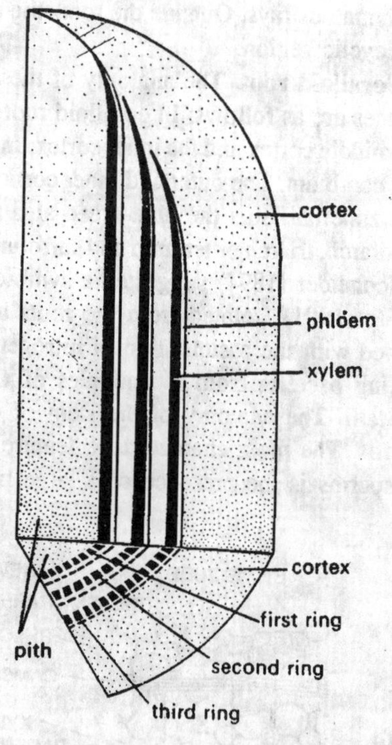

Fig. 2.16. *Cycas revoluta.* Diagrammatic radial and transverse aspects of stem showing concentric rings of vascular bundles.

The leaf traces curving about through the cortex are known as girdles. Each leaf receives two bundles from the stem stele. A leaf gap is produced whenever a leaf-trace is given off. The traces are four in number, two of which are direct and the remaining two pass round the stem through the cortex and often enter a leaf on the opposite side of the stem. The girdles develop in later stages when the plant becomes old. The leaf traces are concentric and endarch when they start, and mesarch when they pass out the cortex.

The presence of leaf gaps and concentric bundles denotes a prominent feature of pteridophytes, which shows a strong affinity between *Cycas* and ferns.

Rachis. The rachis of the leaf of *Cycas* is quite stout and woody. To study the internal structure of the rachis the transverse sections are required. The outermost layer is the single layered epidermis covered by a thick cuticle. Just beneath the epidermis there is a hypodermis which is two or three layers in thickness on the adaxial side and many layered on the abaxial surface. The cells of the hypodermis are sclerenchymatous. The ground tissue which follows

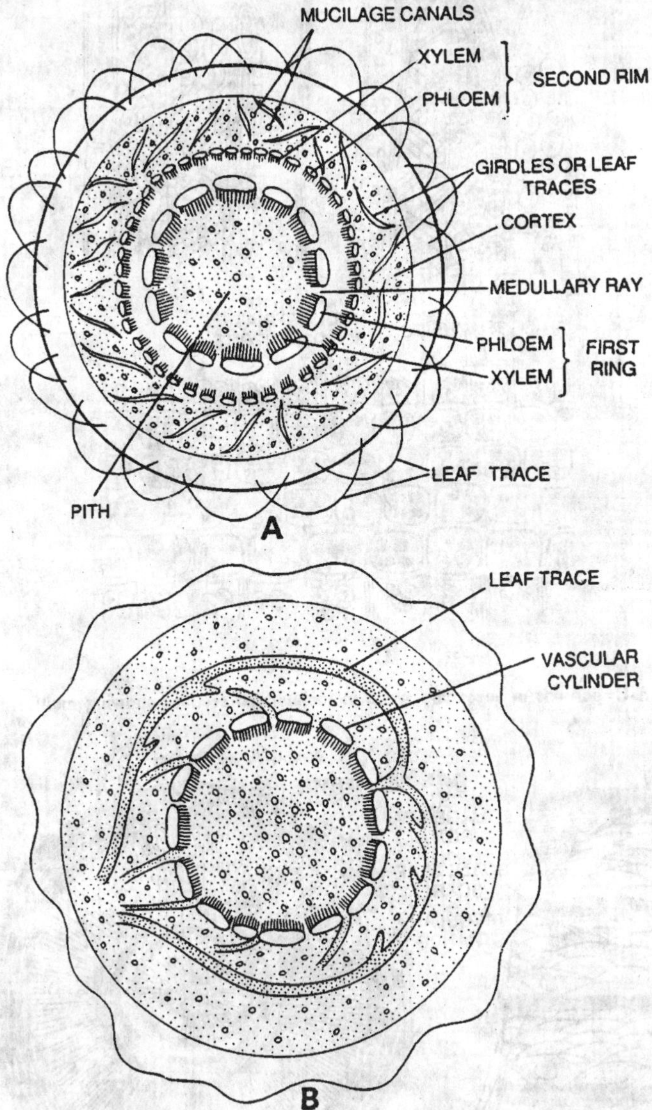

Fig. 2.17. *Cycas* sp. Anatomy of stem. A, T.S. of stem (diagrammatic); B, T.S. of stem showing the origin of leaf traces from vascular cylinder (diagrammatic).

the hypodermis is parenchymatous. The most characteristic feature of the rachis is the arrangement of vascular bundles in a omega-shaped (Ω) outline. The vascular bundles vary in their structure at the base, centre and apex of the rachis. Each vascular bundle is ensheathed within a single layered sclerenchymatous sheath. The vascular bundles are open and collateral. A single layered endodermis and one or multilayered pericycle surround the vascular bundle.

The interesting changes take place at the base of the rachis when the transition from the stem type of xylem takes place into the petiolar arrangement. The bundles are endarch at the base, *i.e.*, the protoxylem remains towards the centre. Here, the xylem develops centrifugally. The phloem lies towards the abaxial side. The cambium is found at the upper surface of the

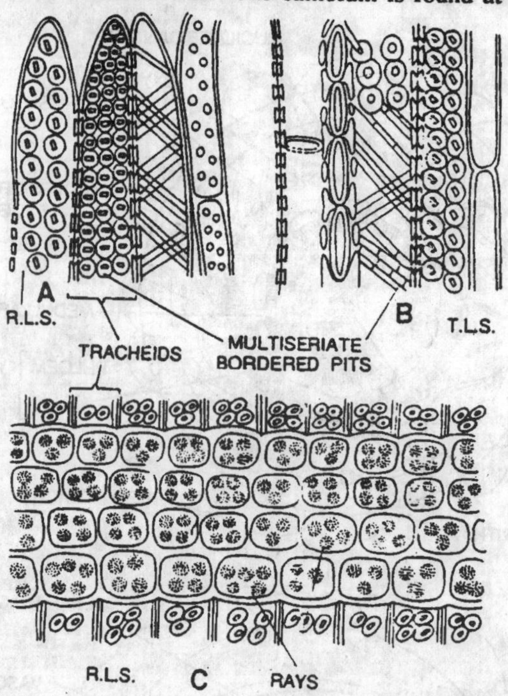

Fig. 2.18. *Cycas*. A-C, portions of secondary wood in longitudinal sections showing multiseriate bordered pits.

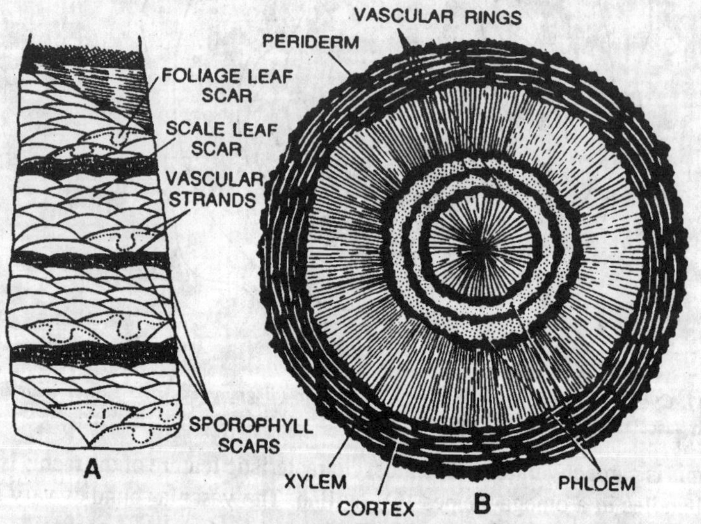

Fig. 2.19. *Cycas revoluta*. A, an old trunk showing different kinds of scars; B, transverse section of an old stem showing polyxylic condition.

centrifugally developed metaxylem. During the upward course of the bundles in the rachis the metaxylem tracheids appear opposite the protoxylem and constitute the centripetally formed metaxylem. Here, the metaxylem is present on both the sides of protoxylem and the bundles are mesarch. Simultaneously, with the appearance of the centripetal metaxylem elements the cambium begins to function on the inner side. In further upward course the centripetally developed metaxylem increases in quantity and on the other hand the centrifugal metaxylem decreases

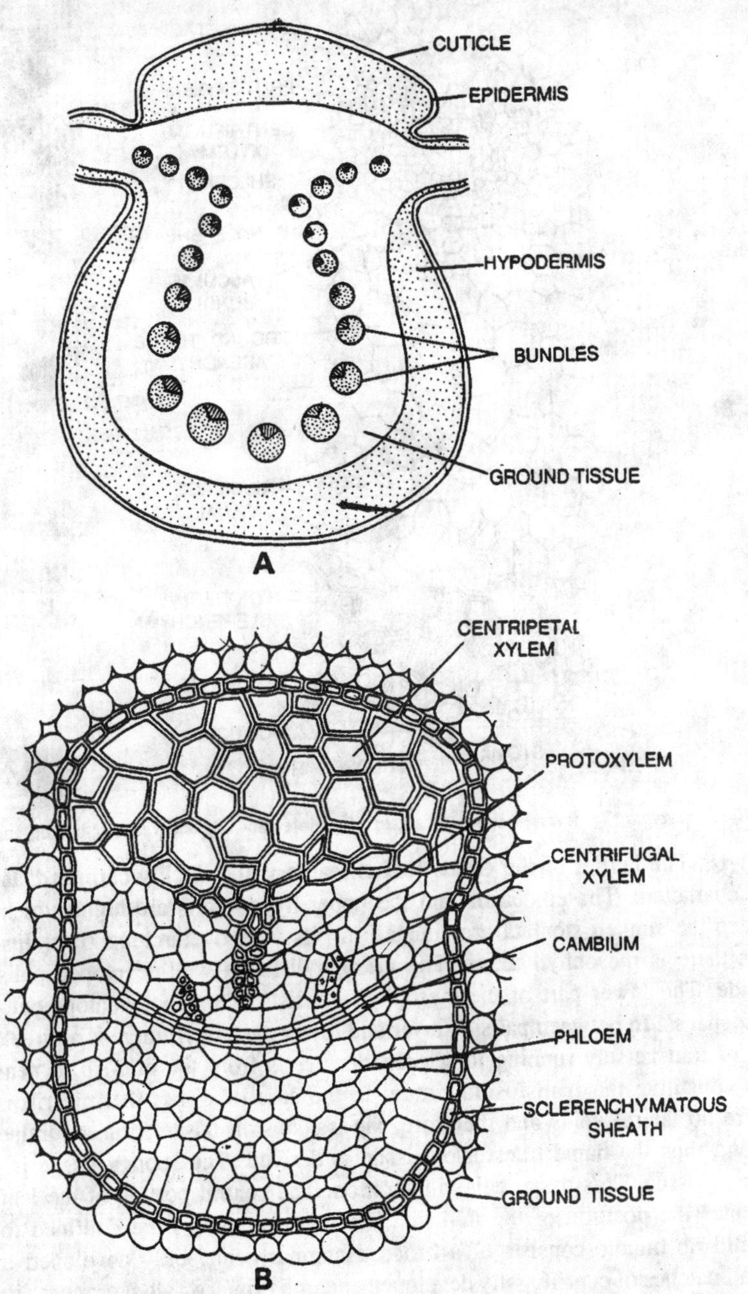

Fig. 2.20. ***Cycas*** **sp. Anatomy of rachis. A, T.S. of rachis (diagrammatic structure); B, detailed structure of a single bundle of the rachis.**

and only two patches of them remain, each consisting of a few tracheids only. All along the phloem remains towards abaxial side. This way from the base to the apex of the rachis the bundles change their position of xylem elements from endarch to exarch.

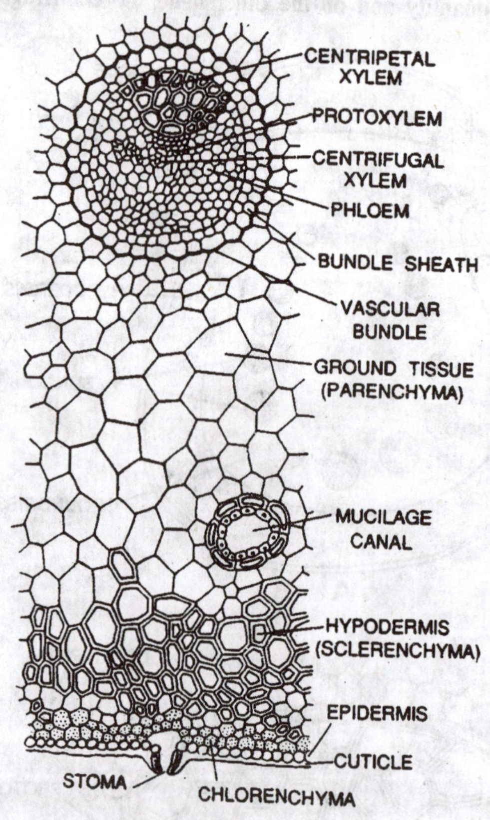

Fig. 2.21. *Cycas revoluta*. Transverse section of a part of petiole showing one vascular bundle and other tissues.

Leaflets. The leaflet of the *Cycas* leaf is leathery and thickly cutinized. It possesses all xerophytic characters. The epidermis and the upper hypodermis are highly thickened. On the lower surface the sunken stomata are situated in pits with overarching rims. Just beneath the hypodermis there is mesophyll tissue. The mesophyll has a well-developed palisade layer on its upper side. The lower part of the mesophyll consists of parenchymatous cells with scanty intercellular spaces. In between paliside layer and lower mesophyll there is a three or four celled thick layer of transversely running long colourless cells from the mid-rib to near the margin. These cells constitute the **transfusion tissue.** In the case of *Cycas* the mid rib is unbranched and there are no lateral veins and therefore, the transfusion tissue form a conducting channel for water, and thus the name transfusion tissue is derived. The spongy tissue is found below the transfusion tissue. This tissue helps in aeration. The central portion of the leaflet is bulged which indicates the position of the mid-rib. The vascular tissues are confined to the mid-rib only. The mid-rib bundle consists of a broad triangular centripetlly developed exarch xylem and two small patches of centrifugally developed endarch primary xylem groups. They are known as centripetal and centrifugal xylem patches. The presence of the diploxylic structure is the characteristic feature of this genus. The phloem is found on abaxial side. The pericycle is parenchymatous. The endodermis is single layered and sclerenchymatous which ensheaths the bundle. In between phloem and centripetal xylem there lies a cambium. The phloem consists of sieve

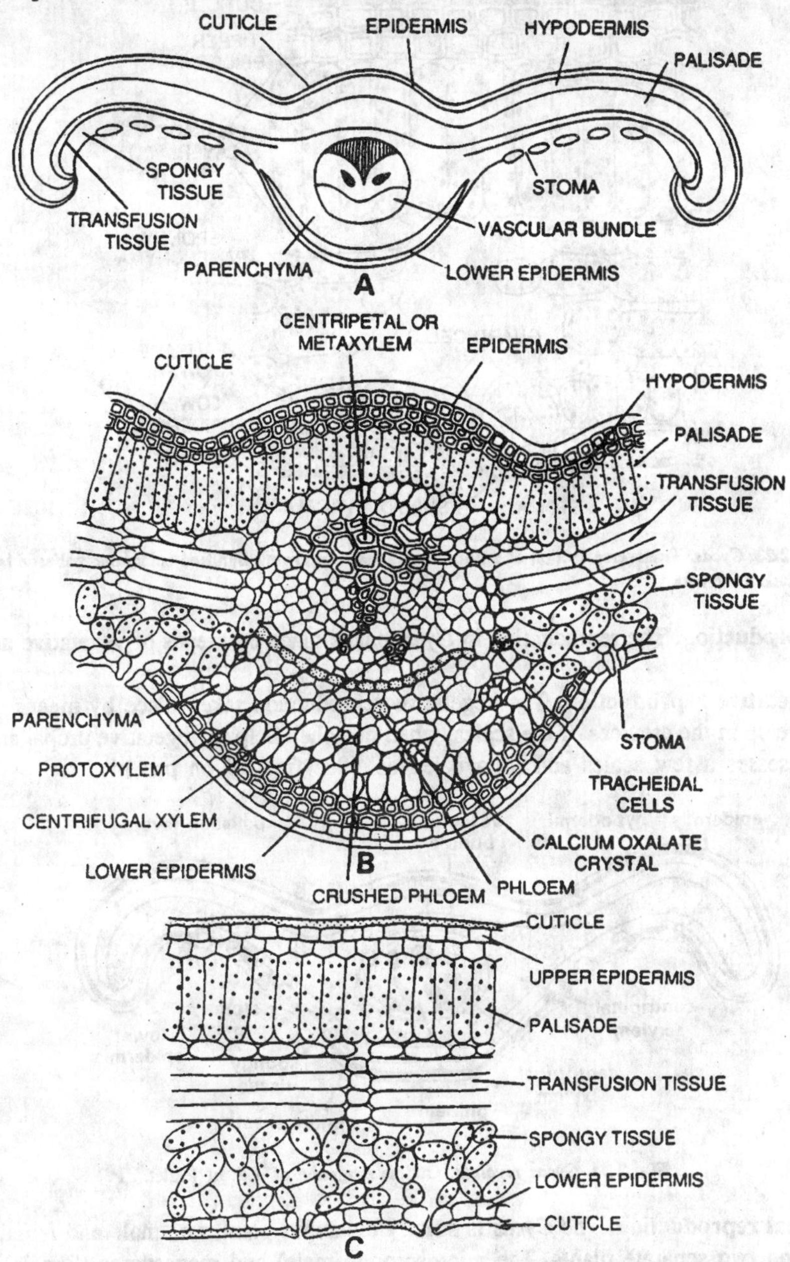

Fig. 2.22. *Cycas* sp. Anatomy of leaflet. A, T.S. of leaflet (diagrammatic structure); B, the detailed structure of mid-rib portion of the leaflet; C, T.S. of a portion of the wing of leaflet.

tubes and phloem parenchyma, the companion cells are altogether absent. Calcium oxalate crystals are found in various cells of the mid-rib. The remaining part of the bundle is occupied by parenchyma. A number of tracheidal cells are found on either side of the vascular bundle.

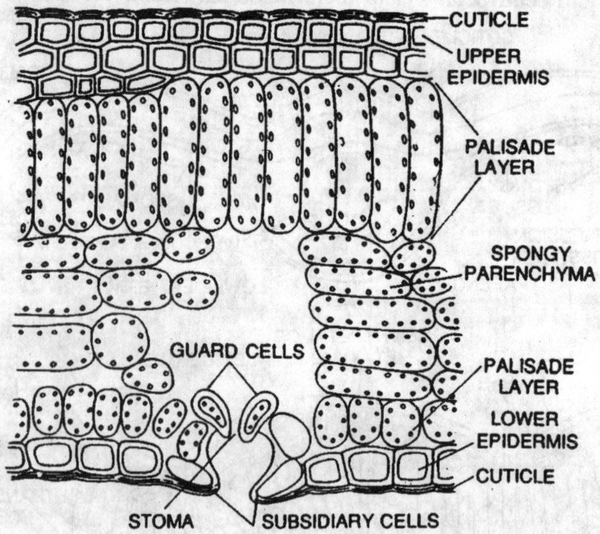

Fig. 2.23. *Cycas.* Transverse section of a portion of a leaflet showing lignified epidermis, palisade layer, spongy tissue and sunken stomata.

Reproduction. The reproduction in *Cycas* takes place by means of vegetative and sexual methods.

Vegetative reproduction. The vegetative reproduction taken place by means of bulbils which develop in the crevices of the scales, when detached help in vegetative propagation. Each bulbil possesses a few scales and foliage leaves. (See fig. 2.9 on pp. 14).

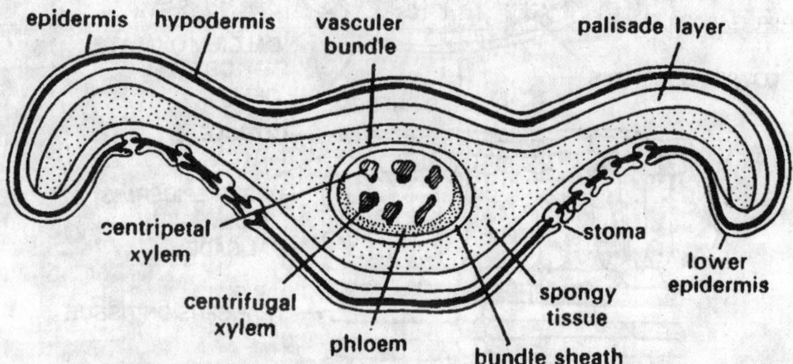

Fig. 2.24. *Cycas revoluta.* Outline diagrame of T.S. of leaflet.

Sexual reproduction. The *Cycas* is strictly dioecious plant, *i.e.*, male and female organs are found on two separate plants. The microsprores (male) and megaspores (female) develop in micro and megasporangia respectively. The male plants of *Cycas* are very rare in our country. The cones or strobili are found at the apex of the trunk concealed at first among the bases of the leaves. The male flowers are found to be arranged in erect cones. Usually a single male cone develops at the apex of male plant. The megasporophylls or female strobili are found in crowded whorls around the apex of the female plant alternating with whorls of foliage leaves.

Male cone. The male cone is about more than 50 cm. in length. Usually a single male cone develops at the apex of the plant. The sporophylls are closely imbricate and spirally arranged in a compact cone. Some of the sporophylls of the base and apex of the cone are being sterile.

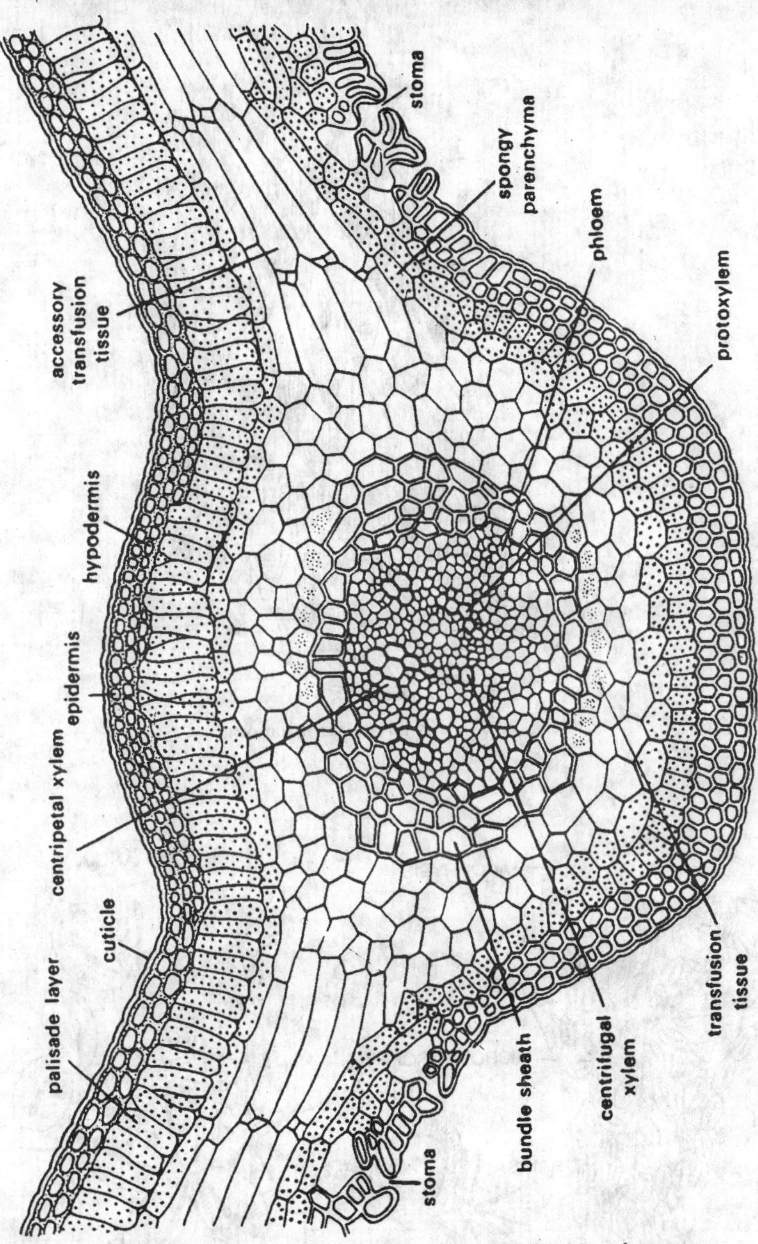

Fig. 2.25. *Cycas cercinnalis.* T.S. leaflet showing detailed internal structure.

Each microsporophyll is narrow below and broad above terminating into a projection, the **apophysis.** The microsporangia are confined to the abaxial (lower) surface of the microsporophyll, the upper surface is sterile. The sporangia may number upto one thousand in certain species. The sporangia occur on two flanks of the microsporophyll differentiated by a median sterile

Fig. 2.26. *Cycas cercinnalis*. Male cone.

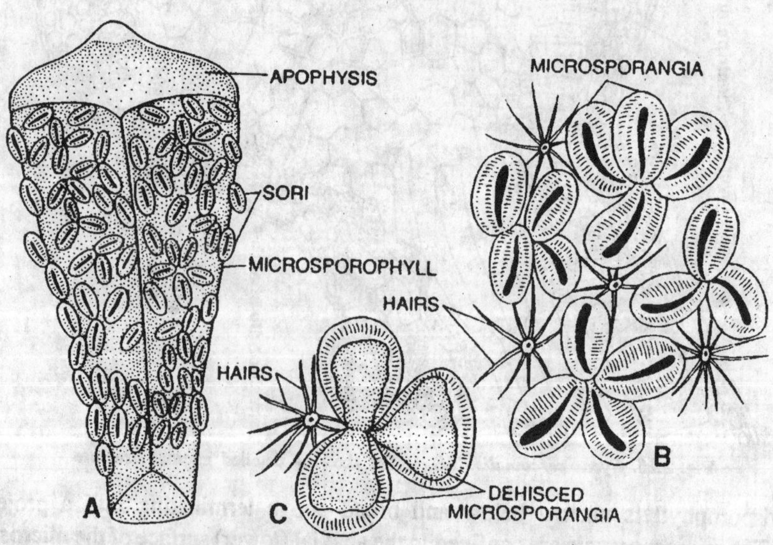

Fig. 2.27. *Cycas*. A, microsporophyll; B, microsporangia; C, dehisced microsporangia.

ridge. They are usually found in definite sori each consisting of two to six sporangia. The sporangia are more or less united at the base. The microsporangia develop from the hypodermal cells.

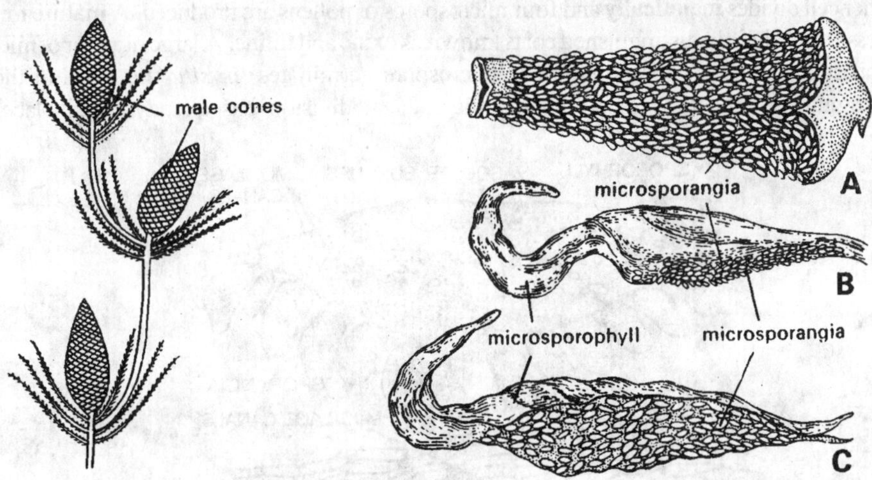

Fig. 2.28. *Cycas cercinnalis.* Mode of branching of male plant.

Fig. 2.29. *Cycas cercinnalis.* A, microsporophyll as seen from above; B-C, microsporophylls as seen in lateral view.

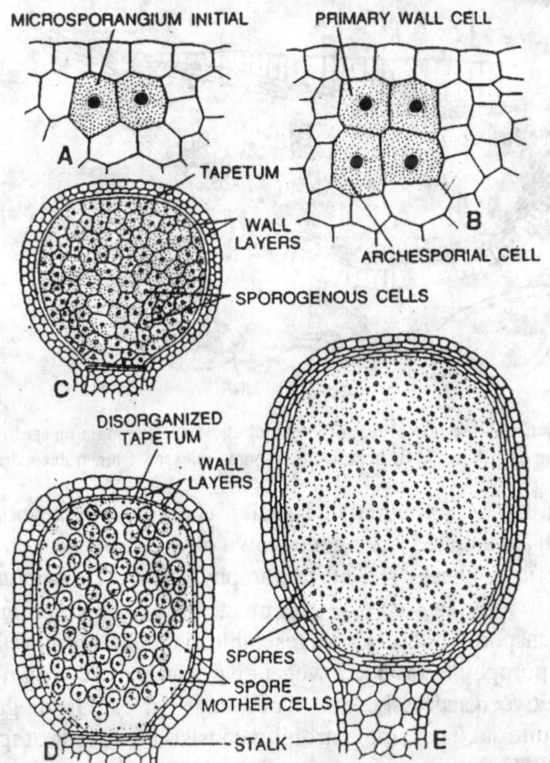

Fig. 2.30. *Cycas revoluta.* A-E, stages in the development of microsporangium.

The wall of sporangium is four to seven layered. There is definite **tapetal** layer developed from the **sporogenous** tissue for the nourishment of the microspores. The son of **microsporangia** are intermingled with hair.

The last generation of the sporogenous **tissue** forms the spore mother cells. Each spore mother cell divides **meiotically** and four microspores or pollens are produced. A mature microspore possesses two sharply distinguished coats known as **exine** and **intine.** A large number of microspores are produced in each sporangium. The microspore germinates *in situ, **i.e.,*** within the micro-sporangium. At the time of pollination the nuclear division has already taken place.

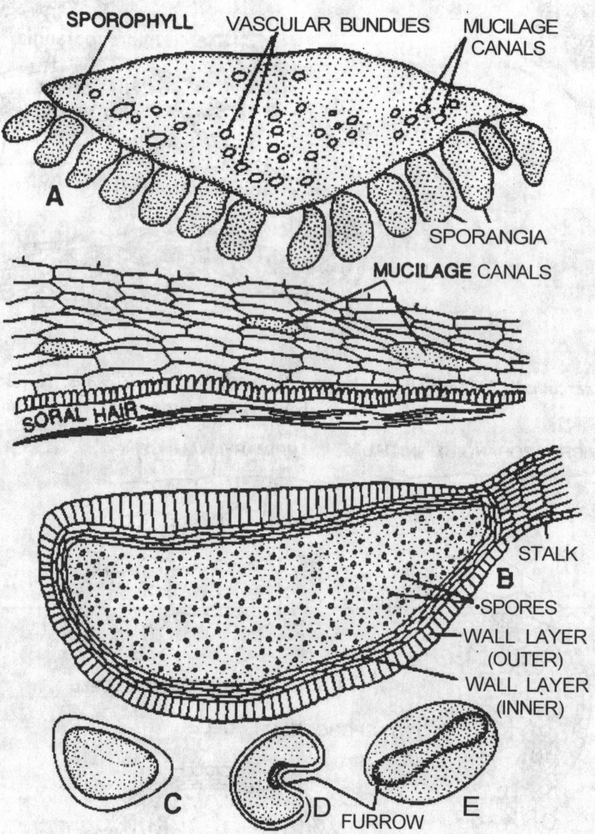

Fig. 231. ***Cycas***. A, transverse section of **microsporophyll** showing sporangia on its lower side; B, enlarged view of **microsporangium** showing its details; **C-E**, mature microspores **released** from **microsporangium.**

Female strobilus. In *Cycas* the female **strobilus** is made up of a whorl of spirally arranged megasporophylls which are situated among the crown of vegetative leaves. The **megasporophylls** are not arranged in definite cones. The **megasporophyll** of *Cycas **revoluta*** is one of Nature's curiosities. It preserves in a living seed plant the image of the seed-bearing fronds of Palaeozoic **pteridosperms.** The megasporophylls **of** *Cycas* resemble to the foliage leaves. The apical **meristem** after producing megasporophylls produces vegetative leaves. The crown of megasporophylls develops each year in between successive crown of foliage leaves. The megasporophylls develop only when the plant is quite old. The female strobilus consists of a rosette of **sporophylls** resembling reduced foliage leaves. Here, the ovules replace the lower pinnae or spines.

The megasporophylls of *Cycas revoluta* resemble too much to the vegetative leaves than in any other species of *Cycas*. In *C. circinnalis*, more reduction taken place but the pinnate

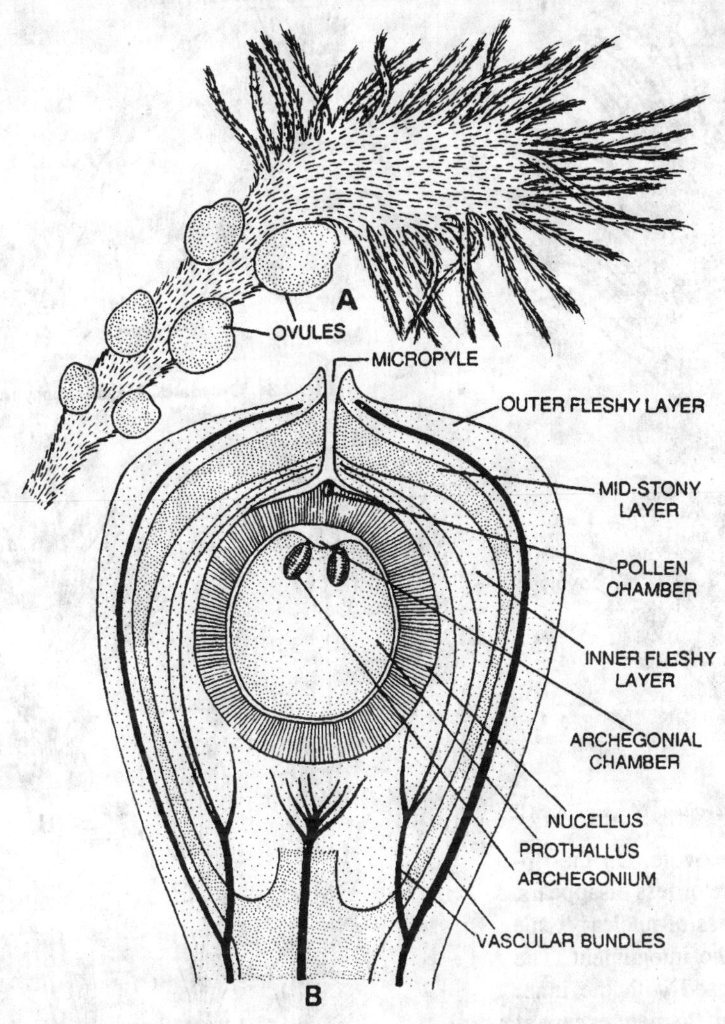

Fig. 2.32. *Cycas revoluta*. A, female strobilus; B,L.S. of ovule.

character of vegetative leaf still persists. In *Cycas revoluta* and *C. circinnalis* several ovules develop on the margins of the megasporophyll whereas in *C. siamensis* only two ovules are found. In *C. rumphii* the magasporophyll bears two to five ovules. The ovules of *Cycas* are quite large. In *Cycas circinnalis*, the largest ovules are formed. Each ovule is 6 cm. x 4 cm. The ovules of *C. circinnalis* are smooth and dark-green, whereas of *C. revoluta* are densely hairy and orange-red in colour.

The ovule. The ovules are sessile and orthotropous in their position. They are ovoid or spherical and vary in their size. The ovule of *Cycas circinnalis* may attain the length of 6 cm. Each ovule has a single thick integument which is differentiated in three layers. The inner and outer fleshy layers of the integument remain separated from each other by a hard stony layer in between the two. The inner fleshy layer remains more prominent in earlier stages of the

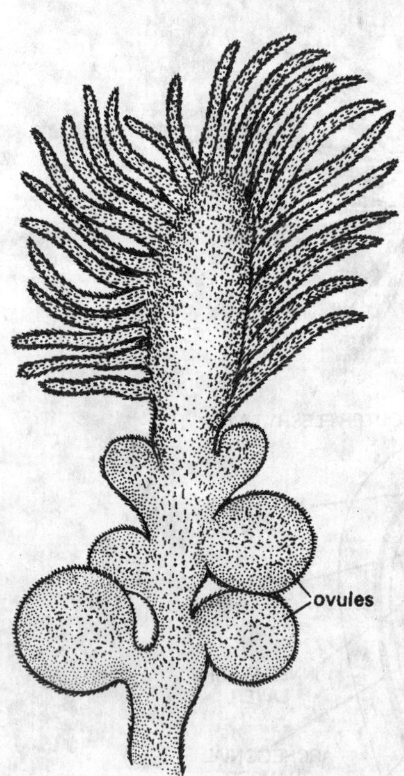

Fig. 2.33. *Cycas revoluta*. Megasporophyll.

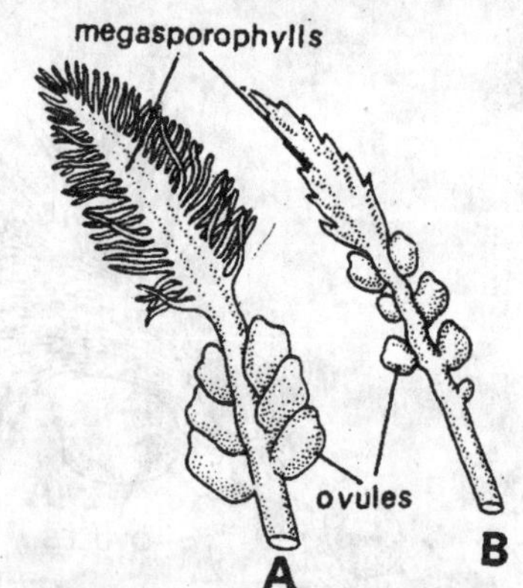

Fig. 2.34. Cycadales. Megasporophylls. A, *Cycas revoluta*; B, *C.circinnalis*.

development of the ovule. On the ripening of seed it more or less disappears.

There is a mass of nuclear tissue of nucleus within the integument. The nucellus remains fused with the inner fleshy layer of the integument except at the tip portion. In the top portion of the ovule the integument forms a long micropyle. In the apical portion the nucellus forms a nucellar beak. The beak forces its way into the micropyle, while in the centre of the beak and below it, cells break down and form the pollen chamber.

The ovule is well supplied with vascular bundles. The bundles serve the base of the ovule. Each vascular bundle

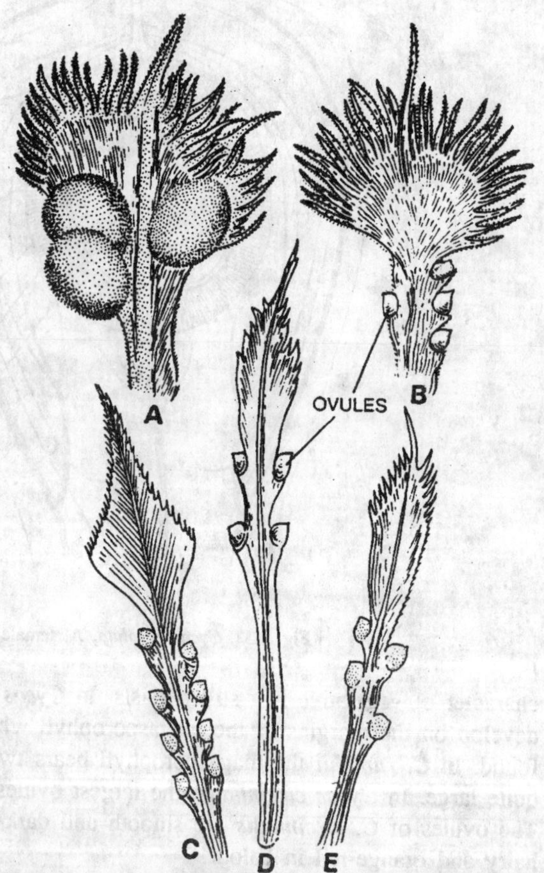

Fig. 2.35 *Cycas* sp. Megasporophylls (female strobili). A, *Cycas pectinata*; B, *C. siamensis*; C, *C. circinnalis*; D, *C. rumphii*; E,*C. beddomei*.

divides into two branches, each serving the outer and inner fleshy layers. These branches further divide for proper vascular supply.

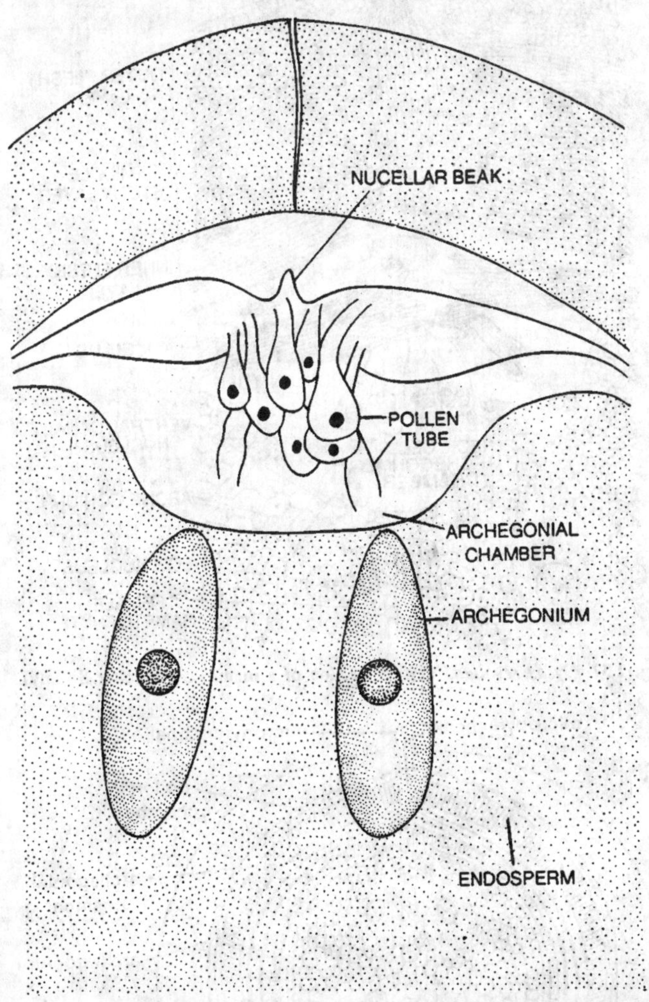

Fig. 2.36. *Cycas* sp. L.S. of the ovule showing female gametophyte with archegonia and pollen tubes.

The ovule arises as a hypodermal mass of meristematic cells. The cells above it divide rapidly and form the nucellus. The surrounding cells also divide rapidly and the nucellus becomes embedded within these cells. A thick integument develops outside the nucellus. A single large megaspore develops deep within the nucellus.

Megaspore. In the beginning all the cells of nucellus are quite-alike. They are thin-walled and parenchymatous. Soon after, one of the cells in the central region of the nucellus becomes prominent. This bigger prominent cell acts as **megaspore mother cell.** The megaspore mother cell divides meiotically and a linear tetrad of four cells is resulted. This way the megaspore mother cell represents the last stage of sporophytic generation. The three upper cells or megaspores of the linear tetrad degenerate and the basal functional megaspore increases in size. The upper

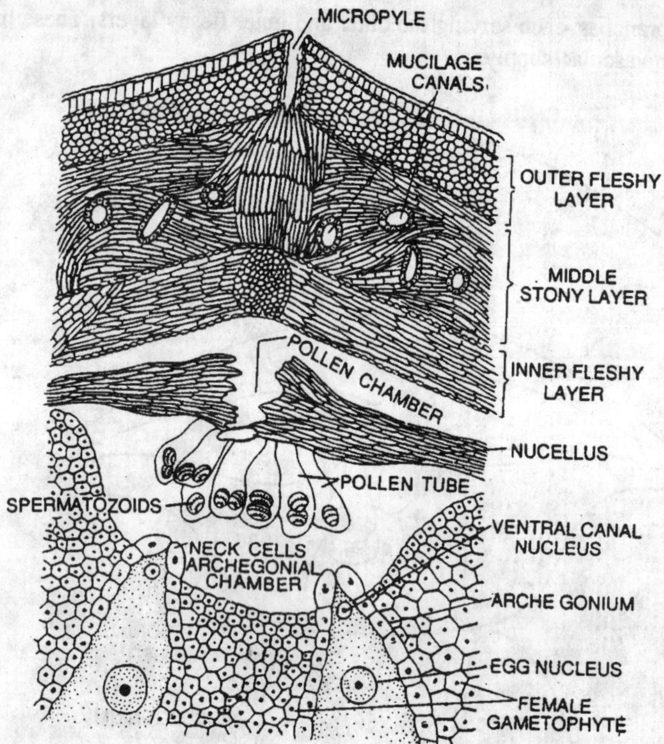

Fig. 2.37. *Cycas revoluta.* Vertical section of a mature ovule (after Swamy).

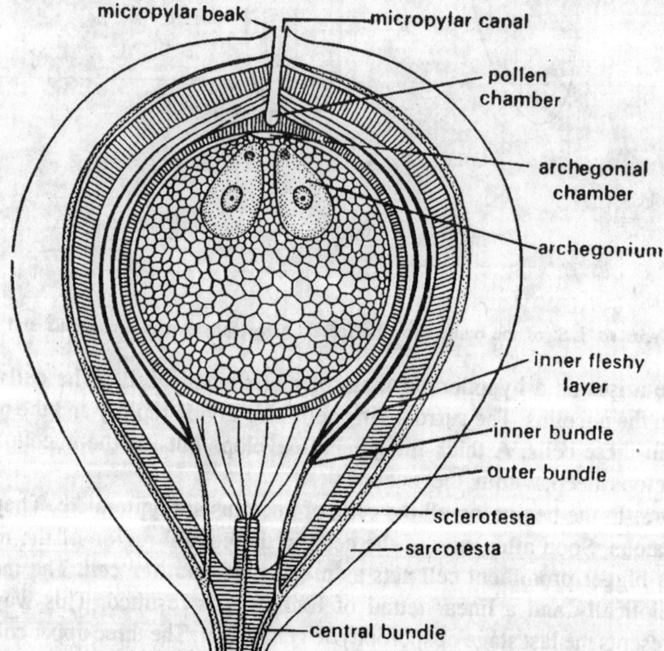

Fig. 2.38. *Cycas revoluta.* V.S. of ovule showing internal structure before fertilization and its vascular supply.

three degenerating megaspores are nutritional in function. They provide the nutrition for the development of the functional megaspore. The megaspore is also known as **embryo-sac.**

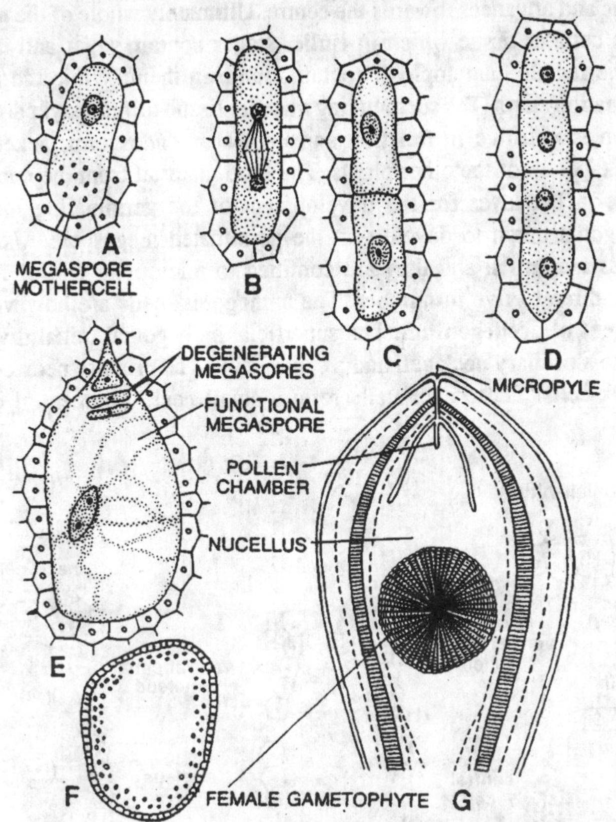

Fig. 2.39. *Cycas*. Development of female gametophyte. A-D, stages in the formation of a megaspore linear tetrad; E, lower cells of the linear tetrad develops into a megaspore; F, free nuclear division; G, cellular female gametophyte.

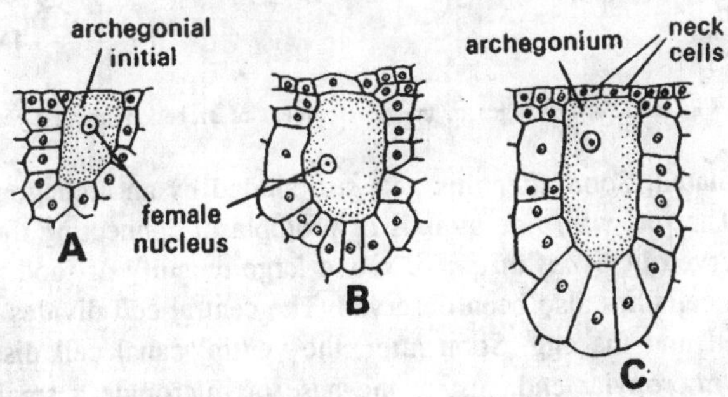

Fig. 2.40. *Cycas*. A-C, stages in the development of an archegonium.

Development of female gametophyte. The megaspore increases in size and gradually continues to absorb some of the neighbouring cells of the nucellus. The surrounding cells of the nucellus form a spongy nutritive tissue around the developing megaspore. Free nuclear division takes place until a thousand free nuclei are formed within the megaspore. A large central vacuole appears and the nuclei are now being pushed towards periphery. After the peripheral shifting of the nuclei, the wall formation begins to take-place. The wall formation begins at the periphery of the embryo sac and advances towards the centre. Ultimately whole of the megaspore of embryo sac fills up with cellular tissue which in earlier stages contain sugar and in later stages starch. Some cells of the female gametophyte contain tannin in them. While the megaspore increases in size, its germination starts. The surrounding nucellar tissue forms a layer around the germinating megaspore which is nutritive in function and known as endosperm jacket. This layer closely resembles to the tapetum of the other plants. The food material from the nucellus passes through this endospermic jacket layer for the development of the germinating megaspore.

The archegonia begin to develop on the germinated megaspore. Usually the number of archegonia is lesser than ten and they are confined to micropylar end of female gametophyte. They are usually three to five in number. The archegonial cells are however, more in number.

Development of archegonium. The superficial archegonial initial divides by a transverse wall resulting into a primary neck cell and a central cell. The primary neck cell divides vertically forming two neck cells. These neck cells form a small neck. The central cell enlarges in size

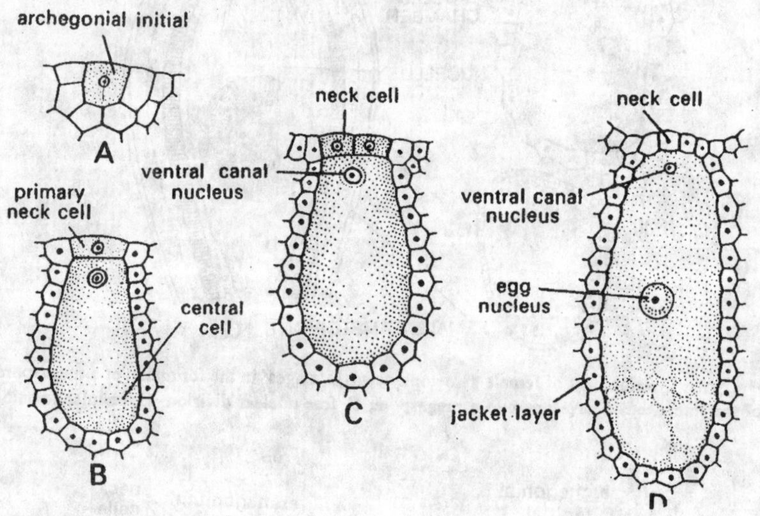

Fig. 2.41. A-D, stages in tne development of archegonium in *Cycas*.

and becomes vacuolated. Soon after this gets surrounded by an archegonial jacket which acts as nutritive layer. The pits with fine strands of protoplasm connecting the jacket cells and the egg have been observed in *Cycas circinnalis*, and large quantity of food passing through these pits into the central cells has also been observed. The central cell divides into two cells known as ventral canal cell and the egg. Soon after, the ventral canal cell disappears.

Towards the micropylar end, just at the base of micropyle a small depression appears at the place where the archegonia are developed. The surrounding endospermic tissue is gradually absorbed and the mature archegonia open into a cavity, the **archegonial chamber.** The overlined nucellar tissue has already been absorbed; the megaspore membrance has already been ruptured and the archegonia open into the archegonial chamber.

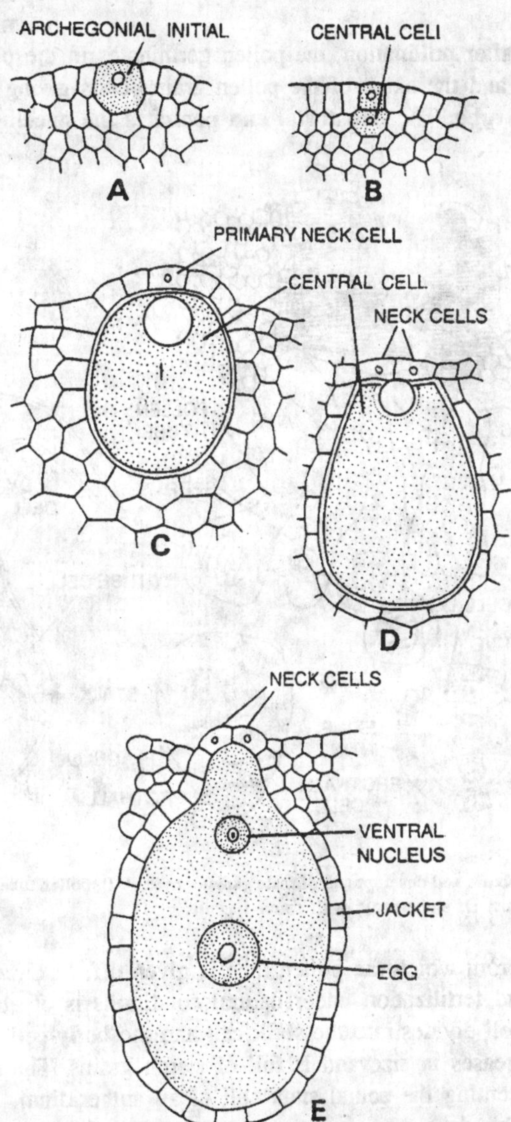

Fig. 2.42. *Cycas* sp. A-D, successive stages in the development of archegonium; E, mature archegonium.

Pollination and development of male gametophyte. The *Cycas* is wind pollinated. At the time of pollination, the nuclear division of the microspore results in the formation of a small persistent prothallial cell and a large antheridium initial. Soon after the nucleus of the antheridium initial divides into two unequal cells. The small cell which remains in close association of the prothallial cell is known as **generative cell,** whereas the other bigger cell is the **tube-cell.** This way, the microspores are three-celled at the time of pollination. At this stage some of the nucellus cells degenerate forming a **pollen chamber**. A mucilaginous pollination drop oozes out from the micropyle. In the liquid, a large number of pollen grains (microspores) are caught and, as it dries up they are drawn into the pollen chamber. Later on the apex of nucellar beak becomes very hard.

About a week after pollination, the pollen germinates in the pollen chamber. The tube cell increases in size and the exine of the pollen grain bursts giving the way to the intine in the form of a **pollen tube.** The **haustorial end** pentrates the nucellus. The haustorial end is

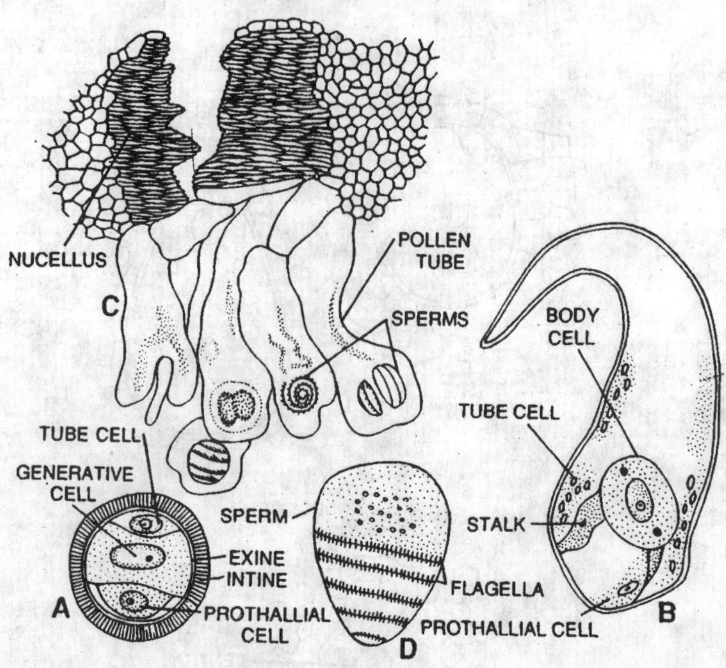

Fig. 2.43. *Cycas.* Structure and development of male gametophyte, A-B, pollen tube development; C, pollen tubes, developing inside the nucellus; D, antherozoid with spiral flagella.

the end of the pollen from which the pollen tube is given off. In *Cycas revoluta,* the time in between pollination and fertilization under natural conditions is of about four months.

The generative cell divides into the **stalk cell** and the **body cell.** The stalk cell does not divide further, but increases in size and is full of starch grains. The stalk cell is sterile cell and regarded as representing the actual stalk cell of an antheridium. The body cell enlarges sufficiently, however, this does not contain starch grains. Two blepharoplasts appear at the two poles of the nucleus of the body cell. The blepharoplasts are small, star-shaped and protoplasmic bodies. They appear a few months before the division of the body cell. The **blepharoplasts** are concerned in the production of cilia; they become vacuolated and break up into a mass of granules which fuse and finally give rise to five or six spiral bands along with numerous cilia are arranged. In the earlier stages the blepharoplasts lie in the plane of long axis, but later on with the enlargement of the pollen tube and body cell they rotate through 90° and come to lie transverse to the long axis of the pollen tube.

Ultimately the body cell divides into two forming two **antherozoids.** The sperms of *Cycas* are the largest among plants. They are easily visible to the naked eye. The movement of the sperms begins even when they are still within the mother cells. The sperms are liberated and they swim actively within the pollen tube. The pollen tube becomes turgid and ultimately ruptures. Ikeno described sperms of *Cycas* to be completely naked. According to Swamy (1948) the pollen tube does not burst but simply an aperture in formed at the end of the tube and the spermatozoids are squeezed through it by amoeboid movements.

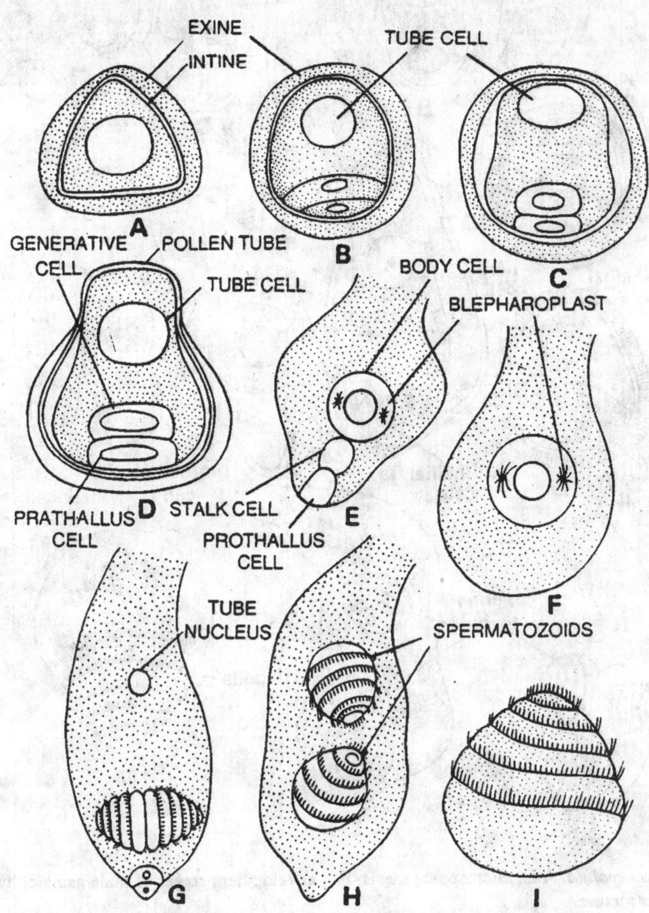

Fig. 2.44. *Cycas* sp. A-G, successive stages in the development of male gametophyte; H, two spermatozoids. I, one mature spermatozoid.

Fertilization. On breaking of the pollen chamber through the base of the nucellar cap, the archegonial chamber and the pollen chamber merge together forming a continuous cavity. The archegonial chamber does not contain any liquid but is moist. On rupturing the pollen tube the liquid is discharged and the sperms swim in this liquid till they enter the necks of archegonia. In earlier stages of the development the pollen tube acts as a haustorium. The spermatoziods swim in the liquid discharged from the pollen tubes and enter the necks of archegonia with a sufficient force. The entire sperm enters the egg but soon after the cytoplasmic sheath and ciliated band are removed and the naked male nucleus moves to the large egg nucleus. The male and female nuclei are united and the fertilization is effected. With the result of union the oospore is formed. The oospore is diploid (2n) and this represents the beginning of the sporopphyte generation. The fertilization which is furnished by means of a pollen tube is known as **siphonogamy.** In *Cycas*, the sperms are motile and, therefore, pollen tube acts as a haustorium and not as a sperm carrier.

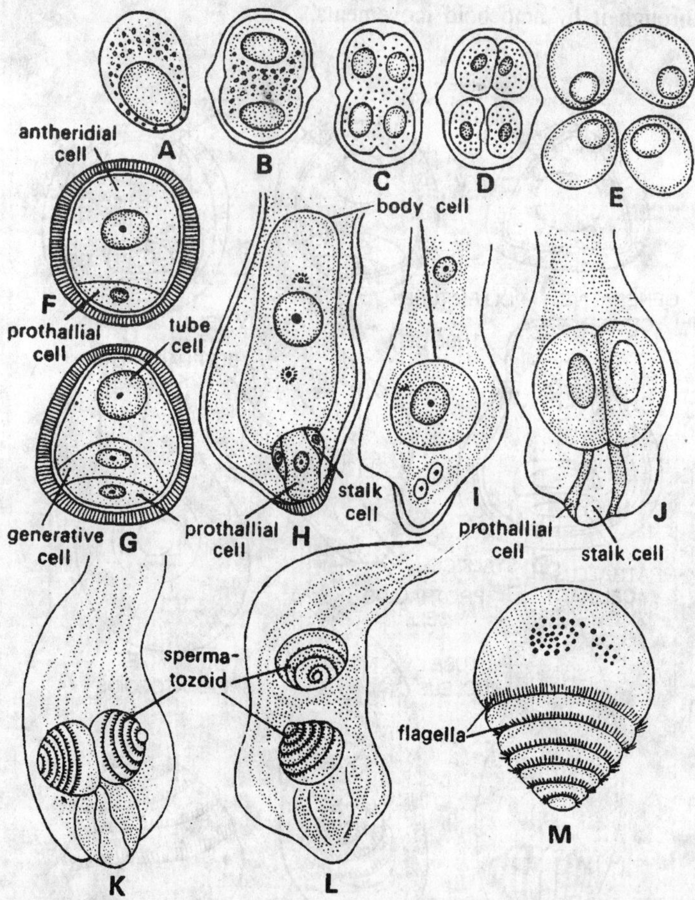

Fig. 2.45. *Cycas revoluta*. A-E, microsporogenesis; F-J, development stages of male gametophyte; K-M, formation of spermatozoids and structure.

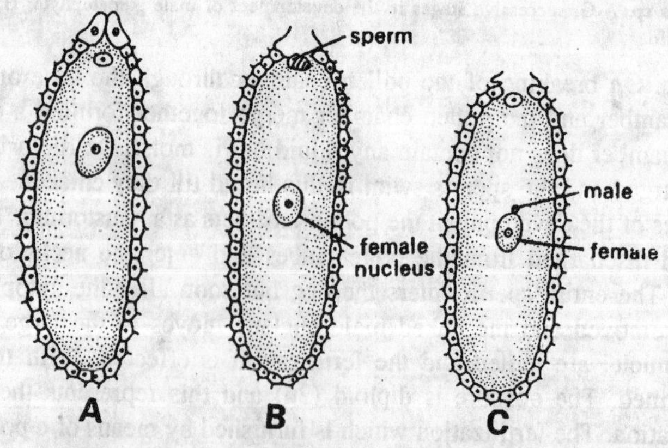

Fig. 2.46. *Cycas* spp. A-C, stage in the process of fertilization.

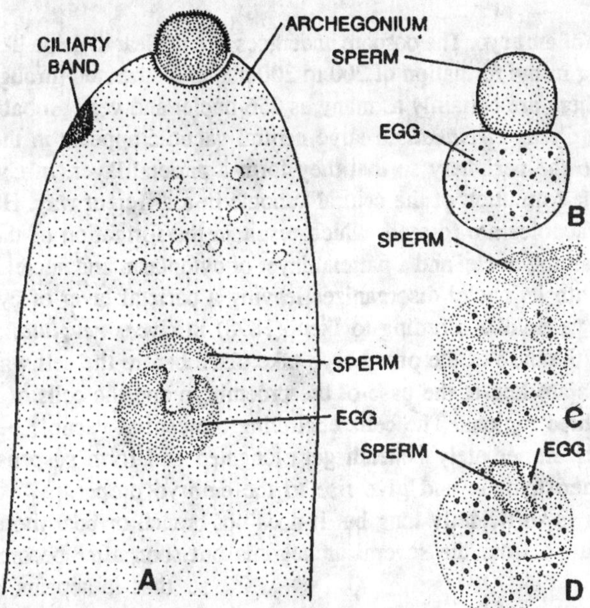

Fig. 2.47. *Cycas revoluta*. Stages of fertilization. A-B, preliminary stages; C-D, gametic fusion.

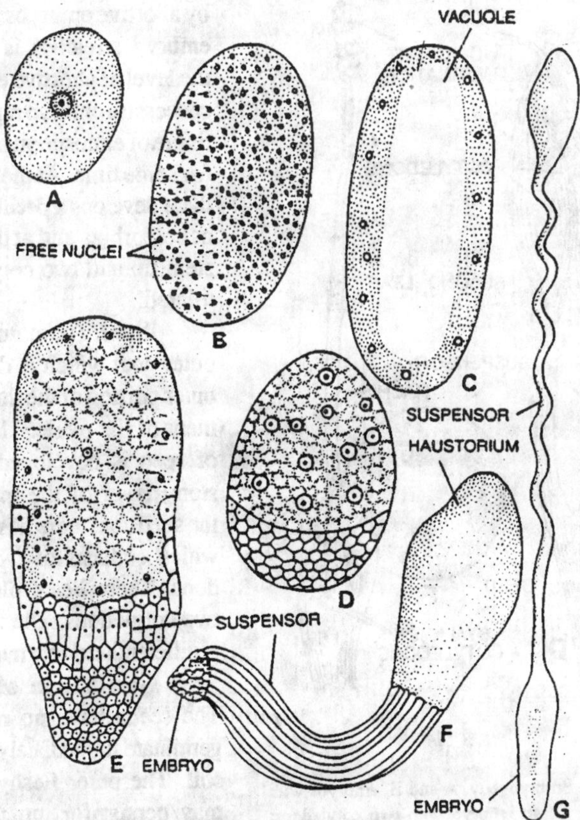

Fig. 2.48. *Cycas* spp. A-F, successive stages in the development of embryo; G, embryo.

Development of embryo. The oospore undergoes free nuclear division like that of germinating megaspore, resulting in the formation of 200 to 300 nuclei distributed throughout the cytoplasm. In *Cycas revoluta* there are certainly as many as 256 nuclei and very probably more. According to Treub (1884), in *Cycas circinnalis* a large central vacuole appeers in the centre and all the nuclei are pushed to the periphery so that they form a parietal layer. In *Cycas revoluta*, Ikeno in 1898 observed the formation of the central vacoule in a different way. He stated that a large number of small vacuoles are formed, which result in the formation of the central portion of the proembryo containing nuclei and a parietal layer of cytoplasm and nuclei. The central portion of the proembryo with its nuclei disorganizes leaving a parietal layer of cytoplasm and nuclei and a large central vacuole. According to Ikeno(1898) in *Cycas revoluta*, simultaneously free nuclear divisions at the base of the proembryo take place before the cell walls begin to appear, permanent cell walls are laid at the base of the proembryo and the cells of this region become differentiated into three regions. The cells encircling the central vacuole become more or less haustorial; the cells immediately beneath get elongated and form suspensor; while the cells at the tip remain meristematic and give rise to the embryo proper.

The suspensor is sufficiently long but it does not branch or bear more than one embryo. The suspensors separate from the several archegonia but soon after become twisted together, so that it appears a single suspensor composed of two, three or more suspensors. The extra suspensors are terminated by abortive embryos; only the successful embryo stretches is suspensor beyond the level of abortive embryos. The single successful embryo rapidly grows into the tissue of endosperm and increases in size for some time. At the base of the embryo there develops a peculiar structure known as coleorhiza and at the opposite end the stem tip and two cotyledons are differentiated.

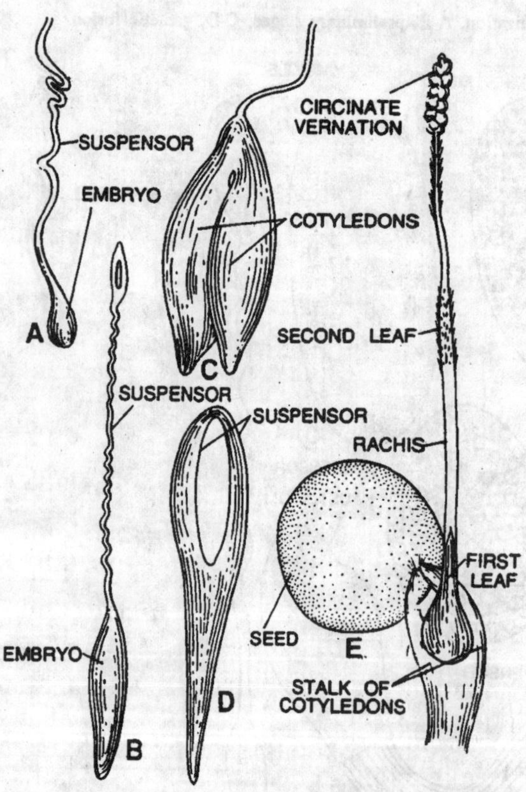

Fig. 2.49. *Cycas*. Embryogeny. A and B, embryos with long and coiled suspensors; C, an embryo with two cotyledons; D, young embryo with two suspensors; E, young seedling.

Seed. The mature seed has a fleshy outer coat which is developed from the outer portion of the three-layered integument of the ovule. The inner stony coat of seed is developed from the middle stony layer of the integument, within the seed coat there lies a straight embryo, which consists of two unequal cotyledons. There lies a plumule in between two cotyledons. The axis of the embryo is differentiated into hypocotyl and radicle.

Gemination of seed (seedling). The seeds have no resting period and germinate immediately if placed on moist soil. The outer fleshy coat of the seed may persist for months even after the seed has started its germination. The seed is large and longer than broad. It falls on the soil with long axis parallel to the earth. It absorbs water from the moist soil and the coleorhiza comes first and the tip of the primary root makes its way through the coleorhiza. The major part of the cotyledons comes out of the seed, but the tips of the cotyledons remain within the seed for considerably a long time till the whole of endosperm is transformed on to the cotyledons. The leaves begin to develop singly on the new plants and it has been observed that for a number of years only one leaf is developed at a time. It is after several years that a crown of leaves begin to develop at the apex of the stem. In the beginning the number of the leaves is quite less and after several years the normal crown of the leaves is formed. In *Cycas*, the leaves during development show circinate vernation like those of ferns.

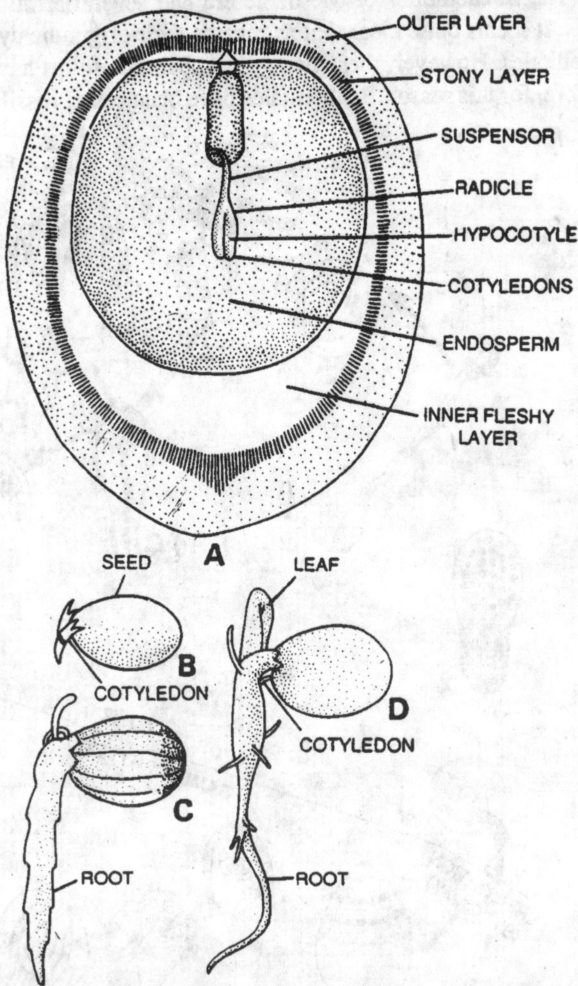

Fig. 2.50. *Cycas* **sp. A, L. S of seed; B-D, successive stages in seed germination.**

Origin and relationsips. According to Arnold (1953), the modern cycads are the remnant of a much larger vascular plant group that during the past ranged far beyond its present limits. Every genus probably evolved independently, within or near its present general range and none arose from any of the others. The genus *Cycas* has been considered the most primitive because of the lack of definite ovulate strobili.

The Cycadales are believed to have evolved from pteridosperms (Cycadofilicales) during the later part of the Carboniferous period, but the particular pteridosperm which served as ancestor has not been identified.

CYCAS IS A LIVING FOSSIL.
(Relationships to other gymnosperms)

Cycadofilicales resembled with the Filicales and Cycadales. On one hand the Cycadofilicales resembled the ferns in their general habit, vascular anatomy and occurrence of sporangia in sori while on the other hand they resembled with the cycads in the structure of their ovule. The ovules in both the cases possess three-layered integument, the two vascular strands traversing the outer and inner fleshy layers and the nucellar beak with a distinct pollen chamber.

The Bennettitales of Mesozoic era had several affinities with Cycadofilicales, ferns and cycads. It seems quite clear that as regards their origin the cycads are connected with the extinct Bennettitales. However, these two gymnospermic groups have so many features in common and therefore, for this reason Bennettitales have been called fossil cycads. The stem of the Cycadales,

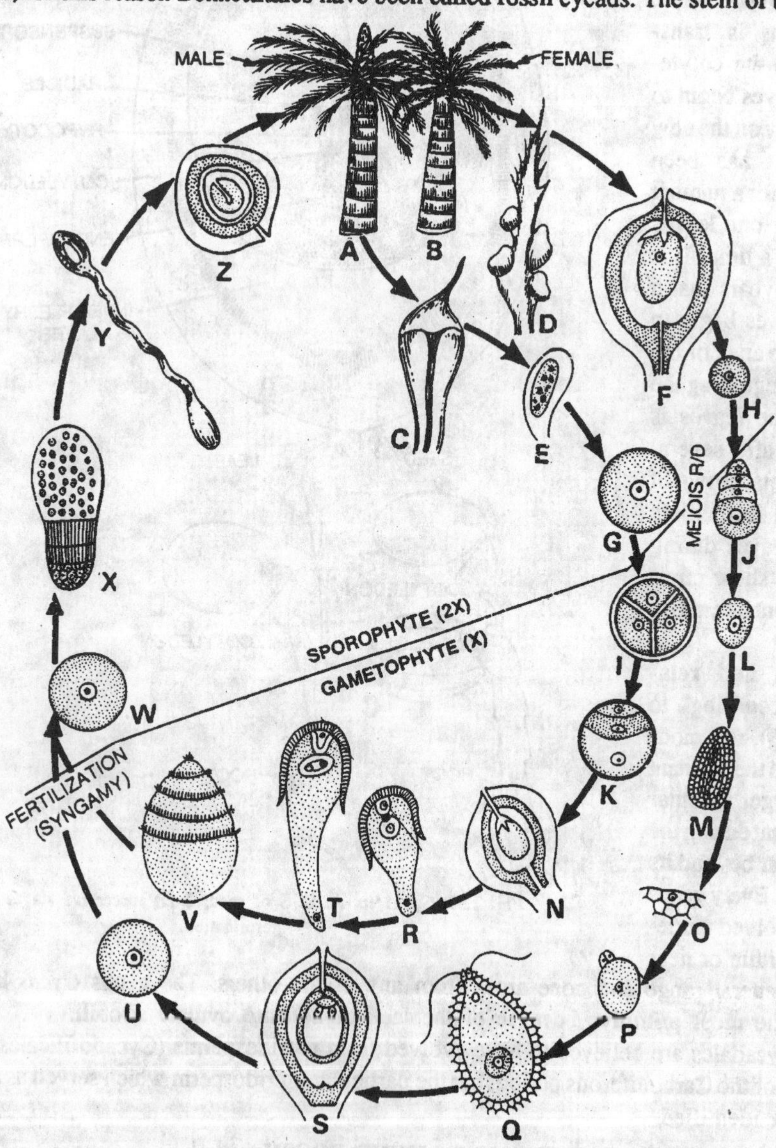

Fig. 2.51. *Cycas* sp. Diagrammatic life-cycle; A, male plant; B, female plant; C. microsporophyll; D, female strobilus; E, microsporangium; F, ovule; G, microspore mother cell; H, megaspore mother cell; I, microspore tetrad; J, megaspore linear tetrad; K, male gametophyte; L, megaspore; M, female gametophyte; N, ovule; O, archegonial initial; P, young archegonium; Q, mature archegonium; R, germinating pollen; S, germinating pollen in pollen chamber; T, germinating pollen; U, egg; V, spermatozoid; W, oospore; X, young embryo; Y, embryo; Z, seed.

like that of the Bennettitales, is an endarch siphonostete with a narrow zone of wood, the xylem in other aerial parts of the plant being mesarch. The dicotyledonous embryo of Cycadales is also suggestive of that of Bennettitales. The Bennettitales began to come in prominance as the

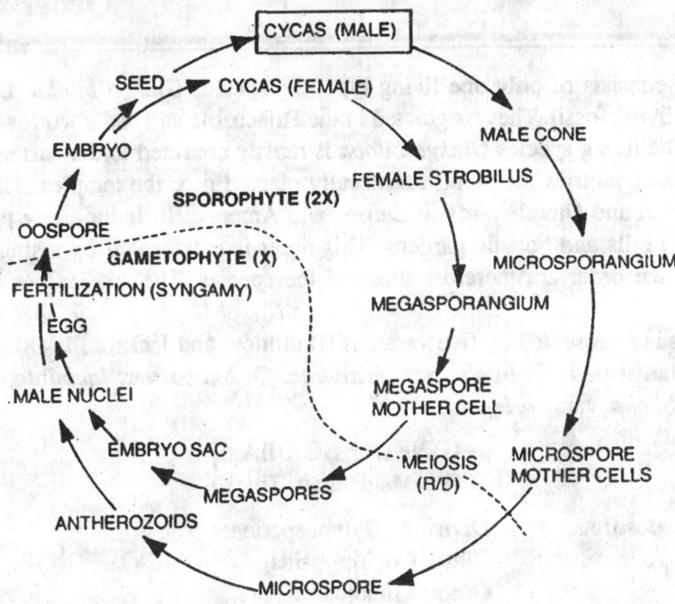

Fig. 2.52. *Cycas*. Graphic life-cycle.

Cycadofilicales declined, with the result Bennettitales carried forward many of the primitive features of the Cycadofilicales.

Cycadales best flourished in Mesozoic. The Cycads resemble the ferns in their general habit, vascular anatomy, structure and venation of the leaves, occurrence of the sporangia in sori, structure of the microsporangia and multiciliate sperms. All the above mentioned characters and the structure of ovule, are common to the Cycadofilicales, Bennettitales and Cycadales. Hence it is thought that three orders are closely related with each other and represent a distinct line of evolution reaching far back into the Palaeozoic. *Cycas* being the most primitive of living Cycadales, on account of its loose and fern like megasporophylls, stand nearest to the fossil seed plant line and therefore, the *Cycas* (cycads) is thought to be a 'living fossil'.

Economic importance. The pith of *Cycas revoluta* yields sago and the fruits can be eaten, being rich in protein and soluble non-nitrogenous substances. In Europe, the leaves of *Cycas revoluta* after silvering and treatment in various ways are made into funeral wreaths, there the leaves are called 'palm leaves'. The leaves of *C. circinnalis* are used in making mats in South India. The young shoots of *C. circinnalis* are eaten, and sago is also obtained from the trunk. For obtaining the sago, the trunk is cut into disks which when dry are pounded into flour and thereafter thrown into water where the starch settles down. The large fruits of *C. circinnalis* yield annually about the same quantity of starch. The hill tribes of Assam eat the seeds and tender fleshy shoots of *Cycas pectinata*. The fleshy stem of *C. pectinata* is pounded and used as a hair wash for diseased root hair. The sago is also extracted from the trunk of *Cycas rumphii*. The cooked fruits of *C. rumphii* are eaten by Andamanese tribes while the uncooked fruits are poisonous.

3

Class-Coniferopsida

Order-Ginkgoales

This order consists of only one living type, the species *Ginkgo biloba*. Like cycads this is also called the living fossil. They originated in the Palaeozoic and had a worldwide distribution in the Jurassic. The living species *Ginkgo biloba* is mostly confined to the eastern part of China and Japan. In these countries the plants are usually planted near the temples. This plant is also grown in the garden and church yards in Europe and America. In India a few plants are found to be cultivated in hills and botanic gardens. This order is represented by a single species, and the characters of the order are therefore those of the species. The species has been discussed here in detail.

The species is represented by five varieties (Dallimore and Jackson, 1948). These varieties are : *G. biloba* var. *Aurea, G. biloba* var. *fastigiata, G. biloba* var. *laciniata, G. biloba* var. *pendula* and *G. biloba* var. *variegata.*

GINKGO BILOBA
(The Maidenhair Tree)

Systematic position.

Division. Gymnospermae
Class. Coniferopsida
Order. Ginkgoales
Family. Ginkgoaceae
Genus. *Ginkgo*

Morphological features. This is a graceful, strictly pyramidal tree, reach nearly 100 feet in China but rarely half that height in Europe. With age it acquires a rounded head by the upgrowth of the branches. The branches are dimorphic, *i.e.*, they consist of long and dwarf (spurs) shoots as found in *Pinus.* There is in fact no hard and fast line between the two types of shoots, and dwarf shoots may sometimes grow into long ones. The long shoot elongates rapidly and bears scattered leaves on it while the dwarf shoot elongates very slowly and bears a crown of leaves above and leaf scars below. They possess long tap root that penetrates deep in the soil.

The leaves are petiolate and large while the scale leaves are small in size. The leaves are bilobed and hence the specific name *biloba.* The leaves are produced in a spiral order but on the spurs the internodes are suppressed, so that, the leaves seem to arise from the same level. These spur leaves are usually undivided. The leaves which spring from long shoots are deeply lobed into two. The leaves of the long shoot that arise from the seedlings are more deeply lobed. Generally the leaf venation is of dichotomous type. The lower margin of the leaf is thickened, the thickness increasing toward the petiole, so that, the blade is really shaped like a flat spoon.

Anatomy of stem. Anatomically the stem of *Ginkgo biloba* resembles with that of *Pinus.* The protoxylem consists of spiral tracheids. The later-formed tracheids are all of the bordered pit type. The pitting is irregular, often only in a single loose row. The pits may also be arranged in two rows either opposite or alternate. The thickenings, known as *Bars of Sanio* are prominently found between the pits in the secondary wood. The trabeculae are also found here and there traversing the interior of the tracheids.

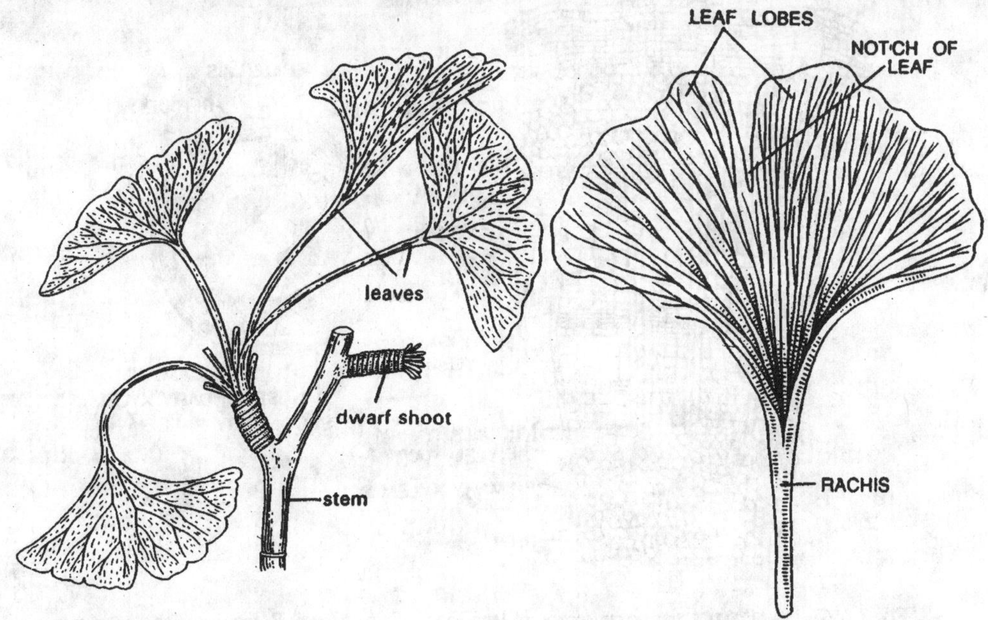

Fig. 3.1. ***Ginkgo biloba*. A dwarf shoot bearing leaves.**

Fig. 3.2. ***Ginkgo biloba*. A foliage leaf showing apical notch, leaf lobes and veins.**

Little secondary thickening is found in the spurs. There is an external phellogen; a heavy periderm is found towards outer side of the phellogen whereas a secondary cortex or phelloderm on the inner side of the phellogen.

The medullary rays are exceptionally small. Each medullary ray is one cell broad and seldom more than three cells high, except in the spurs. In spurs the medullary rays traverse the tracheids they connect to them by large simple pits.

The mucilage cavities are abundantly found throughout the tissues.

The annual rings are also found. The wood is wholly composed of tracheids. A very small amount of xylem parenchyma is also found. Here the cells contain calcium oxalate crystals.

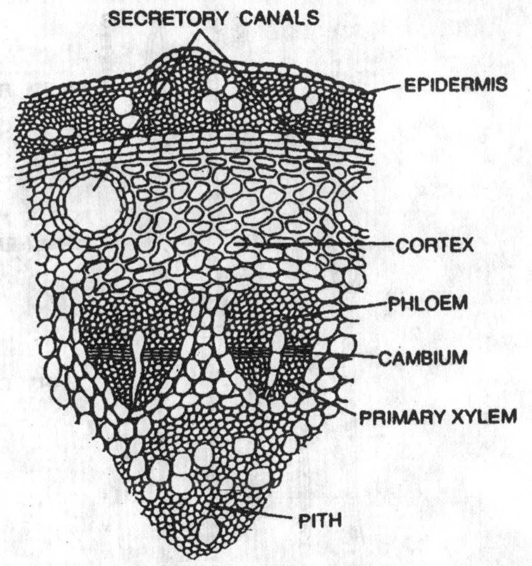

Fig. 3.3. ***Ginkgo biloba*. Transverse section of a young dwarf shoot showing secretory canals in the cortical region and a ring of vascular bundles around the pith.**

Around the vascular cylinder there is an imperfect ring of stone cells forming the pericycle. The pericyclic zone remains accompanied by an inner ring of thickened and gelatinized cells.

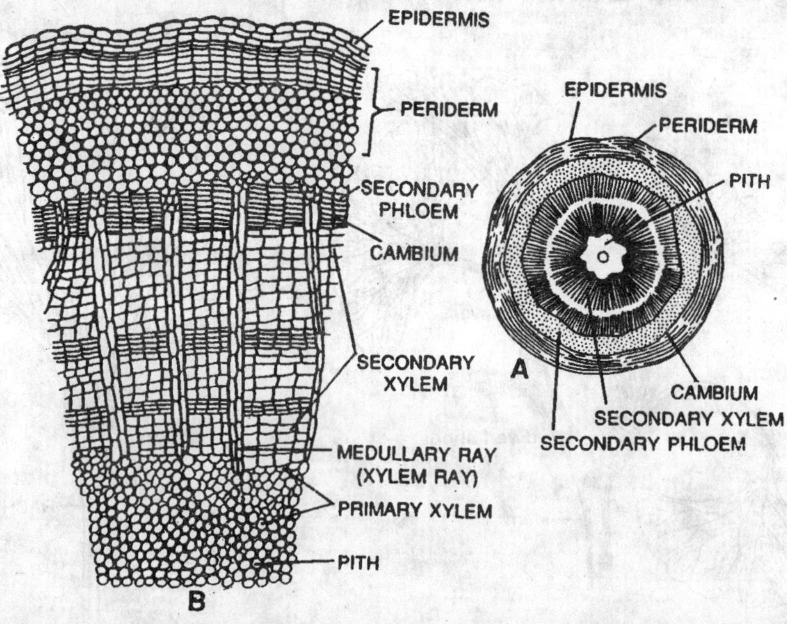

Fig. 3.4. *Ginkgo biloba*. A transverse section of long shoot (diagrammatic) B, transverse section of a long shoot showing secondary growth (detailed structure).

The older stem contains broad phloem with prominent large sieve tubes. The phloem contains discontinuous zones of fibres and a small amount of parenchyma.

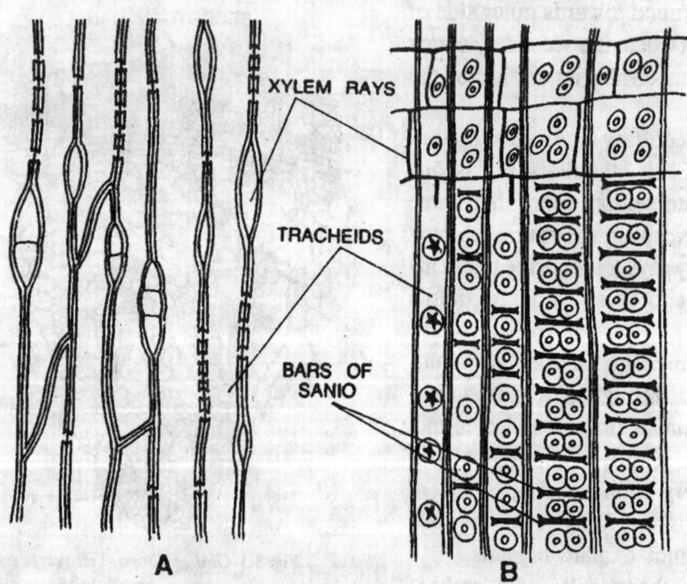

Fig. 3.5. *Ginkgo biloba*. A, T.L.S. of wood showing tracheids and xylem rays; B,R.L.S, of wood showing Bars of Sanio, tracheids and xylem rays.

Anatomy of leaf. The leaf consists mainly of two parts, the petiole and leaf lamina. The anatomy of these parts is an follows:

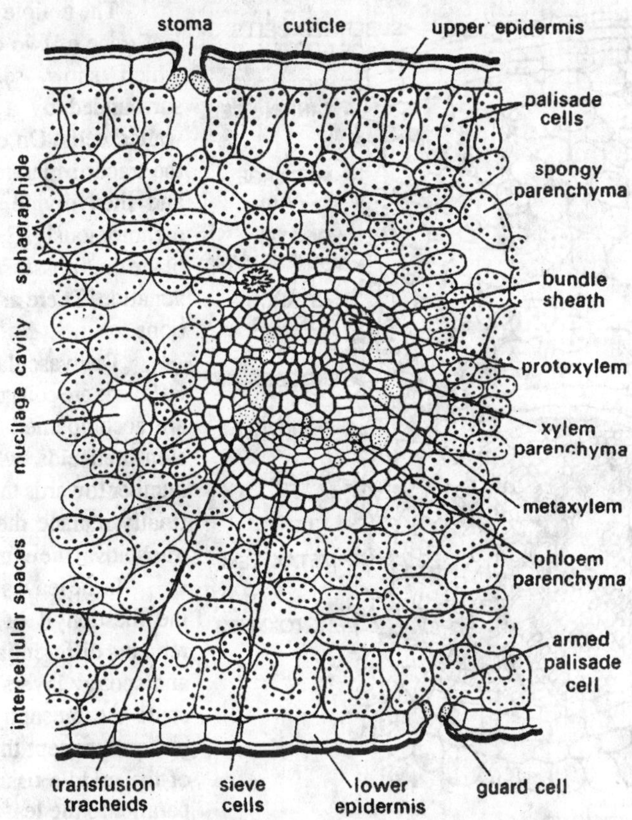

Fig. 3.6. *Ginkgo biloba*. Transverse section of the foliage leaf.

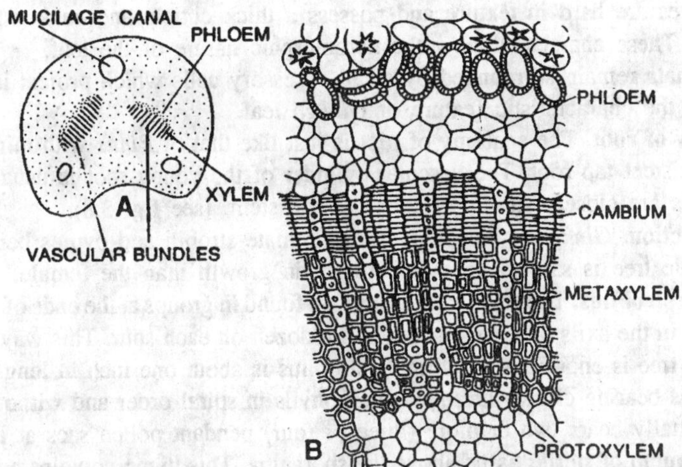

Fig. 3.7. *Ginkgo biloba*. A, T.S. of petiole (diagrammatic); B, T.S of petiole (detailed structure).

The petiole remains surrounded by a single-layered epidermis. In side the epidermis there is well developed parenchyma. The mucilage canals are found here and there in this parenchyma.

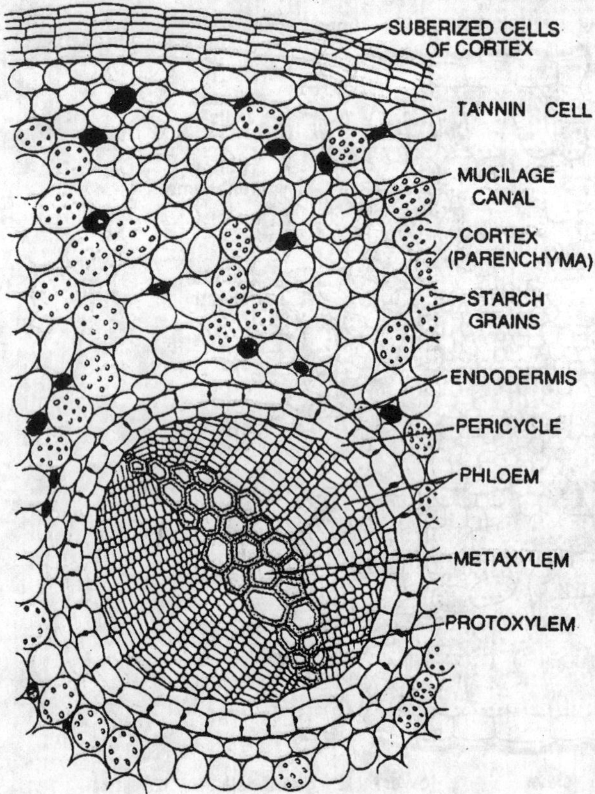

Fig. 3.8. *Ginkgo biloba*. Transverse section of young root.

The petiole contains a double leaf trace of two collateral bundles which show secondary growth, surrounded by a sheath of thick-walled cells. On entering the blade the traces separate to right and left and then divide again and again dichotomously. Each half of the leaf blade thus possesses an independent venation. There are no side connections.

The vascular bundles of the leaf blade are collateral. These bundles are accompanied by short transfusion tracheids which increase in number towards the leaf margin and finally replace the ordinary xylem tracheids at the extremity of the vein.

In the leaves of the long shoots the mesophyll of the leaf lamina remains differentiated into palisade and spongy layers. In the leaves of spurs the mesophyll is practically uniform, except that the loose cells of the middle tissue are drawn out parallel to the leaf surface and may be compared with the transfusion tissue of *Cycas* leaflet. Long mucilage canals may be seen like dark streaks between the veins.

The leaves are hard in texture and possess a thick cuticle and sunken stomata on the lower surface. These characters show the xeromorphic nature of the leaf.

The stomata remain surrounded by small accessory cells which project inward over the stoma. This is the characteristic feature of *Ginkgo* leaf.

Anatomy of root. The anatomy of root is just like that of *Pinus* with minor differences. There is a persistent tap root. The vascular cylinder of the young root is usually diarch. The vascular strands are closely similar to those of the stem. (see fig. 3.8).

Reproduction. *Ginkgo* is dioecious, the staminate strobili and ovules being on separate plants. The male tree is said to be more upright in growth than the female.

The male strobilus. The staminate strobili are found in groups at the ends of short branches. They are borne in the axils of leaves, about half a dozen on each spur. This way, the aggregate number for the tree is enormous. Each male strobilus is about one inch in length and consists of a slender axis bearing closely arranged sporophylls in spiral order and without bracts. Each sporophyll normally bears two or more (three or four) pendant pollen sacs at its apex. There is a prominent hump of sterile tissue above the sporangia. This hump contains a large mucilage cavity. The hump is a protective thickening of the sporophyll which covers the young sporangia during their early development.

The whole strobilus looks very much like a small catkin. Its axis contains a circle of 9-10 vascular bundles which gives off a double trace of small collateral bundles, to each microsporophyll.

The sporangial wall opens by a longitudinal slit. The middle layer of the wall of sporangium, the endothecium, bears cuticularized thickenings and is active in causing dehiscence. In this

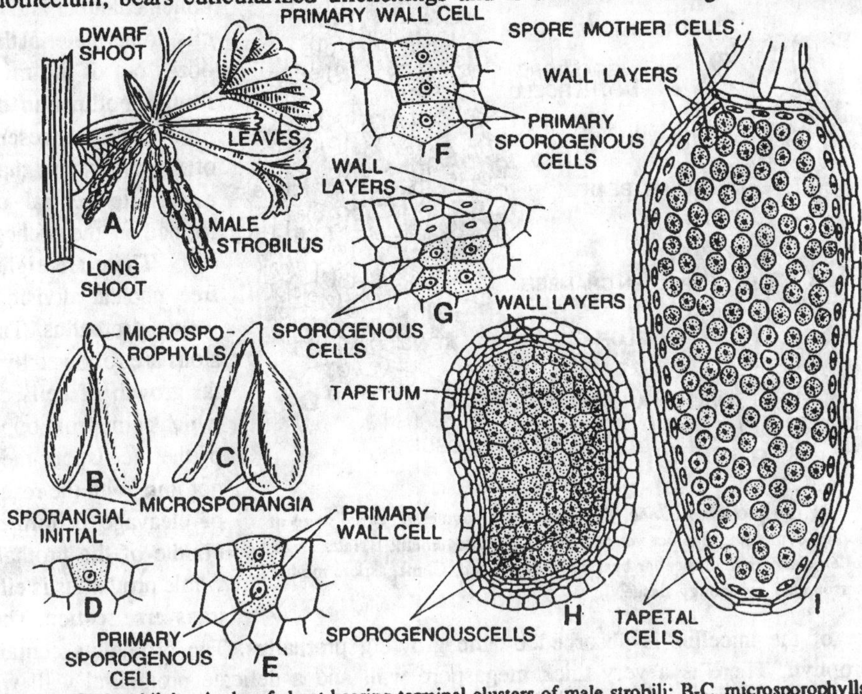

Fig. 3.9. *Ginkgo biloba.* A, dwarf shoot bearing terminal clusters of male strobili; B-C, microsporophylls bearing two microsporangia; D-H, development of microsporangium; I, a microsporangium containing spore mother cells.

respect the genus *Ginkgo* resembles with the angiosperms and differs from other gymnosperms, in which the outer layer or the exothecium of the sporangial wall is thickened, active layer.

The pollen grains are very much like that of *Cycas*. The division of the contents of the pollen grain to form prothallial cells begins before the sporangium opens. The pollen grain germinates to form two prothallial cells. The first prothallial cell aborts. The second prothallial cell persists. The remaining third nucleus divides again producing an antheridial cell in contact with the prothallial cell, and a large tube cell, which does not divide again. The pollen is shed in this four-celled condition.

The ovule. The sessile ovules are borne in pairs at the ends of slender stalks which occur in groups at the ends similar of spurs. The ovule is very similar to that of *Cycas*. The base of ovule remains surrounded by a

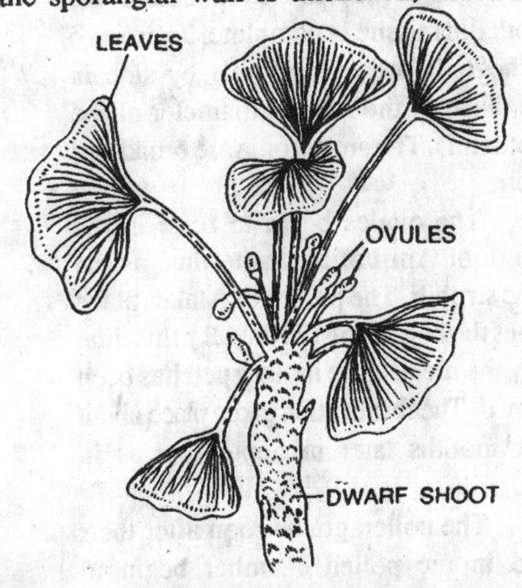

Fig. 3.10. *Ginkgo biloba.* Dwarf shoot with young ovules and a cluster of foliage leaves.

fleshy cup, the **ring or collar**. As in *Cycas*, there is a nucleus surrounded by a single integument. The integument contains three layers. The integument has no vascular system.

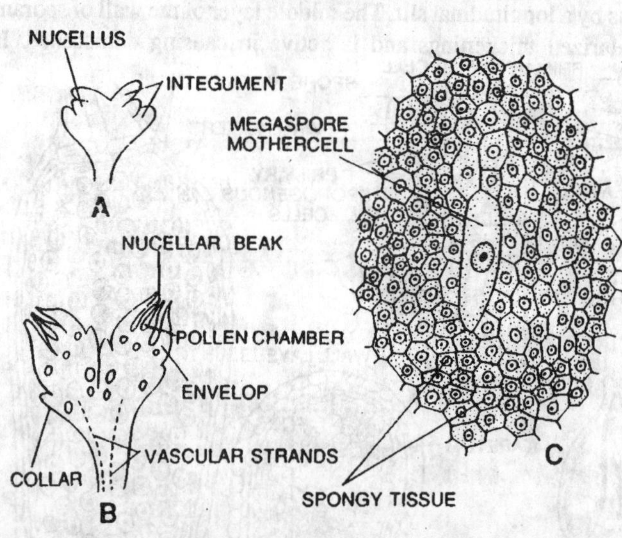

Fig. 3.11. *Ginkgo biloba*. Development of ovule. A, early stage of developing ovule, nucellus not yet covered by the integument; B, later stage, the pollen chamber and nucellar beak have developed; C, megaspore mother cell surrounded by spongy tissue.

The free apex of the nucellus breaks down into a pollen chamber and of the mucilaginous material thus formed oozes out of the micropyle to form a pollination drop.

It closely resembles that of *Cycas*, except that there is a definite ventral canal cell formed in the archegonium.

There are eight or more free nuclear divisions in the young prothallus. These divisions are followed by centripetal growth of cells. The cells which come from opposite side of the megaspore meet but do not join, with the result a place of cleavage is formed in the middle of the prothallus. The whole prothallus is elliptical in transverse section. The spongy tissue of the nucellus is absorbed by the growing prothallus. The prothallus contains some chlorophyll. There is a very thick megaspore wall and a delicate prothallial wall within the prothallus, which remain quite distinct from each other.

At the apex of nucellus there are usually two archegonia. These archegonia begin to develop even before the prothallus tissue is complete. These two archegonia are formed side by side in the plane of the longest diameter of the prothallus. The archegonia are broad and short.

The ovules begin to form in the month of April at the same time as the male strobili. The pollination takes place about the beginning of May. By this time only the megaspore mother cell has been formed. The fertilization takes place about four months later in September as in *Cycas*.

The pollen grains soon after their shed in the pollen chamber begin to germinate. The exine of the pollen grain does not cover the whole cell, and there

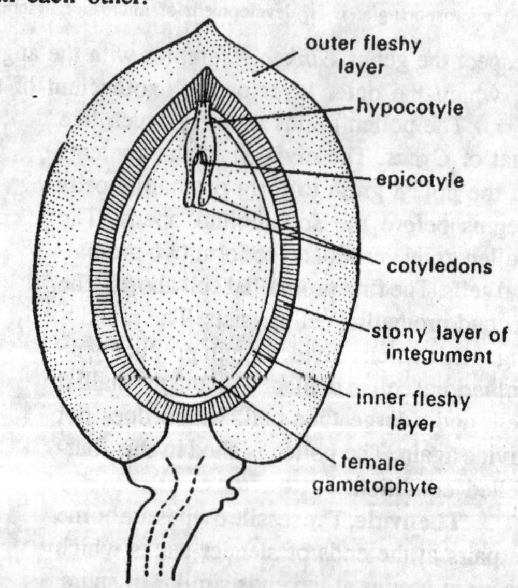

Fig. 3.12. *Ginkgo biloba*. Longitudinal section of the mature ovule possesing female gametophyte with two archegonia.

is a thin portion covered only by the intine. Thin intine protrudes out in the form of a haustorial (pollen) tube burrowing into the nucellus as also found in *Cycas*. This tube becomes enlarged

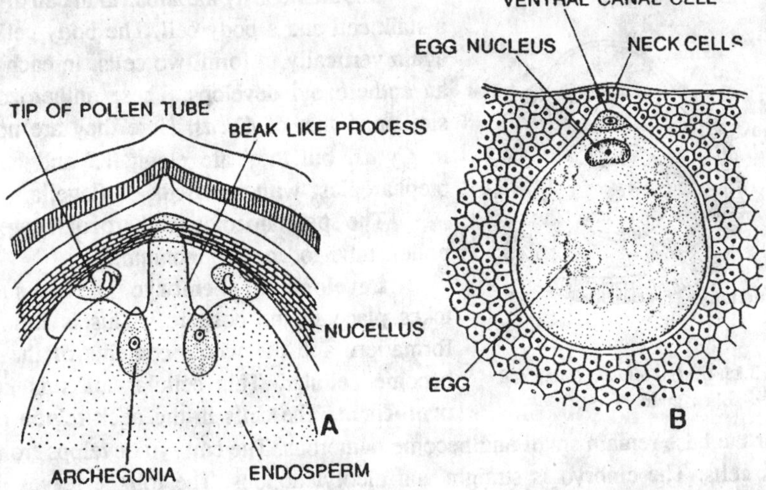

Fig. 3.13. *Ginkgo biloba*. A, longitudinal section of the apical part of the ovule; B, a mature archegonium.

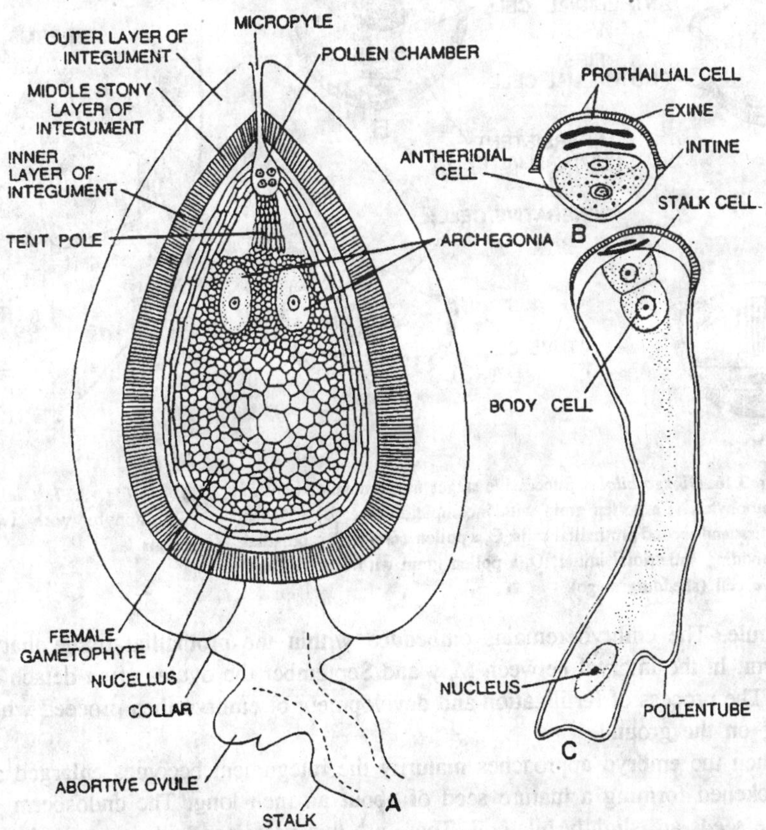

Fig. 3.14. *Ginkgo biloba*, A, ovule in longitudinal, section; B-C, germinating microspore.

and bladder like. This tube pushes the pollengrain down through the floor of the pollen chamber. This way, the floor of pollen chamber disintegrates above the archegonia.

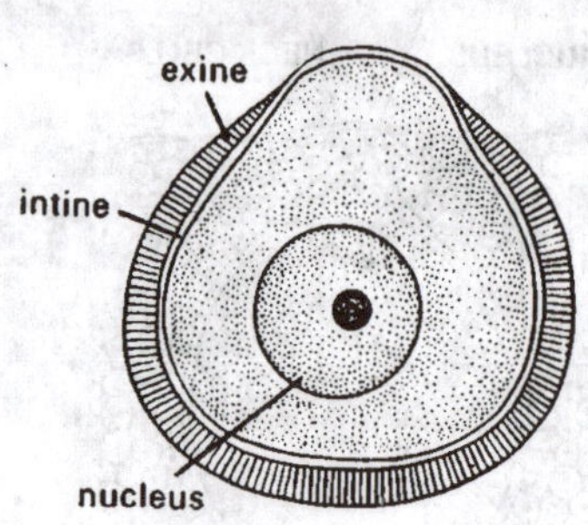

Fig. 3.15. *Ginkgo biloba*. Young and ungerminated pollen grain.

Simultaneously the antheridial cell divides into two, a stalk cell and a body cell. The body cell divides once again vertically to form two cells. In each of these cells an antherozoid develops. These antherozoids are quite similar to that of *Cycas*. Here they are not so large as in *Cycas*, but they are elongated and have the spiral blepharoplast with hundreds of flagella.

The spermatozoids (antherozoids) are carrid by the pollen tube to an archegonium.

Development of embryo. The free nuclear division takes place in the oospore. This is followed by wall formation, and the whole contents of the archegonium become cellular. This cellular mass is known as the **protocorm.** The cells in the neck region enlarge while those near the base remain small and become numerous. The embryo develops from these small numerous cells. The embryo is straight and dicotyledonous. The foliage leaves develop from

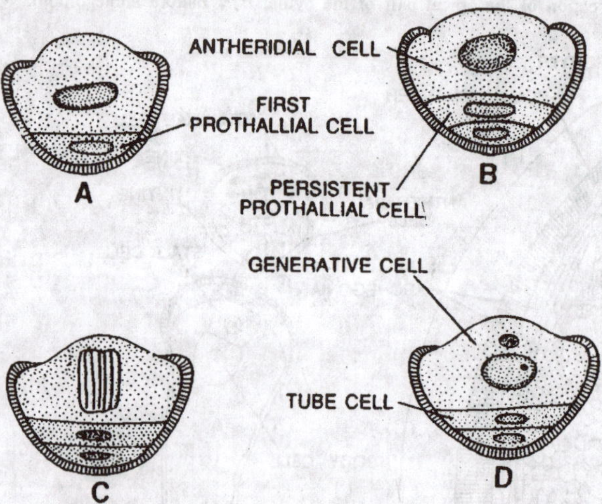

Fig. 3.16. *Ginkgo biloba*. Successive stages in the development of male gametophyte. A, a pollen grain with incomplete exine; B, a pollen grain with first and second prothallial cells; C, a pollen grain with prothallial cells and dividing antheridial initial; D, a pollen grain with tube cell and a generative cell (shedding stage).

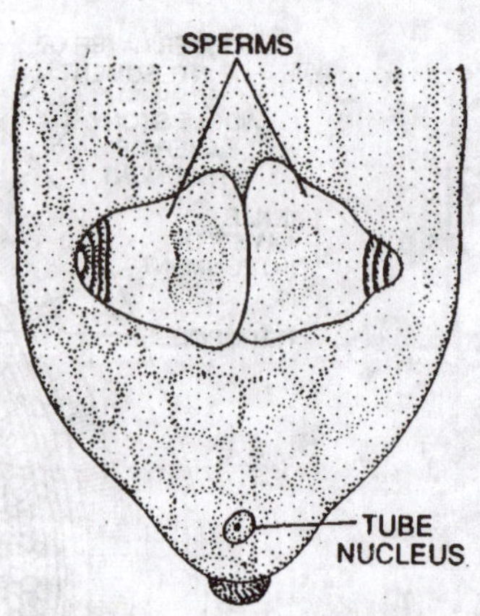

Fig. 3.17. *Ginkgo biloba*. Male gametophyte with two multiciliated sperms.

the plumule. The embryo remains embedded within the prothallial tissue that functions as endosperm. In the interval between May and September the ovules often detach and fall from the tree. The process of fertilization and development of embryo then proceed while the ovules are lying on the ground.

When the embryo approaches maturity the integument becomes enlarged and the outer flesh thickened forming a mature seed of about an inch long. The endosperm is edible.

The seeds are slightly bilateral. There are two prominent ridges on the hard testa of the seed. Generally only one of each pair of ovules develops to maturity. The two cotyledons are somewhat unequal. The large cotyledon bears a small notch at its apex while the smaller cotyledon is clifted nearly into two.

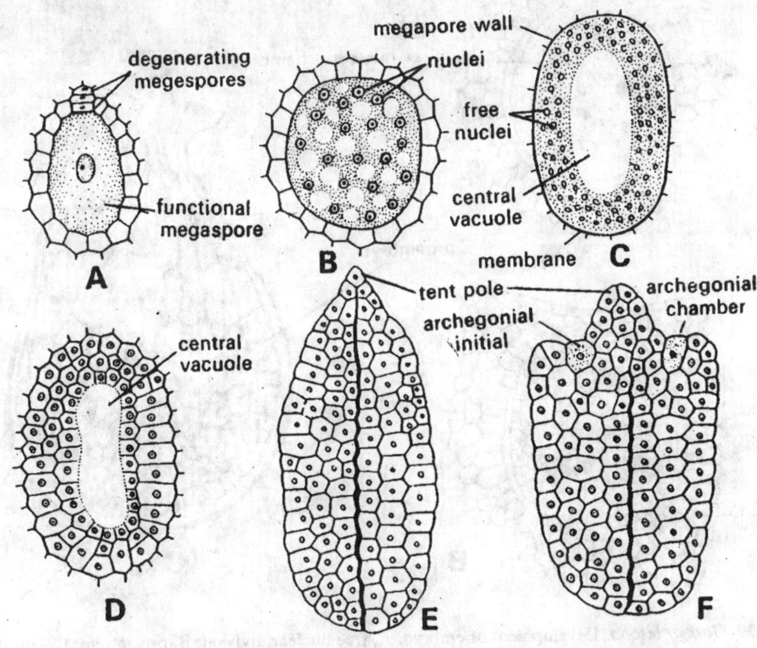

Fig. 3.18. *Ginkgo biloba*. Development of female gametophyte.

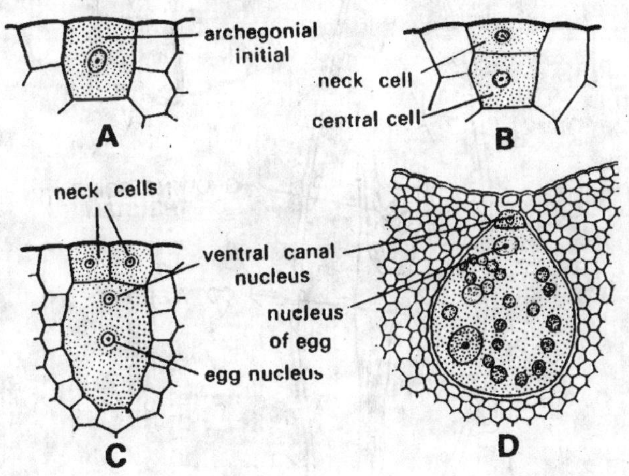

Fig. 3.19. *Ginkgo biloba*. A-D, development of archegonium.

Germination of seed. The cotyledons are hypogeal. They remain embedded in the nutritive endosperm of the seed. There is primary tap root. The young leaves are deeply bilobed.

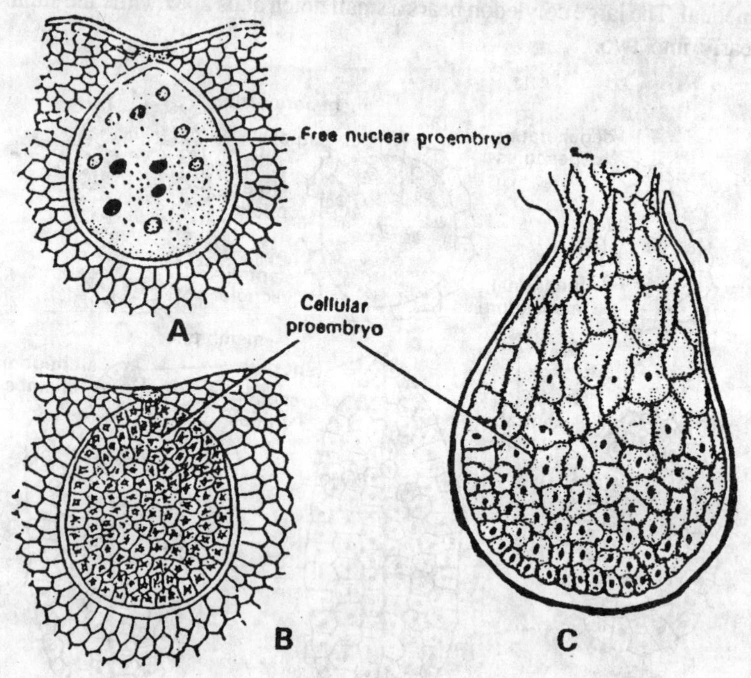

Fig. 3.20. *Ginkgo biloba*. Development of embryo. A, free nuclear division; B, proembryo; C, late stage of cellular proembryo.

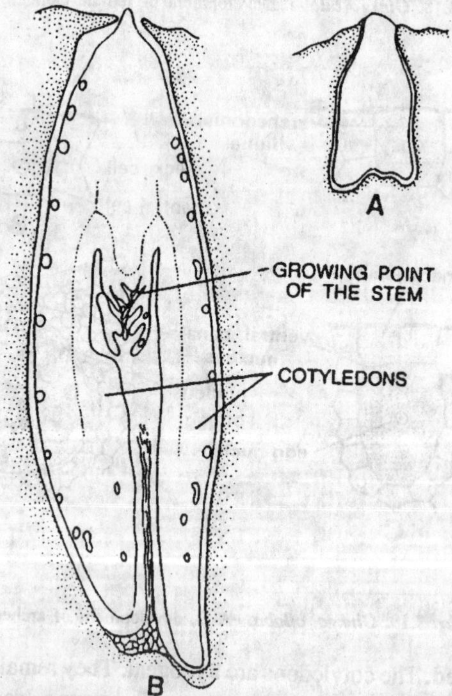

Fig. 3.21. *Ginkgo biloba*. A, late stage in the development of embryo; B, embryo showing two cotyledons.

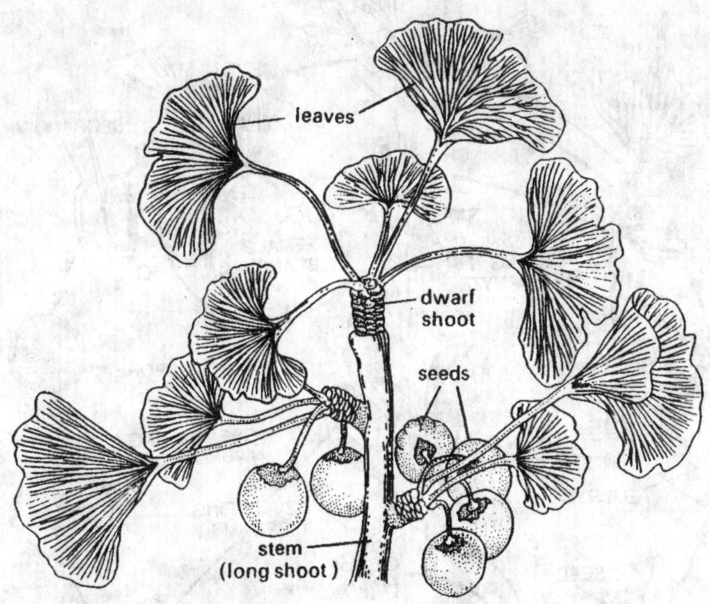

Fig. 3.22. *Ginkgo biloba*. A dwarf shoot with foliage leaves and ripening seeds.

Affinities. *Ginkgo biloba* is a very interesting plant from the view point of its resemblances with Cycadales, Cycadofilicales, Cordaitales and Coniferales. These resemblances may be summarized as follows:

Resemblances with the Cycadales. In both the groups there are several common features, such as, the presence of wide pith, cortex and a great number of mucilaginous ducts in stem, leaf, petiole, root and embryo. In addition to these characters the Ginkgoales resemble the Cycadales in having mesarch xylem in the cotyledonary and adult leaves, eusporangiate type of development of microsporangia, presence of a nucellar beak, pollen chamber, multiflagellate sperms, foliar nature of micro and megasporophylls and free nuclear division during the embryo development. Both the groups are thought to be primitive.

Resemblances with Cycadofilicales and Cordaitales. *Ginkgo biloba* resembles the pteridosperms and Cordaitales in having the common primitive character, such as, the accumulation of abundant reserve food material and lignification of the ovular integument prior to fertilization. The fertilization and development

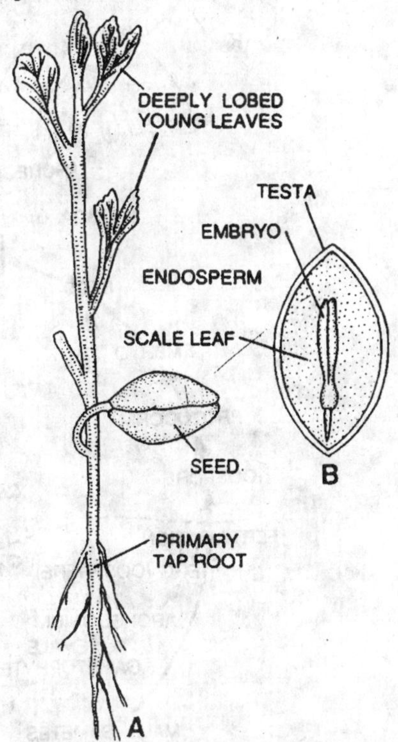

Fig. 3.23. *Ginkgo biloba*. Seed and its germination. A, germinating seed with young seedling and primary tap root; B, seed in longitudinal section.

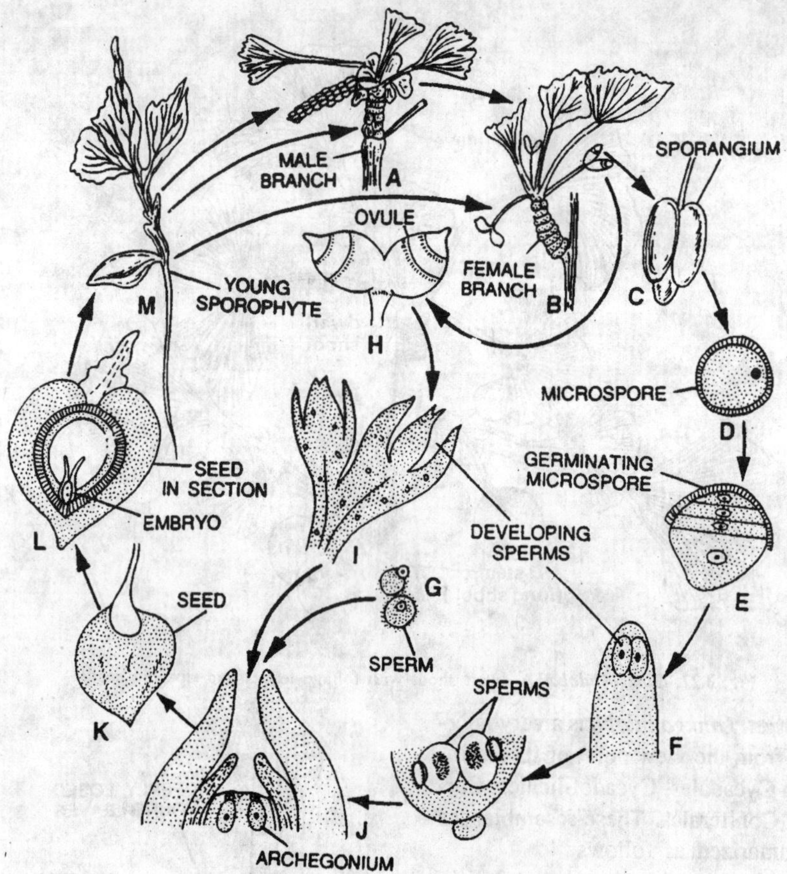

Fig. 3.24. *Ginkgo biloba*. Diagrammatic life-cycle.

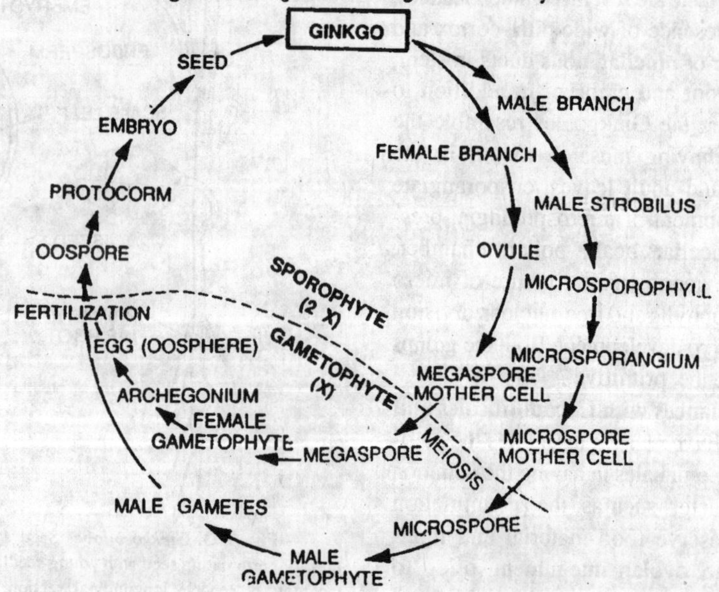

Fig. 3.25. *Ginkgo biloba*. Graphic life-cycle.

of embryo of *Ginkgo* also resembles to some extent which probably also occurred in the pteridosperms and Cordaitales.

Resemblances with the Coniferales. In many characters the Ginkgoales also resemble the Coniferales. These characters are: presence of annual rings in secondary wood of stem and root, narrow xylen rays, tree like habit of the plant, presence of dimorphic shoots and leaves with sunken stomata. Besides, several other characters are those which have been discussed in the text that support the independent and distinct status of Ginkgoales.

4

Class-Coniferopsida

Order-Coniferales

According to Buchholz (1948), the existing Coniferales have 50 genera and 550 species, 30 of the genera being confined to the Northern and 14 to the Southern hemisphere.

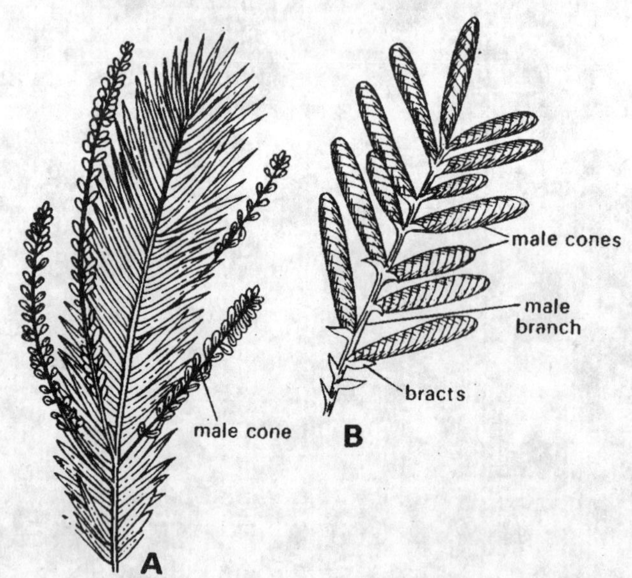

Fig. 4.1. *Podocarpus* **spp. A, shoot bearing male cones; B, male branch with male cones in bracts.**

The Coniferales are well represented in the tropics but are abundant in temperate regions of the world. Many genera, *e.g.*, *Pinus,Picea, Cupressus* and *Juniperus* are common to both Eastern and Western hemispheres. All the Pinaceae except one species of *Pinus* are confined to the Northern hemisphere while Araucariaceae and most of Podocarpaceae belong to the Southern hemisphere. The Taxaceae are mostly found in the Northern hemisphere. The Taxodiaceae, Cupressaceae and some of Podocarpaceae are commonly found in both the Northern and Southern hemispheres of the world. A large number of Coniferales are found abundantly in the western parts of the United States and also in the extratropical regions of Eastern and Central Asia. A very few conifers have been reported from tropical regions of Africa and especially of South Africa. In India, about 13 genera and 25 species of Coniferales are found. *Cephalotaxus* and *Taxodium* have been reported to be found in South India only while the remaining 11 genera are found throughout the country especially in cool hilly places. About 7 conifers have been reported from the vicinity of Simla, *e.g., Cupressus, Pinus, Picea, Abies, Juniperus, Podocarpus* and *Taxus*. In Simla hills, there are five well marked species of Conifers. *Pinus roxburghii* (Syn. *P.longifolia)* is found upto a height of 6,000' and this species is also commonly found in the plains. *P. wallichiana*, (Syn. *P.excelsa)* is found between 5,000' to 7,000' elevation, *Cedrus deodora*, the well known deodar tree is found at 8,000' elevation or so. *Picea morindoides* is found between 7,000' to 9,000' in Simla hills. *Abies pindrow, Taxus* sp. and *Cupressus* sp. are found at about 8,000' elevation in Simla hills.

About all the conifers are evergreen plants with few exceptions, *e.g.*, dwarf junipers and few others. The trees are with a long shaft which may reach to a height of 250 or more and may attain the girth of even 50' or so. The conifers include the biggest trees of the world, *e.g.*,

Sequoia gigantea and *Taxodium mucronatum.* All the conifers are longlived plants and some

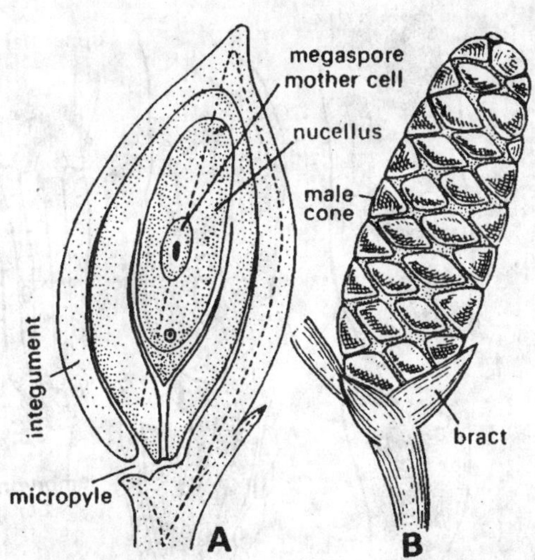

Fig. 4.2. *Podocarpus* spp. A, L.S. of ovule; B, single male cone arising from axil of bract.

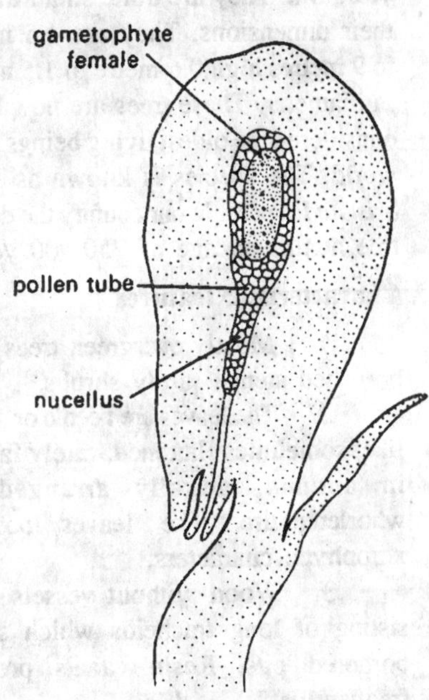

Fig. 4.3. *Podocarpus* spp. L.S. ovule showing micropyle, pollen tube, nucellus, female gametophyte, stalk and bract.

of them may live 1,000 years and few of them like *Sequoia* and *Taxodium* sp. are said to be living for last 4,000 years or more. Lobb, an English explorer discovered these trees *(Sequoia* and *Taxodium)* in 1850 on the Sierra Nevada Mountains in California. Dr. Schulman at the Arizona University,

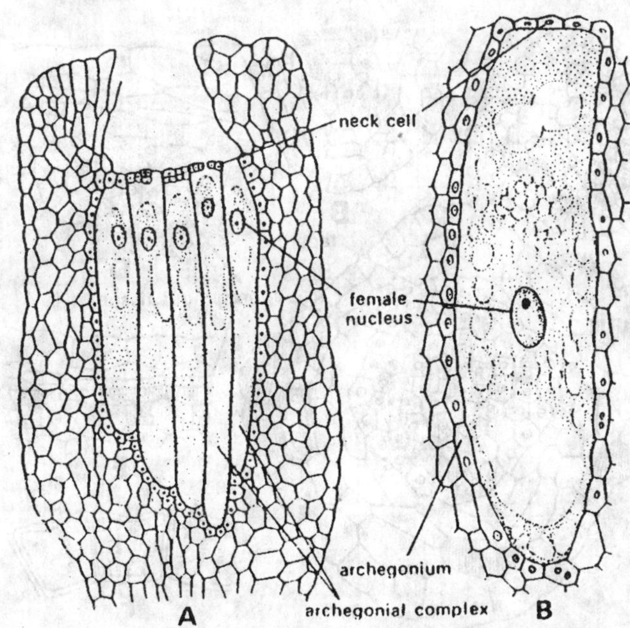

Fig. 4.4. *Podocarpus* spp. A, female gametophyte with archegonial complex in micropylar region (L.S.); B, single archegonium.

discovered the Pine trees called 'bristle cone pines' which he estimates to be more than 4,000 years old. They are quite small trees in their dimensions. They are 4.5 metres to 9 metres high 0.6 metre to 1.5 metres in diameter. These trees are now being considered the oldest living beings in the world. The species is known as *Pinus aristata* Englem. In our country the deodar tree attains an age of 750-900 years.

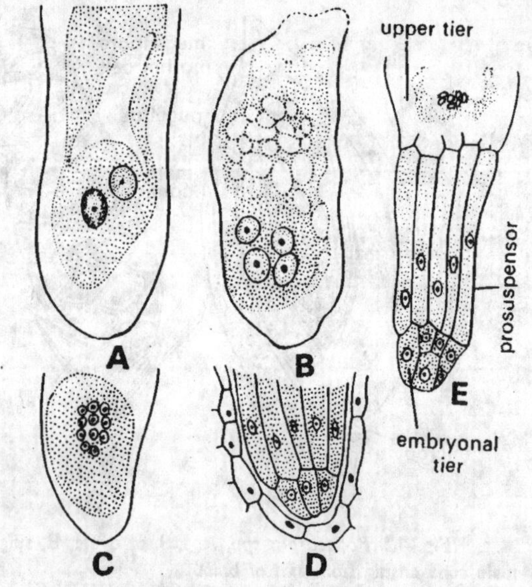

Fig. 4.5. *Podocarpus* spp. Embryogenesis. A-C, 2-16 nucleate proembryo; D, embryonal and suspensor tiers of proembryo; E, three-tiered embryo.

Characteristic features

1. Mostly evergreen trees with branched stems, rarely shrubs.
2. The leaves are needle or scale-like, sometimes flattened, rarely falling in autumn, spirally arranged or whorled,entire. The leaves possess xerophytic characters.
3. Wood without vessels consisting of long tracheids which show bordered pits. Resin canals present frequently.
4. The flowers are monoecious or dioecious, *e.g., Juniperus, Podocarpus,* Taxaceae etc. The female flowers are terminal or lateral and then surrounded by supporting bracts.

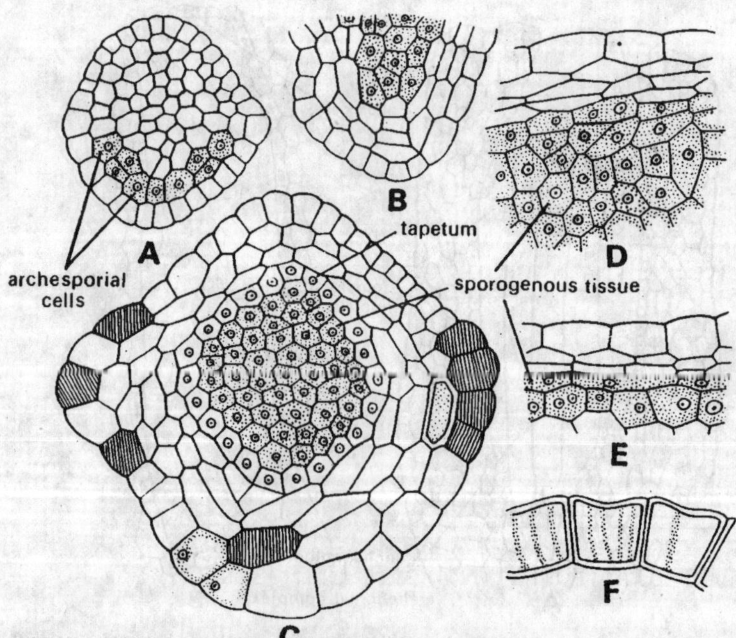

Fig. 4.6. *Cephalotaxaceae. Cephalotaxus* spp. A-F, successive stages in the development of microsporangium.

5. The male flowers consist of a number of stamens arranged in strobili. The stamens are usually many, each with 2 to 20 pollen sacs, connective often produced as an appendage.

6. Pollen grains may be winged, *e.g., Pinus*.

7. The female flowers are arranged in cones or catkins with the exception of Taxaceae, Cephalotaxaceae and Podocarpaceae.

8. Each female flower consists of a bract (sterile) and a scale (fertile). The scale is found above the bract. The ovules develop on the upper surface of ovuliferous scales.

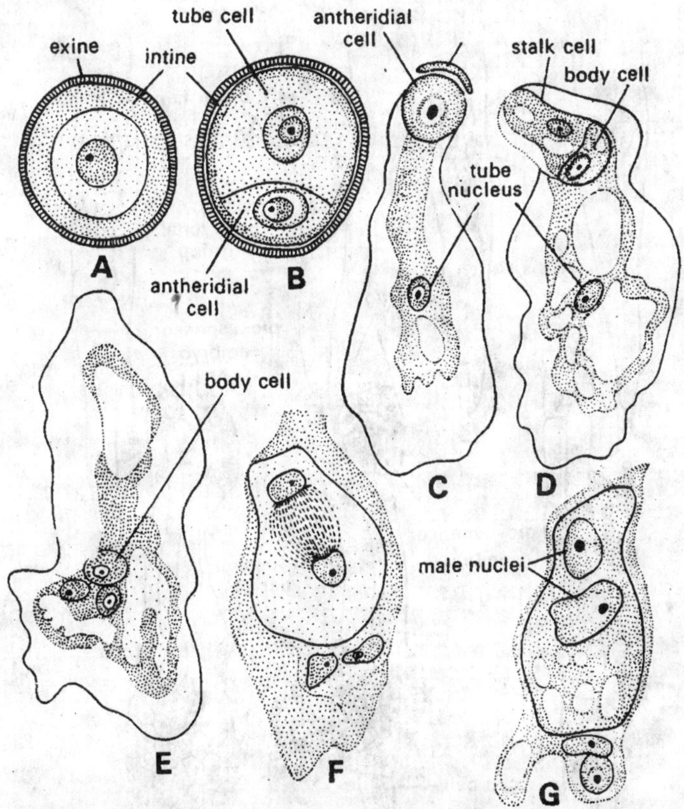

Fig. 4.7. *Cephalotaxus* spp. Male gametophyte. A, pollen grain; B, two-celled stage of pollen grain; C, pollen tube with tube nucleus and antheridial cell; D, antheridial cell formed stalk cell and body cell; E, pollen tube containing body cell stalk and tube nuclei; F, dividing body cell; G, two male nuclei formed.

9. The female cone ripens in 1-3 years and is usually dry on ripening.

10. The seeds are often winged, nut like and with a leathery or woody testa.

11. The cotyledons are epigeal and 2-16 in number.

12. Polyembryony is quite common.

13. They produce non-motile sperms at the time of fertilization.

Classification. The coniferales may be divided into two main series-the Pinades and the Taxades characterized by the presence or absence of an ovulate cone.The classification of conifers depends on the worker's personal stand point. In fact all external features of the various parts, the superficial characters of reproductive structures and the anatomy of the vegetative and reproductive parts are taken into consideration for the study of their taxonomy.

According to Sexton, the structure of male gametophyte,the position of archegonia and the structure and development of the embryo are sufficiently distinctive to be used in the classification of conifers. He, however, recognizes only five families- 1. Pinaceae; 2. Cupressaceae; 3. Araucariaceae; 4. Podocarpaceae and 5. Taxaceae. He does not recognise Taxodiaceae.

Upto 1917 Chamberlain divided the conifers into two main series co-ordinate with each other and to other groups of gymnosperms-Pinaceae and Taxaceae.

Chamberlain in 1935 slightly modified his view and divided the conifers into six families- 1. Abietaceae; 2. Taxodiaceae; 3. Cupressaceae; 4. Araucariaceae; 5. Podocarpaceae and 6.

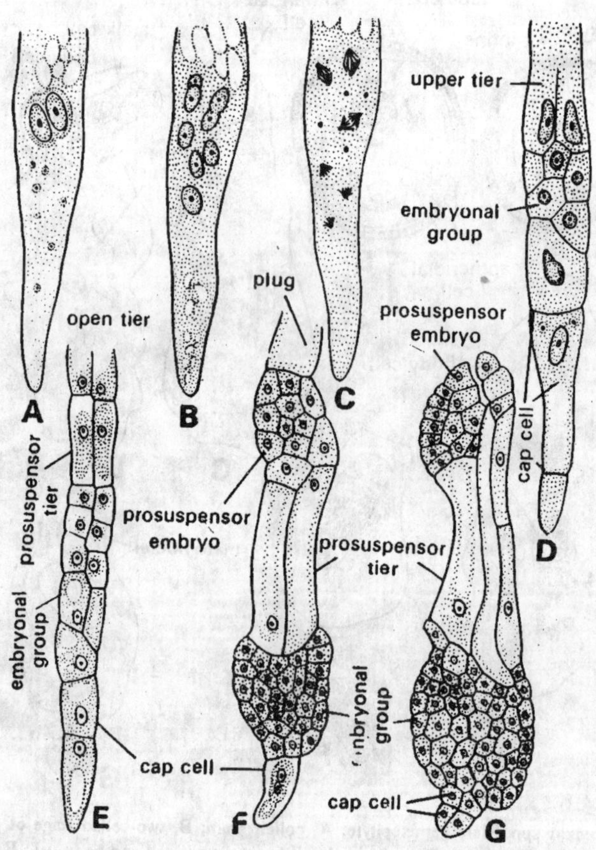

Fig. 4.8. *Cephalotaxus* **spp. Successive stages in the development of embryo. A, two-nucleate stage; B, eight nucleate stage; C, further free nuclear division; D, wall formation in proembryo; E, proembryo with prosuspensor and embryonal tiers; F-G, embryo with cap cells.**

Taxaceae. He included the first four families in his series Pinades and the last two families in series Taxades.

Pulle has proposed a revised classification of the group. It is as Follows:

1. Order - Araucariales
   Family Araucariaceae
2. Order - Podocarpales
   Family - Podocarpaceae

3. **Order - Pinales**
   Family - Pinaceae
4. **Order - Cupressales**
   Families - Cupressaceae
   Taxodiaceae
5. **Order - Taxales**
   Families - Cephalotaxaceae
   Taxaceae.

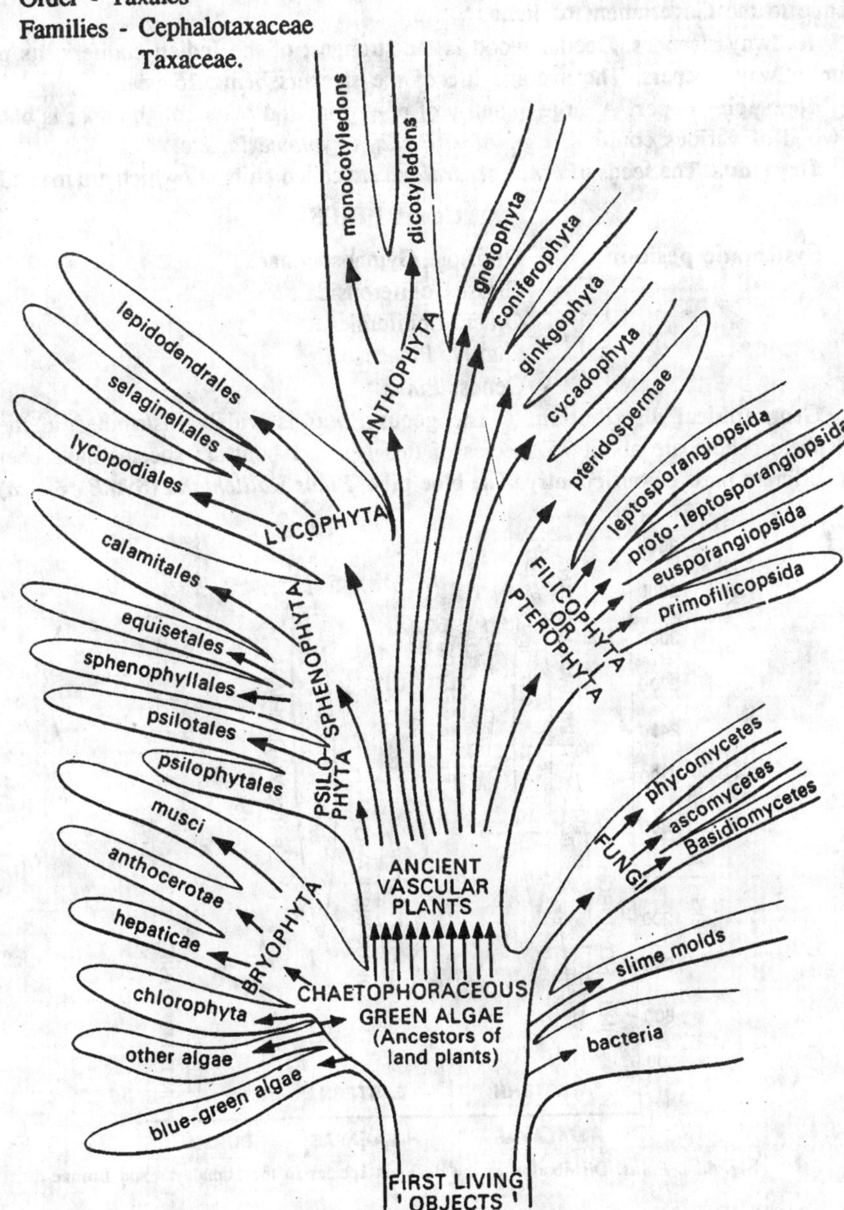

Fig. 4.9. Evolution of higher plants from first living objects.

Economic importance. From the point of their utility, the conifers are the most important.

Timber. Their timber is of universal importance. The timber is recognized for its durability,strength,lightness, elasticity,fineness in grain etc. The conifers are abundantly found, and therefore the timber produced by them is the cheapest and the best.

Resin and turpentine. Resin and turpentine are obtained from the coniferous trees on commercial scale. A well known resin Canada balsam is obtained from *Abies balsamia.* It is used as mounting medium for preparing permanent biological slides. They are a source of immense revenue to the Government of India.

Railway sleepers. Deodar wood is the strongest of the Indian conifers. Its primary use is for railway sleepers. The average life of the sleepers being 15 years.

Newsprint paper. A large quantity of newsprint and other rough paper is obtained from the wood of various conifers, *e.g., Abies,Picea, Cryptomeria,* etc.

Dry fruits. The seeds of *Pinus gerardiana* are called **chilgoza** which are roasted and eaten.

Genus PINUS

Systematic position.

Division. Gymnospermae
Class. Coniferopsida
Order. Coniferales
Family. Pinaceae
Genus. *Pinus.*

Geographical distribution. The genus *Pinus* is widely distributed in the Northern hemisphere. There are about 75 species of this genus. About six species have been recorded from different parts of our country. The blue pine, *Pinus wallichiana* (Syn.*P.excelsa*) is largely

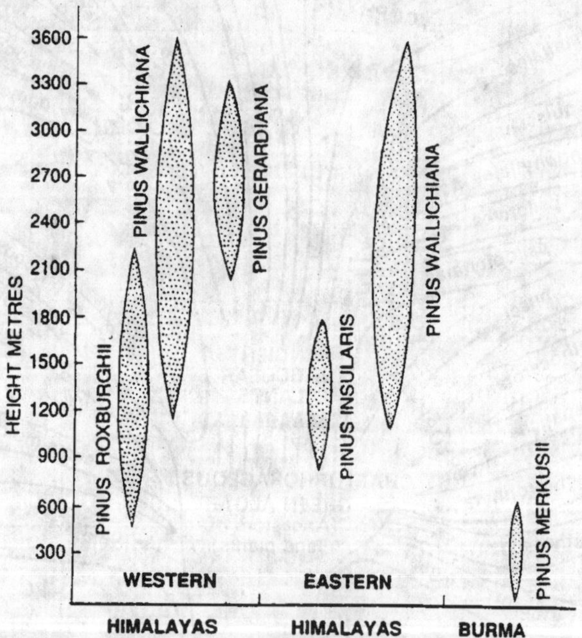

Fig. 4.10. *Pinus.* Distribution of the different species in the Himalayas and Burma.

found in North-West Himalayan region at 1,800 metres to 7000 metres elevation. The chir pine *P. roxburghii* (Syn. *P. longifolia*) occurs from Afghanistan to Bhutan in the outer range of the Himalayas. This species is also commonly found in the Indian plains. The chilgoza pine,

P. gerardiana is found in the inner arid valley of the Himalayas at 1,800 m. to 3,000 m. elevation. The Khasi pine, *P. insularis* is distributed in the Khasi, Naga hills and Manipur region at elevations of 750 m.to 1950 m. Recently a Chinese species *P.armandi* has also been recorded from the north-east hills of Assam. The merkus pine, *P. merkusii* is found in the hills of Burma at

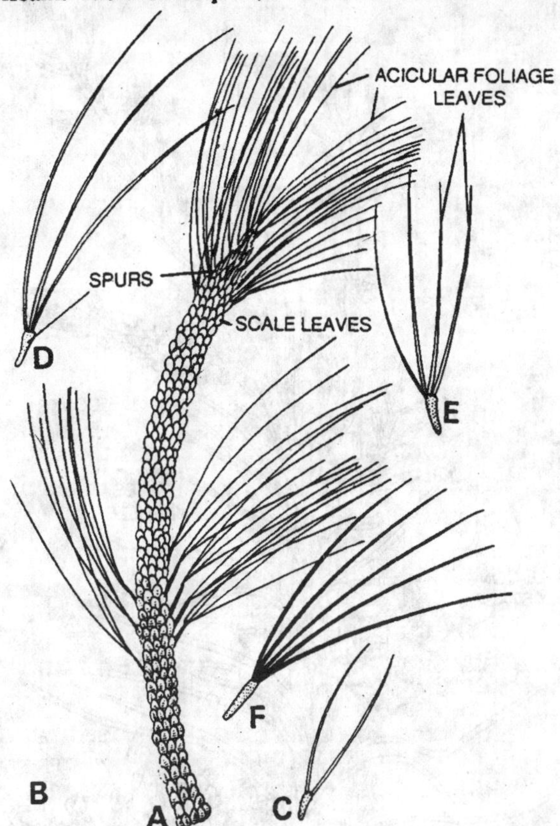

Fig. 4.11. *Pinus*. External morpholgy. A, shoot of unlimited growth of *P. roxburghii;* B, spur of *P. monophylla*; C, spur of *P. sylvestris*; D, spur of *P. roxburghii*; E, quadrifoliar spur; F, pentafoliar spur of *P.wallichiana*.

150 m to 172 m elevations. In addition to these species several exotic pines have been introduced in India, *e.g.*, *P. montana*, *P. laricio*, *P. sylvestris*, etc.

External morphology. The full-grown plant of *Pinus* is a large tree giving rise to a series of widespread branches. In most of pine trees a whorl of branches is produced each year. Sometimes, in young and vigorous trees two such whorls may be produced in one year. The whorls are formed in the axils of scale leaves every year. The main shaft is cylindrical and covered with a rough scaly bark. The branching is confined to the upper part of the stem, giving a pyramid-like appearance to the plant. The branches are dimorphic, the two forms being characterized as **long shoots and dwarf shoots** (spurs). These shoots are also known as the shoots of unlimited growth and shoots oflimited growth respectively. There are two kinds of leaves, the scale leaves and the green acicular foliage leaves which are commonly termed as **needles.** The dwarf shoots or the shoots of limited growth bear the foliage leaves while the long shoots or the shoots of unlimited growth bear scale leaves on them. The dwarf shoots with their cluster of green leaves are known as **'spurs'.** The number of needles in each spur varies from species to species. Each spur of *P. monophylla* is unifoliate; the spurs of *P. sylvestris* and *P. pinaster* are bifoliate; the spurs of *P. roxburghii*, *P. gerardiana* and *P. insularis* are trifoliate while the spurs of *P. wallichiana* are pentafoliate. The pine trees possess tap root. The tap root is elongated and possesses strong lateral roots.

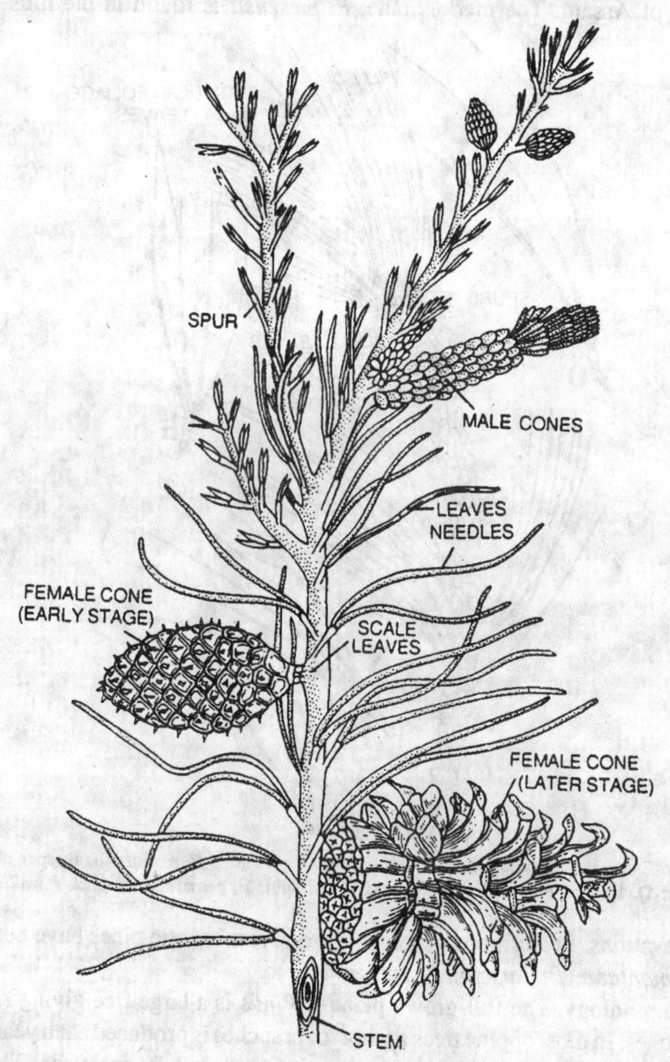

Fig. 4.12. ***Pinus*** **sp. A twig showing male and female cones, spurs, scale leaves and foliage leaves.**

The flowers are monoecious, *i.e.*, male and female strobili (cones) are borne on the same lant. The male flowers are catkin-like (cones) but erect in position. They are found in the ixils of membranous bracts which are spirally arranged on the axis. The young female cones are solitary, paired or whorled at the apex of the current year's shoot and consist of a central woody axis on which the two sets of scales are arranged in spiral way. The female cones are usually found on the shoots which do not bear male cones and take the place of shoots of unlimited growth.

Internal morphology (anatomy)

Root. The internal structure of root resembles to that of a dicotyledonous root. In a transverse section the root shows a piliferous layer bearing unicellular root hairs. The root hairs are found only in the young roots and root tips. In young roots there is fungal growth of ectophytic mycorrhiza. With the appearance of this fungus the root hairs of the root disappear.

Just beneath the piliferous layer there lies a broad cortex which consists of 4 to 5 layers of thinwalled parenchymatous cells. The innermost layer of the cortex is single-layered endodermis consisting of brown suberized cells containing tannin in them. Just below the endodermis there is multilayered pericycle containing tannin and starch grains. Lateral roots are developed from the second layer of the pericycle. The outermost layer of the pericycle helps in the formation of the digestive sac which enables the lateral roots to penetrate through the cortex to the outside.

In the centre of the stele there are two to six Y-shaped xylem bundles, and an equal number of phloem bundles alternating with them. The xylem has no true vessels and consists of tracheids. The phloem consists of sieve tubes and phloem par^nchyma. Companion cells are

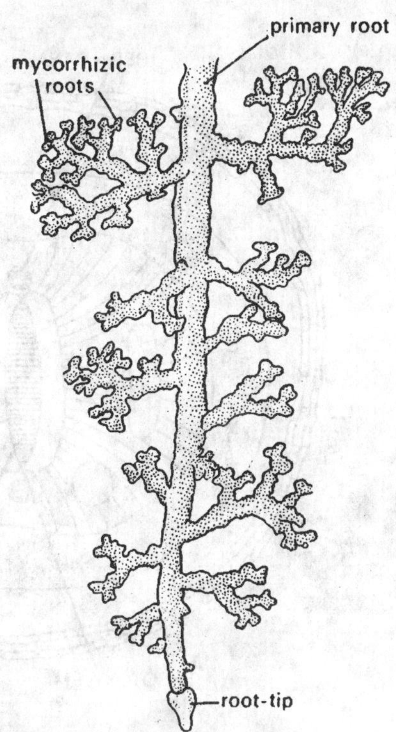

Fig. 4.13. *Pinus roxburghii.* The primary tap root bearing mycorrhizic roots.

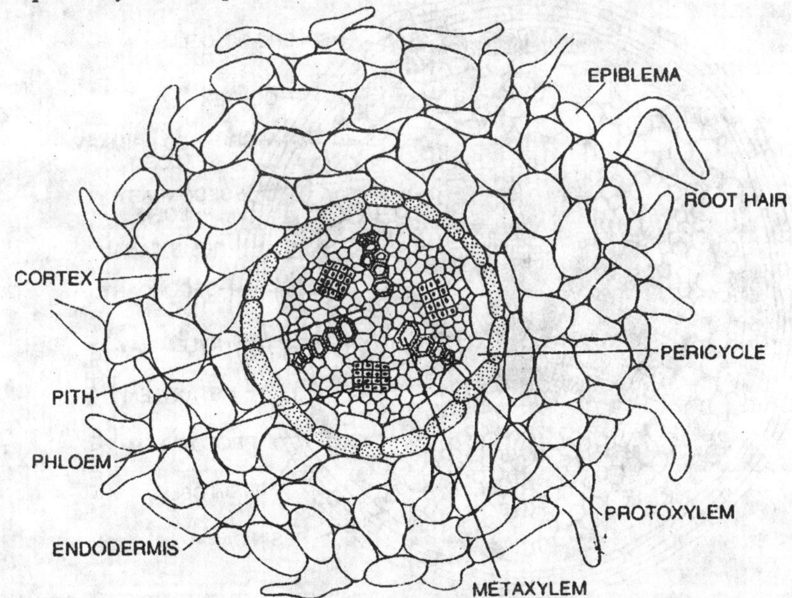

Fig. 4.14. *Pinus.* Transverse section of a young triarch root.

altogether absent. In between the arms of a Y-shaped xylem bundle there lies a resin canal. In the centre there is a small pith.

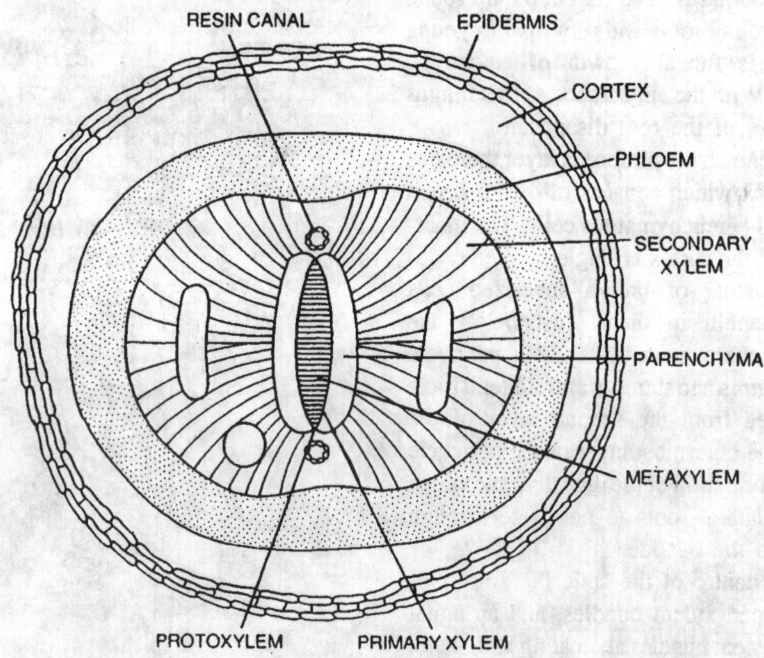

Fig. 4.15. ***Pinus.*** **Anatomy of root. T.S. of diarch root (diagrammatic structure).**

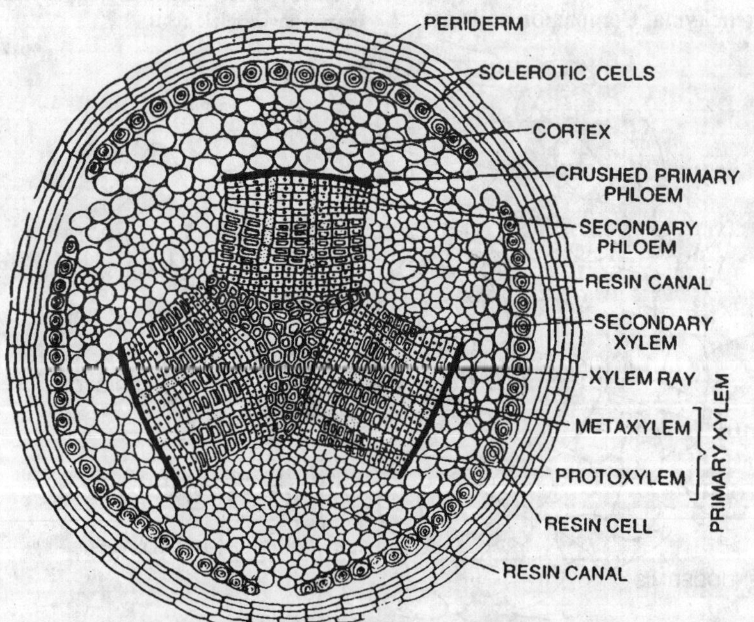

Fig. 4.16. ***Pinus.*** **Transverse section of an old triarch root with secondary growth.**

The secondary growth takes place as in dicotyledonous roots. A cambial strip develops from parenchymatous cells in between the phloem and metaxylem. This cambium cuts secondary xylem towards the inner side and the secondary phloem towards the outside. Simultaneously

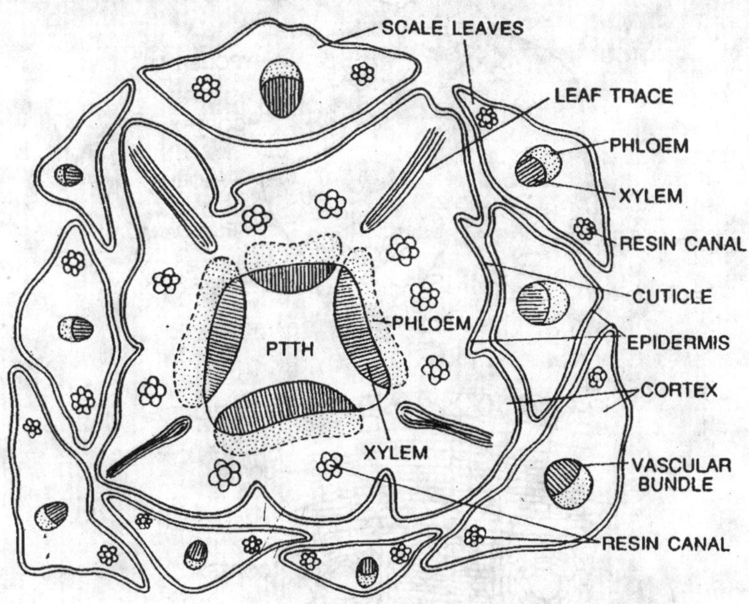

Fig. 4.17. *Pinus roxburghii.* Transverse section of young stem (diagrammatic).

a layer of the pericycle functions as a cork cambium and cuts off a layer of cork cells towards the outside. A thick layer of cork develops which separates the cortex from the stele and because of this barrier of cork the cortex does not get food and dies.

Stem. The internal structure of the stem of *Pinus* resembles to that of a dicotyledonous stem, though on the whole it displays the simpler structure. The young stem of *Pinus* shows the undermentioned structure in transverse section. The young stem is somewhat wavy in outline. It is surrounded by a single-layered cuticularized epidermis. Just beneath the epidermis there is multilayered hypodermis consisting of lignified sclerenchymatous cells. The hypodermis constitutes the outer region of the cortex. Underneath the hypodermis there lies the inner cortex consisting of thinwalled parenchymatous cells containing chloroplasts and resin canals. Each resin canal is surrounded by a layer of glandular epithelial cells. The innermost layer of the cortex may be considered as endodermis but it is not at all clear. The pericycle is also inconspicuous. It is parenchymatous.

The vascular bundles are conjoint, collateral and open forming a ring in the transverse section. The primary bundles are separated from each other by narrow medullary rays. The phloem consists of sieve tubes and phloem parenchyma. The sieve tubes are elongated and possess sieve plates on the radial walls. The companion cells are altogether absent. The xylem consists of tracheids. The protoxylem consists of annular and spiral tracheids.

Secondary growth. The secondary growth in the stem of *Pinus* takes place exactly in the same way as in dicotyledonous stems. A complete ring of vascular cambium develops. It cuts the secondary xylem towards the inner side and secondary phloem towards the outside. After the secondary growth the primary xylem may be seen outside the pith region while the primary phloem is completely obliterated.

Medullary rays. At many places the secondary xylem and secondary phloem zones are traversed by secondary medullary rays. These vascular rays also develop from the cambial cells

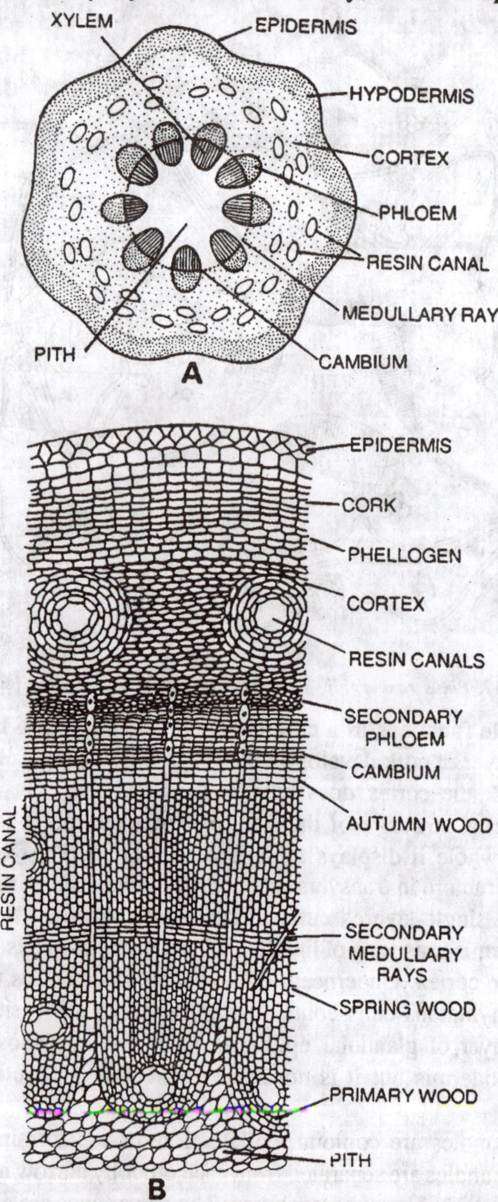

Fig. 4.18. ***Pinus.*** **Anatomy of stem. A, T.S. of young stem (diagrammatic structure); B, T.S. of three year old stem (detailed structure).**

to replace the original primary rays. These rays are usually uniseriate. Each ray is composed of rectangular cells with thick walls and numerous simple pits. The cells contain cytoplasm, a nucleus and many starch grains. The upper and lower margins of the medullary rays are constituted of one or two rows of marginal ray-tracheids which run horizontally and resemble short tracheids of the xylem from which they have been derived. In the cambium and phloem zones instead of marginal ray-tracheids, large thinwalled cells develop which extend upwards and downwards in between the cells of vascular tissue. In the secondary phloem the rays consist partly of starch containing cells and partly of albuminous cells. The medullary rays vary in their size. The smallest

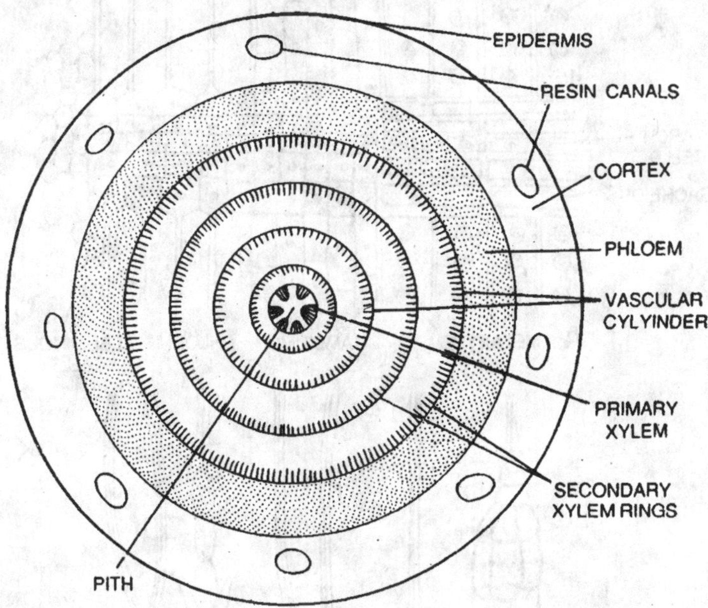

Fig. 4.19. ***Pinus.*** Anatomy of stem. T.S. of stem showing secondary growth (diagrammatic structure).

medullary ray being only two cells high and one cell wide. Usually they are only one cell broad and less than a dozen cells in height.

The secondary wood or metaxylem consists of tracheids with typical bordered pits. The tracheids are about 4 mm. long and pointed at both ends. The bordered pits are found on the radial walls and in certain species of *Pinus* the bordered pits are also found on tangential walls but fewer in number. The protoxylem consists of annular and spiral tracheids. The resin canals are present both in primary and secondary wood. Each resin canal is surrounded by the epithelial cells. The secondary phloem consists of sieve tubes and phloem parenchyma. The sieve tubes are elongated and possess the sieve plates on the radial walls. The companion cells are altogether absent.

As seen in transverse section the secondary growth is not uniform throughout. It shows annual rings. Each annual ring represents the growth for one year. The cambium ceases its activity in the winter season which is renewed in the following spring season. The wood formed in the spring differs from that formed in autumn. In autumn there is less supply of food material and therefore, the tracheids formed in autumn are quite small in size while in spring there is abundant supply of food material and the larger tracheids are formed. By counting the number of annual rings one can estimate the age of the tree. The walls of autumn tracheids are much thicker than those of spring tracheids. In the *Pinus* stem the wood is dense and massive and known as **pycnoxylic wood.**

The cork cambium or phellogen (meristematic layer) originates in the outer cortical region near the surface. This meristematic layer cuts off additional cortical cells (phelloderm) towards the inner side and forming cork towards the outside. This cork constitutes the bark which is impervious to water and protects the stem and other delicate tissues from excessive evaporation and injury.

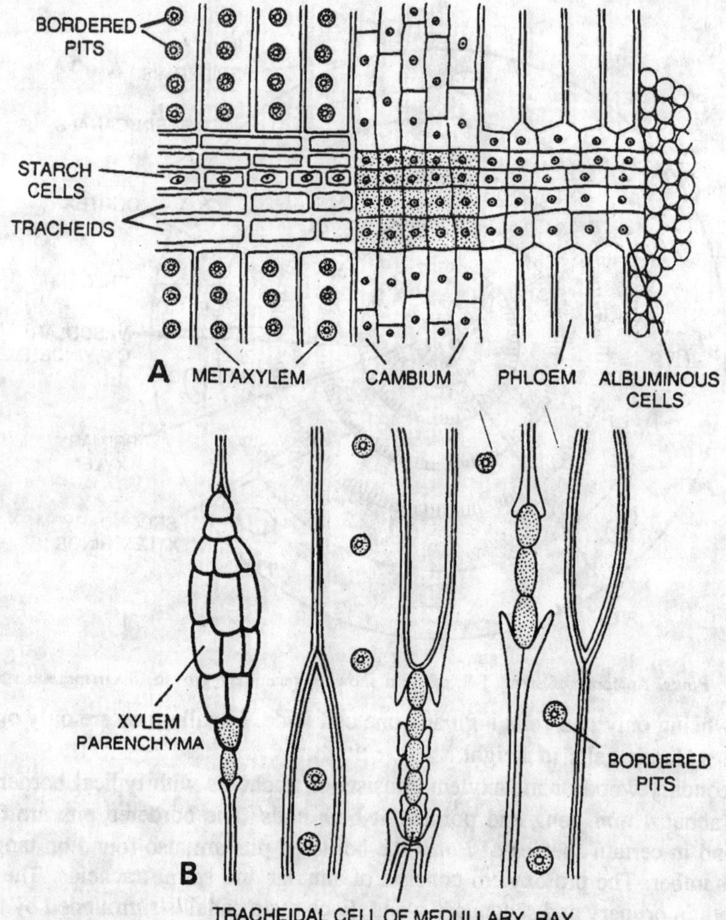

Fig. 4.20. ***Pinus*** **sp. Medullary rays. A, radial longitudinal section of old stem; B, a part of tangential longitudinal section showing the medullary rays and xylem parechyma.**

In *Pinus*, the resin canals are found in the cortical region opposite each primary vascular bundle. The resin canals form an interconnected system throughout the xylem, phloem and the cortex. Each canal is bounded by a layer of glandular secreting epithelial cells, which secrete turpentine. The turpentine acts as an antiseptic in healing the wounds caused by fungi or bacteria. The turpentine is oxidized to solid resin when exposed to air. This solid resin covers the wound until new bark is replaced. Sometimes, the epithelial cells become large and known as tyloses which block the resin canals.

The leaf. The leaf of *Pinus* is xeromorphic. The whole anatomy of the leaf makes it adaptable to withstand the low temperature and scarcity of water supply.

The outline of the needle (foliage leaf) in a transverse section depends on the number of the needles in the dwarf shoot (spur). In *P.monophylla* the spur bears a single needle and, therefore, the outline of the needle is circular. In *P. sylvestris* each spur consists of two needles and the outline of each needle is semi-circular. In *Pinus roxburghii* and *P. wallichiana* each spur consists of three needles and, therefore, the two flat faces of each needle are towards the

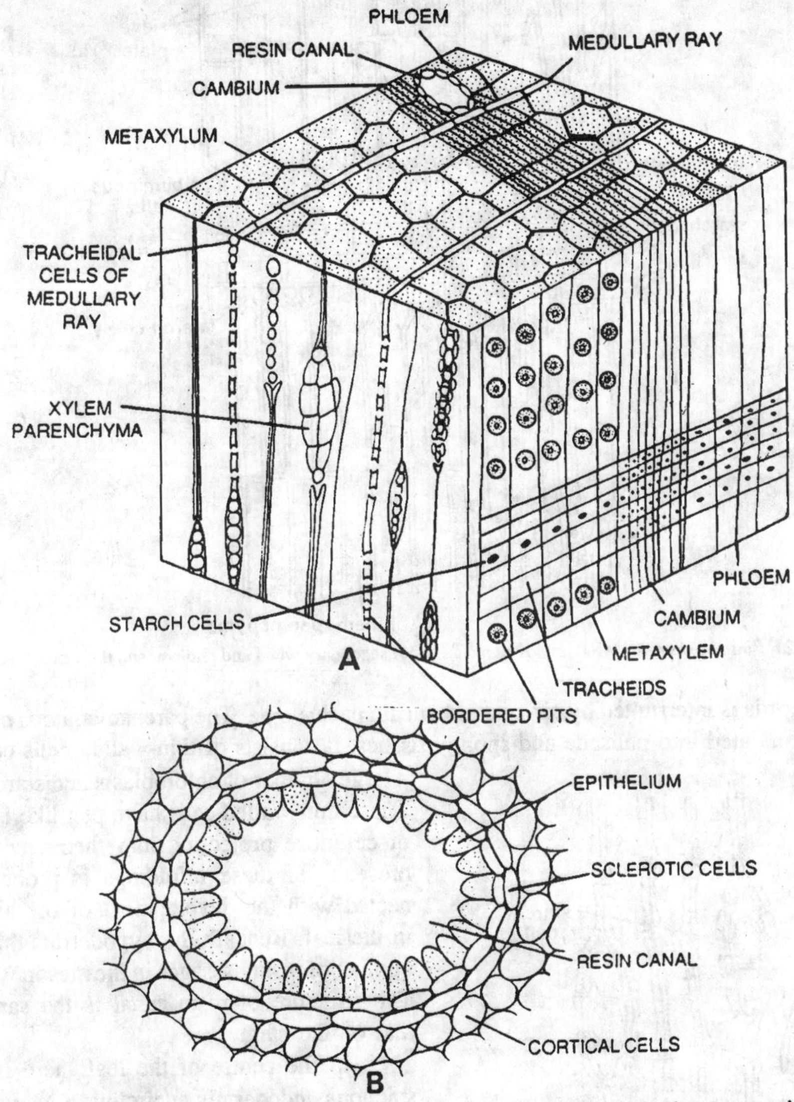

Fig. 4.21. *Pinus* sp. Anatomy of stem. A, three dimensional diagram showing spring and autumn wood; the medullary rays in T.L.S. and R.L.S. are also seen; B, T.S. of a schizogenous resin canal in the cortical region.

inner side and the curved face towards the outside. In these species the outline of the needle is somewhat triangular.

The outermost layer is the epidermis which consists of extremely thick-walled and cuticularized cells. A number of depressions are found over the epidermis. The stomata are developed all over the epidermis in these depressions. The guard cells are sunken in depressions below the level of the epidermis. Just beneath the epidermis there is a hypodermis which is composed of one or two layers of sclerenchymatous cells. The hypodermis is several layered at the corners.

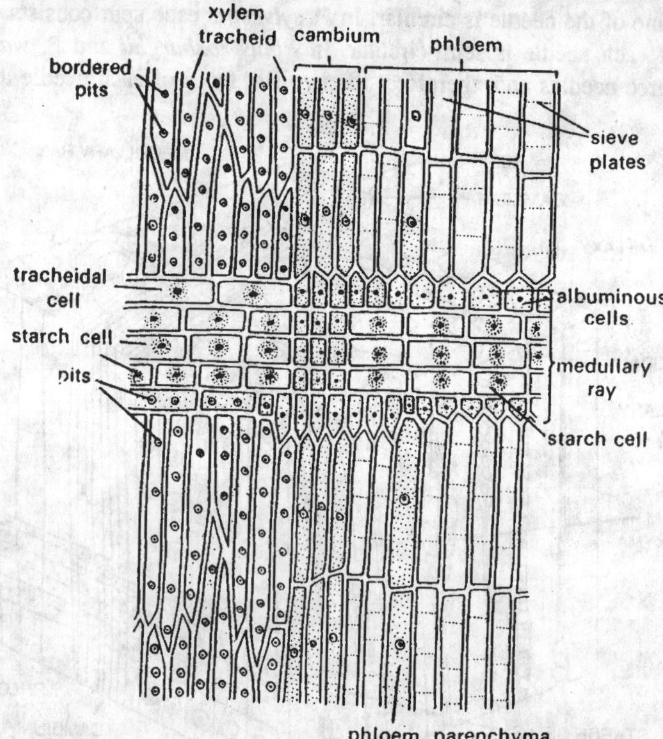

Fig. 4.22. *Pinus roxburghii.* Anatomy of stem; T.L.S. of secondary wood and pholem, and the structure of secondary medullary ray.

The hypodermis is interrupted by air-spaces beneath each stoma. The parenchymatous mesophyll is not differentiated into palisade and spongy tissues. It consists of thin-walled cells containing a large number of chloroplasts and starch grains. These thin-walled cells have peg-like infoldings of cellulose projecting into their cavities. The presence of these infoldings is probably connected with the development of the air spaces in the leaf. Beneath the hypodermis there are a number of resin passages in the mesophyll tissue. The structure of resin canal is the same as in that of the stem

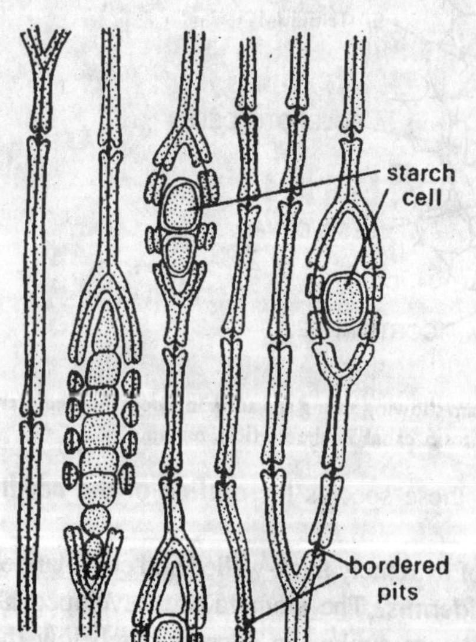

Fig. 4.22 (a). *P.roxburghii.* R.L.S. of secondary wood showing the medullary rays.

In the centre of the leaf there is a conspicuous endodermis encircling a many-layered pericycle. Within the pericycle there are two vascular bundles. The xylem of the bundle is being directed to the inner side and phloem towards outside. The two bundles run parallel and unbranched from base to apex of the leaf. These two vascular bundles arise from a single leaf trace.

The pericycle consists of different types of cells. Firstly, they are parenchymatous cells rich in proteins and known as albuminous cells.

These cells are found above the phloem of vascular bundles. The remaining tissue within the sheath is the **'transfusion tissue'** which consists of two kinds of parechymatous cells- 1. The cells without protoplasm and pitted. They are tracheidal cells which serve to conduct water from the xylem to the mesophyll and they are thought to represent in function an extension

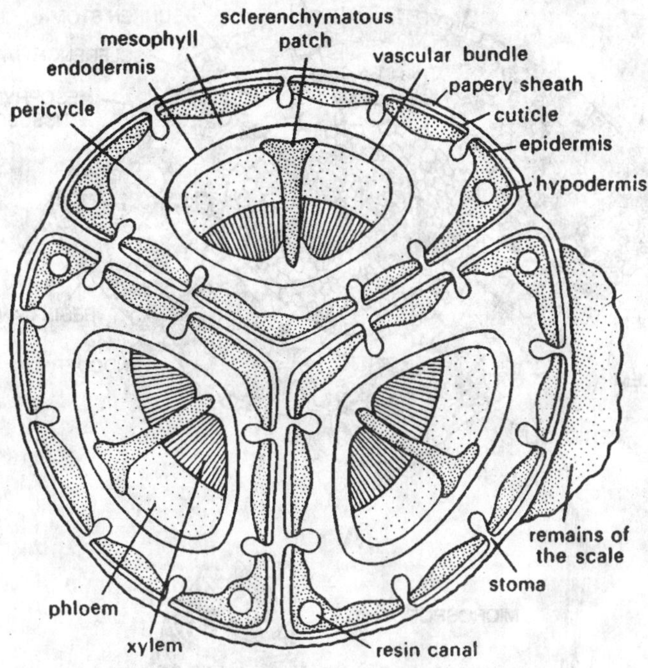

Fig. 4.23. ***Pinus.*** **Leaf anatomy. T.S. of spur of *P. roxburghii* showing three needles (diagrammatic structure).**

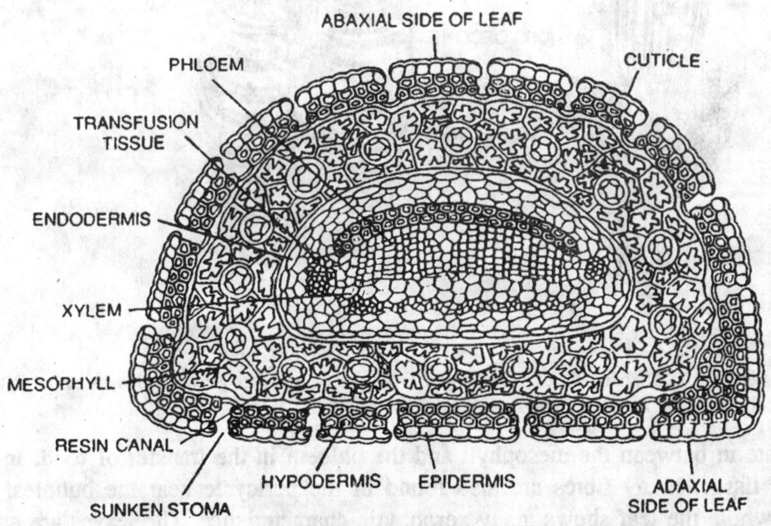

Fig. 4.24. Transverse section of needle (leaf) of *Pinus sylvestris* (detailed structure).

of the tracheidal system. According to Worsdell, these cells have been derived from the centripetal xylem of the ancestral mesarch bundle. 2. The cells with protoplasm are not pitted. These cells

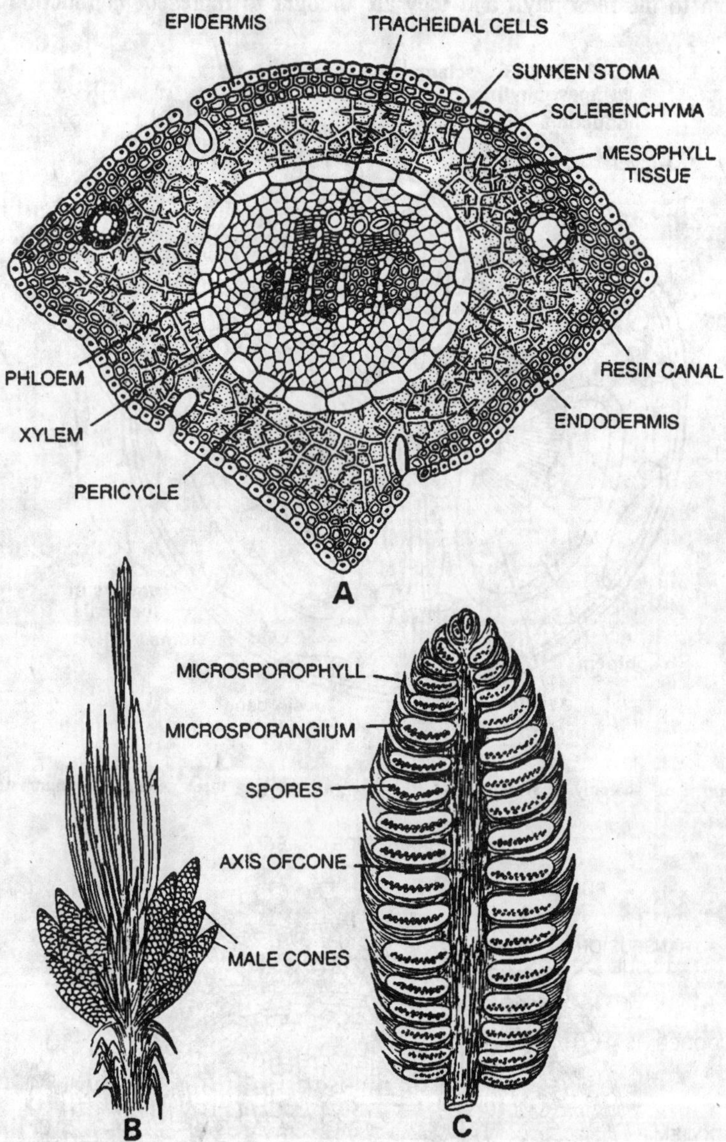

Fig. 4.25. *Pinus* sp. A, T.S. of needle of *P. roxburghii*; B, a shoot of unlimited growth with male cones; C, longitudinal section of male cone showing microsporophylls and microsporangia.

are intermediate in between the mesophyll and the phloem in the transfer of food. In addition to transfusion tissue, many fibres are also found in the pericycle near the bundles.

On the whole the leaf shows many xerophytic characteristics. The leaves are acicular in form. In the transverse section the leaf shows thick cuticle, the sunken stomata, sclerenchymatous hypodermis, the simple vascular system and the peculiar transfusion tissue.

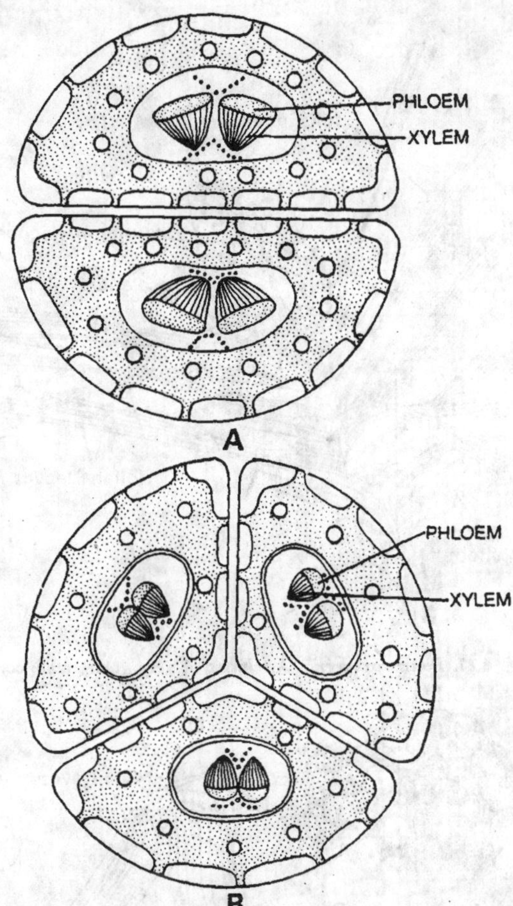

Fig. 4.26. Anatomy of leaf. A, T.S. of spur of *Pinus sylvestris;* B, T.S. of spur of *P. roxburghii.*

Reproduction. The reproduction takes place by means of spores developed in sporangia situated on the sporophylls. The strobili or cones of *Pinus* are monosporangiate, *i.e.*, the two kinds (male and female cones) occur on the same plant. The plants are monoecious. In some of the species bisporangiate strobili are occasionally produced, they have been reported in *P. roxburghii* by Rao (1931), in *P. montana* by Stell (1918) and in *P.maritima* by Goebel.

Male cones. The male cones of *P.wallichiana* reach maturity and shedding of pollen takes place earliest at low elevations and in hot dry weather.

The male cone is produced in the axil of a scale leaf at the base of the current year's young shoot and thus replaces a shoot of limited growth. The cones of *P.wallichiana* are about 7 mm. to 1 cm. long immediately before their ripening. They lengthen to 1-2 cm. and fall soon after ripening. At the base of the cone there is an involucre consisting of a number of small imbricate scales. The cone bears 60 to 100 spirally arranged specialized leaves known as microsporophylls. Each microsporophyll bears two microsporangia or pollen sacs on its underside. The development of the microsporangium is of eusporangiate type. The sporogial wall consists of four layers. It takes place either by a single cell or by a layer of hypodermal cells. Konar and Ramchandani (1959) reported that in *P. wallichiana* the wall of microsporangium

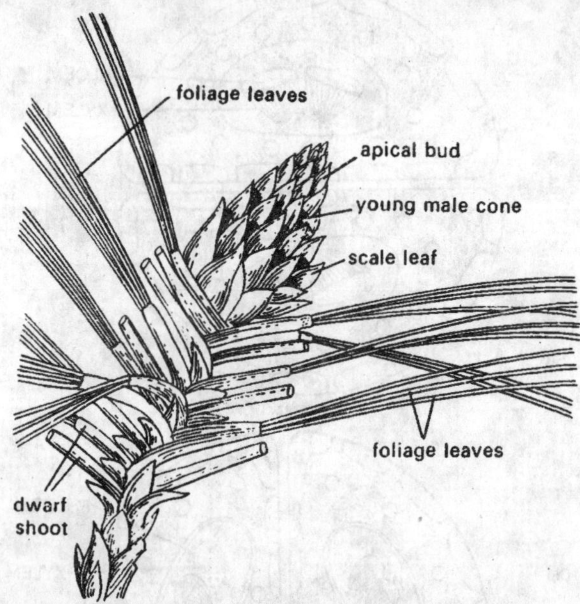

Fig. 4.27. *Pinus roxburghii*. A, cluster of very young male cones.

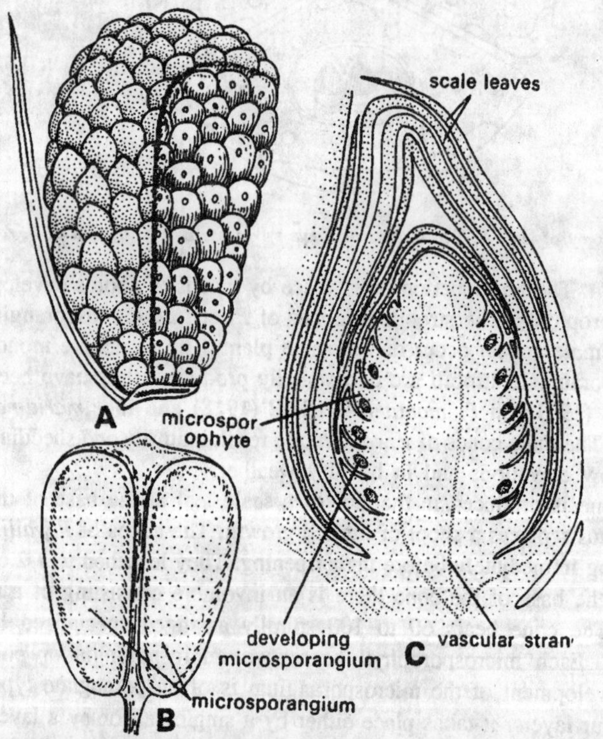

Fig. 4.28. *Pinus roxburghii*. A, fully developed male cone; B, microsporophyll bearing two undehisced sporangia; C, V.S. of a very young male cone showing developing sporangia.

consists of the epidermis, two middle layers and a glandular tapetum. Within young sporangium there is a peripheral tapetum and a central archesporial tissue which forms a number of microspore mother cells. The microspore mother cells represent the last stage of sporophyte generation. Each microspore mother cell divides meiotically producing four haploid (n) microspores. The microspores or pollen grains separate from each other and absorb their nourishment from the disintegrating tapetal cells. According to Sethi (1928), in *P.roxburghii* the meiosis takes place during the last week of January.

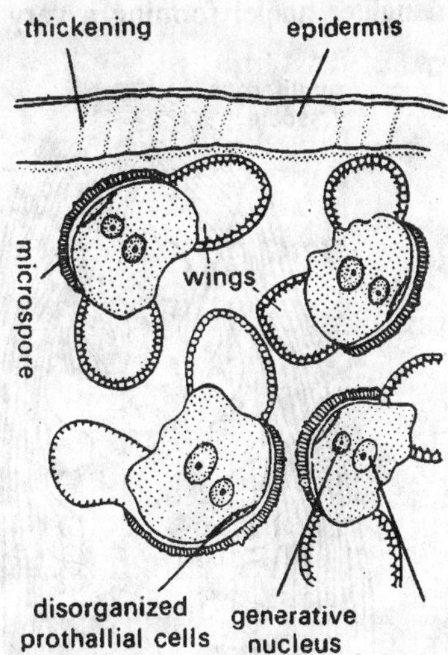

Fig. 4.29. *Pinus* spp. Part of mature microsporangium with winged microspores.

The microspore. The microspore is sorrounded by a three layered wall. The exine is heavily cuticularized and is found only on one side of the microspore. It does not completely cover the mature spore. The exo-intine or the middle layer covers the rest of the spore. The exo-intine is projected outward into two large balloon-like air sacs or wings which make the spore much more buoyant and aid in its dispersal by wind. The intine or inner layer of the spore is very thin.

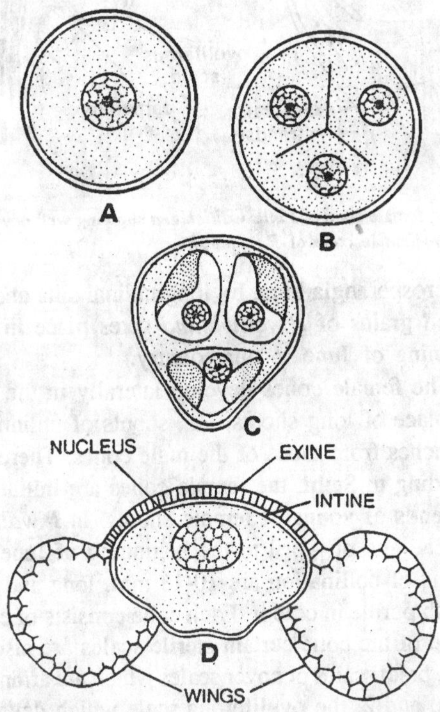

Fig. 4.30. *Pinus* sp. Formation of microspores. A, microspore mother cell; B, microspore mother cell dividing meiotically; C, microspore tetrad; D, a microspore or pollen grain.

On the maturation the spores germinate *in situ, i.e.,* within the microsporangium. Firstly, the nucleus of microspore divides into two and simultaneously a wall develops in between these two daughter nuclei forming a very small flattened prothallial and a large tube cell.

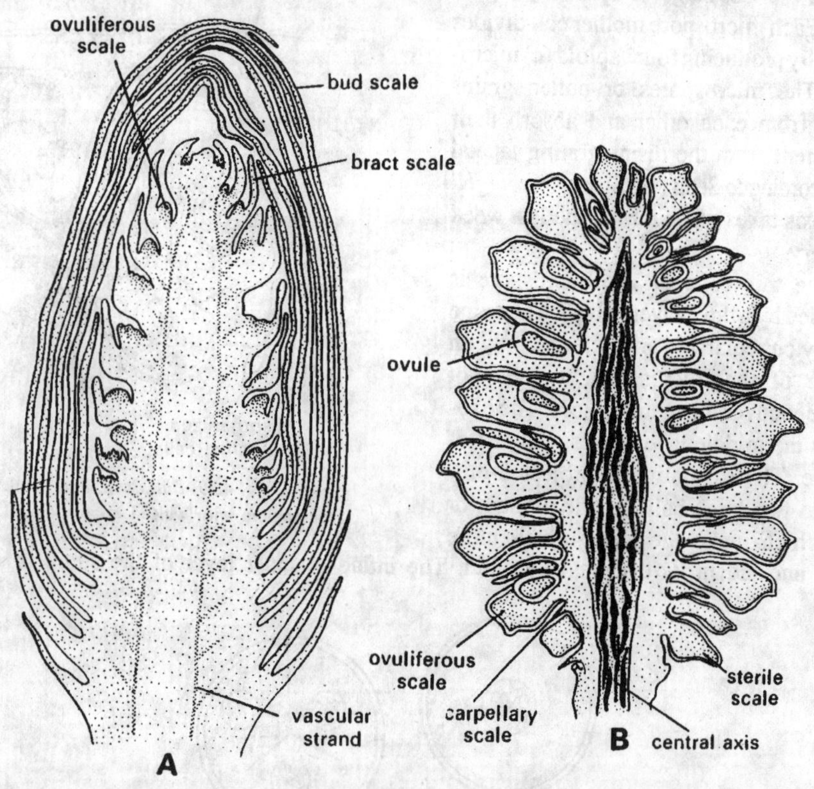

Fig. 4.31. A, L.S. of young female cone of *Pinus wallichiana* showing well developed bract scales and developing ovuliferous scales; B, V.S. mature female cone of *P.roxburghii.*

At this stage the microsporangia burst by longitudinal slits and the microspores are shed. The shedding of the pollen grains of *P.wallichiana* takes place in hot dry weather from the end of April to the beginning of June in our country.

The female cone. The female cones develop laterally in the axil of scale leaves. They are formed in clusters in place of long shoots, *i.e.,* shoots of unlimited growth. They are produced on the different branches from those of the male cones. There may be one to four cones on each long shoot. According to Sethi, the female cones are initiated in January and become visible in February at the ends of youngest female shoots. In *P.wallichiana* the young female cones are pollinated from the end of April to the beginning of June. The young female cones of *P.wallichiana,* at the time of pollination are 10-13 mm. long and 3.8 mm. in diameter. The young cones are dark reddish purple in colour. Each cone consists of central axis bearing spirally arranged scales. At the base of the cone certain sterile scales are also found. Each fertile scale consists of two structures- 1. bract scales or cover scales which are arranged spirally and developed directly from the cone axis, and 2. the ovuliferous scale which develops on the upper surface of the bract scale. The bract scale is leathery while the ovuliferous scale is woody in structure. Each ovuliferous scale bears two ovules on its upper surface. The bract scales are concealed under the ovuliferous scales and, therefore, they are not visible from outside.

The ovulate strobilus of Pinaceae has long been a morphological puzzle. The question arises whether structure of ovulate strobilus is simple or compound. There is much confusion about the double structure of the strobilus. The views of the various workers are as follows:

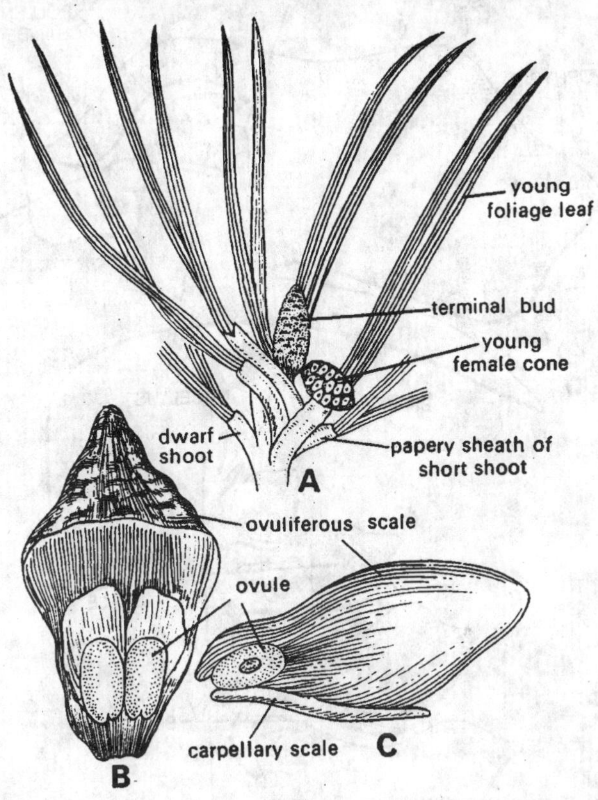

Fig. 4.32. ***Pinus sylvestris.*** **A, long shoot bearing dwarf shoots with two needles, apical bud and female young cone also seen; B, megasporophyll bearing two seeds; C, sectional view of megasporophyll showing ovule.**

Robert Brown regarded the ovuliferous scale as an open carpel. Alexander Braun (1853) stated that the ovuliferous scale represents the first two leaves of an axillary shoot which are fused by their upper margins. This view was accepted by many botanists of the time. According to Sachs and Eichler, the ovuliferous scale is an outgrowth of the bract scale comparable to a ligule or placenta. Kubart and Bessey regarded the ovuliferous scale as a combined outgrowth of the ovules themselves and call an aril or an enlargement of the chalaza of the ovules. According to Delpino the ovuliferous scale develops from two lateral lobes of the bract scale which have been turned inwards and fused together. Hirmer states that the ovuliferous scale and the bract scale are both parts of one structure which has forked vertically in the *Cheirostrobus*. Florin, however, regards the cone as a compound structure. He regards the cone as inflorescence and the ovuliferous scales as short shoots.

The ovule. The two ovules are found side by side on the upper surface of the ovuliferous scale. Each ovule consists of a group of the cells forming a tissue, the nucellus. The ovule is surrounded by a two-lipped covering known as integument. The integument develops up around the ovule. It starts from the abaxial or outer end of the nucellus and grows inwards towards the base of the ovuliferous scale. The integument completely surrounds the nucellus except at the inner and where a wide aperture lies, known as **micropyle.** The integument is fused to the

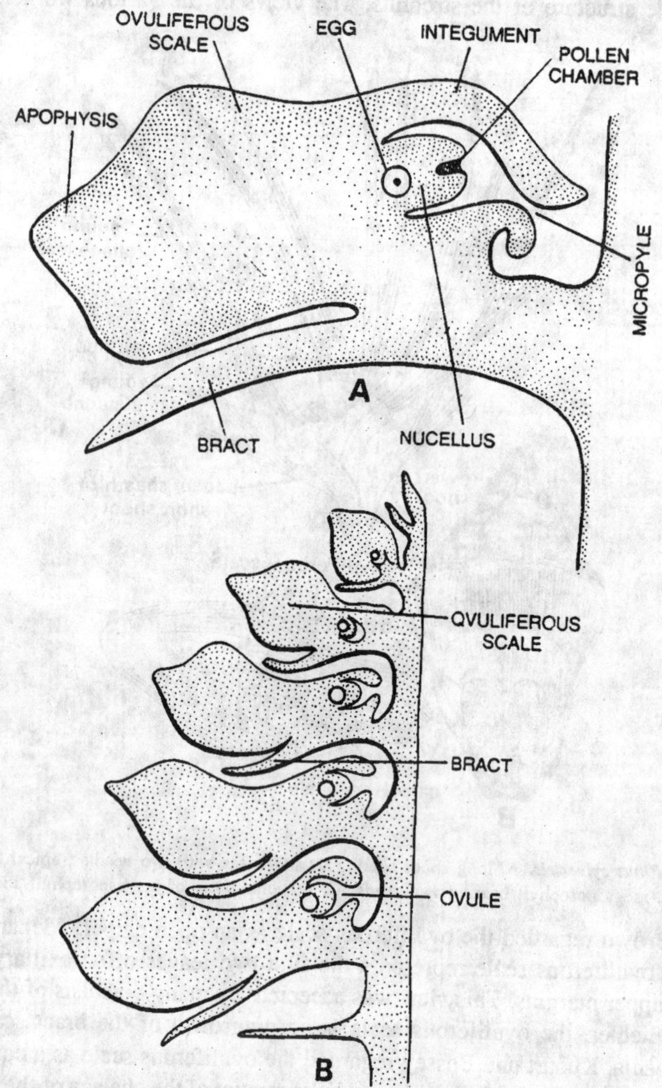

Fig. 4.33. ***Pinus.*** Female cone. A, a single ovuliferous scale showing ovule in longitudinal view; B, longitudinal section of female cone showing bract and ovuliferous scales (diagrammatic).

nucellus except for a short distance near the micropyle. At the apex of the young nucellus a single deep seated hypodermal cell is found. It is known as archesporial cell. The archesporial cell divides forming a tapetal cell and a megaspore mother cell. The megaspore mother cell divides meiotically producing a linear tetrad of four haploid megaspores. The three upper megaspores of the linear tetrad degenerate leaving the lowermost functional megaspore of embryo-sac.

As regards the development the sporogeneous tissue is small and hypodermal. The hypodermal cells cut off the tapetum which forms a more or less nutritive layer around the megaspore. This is described as spongy tissue, which is formed by tapetum surrounding the megaspore mother cell.

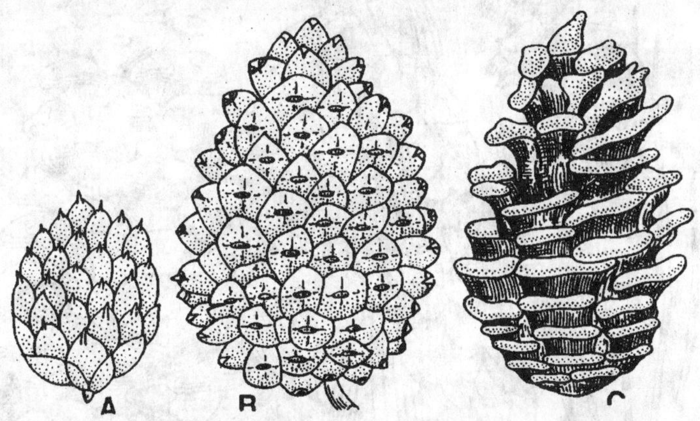

Fig. 4.34. ***Pinus*** **sp. Female cones. A, first year; B, second year and C, third year female cones.**

At this stage the pollination takes place. The development of the embryo-sac takes place after pollination.

Pollination. In *Pinus* the pollination is effected by the wind. The balloon-like wings of the pollen grains aid them to reach the ovule. Usually the pollination occurs towards the end

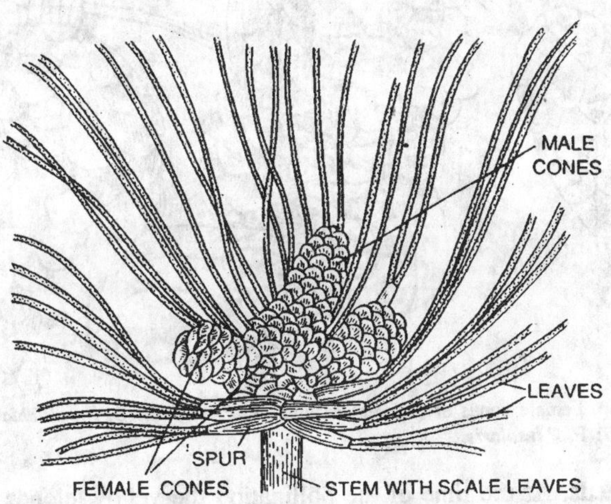

Fig. 4.35. ***Pinus roxburghii.*** **A long shoot with male and female cones.**

of May when the pollen grains are liberated in large quantities from the microsporangia of male flowers. The pollen grains fall on the scales of the female cone. At this time, the edges of the bract scales become incurved separating the ovuliferous scales, which allow the pollen grains to reach to the ovules. Prior to the pollination the scales remain tightly closed around the cone axis and soon after pollination the scales close again and remain in the same condition until

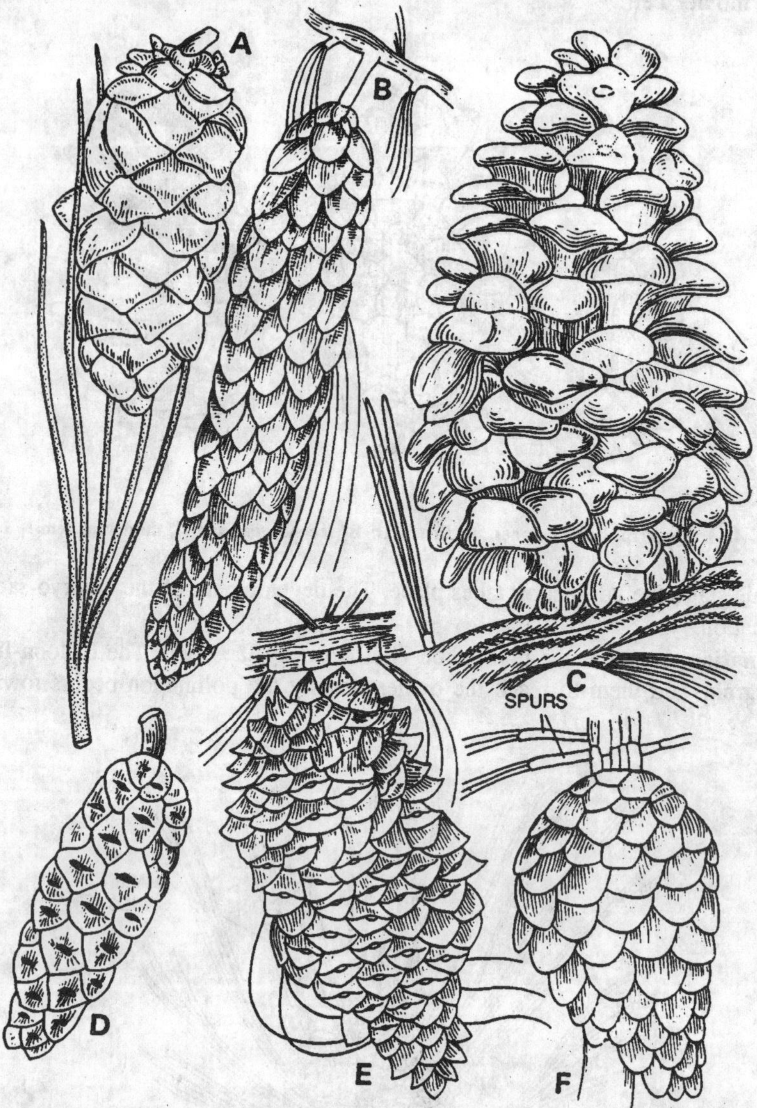

Fig. 4.36. ***Pinus*** **spp. Female cones of different species. A,** ***P.armandi;*** **B,** ***P.wallichiana;*** **C,** ***P.gerardiana;*** **D,** ***P.merkusii;*** **E,** ***P.roxburghii;*** **F,** ***P.insularis.***

the ripening of the seeds. At the time of the pollination the yellow clouds of the microspores may be seen in the atmosphere. Most of the pollen grains go waste. As the ovuliferous scales are separated from each other, the nucellus secretes a mucilaginous fluid. The pollen grains become entangled in this mucilaginous fluid. As the pollination is over this drop of mucilaginous fluid dries up, drawing the pollen grains through the micropyle to the apex of the nucellus, where they germinate. After pollination the scales close, the young cones increase somewhat in size and begin to become hard.

The male gametophyte and its development. At the time of pollination the first nuclear division has already occurred in the pollen grain. The first prothallial cell and the tube cell have also been formed at this stage. The further development of the male gametophyte takes place when it germinates on the nucellus. The tube cell divides forming a small second prothallial

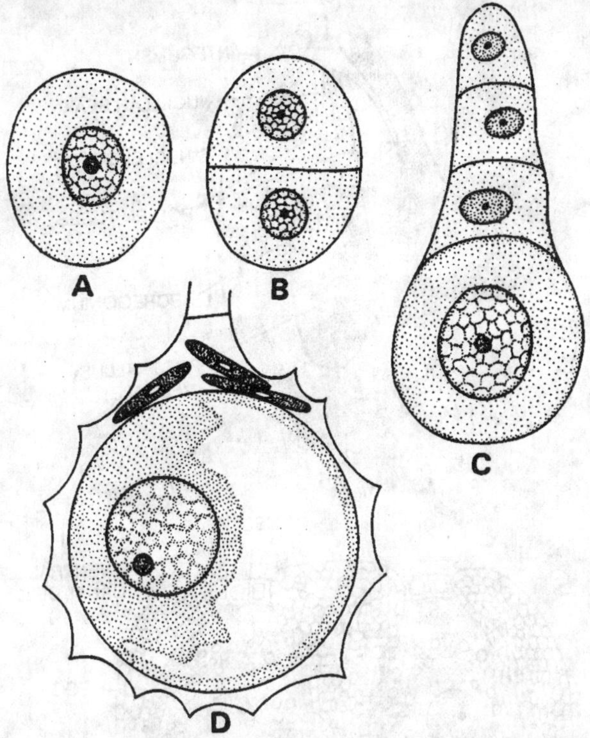

Fig. 4.37. *Pinus* spp. Megasporogenesis. A, megaspore mother cell; B-D successive stages in the formation of single megaspore.

cell and a large tube cell. The first and second prothallial cells soon become flattened and disintegrate. The remaining nucleus divides third time producing a large rounded cell, the antheridial cell on top of the degenerating prothallial cells. The nucleus which remains in the tube cell of the pollen grain is known as tube nucleus. On making the contact with the apex of the nucleus the exo-intine of the microspore or pollen grain breaks between the wings, and the intine grows out in the form of the pollen tune. The tube nucleus passes in the pollen tube. The pollen tube penetrates the nucellar tissue and grows slowly through out the following summer. Now the scales of the female cone become thickened and thus the female cone becomes completely closed. The developing pollen rests there throughout the winter. In the next April, the antheridial cell divides into two cells, forming the **body cell** and **stalk cell**. The stalk cell does not divide further. The body cell divides into two unequal cells with less cytoplasm and large nuclei. These cells are known as **male cells** (see fig. 4.42).

The female gametophyte and its development. During the period between pollination and fertilization many changes take place in the ovule and the cone as a whole. The female cone increases in size and becomes green. This increase in size of the cone takes place due to the enormous growth of the axis and of the ovuliferous scales.

The megaspore is formed when the surrounding nucellus is in earlier stages of development.

Both the megaspore and nucellus continue to grow simultaneously. The nucleus within the megaspore enlarges and begins to divide. Free nuclear division takes place and from 256 to

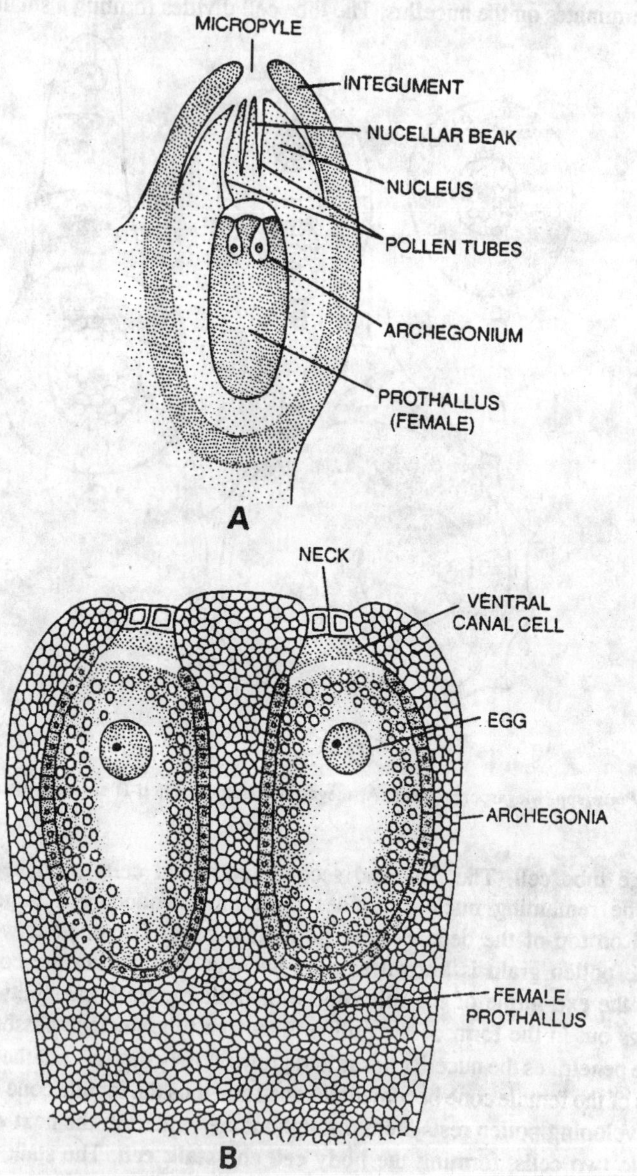

Fig. 4.38. ***Pinus*** **spp.** A, longitudinal section of mature ovule showing pollen tubes and process of fertilization; B, female gametophyte; C, prothallus with two archegonia.

more than two thousand nuclei are formed within the germinating megaspore. After free nuclear division a large vacuole appears in the centre and all nuclei now come to lie near periphery. After this peripheral disposition of the nuclei the wall formation starts. During all this time the megaspore gradually increases in size and its membrane that of the ovule becomes thicker.

The wall formation takes place by the formation of perpendicular walls from the periphery and gradually proceeds towards the centre. As the cell formation proceeds towards the centre, in the beginning the cells remain open, and towards the inner side in the megaspore the walls

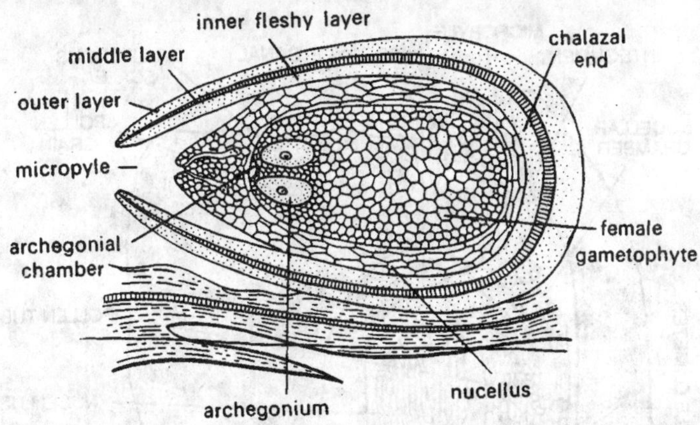

Fig. 4.39. *Pinus roxburghii.* L.S. of mature ovule showing archegonia.

are laid down. In *P.wallichiana*, during the first year the functional megaspore undergoes only a few free nuclear divisions, and in the next year the free nuclear division continues and cell formation begins in the following May.

Thereafter the embryo sac or megaspore is cut into a number of radial spaces in the form of long tubes, containing several nuclei described as **alveoli.** These alveoli project towards the centre of the germinating megaspore. They are open on the inner side in the beginning but later on they become closed and new cross walls are laid down in these tubes. They form the

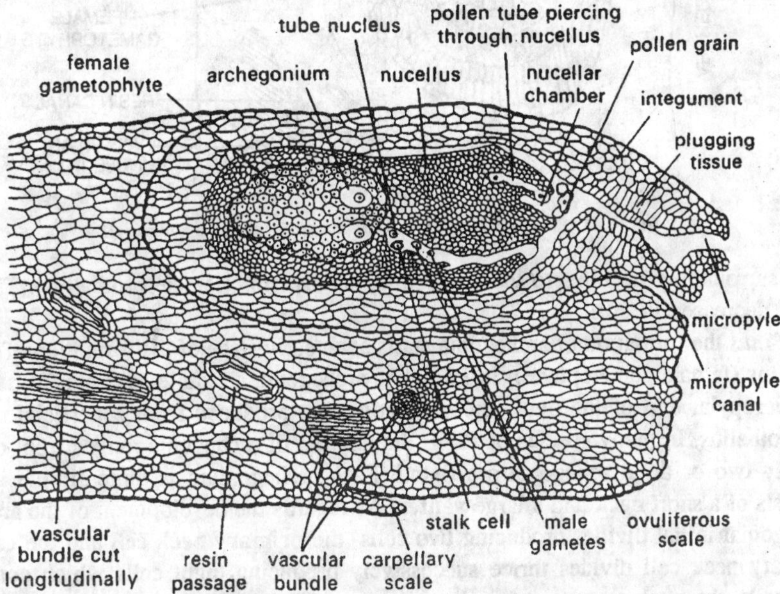

Fig. 4.40. *Pinus* roxburghii. Longitudinal section of mature ovule (cellular).

endosperm or female prothallus (female gametophyte) within the megaspore.

In early stages of the development of megaspore, it is surrounded by a layer of cells formed by tapetum, this layer forms the endosperm jacket. The nutritive matter is gradually passed on

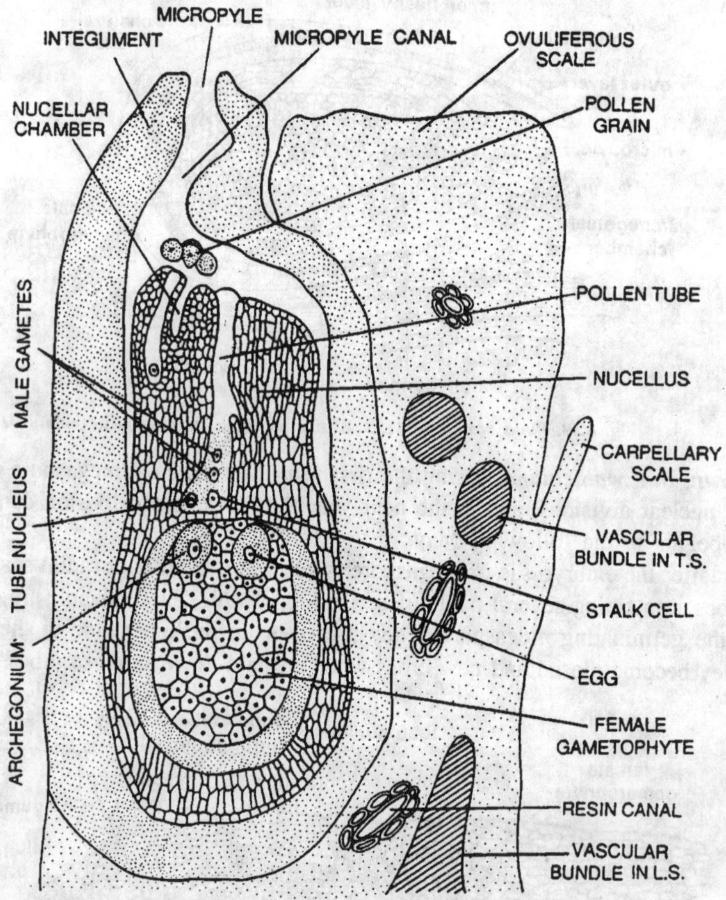

Fig. 4.41. *Pinus roxburghii.* **L.S. of mature ovule (diagrammatic).**

from the surrounding nucellus to the developing endosperm (female prothallus). The nucellus gradually decreases in size.

In *Pinus* the formation of archegonia starts at an early stage of the formation of endosperm or prothallus (female gametophyte). In *Pinus* the archegonia are quite separate from each other. At the micropylar end of the ovule the archegonia are produced from superficial cells of the female prothallus. In many species of *Pinus* the number of archegonia ranges from one to five, but usually two or three achegonia are found. Each archegonium is quite simple in structure and consists of a short **neck** and a large **venter**. As regards the development of the archegonium the archegonial initial divides producing two cells- the **primary neck cell** and the **central cell**. The primary neck cell divides thrice successively producing eight cells which constitute the neck. Usually the neck consists of eight cells arranged in two tiers of four cells each.

The central cell divides to produce an **oosphere** (egg) and the **ventral canal cell**. The cells surrounding the archegonium form a jacket layer which supplies food to the developing egg. The cells of the female prothallus surrounding the archegonium grow more vigorously than the neck cells so that an **archegonial chamber** is resulted.

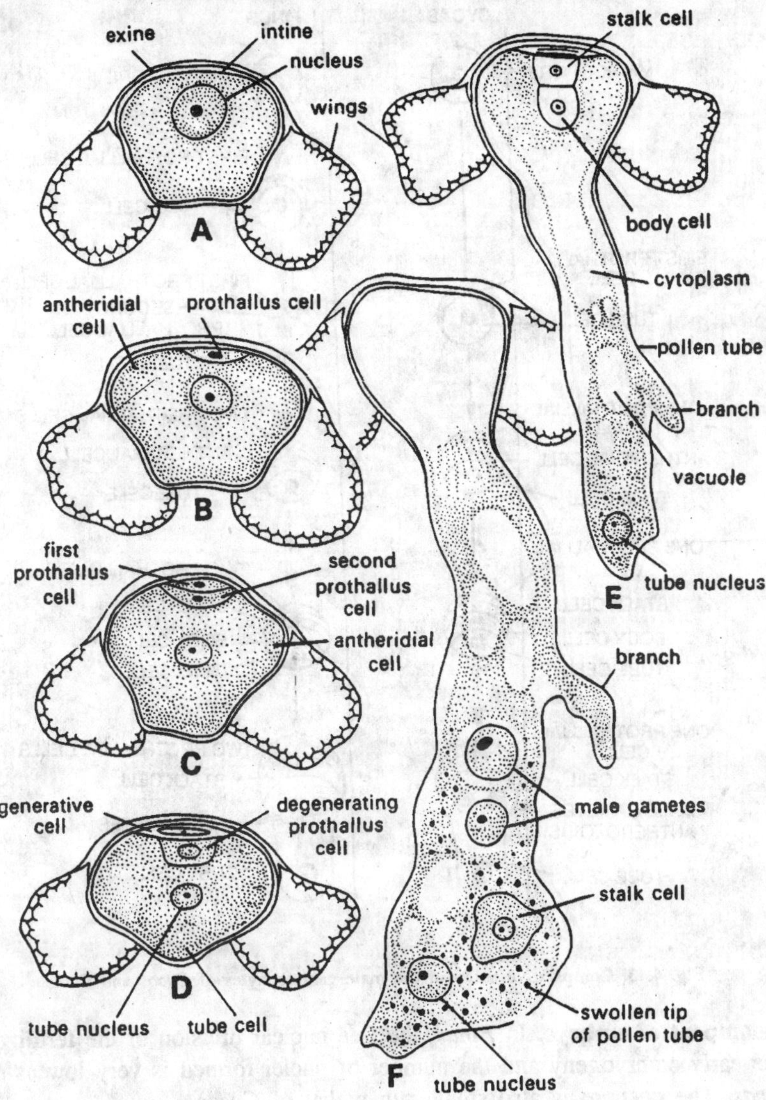

Fig. 4.42. *Pinus* spp. Development of male gametophyte. A, a pollen grain; B-D, successive stages in the development; E, pollen tube with nucleus; F, two unequal male gametes, stalk nucleus, tube nucleus within pollen tube.

Fertilization. In the month of April of the second year the pollen tube completes its resting period and again becomes active and the two unequal male cells are formed from the body cell. The pollen tube elongates until it reaches the neck of an archegonium. The end of the pollen tube enters the neck and bursts discharging the stalk cell, tube nucleus and the two unequal male cells. The larger nucleus enters the egg and fuses with the female nucleus, thus effecting

the fertilization. The other nuclei, however, die and disappear. With the result of fertilization, a thick-walled, diploid (2n) oospore is formed. The fertilization is completed by the end of June.

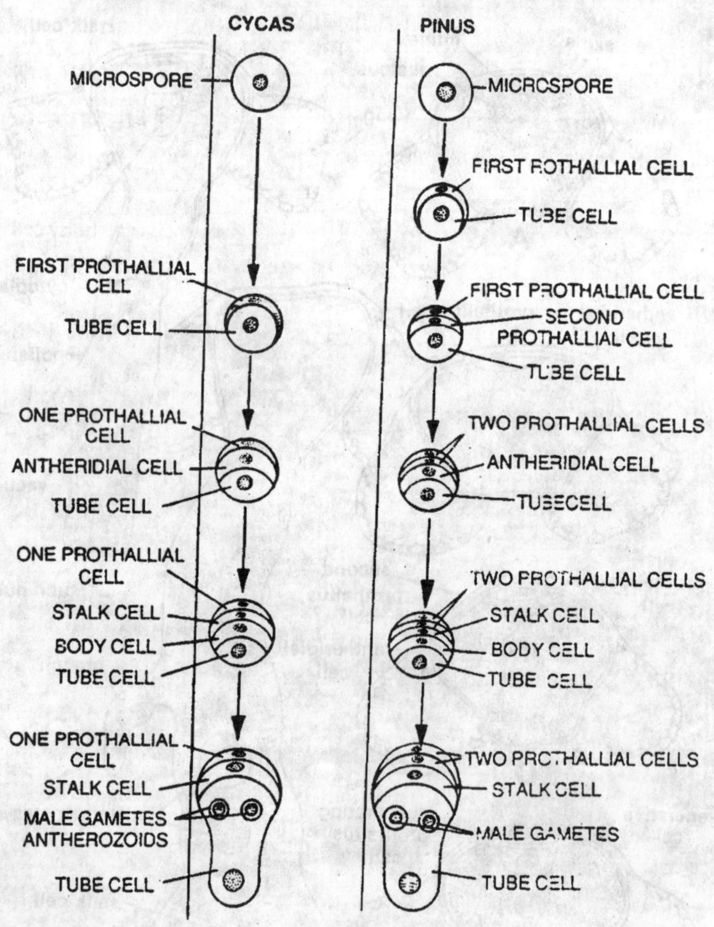

Fig. 4.43. Comparative study of the male gametophytes of *Cycas* and *Pinus*.

Develompent of embryo. In *Pinus*, the free nuclear division of the fertilized egg takes place in the early embryogeny and the number of nuclei formed is very low as compared to that of *Cycas*. The oospore is also smaller than that of *Cycas*.

In *Pinus*, the fertilized diploid nucleus of the oospore divides twice to form four nuclei which come to lie side by side at the bottom of the egg. Thereafter a third nuclear division takes place forming the eight nuclei which are arranged in two tiers of four each.

The cell walls are laid down across the four basal nuclei, and this way four small cells have been produced at the lower end of the oospore. The upper free nuclei are separated only by imperfect walls and they do not take any part in the formation of the embryo. The lower cells again divide and this way, three tiers of four cells each have been formed. Once again the lowermost nuclei divide and four tiers of 4 cells each have been formed. This structure

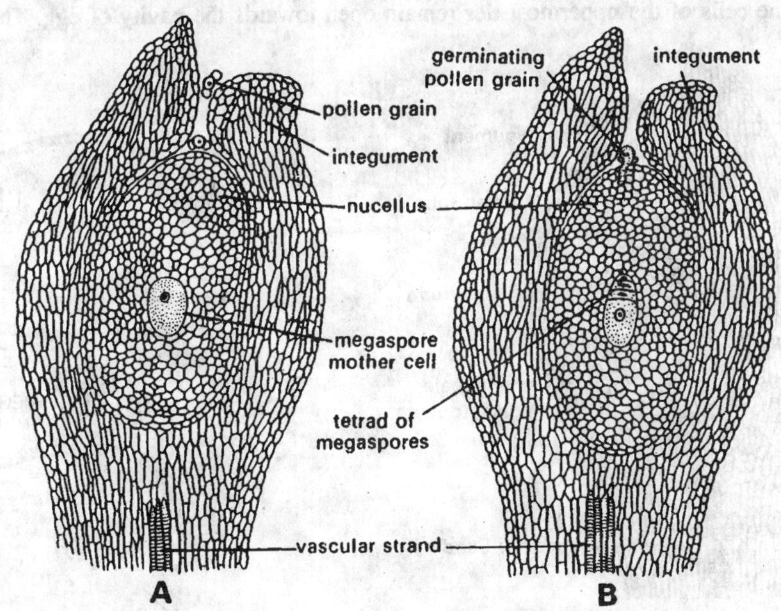

Fig. 4.44. ***Pinus roxburghii.*** **A, V.S. of young ovule showing a megaspore mother cell in the nucellus; B, V.S. of young ovule showing a tetrad of megaspores (haploid n) in the nucellus.**

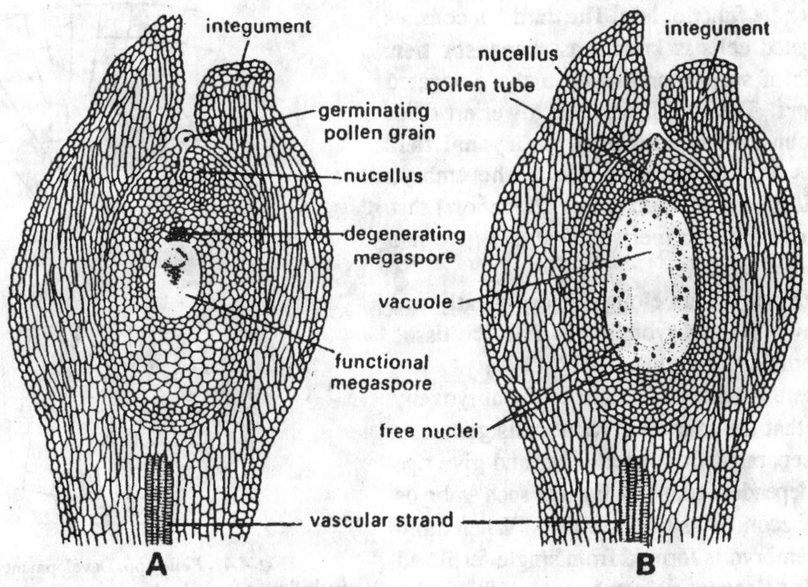

Fig. 4.45. ***Pinus roxburghii.*** A, V.S. young ovule showing enlargement of functional megaspore and degeneration of other three; B, V.S. ovule with female gametophyte in free nuclear stage.

consists of sixteen cells four tiers of four cells each, is known as **proembryo.**

The cells of the uppermost tier remain open towards the cavity of egg. This tier of four

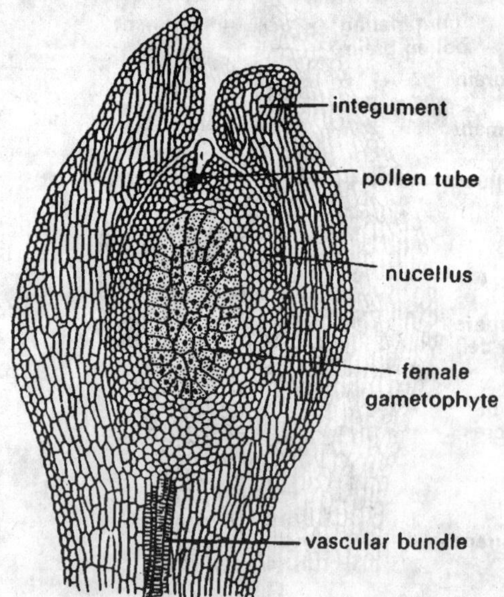

Fig. 4.46. *Pinus roxburghii.* V.S. ovule with young cellular female gametophyte.

cells is known as **open tier**. The second tier of four cells constitutes the **rosette tier**. The cells of this tier are functionless. The third tier consists of elongated cells is known as **suspensor tier**. The cells of suspensor tier form the elongated **suspensors**. The fourth and the lowermost tier of the four cells is known as **embryonal tier**. The cells of this tier give rise to the embryo proper. It is seen in some cases (Buchholz) that rosette cells are also meristematic and give rise to secondary embryos.

The suspensors elongate very greatly and thrust down the embryonal cells into the tissue of the female prothallus.

There is a peculiar feature of embryogeny in *Pinus* that the four embryonal cells generally become separated from each other and give rise to four independent embryos. Each of such embryos develops secondary suspensor cells. When more than one embryo is formed from single fertilized egg is called **cleavage polyembryony.** Ultimately, only one of these embryos survives and becomes mature. In *Pinus*, the cleavage polyembryony can be recognized both by counting the number of embryos produced from each zygote and by

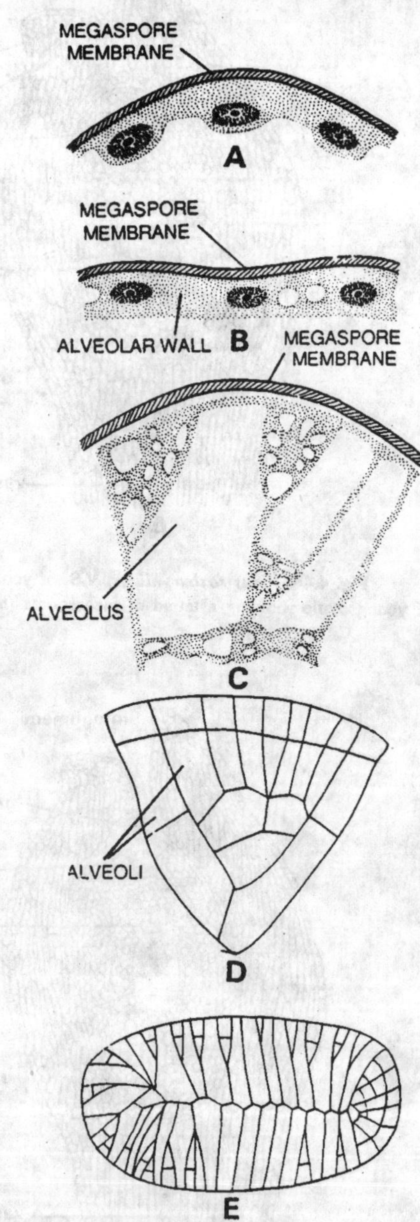

Fig. 4.47. *Pinus* spp. Development of the female prothallus. A and B, alveolation; C, alveoli extending into central vacuole; D, alveoli; E, completion of the formation of alveoli to form the female prothallus.

tracing each embryo back to single-celled suspensor (Buchholz, 1926).

Structure of embryo. The mature embryo consists of a short axis with the **radicle** towards the micropylar end and a small **plumule** downwards. The plumule is surrounded by a number

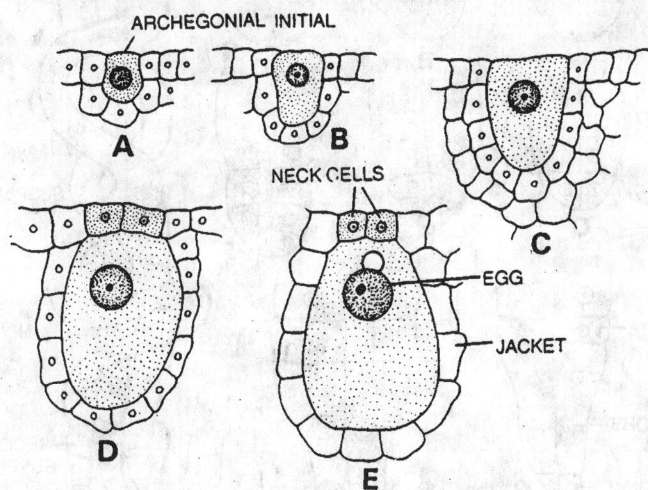

Fig. 4.48. ***Pinus*** **spp. A-D, successive stages in the development of archegonium; E, mature archegonium.**

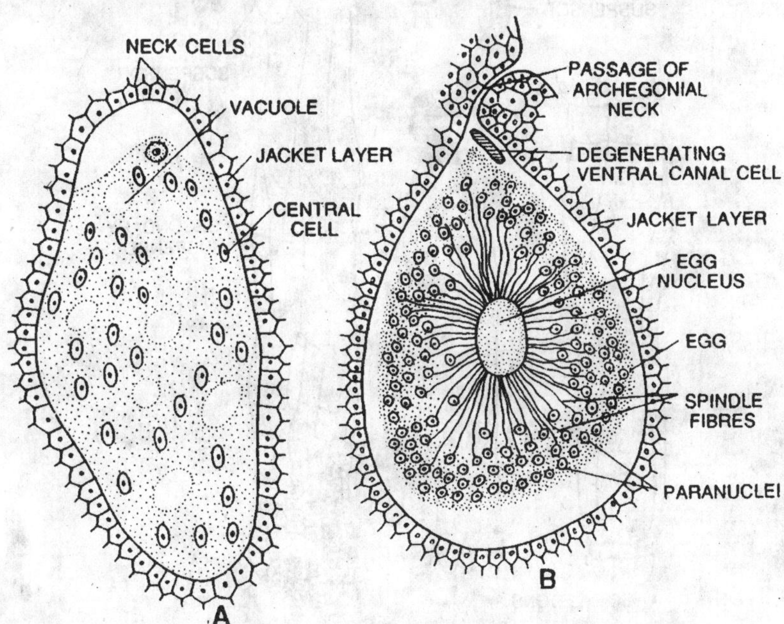

Fig. 4.49. ***Pinus roxburghii.*** A, a young archegonium with two neck cells and a central cell; B, a mature archegonium with degenerating ventral canal cell.

of tiny leaves or **cotyledons.** The cotyledons may be ten or so in number. The suspensor remains attached as a small thin coil to the tip of the radicle which forms a thick cap over it. The mature embryo is embedded in the **endosperm** which is laden with food material. The nucellar

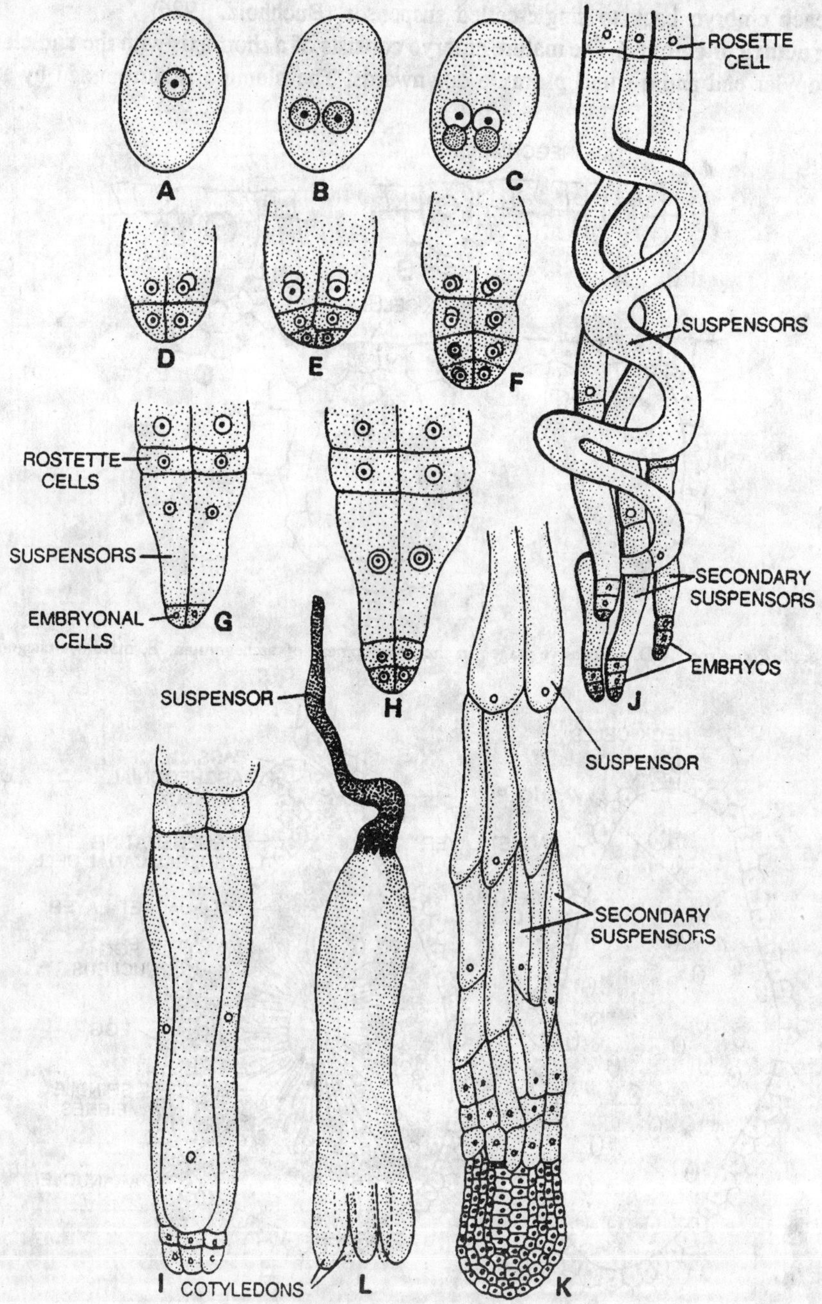

Fig. 4.50. ***Pinus*** **spp.** Development of Embryo, A, oospore; B-H, successive division of the oospore; I, primary suspensors elongate; J, formation of secondary suspensors and embryos; K, formation of single embryo; L, formatior of cotyledons.

tissue is altogether crushed and disorganized because of the expansion of endosperm and embryo, however, the remaining thin layer of nucellus is known as **perisperm**

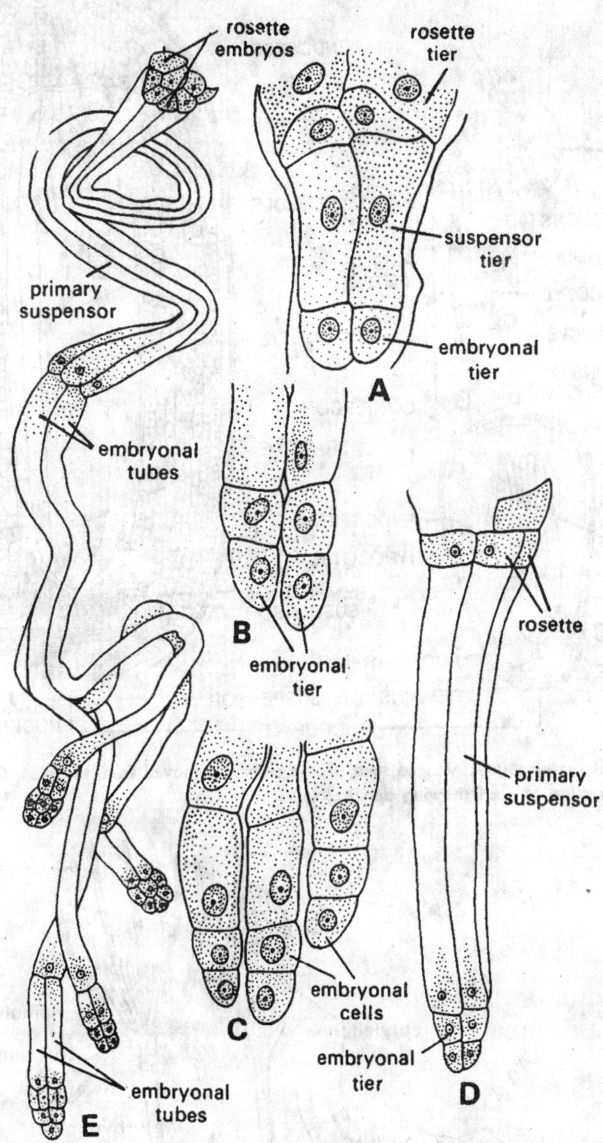

Fig. 4.51. *Pinus walichiana.* A-E, later stages in the development of embryo.

The seed. The seed contains the following structure:

1. **Embryo.** It contains a straight mature embryo. The structure of the mature embryo has already been described in preceding lines.

2. **Endosperm.** The endosperm is laden with food material. The embryo remains embedded in the endosperm and absorbs its nutrition from it.

3. **Perisperm.** A thin layer of crushed nucellus persists which is known as perisperm.

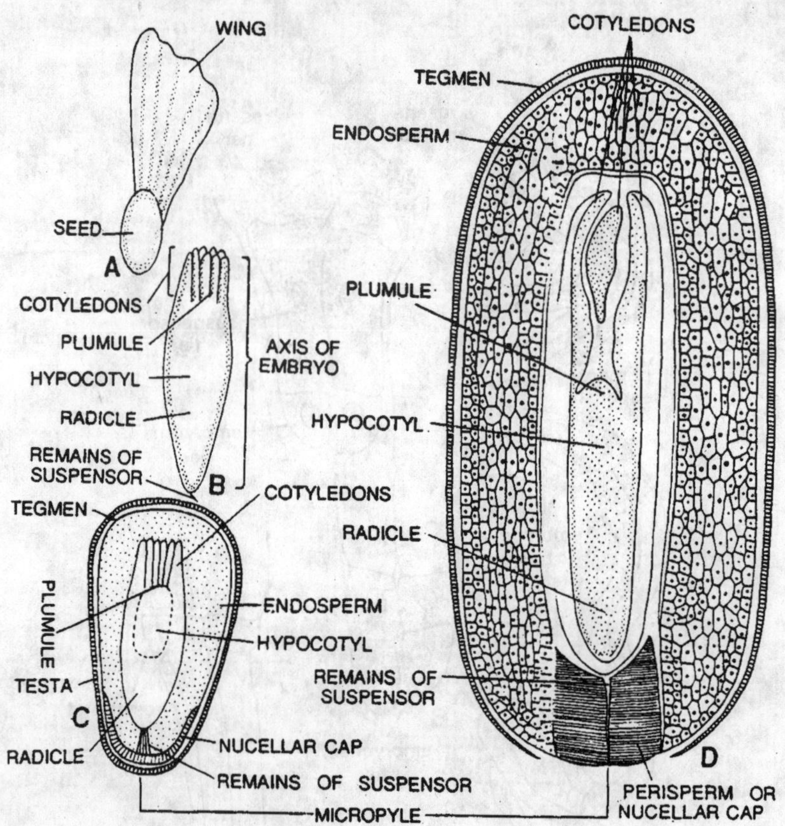

Fig. 4.52. *Pinus roxburghii.* A, winged seed; B, the embryo removed from the seed; C, longitudinal section of seed; D, longitudinal section of seed showing details.

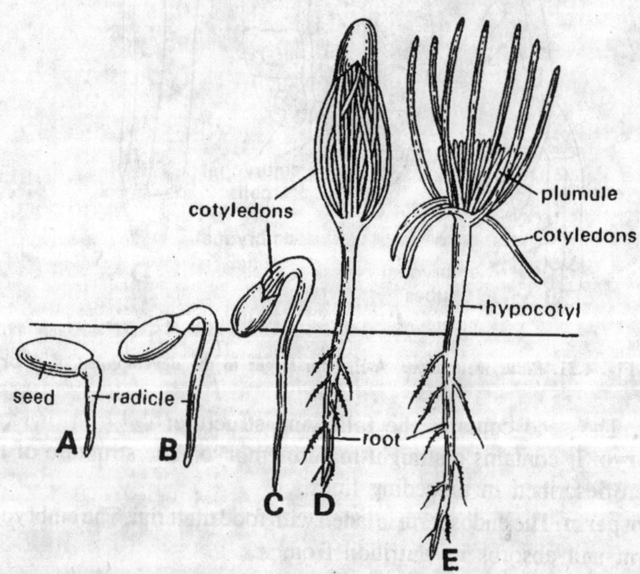

Fig. 4.53. *Pinus roxburghii.* A-E, stages in the germination of seed.

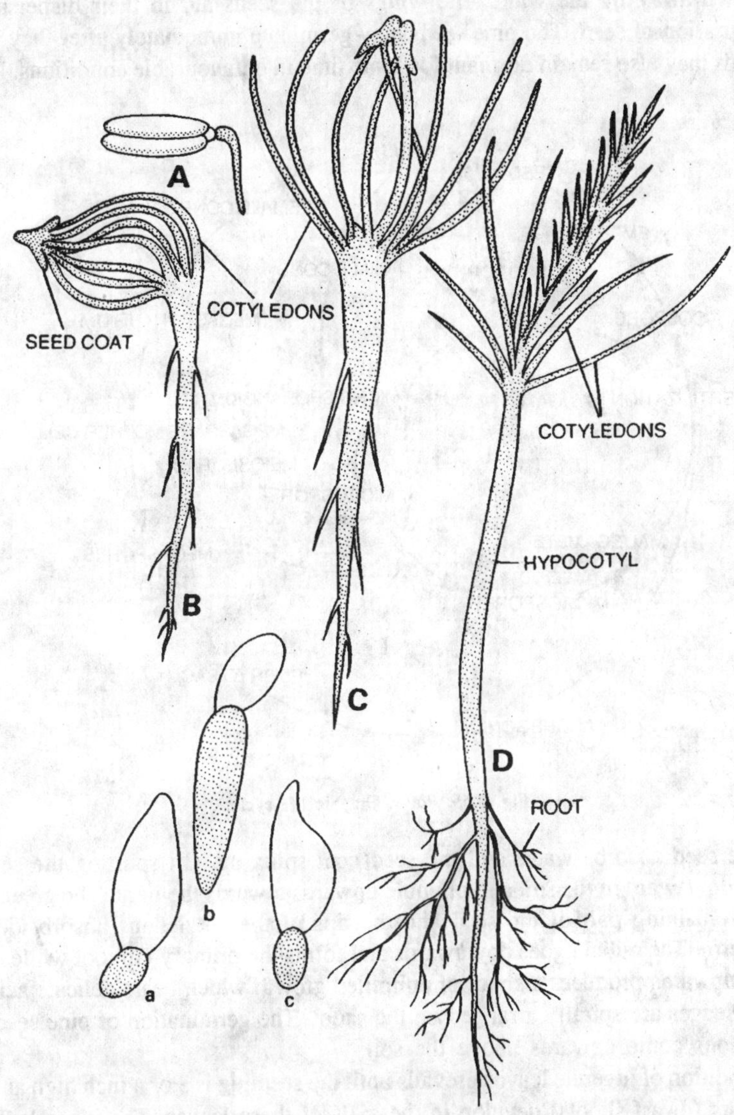

Fig. 4.54. ***Pinus*** **spp. Seed and its germination. A-D, successive stages in the formation of seeding. (a) seed of** ***P. roxburghii*** **(b) seed of** ***P. gerardiana;*** **(c) seed of** ***P. wallichiana.***

4. Testa. The integument of the ovule becomes the seed coat or testa. The testa is hard and stony as it develops from the middle stony layer of the integument.

5. Wing. The seed has a thin membranous wing which is derived from the surface of the ovuliferous scale. The wing helps in the dispersal of the seed.

Dispersal of seeds. In the third year when the female cone reaches maturity the cone becomes dry, brown and woody. Each ovuliferous scale bears two mature seeds placed side by side on its upper surface. When the seeds are mature the axis of cone elongates and the scales are separated from each other leaving spaces among them. The seeds are liberated and they are blown away by the wind. The wings of the seeds aid in their dispersal.

Germination of seed. The pine seeds may germinate immediately after they fall on moist soil. The seeds may also remain dormant for some time in unfavourable conditions. In favourable

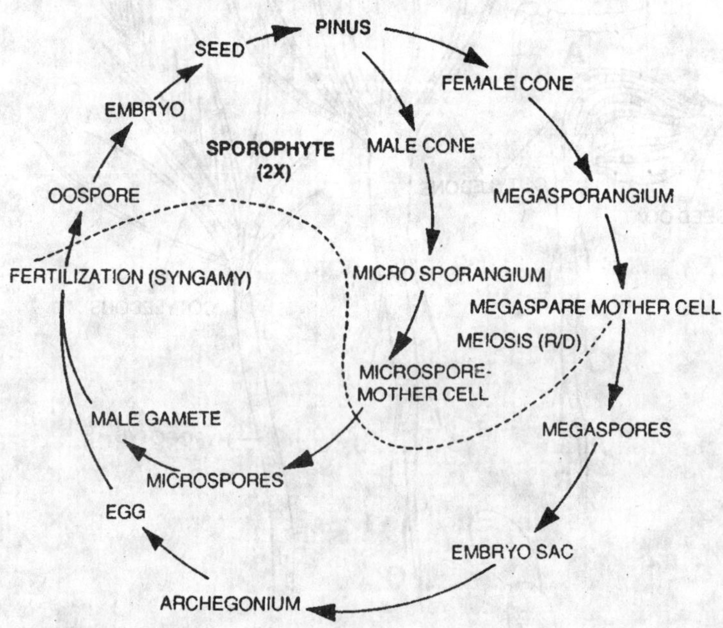

Fig. 4.55. *Pinus*. Graphic life-cycle.

conditions the seed absorbs water and the seed coat splits up. On splitting the seed coat the radicle grows downward in the soil and plumule upwards towards the light. The green cotyledons carry up the remaining part of the seed and the tips of the cotyledons absorb the remainder of the endosperm. The radicle goes downwards and forms the primary tap root while the plumule which passes upwards produces a shoot of unlimited growth which bears delicate acicular green leaves. These leaves are spirally arranged on the shoot. The germination of pine seed is **epigeal** as the cotyledons come upwards above the soil.

This condition of juvenile leaves prevails until the seedling is 3 or 4 inch high and thereafter first foliar spurs (dwarf shoots) develop in the axils of these leaves. The juvenile condition is a primitive feature of *Pinus*. The later formed juvenile leaves of the seedling become smaller and smaller and pass on into the scale leaves of the stem.

Economic importance. Several species of *Pinus* yield wood which is used for building material, furniture, poles, match boxes and other such articles. The wood of *P. wallichiana* (**Kail**) and *P. roxburghii* (**Chir**) yield good timber. The wood is also used as fuel because it catches fire quite easily because of the presence of resin. The old female cones are also used as fuel and decoration pieces.

The pine trees yield large amount of resin which is used to manufacture turpentine. Resin and turpentine are used to make paints, varnishes and medicines.

The seeds of *P. gerardiana* are edible and known as '**chilgoza**'. The seeds of *P. roxburghii* are also edible.

Many species of *Pinus* make cheap sources of cellulose which is used for various purposes. Usually the T.B. sanatoria are situated in the pine forests because of attractive and healthy atmosphere.

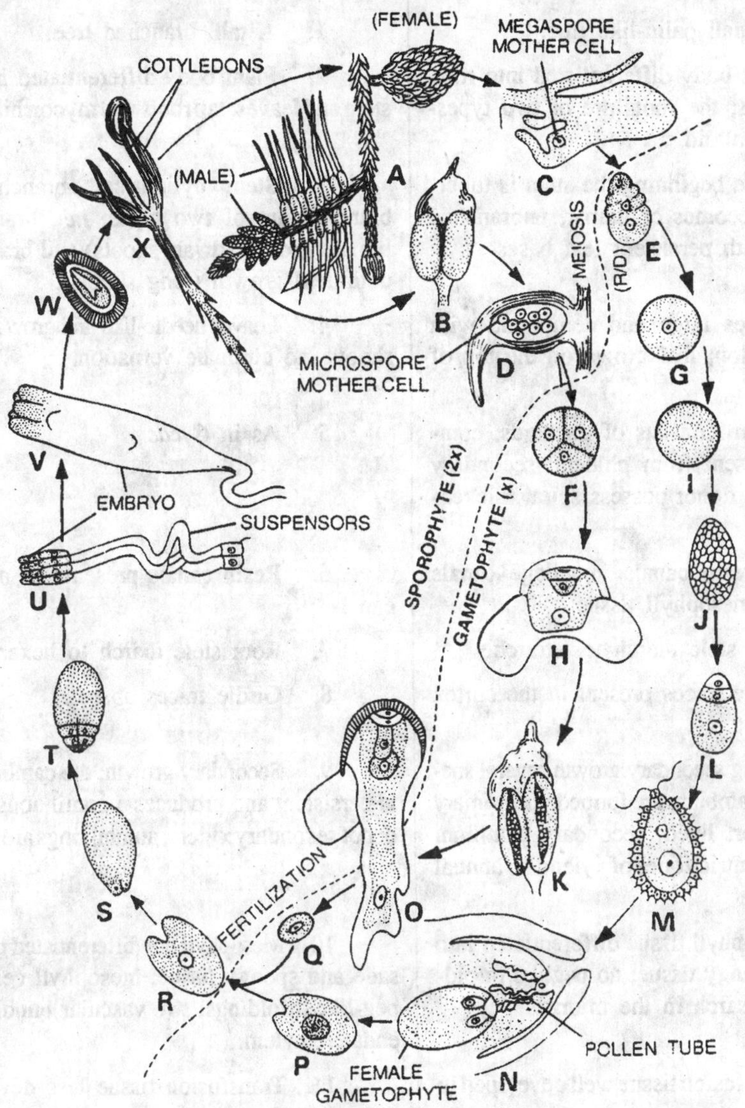

Fig. 4.56. *Pinus* spp. Diagrammatic life-cycle. A, a twig with male and female cones; B, microsporophyll with microsporangia; C, ovuliferous scale with ovule; D, microsporangium with microspore mother cells; E, megaspore linear tetrad; F, microspore tetrad; G, megaspore; H, microspore; I-J, formation of female prothallus; K, dehiscence of microspores; L-M, development of archegonium; N, mature ovule and fertilization; O, male gametophyte; P, egg; Q, male gamete; R, oospore; S-V, successive stages in the development of embryo; W, seed; X, seedling.

In India *P. gerardiana* provides edible seeds, 'chilgoza'; *P. roxburghii* and *P. wallichiana* yield timber; *P. roxburghii* and *P. insularis* yield resin and turpentine.

Comparison of *Cycas* and *Pinus*

Cycas	*Pinus*
1. A small palm-like tree.	1. A tall, branched tree.
2. Plant body differentiated into root, stem and leaves; the roots are of two types-normal and coralloid.	2. Plant body differentiated into root, stem and leaves; tap roots and mycorrhizic roots.
3. In the beginning the stem is tuberous, later on becomes columnar, unbranched and covered with persistent leaf bases.	3. Stem cylindrical, branched; the branches are of two types, *i.e.*, branches of limited growth (dwarf shoots) and branches of unlimited growth (long shoots).
4. Leaves large and compound with circinate vernation; leaf crown on the top of the plant.	4. Leaves needle-like and grow on dwarf shoots; no circinate vernation.
5. Xylem consists of tracheids; companion cells absent from phloem; secondary xylem tracheids donot possess spiral thickenings.	5 As in *Cycas*.
6. No resin canals; mucilage canals present in the mesophyll tissue.	6. Resin canals present; no mucilage canals.
7. Root stele diarch to tetrarch.	7. Root stele diarch to hexarch.
8. Girdle traces present in the cortex of stem.	8. Girdle traces absent.
9. During secondary growth several successive rings of cambium are formed; the primary cambium is short lived; secondary cambium rings form concentric zones of xylem; no annual rings.	9. Secondary growth; the cambium ring is persistent and produces a continuous cyclinder of secondary xylem; annual rings are formed.
10. Mesophyll tissue differentiated into palisade and spongy tissue; no peglike infoldings; xylem mesarch in the midrib.	10. Mesophyll not differentiated into palisade and spongy tissue; mesophyll cells have peg-like infoldings; two vascular bundles with endarch xylem.
11. Transfusion tissue well developed in the leaves.	11. Transfusion tissue less developed.
12. Vegetative reproduction by buds or bulbils.	12. No vegetative reproduction.
13. Plants usually dioecious; cones simple and unisexual.	13. Plants usually monoecious; cones simple and unisexual.
14. Sporophylls differ from foliage leaves, arranged in cones, *i.e.*, male and female cones.	14. Sporophylls arranged in male and female cones.

Cycas	*Pinus*
15. Microsporophyll triangular, woody, not peltate bearing numerous microsporangia.	15. Microsporophyll scaly, not peltate, bearing two microsporangia.
16. Male cone terminal, large bearing numerous microsporophylls.	16. Male cones small, grouped in clusters, replace the dwarf shoot and not main axis.
17. Microspores (pollen) not winged.	17. Microspores (pollen) winged, exine inflated to form balloon-like expansions.
18. Female cone has a whorl of loosely arranged megasporophylls.	18. Female cone compact and corresponds to an inflorescence; ovuliferous scale represents female flower.
19. Each megasporophyll quite large in size, bearing 1-5 pairs of big ovules.	19. Each megasporophyll or carpel is quite simple; each ovuliferous scale bears two ovules on its upper surface near the central axis.
20. Ovule bears 1-2 integuments.	20. Ovule bears single integument.
21. One megaspore (embryosac) in each megasporangium (ovule).	21. As in *Cycas*.
22. A sharp nucellar beak and a pollen chamber.	22. Pollen chambei and nucellar beak absent.
23. Nucellus free in the upper part or through out and not fused with the integument.	23. As in *Cycas*.
24. At the time of fertilization the embryo sac consists of a gametophyte having many food laden cells and 2-3 archegonia.	24. At the time of fertilization the embryosac consists of a gametophyte having many food laden cells bearing 2-3 archegonia.
25. 3-celled pollen are shed from the pollen sacs and pollinated by wind. Pollen grains are caught in the mucilage drop oozing out from the micropyle. As the drop dries the pollen are taken back into the pollen chambers.	25. Wind pollination; pollen grains enter the open ovuliferous scale and get entangled in the oozing mucilage drop from micropyle.
26. Pollen deposited directly on the nucellus during pollination; further development of pollen is completed in pollen chamber.	26. As in *Cycas*.
27. Microspore germinates *in situ* (within microsporangium) while the germination is completed on the nucellus after pollination.	27. As in *Cycas*.
28. The microspore cuts off only one prothallus cell, which is persistent (male gametophyte).	28. The microspore cuts off two prothallial cells towards the pointed end (male gametophyte).
29. Generative cell produces two male gametes while large vegetative cell gives rise to tube cell and stalk cell.	29. As in *Cycas*.

Cycas	*Pinus*
30. The male gametes are large and multi-flagellate.	30. Male gametes small non-motile.
31. The motile sperms move in the pollen tube and reach the neck of the archegonium; here the pollen tube acts as haustorium.	31. The non-motile sperms are carried to the mouth of archegonia through pollen tube; pollen tube acts as sperms carrier and haustorium.
32. The female gametophyte is formed by the free nuclear division of the nuclear embryo sac cell; wall formation starts when numerous (1000) nuclei have been formed within the megaspore.	32. As in *Cycas*.
33. The megaspore germinates inside the megasporangium (ovule).	33. As in *Cycas*.
34. Female gametophyte bears 3-6 archegonia per ovule; neck present; no ventral canal cell; ventral nucleus present.	34. Female gametophyte bears one or two archegonia; no neck canal cell.
35. Female gametophyte parasitic on nucellus.	35. As in *Cycas*.
36. The end of pollen tube bursts, discharges antherozoids in archegonial chamber where they swim and approach the archegonial necks; a spermatozoid enters the neck of archegonium; the male nucleus fuses with the egg nucleus resulting in the formation of oospore.	36. Usually a time interval of one year in between pollination and fertilization; fertilization occurs in the end of April; pollen tube reaches the neck of archegonium; the neck breaks and receives the contents from pollen tube; male nucleus fuses with the egg nucleus forming oospore.
37. Endosperm develops from the vegetative parts of the female gametophyte, before fertilization; it is haploid.	37. As in *Cycas*.
38. The oospore divides by free nuclear division and the wall formation starts when numerous nuclei are formed.	38. Wall formation starts when eight-nucleate stage is attained.
39. Proembryo has no clear tiers.	39. Proembryo has four distinct tiers.
40. Only one embryo is formed; no cleavage polyembryony.	40. Four potential embryos, but only one matures; common occurrence of cleavage polyembryony.
41. Seed has no perisperm; two cotyledons; testa fleshy and coloured; aril absent; endospermic.	41. Perisperm present; numerous cotyledons in each seed; no aril; testa dry; endospermic.
42. Seeds have no wings.	42. Seeds winged.
43. Hypogeal germination of seed.	43. Epigeal germination of seed.

Family-Cupressaceae

This is the largest family of order Coniferales. There are 20 genera and 148 species in this family. About half of the genera are confined to northern hemisphere such as *Thuja, Biota, Cupressus, Juniperus* etc. The remaining genera such as *Callitris, Libocedrus, Papuacedrus* etc., are commonly found in southern hemisphere. The characteristics of genus *Biota* are mentioned here.

Genus BIOTA

The sporophyte is small conical or pyramid-like tree. They are commonly grown in the houses and parks as ornamentals. The leaves are small and scale-like. The leaves are commonly opposite or whorled. The shoot is not dimorphic. Generally the sporophyte is dioecious. The megasporangiate strobili are small and dry. The ovuliferous scales are generally overlapped on each other. The ovuliferous scales are comparatively bigger than the bract scales. Commonly the ovules are found in erect position.

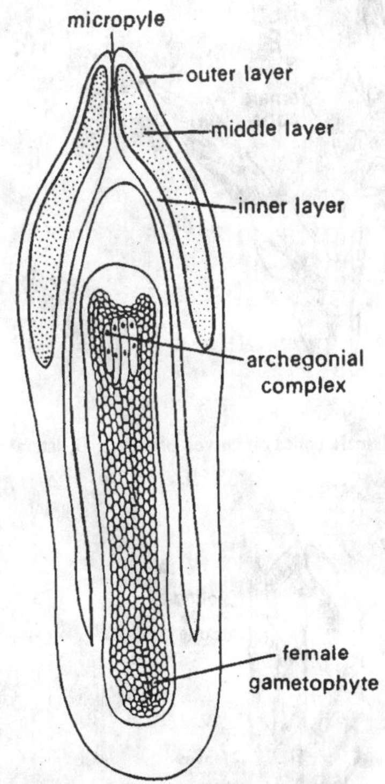

Fig. 4.57. *Biota orientalis*. L.S. of mature ovule.

The microsporangiate strobili are small in size. The whorls of microsporangiophores are found on the central axis of microsporangiate strobili. The microsporangia are somewhat spherical, and they are developed from 2-3 hypodermal cells. The young sporangium contains external wall layer, middle layer and tapetum. The tapetal layer surrounds the group of mother cells. These mother cells divide meiotically and form microspore tetrads. Each microsporangiophore bears 2-6 spherical microsporangia on it. The pollens are wingless. The pollination takes place by means of pollination drop mechanism. The male gametophyte has no prothallial cells. The pollens (pollen grains) are liberated from microsporangia (pollen sacs) when they are two-celled. This two-celled stage consists of one generative cell and the other tube cell.

The microgametes (male gametes) are of same size. The single pollen tube enters the two archegonia by its branching, or it enters the single archegonium without any such branching. The pollens carried away by wind are entrapped in the pollination drop, and gradually become smaller in size. The processes of pollination, fertilization and development of seed are being completed in the same season.

The outermost layer of the sporangium of *Biota orientalis* is cutinized and bears annular thickening. The spore tetrads are either isobilateral or tetrahedral.

The integument remains separated from nucellus. In the young ovule of *Biota orientalis*, the integument near its base is 3-4 celled thick, while at the micropylar end it is single-celled in thickness. After pollination, the integument cells neighbouring the micropyle, elongate radially and close the canal. The differentiation of the archesporial cells within the nucellus takes place at sufficient depth in the tissue, just near the chalaza. Generally they are two in number. The

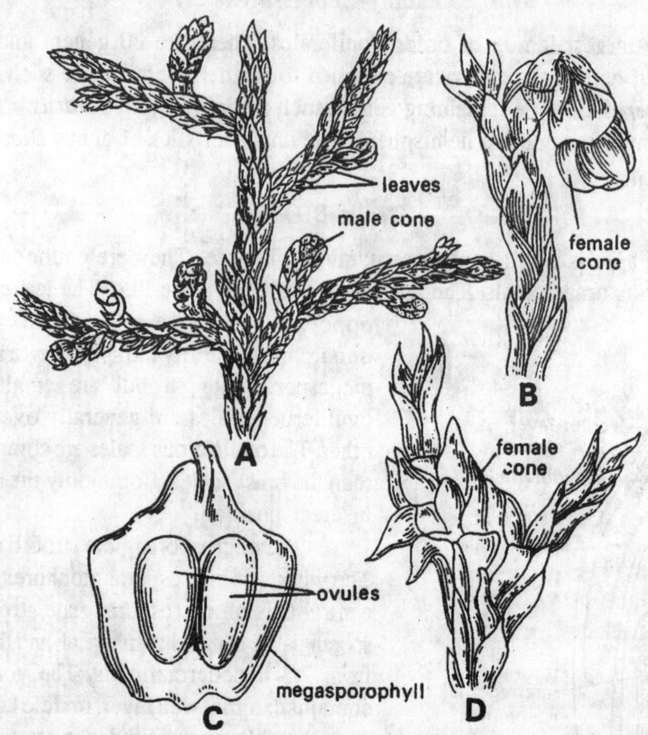

Fig. 4.58. ***Biota orientalis*. A, branch bearing male cones; B-C, female cones on curved branches; D, lateral view of megasporophyll bearing ovules.**

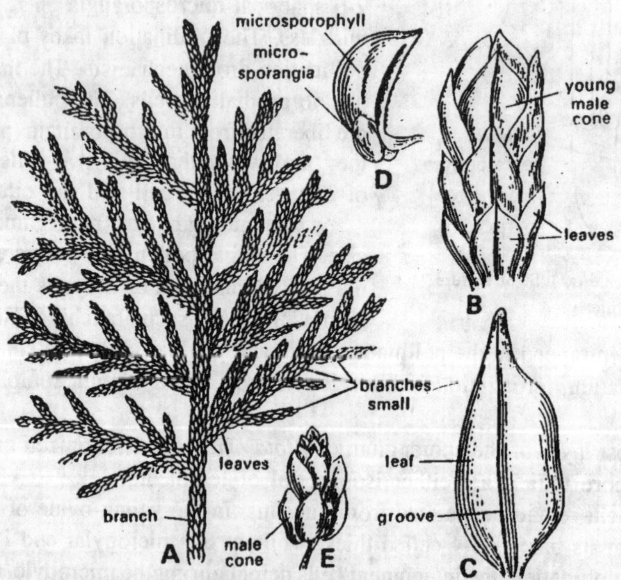

Fig. 4.59. ***Biota orientalis*. A, a branch bearing smaller branches and leaves; B, a young male cone covered by leaves; C, abaxial surface of a green leaf showing a groove running throughout its length; D, a microsporophyll as seen from one side with microsporangia; E, a male cone as seen after dehiscence.**

archesporial cells divide repeatedly, and form a group of sporogenous cells, of which one of the cells becomes megaspore mother cell.

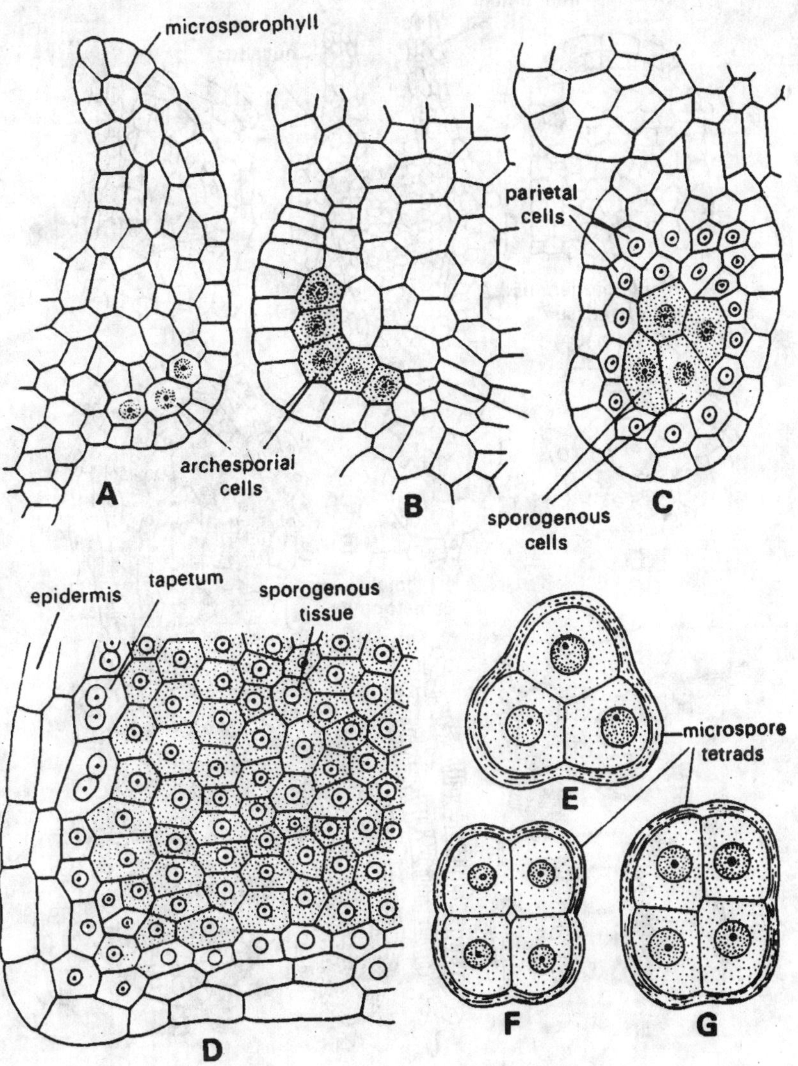

Fig. 4.60. *Biota orientalis.* A-B, L.S. of microsporophyll showing a row of hypodermal archesporial cells that will develop into a microsporangium; C, L.S. of microsporophyll showing sporogenous cells and parietal cells; D, L.S. of young microsporangium showing wall layers, tapetum and spore mother cells; E-G, microspore tetrads.

This megaspore mother cell divides meiotically and a linear tetrad is formed. In *Biota*, this linear tetrad consists of three cells only.

The active megaspore remains surrounded by a conspicuous spongy tissue. The megaspore enlarges in its size. Its nucellus divides repeatedly forming 4,000 independent nuclei (Singh and Oberoi, 1962). These nuclei remain arranged in the peripheral cytoplasm around a big vacuole.

The walls are formed around the nuclei centripetally. The developing gametophyte near about consumes the whole nucellus.

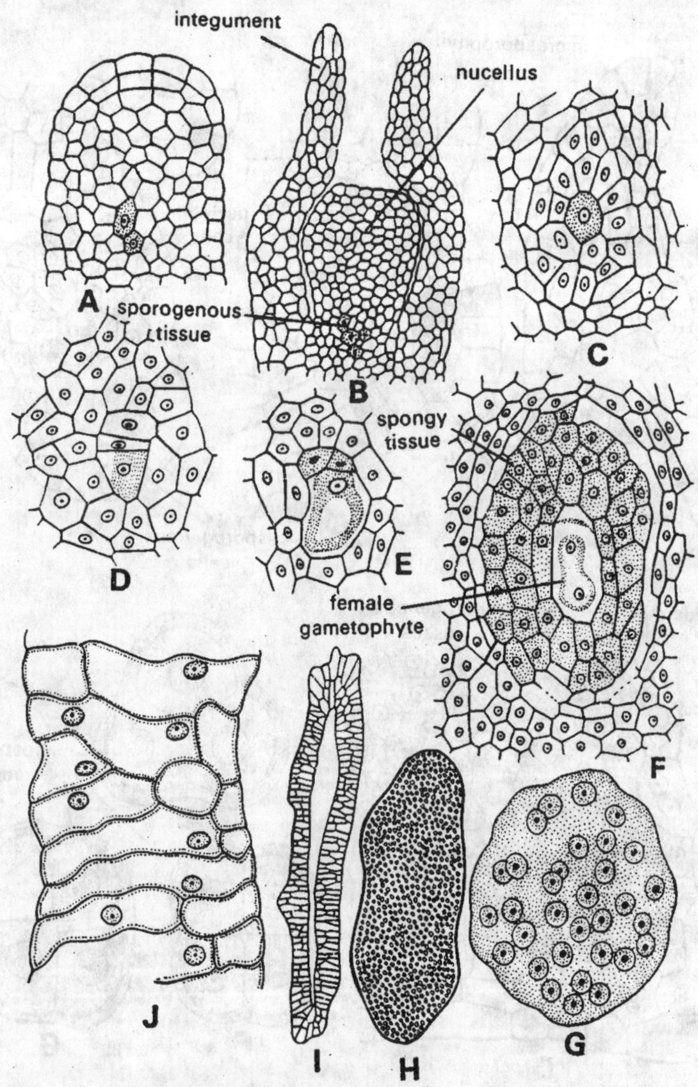

Fig. 4.61. *Biota orientalis*. Stages in the development of female gametophyte (A-J).

The archegonia are developed in groups, and make the micropylar, lateral or chalazal complex. Singh and Oberoi (1962) have reported an inactive chalazal complex in *Biota*. This complex is in addition to a 15-28 archegonial micropylar complex. They also saw one or two supernumerary archegonia just beneath the micropylar archegonial complex. The archegonial neck in *Biota orientalis* consists of 4-celled solitary tier. The neck cells disorganise just before fertilization. Prior to fertilization, the central cell divides and gives rise to a ventral canal cell and an egg.

During the development of embryo zygote nucleus divides, and eight independent nuclei are formed. Four embryonal cells of the lower tier become walled, while the four cells of the

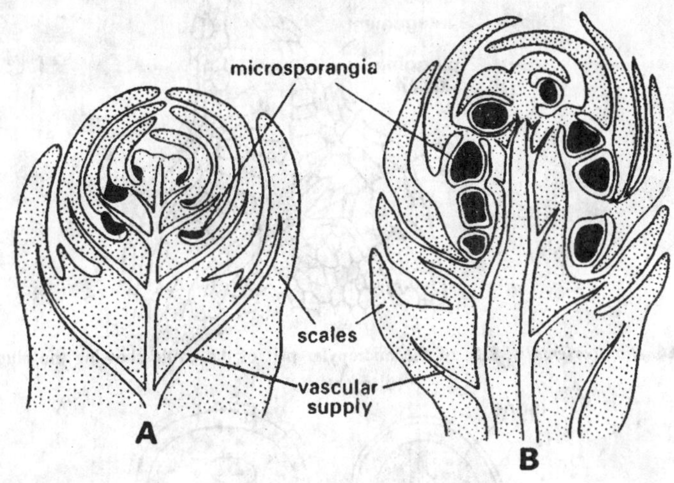

Fig. 4.62. *Biota orientalis*. A, L.S. of a young male cone; B, L.S. of mature male cone showing microsporangia.

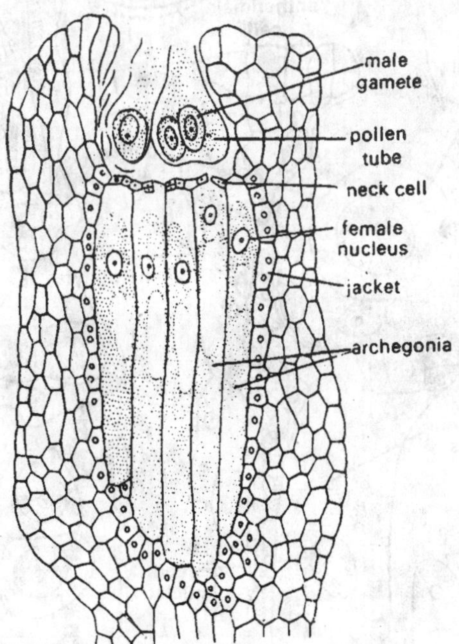

Fig. 4.63. *Biota orientalis*. L.S. of ovule showing female gametophyte and archegonial complex.

upper tier remain open towards the archegonial neck. The cells of the upper tier divide again forming prosuspensor tier and upper open tier. Thus three tiers of four cells each are formed.

In *Biota orientais,* the four primary embryonal cells are being arranged in tetrahedral, isobilateral or T-shaped fashion. The prosuspensor cells become elongated, and push the embryonal

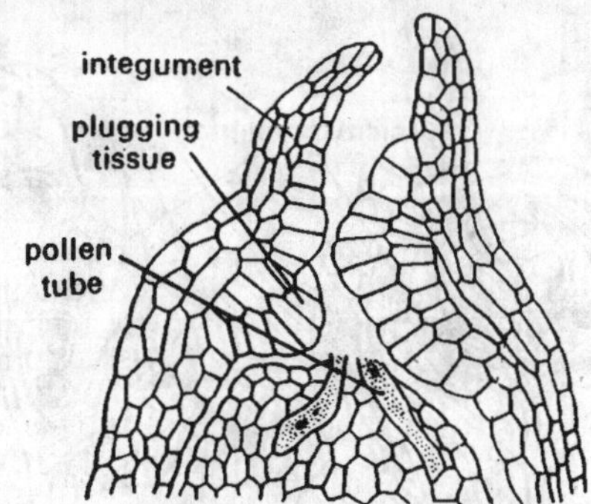

Fig. 4.64. ***Biota orientalis*. L.S. of the micropylar part of the ovule showing the plugging tissue.**

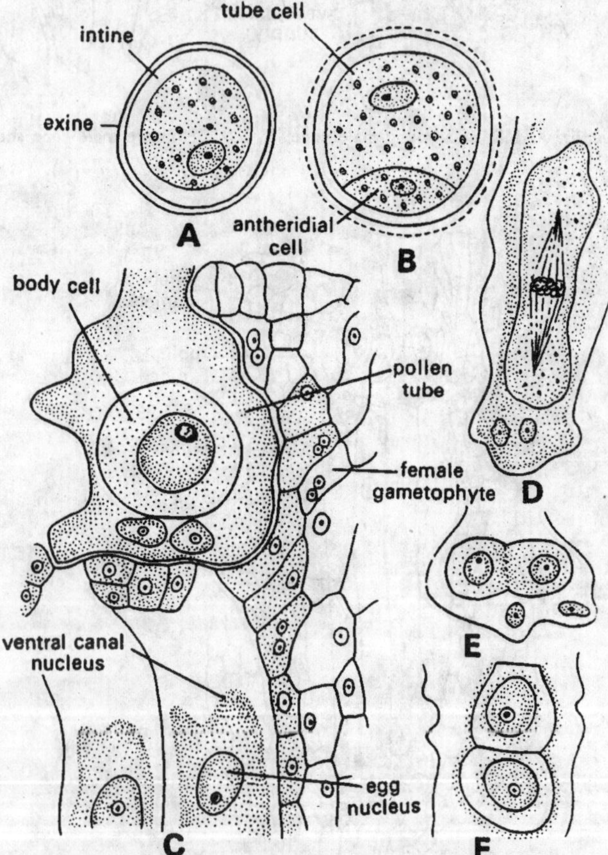

Fig. 4.65. ***Biota orientalis*.** A, microspore with single nucellus; B, two-celled microspore at the ume of dehiscence of microsporangium; C, part of pollen tube containing body cell, stalk nucleus and tube nucleus; D, dividing body cell into two male cells; E-F, part of pollen tube with two male cells.

cells out of the archegoniate complex. Thereafter the primary embryonal cells divide transversely forming the upper tier of suspensor cells, and the lower tier of embryonal cells. Just after this process the suspensor cells begin to elongate and cleavage begins. Thus the **cleavage polyembryony** is found is *Biota.*

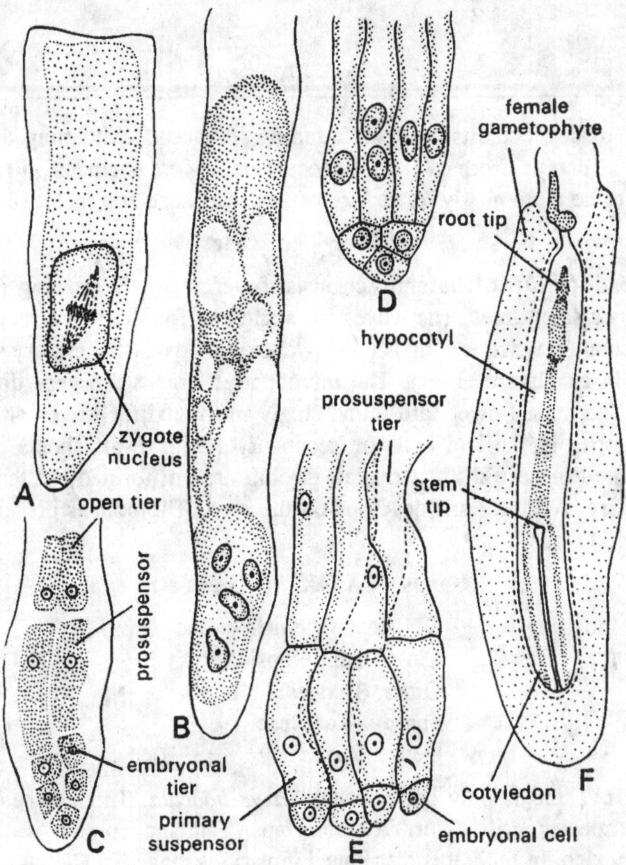

Fig. 4.66. *Biota orientalis*. Stages in the development of embryo. A, division of zygote nucleus showing intranuclear division; B, four nucleate proembryo; D, later stage with divided cells of embryo tier; E, the proembryo showing prosuspensor, primary suspensor and embryonal cells; F, L.S. of female gametophyte showing mature embryo

The mature embryo has two cotyledons. The cotyledons are not too long. The basic chromosome number in *Biota* is 11 (*i.e.*, n=11).

5

Class-Coniferopsida
Order-Taxales

The order Taxales, consists of a single family, Taxaceae, containing three genera, *Taxas, Cephalotaxus* and *Torreya.* Formerly the Podocarpineae were included in this group but now they are thought to be more nearly related to the Araucariineae and included in the Coniferales.

Family-Taxaceae

Characteristic features. The family consists of evergreen slow growing, profusely branched, long-lived shrubs or small trees. The leaves are acute and found to be arranged spirally on the branches. The secondary wood is compact and pycnoxylic possessing tertiary spirals on the walls of tracheids. Resin canals are absent. The reproductive organs are very different from those of true conifers. The male flowers are found singly in the axil of leaves or bracts. the female flowers are developed in the leaf axils on secondary fertile dwarf shoots or on long shoots. In Taxaceae, unlike Pinaceae distinct ovulate strobili are not formed but the ovules are freely exposed and solitary. A fleshy aril develops on the seed. The pollens do not bear wings. The embryo is dicotyledonous.

Genus TAXUS (The Yew)

Systematic position. Division. Gymnospermae
Class. Coniferopsida
Order. Taxales
Family. Taxaceae
Genus. *Taxus*

Distribution. There is only one species, *Taxus baccata.* Though there are several sub-species. These sub-species occur in North America, Japan, England, Europe, Asia Minor, Himalayas, Philippines and Mexico. In India it is distributed in various ranges of Himalayas. It is commonly found above 10,000 feet in the Himalayas.

Habit and morphological features. It is a slow growing evergreen tree seldom reaching a height of 10-12 metres. Sometimes the Yew attains a height of 30 metres and a girth of five metres. The general habit of the plant body of *Taxus* resembles *Pinus.* The branches grow horizontally and form a very dense canopy. There are no foliar spurs. During its growth branching occurs from the base, and the branches fuse together to form the compound columnar trunk which is characteristic of the tree. The trunk remains covered by a reddish brown scaly bark. The branches are spreading and remain covered by brown scales at their bases.

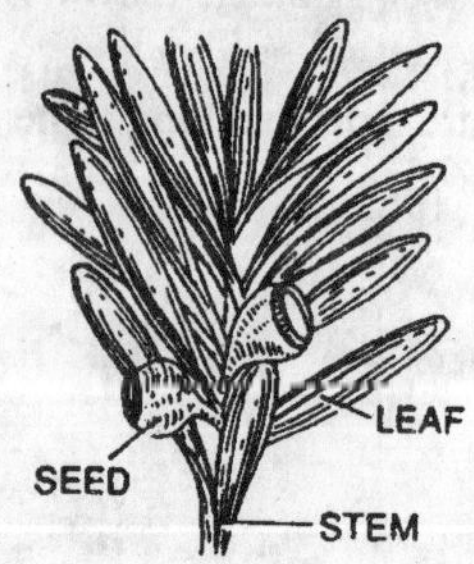

Fig. 5.1. *Taxus baccata.* A twig with leaves and seeds.

The oldest specimen estimated to be 3,000 years old stands in the churchyard of Fortingal in Peshire, Scotland.

Foliage is dark green. The leaves are flattened, flexible, linear, spirally arranged but spreading in two opposite ranks narrowed to a short stalk. The leaves are 2-3 cm. long, narrow and obtuse.

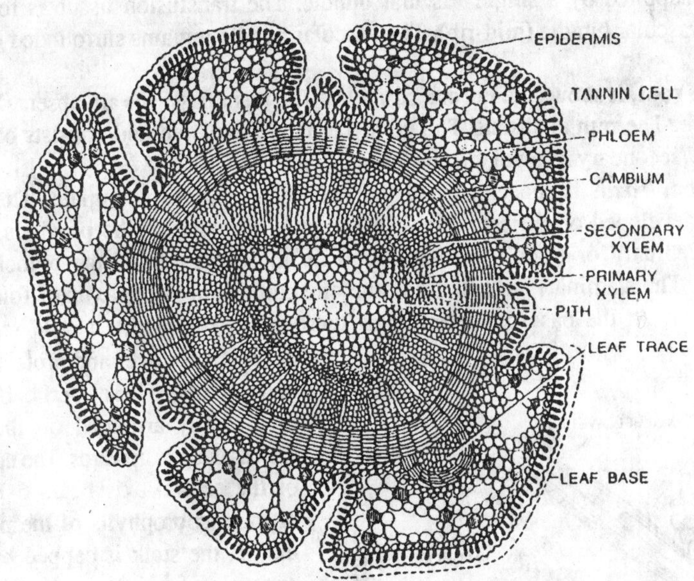

Fig. 5.2. *Taxus baccata*. Transverse section of the stem showing secondary tissues and a single leaf trace.

Anatomy of stem. Anatomically *Taxus* agrees to the general type of *Pinus* except for some minor differences. There is complete absence of resin passages. The cortex and pith are small in comparison of vascular cylinder. The tracheids of the secondary xylem show a spiral thickening, in addition to bordered pits in a single series, as in *Pinus*. The medullary rays are uniseriate. The vascular bundles are conjoint, collateral, endarch and open. The secondary growth is abundant.

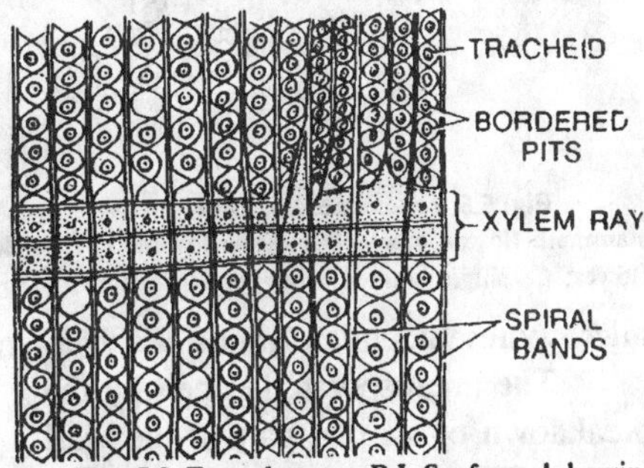

Fig. 5.3. *Taxus baccata*. R.L.S. of wood showing a biseriate xylem ray and tracheids.

Anatomy of leaf. The leaf shows a dorsiventral structure. There is a single layered upper and lower epidermis covered over with

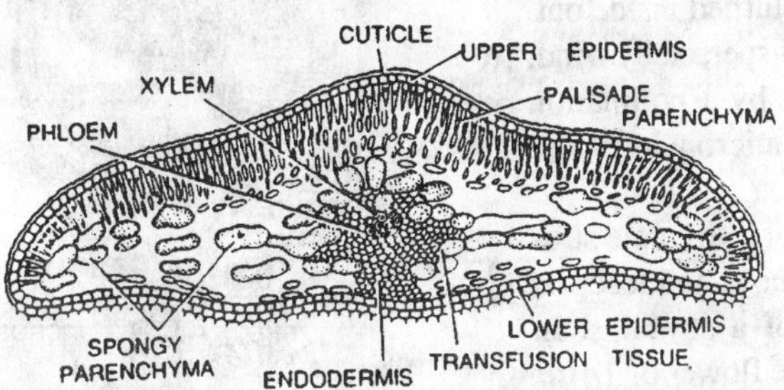

Fig. 5.4. *Taxus baccata*. Transverse section of leaf.

cuticle. The cuticle along the lower epidermis is generally wavy. The sunken stomata are present in the lower epidermis.

The mesophyll is differentiated into the palisade and the spongy parenchyma.

The leaf is supplied by a single vascular bundle. The transfusion tissue is found on the either side of the vascular bundle (mid-rib). The vascular bundle remains surrounded by a bundle sheath.

Anatomy of root. It shows the normal structure. The resin passages are absent. It possesses the diarch structure. The pith is absent. The pericycle is multilayered and consists of parenchymatous cells. The secondary growth occurs as in *Pinus*.

Reproduction. *Taxus* is dioecious and the strobili are monosporangiate. The male and female cones are developed on different trees in the axils of leaves of the previous years. The cones appear in February or March and are found on the underside of the branches.

Male cones. The staminate strobili are borne on short stalks in the axils of foliage leaves of the previous year. At the base of the strobilus there are ten decussate sterile scales which at first cover over and protect the young sporophylls. These sterile bract scales are spirally arranged on the axis and increase in size upwards. The upper portion of the central axis bears 6-14 stamens or microsporophylls of the peltate type that is, the stalk is capped by a peltate expansion from the underside of which are situated the pendant sporangia. The number of sporangia is generally 5-7 and sometimes even nine. The microsporangia (pollen sacs) remain united with one another and attached to the stalk of the sporophyll.

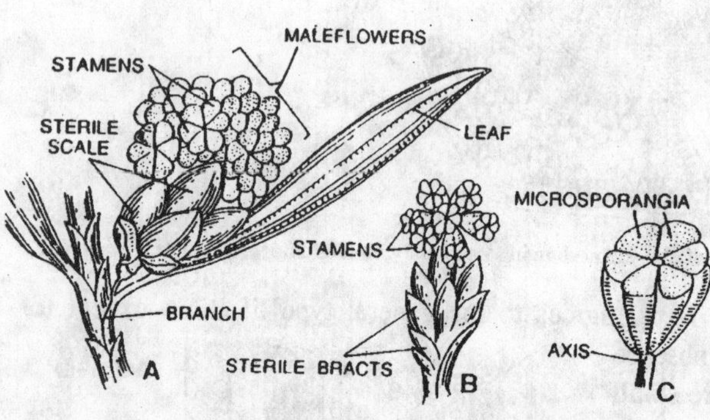

Fig. 5.5. *Taxus baccata*. Reproductive parts. A, a twig with staminate flower, stamens and sterile bracts; B, a detached male flower; C, single stamen with pendant microsporangia.

The microsporangia contain a number of microspore mother cells which divide meiotically to produce haploid microspores or pollen grains. Each microsporangium contains numerous small microspores.

The microsporangia open by the breakdown of all that part of the wall nearest the axis and the curling back of the upper part with the peltate shield, as if the umbrella had turned inside out. The pollen grains are dispersed by wind. The pollen are caught by a pollination drop of liquid at the micropyle of the ovule.

Female flower. The ovulate strobilus is very much reduced and loses its ordinary appearance of a strobilus. In other words, the female flower of *Taxus* is not a cone.

The ovules appear at the same time and in the same axillary position as the male cones, but upon different

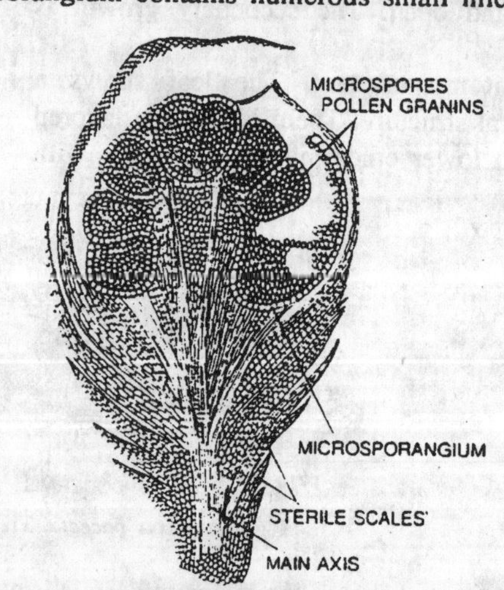

Fig. 5.6. *Taxus baccata*. Male flower in longitudinal section showing dehiscing microsporangia.

trees. There is a short axis, called the **primary axis,** that bears a closely imbricated succession of sterile scales there on its lower portion. In the axil of one of the upper scales there arises a very short lateral fertile shoot, the **secondary axis.** This bears three, crossed pairs of minute scales, the lowest pair standing transversely to the subtending bract and ends in a straight ovule. In the axil of the eighth bract counting downwards from the apex there is a bud which was formerly regarded as the apex of the primary axis, is not known as a secondary shoot which sometimes produces a second ovule.

The ovule. The structure and development of the ovules is similar to that of *Pinus* except minor differences. The ovule is somewhat flattened with two lateral ribs. It is transversely placed to the uppermost pair of bracts. It consists of a nucellus and micropyle. There is a single thick integument, which forms a long micropylar canal above the nucellus. Two vascular bundles supply the integument, running along the lateral ribs. These bundles arise by the fusion of two pairs in the stalk of the ovule.

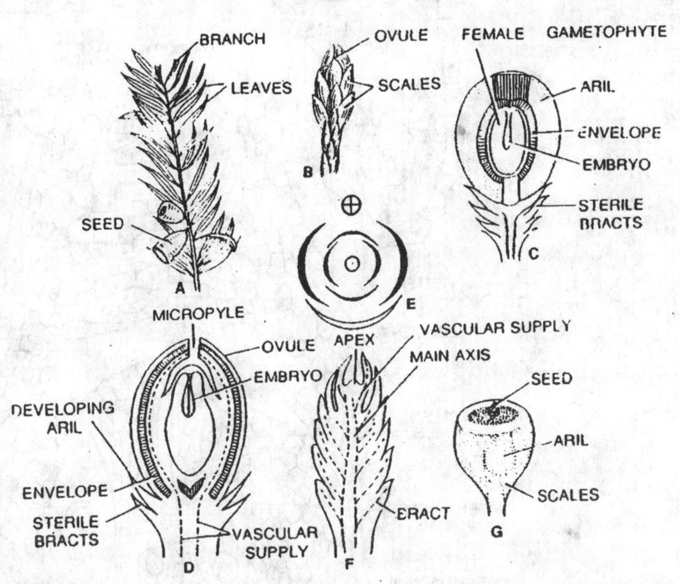

Fig. 5.7. *Taxus baccata.* Seed and female reproductive parts. A, a twig with seeds and leaves; B, a female strobilus with terminal ovule; C, seed in longitudinal section; D, longitudinal section of ovule; E, floral diagram of a female flower; F, longitudinal section of female strobilus showing terminal ovule; G, the seed with aril and scales.

Just below the integument there is a ring-like swelling, the **aril,** which slowly grows and surrounds the whole ovule. This aril is correctly known as the **cupule.**

The integument of the ovule is differentiated into an outer fleshy, a middle very hard stony and an inner fleshy layers. The vascular bundles of the integument enter the stony layer at the base through two pores or **foramina.**

The ovule does not have a nucellar beak or pollen chamber According to Dupler, in very young ovule the nucellus and integument are free from each other. The nucellus becomes conical upward.

On the maturation of the seed the stony layer becomes lignified and a tube-like sheath of cells with thick yellow cuticle develops around the micropyle, which prevents the micropyle from being closed by the growth of the integument.

Microspore and male gametophyte. Each microspore is a spherical structure having two wall layers, the exine and the intine. It contains a mass of protoplasm and a central nucleus. There are no wings or balloon-like expansions.

The pollination and fertilization are effected as in *Pinus,* but in the same year. After pollination the pollen grains begin to germinate at once. The microspore is shed in uninucleate stage. It divides transversely to form a generative cell and a tube cell while the spore rests on the tip of the nucellus. The generative cell divides transversely to form a stalk cell and a body cell. In *Taxus* no prothallial cell is formed. The exine of the spore ruptures and the intine protrudes out in the form of pollen tube. The pollen tubes remain on the nucellus for about two months before fertilization. The tube nucleus and the stalk nucleus pass into the pollen

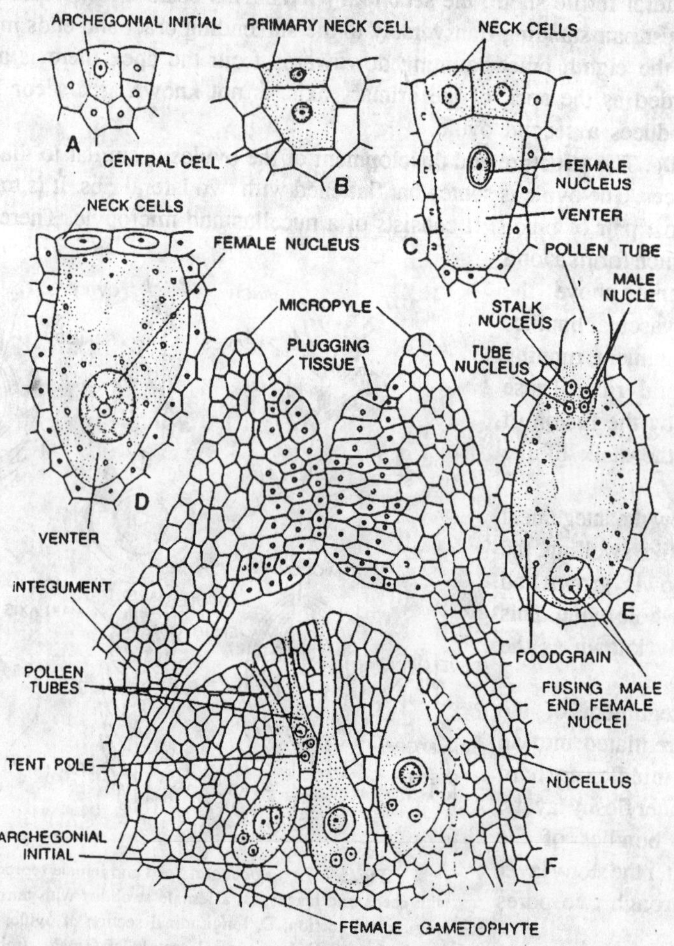

Fig. 5.8. ***Taxus baccata.*** **A-D, stages showing the development of archegonium; E, fertilization-fusion of male and female nuclei; F, vertical section of a part of mature ovule showing plugging tissue and pollen tubes.**

tube. The body cell now divides into unequal male nuclei which also migrate into the pollen tube. The tip of the pollen tube discharges all its contents into the cytoplasm of the egg.

Megaspore and female gametophyte. One or several archesporial initials develop in the hypodermal region at the apex of the nucellus. Each archesporial initial divides periclinally to form two cells, an upper cell, *i.e.*, the **primary wall cell,** and a lower cell, *i.e.*, the **primary sporogenous cell**. The primary wall cell divides again and again periclinally producing a row of cells between the primary sporogenous cell and the surface of the nucellus.

The primary sporogenous cell divides again and again forming a compact mass of sporogenous tissue. One or more of these cells become elongated and differentiated into megaspore mother cells. Each megaspore mother cell divides transversely to form four cells. One of these cells, usually the lowermost, functions as a megaspore and develops into an embryo sac. The remaining three cells are being aborted.

Inside the ovule the megaspore nucleus divides and by a series of simultaneous nuclear divisions a large number (*i.e.*, 256) of free nuclei are produced. These nuclei are arranged round

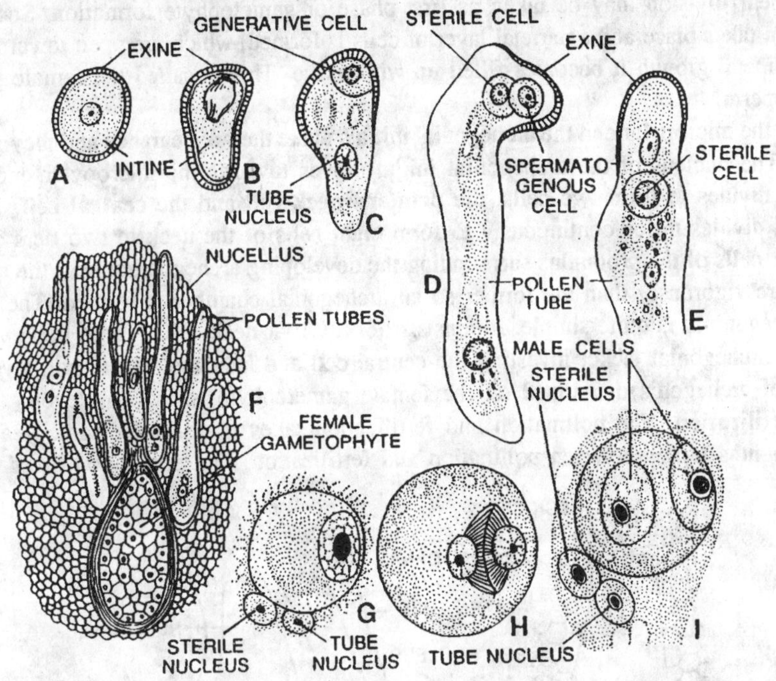

Fig. 5.9. *Taxus baccata.* Development of male gametophyte. A, mature microspore, B, first division; C, formation of generative cell and tube nucleus; D, formation of spermatogenous cell and pollen tube; F, female gametophyte and the pollen tubes within the nucellus; G, a mature spermatogenous cell; H-I, formation of two unequal male cells.

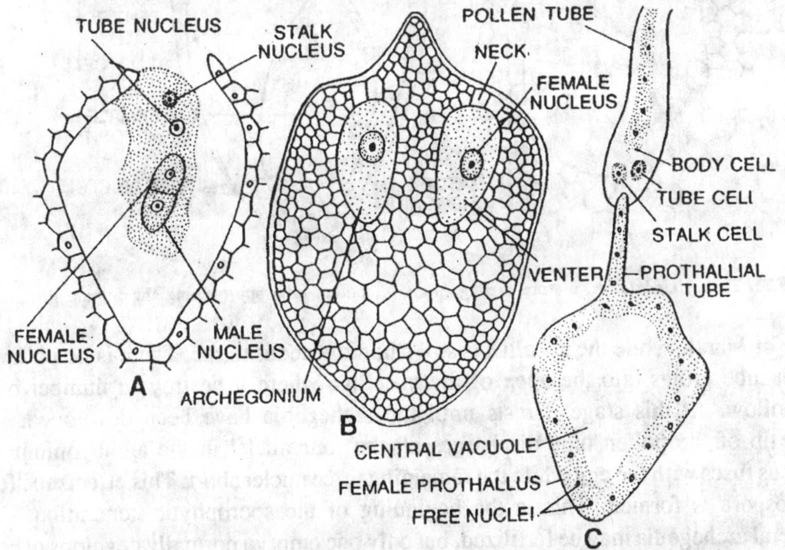

Fig. 5.10. *Taxus baccata.* A, fertilization-fusion of male and female nuclei; B, female gametophyte with two archegonia; C, fertilization in process.

the periphery of the rapidly enlarging cell. A large central vacuole develops in the centre of the cell that restricts the free nuclei to the layer of cytoplasm lining the wall. This period of free nuclear division may be taken as first phase of gametophyte formation. Soon after wall formation takes place and a parietal layer of cells is formed, which are open towards the centre. By centripetal growth it becomes filled up with tissue. This is called the female gametophyte or endosperm.

At the micropylar end the archegonial initials make their appearance and they grow rapidly in size. The nucleus of the archegonial initial moves towards the micropylar end of the cell where it divides to form two cells, the **primary neck cell** and the **central cell.** The primary neck cell divides thrice continuously to form eight cells of the neck in two tiers of four cells each. The cells of the prothallus surrounding the developing archegonium near the apical region grow more vigorously than the former so an archegonial chamber is formed. The structure of each archegonium is quite simple. It consists of only four neck cells. There is no ventral canal cell. The archegonial jacket investing the central cell and later the egg, is not developed. The number of archegonia developed on the female gametophyte is five to eleven.

Fertilization. The pollination and fertilization takes place in the same year. A period of two months lapses between pollination and fertilization. The pollination takes place about

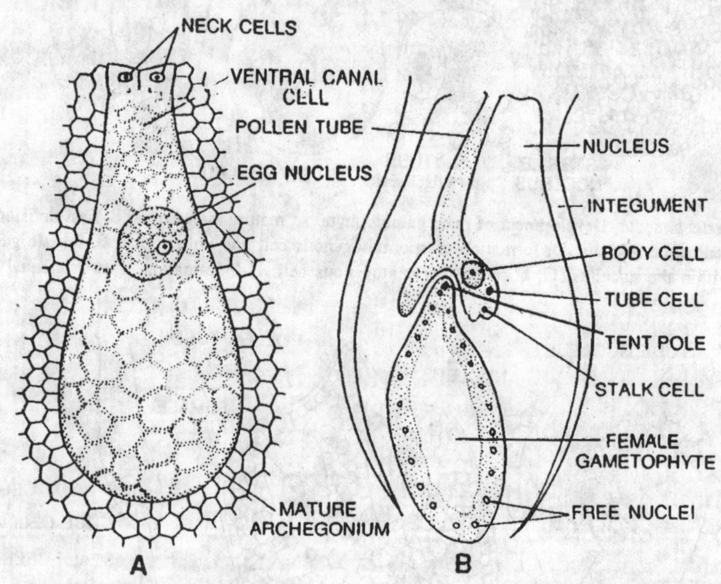

Fig. 5.11. *Taxus baccata.* A, mature archegonium; B, pollen tube approaching the female gametophyte.

the middle of March, while the nucellus is still a homogeneous tissue. There is no pollen chamber. The pollen tube grows into the apex of the nucellus, where it destroys a number of cells and forms a hollow. At this stage it rests until the archegonia have been developed.

The tip of the pollen tube discharges all the four nuclei in the archegonium. The large male nucleus fuses with the egg while the remaining three nuclei abort. This effects the fertilization and the oospore is formed. This is the beginning of the sporophytic generation.

Several archegonia may be fertilized, but only one embryo normally develops to completion.

Development of embryo. The nucleus of oospore undergoes free nuclear division upto thirty-two nuclei being formed. Now the walls around the nuclei are laid down so that a cellular proembryo is formed. The proembryo fills up the whole oospore.

The proembryo has a general cone-like appearance. The cells are being arranged in three tiers. The lowest tier usually consists of a single cell. The cells of the upper and middle tiers

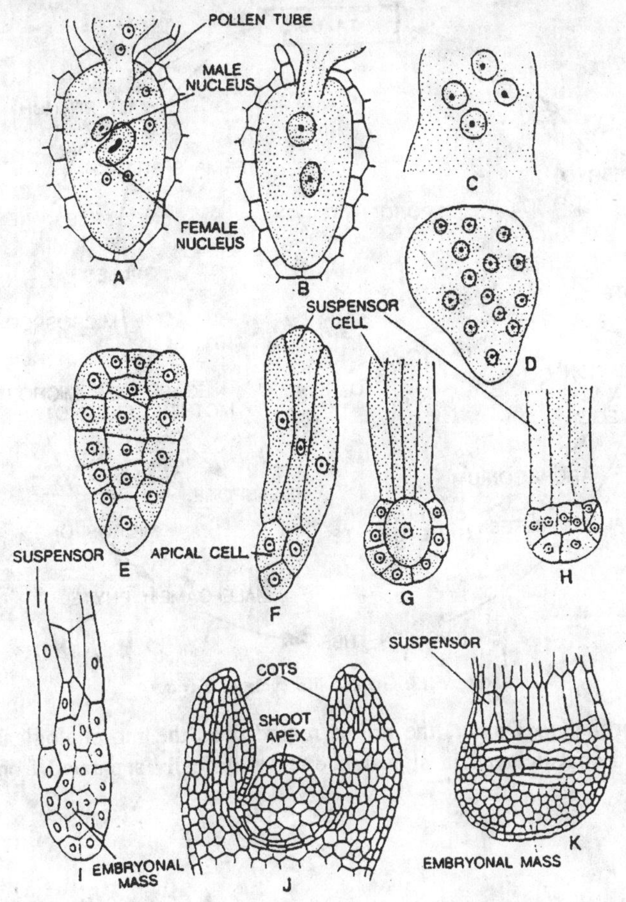

Fig. 5.12. *Taxus baccata*. A-K, stages in the development of embryo (After Saxton and Dupler).

elongate and form the suspensor. The suspensor cells push the embryonal mass out of the archegonium into the endosperm (female gametophyte). Thereafter an apical cell is formed at the apex of the embryonal mass of cells. This cell produces a single straight embryo consisting of two cotyledons, a plumule and a radicle. The embryo consumes all the endosperm during its maturation and hence the seed becomes non-endospermic.

Seed. The ovule ultimately gets transformed into a seed. During the development of the embryo; a cup-shaped structure grows up from the base of the ovule and this is the **aril,** which becomes fleshy and forms the red investment of the seed, the integument becomes the testa which is hard and at first is olivegreen but later becomes dark, in mature seed. The testa consists of three layers-an outer fleshy, the middle stony and the inner fleshy which is reduced to a thin membrance.

The aril of the seed has usually been compared in the past with the epimatium of *Podocarpus* and with the ovuliferous scale of the other conifers. It was perhaps on the basis of this similarity that Podocarps were previously included in the Taxaceae. The morphology of the aril is a controversial topic. Mclean and Cook have suggested that it is misnomer to call the structure, the aril, since this is not considered homologous to the structure bearing the same name in

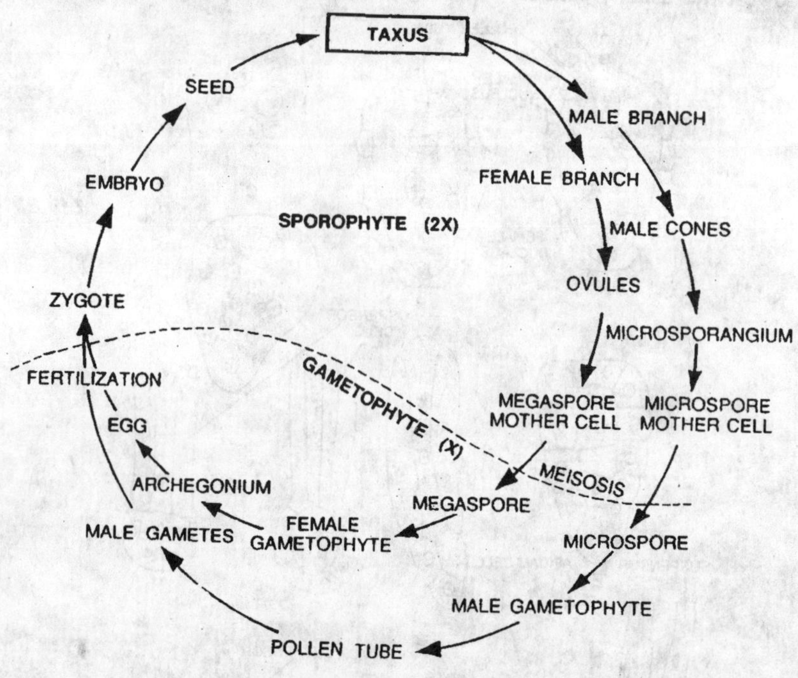

Fig. 5.13. Graphic life cycle of *Taxus*.

angiosperms. According to Dupler, the aril is regarded as the morphological fleshy layer of a three-layered seed coat delayed in appearance and physically separated from the hard stony layer.

6

Class-Gnetopsida

Order-Gnetales

The Gnetales are a small group of plants, comprising ot three genera-*Gnetum, Welwitschia* and *Ephedra.* The importance of this group is because of its forming a connected link between gymnosperms and angiosperms. They possess extraordinary morphology and diverse habit. They are gymnosperm-like in that their ovules are naked and borne on cones, whereas angiosperm-like in that their ovules and microsporangia are borne on somewhat flower-like fertile shoot. The ovules are, however, enclosed within coverings additional to the true integument, which are sometimes considered to be the equivalent of ovary walls. Another angiosperm-like character is the presence of vessels in the xylem.

A common character of the group is the prolongation of the apex of the integument into a long narrow tube with a flattened orifice, by means of which the pollen is collected. They are so advanced in structure that in several respects they resemble the angiosperms. However, there is no direct relationship, and it has been established that the Gnetales were not the ancestors of the angiosperms. They are a parallel line of evolution.

Three genera, *i.e, Gnetum, Welwitschia* and *Ephedra*, were long considered to form a single family Gnetaceae of the order Gnetales, but on the basis of morphological evidences the order has been divided into three families, *i.e.*, Gnetaceae, Welwitschiaceae and Ephedraceae. Now the three orders have been established known as Ephedrales, Welwitschiales and Gnetales.

Whatever the system of classification may be, there are three genera. 1. *Ephedra,* 2. *Welwitschia* and 3. *Gnetum.*

The genus *Ephedra,* consists of shrubby plants with minute leaves. There are about 35 species distributed in the New and the Old Worlds. The species have mostly a restricted range of distribution.

The genus *Welwitschia,* comprises the remarkable living plants. There is only one species, *W. mirabilis,* particularly confined to the desert region of South West Africa. The stem resembles an enormous woody carrot, the top of which is concave and may reach 4 feet in diameter. It is entirely buried in the sand and bears only two long strap-shaped leaves, which persist throughout life.

The genus *Gnetum* includes trees, shrubs or woody climbers. The climbers are predominating. There are about 4 species distributed in the tropical regions of the world, especially West Africa to the Amazon region. They are invariably the jungle plants.

Their vegetative appearance is entirely angiospermic, and their large netveined leaves cannot be distinguished from those of dicotyledons.

Here the detailed account of the genus *Ephedra* has been give.

Genus EPHEDRA

Systematic position. Gymnospermae
Class. Gnetopsida
Order. Gnetales

Family: Gnetaceae
Genus. *Ephedra.*

There are about 35 species of *Ephedra* growing in the arid regions about the Mediterranean, and in tropical to temperate Asia, North America and South America. The species have mostly a restricted range and none of the species is common to both sides of the Atlantic.

A few species are found in our country. They are *Ephedra intermedia, E. gerardiana, E. foliata* and *E. nebrodensis. E. intermedia* occurs in the natural state in the Kashmir and North-West Himalayas at an altitude of about 2500 to 4200 metres above sea level. *E. foliata* occurs in the wide state in the Punjab and certain parts of Rajasthan. This species is commonly met with in the forests near Lahore in Pakistan. *E. gerardiana* and different varieties of this species are found along the length of the Himalayas at an altitude of about 3050 metres above sea level.

Habit. Usually the plants are low straggling shrubs reaching the height of one metre or so. The green, long jointed, slender branches bear minute leaves. The scaly leaves are opposite

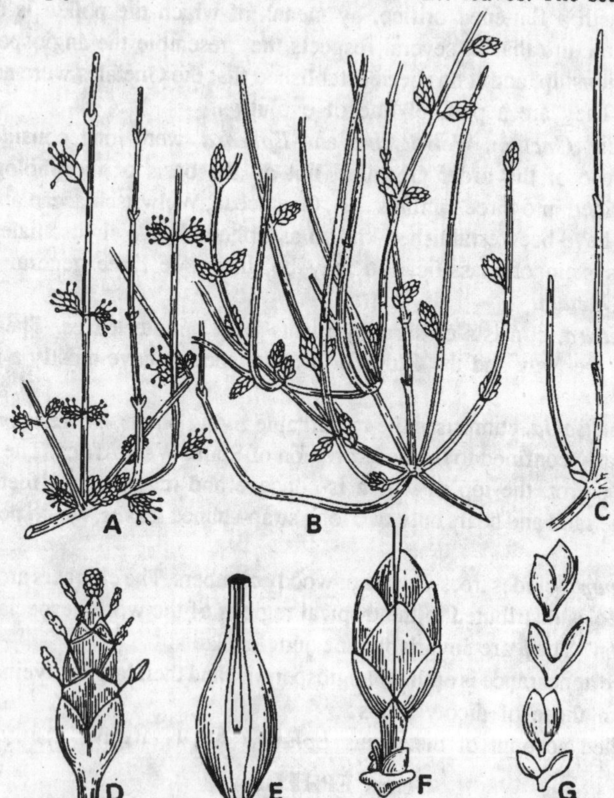

Fig. 6.1. *Ephedra* sp. A, male (staminate) branches; B, female (ovulate) branches; C, sterile branches; D, staminate strobilus; E, L.S. of seed; F, ovulate strobilus; G, bracts of the female strobilus.

and connate in a two-toothed sheath. The whole aspect of the plant is xeromorphic. True foliage leaves are generally lacking and the whole habit is suggestive of a shrubby *Equisetum.* The leaves also occur in threes and rarely in fours.

Growth. The growth of the internodes is independent and takes place by a pale-coloured meristematic zone at the base of each node. The meristem in some species becomes hard and turns into an abscission layer in the end of the growing season, with the result the branches drop off, which are to be replaced by new axillary shoots in the next growing season. The species may be called deciduous and they are supposed to be analogous to the trees which drop their leaves in winter.

Anatomy of the stem. The surface of the stem is ribbed and therefore, the outline of the stem in transverse section is not circular. The epidermis is quite thick and just beneath the

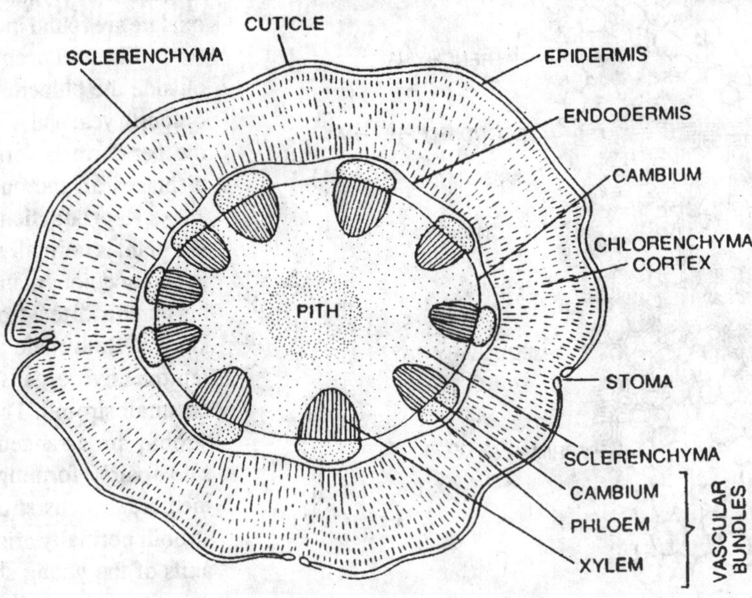

Fig. 6.2. *Ephedra foliata*. Transverse section of young stem showing primary structure (diagrammatic).

ribs there are groups of fibre cells. The stomata occur in furrows between the ribs and each stoma is situated in a circular pit.

The cortex is differentiated into a loose outer palisade layer and an inner spongy tissue. Both of these contain chlorophyll. The innermost layer of the cortex is known as endodermis.

The endodermis surrounds the ring of collateral and endarch vascular bundles. The large and small bundles are grouped into pairs of threes. Each small bundle is represented by a leaf trace. Each vascular bundle runs through two internodes and thereafter passes out in pairs into the leaves. This way, there is a double leaf trace.

The cambium is present and sometimes the older stems become greatly thickened due to secondary growth. In the course of the formation of secondary wood the annual rings are formed. The young stems contain uniseriate medullary rays, whereas in older stems these rays become very broad and long, vertically with the result of their own growth as well as by the fusion of adjacent rays. They are composed of lignified cells and therefore, the wood texture is very hard. The xylem elements, *i.e.*, the tracheids possess single rows of scattered bordered pits arranged on the radial and tangential walls. Bars of Sanio and trabeculae are also present across the tracheids.

The most interesting feature is the presence of vessels, resembling those of angiosperms. The vessels are found in abundance. In the spring or growing season they are largest in size but they decrease in size and number in the autumn wood which completely lacks the vessels.

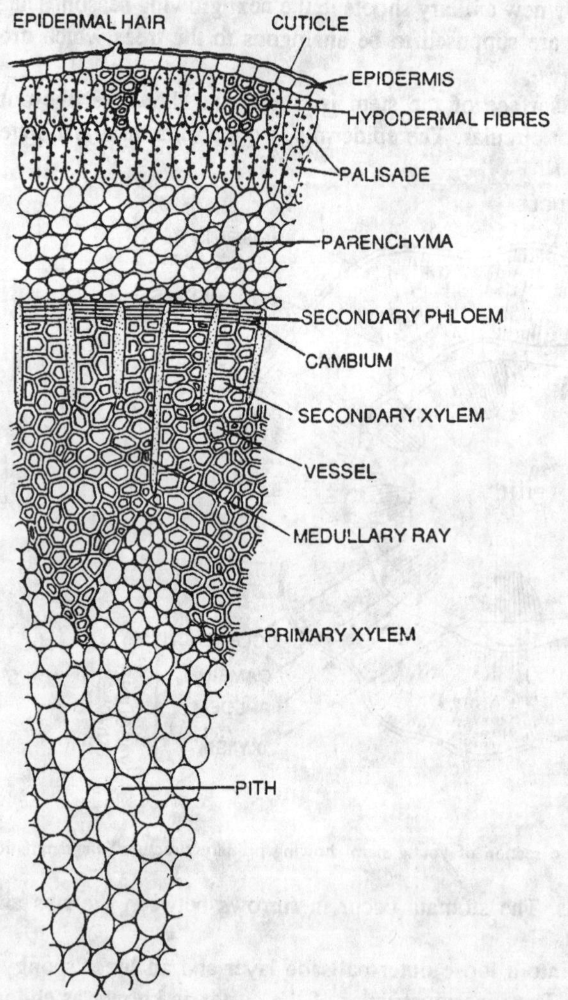

Fig. 6.3. *Ephedra* sp. T. S. of stem showing broad cortex with photosynthetic tissue. Vessels are found in the secondary xylem.

The pith cells of the nodal region become lignified resulting in the formation of transverse plates. The resin ducts are not found, but the cells with large stellate crystals of calcium oxalate are found in abundance.

The phellogen appears just outside the phloem in the third or fourth year and with the result the periderm is formed which displaces all the outer tissues.

Reproduction. The *Ephedra* plant is normally dioecious, but monoecious plants and even hermaphrodite flowers are known. Both male and female reproductive organs are borne on small strobili. The scales of strobili are arranged in decussate pairs forming a short membranaceous strobilus. The strobili normally arise from the axils of the young shoot. They may develop simply or in small sessile groups.

As the rule the plants are dioecious and strictly male or strictly famale flowers are formed. However, P.N. Mehra (1950) reported the occasional occurrence of hermaphrodite flowers in *Ephedra intermedia*. Here, the hermaphrodite flowers are borne on bisexual strobili. According to Eames (1952), the *Ephedra* plants lack totally the hermaphrodite flowers, he regarded these hermaphrodite flowers mere monstrosities. Thibout and Takhtadjan have also reported 'bisexual cones' in *Ephedra*. The bisporangiate strobili produce stamens at the base and ovules above.

The male strobilus. Each male strobilus arises from the axil of a leaf. It consists of a short axis which bears a number of thick and closely arranged bracts found in decussate pair. The pairs of bracts vary from two to twelve in number. A single male flower stands in the axil of each bract. Each male flower consists of a perianth of two scales and a stalked stamen arises from between the two scales of the perianth and bears two to several anthers (sporangia) at its top. The simple stamen bears two sporangia and when it becomes branched it bears several sprangia. The anther is two or three lobed. Each lobe opens by an apical slit and the pollens are released. The pollen are elliptical in shape.

The female strobilus. Each female strobilus is found in the axil of a leaf on the young shoot. The female strobilus though quite similar in structure to the male strobilus yet smaller in size. Each strobilus bears two to four pairs of bracts. Each strobilus ends in a group of female

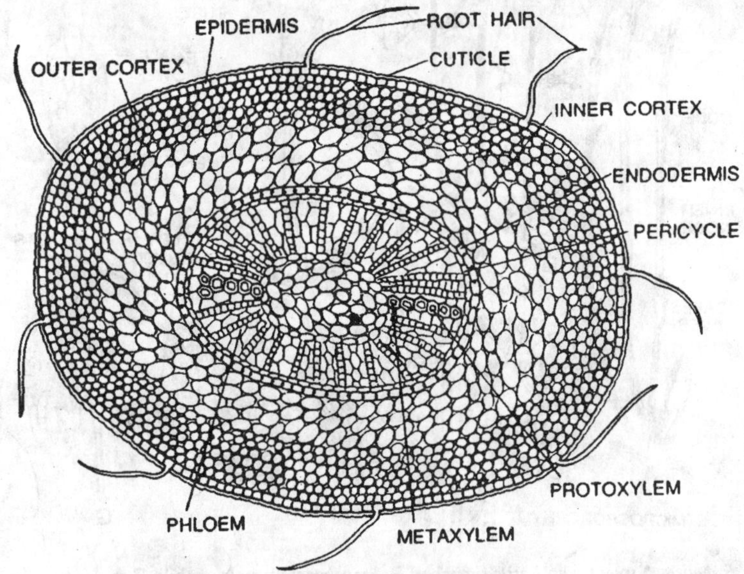

Fig. 6.4. *Ephedra foliata.* Transverse section of young root.

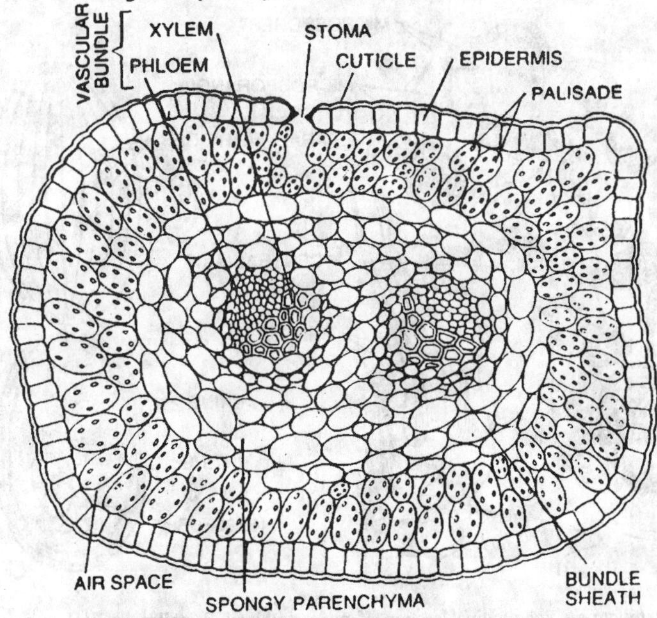

Fig. 6.5. *Ephedra foliata.* Vertical section of leaf.

flowers, one, two or three in number. Each flower consists of a nucellus with two envelopes. The inner envelope consists of two segments while the outer of four segments. The inner envelope is a true ovular integument, and the outer envelope is perianth. The whole structure is found in the axil of one of the bracts of the uppermost pair. The upper half of the nucellus is free

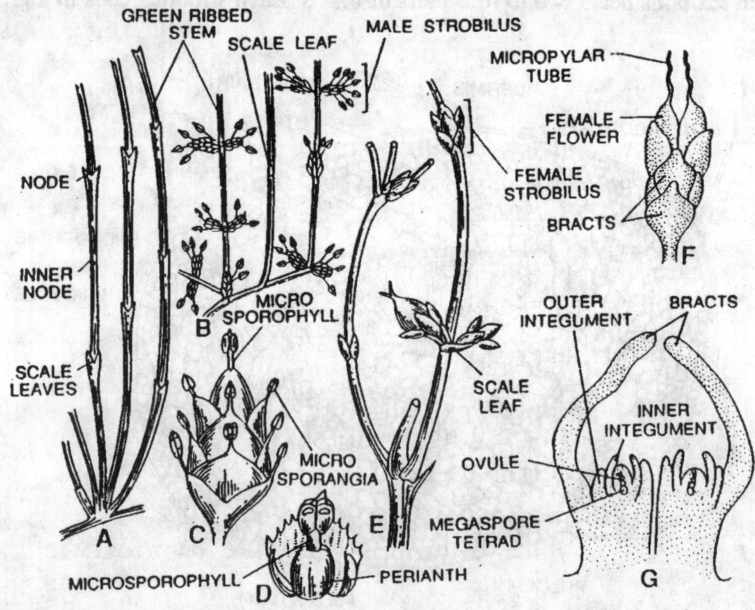

Fig. 6.6. *Ephedra*. A, plant with sterile branches; B, branches with male strobili; C, male strobilus : D, male flower; E, branches with female strobili; F, female strobilus; G, female flower showing ovules and megaspore tetrads.

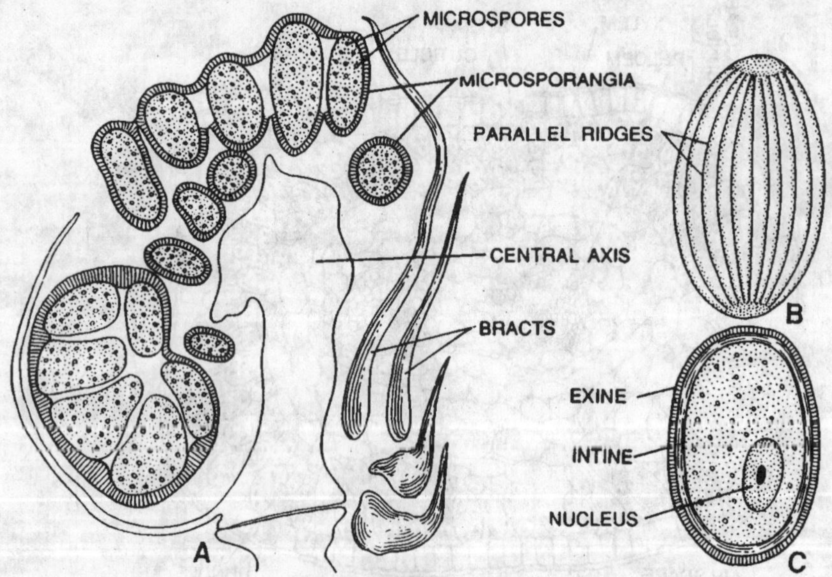

Fig. 6.7. *Ephedra foliata*. A, longitudinal section of male strobilus; B, pollen grain showing parallel ridges on its exine; C, pollen grain in sectional view.

from the inner integument, whereas the inner integument is prolonged out into a thin walled micropylar tube. Sometime, this is found spirally twisted and opening at its tip into a tongue-like extension. The pollination drop extrudes from here by means of which the pollen is drawn into the micropylar tube. Later on the perianth becomes closely fit to the integument so that the structure becomes woody or fleshy developing into the outer coat of the mature seed.

The male gametophyte. The nuclei of the pollen (microspore) of *Ephedra* divide in the similar plan as that of *Pinus*. This indicates the positive sign of linkage of *Ephedra* with conifers. After a short time of the rounding off of the spore, the nucleus divides resulting in the formation of first prothallial cell. A second division gives rise to second prothallial cell and the antheridium initial, which are not separated from each other by any wall. The second prothallial cell and its nucleus become flattened, whereas the nucleus of antheridium initial becomes very much enlarged, and divides producing the generative and tube nuclei. A zone of cytoplasm surrounds the generative nucleus. This zone may be differentiated from the general cytoplasm. The second prothallial cell and tube nuclei are still embedded in the general cytoplasm with no wall of demarcation. The generative cell nucleus divides giving rise to the stalk and body nuclei. These nuclei remain invested in the cytoplasm of the generative cell, and the body cell becomes larger than the other.

The pollination occurs at this stage during the first two weeks of April. At this stage the spore contains five nuclei, *i.e.*, first and second prothallial, stalk, body and tube nuclei.

After reaching the deep pollen chamber the exine of the spore is thrown off and the intine remains free with its contents. The two prothallial cells soon disappear and the body cell nucleus divides forming two equal male nuclei. The tube nucleus moves about the cavity of microspore and in the end rests against the spore wall. Soon after a pollen tube comes out from this very point where the tube nucleus rests. The tube nucleus and the two male nuclei accompanied by the nucleus of the stalk cell pass into the pollen tube.

After about twelve hours of pollination, the fertilization takes place. The pollen tube penetrates between the neck cells of the archegonium and the two male nuclei pass directly into the oosphere. In the case of *Ephedra* the fertilization takes place only twelve hours or so after pollination, whereas there is prolonged delay in other Gymnosperms.

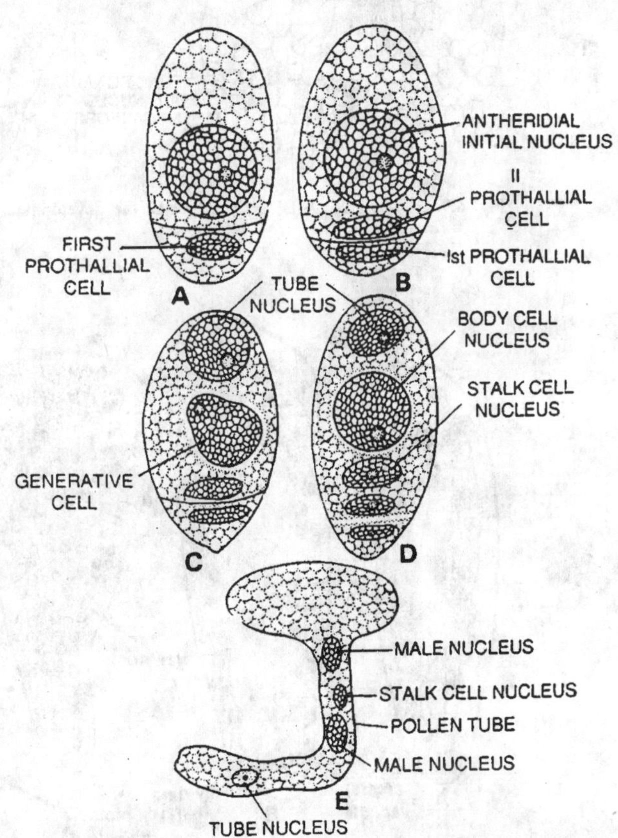

Fig. 6.8. *Ephedra trifurca*. Development of male gametophyte. A-D, successive stages in the germination of the pollen grain (microspore); E, pollen grain with the pollen tube having male muclei 1 and 2

The female gametophyte. The details of the female gemetophyte have been worked out by Land (1904) in *Ephedra trifurca.* In the beginning a row of four megaspores is formed, the

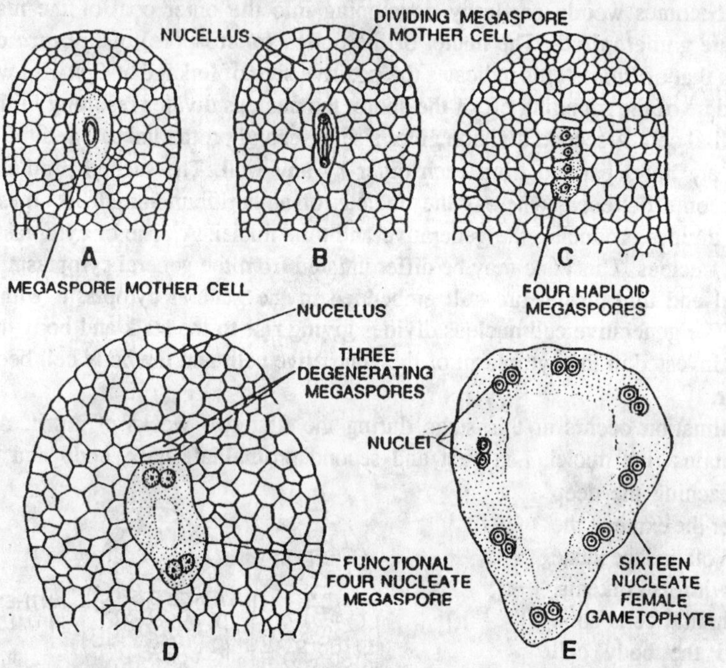

Fig. 6.9. *Ephedra* sp. A-E, early stages in the development of female gametophyte.

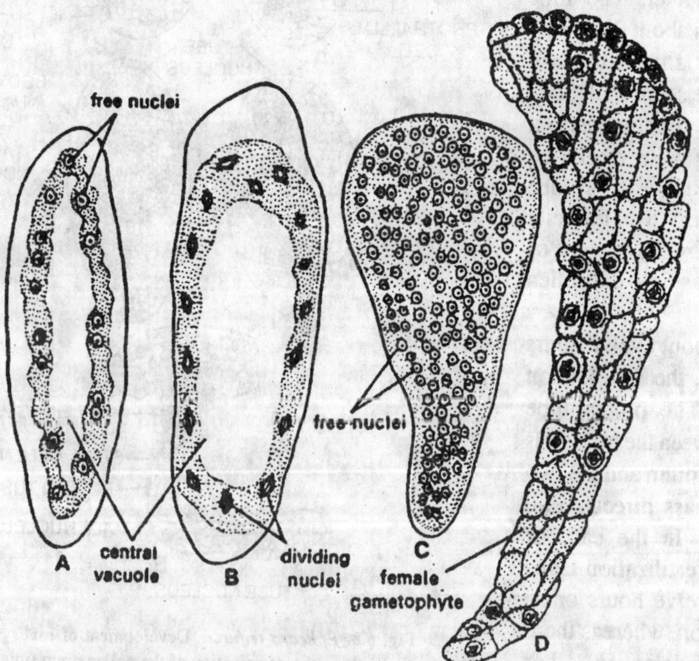

Fig. 6.10. *Ephedra* sp. A-D, late stages in the development of female gametophyte.

lowest megaspore enlarges in size and becomes conspicuous. This megaspore nucleus divides into two nuclei, thereafter these two nuclei divide again forming from nuclei situated at the periphery of the megaspore sac at equidistant positions. Successive simultaneous divisions follow rapidly until about 256 free nuclei are formed by the process of free nuclear division. This process takes place through a period of about twenty days. The cell walls appear in the endosperm,

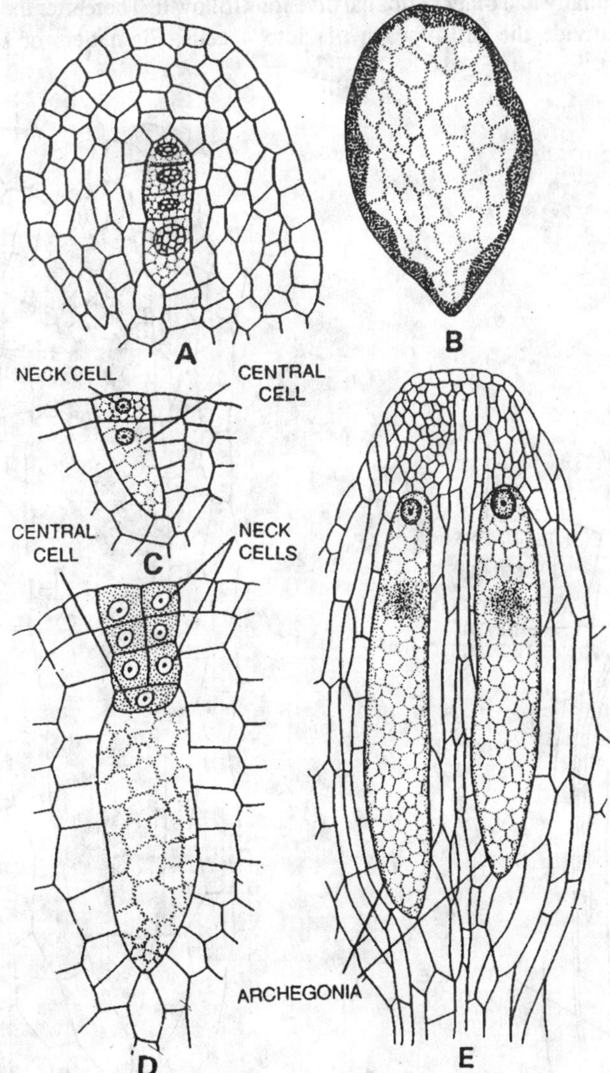

Fig. 6.11. *Ephedra trifurca.* Female gametophyte. A, row of four megaspores, the lower begins to enlarge; B, free nuclear division to form 64 nucleate stage of female gametophyte; C-E. successive stages in the development of archegonium; C, showing neck and central cell; D, showing three tiers of the neck cell, central cell still undivide E, two archegonia, neck not distinguished.

and the latter differentiates into two distinct regions known as micropylar and antipodal regions. The micropylar region consists of loose cells and the antipodal region of compact cells. The antipodal region becomes differentiated into two regions known as storage and haustorial regions. The storage region comprises of the large portion of the endosperm which is full of starch and other foods. The haustorial region consists of one or two outermost layer of endosperm cells situated at the antipodal end. The micropylar region contains two elongated archegonia and consists of elongated thin-walled cells.

Development of archegonia. Usually the two archegonium initials begin to develop in the micropylar region. The initial divides transversely cutting off the primary neck cell. The latter divides periclinally and other periclinal divisions follow it. Thereafter the anticlinal divisions take place which divide the cell of each tier into 4 cells. Eight tiers of such neck cells are

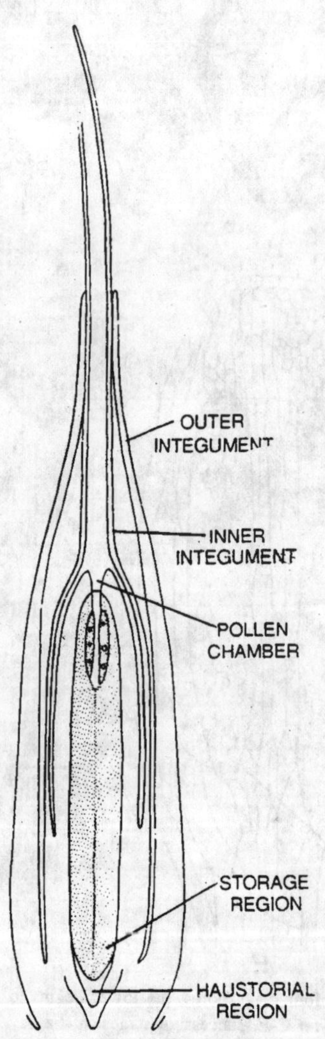

Fig. 6.12. *Ephedra trifurca.* L.S. of ovule showing nucellus with deep pollen chamber, inner and outer integuments, and female gametophyte with archegonia, storage and haustorial region.

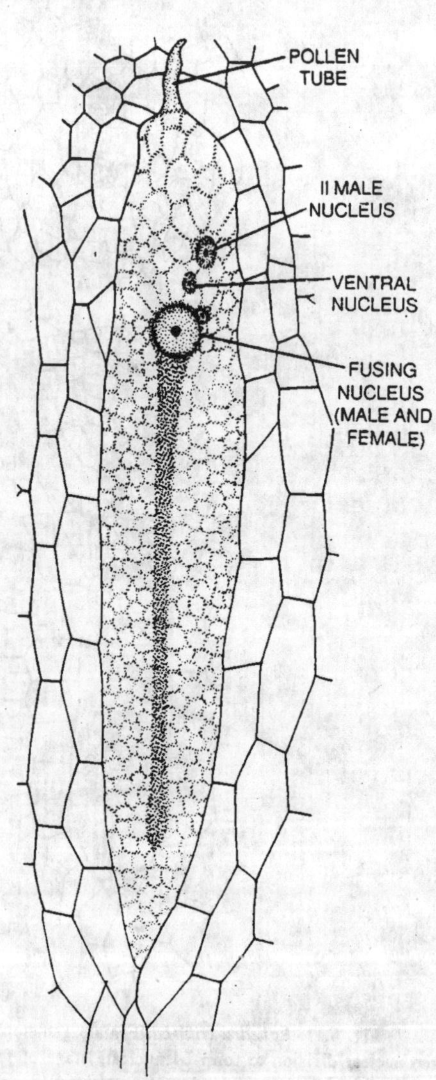

Fig. 6.13. *Ephedra trifurca.* Fertilization, pollen tube forces its way between the neck cells, thereafter male and female nuclei fuse together.

formed. Later on irregular divisions may take place and the appearance of the neck becomes like the ordinary tissue. The minimum number of neck cells is probably thirty-two, but sometimes they may be more. The archegonium neck of *Ephedra* is longest among the gymnosperms and thus associated with the deepest pollen chamber.

Structure of archegonium. Each archegonium is elongated and possesses an ill-defined jacket layer. The neck is at first made up of three tiers of cells but later eight or more tiers may be formed. According to the findings of Narang in some species of *Ephedra* archegonium groups occur at the micropylar end and sometimes an archegonial complex is formed by the aggregation of 4 to 8 archegonia. The oosphere (egg) of the archegonium contains a large nucleus, with a dense plasmatic sphere around it. A large ventral canal nucleus is present but ventral canal cell is absent.

Fertilization. Both the male nuclei are released from the pollen tube into the archegonium, but only the first male nucleus fuses with the nucleus of the egg. The second male nucleus

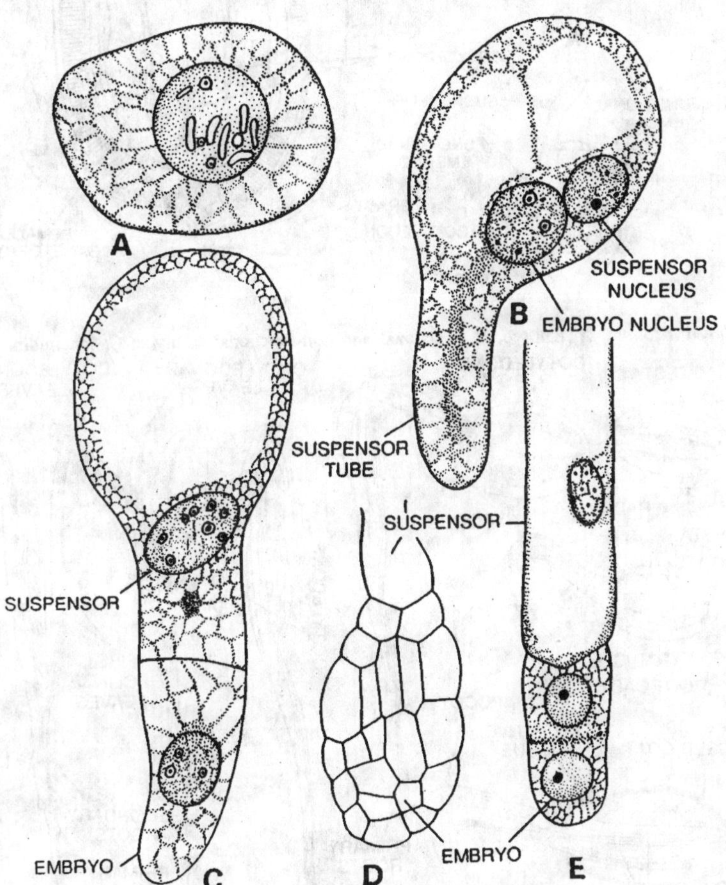

Fig. 6.14. *Ephedra trifurca*. Development of embryo. A, one of the proembryos prior to develop suspensor tube; B, pro-embryo with suspensor tube and two free nuclei; C, later stage with suspensor; D, division of embryo initial cell; E, later stage of embryo.

and the ventral canal nucleus divide forming many nuclei for a number of small cells around the top of the egg whose fate is not clearly understood. This tissue is soon absorbed by the embryo and, therefore, may be regarded as physiological endosperm. The oospore nucleus divides into eight nuclei of which 3 to 5 may give rise to emryos.

Development of embryo (embryogeny). Land (1907) has worked out the details of the formation of the embryo of *Ephedra trifurca* which is as follows: The oospore nucleus gives rise to eight free nuclei, more or less unequal in size. Three to five free nuclei enter into the

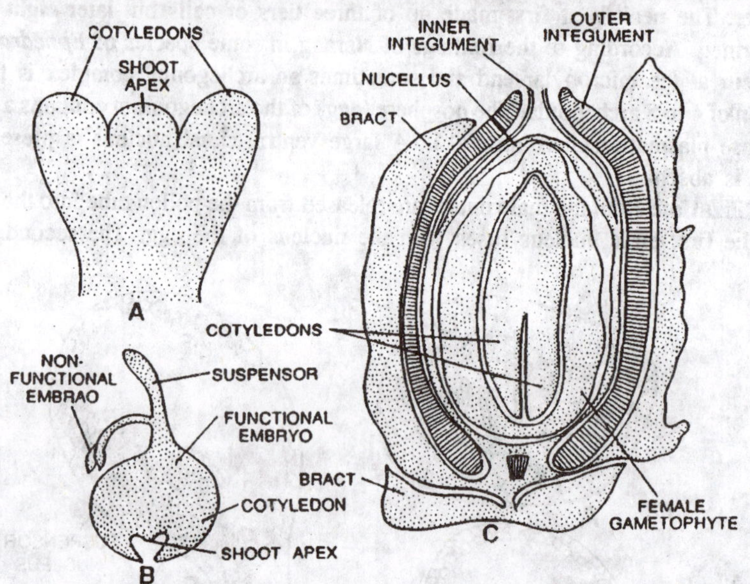

'ig. 6.15. *Ephedra foliata*. A, embryo, B, functional and non-functional embryos; C, longitudinal section of seed.

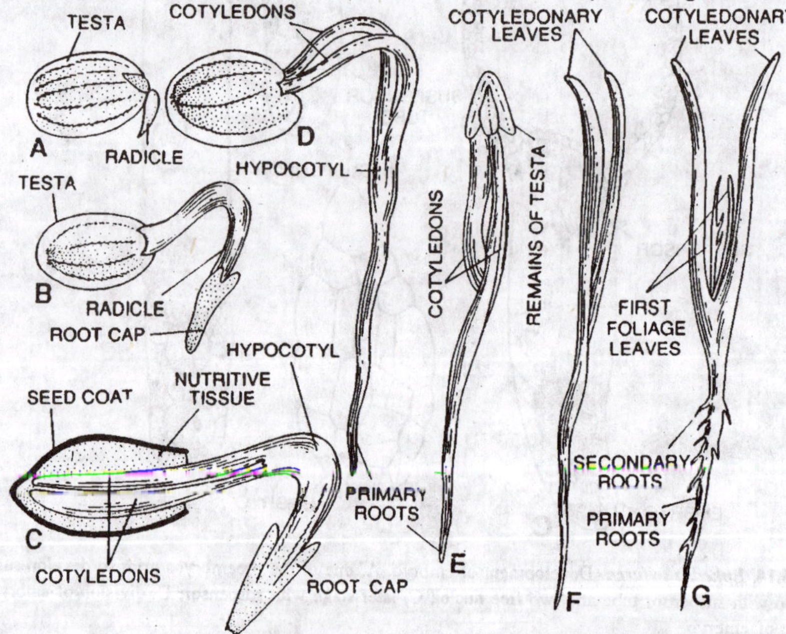

Fig. 6.16. *Ephedra foliata*. A-G, stages of seed germination.

organization of walled cells and produce embryos. The proembryonal cells are isolated from the surrounding cytoplasm by the cleavage formation, and around each nucleus a more or less irregular sheath of cytoplasm is deposited. In usual way, the lower cells successfully produce embryos but sometimes, the upper cells of micropylar region produce embryos. Each functioning proembryonal cell becomes rounded and its nucleus divides producing two daughter nuclei which are equal in the beginning but soon one becomes larger than the other. Due to the formation of the vacuoles, the two nuclei become shifted on one side of the cell. The larger nucleus passes

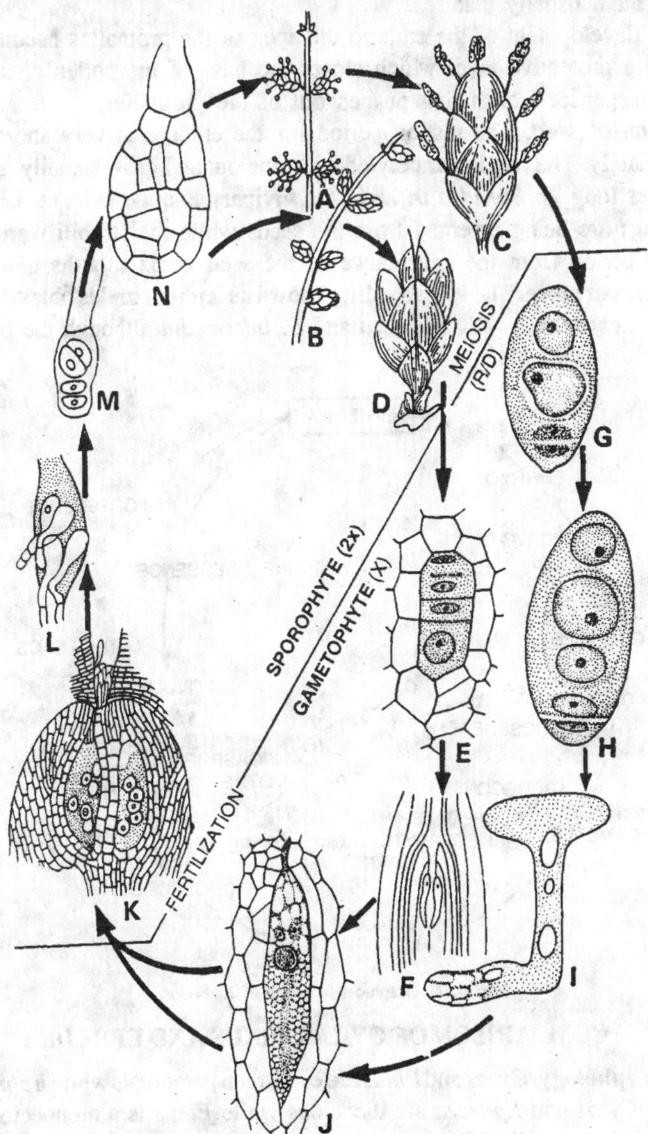

Fig. 6.17. ***Ephedra*****.** Diagrammatic life cycle. A, male or staminate branch; B, female or ovulate branch; C, staminate strobilus : D, ovulate strobilus: E, megaspore formation; F, ovule; G-I, formation of male gametes; J-K, fertilization; L-N, development of embryo.

nearly to the end of the tube and a wall is formed which cuts it off in a terminal embryo forming cell. The outer larger nucleus enters the suspensor and functions as suspensor nucleus. The terminal embryo cell divides and becomes a proembryo. The remainder of the tube forms the suspensor which elongates and pushes the young embryo down in the nutritive tissue at the base of the prothallus. The embryo cell divides transversely, and this is followed by a succession of division in irregular planes producing an ovoid mass of cells. The basal cells of the pro-embryo elongate successively, and form a secondary suspensor, which is not clearly differentiated from the root cap of the embryo proper. Although several embryos may begin to develop in a single ovule but only one of them usually matures.

During the development of the embryo the apex of the prothallus becomes meristematic and gives rise to a protective layer which closes the base of the pollen chamber. This layer also checks the suspensor from being pushed out of the prothallus.

Germination of seed. The resting period for the embryo is very short. The seed may germinate immediately. The two linear cotyledons come out and grow steadily until they become several centimetres long. In *Ephedra trifurca,* the vivipary is quite evident, *i.e.,* in some cases numerous seedlings are being emerged from the seeds while the strobili were attached to the stem. The whole period from the appearance of the seed is six months.

Economic importance. The valuable drug known as **'ephedrine'** is obtained from *E. sinica, E. gerardiana, E. nebrodensis, E. equisetica* and *E. intermedia* although the percentage of the available ephedrine is variable.

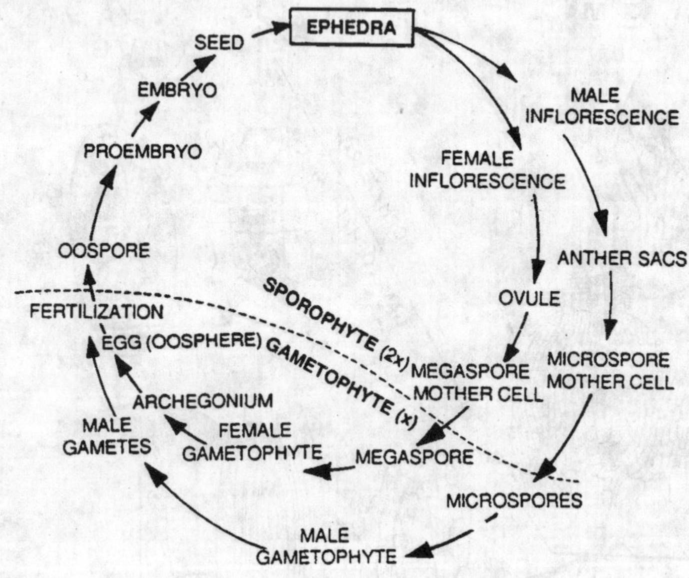

Fig. 6.18. Graphic life cycle of *Ephedra.*

COMPARISON OF CYCAS, PINUS AND EPHEDRA

External morphology. *Cycas* and *Pinus* are evergreen perennials while *Ephedra* is perennial with scaly leaves. *Cycas* and *Ephedra* are dioecious while *Pinus* is a monoecious plant. *Cycas* is 5-15 feet in height (*C. media 20'*), resembling in habit with palms on the one hand and tree ferns on the other. *Pinus* is about 15-45 metres in height while *Ephedra* is a straggling shrub with few species as lianes.

In all the three cases the plant body is distinctly differentiated into root, stem and leaves.

1. Stem

In *Cycas* it is tuberous when young but erect, columnar and unbranched. Sometimes it is branched due to injury or growth of a lateral bud. Male plants show sympodial growth (terminal cone) and female monopodial. The surface of stem is rough due to the presence of armour of persistent leaf bases and scale bases in alternate whorls. In case of *Pinus* and *Ephedra* the growth is monopodial one. In *Pinus*, branches are of 2 types, *i.e.*, long-shoots-(branches of unlimited growth) which are given out in horizontal direction; and short or dwarf shoots (branches of limited growth) present in the axils of scaly leaves on branches of unlimited growth. In *Ephedra* the stem is green, ribbed with distinct nodes and long internodes. The branches arise from axillary buds.

Anatomy of stem. All the three plants show a general pattern like that of dicot stem, *i.e.*, vascular bundles are conjoint, collateral, open with endarch xylem, and are arranged in a ring. All the members show a well developed secondary growth. In case of *Cycas* mucilage canals are present scattered in cortex and pith regions. Each mucilage canal is made up of inner single layer of epithelial cells which secrete mucilage. Besides this girdle-shaped leaf traces are also found in the cortex. They really are the vascular supplies of the leaves given by the main vascular cylinder. Each leaf receives four traces, two of which are direct traces and other two are given out from the opposite side of direct traces. These girdle around the main vascular cylinder through cortex and then enter the upper leaf base on the opposite side. These leaf traces are known as **girdle traces.** In *Pinus,* cortex and pith cells have tannin and resin canals, which secrete resinous substances like turpentine. Each resin canal is internally lined up by epithelial cells surrounded by sclerotic sheath. In *Ephedra,* the cortex is differentiated into outer palisade parenchyma and inner spongy chlorenchyma which are provided with patches of sclerotic cells.

In case of *Cycas* annular rings are formed due to secondary growth by successive cambium rings which are active only for a short while. These cambium rings are formed either from pericycle or cortex. In *C. pectinata* as many as 14 concentric vascular rings are reported. The successive outer rings decrease in thickness. In *Pinus* annual rings are formed due to secondary growth differentiated into spring wood and autumn wood. The cells of the spring wood are thin-walled, polygonal and larger in size while those of autumn wood are thick-walled, squarish and smaller in size. By counting the number of annual rings we can count the age of the plant. This type of differentiation is clearly visible only in trees growing in temperate countries where the winters are severe, but where the winter is not severe, the secondary growth may even continue for more than one year without demarcation into annual rings. In this case they are called as **growth rings** instead of annual rings. In *Ephedra,* secondary growth is very little and here also annual rings are formed with distinct spring wood and autumn wood. A sclerotic layer is also formed outside the secondary phloem. *Ephedra* is characterized by the presence of companion cells in phloem and vessels in the xylem elements while both these are absent in *Cycas* and *Pinus.* In this case sieve tubes and companion cells are developed from the same initial. In the same way vessels are developed from pitted tracheids in *Ephedra* and from scalariform tracheids in angiosperms.

In *Pinus* and *Ephedra* cork cambium develops which cuts off secondary cortex on the inner side and cork on the outer side.

The structure of the secondary medullary rays is very peculiar in *Pinus*. In r.l.s. the height and length of secondary medullary ray are seen. The ray on both sides is surrounded by 2 rows of dead empty elongated cells known as ray tracheids having pits in the field. In phloem region it is made up of starch cells surrounded by elongated dome-shaped albuminous cells. The bordered pits on xylem tracheids are seen in surface view and are arranged in uniseriate manner. In t.l.s. the wood tracheids are cut longitudinally showing longisection of bordered pits. The torus is very distinct, surrounded by two dome flaps of thickening. Medullary rays are cut transversely and show breadth and height. These are either linear (uniseriate) or fusiform (multiseriate). In multiseriate medullary ray, a resin canal is present in the centre. In case of *Pinus* the wood is dense and compact as a result of which it has been termed as pycnoxylic.

2. Leaf

The leaves in gymnosperms show a wide variation in their external morphology. In *Cycas*, it is pinnately compound fernfrond like structure. The size of a frond may be as small as 5 cm. as in *Zamin pygmea* or may be as long as 3 metres, *e.g.*, *Cycas circinalis*. In young stages they show circinate vernation-a prominently pteridophytic character. In Coniferales the leaves are linear and acicular (*Pinus*). In other genera of Coniferales the shape of the leaves is varying. In *Gnetum*, however, the leaves are broader whereas in *Welwitschia* there are only two leaves throughout the life of the plant. In *Ephedra*, only rudimentary leaves are present at the nodes.

A characteristic feature of *Cycas* and conifers is the presence of dimorphic leaves, *i.e.*, scale and foliage leaves. Their distribution is characteristic in different members.

In *Cycas*, scale leaves are arranged spirally alternating with the whorl of foliage leaves on the stem covering stem apex and developing foliage leaves. In *Pinus*, these are present on long and dwarf shoots. In *Ephedra*, scale leaves are present on the nodes of the phylloclade, forming a sheath which may be two or three toothed, rarely four toothed. In *Cycas* scale leaves are persistent protective, and are covered with brown ramental hairs. In *Pinus* they fall off as the branches mature.

Foliage leaves are absent in *Ephedra*. In *Cycas*, leaves are large pinnately compound up to 3 metres long in *C. circinalis*. In *Pinus* they are needle like simple borne on dwarf shoot constituting with the latter the foliar spur. The leaves of *Cycas* are shed leaving leaf bases but in *Pinus* foliar spur is shed. Number of leaflets in *C. revoluta* is more than hundred, in *C. circinalis* 80-100. In *Cycas revoluta*, the leaflets are lanceolate, leathery with entire curved margin and spiny apex. In *C. circinalis* margin is not curved. In *Pinus* foliar spur may be monofoliar, bifoliar, quadrifoliar or pentafoliar, depending upon the number of foliage leaves on the dwarf shoot. Sometimes the leaves are present in groups directly on the main branch or shoot, *e.g.*, in *Cedrus*.

Anatomy. Scale leaves are dry and without vascular tissue in *Pinus* but in *Ephedra* each scale leaf is provided with a midrib which consists of two parallel vascular bundles. In *Pinus*, needle may be circular, semicircular or triangular in its outline. Epidermis consists of thick-walled parenchymatous cells covered with thick cuticle. *Cycas* leaf is hypostomatic and *Pinus* leaf is amphistomatic. Stomata are sunken. A well defined hypodermis is single layered below upper epidermis but is absent above lower epidermis in *Cycas*. In midrib region, however, it is 2-4 layered on both the sides. *Pinus* needle has multilayered hypodermis on all the sides. Mesophyll is differentiated into palisade tissue and spongy parenchyma in *Cycas* but not so in *Pinus*. Resin canals are found in *Pinus*. Transfusion tissue is present in both. Midrib consists of only one vascular bundle in *Cycas* while in *Pinus* two vascular bundles are found. In *Cycas*, vascular bundle is surrouded by a bundle sheath and pericycle is one layered. Both the vascular bundles of *Pinus* needle are separated with 'T' shaped tissue of sclerenchyma, xylem is diploxylic (centripetal and centrifugal) in *Cycas* while in *Pinus* only of one type.

In *Cycas* leaf consists of leaflets and rachis. In T.S. rachis also shows different type of tissues. Outermost layer is epidermis covered with thick cuticle. Epidermis is followed by continuous hypodermis which may be 3 to 4 layered and sclerenchymatous. Hypodermis is followed by chlorenchymatous tissue. In the centre parenchymatous ground tissue is found. Mucilage canals are found in the ground tissue. Numerous vascular bundles are embedded in the ground tissue arranged as omega. Each bundle is conjoint collateral surrounded by a thick-walled fibrous bundle sheath and is open.

The vascular bundles are different in structure from base to apex of the rachis. At the base of the rachis bundles are with endarch xylem. As the vascular bundles are traced upward the centrifugal xylem gradually diminishes and the centripetal xylem appears on the inner side of the protoxylem. Throughout the greater part, the bundles contain a large mass of centripetal xylem with only traces of centrifugal xylem which is separated from the protoxylem by parenchymatous cells. The vascular bundle in this region shows pseudomesarch condition. At the apex the centrifugal xylem completely disappears and is exarch. The gradual disappearance of centrifugal and simultaneous appearance of the centripetal xylem is characteristic of *Cycas*

3. Root

External morphology. *Cycas* possesses a well developed tap root system which grows deep into the soil. Wordsdell (1909), however, believed that tap root of *Cycas* is later replaced by adventitious roots. Coralloid roots-coral like in structure, also develop. These are negativaly geotropic roots and are dichotomously branched.

Development of coralloid roots. These roots develop from normal roots due to some changes brought about in the growing region of the roots. During growth root apex eventually comes in contact with the bacteria. The latter enter the roots through lenticels or even epiblema and growing apex and inhabit the roots. Due to the presenee of these bacteria the cortex gets loose. Soon the elements of blue green algae also enter into it. Due to the presence of these external elements, the physiology of roots changes, they become negatively geotrpic, divide dichotomously repeatedly and never show signs of secondary growth. These roots, as are comparable to corals, are known as coralloid roots. These eventually come out of the earth and help perhaps in respiration and absorption of water.

In *Pinus* there is a tap root system which forms a massive root system due to strongly developed lateral roots. Root hairs are few. Ectotrophic mycorrhiza is developed and as such root hairs soon vanish.

Anatomy. The roots have radial vascular bundles with exarch xylem. It may be diarch, triarch or polyarch. In *Cycas* there are two types of roots, *i.e.*, (1) Normal (2) Coralloid root.

Cycas normal root resembles the dicot root internally. Outermost layer is epiblema (piliferous layer) composed of thin-walled, elongated, compactly arranged parenchymatous cells. Root hairs are present on younger roots. In *Pinus* root hairs disappear due to formation of ectotrophic mycorrhiza. In *Ephedra,* root hairs are present.

Below the epiblema is a many layered cortex composed of loosely arranged parenchyma cells filled with starch. Mucilage canals are present in *Cycas.* The innermost layer of cortex forms the endodermis which is well differentiated in *Pinus.* Tannins are absent in *Ephedra.* In coralloid root of *Cycas,* cortex is divided into three zones, *i.e.*, outer, inner and middle zone. The last zone consists of loosely arranged, radially elongated thin-walled cells with large intercellular spaces. These spaces and the cell lumen are occupied by blue green algae.

Algal zone contains usually blue green alga like *Anabaena cycadae* (Reinke 1872, 1879), *Nostoc punctiforme* (Hariot, 1892) or *Oscillatoria,* diatoms and other algae associated with nitrogen fixation, or a nitrogen fixing bacteria, *e.g.*, *Azotobacter, Pseudomonas radicicola.* It has been suggested that (a) algal zone and root act as symbionts (Schneider, 1894) (b) algal zone as aeration organ (Life 1901). According to Schaedae (1944), there are no bacteria in coralloid root.

Vascular bundle. Radial vascular bundles are surrounded by many layered pericycle. Xylem is diarch to triarch in *Cycas,* diarch to hexarch in *Pinus* and diarch in *Ephedra.* It is exarch but Atwood (1939) reported mesarch in *C. revoluta.* Xylem in *Pinus* is 'Y' shaped and has resin canals. Xylem cnsists of tracheids only in *Cycas* and *Pinus* but tracheal tubes (vessels) are also present in *Ephedra*. Phloem consists of sieve tubes and parenchyma in *Pinus* but only parenchyma in *Cycas*.

Secondary growth. In *Cycas* and *Pinus* secondary growth is due to cambium which develops inner to phloem patches. These patches ultimately become continuous and cut secondary xylem and phloem on inner and outer sides respectively.

4. Reproductive Parts

Sexual reproduction is the only method of reproduction in these except in *Cycas*, which can reproduce vegetatively by means of bulbils. These are formed in the axil of scale leaves throughout the stem. These may germinate either on the stem itself giving the false appearance of dichotomy or fall off on the soil and sprout whenever suitable conditions are available.

Sexual reproduction is basically the same in all the three genera. These are heterosporous, producing two types of spores-megaspores giving rise to female gametophyte and microspores forming male gametophyte, *Cycas* and *Ephedra* are dioecious, but Chamberlain (1935) reported a plant of *C. revoluta* from Australia which at first produced a female strobilus and a few years later, a male strobilus. Similarly *Ephedra intermedia* and *E. foliata* are monoecious. *Pinus* is strictly monoecious and is interesting as male and female cones are borne on separate branches.

Male reproductive parts. The microspores are produced in microsporangia which in turn are present on leaflike microsporophylls. The microsporophylls are aggregated to form the male cone. In *Cycas*, the male cone is present at the top of the main axis and is borne singly. In *Pinus*, groups of male cones are present in the axil of the scale leaves, while in *Ephedra* 2-4 male cones are present at each node in the axil of the scale leaves.

In *Cycas* male cones attain a length of 50-60 cm. Large number of microsporophylls are present spirally arranged around the cone axis. Each microsporophyll is more or less triangular, the apical portion of which extended in a sterile disc known as apophysis. Microsporophylls show traces of pinnate characters. On the ventral surface of each microsporophyll are present large number of stalked sporangia in groups of two to five. Around each sorus are present many multicellular hairs, known as indusial hairs. In *Pinus* cones are only 2-10 cms. in length, and each microsporophyll bears only two sessile sporangia. In this case pollen grains are winged. In *Ephedra* bracts are arranged in opposite and decussate manner and all of them except the lower 1 or 2 pairs are fertile. A stamen is present in the axil of bilabiate perianth. The anthers are 2-5 lobed and are provided with a terminal aperture.

Female reproductive parts. Megasporophylls are the ovule bearing organs. In *Pinus* and *Ephedra* the megasporophylls are forming compact cones while in *Cycas* the megasporophylls are borne in succession of the leaves and are not forming any cone. It is interesting to note that in *Cycas* the megasporophylls are present on the main axis while in *Pinus* and *Ephedra* female cones are present on lateral branches. This has led Meeuse to suggest that each megasporophyll in *Cycas* is equal to a full cone of other genera. In certain species of *Cycas, e.g., C. circinalis, C. normanbyana and C. rumphii*, there is reduction in the number and size of sporophylls. The ovules are borne on the margins of the sporophylls. The largest ovules in the plant kingdom are borne by *Cycas* species (about 6 cm. in diameter). In *Ephedra* all the lower bracts are sterile except the uppermost pair which bears a pair of ovules. In *Pinus* the female cone matures in 3 years. Each seed-scale complex here is differentiated into two parts:

(i) Bract scale. Small, dry, membranous structure attached directly to the cone axis and bearing the ovuliferous scale in its axil.

(ii) Ovuliferous scale. Woody structure borne above the bract scale bearing at its base two ovules with micropyles directed towards the cone axis.

The morphological nature of the female cone of *Pinus* and in particular that of ovuliferous scale has been variously described in the past. Views have been put forward by Robert Brown (1827), Schleiden (1839), Van Tiegham (1869) and Sachs (1868). The most accepted view is, however, that of Florin (1955). According to him, the elongated axis is a peduncle, the bract scale is a real bract and the ovuliferous scale with two ovules is a female flower of very rudimentary type. The entire cone thus represents an inflorescence or a 'compound strobilus' consisting of a number of rudimentary female flowers.

The ovule of *Cycas* is orthrotropous and a chiefly made up of central mass of parenchymatous cells known as nucellus. The nucellus is covered by the integument all around except at the top where a small pore-the micropyle is left. The integument may be differentiated into 3 layers-the outer fleshy, the middle stony and the inner fleshy layer. In the region of micropyle the nucellus protrudes into a nucellar beak which extends on peripheral sides to form a pollen chamber. Embedded in the nucellus there is a functional megaspore with rudiments of non-functional megaspores here and there. The ovule of *Pinus* is similar to that of *Cycas* except that no nucellar beak and pollen chamber are present in *Pinus*. In *Ephedra*, however, a very long micropyle is present which protrudes out of the bracts also. The female flowers of *Ephedra* can be identified morphologically by the presence of long micropyles at the top of the cone.

5. Male Gametophyte

Gametophytic generation starts with the formation of spores, after meiosis in the sporogenous tissue within the sporangium. Gymnosperms are heterosporous, consequently micro- and megaspores are formed inside the micro-and megasporangia respectively. The microspores, also called as pollen grains in case of gymnosperms and angiosperms are the initial stages of male gametophyte. The microspores of *Cycas, Pinus*, and *Ephedra* are built on the same architectural pattern. Their development can be studied in two stages.

(i) Development of male gametophyte before pollination, *i.e.*, inside microsporangium.

(ii) Development of male gametophyte after pollination, *i.e.*, inside ovule.

The germination of pollen grains starts *in situ*. The first division, in all the three genera results in the formation of a small prothallial cell and a bigger cell. The bigger cell directly functions as antheridial initial in *Cycas*, while in *Pinus* and *Ephedra* it cuts off a second prothallial nucleus. In *Pinus* this division of nucleus is followed by cytokinesis resulting in second prothallial cell and antheridial cell. However, in *Ephedra* there is no wall between second prothallial nucleus and the antheridial nucleus. The antheridial cell divides to form generative cell and tube cell. In *Cycas*, pollen is at three celled stage (one prothallial cell, one generative cell and one tube cell). In *Ephedra* the antheridial nucleus divides to form generative nucleus and nucleus. The generative nucleus divides giving rise to body nucleus and stalk nucleus. At the time of shedding the pollen is five nucleate (two prothallial nuclei+one stalk nucleus+one body nucleus+one tube nucleus). Each pollen grain develops a thick exine and thin intine. It is winged in *Pinus*.

Formation of male gametes from the body cell/body nucleus takes place in pollen chamber of ovule. Pollen tube is formed by the prolongation of intine. The tube nucleus migrates into it. The pollen tube in gymnosperms is not the gamete carrier but functions as haustorium absorbing food from the nucellar tissue for the developemtn of male gametophyte. The tube usually divides to ramify into the nucellus. The two prothallial cells of *Pinus* male gametophyte are short lived and degenerate soon. Body cell in *Cycas* divides to form two sperm mother cells, which give rise to motile sperms. The sperm is top shaped, multiciliated and is visible to naked eyes (perhaps biggest sperm in plant kingdom). In *Ephedra* and *Pinus* the male gametes formed from body nucleus are only two nuclei.

6. Female Gametophyte

Megasporogenesis. Megaspore mother cell differntiates in the nucellus. Reduction division of megaspore mother cell nucleus results in the formation of a linear tetrad of megaspores, the chaiazal megaspore out of them is functional and the rest 3 degenerate.

Development of female gametophyte. The development of female gametophyte in *Cycas, Pinus* and *Ephedra* takes place almost in the same manner. The functional megaspore, before its division enlarges by absorbing some of the neighbouring cells of the nucellus tissue. The development of the female gametophyte takes place *in situ.* The nucleus of the megaspore divides by a free nuclear division to form a large number of nuclei as many as 2000. The nuclei thus produced are pushed towards the periphery as a vacuole appears in the centre of this structure. The wall formation starts from periphery and proceeds to the centre. The tissue produced in this manner is called endosperm or female prothallus. The compactly arranged upper cells of this structure are smaller in size than those of the cells of lower region. The latter cells perform the function of nutrition.

Much before the gametophyte has been cellular, one to two cells thick layer endosperm jacket (the nutritive layer) develops contact with the gametophyte. The cells of the jacket, like those of the tapetum pass on the nutritive material from the cells next to it into the growing gametophyte. The cells of the gametophyte contain starch grains. Some of the cells may contain tannin also.

Although the wall formation in *Pinus* is also centripetal but the first wall is perpendicular to the megaspore membrane, and extends to the middle of the megaspore cavity, so that the cells are like long tubes which are called as "alveoli". Cross walls are then formed in such a manner that the alveoli become divided into a number of uninucleate cells.

In *Ephedra* the megaspore mother cell is deep seated because of prolific divisions in the overlying nucellar epidermis. The number of nuclei produced by free nuclear division in megaspore is 256 or more within 20 days. The tissue of the mature female gametophyte is gradually differentiated into:

(i) Micropylar region or upper reproductive zone-consisting of thin-walled loosely arranged cells which function as archegonial initials in further development.

(ii) Antipodal region:

(a) Lower storage region. Greater part of the endospern is of compactly arranged cells which are full of food materials.

(b) Basal haustorial zone. It consists of outer two or more layers of cells at the antipodal end. The cells of this region are also compactly arranged and perform the function of absorption of food material from the surrounding cells of the nucellus.

Development of archegonium. Each archegonium develops from a single archegonial initial situated superficially in the tissue of endosperm on the micropylar end. It enlarges in size and gets differentiated from rest of the cells of the endosperm. It divides by a transverse or periclinal division into an upper primary neck cell, and lower central cell. The primary neck cell divides into two neck cells by a vertical division. In *Pinus* this primary neck cell divides by two successive vertical divisions at right angles to each other followed by a transverse divison forming an eight celled neck arranged in two tiers of four cells each. In *Ephedra* the primary neck cell divides to provide a neck of eight or more tiers with a minimum of 32 cells. In the beginning these divisions are regular but later on become irregular so that the cells of neck can no longer be distinguished from the surrounding gametophyte tissue. The neck of archegonium in *Ephedra* is the largest among Gymnosperms.

The central cell enlarges in size and its nucleus divides into a ventral canal nucleus and egg nucleus. There is no wall formation between two nuclei in *Cycas* and *Ephedra* while there is a definite wall between two nuclei in *Pinus* giving rise to an upper small ventral canal cell and lower larger egg. The egg of *Cycas* and its nucleus are large in living plants.

The neck canal cells are never formed in gymnosperms. Later the ventral canal nucleus or the cell degenerates. During this development of archegonium, the endosperm tissue near micropyle grows upward and forms the archegonial chamber. The archegonial jacket also develops in *Cycas* and *Ephedra* from the surrounding tissue which helps in the transfusion of food to egg. Swamy (1948) has, however, observed pores in the wall of the jacket facing the venter of archegonium. It has been suggested that exchange of food materials takes place through these pores.

Mature archegonium. Each archegonium consists of an egg and a neck formed by two, eight and 32 cells in *Cycas, Pinus* and *Ephedra* respectively. The neck is not very distinct during later stage in *Ephedra,* but well distinct in the rest two cases. A ventral canal cell is present in *Pinus* but in *Cycas* and *Ephedra* there is only ventral canal nucleus. The surrounding cells of the endosperm form the archegonial jacket, around the archegonium in *Cycas* and *Ephedra.*

From evolutionary standpoint, *Cycas* appears to be more primitive than *Pinus* and *Ephedra.* This is illustrated by the presence of largest egg and a massive ovule with integument differentiated into 3 regions.

7. Embryo

Pollination. Like all other gymnosperms *Cycas, Pinus* and *Ephedra* are also anemophyllous. A drop of mucilaginous substance oozes out of the micropyle in which the pollen grains are entangled. Later on the pollen grains are drawn into the pollen chamber due to the drying up of pollination drop.

Fertilization. The act of fertilization is not immediately accomplished, but there is an intervening period between pollination and fertilization which varies in various genera. In *Cycas* it may be 5-6 months, in *Pinus* it is about 9 months whereas in *Ephedra* it may be as short as only 10 hours. Fertilization is siphonogamous. The pollen tube penetrates into the archegonial chamber and bursts to release the male gametes. In *Cycas* the spermatozoids swim for some time and on coming in contact with the neck cell, one of them enters the archegonium and finally penetrates the egg. Here the ciliated band is left at one of the corners of egg and nucleus slips out of it and fuses with the egg nucleus forming zygote. In *Pinus* pollen tube bursts near the egg releasing two male nuclei; one of them degenerates and the other fuses with egg nucleus forming zygote. In *Ephedra* the pollen tube bursts within the archegonium discharging the tube nucleus, stalk-cell nucleus and two sperm nuclei into the cytoplasm of egg. One sperm nucleus fuses with the egg nucleus to form zygote. The bahaviour of other nucleus is extraordinary and it fuses with ventral canal nucleus (Khan 1943).

Embryo formation starts with a prolonged period of free nuclear divisions without resting period. In *Cycas* 64-256 free nuclei have been reported. A large central vacuole is formed by coalescence of smaller ones. As the real embryo of *Cycas* is formed only from a part of this structure it is termed proembryo. Cell formation begins at the base of the proembryo and gradually extends upwards. The proembryo becomes differentiated into three regions. Cells present just below the central vacuole become haustorial; those immediately below this elonagte to form a long, coiled and twisted suspensor while the cells situated at the tip of the suspensor remain meristematic and develop the embrro proper.

The fertilized egg or the zygote in *Pinus* nucellus divides into two and then four nuclei by free nuclear divisions. These nuclei enlarge, migrate to the base of egg cell and arrange themselves in a single plane. The four nuclei now divide simultaneously to form eight free nuclei. It is after this division that the cell wall formation starts so that the eight nuclei become arranged in two tiers of four cells each. The two tiers again divide to form four tiers giving rise to a sixteen celled proembryo. Uppermost tier is open at the top and is believed to be nutritive. The next tier is the rosette tier; the third one is suspensor, and the fourth the embryo proper that divides further. The suspensor cells elongate very much and split into four separate suspensors each bearing at its tip a group of embryonal cells developed from a cell of the embryonal tier. The four separated cells of the embryonal tier give rise to four separate embryos. This phenomenon is known as **cleavage polyembryony.** Apart from this, adventive polyembryony is also found in *Pinus.*

The zygote undergoes three free nuclear divisions in *Ephedra* forming eight nuclei all capable of giving rise to embryos. They are evenly distributed in the cytoplasm. Each of them invests itself with cytoplasm and cell wall thus becoming a proembryo. Out of these, usually 3-6 proembryos begin to develop into embryos. The proembryonal cell divides to form a suspensor cell and an embryo initial which is pushed down into the female prothallus by the elongating suspensor.

8. Seed

Seeds are mature ovules with all their contents like embryo and endosperm and integuments etc. It is to be noted that a seed has condensed within it three successive generations : sporophytic, gametophytic and sporophytic generations, beautifully packed in one. These seeds dominate the life of higher plants to such an extent that these plants are better known as seed plants.

The formation of seed takes longer duration among gymnosperms and these do not have any resting period before germination.

Development of seeds. In *Cycas,* seed formation sets on after about one year since fertilization. The integuments are variously coloured and have three layers. The outermost fleshy layer starts drying up and becomes appressed to middle stony layer and forms seed coat. The inner fleshy layer also starts giving up all cellular contents. They are used up by developing gametophyte and embryo. The embryo increases in size without much differentiation at this stage. The nucellus is also used up leaving behind a small papery cap just below the micropyle. The tissue of female gametophyte modifies into the massive and nutritive tissue-the endosperm. The embryo at this stage has a long twisted suspensor. At the point where suspensor joins embryo proper a hard covering tissue called as coleorhiza is formed; may be that it forms a protective layer for radicle. The embryo usually has two cotyledons but rarely one or three cotyledons may also be present. The cotyledons are unequal in size and sometimes the larger one may show pinnate nature. Between the cotyledons there may be one or two scales and a small leaf arising at right angles to cotyledons. The development of roots is much delayed in *Cycas.* The embryo may be more than one in some cases, *e.g.,* in *C. circinalis.* Shedding of seeds may take place at various stages. Soon after shedding germination starts which is very similar to that in dicotyledons.

In *Pinus* ovules are small in size and consequently the seeds too. Of course number of seeds per cone is large. Each sporophyll bears two seeds. The integuments after fertilization constitute the seed coat. The embryo in *Pinus* is slightly different from that of *Cycas.* It has a whorl of cotyledons which may range up to 14. It has been supposed that large number of cotyledons indicate primitiveness of embryo. The outer fleshy layer of integument dries up and forms the wings of *Pinus* seeds. The inner fleshy layer helps in nutrition of embryo and forms a papery layer. Remains of nucellus are represented by a small cap-like and shrunken layer of cells at apex. At the time of germination the first leaves are needle-like and spurs arise in the axil of these primary leaves. The cotyledons may develop chlorophyll without light also. Due to ecological factors seedlings of *Pinus* do not develop under the canopy of parent tree.

Ephedra ovules are smaller and bitegmic. The outer integument is of doubtful status and is not considered generally as true integument. It is supposed to be made of two closely appressed valves. Moreover, the bracts which are present outside the ovule become fleshy. Perhaps a step forward towards fruit formation from protection and dispersal point of view. The embryo has two cotyledons embedded in female gametophyte tissue and the nucellus. The nucellus results in a papery covering and gametophyte forms endosperm. The outer integument becomes very hard after some time and forms the seedcoat with two layers. The germination may even take place on the parent plant. The polyembryony may also be present in *Ephedra*, but is very rare.

Dispersal and germination of seeds.*Cycas* seeds are very large and variously coloured attracting, birds and animals for dispersal. Seeds of *Pinus* are small and winged, therefore, wind currents also help in their dispersal. Seeds of *Pinus* are good to taste and taken away by animals too. The *Ephedra* seeds may start germinating on parent plant without going away from parents. *Cycas* and *Pinus* seeds, however, germinate after detachment from the parent plants. *Cycas* seeds are very sensitive to injury which may result in failure of germination.

At the time of germination of *Cycas* seed, root initial forms the primary root and pushesd out through the micropyle. Germination is hypogeal. Primary leaf starts arising at right angles to cotyledons and becomes green on coming out.

Pinus seeds show epigeal germination, cotyledons also with primary needle-like leaves comig out of the seed coat and seed coat too being pushed out of the soil.

Ephedra seeds do not require long resting periods and germinate soon after maturity. In fact germination process is alike in *Pinus* and *Ephedra* and is epigeal.

Genus GNETUM

Systematic position. Division. Gymnospermae
Class. Gnetopsida
Order. Gnetales
Family. Gnetaceae
Genus. *Gnetum*

Distribution. There are about thirty-four species in this genus. They are mainly distributed in the tropicl regions of the world. Out of these thirty-four species about nineteen species are common to the Eastern and Western hemispheres. The genus is not found in North America, Europe and Australia. One species, *Gnetum scandens* a large climbing plant is widespread in Asia and in India it is found in the Western Ghats. Another species *G. gnemon*, a small evergreen tree is found in Burma.

Habit. Majority of species of *Gnetum* are climbing or trailing plants. Only few species are either trees or shrubs. The plant is easily mistaken for a dicot unless it is in flower or bearing fruits.

The leaves are net-veined, large, oval and entire. In most of species there are branches of limited and unlimited growth, but usually there is no strong difference between two types of branches as seen in *G. gnemon* which is an evergreen small tree, the shoots of unlimited growth bear scale leaves on them while the foliage leaves are borne on shoots of limited growth.

G. gnemon can reproduce vegetatively by forming buds on leaves just as in *Bryophyllum*. So in external appearance and the form of leaf, the plant is very much like a dicot.

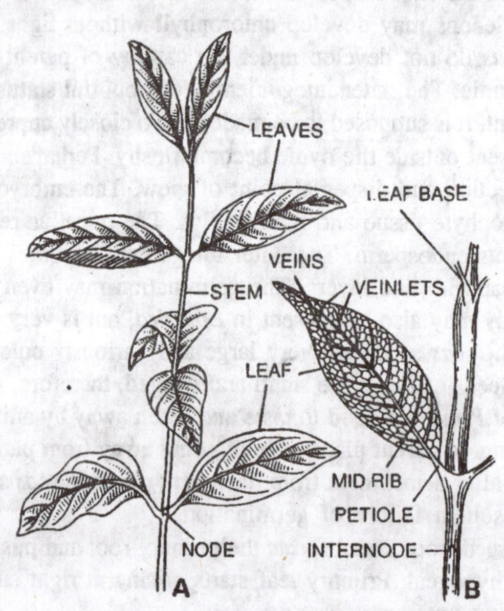

Fig. 6.19. *Gnetum gnemon*. A, a twig with opposite and decussate leaves; B, a leaf showing reticulate venation.

Anatomy of stem. In the internal anatomy also, especially in climbing species the structure is like a dicot climber. There is a ring of collateral endarch bundles separated from one another by broad medullary rays. Towards outside there is a broad cortex. The epidermis is single-layered and its cells are papillate and rectangular with thickened outer walls. The cuticle is thick and the stomata are sunken. The endodermis and pericycle are inconspicuous. The wood is composed of a large number of tracheids and a few vessels. The phloem is composed of sieve cells and phloem parenchyma. The pith consists of round or polygonal parenchymatous cells. Some of the cells of pith in older stems become lignified and pitted.

The secondary growth begins in a normal manner but after a short time an extrafascicular cambium develops and it produces a normally oriented ring of vascular bundles. However, in *G. ula* and *G. africanum* there is anomalous type of secondary growth. Here several successive rings of cambia develop one after the other in different parts

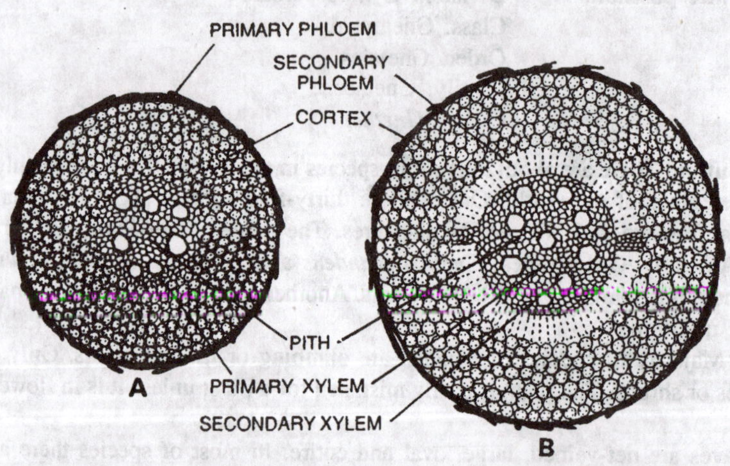

Fig. 6.20. *Gnetum gnemon*. A, T.S. of young root; B, T.S. of old root.

of the cortical region. These rings gradually become incorporated into a continuous xylem cylinder possessing wedge-shaped vascular bundles. The growth of the first ring ceases and thereafter the growth of the second ring begins. This way, the rings are formed successively one after another. Some of the rings remain incomplete and become in eccentric position with regard to pith.

The protoxylem is small in amount and it is made up of spiral vessels. The secondary xylem is made up entirely of vessels which have pitted thickenings of their walls and are of various sizes. These vessels is *Gnetum* are more advanced than seen in *Ephedra*. In *Ephedra* the end walls of tracheids have several perforations while in *Gnetum* there is a single broad perforation in the end wall. In angiosperms the vessels are formed by complete breaking down of end walls. Here in *Gnetum* there is a broad perforation and the end wall does not completely break down. Really the vessels in Gnetales, especially in *Gnetum* have no real genetic connection with those of angiosperms. The phloem in *Gnetum* is of uniform cells made up of the sieve tubes and phloem parenchyma. The companion cells are altogether absent. In some cases small cells have been seen to be present near the sieve tube, but during development it is seen that they are not cut off from sieve tubes, so unlike the phloem of angiosperms there are no real companion cells.

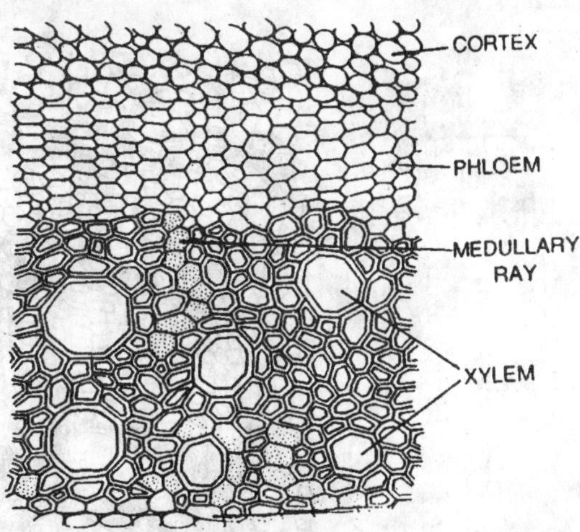

Fig. 6.21. *Gnetum gnemon*. A portion of the T.S. of root.

The primary cambium in *Gnetum* as seen in *G. gnemon* is short lived, successive cambial rings are produced which form concentric series of vascular bundles, some are very close and practically combined to one another. Secondary growth is practically as seen in old stems of cycads.

Fig. 6.22. *Gnetum gnemon*. A portion of the T.S. of young stem.

Anatomy of leaf. The epidermis on the upper and lower surface of the leaf is undulate. It possesses a well-developed cuticle upon its upper side. The stomata are found only in the lower surface. The mesophyll tissue is differentiated into palisade and spongy tissues. Besides these, are present some spicular cells, latex tubes, fibres and sclerotic cells. The vascular bundles are arranged in an arch-like manner in the midrib region. They are conjoint, collateral and endarch; protoxylem is found towards upper side. Each vascular bundle remains surrounded by sclereids on all sides. According to Rodin (1966), in petioles the vascular bundles are associated with transfusion tissue.

Anatomy of root. The root shows diarch structure and generally resembles with the root of angiosperms. The cortex is sufficiently broad and consists of parenchymatous cells which

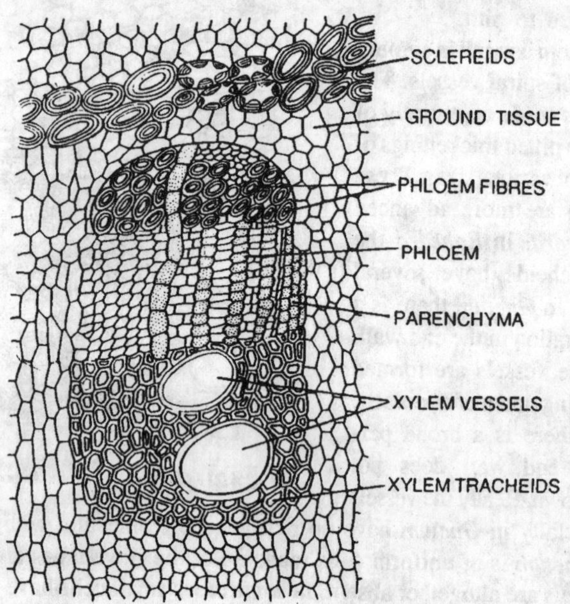

Fig. 6.23. *Gnetum* sp. A portion of the T.S. of stem showing xylem, vessels, xylem tracheids, phloem and sclereids.

are polygonal or oval and full of starch grains. Sometimes thickwalled fibres are also present in the cortex. There is a single layered endodermis around the pericycle. The primary xylem

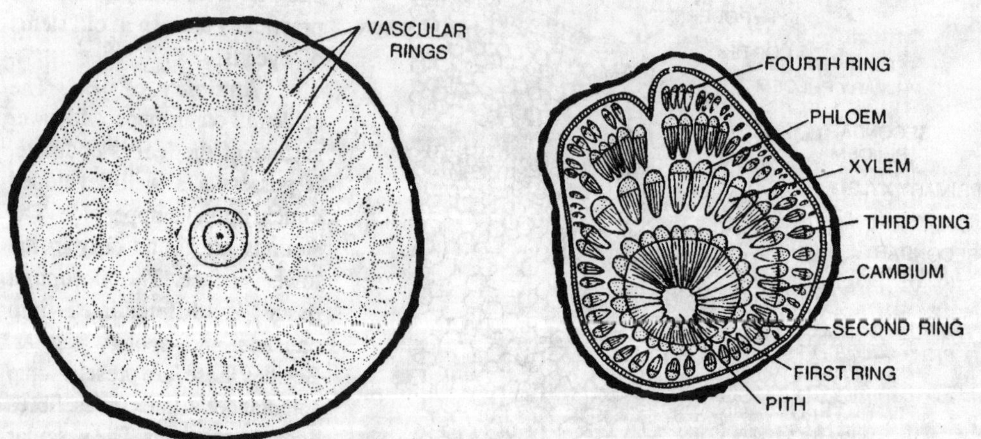

Fig. 6.24. *Gnetum* sp. T.S. of an old stem showing vascular rings.

Fig. 6.25. *Gnetum* sp. T.S. of stem showing eccentric secondary growth.

cannot be differentiated after scondary growth. Secondary growth is of normal type. The secondary xylem consists of tracheids possessing uniseriate bordered pits. The vessels are also present.

The pits which occur on the vessels may be bordered or simple, small and multiseriate. Commonly crassulae are absent. The phloem is composed of sieve cells and parenchyma. Medullary rays

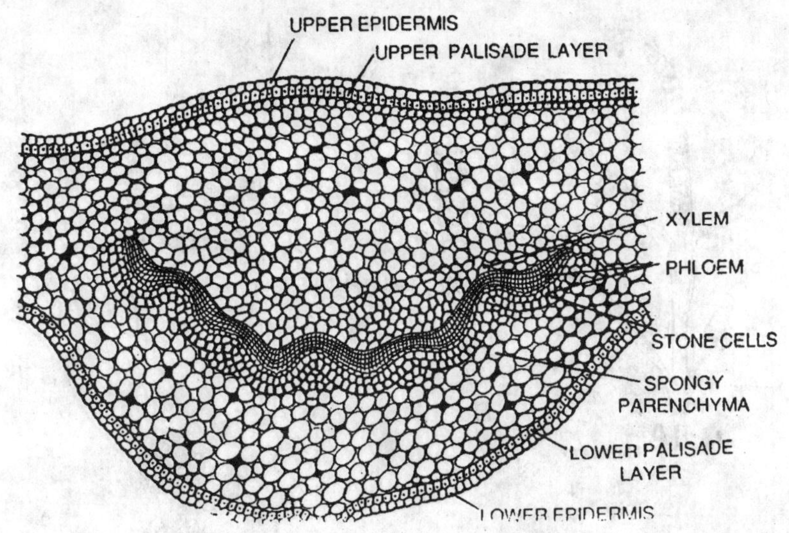

Fig. 6.26. *Gnetum* sp. T.S. of leaf.

are broad and composed of thinwalled parenchymatous cells. These cells are impregnated with calcium oxalate cystals.

Reproduction. The plants are dioecious and the strobili both male and female are compound. They may be axillary or terminal in position, and they may occur singly or in groups. In the case of male strobilus the whorls of male flowers are very close to one another but not overlapped. In female they are closely arranged to one another when young but when strobilus becomes ripe and reaches the fruiting stage the whorls are at a distance from one another.

Male strobilus. It has a long slender axis which bears numerous decussating pairs of bracts. These bracts are practically fused along their margins, so round the axis as short intervals there are cup-shaped structures formed by the fusion of these bracts. In the axils of these bracts there are whorls of staminate flowers, 3-5 in each whorl. Each one of these male flowers is made up of a stalk with two anthers at its tip and a small sheathing perianth at the base of the stalk. The anthers themselves are unilocular. Sometimes it is seen that in the place of stamen a female flower may be developed but later on such female flowers abort but few cases have been noted in which such female flowers form seeds, so here it is not case of bisporangiate flower but it is that male and female flowers may be developd in the axil of same bract. Here the male flower is represented by a single stamen whose place sometimes being taken by a female flower. On maturation the stalk of the anther elongates rapidly pushing the anther through a slit in the perianth beyond the collars of the strobilus.

Female strobilus. Like the male strobilus the female strobilus also possesses an axis which bears collars arranged one above the other. Above each collar about 4-10 ovules are found to be arranged in a ring. The ovules are dimorphic. The smaller ovules ultimately degenerate. Each female flower is represented by a single ovule.

In *G. gnemon* the female strobilus is made up of pairs of opposite sheathing bracts in the axil of which there are 5-6 whorls of ovules with 5-7 ovules in each whorl. In the young strobilus the decussating bracts form the cups round the axis of the strobilus. In the axils of these bracts there are whorls of ovules present. Here also in each whorl there are 5-7 ovules present.

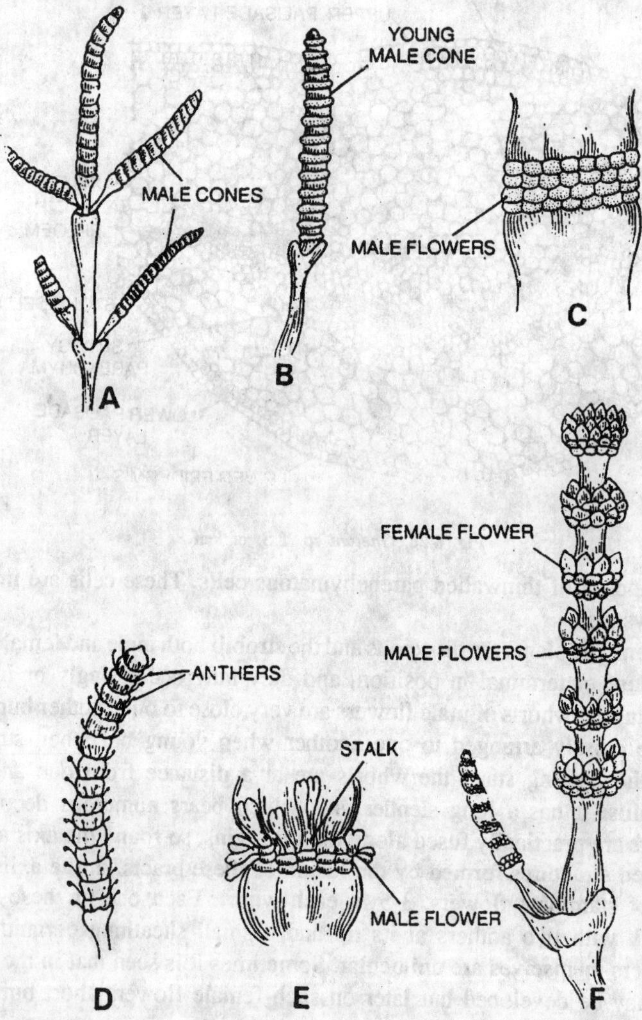

Fig. 6.27. ***Gnetum*** **sp. Structure and development of male strobili. A, a twig with male cones; B, a young male cone; C, a portion of male cone showing male flowers; D, mature male cone; E, a portion of male cone with dehiscing flowers; F, enlarged cones with male flowers and ovular ring.**

Female flower. Each of the ovule represents a female flower. The ovule is borne on a very short stalk. Round the ovule there are three investing coverings. The inner two are the integuments while the outermost one is the perianth corresponding to the perianth of male flower which here has become very large in size and encloses the ovule so the ovule is completely invested by three coverings. The innermost, *i.e.*, the inner integument extends far beyond the ovule and forms a long micropyle. The middle covering is the outer integument in the ripe ovule which seems to become differentiated into an inner stony and outer fleshy layer. Two sets of vascular bundles are found in the outer fleshy layer of the outer integument, and one set in the inner integument is seen to be running along the length of integuments. The outermost covering which represents the perianth has its own vascular supply. It is seen that during

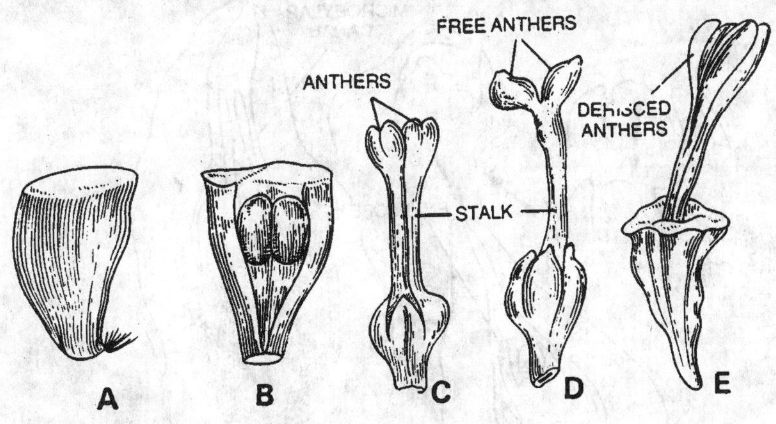

Fig. 6.28. *Gnetum* sp. Male flowers. A, B and D, male flowers of *G. ula* in different views; C and E, male flower of *G. gnemon*.

development of ovule that these three coverings appear in acropetal succession. The outermost develops rapidly to surround the young ovule. The innermost projects beyond the body of ovule

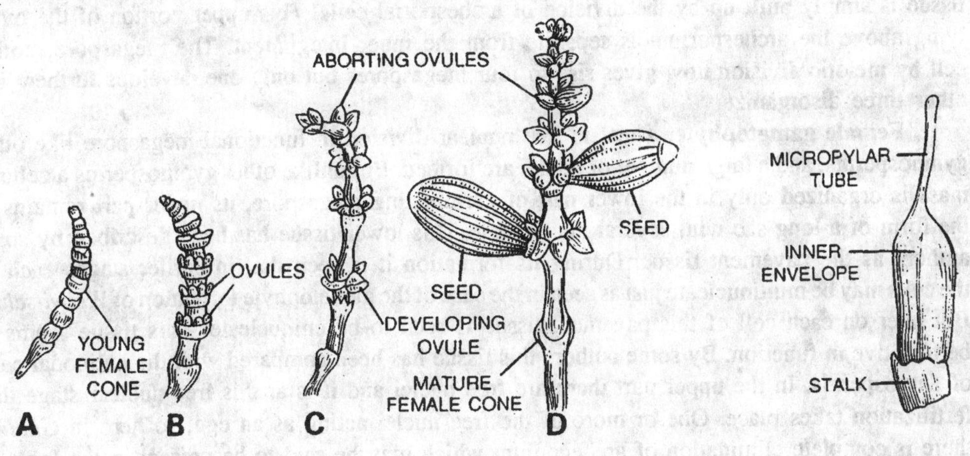

Fig. 6.29. *Gnetum* sp. Structure and development of the female strobili and seeds. A, young female cone; B, old female cone; C-D, still older female cones with seeds.

Fig. 6.30. *Gnetum* sp. Structure of ovule.

forming the micropylar tube. The outer covering becomes finally orange red in colour, and the middle covering hard and sclerotic with a ridged surface.

Megasporogenesis. At the apex of the nucellus of the ovule there lies a small cavity, the rudimentary pollen chamber. In early stages of development the ovule shows one to three hypodermal archegonial cells in the nucellus. Two to four archesporial initials are differentiated in the hypodermal region of the nucellus. These archesporial cells divide periclinally forming parietal and sporogenous cells divide to form 8-16 megaspore mother cells. There is small amount

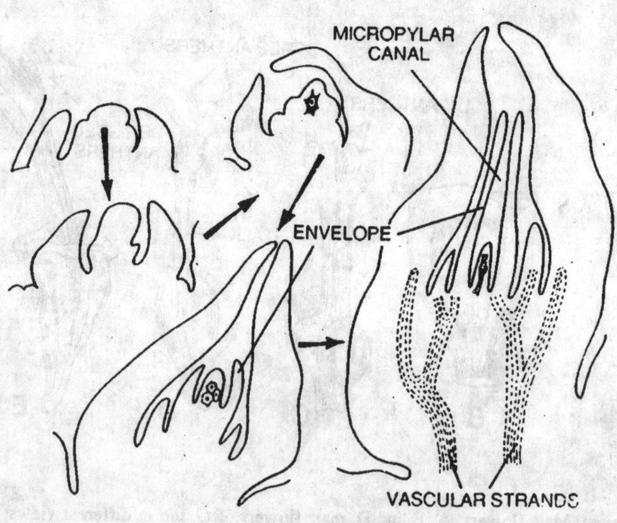

Fig. 6.31. *Gnetum* sp. Successive stages in the development of ovule. The ovule shows different layers of the envelope and micropylar canal.

of nucellar tissue lying above the megaspore. Some authors are of opinion that the overlying tissue is simply built up by the division of archesporial cells. This upper portion of the ovule lying above the archesporium is separate from the inner integument. The megaspore mother cell by meiotic division now gives rise to four megaspores but only one develops further. The other three disorganize.

Female gametophyte. There is free nuclear division in functional megaspore like other gymnosperms, and a large number of nuclei are formed. But unlike other gymnosperms a cellular mass is organized only in the lower part of germinating megaspore, its upper part remains in the form of a long sac with several nuclei in it. This lower tissue has been described by some authors as the pavement tissue. During its formation it is seen that in earlier stages each of the cells may be multinucleate just as seen in the case of the gametophyte formation of *Welwitschia*. But later on each cell of this pavement tissue is seen to be uninucleate. This tissue seems to be nutritive in function. By some authors this tissue has been compared with the antipodal cells of angiosperms. In the upper part there are free nuclei and it is at this free nuclear stage that fertilization takes place. One or more of the free nuclei acting as an egg, so here in *Gnetum* there is complete elimination of archegonium which may be said to be present in the form of an archegonial initial represented by prothallial tubes in the case of *Welwitschia*.

The nucellus at the four nucleate stage of the gametophyte is masive and, therefore, the gametophyte gets deeply buried in the nucleus. The pavement tissue is generally consumed during the development of the gametophyte and thought to be nutritive in function.

Male gametophyte. The pollen grains in *Gnetum* are shed at trinucleate stage. There is even no prothallus nucleus. At the time of shedding the pollen grain contains three nuclei representing the body, tube and the stalk nucleus, so the condition is exactly the same as seen in angiosperms. The pollination takes place, and the cases observed in which the pollen grains germinate while they are within the micropylar tube and a sufficient distance from the main

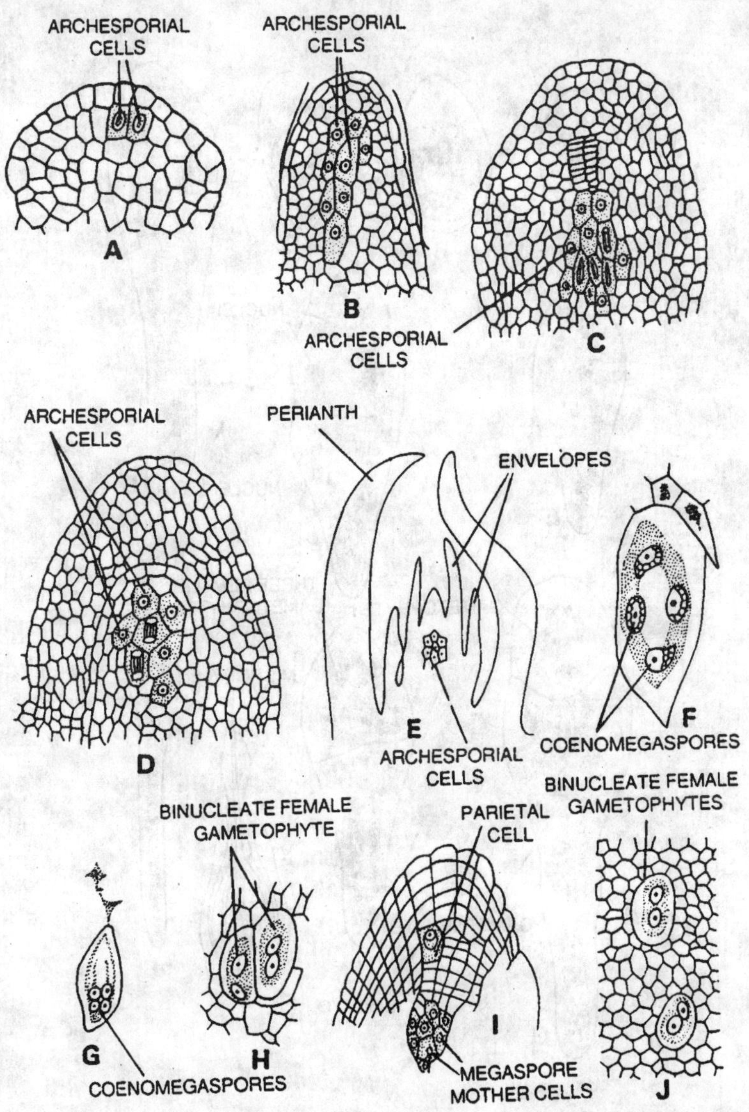

Fig. 6.32. ***Gnetum*** **sp. A-J successive stages in the development of coenomegaspores, megaspore mother cells and female gametophyte.**

body of the ovule, this is again another angiospermic feature. In all gymnosperms excepting *Gnetum* it is seen that at the micropylar end of ovule a small or larger depression, the pollen chamber is always present. In some species of *Gnetum* there is a very small depression or we may say that the micropylar end of ovule is somewhat flattened where pollen grains lie. After pollination the exine of the pollen grain is discarded and the intine grows into a pollen tube. A partition wall is laid down round the body nucleus, and this cell along with the tube nucleus passes down into the pollen tube. The stalk nucleus left behind in the pollen itself. The pollen tube grows towards the ovule pushed down through any of the nucellar cells and during this passage the body tube within pollen tube divided to form two male cells. Some workers think

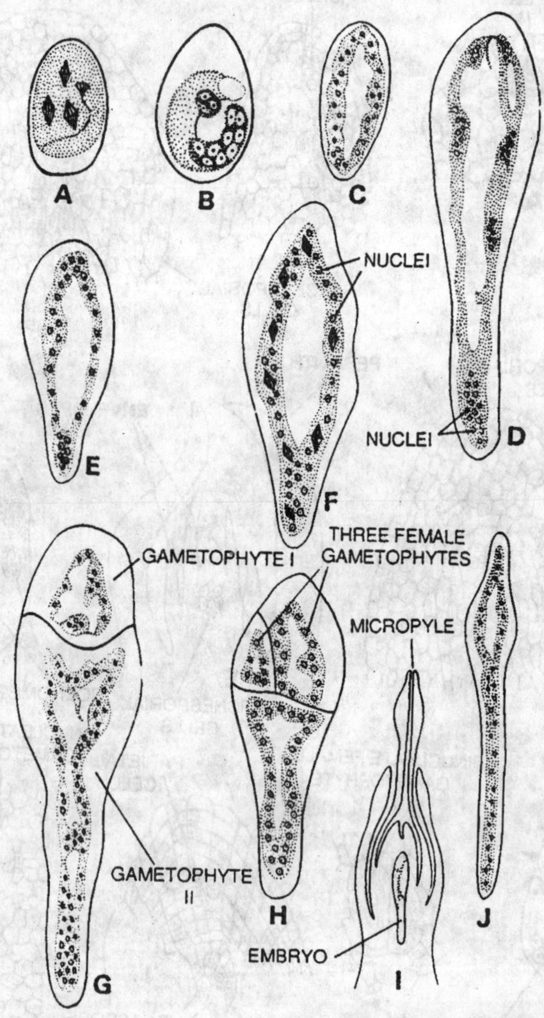

Fig. 6.33. *Gnetum* sp. Development of female gametophyte. A-F, free nuclear division stages; G-H, development of two and three female gametophytes respectively; I, V.S. of ovule; J, an abnormal female gametophyte.

that these two male cells are of unequal size, the large one acts as male cell while smaller one degenerates.

Fertilization. The outstanding fact is that no archegonia are formed and that apparently any of the nuclei of the embryo sac may be fertilized. The fertilization takes place when about 700-800 free nuclei are formed in the female gametophyte. One to four pollen tubes are seen in the female gametophyte. Usually two, rarely one or three nuclei of female gametophyte enlarge. These are egg nuclei and have dense contents. The egg nucleus may attain a size 3 or 4 times larger than that of a normal vegetative nucleus. The pollen tube comes in contact with the megaspore. At the point of contact the wall disorganises and both male cells from each of pollen

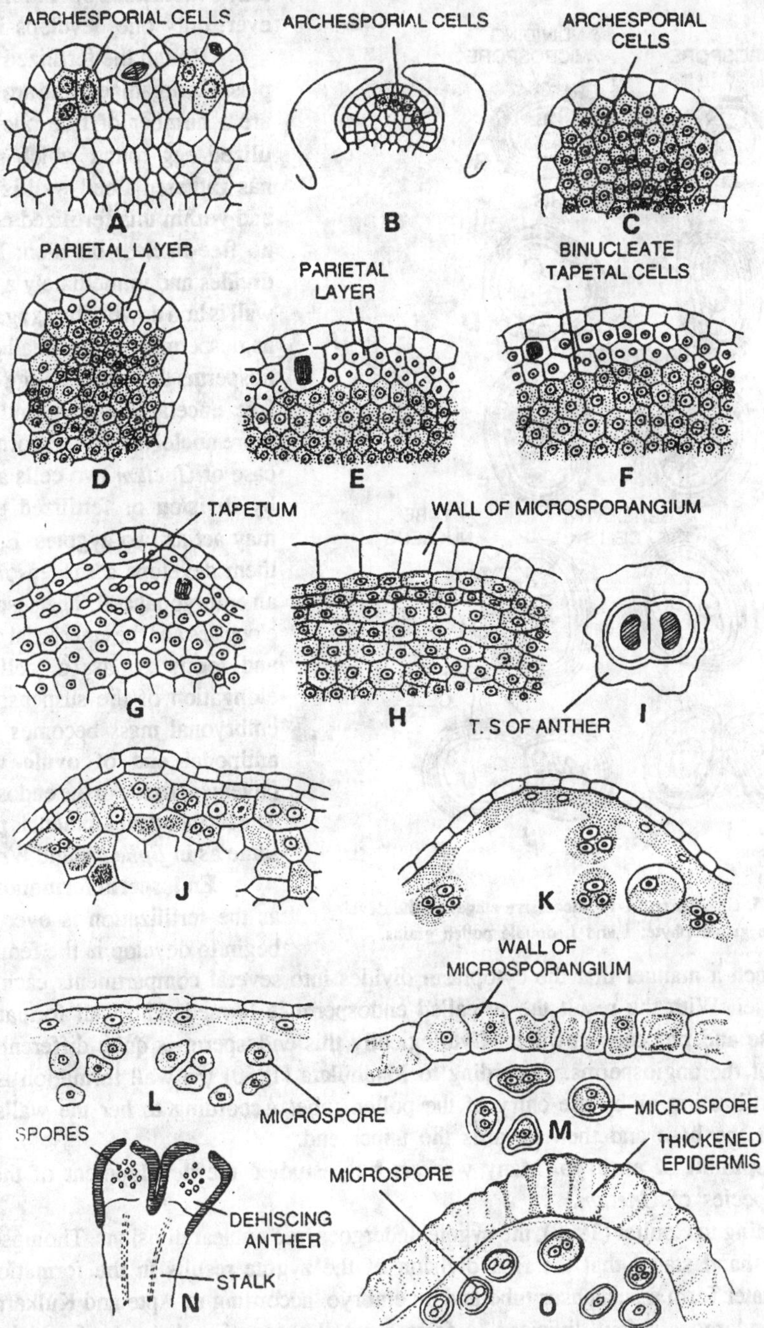

Fig. 6.34. *Gnetum* sp. Development of microsporangium and dehiscence of anther. A-H, successive stages in the development of archesporial cells; I, T.S. of anther; J-O, structure of anther and its dehiscenee.

tube are discharged into the free nuclear part of germinated megaspore. Fertilization takes place and it is seen that the fertilized eggs come to lie in pairs. More than one egg in each ovule may be fertilized but ultimately, however, only one develops in embryo.

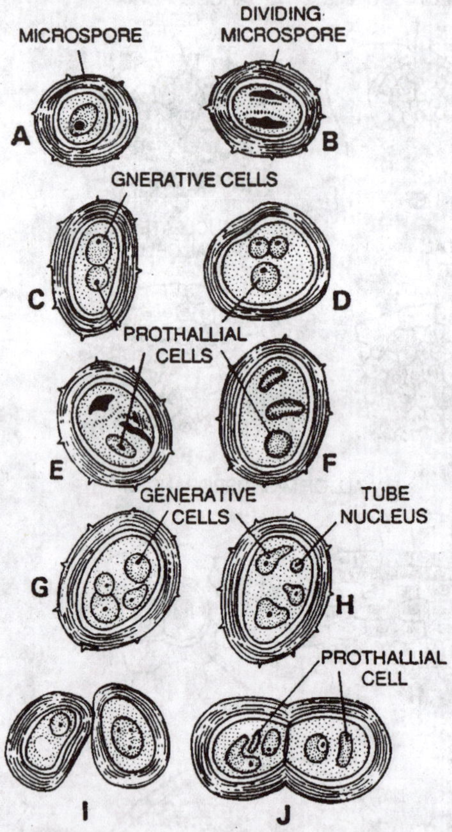

Round the fertilized egg protoplasm is organized, and around it there are a number of free nuclei. In fertilized egg round which protoplasm has gathered a cell wall is laid down and within this fertilized cell there is no free nuclear division. Its nucleus divides and immediately a separating wall is laid down. This is again another angiospermic feature. In all other gymnosperms the fertilized egg undergoes free nucelar division, and a few or more nuclei are always formed. In the case of *Gnetum* two cells are formed by division or fertilized egg which may act as two zygotes. Each one of them develops a suspensor tube and an embryonal cell. The suspensor cell elongates and the embryonal cell divides and forms a mass of cells. By the elongation of the suspensor cell the embryonal mass becomes pushed to antipodal end of ovule where the pavement tissue or the endosperm lies. The further development is practically same as in *Ephedra* and *Welwitschia*.

Fig. 6.35. *Gnetum* sp. A-H, successive stages in the development of male gametophyte; I and J, double pollen grains.

Endosperm formation. As soon as the fertilization is over the walls begin to develop in the female gametophyte in such a manner that the cytoplasm divides into several compartments each with one or more nuclei. With the result the so-called endosperm is formed. The wall formation starts from the base and proceeds upwards. Cytologically this endosperm is quite different from the endosperm of the angiosperms. According to Madhulata (1960) the wall formation is initiated either by fertilization or by the entry of the pollen tube. According to her the walls are first laid down at the base and then towards the upper end.

Development of embryo. Many workers have studied the development of the embryo in various species of *Gnetum*.

According to Coulter (1908), the zygote undergoes free nuclear divisions. Thompson (1916) on the other hand stated that the first division of the zygote results in the formation of two cells which later form a suspensor tube and an embryo. according to Apte and Kulkarni (1953), the zygote undergoes a few divisions to form a small mass of cells each of which develops into a suspensor tube. According to Waterkeyn (1954), as a result of successive divisions of zygote a row of several cells may be produced and each cell produces a tube.

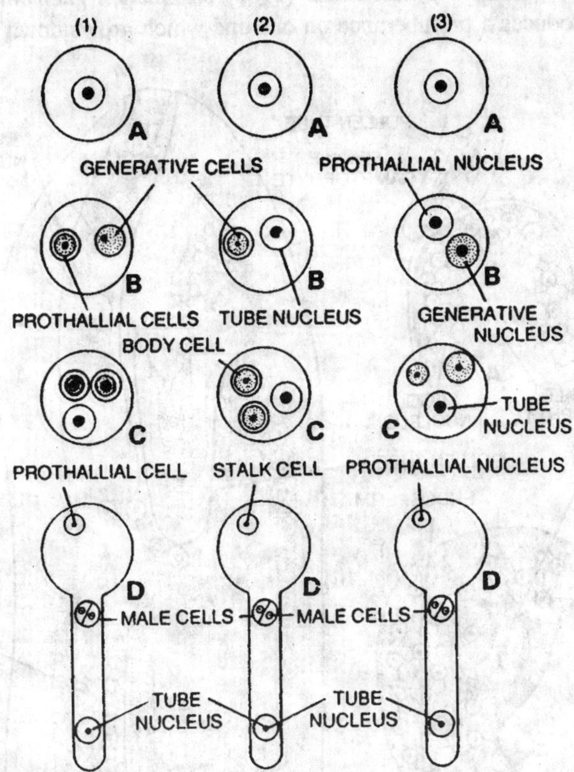

Fig. 6.36. ***Gnetum*** **sp.** Old and new concepts about the development of male gametophyte I, Thompson's view; 2-3, Negi and Madhulata's view.

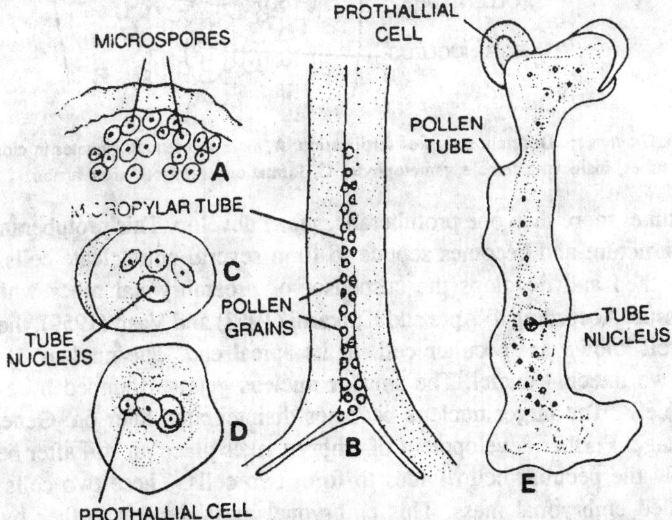

Fig. 6.37. ***Gnetum*** **sp. Germination of pollen grains. A, portion of anther with microspores; B, pollen grains germinating in micropylar canal; C-D, germinating pollen grains; E, germinated pollen grain showing pollen tube, prothallial cell and tube nucleus.**

Recent studies made my Madhulata (1960) on *Gnetum gnemon* have revealed the fact that the zygote produces a protuberance on one end which grows down towards the endosperm

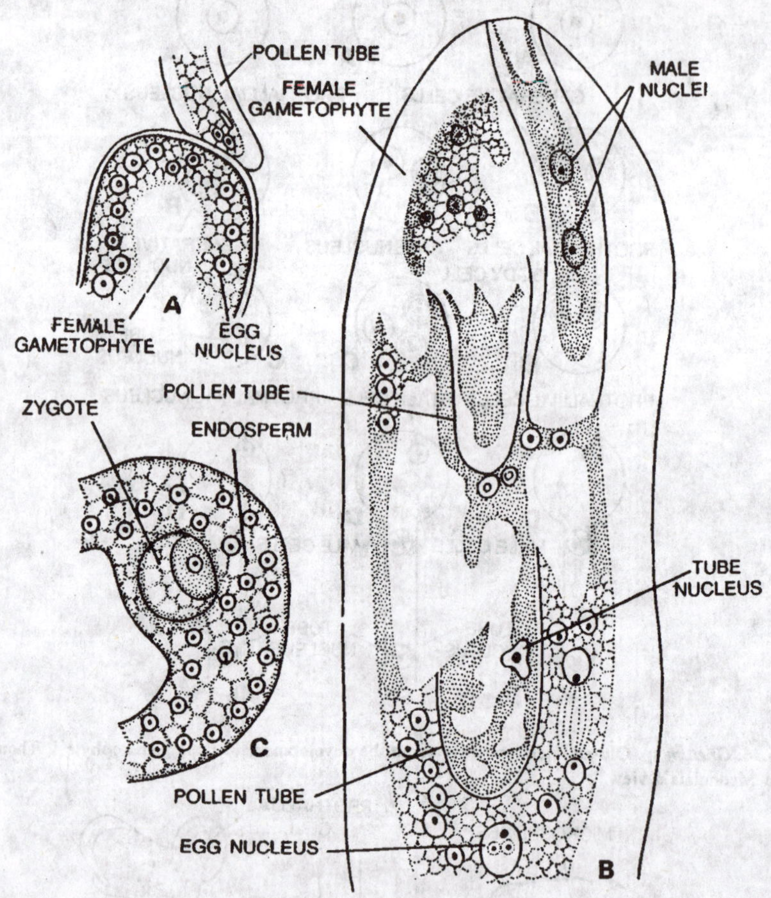

Fig. 6.38. ***Gnetum*** **sp. Different stages of fertilization. A, male and female gametes in close contact; B, elongation of many pollen tubes, inside the female gametophyte; C, formation of zygote after fusion.**

tissue. Sometimes more than one protuberances may develop. This protuberance further develops in a tubular structure and becomes septate to form several uninucleate cells. Later on this tube becomes branched and develops the suspensor or proembryonal tubes which go towards the endosperm tissue. According to Apte and Kulkarni (1953) and Vasil (1959), the primary suspensor tube cuts a cell known as 'peculair cell' at its apical end. The nucleus of the suspensor cell divides into two unequal nuclei. The smaller nucleus gets surrounded by a wall and becomes the 'peculair cell'. The larger nucleus becomes disintegrated later on. Generally the seeds are shed at this stage. Further development of embryo takes place on soil after being detached from the plant. Now the peculiar cell divides to form two cells. These two cells divide once again to form 4-celled embryonal mass. This embryonal mass divides further by a transverse wall to form eight-celled embryo. Thereafter irregular divisions take place to form many cells. As the embryonal mass increases in size some of the cells found towards the primary suspensor tubes become elongated and develop into secondary suspensors. The cells that are situated at the lower end of the secondary suspensors form the embryo proper later on.

The growth of the primary and secondary suspensors pushes the embryo deep into the endosperm tissue. As far as the structue of embryo is concerned it assumes a conical shape.

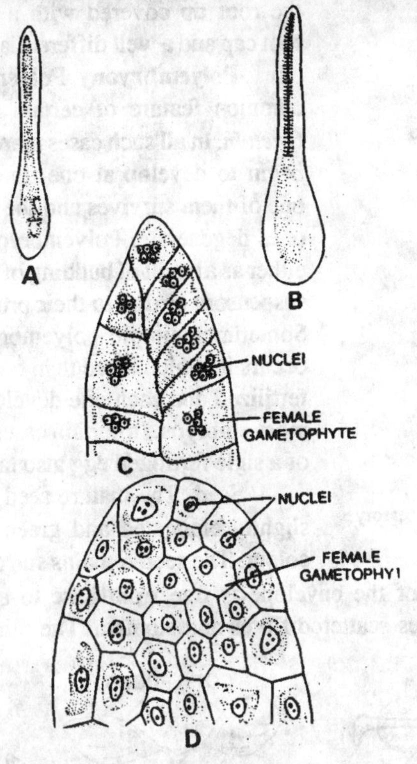

Fig. 6.39. *Gnetum* sp. Development of endosperm. A-B, female gemetophyte in longitudinal sections; C, a portion of multinucleate endosperm; D, uninucleate and multinucleate chambers.

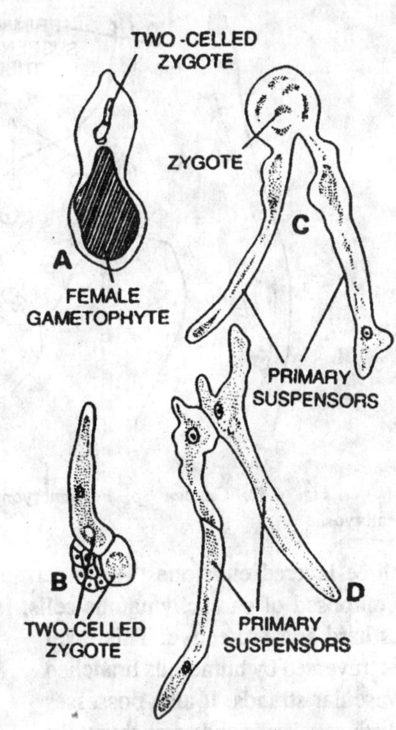

Fig. 6.40. *Gnetum* sp. Development of embryo (early stages). A, vertical section of female gametophyte; B, two celled zygote; C, a zygote with two primary suspensors; D, two zygotes developing together.

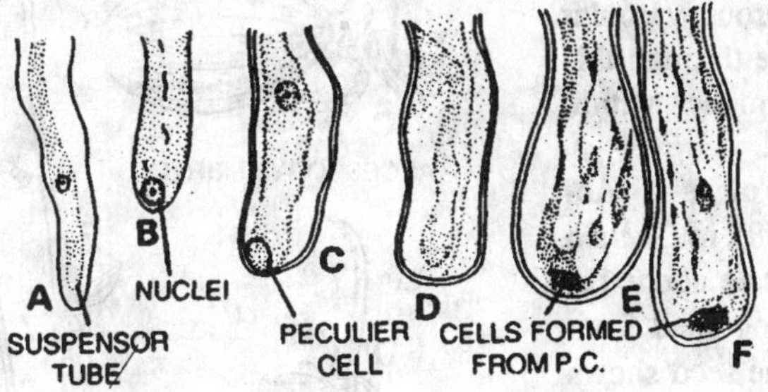

Fig. 6.41. *Gnetum* sp. A-F, formation of peculiar cell and its further division.

There lies a centrally placed axis at the apex of the embryonal mass. This axis is future stem. The axis remains surrounded by two lateral cotyledons. The root tip with its prominent root cap gets differentiated on the other side of the axis. Soon after the development of stem and root, a lateral protuberance called the 'feeder' becomes prominent. The feeder develops an epidermis, cortex, vascular bundles and pith like that of hypocotyl. This way, the embryo consists of a stem tip with two lateral cotyledons, the root tip covered with a prominent root cap and a well differentiated feeder.

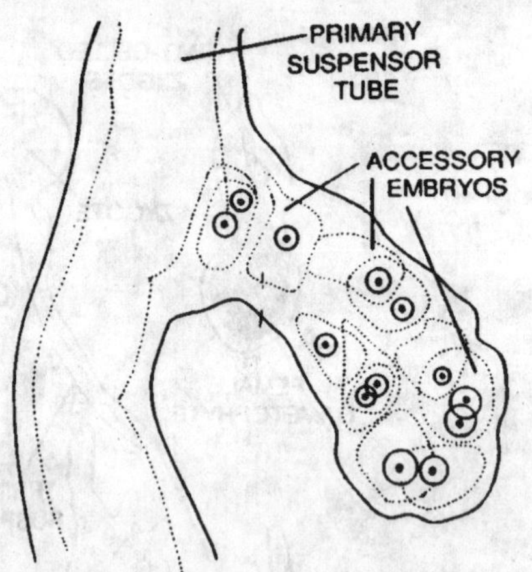

Fig. 6.42. *Gnetum* sp. Polyembryony with accessory embryos.

Polyembryony. Polyembryony is common feature of certain species of *Gnetum*. In all such cases many embryos begin to develop at one time but only one of them survives and the remaining ones degenerate. Polyembryony occurs either as a result of budding of secondary suspensors or due to their proliferation. Sometimes simple polyembryony also occurs in which more than one egg gets fertilized and each one develops one or more embryos. In certain cases cleavage of a sigle fertilized egg also takes place.

Seed. The mature seed is oval or slightly elongated and green to red in colour. The seed remains surrounded by three-layered envelope. The outermost layer of the envelope is free from base to apex. It is composed of parenchymatous cells. It possesses scattered fibres and stomata. The middle layer is hard and protective. This layer is traversed by numerous branched vascular strands. It also possesses distinct palisade and parenchymtous cells. The innermost layer of the envelope is composed of parenchymatous cells and remains united with the nucellus upto two-third of its length. It projects out beyond the nucellus. Numerous vascular bundles also traverse through this layer. The endosperm lies within the innermost layer.

Germination of seed. According to Negi (1958), in *Gnetum ula* the seeds germinate about 11-12 months after being shed from the mother plant. The seed shows epigeal germination. First of all the root along with its cap comes out and gradually develops in a tap root system. Soon after the hypocotyl elongates and pushes the two cotyledons out of the seed. The stem

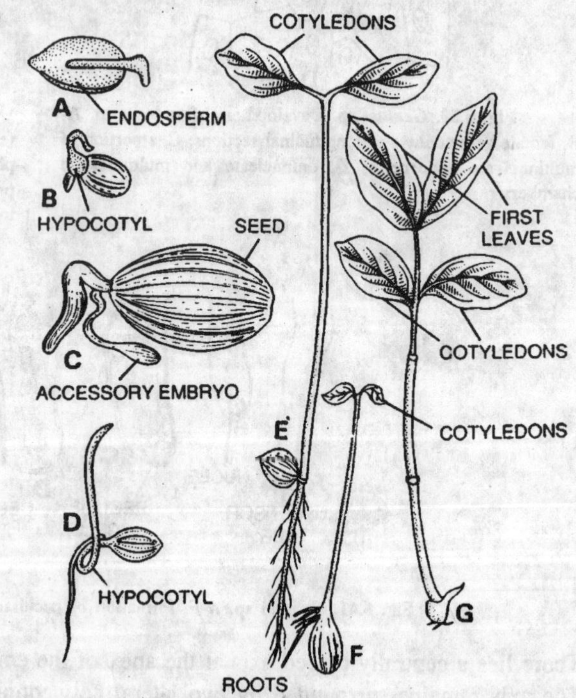

Fig. 6.43. *Gnetum* sp. A-G, successive stages in seed germination

tip becomes enlarged and develops a pair of plumular leaves. The 'feeder' remains inside the seed.

Economic importnce. The young leaves and strobili of *Gnetum* are used as vegetables. The seed kernels of *G. gnemon* are moulded into cakes. The seeds are also eaten after boiling or roasting. The seeds of *G. ula* yield an edible oil. This oil also possesses antirheumatic properties. The oil obtained from *G. montanum* is reported to possess piscidal properties. The bark of *G. latifolium* yields a strong fibre that is used for making nets and ropes. Paper is manufactured from fibre pulp.

Relationships (Affinities). *Welwitschia* and *Gnetum* are more closely related to one another, though they are widely differentiated from each other in form and habitat. The more important common features between two are their general anatomy, the form of male and female strobili, their gametophytes and especially in the development of embryo. These two genera cannot be separated from *Ephedra,* for all the three have many features in common. The general anatomy of the three, especially presence of vessels, presence of spicular cells, the compound strobili of both male and female and the presence of a long micropyle in all three genera are strong common features. *Welwitschia* and *Gnetum,* especially *Gnetum* has made much advance, especially in having completely eliminaated archegonial initial cell.

These three genera, and more specially *Ephedra* have evident relationships with other gymnosperms. The presence of naked ovule in all three, the general structure of ovule, the presence of both male and female gameophytes both in *Ephedra* and *Welwitschia* and the presence of archegonia in *Ephedra* and archegonial initial in *Welwitschia* are all gymnospermic features.

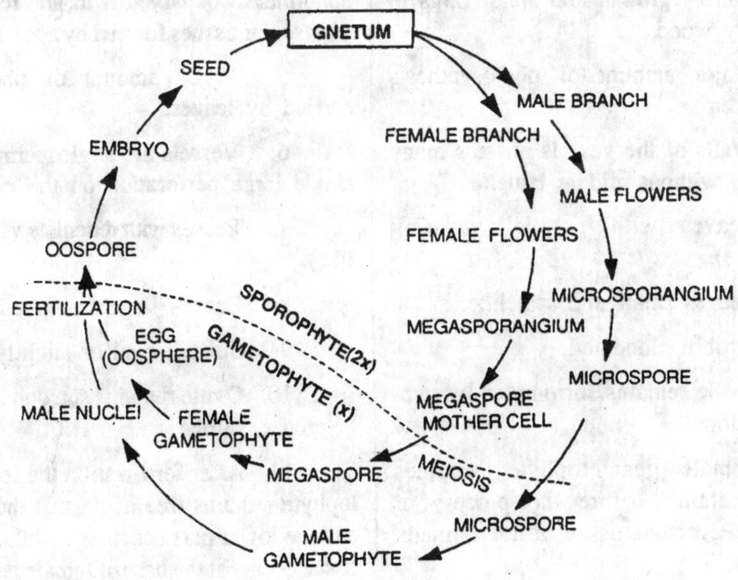

Fig. 6.44. *Graphic* life cycle of *Gnetum.*

In spite of above mentioned resemblances the group has decidedly made an advance and developed angiospermic features. These angiospermic features are-the leaves are broad and green possessing reticulate venation. Vessels similar to those of angiosperms occur though their formation is different. In *Ephedra* archegonia are present but there is no separation wall between egg and venter canal cell. In *Welwitschia* there is only one archegonial initial cell and in *Gnetum* archegonia are altogeter absent. The development of female gametophyte in *Gnetum* is tetrasporic. Free nuclear divisions similar to that of angiosperms occur in the embryosac. In *Gnetum* the endosperm completes its development after fertilization. The presence of small bracts at the base of stamens or as ovule of *Gnetum* are considered the perianth of angiospermic flower. All these angiospermic features in Gnetales were considered to mean a genetic relationship between Gnetales and angiosperms. Some workers were of opinion that the Gnetales considered to be a group from which angiosperms originated but other workers did not support this view. According to this view the Gnetales are not ancestors of angiosperms but both of them arisen from a common stalk and so much common ancestry would explain the resemblances between the groups.

Comparison of *Ephedra* and *Gnetum*

Ephedra	*Gnetum*
1. It grows in warms dry regions and at high altitudes on the hills.	1. Usually grown in humid tropical and sub-tropical regions.
2. A bushy shrub, rarely a climber or tree.	2. Usually a big-sized tree.
3. A xerophyte, bushy in appearance.	3. Angiosperm-like appearance.
4. The internal structure of stem bears normal secondary growth and annul rings in the secondary wood	4. The internal structure of stem shows abnormal secondary growth, and eccentric rings of vascular tissues formed by accessory cambia.
5. Major amount of photosynthesis carried by stem.	5. Major amount of photosynthesis carried by leaves.
6. Walls of the vessels possess many bordered pits without middle lamella.	6. Vessels are angiosperm-like with a single large perforation on the end walls.
7. Leaves with parallel venation (monocot like).	7. Leaves with reticulate venation (dicot like).
8. Leaves small and scalelike.	8. Leaves large nad photosynthetic.
9. Strobili elongated.	9. Strobili oval or slightly elongated.
10. Ovule remains surrounded by two-layered envelope.	10. Ovule remains surrounded by three-layered envelope.
11. Female gaametophyte becomes completely cellular before the process of fertilization. Pavement tissue is not formed.	11. After fertilization the female gametophyte remains free nuclear in the upper part and the lower part becomes cellular. Pavement tissue occurs at the base of female gametophyte.
12. Each male flower bears 1-8 microsporangia.	12. Each male flower bears two unilocular microsporangia.
13. Microspores shed at 5-nucleate stage (a prothallial cell, a prothallial nucleus, a stalk nucleus, a body nucleus and a tube nucleus).	13. Microspores shed at 3-celled stage (a prothallial cell, generative cell and a tube nucleus).

Ephedra	*Gnetum*
14. Archegonium with longest neck among Gymnosperms. Number of archegonia 2-3 per ovule.	14. There is no archegonium but any of the cell of upper cellular part of female gametophyte towards the micrcpylar end may act as an egg.
15. In certain cases the second male cell fuses with the ventral nucleus causing 'double fertilization'.	15. Only one male cell is functional and no double fertilization.
16. Upto 8-celled stage of the embryo there occur few nuclear divisions in the zygote.	16. Here even the first division of the zygote is followed by wall formation.
17. Zygote divides to form about 8 proembryos which later develop into embryos each with its own primary and secondary suspensor.	17. No proembryos formed; the zygote directly gives rise to several primary suspensor cells at the tip of which develops the embryonal mass.
18. There occurs monosporic embryo sac.	18. There occurs tetrasporic embryo sac.
19. Feeder absent.	19. Embryo develops feeder which persists for a long time and helps in nutrition as well as at the time of seed germination.
20. No rest period during seed germination.	20. Seed germination is delayed after being shed from the plant.

7

Methods of Fossil Study

PROCESSES OF PLANT FOSSILIZA TION

Palaeo-botany or the study of fossil plants, *i.e.*, the plants existed in the past and now are entirely extinct. This is the difficult branch of Botany in respect that the fossil plants are difficult to obtain and they are rather scarce. Whenever the fossil plants are found, they are in parts which are to be co-ordinated. This is a tough process of the study. The fossils are cut in sections with a great difficulty and thereafter the preparations are made which require great labour, time and technique. The study of fossils is useful academically as well as economically. The academic interest lies in that their study clears up to a great extent the inter-relationships and evolution of the ancient groups of the plants.The economic interest lies in that some fossils are confined to definite strata of earth crust and they are associated with petroleum, coal and similar other things of economic value. Actually some coal fields were discovered only on account of the presence of certain fossils just above the coal mines. Another point of view of academic interest is that the fossils help in the determination of the climate of ancient time in different regions. During coniferous period the earth was covered by very luxuriant forests and it is assumed that climate in those days was very uniform because the plants flourished in those times (*i.e.*, Carboniferous period) were greatly the same on the whole surface of the earth.

According to one school of thought, the outer part of earth crust is solid and comparatively thinner than interior of earth which is supposed to be in molten condition. Some workers are of opposite view. They believe that the interior portion of the earth is firmer and more solid while the outer crust of the earth is of light matter. They also presume that the interior crust of the earth consists of many heavy metals like iron and lead. Whatsoever might have been the position but it is very clear that the first crust which formed the first surface much have been very uniform.

Later on the water and air appeared and with the appearance of these two destructive factors the surface of the earth began to change. Rain and air affected the surface of the earth to great extent and at certain places the surface rose up in high mountains by bursting of interior of matter and at other places it sank down and formed seas and oceans. The rocks withered off, the volcanic eruptions and other changes took place and the general topography of the earth was completely changed. Rivers coming down from mountains bring down pieces of rocks and large quantity of sand when they flow in the plains. This sand settles down at the bottom of the water. Along with this sand the parts of the plants and animals which get into that sand would have a chance of being preserved. This preservation takes place in different forms.

The fossils are mainly found in the sand which has been brought down by the rivers from the mountains. This sand has become compressed and sedimentary rocks have been formed in the bottom of water. They contain the pieces of plants or entire bodies or bodies of animals which have been surrounded by mud and salt and they got the chance of being preserved.

This preservation takes place in two forms:

1. **Impressions.** These impressions may be of the surface markings of the plant material or be the markings of the internal cavities in the plant material. The surrounding mud gets deposited on the surface of the plant material or gets into the cavity and their impressions are left. In some fine fossils even details of the venation, structure of stomata and epidermal hairs have been studied. In such impressions the plant material itself has decayed or turned into coal. However, the surface impressions cannot be relied upon as we know that the plants, *e.g.*, *Equisetum* (a pteridophyte), *Ephedra* (a gymnosperm) and *Casuarina* (an angiosperm) gave same type of impressions, but all of them are widely differentiable and have no affinities and relationships with each other.

2. **Petrifaction.** Another method of preservation of great importance is that the plant material when lying in water has become infiltrated with mineral matter such as lime, silica, magnesium salts and other similar substances. These salts penetrate the minute cells and the organic matter is being replaced by mineral salts. The original organic matter may be completely or partially changed but its outline and general structure is left in the form of these infiltrated substances.

In coal mines generally the coal balls are found which are formed by the infiltration of calcium bicarbonate.

In another method of preservation in which rare and small parts of plant material such as seeds, pollen grains, spores etc., have been preserved. In this method the plant material happens to fall in resin or amber extruded by some plants, such as pine trees and others.

Technique of Fossil Study

The different methods of the study of fossil plants are as follows:

It is a laborious process and requires sufficiently great time. Usually the petrified specimens are cut in serial sections which give an idea of the actual structure of the fossil plant. These petrified pieces are cut into very fine slices by different methods. In one method each such piece is attached to glass plate and grounded to sufficient thinness and thereafter studied under the microscope.

Another improved method of the study of these petrified specimens is to prepare the films of the material by special techniques. The method of preparing thin films is as follows: First of all the surface of the section of the petrified material is made smooth. If the material consists of calcium carbonate, then on the smooth surface of the slice a film of 5 percent hydrochloric acid is allowed to act for five minutes. If the slice of the petrified material is silica then the film of 10 percent hydrofluoric acid (HF) is allowed to act on the smooth surface for ten minutes so that the silica is dissolved. The surface of the petrified section by the action of these acids becomes rough on account of the dissolution of the mineral matter. If any organic matter remains on this surface, now put hot gelatin on the surface. As soon as the things dry up, they are removed and studied under the microscope. This process may be successful only in the case when organic matter is left in petrified specimens. In cases where organic matters are already decayed, such preparations are never good.

The fossils naturally would be pieces of plants. It is very rare that entire plant could have preserved. This way, only pieces can be studied. In such type of study the individual pieces are given botanical names, just as in living plants. The botanical names of the fossil plants are not so significant as those of living ones. As they are represented by the pieces of the plants and, therefore, their generic names would be according to stem, leaf and root or any reproductive structure. The stems are usually given the generic names which end with **'dendron'** (tree) or

'xylon' such as, *Lygenodendron* or *Cladoxylon*. The leaves end with **'pteris'** or **'phyllum'** and reproductive parts end with **strobilus**. This way, the palaeobotany is the study of the parts of fossil plants and in certain cases marvelious results have been obtained.

In the case of *Lygenopteris*, one of the Cycadofilicales of Carboniferous period which was found in pieces and later on the palaeobotanists supported that all the pieces belonged to one particular plant. Later on after few years the complete intact plant was found.

Important Strata

The Palaeozoic and Mesozoic strata are very important from the study point of view of the fossil plants. It is in the upper middle Palaeozoic, *i.e.*, the Devonian strata, we come across with the first land plants such as lycopods belonging to Lycopodiales, Equisetales, the seeded ferns, the primitive gymnosperms, the pteridophytes etc. In the late Palaeozoic, *i.e.*, both in the upper and lower, Carboniferous strata the earth was covered up by the very luxuriant forests. These forests were formed by lycopods, horsetails, seeded ferns and later on with primitive gymnosperms. The Carboniferous strata is most important. The coal mines are situated in this strata. The coal mines are the result of dense forests having got submerged in those times. The Mesozoic is also very important from the point of view that the first angiosperms made their appearance, otherwise the higher gymnosperms formed luxuriant forests in those times. In the later Mesozoic some of the gymnosperms disappeared. Majority of the Cycads disappeared and only a few forms have been left upto the present day.

In India the most important strata is described technically the **'Gondwana system'** named after Gondwana Kingdom. For the first time the rocks of this period were discovered near Narmada river.

Work on Fossils in India

The firstmost work on Indian fossils was done by Fiestmental, the Director of Geological Survey of India. He published a monograph of Indian fossil forms. Thereafter Seward and Bencroft made a number of additions. From 1895 to 1909 most of the work on Indian fossils was carried on in Europe. It had been since 1919 till the death of Dr. Birbal Sahni, *i.e.*, 1949, that work of Indian fossils had been added to a great deal. He first revised the known Indian fossils and then published a number of monographs describing numerous fossil forms. The most important Indian fossil of Cycads is *Williamsonia sewardiana*. In India it is only in few places that there is a good collection of fossil plants, *e.g.*, Museum of Calcutta, the Lucknow University Museum; Presidency College, Madras, Prince of Wales College, Jammu. The collections are made from Raniganj coal field, in Bengal, places near about Nagpur, Reeva state, Kotah state, in some parts of Kashmir near about the Satlaj river, salt range in Punjab (now in Pakistan), and Rajmahal hills, Bihar.

GEOLOGICAL PERIODS

GEOLOGICAL PERIODS IN THE NORTHERN HEMISPHERE

Era	Period	Age in million years	Type of vegetation
CENO ZOIC	Quaternary	1	Modern
	Upper Tertiary, Pliocene Miocene	10 20	Modern

Era	Period	Age in million years	Type of vegetation
MESOZOIC	Lower Tertiary, Oligocene	35	Modern, with tropical plants in Europe
	Eocene	50	
	Upper Cretaceous	75	
	Lower Cretaceous	100	Gymnosperms dominant
	Upper Jurassic	130	(Conifers and Bennettitales)
	Lower Jurassic (Liassic)	140	Luxuriant forests of Gymnosperms and Ferns
	Upper Triassic (Rhaetic)	160	
	Lower Triassic (Bunter)	180	Sparse desert flora with Gymnosperms
	Upper Permian	190	(Conifers and Bennettitales)
PALAEOZOIC	Lower Permian Upper Carboniferous (Coal Measures)	200	Tall swamp forests with early Gymnosperms, Tree Lycopods *Calamites* and Ferns.
	Lower Carboniferous	250	Early Gymnosperms, large tree Lycopods and Ferns
	Upper Devonian	260	
	Middle Devonian	275	*Rhynia* vegetation in marshy localities.
	Lower Devonian Upper Silurain	300	Herbaceous marsh plants (*Psilophyton* and *Zosterophyllum*) and some small shrubs.
	Silurian	350	Marine algae
	Ordovician	425	Marine algae
	Cambrian	500	Marine algae, but some evidence of land plants too.
PRECAMBRIAN		4500 ?	Fungi and bacteria reported to have occurred 2,000 million years ago.

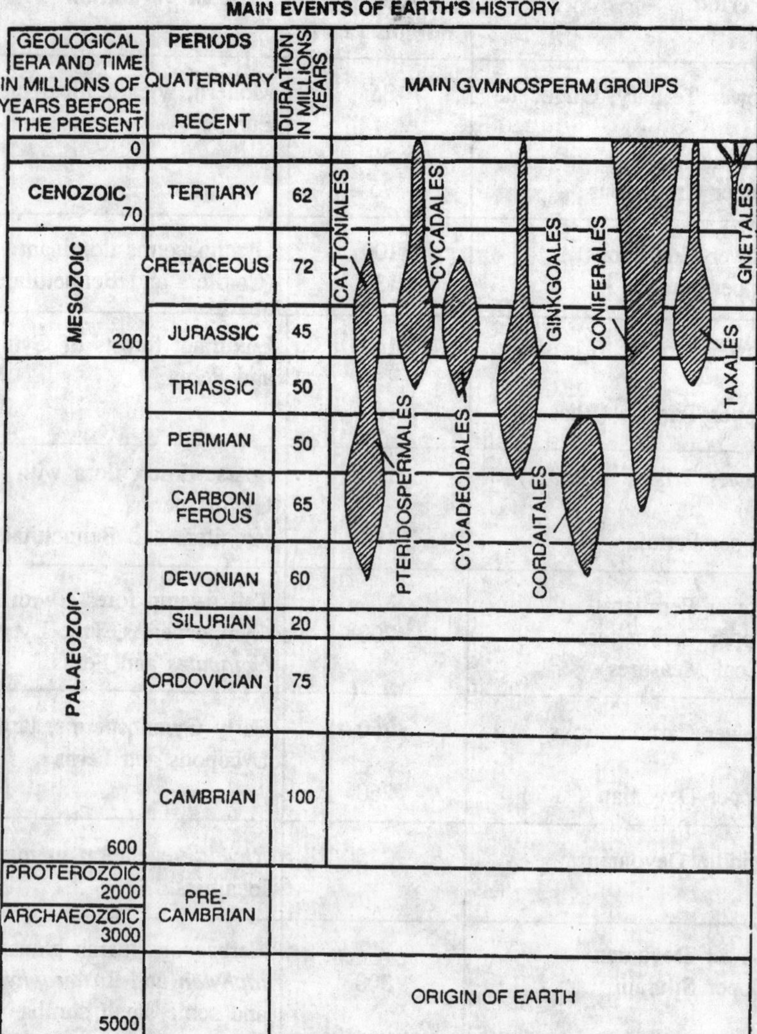

Fig. 7.1. Geological history of the Gymnosperms.

8

Division-Psilophyta
Class-Psilotopsida

Order-Psilophytales

There are about 25 known genera in this order. All genera are fossil. The fossil record indicates that the Psilophytales became extinct during the Upper Devonian. They however, lived from the Middle Silurian to the Upper Devonian. The fossil remains of the Psilophytales have been found in widely scattered regions, such as Canada, Scotland, France, United States of America, Australia etc.

The characteristic features of the order are as follows:

The plant body of the sporophyte was rootless, dichotomously branched and was differentiated into a prostrate subterranean and an erect aerial portion. The aerial portion was leafless or with small simple leaves.

In majority of the genera, the sporangia were borne at the tips of the branches. However, a few genera had the sporangia in clusters, at the tips of their branches.

The stele was simple and protostelic type.

Classification. The order Psilophytales has been divided into nine families. Some of the families are-Rhyniaceae, Psilophytaceae, Pseudoporochnaceae, Zosterophyllaceae, Asteroxylaceae etc. Here the genus *Rhynia* of Rhyniaceae has been described.

Genus RHYNIA

Systematic position.

Pteridophyta
Division. Psilophyta
Class. Psilophytopsida
Order. Psilophytales
Family. Rhyniaceae
Genus. *Rhynia*

Till now only two species of *Rhynia* named as *Rhynia major* and *R. gwynne-vaughani* have been described by Kidston and Lang (1917,1921) from the Old Red Sandstone Beds of Middle Devonian. As the genus has been derived from the place *Rhynie* situated in Scotland and, therefore, it is named after the name of the place of its origin and known as *Rhynia.* A number of specimens of these two species have been found at Rhynie. The specimens are so well preserved that they give detailed information about the form and structure of the very primitive vascular plant.

Kidston and Lang (1921) thought and supported by evidences that in those times the plants grew in swampy marshes near the volcanoes.

Both the species of *Rhynia, i.e., R. major* and *R. gwynne-vaughani* were herbaceous plants. Comparatively the *Rhynia major* was larger in size. The plant body of the species consisted of dichotomously branched horizontal rhizome and the erect, aerial dichotomously branched stem. The aerial stems of *R. major* were about 50 cms. high and 1.5 mm. to 6 mm. in diameter, whereas, the corresponding structures of *R. gwynne-vaughani* were 20 cms. in height and 1 to

3 mm. in diameter. The aerial branches were leafless shoots. The plants were present in patches on the underside of the rhizome. The aerial branches were naked, leafless, cylindrical, dichotomously forked and tapering at their apices.

The terminal, elongated sporangia were found on the tapering vegetative apices.

The stems of *R. gwynne-vaughani* bore adventitious branches at the bases of aerial shoots and rhizome. It is thought that these branches got detached readily from the parent plant and helped in vegetative propagation.

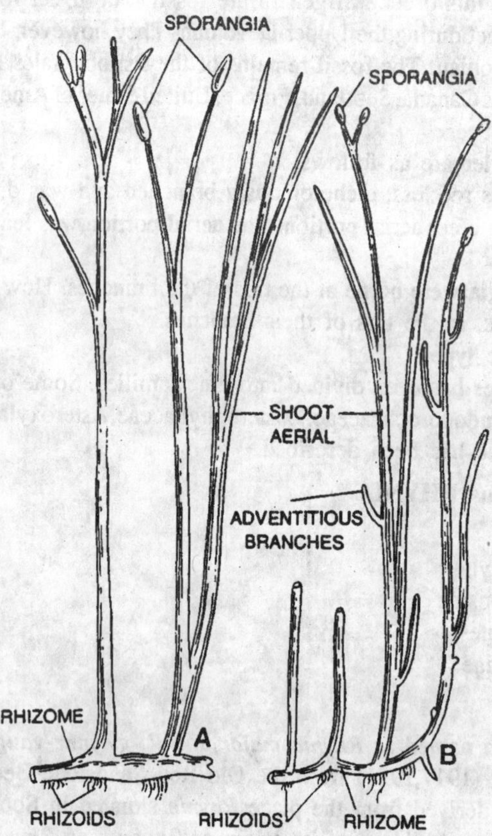

Fig. 8.1. *Rhynia*. A, reconstruction of *R. major;* B, reconstruction of *R. gwynne-vaughani* (After Kidston and Lang).

Anatomy. The internal structure of the rhizome and aerial stem was quite similar to each other. This was differentiated into epidermis, cortex and stele. The details of the structure were as follows:

1. **Epidermis.** The epidermis was one cell in thickness with well-defined thick outer wall and conspicuous cuticle on the outer side. The cells of the epidermis were broadly fusiform from the surface view. The aerial stems bore the stomata on the epidermis here and there with pairs of guard cells. However, the rhizome did not possess any stomata. The rhizome possessed unicellular rhizoids on its surface.

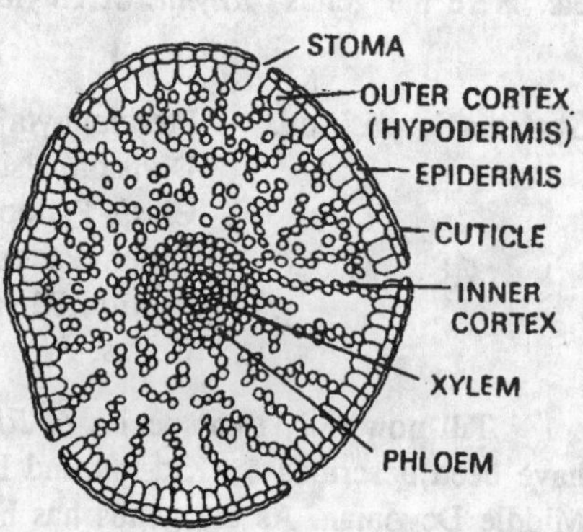

Fig. 8.2. *Rhynia gwynne-vaughani*. T.S. of an aerial branch.

2. **Cortex.** The cortex consists of two zones (a) the outer cortex (hypodermis) and the (b) inner cortex.

(a) Outer cortex or hypodermis. Immediately beneath the epidermis there was a zone one to four cells in thickness consisted of angular parenchymatous cells having no intercellular spaces known as hypodermis. This zone was also known as the outer cortex.

(b) Inner cortex. Inner to the hypodermis there was broad zone of inner cortex. In comparison of the central stele the cortex was ten fold. The inner cortex consists of rounded cells separated from each other by large intercellular spaces which communicate to the stomata through the gaps in hypodermis. It is also believed that this zone of the inner cortex acted as the chief photosynthetic tissue.

The endodermal and pericyclic layers were altogether absent.

3. **Stele.** There was a protostelic type (hapostele) of stele. The central xylem core was completely surrounded by phloem layer four or five cells in thickness. The xylem consists of tracheids only. In majority of the specimens the xylem had imperfectly preserved. According to Kidston and Lang (1917), the tracheids had annular thickenings at their walls. Kidston and Lang (1920) have reported that in some of the specimens the tracheids were isodiametric, whereas in some other specimens the central tracheids were smaller in diameter and the peripheral ones large in diameter

The phloem consists of elongated thin-walled cells with oblique end walls.

The reproductive structures (the sporangia). The elongated or oval sporangia were terminally developed at the apices of the branches of the aerial shoot. It is not clear whether the sporangia were borne on all the apices of the branches or on certain branches of the plant. Each apex bore a single sporangium.

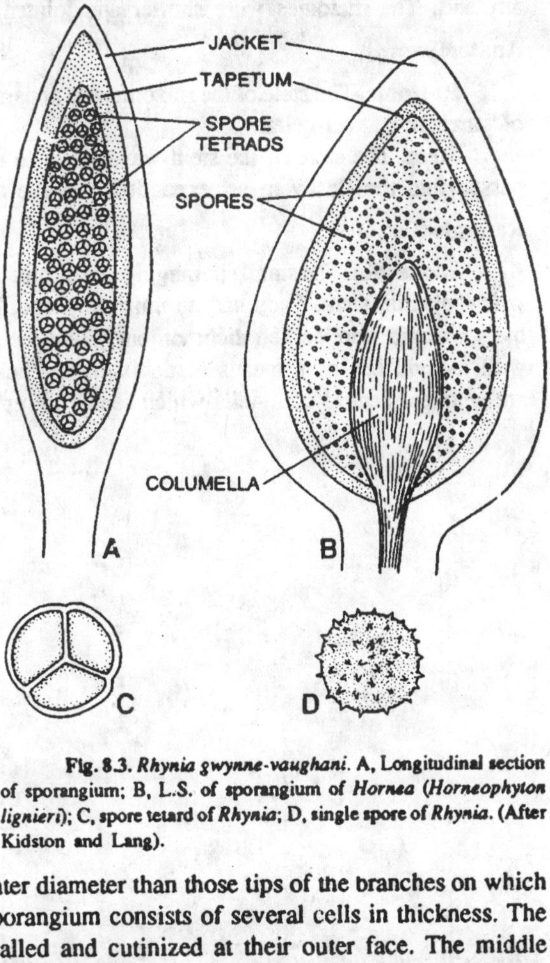

Fig. 8.3. *Rhynia gwynne-vaughani.* A, Longitudinal section of sporangium; B, L.S. of sporangium of *Hornea* (*Horneophyton lignieri*); C, spore tetard of *Rhynia*; D, single spore of *Rhynia*. (After Kidston and Lang).

As regards the shape the sporangia were cylindrical and they possessed the greater diameter than those tips of the branches on which they were found. The jacket layer of the sporangium consists of several cells in thickness. The outermost cells of this layer were thick-walled and cutinized at their outer face. The middle layer or the sporangial wall possessed thin-walled cells and innermost layer consists of thin-walled rounded cells which probably act as tapetal layer. The sporangium has no columella within it. The cavity contained many spores of the same size. The sporangia were indehiscent and it is thought that the spores were released only on the decay and disintegration of the thick sporangial wall.

The spores are spherical and cuticularized. They are found arranged in spore tetrads.

Genus HORNEOPHYTON

Systematic position. Pteridophyta
Division. Psilophyta
Class. Psilophytopsida
Order. Psilophytales
Family. Rhyniaceae
Genus. *Horneophyton*

Habit

Horneophyton formerly known as *Hornea* was similar to *Rhynia*, but was a smaller more slender plant with somewhat more freely branched stems, which were without appendages of any sort. The rhizomes were shorter and jointed, with thick tuberous segments.

Anatomy

Rhizome. The stele of the rhizome was apparently without vascular tissue consisting wholly of parenchymatous cells.

Stem. The stele of the stem was similar to that of *Rhynia*. The steles of the aerial stems passed into the fleshy mass, expanded bell-like and faded out.

Reproduction

Sporangia. The small sporangia were terminal on the stem tips and sessile, being continuous with the aerial stem. They had no uniform shape, being merely fertile stem tips. In many cases the sporangia were forked dichotomously like the stem. The wall of sporangium was thick and without evidence of a means of dehiscence. The spores lay in a dome-shaped cavity around and above a central columella which seems to represent a continuation of the phloem of stele.

9

Class-Cycadopsida

Order-Pteridospermales (Cycadofilicales)

The Pteridospermae or the Cycadofilicales are fossilized plants that first appeared in the upper Devonian. They were very common throughout the Corboniferous and extended upto Permian strata. After the Fermian they began to decline and disappeared altogether. It is in the early Devonian rocks that the members of some very primitive type of land plants disappeared. These plants were probably the simplest pteridophytes. They had no roots, some without leaves and some had small tiny outgrowths which represented the leaves.This primitive group of land plants was different from pteridophytes. Their stem had well formed xylem and phloem tissue and the tips of the branches bore sporangia on them. It is from the Upper Devonian through the Carboniferous and upto end to some extent into Permian strata also that we come across with a large number of fossilized plants belonging to Bryophytes, Pteridophytes and Gymnosperms. The Carboniferous strata is very rich in plant remains. It appears that in those times the whole surface of earth was covered up by very extensive and luxurious forests. It seems that the climate in those times on the whole surface of earth was very uniform. This view is based because identical fossils have been found in very widely separated areas. It appears that during this Corboniferous which is also described as coal age there were general and sudden changes in the topography of earth. It appears that very extensive forests got submerged and it is the submerged forests from which we get the extinct coal supply of present day. During this Carboniferous which has been sometimes described as age of ferns, very large sized Lycopods and Equisetales were common along with these fossil pteridophytes. There were also present seeded plants which developed seeds of Cycadian type on them. Those plants which bore seeds were formerly described as Cycadofilicales. English workers (Oliver and Scott) described them as **Pteridosperms** while American workers call them **Cycadofilicales.**

These seeds were known to previous workers also but their association with fern like leaves was recognized in 1903. It was during this year that some of these Carboniferous seeds were found in organic continuity with the leaves of fern like plants. These seeded fossil plants have definite relation with gymnosperms of which American workers think they form an extinct class. They show definite relationships with ferns on one side and cycads on the other. They were distinctly spermous plants for their seeds were not enclosed in closed carpel as in angiosperms. This group of seeded plants was dominant during Carboniferous, they probably began in the Upper Devonian and began to decline in the Permian after which they disappeared altogether. They were very widespread. They have been found in all countries in Northern Hemisphere and in Southern Hemisphere also they have been found wherever they have been searched for.

One of the best known plants is *Lyginopteris* which is best known of all Carboniferous plants. Before 1903 different parts of the plants were known, their stems, roots, leaves and also the seeds. On the whole surface of seeds capitate glands were present and it was the presence of these capitate glands that gave rise to idea that all these different parts of plants might be of one or same plant. Later on the entire plant was found in one place in organic continuity from roots, stems and leaves.

A number of stem genera have been described from time to time. These stem genera may be grouped under a number of families. The more important families are -Lyginopteridaceae, Medullosaceae, Calamopityaceae, Peltaspermaceae, Corystospermaceae.

Pant (1957) makes a distinct class Pteridospermopsida for these plants. He divided this into six orders. These orders are-Lyginopteridales, Medullosales, Glossopteridales, Peltaspermales, Corystopermales and Caytoniales.

Genus LYGINOPTERIS (CALYMATOTHECA)

Systematic position. Gymnosperms
Class. Cycadopsida
Order. Pteridospermales
Family. Lyginopteridaceae
Genus. *Lyginopteris*
(*Calymatotheca*)

The genus *Lyginopteris* is also known as *Lyginodendron*.

Morphological features. The stem *Lyginopteris* was slender and covered with large scaly leaves. Near the base of the plant adventitious roots developed. The plant seems to have been a climber.

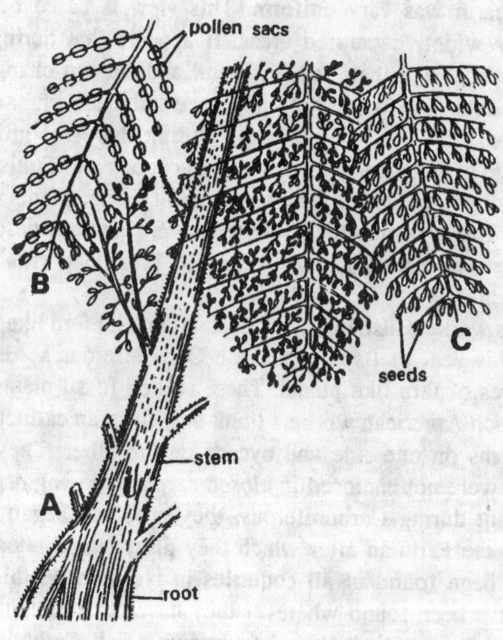

Fig. 9.1. *Lyginopteris oldhamia.* Restoration showing external characters. The frond on the left bears the pollen sacs on peltate leaflets (*Crossotheca*); That on the right bears seeds. Stem and roots also present.

Lyginopteris oldhamia also known as *Calymatotheca hoeninghausi* was described in detail by Williamson, Scott, Brongniart, Binney, Potonie, and Oliver and Scott. It was found abundantly in the coal ball horizon of Lancashire and Yorkshire

Anatomical features. The Primary structure was an ectophloic siphonostele with large pith round which a number of primary mesarch bundles. Older plants showed normal secondary growth. In some specimens, however, the xylem portion of primary vascular bundles was in a continuous ring. In some there was an abnormal type of secondary growth.

This abnormality was of two forms, either there was an inner ring of secondary phloem developed or it was that the cambium appeared in strips found separately in vascular bundles giving rise to a polystelic appearance.

Reproductive structure. Some of these Palaeozoic leaves bore microsporangia on them. The fertile pinnules were more or less peltate in form and on their underside they bore usually six sporangia. These sporangia are usually bilocular. Such a type has been described as *Crossotheca* type. The microspores seem to have formed a male prothallus. The sperms seem to have been of like those of presnt-day cycads.

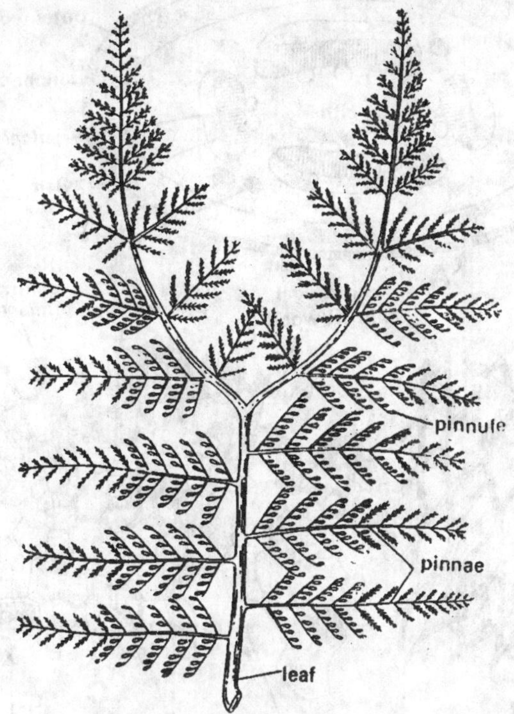

Fig. 9.2. ***Lyginopteris oldhamia.*** **A frond.**

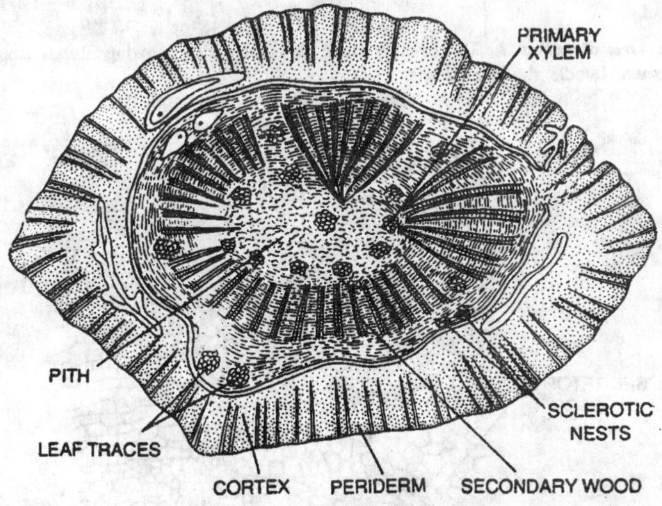

Fig. 9.3. ***Lyginopteris oldhamia.*** **T.S. of the stem showing well-developed secondary wood and leaf traces.**

Seed. The best known seed has been described under the name *Lagenostoma*. These seeds were small in size, only about 1/4" but they were highly organized. It was barrel shaped and

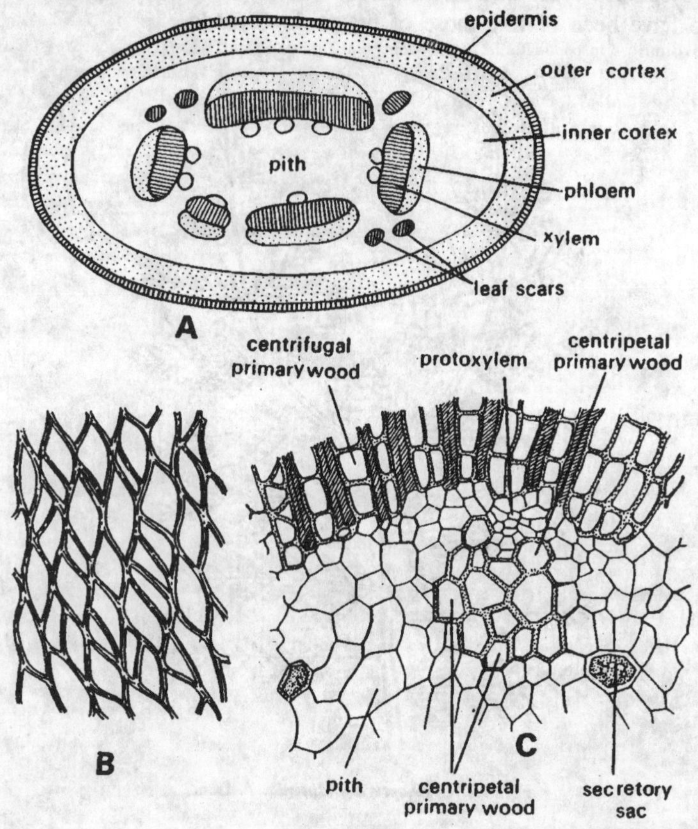

Fig. 9.4. *Lyginopteris oldhamia*. A, T.S. primary stem; B, outer cortex showing fibrous strands as they appear in L.S.; C, primary vascular bundle showing details.

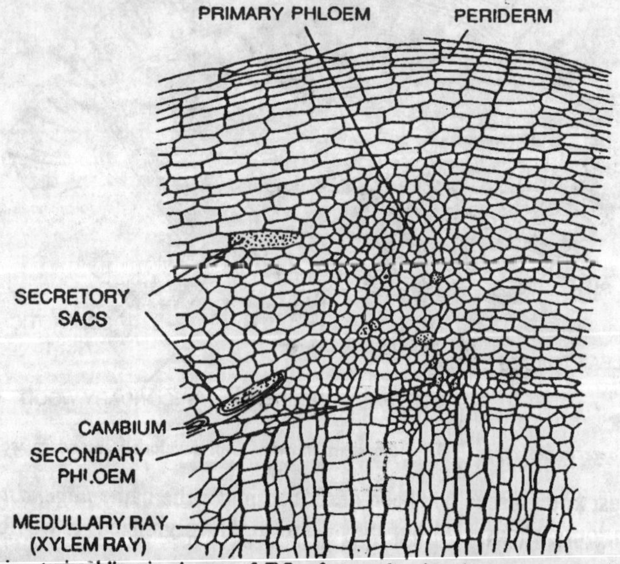

Fig. 9.5. *Lyginopteris oldhamia*. A part of T.S. of stem showing the outer portion of the stele.

whole seed enclosed in cupule. This cupule opened out when seed was mature.

Each seed was borne at the tip of stalk. The cupule rose from the base of the seed but not fused with it. The cupule was in three main lobes. These lobes were divided in the upper

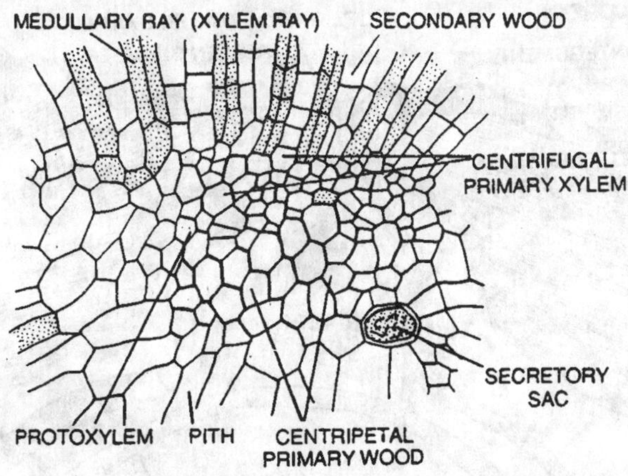

Fig. 9.6. *Lyginopteris oldhamia.* **A part of T.S. of stem primary and secondary xylem, medullary rays and pith.**

parts of the seeds. The seed or ovule was orthrotropus and of cycadian type. It was radially symmetrical. The cupule was separated from the seed along its entire length. The seed itself has an integument which surrounded the nucellus. The integument and nucellus were fused except at the top. The integument formed nine projections. In each of these projections there was a vascular bundle present. These projections surrounded the nucellar beak.

The pollen grains then came to lie in pollen chamber formed by disorganization of some of nucellar tissue round the base of nucellar beak.

In some of preparations the megaspore membrane is very well seen. In the centre of the seed there was a tissue but so far neither any archegonium nor any embryo has been found in these Palaeozoic seeds.

The seeds were borne at the tips of the stalks. They were not organized to form cover. In addition to this *Lagenostoma* there are many other Palaeozoic seeds which have been described some simpler other more complex than *Lagenostoma.*

A peculiar feature of all Palacozoic seeds which so far has not been explained is the absence of any embryo in them. Pollen grains have been found in pollen chambers. Some of them had even

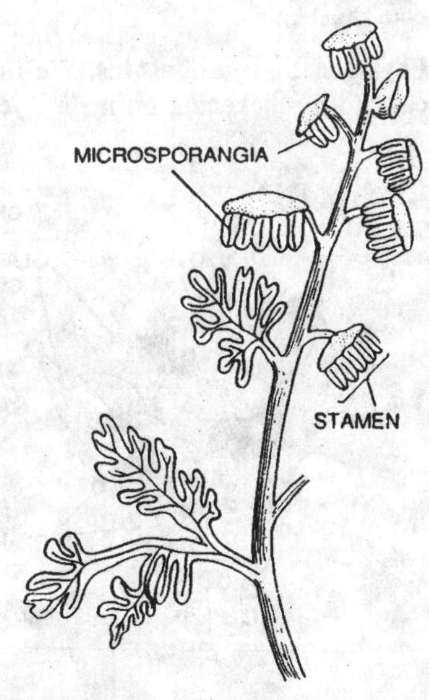

Fig. 9.7. *Crossotheca* **sp. The male reproductive structures of** *Lyginopteris oldhamia* **with few stamens and pendant sporangia.**

showed their germ tube, but so far no seed is discovered in which embryo was developed. It is possible that all the seeds described might have not preserved or they preserved before embryo formation.

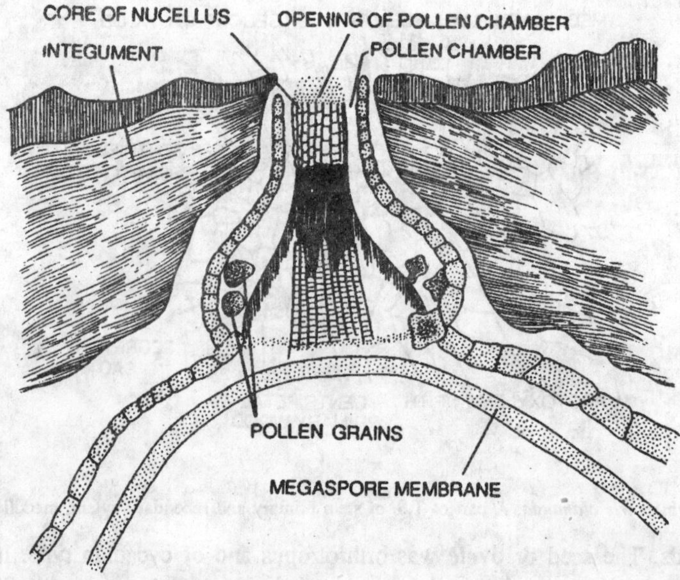

Fig. 9.8. *Lagenostoma lomaxi*. L.S. through the apex of the seed of *Lyginopteris oldhamia* showing pollen grains in the pollen chamber.

Phylogenetic relationships. We find that the xylem in the case of pteridophytes show only scalariform thickening on it. In Cycads we find the pitted type of thickening on it. It is

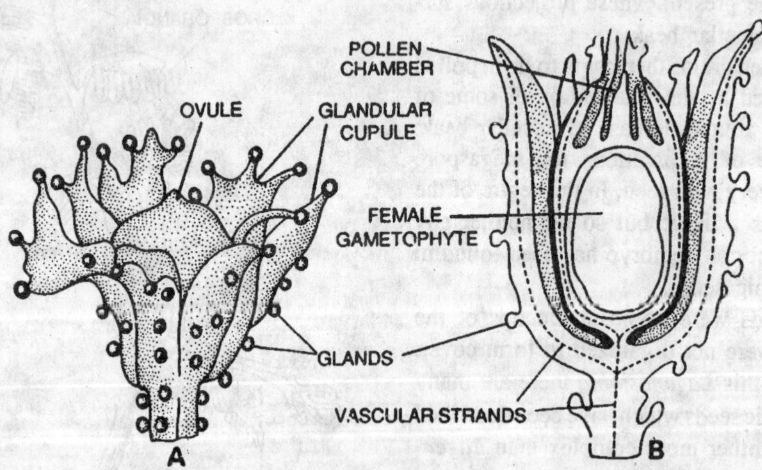

Fig. 9.9. *Lagenostoma lomaxi*. A, the seed of *Lyginopteris oldhamia* enclosed in a glandular cupule; B, L.S. of seed showing details of cupule.

seen that the xylem of some cycads passes through scalariform stage and one recent cycad *Stangeria*, here the xylem shows only scalariform thickening in tracheids. There were no pits. On account of the absence of pitted thickening the genus *Stangeria* was for a long time included among ferns, and it was only later when a plant was found bearing seeds on it that it was shifted from ferns to Cycadales.

The stem of Cycadofilicales has combined structure of both ferns and cycads. In Cycadofilicales arranged from protostelic to polystelic structure and there is found only in some of lower types like *Marattia, Angiopteris, Ophioglossum* and *Botrychium*. These Cycadofilicales in appearance were like the ferns. In Carboniferous times the vegetation was very much like most of present day ferns. It was due to the strong resemblance in the external features that the group was described as Cycadofilicales. It was later on when seeds of *Lagenostoma* were found growing on leaves of *Lyginopteris* that fern connection of the group came to an end.

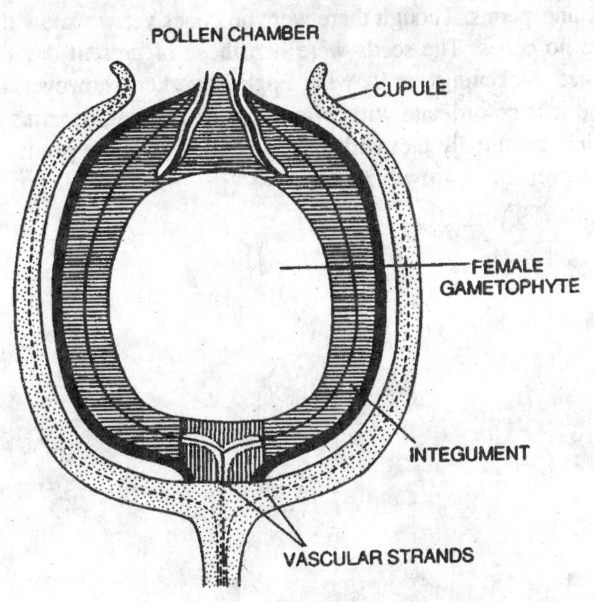

Fig. 9.10. *Lagenostoma lomaxi.* **Longitudinal section of the ovule with the investing cupule.**

English workers (Oliver and Scott) call the group Pteridosperms. They are of opinion that they belong to an extinct group of pteridophytes which has developed seeds on them. American workers called the group as Cycadofilicales. They think that the group has distinct affinities with ferns on one hand and Cycadales on other hand also. The fossil leaves found in Carboniferous strata cannot be suitably assigned either to the ferns or to Cycadofilicales. With more advancing knowledge more and more of these Corboniferous leaves are coming to be associated with Cycadofilicales, and some workers have even gone so far to say that during the Carboniferous the ferns were not at all common and they did not exist in those times. Some of the more primitive type of ferns probably did exist in those times but so far known all those types were homosporous. Heterospory probably developed only in Cycadofilicales. These Cycadofilicales are of course definitely related to ferns in the general appearance of plant, the form of leaves, the general anatomy of stem and particularly in the form of microsporangia which in the two groups are very much alike the features which separate the Cycadofilicales from the ferns in a more complex stem anatomy with secondary growth and the presence of seed structure. The origin of this seed structure has not been explained so far. There is not direct evidence with regard to the evolution of structure. There is only guess work, this seed habit must have preceded heterospory. It must have from some heterosporous plants that the Cycadofilicales had originated. We know all bryophytes and primitive ferns are homosporous. It is only a few types of present day pteridophytes like *Selaginella* and heterosporous ferns that heterospory is seen, it appears that the two groups pteridophytes and Cycadofilicales have originated from some common ancestral group in which heterospory had come into existence or it may be that homosporous ancestral ferns may have given rise to pteridophytes and the heterosporous ones to the Cycadofilicales. So clearly the Cycadofilicales must have come from some heterosporous ancestors, but so far no such fossil has been discovered. Chamberlain and other American workers are of opinion that Cycadofilicales form the most primitive group of gymnosperms. Though there were no cones yet we know that in *Cycas* itself in the female there are no cones. The seeds were like those of present day cycads, and like them the xylem had pitted thickenings on its wall. English workers, however, think that the group is an extinct one and it is co-ordinate with gymnosperms and angiosperms and they call it Pteridosperms. They think, commonly met within the stem of *Cycas* the primary structure was polystelic. Each stem has secondary growth round it.

10

Class-Cycadopsida

Order-Cycadeoideales (Bennettitales)

Introduction. The Cycadofilicales, they formed the dominant fossil plants during Palaeozic age. The Cycadofilicales have of course definite affinities with the cycads on one side and ferns on the other, but they had no cones either in the male or in the female part of the plants, so some workers think that the Cycadofilicales form a separate group quite distinct from gymnosperms.

In the Mesozoic times, however, we came across fossil plants which had cones and were definitely related to gymnosperms. So in Mesozoic the Cycadofilicales were replaced by true gymnosperms which formed strobili, and the seeds had a naked dicotyledonous embryo in them. The ovule or the seed was never enclosed in a closed carpel. The Mesozoic gymnosperms can be placed into two separate groups-1. Cycadeoideales (Bennettitales) and 2. Cycadales.

Pant (1957) has placed the cycadeiods in a distinct class, the Cycadeoideopsida of the division Cycadophyta.

The Cycadeoideales (Bennettitales) first appeared in the Permian they reached their highest range during the Jurassic period, after which they disappeared altogether.

The second group Cycadales had a world-wide distribution during the Mesozoic period. Majority of them had altogether disappeared, only a few types have been left which are confined to special parts of the East. The present day cycads are only the remnants of very large dyeing out group, *i.e.*, they are sometimes described as **living fossils,** because they are on their way to extinction.

The Cycadeoideales (Bennettitales) were very much like the cycads in their general appearance, and as the Mesozoic had these two prominent groups of gymnosperms, so that period sometimes described as age of cycads. These Cycadeoideales are closely related to the Cycadofilicales on one side and to cycads on the other but they have their own characteristic features which distinguish them from all other gymnosperms except the Gnetales. This important feature which separates the Cycadeoideales from other gymnosperms is the presence of bisporangiate strobili.

The plants of this group were diversified in their habit. Some types had short columnar stems like most of the living cycads. The short columnar stem was usually unbranched and at the apex of the plant there was a terminal crown of leaves which in most cases pinnate. Some other forms had branched stems with multiple crown. In present day cycads we know that young leaves and megasporophylls are covered up by unicellular hairy outgrowths known as **ramenta.** In Cycadeoideales (Bennettitales) these ramenta were not unicellular; they were scale like, flattened and were several cells in breadth. Like cycads the plants had well organized strobili or cones, but in cycads they are monosporangiate whereas in Cycadeoideales they were usually bisporangiate and they were either terminal or axiarlly in position. Majority of Cycadeoideales (Bennettitales) seem to have flowered only once in their life and after flowering the plant died out as we find in some of present day angiosperms.

Classification. According to Sporne (1965), the order Cycadeoideales (Bennettitales) has been divided into three families. They are-

1. Cycadeoideaceae.
2. Williamsoniaceae, and
3. Wielandiellaceae.

Here *Cycadeoidea* (*Bennettites*) of Cycadeoideaceae and *Williamsonia* of Williamsoniaceae have been discussed in detail.

CYCADEOIDEA (Bennettites)

Systematic position. Gymnosperms
Class. Cycadopsida
Order. Cycadeoideales
Family. Cycadeoideaceae
Genus. *Cycadeoidea.*

Bennettites, by American workers has been described as *Cycadeoidea.*

Morphological features. In *Cycadeoidea* the stem was unbranched with a single crown of pinnate leaves at the tops, but some species had branched stem with a multiple crown. In

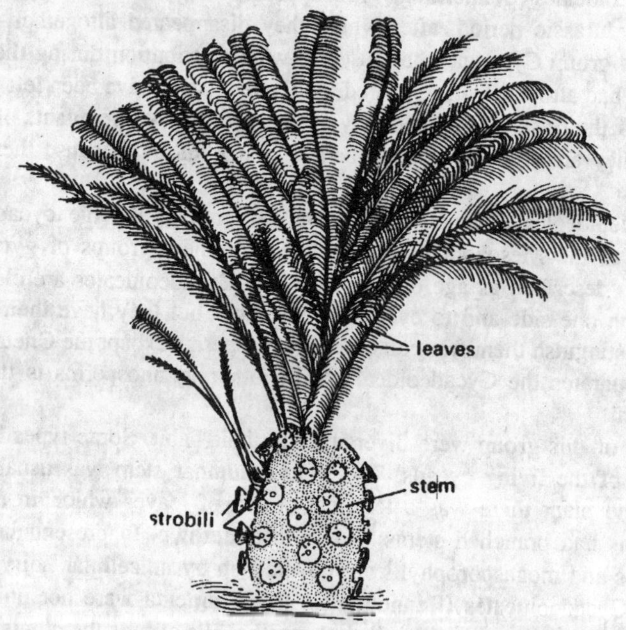

Fig. 10.1. *Cycadeoidea dacotensis.* External features.

some the stem was tuberous. In all cases the stem was covered up by persistent leaf bases as we find in *Cycas.*

Anatomical features. In structure the stem usually had large pith and thin vascular cylinder in which the protoxylem was endarch, thick cortex with a number of gum canals in it. There was small amount of secondary growth. Growth rings were only in few cases where the cambium persisted and was more active, so on the whole the stem anatomy was like those of present day cycads. *i.e.*, with large pith, broad cortex and narrow vascular cylinder. In some few cases,

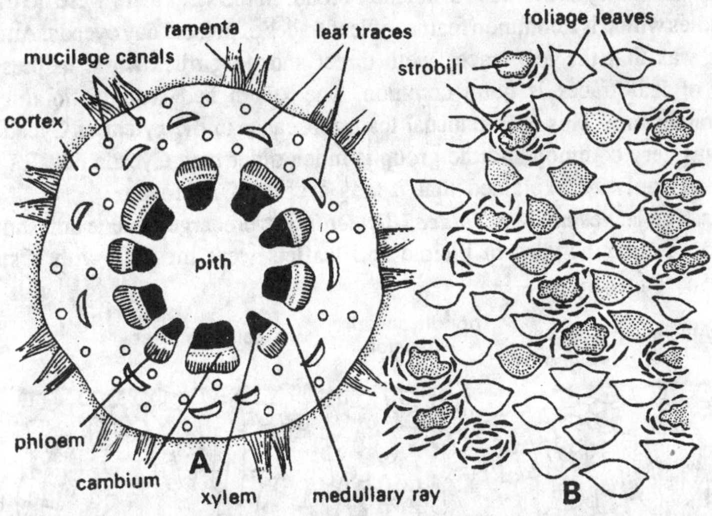

Fig. 10.2. *Cycadeoidea*. A, T.S. of stem showing primary structure; B, tangential section of a trunk.

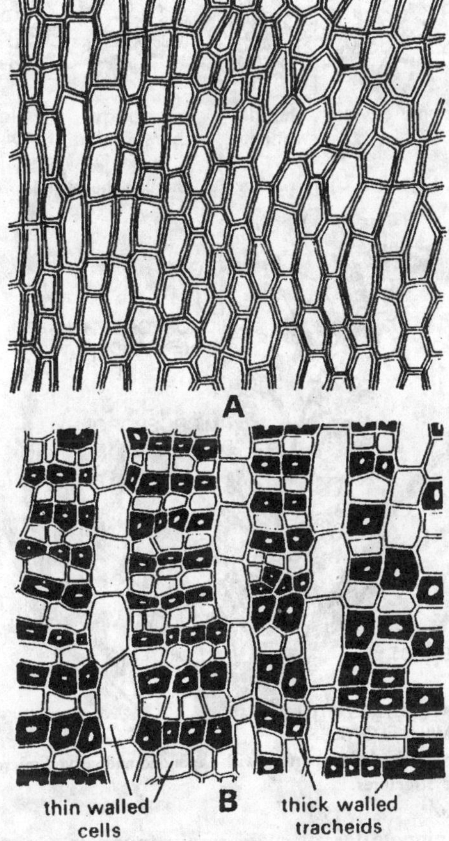

Fig. 10.3. *Cycadeoidea*. A, T.S. of secondary phloem; B, T.S. of secondary wood.

however, the vascular cylinder was sufficiently broad. In the stem there were no traces of mesarch vascular bundles which is a common feature of leaf traces of present day cycads. Another distinction from cycads was that the leaf traces were direct and no girdles while in present day cycads the girdling of leaf traces is quite common. The xylem had scalariform thickenings; pitted thickenings rather rare. This is an unusual feature because in the xylem of Cycadofilicales pitted thickening was very common and the group is much older than Cycadeoideales (Bennettitales). The feature can only be explained that it was a case of reversion.

Leaves. The leaves in *Cycadeoidea (Bennettites)* were large pinnate and showed xerophytic features. The vascular bundles in petiole and leaflets were mesarch with a strong sheath of

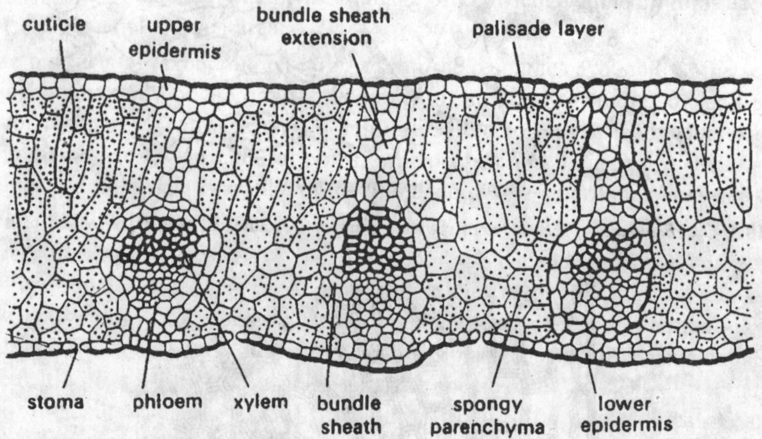

Fig. 10.4. *Cycadeoidea.* T.S. through the pinnule.

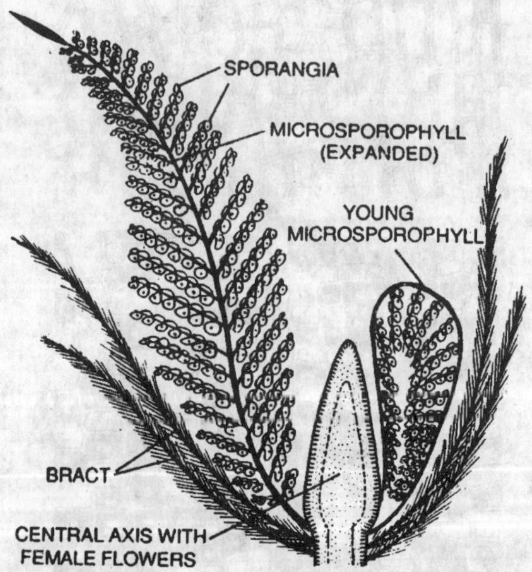

Fig. 10.5. *Cycadeoidea dacotensis.* Apical portion with expanded and curved microsporophylls; the conical central axis possesses female reproductive structures.

sclerenchyma around it. Bipinnate leaves were rarely found in *Cycadeoidea* so the form and structure of leaf is practically like that of living cycads.

Fructifications. The fructification in *Cycadeoidea* was bisporangiate. The strobili were developed in the upper part of the plant in large numbers. In some cases each leaf seems to have an axillary strobilus. The whole of the strobilus and the bases of leaves were covered up by large sized scales which were several cells in breadth and sometimes more than one cell in thickness; strobili so were axillary and borne at the tip of axillary stalk or peduncle and therefore, these strobili can be described as dwarf branches. Each strobilus was made up of

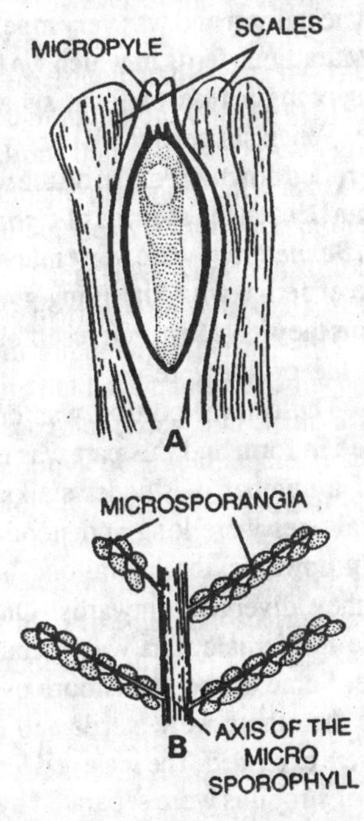

Fig. 10.6. *Cycadeoidea dacotensis*. A, longitudinal section of the seed; B, portion of a male reproductive organ with pendant microsporangia.

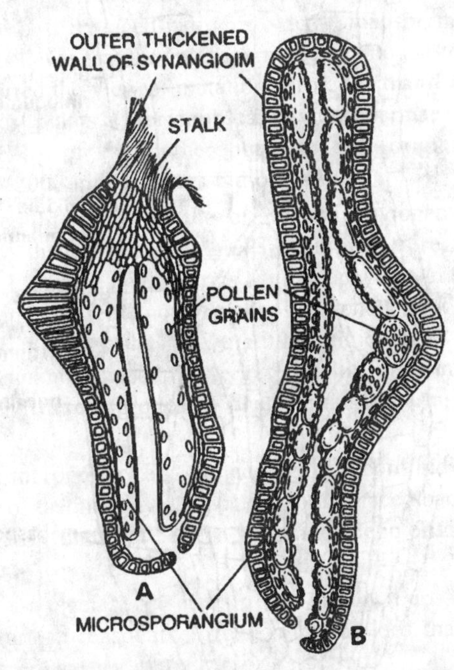

Fig. 10.7. *Cycadeoidea dacotensis*. A, central structure of a synangium showing stalk and two microsporangia in longitudinal section; B, T.S. of the synangium showing thickened outer wall and inner parenchymatous layer.

a number of heavy imbricate reduced leaves or bracts. These bracts completely surrounded the strobilus when it was the earliest stage of development where the strobilus developed these imbricate bracts separated and the inner part of the strobilus exposed. The second whorl was made up of a number of leaf like microsporophylls, all of which were united at the base to form a cup-shaped structure round the central part of strobilus. The third central portion was hemispherical or domeshaped in appearance. The central part was made up of a number of ovulate sporophylls. These megasporophylls were simply stalked. At the tip of the stalk was developed an ovule. Some stalks were sterile and the tips of sterile stalks were flattened. The central stalks stood up vertically upward and they were longer in length. The lateral ones continue to decrease in size from above downwards and stood projecting from lateral side, so the ripe female portion of strobilus hemispherical or dome-shaped in appearance. It is clear that in this bisporangiate strobilus the stamens or microsporophylls ripe first at which time the ovules were immature. When the ovules matured the stamens were shed, so in the strobilus in which the ovules were ripe the microsporophylls were absent but in young strobilus both were present. The microsporophylls or the stamens were 10 or 20 in number. These microsporophylls or stamens were all united at the base and each stamen was pinnate in form and on each stamen there were about twenty slender pinnae on either side, under the pinnae were developed two rows of fused sporangia or synangium had a short stalk and two pollen sacs in it; so each stamen was pinnate in form and was very much like the Marattious ferns in which we know the sporangia fuse to form synangia.

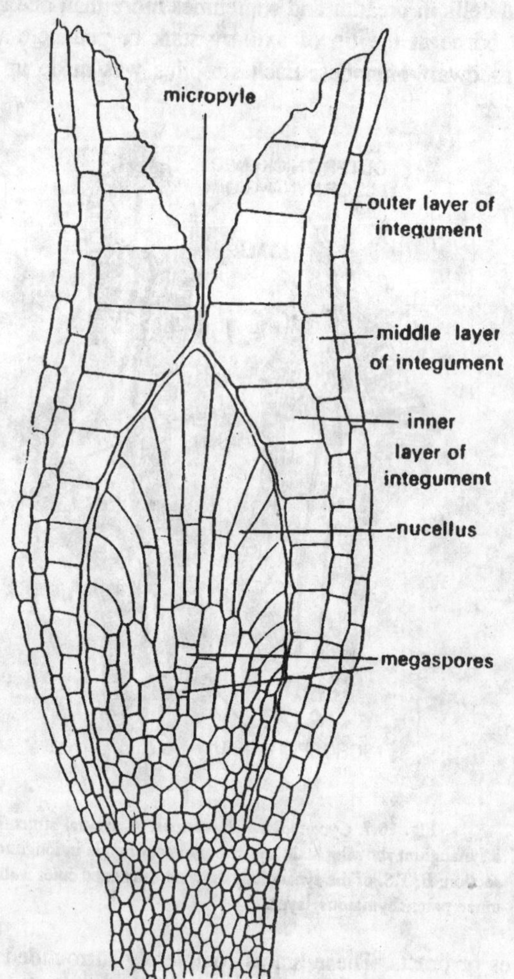

Fig. 10.8. *Cycadeoidea.* V.S. through young ovule.

When stamens were very young they rolled downwards; so on the whole we can say that the stamens of *Cycadeoidea (Bennettites)* were very much like those of ferns, while in living gymnosperms they have lost their resemblance with the ordinary ferns.

The central portion was dome-shaped in form and this part was made up of a number of slender stalks, the central ones were long and stood vertically upwards, the lateral ones short and they diverged outwards. On the whole the female part was oblong in shape. Some of the stalks bore ovules while the others were sterile and their tips were expanded. The male and female parts of strobilus were separated by the presence of some sterile bracts or scales. The ovule was orthrotropous and terminal. It was small in size and was surrounded by three-layered testa like that of *Cycas*. The nucellus was separate from integuments only in the upper part the integument projects forward to form a long micropyle. At the base of the micropyle the nucellus projected to form nucellus beak and round the base of nucellus beak there was a depression, the pollen chamber. On the outside of the ovule there was a small basal cup which suggests the cupule of *Lagenostoma*, but it was much reduced in size and never surrounded the whole of ovule or seed at any time.

The ovule seems to fertilize by swimming sperms. After fertilization the ovule developed into a dicotyledonous embryo which was non-endospermic. When the seeds were developed the whole of the female part of strobilus became fleshy and formed a fruit. In ripe fruit holes were present on the surface and at the base of these holes was dicot embryo while the fleshy portion of fruit was formed by stalk and interspersed sterile scales.

During the development of seed it appears that there was no suspensor developed. It might have been developed in the earlier stages of development of seed, but it is doubtful, so Cycadeoideales

(Bennettitales) differ from other gymnosperms in these two important respects- 1. The non-formation of suspensor and 2. The presence of a non-albuminous dicot embryo.

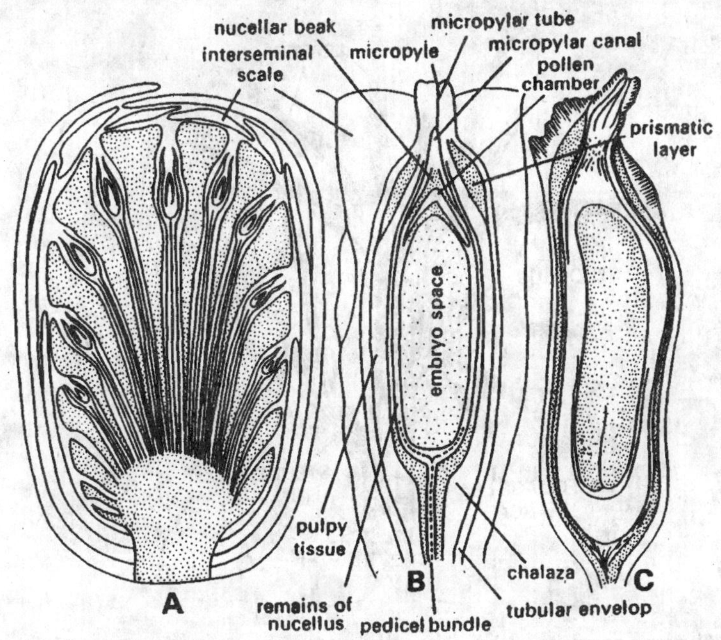

Fig. 10.9. *Cycadeoidea*. A, V.S. of gynoecium showing convex thalamus; B, V.S. of ovule; C, L.S. of seed.

WILLIAMSONIA

(*Williamsonia sewardiana* Sahni)

Systematic position.
Gymnosperms
Class. Cycadopsida
Order. Cycadeoideales
Family. Williamsoniaceae
Genus. *Williamsonia*

The second type of Cycadeoideales (Bennattitales) are known as Williamsonias cycads. This group of plants was more ancient than *Cycadeoidea* (*Bennettites*). The group differs both from *Cycadeoidea* and cycads in some important features, in the stem anatomy and also in the form and structure of fructification.

An Indian type *Williamsonia sewardian*a Sahni has been described from the Jurrasic period. It was found in a silicified condition in Rajmahal Hills (Bihar) which is made up of upper Gondwana rocks.

The stem was erect; cylindrical and branched. The branches were covered up by scales and also the leaf bases.

The stem anatomy of this type was more like those of conifers. The pith, however, was large, dense compact mass of xylem and phloem and the cortex was comparatively narrow.

The strobili were borne on short branched lateral shoots. They differ from those of *Cycadeoidea* in that they were monosporangiate. In this *Williamsonia sewardiana* only the female has so far been described but Dr. Sahni himself was of opinion that this Indian form might have

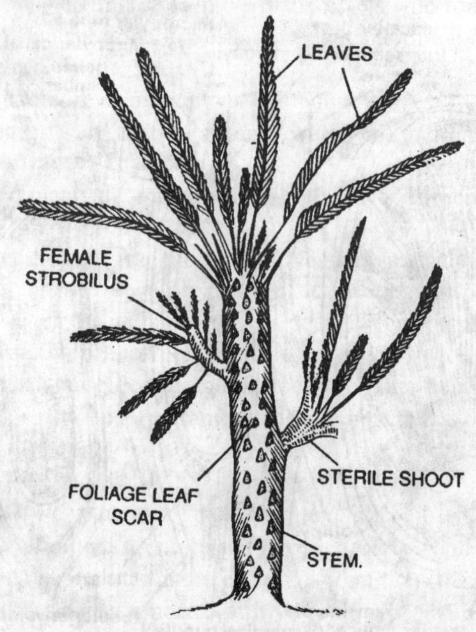

Fig. 10.10. *Williamsonia sewardiana* Sahni. Reconstruction showing a female strobilus and a sterile shoot.

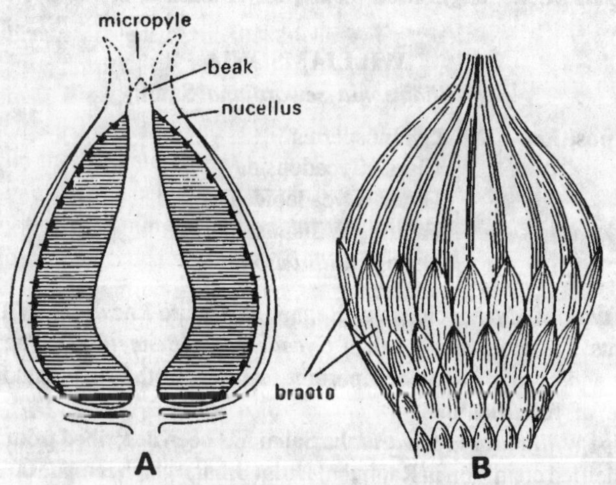

Fig. 10.11. *Williamsonia gigas*. A, V.S. of ovuliferous flower; B, floral bud.

been bisporangiate. The specimen got separated from the strobilus. The structure of female portion was practically like that of *Cycadeoidea (Bennettites)*. The ovules were borne on short stalks which had interstaminal scales in between. The ovule structure was again that of *Cycadeoidea (Bennettites)*.

The male strobilus rather which has been described in some other species of *Williamsonia* had a number of stamens which fused together to form a cup or discshaped structure. The free parts of stamens had two rows of slender leaflets were absent and only stalks were present.

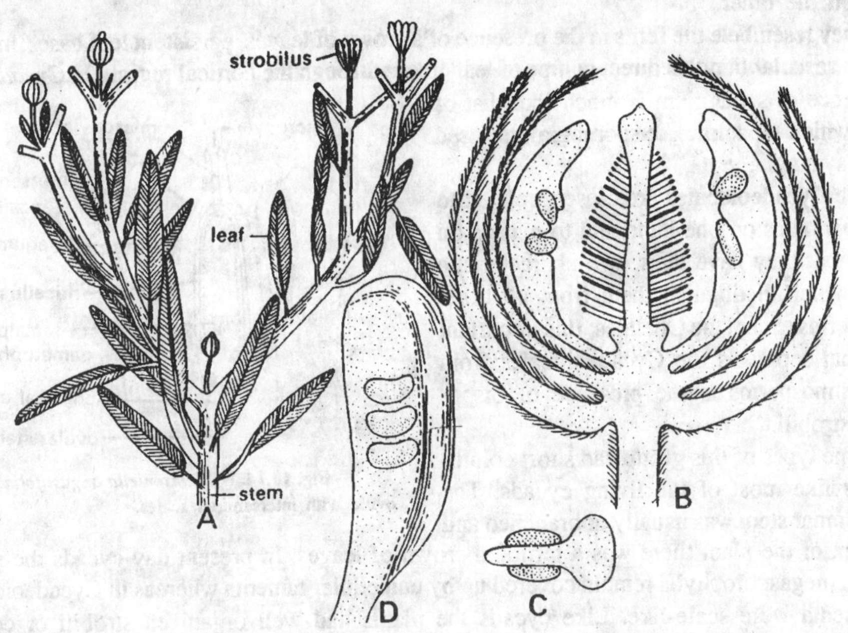

Fig. 10.12. *Williamsoniella coronata*. A, reconstruction of a portion of plant showing branched stem, simple leaves and strobili; B, V.S. of bisexual flower; C-D, microsporophylls.

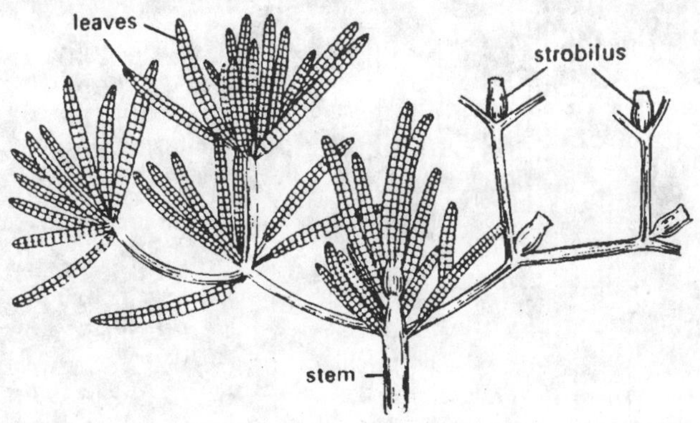

Fig. 10.13. *Williamsoniella angustifolia*. Reconstruction of a portion of plant showing branched stem, simple leaves and strobili.

So either on leaflets or on stalks there were two rows of pollen sacs. This way, the male portion of *Williamsonia* was like that of *Cycadeoidea*. A large number of male strobili have been described, but no trace of female portion in them has been found, so these male strobili were evidently unisexual while the female ones might have been bisexual.

In other genus *Williamsoniella*, the strobilus was bisexual but round it there were no basal bracts, so it differed from *Cycadeoidea* only in the absence of bracts.

Affinities. In many respects the Cycadeoideales resemble the ferns on one hand and the cycads on the other.

They resembele the ferns in the presence of a crown of leaves, persistent leaf bases, mesarch nature of vascular bundles direct course of ieaf-traces through the cortical region. In *Cycadeoidea* the presence of synangium is much like that of ferns in which we know the sporangia are fused to form synangia.

The Cycadeoideales are closely related to Pteridospermales on one hand and to cycads on the other but they have their own characteristic features which distinguish them from all other gymnosperms except the Gnetales; this important feature that separates the Cycadeoideales from other Gymnosperms is the presence of bisporangiate strobili.

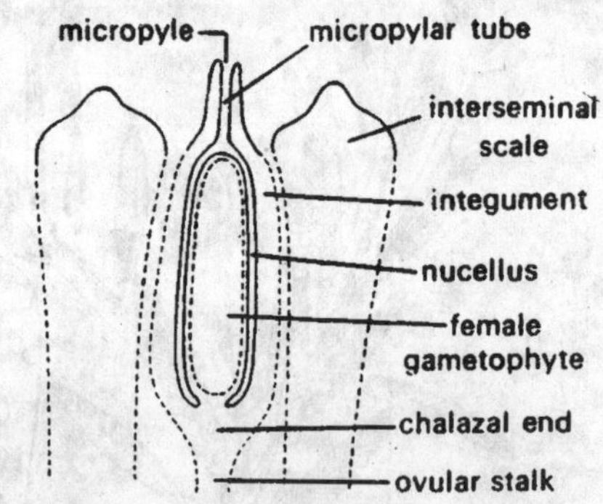

Fig. 10.14. *Williamsoniella angustifolia*. L.S. of ovule with interseminal scales.

Some types of this group had short columnar stems like most of the living cycads. The short columnar stem was usually unbranched and at the apex of the plant there was a terminal crown of leaves. In present day cycads the young leaves and megasporophylls remain covered up by unicellular ramenta whereas in Cycadeoideales these ramenta were scale-like. Like cycads the plants had well organized strobili or cones.

Majority of Cycadeoideales seem to have flowered only once in their life and after flowering the plant died out as we find in some of present angiosperms.

11

Class-Coniferopsida

Order-Cordaitales

The Cordaitales is another fossil group which lived side by side with the Cycadofilicales in Palaeozoic times. Both the groups were very common, the Cordaitales being then the prevailing gymnospermous forest plants. The group extends from Devonian through Carboniferous upto the Permian times, and in India they had been found even in the lower Mesozoic, *i.e.*, in Lower Gondwana strata of India. The plants of this group were trees some even more than ninety feet in height and three feet in diameter. The plant branched at only tips and bore generally narrow and simple leaves in clusters at top. According to Scott, the order may be divided into three groups– 1. Poroxyleae. 2. Pityeae and 3. Cordaiteae.

The details, especially of first two groups are not fully known. Their separation from Cordaiteae is based on structure of vegetative parts.

1. **Poroxyleae.** This had a single genus known as *Poroxylon*. Here the stem was slender with long internodes and within the stem were collateral exarch primary bundles, the pith was large, and double leaf traces for each leaf which had thick parallel venation in it.

2. **Pityeae.** It included a single genus *Pitys* where the bundles were mesarch, the wood was with multiciliated pits, so it differed from Cordaiteae having mesarch bundles.

3. **Cordaiteae.** It included a number of genera. The best known genus was *Cordaites.*

Genus CORDAITES

Systematic position. Division. Gymnospermae
Class. Coniferopsida
Order. Cordaitales
Family. Cordaitaceae
Genus. *Cordaites.*

Morphological features. The plant was tall and slender. It had a crown of branches near the top, leaves were large simple and pendulate and not found on any living gymnosperms.

Anatomical features. The internal structure of stem in *Cordaites* showed a combination of *Cycas* and conifers characters. Pith like cycads layer but discoid in form, *i.e.*, it was in form of plates at intervals. The vascular cylinder was like those of conifers. It was made up of thick collateral endarch primary bundles which sometimes however, seem to have been mesarch as in *Mesoxylon*. There was good deal of normal secondary growth. Secondary xylem was made up of living pitted tracheids with pits on radial walls. In fact it is clear that majority of plants found in Carboniferous period belonging to Cycadofilicales, Cordaitales and a number of fossil pteridophytes. The stems had good deal of secondary growth in them, but in these no annular rings. The stem anatomy of Cordaitales shows variations from exarch to endarch collateral primary vascular bundles with good deal of secondary growth in stem. So far anatomical structure is concerned the group is related to Cycadofilicales in having exarch and mesarch bundles and to gymnosperms especially to conifers in having developed endarch bundles.

Reproductive structures. The plants were monoecious or dioecious but the strobili were always monosporous and never bisporangiate like Cycadeoideales (Bennettitales) or living Gnetales.

Cordaitanthus was the reproductive shoot of *Cordaites* itself. It was not definitely organized to form a cone but it was more or less a strobilus. The male and female strobili were borne separately in lateral sides of short simple axis. The strobili were enclosed when young by bracts so outwardly they could not be distinguished unless sections were cut and examined. Both were usually of small size.

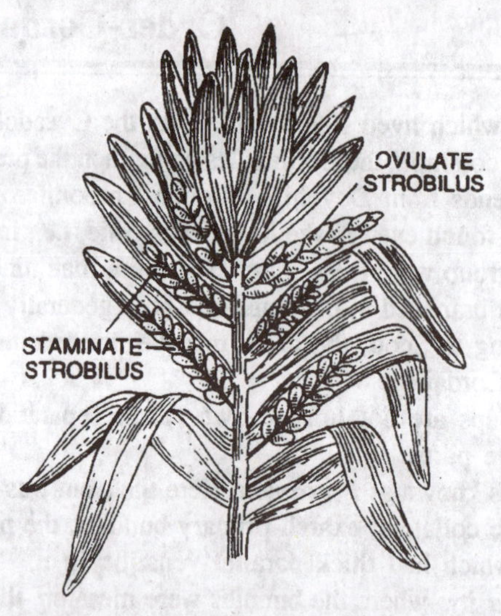

Fig. 11.1. *Cordaites*. Apical portion of the stem with leaves and male and female strobili.

Male strobilus. The male strobilus had a thick axis bearing spirally arranged bracts with stamens on them, the stamens themselves were either solitary or grouped near the apex and in between the stamens there were sterile bracts present. These bracts might have been sterile microsporophylls. Each individual stamen was peculiar in form, in that it had a cluster of three to six terminal sporangia similar types of stamens have also sometimes described belonging to Cycadofilicales, but the important difference was that at the tips were borne synangia and separate or individual sporangia. If in this male catkin of *Cordaites* there were no sterile bracts the structure would

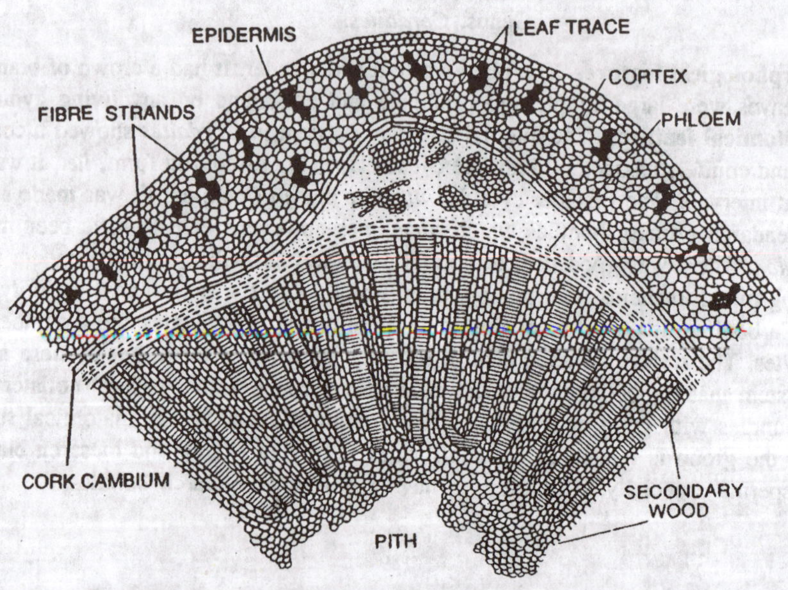

Fig. 11.2. *Cordaites*. Portion of transverse section of stem showing well developed secondary wood.

be very much like the staminate strobilus of *Ginkgo* but there is a very important difference and that in the *Cordaitanthus* the sporangia were terminal while in *Ginkgo* they were pendant

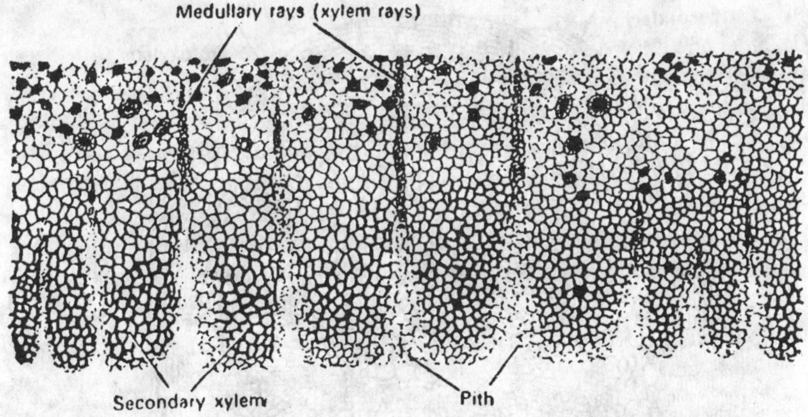

Fig. 11.3. *Cordaites.* Inner part of secondary wood in T.S. showing narrowing of wood towards pith

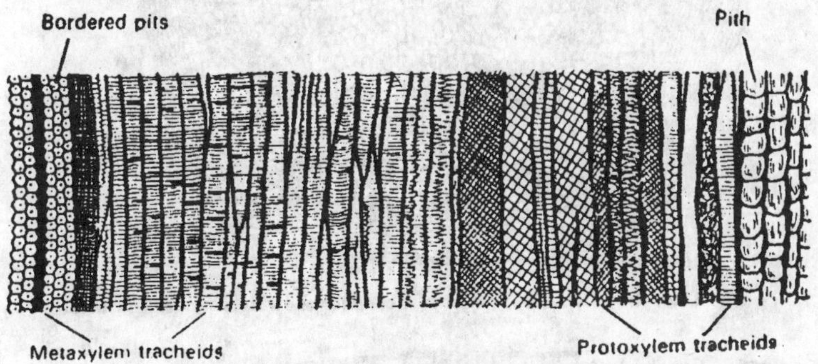

Fig. 11.4. *Cordaites.* R.L.S. of secondary wood showing various type of thickenings of tracheids.

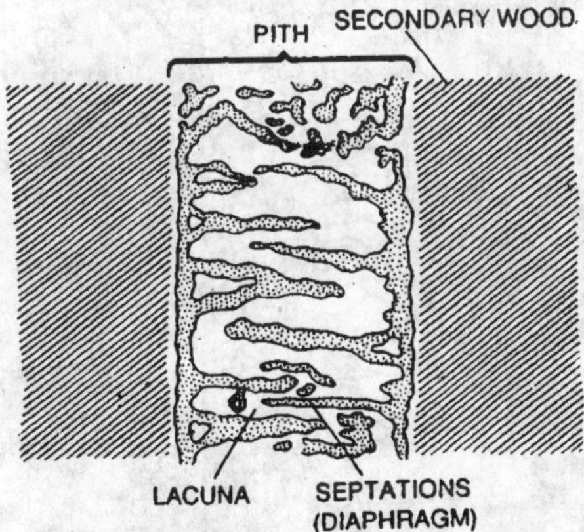

Fig. 11.5. *Cordaites.* R.L.S. of stem showing secondary wood and discoid pith.

and only two in number. In fact the morphology of this male strobilus of *Cordaites* is not quite clear. Different interpretations are put on male catkin of *Cordaites*. Some think that whole of

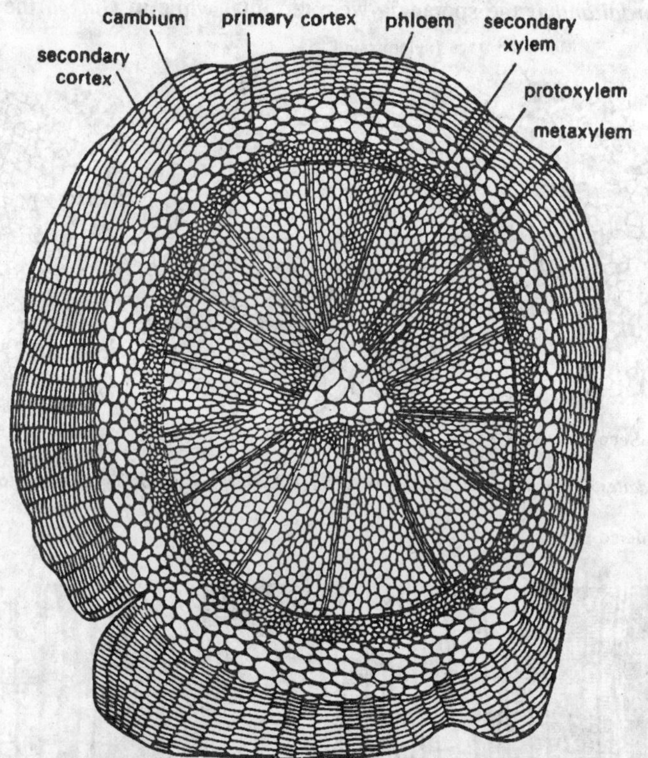

Fig. 11.6. *Amyelon*. T.S. of root of *Cordaites* showing various structures.

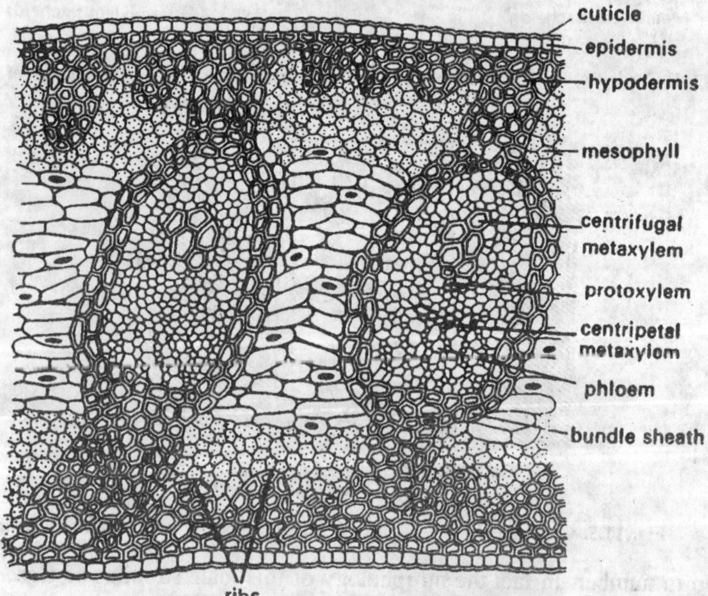

Fig. 11.7. *Cordaites angulostriatus*. V.S. of leaf.

strobilus acting as single male flower in which there are large number of stamens, the stalk being the filamentous bearing a number of pollen sacs were simply sterile filaments. Other workers

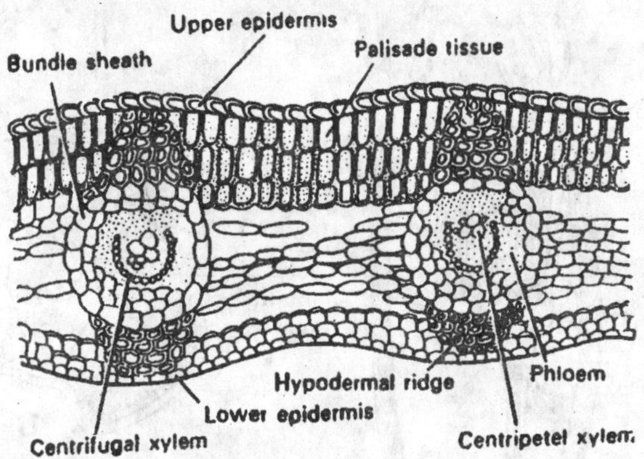

Fig. 11.8. *Cordaites lingulatus.* T.S. of leaf showing various structures.

described this male catkin as inflorescence, each catkin bearing a flower itself, the filament is stalk of flower, the pollen sacs the stamens.

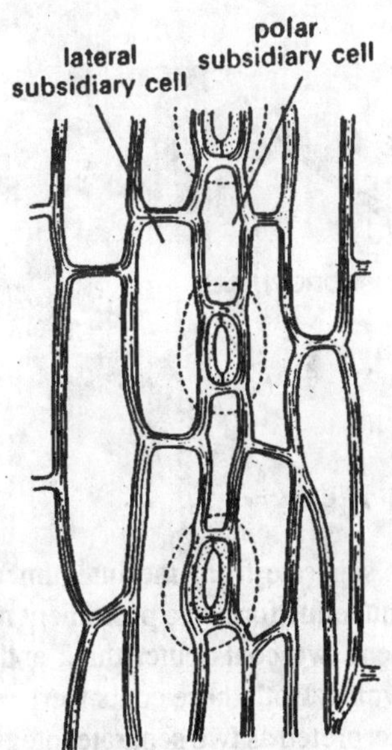

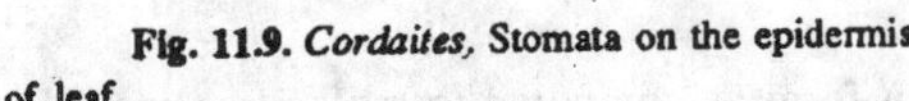

Fig. 11.9. *Cordaites,* Stomata on the epidermis of leaf.

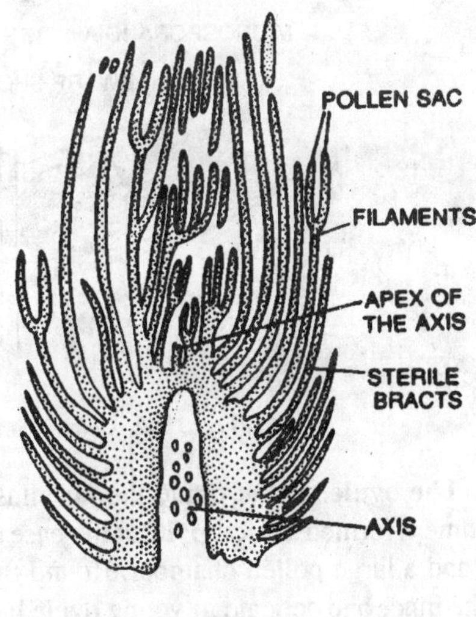

Fig. 11.10. *Cordaitanthus.* The staminate strobilus of *Cordaites*, showing sterile bracts and stalked stamens.

Female strobilus. The female strobilus when young was enclosed by bracts. It had a thick conical axis bearing a number of spirally arranged bracts. In the axils of some of these bracts there were dwarf shoots present. These dwarf shoots in turn had minute bracts. On the tip of dwarf shoot there was a terminal ovule.

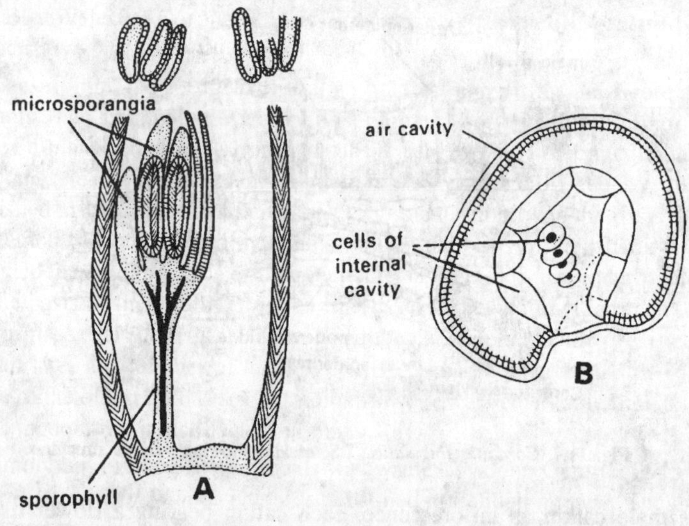

Fig. 11.11. *Cordaitanthus*. A, microsporophylls and microsporangia; B, pollen grain.

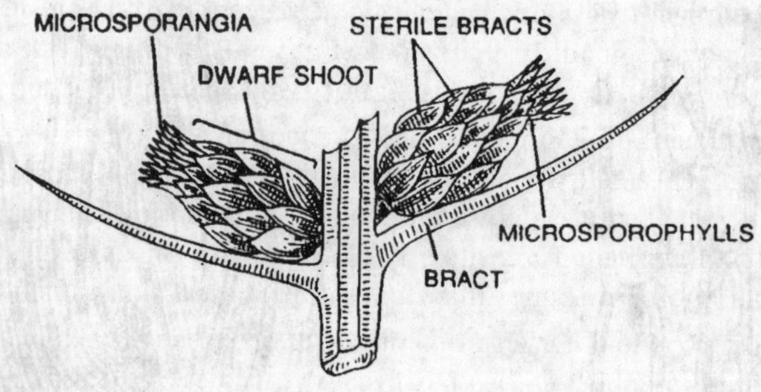

Fig. 11.12. *Cordaitanthus*. Two male strobili of *Cordaites*.

The ovule. In the ovule the nucellus was completely separate from the integument. The integument formed a micropyle; at the base of micropyle the nucellus formed a prominent nucellar beak and a large pollen chamber. Round the ovules there were two coats outer thick and fleshy and the inner one delicate in young ovule but hard in older ovules. Both these coats were separate below but more or less united above. So these coats may be interpreted as two separate integuments which are formed at upper part or a case of single integument which differentiated into three coats one of which not preserved.

The female strobilus is distinctly compound in form as compared to male stroblius. The ovule bearing stalks were distinctly axillary in position just like those of present day conifers. In Cycadofilicales, Cycadeoideales (Bennettitales) and the living cycads the ovule bearing structures are not axillary in position and the ovules in these groups were borne on leaves while in Cordaitales and the Coniferales the ovules are borne on stem structure. Dr. Sahni according to the position of ovules recognized two groups.

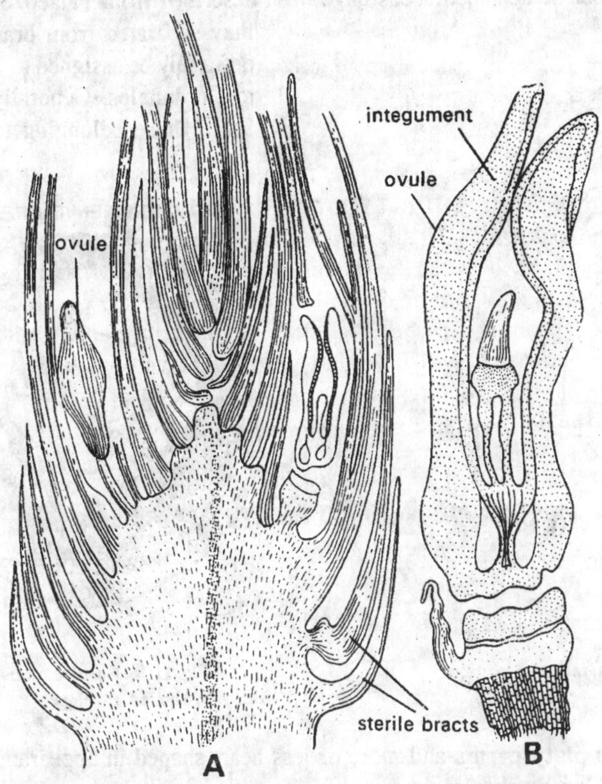

Fig. 11.13. *Cordaitanthus*. The female reproductive organs of *Cordaites*. A, longitudinal section of female strobilus; B, longitudinal section of an ovule.

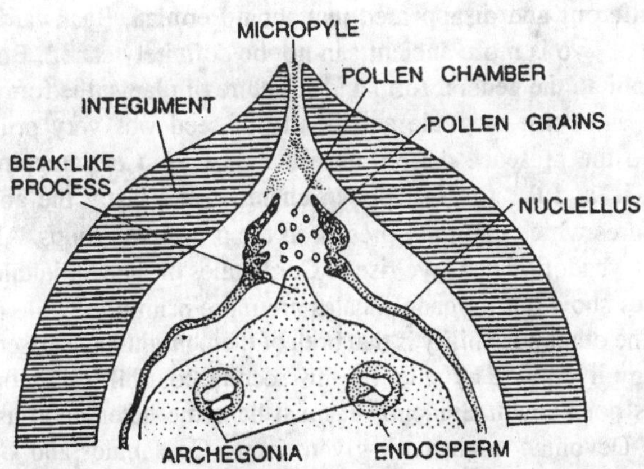

Fig. 11.14. *Cycadinocarpus*. Longitudinal section of the apical portion of a mature ovule showing various parts and pollen grains within pollen chamber.

1. Phyllosperms and 2. Stachysperms.

1. **Phyllosperms.** The ovules borne on leaves as seen in pteridosperms and *Cycas*.
2. **Stachysperms.** The ovules as seen in Cordaitales, *Ginkgo* and the Coniferales.

A large number of detached seeds have been described from Palaeozoic times and when they separated from branches they can not definitely be assigned to Cycadofilicales and to Cordaitales. Generally according to Dr. Sahni those belonging to Cordaitales were flat and described as **platysperms** and more or less heart shaped in appearance and no embryo in them like those Cycadofilicales.

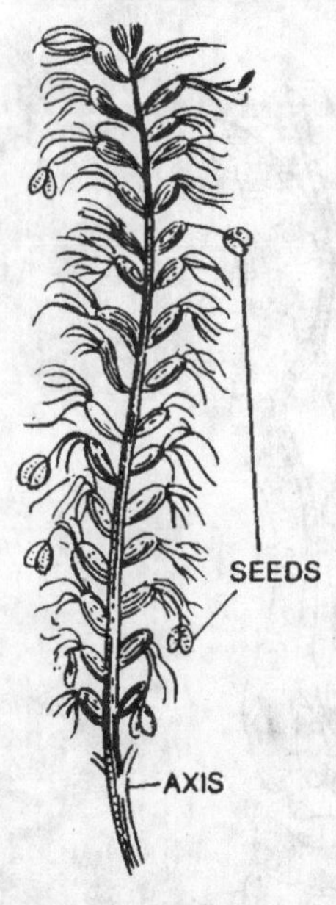

Fig. 11.15. *Cordaitanthus*. A portion of axis bearing seeds.

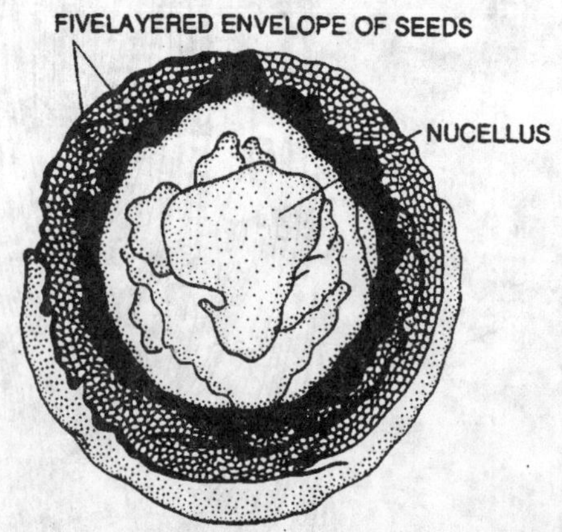

Fig. 11.16. *Cardiocarpus*. The seed of *Cordaites* with five layered envelope.

AFFINITIES

Cordaitales and Cycadofilicales lived side by side in Palaeozoic time, both were very abundant in Carboniferous and disappeared just about Permian. Backwards both groups extend to Devonian, which of two is more ancient can not be definitely stated. Both seem to be related is clear from their habitat, the general form and structure of plants, the form of strobili especially male one, structure and form of ovule and finally of seed was very primitive in the two. In the ovule of the two the presence of free nucellus. The form of integument with two sets of vascular bundles in it, and the form of pollen chamber and lastly the absence of embryo are really primitive features which naturally present in the primitive groups. All these resemblances mean that either the Cycadofilicales give rise to Cordaitales or the Cordaitales to Cycadofilicales. All combined features show that Cycadofilicales are more primitive while the Cordaitales show advanced features. The other possibility is that both of them might have arisen from some common ancestral form as they lived side by side. So this second possibility is more accepted but what the ancestral form is not known. Evidently it must be some heterosporous pteridophyte which lived near about the Devonian might have given rise to Cordaitales and Cycadofilicales which is clear from the stem anatomy of Cordaitales in which endarch siphonostele was more common.

The strobili of Cordaitales are better organized. The sporophylls here completely lost their leaf like appearance. The female strobilus was more complicated in that ovule bearing branch was always axillary in position. In Cycadofilicales the strobilus was simple and ovules were borne on leaves. So these Cycadofilicales and Cordaitales can be said to represent two different lines of evolution; the Cycadofilicales was left behind with more primitive features in it and before it disappeared it gave rise to two branches of Cycadophyta line the fossil Cycadeoideales (Bennettitales) and living Cycadales. The Cordaitales made some advance both in anatomy of stem organization and complexity of its reproductive parts. Being a primitive group naturally it had resemblances with *Cycas* which is seen in the presence of large pith, general leaf anatomy, general structure of seed and ovule and presence of sperms.

Chamberlain is of opinion that Cordaitales represents one line of Gymnosperms, the Coniferophyte line characterized by the presence of large sized plant with profused branched stem and simple leaf, small pith, abundant wood and narrow cortex. The reproductive structure organized to form more or less cones. The Cordaitales seems to have given rise to Ginkgoales and conifers before it disappeared in or about in Permian times. In Ginkgoales there are good deal of resemblances. In both the plants are lofty and branched. The stem anatomy, the structure of ovule and in presence of sperms two groups seem to be more or less closely related. There is great similarity in the outward appearance of male strobili of *Ginkgo* and that of Cordaitales.

Glossary

Abaxial–The surface of any structure which is turned away from the axis.

Acicular–Needle like in shape, sharp pointed.

Acotyledon–A plant without a cotyledon.

Acrogenous–Increasing in growth at apex.

Acropetal–Ascending, *e.g.*, leaves, flowers or roots, developing successively from an axis so that youngest arise at apex.

Actinostele–Stele with xylem radiating outwards and forming ridges, *e.g.*, ferns.

Adaxial–Turned towards the axis.

Adherent–Attached to some substratum.

Adventitious–Found is an unusual place, *e.g.*, Adventitious roots.

Aerenchyma–Aerating cortical tissue in floating plants.

Alternation of generations–The occurrence in one life history of two or more different forms differently produced, usually an alternation of a sexual with an asexual form.

Alveolus–A small pit or depression.

Amphibian–Adapted for life either on land or water.

Amphicribral–Amphiphloic.

Amphiphloic–With phloem both peripheral and central to xylem.

Anacrogynous–Certain liverworts in which female reproductive bodies do not arise at or near apex of shoot.

Anatomy–The science which treats of the structure of plants.

Androcyte–A cell arising by growth from an antheridium and giving rise to antherozoid.

Annulus–Any ring like structure; special ring in fern sporangium, by action of which sporangium bursts.

Anterior–Nearer head end.

Antheridia–Plural of antheridium.

Antheridial cell–The cell representing an antheridium, as in Gymnospermae.

Antheridiophore–A gametophore bearing antheridia.

Antheridium–Male gametangium in which male sexual cells are produced in many cryptogams.

Antherozoids–Male sexual cells in antheridia.

Anticlinal–Line of division of cells at right angles to surface of apex of a growing point.

Apical–Cell at tip of growing point.

Apogamy–Reproduction without intervention of sexual organs.

Apophysis–Small protuberance at apex of ovuliferous scale in pine.

Apospory–Production of a diploid gametophyte from a sporophyte without intervention of spore formation.

Aquatic Living in water.

Archegoniophore–Gametophore bearing archegonia, *e.g.*, *Marchantia*.

Archegonium–A female gametangium in which oospheres are formed, and in which the young plant begins development.

Archesporium–A cell or mass of cells dividing to form spore mother cells and elater forming cells.

Asexual–Having no sexual organs.

Asymmetrical–Having two sides disproportionate.

Autoecous–Bearing antheridia and archegonia on the same plant but different branches, *e.g., Funaria.*

Axial–Pertaining axis or stem.

Axillary–Growing in axil, as buds.

Basal–Near the base.

Basipetal–An order of development of organs on which the youngest structures are at the base and the oldest at the apex.

Biennial–Lasting for two years.

Biflagellate–Having two flagella.

Bifoliar–Having two leaves.

Blepharoplast–A basal granule which gives rise to the cilia of antherozoid.

Body cell–An antheridial cell.

Bordered pit–A form of pit developed on walls of tracheids and wood vessels, with overarching border of secondary cell wall.

Botany–The branch of biology dealing with plants.

Bract scales–Small scales developed directly on axis of cones, *e.g.*, ovuliferous scales.

Bulbous–Like a bulb.

Calyptra–Tissue enclosing developing sporogonium in liverworts.

Cambium–The tissue from which secondary growth arises in stem.

Capsule–A sac containing spores; sporogonium in Bryophyta.

Carinal–Canals in protoxylem beneath ridges of stem in Equisetales.

Casparian band–Casparian strip; a strip encircling radial walls of endodermis cells.

Cauline–Belonging to a stem.

Centrifugal–Turning away from axis.

Chloroplast–A plastid containing chlorophylls a and b.

Columella–A prolongation of stalk into sporogonium.

Cone–The female flower of Coniferae, with woody axis and spirally arranged carpels.

Coralloid–Braching like a coral, *e.g., Cycas* roots.

Cortex–The extrastelar fundamental tissue of the sporophyte.

Cosmopolitan–Worldwide in distribution.

Cotyledon–The seed-leaf, primary or first leaf of a embryonic sporophyte.

Cryptogam–A plant without apparent reproductive organs.

Cuneate–Wedge shaped.

Cupule–The gemmae bearing cup of *Marchantia.*

Cushion–The central thick region in prothallus of fern.

Decumbent–Lying on the ground but rising at apex.

Dehiscenc–The spontaneous opening of an organ releasing spores etc.

Development–The changes undergone by an organism from its beginning to maturity

Diarch–With two xylem and two phloem bundles.

Dichotomy–Repeated forking.

Dictyostele–A network formed by meristeles.

Dioecious–Having male and female sexes on different individuals.

Diploid–Having a double set of chromosomes (*2x*).

Dorsal–Upper surface of thallus or prothallus.

Dorsiventral–With upper and lower surfaces distinct.

Ectophloic–With phloem surrounding xylem.

Egg–Oosphere, mature female germ cell.

Elater–One of the cells with a spiral tkickening which assists in dispersing sporés from capsule in liverworts, *e.g., Marchantia.*

Elaterophore–Tissue bearing the elaters, in some liverworts, *e.g., Pellia.*

Embryo–A young organism in early stages of development.

Endodermis–Innermost layer of cortex in plants.

Endosperm–The nutritive tissue of certain seeds.

Endospore–Inner coat of spore; intine.

Endosporium–Inner coat of spore wall.

Endothecium–The central region of an epibasal octant of oospore of liverwords and mosses.

Entire–With continuous margin.

Epidermis–The outermost protective layer of stem.

Epithelium–Any cellular tissue lining a cavity, *e.g.*, resin canal.

Erect–Directing upwards.

Evolution–The gradual development of organisms from pre-existing organisms since the dawn of life.

Exarch–With protoxylem strands outside metaxylem towards periphery.

Exine–Outer layer of spore wall; exospore, exosporium.

Fertilization–The union of male and female nuclei.

Foot–An embryonic structure in vascular cryptogams through which nourishment is obtained from prothallus or thallus; basal portion of sporophyte in mosses.

Fossil–Petrified plant portion found in rocks.

Frond–A leaf especially of fern.

Gametangium–A structure producing sexual cell.

Gametes–Sexual cells.

Gametogenesis–Gamete formation.

Gametophyte–The gamete forming phase in alternation of plant generations; sexual generation of plants.

Gemma–A bud or outgrowth of a plant which develops into a new organism, *e.g., Marchantia, Funaria* etc.

Gemma-cup–Cupule containing gemmae, *e.g., Marchantia.*

Genera–Plural of genus.

Genus–A group of closely related species, in classification of plants.

Germ tube–Short filamentous tube put forth by a germinating spore.

Germination–Process of initial development.

Glossopodium–The sheathing leaf base of the ligule of *Selaginella.*

Groove–Furrow or depression.

Ground tissue–Conjunctive parenchyma.

Growing point–A part of plant body at which cell division is localized, generally terminal and composed of meristematic cells.

Guard cells–Two cells surrounding stomata of aerial epidermis of plant tissue.

Gymnosperms–The plants having seeds not enclosed in true ovary, *e.g., Cycas, Pinus.*

Habitat–The locality or external enviroment in which a plant lives.

Haploid–Having *x* number of chromosomes.

Haplont–An organism having haploid somatic nuclei.

Hepatic–Liverworts.

Heterosporous–Producing two kinds of spores, *e.g., Selaginella, Marsilea* etc.

Heterospory–The condition of being heterosporous; the production of megaspores and microspores.

Homosporous–Bearing spores of one kind, *e.g.*, Fern.

Hypobasal–The lower segment of a developing zygote.

Indusium–An outgrowth of plant epidermis covering and protecting a sorus, *e.g.*, Ferns.

Integument–Coat of ovule.

Internode–The part between two successive nodes.

Intine–The inner covering membrane of a pollen grain or of a spore.

Karyogamy–Union of nuclei after cytoplasmic fusion (plasmogamy).

Lacuna–A space between cells.

Lanceolate–Lance-shaped.

Leaf gap–Gap in vascular cylinder of stem, a parenchymatous region associated with leaf traces.

Leaflet–Individual unit of a compound leaf; pinna.

Leptosporangiate–With sporogenous tissue developing from outer cell of periclinal division.

Life-cycle–The various phases through which an individual species passes to maturity.

Ligule–Small scale on upper surface of leaf-base in Selaginellaceae and *Isoetes*.

Macrospore–Macrospore; a larger spore of heterosporous plants.

Medullary rays–A number of strands of connective tissue extending between pith and pericycle.

Megagametophyte–The female gametophyte developed from a megaspore, *Selaginella*, *Marsilea* etc.

Megaspore–Larger spore of heterosporous plants, regarded as female, *e.g.*, *Selaginella*, *Marsilea*.

Megasporophyll–A spore bearing leaf developing megasporangia, *e.g.*, *Selaginella*.

Mesarch–Xylem having metaxylem developing in all directions from the proroxylem, characteristic of ferns.

Mesophyll–The internal parenchyma of a leaf; chlorophyllous cells.

Metamorphosis–Change of form and structure, *e.g.*, the formation of antheozoid from an androcyte.

Microgametophyte–The male gametophyte developed from a microspore, *e.g.*, *Selaginella*, *Marsilea*.

Micropyle–Aperture for admission of pollen tube at ovule apex.

Microsporangium–A sporangium bearing a number of microspores, *e.g.*, *Selaginella*, *Marsilea*.

Microspore–The spore developed in a microsporangium of heterosporous plant.

Mid-rib–The large central vein of a leaf.

Monoecious–With sex organs on one gametophyte; hermaphrodite.

Morphology–The science of form and structure of plants.

Mycorrhiza–Association of fungal mycelium with roots of a higher plant.

Neck-canal–The hollow within the elongated part of an archegonium.

Node–The joint of a stem at which leaves arise.

Nucellus–Parenchymatous tissue between ovule or megaspore and its inner integument, and extending from chalaza to micropyle.

Nurse cells–Cells providing nourishment to the developing spores in the sporogonium.

Octant–One of the units in eight celled stage in segmentation of ovum.

Oosphere–An egg before fertilization.

Oospore–The zygote or fertilized egg cell.

Operculum–A lid or covering flap at apex of capsules of mosses.

Order–In classification, group of organisms closely allied, ranking between family and class.

Ovule–The megasporangium of seed plant.

Ovuliferous–Containing ovule; scales, each bearing one or more ovules, developed on bract scales, as in Coniferae.

Palaeobotany–Botany of fossil plants and plant impressions.

Palisade–Arrangement of apposed elongated cellular structures.

Paraphysis–A slender filamentous epidermal outgrowth occurring among sex organs, *e.g., Funaria.*

Parthenogenesis–Reproduction without fertilization by a male element.

Peduncle–Stalk.

Perichaetium–One of membranes enveloping archegonia of bryophytes, *e.g., Marchantia.*

Periclinal–System of cells parallel to surface of apex of a growing point.

Pericycle–The external layer of stele, the layer between endodermis and conducting tissues.

Perigynium–Membranous envelope of archegonium in liverworts, *e.g., Marchantia.*

Peristome–The region surrounding mouth; term used in connection with moss capsule.

Pinna–A leaflet of a pinnate leaf.

Pinnate–A compound leaf, having leaflets on each side of an axis or mid-rib.

Pit–A depression formed in course of cell-wall thickening in plant tissue.

Pollen–The powder produced by anthers, consisting of pollen grains.

Pollen chamber–Pit formed at apex of nucellus below micropyle.

Pollen grain–The haploid microspore of seed plants.

Pollen-tube–A tubular process developed from pollen grains, represents male gametophyte.

Pollination–Transference from anther to ovule.

Pollination-drop–Mucilaginous drop exuded from micropyle and which detain pollen grains, as in gymnosperms.

Polyembryony–Formation of several embryos in one ovule, *e.g., Pinus.*

Polystelic–With several steles.

Posterior–Situated behind.

Primitive–Of earliest origin

Prothallus–A small thin structure, the gametophyte or haploid sexual generation of pteridophytes.

Protonema–The filamentous thallus of mosses from which the moss plant buds.

Pseudoelater–One of the chains of cells of sporogonium of liverworts.

Pteridology–The branch of botany dealing with ferns.

Pteridophyte–A vascular cryptogam, any of the ferns, club mosses or horse tails.

Quadrifoliate–Four leaved, compound palmate leaf, with four leaflets arising at a common point, *e.g., Marsilea.*

Rachis–The axis bearing leaflets.

Ramenta–Brown scale-like structures found on fern leaves.

Receptacle–A sporophore, terminal disc of mosses.

Resin–An acidic excretion product of certain plants.

Resin canals–Ducts in bark, wood, mesophyll, etc., particularly of conifers, lined with glandular epithelium excreting essential oils.

Rhizoid–A root-like outgrowth of thallus, *e.g.,* liverworts, mosses, ferns, unicellular hairs on lower side of prothallus.

Rhizome–A thick horizontal stem, partly along and partly underground, sending out shoots above and roots below, *e.g.,* Ferns.

Rhizophore–A naked branch which grows down into soil and develops roots from apex, *e.g., Selaginella.*

Root–Descending portion of plant, fixing it in soil, and absorbing moisture and nutrients.

Secondary growth–Development of secondary meristem or cambium producing new tissue on both sides.

Seed–A mature fruit containing an embryo ready for germination under suitable conditions.

Segment–A division formed by cleavage of an ovum.

Septa–Plural of septum.

Septum–A partition separating two cavities of tissue,

Seta–Sporophore of liverworts and mosses.

Sexual cell–Ovum or sperm.

Shoot–Stem of a vascular plant derived from the plumule.

Sieve tubes–Phloem vessels, long slender structure consisting of elongate cells placed end to end.

Siphonostele–The hollow vascular cylinder of a stem, which may contain pith.

Solenostele–A stage after the siphonstele in fern stem development.

Sori–Plural of sorus.

Sorophore–Stalk bearing sori, *e.g., Marsilea.*

Sorus–A collection of small stalked sporangia on under surface on fern pinnule.

Spermatozoid–An antheozoid; a free-swimming male gamete.

Spike–Axis bearing sporophylls, *e.g., Selaginella.*

Sporangia–Plural of sporangium.

Spore–A highly specialized reproductive cell of plants.

Spore-mother-cells–Sixteen cells produced by repeated division of an archesporium, each in turn dividing into four spores.

Sporocarp–Special spore bearing structure of Marsileales.

Sporogenesis–Spore formation.

Sporogenous–Spore producing.

Sporophyll–A sporangium-bearing leaf.

Sporophyte–The diploid ($2x$) spore-producing phase in alternation of plant generations.

Spur–Dwarf shoot, *e.g., Pinus.*

Stalk-cell–The barren cell of two into which the antheridial cell of gymnosperms divides.

Stele–A bulky strand or cylinder of vascular tissue contained in stem and root of plants.

Stomium–Group of thin-walled cells in fern sporangium where it ruptures on maturity.

Suspensor–A modified structure developing in the embryo development of *Selaginella* and *Pinus.*

Syngamy–Sexual reproduction; fertilization.

Tapetum–Special nutritive layer investing sporogenous tissue of sporangium.

Terminal–Situated at the end.

Tetrad–A group of four spores formed by meiotic division of spore mother cells.

Thallus–A combination of cells presenting no differentiation of leaf and stem.

Theca–A spore-case.

Trabeculae–Plates of sterile cells extend across cavities of pteridophytes.

Tracheid–One of the cells with spiral thickening or bordered pits, conducting water and solutes, and forming woody tissue.

Tracheidal cells–Pericycle cells resembling tracheids.

Vallecular canal–One of canals in cortical tissue of stem of horse-tails (*Equisetum*).

Vascular bundle–A group of special cells consisting of two parts, xylem and phloem.

Vertical–Standing upright; lengthwise in direction of axis.

Villose–Pubescent.

Wing–Broad membranous expansion of pollen grain of *Pinus.*

Xylem–Lignified portion of vascular bundle.

Zone–An area characterized by similar flora.

Zygote–Cell formation by union of two gametes or reproductive cells.

QUESTION BANK

Long Answer Type

1. Compare the male gametophytes of *Cycas*, *Pinus* and *Gnetum*.
2. Compare the male gametophytes of *Cycas*, *Pinus*, *Gnetum* with that of Angiosperm.
3. Enumerate the stages in development of the male gametophyte of *Gnetum*.
4. Compare the developmental stages in the female gametophyte of *Gnetum* with that of an Angiosperm.
5. Enumerate the advanced features seen in the male flower of *Gnetum* when compared with other gymnosperms.
6. Enumerate the angiosperm-like features of *Gnetum*. Give reasons for and against considering this genus as ancestral to the angiosperms.
7. Enumerate the stages in the development of the male gametophyte of *Pinus*.
8. Explain the structure of male cone of *Pinus* with the help of a labelled diagram.
9. Enumerate the stages in the development of the male gametophyte of *Cycas*.
10. Compare the secondary thickening in the stems of *Cycas* and a Dicot.
11. How will you distinguish a *Cycas* plant from a palm? Draw a well labelled diagram to illustrate the structure of the ovule of *Cycas*. (Kanpur, 1981).
12. Compare the male gametophyte of *Cycas* with that of *Pinus*. (Kanpur, 1981).
13. Write a comparative account of the pollen bearing structures of *Cycas* and *Pinus*. (Kanpur, 1982).
14. Describe some important processes of fossilization in nature. (Kanpur, 1983).
15. Write a comparative account of the ovule bearing structures of *Cycas* and *Pinus*. (Kanpur 1983, 1985).
16. Give an illustrated account of the anatomy of *Ephedra* stem and comment upon the features of special interest found in the resin. (Kanpur, 1983).
17. Compare the internal structure of coralloid root of *Cycas* with that of its normal root. (Kanpur, 1983).
18. Give an account of female gametophyte of *Ephedra*. (Kanpur, 1983)
19. Give a comparative account of the male and female gametophytes of *Pinus* and *Cycas*. (Kanpur, 1981)
20. Compare and contrast *Cycas* and *Pinus* on the basis of (a) post-fertilization changes in ovule and (b) young seedlings. (Kanpur, 1982, 1985)
21. Give a comparative account of development of male gametophytes of *Cycas*, *Pinus* and *Ephedra*. (Kanpur, 1983).
22. Compare the female gametophytes of *Cycas* and *Ephedra* with the help of suitable diagrams. (Kanpur, 1984)
23. Illustrate the development of female gametophyte of *Cycas* and compare it with developed female gametophyte of *Pinus*. (Kumaon, 1987).
24. Give an account of female gametophyte of *Cycas*. (Kanpur, 1984)
25. Give a comparative account of the male cone of *Pinus* and *Ephedra*. (Kanpur, 1984).
26. Compare the morphology of megasporophyll of *Pinus* with that of *Cycas*. (Kanpur, 1984, 1986).
27. Write an outline of any recent classification of Gymnosperms. What is the criterion of this classification? (Kanpur, 1985).
28. Describe the economic importance of *Cycas* and *Pinus*. (Kanpur, 1986).
29. Give the distribution of *Cycas* and *Pinus* in India. How would you distinguish a *Cycas* plant from a palm? (Kanpur, 1987)
30. Describe the structure and development of the male gametophyte of *Pinus*. (Kanpur, 1987).
31. What is fossil? Discuss the different types of fossils studied by you. (Kanpur, 1988).
32. Describe in short the primary structure of *Cycas* stem and explain the secondary growth of stem with the help of well labelled diagrams. (Kanpur, 1988).
33. Give the structure and development of male and female gametophytes of *Cycas*. (Rohilkhand, 1979)
34. Write an illustrated account of the male and female cones of *Pinus*. (Rohilkhand, 1979).
35. Describe the structure of male cones of *Cycas* and *Pinus*. (Rohilkhand, 1980, 1981).
36. Discuss the secondary growth in *Cycas* and *Pinus*. (Rohilkhand, 1981).
37. Give an illustrated account of the post-fertilization changes which occur in the ovule of *Pinus*. (Rohilkhand, 1982)
38. Describe the process of pollination and fertilization in *Cycas*. (Rohilkhand, 1983).
39. Describe the female cone of *Pinus*, and discuss the morphological nature of ovuliferous scale. (Rohilkhand, 1984).

40. Give an illustrated account of structure of ovule of *Cycas* at the time of pollination. (Rohilkhand, 1985)

41. Give a general account of different processes of fossilization in plant kingdom. Write full name of the world fame institute of palaeobotany situated in Uttar Pradesh. (Rohilkhand, 1986).

Diagrammatic Type

1. Illustrate the internal structure of the normal roots of *Cycas* and *Pinus* with the help of labelled diagrams only. (Kanpur, 1981).

2. Draw labelled diagrams only to show the structure of the following: (i) T.S. dwarf shoot of *Pinus*, (ii) T.S. stem of *Cycas*, (iii) L.S. ovule of *Ephedra*. (Kanpur, 1981).

3. With the help of labelled diagrams only, depict The life-history of *Ephedra*. (Kanpur, 1981, 1986, 1988).

4. Draw well-labelled diagrams of the following: (a) T.S. of *Cycas* petiole, (b) R.L.S. of *Pinus* stem, (Kanpur, 1982).

5. Draw well-labelled diagrams of the following: (a) T.S. *Pinus* stem, (b) T.S. *Pinu* needle. (Kanpur, 1983).

6. Draw well-labelled diagrams of the following: (a) T.S. of *Pinus* stem, (b) T.L.S. of *Pinus* stem, (c) R.L.S. of *Pinus* stem. (Kanpur, 1984).

7. Draw well-labelled diagrams of the following: (a) T.S. of *Cycas* rachis, (b) T.S. of *Pinus* stem. (Kanpur, 1985)

8. Draw well-labelled diagram of the following: (i) T.S. of *Pinus* dwarf shoot, (ii) T.S. of *Ephedra* stem, (iii) T.S. of *Cycas* leaflet. (Kanpur, 1986).

9. Draw well-labelled diagrams of the following: (i) T.S. rachis of *Cycas*, (ii) L.S. male cone of *Pinus*, (iii) L.S. of *Ephedra* ovule, (iv) R.L.S. and T.L.S. of *Pinus* wood. (Kanpur, 1987).

10. With the help of labelled diagrams only depict the life-cycle of *Cycas*. (Kanpur, 1987).

11. Draw well-labelled diagrams of the following: (a) T.S. of coralloid root of *Cycas*, (b) T.S. *Pinus* needle, (c) T.S. of *Ephedra* stem. (Kanpur, 1988).

12. Give the diagrammatic representation of the life-cycle of *Cycas* and *Pinus*. (Rohilkhand, 1979).

13. Give the diagrams of any two of the following:
(i) radial and tangential longitudinal sections of *Pinus* stem, (ii) T.S. of *Cycas* petiole, (iii) T.S. of *Pinus* needle. (Rohilkhand, 1980).

14. Draw labelled sketches of the following: (i). T.S. rachis *Cycas*, (ii) T.S. coralloid root of *Cycas*. (Rohilkhand, 1982).

15. Draw labelled cellular diagrams of any four of the following. No description is required. (1) *Cycas* leaflet, (2) *Cycas* rachis, (3) coralloid root of *Cycas*, (4) *Pinus* dwarf shoot, (5) R.L.S. *Pinus* stem. (Kumaon, 1983).

16. Draw labelled diagrams of any five of the following. No description is required. (i) coralloid root of *Cycas*, (ii) L.S. male strobilus of *Cycas*, (iii) *Cycas* megasporophyll, (iv) T.S. *Pinus* stem, (v) T.L.S. *Pinus* stem, (vi) *Pinus* dwarf shoot. (Kumaon, 1984).

Short Answer Type

1. Distinguish between microsporophylls of *Cycas* and *Pinus*.
2. Distinguish between transverse sections of coralloid root of *Cycas* and mycorrhizal root of *Pinus*.
3. Compare the nature of embryo of *Pinus* with that of *Cycas*.
4. Write explanatory note on albuminous cells.
5. What is a manoxylic type of stem, and how does it differ from the pycnoxylic type?
6. On the basis of morphology and anatomy, how would you distinguish the components in the pollen grains of *Cycas* and *Ephedra*.
7. Write a short note on transfusion tissue.
8. List the xeromorphic characters of gymnosperms.
9. Write explanatory note on mycorrhiza.
10. List the characters common to cycads and ferns.
11. What is characteristic of nodal anatomy of *Ephedra*.
12. The mature pine cone is a closed structure. Why is *Pinus* then placed among the gymnosperms?
13. Describe the *Lyginopteris*.
14. Write note on ovule of *Taxus*.
15. Write notes on centripetal and centrifugal xylem.
16. Write brief note on phloem in gymnosperms.
17. How does the behaviour of the pollen tube differ in *Cycas*, *Pinus* and *Gnetum*.

18. How would you distinguish the wood of *Gnetum* from that of *Pinus* in T.L.S.

19. Write explanatory note on bars of sanio.

20. Would you consider the motility of sperms in gymnosperms an advanced or a primitive character? Explain.

21. Write explanatory note on gymnosperms of the Himalayas.

22. Write brief account on distribution of gymnosperms (studied by you) in India.

23. Write brief account of the development of the male gametophyte in *Pinus*?

24. Write note on the L.S. of *Cycas* ovule at the time of fertilization.

25. Write brief account of the male cones of *Ephedra*.

26. Write brief account of the stem of *Cycas*.

27. What are the characteristic features of Bennettitalean flowers.

28. How can the xylem of *Gnetum* be distinguished from that of *Pinus*.

29. With the help of labelled diagrams, compare the internal structure of the leaflet of *Cycas* with that of the needle of *Pinus*.

30. Explain how the development of endosperm in angiosperms differs from that of gymnosperms.

31. Compare the male cone of *Pinus* with that of *Gnetum*.

32. Explain how the female gametophyte of *Gnetum* shows advanced features than that of *Cycas* and *Pinus*.

33. Compare the ovules of *Cycas*, *Pinus* and *Gnetum* with reference to internal structure.

34. Mention any three common characters shared by all gymnosperms.

35. Write short notes an any four of the following: (i) coralloid roots of *Cycas*, (ii) ovuliferous scale, (iii) periderm formation in *Cycas*, (iv) vessels in *Ephedra*, (v) polyembryony in *Pinus*, (vi) dwarf shoots of *Pinus*. (Kumaon, 1981, 1985)

36. (a) What are the remarkable features in the embryogeny of gymnosperms unknown either to pteridophytes or gymnosperms. (b) Explain simple and cleavage polymbryony. (c) Describe the integument of a typical gymnospermous seed. (d) What is free-nuclear division? When does it occur in the life-cycle of gymnospermous plants? (Kumaon, 1982).

37. Write short notes on any four of the following: (i) coralloid root, (ii) ovuliferous scale, (iii) pollen tubes, (iv) economic importance of gymnosperms, (v) transfusion tissue, (vi) leaves of *Ephedra*. (Kumaon, 1986).

38. Write short notes on any four of the following: (i) dwarf shoot, (ii) pollen chamber, (iii) ovuliferous scale, (iv) megasporophyll of *Cycas*, (v) girdle trace. (Rohilkhand, 1980).

39. Write short notes on the following : (i) centripetal xylem, (ii) coralloid root, (iii) ovuliferous scale, (iv) bordered pits. (Rohilkhand, 1981).

40. Write short notes on: (i) resin canal, (ii) bars of Sanio. (iii) pycnoxylic wood. (Rohilkhand, 1985).

41. Write a brief account on: (i) palaeobotany, (ii) polyembryony. (Rohilkhand, 1985).

42. Explain the following: (a) *Cycas* is living fossil, (b) Plants of *Pinus*, in India, are found in plenty on the Himalayan hills only. (Rohilkhand, 1986).

43. Write brief account on: (a) Polyebryony, (b) fossilization. (Rohilkhand, 1987).

44. Write notes on the following: (i) economic importance of *Cycas*, (ii) bordered pits in *Pinus*. (Kanpur, 1981).

45. Write short notes on: (i) economic importance of gymnosperms, (ii) polyembryony in *Pinus*. (Kanpur, 1983).

46. Describe the economic importance of *Cycas* and *Pinus*. (Kanpur, 1986).

47. Write short notes on: (a) ovuliferous scale of *Pinus*, (b) ciliated sperms, (c) resin canal. (Kanpur, 1987)

48. Answer the following: (a) explain why all *Cycas* plants do not produce seeds, (b) are the coralloid roots of *Cycas* useful for aeration or for nitrification, (c) mention two common characters shared by all gymnosperms, (d) *Pinus* wood has more economic value than *Cycas* wood. (Kanpur, 1987).

49. Give the differences between the following: (a) pollen grains of *Cycas* and *Pinus*, (b) ovule of *Cycas* and *Pinus*, (c) Root of *Cycas* and *Pinus*. (Kanpur, 1988).

50. How would you differentiate the following: (i) root of *Cycas* and *Pinus*, (ii) microsporophyll of *Cycas* and *Pinus*, (iii) ovule of *Cycas* and *Pinus*. (Kanpur, 1989).

Multiple Choice Type

1. The characteristic feature of the stomata in gymnosperms is: (i) that they are deeply sunken in pits, (ii) that there are subsidiary cells, (iii) that the walls of the guard cells are completely free of lignin. (iv) that the walls of the guard cells and subsidiary cells are partly free of lignin.

2. Stalked ovules are seen in: (i) Bennettitales, (ii) Cycadales, (iii) Cycadofilicales, (iv) Coniferales.

3. *Gnetum*, when not in flowering can be easily mistaken for a: (i) tree fern, (ii) monocot plant, (iii) dicot plant. (iv) none of the above.

4. The ovuliferous scale of *Pinus* is a part of: (i) megasporophyll, (ii) microsporophyll, (iii) ovule, (iv) dwarf shoot.

5. The wing on the seed of *Pinus* is developed from: (i) integument, (ii) nucellus, (iii) ovuliferous scale, (iv) carpellary scale.

6. Spermatozoids large enough to be seen by naked eye are produced in: (i) *Cycas revoluta*, (ii) *Pinus roxburghii*, (iii) *Gnetum ula*, (iv) *Pinus wallichiana*.

7. The anatomical features of the *Cycas* leaflet indicate that *Cycas* is a : (i) xerophyte, (ii) mesophyte, (iii) gymnosperm, (iv) conifer.

8. What could be the best function attributable to the transfusion tissue seen in *Cycas* leaflets: (i) mechanical, (ii) photosynthetic, (iii) conduction, (iv) storage.

9. Which of the following is not the characteristic feature of *Cycas* : (i) circinate vernation of foliage leaves, (ii) presence of arm parenchyma, (iii) presence of motile sperms, (iv) absence of vessels in the xylem.

10. The mesophyll tissue in *Pinus* is called : (i) armed parenchyma, (ii) Chlorenchyma, (iii) transfusion tissue, (iv) spongy parenchyma.

11. The wood of *Cycas* consists of : (i) tracheids only, (ii) vessels only, (iii) equal number of tracheids and vessels, (iv) more of tracheids and less of vessels. (C.P.M.T. 1978).

12. Gymnosperm differs from the angiosperm: (i) having seeds, (ii) in being smaller in size, (iii) in showing heterospory, (iv) in having naked ovules. (C.P.M.T.1978).

13. *Cycas* differs from *Pteris* in having: (i) tracheids, (ii) motile sperms, (iii) pollen tubes, (iv) archegonia. (C.P.M.T.1980).

14. In *Cycas*, the pollen grains are shed at the: (i) 4-celled stage, (ii) 3-celled stage, (iii) 2-celled stage, (iv) 1-celled stage. (C.P.M.T.1981).

15. In which of the following sexual reproduction is possible without water: (i) *Ulothrix*, (ii) *Funaria*, (iii) *Dryopteris*, (iv) *Cycas*. (C.P.M.T.1982).

16. The sperms of *Cycas* are: (i) very large and have numerous spirally arranged cilia, (ii) very small and have no flagella, (iii) large and nonciliate, (iv) small and nonmotile. (C.P.M.T.1982).

17. In *Cycas*: (i) male strobilus (cone) and megasporophylls occur on separate plants, (ii) male strobilus (cone) and megasporophylls occur on same plant, (iii) The same cone contains both microsporangia and ovules, (iv) The same sporophyll bears microsporangia and ovules (C.P.M.T.1982).

18. Which of the following conditions found in *Cycas* is also found in *Pteris*: (i) naked ovules, (ii) circinate vernation of leaves, (iii) presence of coralloid roots, (iv) presence of secondary growth. (C.P.M.T. 1983).

19. Palaeobotany is a branch of botany which deals with: (i) living plants, (ii) fossil plants, (iii) none of plants, (iv) all of the above. (C.P.M.T. 1984-A).

20. Fruits are not formed in gymnosperms because (i) they are seedless plants, (ii) they are not pollinated, (iii) they have no ovary, (iv) the process of fertilization does not take place in them. (C.P.M.T. 1984-B).

21. Which of the following is a living fossil: (i) *Pinus*, (ii) *Cycas*, (iii) *Metasequoia*, (iv) *Ephedra*. (C.P.M.T. 1984).

22. In which of the following features do gymnosperms resemble angiosperms: (i) presence of vessels in wood, (ii) mode of fertilization, (iii) nature ot endosperm, (iv) presence of ovule. (C.P.M.T. 1984).

23. A blue-green alga lives in : (i) *Riccia* thallus, (ii) *Marchantia* thallus, (iii) *Cycas* roots, (iv) *Pisum* roots. (C.P.M.T. 1984-B).

24. Which of the following statements is not true for gymnosperms: (i) leaves are compound, (ii) naked seeds are formed, (iii) xylem is made up of tracheids, (iv) xylem is made of vessels. (C.P.M.T. 1984-B).

25. In the plant *Cycas* the male cone lacks: (i) microspores, (ii) microsporophyll, (iii) microsporangium (iv) nucellus. (C.M.P.T 1984-B).

26. The trachieds of *Cycas* are: (i) with uniseriate bordered pits, (ii) with biseriate bordered pits, (iii) with multiseriate bordered pits, (iv) without bordered pits. (C.P.M.T. 1985).

27. The smallest gymnosperm known is (i) *Cycas revoluta*, (ii) *Zamia pygmia*, (iii) *Cupressus* sp, (iv) *Pinus roxburghii*. (C.P.M.T. 1986).

28. Gymnosperms differ from angiosperms in : (i) presence of naked seeds and absence of vessels, (ii) presence of seeds in vessels, (iii) having abortive seeds, (iv) having tracheids. (C.M.P.T. 1986).

29. Coralloid roots are found in: (i) *Cycas*, (ii) *Dryopteris*, (iii) *Pinus*, (iv) *Lycopodium*. (C.P.M.T. 1987).

30. Xylem of gymnosperm lacks: (i) tracheids, (ii) xylem parenchyma, (iii) xylem fibres, (iv) vessels. (C.M.P.T. 1987).

31. Endosperm in gymnosperm is formed: (i) at the time of fertilization, (ii) before fertilization, (iii) after fertilization, (iv) along with development of embryo. (C.P.M.T. 1987).

32. Red wood tree is: (i) *Cedrus*, (ii) *Pinus*, (iii) *Sequoia*, (iv) *Metasequoia*. (C.P.M.T. 1987).

33. The resin duct of a gymnospermous stem is an example of: (i) intercellular space, (ii) schizogenous cavity, (iii) lysigenous cavity, (iv) big vacuole. (C.M.P.T. 1989).

34. The pollination in *Cycas* takes place through: (i) ant, (ii) water, (iii) bat, (iv) wind. (C.P.M.T. 1990).

35. *Cycas* ovule is: (i) anatropous, (ii) orthrotropous, (iii) hemianatropous, (iv) campylotropous.

Answers

1(i), 2(iii), 3(iii), 4(i), 5(iii), 6(i), 7(i), 8(iii), 9(ii), 10(i), 11(i), 12(iv), 13(iii), 14(ii), 15(iv), 16(i), 17(i), 18(ii), 19(ii), 20(iii), 21(ii), 22(iv), 23(iii), 24(iv), 25(iv), 26(iii), 27(ii), 28(i), 29(i), 30(iv), 31(ii), 32(iii), 33(ii), 34(iv), 35(ii).

INDEX

ANGIOSPERMS

1

Angiosperms: Introduction

Characteristic Features of Angiosperms

Angiosperms (Gr., *angios* = vessel or receptacle; *sperma* = seed).

1. The sporophyte which is the dominant plant in the life-cycle is differentiated into roots, stem and leaves.

2. The highest degree of perfection of the vascular system with true vessels in the xylem and companion cells in the phloem.

3. The organisation of the **microsporophylls** (stamens) and **megasporophylls** (carpels) into a structure called the **flower**, which is typical only of the angiosperms.

4. The presence of four **microsporangia** (pollen sacs) per microsporophyll (stamen).

5. The **ovules** are always enclosed in an ovary which is the basal region of the megasporophyll.

6. Production of two kinds of spores, **microspores** (pollen grains) and **megaspores.** Angiosperms thus are **heterosporous.**

7. Presence of single functional megaspore which is permanently retained within the **nucellus** or **megasporangium.**

8. Adaptation of flower to insect pollination.

9. Pollination consists in the transference of pollen grains from anther to stigma.

10. Spore dimorphisim having resulted in the production of gametophytes, male and female.

11. Extreme reduction in size, duration of existence and complexity of the structure of the gametophytes which are entirely parasitic.

12. The male **gametophyte** has reached the limits of reduction. It consists only of the pollen grain and the pollen tube contains the tube nucleus and two male gametes or nuclei. The male cells (gametes) are non-ciliated.

13. The female **gametophyte** lacks any extensive development of vegetative tissue. It consists of three egg apparatus cells, three antipodal cells and two polar nuclei in the centre of the emryo sac.

14. The non-motile male cells or nuclei are carried bodily to the neighbourhood of egg apparatus by the pollen tube.

15. The seed or seeds remain enclosed in the ripened ovary called the fruit.

16. The phenomenon of **double fertilization** or **triple fusion** is the characteristic of the angiosperms.

17. The endosperm develops after fertilization. It is triploid.

18. The angiosperms are completely adapted to life on land.

Evolution of Angiosperms

The evolutionary high 'status' of angiosperms over other groups of the plant kingdom is due to their :

(*i*) diversified habit and vegetative forms;

(*ii*) higher degree of perfection of the vascular system; the xylem in addition to the tracheids, contains wood vessels, and the phloem possesses companion cells;

(*iii*) successful invasion of all habitats;

(*iv*) adaptation of flower to insect pollination;

(*v*) bisexual flower, the bisexuality ensures self-pollination if cross-pollination fails, hence the seeds are formed in any case;

(*vi*) development of ovules within the ovary which ensures proper protection to the developing ovules and seeds;

(*vii*) efficient and effective dispersal by insects, birds, other animals, wind, water and by several other methods;

(*viii*) efficient and varied means of vegetative propagation, which result in rapid multiplication;

(*ix*) their utmost economic importance, and

(*x*) besides the above listed reasons, there must be some hereditary causes such as better equipment of gene potential and useful gene mutations, which enable them to encounter variations in temperature and other environmental changes and led to their conquest over other plant groups.

CONCEPT OF SYSTEMATICS

What is Systematic Botany? Systematic botany is the science which deals with the classification and naming of the plants. The science of classifying the plants is said to be **plant taxonomy** and lays emphasis upon phylogenetic relationships. The naming of the plants is known as **nomenclature** and provides each plant with a name. This way, the systematic botany consists of classifying and naming of the plants. The important task included in the field of systematic botany is the collection of the plant specimens from all over the world and building the great herbaria. The plants are being classified chiefly on the basis of the study of their comparative morphology. In modern days the study of systematic botany needs a good background in general botany, cytology, genetics, ecology, plant geography and palaeobotany; otherwise, the origin of the plants cannot be traced out.

Nomenclature makes the important part of systematic botany. It deals with names which may or may not indicate relationships. The science of plant taxonomy classify the plants on the basis of similarities and differences, which are now known as "phylogenetic relationships". Without naming the plants, they cannot be classified and therefore, nomenclature makes a very important part of systematic botany.

Ecology and systematic botany are interrelated to each other. Certain species are definitely hydrophytic, xerophytic, mesophytic or halophytic. The soil characters are being indicated by the vegetation growing upon. The character of the soil may be indicated by its vegetation.

Scope of Systematic Botany–As already mentioned, the study of systematic botany consists of two important steps–(1) plant taxonomy–which establishes the phylogenetic relationships that exist naturally between many groups of plants, (2) nomenclature, *i.e.*, giving of names to all kinds of plants. There are millions of plants all over the world, and it is very clear, that no two individuals are **exactly** alike. Most of the individuals are so different from each other that hardly they show any bond of relationship. On the basis of the comparative study of their morphology, the taxonomists have arranged the individual plants in a fairly orderly system. Thousands of specimens have been provided with their names and descriptions. The group relationships, distribution, properties, etc., of the plants have been studied. This way, systematic botany prepares foundation for all sciences concerning plants.

The systematic botany is of great value in forestry. Every type of tree of the forest must be named and classified. At the same time, the characteristic features, distribution and abundance of the plants must be learned.

The study of systematic botany is equally essential in agriculture and horticulture. To get resistant varieties, etc., the crossing is needed, which is furnished by foreign seeds and plants. The foreign seeds and plants are introduced by the men who have sufficient knowledge of systematic botany.

The work of soil conservation needs the good study of systematic botany. The ecologists who are well trained in systematic botany choose the native grasses and plants which act as sand soil binders.

To study the plant ecology, the knowledge of systematic botany becomes essential. The plant ecologist must be knowing the names of the plants and their relationships to habitat and environment. The character of the soil of a land area may easily be judged by the vegetation growing there.

The study of systematic botany is of great use in tracing the origin of plants. The fossils, scattered here and there in fragments are found to be very useful in tracing out the characters of the ancestral forms of the plants. The study of these fossils requires the sound knowledge of systematic botany.

Objectives of Systematic Botany–The study of systematic botany fulfils four main objectives. They are as follows :-

1. The first objective of systematic botany is to know the various kinds of plants on the surface of earth with their names, affinities, geographical distribution, habit characteristics and their economic importance.

2. The second objective of this science is to have a reference system for plants, where the scientists can work with named entities such as species, genus and family.

3. The third objective is to demonstrate the manifold diversities of plant kingdom and their relation to evolution of plants. A systematic reconstruction of plant kingdom can be made only after the complete knowledge of the individual plants. After making an inventory of the components of plant kingdom, the different facts of evolutionary knowledge with an accurate phylogenetic scheme can be obtained.

4. The fourth objective is to ascertain nomenclature. To every plant a binomial name is given, *e.g., Solanum tuberosum* Linn. The first name refers to its genus, the second to its species and Linn., or L. for Linnaeus, who first observed and reported the plant.

ANGIOSPERMS OR FLOWERING PLANTS

Angiosperms or flowering plants form the largest group of plant kingdom, including about 300 families (411 families, Hutchinson), 8,000 genera and 300,000 species. They are considered to be highest evolved plants on the surface of the earth. From Cretaceous age, the angiosperms eclipsed all other vegetation and now they are dominant. They are found almost everywhere in each possible type of habitat and climate. They occur in deep lakes, deserts, in beds of seas and even on high peaks of mountains. The species of *Opuntia* (Cactaceae) can survive without water in acute desert conditions, whereas on the other hand the species of *Hydrilla* (aquatic plant) are extremely sensitive to drought conditions. Some species are found on rocks, some in waterfalls and also some are marine. The species of *Rhizophora*, popularly known as 'mangrove vegetation' are found near the water of the sea. The epiphytes, parasites, saprophytes, symbionts and even insectivorous plants are also not uncommon.

They may be annual, biennial or perennial herbs, shrubs, trees, climbers, twiners and lianes. On one hand the angiosperms may be as minute in size as a pin head, *e.g.*, *Wolffia microscopica*, on the other extremity like eucleptiles of Australia may reach upto 300 feet in height.

ALTERNATION OF GENERATIONS: LIFE CYCLE OF ANGIOSPERMS

In the life-cycle of angiosperms, there is alternation of nutritionally independent and more complex sporophyte with the inconspicuous, reduced and parasitic gametophytes. The sporophyte, which may be a **herb**, **shrub** or a **tree** is differentiated into roots, stem and leaves each with a vascular tissue with the highest degree of perfection. In addition, it bears floral leaves or

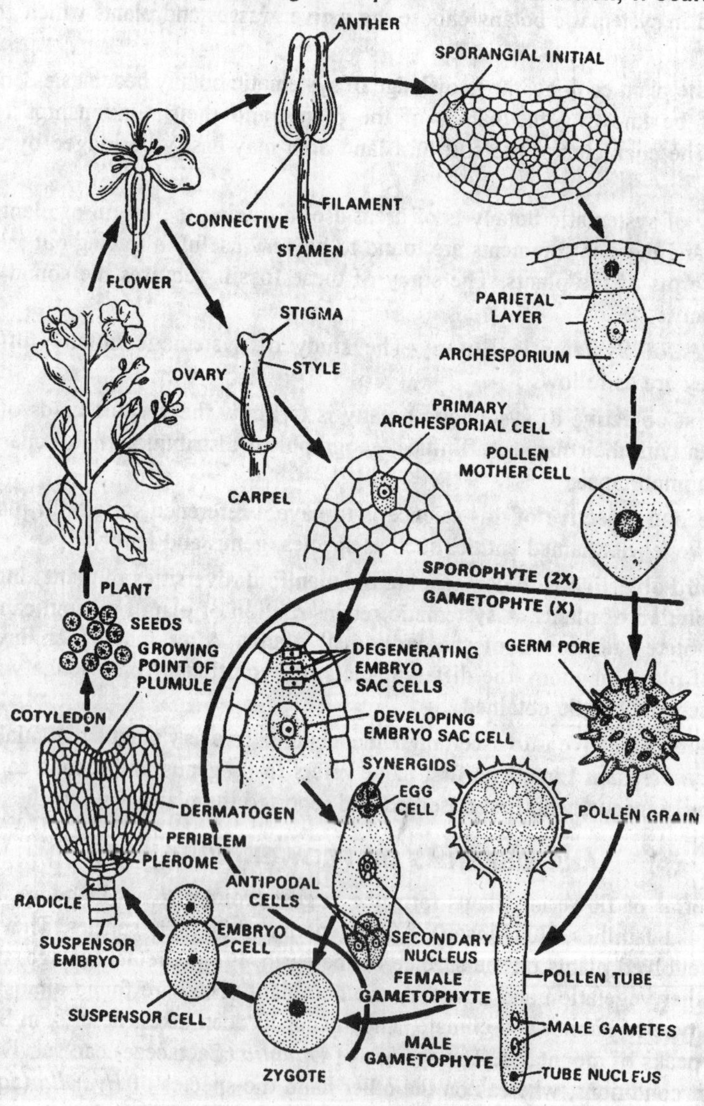

Fig. 1.1. Life-history of a typical angiosperm

sporophylls organised into a structre called the **flower.** The sporophylls are of two kinds : **microsporophylls** and **megasporophylls.** Each microsporophyll bears four **microsporangia.** The latter contains **microspores** which are differentiated by meiosis from the diploid **microspore mother cells.** The megasporophyll in a flower part, which is called the **ovary,** contains one or more **ovules.** The ovule consists of one or two integuments enclosing the **nucellus** or **megasporangium.** Towards the micropylar end of the megasporangium is differentiated, by meiosis, a single functional megaspore which is haploid. The **microspores** and **megaspores** are the first structures of the gametophyte generation. The **sporophyte generation** in the angiosperms, thus consists of the sporophyte plant, flowers, microsporophylls and megasporophylls, microsporangia, megasporangia, microspore mother cells and the megaspore mother cells.

With meiosis the sporophyte phase switches on the gametophyte phase. The first structures of the gametophyte phase are the **microspores** and the **megaspores.** On germination, they produce the alternate structures in the life-cycle which are the male and the female gametophytes and not the sporophyte plant. The male and the female gametophytes are extremely reduced and parasitic. The female gametophyte consists only of six cells and two nuclei. Three cells at the micropylar end form the **egg apparatus.** The other three at the chalazal end form the **antipodal group.** There are two **polar nuclei** in the centre. The male gametophyte consists of a pollen tube containing the *tube nucleus* and two **male nuclei.** The eggs and the male nuclei represent the gametes which are the last structures of the gametophyte phase. With the fusion of one of the male nuclei with the egg, the gametophyte phase ends. **Fertilization** thus is the second critical point in the life cycle. With it the gametophyte generation switches on to the sporophyte generation. The fertilized egg or **oospore** is the first structure of the next sporophyte. By segmentation it produces the embryo (*i.e.*, the baby plant in the seed); the ovule to the seed and the ovary as a whole to the fruit. The embryo lies dormant in the seed and the latter lies embedded in the fruit. The seed and the fruit give adequate protection to the embryo, store up food material for it, and are often well adapted for dispersal. Sooner or later as the seed germinates the embryo grows into a **seedling** which gradually grows into a mature plant.

Taxa or Taxon

Taxa or Taxon – Taxonomic group of any rank or unit, *viz.*, family, class, order, genus, species, etc., is called *taxon* (plural : *taxa*). For example, the genus *Solanum* (the name of the genus begins always with a capital letter) consists of many species (the name of the species begins always with a small letter) such as *tuberosum*, *melongena*, *nigrum*, etc., but certain common characters of the genus *Solanum* can separate this genus from all other genera of the family Solanaceae.

Higher to genus is the family, where one or many genera are correlated between greater number of morphological characters are grouped together. The name of the family ends in *aceae*, *e.g.*, Brassicaceae, Poaceae, etc.

The family represents a more natural *taxon* or unit than any of the higher categories, e.g., Brassicaceae, Palmaceae, Solanaceae, Rosaceae are identified as natural *taxa* by certain characters peculiar to them. The angiospermic families may be separated from one another by floral characters, inflorescence, type, nature of perianth, ovary position, number of carpels, placentation, etc.

The families which are allied to one another are grouped into an order which makes a natural taxon. For example, the order Ranales comprises of Ranunculaceae, Nymphaeceae, etc. Here the families are related to each other in general floral constitution.

What is a Species

According to new definition the species are biological units that have evolved from a series of often identifiable ancestors, by which the smaller taxa of plants (genera, species, and their subdivisions) may be circumscribed by a combination of genetical, morphological and ecological criteria. According to this view, a species is considered as an objective definitive unit.

TAXONOMY AND ITS SIGNIFICANCE

The study of taxonomy has among its objectives the learning of the kinds of plants on earth and their names, of their distinctions and their affinities, their distributions and characteristics of habitats, and the correlation of these facets of knowledge with pertinent scientific data. A secondary objective of taxonomy is the assemblage of knowledge gained. Floras are published to account for the plants of a given area.

All the products of taxonomic research add to the resources available to the scientist. They are essential to any study of the natural resources of an area, to studies of land potentials, to evaluation of resources of raw materials possible suited to man's needs (for example, forest products, medicines, food, agricultural crops, ornamentals and industry). A third important and scientific objective is the demonstration of the tremendous diversity of the plants and their relation to man's understanding of evolution.

Interrelationships with allied sciences. Taxonomy is dependent on many other sciences and they in turn are equally dependent on it. One who studies taxonomy must have a knowledge of morphology, embryology, floral anatomy, ontogenetical development and teratological variations of the plants with which he works. Modern taxonomists place considerable value on the importance of cytogenetic findings as criteria in delimiting the species and its elements. In addition to an appreciation and understanding of the contributory value of morphological, anatomical and cytogenetical findings, modern taxonomic studies reflect the significance of distributional patterns and of more detailed data of the extent of normal variation and its causes. All these viewpoints demonstrate the increasing dependence of taxonomy on the findings of related sciences; the product of modern taxonomic research is rapidly becoming one of synthesis rather than of individual conclusions.

Problems in taxonomy. During the eighteenth and early nineteenth centuries the study of taxonomy dominated the field of botanical activity. Today the taxonomist is interested in the problems associated with the distribution of the plants. Knowledge of plant distribution is relevant to the determination of geographic areas of origins of species, of genera, and often of families. These studies in distribution and geography bring taxonomy into the field of phytogeography, the inquiry into why a group occupies that area, how long it has been there, how rapidly it is migrating, and what evolutionary trends it is showing. Studies with this wider viewpoint represent a synthesis of ecologic, genetic and taxonomic aspects leading to a better understanding of a series of common problems.

ORIGIN OF ANGIOSPERMS

The angiosperms appeared suddenly in Cretaceous age about 65 million years back. **Charles Darwin** described this sudden appearance of angiosperms in lower or upper Cretaceous as an '*abominable mystery*'. When angiosperms appeared for the first time in lower or upper Cretaceous, they were full fledged like the trees and the herbs of today. In support of this view *Prof. Knowlton* advocates in his '*Plant of the past*', "*from the time of their appearance they did not progress at all due to their full-fledged appearance in the Cretaceous*".

The fossil records of the angiosperms also support their appearance full-fledged in lower or upper Cretaceous. The fossils of that age are so characteristic and modern in appearance that most of them can be referred unmistakably to living families, general and even to some species. *Prof. Knowlton* stated in his book, *'Plant of the past' "if a student of present day trees and shrubs, could have wandered over the hills and vales in those days, he would have found himself quite at home among the trees and shrubs growing there"*. The forms of cycads and conifers, which long dominated the universe were already pushed background and the earth had become infact the earth of flowering plants. Charles Darwin has called this sudden appearance of angiosperms as an *"abominable mystery"*.

However, some workers do not agree with the doctrine of 'abominable mystery'. According to **H.H. Thomas** (1936), the angiosperms of the past replaced many of older gymnosperms in asturine and marshy waters. **Graud Eury** (1906) believes that the angiosperms came into existence through mutation. **Guppy** (1919) however, supported the view of mutation. **Prof. Bertrand** is of opinion that all the great groups of vascular plants (Pteridophyta, Gymnospermae and Angiosperms) not only arose quite independently of each other but also they originated simultaneously as far back in the Archian period (2000 million years old-oldest).

There is a very considerable but scattered literature on the origin and phylogeny of angiosperms. The palaeobotanical evidence shows that there seems three possibilities as regards the origin of angiosperms. These possibilities are–1. that the angiosperms are monophyletic in their origin but have had a very much longer history than at present known, perhaps stretching back into Palaeozoic times and with a whole series of missing links; 2. that the angiosperms are monophyletic but that the first and at present unknown group diverged quickly in terms of geological time, into a considerable number of different groups; 3. that the angiosperms are polyphyletic.

According to Campbell, "both comparative morphology and the geological record indicate that the existing angiosperms represent a number of distinct phyla which cannot be traced back to a single ancestral type". This statement shows that he does not believe in monophyletic origin of angiosperms.

According to Thomas, the evolutionary tendencies detected in the three groups, *i.e.*, Caytoniales, Bennettitales and Pteriodsperms furnish reasonable grounds for the idea that the angiosperms were derived from various pteridosperms early in the Mesozoic period.

Parkin also argues for the monophyletic origin of the angiosperms.

In conclusion we can say, that the history of the angiosperms is still almost as great a mystery as it was in the time of Darwin. We do not know, when, where or from what the presumably most recent and now dominant large group of existing terrestrial plants arose.

Lotsy says, that hybridization is the key to evolution of angiosperms. This view has been supported by Anderson. He has suggested on the basis of cytological investigations that the angiosperms may have arisen as a result of hybridization between two gymnosperms.

According to Hagerup, the origin of some angiosperms took place from the Coniferales through Gnetales.

According to Eames, Sinnott and Bailey the more primitive angiosperms were arboreal in habit and the herbaceous angiosperms have been evolved from them.

According to Hutchinson, in certain groups, trees and shrubs are probably more primitive than herbs.

Evidently views in this respect are divergent and speculative, the available data being meagre, fragmentary and isolated. Some of the theories proposed from time to time in this connection, are as follows:

1. **Bennettitales-Ranales Theory.** (a) Hallier's view (1906) regarding the origin of angiosperms is that Ranales, (*e.g.*, *Magnolia*) seems to be related to Bennettitales and may have

been derived from *Cycadeoidea*, and that the monocotyledons are an offshoot of the dicotyledons. The elongated floral axis of Ranales with spirally arranged male and female sporophylls and the cone of *Cycadeoidea* are definitely alike. Both the groups were also abundant in the Cretaceous. Ranales is regarded as the earliest stock from which the polycarpic families of dicotyledons have arisen, and also the monocotyledons as an offshoot. But the differences in the anatomical structure of the wood, types of sporophylls, nature and position of ovules, etc., in the two groups, *i.e.*, Bennettitales and Ranales, have led to difficulties in accepting this view. It is more probable that both the groups have evolved from a common ancestry and developed in unrelated parallel lines.

(b) Arber and Parkin (1907), proposed the existence of an imaginary group (Hemiangiospermae) having cycadeoid type of flowers linking the above two groups. From this imaginary group the Ranalian type of flowers might have originated, and from this Ranalian type all other angiosperms had sprung up. They were of the opinion that the earliest monocotyledons were more primitive than the dicotyledons. They also supported the Hallier's view.

(c) Hutchinson (1925) considered the origin of angiosperms as *monophyletic*, and supported the views of Hallier (1906) and Arber and Parkin (1907). He believed in the Bennettitalean origin of angiosperms and stressed on two parallel evolutionary lines for the primitive dicotyledons: a woody (arborescent) line, called *Lignosae*, starting from Magnoliales, and a herbaceous line, called *Herbaceae* starting from Ranales. He further held the view that the monocotyledons were derived from a primitive dicotyledonous order, The Ranales.

2. Coniferae–Amentiferae Theory. Engler and Prantl (1924) rejected the Cycadeoidean origin of angiosperms, as proposed by Hallier (1906) earlier. They held the view that dicotyledons and monocotyledons had arisen independently from a hypothetical group of extinct gymnosperms (allied coniferae) with unisexual strobilus which developed in the Mesozoic. Thus, according to them, the angiosperms had a *polyphyletic* origin, and evolution took place on several parallel lines from the beginning. They considered the monocotyledons to be more primitive than the dicotyledons. The unisexual naked (without perianth) condition of the angiospermic flowers, such as Pandanales (monocotyledons) and Amentiferae (catkin-bearing dicotyledons, *e.g.*, *Casuarina, Salix, Betula*, etc.) was most primitive. But according to the modern classification these orders are regarded more advanced.

3. Gnetales-Casuarinales Theory. Wettstein (1935) held the view that angiosperms of *Casuarina* type evolved from Gnetales, (particularly *Ephedra*), a highly advanced group of gymnosperms, which branched off from the main gymnospermic line. However, there are some angiospermic features in Gnetales but this group has meagre fossil records, and not gone earlier than Tertiary. Fagerland (1947) was of the opinion that both Gnetales and Proangiosperms had a common ancestor and the modern angiosperms evolved from the Proangiosperms in Polyphyletic lines.

4. Caytoniales-Angiosperm Theory. Thomas (1925) suggested that Caytoniales, a small group of angiosperm-like plants, discovered in the Jurassic rocks of Yorkshire (England), might be the ancestor of angiosperms. However, Harris (1932-33) opposed this view. Knowlton (1927) expressed the view in his book *Plants of the Past* that both Caytoniales and angiosperms evolved from the large extinct Palaeozoic group of Pteridosperms. Arnold (1948), expressed the view that Caytoniales were allied to the Pteridosperms rather than to the angiosperms.

5. Pteridosperm-Angiosperm Theory. Andrews (1947) was of opinion that seed ferns or Pteridosperms, an ancient group of the Palaeozoic, might be the starting point for the angiospermic plants.

Origin of Dicotyledons

There are two large groups of angiosperms (1) Dicotyledons and (2) Monocotyledons. The dicots are more important and they are supposed to have originated before the monocotyledons. It is thought that for the first time dicotyledons appeared in the early Mesozoic era, or perhaps the late Palaeozoic. The oldest known fossils of dicots are from the lower Cretaceous rocks.

It is generally believed and agreed that the dicotyledons have been originated from gymnosperms of a type somewhat different from present day forms of gymnosperms. According to some similarities noted in these two groups, the development seems to be parallel, or in some respects convergent development of the two groups.

Origin of Monocotyledons

This is the subject of great controversy. For some time this was thought that the monocotyledons were more primitive than the dicotyledons and probably they have given rise to the dicotyledons. Now, this belief has been totally given up. It is now generally thought that the dicotyledons are more primitive and they have given rise to the monocotyledons. This is also believed, that the monocotyledons were an offshoot of the primitive dicotyledons back in the early Mesozoic era. They are thought to be monophyletic (*i.e.*, of one origin) in their origin. According to another conception the monocotyledons are not monophyletic but polyphyletic in their origin.

PRINCIPLES OF TAXONOMY AND PHYLOGENY OF ANGIOSPERMS

The problem of classifying the angiospermic plants in a systematic way may either be termed as, **'taxonomy of angiosperms'** or **'systematic botany'**. This is a functional science and deals with identification, nomenclature and classification of the plants found all over the world. The angiosperms are widely distributed with so many morphological variations that sometimes it seems almost impossible to arrange them in systematic order. Since the prehistoric times, people were interested in placing the plants in a systematic way and for the first time a few plants were classified according to their medicinal and food value and thus the study of taxonomy of flowering plants began. In nineteenth century, the formation of the principles of taxonomy began and several principles were formed, which are still very helpful in arranging the plants in a systematic order.

Swingle has proposed the principles, which have been uniformly accepted by the botanists. They are as follows:

1. Plant relationships are up and down genetic lines and these must constitute the framework of phylogenetic taxonomy. This will naturally form a branching but not reticulate structure.

2. Some evolutionary processess are progressive while others are regressive.

3. Evolution does not necessarily involve all organs of the plant at one time or in the same direction. One organ may be advancing while another is stationary or retrogressing

4. Evolution has generally been consistent and when a particular progression or regression has set in, it is persisted into the end of the phylum.

5. In any phylum the chlorophyll bearing plants precede the chlorophyll less ones. Saprophytes are derived from independent forms and parasites usually from the saprophytes among the lower plants, and from independent forms among the flowering plants.

6. Usually structures with many similar parts are more primitive, and those with fewer and dissimilar parts are more advanced.

7. Among seed plants the stem structure with collateral bundles arranged in a cylinder is more primitive than that with scattered bundles.

8. In most groups of seed plants woody members have preceded the herbaceous ones.

9. In most groups of seed plants erect members have preceded the vines.

10. Perennials are more primitive than biennials and biennials are usually more primitive than annuals.

11. Scalariform vessels are more primitive than vessels with round pits.

12. The spiral arrangement of the leaves on the stem and of the floral leaves precedes that of opposite and whorled types.

13. Simple leaves are more primitive than compound leaves.

14. Historically leaves were first persistent (evergreen) and later deciduous. This way, the deciduous leaves show advanced character.

15. Among the seed plants the netted venation of leaves is more primitive than the parallel venation.

16. The many-parted flower is the more primitive, the type with few parts being derived from it, and the change is accompanied by a progressive sterilization of sporophylls.

17. A condition in which the perianth consists of like segments is more primitive than one in which sepals and petals are unlike each other.

18. Flowers with petals preceded apetalous ones, the latter being derived by reduction.

19. Polypetalous flowers are more primitive than gamopetalous ones, the latter being derived from the former by symphysis.

20. Regular flowers are more primitive than irregular ones.

21. Spirally imbricate floral parts are more primitive than those that are whorled and valvate.

22. Hypogyny is the primitive structure and from it perigyny and epigyny have been derived.

23. Numerous carpels represent a more primitive condition (than united carpels).

24. Separate carpels represent a more primitive condition than united carpels.

25. Axile placentation preceded parietal and central placentation of ovaries.

26. The presence of numerous stamens indicates a more primitive condition than that of a few stamens.

27. Separate stamens preceded united stamens.

28. Evolution in angiosperms is believed to have proceeded from seeds with two seed coats to those with only one.

29. The primitive seed contains endosperm and a small embryo, the advanced type has little or no endosperm, with the food stored in a large embryo.

30. A straight embryo is usually more primitive than a curved one.

31. The solitary flower is more primitive than the inflorescence.

32. Bisexual flowers preceded unisexual flowers.

33. The monoecious condition is primitive than dioecious.

34. Simple and aggregate fruits preceded multiple fruits.

35. The same evolutionary phenomena have often been repeated as separate occurrences in different parts of the plant kingdom, *e.g.*, loss of chlorophyll, loss of petals, stamens and carpels, acquisition of fleshy texture in fruits and of various types of thorns, change from simple to compound leaves, from erect to prostrate habit and from hypogynous to perigynous or epigynous insertion of floral parts, and lateral union (symphysis) of petals, stamens and carpels.

36. In determining the closeness of relationship between two families or other groups, it is usually best to compare with each other the most primitive, or basal members of each group, rather than those that are simplified by reduction or those that are most highly specialized.

Besides the above mentioned principles proposed by Swingle, the following parallel principles of taxonomy proposed by C.E. Bessey (1915) are of equal importance and are used by all modern taxonomists of twentieth century. These are known as Besseyan principles. They are as follows:

1. There is progressive evolution, *i.e.*, life advances from simplicity to complexity.

2. The existing simpler forms resemble more to their simple ancestors than to the present complex ones.

3. The evolved forms never become like ancestors.

4. The herbaceous habit of plants is more advanced than arboreal habit.

5. Annuals are more advanced than biennials and biennials are more advanced than perennials.

6. Hydrophytes, epiphytes, saprophytes and parasites are more advanced than ordinary terrestrial forms.

7. Compound leaves are more advanced than simple leaves.

8. Bisexual flowers are more primitive than unisexual ones.

9. Dioecious plants are more advanced than monoecious ones.

10. Inflorescence is more advanced than solitary flowers.

11. Gamopetalous flowers are more advanced than polypetalous flowers. The gamopetaly has been derived from polypetaly by the fusion of the petals.

12. Flowers lacking petals (apetaly) are more advanced.

13. Actinomorphic flowers are more primitive than zygomorphic ones.

14. Hypogyny precedes epigyny.

15. Apocarpous condition is more primitive than syncarpous condition.

16. Exalbuminous seeds are more advanced than albuminous seeds.

17. Polyandrous stamens precede jointed condition of stamens.

18. Simple fruits are more primitive than compound fruits.

Turril states, that "the angiosperms were the last of the great groups (Thallophyta, Bryophyta, Pteridophyta, Gymnosperms, Angiosperms) to appear. Hence, changes towards characters peculiar to the angiosperms may with some justification be read as progressive, the more so that they are now the dominant group ecologically over much of the land surface of the globe."

The 'conservative' characters have been supposed by many taxonomists to be most useful and valuable in phylogenetic studies. For plants, the reproductive organs are supposed to be more conservative than the vegetative organs. For example, the morphogenetic evolutionary changes in carpels have been far greater than in any vegetative organ considered throughout Spermophyta.

According to Tuzson, the monocots form an older group than the dicots, because the smaller groups of monocots are separable into series and families which show greater gaps on the whole than those in dicots.

Asa Gray is of opinion that a natural system of classification of plants aim to arrange all the known plants of the plant kingdom, in a series of grades according to their resemblances, in all respects, so that each species, genus, tribe, family, order, etc., shall stand next to those which it most resembles in all respects, or rather in the whole plan of structure. For example, two plants may be very much alike in external appearance, yet very different in their principal structure.

According to Sprague the monophyletic groups are regarded as the only natural ones at and above the family level in existing angiosperms. On the other hand, Hutchinson in his phylogenetic scheme and classification definitely retains groups he considers polyphyletic, *e.g.*, Asterales and Euphorbiales.

Hutchinson's principles of taxonomy and phylogeny of Angiosperms are most acceptable to the present taxonomists. They are as follows —

Hutchinson's General Principles Adopted for the Classification of Flowering Plants:

Other things being equal, it may be stated that:

1. Evolution is both (1) upwards and (2) downwards. The latter involving degradation and degeneration; examples : (1) towards the sympetalous condition; epigyny; (2) towards the apetalous state of many flowers; unisexuality in flowering plants.

2. Evolution does not necessarily involve all organs of the plant at the same time; and one organ or set of organs may be advancing while another set is stationary or retrograding. For example in the family Crassulaceae the carpels have remained free while some genera are sympetalous.

3. Evolution has generally been consistent, and when a particular progression or retrogression has set in, it is persisted into the end of the phylum; examples: the strong tendency to zygomorphy of the corolla coupled with the reduction in the number of stamens in Engler's hypogynous *Metachlamydeae* : the great tendency to perigyny and epigyny in the *Archichlamydeae* and *Metachlamydeae* as exhibited in the families *Apiaceae* and *Rubiaceae* respectively.

Relating to the General Habit of Plants:

4. In certain groups, trees and shrubs are probably more primitive than herbs; examples:Mimosaceae and Caesalpiniaceae (trees and shrubs) as compared with the family Fabaceae (Papilionaceae) (becoming herbaceous).

5. Trees and shrubs are older than climbers, the latter habit having been acquired through particular environment.

6. Perennials are older than biennials, and from them annuals have been derived; note the extraordinarily few annuals in the primitive family Ranunculaceae and in Rosaceae; the great number in the more advanced and natural family Brassicaceae.

7. Aquatic phanerogams are as a rule more recent than terrestrial (at any rate in the members of the same family or genus), and the same may be said of epiphytes and parasites.

Relating to the General Structure of Flowering Plants:

8. Plants with collateral vascular bundles arranged in a cylinder (Dicotyledons) are more primitive in origin than those with scattered bundles (Monocotyledons), though it does not necessarily follows that the latter have been directly derived from the former.

9. The spiral arrangement of leaves on the stem and of the floral leaves precedes that of the opposite and whorled type.

10. As a rule simple leaves precede compound leaves.

Relating to the Flowers and Fruits of Plants:

11. Bisexual (hermaphrodite) precede unisexual flowers, and the dioecious is probably more recent than the monoecious condition.

12. The solitary flower is more primitive than the inflorescence, the highest forms of the latter being the umbel and capitulum; examples of the latter: Apiaceae and Asteraceae respectively.

13. Spirally imbricate floral parts are more primitive than whorled and valvate; examples : *Magnolia* and *Clematis*.

14. Many-parted flowers (polymerous) precede, and the type with few parts (oligomerous) follows from it, being accompanied by a progressive sterilisation of reproductive parts (sporophylls); examples : *Magnolia* and *Cherianthus*.

15. Petaliferous flowers precede apetalous ones, the latter being the result of reduction.

16. Free petals (polypetaly) are more primitive than connate petals (sympetaly).

17. Actinomorphy (regularity) of the flower is an earlier type than zygomorphy (irregularity); examples: *Caltha* and *Delphinium*.

18. Hypogyny is the primitive structure, and from it perigyny and epigyny were derived later.

19. Free carpels (apocarpy) are more primitive and from them connate carpels resulted; sometimes however, when the carpels have remained loosely united during evolution they may again become quite free, especially in fruit; example : Asclepiadaceae.

20. Many carpels (polycarpy) precede few carpels (oligocarpy); example: *Ranunculus* and *Nigella*.

21. The endospermic seed with small embryo is primitive and the non-endospermic seed more recent; examples: Ranunculaceae and Rosaceae.

22. In primitive flowers there are many stamens, in more advanced flowers few stamens; examples: *Ranunculus* and *Cherianthus*. This condition may, however, be reversed within the confined of a single family like Papaveraceae, where bees feed on the pollen.

23. Separate stamens precede connate stamens; Examples : Tiliaceae and Malvaceae; Campanulaceae and Lobeliaceae.

24. Aggregate fruits (fruits formed several separate flowers) are more recent than single fruits, and as a rule the capsule.

FOSSIL ANGIOSPERMS

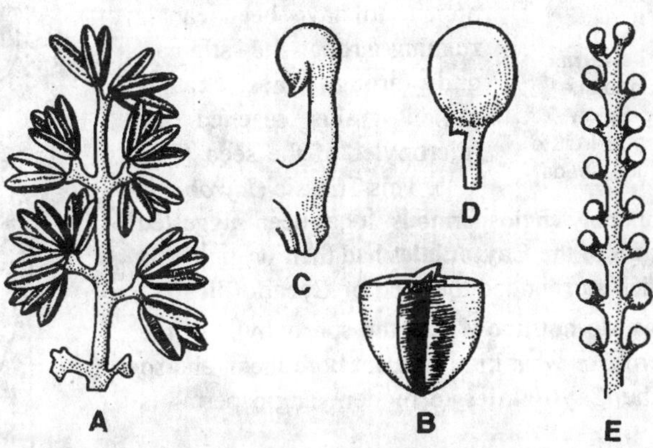

Fig. 1.2. *Gristhorpia nathorsti* A, sporophyll with anthers; B, cross section of anthers showing four locules C, very young carpel D, seed; E, sporophyll with carpels.

Oldest Angiosperms

Caytoniales. In Jurassic rocks have been found the oldest known plants which were angiospermous in the true sense of the term; that is, in having seed enclosed in a carpel. These plants represent two closely-related genera of the order *Caytoniales.*

The carpels were borne on sporophylls. Each sporophyll consisted of a central stalk with pinnately arranged short side branches each of which was terminated by a carpel or fruit. The carpel was completely closed, and the tip of the portion of the pinnule which bent over became the stigma. Pollen grains have been found attached to the stigmas. The seed were borne within the closed carpel (Fig. 1.3). The integument of the seed is rather strikingly similar to that of certain seed of seed-ferns. The fruits were fleshy, so the seeds may have been scattered by animals that ate the fruit.

The anthers are found on sporophylls similar in general outline to those which bore the carpels. The branches of the sporophylls divided, and the tips of the divisions bore groups of stamens (Fig. 1.2). The anthers were sessile or at the ends of short filaments. They had a four-winged form, and each wing seems to have contained a pollen sac. Thus they had the same general form as the stamen of a modern angiosperm.

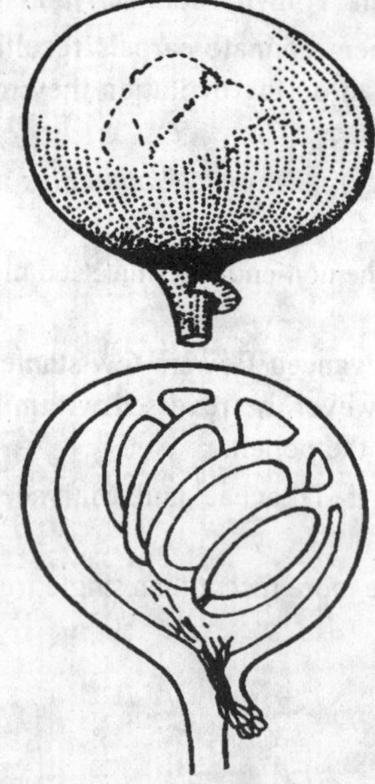

Fig. 1.3. ***Caytonia sewardi.*** Above, fruit with flange-like stigma on the stalk. Below, diagrammatic section of a fruit to show arrangement of the seeds.

The leaves which appear to belong to the known **Caytoniales** are of a type which was very widespread during the Jurassic period and extended from Triassic to Cretaceous times. The venation was netted, but similar to that of *Glossopteris,* which is generally regarded as a seed-fern.

The general character of the leaves, sporophylls, and seed indicate that the **Caytoniales** were derived from the seed-ferns.

In the **Caytoniales** from the Jurassic the pollen seems to have been caught on a stigma, and the ovules are enclosed in ovaries, but in a form from the upper Triassic pollen grains are found in the micropyles of the seed. In this form there appear to have been canals running through the "stigma," and through these canals pollen grains reached the micropyles of the seed (Fig. 1.4). This Triassic *Caytonia* is evidently not yet completely angiospermous. It has been suggested that the walls of the fruit of the **Caytoniales** had their origin in the fusion of cupules which surrounded the seed of **Cycadofilicales.**

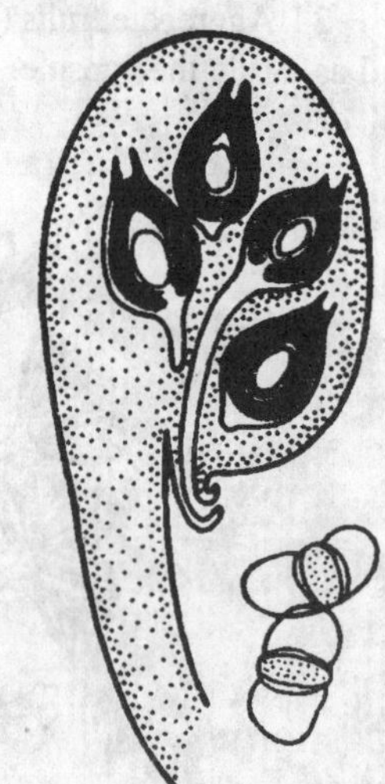

Fig. 1.4. ***Caytonia thomasi.*** Left, longitudinal section of fruit showing seed. In one case the canal leading to a seed is shown. Right, pollen grains.

Unfortunately we do not know how the sporophylls of the **Caytoniales** were borne, or what kind of plant bore them, and so the relationship of the **Caytoniales** to modern angiosperms is obscure.

ANCESTRY OF ANGIOSPERMS

The ancestry of angiosperms has long been a moot question. There is not enough evidence to reach a definite decision, and there is much disagreement as to theory.

Among living gymnosperms the group which is most similar to the angiosperms is the **Gnetales.** While there is great dissimilarity between the **Gnetales** and the angiosperms, still there is enough similarity to convince many that either the angiosperms are descended from the **Gnetales,** perhaps from extinct forms, or else the two groups are closely related and have a common ancestry. Some regard the **Gnetales** as being intermediate between the conifers and angiosperms, and so are inclined to a belief in a coniferous ancestry for angiosperms. This belief is based in part on similarity in wood structure and on the fact that in conifers, Gnetales, and angiosperms the fertilization is by male nuclei and not by spermatozoids as in all other living groups of land plants.

A rather complete understanding of the "flower" of *Cycadeoidea* was followed by an extensive discussion of a cycadophyte ancestry for angiosperms; not from *Cycadeoidea* with its highly specialized stem, rather from one more nearly related to *Williamsonia* or *Williamsoniella*, but still more primitive. It was pointed out that, as previously mentioned, the bracts, stamens, and ovuliferous cone of *Cycadeoidea* were in the same relative positions as the perianth, stamens, and carpels of the angiosperms. Also, the embryo had two cotyledons, while the seeds were almost enclosed by the interseminal bracts.

The discovery of the **Caytoniales**, first in the Jurrasic and later in the late Triassic, has been taken by some as indicating that the angiosperms, through forms related to the **Caytoniales**, are descended direct from the seed-ferns. A recent theory is that the carpels of angiosperms were derived through a modification and fusion of cupules which surrounded the seeds of **Cycadofilicales**, and that the ovaries of the **Caytoniales** represent an intermediate condition between the seed-ferns and modern angiosperms.

One feature which is common to practically all theories as to the origin of the angiosperms is that they go far back into early Mesozoic or latter Palaeozoic times; also that their ancestors were generalized forms and the not such specialized ones as modern conifers or cycads or the late Mesozoic **cycadeoideas**. A conservative suggestion is that they were derived from somewhere in the general gymnospermous complex, from a line in which the marked peculiarities of more modern groups had not become so pronounced as they appear in the well-known specialized types.

Our ignorance as to the ancestry of the angiosperms is not surprising when we remember how scanty is our knowledge of plant floras. If, as seems probable, the angiosperms evolved in Arctic regions, much of the record may be thickly covered with an icy mantle and inaccessible to us. Also, the record of former vegetation is largely that in and around swamps, about lakes, and in similar situations; and any trace of much of that which flourished on higher ground is forever lost. However, in comparatively recent years a great deal of evidence as to the history and relationships of land plants has been discovered; so we may hope that perhaps some day we may have a fairly connected account of the development of the angiosperms.

2

Evolution of Classification

In India, botanical science had its origin in a very remote age with the development of agriculture, when Indians of the Vedic Period (2000 B.C. - 800 B.C.) cultivated various food crops. In the Vedic literature several technical terms are used in the description of plants and plant parts. Both the external features and the anatomy were studied. One of the most ancient works dealing with plant life in scientific manner is the *Vrikshayurveda* (science of plants and plant life), compiled by Parasara even before the beginning of the Christian era, and which formed the basis of botanical teaching and the medical studies in ancient India. This ancient work deals with the morphology and anatomy of the plant organs, nature and properties of soil, description and distribution of forests in the country, etc. According to Parasara there are innumerable cells (*rasakosa*) which serve as storehouse of sap derived from the soil, through the roots by transporting system (*syandani*) and that the sap is digested with the help of chlorophyll (*ranjakena pacyamanat*) into nutritive substances and by-products. A system of classification was formulated, and the plants were divided into *eight* groups : 1. *Vriksha*; 2. *Vanaspati*; 3. *Guccha*; 4. *Gulma*; 5. *Trina*; 6. *Aushadhi*; 7. *Valli* and 8. *Kashtha* Charaka was devoted for medicinal plants, and he wrote the famous work *Charaka Samhita*. The ancient system of classification was formulated based on the study of the comparative morphology of plants, which according to Majumdar was more advanced than any system developed in Europe before the eighteenth century. Many families (*ganas*) were clearly distinguished as to be recognizable today. The present day family Cruciferae was called *Swastikaganiyam* because the calyx resembles a *Swastika*, and was characterized by having superior ovary, four free sepals, four free petals, six stamens two of which are shorter and two carpels fused and forming a two-locular fruit. The present day Cucurbitaceae was known as *Tripusaganiyam* and was characterized by having flowers which were epigynous, sometimes bisexual with five sepals, five fused petals, three stamens and a trilocular ovary with three rows of ovules.

In middle ages there was an extraordinary rise of Islam among the Arabians who spread their culture from Spain to Bokhara. Islamic scholars contributed a lot to the botanical science. A Persian writer Abu Mansur wrote a book on medicinal plants, also referred Indian plants. Ibn Sina (Avicenna) (980 - 1037A.D.) also a Persian, compiled a book *Canon of Medicine* containing numerous medicinal plants, which was much valued and had more than twenty editions in the sixteenth century.

The problem of classifying the heterogenous groups of angiosperms is very interesting and as well as too tough for taxonomists. The angiosperms are widely distributed with so many variations that sometimes it seems almost impossible to arrange them in systematic order. Since the prehistoric times the people were interested in the solution of the problem, and for the first time a few plants were classified according to their medicinal and food value and thus the taxonomy of flowering plants originated. After the discovery of some useful plants, the mind of the people of the past diverted towards the cultivation of food producing and other beneficial plants. They prepared fields, sown the seeds and harvested the crops which were assumed to be God given gifts.

There can be three different possible types of systems of classification, *the aritificial, the natural* and *the phylogenetic* classifications:-

1. **Artificial System**– In artificial system of classification only a few characters of the plants are being considered, for example, the grouping of plants into *herbs, shrubs* and *trees* or the sexual system of Linnaeus based on number of stamens and styles. There are so many drawbacks in this system the most important one is that the plants closely resembling each other are very often placed in widely separated groups, while those quite different from each other are being placed in the same group. This system can be compared to the alphabetical arrangement of words in a dictionary, where two words arranged one after the other have nothing to do with each other. The artificial systems are not commonly in current use. The best known artificial system is of Linnaeus published in 1735.

2. **Natural System**–In the natural system all the important characters of the plant are being considered, and the plants are classified according to their related affinities. This system reflects the situation as it is thought to exist under natural conditions, *e.g.*, Bentham and Hooker's system (1862-83).

3. **Phylogenetic System**–In phylogenetic system, the plants are classified according to their evolutionary and genetic affinities. But it seems a bit difficult to clasify the plants perfectly on the basis of evolutionary tendencies due to imperfection of fossil records, and therefore, at present the plants are classified partly according to natural and partly according to phylogenetic basis. The systems of classification proposed by Engler in 1886, by Hutchinson in 1926, and by Tippo in 1942 are phylogenetic. The plant taxonomy begins with the Greeks.

ANCIENT SYSTEMS OF CLASSIFICATION

From Theophrastus to Linnaeus

Theophrastus (370-285 B.C.), the father of Botany, was a pupil of Plato (426-347 B.C.) and friend of Aristotle (384-322 B.C.) classified plants on the basis of form and texture. In his *Historia Plantarum* he had classified and described about 450 cultivated plants. He had classified them as herbs, under-shrubs, shrubs and trees. Theophrastus considered the trees to be highest evolved while herbs very primitive. He also distinguished between annual, biennial and perennial plants. He also noted the difference between centripetal and centrifugal inflorescences. His classification was strictly artificial.

Dioscorides (64 A.D.) of Christian era prepared a '*Materia Medica*' and described many plants of medicinal value, but his descriptions were imperfect and classification of little importance.

Pliny (23-79 A.D) also a botanist of Christian era had written seventeen volumes on the subject and gave many details of the cultivation of gum and spices yielding plants. He had described medicinal plants in three volumes of his great work, '*Historia Naturalis*' (77 A.D.), largely compiled from Greek authors.

Albertus Magnus (1193-1280), for the first time distinguished between monocotyledons and dicotyledons on the anatomical basis with the help of crude lenses available at that time. In his *De vegetalis* he had described many plants and methods of gardening and orcharding.

Otto Brunfels (1464-1534), could recognise for the first time between perfecti and imperfecti groups of plant kingdom on the basis of the presence and absence of flowers Sometimes it is also said that modern study of systematic botany has started from the time of Brunfels.

Jerome Bock (1498-1554), classified the plants as trees, shrubs, and herbs but he tried to follow the natural system of classification and brought many plants close to each other according to their related affinities.

William Turner (1512), is the acknowledged 'Father of English Botany' and was the author of several botanical works. He wrote the first *English Herbal*. In this work the arrangement is alphabetical and the woodcuts are inferior to those of Brunfels. The work was published in three parts.

The Herbal of Valerius Cordus (1561) of Wittenberg showed a great advance and is a landmark in descriptive botany.

Andrea Caesalpino (1519-1603), an Italian botanist and physician, wrote a set of sixteen volumes on plants. The first volume was philosophical describing some minor differences of cultivated and wild plants, trees and shrubs, shrubs and herbs and some medicinal plants. In the other fifteen volumes he divided the plant kingdom into fifteen heterogeneous groups. In his work *De plantis* published in 1583, he classified about 1500 plants chiefly on the basis of the characters of the seed and embryo. He classified the plants first on the basis of their habit (trees and herbs) and then subdivided them on the basis of their characters of the seed and embryo. He classified the plants first on the basis of their habit (trees and herbs) and then subdivided them on the basis of their characters of the seeds and fruits. He also used the characters like the presence or absence of bulbs, presence of latex and number of locules, etc. He prepared a herbarium of 768 plants which still exists as one of the oldest herbaria. Caesalpiono's work was remarkable in that time to classify the whole plant kingdom but it also lacked in many ways.

The most important of the British herbals was written by John Gerard (1545-1612) entitled *Herbal or General Historie of Plante*. This was published in London in 1597. The groups of plants recognised are based on well marked characters of general form, manner of growth and economic uses. They started with supposed simpler forms, grass-like plants with narrow leaves, and advanced through the broader-leaved bulbous and rhizomatous monocotyledons to dicotyledonous herbs, culminating in shrubs and trees, the latter being regarded as the most perfect.

Jean Bauhin (1541-1613), a Swiss physician, wrote an illustrated *Historia Plantarum Universalis* which was published after his death in 1650-51 in three volumes, which contained the descriptions of about 5,000 plants with 3,500 illustrations.

G.Bauhin (1560-1624), a Swiss botanist and brother of Jean Bauhin wrote '*An Introduction to Botany*' in which he described about 6,000 species of plants. He also made considerable advance in the frequent use of binomial nomenclature. He described many plants with generic and specific names. His great work '*Phytopinax*' of 1596 is great contribution of Botany. Many of the plant descriptions of Bauhin's work are very much alike modern description of species.

M. de Lobelius (1538-1616), he differentiated between monocotyledons and dicotyledons chiefly on the basis of leaves. He clearly recognized certain natural affinities in plants. His great work '*Kruydtboeck*' was published in 1581, which contains many excellent illustrations.

Robert Morison (1620-1683), published his *Plantarum Historia Universalis* at Oxford in 1680. His system much resembles that of Caesalpino. Woody plants are kept distinct from herbaceous, and the latter are divided into fifteen sections based partly on habit and partly on characters derived from the fruit and seed. Morison had the distinction of publishing the first systematic monograph of a limited group. This was his *Plantarum Umbelliferarum Distributio Nova*, issued in Oxford in 1672. The arrangement is based on the form of the fruit, which is still employed for the classification of Umbelliferae.

John Ray (1628-1705), an English botanist contributed much and the foundation of modern Botany was laid. His most important work was the '*Historia Plantarum*' which was published in three volumes between 1686 and 1704. He clearly showed the importance of the monocotyledonous and dicotyledonous nature of the embryo from the point of view of related affinities and classification. He divided the plant kingdom in two main groups, A. Herbae, B.

Arbores. For the first time he used the terms *Dicotyledons* and *Monocotyledons*. In descriptions he used many binomial names. His classification was as follows –

Herbae: (*i*) Imperfecti (flowerless)

(*ii*) Perfecti (flowering)

(*a*) Dicotyledons (with two cotyledons),

(*b*) Monocotyledons (with one cotyledon).

Arbores: (*i*) Dicotyledons (with two cotyledons).

(*ii*) Monocotyledons (with one cotyledon).

J.P. de Tournefort (1656-1708), is said to be the founder of the concept of modern genera. Though generic names were used before, yet Tournefort was the first to provide genera with descriptions and placed them definitely apart from species. His classification was inferior in many ways to John Ray's classification. He never distinguished between Cryptogams and Phanerogams whereas monocots and dicots. He classified the plants on the basis of presence and absence of petals. He called the plants with petals as *Petalodes* and without petals as *Apetali*. His famous work *Elements de Botanique* was published in 1684 which was later translated into *Latin*.

Camerarius (1665-1721), a German botanist could demonstrate for the first time the sexuality in the flowering plants on the experimental basis. He proved experimentally that pollen is absolutely necessary for fertilization and seed formation in the life history of plants.

MODERN SYSTEMS OF CLASSIFICATION

From Linnaeus to Hutchinson

Carolus Linnaeus (1707-1778), a great Swedish botanist and said to be the father of modern botany was born in 1707. Only due to this great man, botany and other sciences got inspiration and expanded day by day. From the time of Linnaeus actually the concept of modern botany began. He followed the binomial system of nomenclature and described hundreds of plants from different parts of the world. He also revised many older genera and gave great stability to species. He published a number of important works, *e.g.*, *Genera Plantarum, Species Plantarum, Flora Lapponica*, etc., in which he described hundreds of plants known in that time. Linnaeus proposed an *artificial sexual system* of classification containing twentyfour classes. The plants were classified on the basis of number, cohesion, length and certain other characters of stamens. The outline of the classes of his system was published in his *Systema Naturae* in 1735, and again in *General Plantarum* in 1737 along with all the known plants of that time arranged systematically. The outline of the classification followed by him is as follows:

1.	**Monandria**	one stamen	*e.g., Canna, Lemna, Salicornia, Scirpus.* etc.
2.	**Diandria**	two stamens	*Veronica, Lepidium.*
3.	**Triandria**	three stamens	*Iris, Galium, Grasses.*
4.	**Tetrandria**	four stamens	*Protea, Ulmus.*
5.	**Pentandria**	five stamens	*Ipomoea, Primula.*
6.	**Hexandria**	six stamens	*Lilium, Alisma.*
7.	**Heptandria**	seven stamens	*Trientalis, Aesculus.*
8.	**Octandria**	eight stamens	*Vaccinium, Dirca.*
9.	**Enneandria**	nine stamens	*Rheum, Ranunculus.*
10.	**Decandria**	ten stamens	*Linum, Acer, Oxalis.*
11.	**Dodecandria**	11-19 stamens	*Reseda, Asarum, Calla.*

12.	**Icosandria**	20 or more stamens attached to the calyx	*Rosa, Rubus, Spiraea.*
13.	**Polyandria**	20 or more stamens attached to the receptacle	*Tilia, Papaver.*
14.	**Didynamia**	Didynamous stamens	Labiatae
15.	**Tetradynamia**	Tetradynamous stamens	Cruciferae
16.	**Monadelphia**	Stamens in one bundle	*Linum,* Malvaceae.
17.	**Diadelphia**	Stamens in two bundles	Papilionatae.
18.	**Polyadelphia**	Stamens in many bundles	*Tilia, Hypericum.*
19.	**Syngenesia**	Syngenesious stamens	Compositae, *Solanum.*
20.	**Gynandria**	Stamens grown to the pistil	Orchidaceae
21.	**Monoecia**	Plants monoecious	*Morus, Typha, Betula.*
22.	**Dioecia**	Plants dioecious	*Salix, Juniperus.*
23.	**Polygamia**	Plants polygamous	Compositae, *Acer.*
24.	**Cryptogamia**	Hidden flowers	Ferns, Mosses, Algae, Fungi, etc.

Though system was artificial, yet due to its simplicity and easy way of identification of plants, the system continued for more than one hundred years after the death of Linnaeus. He recognised the weakness of his system and was fully aware of the shortcomings. He attempted once again a more natural system but anyhow he could not complete it. On the whole, whatever Linnaeus presented to the scientific world is undoubtedly remarkable and of great value.

B.de Jussieu (1699-1777) and **A.L.de Jussieu** (1748-1836). The two French botanists Bernard de Jussieu and his nephew A.L. de Jussieu at the end of eighteenth century proposed a scheme of classification which was based upon the sexual system of Linnaeus but in an improved form. B. de Jussieu made many changes in the classification of Linnaeus and more or less gave a new form to it, but this was never published. His nephew, A.L. de Jussieu modified his uncle's plan with many changes and published it in his own way, in his famous work "*Genera Plantarum*" in 1789. He included fifteen classes and more than one hundred orders (now known as families) in his system. His system was chiefly based on the number and position of cotyledons and adhesion of petals. His system of classification can be claimed to be a natural one. The outline of his plan is as follows :

Acotyledones	Class 1
Monocotyledones	
Stamina hypogyna	Class 2
Stamina perigyna	Class 3
Stamina epigyna	Class 4
Dicotyledones	
Apetalae	

Stamina epigyna	Class 5
Stamina perigyna	Class 6
Stamina hypogyna	Class 7
Monopetalae	
Corolla hypogyna	Class 8
Corolla perigyna	Class 9
Corolla epigyna	
Antheris connatus	Class 10
Antheris distinctis	Class 11
Polypetalae	
Stamina epigyna	Class 12
Stamina hypogyna	Class 13
Stamina perigyna	Class 14
Diclines Irregulares	Class 15

Michel Adanson (1727-1806), published his *Families des plantes* in Paris in 1763. This contains an exhaustive account of previous systems, and an arrangement of genera in fifty-eight families, which are named and characterized.

A.P. de Candolle (1778-1841), a great French botanist proposed a very reasonable natural system of classification. In his younger days, he spent much of his time at Paris but after 1816 he shifted to Geneva, Switzerland, where he carried out several researches and made the city centre of botanical research. He wrote many monographs concerning classification and plant families, he was also interested in Physiology and Plant Geography. His famous work "*Theorie elementaire de la botanique*" was published in 1813, in which he proposed a classification completely based on natural system. He declared certain principles which later provided the base of origin of Bentham and Hooker's classification. His system contains 161 orders or families. The last edition of his "*Theorie elementaire*" was published in 1844 by his son Alphonse de Candolle in which he gave 213 orders or families.

The outline of de Candollean classification is as follows –

Vasculares – Plants with vascular bundles or plants with cotyledons.

Class 1. Exogenae – Dicotyledoneae, embryo with two cotyledons and vascular bundles arranged in a ring.

A. Diplochlamydeae – Flowers with two whorls of perianth, *i.e.*, calyx and corolla both present.

(*a*) Thalamiflorae – Flowers polypetalous and hypogynous.

Cohort 1. Carpels many and stamens opposite the petals, orders 1–8 Ranunculaceae, Berberidaceae, etc.

Cohort 2. Carpels solitary or joined; placenta parietal, order 9-20. Cruciferae, Violaceae, etc.

Cohort 3. Ovary solitary; placenta central, orders 21-44. Caryophyllaceae, Malvaceae, etc.

Cohort 4. Fruit gynobasic, orders 45-46.

(*b*) Calyciflorae – Flowers perigynous or epigynous, stamens inserted on the calyx tube, orders 47-84.

(*c*) Corolliflorae – Flowers gamopetalous and hypogynous, stamens epipetalous, orders 85-108.

B. Monochlamydeae – Flowers with single whorl of perianth, orders 109-128.

Class 2. Endogenae – Monocotyledoneae, embryo with single cotyledon; vascular bundles scattered.

A. Phanerogamae – Flowers present, *i.e.*, visible and regular, orders 129-150.

B. Cryptogamae – Flowers absent, orders 151-155.

Cellulares – Plants without vascular bundles and cots.

Class 1. Foliaceae – Leafy; sexuality known, orders 156-157 (mosses and liverworts).

Class 2. Aphyllae – Not leafy; sexuality unknown, orders 158-161 (algae, fungi, lichens).

A.P. de Candolle's system was definitely superior to Jussieu's system of classification including about 161 orders (families) of plants as compared with 167 families of Jussieu and Linnaeus respectively. The last edition of de Candolle's "*Theorie elementaire*" was published by his son Alphonse de Candolle in 1844, which contained 213 orders or families.

Robert Brown (1773-1858)–The name of Robert Brown seems fairly large in the chronicles of systematic botany, because he has often been referred to as *botanicorum facile princeps*. Brown was the first keeper of the Botanical Department of the British Museum. In 1810 Brown published his *Prodromus Florae Noval Hollandiae*, containing descriptions of Australian plants collected by Banks and Solander. In 1827 Brown published a paper entitled, '*The female flower in Cycadaceae*' and *Coniferae*, and in this he announced the important discovery of the distinction between Angiosperms and Gymnosperms.

John Lindley (1799-1865), an English botanist, who was Professor of Botany at the University College, proposed a system of classification and published "*Introduction to the Natural order of Plants*". His system was based on de-Candollean system with few remarkable improvements.

Stephen Endlicher (1804-1849), was a prominent systematist of the first half of the nineteenth century. His scheme of classification was published in "*Genera Plantarum Secundum Ordines Disposita*" in 1836-40 which was widely used in the continental countries. He divided the plant kingdom into two main groups 1. *Thallophyta*, 2. *Cormophyta*. In Thallophyta he included algae, fungi and lichens while in *Cormophyta* mosses, ferns and seed plants were included. In his published work "*Genera plantarum*" he described 6,235 vascular plants.

Henri Baillon (1827-1895), Henri Baillons' *Histoire des plantes* (1867-1895), a work of thirteen volumes, contain useful descriptions of the families and genera of flowering plants. Its chief merits, however, are the splendid illustrations with accurate dissections. He also published monographs on Euphorbiaceae and Buxaceae, etc. accompanied by beautifully reproduced plates.

In 1851, **Hofmeister** carried out many brilliant researches and established the existence of alternation of generations in lower as well as in higher plants. Following the researches of Hofmeister in 1859, Charles Darwin published his famous doctrine of evolution "*Origin of Species*". The doctrine of evolution was known to every educated man in the world, which brought revolutionary changes in solving the problem of classification of plants and animals.

Bentham and Hooker, the well known English systematists, established an important system of classification which appeared in the last half of the nineteenth century. George Bentham (1800-1884) and Sir Joseph Dalton Hooker (1817-1911), two English botanists, whose researches were carried over at the great herbarium of the Royal botanical garden at Kew. After the death of his father Sir W.J. Hooker, Sir Joseph Dalton Hooker was director of the Kew gardens for twenty years. Bentham and Hooker published a joint system of classification, containing 202 orders (families) in a monumental work "*Genera Plantarum*" (1862-1883) which became most popular classification in European countries except France for many years. The system of classification by Bentham and Hooker is based on that of De Candolle's and Jussieu's systems, but it is a great improvement over those and other previous schemes. An outline of Bentham and Hooker's classification is given below:

SYSTEM OF CLASSIFICATION BY BENTHAM AND HOOKER (1862-1883)

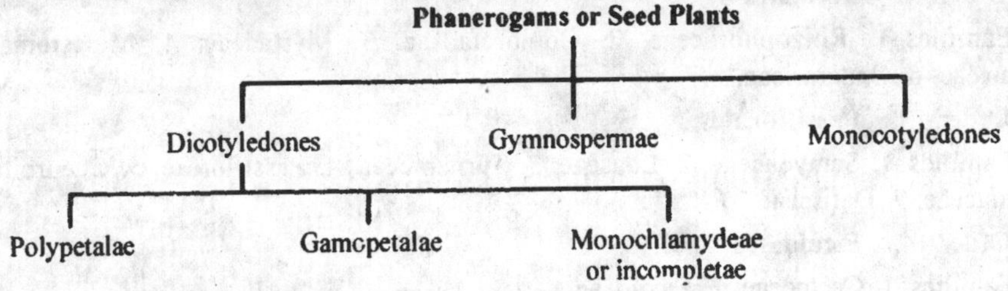

1– Polypetalae – Petals separate.

Series (A) Thalamiflorae—Hypogynous, stamens and pistils many, indefinite, mostly free, distinct sepals free from ovary.

Order 1. **Ranales**

Families –1. Ranunculaceae, 2. Dilleniaceae, 3. Calycanthaceae, 4. Magnoliaceae, 5. Anonaceae, 6. Menispermaceae, 7. Berberidaceae, 8. Nymphaeaceae.

Order 2. **Parietales.**

Families –1. Sarraceniaceae, 2. Papaveraceae, 3. Cruciferae, 4. Capparidaceae, 5. Resedaceae, 6. Cistineae. 7. Violaceae, 8. Canellaceae, 9. Bixineae.

Order 3. **Polygalineae**

Families–1. Pittosporeae, 2. Tremandreae, 3. Polygaleae, 4. Vochysiaceae.

Order 4. **Caryophyllineae**

Families–1. Frankeniaceae, 2. Caryophyllaceae, 3. Portulacaceae, 4. Tamariscineae.

Order 5. **Guttiferales**

Families–1. Elatineae, 2. Hypericineae, 3. Guttiferae, 4. Ternstroemiaceae, 5. Dipterocarpeae, 6. Chlaenaceae.

Order 6. **Malvales**

Families–1. Malvaceae, 2. Sterculiaceae, 3. Tiliaceae.

Series (B) Disciflorae. Hypogynous, calyx consists of free or united sepals which are usually free from ovary. A prominent cushion shaped disc present below the ovary. Ovary free or embedded in the receptacular disc. Stamens usually definite in number.

Order 7. **Geraniales**

Families–1. Lineae, 2. Humiriaceae, 3. Malpighiaceae, 4. Zygophylleae, 5. Geraniaceae, 6.Rutaceae, 7. Simarubeae, 8. Ochnaceae, 9. Burseraceae, 10. Meliaceae, 11. Chailletiaceae.

Order 8. **Olacales**

Families–1. Olacineae, 2. Ilicineae, 3. Cyrilleae.

Order 9. **Celastrales**

Families–1. Celastrineae, 2. Stackhousieae, 3. Rhamneae, 4. Ampelideae.

Order 10. **Sapindales**

Families–1. Sapindaceae, 2. Sabiaceae, 3. Anacardiaceae, 4. Coriarieae, 5. Moringeae.

Series (C) Calyicflorae–Calyx consists of united sepals rarely free and adnate to ovary. Perigynous or epigynous. Ovary enclosed by the axis sometimes inferior.

Order 11. **Rosales**

Families–1. Connaraceae, 2. Leguminosae, 3. Rosaceae, 4. Saxifrageae, 5. Crassulaceae, 6. Droseraceae, 7. Hamamelideae, 8. Bruniaceae, 9. Halorageae.

Order 12. **Myrtales**

Families–1. Rhizophoraceae, 2. Combretaceae, 3. Myrtaceae, 4. Melastomaceae, 5. Lythrarieae, 6. Onagrarieae.

Order 13. **Passiflorales**

Families–1. Samydaceae, 2. Loaseae, 3. Turneraceae, 4. Passifloreae, 5. Cucurbitaceae, 6. Begoniaceae, 7. Datisceae.

Order 14. **Ficoidales**

Families–1. Cactaceae, 2. Ficoideae.

Order 15. **Umbellales**

Families–1. Umbelliferae, 2. Araliaccae, 3. Cornaceae.

2. Gamopetalae : Petals united.

Series(A) Inferae–Ovary inferior, stamens as many as corolla lobe s or fewer.

Order 1. **Rubiales**

Families–1. Caprifoliaceae, 2. Rubiaceae.

Order 2. **Asterales**

Families–1. Valerianeae, 2. Dipsaceae, 3. Calycereae, 4. Compositae.

Series(B) Heteromerae–Ovary usually superior, carpels more than two, stamens as many as corolla lobes or fewer, epipetalous or free.

Order 4. **Ericales**

Families–1. Ericaceae, 2. Vaccinieae, 3. Monotropeae, 4. Epacrideae, 5. Diapensiaceae, 6. Lennoaceae.

Order 5. **Primulales**

Families–1. Plumbagineae, 2. Primulaceae, 3. Myrsineae

Order 6. **Ebenales**

Families–1. Sapotaceae, 2. Ebenaceae, 3. Styraceae.

Series (C) Bicarpellatae–Ovary usually superior, carpels usually two, rarely 1 to 3, stamens alternate with the corolla lobes and equal in number or fewer.

Order 7. **Gentianales**

Families–1. Oleaceae, 2. Salvadoraceae, 3. Apocynaceae, 4. Asclepiadaceae, 5. Loganiaceae, 6.Gentianaceae.

Order 8. **Polemoniales**

Families–1. Polemoniaceae, 2. Hydrophyllaceae, 3. Poragineae, 4. Convolvulaceae, 5. Solanaceae.

Order 9. **Personales**

Families–1. Scrophularineae, 2. Orobanchaceae, 3. Lentibularieae, 4. Columelliaceae, 5. Gesneraceae, 6. Bignoniaceae, 7. Pedalineae, 8. Acanthaceae.

Order 10. **Lamiales**

Families–1. Myoporineae, 2. Selagineae, 3. Verbenaceae, 4. Labiatae, 5. Plantagineae.

3. Monochlamydeae or Incompletae–Flowers with simple perianth which is usually sepaloid and sometimes altogether absent.

Series 1. Curvembryae

Families–1. Nyctagineae. 2. Illecebraceae, 3. Amarantaceae, 4. Chenopodiaceae. 5. Phytolaccaceae, 6. Batideae, 7. Polygonaceae.

Series 2. Multivulatae Aquaticae

Families–1. Podostemaceae

Series 3. Multivulatae Terrestres

Families-1. Nepenthaceae, 2. Cytinaceae, 3. Aristolochieae.

Series 4. Micrembryae

Families–1. Piperaceae, 2. Chloranthaceae, 3. Myristiceae, 4. Moniaceae.

Series 5. Daphnales

Families–1. Laurineae, 2. Proteaceae, 3. Thymelaceae, 4. Penaeaceae, 5. Eleagnaceae.

Series 6. Achlamydosporeae

Families–1. Loranthaceae, 2. Santalaceae, 3. Balanophoreae.

Series 7. Unisexuales

Families–1. Euphorbiaceae, 2. Balanopseae, 3. Urticaceae. 4. Plantanaceae, 5. Leitnerieae, 6.Juglandeae, 7. Myricaceae, 8. Casuarineae, 9. Cupuliferae.

Series 8. Ordines anomali (*Anomalous Families*)

Families–1. Salicaceae, 2. Lacistemaceae, 3, Empetraceae, 4. Ceratophylleae.

Gymnospermae :

Families–1. Gnetaceae, 2. Coniferae, 3. Cycadaceae.

Monocotyledones :

Series 1. Microspermae : Ovary superior tricarpellary. Seeds very minute and many. Inner perianth leaves petaloid.

Families–1. Hydrocharideae, 2. Burmanniaceae, 3. Orchideae.

Series 2. Epigynae : Ovary inferior.

Families–1. Scitamineae, 2. Bromellaceae, 3. Haemodoraceae, 4. Irideae, 5. Amaryllideae, 6. Taccaceae, 7. Dioscoreaceae.

Series 3. Coronarieae : Ovary superior, albuminous.

Families–1. Roxburghiaceae, 2. Liliaceae, 3. Pontederiaceae, 4. Philydraceae, 5. Xyrideae, 6. Mayacaceae, 7. Commelinaceae, 8. Rapateaceae.

Series 4. Calycineae : Ovary superior ; albumen copious; perianth inconspicuous, sepaloid, stiff or herbaceous.

Families–1. Flagellarieae, 2. Juncaceae, 3. Palmae.

Series 5. Nudiflorae : Ovary superior; perianth absent or reduced to scales.

Families–1. Pandaneae, 2. Cyclanthaceae, 3. Typhaceae, 4. Aroideae, 5. Lemnaceae.

Series 6. Apocarpae : Ovary superior, carpels separate or single.

Families–1. Triurideae, 2. Alismaceae, 3. Najadaceae.

Series 7. Glumaceae : Ovary single celled, single ovuled. Flowers in spikelets or heads. Perianth scale-like.

Families–1. Eriocauleae, 2. Centrolepideae, 3. Restiaceae, 4. Cyperaceae. 5. Gramineae.

Merits and Demerits of Bentham and Hooker's System of Classification

Merits: Though this system is not very natural, yet it is very easily workable, and is important from the point of view of its field applications. The system has been worked out as a result of very careful comparative examination of all the known genera of phanerogams. There are 202 orders (now known as families) beginning with Ranunculaceae and ending in Gramineae. This system was accepted by entire British Empire, U.S.A. and other European countries. Even today many botanists follow this classification because of its simplicity. This classification makes the basis for the arrangement of plants in Kew Herbarium and other important herbaria of Commonwealth Countries.

A special feature of this system is the addition of the *Disciflorae* and a curious arrangement of dividing certain groups on the basis of aquatic or terrestrial characteristics.

In this system, the greater emphasis has been given on the contrast between *free* and *fused* petals, *e.g.*, the Dicots are divided into the three groups –*Polypetalae, Gamopetalae* and *Monochlamydeae*.

In Monocots, the stress is being given to the relative position of ovary and perianth characteristics.

Demerits: The greatest disadvantage of this system is the retention in the group Monochlamydeae, a number of orders which show affinities with those in which a biseriate perianth is the rule.

The position of Gymnosperms between Dicots and Monocots is only for convenience rather than an indication of affinities.

This system of classification is based mainly on single and mostly artificial characters, with the result the closely related families are widely separated from each other.

The extreme simple nature of the parts of certain orders such as Salicineae, Cupuliferae, etc. indicate an affinity with an early group or groups which are now extinct. Such orders would therefore have no near allies in existing orders with an elaborate arrangement of the flower parts.

In Polypetalae, the orders (families) with inferior ovary are placed afterwards, whereas in Gamopetalae they are treated before than those with superior ovary.

In Monocots, much stress is being given to the relative position of the ovary and the perianth characters in determining the affinities, than seems justified by the comparative study of the orders. For example, the families Iridaceae and Amaryllidaceae exhibit greater affinities to Liliaceae than to Scitamineae and Bromeliaceae with which they are allied in this system because of common characters of epigyny. Thus the treatment of Monocots in general, except for position of Glumaceae is not proper.

Eichler's System of Classification

In 1883, a system of classifying the flowering plants was proposed by A.W. Eichler (1839–1887) is also remarkable because of its resemblance with widely accepted Englerian system. An outline of Eichler's system is given below:

A. Cryptogamae.

B. Phanerogamae.

1. Gymnospermae.

2. Angiospermae.

1. *Class:*	**Monocotyleae**
Series 1.	**Liliflorae**
Series 2.	**Enantioblastae**
Series 3.	**Spadiciflorae**
Series 4.	**Glumiflorae**
Series 5.	**Scitamineae**

Series 6.	**Gynandrae**
Series 7.	**Helobiae**
2. *Class:*	**Dicotyleae**
Sub-class :	**Choripetalae**
Series 1	**Amentaceae**
Series 2.	**Urticineae**
Series 3.	**Polygoninae**
Series 4.	**Centrospermae**
Series 5.	**Polycarpicae**
Series 6.	**Rhoeadinae**
Series 7.	**Cistiflorae**
Series 8.	**Columniferae**
Series 9.	**Gruinales**
Series 10.	**Terebinthinae**
Series 11.	**Aesculinae**
Series 12.	**Frangulinae**
Series 13.	**Tricoccae**
Series 14.	**Umbelliflorae**
Series 15.	**Saxifraginae**
Series 16.	**Opuntinae**
Series 17.	**Passiflorinae**
Series 18.	**Myrtiflorae**
Series 19.	**Thymelaeinae**
Series 20	**Rosiflorae**
Series 21.	**Leguminosae**
Anhang :	**Hysterophyta**
Sub-class:	**Sympetalae**
Series 1.	**Bicornes**
Series 2.	**Primulinae**
Series 3.	**Diospyrinae**
Series 4.	**Contortae**
Series 5.	**Tubiflorae**
Series 6.	**Labiatiflorae**
Series 7.	**Campanulinae**
Series 8.	**Rubiinae**
Series 9.	**Aggregatae**

The "series' in Eichler's system can easily be compared with modern 'order' and whereas the 'orders' of Eichler's system may be referred to modern families. The inclusion of the Hysterophyta as an "anhang" in the sub-class Choripetalae is the indication that in those days the position of angiospermic parasites was not at all clear. This group includes Loranthaceae, Rafflesiaceae and the Balanophoraceae. Eichler also included one non-parasitic family Aristolochiaceae in this group. Eichler's system also separates the Gymnosperms from their usual position between the Dicotyledons and Monocotyledons (as proposed by Bentham and Hooker) and raised the group to equivalent group with the angiosperms and lower than the monocotyledons.

ENGLER AND PRANTL'S SYSTEM OF CLASSIFICATION

Adolf Engler (1844-1938), a German botanist who served as Professor of botany in the University of Berlin for thirty years and director of Botanical Gardens from 1889 to 1921. His phylogenetic system of classification was first published as a guide to the botanical garden of Breslau in 1892. Later on the system expanded in a monumental work called "*Die Naturlichen Pflanzebfamilien*"! with means for the identifiction of the genera of whole plant kingdom. This publication continued with numerous volumes, many supplements, syllabi and revisions from 1895 to the present day. This system has been the dominant one of the plant classification in most of the scientific world since 1900. Most of the prominent herbaria of the world are arranged according to this system. This great work was completed in the collaboration of his associate worker Eugen Prantl (1849-1893).

According to this system the families were arranged in accord to the increasing complexity of the flower, fruit and seed development. The general outline of the system proposed by Engler and Prantl is given below—

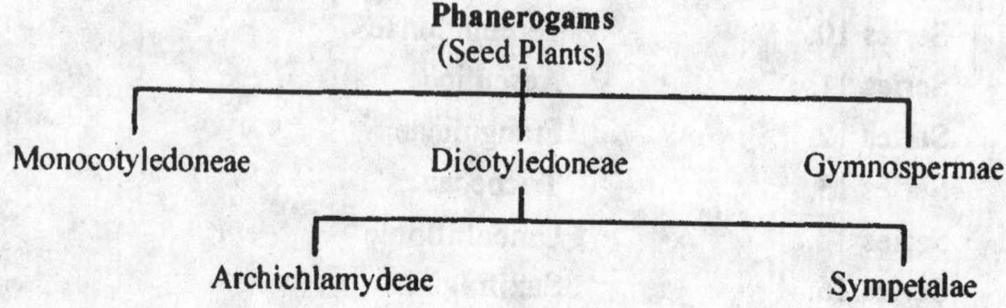

Class 1. MONOCOTYLEDONEAE

Order 1. **Pandanales**

Families–1. Typhaceae, 2. Pandanaceae, 3. Sparganiaceae.

Order 2. **Helobiae**

Families–1. Potamogetonaceae, 2. Najadaceae, 3. Aponogetonaceae, 4. Scheuchzeriaceae, 5. Alismaceae, 6. Butomaceae, 7. Hydrocharitaceae.

Order 3. **Triuridales**

Family–1. Triuridaceae.

Order 4. **Glumiflorae**

Families–1. Gramineae, 2. Cyperaceae.

Order 5. **Principes**

Family–1. Palmae.

Order 6. **Synanthae**

Family–1. Cyclanthaceae.

Order 7. **Spathiflorae**

Families–1. Araceae, 2. Lemnaceae.

Order 8. **Farinosae**

Families–1. Flagellariaceae, 2. Restionaceae, 3. Centrolepidaceae, 4. Mayacaceae, 5. Xyridaceae, 6. Eriocaulaceae, 7. Thurniaceae, 8. Rapateaceae, 9. Bromeliaceae, 10. Commelinaceae, 11. Pontederiaceae, 12. Cyanastraceae, 13. Philydraceae.

Order 9. **Liliflorae**

Families–1. Juncaceae, 2. Stemonaceae, 3. Liliaceae, 4. Haemodoraceae, 5. Amaryllidaceae, 6. Velloziaceae, 7. Taccaceae, 8. Dioscoriaceae, 9. Iridaceae.

Order 10. **Scitamineae**

Families–1. Musaceae, 2. Zingiberaceae, 3. Cannaceae, 4. Marantaceae.

Order 11. **Microspermae**

Families–1. Burmanniaceae, 2. Orchidaceae

Class 2. DICOTYLEDONEAE

Sub-class A. Archichlamydeae

Order 1. **Verticillatae**

Family–1. Casuarinaceae.

Order 2. **Piperales**

Families–1. Saururaceae, 2. Piperaceae, 3. Chloranthaceae, 4. Lacistemaceae.

Order 3. **Salicales**

Family–1. Salicaceae.

Order 4. **Garryales**

Family–1. Garryaceae.

Order 5. **Myricales**

Family–1. Myricaceae.

Order 6. **Balanopsidales**

Family–1. Balanopsidaceae.

Order 7. **Leitneriales.**

Family–1. Leitneriaceae.

Order 8. **Juglandales**

Family–1. Juglandaceae.

Order 9. **Batidales**

Family–1. Batidaceae.

Order 10. **Julianales**

Family–1. Julianaceae.

Order 11. **Fagales**

Families–1. Betulaceae, 2. Fagaceae.

Order 12. **Urticales**

Families–1. Ulmaceae, 2. Moraceae, 3. Urticaceae.

Order 13. **Proteales**

Family–1. Proteaceae.

Order 14. **Santalales**

Families–1. Myzodendraceae, 2. Santalaceae, 3. Opiliaceae, 4. Grubbiaceae, 5. Olacaceae, 6. Octoknemataceae, 7. Loranthaceae, 8. Balanophoraceae.

Order 15. **Aristolochiales**

Families–1. Aristolochiaceae, 2. Raflessiaceae, 3. Hydnoraceae.

Order 16. **Polygonales**

Family–1. Polygonaceae.

Order 17. **Centrospermeae**

Families–1. Chenopodiaceae, 2. Amarantaceae, 3. Nyctaginaceae, 4. Cynocrambaceae, 5. Phytolaccaceae, 6. Aizoaceae, 7. Portulacaceae, 8. Basellaceae, 9. Caryophyllaceae.

Order 18. **Ranales**

Families–1. Nymphaeaceae, 2. Ceratophyllaceae, 3. Trochodendraceae, 4. Cercidiphyllaceae, 5.Ranunculaceae, 6. Lardizabalaceae, 7. Berberidaceae, 8. Menispermaceae, 9. Magnoliaceae, 10.Calycanthaceae, 11. Lactoridaceae, 12. Anonaceae, 13. Eupomatiaceae, 14. Myristicaceae, 15.Gomortegaceae, 16. Monimiaceae, 17. Lauraceae, 18. Hernandiaceae.

Order 19. **Rhoeadales**

Families–1. Papaveraceae. 2. Capparidaceae, 3. Cruciferae, 4. Tovariaceae, 5. Resedaceae, 6.Moringaceae.

Order 20. **Sarraceniales**

Families–1. Sarraceniaceae, 2. Nepenthaceae, 3. Droseraceae.

Order 21. **Rosales**

Families–1. Podostemaceae, 2. Tristichaceae, 3. Hydrostachnaceae, 4. Crassulaceae, 5. Cephalotaceae, 6. Saxifragaceae, 7. Pittosporaceae, 8. Brunelliaceae, 9. Cunoniaceae, 10. Myrothamnaceae, 11. Bruniaceae, 12. Hamamelidaceae, 13. Eucommiaceae, 14. Platanaceae, 15. Crossosomataceae, 16.Rosaceae, 18. Leguminosae.

Order 22. **Pandales**

Family–Pandaceae.

Order 23. **Geraniales**

Families–1. Geraniaceae, 2. Oxalidaceae, 3. Tropaeolaceae, 4. Linaceae, 5. Humiriaceae, 6.Erythroxylaceae 7. Zygophyllaceae 8. Cneoraceae, 9. Rutaceae, 10. Simarubaceae, 11. Burseraceae, 12. Meliaceae, 13. Malpighiaceae, 14. Trigoniaceae, 15. Vochysiaceae. 16. Tremandraceae, 17.Polygalaceae, 18. Dichapetalaceae, 19. Euphorbiaceae, 20. Callitrichaceae.

Order 24. **Sapindales**

Families–1. Empetraceae, 2. Coriariaceae, 3. Limnanthaceae. 4. Anacardiaceae, 5. Cyrillaceae, 6. Buxaceae, 7. Pentaphylacaceae, 8. Corynocarpaceae, 9. Aquifoliaceae, 10. Celastraceae, 11.Hippocrateaceae, 12. Salvadoraceae, 13. Stackhousiaceae, 14. Staphyleaceae, 15. Icacinaceae, 16. Aceraceae, 17. Hippocastanaceae, 18. Sapindaceae, 19. Sabiaceae, 20. Melianthaceae, 21. Balsaminaceae.

Order 25. **Rhamnales**

Families–1. Rhamnaceae, 2. Vitaceae.

Order 26. **Malvales**

Families–1. Elaeocarpaceae, 2. Chlaenaceae, 3. Gonystilaceae, 4. Tiliaceae, 5. Malvaceae, 6. Bombacaceae, 7. Sterculiaceae, 8. Scytopetalaceae.

Order 27. **Parietales**

Families–1. Dilleniaceae, 2. Eucryphiaceae, 3. Ochnaceae, 4. Caryocaraceae, 5. Marcgraviaceae, 6. Quinaceae, 7. Theaceae, 8. Guttiferae, 9. Dipterocarpaceae, 10. Elatinaceae, 11. Frankeniaceae, 12.Tamaricaceae, 13. Fouquieriaceae, 14 Cistaceae, 15. Bixaceae, 16. Cochlospermaceae, 17. Winteraceae, 18. Violaceae, 19. Flacourtiaceae, 20. Stachyuraceae, 21. Turneraceae, 22. Malesherbiaceae, 23. Passifloraceae, 24. Achariaceae, 25. Caricaceae, 26. Loasaceae, 27. Datiscaceae, 28. Begoniaceae, 29. Ancistrocladaceae.

Order 28. **Opuntiales**

Family–1. Cactaceae.

Order 29. **Myrtiflorae**

Families–1. Geissolomataceae, 2. Penaeaceae, 3. Oliniaceae, 4. Thymelaceae, 5. Elaeagnaceae, 6. Lythraceae, 7. Sonneratiaceae 8. Punicaceae, 9. Lecythidaceae, 10. Rhizophoraceae, 11. Nyssaceae, 12. Alangiaceae, 13. Combretaceae, 14. Myrtaceae, 15. Melastomaceae, 16. Onagraceae, 17. Haloragidaceae, 18. Hippuridaceae, 19. Cynomoriaceae.

Order 30. **Umbelliflorae**

Families–1. Araliaceae, 2. Umbelliferae, 3. Cornaceae.

Sub class B. Sympetalae

Order 1. **Ericales**

Families–1. Clethraceae, 2. Pyrolaceae, 3. Lennoaceae, 4. Ericaceae, 5. Epacridaceae, 6. Diapensiaceae.

Order 2. **Primulales**

Families–1. Theophrastaceae, 2. Myrsinaceae, 3. Primulaceae.

Order 3. **Plumbaginales**

Family–1. Plumbaginaceae.

Order 4. **Ebenales**

Families–1. Sapotaceae, 2. Ebenaceae, 3. Symplocaceae, 4. Styracaceae.

Order 5. **Contortae**

Families-1. Oleaceae, 2. Loganiaceae, 3. Gentianaceae, 4. Apocynaceae, 5. Asclepiadaceae.

Order 6. **Tubiflorae**

Families–1. Convolvulaceae, 2. Polemoniaceae, 3. Hydrophyllaceae, 4. Boraginaceae, 5. Verbenaceae, 6. Labiatae, 7. Nolanaceae, 8. Solanaceae, 9. Scrophulariaceae, 10. Bignoniaceae, 11.Pedaliaceae, 12. Martyniaceae, 13. Orobanchaceae, 14. Gesneriaceae, 15. Columelliaceae, 16.Lentibulariaceae, 17. Globulariaceae, 18. Acanthaceae, 19. Myoporaceae, 20. Phrymaceae.

Order 7. **Plantaginales**

Family–1. Plantaginaceae.

Order 8. **Rubiales**

Families–1. Rubiaceae, 2. Caprifoliaceae, 3. Adoxaceae 4. Valerianaceae. 5. Dipsaceae.

Order 9. **Cucurbitales**

Family–1. Cucurbitaceae

Order 10. **Campanulatae**

Families–1. Campanulaceae, 2. Goodeniaceae, 3. Brunoniaceae, 4. Stylidiaceae, 5. Calyceraceae, 6. Compositae.

Merits and Demerits of Engler and Prantl's System of Classification

Merits–According to this system, the large artificial group of Bentham and Hooker's system, the *Monochlamydeae* has been completely abolished, and its families have been distributed among the related forms with free petals in the large series of this (Engler's) system, the *Archichlamydeae*.

The Sympetalae of this system corresponds to the Gamopetalae of Bentham and Hooker's system.

In this system the Gymnosperms are treated separately.

The families with inferior ovary have been treated in the last, both in Archichlamydeae and Sympetalae. The advancement is marked from the hypogyny to complete epigyny.

Engler considered the orchids to be more highly evolved than the grasses.

Demerits–In this system the Amentiferae or catkin bearers, (*e.g.*, Salicaceae, Juglandaceae, Betulaceae, etc.) have been regarded as most primitive and precede petaliferous families, (*e.g.* Ranunculaceae and Magnoliaceae). The Amentiferae are a reduced rather than a primitive group. According to Bessey and others the polypetaly was earlier, and apetaly was derived from it through modification.

The acceptance of the derivation of dichlamydeous flowers (perianth in two series) from monochlamydeous ones (perianth in single series) is objectionable.

The other demerits of this system (Engler's) are :

Derivation of parietal placentation from axile placentation.

Derivation of free-central placentation from parietal placentation.

Derivation of bisexual flowers from unisexual flowers.

Derivation of entomophily from anemophily.

In this system Monocots have been considered to be more primitive than Dicots, which does not correspond to the present day knowledge.

C.E. Bessey, an American botanist (1845–1915) proposed a system of classification based on Bentham and Hooker's system. According to Bessey the Ranales were the primitive angiosperms from which two branches were given out which gave birth to Monocots and Dicots. The system proposed by C.E. Bessey has never been accepted and adopted to any great degree in the world.

Richard von Vettstein (1862–1931), an Austrian botanist also proposed a system of classification parallel to Engler and Prantl's system with minute differences. In this system the unisexual and naked flowers are considered more primitive than bisexual and flowers with perianth. He considered dicots to be more primitive than monocots.

Hans Hallier (1868–1932), a German botanist proposed a system based on phylogenetic relationships parallel to Besseyan system with minor alternations. He considers dicots to be older and more primitive than monocots.

A.B.Rendle (1865–1938), an English botanist has also classified the flowering plants in his two-volumes "*The classification of flowering plants*", Vol. I contains Monocots and Vol. II contains Dicots. His system resembles Englerian system with very minor differences. According to him the Dicots are arranged in three grades: 1. Monochlamydeae, 2. Dialypetalae, and 3. Sympetalae. He considers the family Compositae (Dicot) to be highest evolved.

Oswald Tippo, of the University of Illinois published an outline (1942) of a projected phylogenetic classification. This system, however, was not his original, but was a compilation, a synthesis of several principles of Smith and Eames. The classification of non-vascular plants was that followed by G.M. Smith (1938) and the basic arrangement of vascular plants was that proposed by A.J. Eames (1936). Tippo studied and synthesised the palaeobotanical data and compared these data with the data obtained through the study of living plants. He emphasized the dependence on the vegatative characters such as plant form, branching and vascular anatomy in arriving at the natural grouping of major groups. This system emphasises that the Pteridophyta is not a homogeneous group and that there is no distinct line of differentiation between Pteridophyta and Spermatophyta. The outline of this classification is as follows:

Kingdom: Plantae

Sub-kingdom: Thallophyta

Phyla. 1.Cyanophyta; 2. Chlorophyta; 3. Chrysophyta; 4. Pyrrophyta; 5. Phaeophyta; 6.Rhodophyta; 7. Schizomycophyta; 8. Myxomycophyta; 9. Eumycophyta.

Sub-kingdom: Embryophyta

Phylum : Bryophyta

Class. 1. Musci; 2. Hepaticae; 3. Anthocerotae

Phylum : Tracheophyta

Sub-phylum : Lycopsida

Class: Lycopodineae

Orders: 1. Lycopodiales; 2. Selaginellales; 3. Lepidodendrales; 4. Pleuromeiales; 5. Isoetales.

Sub-phylum : Sphenopsida.

Class: Equisetineae

Orders: 1. Hyeniales; 2. Sphenophyllales; 3. Equisetales.

Sub-phylum: Pteropsida

Class: Filicineae

Orders: 1. Coenopteridales; 2. Ophioglossales; 3. Marattiales; 4. Filicales

Class: Gymnospermae

Sub-class: Cycadophytae

Orders: 1. Cycadofilicales; 2. Bennettitales; 3. Cycadales.

Sub-class: Coniferophyta

Orders: 1. Cordaitales; 2. Ginkgoales; 3. Coniferales; 4. Gnetales.

Class: Angiospermae

Sub-class: 1. Dicotyledoneae; 2. Monocotyledoneae

John Hutchinson (1884-1972) an English botanist has given the latest information about the phylogenetic classification of angiosperms which has been published in his famous work, "*Families of Flowering Plants*" recently in 1959. Formerly this work was published twice in 1926 and 1934. Hutchinson's classification is more closely related to that of Bentham and Hooker's and Bessey's systems of classification than that of Engler and Prantl's system. According to Hutchinson, the primitive polypetalous forms have been diverged from the very beginning along two separate lines. One of the lines still retains the arboreal habit whereas the other one has adopted the herbaceous habit. The first order of his arboreal line is Magnoliales and the first order of herbaceous line is Ranales. Hutchinson has divided the angiosperms into two large groups 1.Herbaceae and 2. Lignosae. The group Herbaceae contains the herbaceous representatives and the group Lignosae contains the woody representatives. He has proposed that the families having apetalous flowers have been derived partly through the Magnoliales and partly through the Ranales.

Hutchinson has divided the seed plants into two phyla–1. *Gymnospermae* and 2. *Angiospermae*. The phylum Angiospermae has been further divided into two sub-phyla–1. *Dicotyledones* and 2. *Monocotyledones*. The sub-phylum Dicotyledones has been divided into two divisions–1. *Lignosae* and 2. *Herbaceae*. The Lignosae, a woody group and Herbaceae, a herbaceous group. The sub-phylum Monocotyledones has been divided into three groups–1. *Calyciferae*, 2. *Corolliferae* and 3. *Glumiflorae*. The flowers of group Calyciferae possess distinct calyx and corolla; the flowers of group Corolliferae possess more or less similar calyx and corolla; the flowers of group Glumiflorae possess much reduced perianth or lodicules.

Phylum Angiospermae–Ovules enclosed in an ovary usually crowned by a style and stigma, the latter receiving the pollen grains mainly through the agency of insects becoming wind-pollinated when much reduced. Wood when present consisting of true vessels. A more recently evolved phylum than the *Gymnospermae*, and constituting the bulk of the present world vegetation, yielding valuable timbers and practically all food, forage and medicinal plants.

Sub-Phylum I. **Dicotyledones**–Embryonic plant with two cotyledons. Vascular bundles of the stem usually arranged in a circle (except in a few genera of the lower herbaceous families which have scattered bundles). Leaves typically net veined, opposite or alternate. Flowers usually pentamerous or tetramerous.

Division I. **Lignosae**–Trees and shrubs and some herbs clearly derived from and related to other woody plants-a fundamentally woody group.

Division II. **Herbaceae**–Herbs, or rarely shrubby plants related to and derived from herbaceous stocks—a fundamentally herbaceous group.

Sub-Phylum II. **Monocotyledones**-Embryonic plant with only one cotyledon; vascular bundles of the stem closed and scattered. Leaves typically parallel nerved, alternate, often sheathing at the base. Flowers usually trimerous.

Primitive and advanced characters which formed the basis of his classification are : polypetalous condition is more primitive than gamopetalous condition; unisexual flowers are more advanced than bisexual flowers; apetalous forms have originated from petaliferous stock, and represent reduction and advance; free parts on the whole are considered more primitive than connate or adnate parts; thus numerous free stamens are earlier than few or connate stamens; similarly apocarpous pistil is more primitive than syncarpous pistil; spiral arrangement of parts is more primitive than cyclic.

The outline of Hutchinson's system of classification is as follows:

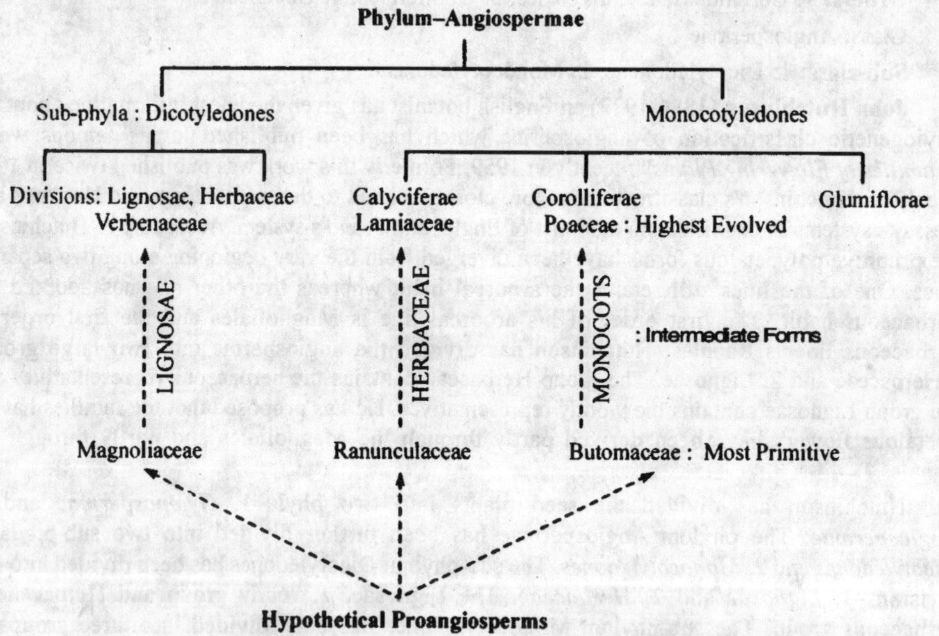

HUTCHINSON'S SYSTEM OF CLASSIFICATION, 1959

Phylum – *Angiospermae*

Sub-Phylum – *Dicotyledones*

Division I – *Lignosae*

Order 1. **Magnoliales**

Families–1. Magnoliaceae, 2. Illiciaceae, 3. Winteraceae, 4. Canellaceae, 5. Schisandraceae, 6.Himantandraceae, 7. Lactoridaceae, 8. Trochodendraceae, 9. Cercidiphyllaceace.

Order 2. **Annonales**

Families–10. Annonaceae, 11. Eupomatiaceae.

Order 3. **Laurales**

Families–12. Monimiaceae, 13. Austrobaileyaceae, 14. Trimeniaceae, 15. Lauraceae, 16. Gomortegaceae, 17. Hernandiaceae, 18. Myristicaceae.

Order 4. **Dilleniales**

Families–19. Dilleniaceae, 20. Connaraceae, 21. Crossosomataceae. 22. Brunelliaceae.

Order 5. **Coriariales**

Family–23. Coriariaceae.

Order 6. **Rosales**

Families–24. Rosaceae, 25. Dichapetalaceae, 26. Calycanthaceae.

Order 7. **Leguminales**

Families–27. Caesalpiniaceae, 28. Mimosaceae, 29. Papilionaceae.

Order 8. **Cunoniales**

Families–30. Pterostemonaceae, 31. Cunoniaceae; 32. Philadelphaceae, 33. Hydrangeaceae, 34. Grossulariaceae; 35. Oliniaceae, 36. Greyiaceae, 37. Escalloniaceae, 38. Baueraceae. 39. Crypteroniaceae.

Order 9. **Styracales**

Families–40. Lissocarpaceae. 41. Styracaceae, 42. Symplocaceae.

Order 10. **Araliales**

Families–43. Cornaceae, 44. Alangiaceae, 45. Garryaceae, 46. Nyssaceae, 47. Araliaceae, 48. Caprifoliaceae.

Order 11. **Hamamelidales**

Families–49. Tetracentraceae, 50. Hamamelidaceae; 51. Myrothamnaceae, 52. Platanaceae, 53. Stachyuraceae 54. Buxaceae, 55. Daphniphyllaceae, 56. Bruniaceae.

Order 12. **Salicales**

Family–57. Salicaceae.

Order 13. **Leitneriales**

Family–58. Leitneriaceae.

Order 14. **Myricales**

Family–59. Myricaceae.

Order 15. **Balanopsidales**

Family–60. Balanopsidaceae.

Order 16. **Fagales**

Families–61. Betulaceae, 62. Fagaceae, 63. Corylaceae.

Order 17. **Juglandales**

Family–64. Rhoipteleaceae, 65. Juglandaceae, 66. Pictrodendraceae.

Order 18. **Casuarinales**

Family–67. Casuarinaceae.

Order 19. **Urticales**

Families–68. Ulmaceae, 69. Cannabiaceae, 70. Moraceae, 71. Urticaceae, 72. Barbeyaceae, 73. Eucommiaceae.

Order 20. **Bixales**

Families–74. Bixaceae, 75. Cistaceae, 76. Flacourtiaceae, 77. Cochlospermaceae, 78. Hoplestigmataceae, 79. Achatocarpaceae, 80. Lacistemaceae.

Order 21. **Thymelaeales**

Families–81. Gonystylaceae, 82. Aquilariaceae, 83. Geissolomataceae, 84. Penaeaceae, 85. Thymelaeaceae, 86. Nyctaginaceae.

Order 22. **Proteales**

Family–87. Proteaceae.

Order 23. **Pittosporales**

Families–88. Pittosoporaceae, 89. Byblidaceae, 90. Stegnospermaceae, 91. Vivianiaceae, 92. Tremandraceae.

Order 24. **Capparidales**

Families–93. Capparidaceae, 94. Moringaceae, 95. Tovariaceae.

Order 25. **Tamaricales**

Families–96. Frankeniaceae, 97. Tamaricaceae, 98. Fouquieriaceae.

Order 26. **Violales**

Family–99. Violaceae.

Order 27. **Polygalales**

Families–100. Polygalaceae, 101. Krameriaceae, 102. Trigoniaceae, 103. Vochysiaceae.

Order 28. **Loasales**

Families–104. Turneraceae, 105. Loasaceae.

Order 29. **Passiflorales**

Families–106. Malcsherbiaceae, 107. Passifloraceae, 108. Achariaceae.

Order 30. **Cucurbitales**

Families–109. Cucurbitaceae, 110. Begoniaceae, 111. Datiscaceae, 112. Caricaceae.

Order 31. **Cactales**

Family–113. Cactaceae.

Order 32. **Tiliales**

Families–114. Dirachmaceae, 115. Scytopetalaceae, 116. Tiliaceae, 117. Sterculiaceae, 118. Peridiscaceae, 119. Bombacaceae.

Order 33. **Malvales**

Family–120. Malvaceae.

Order 34. **Malpighiales**

Families–121. Ixonanthaceae, 122. Malpighiaceae. 123. Humiriaceae, 124. Linaceae, 125. Irvingiaceae, 126. Huaceae, 127. Ledocarpaceae, 128. Erythroxylaceae, 129. Ctenolophonaceae, 130. Lepidobotryaceae, 131. Balanitaceae, 132. Zygophyllaceae.

Order 35. **Euphorbiales**

Family–133. Euphorbiaceae.

Order 36. **Theales**

Families–134. Bonnetiaceae, 135. Theaceae, 136. Sauraniaceae, 137. Actinidiaceae, 138. Pellicieraceae, 139. Pentaphylacaceae, 140. Tetrameristaceae, 141. Marcgraviaceae, 142. Caryocaraceae, 143. Medusagynaceae.

Order 37. **Ochnales**

Families–144. Strasburgeriaceae, 145. Ochnaceae, 146. Rhodolaenaceae, 147. Sphaerosepalaceae, 148. Dipterocarpaceae, 149. Ancistrocladaceae.

Order 38. **Ericales**

Families–150. Clethraceae, 151. Pyrolaceae, 152. Ericaceae, 153. Epacridaceae, 154. Diapensiaceae, 155. Monatropaceae, 156. Lennoaceae, 157. Vacciniaceae.

Order 39. **Guttiferales**

Families–158. Hypericaceae, 159. Clusiaceae; 160. Eucryphiaceae, 161. Quiinaceae.

Order 40. **Myrtales**

Families–162. Myrtaceae, 163. Lecythidaceae, 164. Rhizophoraceae, 165. Sonneratiaceae, 166. Punicaceae, 167. Combretaceae, 168. Melastomataceae.

Order 41. **Celastrales**

Families–169. Pandaceae, 170. Aquifoliaceae, 171. Salvadoraceae, 172. Koeberliniaceae, 173. Cneoraceae, 174. Cardiopteridaceae, 175. Cyrillaceae, 176. Icacinaceae, 177. Empetraceae, 178. Aextoxicaceae, 179. Pentadiplandraceae, 180. Celastraceae, 181. Corynocarpaceae, 182. Stackhousiaceae, 183. Goupiaceae, 184. Hippocrateaceae, 185. Erythropalaceae, 186. Capusiaceae, 187. Scyphostegiaceae.

Order 42. **Olacales**

Family–188. Olacaceae, 189. Opiliaceae, 190. Octoknemaceae, 191. Aptandraceae, 192. Dipentodontaceae, 193. Medusandraceae.

Order 43. **Santalales**

Families–194. Loranthaceae, 195. Grubbiaceae, 196. Santalaceae, 197. Myzodendraceae, 198. Balanophoraceae.

Order 44. **Rhamnales**

Families–199. Heteropyxidaceae, 200. Elaeagnaceae, 201. Rhamnaceae, 202. Vitaceae.

Order 45. **Myrsinales**

Families–203. Myrsinaceae, 204. Theophrastaceae, 205. Aegicerataceae.

Order 46. **Ebenales**

Families–206. Ebenaceae, 207. Sapotaceae, 208. Sarcospermaceae.

Order 47. **Rutales**

Families–209. Rutaceae, 210. Simaroubaceae, 211. Burseraceae, 212. Averhoaceae.

Order 48. **Meliales**

Family–213. Meliaceae.

Order 49. **Sapindales**

Families–214. Melianthaceae, 215. Sapindaceae, 216. Podoaceae, 217. Sabiaceae, 218. Anacardiaceae, 219. Aceraceae, 220. Hippocastanaceae, 221. Staphyleaceae, 222. Akaniaceae, 223. Julianiaceae, 224. Didiereaceae.

Order 50. **Loganiales**

Families–225. Potaliaceae, 226. Loganiaceae, 227. Buddleiaceae, 228. Antoniaceae, 229. Spigeliaceae, 230. Strychnaceae, 231. Oleaceae.

Order 51. **Apocynales**

Families–232. Plocospermaceae, 233. Apocynaceae, 234. Periplocaceae, 235. Asclepiadaceae.

Order 52. **Rubiales**

Families–236. Dialypetalanthaceae, 237. Rubiaceae.

Order 53. **Bignoniales**

Families–238. Cobaeaceae, 239. Bignoniaceae, 240. Pedaliaceae, 241. Martyniaceae.

Order 54. **Verbenales**

Families–242. Ehretiaceae, 243. Verbenaceae, 244. Stillbeaceae, 245. Chloanthaceae, 246. Phrymaceae.

Subphylum ... *Dicotyledones*

Division II ... *Herbaceae*

Order 55. **Ranales**

Families–247. Paeoniaceae, 248. Helleboraceae, 249. Ranunculaceae, 250. Nymphaeaceae, 251. Podophyllaceae, 252. Ceratophyllaceae, 253. Cabombaceae.

Order 56. **Berberidales**

Families–254. Sargentodoxaceae, 255. Lardizabalaceae, 256. Menispermaceae, 257. Nandinaceae, 258. Circaeasteraceae, 259. Berberidaceae.

Order 57. **Aristolochiales**

Families–260. Aristolochiaceae, 261. Hydnoraceae, 262. Cytinaceae, 263. Nepenthaceae.

Order 58. **Piperales**

Families–264. Piperaceae, 265. Saururaceae, 266. Chloranthaceae.

Order 59. **Rhoeadales**

Families–267. Papaveraceae, 268. Fumariaceae,

Order 60. **Cruciales (Brassicales)**

Family–269. Cruciferae (Brassicaceae).

Order 61. **Resedales**

Family–270. Resedaceae.

Order 62. **Caryophyllales**

Families–271. Elatinaceae, 272. Molluginaceae, 273. Caryophyllaceae, 274. Ficoidaceae, 275. Portulacaceae.

Order 63. **Polygonales**

Families–276. Polygonaceae, 277. Illecebraceae.

Order 64. **Chenopodiales**

Families–278. Barbeniaceae. 279. Phytolaccaceae, 280. Gyrostemonaceae, 281. Agdestidaceae, 282. Petiveriaceae, 283. Chenopodiaceae, 284. Amaranthaceae, 285. Cynocrambaceae, 286. Batidaceae, 287. Basellaceae.

Order 65. **Lythrales**

Families–288. Lythraceae, 289. Onagraceae, 290. Trapaceae, 291. Haloragidaceae, 292. Callitrichaceae.

Order 66. **Gentianales**

Family–293. Gentianaceae, 294. Menyanthaceae.

Order 67. **Primulales**

Families–295. Primulaceae, 296. Plumbaginaceae.

Order 68. **Plantaginales**

Family–297. Plantaginaceae.

Order 69. **Saxifragales**

Families–298. Crassulaceae, 299. Cephalotaceae, 300. Saxifragaceae, 301. Eremosynaceae, 302. Vahliaceae, 303. Francoaceae, 304. Donatiaceae, 305. Parnassiaceae, 306. Adoxaceae,

Order 70. **Sarraceniales**

Families–307. Droseraceae, 308. Sarraceniaceae.

Order 71. **Podostemales**

Families–309. Podostemaceae, 310. Hydrostachyaceae.

Order 72. **Umbellales**

Family–311. Umbelliferae (Apiaceae).

Order 73. **Valerianales**

Families–312. Valerianaceae, 313. Dipsacaceae, 314. Calyceraceae.

Order 74. **Campanales**

Families–315. Campanulaceae, 316. Lobeliaceae,

Order 75. **Goodeniales**

Families–317. Goodeniaceae, 318. Brunoniaceae,319. Stylidiaceae.

Order 76. **Asterales**

Family–320. Compositae (Asteraceae).

Order 77. **Solanales**

Families–321. Solanaceae, 322. Convolvulaceae, 323. Nòlanaceae.

Order 78. **Personales**

Families–324. Scrophulariaceae, 325. Acanthaceae, 326. Gesneriaceae, 327. Orobanchaceae, 328. Lentibulariaceae, 329. Columelliaceae.

Order 79. **Geraniales**

Families–330. Geraniaceae, 331. Limnanthaceae, 332. Oxalidaceae, 333. Trapaeolaceae, 334. Balsaminaceae.

Order 80. **Polemoniales**

Families–335. Polemoniaceae, 336. Hydrophyllaceae, 337. Cuscutaceae.

Order 81. **Boraginales**

Family–338. Boraginaceae.

Order 82. **Lamiales**

Families–339. Myoporaceae, 340. Selaginaceae, 341. Globuriaceae, 342. Labiatae (Lamiaceae).

Subphylum – ***Monocotyledones***

Division I – ***Calyciferae***

Order 83. **Butomales**

Families–343. Butomaceae, 344. Hydrocharitaceae.

Order 84. **Alismatales**

Family–345. Alismataceae, 346. Scheuchzeriaceae, 347. Petrosaviaceae.

Order 85. **Triuridales**

Family–348. Triuridaceae.

Order 86. **Juncaginales**

Families–349. Juncaginaceae, 350. Lilaeaceae (Heterostylaceae), 351. Posidoniaceae.

Order 87. **Aponogetonales**

Families–352. Aponogetonaceae, 353. Zosteraceae.

Order 88. **Potamogetonales**

Families–354. Potamogetonaceae, 355. Ruppiaceae.

Order 89. **Najadales**

Families–356. Zannichelliaceae, 357. Najadaceae.

Order 90. **Commelinales**

Families–358. Commelinaceae, 359. Cartonemataceae, 360. Flagellariaceae, 361. Mayacaceae.

Order 91. **Xyridales**

Families–362. Xyridaceae, 363. Rapateaceae

Order 92. **Eriocaulales**

Family–364. Eriocaulaceae.

Order 93. **Bromeliales**

Family–365. Bromeliaceae.

Order 94. **Zingiberales**

Families–366, Musaceae, 367 Sterlitziaceae, 368. Lowiaceae, 369. Zingiberaceae, 370. Cannaceae, 371. Marantaceae.

Division II – *Corolliferae*

Order 95. **Liliales**

Families–372, Liliaceae, 373. Tecophilaceae, 374. Triliaceae, 375. Pontederiaceae, 376. Smilacaceae, 377. Ruscaceae.

Order 96. **Alstroemeriales**

Families–378. Alstroemeriaceae, 379. Petermanniaceae, 380. Philesiaceae.

Order 97. **Arales**

Families–381. Araceae, 382. Lemnaceae.

Order 98. **Typhales**

Family–383. Sparganiaceae, 384. Typhaceae.

Order 99. **Amaryllidales**

Families–385. Amaryllidaceae.

Order 100. **Iridales**

Family–386. Iridaceae.

Order 101. **Dioscoreales**

Families–387. Stenomeridaceae, 388. Trichopodaceae, 389. Roxburghiaceae, 390. Dioscoreaceae.

Order 102. **Agavales**

Family–391. Xanthorrhoeaceae, 392. Agavaceae.

Order 103. **Palmales**

Family–393. Palmae (Palmaceae, Arecaceae).

Order 104. **Pandanales**

Family–394. Pandanaceae.

Order 105. **Cyclanthales**

Family–395. Cyclanthaceae.

Order 106. **Haemodorales**

Families–396. Haemodoraceae, 397. Hypoxidaceae, 398. Velloziaceae, 399. Apostasiaceae, 400.Taccaceae, 401. Philydraceae.

Order 107. **Burmanniales**

Families–402. Burmanniaceae, 403. Thismiaceae, 404. Corsiaceae.

Order 108. **Orchidales**

Family–405. Orchidaceae.

Division III – *Glumiflorae*

Order 109. **Juncales**

Families–406. Juncaceae, 407. Thurniaceae, 408. Centrolepidaceae, 409. Restionaceae.

Order 110. **Cyperales**

Family–410. Cyperaceae.

Order 111. **Graminales**

Family–411. Gramineae (Poaceae)

Merits and Demerits of Hutchinson's System

Merits. 1. Most taxonomists are of opinion that this system has given a much better idea of phylogenetic conception and has stimulated phyletic rethinking to a greater extent. 2. Primitive or basic orders are Magnoliales representing arborescent families, and Ranales representing herbaceous families, giving rise to woody and herbaceous forms respectively on parallel lines; 3. Bisexual and polypetalous flowers precede unisexual and gamopetalous flowers; 4. Amentiferae(catkin-bearing families with unisexual, apetalous flowers), *e.g.*, Fagaceae–oak, Betulaceae–birch, Juglandaceae–walnut, Salicaceae–willow and poplar, etc., is regarded as advanced (and not primitive as considered by Engler) and thus transferred (including Urticales) to a new phyletic position close to Rosales and Leguminales; their apparent simplicity represents reduction and specialization (and not primitiveness); 5. Casuarinaceae has been assigned an advanced position and placed at the top of the Amentiferae; 6. Several big orders have been split up into distinct small families, thus simplifying matters, *e.g.*, Rosales, Parietales, Malvales, etc. 7. Many families have been raised to the rank of orders, *e.g.*, Leguminosae to Leguminales, Saxifragaceae to Saxifragales, Podostemaceae to Podostemales, etc. 8. Reshuffling of several orders and families, *e.g.*, Cactales (Opuntiales of Engler) placed close to Cucurbitales and Passiflorales; 9. Origin of monocotyledons from dicotyledons at an early stage of evolution, the point of origin being the Ranales. 10. Splitting up of Helobiae (of Engler) into separate orders and considering the order Butomales (Butomaceae and Hydrocharitaceae) as the starting point of monocotyledons. 11.Reshuffling of genera of Liliaceae and Amaryllidaceae on the basis of inflorescence characters. 12.Rearrangement of several orders, finally ending in Cyperales and Graminales.

Demerits. 1. Many taxonomists do not agree with his rigid bifurcation of dicotyledons into Lignosae (woody plants) and Herbaceae (herbaceous types). 2. Many taxonomists hold different views regarding the relationship between various orders and families. 3. Monophyletic view regarding the origin of angiosperms is not universally accepted. 4. Monophyletic origin of Monocotyledons from the Ranales, as against the polyphyletic (diphyletic) views of Lotsty (1911) and Hallier (1912). 5. Urticales, Umbellales, Euphorbiales, etc., as originating from different ancestors.

TABLE NO 2.1 COMPARISON OF THREE SYSTEMS

Bentham & Hooker (1862) *Genera Plantarum*	**Engler and Prantl (1931)** *Die naturlichen Pflanzenfamilien*	**Hutchinson** *The Families of Flowering Plants*
Flowering plants divided into a few large groups based mainly on single and mostly artificial characters. Closely allied families often widely separated in consequence.	A phylogenetic system proposed subsequent to the acceptance of the theory of descent, Based on Eichler's system, and on the assumption that in fiowering plants, the *absence of a perianth is a primitive* feature.	A phylogenetic system based on the assumption that plants with sepals and petals, associate with other floral and anatomical characters regarded as also primitive are more ancient phylogenetically than plants without sepals or petals.
This was never claimed to be a phylogenetic system but for practical purposes has served a very useful term.	Thus families such as Casuarinaccae, Piperaceae, Salicaceae, Betulacec, Fagaceae, Urticaceae, etc., are regarded as primitive and are placed before families such as Magnoliaceae and Ranunculaceae.	This seems a logical interpretation of the theory that the parts of an angiospermous flower are modified leaves. Free parts are regarded as primitive, and connate or adnate parts as more recent. The spiral arrangement of parts is more primitive than the cyclic (whorled) and numerous free stamens are earlier than the few or connate. Also the bisexual precedes the unisexual flower.
Based on the earlier de Candolle's System (1818), itself an elaboration and slight, modification of Jussieu's arrangement proposed as early as 1779. This system was therefore cenceived under the influence of the old dogma of the fixity of species, and long before the theory of descent was proposed	Disregards the fact that many of these families without a perianth are characterized by having a gynoecium composed of the union of two or more carpels, a condition at variance with the universally accepted theory that the carpel is a modified leaf. With such an accepted view as to the origin of the carpel a syncarpous ovary could hardly have preceded an apocarpous one. Moreover a syncarpous ovary is characteristic of nearly all the highly placed of Engler's Archichlamydeae and of Metachlamydeae (Gamopetalae), the latter group accepted by all authors as being the most highly evolved of the Dicotyledons.	In this system the groups (orders of cohorts) are smaller and families are associated which show close general relationship Resemblances emphasized rather than differences. The monochlamydeae and Gamopetalae of Bentham and Hooker are distributed amongst the Polypetalae according to their relationships.
As a *Genera Plantarum* will ever remain a classic	Monocotyledons precede the Dicotyledons. Recent work on the fossil records gives little support to this system.	The monocotyledons are placed after the Dicotyledons, from which they were derived at an early stage, the point of origin being the Ranales.

List of Families of Dicotyledones in the Present Text with certain more or less constant characters.

Leaves : Opposite (or verticillate) leaves constant

Families with constantly opposite (or verticillate) leaves.

Caryophyllaceae; Casuarinaceae; Rubiaceae.

Leaves : Opposite (or verticillate) leaves predominant.

Families in which opposite (or verticillate) leaves are predominant.

Acanthaceae ; Apocynaceae ; Asclepiadaceae ; Loranthaceae ; Myrtaceae; Nyctaginaceae ; Pedaliaceae ; Rutaceae ; Scrophulariaceae.

Leaves : Compound : leaves predominant

Families in which compound leaves are predominant.

Umbelliferae (Apiaceae); Caesalpiniaceae; Papilionaceae (Fabaceae); Mimosaceae; Rosaceae; Rutaceae.

Leaves : Sometimes Compound

Families in which compound leaves sometimes occur.

Convolvulaceae; Cucurbitaceae; Euphorbiaceae; Ranunculaceae; Sterculiaceae; Verbenaceae.

Stipules : Leaves always stipulate.

Stipules are constant in the following families.

Caesalpiniaceae; Papilionaceae (Fabaceae); Magnoliaceae; Malvaceae; Mimosaceae; Polygonaceae; Rubiaceae.

Stipules : Leaves mostly stipulate

Families in which the leaves are mostly stipulate.

Linaceae; Rosaceae; Sterculiaceae.

Glandular Leaves

Families in which glandular or pellicid dots occur or sometimes occur in the leaves.

Acanthaceae; Anacardiaceae; Annonaceae; Asteraceae (Compositae); Combretaceae; Euphorbiaceae; Magnoliaceae; Meliaceae; Myrtaceae; Nymphaeaceae; Piperaceae; Polygonaceae; Rubiaceae; Rutaceae; Urticaceae; Verbenaceae; Violaceae.

Anther opening by terminal pores

Loranthaceae (some); Malvaceae; Solanaceae.

Inferior Ovary

Apiaceae (Umbelliferae); Asteraceae (Compositae); Cactaceae; Combretaceae; Cucurbitaceae; Loranthaceae; Myrtaceae; Nymphaeaceae (some); Rosaceae (some); Rubiaceae; Saxifragaceae (some).

CURRENT SYSTEMS

Cronquist's System

Arthur Cronquist (1919—) of U.S.A. presented a comprehensive system of classification in 1968 and thereafter its revised version in 1981. His system is parallel to that of Takhtajan's system in so many respects.

He divided the angiosperms in two classes-Magnoliopsida (=Dicotyledons) and Liliopsida (=Monocotyledons). Magnoliopsida is further divided into six sub-classes, *viz.*, Magnoliidae, Hamamelidae, Caryophyllidae, Dilleniidae, Rosidae and Asteridae; Liliopsida has five sub-classes, *viz.*, Alismatidae, Arecidae, Commelinidae, Zingiberidae and Liliidae.

The division Magnoliophyta has two classes, *viz.*, Magnoliopsida (= Dicotyledons) and Liliopsida (= Monocotyledons). There are six sub-classes, sixty four orders, 318 families, and about 165,000 species in Magnoliopsida (=Dicotyledons) whereas in Liliopsida (= Monocotyledons) there are five sub-classes, nineteen orders, 65 families and about 50,000 species. The outline of Cronquist's (1981) system of classification is as follows :

CLASS A. MAGNOLIOPSIDA (=DICOTYLEDONES)

Sub–class I Magnoliidae
Order 1. Magnoliales
" " 2. Laurales
" " 3. Piperales
" " 4. Aristolochiales
" " 5. Illiciales
" " 6. Nymphaeales
" " 7. Ranunculales
" " 8. Papaverales

Sub–Class II. Hamamelidae
Order 1. Trochodendrales
" " 2. Hamamelidales
" " 3. Daphnophyllales
" " 4. Dedymelales
" " 5. Eucommiales
" " 6. Urticales
" " 7. Leitneriales
" " 8. Juglandales
" " 9. Myricales
" " 10. Fagales
" " 11. Casuarinales.

Sub–class III Caryophyllidae
Order 1. Caryophyllales
" " 2. Polygonales
" " 3. Plumbaginales

Sub–class IV. Dilleniidae
Order 1. Dilleniales
" " 2. Theales
" " 3. Malvales
" " 4. Lecythidales
" " 5. Nepenthales
" " 6. Violales
" " 7. Salicales
" " 8. Capparales
" " 9. Batales
" " 10. Ericales
" " 11. Diapensiales
" " 12. Ebenales
" " 13. Primulales

Sub–Class V. Rosidae
Order 1. Rosales
" " 2. Fabales
Order 3. Proteales
" " 4. Podostemales
" " 5. Haloragales
" " 6. Myrtales
" " 7. Rhizophorales
" " 8. Cornales
" " 9. Santalales
" " 10. Rafflesiales
" " 11. Celastrales
" " 12. Euphorbiales
" " 13. Rhamnales
" " 14. Linales
" " 15. Polygalales
" " 16. Sapindales
" " 17. Apiales

Sub–class VI. Asteridae
Order 1. Gentianales
" " 2. Solanales
" " 3. Lamiales
" " 4. Callitrichales
" " 5. Plantaginales
" " 6. Scrophulariales
" " 7. Campanulales
" " 8. Rubiales
" " 9. Dipsacales
" " 10. Calycerales
" " 11. Asterales

CLASS B. LILIOPSIDA (=MONOCOTYLEDONS)

Sub–class I. Alismatidae
Order 1. Alismatales
" " 2. Hydrocharitales
" " 3. Najadales
" " 4. Triuridales

Sub–class II. Arecidae
Order 1. Arecales
" " 2. Cyclanthales
" " 3. Pandanales
" " 4. Arales

Sub–class III. Commelinidae
Order 1. Commelinales
" " 2. Eriocaulales

Order 3.	Restionales		Sub–class IV.	Zingiberidae	
" " 4.	Juncales		Order	1.	Bromeliales
" " 5.	Cyperales		" "	2.	Zingiberales
" " 6.	Hydatallales		Sub–Class V.	Liliidae	
" " 7.	Typhales		Order	1.	Liliales
			" "	2.	Orchidales

Takhtajan's System

Armen Takhtajan, an academician of Leningrad, Russia, presented a system of classification of angiosperms for the first time in 1942, and thereafter its revised versions in 1954, 1966, 1969 and 1980 respectively. Takhtajan's approach is quite similar to that of Cronquist. According to him the angiosperms are of monophyletic origin and evolved from some ancient group of gymnosperms.

The evolutionary tendencies followed by Takhtajan are as follows:

Woody plants precede herbaceous plants.

Sparingly branched trees gave rise to the trees with numerous slender branches.

Evergreen trees precede deciduous woody plants.

Simple pinnately–nerved leaves precede pinnately lobed, pinnatifid and pinnatisect leaves with pinnate venation.

Simple leaves precede compound leaves and *vice versa.*

Pinnate venation precedes palmate venation. Reticulate venation precedes parallel venation.

Alternate leaf arrangement precedes opposite and verticillaster arrangements.

Stomata with subsidiary cells are primitive and without subsidiary cells are advanced.

Tripentalacunar type of nodal structure is primitive, and unilacunar type is advanced.

The vessel members with scalariform perforations precede the vessel members with simple perforations.

Tracheids gave rise to fibre tracheids, and then to libriform fibres.

Cymose inflorescence is primitive, and racemose is derived one.

Primitive flowers are spiral whereas advanced are cyclic in arrangement of their floral organs.

The primitive stamens were leaf–like, and with marginally situated microsporangia.

The pollen is derived from monocolpate to tricolpate and from tricolpate to triporate. The exine of primitive pollen was without external sculpturing whereas in advanced pollen there was sculpturing of various types.

Apocarpy precedes syncarpy.

Bitegmic ovules are primitive, whereas unitegmic are advanced.

Crassinucellate ovules precede tenuinucellate ovules. Anatropous ovule is primitive; the other types are derived from it.

Entomophily gave rise to anemophily and *vice versa.*

Polygonum-type of embryo sac is primitive, the other types are derived from it.

Endospermic seeds are primitive, whereas non–endospermic seeds are advanced. The monocot embryo is evolved from dicot embryo.

Many seeded follicle is most primitive type of fruit, from this other types are evolved.

Takhtajan has divided his Division–Magnoliophyta (=Angiospermae) into two classes–Magnoliopsida (= Dicotyledones) and Liliopsida (=Monocotyledones). He has further divided–Class Magnoliopsida into seven sub–classes and twenty superorders, and Liliopsida into three sub–classes and eight superorders.

According to him the order Magnoliales is most primitive among the flowering plants. All other angiosperms have been derived from the Magnoliales.

The order Alismatales has been supposed to be most primitive among monocotyledons and the members of the other monocotyledonous orders have been derived from them.

According to him the highest evolved order among dicotyledons is Asterales; and Arales among monocotyledons.

The outline of Takhtajan's (1980) system is as follows:

DIVISION MAGNOLIOPHYTA (=ANGIOSPERMAE)
CLASS A. MAGNOLIOPSIDA (= DICOTYLEDONS)

Sub–class I	Magnoliidae
Superorder	I. Magnolianae
Order	1. Magnoliales (Annonales)
" "	2. Illiciales
" "	3. Laurales
" "	4. Piperales
" "	5. Aristolochiales
Superorder II.	Rafflesianae
Order	6. Rafflesiales
Superorder III.	Nymphaeanae
Order	7. Nymphaeales
" "	8. Nelumbonales
Sub-calss B.	Ranunculidae
Superorder IV.	Ranunculanae
Order	9. Ranunculales
" "	10. Papaverales
" "	11. Sarraceniales
Sub-class C.	Hamamelididae
Superorder V.	Hamamelidanae
Order	12. Trochodendrales
" "	13. Cercidiphyllales
" "	14. Eupteleales
" "	15. Didymelales
" "	16. Hamamelidales
" "	17. Eucomuniales
" "	18. Urticales
" "	19. Barbeyales
" "	20. Casuarinales
" "	21. Fagales
" "	22. Balanopales
" "	23. Leitneriales
Superorder VI.	Juglandanae
Order	24. Myricales
" "	25. Juglandanles
Sub-class D.	Caryophyllidae
Superorder VII.	Caryophyllanae
Order	26. Caryophyllales
" "	27. Polygonales
Super order VIII.	Plumbaginanae
Order	28. Plumbaginales
Sub-class E.	Dilleniidae
Order	29. Dilleniales
" "	30. Paeoniales
" "	31. Theales
" "	32. Violales
" "	33. Bigoniales (Datiscales)
" "	34. Capparales
" "	35. Tamaricales
" "	36. Salicales
Superorder X.	Ericanae
Order	37. Ericales
" "	38. Ebenales
" "	39. Primulales
Superorder XI.	Malvanae
Order	40. Malvales
" "	41. Euphorbiales
Sub–Class F.	Rosidae
Superorder XII	Rosanae
Order	43. Saxifragales
" "	44. Rosales
" "	45. Fabales
" "	46. Connarales
" "	47. Podostemales
" "	48. Nepenthales
Superorder XIII.	Myrtanae
Order	49. Myrtales
Superorder XIV.	Rutanae

Order 50. Rutales
" " 51. Sapindales (Acerales)
" " 52. Geraniales
" " 53. Polygalales
Superorder XV. Aralianae
Order 54. Cornales
" " 55. Araliales (Apiales)
Superorder XVI. Celastranae
Order 56. Celastrales
" " 57. Santalales
" " 58. Balanophorales
" " 59. Rhamnales
" " 60. Elacagnales
Superorder XVII. Proteanae
Order 61. Proteales
Sub–class G. Asteridae

Superorder XVIII. Gentiananae
Order 62. Gentianales
" " 63. Oleales
" " 64. Dipsacales
" " 65. Loasales
Superorder XIX. Lamianae
Order 66. Polemoniales
" " 67. Lamiales
" " 68. Scrophulariales
Superorder XX. Asteranae
Order 69. Campanulales (including) Goodeniales)
" " 70. Calyceralesz
" " 71. Asterales

CLASS–LILIOPSIDA (=MONOCOTYLEDONES)

Sub-class A. Alismatidae
Superorder I. Alismatanae
Order 1. Alismatales
" " 2. Najadales (including Aponogetonales, Potomogetonales and Zosterales)
Sub–class B. Lilidae (including Commelinidae and Zingiberidae)
Superorder II. Triuridanae
Order 3. Triuridales
Superorder III. Lilianae
" " 4. Liliales
" " 5. Smilacales
Order 6. Burmanniales
" " 7. Orchidales

" " 8. Bromeliales
Superorder IV. Juncanae
Order 9. Juncales
" " 10. Cyperales
Superorder V. Commelinanae
Order 11. Commelinales
" " 12. Eriocaulales
" " 13. Restionales
" " 14. Hydatellales
" " 15. Poales
Superorder VI. Zingiberanae
Order 16. Zingiberales
Sub–class C. Arecidae
Superorder VII. Arecanae
Order 17. Arecales
" " 18. Cyclanthales
" " 19. Pandanales
" " 20. Typhales

Superoder VII. Aranae
Order 21. Arales

Dahlgren's System

Rolf Dahlgren, a botanist of Copenhagen, Denmark published a system of classification of angiosperms in 1975, and thereafter a revised and improved version in 1980, 1981 and 1983 respectively. He has taken into consideration the utmost possible information. He has taken into consideration the following : apocarpy, syncarpy, and monocarpy; choripetaly and sympetaly; monoaperturate pollen grains; sclereid idioblasts; microsporogenesis types; bi–and unitegmic ovules, tenuinucellate, pseudocrassinucellate and crassinucellate ovules; bi–and trinucleate pollens; oxalate raphides, silica bodies; iridoid compounds; ellagic acid and ellagitannins; benzylisoquinoline alkaloids, tropane alkaloids, pyrrolizidine alkaloids; polyacetylene and several flavonoids. This shows that Dahlgren has used chemical characters extensively in his system of classification.

According to him the angiosperms have evolved in one particular line of gymnosperms and thus treated of monophyletic origin. He is not of opinion that Alimatiflorae are closely connected with the monocotyledonous ancestors. According to him "the appearance of the monocotyledons may have its counterpart in the evolution of the mainly herbaceous Aristolochiaceae, which also evolved out of early Magnoliales with trimerous flowers and two perianth whorls Aristolochiaceae share with the monocotyledons in possession of cuneate crystalloids in the sieve element plastids." According to him the "Amentiferae" have been interpreted as derived from trimerous ancestral type.

Dahlgren is in agreement that the discontinuity between monocotyledons and dicotyledons is not particularly marked. He is of opinion that the monocotyledons can be clearly circumscribed on the combination of a single–cotyledoned embryo and sieve element plastids. This system has also been advocated by serological studies. In Dahlgren's classification the angiosperms are made up of a class, Magnoliopsida. It is as follows:

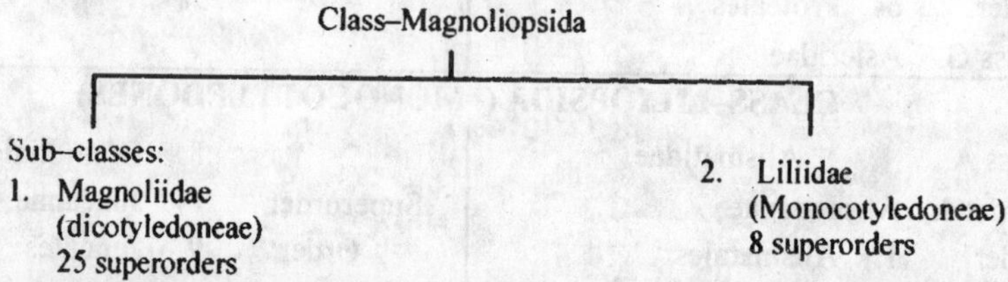

The superorders end with the suffix–florae.

According to Dahlgren the superorder magnoliiflorae have been treated to be most primitive among dicotyledons. They possess the following characters: woody habit, primitive wood; P–type plastids in the sieve elements; alternate, exstipulate leaves, cells with essential oils in leaves; benzylisoquinoline alkaloid present, trimerous tepals; flat and leaf–like stamens with sporangia below the apex, pollengrains sulcate, 2–nucleate; gynoecium apocarpous, or monocarpellary; endosperm formation cellular. The order Nelumbonales is included in superorder Magnoliiflorae and not in Ranunculiflorae.

The superorder Alismatiflorae has the following characteristics which are generally unusual for monocotyledons : intervaginal squamules present; vessels absent in the stem and often in the roots; root hairs attached to short epidermal hairs, amoeboid tapetum; 3–nucleate pollen grains; apocarpous or monocarpellary condition of gynoecia; embryo formation of caryophyllad type; oxalate raphides and silica crystals absent.

The outline classification is as follows:

CLASS–MAGNOLIOPSIDA (ANGIOSPERMAE)

Sub–class MAGNOLIIDAE (=DICOTYLE-DONEAE)		Order	–Illiciales
		" "	–laurales
Superorder	–Magnoliiflorae	" "	–Nelumbonales
Order	–Annonales	Superorder	–Nymphaeiflorae
" "	–Aristolochiales	Order	–Piperales
" "	–Rafflesiales	" "	–Nymphaeales
" "	–Magnoliales	Superorder	–Ranunculiflorae
Order	–Lactoridales	Order	–Ranunculales
" "	–Chloranthales	" "	–Papaverales

Superorder	–Caryophylliflorae
Order	–Caryophyllales
Superorder	–Polygoniflorae
Order	–Polygonales
Superorder	–Plumbaginiflorae
Order	–Plumbaginales
Superorder	–Malviflorae
Order	–Malvales
" "	–Urticales
" "	–Euphorbiales
" "	–Thymelaeales
" "	–Rhamnales
" "	–Elaegnales
Superorder	–Violiflorae
Order	–Violales
" "	–Cucurbitales
" "	–Salicales
" "	–Tamaricales
" "	–Capparales
" "	–Salvadorales
Superorder	–Theiflorae
Order	–Dilleniales
" "	–Paeoniales
" "	–Theales
Superorder	–Primuliflorae
Order	–Primulales
" "	–Ebenales
Superorder	–Rosiflorae
Order	–Trochodendrales
" "	–Cercidiphyllales
" "	–Hamamelidales
" "	–Geissolomomatales
" "	–Balanopales
" "	–Fagales
" "	–Juglandales
" "	–Myricales
" "	–Casuarinales
" "	–Buxales
" "	–Cunoniales
" "	–Saxifragales
" "	–Droserales
" "	–Gunnerales
Order	–Rosales
Superorder	–Podostemiflorae
Order	–Podostemales
Superorder	–Proteiflorae
Order	–Proteales
Superorder	–Myrtiflorae
Order	–Haloragales
" "	–Rhizophorales
" "	–Myrtales
" "	–Chrysobalanales
Superorder	–Fabiflorae
Order	–Fabales
Superorder	–Rutiflorae
Order	–Sapindales
" "	–Rutales
" "	–Polygalales
" "	–Geraniales
" "	–Balsaminales
" "	–Tropaeolales
Superorder	–Santaliflorae
Order	–Celastrales
" "	–Vitales
" "	–Santalales
Superorder	–Balanophoriflorae
Order	–Balanophorales
Superorder	–Araliflorae
Order	–Pittosporales
" "	–Araliales
Superorder	–Asteriflorae
Order	–Campanulales
" "	–Asterales
Superorder	–Solaniflorae
Order	–Solanales
" "	–Boraginales
Superorder	–Corniflorae
Order	–Fouquieriales
" "	–Ericales
" "	–Eucommiales
" "	–Sarraceniales
" "	–Cornales
" "	–Dipsacales
Superorder	–Losaliflorae

Order	–Losales	Order	–Liliales
Superorder	–Gentianiflorae	" "	–Burmanniales
Order	–Goodeniales	" "	–Orchidales
" "	–Oleales	Superorder	–Bromeliiflorae
" "	–Gentianales	Order	–Velloziales
Superorder	–Lamiiflorae	" "	–Bromeliales
Order	–Scrophulariales	" "	–Haemodorales
" "	–Hippuridales	" "	–Philydrales
" "	--Lamiales	" "	–Pontederiales
" "	–Hydrostachyales	" "	–Typhales
Sub-class	–LILIIDAE	Superorder	–Zingiberiiflorae
(=MONOCOTYLEDONEAE)		Order	–Zingiberales
Superorder	–Alismatiflorae	Superorder	–Commeliniflorae
Order	–Hydrocharitales	Order	–Commelinales
" "	–Alismatales	" "	–Hydatellales
" "	–Zosterales	" "	–Juncales
Superorder	–Triuridiflorae	" "	–Cyperales
Order	–Triuridales	" "	–Poales
Superorder	–Ariflorae	Superorder	–Areciflorae
Order	–Arales	Order	–Arecales
Superorder	–Liliiflorae	" "	–Cyclanthales
Order	–Dioscoreales	" "	–Pandanales
" "	–Asparagales		

Thorne's system

Robert F. Thorne of California U.S.A., published a system of classification in 1968. Thereafter a revised outline of this system was published in 1981 and 1983 respectively. This system has been used to express the phylogenetic relationships among the higher taxa of flowering plants. The modern trends of taxonomy have been taken into consideration in the depiction of more accurate relationships of different taxa. He has laid much emphasis on phytochemical approach. The other modern approaches such as palynology, seed morphology, anatomy, embryology, host-parasite relationships, plant geography, palaeobotany, microstructure and cytology have been utilized in the differentiation of taxa. According to Thorne the Angiospermae are monophyletic in origin. He has divided the flowering plants into two sub-classes (1) Dicotyledoneae (Annonidae) and (2) Monocotyledoneae (Liliidae). There are 19 superorders in Dicotyledoneae and 9 superorders in Monocotyledoneae. The outline of Thorne's system is as follows:

CLASS–ANGIOSPERMAE (ANNONOPSIDA)

Sub–classes : (1)DICOTYLEDONEAE (ANNONIDAE) 19 Superorders

(2) MONOCOTYLEDONEAE 9 Superorders

Sub-Class=DICOTYLEDONEAE (ANNONIDAE)

Superorder	–Annoniflorae
Order	–Annonales
" "	–Nelumbonales
" "	–Paeoniales
" "	–Berberidales
Superorder	–Nymphaeiflorae
Order	–Nymphaeales
Superorder	–Rafflesiiflorae
Order	–Rafflesiales
Superorder	–Theiflorae
Order	–Theales
" "	–Ericales
" "	–Ebenales
" "	–Primulales
" "	–Polygonales
Superorder	–Chenopodiiflorae
	–(Centrospermae)
Order	–Chenopodiales
Superorder	–Geraniiflorae
Order	–Geraniales
Superorder	–Santaliflorae
Order	–Celastrales
" "	–Santalales
" "	–Balanophorales
Superorder	–Violiflorae
Order	–Violales
" "	–Caparales
Superorder	–Malviflorae
Order	–Malvales
" "	–Urticales
" "	–Rhamnales
" "	–Euphorbiales
Superorder	–Rutiflorae
Order	–Rutales
Superorder	–Proteiflorae
Order	–Proteales
Superorder	–Hamamelidiflorae
Order	–Hamamelidales
" "	–Casuarinales
" "	–Fagales
Superorder	–Rosiflorae
Order	–Rosales
" "	–Pittosporales
Superorder	–Loasiflorae
Order	–Loasales
Superorder	–Myrtiflorae
Order	–Myrtales
Superorder	–Gentianiflorae
Order	–Oleales
" "	–Gentianales
" "	–Bignoniales
" "	–Lamiales
Superorder	–Solaniflorae
Order	–Solanales
" "	–Campanulales
Superorder	–Corniflorae
Order	–Cornales
" "	–Araliales
" "	–Dipsacales
Superorder	–Asteriflorae
Order	–Asterales
SUB–CLASS MONOCOTYLEDONEAE (LILIIDAE)	
Superorder	–Liliiflorae
Order	–Liliales
Superorder	–Triuridiflorae
Order	–Triuridales
Superorder	–Alismatiflorae
Order	–Alismatales
" "	–Zosterales
" "	–Najadales
Superorder	–Ariflorae
Order	–Arales
Superorder	–Cyclanthiflorae
Order	–Cyclanthales
Superorder	–Pandaniflorae
Order	–Pandanales
Superorder	–Areciflorae
Order	–Arecales
Superorder	–Typhiflorae
Order	–Typhales
Superorder	–Commeliniflorae
Order	–Commelinales
" "	–Pontederiineae
" "	–Juncineae
" "	–Zingiberales

FIVE KINGDOM CLASSIFICATION

Today, many biologists believe that a five-kingdom classification system is best. The new five-kingdom classification of organisms was proposed by Whittaker in 1969, to replace the old two-kingdom system. At present, the following three traits are used to place organisms into one of five kingdoms :

1. The kind of cells that make up the organism.

2. The number of cells that make up the organism.

3. The method of which the organism obtains its food.

The five kingdoms are Monera, Protista, Plantae, Fungi and Animalia.

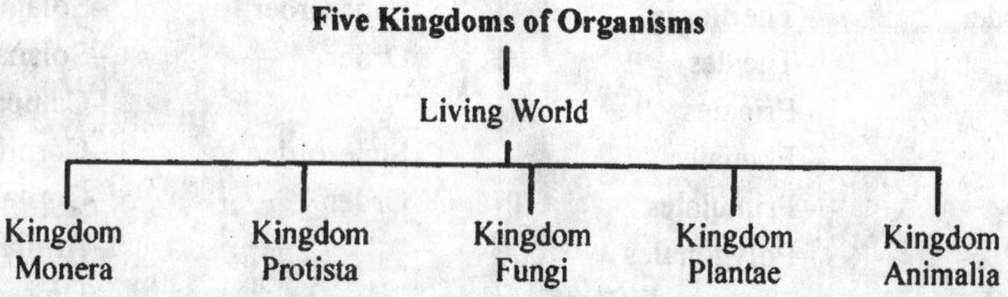

Kingdom Monera — The Kingdom of Procaryotes

Most organisms in this kingdom are made up of a simple, **procaryotic** cell.

Some monerans make their own food, while others take in food from an outside source.

The monerans include the smallest, the simplest, and the most widespread groups of organisms on Earth. Except for one species of bacteria (*i.e., Escherichia coli*) that lives in our intestines less is known about this kingdom than all of the others. There are two major phyla in this kingdom.

Bacteria. They are the smallest organisms in the five kingdoms. Most bacteria are either round (cocci), rod-shaped (bacilli), or spiral-shaped (spirilli). Some kinds form colonies of chains, others form clusters. Some bacteria are capable of dividing in half every 20 minutes. At this rate one bacterium can give rise to one million bacteria in just six hours.

Bacteria are omnipresent, and found almost everywhere on Earth. They are found even on the surface of human body and inside the alimentary canal. Most bacteria feed on dead organisms and cause them to rot. Bacteria are used to make inudstrial chemicals, to treat sewage, and to make foods like yogurt, cheese, butter and curd. However, certain bacteria are also the cause of many human infections such as tooth decay, pneumonia, tuberculosis, etc.

Cyanobacteria. They are also called blue-green bacteria. They are bigger than most bacterial cells. They also contain pigments in them which absorb sunlight and enable the organisms to make their own food. Blue-green bacteria are commonly found growing in water and on wet surfaces. Commonly, they reproduce in large numbers. The Red Sea was named after a blue-green bacteria containing red pigments. When these bacteria increase in great numbers, the water looks red in colour.

Other forms. The Actinomycetes (filamentous bacteria) and Archaebacteria are also included in the kingdom Monera. The Archaebacteria thrive under extreme environmental conditions such as high temperature (80°C), absence of oxygen, high salt concentration and acidic pH. They are thought to be early forms of life.

The members of kingdom Monera are the important decomposers and mineralisers in the biosphere.

Kingdom Protista — The Kingdom of Unicellular Eucaryotes

The protists are much more complicated than monerans. A protist cell is somewhat like the whole body of a plant or an animal stuffed into a single cell. They are mostly unicellular and aquatic eucaryotes. They contain typical eucaryotic cell organelles such as nucleus, mitochondria, plastids, endoplasmic reticulum and Golgi bodies. Very commonly they are flagellate with 9 + 2 internal microtubular structure of flagella. Protist cells are so complex that they are often thousands of times larger than human body cells. The main groups of this kingdom are—Diatoms, Dinoflagellates, Euglena, Sarcodines, Ciliates, Flagellates and Sporozoans.

Diatoms. The bodies of such protists have two shells which fit together like a pill box. They float on the surface of fresh and salt water.

Dinoflagellates. These protists have two whiplike tails for movement. Like diatoms they also make their own food.

Euglena. For a long time, biologists could not decide whether these protists were plants or animals. They contain green pigments and make their own food.

Sarcodines. They resemble small masses of cloudy gelatin. They move by pushing out parts of their cell membrances.

Ciliates. They are covered with short, hairlike structures called cilia. The cilia contract in waves to push the organisms along.

Flagellates. They move by using one or two whiplike projections or flagella. The flagella possess 9 + 2 internal microtubular structure. Most of these protists live independently, but some live inside other organisms. The causal organism of African sleeping sickness belongs to this group.

Sporozoans. They live inside other organisms. Most have no structures for moving about. They depend on their hosts for transportation, *e.g.*, malaria protozoans in the human blood stream.

Kingdom Fungi — The Kingdom of Multicellular Decomposers

More than 100,000 kinds of fungi have been described and named. Biologists estimate that at least another 10,000 remain to be discovered. The members of this kingdom are diverse kinds of eucaryotic, predominantly multicellular hetetrophic organisms. They feed by absorbing the dead or living material of other organisms. They are termed **saprobes** when they live on decaying organic matter, and **parasites** when they assimilate tissues of living things. They produce thread like structures (mycelia) that grow deep into their food materials. Chemicals from the mycelia dissolve the food. The fungi then absorb the dissolved food through these mycelia. When the food is part of a living organism, the fungus may cause disease to its host, such as atheletes foot in humans and rusts in cereals.

However, the yeasts have a single-celled body. They are involved in fermentation. Their mode of sexual reproduction shows their relationship with multicellular fungi.

The fragmentation of hypha helps in multiplication where each fragment grows into new mycelium. The fungi also reproduce by asexual spores such as zoospores, aplanospores, conidia, gonidia, ascospores and basidiospores. Sexual reproduction takes place by the fusion of similar gametes (isogamy) and dissimilar gametes (oogamy). Fungi are grouped into following important classes :

Class Myxomycetes. They are frequently found in the rainy season in cool, shady moist places in the woods, on manure or other decomposing plant material which holds abundant moisture.

These organisms are naked during all stages of development except the spore stage. This naked vegetative phase consists of a naked amoeboid multinucleate free living plasmodium. They ingest particles of food and reject the portions remaining after digestion. The plasmodia and fructifications of many species of slime molds are very colourful. The spores are provided upon or within aerial sporangia and are more or less dispersed by wind. The asexual spores called zoospores produced in sporangia are also means of reproduction. Sometimes these motile spores also act as compatible gametes which fuse and form zygote. The thalloid body called plasmodium represents a diploid (2*n*) stage.

Class Phycomycetes. They are frequently found in damp places. They reproduce asexually by biflagellate zoospores. The zoospores are borne in zoosporangia of various types. Sometimes, aplanospores are produced within sporangia. The sexual reproduction takes place by fusion of similar (isogamy) or dissimilar (oogamy) gametes.

The mycelium is coenocytic (multinucleate) and aseptate.

The main examples are — *Saprolegnia, Mucor, Albugo,* etc.

Class Ascomycetes. They are known as **sac fungi**. They may be unicellular (*e.g.*, yeasts) and multicellular (*e.g.*, *Aspergillus*) fungi. The asexual reproduction takes place by means of **conidia**. The conidia are generally produced in chains. The spores are formed exogenously, *i.e.*, outside sporangium. The conidia germinate immediately after their detachment from the conidiophores when they fall upon suitable substrata in appropriate atmospheric conditions producing germ tubes.

Sexual reproduction is through ascospores which are formed endogenously within an **ascus**. The gametes involved in sexual reproduction are non-motile and represented by + and – strains. The fusion of gametes is followed by meiosis that produces haploid (*n*) ascospores. The fruiting body called **ascocarp** is generally cup shaped and contains spore bearing structures the asci. The examples of this class are — yeast, *Aspergillus, Penicillium, Claviceps,* etc.

Class Basidiomycetes. They are also called club fungi. The Basidiomycetes include the fungi which bear their spores exogenously on **basidia** (singular **basidium**). The basidia may occur singly or be variously arranged by hymemia. These fungi have septate mycelia. The asexual spores called basidiospores are generally four in number. They are produced exogenously on basidium. Two compatible nuclei fuse to form zygote, which undergoes meiosis and forms four basidiospores. The fruiting body containing basidia is a multicellular structure called **basidiocarp.** Common examples are — rusts (*Puccinia*), smuts (*Ustilago*), mushrooms (*Agaricus*), etc. Some mushrooms are edible and some are deadly poisonous.

Kingdom Plantae — The Kingdom of Multicellular Producers

This kingdom contains mostly green, multicellular organisms. Their cells are of various types and form special organs. Most of these organs help in the production of food and in the reproduction of the plant. The plants are found on land, sea shores, in lakes and streams. The main groups of kingdom Plantae consist of complex red, brown and green algae, liverworts, mosses, ferns, gymnosperms and angiosperms.

The plant cells possess cellulose cell wall, and therefore, cannot contract and relax like animal cells. The plants are immobile. They are primary producers on land, and along many sea shores. Sometimes, they make dominant vegetation of lakes and ponds. They synthesize all their organic constituents from water, carbon dioxide, and inorganic forms of essential elements, using light energy trapped by chlorophyll and accessory pigments. This great phenomenon is photosynthesis. Certain flowering plants are parasitic and absorb their food partially or wholly from host plants. Some plants are insectivorous and absorb their nutrition by trapping and ingesting the small insects. The main phyla of this kingdom are: Brown algae, Red algae, Green algae, Bryophytes, and Tracheophytes (*i.e.*, ferns, gymnosperms and angiosperms).

Brown algae. The members of this phylum are commonly found in cool sea water They contain golden brown and green pigments that give them their brown colour. They possess stemlike organs called **stipes** and leaf-like organs called **blades.**

Red algae. The red algae are usually found in urban tropical seas. Their colour is due to a red pigment. The red pigment absorbs the blue light that penetrates deep water. They use the energy in this blue light to make food. Gelatin substitutes and other foods are obtained from them.

Green algae. They are found both in sea and fresh water. On land, they cover most of the surfaces that receive sunlight and stay damp. They are green, store food and possess cell walls.

Bryophytes. They are small green (*e.g.*, liverworts and mosses) that live only in moist places.

Tracheophytes. The members of this phylum have vessels that transport food, water and minerals to all parts of the plant body. This phylum is divided into three classes according to the kinds of seeds each plant forms. **Ferns** (*i.e.*, pteridophytes) have no seeds; **cone bearing** plants (*i.e.*, gymnosperms) have uncovered (naked) seeds, and flowering plants (*i.e.*, angiosperms) possess covered seeds that are enclosed in a fruit.

Kingdom Animalia—The Kingdom of Multicellular Consumers

The animal kingdom is the most complex of all five kingdoms. They are multicellular holozoic eucaryotes, and also known as **Metazoa**. They obtain food from outside sources and move from place to place. They swallow the food, and digest it inside the body. Ecologically they are called **consumers**, forming links in different food chains and complex food webs through which a part of matter and energy flows from the producer to the decomposer.

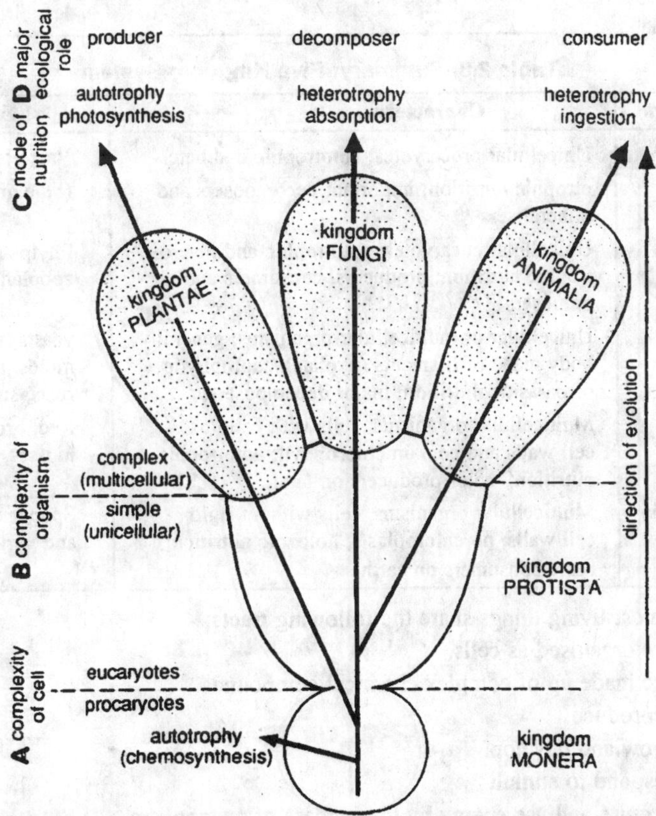

Fig. 2.1. The five Kingdoms of life. They are being arranged to show the probable phylogenetic relationships among the kingdoms.

Table 2.2. Characteristics of Five Kingdoms

Kingdom	Characteristic features	Examples
Monera	Unicellular, procaryatic autotrophic, asexual reproduction, multiplication by amitosis.	Archaebacteria, eubacteria and cyanobacteria.
Protista	Unicellular, eucaryotic, aquatic, autotrophic and heterotrophic, asexual reprouction by division into two and sexual reproduction by fusion.	Protozoans, slime molds and water molds.
Fungi	Multicellular eucaryotic, cell wall of chitin, heterotrophic, saprophytic, parasitic, asexual and sexual reproduction by spores and gametes respectively.	Bread mold, yeast, rusts, smuts and mushrooms.
Plantae	Multicellular, non-motile eucaryotic cellulose cell wall, well developed tissue system, autrotrophic, photosynthesis, asexual reproduction by multiplication, sexual reproduction by distinct sex organs, life cycle exhibits alternation of generations.	Algae, bryophytes, pteriodophytes, gymnosperms and angiosperms.
Animalia	Multicellular, motile forms, eucaryotic, well developed tissues, heterotrophic, sexual reproduction by distinct sex organs, control and coordination system, distinct embryological development.	Sponges, coelenterates, worms, insects, molluscs, star-fish, amiphibians, reptiles, birds, and mammals.

Table 2.3. Summary: Five Kingdom System

Kingdom	Characteristics	Examples
1. **Monera**	Unicellular procaryotes; autotrophic and heterotrophic nutrition; important decomposers and mineralisers	Bacteria, cyanobacteria (blue green algae)
2. **Protista**	Unicellular eucaryotes; autotrophic and heterotrophic nutrition; producers, consumers and decomposers	Phytplankton, zooplankton, protozoans
3. **Fungi**	Unicellular or multicellular organisms; generally body made of filaments (hyphae); saprophytic or parasitic nutrition; major decomposers.	yeasts (unicellular), molds, mushrooms, rusts, smuts, etc.
4. **Plantae**	Multicellular organisms; cells with cellulose cell walls and contain chlorophyll; autotrophic nutrition; major producers on land.	Red, brown, green algae, mosses, ferns, seed plants
5. **Animalia**	Multicellular organisms; cells without rigid cell walls; no chloroplasts; holozoic nutrition; basic consumers on earth.	Sponges, invertebrates and vertebrates.

However, most living things share the following tracts:

1. They are organised as cells.
2. They are made up of complex chemical compounds.
3. They reproduce.
4. They grow and develop.
5. They respond to stimuli
6. They acquire and use energy by the process of metabolism.

The millions of kinds of organisms are grouped by most biologists into five kingdoms: (*i*) monerans (*ii*) protists (*iii*) fungi (*iv*) plants and (*v*) animals. The merits and demerits of five kingdom classification are as follows:

Merits and Demerits

At first biologists grouped all organisms into two kingdoms. Organisms either belonged to the plant kingdom or to the animal kingdom. Basically the plants are autotrophic and animals are heterotrophic.

Later on this was realised that some groups did not fit well into either kingdom. For example, organisms such as bread mold do not make their own food as plants do but they are not animals.

Biologists also discovered that two kinds of one-celled organisms exist. One group has a nucleus without a covering or membrane around it and the nuclear material floats freely in the cell. Such a cell is called a **procaryote**. Whereas organisms in the other group have an organised nucleus surrounded by a membrane and the structures inside these cells are more complex. This type of cell is called **eucaryote.**

As a result of these findings this was decided that other kingdoms should be added to distinguish the different kinds of organisms.

With the result, many biologists believe that a five-kingdom classification system is best.

Identification of Species and Use of Key

A taxonomist makes it possible for other biologists to identify the species revised by him. This can be done in three ways (*i*) by placing specimens of species classified by him (taxonomist) in a museum or herbarium, (*ii*) by publishing descriptions and illustrations of the species, and (*iii*) by forming taxonomic keys.

Indented key

Character		**Genus**
Carpel single-	ovuled, fruit achene, leaves opposite, compound petals absent, leaves without tendril	*Clematis*
	Petals present third or terminal leaflet modified into tendril	*Naravelia*
	Leaves alternate or radical	*Anemone*
Carpel many-	ovuled; fruit follicle, carpels united at base; flowers regular	*Nigella*
	Carpels free at base; flowers irregular	*Aconitum*

Bracketed Key

Character	**Genus**
1. Carpel single-ovuled; fruit achene	2
1. Carpel many-ovuled; fruit follicle	4
2. Leaves opposite; compound	3
2. Leaves alternate; radical	*Anemone*
3. Petals absent, leaves without tendril	*Clematis*
3. Sepals as many as petals; third or terminal leaf modified into tendril ..	*Naravelia*
4. Carpels united at the base; flowers regular	*Nigella*
5. Carpels free; flowers irregular	*Aconitum*

N.B. In this key the number on the right indicates the next choice of paired contrasting statements.

Other taxonomic aids. (i) Monographs, which give comprehensive account of complete compilation of available information of any one family or genus at a given time.

(*ii*) **Manuals,** which contain compiled information about area covered, keys, description of families, genus and species.

(*iii*) **Other publications,** such as periodicals and dictionaries are brought out to give us information about new additions and updated information.

Utility of Systematics

Systematics is an observational science and forms a part of natural history. This gives us an idea of organic diversity, its origin, and evolution. The role of systematics is obvious in the study of population biology, population genetics, evolutionary biochemistry, immunobiology, etc.

Systematics is of value in the identification and breeding of a number of economically important crop plants. It gives the correct identity of the plants. In our country, the epidemiological problem is important. A few species of mosquitoes transmit malaria, filaria, dengue, kala azar, etc. The identification of these, their feeding and breeding biology and their behaviour have led to their eradication and control. Systematics provides the basic information needed in many applied disciplines and thus becomes complementary to all branches of basic and applied biology.

3

Modern Trends in Taxonomy

It has now been generally accepted by taxonomists that morphological characters alone should not be regarded in systematical classification of plants. Other disciplines like cytology, genetics, physiology, geographical distribution, anatomy, embryology, ecology, palynology, phenology, biochemistry, numerical taxonomy and transplant experiments have been found to be especially useful in solving some of the taxonomical problems by providing additional characters. The cytogenetical data may also indicate the origin and relationship of different taxa. The impact of above mentioned disciplines on the present day taxonomy has changed the face of it from alpha (classical) to omega (modern kind). New systematics (taxonomy) thus has become a very interesting subject, claiming to have solved many problems of classification and to have evolved a better taxonomy in which relationships of all kinds are knit together. However, it has also created confusion to the discomfort of classical taxonomists (Chennaveeraiah, 1970). There are amongst the taxonomists lumpers as well as the splitters. In such considerations morphology must still occupy a prime place although amongst the other disciplines cytology and genetics have played a very significant role (P.N. Mehra, 1970).

1. GROSS MORPHOLOGICAL

Turril says, the morphology, either of plants or animals, now a days includes a good deal more than the mere study of shape. The term morphology' is very wide and includes anatomy, and much of cytology, ontogeny, embryology, palaeontology and genetics. He says, "even within morphology in a very strict sense the taxonomist is forced to consider matters which involve physiology."

Turril says, that some one or other morphological classification provides the best basis for a general classification of the widest possible use, because morphological characters are, in general easier to use than others. According to him the great value of a classification based essentially on gross morphology is that it enables correlations of the characters to be easily determined.

2. ANATOMICAL

The study of morphology with a compound microscope is known as anatomy (histology). According to some anatomists the anatomical features are more conservative than those of gross morphology and are, therefore, of greater use in tracing phylogeny or organogenesis. Vesque has given much emphasis upon the value of anatomy in taxonomic and phylogenetic research. Tippo has proposed the lines of evolution of stem structure, more particularly for the secondary xylem of angiosperms. The view has been accepted by most plant anatomists. Stem anatomy shows that the Magnoliales are a relatively primitive of the angiosperms.

Chalk advocates that the structure is probably more conservative than that of the flower and the minor floral differences by which many species are distinguished from one another are not reflected in the wood; thus the wood of different species of a genus are often indistinguishable. According to Chalk there is one danger in using wood anatomy in classification is that wood usually means secondary wood, and the stressing of criteria derived, atleast, mainly from the secondary wood of trees and large shrubs may give results statistically biased against herbaceous parts of woody groups.

The following anatomical characters are of taxonomic significance:

1. Epidermal appendages, trichomes and emergences of all kinds.
2. Structure of stomata.

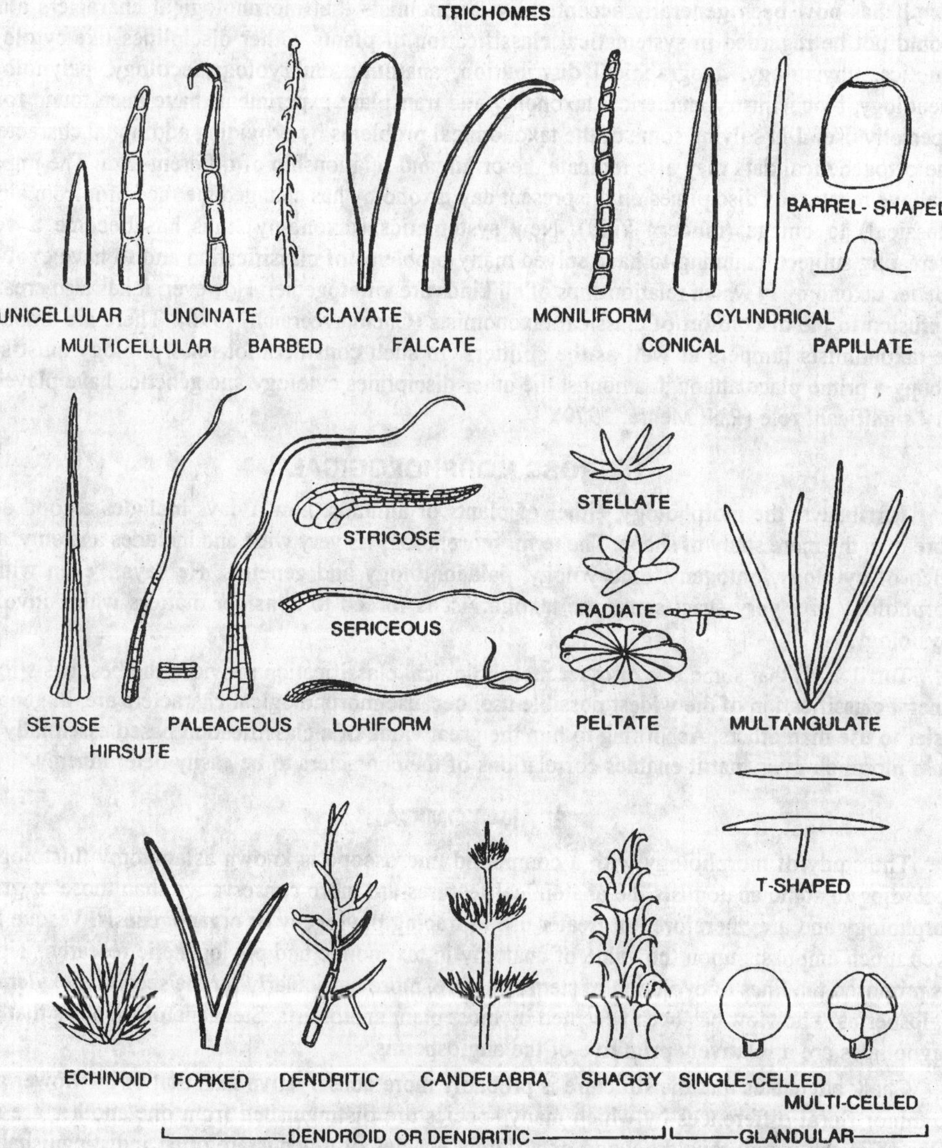

Fig. 3.1. Trichomes of different types

3. The wood anatomy including the elements of the secondary xylem; this is very useful in modern taxonomic studies, particularly in establishing the phylogenetic relationships among taxa, and in many cases serves as an aid in the identification of orders.

4. The nodal anatomy of the axis, the leaf trace nature.

5. Floral anatomy.

Epidermal emergences of all kinds: All unicellular and multicellular appendages are known as **trichomes;** more massive structures, *e.g.,* warts, spinous bodies both epidermal and subepidermal are called **emergences.** The use of trichomes in taxonomic studies are of much value, and has been known for many years. In identification of some families of angiosperms, the presence of a particular type of trichomes helps a lot, *e.g.,* simple unicellular or multicellular uniseriate trichomes occur in many taxa; similarly the occurrence of glandular trichomes of various kinds forms an aid to taxonomy in some respects.

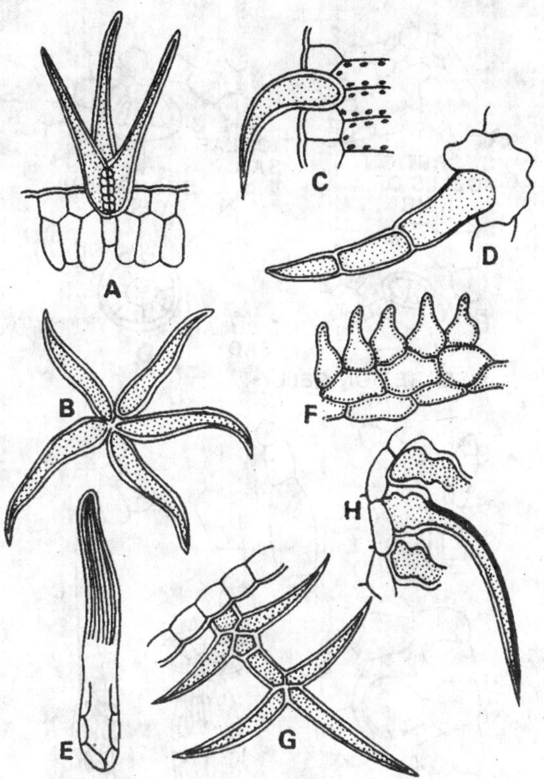

Fig. 3.2. Various kinds of trichomes. A and B, leaf of *Hamamelis;* C, leaf of *Cassia;* D, leaf of *Datura;* E, testa of *Strychnos nuxvomica;* F, petal of *Arnica;* G, leaf of *Verbascum;* H, testa of *Strophanthus.*

Structure of stomata: Stomata which also belong to the epidermal tissue system have been the subject matter of studies, which is very useful in the fields of taxonomy. Stomata have been classified according to the position of subsidiary cells, guard cells in relation to the aperture. They are of the following types :

1. **Ranunculaceous or anomocytic. Type A.** (*Anomocytic* = irregular celled). This type is characterised by having a limited number of subsidiary cells which are quite alike the remaining epidermal cells; the accessory or subsidiary cells may be four or five in number. In most of the cases, these subsidiary cells are just like the other epidermal cells. This type of stomata occur in Ranunculaceae, Capparidaceae, Malvaceae and some other families.

2. **Cruciferous or anisocytic. Type B.** (*Anisocytic* = unequal celled). This type of stomata occur in Cruciferae (Brassicaceae) and many genera of Solanaceae. In this type, each stoma remains surrounded by three subsidiary cells of which one is distinctly smaller than the other two.

3. **Rubiaceous or paracytic. Type C.** (*Paracytic* =parallel celled). In this type, the stoma remains surrounded by two subsidiary or accessory cells which are parallel to the long axis of the pore and guard cells. This type of stomata occur in Rubiaceae and allied families.

4. **Caryophyllaceous or diacytic. Type D.** (*Diacytic* = cross celled). In this type each stoma remains surrounded by a pair of subsidiary or accessory cells and whose common wall is at right angles to the guard cells. This type of stomata occur in Caryophyllaceae and allied families.

5. **Gramineous type.** The gramineous stoma possesses guard cells of which the middle portions are much narrower than the ends so that the cells appear in surface view like dump-bells. They are commonly found in Gramineae (Poaceae) and Cyperaceae of monocotyledons. (see fig. 3.6).

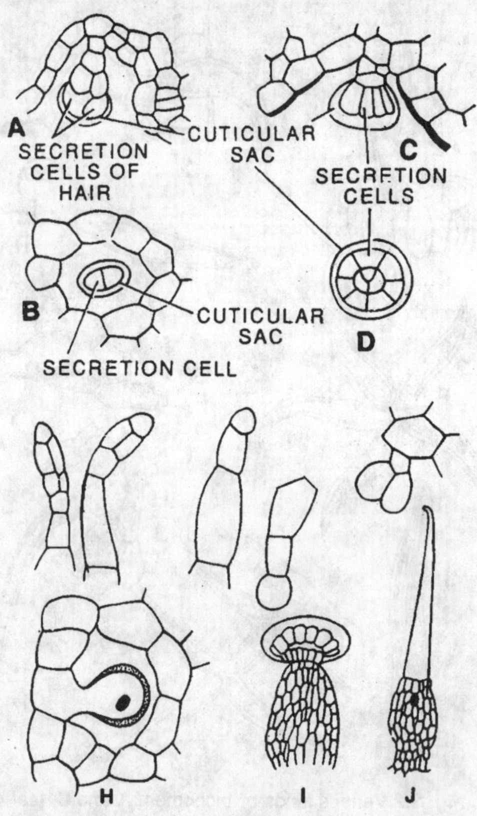

Fig. 3.3. Types of glandular A and B, section and surface view, *Artemisia;* C and D section and surface view, *Thymus;* E, *Hyascyamus*; *F, Atropa; G, Digitalis*; *H, Dryopteris filix-mas (rhizome)*; 1, *Cannabis;* J, *Urtica dioica.*

Wood anatomy: The plant anatomists have established a number of lines of phylogenetic specialisation in the stelar structure on the evolutionary basis and have applied the same in establishing data regarding the primitiveness of taxa and relationships between them and ultimate determination of taxon position.

These lines of anatomical specialisation are as follows:

1. The protostele is more primitive than siphonostele or dictyostele.

2. In angiosperms the woody axis of trees and shrubs is more primitive than the stem of herbs.

3. Vessel elements with scalariform perforation plates are more primitive than vessels with single opening in the perforation plate (*i.e.*, simple perforation).

4. The type with numerous bars and narrow openings is more primitive than the type with few bars and wide openings.

5. Vessel elements, long, small in diameter, angular in cross section are more primitive than those with short ones having broad circular cross sections; also the vascular elements with long sloping end walls are more primitive than those having non-sloping transverse walls.

6. Scalariform type of pitting on the lateral walls of the vessel is more primitive; opposite and alternate pitting are advanced.

7. Solitary or single pores occurring in vessel arrangement throughout the wood (*i.e.*, solitary pores) is more primitive than aggregate groupings of various types, such as pore multiples, pore fascicles, pore chains, etc.

8. Diffusc porous conditions of wood is more primitive than ring porous condition.

9. Evolutionary line of origin runs as follows:

Tracheids-to fibre-tracheids- to libriform wood fibres-to scalariform tracheids-to circular bordered pitted tracheids.

10. Diffuse wood parenchyma is more primitive than various aggregate arrangement. Apotracheal wood parenchyma is more primitive than paratracheal wood parenchyama.

Heterogeneous rays, *i.e.*, rays with both the vertically as well as radially elongated cells are less specialised and primitive than homogeneous rays in which cells are only radially elongated.

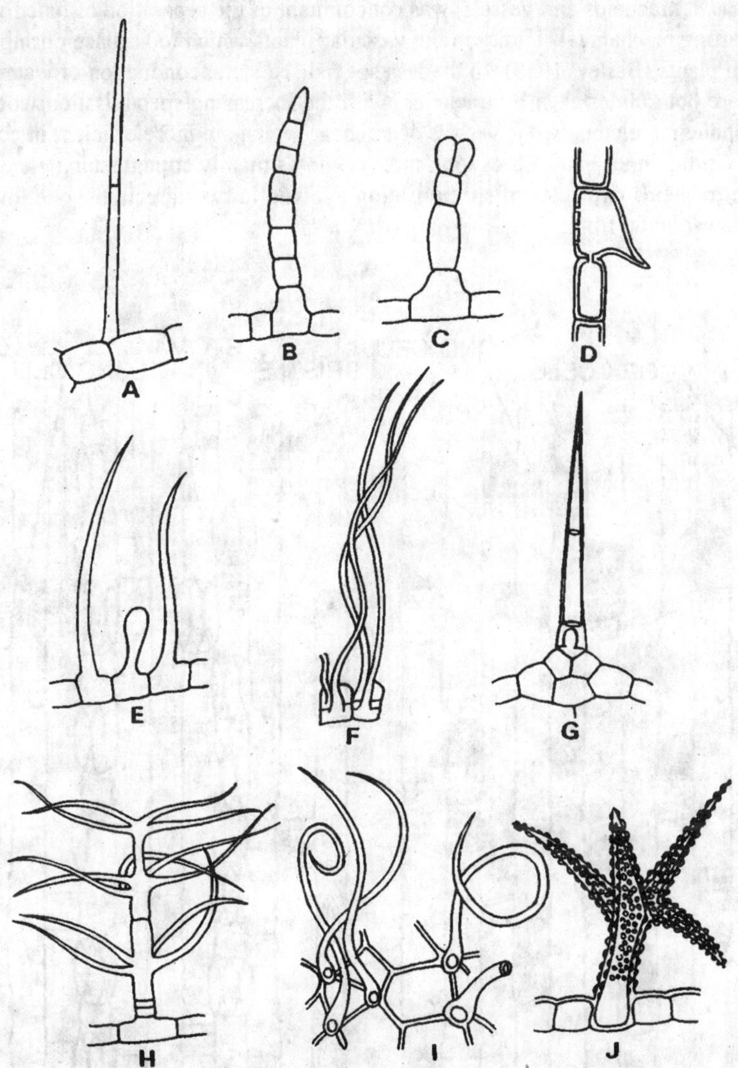

Fig. 3.4. Various types of trichomes (hairs)—A, from petal of *Epigaea;* B, from leaf of *Coreopsis;* C, from petal of *Phryma;* D, from leaf of *Avena;* E, from sepal of *Heliotropium;* F, from stem of *Onopordum;* G, from *Cucumis* leaf; H, from *Platanus* leaf; I, from *Rubus* fruit; J, from *Aubrietia* stem. (After E. & M.)

Schemes based on floral morphology and together with anatomical characters as mentioned above must go together in determination, of taxonomic position, generic identification, etc.

For example, Amentiferae according to Engler is most primitive in the Dicotyledons, but the wood anatomy shows that this taxon is very advanced in comparison to Magnoliales-Ranales.

Amentiferae : Simple perforation plates, *i.e.*, an advanced character; vessels are ring porous, short, round in cross section, large in diameter with alternate or opposite pitting, *i.e.*, an advanced character; libriform wood fibres prevail, an advanced character.

Phylogenetic specialisation of tracheary and fibre cells: The specialisation of tracheary elements (*i.e.*, tracheids and vessels) was concomitant of the separation of functions of conduction and imparting mechanical rigidity in the vascular plants which took place during the evolution of terrestrial plants (Bailey, 1953). In the less specialised state, conduction of water and mechanical support are both undertaken by tracheids; with the increasing specialisation woods evolved with better conducting elements, the vessels or tracheae which are more efficient in conduction than in support. On the other hand fibres took their origin as mainly strengthening elements. Thus, from primitive tracheids two lines of specialisation evolved in two directions, one towards the vessel, another towards the fibres.

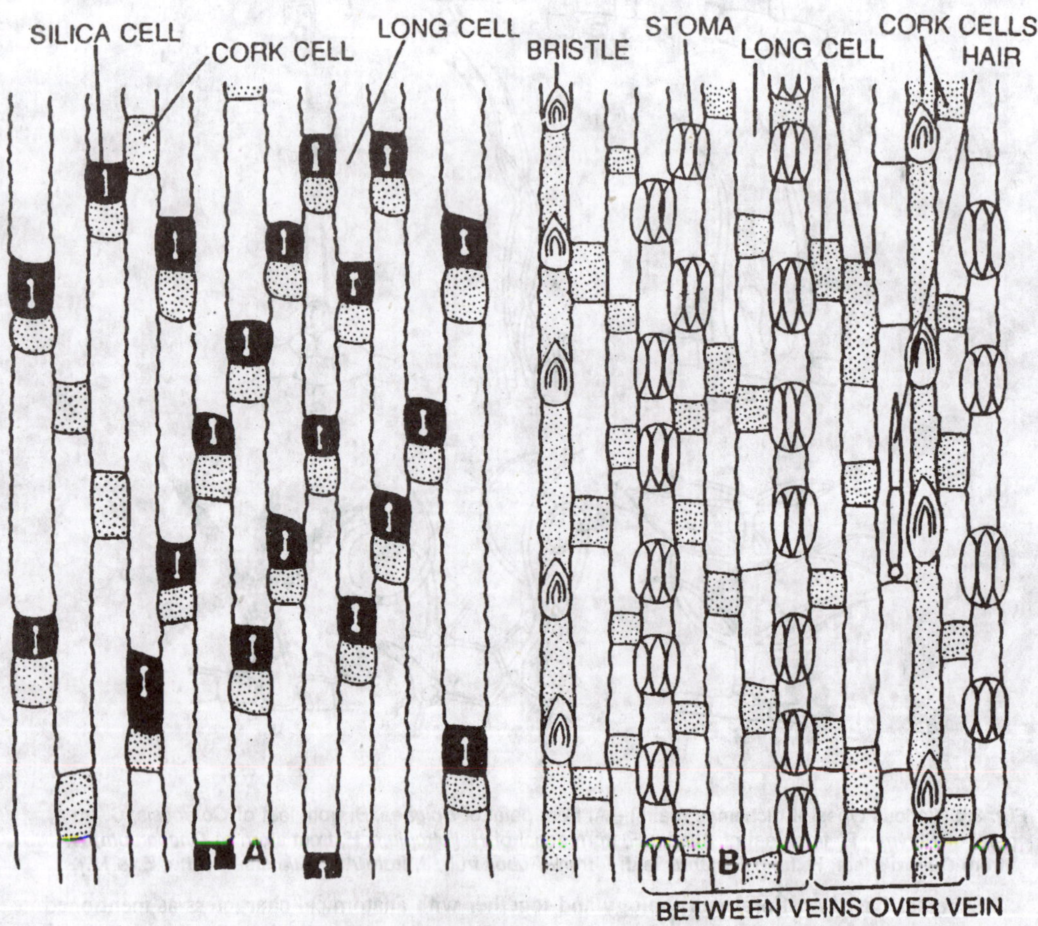

Fig. 3.5. Epidermis. A, epidermis of sugarcane stem showing alternation of long cells with pairs of short cells, the cork cells and silica cells; B, lower epidermis from leaf blade of sugarcane, showing distribution of stomata and their epidermal cells.

Vessels arose separately in several groups (taxa) among Tracheophyta, *i.e.*, vascular plants. The common accepted idea is that the vessels evolved independently in angiosperms. Evidence supports that in the Dicotyledons vessels originated and undergone specialisation first in the secondary xylem, then in metaxylem and were lost in the early primary xylem, *i.e.*, protoxylem.

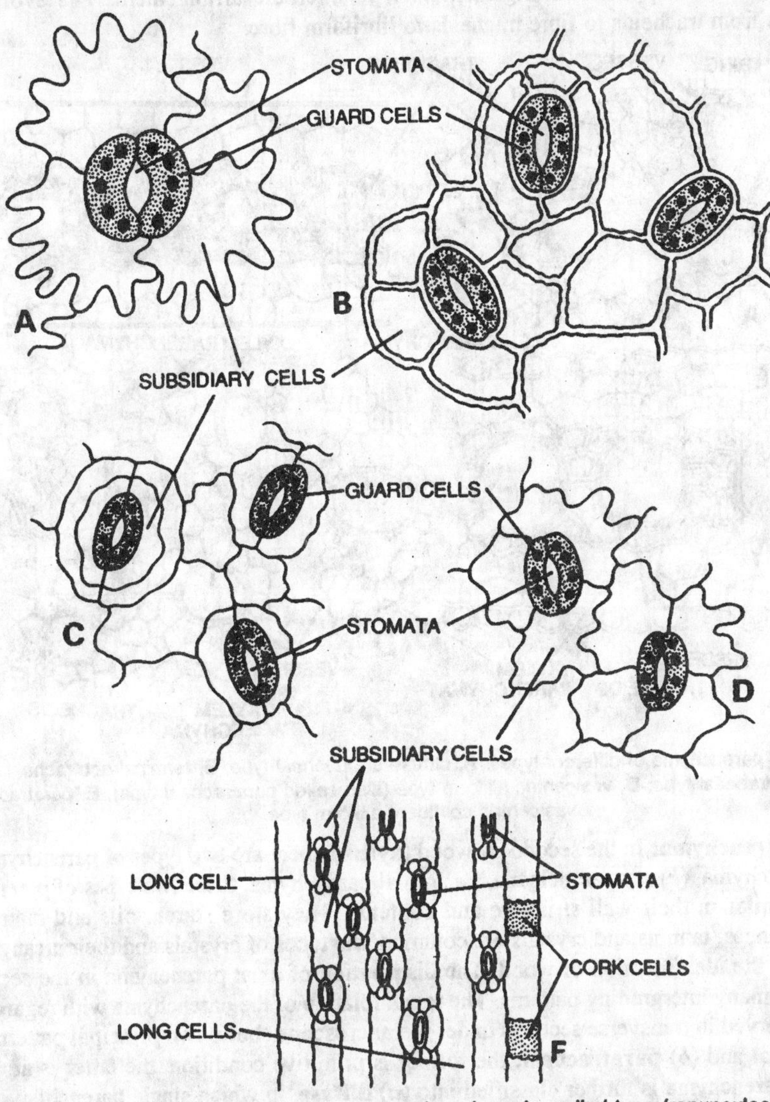

Fig. 3.6. Stomata—types of stomata. A, anomocytic or irregular celled type (ranunculaceous type); B, anisocytic or unequal celled type (cruciferous type); C, paracytic or parallel celled type (rubiaceous type); D, diacytic or cross celled type (caryophyllaceous type); E, gramineous type.

The evolutionary sequence of the vessel members of the secondary xylem of the Dicotyledons began with long scalariform pitted tracheids (primitive character) as found in the very primitive dicotyledons, such as Winteraceae, Trochodendraceae, Tetracentraceae, Chloranthaceae, etc. These tracheids were succeeded by vessel members of long and narrow form with tapering ends. The cells became short progressively became wider and less inclined and finally transverse. In very primitive state the perforation plate was scalariform with numerous bars resembling a wall with scalariformly arranged pits without pit memberanes; further specialisation went on, as a result number of bars decreased and finally total elimination of bars took place resulting in the simple perforation (advanced). The pitting on vessel walls also underwent change in course of evolution. In intervessel pitting bordered pit pairs in opposite arrangement were changed into smaller bordered pit pairs in opposite arrangement then to alternate arrangement. The evolutionary sequence was from tracheids to fibre tracheids to libriform fibre.

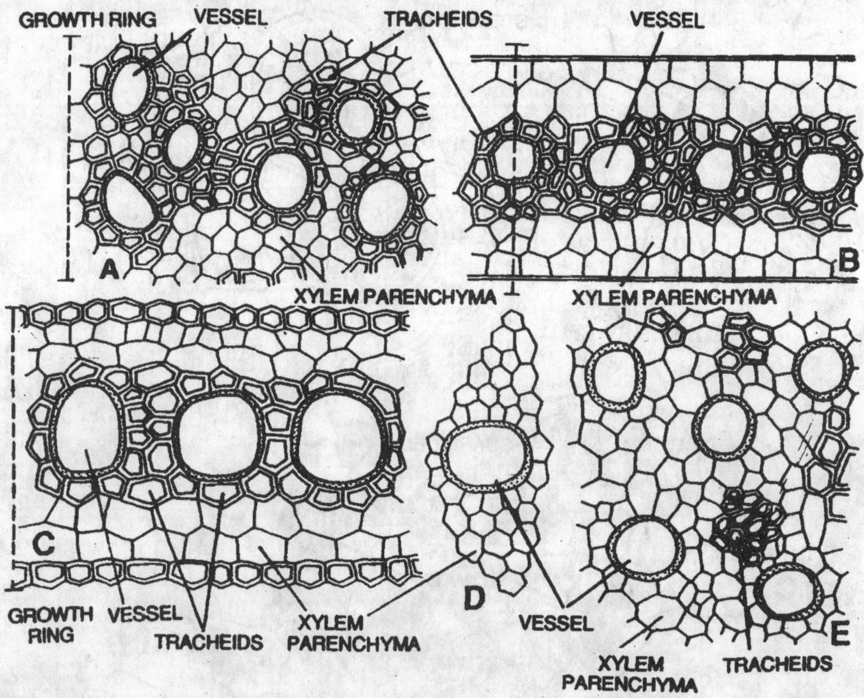

Fig. 3.7. Xylem parenchyma of different types. A, diffuse apotracheal type; B, terminal apotracheal type; C, banded apotracheal type; D, vasicentric ailiform type (also called paratracheal type); E, paratracheal vasicentric confluent ailiform type.

Wood parenchyma: In the secondary wood (xylem), there are two types of parenchyma, *i.e.*, (*a*) axial parenchyma, (*b*) ray parenchyma, *i.e.*, radial parenchyma, these two types of parenchyma are almost similar in their wall structure and contents. They store starch, oils and many other ergastic substances; tannins and crystals are common, the types of crystals and their arrangements serve much in the identification of wood. The distribution of axial parenchyma in the secondary wood exhibits many intergrading patterns. The touch relation of the parenchyma with regard to the vessels as observed in transverse sections under the microscope shows two principal patterns, *e.g.*, (*a*) **apotracheal** and (*b*) **paratracheal;** the former is primitive condition, the latter is advanced. Apotracheal parenchyma is further classified into (*a*) **diffuse**, in which single parenchyma cell or parenchyma strand scattered among fibres and (*b*) **banded** apotracheal in which parenchyma cells in a line form band (not in touch) over and under the vessels. The paratracheal parenchyma presents two types also, *e.g.*, (*a*) **vasicentric** (very advanced), it forms a complete sheath in touch and round the vessel and (*b*) **confluent** in which coalesced parenchyma form irregular or tangential or diagonal bands.

Structure of the rays: In contrast to the predominantly uniseriate rays of *Pinus* (Gymnosperm-conifer) those of the Dicotyledons may be one to many cells wide, *i.e.*, the rays (secondary rays) may be uniseriate and multiseriate. The structure of rays also play an important role in helping to identify the primitiveness of the wood anatomy. Rays may range in height from one to many cells. The appearance of rays in radial and tangential sections can be used as basis for classification.

Arrangement of vessels: The arrangement of vessels in the secondary xylem of Dicotyledons is a characteristic feature and is used in the identification of species. When the vessels are more or less equal in diameter and are uniformly distributed throughout the wood or when there is gradual change in size and distribution throughout the growth ring the wood is known as "**diffuse porous**", as found in *Salix, Populus, Acacia*, etc. When the wood consists of vessels of different diameters and in which those produced at the beginning of the season are distinctly larger than the late wood, the wood is called "**ring porous**", as found in *Fraxinus, Quercus*, etc. From the phylogenetic point of view diffuse porous wood is primitive than ring porous wood.

Wall thickenings of the vessels: In many vascular plants (Tracheophyta) the secondary wall thickenings of the protoxylem are annular or spiral. From an ontogenetic viewpoint the annular thickenings develop prior to the spiral. In later formed tracheary elements the spiral bands became joined in certain areas giving rise to scalariform thickenings. In tracheary elements developed at later ontogenetic stage, the thickening of wall becomes reticulate; when the openings in the secondary wall of such a network are elongated in a direction perpendicular to the longitudinal axis of the element, the thickening is known as scalariform-reticulate. Ontogenetically the most advanced type of thickening is pitted which occurs in the late primary and secondary xylem (wood). The pittings according to their arrangements are known as (*a*) scalariform, (*b*) opposite and (*c*) alternate. When the pits are transversely elongated and are consequently arranged in ladder-like manner, it is termed scalariform; when circular pits are arranged in horizontal or transverse rows they are known as opposite; when in diagonal series, known as alternate. The different types of pitting are of taxonomic importance.

Vessels and the structure of perforation plates: Vessels have limited length and in those vessel members which end in a tube bearing perforation at the terminal end, sometimes perforations also are present on the side wall. The end or side wall bearing perforation is known as perforation plate. The perforation plate consisting of a single large pore is known as simple perforation. In case of end wall containing numerous perforations they may be scalariform perforation plate in which perforations are arranged in parallel series in ladder-like fashion, reticulate perforation plate is distinguished by the arrangement of perforations in reticulate or net-like fashion; when end wall has many circular perforations arranged in scattered fashion, it is known as **foraminate** type characteristic of many gymnosperms and a few angiosperms.

The following structural features of angiosperm tracheary elements are used as the basis for the study of evolution:

1. The length of the element, *e.g.*, tracheids are long cells, the average length in *Trochodendron* is 4.35 mm., in the monocots the same is 5.07 mm.

2. Diameter of tracheary elements (vessels).

3. The perforation plates; the scalariform oblique perforation plates with numerous perforations are most primitive; simple horizontal one is the most advanced.

4. Shape of the cross-section of the vessel; angular primitive; circular recent and advanced.

Origin and specialisation of vessels: In the Dicotyledons the evolution of vessels took place first in the woody plants, apparently they arose independently in taxa with vessels which occur in most existing dicotyledons. As a result of the facts collected, it has been assumed that the vessels arose first in the secondary xylem, and later in metaxylem of the primary xylem, the specialisation and consequent evolutionary steps gradually advanced from secondary to primary xylem. It could also be assumed that herbaceous plants have evolved from woody plants as a result of reduction of cambial activity only after obvious development of vessels had taken place in the woody ancestral plants. In some specialised Dicotyledons, *e.g.*, in certain members of Cactaceae the secondary xylem lacks vessels which are replaced by so-called vascular tracheids, here the lack of vessels is the result of secondary reduction (Bailey, 1957).

From the evolutionary point of view, vessels in the Monocotyledons first appeared in roots and later in stems and leaves. The specialization of the vessels followed in the same pattern. Phylogenetically the vessels first appeared and became specialised in late formed metaxylem and advanced gradually in early formed metaxylem and finally into protoxylem (Cheadle, 1944). In the existing Monocotyledons of present day the last formed metaxylem roots contain vessels having scalariform perforation plates. A few aquatic Monocotyledons lack vessels in their xylem of roots, stems and leaves; their character is probably secondary one, *i.e.*, reduction stage and advanced.

Floral anatomy — Floral anatomy for delimiting taxa has been utilized by B. Tiagi and G. Dixit (1965) in Asclepiadaceae; Y.D. Tiagi (1964), B. Tiagi (1965, 1966) in *Opuntia, Cuscuta* and *Evolvulus.* V. Puri (1970) has reviewed the opinions concerning the nature of ovule in angiosperms and suggested that the angiospermic ovule has possibly arisen independently of the gymnosperms and the two, due to lack of any direct link, should not be homologised. Y.D. Tiagi (1961, 1963, 1968) on the basis of evidence derived from floral morphology, anatomy, embryology and cytology in Cactaceae considers the family as closely related to Calycanthaceae. The floral anatomy of two species of *Cuscuta* has been examined by B. Tyagi (1966) and its retention in the family Convolvulaceae has been suggested by C.M. Govil (1968). V. Singh (1977) has studied the primary pattern of floral development in all the taxa of the Alismatales as trimerous. Establishment of Hydrangeaceae and Escalloniaceae of sub-families Hydrangeoideae and Escallonioideae of Saxifragaceae, as separate families has been supported by N.P. Saxena (1966) on anatomical and embryological grounds. Eames (1953) has suggested the removal of *Paeonia* from Ranunculaceae and elevation of a new family Paeoniaceae on the basis of floral anatomy.

Anatomy of Leaf — B.C. Kundu and B.Gupta (1963, 1964 a, b) have suggested the use of quantitative microscopy of leaves in distinguishing taxa which is found useful in Solanaceae.

Anatomy of Axis — This has widely been used for comparing taxa, in *Corchorus* (R.M. Datta and K.Roy, 1963), in *Ipomoea* (C.M. Govil, 1971), *Phaseolus* (P.C. Datta and A.Saha, 1970).

Epidermal and Cuticular Anatomy — This has also been used for taxonomic delimitation of medicinal plants (P. Singh and B.C. Kundu, 1962, Y.N. Pandey, 1970; K.J. Ahmad, 1964 *a*, *b*).

Seed Coat Anatomy — Anatomy of seed coats of pulses has been utilised for delimitation of taxa by P.C. Datta and R.K. Maiti (1968). Chuang and Lawrence (1972) studied the anatomy of seed coat in the genus *Cordylanthus* of Scrophulariaceae.

3. POLLEN MORPHOLOGY OR PALYNOLOGICAL

The science of palynology or study of pollen morphology deals with the detailed study of pollen grains, *i.e.*, the microspores of Spermatophyta, particularly of Angiosperms. It mainly deals with the structure, walls, etc., of pollen grains. Palynological studies help in the confirmation of relationship and affinities between the related taxa. The pollen grains of related families and genera are usually of more or less of same type. The sculpture and pattern of the outer wall exine, number of apertures on the wall, size and the shape of the pollen grains, etc., help in taxonomy.

Mature pollen grains usually have two walls, the outer *exine*, and the inner *intine*. The intine is thinner and more tender than exine. The exine at first appears within the special wall as a thin membrane. In further development the exine thickens considerably and two layers can be distinguished on it, the outermost of which is the *sexine* and the inner is *nexine*. The sexine is thin and has high refractive index, and therefore is not easily visible; the surface of sexine is at first smooth, but later on after the formation of nexine many types of projections may develop on it. In the aperture regions of characteristic shapes from which pollen tube comes out, the exine may be completely wanting or it may be represented by nexine (Erdtman, 1952). The nexine is relatively thick and cutinized. The specific cutin is called *sporopollenin* (Frey-Wyssling, 1959). This substance is more stable than cutin. The intine is not of constant thickness, it is thicker at apertures. Numerous characteristic projections and sculpturings are formed on sexine, in addition to various types of sculpturings of the sexine there are other morphological characters which are used in the classification of the pollen and also imparts great value in angiosperm taxonomy.

Pollen grains are aperturate, *i.e.*, provided with pores or apertures. The position, shape, structure and number of apertures are of taxonomic significance. As a result of research on palynology based on the variability of the grains, a detailed nomenclature of the structural characteristics of pollen has been developed by which exact morphological descriptions of various pollen grains can be done accurately. The categories of the structural morphology of pollen that help in plant taxonomy are as follows:

(*i*) Size of pollen grain to be measured in μ (μ = 001 mm); (*ii*) structure of pollen grain, *i.e.*, oval, ellipitical, triangular, etc.; (*iii*) whether grains are free or in pollinia, or in tetrads; (*iv*) the nature of sculpturing of the pollen grains, the nature of exine, *e.g.*, smooth, spiny, warty, pitted, etc.; (*v*) sulcus–an elongated furrow perpendicular to the long axis at the pole of the grain; (*vi*) colpa–this is an elongated furrow at right angle to the equatorial plane, the ends of the furrow are directed towards the poles of the grain; (*vii*) ruga—an elongated furrow, the direction of which differs from both of the above types; (*viii*) pore– a pore is a circular aperture, when the number of pores is small, they are restricted in the equatorial region; when large they occur on all over the surface.

The comparative study of the structure of pollen grains of living and fossil plants plays as important role in the determination of classification of the plants. According to Wodehouse the pollen grain characters of gymnosperms are primitive to those of angiosperms. Wodehouse says, that the one furrowed grain is a primitive character in Saururaceae, Piperaceae, Chloranthaceae and Magnoliaceae all of which families have been considered by various authors as primitive.

According to P.K.K. Nair, pollen morphology of primitive angiosperms has indicated that there are two primary dicot stocks namely the "Magnolian" stock (Magnoliaceae and its allies), and the ``Ranalian" stock (Ranunculaceae and its allies), each stock being characterized by monocolpate and tricolpate pollen respectively.

Cronquist, A. (1968) has proposed that the two subclasses Ranunculideae and Magnoliideae, should be considered to represent the two palynological stocks, and should be raised to higher taxa levels.

4. CHEMOTAXONOMY

The science of chemical taxonomy (chemotaxonomy) is based on the classification of plants on the basis of their chemical constituents which are deeply concerned with the molecular characteristics. The method of chemical taxonomy is simple in principle and is based on the investigation of the distribution of chemical compounds, or groups of biosynthetically related compounds in series of related plants. Different plants sometimes contain substances which although belong to different chemical compounds appear to be biosynthetically analogous. Such plants may contain similar enzyme system, and the compounds produced by such enzymes infer that relationships exist between related plants. However, the chemotaxonomic studies include the investigations of the patterns of the compounds existing in plants, and in all the individual parts of the plants, such as bark, wood, leaves, roots, etc. The chemical constituents usually differ much in different organs. Such investigations are needed for obtaining real evidence for alliance and non-alliance of plants.

According to Turril, the physiological aspects, to some extent, lead the taxonomist towards the determination of phylogeny. It has been biochemical rather than directly functional criteria that have been used in taxonomy. The presence or absence of essential oils, resins, latex, glucosides, alkaloids, etc., has been used in classification. According to Molisch, a definite chemical substance may appear in a single species, in several species of the same genus, in a single genus, in several genera of one family, in an entire family, in two to many related families, or in large divisions of the plant kingdom.

Mc Nair showed the distribution of certain chemical substances in the families of angiosperms and its relation to climate. Climatic conditions have a major influence on the distribution of plants containing certain substances, *e.g.*, fats, volatile oils, alkaloids, etc. He concluded, that for tropics, and perhaps for all climates, the chemical products are highly organized. According to Reichert, it is possible to identify many plants by their starch grains.

Alston and Turner (1963) in their work "**Biochemical Systematics**"; Hegnauer (1965) in his "**Chemotaxonomic der Pflanzen**"; Leone (1964) in his "**Taxonomic Biochemistry and Serology**"; Swain (1963) in his "**Chemical Plant Taxonomy**" and Harborne (1967) have given a review of phytotaxonomical chemistry and emphasized on the role of various chemicals in plant taxonomy. Molery (1967) has stressed on the importance of β-cyanins and β-xanthins in plant taxonomy. β-cyanins are commonly met with in the families of Order Centrospermae. The other chemicals are also found specifically in particular orders or families of flowering plants. For example, **isoquinoline** (alkaloids) is found in the families of Ranales; **retanone** in the families of Leguminales; **biflavonyls** in *Casuarina equisetifolia* (Casuarinaceae). The presence of such chemicals in different groups of plants is of great taxonomic significance.

The choice of host plants by parasites is a part determined by the chemical constitution of the host. Thus, many animal parasites choose plants with similar chemical characters–mustard oil, glucosides, HCN glucosides, alkaloids, organic acids (especially oxalic), salts, etc. In correlation insect larvae which, when free to choose, limit themselves to one plant family or to such families as have similar chemical substances.

Applications and methods. There are a few angiosperamic taxa which are characterized by specific compounds of general occurrence. For example, leaving aside the family Caryophyllaceae, the rest of the families, such as Chenopodiaceae, Amaranthaceae, Aizoaceac, etc., of the taxon Caryophyllales (Centrospermae) contain *Betacyanin*, a colured substance, but differs from anthocyanin. It appears that except Caryophyllaceae, the remaining families are closely related, and therefore, the Caryophyllaceae may be isolated which may form a separate monotypic taxon. Betacyanin also occurs in Cactaceae, and therefore, the members of Caryophyllales are phylogenetically related. There are certain other chemical connections between Cactaceae and members of Caryophyllales, *e.g.*, common presence of isoquinole alkaloid in *Salsola* of Chenopodiaceae and Cactaceae. Another example may be cited, *i.e.*, in the family Cruciferae unsaturated acid, erucic acid is prominent; in *Tropaeolum* erucic acid is also present, it indicates the relationship between Geraniales and Rhoeadales. In Umbelliferac and Araliaceae a structural isomer of oleic acid, *i.e.*, petroselinic acid occurs, these two families are related and belong to the same order. The other examples are from Magnoliales–Ranales taxa, where it shows that in Magnoliaceae Lauraceae, Ranunculaceae, Annonaceae, the alkaloid isoquinoline is present; this supports that these families are closely related. On the other hand, Asclepiadaceae and Gentianaceae are allied due to the common occurrence of *pyridine*. The Liliaceae and Amaryllidaceae are closely associated, and this is supported by the presence of *isoquinoline* in both.

5. SEROTAXONOMY

Serotaxonomic test and consequent phylogenetic relationships between the taxa of angiosperms were established in Germany by Professor K.C. Mez (1866-1944) at the University of Koenigsberg in 1926, and this was modified in 1936. He established that relationship between larger group of angiospermic plants could be determined by serological test; and the closely related taxa and plants could be arranged accordingly. Serotaxonomy consists of the study and analysis of protein reaction of plants of different families with the blood serum of either rabbit or guineapig.

The serum test is done as follows:

For example, two plants A and B of two different families, of controversial relationship are taken for study. The protein extract of plant A, *i.e.*, extracted from tender shoot bud, is injected into the body of a rabbit, then after sometime, blood serum is collected from the injected rabbit in a sterilized test tube; protein extract of plant B is also collected from the tender shoot bud and is added to the rabbit serum having A protein; if there is precipitation then A and B plants are closely related and may be placed taxonomically close to each other; if on the other hand, no precipitation taken place, the plants are not closely related. The precipitation result is known as **positive reaction**, and non-precipitation result is known as **negative reaction**. For example, protein extract of *Passiflora* spp., (plant A) injected into the body of a rabbit, the serum then collected in a sterilized test tube and to it the protein extract of *Cucurbita* spp., (plant B) is added; the result is non-precipitation, *i.e.*, negative reaction, and therefore, it is considered that the families Passifloraceae and Cucurbitaceae are not related at all. If degree of precipitation is much, the families are closely related, if less the families are distantly related.

6. PALAEONTOLOGICAL

Darrah has proposed the following palaeontological evidences in support of the phylogenetic evolution of the plants.

1. The invasion of land by an undifferentiated thalloid plant took place not later than Silurian.
2. The simple upright undifferentiated and protostelic axis formed in Silurian.
3. Enlargement of the plant body with specialization towards a division of labour-sterile and fertile, took place apparently in late Silurian.
4. Origin of the photosynthetic leaf took place in Devonian.
5. Specialization in the service of support, with the resultant secondary body took place in Devonian.
6. Development of the strobilus also took place in Devonian.
7. The development of heterospory took place in upper Devonian if not earlier.
8. Retention of the gametophyte within megaspore, within the megasporangium, where fertilization takes place with the ultimate attainment of the seed took place probably in upper Devonian.
9. Evolution of the pollen grain took place in Carboniferous if not earlier.
10. Evolution of the bisporangiate flower took place in Triassic period.
11. Attainment of the condition of angiospermy took place in Jurassic and Triassic periods.
12. Evolution of the herbaceous habit of the angiosperms took place in Cretaceous (chiefly Cenozoic).

7. ONTOGENETICAL

According to Gunderson the evolution of the flowering plants is based on floral structure and presumed **organogenesis**. He points out that the flowering tendencies are widely accepted as progressive in organogenesis and agree with the findings in development.

Petals — from separate to united.
Petals — from actinomorphy to zygomorphy.
Sepals — from separate to united.
Ovary — from superior to inferior.
Carpels — from separate to united.
Placentation — from parietal to axile.

8. CYTOGENETICS AND BIOSYSTEMATICS

In modern days cytology has played an important role in taxonomy to achieve a classification which represents mutual relationships and is useful in indexing plants. Modern taxonomy takes up cytological evidence based on chromosome numbers, chromosome morphology, chromosome behaviour in meiosis and in aberrant forms of reproduction. The branch of taxonomy principally based on cytology is known as **'cytotaxonomy'**, it is a part of experimental taxonomy. It includes cytological aspects, study of cytogenetics and phenomena together with consideration of classical aspects of taxonomy. Application of data from the fields of cytology and cytogenetics have become very much helpful in taxonomic studies, particularly in determining categories of genus, species and sub-species in controversial cases. In this approach emphasis has been laid on cytogenetics and other aspects of cytotaxonomy supplemented by classical aspects of morphology, and therefore, the cytotaxonomy is the integration of cytology and taxonomy in the attempt of better understanding to solve problems of plant affinities.

The first step in cytotaxonomy research is the thorough sampling of the taxon (it may or may not be species) and its populations and the consequent cytological studies of chromosomes, *i.e.*, the number of chromosomes, chromosome morphology, chromosome behaviour, etc., of many populations within geographic races, species, genera and so on. Difference in chromosome number, chromosome morphology and chromosome behaviour at meiosis usually indicate generic or specific differences of taxonomic significance.

The second step consists in the capacity of different populations to hybridize and a study of vigour and fertility of hybrids. This enables to know the presence or absence of breeding barriers between groups and is of taxonomic value as indicating the natural limits of the taxa of various categories.

The third and last step includes the study of homologies of the chromosomes in the hybrids as determined in the meiosis, this is significant indicator in the degree of genetic relationship. Informations obtained from the above mentioned three steps are compared with the facts obtained from comparative morphology and geographical distribution. The resultant classification of the taxa, *i.e.*, the category of genus and taxa of lower rank, to which it is applied has an increased objective over one obtained from the consideration of morphology.

For example, Gundersen's (1950) system of classification is based on basic chromosome numbers together with anatomical and morphological characters. He has shown in primitive taxon, such as Magnoliflorae the basic number of chromosomes are much less than advanced taxon Rubiflorae and others.

The relationship between genera and the relative primitiveness of the constituent genera in the family can be determined on cytological studies, *e.g.*, cytological studies on Dipterocarpaceae

(Roy and Jha, 1965) established such relationship, the taxonomic position and the probable evolution of the taxon. Roy and Jha, 1965, estimated the chromosome numbers in different genera of the different tribes of Dipterocarpaceae. The number of chromosomes estimated as such 2n = 22, n = 11; 2n = 14, n = 7; 2n = 12, n = 6 according to different genera, *e.g.*, the tribe Vaticae and Dipterocarpae have some basic chromosome numbers as 2n = 22, n = 11; *Shorea* has 2n = 14, n = 7, so Dipterocarpae and Vaticae are more primitive and more closely related than those of others in the taxon.

Synthetic Biosystematic Approach: Taxonomy is an unending synthesis whose basis is becoming broadened decade after decade. It is therefore highly essential to bring about a coordinated synthesis between the "taxonomist" and "biosystematist"–the former posing the various ambiguities wherever located in the understanding of taxa while revising the families and preparing basic frame structure, and the latter working out such problems with the sophisticated technique utilising the relevant data and material supplied by the taxonomists. Such co-ordination between the Government Scientific Institutes and the University Departments would go a long way in solving several taxonomic problems of the Indian Flora (Rolla Seshagiri Rao and R. Sundra Raghavan, 1970).

Cytogenetics and Biosystematics in Synthetic Taxonomy—During modern times, there has been spectacular extension of synthetic taxonomy. Taxonomists, unlike in the past, are not only busy collecting data but basic problems are discussed at the outset as to why should data be collected; how should it be collected and what information to get from them. In all scientific experiments, howsoever, comprehensive and sophisticated it may be, we need some check to maintain balance between technique and concepts derived therefrom. In this context cytogenetics, genecology, population genetics, plant breeding and cytotaxonomy have a special role to play. Phylogenetic relationship at the specific and generic level and origin of species in many cases have been convincingly worked out by cytogeneticists. This approach has brought out new realizations and opened up new lines to study taxonomy (R.P. Roy and R.P. Sinha, 1970).

The cytogenetical data also indicate the origin and relationship of different taxa. Such studies include crossability between taxa and cytological analysis of artifical hybrids along with their parental forms. Similar studies can be carried on natural hybrids. Variation in chromosome number is of special interest to the evolutionist as the new group may be biologically isolated by incompatibility or hybrid sterility and may have been formed as a result of sudden doubling of chromosome number. Types formed by duplication of the same genome are difficult to distinguish from their progenitors as the differences are often in degree of expression of same characters. As a general rule, alloploids are distinct in their morphological characteristics. The detection of more than one chromosome number within any one species may therefore demand its detailed study as the species may be a collective one and needs a taxonomic revision (S.L. Tandon and G.R. Rao, 1970).

9. EMBRYOLOGY AND TAXONOMY

The comparative or phylogenetic embryology deals with the embryological data, which are used as a tool for ascertaining inter-relationships and taxonomic positions. In a symposium on comparative Embryology of Angiosperms held at Delhi in 1967 the systematic positions of various families were discussed in the light of embryological data. This symposium enabled to bring together in a comprehensive manner the scattered embryological literature on various families.

P. Maheshwari, B.M. Johri and S.N. Dixit (1957) pointed out that Loranthoideae and Viscoideae should be raised to the status of families. R.N. Kapil (1970) has pointed out several similarities between Crossosomataceae and Dilleniaceae like numerous centrifugal stamens, 2-3 middle layers, secretory tapetum, bitegmic crassinucellate ovules, micropyle formed by both the integuments. *Polygonum* type of embryo sac, nuclear endosperm showing aggregation of nuclei at chalazal end and reniform arillate seeds. These show that the two families are related.

After a comparative study on the embryology of Magnoliaceae and *Schizandra* R.N. Kapil and S. Jalan (1964) supported the view that *Schizandra* should be raised to a family rank. According to K. Subramanyam (1970) Crassulaceae shows close affinity with Saxifragaceae. For example *Crassula aquatica* has a mode of habit similar to Podostemaceae and it has the most reduced endosperm in Crassulaceae. This taxon according to Subramanyam (1962), forms a connecting link between Crassulaceae and Podostemaceae thus supporting Maheshwari (1945) that Podostemaceae are much reduced apetalous derivatives of the Crassulaceae.

V.K. Sharma (1968) has compared the embryological feature of Coriariaceae and allied families and suggested that it may represent an isolated stock distantly related to the Sapindales.

Dnyansagar (1970) stated that the sub-family Mimosoideae, Caesalpinoideae and Papilionatae should be raised to family levels.

10. NUMERICAL TAXONOMY

Numerical plant taxonomy may be defined as the science in which for the purpose of classifying the plants, mathematical methods are used. In other words, the application of simple mathematical principles or techniques or methods in taxonomical studies of plants may be defined as numerical *taxonomy*. Heywood (1967) defines numerical taxonomy as "the numerical evaluation of the similarity between groups of organisms and the ordering of these groups into higher-ranking taxa on the basis of these similarities." According to Sneath and Sokal (1973) "numerical taxonomy aims to develop methods that are objective, explicit and repeatable, both in evaluation of taxonomic relationships and in the erection of taxa." Thus the establishment of this branch of taxonomy is mainly based on the fact that the taxonomists are becoming now a days more susceptible to more clear, more closure and more reasonable criteria and principles.

Taxometrics or numerical taxonomy is mainly concerned with procedural and operational problems, *i.e.*, with the actual methods we take up in classifications. This science is based on evidence of similarities exhibited by observed and recorded characters of taxa, not on phylogenetic probabilities. It is not concerned with producing new data but methods of handling them by means of electronic computers so as to reduce subjective element involved in comparing sets of data. The necessity of mathematical applications in taxonomy has been realized by J.P.M. Brenon (1968) in Royal Botanic Garden, Kew (England). He has given the example of Kew herbarium with nearly four to five million specimens. The proper arrangement of such a huge number of plants is only possible by computerization of the arranging system.

J. Cullen (1968) has precisely revised the botanical problems of numerical taxonomy and has compared the numerical taxonomical methods with the orthodox taxonomical methods. In the orthodox taxonomy, the meaning of "*character*" is quite different from the meaning of "*character*" in numerical taxonomy. In the former thc *character* stands for any attribute referring to form, structure and behaviour which a taxonomist separates from the whole organism for a particular purpose such as comparison or interpretation, while in latter the *character* means a taxonomic character of two or more states, which within this study cannot be sub-divided any further. This is known as *unit character* (operational taxonomic unit, OTU), which makes a basic unit of numerical plant taxonomy.The character concept is combined with the use of average individuals or some other forms of selecting samples to represent the basic taxa (OTU) of the group under consideration.

Another botanical problem of numerical taxonomy is the use of weighing technique. On the basis of numerical taxonomy, the new reasonable phylogenetic classification may be worked out.

The logical steps involved in numerical taxonmy are as follows:

1. Choice of units to be studied. The first task is to decide what kinds of units are to be studied. These units may be comprised of individuals, strains, species, etc. The units should be as representative as possible of the organisms being considered. The entities of the lowest rank in particular study used are known as *operational taxonomic units*, (OTU's).

2. Character selection. A wide selection of the characters of OTU's is required. Atleast 50 and preferably 100 or more characters are needed to produce fairly stable classification.

3. Measurement of similarity. Overall similarity(s) is calculated by comparing OTU with every other and expressed as percentage, 100% similarity(s) for identity and 0% similarity(s) for non-resemblance. A similarity table may be worked out tabulating similarity coefficients and one for each operational taxonomic unit (OTU).

4. Cluster analysis. The similarity in matrix is now arranged, so as to bring together into clusters OTU'S whose members have highest similarity.

5. Discrimination. This is being done in the last, according to the degree of similarity and tabulated thereon.

In numerical taxonomy, the high-speed electronic computers and the powerful techniques of mathematics are employed to bring out a method of taxonomic analysis. These methods provide results that agree with establised classification of plants. These computerized techniques have highly relevant and increasing repercussions on many of the procedures and concepts of various fields of taxonomy.

PHYLOGENY AND TAXONOMY

Phylogeny is the evolutionary history of a taxon, and attempts to account for its origin and development. The term phylogeny is the autonym of ontogeny. Ontogeny differs from phylogeny in that it accounts for the life history of the individual plant from its development from the zygote to the production of its own gametes. A primary objective of phylogenetic studies in botany is the determination of origins and reletionships of all taxa of both extinct and living plants.

Significance of Phylogeny to Taxonomy–Phylogeny deals with the evolutionary history of all taxa. It is a function of taxonomic research at all levels of classification. The ultimate goal of phylogenetic research is the production of a phylogenetic system of classification. The phylogenetic system shows the genetic and time relationship of any one taxon to another. According to Turril (1942), "taxonomy is based on characters, phylogeny on changes of characters." A phylogenetic system of classification for plants would provide the answer to questions of (*i*) their origin (*ii*) to their modes of evolution, (*iii*) to problems of monophyleticism vs. polyphyleticism, (*iv*) the identity of primitive and advanced characters, etc. It would conclude in a single stable classification of relationships.

Diversity of Phyletic Concepts–The construction of the ultimate phylogenetic classification must be based on established facts regarding the characteristic features of the ancestors of every taxon level. These ancestors existed in remote prehistoric geological time, and because of their relative simplicity their characteristics are known as ***primitive*** which those of their present day descendants are known as *advanced*. For example, Eichler, Wettstein, Rendle and others considered the unisexual apetalous cyclic flower to be primitive, whereas Bessey, Hutchinson and others treated it as advanced and considered the bisexual polypetalous flower with spiral arrangement of parts to be primitive.

The formulation of phylogenetic classifications requires the teamwork and collaboration of botanists of all disciplines and the considered evaluation of data.

"Opinions have differed as to whether the monocotyledonous plants are more advanced or more primitive than the dicotyledonous plants. If the view is accepted that woody habit preceded the herbaceous, then, since the vast majority of ancient and modern monocots are herbaceous, it would point to the monocots been derived from one or more dicot ancestors. Botanists accepting this view treat the dicots as the more primitive. Ms. Arber has held that the monocots preceded the dicots. Current criteria from morphological and anatomical studies of both vegetative and reproductive structures would place the monocots as probably having been derived from the dicots. There is evidence for the belief that the monocots are more likely to be monophyletic in origin than is true for the dicots."

TAXONOMIC LITERATURE

Anonymous 1948-1976. *The Wealth of India*. Raw Materials, 11 vols. CSIR, New Delhi.

Babu, C.R. 1977. *Herbaceous Flora of Dehradun*. CSIR, New Delhi.

Bhandari, M.M. 1978. *Flora of the Indian Desert*. Scientific Publishers, Jodhpur.

Blatter, E. and C.McCann, 1935. *The Bombay Grasses*. Govt. of India Press, Delhi.

Blatter, E., C.McCann and T.S.Sabnis, 1929. *The Flora of Indus Delta*. Madras.

Blatter, E. and W.S.Millard, 1929. *Some Beautiful Indian Trees*. J.Bom. Nat.Hist.Soc. 33:624-635.

Bor, N.L. 1941. *Common Grasses of the United Provinces*. Ind. For. Rec. 2:1-220.

Bor, N.L. 1943. *Manual of Indian Forest Botany*. London.

Bor, N.L. 1960. *The Grasses of Burma, Ceylon, India and Pakistan*. Oxford.

Bor, N.L. and M.B.Raizada, 1954. *Some Beautiful Indian climbers and shrubs*. Bom. Nat. Hist. Soc., Bombay.

Bourdillon. T.F. 1908. *The Forest trees of Travancore*.

Chopra, R.N., Nayer, S.L. and Chopra, I.C. 1956. *Glossary of Indian Medicinal Plants*. C.S.I.R., N.D.

Cliford, H.T. and W. Stephenson, 1975. *An Introduction to Numerical Taxonomy*. Academic Press, N.Y.

Cole, A. J. (Ed.) 1969. *Numerical Taxonomy*. Academic Press, N.Y.

Collett, H. 1902. *Flora Simlensis*.

Cook, C.D.K. 1974. *The Water Plants of the World*. The Hague.

Cooke. T. 1901-1908. *The Flora of the Presidency of Bombay*.

Corven, D.V. 1952. *Flowering Trees and Shrubs in India*. Bombay.

Cronquist, A.1968. *The Evolution and Classification of Flowering Plants*. Thomas Nelson and Sons, Ltd., London.

Dastur, J. F. 1951. *Useful Plants of India and Pakistan*, Bombay.

Davis, P. H. and V. H. Heywood, 1963. *Principles of Angiosperm Taxonomy*. Oliver & Boyd, London.

Duthie, J.F. 1903-1922. *Flora of the Upper Gangetic Plain and of the Adjacent Siwalik and Sub-Himalayan Tracts*, 3 vols.

Duthie, J.F. 1906. *The Orchids of the North-Western Himalayas*. Ann. Roy. Bot. Gdn., Calcutta, 9:81-211.

Elliot, W.1859. *Flora Andhrica*.

Fyson, P.F. *The Flora of the Nilgiri and Pulney Hill-Tops*. 3 Vols. Govt. Press, Madras.

Fyson, P.F. *The Flora of the South Indian Hill Stations: Ootacamund, Coonoor, Kotagiri, Kodaikanal, Yercaund and the Country Round*, 2 vols. Govt. Press, Madras.

Gamble, J.S. 1878. List of *Trees, Shrubs and Large Climbers Found in the Darjeeling District, Bengal*, Calcutta.

Gamble, J.S. and C.E.C. Fischer 1915-1936, *Flora of the Presidency of Madras*. 11 parts.

Gamble, J.S. 1922. *A Manual of Indian Timbers*, London.

Graham, J. 1839. *A Catalogue of the Plants Growing in Bombay and its Vicinity*, Bombay.

Haines, H.H. 1910. *A Forest Flora of Chota Nagpur*.

Haines, H.H. 1921-1925. *The Botany of Bihar and Orissa*, 6 parts.

Hawkes, J.G. (Ed.) 1968. *Chemotaxonomy and Serotaxonomy*. Acad. Press, London.

Heywood, V.H. 1967. *Plant Taxonomy*, London.

Heywood, V.H. (Ed.) 1968. *Modern Methods in Plant Taxonomy*. Acad. Press, London.

Hill, A.F. 1952. *Economic Botany*. N.Y.

Hooker, J.D. 1875-1897. *The Flora of British India*. 7 Vols.

Hutchinson, J.1959. *The Families of Flowering Plants*. 2 Vols, Oxford.

Hutchinson, J. *Evolution and Phylogeny of Flowering Plants*. Acad. Press, London.

Kanjilal, P.C. 1933. *A Forest Flora of Pilibhit, Oudh, Gorakhpur and Bundelkhand*.

Kanjilal, U. 1928. *Forest Flora of the Chakrata, Dehradun and Saharanpur Forest Divisions, United Provinces*.

Kanjilal, U., A. Das, P. C. Kanjilal and R.N. De 1934-1940. *Flora of Assam*. 5 Vols.

Kirtikar, K.R. and Basu, B.D. 1933. Indian Medicinal Plants.

Lawrence, G.H.M. 1951. *Taxonomy of Vascular Plants*. N.Y.

Lawrence, G.H.M. 1955. *An Introduction to Plant Taxonomy*. N.Y.

Maheshwari, J.K. 1963. *The Flora of Delhi*.

Maheshwari, P. and R.N. Kapil, 1963. *Fifty Years of Science in India: Progress of Botany*. Ind. Sc. Cong. Assn., Calcutta.

Maheshwari, P. and U. Singh, 1965. *Dictionary of Economic Plants in India*. ICAR, New Delhi.

Mayurnanthan, P.V. 1929. *The Flowering Plants of Madras City and its Immediate Neighbourhood*.

McCann, C. 1959. 100 *Beautiful Trees of India*. Bombay.

Mooney, H.F. 1950. *A Supplement to the Botany of Bihar and Orissa*. Govt. Press, Ranchi.

Nadkarni, K.M. 1927. *Indian Materia Medica*. 2 Vols. Bombay.

Osmaston, A.E. 1927. *A Forest Flora for Kumaon*.

Pal, B.P. 1960. *Beautiful Climbers of India*, ICAR, New Delhi.

Pandey, B.P. 1980. *Economic Botany*. S. Chand & Co., Ltd., New Delhi.

Pandey, B.P. 1978. *Taxonomy of Angiosperms*. S. Chand & Co., Ltd., New Delhi.

Parker, R.N. 1918. *A Forest Flora for the Punjab with Hazara and Delhi*.

Parkinson, C.E. 1923. *A Forest Flora of the Andaman Islands*.

Poster, C.L. 1959. *Taxonomy of Flowering Plants*.

Prain, D. 1903. *Bengal Plants*. 2 Vols.

Puri, V. 1952. *Floral Anatomy in Relation to Taxonomy*, Agra Univ. J. Res. (Sci.) **1**: 15-35.

Puri, V. 1958. *Floral Anatomy and Taxonomy*. Ind. Bot. Soc. Mem. **1**: 15-18.

Puri, V. 1962. *Floral Anatomy in Relation to Taxonomy*. Bull. Bot. Sur. India, 4: 161-165.

Raizada, M.B. 1958. *Name Changes in Common Indian Plants Ind.* For. **85:** 467-538.

Raizada, M.B. 1975. *Supplement to the Duthie's Flora of the Upper Gangetic Plain.*

Raizada, M.B., R.C. Bhardwaja and S.K. Jain, 1961, 1966. *The Grasses of Upper Gangetic Plain.*

Raizada, M.B. and H.O. Saxena, 1978. *Flora of Mussoorie.*

Randhawa, M.S. 1957. *Flowering Trees in India.* ICAR, New Delhi.

Ranga Achariyar, K. 1921. *A Hand-book of Some South Indian Grasses.*

Rau. M.A. 1975. *High Altitude Flowering Plants of Western Himalayas*. BSI, Calcutta.

Rendle, A.B. 1925. *The Classification of Flowering Plants*. 2 Vols. London.

Royle, J.F. 1833. *Illustrations of the Botany and other Branches of the Natural History of the Himalayan Mountains and of the Flora of Kashmir.*

Santapau, H. 1953. *The Flora of Khandala on the Western Ghats of India.*

Santapau, H. 1957. *The Flora of Purandhar.*

Santapau, H. 1958. *History of Botanical Research in India, Burma and Ceylon Part II Systematic Botany of Angiosperms.*

Santapau, H. 1962. *The Flora of Saurashtra.*

Santapau, H. and Z. Kapadia, 1966. *The Orchids of Bombay.*

Singh, V. and D.K. Jain, 1981. *Taxonomy of Angiosperms*. Rastogi Publications, Meerut.

Smith. P.M. 1976. *The Chemotaxonomy of Plants*, London.

Sneath, P.H. A. and R.R. Sokal, 1973. *Numerical Taxonomy.*

Sporne, K.R. 1956. *The Phylogenetic Classification of the Angiosperms*. Biol. Rev. **31:** 1–29.

Stewart, J.L. 1869. *Punjab Plants.*

Subramaniam, K. 1962. *Aquatic Angiosperms*. CSIR. New Delhi.

Sundararaj, D.D. and Balasubramanyam, G. 1959. *Guide to the Economic Plants of South India.*

Swain, T. 1963. *Chemical Plant Taxonomy*. Acad. Press, London.

Swingle, D.B. 1946. *A Text-book of Systematic Botany*. Mc Graw Hill Book Co., New York.

Takhtajan, A. 1969. *Flowering Plants: Origin and Dispersal.*

Talbot, W. A. 1909-1911. *Forest Flora of the Bombay Presidency and Sind.* 2 Vols.

Turrill, W.B. 1942. *Taxonomy and Phylogeny*. Bot. Rev. **8:** 247–270.

Watson, J.F. 1824. *Plants of Kumaon.*

Watt, G. 1889. *A Dictionary of the Economic Products of India*. 6 Vols.

Willis, J. C. 1973. *A Dictionary of the Flowering Plants and Ferns*. Cambridge Univ. Press, London.

Wodehouse, R.P. 1936. *Pollengrains In the Identification and Classification of Plants*. Am. J. Bot. **23:** 535.

4

Nomenclature of Plants

Historical: The present day binomial system of nomenclature has a long evolutionary development. In near about 200 B.C. Cato in his monumental work `De Re Rustica' used two names for plants, although he lacked the modern concept of genera and species. Later on the descriptive Greek nouns used for genera were translated into Latin, where it took the form of two words, *i.e.*, binary generic names. According to another tendency a descriptive phrase was given for the specific name. These methods resulted in the so called polynomials. In the middle of sixteenth century, Brunfels changed many binary generic names to single ones. A few years later Dodonaeus followed the binomial usage similar in principal to that of today's binomial system of nomenclature. Gaspard Bauhin wrote his monumental work 'Pinax' in 1623, in which he has given a list of about 6,000 plants. He used the binomial system extensively to give names to the plants. None of these workers could make a permanent impression upon the botanical world of that time and works of the writers of those days continued to appear having mixed usage—monomial, binomial, trinomial and polynomial. In those times the names of the plants were composed of several words in a series, constituting more or less a description of the plants. In 1753 Linnaeus published his great work '*Species Plantarum*' where he employed the binomial system to name the plants in true sense. Augustin de Candolle's *Theorie elementaire de la botanique* published in 1813 provided the first significant work since the publications of Linnaeus on nomenclature and morphology of the plants. In this work for the first time, a complete and detailed set of rules on plant nomenclature was given. In 1821, Stendel's *Nomenclator botanicus* was published, which comprised a list of the Latin names of all flowering plants known by then. The second edition of this work appeared in 1840. *Nomenclator* was used universally by the botanists, and formed the basis of the *Index Kewensis*.

Common Names of the Plants : The plants have been provided with names by the different people of the different regions of the world in their own languages. These common or local names of the plants have their own weaknesses. There are three main defects in common names–(*i*) They may be quite indefinite, *e.g.*, Pansy has been provided with fifty English names. (*ii*) They are restricted to the people of one language or even one section of a country. (*iii*) They are not regulated by any constituted authority. These common names give great trouble to a plant collector in a foreign country. There he feels the local names to be more difficult than the botanical names.

Botanical or Binomial Names of the Plants: At the present time each kind of plant is given a generic name followed by a specific name or epithet. The generic name is a noun, and the specific epithet is an adjective indicating which of the several members of the genus has been considered. The genus name is always capitalized. The modern tendency is not to capitalize the specific name regardless of its derivation. Binomials are frequently descriptive of the plants and are usually derived from Greek or Latin, since these languages are internationally known by scholars. For example *Coffea arabica* Linn., stands for coffee; *Ficus elastica* Roxb., for India rubber, *Helianthus annuus* Linn., for sunflower, etc.

Binomials when written should always be underlined and when printed they must be in italicized type. The name or the abbreviated name of the scientist who first described the species follows the binomial–such as *Coffea arabica* Linn. ; *Mangifera Indica* Linn. ; *Rosa alba* Linn.; *Santalum album* Linn., etc. Linn., stands for the name of the great taxonomist Linnaeus; *Albizia procera* Benth. ; *Aloe barbadensis* Mill. ; *Kochia indica* Wight; *Mussaenda glabrata* Hutch.; *Solanum grandiflorum* Ruiz and Pav., here each binomial is followed by the abbreviated name of the scientist who first described the species. Some binomials are followed by two names, the first of which is within the brackets, thus *Albizia lebbeck* (Linn.) Benth. The name within brackets is that of the person who first described the species. The name which follows the parenthesis is that of the person who is responsible for the currently recognized binomial. Accordingly, Linnaeus described the plant for the first time and Bentham corrected the name later on. The other such examples are – *Albizia chinensis* (Osbeck) Merr.; *Alocacia indica* (Roxb.) Schott.; *Amorphophallus campanulatus* (Roxb.) Blume ex Dcne ; *Brassica juncea* (Linn.) Czern. and Cors. var. *Cuneifolia* Roxb. ; *Butea monosperma* (Lamk.) Taubert; *Feronia limonia* (Linn.) Swingle, etc. The reason for such author citations aids the taxonomist in finding the original and subsequent descriptions of a plant when necessary, and to avoid confusion when different workers use the same binomial accidentally to name different species. To aid the taxonomist further, the year in which the plant was described is sometimes written after the author's name following the binomial.

International Rules of Botanical Nomenclature : In 1950, the seventh international botanical congress was held at Stockholm and the rules of nomenclature were the subject of lengthy and considered discussions. Many minor changes and refinements were legislated.

Thereafter in 1956 the rules have been propounded by Lanjouw as the result of thorough study of many years.

They are as follows : 1. Long generic names with difficult pronunciation must be avoided; 2. A generic and a specific name is given to a perfectly named plant; 3. The specific name must differentiate a plant from all allied species; 4. The species is not distinguished by its size; 5. The specific differences of a plant have no concern with the original place of the plant; 6. colour should not be treated as a specific difference; 7. Each plant species must possess a generic name; 8. The generic name must be followed by a specific name; 9. The shorter specific name is always well and good.

The above Linnean principles are quite simple, the role of codifying such rules has been taken up by various International Congresses, which have clarified some rules, modified others, and when necessary have established new ones. In all these changes the main aim has been to produce a uniform system.

The 1956 edition of the code gives detailed instructions on the steps to be taken when any change appears necessary. Such a change is first proposed to the various nomenclature committees elected by the International Botanical Congress and these committees make their recommendations on the subject in preliminary way. The recommendations are then submitted to a vote by post from all the members of the International Association for Plant Taxonomy, from the authors themselves, and from the members of the committees. During the International Botanical Congress all officially enrolled members of the Section cast their final vote. After the inception of the journal '*Taxon*', every change that has been proposed is fully discussed in the journal before any decision is taken on the subject. Any proposal for a change receives full consideration before it is put up for confirmation or rejection during the session of the congress. It is to be noted that during the last three or four International Botanical Congresses there have been but few changes, except perhaps in the rearrangement of the code and in some less important points of nomenclature.

Main Points of the International Code

The code is divided into three parts, (*i*) **principles,** (*ii*) **rules** and (*iii*) **recommendations**. In addition there are few interesting appendices. The **principles** are basic points on which the code is based. They do not give detailed rules about nomenclature, but show the main ideas that have guided the compilers of the Code, and should be kept in view by any botanist attempting to publish a new *taxon*. According to this section of the code, every *taxon* or group of plants can bear only one correct name and *vice versa* a name could be applied to one group of plants. The *rules* give detailed prescriptions on all the points connected with the naming of plants. The *recommendations* are practical applications of the rules.

Some More Important Rules

1. The rank of taxa. In the 1956 edition of the code, the word '*taxon*' has been introduced for the first time, to signify "*taxonomic groups of any rank*" (Article 1). The rank of *species* is basic; one or more species make up a genus; one or more genera make up a *family*; one or more families form an *order*; one or more orders make up a *class*; one or more classess make up a *division*; several divisions make up the *Vegetable Kingdom*. It is to be noted that the Code does not recognize "*Natural Order*", a term often used in older books for the modern equivalent of "*family*" of the code.

The code has special provisions for the naming of the various *taxa*. The names of families must ordinarily end in – *aceae*, those of orders in – *ales*, and both should be treated as feminine plural nouns. Accordingly it is wrong to say, *e.g.*, "The Amaranthaceae consists of ten genera." The correct way is to say, "The Amaranthaceae consist of ten genera" or the Papilionaceae are a family of the Leguminosae, which include two other families, Caesalpiniaceae and Mimosaceae." Names of sub-families, *e.g.*, Amaranthoideae; of tribes, *e.g.*, Amarantheae; of sub-tribes, *e.g.*, Amaranthinae, are all plural names and should be treated like the name of the family in this respect.

2. The type method. This is a modification introduced in the code. According to Principle II, "The application of name of taxonomic groups is determined by means of nomenclatural types." This means that when a species is described as new, the author must indicate which is the type of specimen on which the new species is based. In the case of species, the type is an individual specimen, which is the *type specimen* on which the new species is based. In the small herbaceous plants all mounted on one herbarium sheet, the whole sheet may be marked as the type. When a specimen cannot be preserved, an illustration or figure or a description may be a type. The type of a genus is a species; that of a family is a genus.

In the *nomenclature of types*, several terms are used which are as follows :

The *type* or *holotype* is the one element chosen by the author as the nomenclatural type. An *isotype* is a duplicate of the type or holotype, *e.g.*, if several branches of a tree are collected at the same time, one specimen may be chosen as the *type*, the rest are *isotypes*. Similarly, if a number of small herbaceous plants are collected at the same time, all belonging to the same new species, one may be selected as the *holotype*, the rest are *isotypes*. If they are mounted on the same herbarium sheet, they may be all part of the holotype (Article, 10).

A '*paratype*' is a "specimen cited with the original description other than the holotype or isotype(s)." A paratype usually is a specimen from a collection other than the type on which the description of the new *taxon* has been based. Thus, there can be only one sheet of holotype, whereas a number of sheets of isotypes and paratypes. When an author describes a new species and cites two or more specimens, they become the *syntypes*, if none of them individually has been designated the type or if more than one has been so designated.

"A *lectotype* is a specimen or other element selected from the original material to serve as nomenclatural type when the holotype was not designated at the time of publication or for so long as it is missing." When two or more specimens have been designated as types by the author, they become syntypes, and one must be chosen as a *lectotype*. However, when all the material, on which the description of a new species was based, is missing, then a *neotype* must be selected, and the selection once made must be followed by subsequent taxonomists. This seems to be the case with many of the new taxa described by Roxburgh (1814), for which no actual specimen was either designated or preserved.

The fixing of the type specimen for a species or of the type in general for *taxa* below the rank of order is considered so important in the Code that Article 35 prescribes that "publication on or after January 1, 1958, of the name of a new *taxon* of recent plants of the rank of order or below is valid only when the nomenclatural type is indicated." (Santapau, 1959)

3. Priority. With a few exceptions, where alternative names are permitted, each *taxon* of plants can bear only one correct name; "the correct name is the earliest legitimate one validly published; except in cases of limitation of priority by conservation."

4. Tautonyms. "The name of a species is a binary combination consisting of the name of the *genus* followed by a single *specific epithet*" (Article, 23). The Code gives to authors very great latitude in the selection of names, always provided that the specific name has not been used previously in the genus. There is one case, however, which is specifically condemned as illegitimate and that is the use of *tautonyms*, *i.e.*, when the specific name repeats unaltered the generic name, *e.g.*, *Malus malus* Britt. This method is accepted in zoological nomenclature, where such names are coined, *e.g.*, *Turdus turdus* or *Corvus corvus*, etc. In botanical nomenclature, however, this is condemned, Art., 70(4). Thus, doubts have arisen about the validity of such names as *Sesbania sesban* or *Samanea saman*, etc. Some authors have expressed strongly against such names, while on the other hand, Merrill (1912, 1916) accepts them as valid on the ground that a tautonym is only a name that repeats *unaltered* the generic name, and the names mentioned above do not repeat the generic name in full. Thus, such names can be taken as valid until a further, International Botanical Congress decides the matter.

5. Names of families. The name of a family is a plural adjective and is formed by adding the suffix-*aceae* to the stem of a legitimate name of an included genus, *e.g.*, Rosaceae from *Rosa*, Linaceae from *Linum* and Cucurbitaceae from *Cucurbita*. The new names of the following eight families are :

Palmae	=	Arecaceae
Gramineae	=	Poaceae
Cruciferae	=	Brassicaceae
Leguminosae	=	Fabaceae
Guttiferac	=	Clusiaceae
Umbelliferae	=	Apiaceae
Labiatae	=	Lamiaceae
Compositae	=	Asteraceae

6. Names of infraspecific taxa. There is one case where repetition of the names is not only allowed, but ordered. Article 26 prescribes that the name of an infraspecific taxon (*i.e.*, sub-species, variety or form) must be the same as that of the higher taxon, if the lower taxon includes the type of the higher one. Thus, if a species is divided into varieties, the one that contains the type of the species must be called by the same name as the species and not var., *genuina* or var., *typica* (Art., 71), as often used to be called before. For example, *Carissa congesta* Wt., of Khandala on the Western Ghats has two varieties, (*i*) the typical plant, var., *congesta* (without name of author after the variety) and (*ii*) var., *albida* Sant. If the species has been divided into sub-species, varieties, forms, etc., then a plant is called *Carissa congesta* Wt., subsp., *congesta*, var., *congesta*, f., *congesta*; if the type of each higher rank is included in the lower division, the name *Andropogon sorghum* subsp., *halepensis* (L) Hackel var., *halepensis* is legitimate since the subspecies and the variety have the same type and the epithet must be repeated under (Art., 26; Art. 27, Santapau).

7. Effective and valid publication of new taxa. Publication is effective when one can expect that the general botanical public may come to know of the publication (Santapau). According to Art., 29, publication is effective only when printed matter is distributed to the general public or atleast to botanical institutions with libraries accessible to botanists. It is not effective by communication of a new names at a public meeting such as during the annual session of the Indian Science Congress, nor by placing of names in collections or gardens open to the public as for instance by labelling one's specimens and placing them in public herbaria, where visitors may see the names. The date of the effective publication is the date on which the printed matter became available.

The following are the conditions attached to the publication of new *taxa* for validity.

1. Publication must be effective.

2. There must be a description of the new *taxon*, or a reference to a previously and effectively published description.

3. A combination is not validly published unless the author actually makes it (Art., 32).

4. Publication must be done in Latin (Art., 34) or with reference to a previously published Latin description.

5. When all the conditions for valid publication are not fulfilled at one and the same time, publication becomes valid when the last condition has been fulfilled. Thus, a publication is invalid if a botanist forgets to indicate the type of his new taxa, but it becomes valid when the type of the taxon has been indicated. The date of such publication will be when the type was indicated.

8. Changes in name when taxa are divided, fused or altered. When a genus is divided into two or more genera, or a species is split into two or more species, the original generic or specific name must be retained for the new *taxon* containing the type; this applies also to infra-specific *taxa*.

When a section of a genus or species is transferred to another genus or species *without alteration in rank*, the original name must be retained whenever possible.

When the rank of a genus or infra-generic *taxon* is changed, the correct name or epithet is the earliest legitimate one available in the new rank.

When *taxa* of the same rank are united into one, the oldest legitimate name must be used for the new combined *taxon*; if the names are of the same date, the author who first unites them has the right to choose one of the names and his choice must be followed by subsequent botanists.

9. Form and spelling of names. As per Code, authors have full liberty to coin their names, Authors are recommended not to make such names too long, nor too difficult to pronounce. In any case such names should be treated as Latin names. When the name is actually taken from the Latin language, it retains the original gender and spelling. When the name is taken from the vernaculars, the author assigns a gender to it, and this must be followed by subsequent botanists. Names of genera and higher *taxa* are written with a capital initial letter Specific names should be written with a small initial letter.

Advantages of Binomial System: The binomials are more definite and precise than the common names. They are usually derived from Greek or Latin, since these languages are internationally known by scholars. They are frequently descriptive of the organism or plants. They aid the taxonomists in finding the original and subsequent descriptions which indicate the systematic relationships.

5

Plant Collection and Preparation of Herbarium

What is Herbarium: A herbarium is a collection of plants, which have been dried, pressed, mounted on herbarium sheets, identified and classified according to some approved system of classification. The succulent plants are usually preserved as such in some suitable preservative, such as 4% formaline solution or F.A.A. Many types of dry fruits; cones of gymnosperms and inflorescences of palms, etc., are dried and kept in large containers.

About 1550, Cesalpini and his co-workers began quite definitely to preserve the materials they studied, and since then herbarium making became a great and interesting feature of botanical work. All civilized countries possess their own plant collections (herbaria). The greatest herbarium of the world is at the Royal Botanic Gardens, Kew, England, possessing about six million specimens. A few good herbaria are there in our country. The biggest herbarium of our country is at the Indian Botanic Garden, Calcutta, possessing about one million specimens. The herbarium of the Forest Research Institute, Dehradun has about 3,00,000 specimens. The herbaria of Agricultural College and Research Institute, Coimbatore and National Botanical Gardens, Lucknow, have about 200,000 and 40,000 specimens respectively. There are about 25,000 specimens in the herbarium of the Divisions of Mycology and Plant Pathyolog at Indian Agricultural Research Institute, New Delhi. The herbarium of the Division of Botany at I.A.R.I. New Delhi, contains about 3000 specimens.

Some Important Indian Herbaria

1. Forest Research Institute (FRI), Dehradun. The botany branch, one of the first to be created today comprises of three sections, *viz.*, systematic botany, wood anatomy and plant physiology. The division today maintains a botanical garden, an arboretum, having one of the richest live collections of both indigenous and exotic tree species, and a bambusetum, containing germ plasm of forty species of indigenous and exotic bamboos. All this provides a rich source of material for research, teaching and training.

Systematic Botany and Herbarium. This section has been largely responsible for the collection and indentification of a large portion of the floristic diversity of the Indian sub-continent. Started by Gamble (1890), the herbarium of the FRI has grown to become one of the largest of its kind in Asia. It today houses 3,25,000 authenticated plant specimens, including 1300 type specimens, as well as a carpological collection. Some of the notable publications of the division, based primarily upon the herbarium collection, are 'Manual of Indian Forest Botany' 'World Monograph on the genus *Toona*', 'Forest Flora of Punjab', Kumaon, Chakrata, Andamans, Orchids of north-western India and food from Forests.

2. Herbarium of the Indian Botanic Gardens, Calcutta. Foundation, 1787; Status, directed by the State of West Bengal, Department of Agriculture, Animal Husbandry and Forests; Number of specimens, more than one million Herbarium. World-wide, mainly phanerogams and ferns of

India and neighbouring countries of South and South East Asia. In addition to these there are some authentic collections of Cryptogams, Associated Botanic Gardens: Indian Botanic Garden, Calcutta, and Lloyd Botanic Garden, Darjeeling.

3. Herbarium of the National Botanic Gardens, Lucknow. Foundation, 1948; in 1953 taken over by the Council of Scientific and Industrial Research (CSIR), New Delhi, Government of India. Status, C.S.I.R., Govt. of India. Number of specimens, about half million. The garden has been established by C.S.I.R., as a Central Garden for India.

4. Madras Herbarium, Agricultural College and Research Institute, Coimbatore. Status, Department of Agriculture, Government of Tamil Nadu. Foundation, 1874. Number of specimens, about half million. Herbarium, world-wide, mainly Phanerogams. Important collections, collections of the flora of Madras. Associated Botanic garden, Botanic Garden and Research Institute, Coimbatore.

5. Herbarium of the Division of Botany, Indian Agricultural Research Institute, (I.A.R.I.), New Delhi. Status, maintained by Government of India. Number of specimens about 10,000. Herbarium, North India, introduced plants of economic value, wild relatives of crop plants and associated botanic garden.

Purposes of Herbaria : The drawings, photographs and descriptions of the plants cannot show, what is desired to show by them about individual variation and details of structure and development. In most cases it becomes necessary to compare descriptions with actual specimens which are found in widely separated regions. Most comparative studies for taxonomic purposes are made in the herbarium and laboratory. The plant collections are much used for comparison with new material. The newly collected specimens are identified by the comparison of their morphology and the morphology of the plants collected in the herbarium.

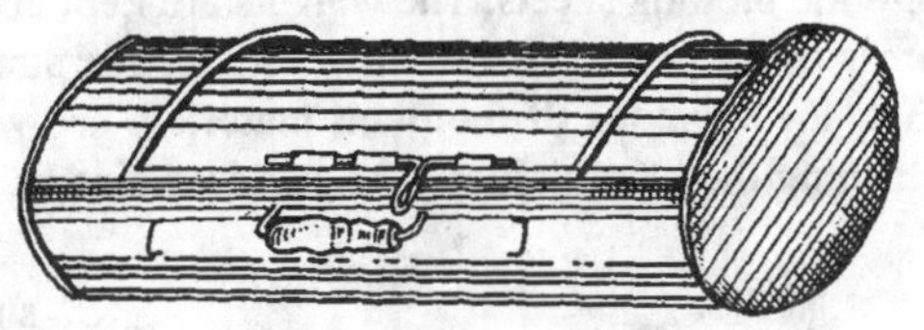

Fig. 5.1. A typical vasculum for botanical collections.

Methods of Plant Collection: The specimens must be collected in every stage of their growth and reproduction and from different localities and habitats. The single specimen may be collected in the late flowering stage having both flowers and young fruits. A complete specimen possesses all parts of the plant including the root system. In certain cases, the roots and other underground parts of the plant are necessary for identification. For collection of the plants one should go out on excursion several times in a season. A list of instruments required by a plant collector, while on excursion is given here :

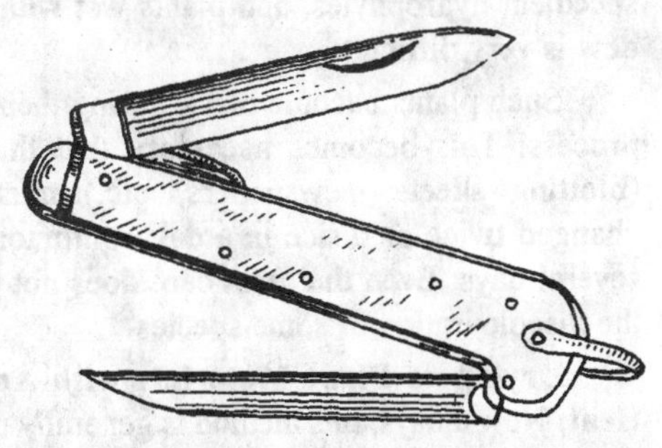

Fig. 5.2. Gardener's knife–one should carry such knife in field trips.

1. A pair of secateurs for cutting woody twigs; 2. A khurpi for digging up roots and underground stems; 3. A knife; 4. A pair of forceps; 5. A vasculum for keeping the collected plants and their twigs; 6. A wooden plant press; 7. Blotting papers and

newspapers; 8. A field book; 9. A pocket lens; 10. A small diary to write notes about plants and their habitat.

Pressing of the Plant Materials: Generally the plants are pressed flat. To get good specimens, the plants must be pressed before their wilting. Generally the plants are collected in a cork lined tin vasculum, and then they are pressed after the day's trip. To get the finest specimens the plants must be pressed directly in the plant press, then and there, as soon as they are collected in the field.

Methods of Pressing the Plant Materials: The plants are pressed in between the sheets of newspaper. These sheets are alternated between sheets of blotting paper. The wooden presses are too heavy to carry them in the fields, and therefore, while going on a collection trip the presses with aluminium covers are preferred. The field press remains tied with the leather straps. The plants become dried by transferring their moisture into the blotting sheets. The weights are kept upon the press or the press is tied tightly with leather straps, so that, the wrinkles of the plant materials are without crushing the plant tissues. The standard size of the press is 12" × 18".

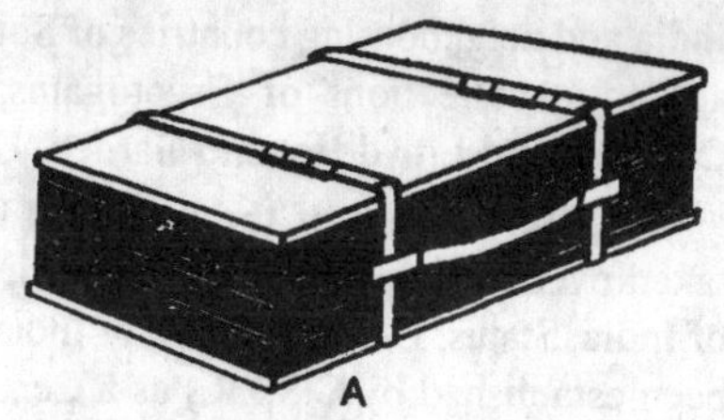

Fig. 5.3. Pressing of specimens in press boards; A, the right way of packing the press; B, the wrong way of packing the press.

Drying of Plant Materials : The blotting sheets or newspaper sheets which have been used for pressing the plant materials must frequently be changed after every 24, 48 or 72 hours. The changing depends very much upon the weather conditions. Changing every 24 hours for two or three days and then two or three more times at longer intervals is a good rule, but this is too much for the plants of arid region. The succulent plants require changing more frequently, and they require a long time to become dry. The drying of some plants such as saprophytes (*Monotropa* sp.), succulent hydrophytes, and plants wet with rain or dew is very difficult.

Such plants become black during their drying process. This becomes necessary that the driers (blotting sheets, newspapers, etc.) must have changed twice or thrice in a day continuously for several days. Even the great care does not prevent the discolouration of some species.

Drying of Plant Materials with Artificial Heat: Now a days, this method is generally used for drying of plants by plant collectors. Usually this method is employed in the damp climates especially in hilly regions during rainy season. The method is

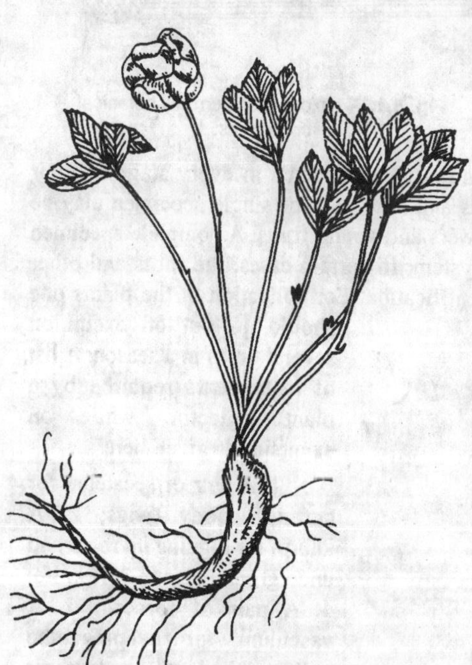

Fig. 5.4. The wrong method of pressing plants. Here the parts of the plant are folded.

as follows–The specimens are pressed in the field press for 24 hours in usual way. The plants are then rearranged within the finely corrugated aluminium sheets. The press is then strapped and thereafter, the heated dry air is passed over through the corrugations of the aluminium sheets. These presses with plant specimens are placed in specially constructed ventilated boxes fitted with electrical incandescent light bulbs. The strapped press may also be placed in a hot air oven at desired temperature. A small electric fan also be placed in the oven, so that the warm air may be forced through the corrugations of the aluminium sheets. Instead of electrical appliances, kerosene lanterns and stoves may also be used for the purpose.

Small and rare specimens may be dried quickly with an electric iron by keeping them in between layers of thick cloth or paper. The time required for drying the plant specimens by artificial heat ranges from 6 to 24 hours. Generally the plants are taken out from the press after 24 hours, and then they are transferred to an ordinary press.

Mounting of Specimens: After drying, the plant specimens are mounted on herbarium sheets. The sheets must be of heavy white paper to give support to the plants in handling. The standard size of the herbarium sheets must be of good quality, so that it will not turn yellow or brittle with age. The specimens are fastened to the sheets by glue or gummed cellophane tape. Usually one specimen, no matter how small it is, should be mounted on one herbarium sheet.

Labelling of the Specimen: A label must contain as much information as possible about the specimen. It is then attached to the lower right hand corner on the herbarium sheet. The label must contain the following information.

Flora of ... ...
Collector's number ... ...
Date of collection ... ...
Botanical name of the plant ... ...
Local name of the plant ... ...
Family... ... ...
Locality... ... ...
Attitude... ... ...
Slope... ... ...
Soil... ... ...
Habit... ... ...
Use... ... ...
Distribution... ... ...
Abundance... ... ...
Remarks... ... ...
Collector's name ... ...

Now-a-days the ecologists show the effects of habitat on morphology and also the distribution of species according to habitat, and such data are only possible when the plant specimens are labelled very properly.

The Field Data : There are two important methods for recording field data. They are as follows– 1. While collecting the plants a plant collector keeps a field book. The field book contains the numbers, which correspond to the numbers of specimens collected. The ecological and other data about the collected plants must also be given in the same field book. 2. According to this method the printed forms with headings indicating the required information are put in convenient sized pads. While making collections each printed form duly filled is attached with a specimen, and then the specimen is pressed in the field press.

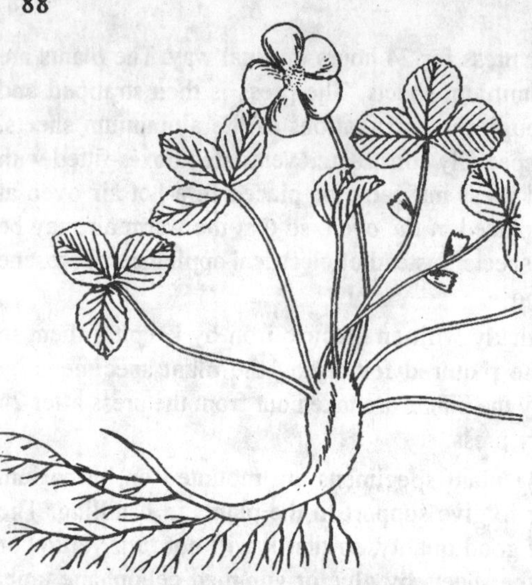
Fig. 5.5. The right method of pressing plants. Here every part of the plant is unfolded.

Storage of Herbarium Sheets: After mounting the plant specimens on herbarium sheets they are usually kept in specially constructed herbarium cases. The wooden or steel cabinets with shelves slightly bigger than the size of the herbarium sheets are used for keeping the mounted plant specimens. The keeping of mounted plant specimen for further record is known as 'filling the specimen.' In comparison to wooden cabinets, the steel almirahs are being preferred as they protect the plants from being damaged by the insects and fire. The specimens are arranged in their cases according to any well-known taxonomic system of classification. In our country and most of Commonwealth countries and British colonies the Bentham and Hooker's system of classification is being used for this purpose. Engler and Prantl's system of classification is also used to classify the plants in many countries. The plants are being arranged familywise and an index card of the families is being prepared. Within the families, the genera are being arranged alphabetically. The specimens of one genus are placed in a folded genus cover. The species are also arranged alphabetically within the genus.

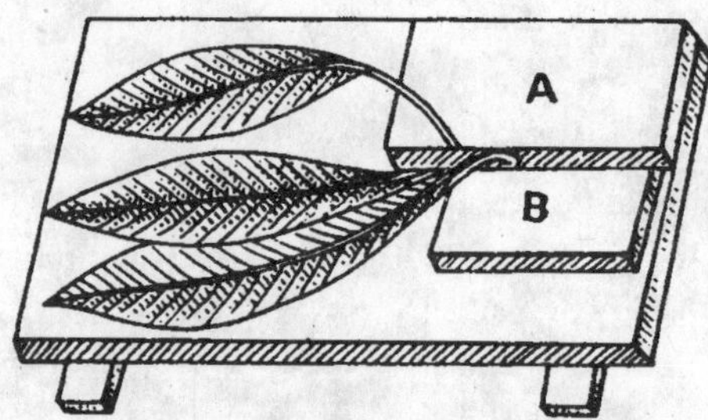

Fig. 5.6. Pressing of specimens with stout branches; A and B-newspaper cushions supporting the stout branch.

Protection of Herbarium Sheets against Molds and Insects: The mold fungi do sufficient harm to the plant specimens, if they are not well dressed by some effective chemical. For this purpose the plant specimens are sprayed with 2% solution of mercuric chloride which is highly poisonous. The insects do much harm to the specimens. To overcome this difficulty, various chemicals have been applied to the specimens. The moth balls, and naphthalene flakes are being placed in the shelves of herbarium to check the insects. These act as repellents to the insects. The mixture of paradichlorobenzene acts as an effective repellent to the insects.

The other chemicals which are used as insect repellents are D.D.T. (*i.e.*, dichlorodiphenyl-trichloroethane), carbon disulphite gas, (insecticidal fumigant) etc. The heat is also being employed to kill the insects. If the mounted sheets are being placed in suitably designed steel almirahs for four to six hours at 60°C, all the insects and their eggs are being destroyed.

Preservation of Type Specimens: The type specimens are those on which the name of some taxon has been based. These specimens are those which are used by the author of a name as a basis for the name and description of a species and designed by him as a type specimen, and therefore, they are supposed to be so valuable that they are always preserved with special care. The type specimens are handled with care and are kept away from regular specimens in separate steel cases by the curators. Usually such specimens are kept in separate cellophane envelopes.

Role of Herbaria in Modern Taxonomic Research: Herbaria and herbarium taxonomists have contributed so significantly to our understanding of Natural History that there is no need to

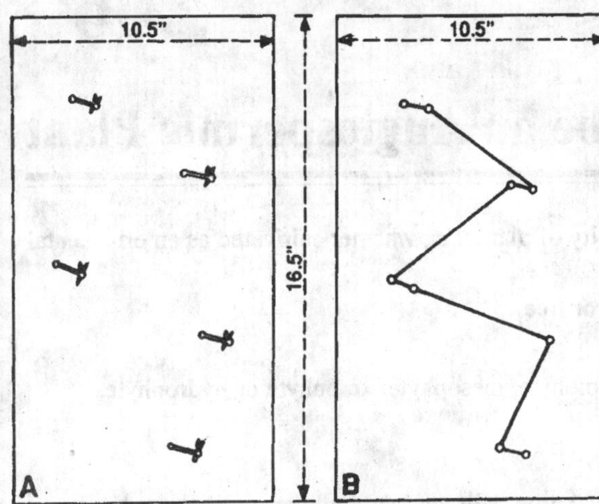

Fig. 5.7. Sticking of specimen on herbarium sheet; A, the right way (back view); B, the wrong way (back view).

defend their existence. There is an urgent need to undertake extensive and intensive explorations of all parts of the country. This is possible and could be carried out very effectively if the Universities start taking an interest in developing herbaria.

Frequently the herbaria are called upon to perform certain duties. While public supported herbaria may be approached for identification of plant specimens and supply of information not easily available from public libraries. Without active basic research identification of specimens may be often difficult or impossible. Hence the herbarium based botanists (taxonomists) should devote at least some time to carry out revisions and monographic studies of difficult groups of plants (K. Subramanyam and C.P. Sreemadhavan, 1970).

Designing of Modern Flora–The species concept and the recognition of intraspecific categories need particular examination; in fact a rational process of taxonomic thought should pervade the entire work. It should reflect adequately the current trends in taxonomic thought and practice and at the same time, keep itself within the limits of a well balanced, integrated botanical work–a creative work meant to inspire botanists of all disciplines and thus help to elevate systematics to its deserved place among the more sophisticated branches of Botany (M.A. Rau, 1970).

Fig. 5.8. Holding of specimen for poisoning.

Phylogeny is the evolutionary history of a taxon, and attempts to account for its origin and development. The term phylogeny is the autonym of ontogeny. Ontogeny differs from phylogeny in that it accounts for the life history of the individual plant from its development from the zygote to the production of its own gametes. A primary objective of phylogenetic studies in botany is the determination of origins and relationships of all taxa of both extinct and living plants.

Significance of Phylogeny to Taxonomy – Phylogeny deals with the evolutionary history of all taxa. It is a function of taxonomic research at all levels of classification. The ultimate goal of phylogenetic research is the production of a phylogenetic system of classification. The phylogenetic system shows the genetic and time relationship of any one taxon to another. According to Turril (1942), "taxonomy is based on characters, phylogeny on changes of characters." A phylogenetic system of classification for plants would provide the answer to questions of (*i*) their origin (*ii*) to their modes of evolution, (*iii*) to problems of monophyleticism vs., polyphyleticism, (*iv*) the identity of primitive and advanced characters, etc. It would conclude in a single stable classification of relationships.

6

How to Describe an Angiospermic Plant

1. Habitat. The natural abode or locality of plant, *i.e.*, whether cultivated as an ornamental plant, a food crop or occurs in a wild state.

2. Habit. (a) Herb, undershrub, shrub or tree.

(b) Annual, biennial or perennial.

(c) Any special feature, *e.g.*, parasite, epiphyte, mesophyte, xerophyte or hydrophyte.

3. Root. (a) Tap or adventitious.

(b) Branched or unbranched.

(c) Any special modification, *i.e.*, fibrous, prop, stilt, aerial, climbing, respiratory, tuberous, parasitic, conical, fusiform, napiform, fasciculated, beeded, nodulated, etc.

4. Stem. (a) Nature of stem, *i.e.*, erect, prostrate, twining or climbing; if climbing, write mode of climbing whether by leaf tendrils, leaflet tendrils, stem tendrils or stipular tendrils or by hooks or some other means.

(b) Any special modification, *i.e.*, rhizome, bulb, corm, tuber, offset, stolon, runner, sucker, phylloclade, etc.

(c) Branched or unbranched. If branched, write the mode of branching, *i.e.*, whether racemose or cymose. If cymose, whether uniparous or multiparous.

(d) Texture, *i.e.*, herbaceous.

(e) Solid or fistular.

(f) Cylindrical (terete), angular flattened or reduced.

(g) Hairy, glabrous, waxy or spiny.

(h) Colour. Whether green, grey, etc.

5. Leaf. (a) Insertion. Radical, cauline or ramal.

(b) Arrangement. Alternate, opposite or whorled (verticiilate). If opposite, whether superposed or decussate.

(c) Petiolate, sessile or subsessile.

(d) Stipulate or exstipulate. If stipulate, describe the nature of stipules, *i.e.*, leafy (foliaceous), spiny, tendrillar, petiolar, etc.

(e) Leaf base. Sheathing connate or ligulate or amplexicaul.

(f) Kind. Simple or compound.

(g) If simple, write about form of lamina, *i.e.*, acicular, linear, lanceolate, oval, ovate, obovate, cordate, obcordate, spathulate, orbicular, rotund, reniform, sagittate, hastate, lyrate, etc.

(h) If compound, whether pinnate or palmate, paripinnate, imparipinnate, bifoliate, trifoliate, etc.

(i) Incision of lamina. Pinnatified or palmatified, pinnatisect or palmatisect.

(j) Margin of lamina. Entire, denate, serrate, hairy, spiny, convolute, etc.

(k) Apex. Acute, obtuse, acuminate, emarginate, truncate, mucronate, cuspidate, etc.

(l) Surface. Hairy, glaucous, spiny.

(m) Venation. Reticulate or parallel, unicostate or multicostate.

(n) Texture. Coriaceous, fleshy, succulent, etc.

(o) Colour. Green or pigmented.

Note. In a compound leaf, the leaflets should be described in the manner as a simple leaf.

6. Inflorescence. Simple mixed, compound and special. If simple, racemose or cymose. If racemose, whether raceme, corymb, spike, catkin, spadix, umbel, hypanthodium, capitulum, etc. If cymose, uniparous, biparous or multiparous cyme, helicoid or scorpioid, or solitary axillary. Compound and mixed, compound spike, compound umbel, compound corymb, spike of spikelets, panicle and panicle of spikelets, etc. Special verticillaster, cyathium, etc.

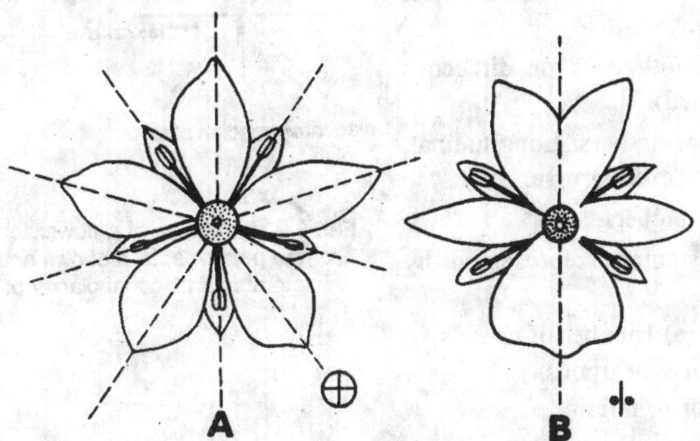

Fig. 6.1. Symmetry of flower. A, regular or actinomorphic; B, Zygomorphic or irregular.

7. Flower. (a) Pedicellate or sessile.

(b) Bracteate or ebracteate; if bracteate, the nature of the bract should be described.

(c) Actinomorphic or zygomorphic.

(d) Unisexual or hermaphrodite (bisexual).

(e) Complete or incomplete.

(f) Regular or irregular.

(g) Hypogynous, perigynous or epigynous.

(h) Colour of the flowers.

8. Calyx. (a) Number of sepals.

(b) Polysepalous or gamosepalous. If polysepalous, describe the nature and form of sepals, *i.e.*, deciduous, caducous, persistent, etc. If gamosepalous, give special forms, *e.g.*, tubular, campanulate, bilabiate, etc.

(c) Sepaloid (green) or petaloid (coloured).

(d) Inferior or superior.

(e) Aestivation of the calyx, *e.g.*, imbricate, valvate, quincuntial, etc.

9. Corolla. (a) Number of petals.

(b) Poly or gamopetalous, if polypetalous, one should describe the form and nature of petals, *i.e.*, rosaceous, papilionaceous, etc. If gamopetalous, give special form, *e.g.*, campanulate, tubular, bilabiate, spurred, ligulate, etc.

(c) Colour of the petals.

(d) Inferior or superior.

(e) Aestivation of the corolla.

10. Androecium. (a) Number of stamens; 1 to 10 or indefinite.

(b) Free or fused; if fused, adelphous or syngenesious; if adelphous, mono, di or polyadelphous. If free, any special form, *e.g.*, didynamous, tetradynamous, etc.

(c) Epipetalous, epiphyllous or free from petals or tepals.

(d) Filament long, short or flattened.

(e) Fixation of anthers. Adnate, basifixed, dorsifixed or versatile.

(f) Anthers. Monothecous or dithecous (single-celled or bicelled).

(g) Dehiscence of anthers. Longitudinal, transverse, porous, extrorse, introrse.

(h) Colour of the anthers.

(i) Sometimes stamens represented by staminodes.

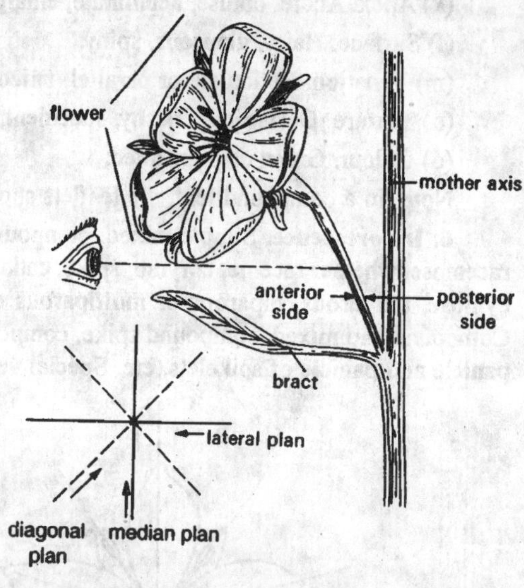

Fig. 6.2. A, relation of a flower to the mother axis or the peduncle; B, it shows antero-posterior, lateral and diagonal planes of the flower.

11. Gynoecium. (a) Number of carpels.

(b) Apocarpous or syncarpous.

(c) Ovary superior or inferior.

(d) Number of loculi.

(e) Number of ovules in each loculus.

(f) Placentation. Marginal, parietal, axile, free central, basal, superficial, etc.

(g) Nature of style, *i.e.*, short, slender, terminal, divided, etc.

(h) Nature of stigma, *i.e.*, simple or lobed, feathery, bearded, hairy, capitate, curved, etc.

(i) Any disc present below the ovary should be described.

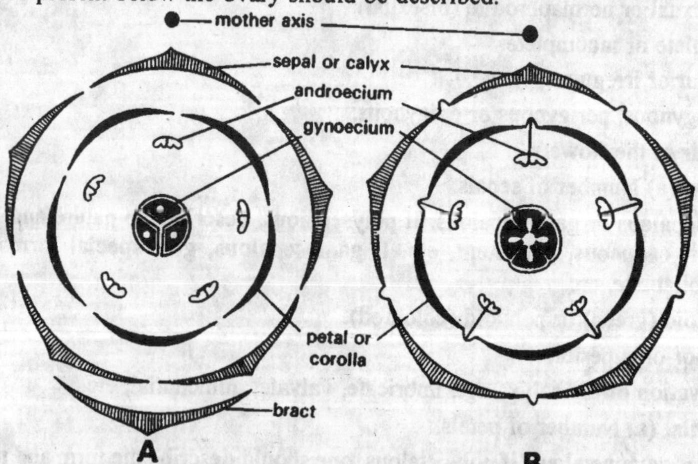

Fig. 6.3. Floral diagrams of polypetalous (A) and gamopetalous (B) flowers, with apocarpous and syncarpous gynoecia respectively.

12. Fruit. Whether it is simple, aggregate, multiple; pod, legume, follicle, lomentum, capsule, siliqua, silicula, achene, cypsela, cremocarp, caryopsis, samara, nut, pome, drupe, berry, carcerule, hesperidium, etaerio of achenes, drupes, follicles or berries, sorosis, syconus.

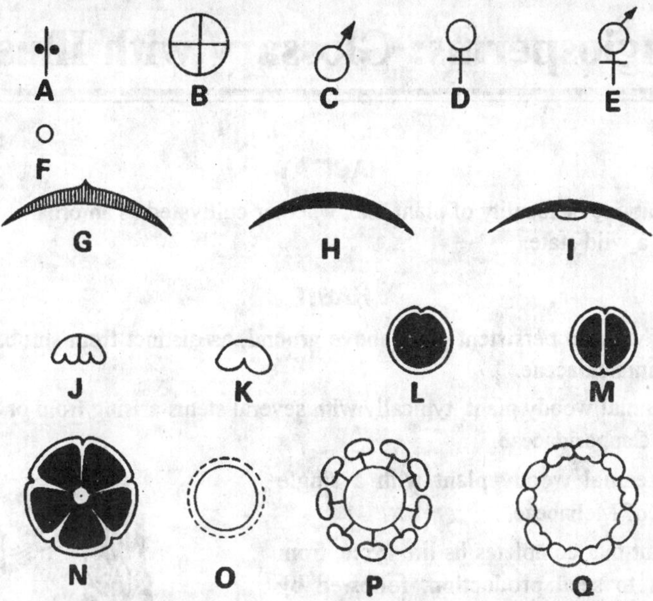

Fig. 6.4. Various symbols used in writing a floral formula and floral diagram. A, zygomorphic; B, actinomorphic; C, male; D, female; E, bisexual; F, mother axis; G, calyx; H, corolla; I, Corolla with pocket nectary; dithecous stamen; K, monothecous stamen; L, monocarpellary gynoecium; M, bicarpellary gynoecium; N, polycarpellary syncarpous; O, nectar secreting disc around carpel; P, monadelphous androecium.

13. Seed. (a) Albuminous or exalbuminous.

(b) Single or two cotyledons.

(c) Position, shape and size of the embroyo.

(d) Cotyledons straight or curved.

14. Floral formula. The floral formula must be given in the end of plant description, *e.g.*, *Solanum*: $\oplus$ ⚥ K(5), $\overline{C(5), A\,5}$, G $(\underline{2})$.

15. Floral diagram.

16. Generic characters. The distinguishing characters of the genus should be underlined in the description.

17. Systematic position. In the end of the description the systematic position of the genus should be given with the help of key.

7

Angiosperms: Glossary with Illustrations

HABITAT

The natural abode or locality of plant, *i.e.*, whether cultivated as an ornamental plant, a food crop or occurs in a wild state.

HABIT

Herb. Plant with no persistent parts above ground, as distinct from shrubs and trees, *e.g.*, *Ranunculus* of Ranunculaceae.

Shrub. Perennial woody plant, typically with several stems arising from or near the ground, *e.g.*, *Capparis* of Capparidaceae.

Tree. A perennial woody plant with a single trunk, *e.g.*, *Melia* of Meliaceae.

Annual. Plant that completes its life-cycle, from seed germination to seed production, followed by death within a single season, *e.g.*, *Brassica* of Cruciferae.

Biennial. Plant that continues its growth from year to year. In *herbaceous* perennials serial parts die away in autumn, replaced by new shoots in the following year from underground structures, *e.g.*, *Delphinium*; in *woody* perennials, permanent woody stems above ground from starting point for each new year's growth, a characteristic that enables some of them to reach a large size, *e.g.*, shrubs and trees.

Parasite. Plant living in or on another plant (its host) from which it obtains food, *e.g.*, *Cuscuta* of Convolvulaceae.

Epiphyte. Plant attached to another plant, not growing parasitically upon it but merely using it for support, *e.g.*, *Vanda* (an orchid) of Orchidaceae.

Fig. 7.1. Saprophyte. *Monotropa*.

Mesophyte. Plant growing under average conditions of water supply.

Hydrophyte. Plant whose habitat is water or very wet places, *e.g.*, *Ranunculus aquatilis* of Ranunculaceae; *Neptunia oleracea* of Mimosoideae.

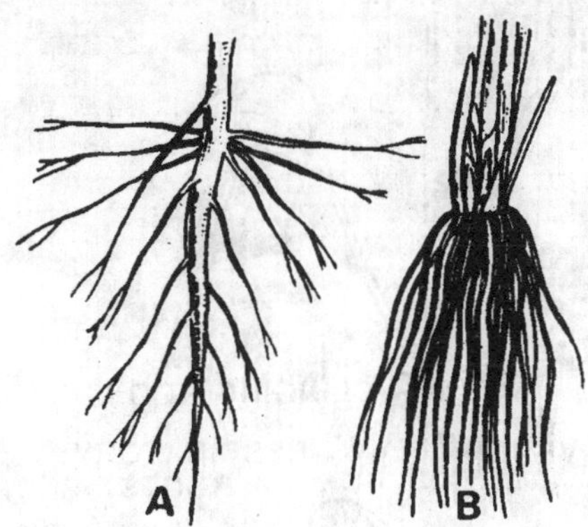

Fig. 7.2. Roots. A, tap and lateral roots of a dicotyledon; B, fibrous roots of a monocotyledon.

Adventitious. Root developing from part of plant other than roots, *e.g.*, from stem or leaf cutting; of buds, developing from part of plant other than in axil of leaf, *e.g.*, from root, most of monocots.

Branched. Roots possessing branches.

Unbranched. Roots without branches.

Fibrous. A fibrous root system consists of a tuft of adventitious roots of more or less equal diameter, arising from stem base or hypocotyl and bearing smaller lateral roots, *e.g.*, wheat, strawberry.

Prop. Hanging aerial roots of *Ficus*.

Stilt. Such roots arise from the nodes of the stem above the soil, *e.g.*, *Saccharum*.

Aerial. Hanging aerial roots, *e.g.*, *Orchids*.

Climbing. The roots that help in climbing of plants, *e.g.*, *Tecoma, Piper, Pothos*.

Respiratory. Spongy roots, *e.g.*, *Jussiaea*.

Pneumatophores. Special root branch produced in large numbers by some vascular plants growing in water or

Xerophyte. Plant of dry habitat able to endure conditions of prolonged drought, *e.g.*, *Capparis decidua* of Capparidaceae.

Saprophyte. Plant which obtains organic matter in solution from dead and decaying tissues of plants (or animals), *e.g.*, *Monotropa*.

Root. That part of vascular plants that usually grows downwards into the soil, anchoring plant and absorbing water and nutrient salts.

Tap. Root system with a prominent main root, directed vertically downwards and bearing smaller lateral roots, *e.g.*, most of dicots; *Cajanus* of Papilionatae.

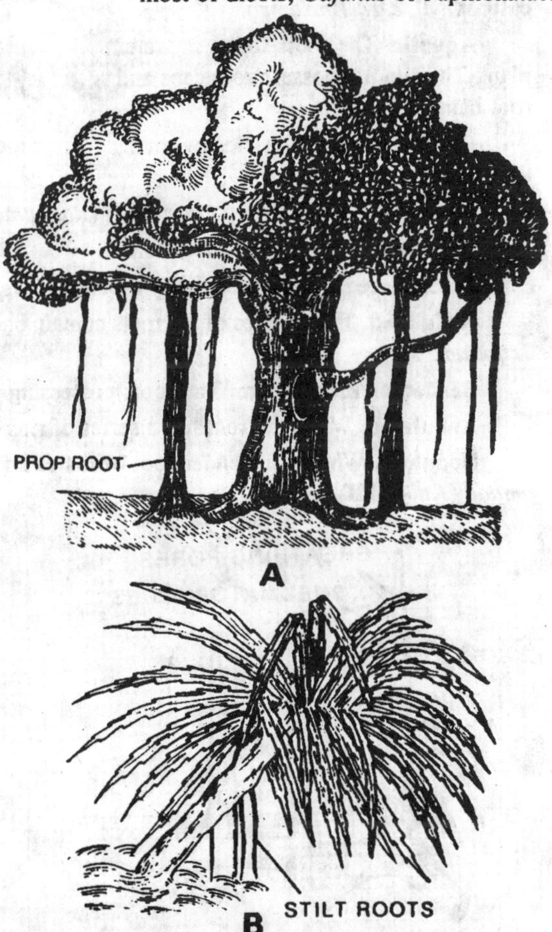

Fig. 7.3. Adventitious modified roots. A, prop roots of banyan; B, stilt roots of Keora.

in tidal swamps, *e.g.*, Mangrove; grows erect, projecting into the air above and contains well developed intercellular system of air spaces in communication with atmosphere through pores on aerial portion.

Tuberous. Irregularly swollen roots laden with starch, *e.g.*, *Ipomoea batatas* of Convolvulaceae.

Parasitic. The sucking roots of parasites, *e.g.*, *Cuscuta, Dendrophthoe, Viscum, Orobanche.*

Epiphytic. Hanging aerial roots of Orchids.

Assimilatory. The aerial roots with chlorophyll, *e.g.*, *Tinospora.*

Aquatic. The roots found in water plant. They do not possess root caps and root hairs, *e.g.*, *Pistia.*

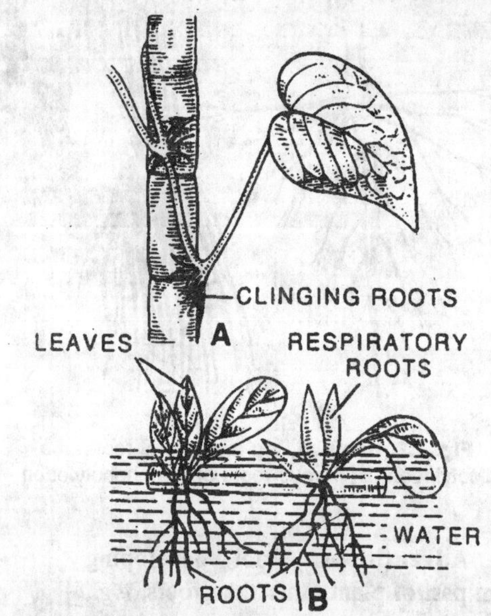

Fig. 7.4. Adventitious root modified roots. A, climbing roots of betel, B, respiratory roots of *Jussiaea*.

Conical. Cone like fleshy roots, *e.g.*, carrot.

Fusiform. *e.g.*, fleshy root of radish (*Raphanus sativus* of Cruciferae).

Napiform. *e.g.*, fleshy root of turnip.

Fasciculated. Clusters of fleshy roots, *e.g.*, *Asparagus.*

Nodulated. The nodules of the roots contain bacteria, *e.g.*, Leguminous roots; *Cicer, Arachis, Trifolium.*

Beaded or Moniliform. The roots possessing beaded structures, *e.g.*, *Vitis.*

Annulated. When the root has a series of ring-like swellings on its body, *e.g.*, ipecacuanha.

Nodulose. When the slender root becomes suddenly swollen near the apex, *e.g.*, *Curcuma amada* **(Am-haldi).**

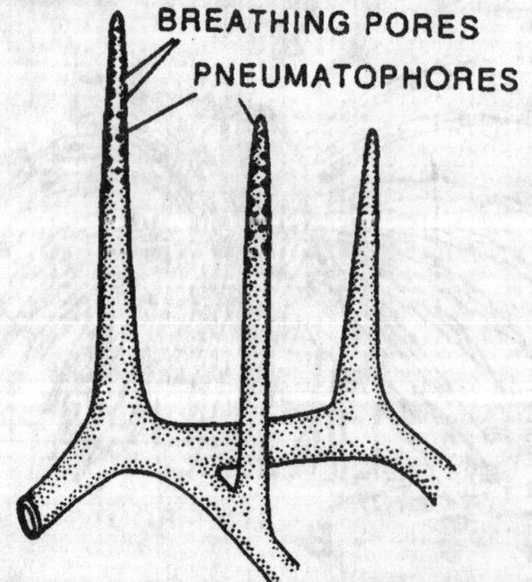

Fig. 7.5. Roots. Pneumatophores arising vertically upwards from an underground root.

STEM

Stem. Normally aerial part of axis of vascular plants, bearing leaves and buds at definite positions (nodes) and reproductive structures, *e.g.*, flowers.

Erect. Rigid, strong and upright stem.

Prostrate. Trailing stem lying flat on the ground, *e.g.*, *Portulaca.*

Twiner. Long, slender and branched stem climbing by twining its body round the support, *e.g.*, *Cuscuta.*

Climbers. A weak stem attaching itself to any neighbouring support by means of special structures such as rootlets, hooks, leaf tendrils, stem tendrils, stipular tendrils, etc.

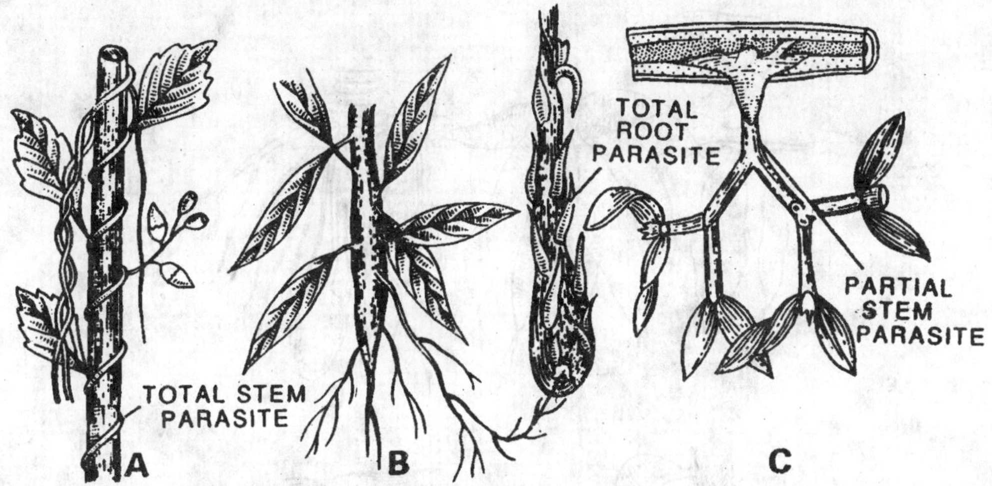

Fig. 7.6. Parasite. A, total stem parasite of *Cuscuta*; B, total root parasite of *Orobanche*; C, partial stem parasite of *Viscum*.

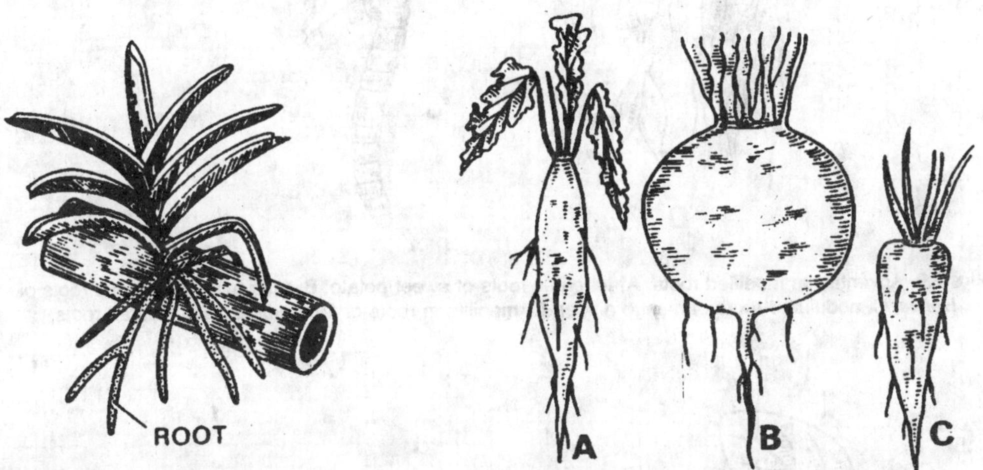

Fig. 7.7. Epiphytic roots of an orchid (*Vanda*).

Fig. 7.8. Modified roots. A, fusiform roots of radish; B, Napiform root of turnip; C, conical root of carrot.

Rhizome. Underground stem, bearing buds in axils of reduced scale-like leaves; saving as a means of perennation vegetative propagation, *e.g.*, *Zinger, Mentha.*

Bulb. Modified shoot consisting of very much shortened stem enclosed by fleshy, scale leaves, *e.g.*, *Allium cepa.*

Corm. Swollen stem base containing food material and bearing buds in the axils of scale-like remains of leaves of previous season's growth, *e.g. Colocasia, Amorphophallus, Gladiolus.*

Tuber. Swollen end of underground stem bearing buds in axils of scale-like rudimentary leaves (stem tuber), *e.g.*, potato.

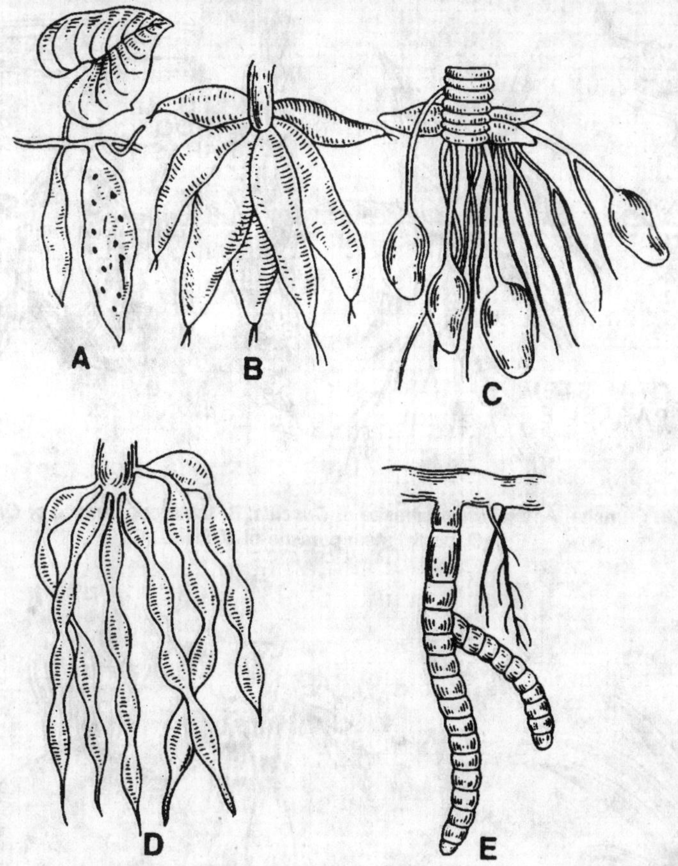

Fig. 7.9. Adventitious modified roots. A, tuberous roots of sweet potato; B, fasciculated tuberous roots of *Dahlia*; C, nodulated roots of mango ginger; D, moniliform roots of *Momordica;* E, annulated roots.

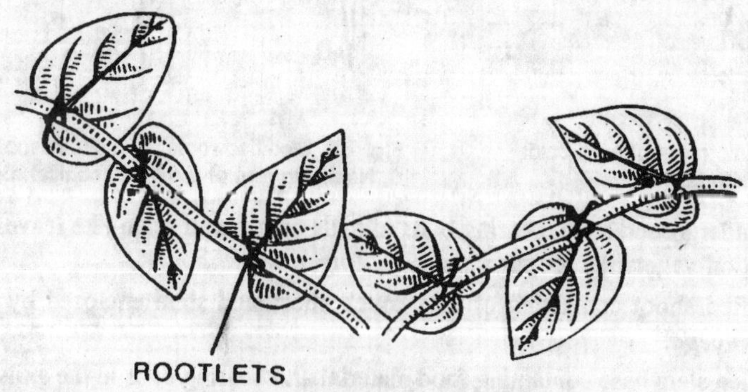

Fig. 7.10. Rootlet climbers of *Ficus pumila*.

Offset. A horizontal, short, more or less apex a tuft of leaves above and a cluster of thickened, prostrate branch producing at the small roots beneath, *e.g.*, *Pistia*.

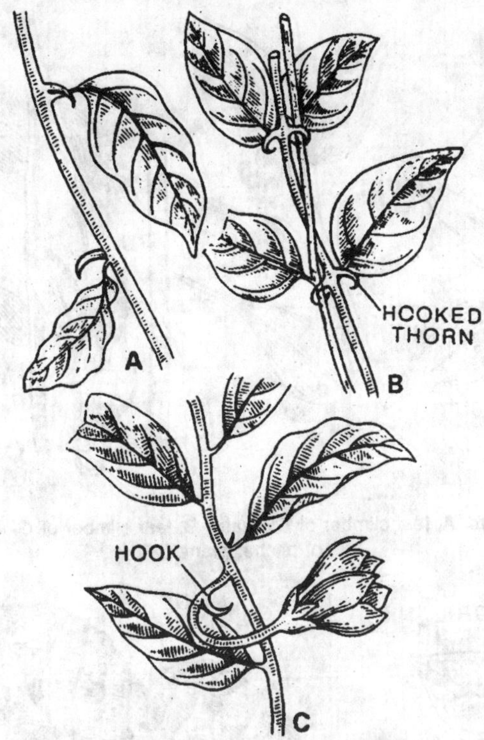

Fig. 7.11. Stem. A, thorn climber of *Bougainvillaea*; B, hooked thorn climber of *Uncaria*; C, hook climber of *Artabotrys*.

Fig. 7.12. Stem. A, prickle climber of cane; B, prickle climber of rose.

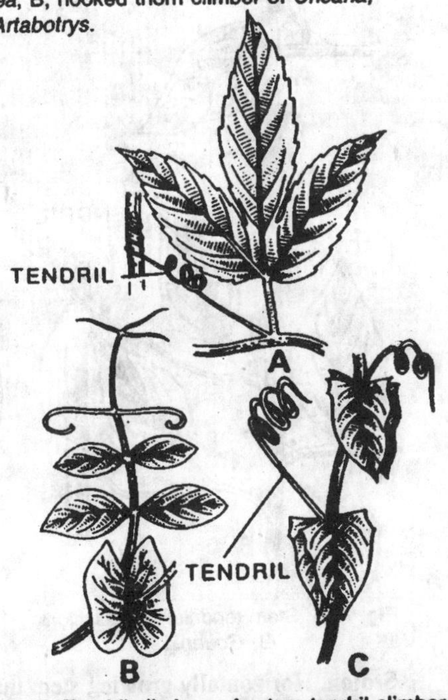

Fig. 7.13. Tendril climbers. A, stem tendril climber of passion flower; B, leaf tendril climber of pea; C, leaf tendril climber of *Lathyrus*.

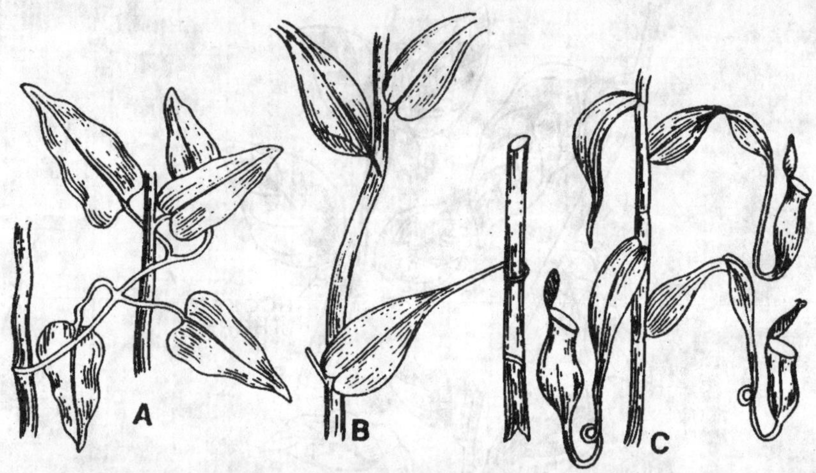

Fig. 7.14. Leaf climbers. A, leaf climber of *Clematis;* B, leaf climber of *Gloriosa*; C, leaf climber of pitcher plant.

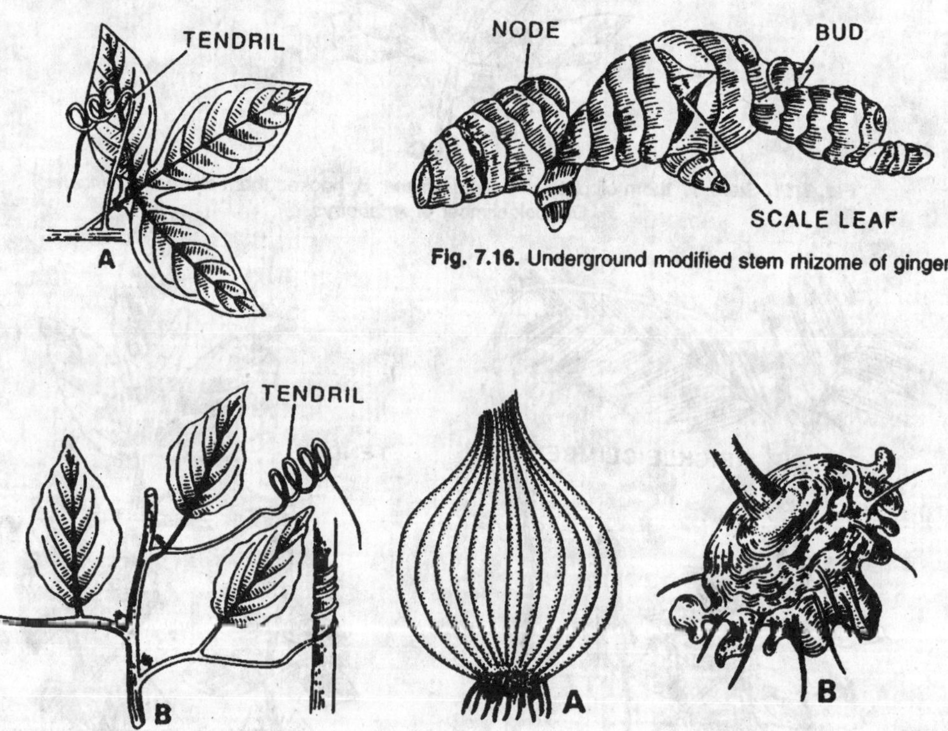

Fig. 7.16. Underground modified stem rhizome of ginger.

Fig. 7.15. Stem tendrils. A, *Passiflora*; B, *Gouania*.

Fig. 7.17. Underground modified stem. A, bulb of onion; B, corm of *Amorphophallus*.

Stolon. Horizontally growing stem that roots at nodes, *e.g.*, strawberry runner, *Colocasia.*

Runner. Stolon that roots at tip forming new plant that eventually is freed from connection with parent by decay of runner, *e.g.*, *Oxalis*.

Fig. 7.18. Underground modified stem; tuber of potato.

Fig. 7.21. Sub-aerial modification of stem. Runner of *Oxalis*.

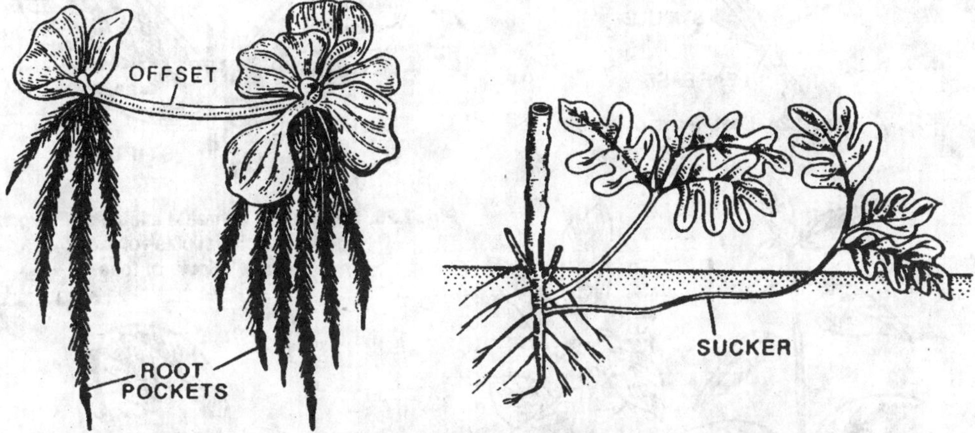

Fig. 7.19. Sub-aerial modification of stem. Offset of *Pistia*.

Fig. 7.22. Sub-aerial modification of stem. Suckers of *Chrysanthemum*.

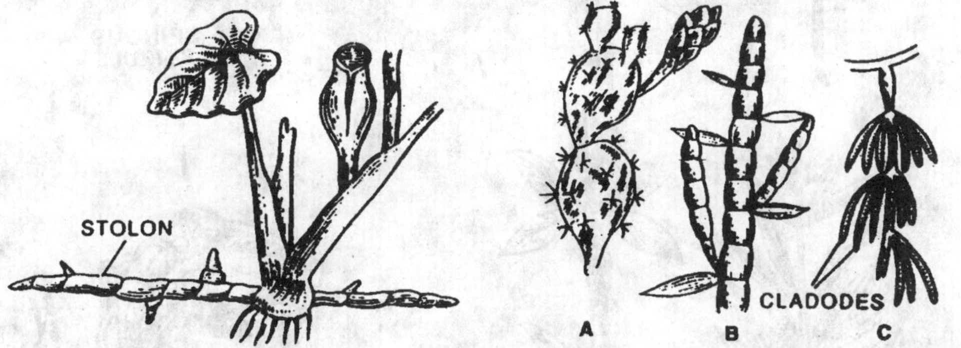

Fig. 7.20. Sub-aerial modification of stem. Stolon of *Colocasia*.

Fig. 7.23. Aerial modification of stem. A, phylloclade of *Opuntia*; B, phylloclade of Cocoloba; C, cladodes of *Asparagus*.

Sucker. A creeping stem but growing obliquely upwards directly giving rise to a leafy shoot, *e.g.*, *Chrysanthemum*.

Phylloclade. Modified stem having appearance and function of a leaf, ***e.g.***, ***Ruscus***, Cocoloba.

Cladode. A phylloclade of single internode, ***e.g.***, ***Asparagus***.

Branched. Stem possessing branches, ***e.g.***, Nim tree.

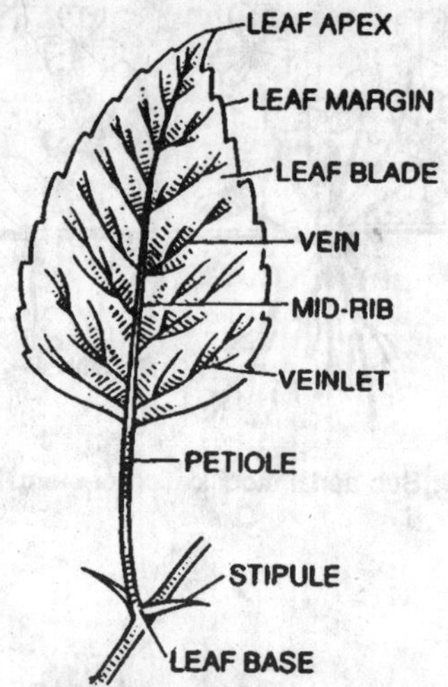

Fig. 7.24. Leaf, Parts of a leaf.

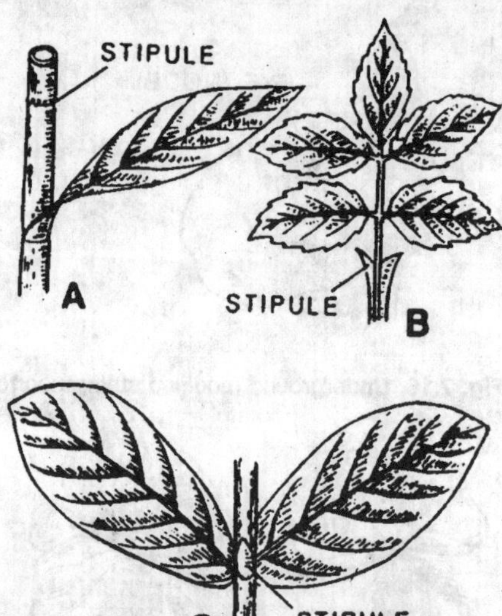

Fig. 7.26. Stipules. A, ochreate stipule of *Polygonum*; B, interpetiolar stipule of *Ixora*; C, adnate stipule of rose.

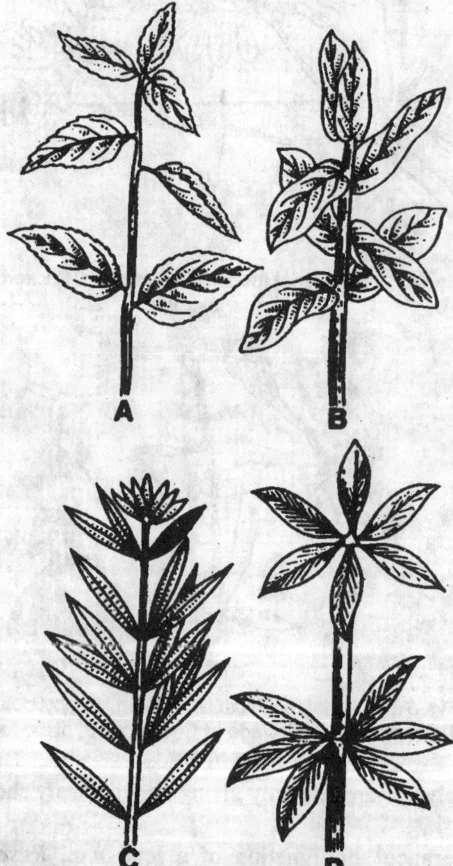

Fig. 7.25. Phyllotaxy of leaves. A, alternate; B, opposite decussate; C, whorled leaves of *Nerium*; D, whorled leaves of *Alstonia*.

Fig. 7.27. Stipules, A, spinous stipules of *Zizyphus*; B, tendrilar stipules of *Smilax*.

Unbranched. Stem having no branches, *e.g.*, Palm.

If branched whether racemose or cymose type of branching. If cymose whether uniparous, biparous or multiparous.

Herbaceous. Having the characters of a herb, *e.g.*, *Ranunculus*.

Woody. Having the characters of a shrub and tree, *e.g.*, *Capparis*, *Melia*.

Solid. Interior portion of the stem is filled up with matter.

Fistular. A stem having hollow interior, *e.g.*, wheat, bamboo.

Cylinderical (terete). Circular stem as seen in T.S.

Angular. A stem shows many angles in T.S., *e.g.*, *Cucurbita*.

Flattened. Flat stem in T.S.

Hairy. Stem possessing hairs on its surface.

Glabrous. Smooth stem.

Waxy. Stem having wax coating, *e.g.*, *Calotropis*.

Spiny. Stem having spines.

Colour. Whether green, grey, etc.

LEAF

Radical. Proceeding from or near the root, *e.g.*, Onion.

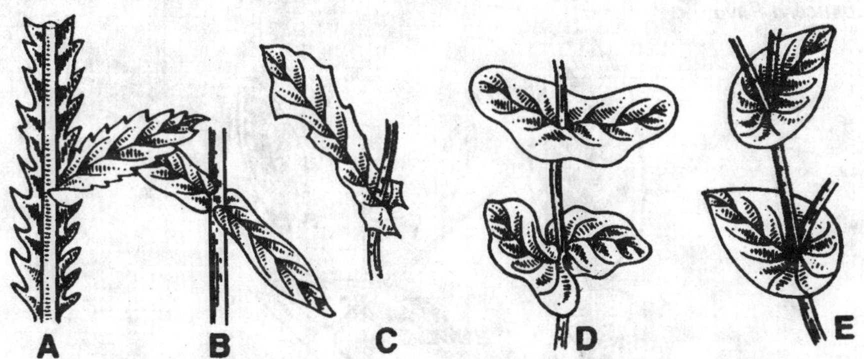

Fig. 7.28. Leaf base of sessile leaves. A, decurrent leaf of *Laggera;* B, auriculate leaf of *Calotropis;* C, amplexicaul leaf of *Emilia;* D, connate leaf of *Lomicera*; E, perfoliate leaves.

Cauline. Pertaining to the stem, *e.g.*, Palms.

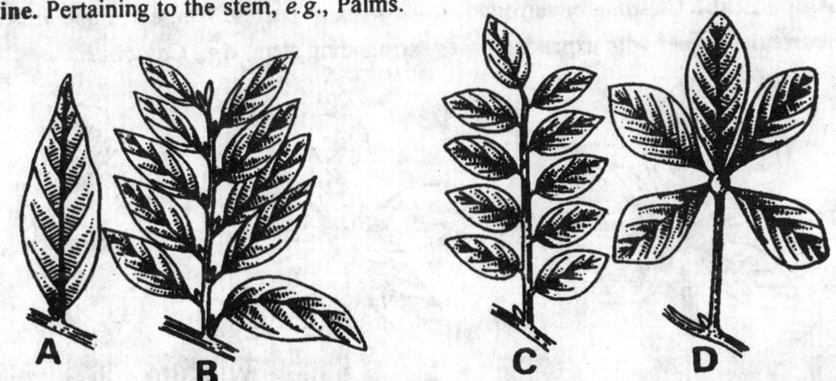

Fig. 7.29. Simple and compound leaves. A, simple leaf; B, a branch; C, pinnately compound leaf; D, palmately compound leaf.

Cauline and ramal. Pertaining to the main stem as well as its branches, *e.g.*, Mango.

Alternate. A single leaf arising at each node, *e.g., Hibiscus rosa-sinensis.*

Opposite. On different sides of the axis with the bases at the same level.

Opposite decusate. In pairs at right angles to one another, *e.g., Calotropis.*

Opposite superposed. A pair of leaves that stands directly over the lower pair in the same plant, *e.g.*, Guava.

Whorled. More than two leaves arranged in a circle round an axis, *e.g., Spergula.*

Petiolate. The leaf blade is situated on the petiole.

Sessile. Without a petiole or stalk.

Sub-sessile. Having short petiole.

Stipulate. With stipules, *e.g.*, rose.

Extipulate. Having no stipules, *e.g., Ipomoea.*

Stipules. An appendage of the leaf. Normally two stipules are developed at base of a leaf petiole, they may be leafy (foliaeous), *e.g., Lathyrus*; free lateral, *e.g.*, China rose; adnate, *e.g.*, rose; interpetiolar, *e.g., Ixora*; *Spergula*; spiny, *e.g., Acacia*; tendrilar, *e.g., smilax.*

Leaf Base

Connate. Two sessile opposite leaves meeting each other across the stem and fusing together, *e.g., Lomicera flava.*

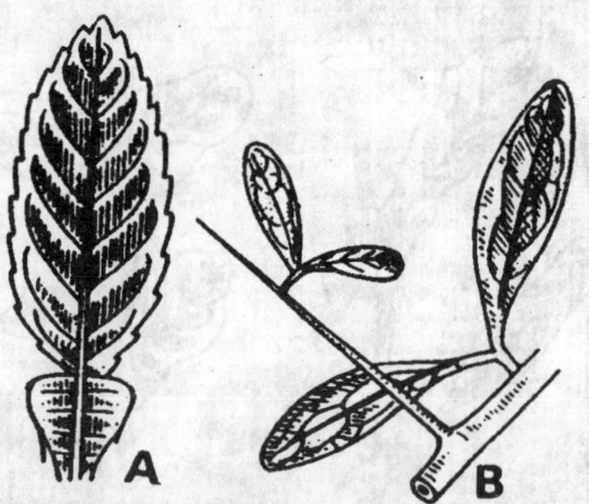

Fig. 7.30. Compound leaf. A, unifoliate leaf (*Citrus*); B, bifoliate leaf (*Prinsepia*).

Amplexicaul. Clasping or surrounding the stem, as base of leaf, *e.g., Sonchus.*

Auriculate. Leaf with expanded bases surrounding stem, *e.g., Calotropis.*

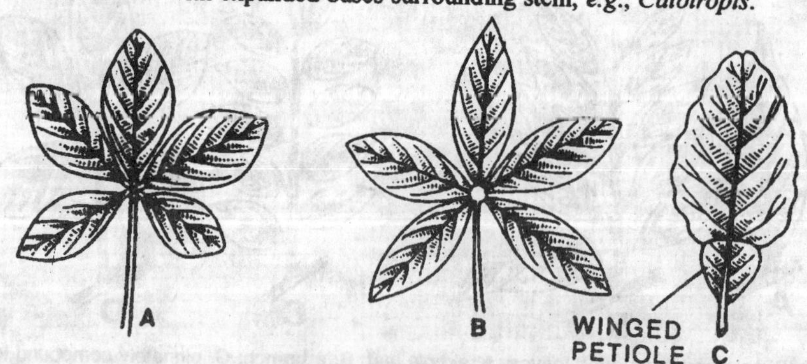

Fig. 7.31. Palmate compound leaves. A, digitate leaf of *Gynandropsis;* B, digitate leaf of *Bombax;* C, unifoliate compound leaf of *Citrus* with winged petiole..

Decurrent. Having leaf base prolonged down stem as a winged expansion or rib, *e.g.*, *Laggera pterodonta.*

Perfoliate. A leaf with basal lobes so united as to appear as if stem ran through it, *e.g.*, *Aloe perfoliata.*

Simple. A leaf which may be entire or incised to any depth, but not down to the midrib or petiole.

Compound. A leaf made up of two or more leaflets, *e.g.*, pea and several other members of Leguminosae.

Unifoliate. Having one leaflet only, *e.g.*, *Citrus.*

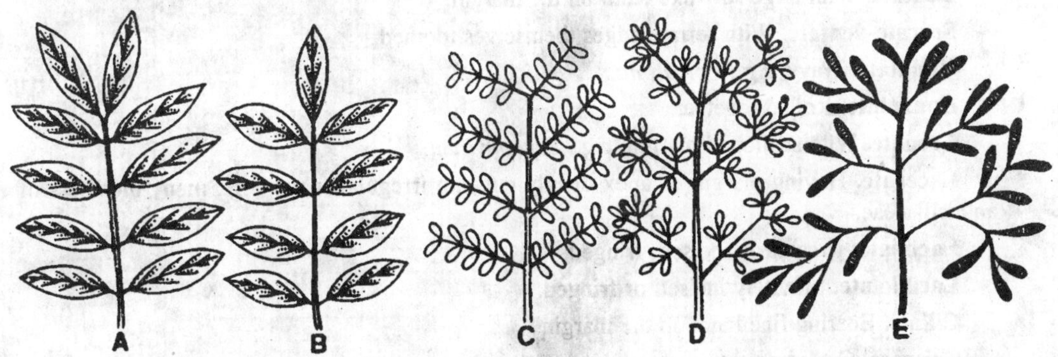

Fig. 7.32. Pinnately compound leaves. A, unipinnate (paripinnate); B, unipinnate (imparipinnate); C, bipinnate; D, tripinnate; E, decompound leaf of coriander.

Bifoliate. Palmate compound leaf with two leaflets, *e.g.*, *Prinsepia, Balanites.*

Trifoliate palmate. Compound leaf having three leaflets growing from same point, *e.g.*, *Oxalis.*

Quadrifoliate. Compound palmate leaf, with four leaflets arising at a common point, *e.g.*, *Marsilea.*

Multifoliate. Compound palmate leaf with five or more leaflets arising at a common point, *e.g.*, *Gynandropsis pentaphylla.*

Pinnate. A compound leaf having leaflets on each side on an axis or mid-rib.

Unipinnate. Having leaflets on each side of an axis, *e.g.*, *Cassia.*

Bipinnate. The central axis produces secondary axis which bears the leaflets, *e.g.*, *Acacia.*

The secondary axes produce the tertiary axes which bears the leaflets, *e.g.*, *Acacia.*

The secondary axes produce the tertiary axes which bear the leaflets, *e.g.*, *Moringa.*

Decompound. More than thrice pinnate, *e.g.*, old leaves of coriander.

Paripinnate. Pinnately compound without a terminal leaflet, *e.g.*, *Cassia.*

Imparipinnate. Pinnately compound leaf with an odd terminal leaflet, *e.g.*, Pea.

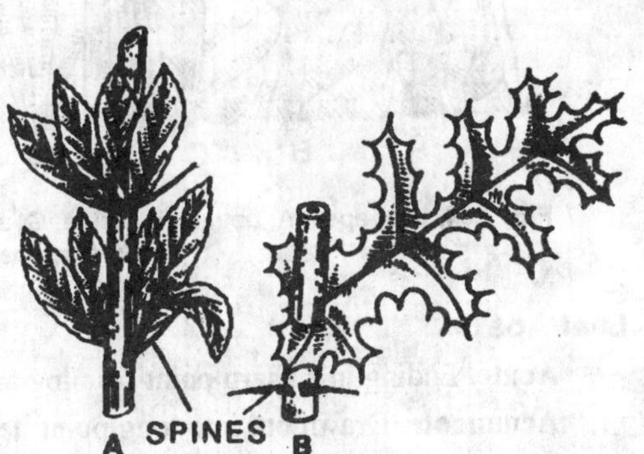

Fig. 7.33. Modified leaves. A, leaves modified into spines; B, leaf of *Argemone* with spines.

Incision of Lamina

Pinnatifid. Leaves lobed half–way to midrib.

Pinnatisect. With leaves lobed almost to base or midrib.

Palmatifid. Leaves divided into lobes to about the middle, at acute angles to each other, *e.g.*, Castor.

Palmatisect. Palmate with division nearly to base.

Margin of Lamina

Entire. With continuous margin, *e.g.*, *Psidium*.

Dentate. With large saw-like teeth on the margin.

Serrate-dentate. With serrate edges themselves toothed.

Undulate. Wavy, *e.g.*, *Polyalthia*.

Convolute. Rolled together.

Crenate. With scalloped margin, *e.g.*, *Bryophyllum*.

Lacerate. Having margin or apex deeply cut into irregular lobes, *e.g.*, many members of Ranunculaceae.

Laciniate. Irregularly incised; fringed.

Laciniolate. Minutely incised or fringed.

Ciliate. Bearing fine hairs on the margin.

Crispate. Curled or extremely undulate margin.

Spinous. Bearing many spines, *e.g.*, *Argemone*.

Pectinate. Comb-like.

Lobed. Leaf margin divided into many lobes, *e.g.*, *Ranunculus*.

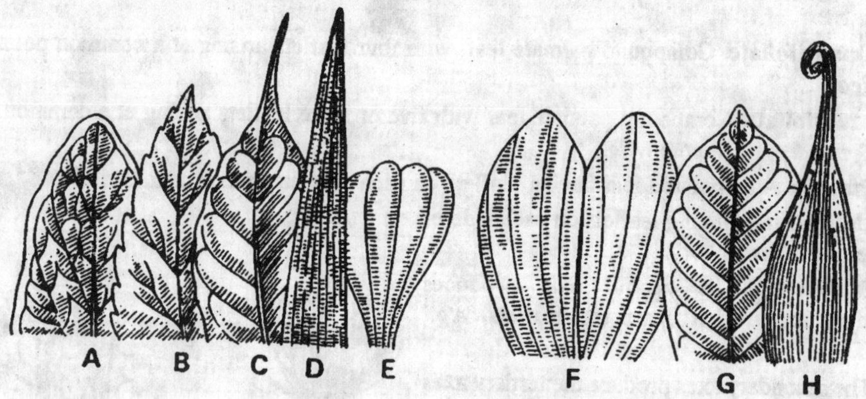

Fig. 7.34. Leaf apex A, obtuse; B, acute; C, acuminate; D, cuspidate; E, retuse; F, emarginate; G, mucronate; H, cirrhose.

Leaf Apex

Acute. Ending in a sharp point forming an acute angle, *e.g.*, mango.

Acuminate. Drawn out into long point; tapering; pointed, *e.g.*, *Ficus religiosa*.

Obtuse. With blunt or rounded end, *e.g.*, Banyan.

Emarginate. Having a notch at apex, *e.g.*, *Bauhinia*.

Truncate. Terminating abruptly, as if tapering end were cut off, *e.g.*, *Caryota urens*.

Mucronate. Abruptly terminated by a sharp spine, *e.g.*, apex of leaflet of *Cassia obtusifolia*.

Cuspidate. Terminating in a point.

Aristate. Provided with awns, or with a well developed bristle.

Retuse. Obtuse with a broad shallow notch in middle, *e.g., Oxalis*.

Cirrhose. Leaf with prolongation of midrib forming a tendril, *e.g., Gloriosa.*

Apiculate. Forming abruptly to a small tip, *e.g., Dalbergia*.

Leaf Surface

Hairy. Leaf surface covered with fine hairs. The hairs are of various types.

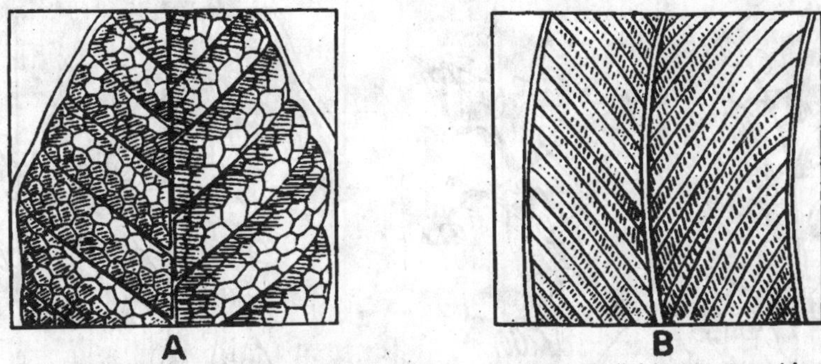

Fig. 7.35. Leaf venation. A, reticulate in a dicot leaf; b, parallel in a monocot leaf.

Glabrous. With a smooth even surface, without hairs, *e.g.*, China rose.

Glaucous. Shiny green, *e.g., Citrus*.

Spiny. Covered with spines, *e.g., Argemone*.

Venation. System or disposition of veins.

Reticulate (net veined). Like net work, *e.g.*, most of dicots.

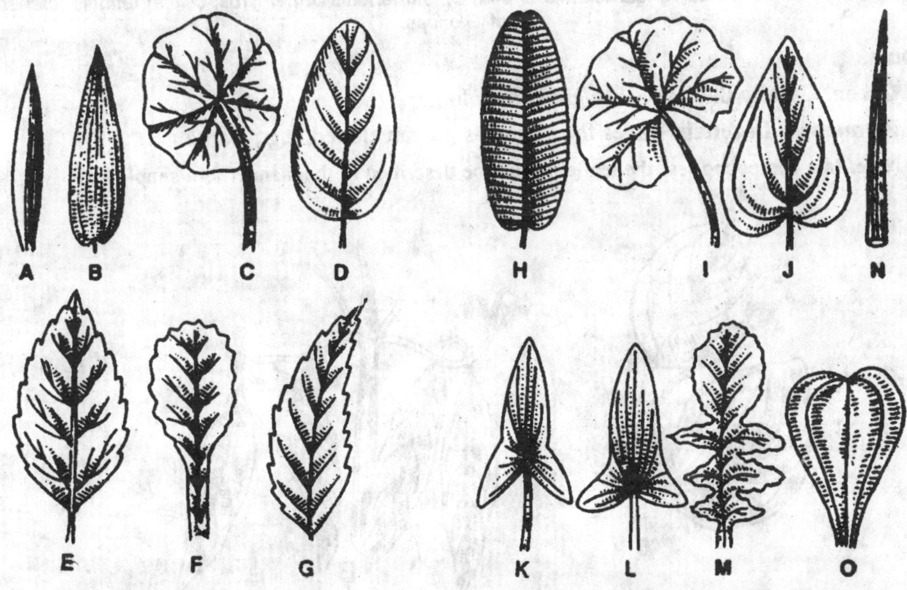

Fig. 7.36. Shape of leaf. A, linear, B, lanceolate; C, rotund; D, oval or elliptical; E, ovate; F, spathulate; G, oblique; H, oblong; I reniform; J, cordate; K, sagittate; L, hastate; M, lyrate; N, acicular; O, cuneate.

Parallel. Parallel veined, *e.g.*, most of monocots.

Unicostale. Having only one principal vein.

Multicostate. Having many principal veins.

Texture

Coriaceous. Leathery, *e.g.*, *Calotropis*.

Fleshy. Soft and thick, *e.g.*, *Spergula*.

Succulent. Full of juice or sap, *e.g.*, *Aloe*.

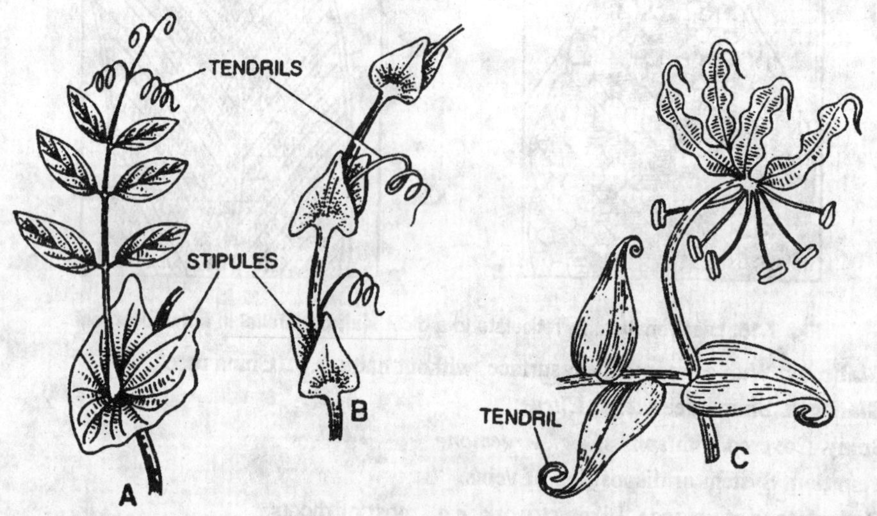

Fig. 7.37. Modified leaves A, leaf tendrils of pea; B, leaf tendrils of *Lathyrus;* C, leaf tendrils (cirrhose) of *Gloriosa*.

Colour

Green. Usually the leaves are green in colour.

Pigmented. In certain leaves the pigments are developed, *e.g.*, *Aerva*.

Note. In a compound leaf the leaflet should be described in the manner as a simple leaf.

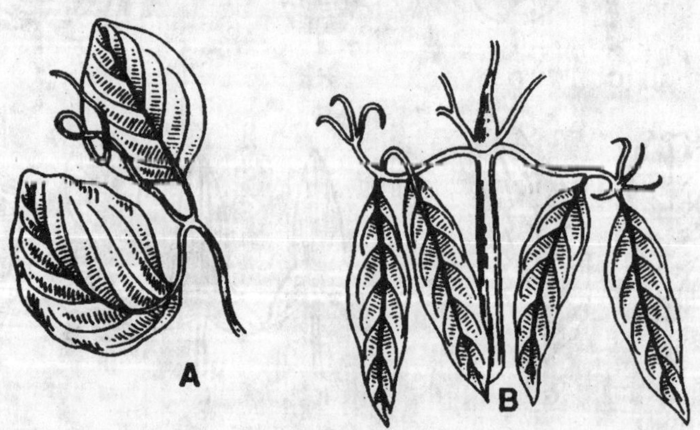

Fig. 7.38. Modified leaves. A, leaf of *Naravelia* with the terminal leaflet modified into a tendril. B, leaflet modified into hooks (*Bignonia*).

Shape of the Leaf

Linear. Long and narrow leaf, *e.g.*, many grasses.

Lanceolate. Lance-shaped leaf, *e.g.*, bamboo, *Nerium*, etc.

Round or orbicular. Leaf with a circular leaf blade, *e.g.*, lotus, garden nasturtium, etc.

Elliptical. An ellipse-shaped leaf, *e.g.*, guava, jack, etc.

Ovate. Leaf with an egg-shaped leaf blade, *i.e.*, slightly broader at the base than at the apex, *e.g.*, banyan, China rose, etc.

Spathulate. Spathula-shaped leaf, *i.e.*, broad and round at the top and narrower towards the base, *e.g.*, *Calendula* and *Drosera.*

Oblique. Leaf with two unequal halves, *e.g.*, *Begonia.*

Oblong. Leaf with wide and long leaf blade. Here the two margins run more or less straight up, *e.g.*, banana.

Reniform. Kidney-shaped leaf, *e.g.*, Indian pennywort.

Cordate. Leaf with heart-shaped leaf blade, *e.g.*,betel (inversely heart-shaped leaf called as obcordate, *e.g.*, wood-sorrel).

Sagittate. Leaf with an arrow-shaped leaf blade, *e.g.*, arrowhead and some aroids.

Hastate. Sagittate leaf with its two lobes directed outside, *e.g.*, water bindweed and *Typhonium*.

Lyrate. Lyre-shaped leaf, *i.e.*, with a large terminal lobe and some smaller lateral lobes, *e.g.*, radish, mustard, etc.

Acicular. Long narrow and cylindrical leaf, *i.e.*, needle shaped, *e.g.*, pine.

Cuneate. Wedge shaped leaf, *e.g.*, water lettuce.

Modified Leaves

Leaves of many plants are often modified into some specified structures to do various specialized functions. Leaf tendril is one of these structures.

Leaf-tendrils. Leaf or part of the leaf modified as a slender, branched or unbranched, thread like structure, used by many climbing plants for attachment to a support in various ways.

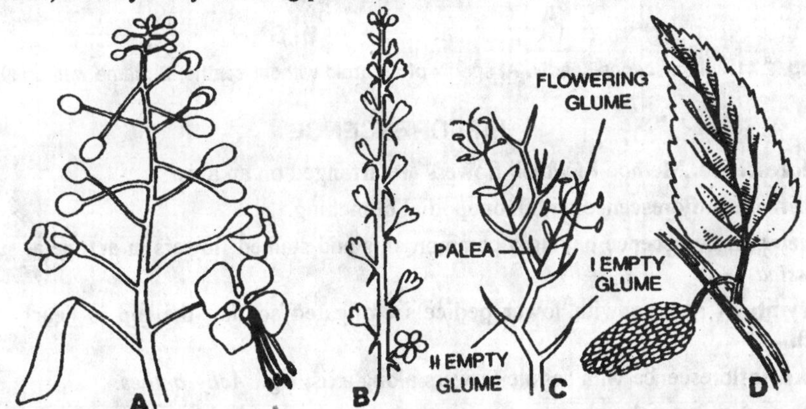

Fig. 7.39. Inflorescence-racemose. A, raceme of gul-mohur; B, spike; C, spikelet of a grass; D,female catkin of mulbery.

In the leaf of pea upper leaflets modify into tendrils while in *Gloriosa* leaf-apex modifies into a tendril. In *Naravelia* the terminal leaflet modifies into a tendril. Likely in *Bignonia unguis-cati* these leaflets are modified into the hooks. *Lathyrus aphaca* is an example of whole leaf modification into a single tendril.

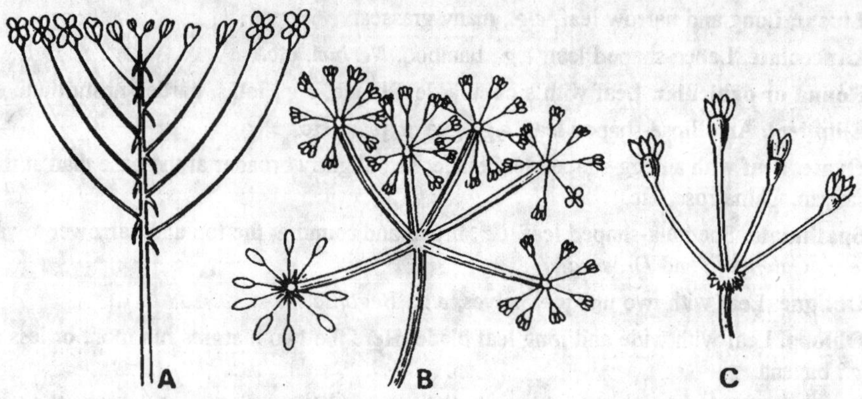

Fig. 7.40. Inflorescence. A, corymb; B, compound umbel (umbel of umbels); C, simple umbel.

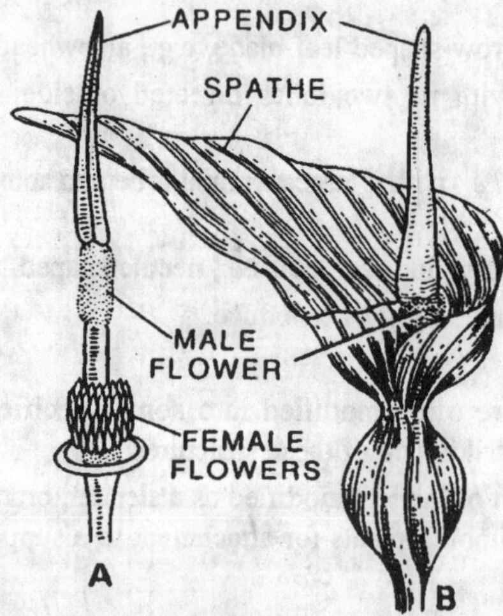

Fig. 7.41. Inflorescence-spadix. A, spadix of an aroid without spathe; B, same with spathe.

INFLORESCENCE

Inflorescence. Method in which flowers are arranged on an axis.

Racemose. Inflorescence with monopodial branching.

Raceme. Inflorescence having a common axis and stalked flowers in acropetal succession, *e.g., Brassica.*

Corymb. A raceme with lower pedicels elongated so that the top is nearly flat, *e.g.,* Candytuft.

Spike. Inflorescence with sessile flowers along axis, *e.g., Achyranthes.*

Catkin. A spike with unisexual flowers and pendulous rachis, *e.g., Morus.*

Umbel. An arrangement of flowers springing from a common centre and forming a flat or rounded cluster, *e.g., Coriandrum, Foeniculum,* etc.

Spadix. A racemose inflorescence with elongated axis, sessile flowers, and an enveloping spathe, *e.g., Colocasia, Alocasia,* Palm, *Musa,* etc.

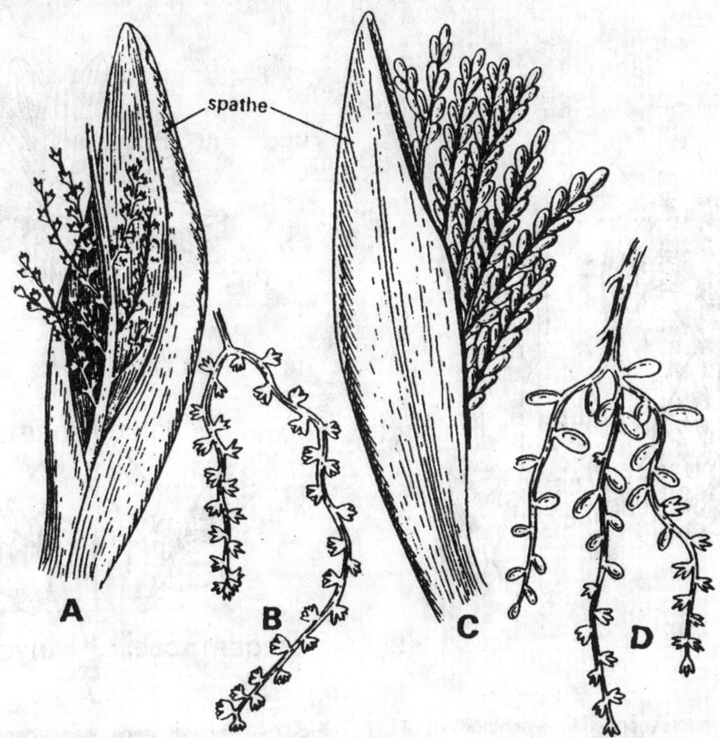

Fig. 7.41. (a). Inflorescence. Spadix of spadices (compound spadix) in coconut palm; A, male spadix of spadices; B, male flowers on the branched peduncle; C, female spadix of spadices; D, female flowers on branched peduncles.

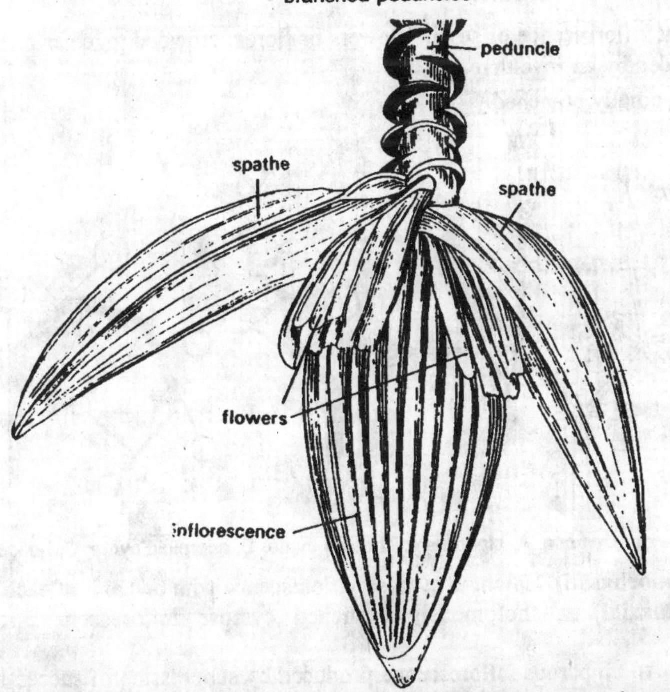

Fig. 7.41. (b). Inflorescence-spadix. Spadix of spadices (compound spadix) in ***Musa***.

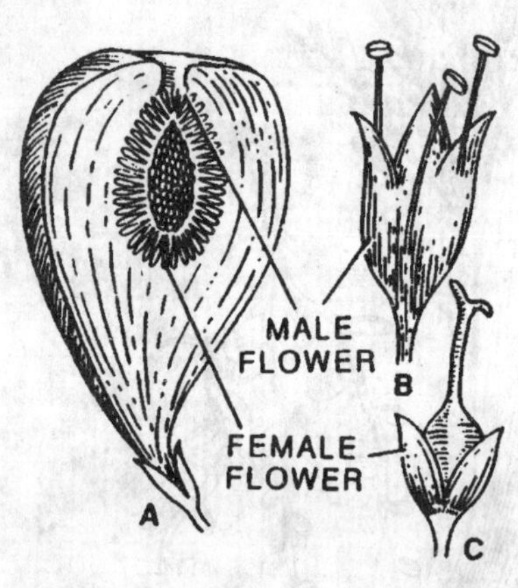

Fig. 7.42. Inflorescence. A, hypanthodium of *Ficus*; B, male flower; C, female fower.

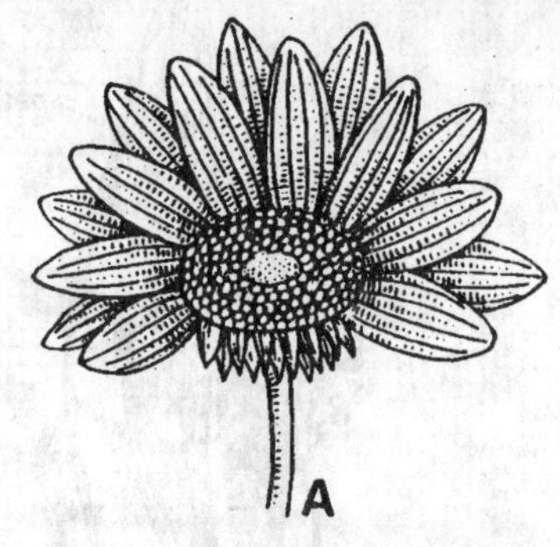

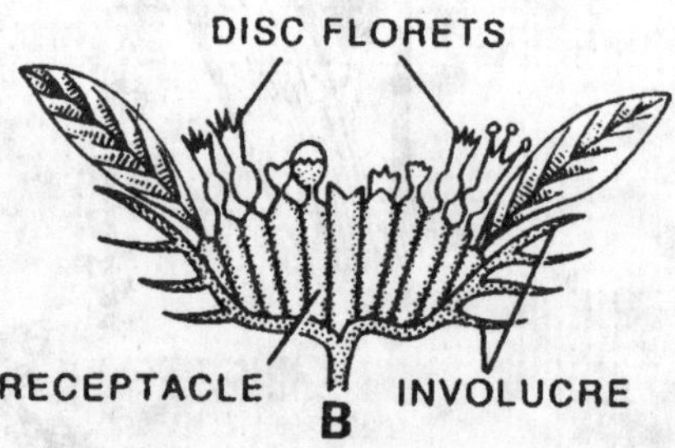

Fig. 7.43. Inflorescence. Head (capitulum). A, a head B, a head in L.S.

Hypanthodium. An inflorescence with concave capitulum on whose walls the flowers are arranged, *e.g.*, *Ficus*.

Capitulum. An inflorescence of sessile flowers or florets crowded together on a receptacle and usually surrounded by an involucre.

Cymose. Sympodially branched.

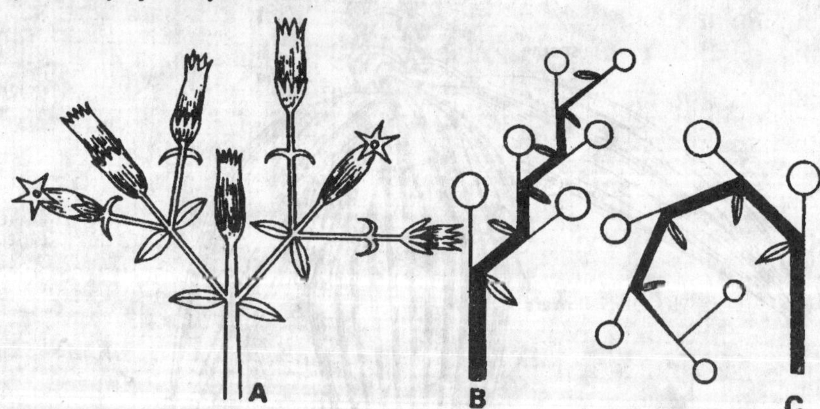

Fig. 7.44. Inflorescence. Cymose. A, biparous (dichasial) cyme; B, scorpioid cyme; C, helicoid cyme.

Uniparous (monochasial). Having a cymose inflorescence with one axis at each branching.

Biparous (dichasial). Dichotomously branched cymose inflorescence, *e.g.*, *Ixora, Saponaria*.

Helicoid cyme. An uniparous inflorescence produced by suppression of successive axes on same side, thus causing the sympodium to be spirally twisted, *e.g.*, *Begonia, Juncus*.

Scorpioid. An uniparous inflorescence in which the lateral branches develop on alternate sides evidently forming a zigzag, *e.g.*, *Heliotropium*.

Special Types

Cyathium. The peculiar inflorescence in *Euphorbia*, a cup-shaped involucre with stamens and stalked gynoecium, each stamen and the gynoecium being a separate flower.

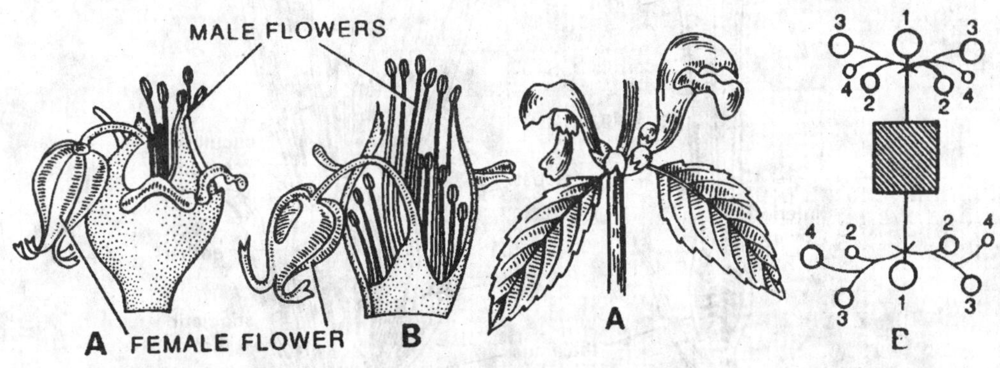

Fig. 7.45. Inflorescence. Special. Cyathium. A, cyathium of *Euphorbia;* B, L.S. of same.

Fig. 7.46. Inflorescence. Special verticillaster. A, verticillaster of Labiatae; B, diagram of same.

Verticillaster. A much condensed cyme with appearance of whorl, but in reality arising in axils of opposite leaves, *e.g.*, characteristic of Labiatae.

Panicle. A compound branched raceme, *e.g.*, Mango, *Melia*.

Spikelet. A secondary spike of grasses, bearing few flowers.

Compound umbel. Umbel of umbels.

Compound corymb. Corymb of corymbs.

Solitary flower. Single flower.

Solitary axillary. Single flower in the axil of leaf, *e.g.*, *Cucurbita*.

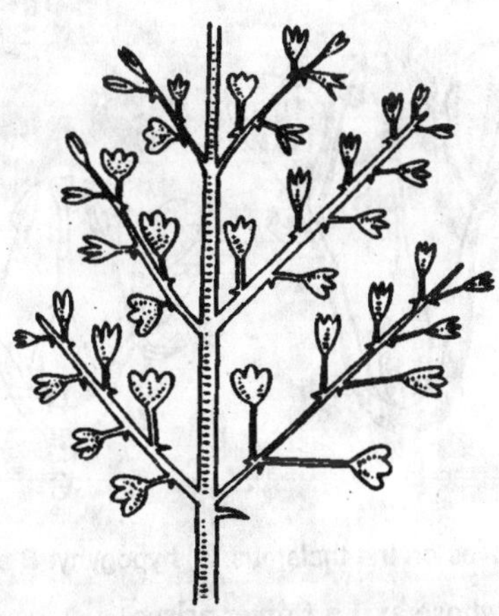

Fig. 7.47. Inflorescence.Panicle.

Solitary terminal. Single flower at the apex, *e.g.*, Poppy.

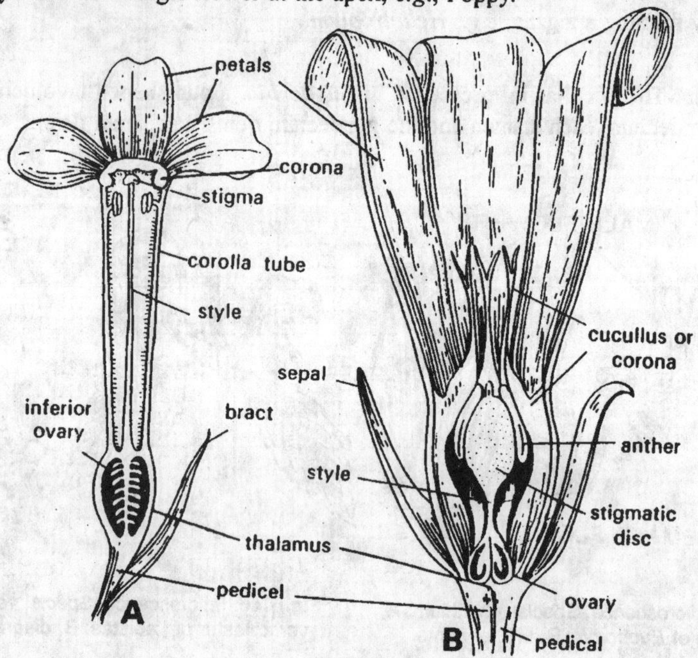

Fig. 7.48. Flower. A, Parts of a flower with inferior ovary; B, L.S. of flower showing position of whorls on the thalamus with superior ovary.

FLOWER

Flower. The blossom of a plant, comprising generally sepals, petals, stamens and pistil a leafy shoot adapted for reproductive purposes.

Pedicel. Stalk of a flower.

Pedicellate. Supported by a pedicel, *e.g.*, Mustard.

Sessile. Without pedicel, *e.g.*, *Achyranthes*.

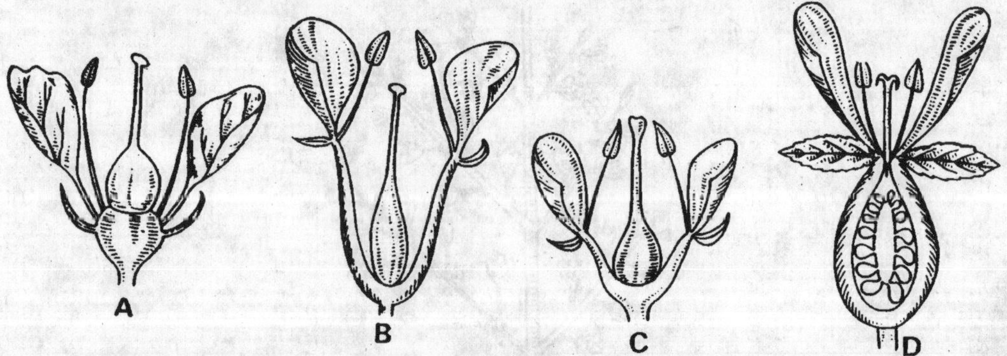

Fig. 7.49. Position of floral leaves on the thalamus. A, hypogyny; B and C, perigyny, D, epigyny.

Bract. A modified leaf in whose axil a flower arises.

Bracteole. Secondary bract at the base of an individual flower.

Bracteate. Flowers having bracts, *e.g.*, *Adhatoda*

Ebracteate. Flowers without bracts, *e.g.*, *Solanum*.

Bracteolate. Flowers with bracteoles, *e.g.*, *Ruellia*.

N.B. Nature of bract or the bracteole should be described separately.

Actinomorphic (regular). Radially symmetrical flower; the flower can be divided into two equal halves along more than one longitudinal plane, *e.g.*, *Vinca*.

Zygomorphic (irregular). Bilaterally symmetrical, with only one plane of symmetry, *e.g.*, *Pisum*.

Unisexual. Flower of one or other sex; distinctly male or female, *e.g.*, *Morus*.

Hermaphrodite (bisexual). The flower having both male and female reproductive organs, *e.g.*, *Brassica*.

Complete. The flower with all the four whorls, *i.e.*, calyx, corolla, androecium and gynoecium, *e.g.*, *Hibiscus*.

Incomplete. The flower lacking any one of the four whorls, *e.g.*, *Lagenaria, Luffa, Cucurbita*.

Regular. Actinomorphic, *e.g.*, *Brassica*.

Irregular. Zygomorphic, *e.g.*, Pea.

Hypogynous. The other whorls inserted below the gynoecium, and not adherent; ovary superior, *e.g.*, *Solanum*.

Perigynous. Having sepals, petals, stamens, round the gynoecium; ovary half superior, *e.g.*, *Prunus*.

Epigynous. Having the various whorls adnate to ovary, thus apparently inserted in ovary; inferior ovary, *e.g.*, *Hamelia*; *Coriandrum*.

Acyclic. Flowers with floral leaves arranged in a spiral, *e.g.*, *Ranunculus*.

Cyclic. Having parts of flower arranged in whorls, *e.g.*, *Solanum*.

Colour. Variously coloured.

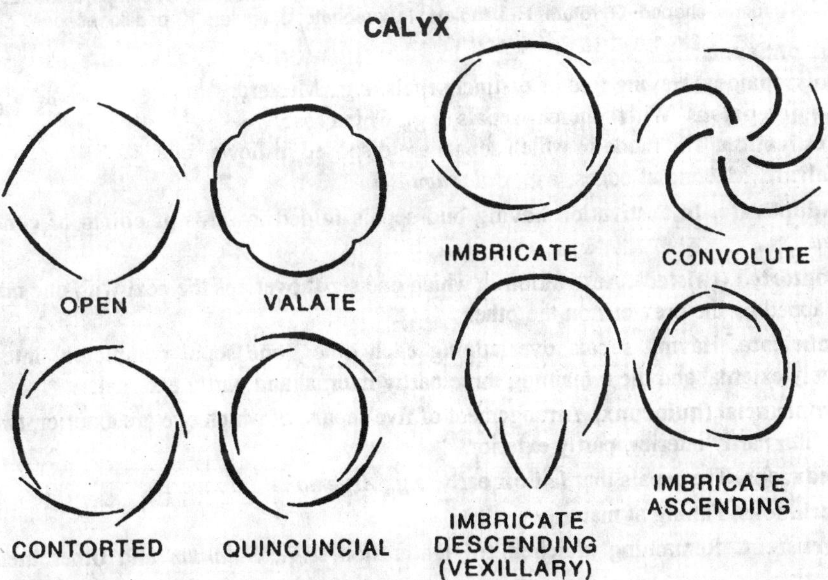

Fig. 7.50. Aestivation of corolla.

Calyx. The outer whorl of floral leaves (sepals).

Sepal. A leaf-like division of calyx.

Sepaloid. Like a sepal; green in colour.

Petaloid. Like a petal; coloured.

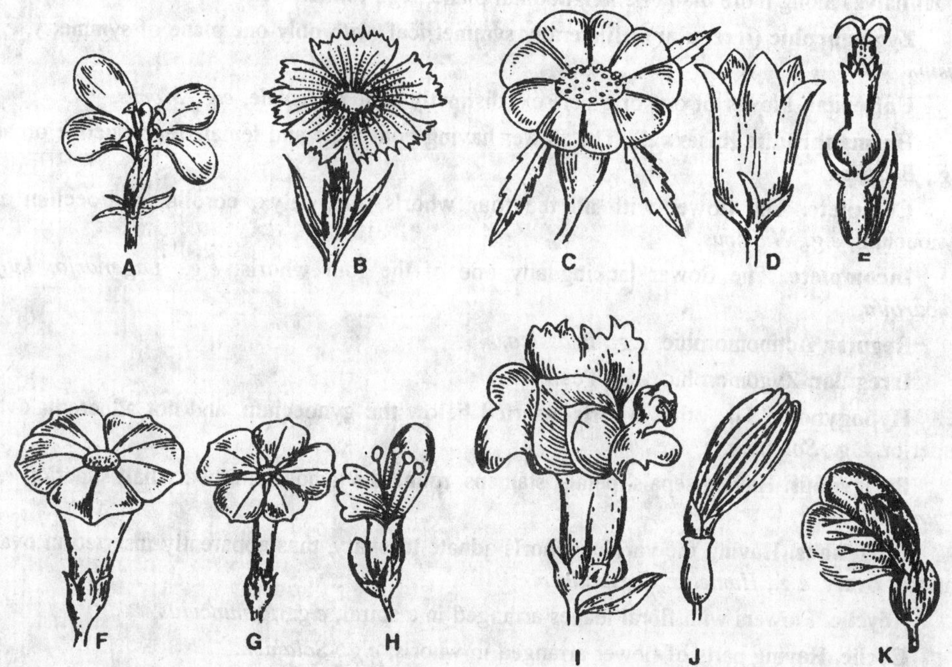

Fig. 7.51. Forms of corolla. A, cruciform; B, caryophyllaceous; C, rosaceous; D, campanulate; E, tubular; F, funnel-shaped; G, rotate; H, bilabiate; I, personate; J, ligulate K, papilionaceous.

Number of Sepals

Polysepalous. Having free or distinct sepals, *e.g.*, Mustard.

Gamosepalous. With coherent sepals, *e.g.*, China rose.

Aestivation. The mode in which sepals are disposed in flower bud.

Valvate. Meeting at edges, *e.g.*, *Solanum*.

Induplicate. In aestivation having bud sepals folded inwards at points of contact, *e.g.*, *Ipomoea*.

Contorted (twisted). Aestivation in which one sepal overlaps the next with one margin, and is overlapped by the previous on the other.

Imbricate. Having sepals overlapping each other, one sepal completely internal, one completely external and the remaining three partly internal and partly external.

Quincuncial (quincunx). Arrangement of five sepals, of which two are exterior, two interior and the fifth partly interior, partly exterior.

Caducous. The sepals that fall off early, *e.g.*, *Argemone, Papaver*.

Deciduous. Falling at maturity.

Persistent. Remaining attached till maturation, *e.g.*, *Solanum* and other members of Solanaceae.

COROLLA

Corolla. The whorl of petals.

Petal. One of the floral leaves of corolla.

Polypetalous. With free petals, *e.g., Brassica, Rosa.*

Gamopetalous. United petals, *e.g, Ipomoea, Petunia.*

Aestivation. The arrangement of the parts of the floral envelopes in bud.

Valvate. An aestivation when the segments of corolla are so placed that their edges touch each other, not overlap, *e.g., Solanum.*

Valvate induplicate. Valvate aestivation with the margins of the petals folded inwards, *e.g., Ipomoea.*

Twisted. Also known as contorted; one margin of the petal overlaps that of the next one, and the next margin overlaps the third one.

Imbricate. A mode of aestivation in which one member of whorl is outside all the others (*i.e.*, its margins are free) and one inside all the others (*i.e.*, both margins are overlapped), the others overlap by one margin only.

Quincuncial. An imbricate (aestivation with 5 petals out of which two are exterior, two interior and the fifth has one margin exterior and one interior.

Vexillary. An imbricate (descending imbricate), in which out of the five petals the posterior one is the largest and covers the two lateral petals, and the lateral petals (wings) overlap the two anterior and smallest petals (keels), *e.g.*, in Papilionaceae.

Shape of Corolla

Cruciform. The corolla with four free petals arranged in the form of a cross; each petal is being differentiated into a *claw* and a *limb, e.g., Brassica.*

Caryophyllaceous. The corolla with 5 free petals; the petals are with long claws and with limbs placed at right angles to the claws, *e.g., Dianthus.*

Rosaceous. The corolla with five or more free petals, not distinguished into limbs and claws and spreading regularly outwards, *e.g., Rosa.*

Campanulate. Bell-shaped; the petals are fused to each other, *e.g., Ipomoea, Cuscuta.*

Infundibuliform. Funnel-shaped corolla; *e.g., Ipomoea, Petunia.*

Salver-shaped (hyporcrateriform). A gamopetalous corolla with a long tube horizontal limb, *e.g., Mussaenda.*

Rotate. Wheel-shaped gamopetalous corolla, *e.g., Solanum*; the gamopetalous corolla with a flat and circular limb at right angles to the short tube.

Papilionaceous. Butterfly-like, as corolla of pea family; the corolla possesses one large posterior *standard,* two lateral *wings* and two innermost and smallest *keels.*

Bilabiate. Two lipped; zygomorphic gamopetalous corolla, *e.g., Justicea, Ocimum, Salvia.*

Personate. Zygomorphic, gamopetalous corolla with two lips, as in *Antirrhinum.*

Ligulate. Zygomorphic, gamopetalous corolla forming a short, narrow tube below and ligule-like flat structure above, *e.g., Sonchus.*

Tubular. Tube-like, gamopetalous corolla, *e.g., Helianthus, Ageratum.*

Spur. A sac-like or tubular projection of a petal, *e.g., Viola, Tropaeolum.*

Nectary. A nectar-secreting gland in petals, *e.g., Salvia, Ranunculus.*

Corona. The appendages found in between corolla and stamens, or on the corolla, *e.g., Calotropis, Asclepias.*

PERIANTH

Perianth. The floral envelopes including both calyx and corolla, no differentiation, into calyx and corolla.

Tepal. A segment of perianth.

Polytepalous. With free tepals.

Gamotepalous. With united tepals.

Sepaloid. Green resembling sepals.

Petaloid. Coloured resembling petals.

Aestivation. The arrangement of the tepal in bud.

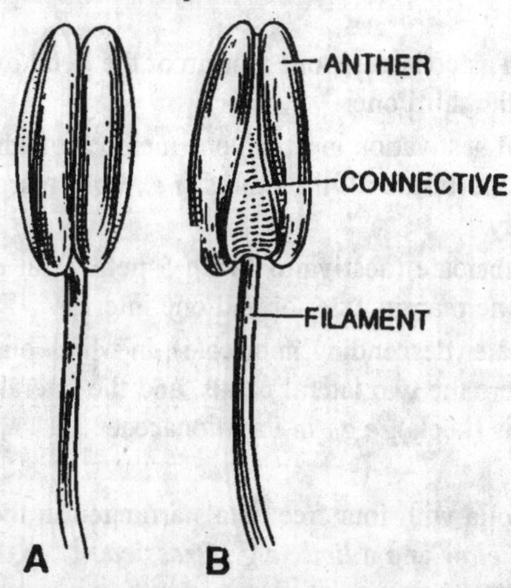

Fig. 7.52. Stamens. A, face; B, back.

ANDROECIUM

Androecium. Collective term for all the stamens and staminodes.

Androphore. A stalk bearing the androecium.

Stamen. A modified leaf in flowering plants bearing pollen sacs. It consists of filaments and anthers.

Staminode. Rudimentary stamens which do not bear fertile pollen, *e.g.*, *Stellaria.*

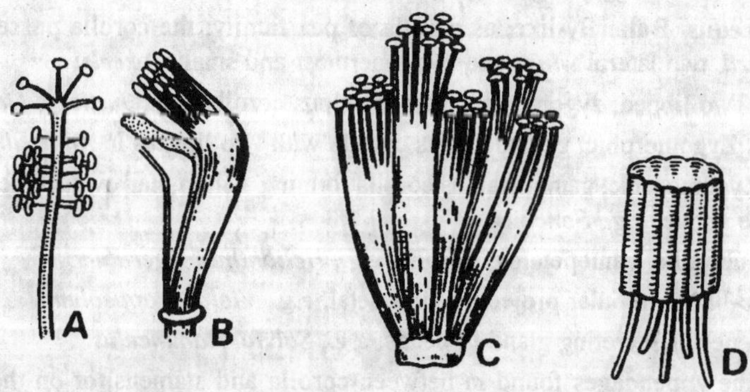

Fig. 7.53. Cohesion of stamens. A, monadelphous; B, diadelphous; C, polyadelphous; D, syngenesious.

Polyandrous. Androecium that consists of free stamens, *e.g.*, *Poppy*.

Monadelphous. United into one bundle by the filaments, *e.g.*, *Hibiscus*.

Diadelphous. Stamens arranged in two bundles, *e.g.*, many members of Papilionaceae.

Polyadelphous. Stamens united in many bundles, *e.g.*, *Citrus*, *Bombax*.

Syngenesious. With anther cohering in a ring, *e.g.*, many members of Compositae, (*e.g.*, *Helianthus*).

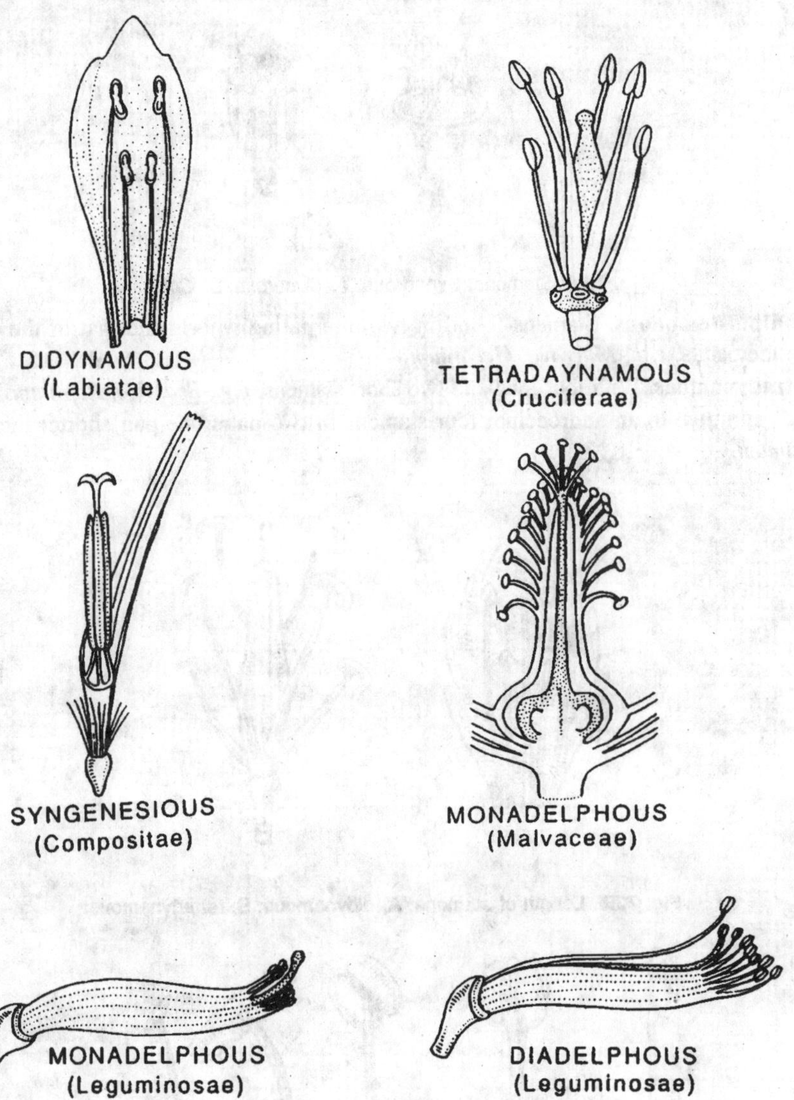

Fig. 7.53 (a). Cohesion of stamens.

Epipetalous. Stamens born on the petals or corolla tube, *e.g.*, *Convolvulus, Ipomoea, Justicea, Solanum*, etc.

Epitepalous. Stamens born on the tepals, *e.g.*, *Asphodelus*.

Gynandrous. Stamens adhering to the carpels, *e.g.*, *Calotropis*.

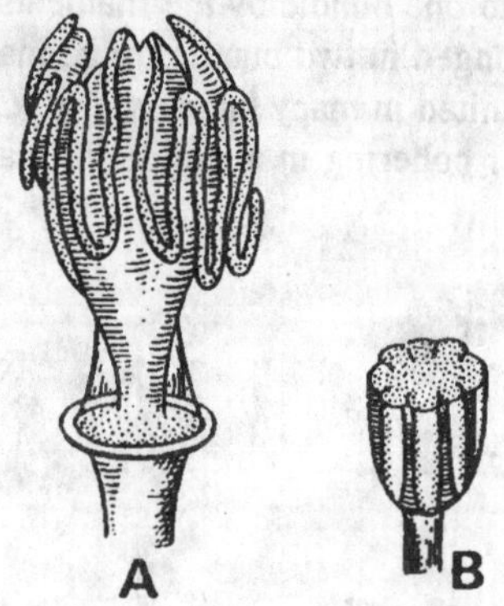

Fig. 7.54. Stamens. Synadrous. A, *Cucurbita;* B, *Colocasia.*

Obdiplostemonous. Stamens found in two alternating whorls and that of the outer whorl opposite the petals, *e.g., Murraya, Geranium.*

Tetradynamous. With four long and two short stamens, *e.g., Brassica, Raphanus, Eruca*, etc.

Didynamous. In an androecium four stamens in two pairs one pair shorter than the other, *e.g., Ocimum.*

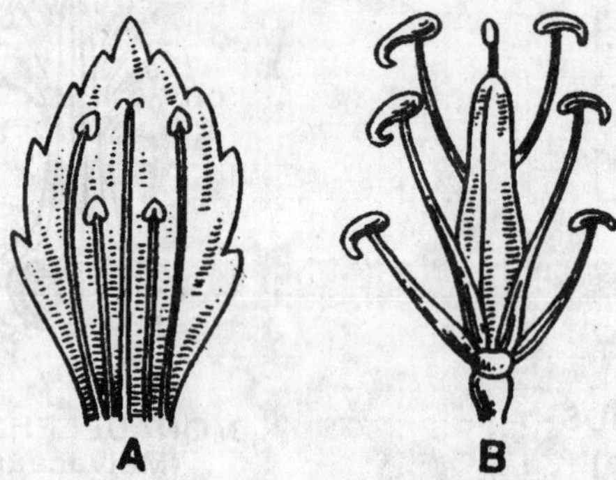

Fig. 7.55. Length of stamens. A, didynamous; B, tetradynamous.

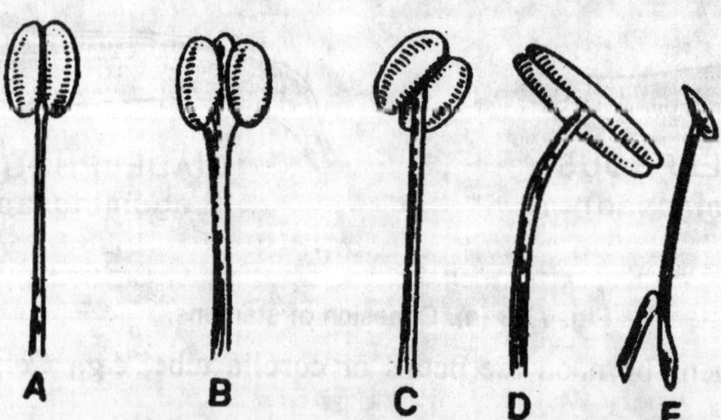

Fig. 7.56. Attachment of the filament to the anther. A, basifixed; B, adnate; C, dorsifixed; D, versatile; E, elongated connective of *Salvia*, separating the two anther lobes (upper fertile, lower sterile).

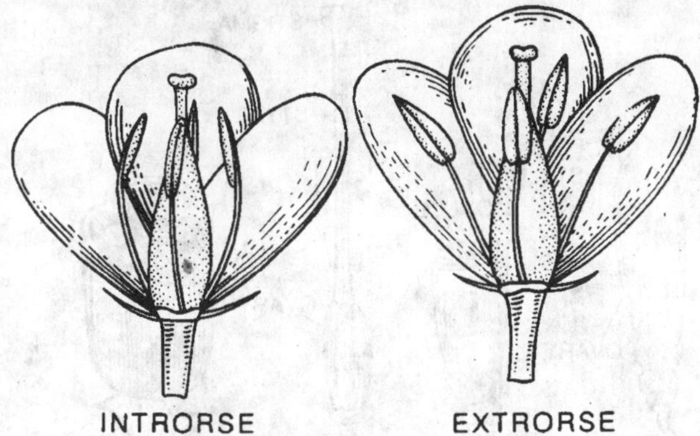

Fig. 7.56. (a). Facing of stamens.

Inserted. Stamens shorter than the corolla tube, *e.g., Ixora, Mussaenda.*

Exerted. The stamens projecting beyond the petals, *e.g.*, passion flower.

Monothecous. Single-celled anther, *e.g., Hibiscus.*

Dithecous. Two-celled anther, *e.g., Papaver.*

Basifixed. Fixed to the filament (stalk) at the base, *e.g., Brassica, Papaver.*

Dorsifixed. Filament attached to the dorsal side of the anther, *e.g., Bauhinia.*

Versatile. The anther can move in any way, *e.g.*, Gramineae.

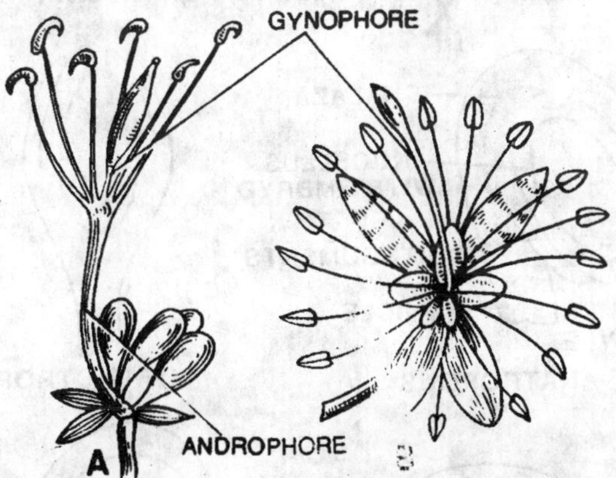

Fig. 7.57. Thalamus. A, flower of *Gynandropsis* showing gynophore.

Adnate. The fusion of unlike parts (anther and filament), *e.g., Michelia.*

Introrse. Anthers facing inwards, *e.g., Hibiscus, Solanum, Stellaria*, etc.

Extrorse. Facing outward from the centre of the flower referred for anthers, *e.g., Papaver, Argemone.*

GYNOECIUM

Gynoecium. The carpel or assemblage of carpels; collective name for the carpels of a flower.

Gynophore. Elongated thalamus in between androecium and gynoecium, *e.g., Cleome.*

Androgynophore. An axis or stalk bearing both stamens and pistil above the point of perianth attachment, *e.g., Gynandropsis.*

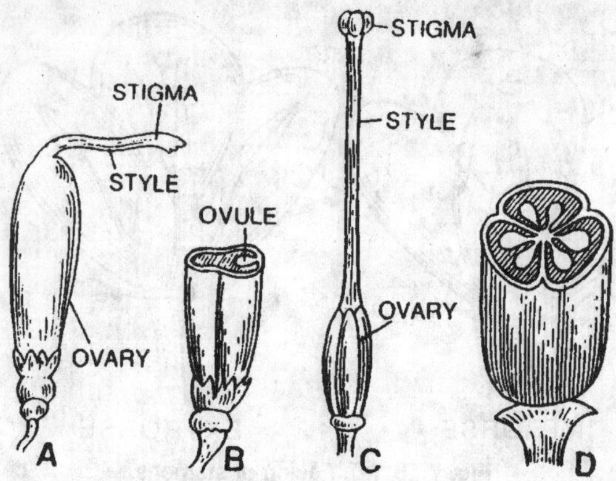

Fig. 7.58. Pistil. A, simple pistil; B, unilocular ovary of simple pistil; C, syncarpous pistil; D, trilocular ovary of syncarpous

Carpel. The modified leaves which bear the ovules, a simple pistil or one of the segments of a compound pistil.

Pistil. A collective term for ovary, style and stigma.

Compound pistil. Where many pistils are aggregated together; the condition many be bicarpellary; *e.g.*, *Brassica*; Tricarpellary, *e.g.*, Lillaceae; tetracarpellary, *e.g.*, *Datura*; pentacarpellary, *e.g.*, *Melia*; polycarpellary, *e.g.*, *Sida, Papaver*, etc.

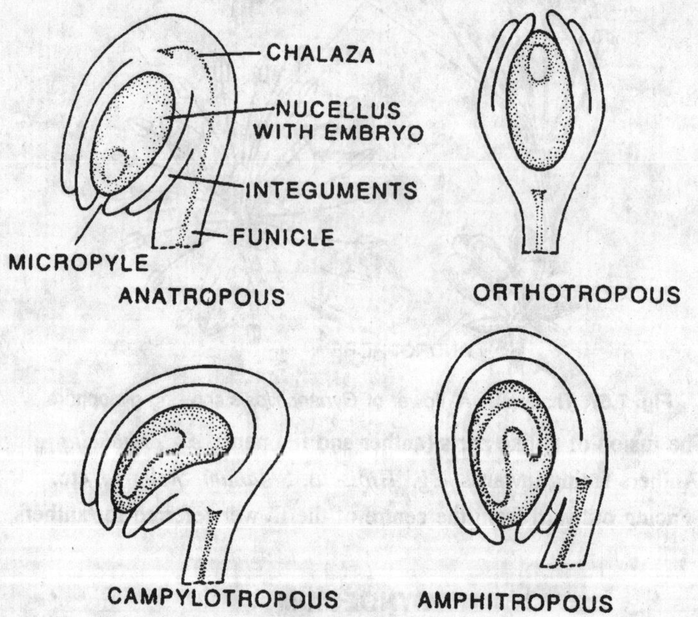

Fig. 7.58 (a). Ovules.

Apocarpous. Carpels in a flower free from each other, *e.g.*, *Ranunculus, Nigella, Clematis*, etc.

Syncarpous. United carpels, compound ovary, *e.g.*, *Citrus*.

Superior. Situated above another member, a superior ovary has its base above the insertion of calyx; a superior calyx or corolla is inserted above the ovary, *e.g.*, *Brassica*.

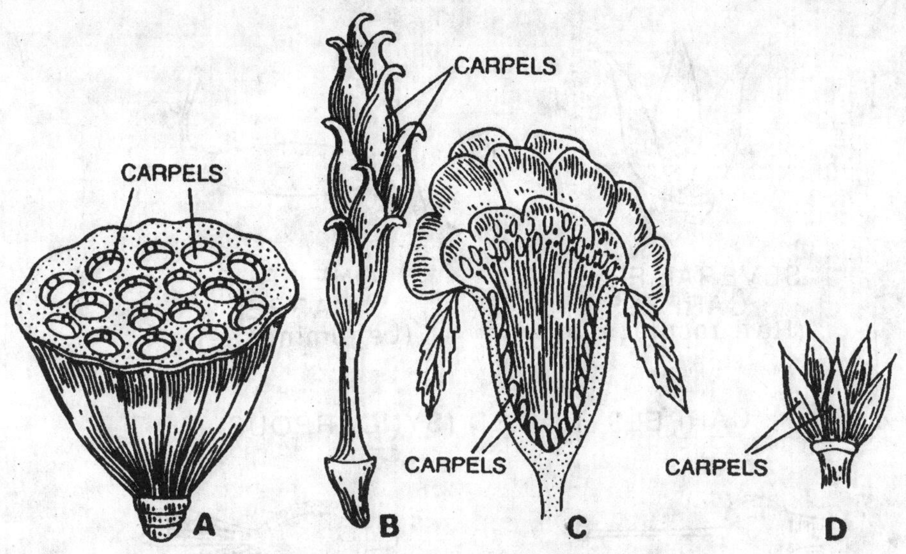

Fig. 7.59. Apocarpous pistil. A, Lotus; B, *Michelia;* C, Rose; D, Sedum.

Inferior. Epigynous condition; the ovary is borne below attachment of all other floral envelopes and adnate to them, *e.g.*, *Coriandrum, Ixora, Pyrus*, etc.

Semi-inferior. Intermediate condition between superior and inferior, *e.g.*, Pea.

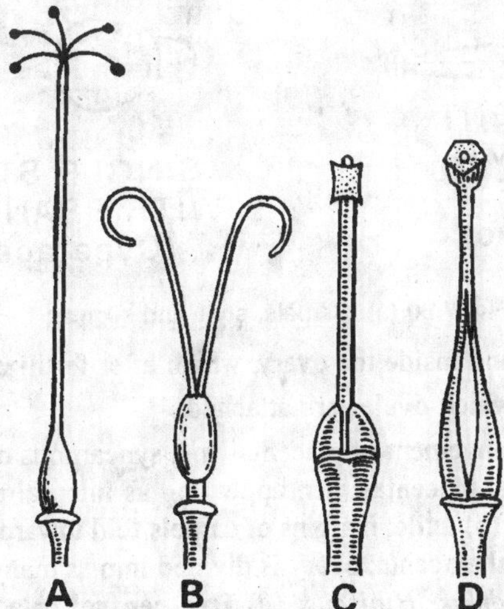

Fig. 7.60. Cohesion of carpels. A, pistil of *Hibiscus rosa-sinensis* with free stigmas; B, pistil with free styles; C, pistil of *Nerium* with free ovaries; D, pistil of *Calotropis* with free ovaries and styles.

Locule. Chamber; the ovary may be unilocular (single chambered), *e.g.*, *Delphinium*; bilocular (two-chambered), *e.g.*, *solanum*; trilocular (three-chambered), *e.g.*, *Allium, Asphodelus*; tetralocular (four-chambered), *e.g.*, *Ocimum*; pentalocular (five-chambered), *e.g.*, *Hibiscus rosa-sinensis*; multilocular (many chambered), *e.g.*, *Citrus*.

CARPELS FREE (APOCARPOUS)

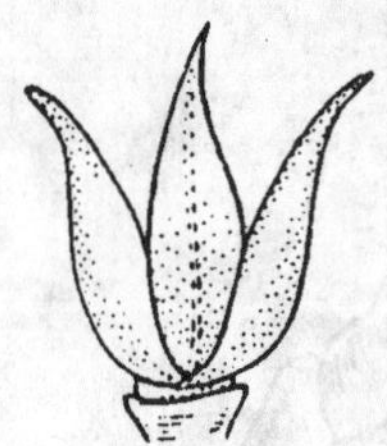

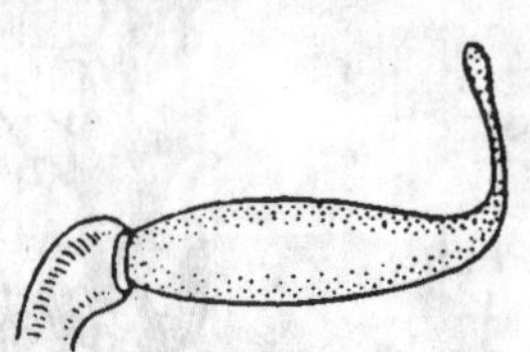

SEVERAL FREE CARPELS (Ranunculaceae)

ONE FREE CARPEL (Leguminosae)

CARPELS UNITED (SYNCARPOUS)

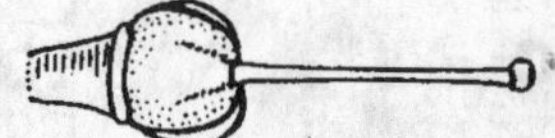

SINGLE STYLE WITH CAPITATE STIGMA (Primulaceae)

STYLES FREE (Linaceae)

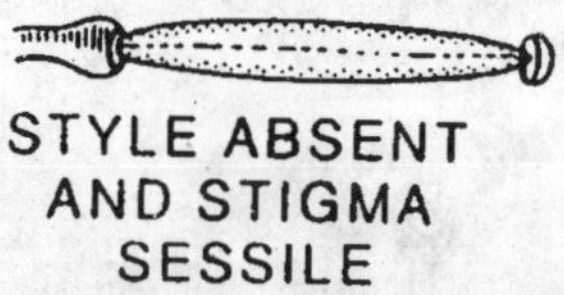

STYLE ABSENT AND STIGMA SESSILE (Cruciferae)

SINGLE STYLE WITH BRANCHES (Cyperaceae)

Fig. 7.60 (a). Carpels, style and stigma

Ovule. Minute bodies found inside the ovary, which after fertilization develop into seeds.

Placenta. The organs to which ovules are attached.

Placentation. Type of arrangement of placentas in a syncarpous ovary; (a) **parietal**; carpels are fused only by their margins, placentas then appearing as internal ridges on ovary wall, *e.g.*, *Argemone, Viola, Brassica*, etc. (b) **axile**; margins of carpels fold inwards, fusing together in centre of ovary to form a single, central placenta; ovary is divided into as many compartments (loculi) as there are carpels, *e.g.*, *Asphodelus, Hibiscus*; (c) **free-central**; placenta arising as a central upgrowth from ovary base, *e.g.*, *Stellaria*; (d) **marginal**; placenta develops along the junction of two carpels, in a unilocular ovary, *e.g.*, pea; (e) **basal**; the ovules are few or reduced to one and are borne at the base of ovary, the ovules when solitary often filling the cavity, the ovary is unilocular, *e.g.*, Compositae; (f) **superficial**; carpels numerous, the placentae develop all round the inner surface of the partition wall, ovary multilocular, *e.g.*, *Nymphaea*; (g) **lamellate**; this is modified parietal placentation, *e.g.*, *Papaver* (Fig. 7.61).

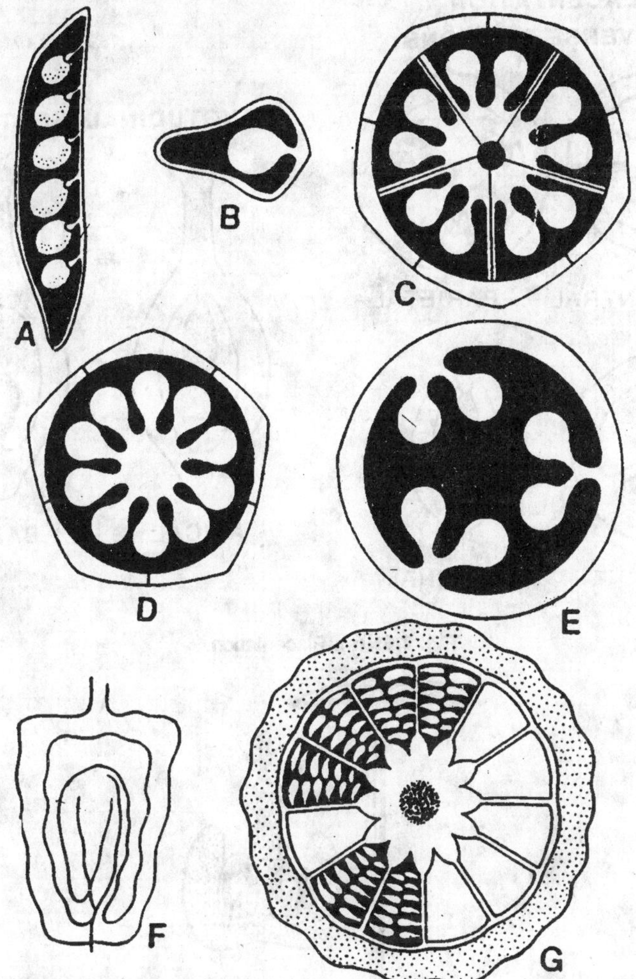

Fig. 7.61. Types of placentation. A, marginal in L.S.; B, marginal in T.S.; C, axile; D, free-central; E, parietal; F, basal; G, superficial.

Style. Prolongation of carpel supporting stigma: (a) terminal-style lying in the same straight line with the ovary, *e.g.*, China rose; (b) lateral-style that arises from the side of the ovary, *e.g.*, *Potentilla*; (c) Gynobasic-style arises from the depression in the centre of ovary or from the thalamus, *e.g.*, *Ocimum, Salvia.*

Stigma. Terminal expansion of style, surface of the carpel which receives pollen; the stigma may be capitate, feathery (*e.g.*, Gramineae), discoid, dumb-bell shaped, bifid to multifid, sticky.

FRUIT

Fruit. Ripened ovary of the flower, enclosing seeds.

Simple fruit. When a single fruit develops from the ovary of a flower with or without accessory parts, it is said to be a simple fruit.

Legume or Pod. A dry monocarpellary fruit developing from a superior, one chambered ovary and dehiscing by both the sutures, *e.g.*, pea, bean, pulses.

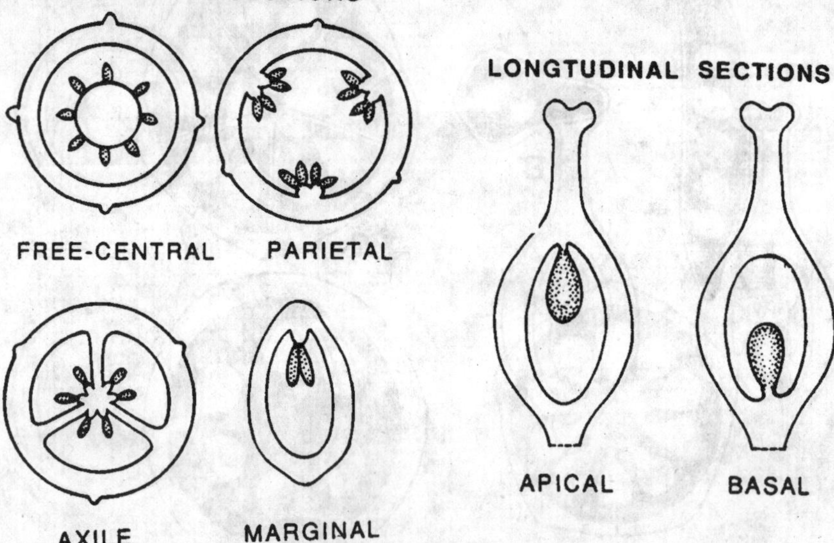

Fig. 7.61 (a). Placentation.

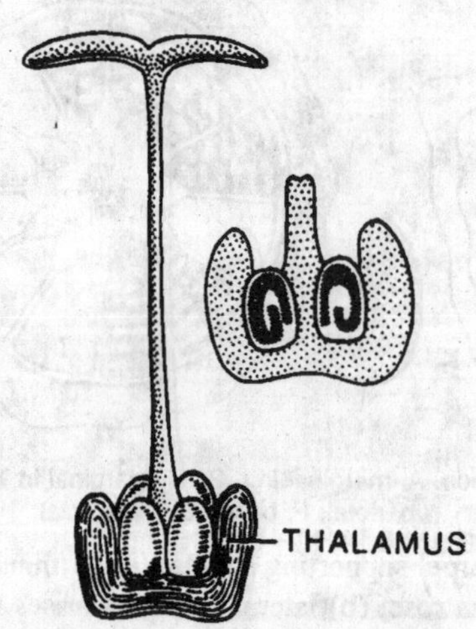

Fig. 7.62. Gynobasic style of *Ocimum*. A, entire gynoecium on thalamus; B, the same in L.S.

Follicle. A dry, monocarpellary, superior one-chambered fruit like the legume, but it dehisces by one sutre only, *e.g., Calotropis, Asclepias, Vinca, Michelia.*

Siliqua. A long, narrow many seeded fruit developing from a superior, bicarpellary ovary with two parietal placentae, dehiscing from below upwards by both the sutures, *e.g., Brassica, Raphanus.*

Silicula. When a siliqua is much shorter and contains only a few seeds, *e.g., Capsella,* candytuft.

Capsule. A many-seeded, uni or multilocular fruit developing from a superior, bi or polycarpellary ovary, and dehiscing in various ways, a capsule may dehisce by pores, as in poppy; or transversely; as in cock's comb; or loculicidally, as in cotton, lady's finger and *Hibiscus mutabilis.*

Caryopsis. A very small, dry one-seeded fruit developing from a superior, monocarpellary ovary, with the pericarp fused with the seedcoat, *e.g.*, Gramineae.

Achene. A small, dry, one-chambered and one-seeded fruit developing from a superior, monocarpellary ovary; but unlike the previous one, the pericarp is free from the seed coat; commonly developing an apocarpous pistil and, therefore, a single flower produces an aggregate of achenes, *e.g.*, *Clematis, Naravelia.*

Cypsela. A dry, one-chambered and oneseeded fruit developing from an inferior, bicarpellary ovary with the pericarp and the seed-coat free, *e.g.*, Sunflower, marigold, *Cosmos.*

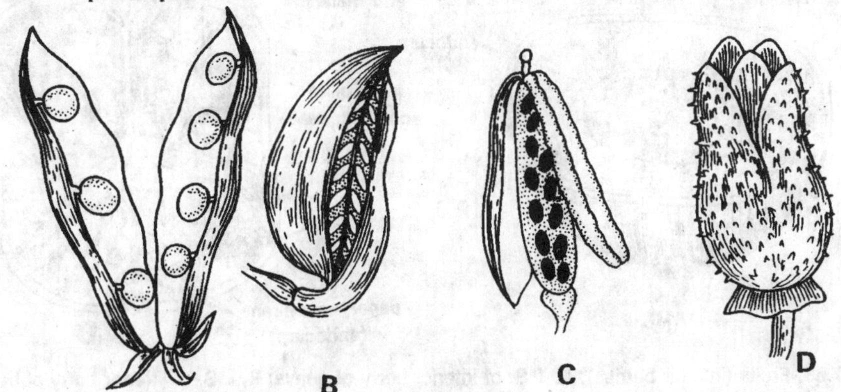

Fig. 7.63. Fruits. A, legume or pod of pea; B, follicle of *Calotropis;* C, siliqua of mustard; D, capsule of *Datura.*

Nut. A dry, one chambered and one-seeded fruit developing from a superior, bi or polycarpellary ovary, with the pericarp hard and woody, *e.g.*, chestnut, oak, etc.

Samara. A dry, indehiscent, one or two-seeded fruit developing from a superior, bi-or tricarpellary ovary, with flattened wing-like outgrowths, *e.g.*, *Hiptage.*

Lomentum. The legume is constricted or partitioned between the seeds into a number of one-seeded parts, *e.g.*, *Acacia.*

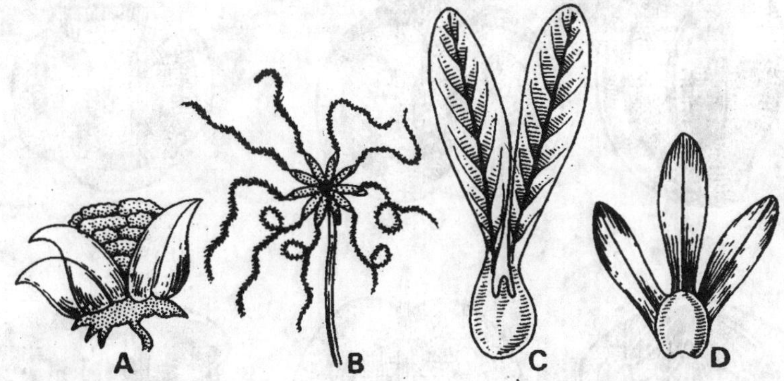

Fig. 7.64. Fruits. A, capsule of cotton; B, achenes of *Naravelia* with persistent feathery styles; C, samara of *Dipterocarpus;* D, samara of *Hiptage.*

Cremocarp. A dry, indehiscent, two-chambered fruit developing from an inferior, bicarpellary ovary; when ripe splits into two, indehiscent, one-seeded mericarps which remain attached to the prolonged end (carpophore) or the axis, *e.g.*, Umbelliferae.

Regma. A dry schizocarpic fruit developing from a tricarpellary, syncarpous, superior ovary and splitting at maturity into three cocci, *e.g.* *Ricinus.*

Carcerulus. The fruit develops from a bicarpellary pistil with a superior ovary which becomes quadrilocular owing to the formation of two false septa; at maturity the four mericarps separate from each other towards the middle, *e.g.*, Labiatae (*Ocimum*) and Boraginaceae. In the carcerulus of *Malva* the superior ovary of the polycarpellary pistil splits into a large number of mericarps.

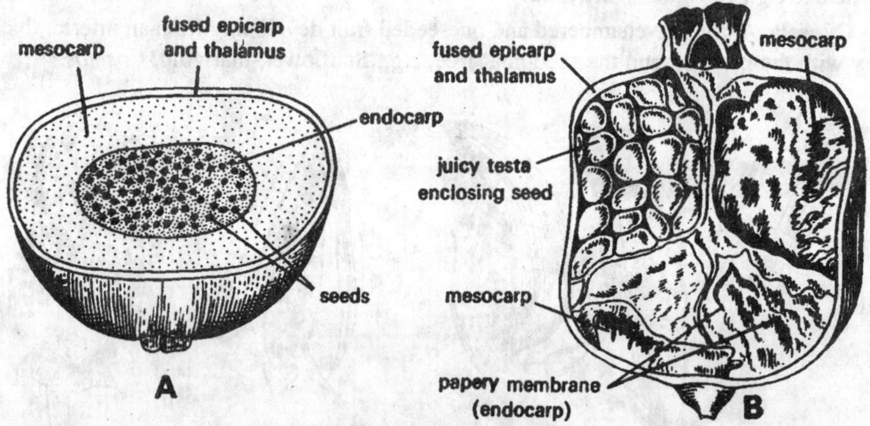

Fig. 7.65. Fruits (inferior berries), A, T.S. of inferior berry of guava; B, L.S. of inferior berry of balusta of pomegranate (*Punica granatum*)

Double samara. It consists of two, three or four samaras, *e.g.*, *Acer* and Sapindaceae.

Drupe. A fleshy, one or more-chambered and one or more-seeded fruit developing from a monocarpellary or syncarpous pistil, with the pericarp differentiated into the *epicarp* which forms the skin of the fruit, the *mesocarp* which is often fleshy, and the *endocarp* which is hard and stony, *e.g.*, mango, peach, plum, coconut-palm, plamyra-palm, almond, etc.

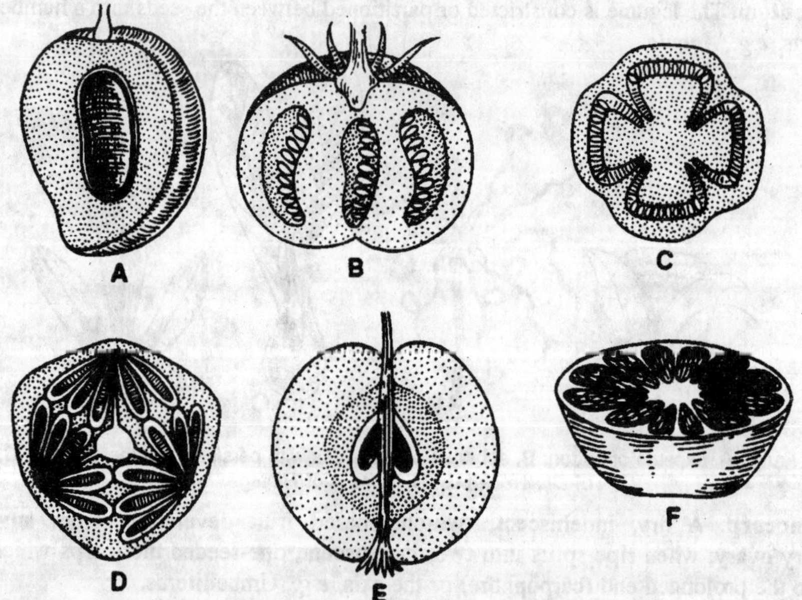

Fig. 7.66. Fruits, A, drupe of mango, B, berry of tomato, C, same in T.S., D, pepo of cucumber in T.S., E, pome of apple, F, hesperidium of orange.

Berry. A superior (rarely inferior) indehiscent, usually many-seeded, fleshy or pulpy fruit developing from a single carpel or more commonly from a syncarpous pistil, with axile or parietal placentation, *e.g.*, tomato, grapes, brinjal, guava.

Pepo. A fleshy, many-seeded fruit developing from an inferior, one-celled or three-celled, syncarpous pistil with parietal placentation, *e.g.*, Cucurbitaceae.

Pome. An inferior, two or more-celled fleshy syncarpous fruit surrounded by the thalamus, *e.g.*, Apple, Pear.

Hesperidium. A superior, many-celled, fleshy fruit developing from a syncarpous pistil with axile placentation; the endocarp projecting inwards forming distinct chambers and the epicarp and mesocarp, fused together, forming the separable rind of the fruit, *e.g.*, *Citrus*.

Aggregate fruit. It develops from a single flower with an apocarpous pistil, carpels being free, each of them develops into a simple fruitlet; an aggregate fruit, therefore, consists of a collection of simple fruitlets as may as the number of free carpels, borne by a single flower; an aggregate of simple fruitlets born by a single flower is known as an *etaerio*. Each fruitlet of an etaerio may be a follicle, an achene, a drupe or a berry.

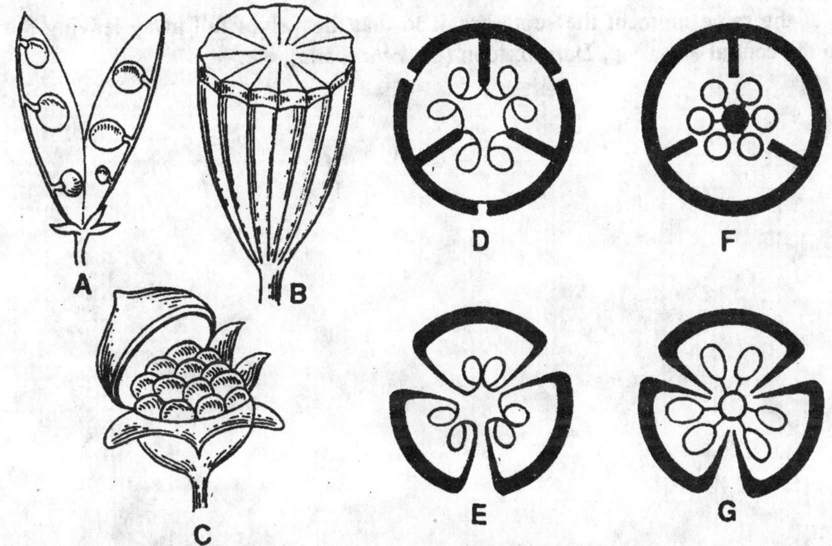

Fig. 7.67. Dehiscence of fruits. A, sutural, B, porous, C, transverse, D, loculicidal, E, septicidal, F, and G, septifragal.

Etaerio of follicles. Each etaerio consists of two or more follicles, *e.g.*, *Calotropis*, *Asclepias*, *Aconitum*, *Michelia*.

Etaerio of achenes. Aggregate of achenes, *e.g.*, *Clematis, Naravelia*, strawberry, etc.

Etaerio of drupes. A number of small drupelets, developing from separate carpels of a flower, are aggregated together on a fleshy thalamus, *e.g.*, raspberry (*Rubus*).

Etaerio of berries. Developing from an apocarpous pistil and lying embedded in the fleshy thalamus, *e.g.*, custard apple; in *Artabotrys* a group of distinct and separate berries is produced.

Multiple of composite fruit. A fruit which develops from a number of flowers juxtaposed together, from a inflorescence.

Sorosis. A multiple fruit developing from a spike or spadix; flowers fuse together by their succulent sepals and the axis bearing them grows and becomes fleshy or woody, and the whole inflorescence becomes a compact mass, *e.g.*, pineapple, jack-fruit, mulberry.

Syconus. Develops from a hollow, pearshaped, fleshy receptacle which encloses a number of minute, male and female flowers; the recepatacle becomes fleshy and forms the so called fruit; it really encloses a number of true fruits or achenes, *e.g.*, banyan, peepul.

Dehiscence of Fruits:

(A) Sutural. Opening of monocarpellary fruit by sutures, either ventral, *e.g.*, madar or dorsal, e.g., *Magnolia* or by both sides, *e.g.*, in Leguminosae, pea, bean, etc.

(B) Porous. Rupturing of fruit by numerous little pores, *e.g.*, Poppy, bath sponge, etc.

(C) Transverse. Bursting of fruit transversely in the way of unhinged lid of a box, *e.g.*, in *Celosia, Portulaca*, etc.

(D) Loculicidal. Splitting of monocarpellary fruit through the back of the loculus (or chamber). Here valves separate from the axis, *e.g.*, *Hibiscus mutabilis* and Acanthaceae, *Adhatoda, Ruellia*, etc.

(E) Septicidal. Dehiscence of monocarpellary fruit through septa, *e.g.*, in mustard family or Cruciferae, radish, mustard, etc.

(F) Septifragal. Dehiscence of a multilocular, monocarpellary fruit loculicidally.

Here at the same moment the septa break so that the valves fall away leaving the seeds attached to the central axis, *e.g.*, *Datura*, toon (*Cedrela toona*), etc.

8

Glossary of Botanical Terms

Abaxial, the side of an organ away from the axis or centre of the axis; dorsal.

Accrescent, continuing to grow.

Achene, a small, dry, hard, one-chambered, one-seeded fruit.

Achlamydeous, flowers without perianth.

Acicular, needle like.

Actinomorphic, a flower having radial symmetry, regular.

Acuminate, ending in a pointed tapering apex.

Acute, tapering more broadly than acuminate.

Adaxial, the side toward the axis; ventral.

Adelphous, combined in groups.

Adnate, the fusion of unlike parts.

Adventitious, appearing in other than the usual place subsequently and irregularly.

Aestivation, the arrangememnt of the parts of the floral envelopes in bud.

Albumen, endosperm, nutritive tissue in the seed outside the embryo.

Albuminous, containing albumen.

Alternate, arranged in a *zigzag* manner, neither opposite nor whorled.

Amphitropous, curved ovule, one end comes nearer to the other.

Amplexicaul, the petiolar base surrounds the stem half way.

Anatropous, an ovule inverted on the stalk in such a way that the nucellus remains straight and the micropyle is directed towards the point of origin of the stalk or funicle.

Androecium, collective term for all the stamens and staminodes.

Androgynophore, an axis or stalk bearing both stamens and pistil above the point of perianth attachment.

Androphore, a stalk bearing the androecium.

Anemophilous, wind pollinated.

Angiosperm, plants having their seed enclosed inside the ovary.

Annual, of one year's growth.

Annular, in a ring or arranged in circle.

Anterior, the lower one (sepal or petal).

Anther, pollen bearing part of the stamen.

Apetalous, a flower without petals.

Apex, the tip or distal end.

Apocarpous, carpels in a flower free from each other.

Appendage, an attached subsidiary or secondary part, as a projecting part, or a hanging part or supplement.

Arborescent, of tree like habit.

Areole, a small pit or raised spot, often bearing a tuft of hairs, glochids or spines.

Aril, a small appendage around the hilum of a seed.

Arista, an awn.

Aristate, awned.

Ascending, becoming erect from a prostrate base.

Attenuate, drawn out into a long slender point.

Auricle, a claw-like appendage.

Awn, a hair or bristle-like terminal appendage.

Axile, the angle formed by an axis and a lateral member (*e.g.*, stem and leaf).

Axil, situated round an axis.

Axillary, situated in an axil.

Axis, any member which bears lateral members may be called the axis of the lateral members.

Baccate, berry like.

Basifixed, fixed to the filament (stalk) at the base.

Beaked, provided with a solid, narrow, tubular, beak-like prolongation.

Berry, a fleshy indehiscent fruit with many seeds. Sometimes seeds are few. The fruit contains thin epicarp, fleshy mesocarp and hard endocarp.

Biennial, the plants which complete their life cycle in two years.

Bifid, divided into two parts, bifurcated.

Bifurcate, forked, divided into two.

Bilabiate, two lipped.

Bilocular, two celled.

Bipinnate, twice pinnate (compound leaf).

Bisexual, containing both stamens and carpels.

Blade, the expanded part (lamina) of a leaf.

Bract, a leaf-like structure at the base of a flower.

Bracteate, with bracts.

Bracteole, small bracts occurring on the pedicel of flower.

Bracteolate, possessing bracteoles.

Bristly, bearing stiff strong hairs or bristles.

Bulb, a thick part in a resting state and made up of scales on a much shortened axis.

Bulbil, a small bulb, usually in leaf axils.

Caducous, falling off very quickly.

Calyx, the outermost green floral leaves.

Calyx-tube, the tube formed by the cohesion of the calyx leaves (sepals).

Campanulate, bell-shaped.

Campylotropous, the ovule bent so as to bring the hilum micropyle and chalaza close together.

Capitate, knob like (swollen).

Capitulum, a head of flowers.

Capsule, dry dehiscent fruit, when ripe opens by two or more valves.

Carpel, the modified leaves which bear the ovules, a simple pistil or one of segments of a compound pistil.

Caryopsis, the grain (fruit) of cereals and grasses one-seeded, indehiscent.

Catkin, a scaly bracted, usually flexuous spike or spike-like inflorescence.

Cauline, belonging to the stem.

Ciliate, possessing hairs.

Claw, the narrow stalk of some petals.

Coccus, a part of a schizocarpic fruit, each of which is usually derived from a single carpel of the ovary.

Compound, composed of two or more similar parts, generally referred to a compound leaf, compound inflorescence and compound ovary.

Connate, united similar part (*e.g.*, sepal to sepal).

Connivent, the bending towards each other of two or more similar organs.

Convolute, rolled up from one or both margins.

Cordate, heart shaped usually referred for leaves.

Coriaceous, leathery.

Corm, an underground stem with sufficient starch.

Corolla, the whorl of petals.

Corymb, a form of racemose inflorescence in which the several pedicels arising at different levels reach more or less the same level at the top.

Corymbose, corymb like arrangement.

Cristate, ribbed.

Culm, the erect stem in Gramineae.

Cuneate, wedge shaped.

Cuspidate, ending in a sharp rigid point.

Cyclic, parts arranged in whorls.

Cyme, the main axis terminates in a flower.

Deciduous, falling off.

Decompound, branched repeatedly.

Decumbent, having the lower parts prostrate.

Decurrent, prolonged downwards from the base.

Dentate, with sharp teeth.

Diandrous, flowers having two stamens.

Diadelphous, in two bundles.

Diclinous, bearing the androecium and gynoecium in separate flowers.

Didynamous, in an androecium, for stamens in two pairs, one pair shorter than the other.

Digitate, finger like division.

Dimorphic, occuring in two different forms.

Dioecious, male and female flowers on different plants.

Disc, a glandular swelling below the gynoecium.

Distichous, arranged alternately in two opposite rows.

Dissected, divided into many segments.

Dorsifixed, attached to the dorsal side.

Drupaceous, more or less drupe like.

Drupe, a stone fruit, the pericarp is usually differentiated into three layers, the epicarp thin and membranous, the mesocarp thick and fleshy, and the endocarp stony, inside which is the seed, *e.g.*, plum.

Ebracteate, without bracts.

Echinate, with long spines.

Emarginate, noted at the apex.

Embryo, dormant plant inside the apex.

Endocarp, the innermost layer of the pericarp.

Endosperm, the nutritive tissue with albumen of the seed.

Entire, without any teeth at the margin.

Entomophilous, insect pollinated.

Epicalyx, a whorl of bracts just below the calyx.

Epicarp, the outermost layer of the pericarp of the fruit.

Epigynous, borne upon, or on top of the gynoecium, with inferior ovary.

Epipetalous, borne-upon on the petals or corolla tube, usually referred for stamens.

Exalbuminous, without endosperm or albumen.

Exerted, the stamens projecting beyond the petals.

Exstipulate, without stipules.

Extra-axillary, situated away from the axil of the leaf.

Extrorse, facing outward from the centre of the flower referred for anthers.

Falcate, somewhat curved.

False septum, not true septum.

Fascicled, closely aggregated.

Female, pistil bearing (pistillate).

Fid, divided, lobed.

Filament, the stalk of an anther or the lower part of the stamen.

Filiform, thread or filament like.

Fimbriate, fringed along the edge.

Fistular, hollow throughout its length.

Fleshy, thick and of leathery or succulent texture.

Flower, a collection of sporophylls or spore bearing leaves with receptacle and perianth.

Foliaceous, leaf like.

Foliate, referring leaflets.

Follicle, the fruit produced from a single carpel and opening by ventral suture.

Free, not united.

Fruit, the ovary and its contents after the fertilization of the ovules.

Frutescent, becoming shrubby.

Fruiticose, shrubby.

Fugacious, falling off rapidly.

Funicle, stalk of ovule.

Fusiform, spindle shaped.

Gamopetalous, united petals (corolla).

Gamophyllous, united tepals (perianth).

Gamosepalous, united sepals (calyx).

Gibbous, swollen on one side.

Glabrous, not hairy, smooth.

Glandular, supplied with glands.

Glaucous, of a blue green colour.

Globose, nearly spherical.

Glumaceous, resembling the glumes of Gramineae, chaffy.

Glume, the bracts and bracteoles of the spikelets of Gramineae.

Gynoecium, the carpel or assemblage of carpels.

Habit, general appearance of a plant.

Habitat, the kind of locality in which plant grows.

Hairy, clothed with hairs.

Head, a compact (capitulum) inflorescence.

Herbaceous, not woody.

Hermaphrodite, bisexual, flower having both androecium and gynoecium.

Heterogamous, bearing two kinds of flowers. In Compositae when ray and disc florets are situated on the same head.

Hispid, with short scattered stiff bristles.

Homogamous, a head of Compositae bearing only one type of flowers.

Homologous, of similar morphological significance.

Hypanthium, tubular or flask-shaped zone of the floral axis which grows above the level of the ovary and bears on its margin the floral leaves and androecium.

Hypogynous, situated below the gynoecium or ovary referring to stamens, petals and sepals.

Imbricate, a mode of aestivation in which one member of the whorl is outside all the others (*i.e.*, its margins are free) and one inside all the others (*i.e.*, both margins are overlapped), the others overlap by one margin only.

Incised, deeply cut.

Incurved, with the ends curved inward.

Indefinite, numerous.

Indehiscent, not opening by valves or pores.

Induplicate, both sides rolled inwards.

Inferior, epigynous condition.

Inflorescence, arrangement of the flowers on the floral axis.

Infundibular, funnel shaped.

Internode, space between two successive nodes.

Introrse, facing inwards usually referred for anthers.

Involucre, whorl or bracts situated below the flower or inflorescence.

Irregular, unsymmetrical, zygomorphic.

Keel, the anterior petal in the Papilionaceae.

Labiate, lipped.

Lanceolate, lance-head shaped, a leaf broad below the middle and tapering towards apex.

Lamina, leaf blade.

Latex, milky juice.

Leaflet, one of the blades of compound leaf.

Legume, the pod of Pea Family, dehiscing along both sutures.

Leguminous, resembling legume.

Ligule, a membranous outgrowth at the junction of leaf blade and leaf sheath in grasses.

Ligulate, with ligule.

Limb, the broader or expanded part of a gamopetalous corolla.

Linear, long and narrow.

Lip, one of the two divisions of a bilabiate corolla.

Lobed, cut less than half way.

Loculicidal, fruits which open by the dorsal suture of the carpels.

Locular, chambered.

Lodicule, rudimentary membranous perianth at the base of the ovary in Grass Family.

Lomentum, a pod or legume constricted between the seeds.

Male flower, androecium bearing (staminate) flower.

Monadelphous, united into one bundle by the filaments.

Monoecious, male and female flowers on the same plant.

Mucronate, tipped with an abrupt point.

Nectary, a nectar secreting gland.

Node, the place of insertion of a leaf on the stem.

Nut, a hard dry one seeded, indehiscent fruit.

Oblong, longer than broad with parallel sides.

Obovate, inversely ovate.

Obtuse, rounded and blunt.

Opposite, arranged on different side of the axis but on the same level.

Orthrotropous, an straight ovule with the micropyle opposite to the chalaza.

Oval, broadly elliptical.

Ovary, part of gynoecium containing the ovules

Ovate, egg shaped, broader end towards the base.

Ovule, minute bodies found inside the ovary, which after fertilization develop into seeds.

Pale, Palea, the scaly bracts within the glumes in the flowers of the Grass family.

Palmate, with palm-like spreading segments.

Panicle, a compound racemose inflorescence.

Paniculate, bearing panicle type inflorescence.

Papilionaceous, butterfly like, as corolla of Pea family.

Pappus, modified calyx into bristles, *e.g.*, in many Composites.

Parietal, borne or peripheral region reference for placenta.

Pedicel, stalk of flower.

Pedicellate, the flower with pedical.

Peduncle, stalk of an inflorescence.

Pentamerous, with five members in each whorl.

Perianth, the floral envelopes including both calyx and corolla no differentiation into calyx and corolla.

Perennial, can live for more than two years.

Perigynous, sepals, petals, and stamens inserted above the hypanthium.

Persistent, not falling off.

Petal, of the floral leaves of the corolla.

Petaloid, white or coloured.

Petiole, the stalk of the leaf.

Petiolate, the leaf with petiole.

Pinna, the leaflets of a bipinnate leaf.

Pinnate, a compound leaf in which the leaflets are arranged on the two sides of the rachis.

Pistil, a collective term for ovary, style and stigma.

Placenta, the organs to which ovules are attached.

Plumose, feathered.

Pod, a dry fruit developed from a monocarpellary ovary.

Pollination, transference of pollen grains from anthers to stigma.

Polypetalous, bearing uni and bisexual flowers on the same plan.

Polypetalous, with free petals.

Polysepalous, with free sepals.

Polymorphic, with many forms.

Pome, most of the edible part develops from the thalamus, *e.g.*, apple.

Prostrate, lying horizontally on the ground.

Pubescent, densely clothed with hairs.

Pulvinus, the swollen base of the petiole.

Quincuncial, an imbricate aestivation, with 5 petals out of which two are exterior, two interior and the fifth has one margin exterior and one interior.

Raceme, an inflorescence with stalked flowers, in which the main axis continues to grow and the lowest flowers are the oldest and open first (arranged in basipetal succession).

Racemose, an inflorescence in which the main axis continues to grow and remains stronger than the laterals which arise from it, the youngest lateral come out from nearest the apex.

Rachis, the axis of an inflorescence or of a compound leaf.

Rachilla, the axis of the spikelet of the grasses.

Ray florets, the outer flowers of a composite head.

Receptacle, the portion of the axis (convex, concave or flattened) of a head on which the florets are arranged.

Regular, actinomorphic, with radial symmetry.

Reniform, kidney shaped.

Reticulate venation, net veined.

Rhizome, an elongated, horizontal underground stem with much quantity of starch.

Root stock, resembling rhizome.

Rotate, wheel shaped usually referred to a gamopetalous corolla, *e.g., Solanum*.

Runner, an elongated prostrate weak stem with adventitious roots at nodes.

Salver-shaped, with a long tube and horizontal limb usually referred to corolla.

Scabrid, clothed with small and stiff hairs.

Scale, a thin membranous outgrowth.

Scaly, with scale.

Schizocarp, a fruit opening in two or more distinct portions each with its separate wall.

Seed, the ovule after fertilization and the development of the embryo.

Sepal, a segment of calyx.

Sepaloid, green.

Septicidal, capsule dehiscing by ventral sutures.

Serrate, toothed with teeth inclined upwards.

Sagittate, arrow shaped with the basal lobes directed backwards.

Sessile, without a stalk.

Simple, not compound.

Sheath, the tubular basal portion of a leaf, usually which encircles the stem.

Sheathing, the leaf having expanded base of the petiole encircling partially or completely the stem.

Sigmoid, curved, in zigzag fashion.

Spathe, a large showy bract usually covering the whole inflorescence.

Spadix, a type of inflorescence with succulent axis possessing male and female flowers, *e.g.*, the Aroid family.

Spike, a racemose elongated inflorescence bearing sessile flowers.

Spikelet, the laterals of the inflorescence of grasses, the small spikes.

Spine, a sharp pointed hard body.

Stamen, a modified leaf in flowering plants bearing pollen sacs. It consists of filaments and anthers.

Staminate, having stamens, male.

Staminode, rudimentary stamens which do not bear fertile pollen.

Standard, the broad upper petal of Papilionaceae.

Stigma, the pollen receiving part of the carpel, usually it is lobed and sticky.

Stipule, a leaf-life outgrowth at the base of the petiole, usually in parts.

Stipulate, having stipules.

Style, a slender outgrowth of a carpel which bears stigma.

Sucker, a shoot of subterranean origin.

Sulcate, grooved.

Superior, situated above another member. A superior ovary has its base above the insertion of calyx; a superior calyx or corolla is inserted above the ovary.

Suture, the line of dehiscence of dry fruits.

Syncarpous, united carpels, compound ovary.

Syngenesious, with anthers cohering in a ring.

Tap root, primary root, deeply penetrating the soil.

Tassel, the male inflorescence of maize.

Tendril, they are either modified branches or leaves. They are filament like usually coiling round the support.

Tepal, a segment of perianth.

Terete, cylindrical.

Tetradynamous, with four long and two short stamens.

Trifoliate, with three leaflets.

Truncate, cut off at the end, square or so.

Tuber, a short underground stem with starch.

Turgid, swollen.

Twiner, a climber, which climbs by spirally coiling stems.

Umbel, an inflorescence having aerial branches of equal length from a common point.

Umbellate, umbel like.

Unilocular, one chambered ovary.

Valvate, an aestivation with the segments of calyx or corolla are so placed that their edges touch each other not overlap.

Venation, the arrangement of veins in a leaf.

Ventricose, suddenly bulged.

Versatile, can move in any way.

Verticillate, whorled.

Viscid, with sticky secretion.

Whorled, verticillate.

Xerophyte, a desert plant.

Zygomorphic, asymmetrical, irregular.

9

THE DICOTYLEDONES

Order-Rosales Family-Rosaceae

According to Hutchinson this is the sixth order of the phylum Angiospermae, subphylum Dicotyledones and division Lignosae. The order comprises of three families, *i.e.*, Rosaceae, Dichapetalaceae and Calycanthaceae, of which the Rosaceae has been described here in detail. Bentham and Hooker have placed this family together with other five families in their eleventh order–the Rosales of class Dicotyledones, division Polypetalae and series Calyciflorae. Engler and Prantl have also included the family Roscaceae with other seventeen families in their twenty first order–the Rosales of class Dicotyledoneae and sub-class Archichlamydeae.

Perigynous to epigynous, mostly apocarpous to syncarpous with central placentation; no endosperm; leaves simple to compound, stipulate.

Family-Rosaceae

This family consists of about 115 genera and 3,200 species (Lawrence).

Distribution : Though the members belonging to this family are cosmopolitan, yet they are mainly distributed in temperate and colder regions. In India most of the genera have been reported from hilly regions. The family is represented in our country by the following few important genera, *e.g., Prunus, Pyrus, Rubus, Rosa, Eriobotrya, Fragaria, Potentilla*, etc.

Habit : There are great variations in the habit of the plants included in this family. They may be trees, shrubs or herbs. Very often they are thorny and sometimes climbing. Most of the species are herbaceous and wild, while on the other hand some species are grown for their delicious fruits, and as ornamentals.

Root : Tap and branched.

Stem : The stem is erect or creeping, herbaceous or hard and woody, cylindrical, branched; many shrubs are spiny (*e.g., Rosa*); in *Rosa* and *Rubus* these spiny outgrowths are cortical and arise from both nodes and internodes. Vegetative propagation takes place by means of runners, suckers and cuttings. The bud grafting is very common in this family.

Leaves : The leaves are alternate (rarely opposite, *e.g.*, in seedling stage of *Prunus*), simple or pinnately compound, petiolate, usually stipulate, the stipules are adnate to the petiole; the leaf base is conspicuous, small spines are often present on the rachis; the margins of the leaves are either serrated or entire.

Inflorescence : There are great variations in different species. It may be corymb, corymbose, umbellate, racemose, or the flowers may be solitary or in the small groups in the twos or threes. Sometimes the inflorescence is compound raceme (panicled flowers).

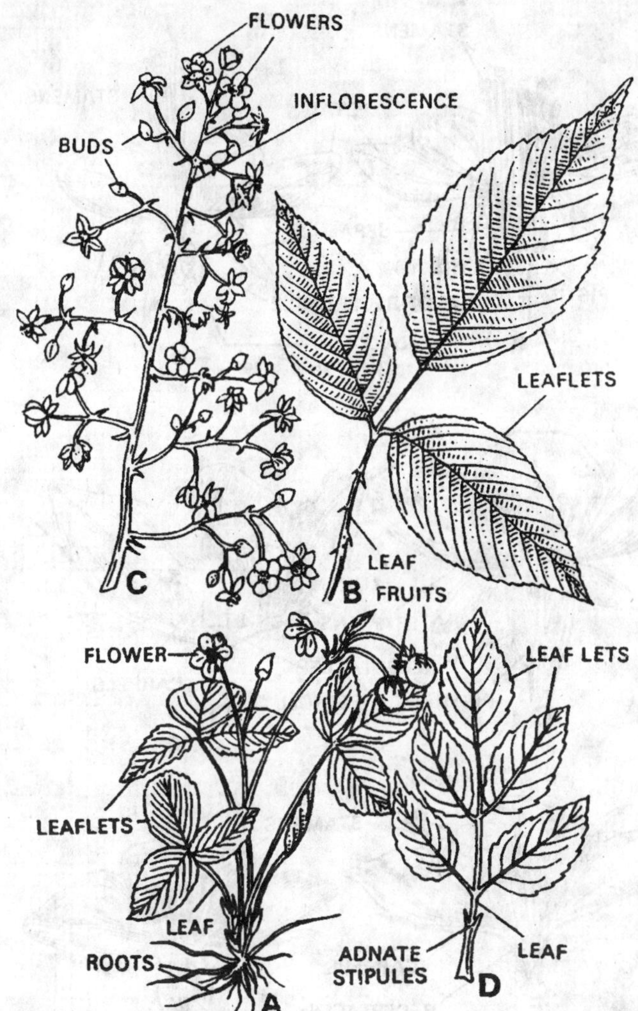

Fig. 9.1. Rosaceae–A, a plant of *Fragaria vesca*: B, a leaf of *Rubus* sp., C, inflorescence of *Rubus* sp.; D, leaf of *Rosa* sp. showing adnate stipules.

Flower: Usually hermaphrodite (bisexual) rarely unisexual (*e.g.*, *Spiraea aruncus*, where the plants are dioecious), actinomorphic, regular (irregular in *Parinarium* and *Parastemon*), bracteolate.

It may be hypogynous (*e.g.*, in *Fragaria*), perigynous (*e.g.*, in *Prunus*) or epigynous (*e.g.*, in *Pyrus*).

Calyx : It consists of 5 sepals, which are connate at the base (gamosepalous); basal portions are usually adnate forming a hypanthium; the lobes of the calyx are free, valvate or imbricate and green; sometimes the epicalyx is formed by bracteoles; the lobes of epicalyx remain alternate to the sepals.

Corolla: It consists of 5 petals (rarely four petals) which are polypetalous; the petals usually arise from the rim of hypanthium; imbricate, very often the hypanthium bears a nectariferous glandular disc. The number of petals increases sufficiently in cultivated varieties because of the conversion of the stamens into petals. They are variously coloured. Sometimes the corolla is altogether absent (*i.e.*, in *Alchemilla*, *Poterium*, etc.).

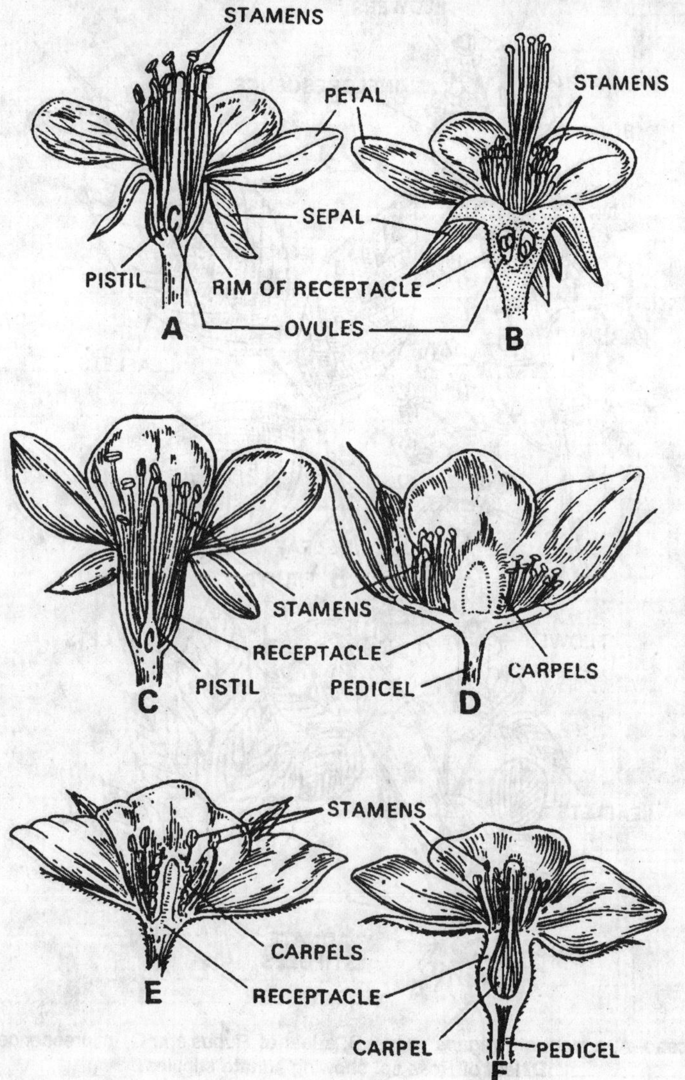

Fig. 9.2. Rosa–flowers cut lengthwise; A, *Prunus* sp.; B. *Malus sylvestris;* C. *Prunus avium* (sweet cherry); D. *Rubus* sp. (raspberry); E. *Fragaria vesca*; F. *Rosa* sp.

Androecium: Usually the number of stamens is indefinite (15-60), sometimes they are 5-10, the stamens are generally arranged in one to many whorls of five each; they are perigynous around the gynoecium and arise from the hypanthium, free; the anthers are small, dithecous (two-celled), introrse and dorsifixed; the dehiscence takes place by means of longitudinal splits; the filaments are usually incurved in bud.

Gynoecium: The number of carpels is one to many; the gynoecium consists of either one compound carpel (syncarpous) or many simple carpels (apocarpous) arranged in cyclic or spiral way. The carpels are usually situated within the hypanthium or the hypanthium remains adnate to the compound ovary; the ovary is either superior or inferior or half superior half inferior (*i.e.*, perigynous condition), when syncarpous, 2-5 locules are found, the placentation is axile, and the stigmatic lobes are as many as the number of carpels. The placentation is basal when one carpel is present (apocarpous); the ovules are one to many in each carpel. The style is free or connate, the stigma is simple, lobed or capitate.

Fruit: They may be dry or fleshy. It may be a pome (*e.g.*, in *Pyrus*), pyriform berry (*e.g.*, in *Eriobotrya*), an etaerio of achenes (*e.g.*, in *Rubus*), or one seeded drupe (*e.g.*, in *Prunus*), etc.

Seeds: The seeds are exalbuminous, generally with small embryo.

Pollination: The pollination takes place usually through the agency of insects (*i.e.*, entomophily), but sometimes there is wind pollination (*i.e.*, anemophily)

Floral Formula : Br ⊕ ⚥ K(5), C5, A ∞, G ∞ or 1 or $\underline{(2-5)}$

Bentham & Hooker(1862)	**Engler & Prantl(1931)**	**Hutchinson (1959)**
Phanerogams	Phanerogams	Angiospermae
Dicotyledones	Dicotyledoneae	Dicotyledones
Polypetalae	Archichlamydeae	Lignosae
Calyciflorae	Rosales	Rosales
Rosales	Rosaceae	Rosaceae
Rosaceae		

Description of some important plants in semitechnical language:-

Prunus persica ; Verna. **Aru.** (See fig. 14.3 on page 320).

Habit: Either shrubs or small trees, usually grown in the gardens for their delicious fruits.

Root: Tap and branched.

Stem: Erect, branched, woody, cylindrical and glabrous.

Leaf: Alternate, simple, stipulate, petiolate, entire or toothed, very frequently two glands present on the petiole. In *Prunus persica* the leaves are oblong–lanceolate or oblong oblanceolate, 4 to 6 inches long, finely serrate, short petioles, fimbriate stipules.

Inflorescence: Flowers are solitary or in fascicles, corymbose or racemose.

Flcwer: Bisexual (hermaphrodite), pedicellate (small pedicels), actinomorphic, regular, perigynous, complete, pink.

Calyx: Five sepals, gamosepalous, sepaloid, imbricate, deciduous in fruit, calyx-tube bell shaped, 5-lobed, inferior.

Corolla: 5 polypetalous (petals free), petals large, arranged on hypanthium. conspicuous, pink, inferior.

Androecium: Stamens indefinite (15-60), perigynous, inserted on the hypanthium, anthers bi-celled, dorsifixed, reddish or brown, dehiscence by longitudinal split, introrse, filaments free.

Gynoecium: One carpel (monocarpellary), ovary situated within the hypanthium (calyx tube), ovary unilocular, with two pendulous ovules, style terminal, stigma broad.

Fruit: Single seeded drupe, fleshy, enclosing an indehiscent or two valved furrowed stone.

Seeds: Exalbuminous, pendulous.

Generic Characters : Ripe carpels not enclosed within the calyx tube (hypanthium), carpels free; carpel one, fruit a drupe with a hard furrowed stone; shrubs or trees with simple leaves; calyx 5 lobed, petals large.

Floral Formula: ⊕ ⚥ K (5), C5, A ∞, G1–

***Eriobotrya japonica*:** Verna-**Lokat.**

Habit: Small trees, grown in the gardens for their edible fruits.

Root: Tap and branched.

Stem: Erect, woody branched with stout branches, cylindrical.

Leaves: Simple, alternate, entire, serrate or crenate, coriaceous; in *Eriobotrya japonica*, the leaves are large, 6-12 inches long, subsessile, lanceolate, or oblanceolate, acuminate, dentate or serrate, net veined, and wooly below.

Inflorescence: Flowers arranged in dense panicles, the inflorescence is wooly.

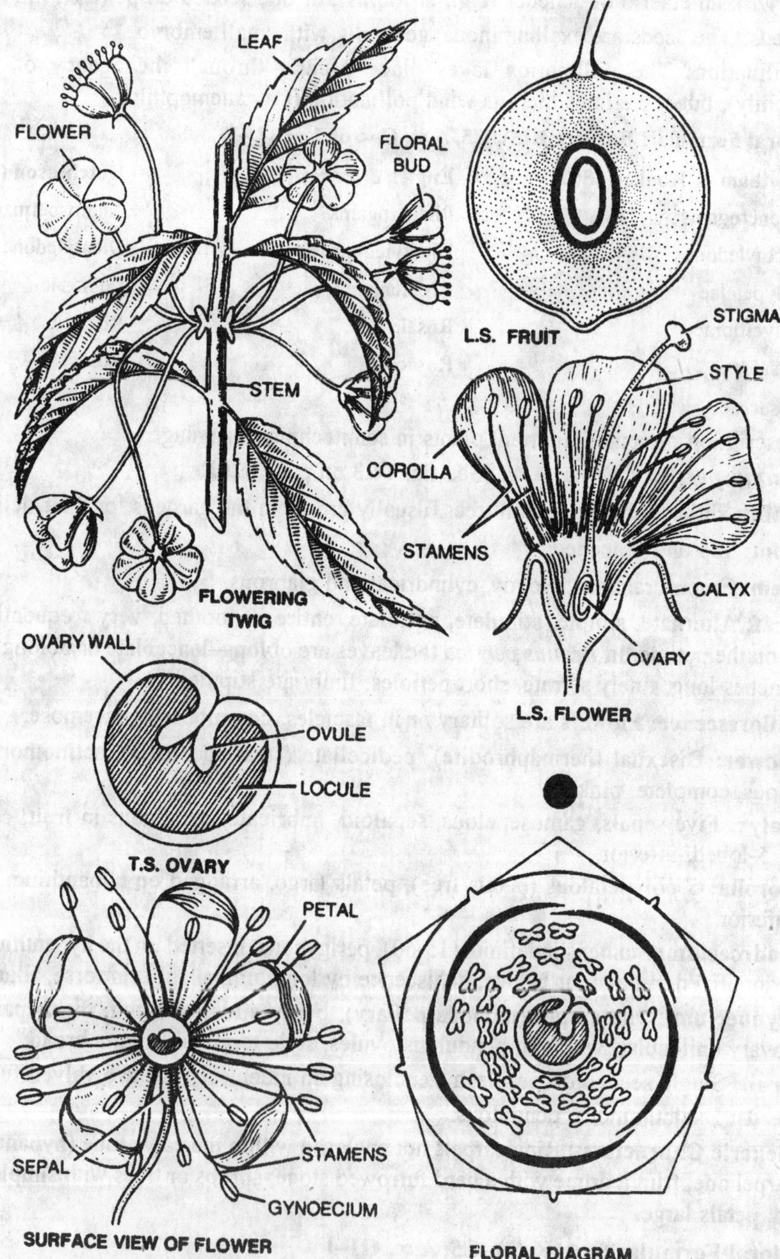

Fig. 9.3. Rosaceae–*Prunus persica* (L.) Stokes; (Verna. Aru).

Flower: Pedicellate (short pedicels), actinomorphic, regular, hermaphrodite (bisexual), epigynous, yellowish white in colour, bracteate and bracteolate.

Calyx: 5 sepals, gamosepalous, small persistent, arranged on the turbinate or clavate hypanthium, sepals also wooly.

Corolla: 5 petals, free, polypetalous contorted or imbricate in bud, obovate or orbicular, usually oblique and notched at margins, claw glabrous and wooly, yellowish white in colour.

Androecium: Indefinite in number, stamens curved inward in the bud, filaments free, anthers 2-celled, dorsifixed, dehiscing longitudinally, introrse.

Gynoecium: Carpels enclosed in the hypanthium, 2-5 celled ovary, with connate and wooly below styles, each cell contains two ovules, axile placentation.

Fruit: Succulent or dry; in *E. japonica*, the fruit is ovoid, yellow, succulent and pyriform berry, 1–1.5 inches long. 1–5 seeded.

Seed: Exalbuminous, large and angled with thick cotyledons.

Generic Characters: Carpels definite, 2-ovuled, enclosed in the hypanthium, and adnate to its walls, more or less connate with one another to form a syncarpous ovary; fruit a berry or pome; flowers arranged in panicles.

Floral Formula: Br Brl ⊕ ⚥ K(5), C5, A ∞, $G_{\overline{(2-5)}}$

Malus sylvestris; Verna. **Seb.**

Habit: Small tree or large shrub.

Stem: Erect, woody, branched, cylindrical, solid.

Leaves: Simple or pinnate, deciduous, petiolate, stipulate (stipules deciduous), glabrous, acute, ovate, unicostate reticulate venation

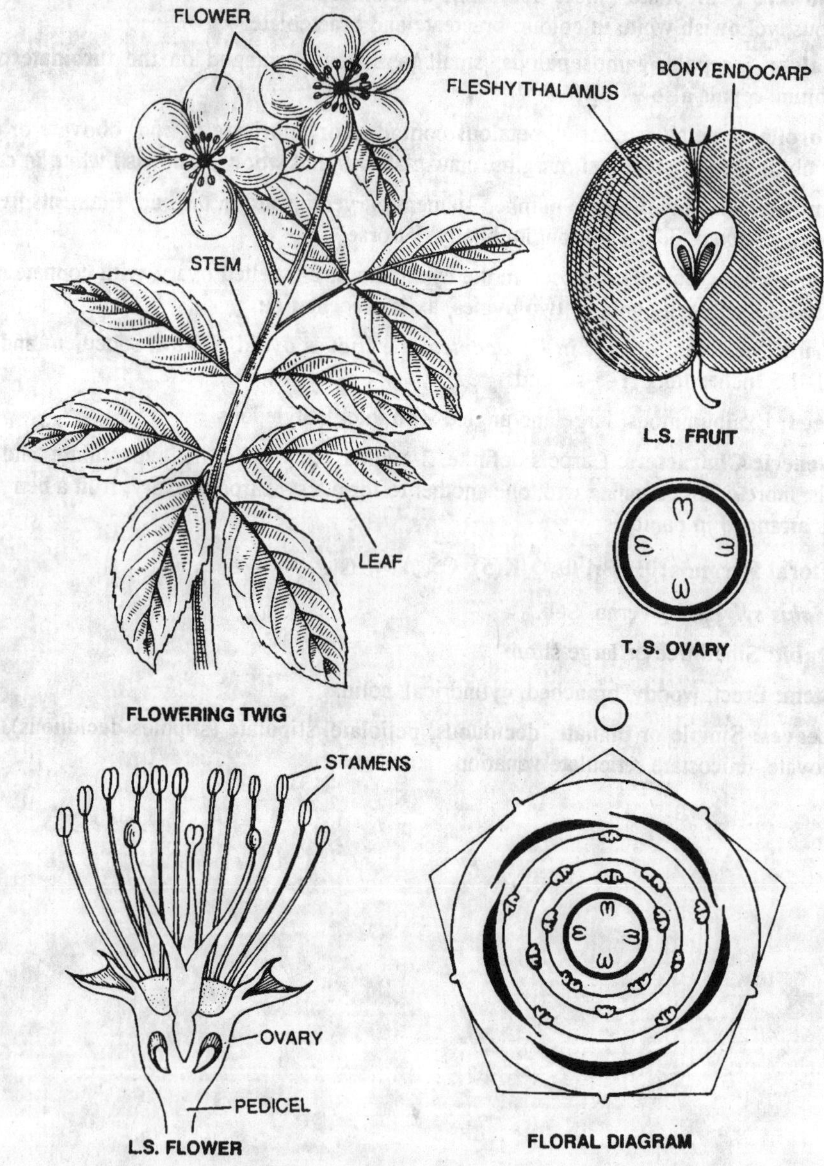

Fig. 9.4. Rosaceae–*Malus sylvestris* (L.) Mill.; syn. *Pyrus malus* Linn. (Verna, *Seb;* Eng. Apple).

Inflorescence: Cymose, terminal cymes or corymbs.

Flower: Pedicellate, bracteate (bracts linear), bracteolate (two bracteoles), hermaphrodite, actinomorphic; complete, epigynous.

Calyx : Five sepals, gamosepalous, calyx-tube obconical lobes erect or reflexed, persistent or deciduous.

Corolla: Five petals, polypetalous, quincuncially imbricate in bud.

Androecium: Indefinite stamens (20 or more), filaments long, connate at the base, anthers dorsifixed, introrse; dehisce longitudinally.

Gynoecium: Two to five carpels, connate and adnate to calyx-tube, syncarpous, ovary inferior, 2-5 locular, axile placentation, styles 2-5, free or connate below, stigmas truncate; ovules 2 in each loculus.

Fruit: A pome, fleshy.

Floral formula: Br Brl ⊕ ⚥ **K**(5), C5, A∝, $G_{\overline{(2-5)}}$

Potentilla supina.

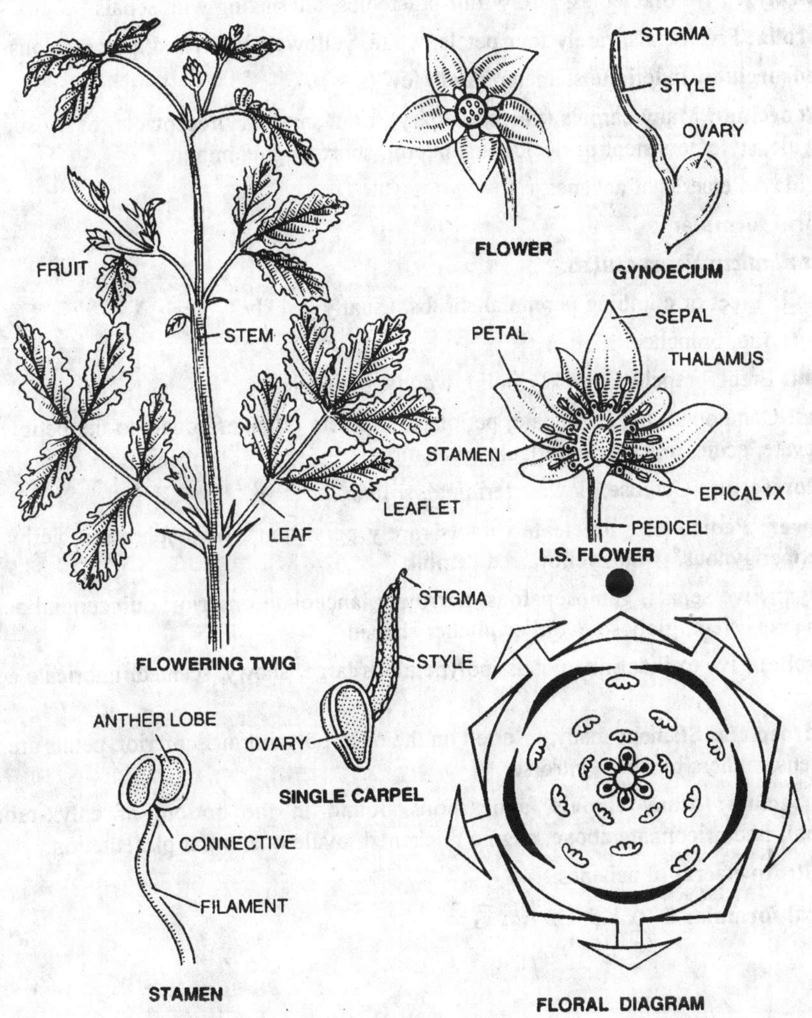

Fig. 9.5. Rosaceae–*Potentilla supina* Linn.

Habit: Perennial herbs, rarely shrubs.

Stem: Erect or diffusely spreading, branched, woody or herbaceous, pubescent, cylindrical, solid.

Leaves: Compound, petiolate stipulate (stipules adnate to the petiole), leaflets mostly serrate, reticulate venation.

Inflorescence: Cymose, solitary or corymbose cymes.

Flower: Pedicellate, hermaphrodite, actinomorphic, complete, hypogynous, white or yellow rarely red.

Calyx: Five sepals, rarely four sepals, persistent, valvate in bud.

Epicalyx: Five bracteoles, rarely four bracteoles, alternating with sepals.

Corolla: Five petals, rarely four petals, white, yellow or rarely red, polypetalous.

Androecium: Indefinite stamens, rarely few (5 - 10).

Gynoecium: Many carpels (polycarpellary), on a small dry receptacle, ovary superior with single ovule, style persistent or deciduous, lateral, short, stigma minute.

Fruit: An etaerio of achenes.

Floral formula:

Rosa indica; Verna-**Gulab**.

Habit: Erect or climbing perennial shrubs, usually prickly.

Root: Tap, branched.

Stem: Erect, branched, prickly, solid, woody, cylindrical.

Leaf: Compound, imparipinnate, petiolate, stipulate (stipules adnate to the petiole), leaflets serrate, ovate, acute, unicostate, reticulate venation.

Inflorescene : Cymose, flowers terminal, solitary.

Flower: Pedicellate, bracteate (bracts rarely persistent), hermaphrodite, actinomorphic, complete, perigynous, white, yellow, red or pink.

Calyx: Five sepals, gamosepalous, narrowly lanceolate, inferior, quincuncial aestivation, calyx-tube persistent, globose ovoid or pitcher shaped.

Corolla: Five to indefinite petals, polypetalous large, showy, scented, imbricate aestivation in bud.

Androecium: Stamens many, inserted on the disk, polyandrous, inferior, petals are modified into stamens, anthers bicelled, introrse.

Gynoecium: Many carpels, apocarpous, found in the bottom of calyx-tube, styles subterminal, free or connate above, stigma thickened, ovule one, basal placentation.

Fruit: An etaerio of achenes.

Floral formula: ⊕ ⚥ K5, C5 A∞ $\overline{G}\infty$

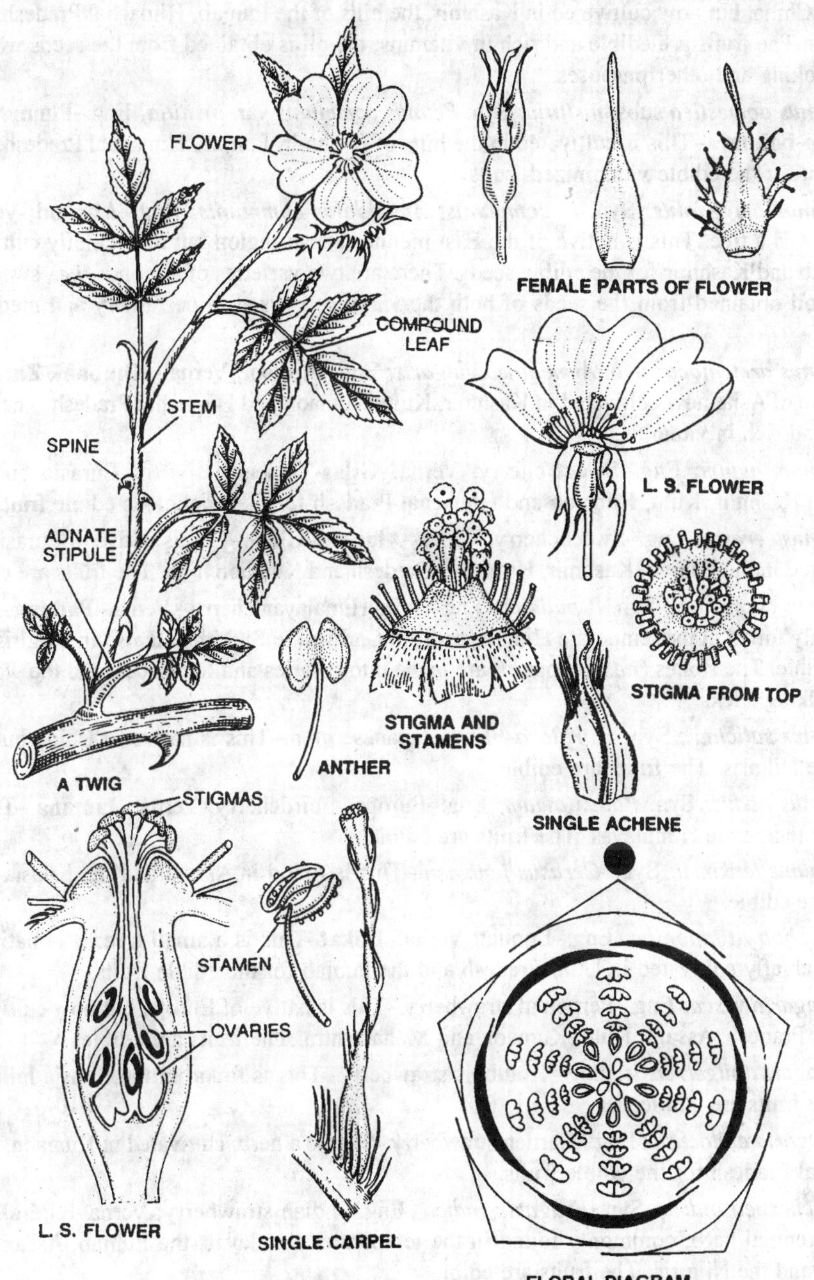

Fig. 9.6. Rosaceae–*Rosa indica*, Eng., rose; Verna. **Gulab, Jangligulab.**

Economic Importance of the Family

Economically this family is of much importance, containing various types of valuable fruit plants, Several other plants are ornamental and are planted in the gardens, public places and houses for their beautiful and fragrant flowers. A list of some important plants is given here–

1. *Prunus persica;* Syn. *Amygdalus persica; Persica vulgaris;* Eng.–Peach; Verna–**Aru**–This is native of China, but now cultivated in Kashmir, the hills of the Punjab, Himachal Pradesh, Kulu and Kumaon.The fruits are edible and rich in vitamins; the oil is obtained from the seeds which is used for cooking and other purposes.

2. *Prunus domestica* subs. *insititia;* Syn. *Prunus communis* var. *insititia;* Eng.–Plum; Verna-**Alucha, Alu-bokhara**–This is cultivated in the hills of Kashmir, Punjab, Himachal Pradesh, Kulu and Kumaon for the edible vitaminized fruits.

3. *Prunus amygdalus;* Syn. *P. communis; Amygdalus communis;* Eng.–Almond; Verna.–**Badam**–This is a tree. This is native of the East mediterranean region but now chiefly cultivated in the Punjab and Kashmir for the edible seeds. There are two varieties of almonds, *i.e.*, sweet and bitter. The oil obtained from the seeds of both the varieties is used in perfumery and medicinal purposes.

4. *Prunus armeniaca;* Syn. *Armeniaca vulgaris;* Eng.–Apricot; Verna.–**Khubani, Zardalu**–This is native of Asia, now cultivated in Kashmir, Kulu, Kumaon and Himachal Pradesh. The fruits are edible and rich in vitamins.

5. *Prunus avium*; Eng.–Sweet cherry; Verna.–**Gilas**-This is native of Eurasia but now cultivated in Kashmir, Kulu, Kumaon and Himachal Pradesh for their delicious edible fruits.

6. *Prunus cerasus;* Eng.–Sweet cherry; Verna.–**Alubalu, Gilas**.–This is native of Eurasia, but now cultivated in the Punjab, Kashmir, Himachal Pradesh and Kumaon hills. The fruits are edible.

7. *Prunus cerasoides;* Syn. *P. puddum;* Eng.–Wild Himalayan cherry ; Verna.–**Paddam**–They are commonly found in the temperate Himalayas from Garhwal to Sikkim and also in Nilgiris. The fruits are edible. The stones (*i.e.*, endocarp) are made into rosaries and necklaces, and the wood is used for walking sticks.

8. *Prunus salicina ;* Syn. *P. triflora ;* Eng.–Japanese plum–This is native of China but now grown in the Nilgiris. The fruits are edible.

9. *Prunus padus:* Syn. *Padus avium;* Eng.–European birdcherry; Verna.–**Jamana**.–This is found in the temperate Himalayas. The fruits are edible.

10. *Prunus jenkinsii;* Syn.–*Cerasus jenkinsii.*–This is found in Assam and the Khasia hills. The fruits are edible.

11. *Eriobotrya japonica;* Eng.–Loquat; Verna.–**Lokat**–This is a small tree. It is native of China, now chiefly cultivated in Uttar Pradesh and the Punjab for the edible fruits.

12. *Fragaria vesca;* Eng.–Perpetual strawberry.–This is native of Europe but now cultivated in Himachal Pradesh, Assam, Kulu, Kumaon and Maharashtra. The fruits are edible.

13. *Fragaria nilgerrensis;* Eng.–Houtbois strawberry.–This is found in the Khasia hills and Nilgiris. The fruits are edible .

14. *Fragaria chiloensis*; Eng.–Garden strawberry.–This is a herb, cultivated in Kumaon, Kulu and Himachal Pradesh for the edible fruits.

15. *Duchesnea indica*; Syn. *Fragaria indica;* Eng.–Indian strawberry; Verna–**Kiphaliya**.–This is a perennial herb, commonly found in the temperate Himalayas, the Punjab, Assam, the Khasia hills and the Nilgiris. The fruits are edible.

16. *Pyrus communis;* Eng.–Pear; Verna.–**Nakh**,–This is a small tree. It is native of Eurasia but now cultivated in Kashmir, Kulu, Kumaon and Himachal Pradesh. The fruits are edible.

17. *Malus sylvestris;* Syn. *Pyrus malus; Malus pumila;* Eng.–Apple; Verna–**Seb**.–This is a small tree; native of Eastern Europe and Western Asia but now cultivated in Kashmir, Himachal Pradesh, Kulu, Kumaon, Assam and in the Nilgiris. The fruits are edible, delicious, rich in iron and vitamins.

18. *Malus baccata;* Syn. *Pyrus baccata;* Eng.–Siberian crab apple; Verma **Jangli Seb**–This is a small tree, cultivated in Kashmir and the Khasia hills. It is used as a stock for apple. The fruits are eaten as vegetable.

19. *Docynia indica;* Syn. *Pyrus indica*; Eng.–Indian crab apple.–This is a tree found in the Eastern Himalayas and the Khasia hills. The fruits are edible.

20. *Pyrus pashia*; Verna.–**Mehal**–This is commonly found in the Western Himalayas. It is used as a root stock for pear. The wood is used in making walking sticks and tobacco pipes.

21. *Pyrus pyrifolia* var. *culta*; Eng.–Chinese pear; Verna.–**Nashpati**–This is native of China but now cultivated in Kashmir, Assam, Kumaon, Darjeeling, the Nilgiris, Himachal Pradesh and Delhi for the edible fruits.

22. *Pyrus vestita;* Verna.**Mauli**–This is found in the Eastern Himalayas and the Khsia hills. The fruits are edible.

23. *Sorbus aucuparia;* Syn. *Pyrus aucuparia;* Eng. European mountain ash; Verna.–**Battal**–This is a tree. It is native of Eurasia but now found in the temperate Himalayas from Kashmir to Kumaon. The fruits are edible.

24. *Spiraea cantoniensis ;* This is a native of China and Japan. This is a small herb grown as an ornamental.

25. *Spiraea corymbosa*–This is a small shrub, grown in the gardens as an ornamental.

26. *Rubus fruticosus;* Eng. European black berry; Verna.–**Vilayati–anchu**–This is cultivated in Kashmır, Assam and Conoor for the edible fruits.

27. *Rubus ellipticus*; Eng. Himalayan yellow raspberry; Verna.–**Lal anchu, Hisalu**–This is a shrub found in the Western Himalayas, South India, the Western ghats and the Khasia hills. The fruits are edible.

28. *Rubus rosaefolius;* Eng. Mauritius raspberry; Verna.–**Yeshul**–This is cultivated in Kumaon, the Khasia hills and Manipur. The fruits are edible.

29. *Rubus lanatus;* Verna.–**Hisalu**–This is found in Kumaon. The fruits are edible.

30. *Rubus lasiocarpus;* Eng. Ceylon Raspberry; Verna.–**Kala anchu, Kala hisalu**–This is found in the Western Himalayas, Sikkim, the Khasia hills, the Western Ghats, Karnataka and the Pulney hills. The fruits are edible.

31. *Rubus moluccanus;* Eng. Black cherry; Verna–**Katsol**.–This is found in Assam, the Khasia hills the Nilgiris and Travancore. The fruits are edible.

32. *Rosa alba;* Eng. Indian white rose ; Verna–**Gulab, Gulchini**–This is a herb, grown as an ornamental in the gardens.

33. *Rosa chinensis;* Syn. *Rosa indica* ; Eng. Edward rose; Verna.–**Cheenia gulab, Desi Gulab**–This is chiefly cultivated in the Kanauj, Kanpur and Hathras regions of Uttar Pradesh. The petals are the sources of rose water, essential oil, pankhuri and gulkand.

34. *Rosa centifolia;* Eng. Cabbage rose; Verna–**Gulab**–This is a herb or small shrub. This is cultivated mainly in the Uttar Pradesh and Bihar. The petals yield an essential oil which is used in perfumery and cosmetics.

35. *Rosa damascena;* Eng. Damask rose; Verna–**Gulab**–They are cultivated chiefly in Aligarh, Ghazipur and Kanauj of Uttar Pradesh. An essential oil known as 'otto of roses' or 'oil of roses' obtained from the petals is used in perfumery. The other important rose products are-rose water, attar, gulkand, pankhuri. Gulkand is used as a tonic and laxative. The dried rose petals are known as pankhuri which are used for preparing cold drinks.

36. *Rosa banksiae;* Eng. Banksia rose–This is climbing herb. This is native of China. It is grown as an ornamental in the gardens.

37. *Crataegus oxycantha;* It is found in the Himalayas from 6,000 feet to 9,000 feet elevation a.s.1. The wood is hard and, therefore, is used in making tool handles.

38. *Potentilla fruticosa*–This is a small herb. The leaves possess medicinal properties and are used as substitute for tea.

39. *Potentilla nepalensis*–This is medicinal plant; used as a remedy of burns.

40. *Potentilla reptans*–The infusion of the herb is used as a remedy of diarrhoea.

41. *Cotoneaster acuminata;* Verna–**Riu**–This is a shrub found in Himalayas. The wood is used for making walking sticks.

42. *Chrysobalanus icaco;* Eng. Coco plum–This is a shrub or small tree, introduced into Malabar. The fruits are edible.

43. *Cydonia oblonga;* Syn. *Cydonia vulgaris*; Verna–**Bihi**–This is a small tree. It is native of West Asia, now grown in Punjab, Kashmir and the Nilgiris for the edible fruits. It is used as a root-stock for pear.

Systematic Relationships : According to Engler the order Rosales consists of about twenty families. Among these families there are very few points of similarity. Hutchinson has splitted the order into a number of distinct orders.

10

Order–Leguminales, Family-Fabaceae

According to Hutchinson this is the seventh order of the phylum Angiospermae, sub-phylum Dicotyledones and division Lignosae. This order consists of only three families, *i.e.*, Caesalpiniaceae, Mimosaceae and Papilionaceae. The Papilionaceae is the highest evolved family within the order. All the three families have been described in detail in this text. Bentham and Hooker have treated these families as sub-families of the family Leguminosae. They have placed the Leguminosae in the eleventh order — the Rosales, of class Dicotyledones, division Polypetalae and series Calyciflorae. Engler and Prantl have also adopted Bentham and Hooker. They have placed the family Leguminosae with its three sub-families, *i.e.*, Caesalpinioideae, Mimosoideae and Papilionatae in the twentieth order the Rosales of class Dicotyledonae and sub-class Archichlamydeae.

Progressive group with increasing zygomorphy of the flower; free to variously united stamens; carpel one; no endosperm, the embryo filling the seed.

Family–Caesalpiniaceae

There are about 135 genera in the family Caesalpiniaceae.

Distribution: Most of the members belonging to this family are tropical and sub-tropical in distribution. In India the family is represented by many genera, *e.g.*, *Cassia, Bauhinia, Tamarindus, Saraca, Poinciana, Parkinsonia, Caesalpinia,* etc.

Habit: The plants show great variations in their habit, they may be trees or shrubs and very rarely herbs. Mostly they are mesophytes, but xerophytes (*e.g., Parkinsonia*) are also reported. Most of them are wild, but many are cultivated for their beautiful flowers and timber. Some plants are woody climbers (*e.g., Bauhinia* sp.).

Root: Tap and branched.

Stem: Erect, woody cylindrical, solid, branched, sometimes herbaceous or climbing. In many species tannin sacs and gum passages are found.

Leaves: The leaves may be simple or compound. If compound they may be pinnate or bi-pinnate. Usually the pinnate leaves are arranged in pairs; petiolate, pulvinus present at the base of the petiole; the pinna ovate or obovate, glabrous, net veined, entire. Usually exstipulate, sometimes minute caducous stipules present.

Inflorescence: Usually racemose, raceme, sometimes pendulous.

Flower: Pedicellate, zygomorphic rarely actionomorphic, hermaphrodite, hypogynous, complete, variously coloured, showy, large or small.

Calyx: Five sepals, free or fused, often petaloid, aestivation imbricate or valvate.

Corolla: Five petals, polypetalous(free); ascending imbricate aestivation, *i.e.*, the posterior petal innermost in bud; inferior, spathulate, showy.

Androecium: Stamens usually ten, free, rarely connate, but, sometimes reduced to staminodes or altogether abortive. In *Cassia* 3 to 5 stamens reduced to staminodes, in *Tamarindus* sp. 3 stamens are well developed, while rest are reduced into staminodes.

Gynoecium: Carpel one (monocarpellary); ovary superior, unilocular; marginal placentation; style long; stigma simple.

Fruit: Pod or lomentum, sessile or stalked, dehiscent or indehiscent.

Seed: Exalbuminous.

Pollination: Entomophilous.

Floral Formula: ⊕ or ·|· ⚥ K5, or (5) C5, A 10 $\underline{G1}$.

Bentham & Hooker (1862)	**Engler & Prantl (1931)**	**Hutchinson(1959)**
Phanerogams	Phanerogams	Angiospermae
Dicotyledones	Dicotyledones	Dicotyledones
Polypetalae	Archichlamydeae	Lignosae
Calyciflorae	Rosales	Leguminales
Rosales	Leguminosae	Caesalpiniaceae
Caesalpinioideae		

1. *Cassia tora* Linn. Verna. **Pamaar, Chakunda;** Eng. Sickle senna.

Habit: A medium sized tree or shrub.

Root: Tap, branched.

Stem: Erect, aerial, woody, branched, cylinderical, glabrous, solid.

Leaves: Cauline and ramal, compound, alternate, paripinnate, pulvinus at the base, petiolate, stipulate (stipules caducous), leaflets 4-8 pairs, ovate, entire, acute, glabrous, venation unicostate reticulate.

Inflorescence: Racemose, typical raceme.

Flower: Pedicellate (long pedicels), bracteate (bracts minute and caducous) or ebracteate, hermaphrodite, actinomorphic, complete, hypogynous, yellow, pentamerous, cyclic.

Calyx: Five sepals, polysepalous, more or less petaloid (yellowish green), inferior, quincunicial aestivation, odd sepals anterior.

Corolla: Five petals, polypetalous, yellow, clawed, ascending imbricate aestivation, inferior.

Androecium: Ten stamens, polyandrous, unequal in length, three posterior stamens reduced to staminodes, basifixed anthers with abortive and indehiscent lobes, dithecous, introrse.

Gynoecium: One carpel (monocarpellary), ovary superior, unilocular, marginal placentation, style short, stigma terminal, hairy.

Fruit: A legume, black when ripe.

Floral formula: Br ·|· ⚥ K5, C5, A7 + 3, G $\underline{1}$.

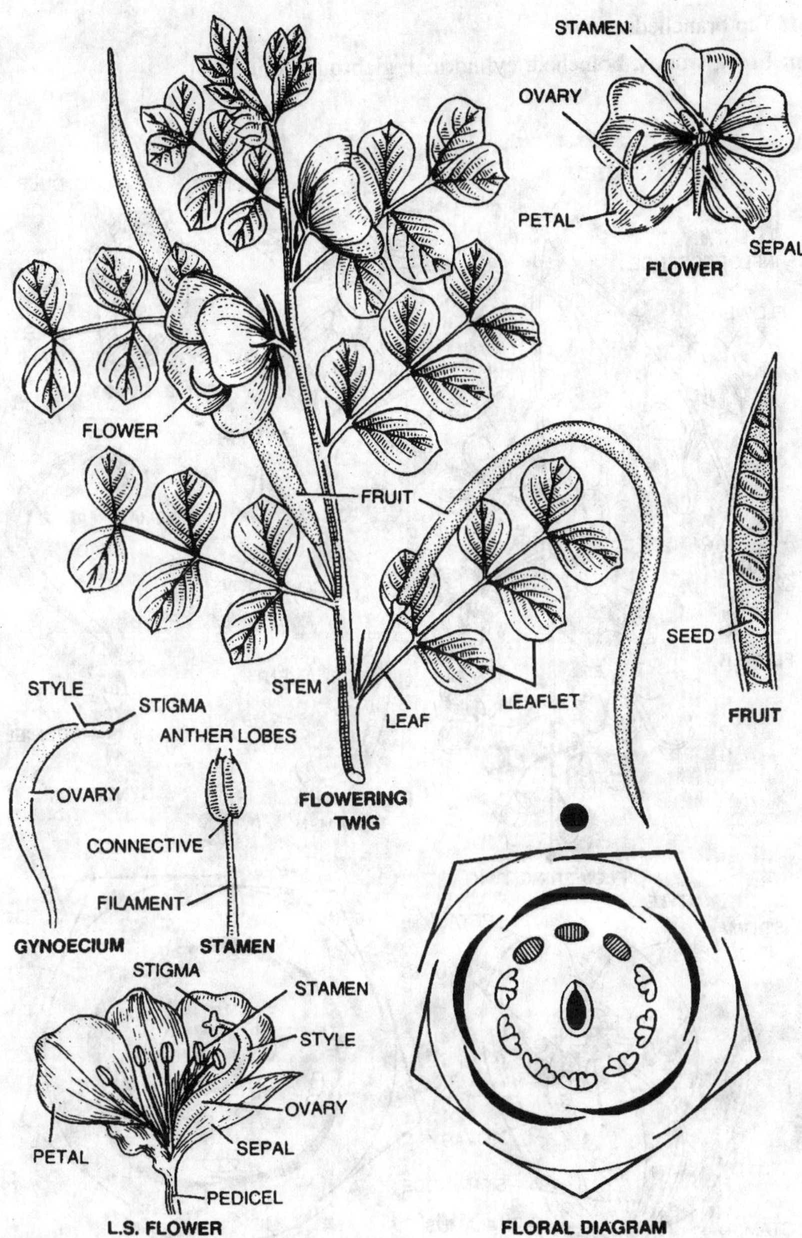

Fig. 10.1. Caesalpiniaceae (Leguminosae). *Cassia tora* Linn.; Eng. Sickle senna; Verna. **Pamaar, chakunda.**

Economic value. The seeds are used in the preparation of sweets and the tender young leaves are used as vegetable. Decoction of leaves is laxative. Leaves and seeds are used in skin diseases, for ringworm and itch. Root is used in snake-bite.

Cassia fistula; Verna–**Amaltas**.

Habit: A medium sized tree.

Root: Tap branched,

Stem: Erect, woody, branched, cylindrical, glabrous, solid.

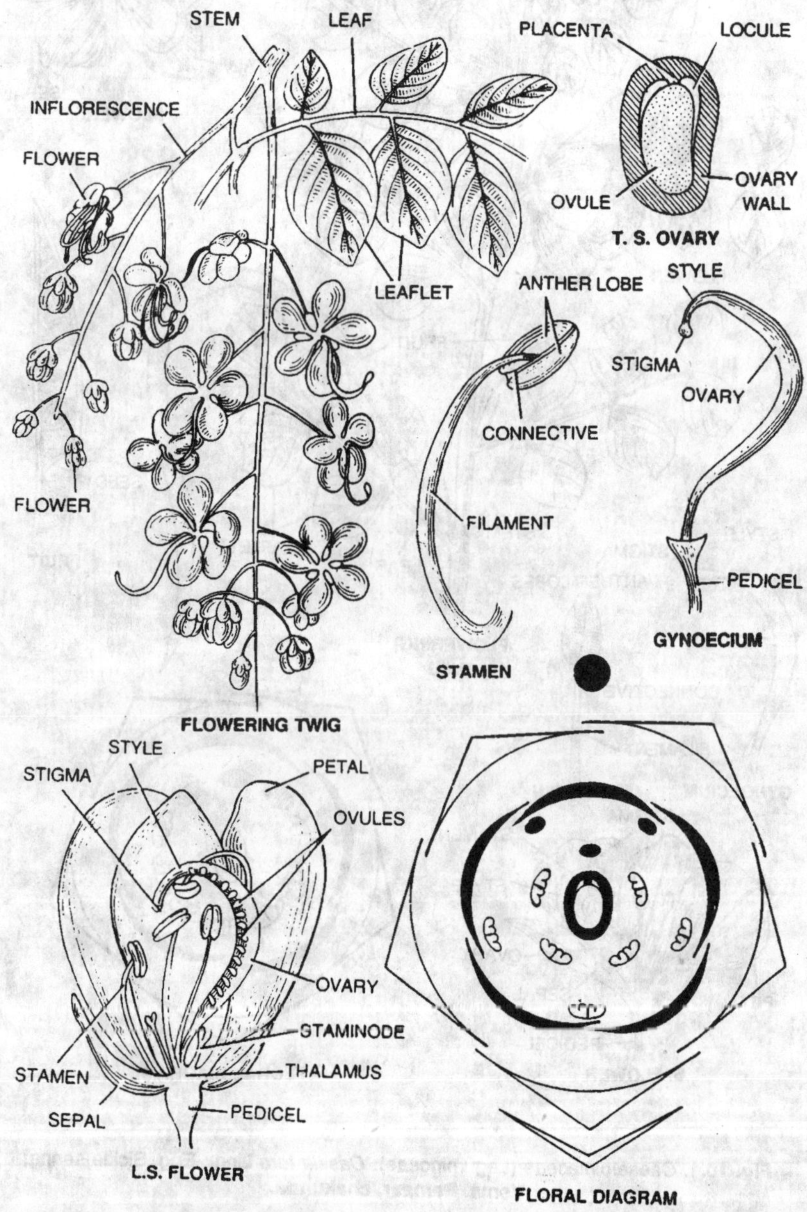

Fig. 10.2. Caesalpiniaceae—*Cassia fistula* Linn; Eng., golden shower; Verna., **amaltas.**

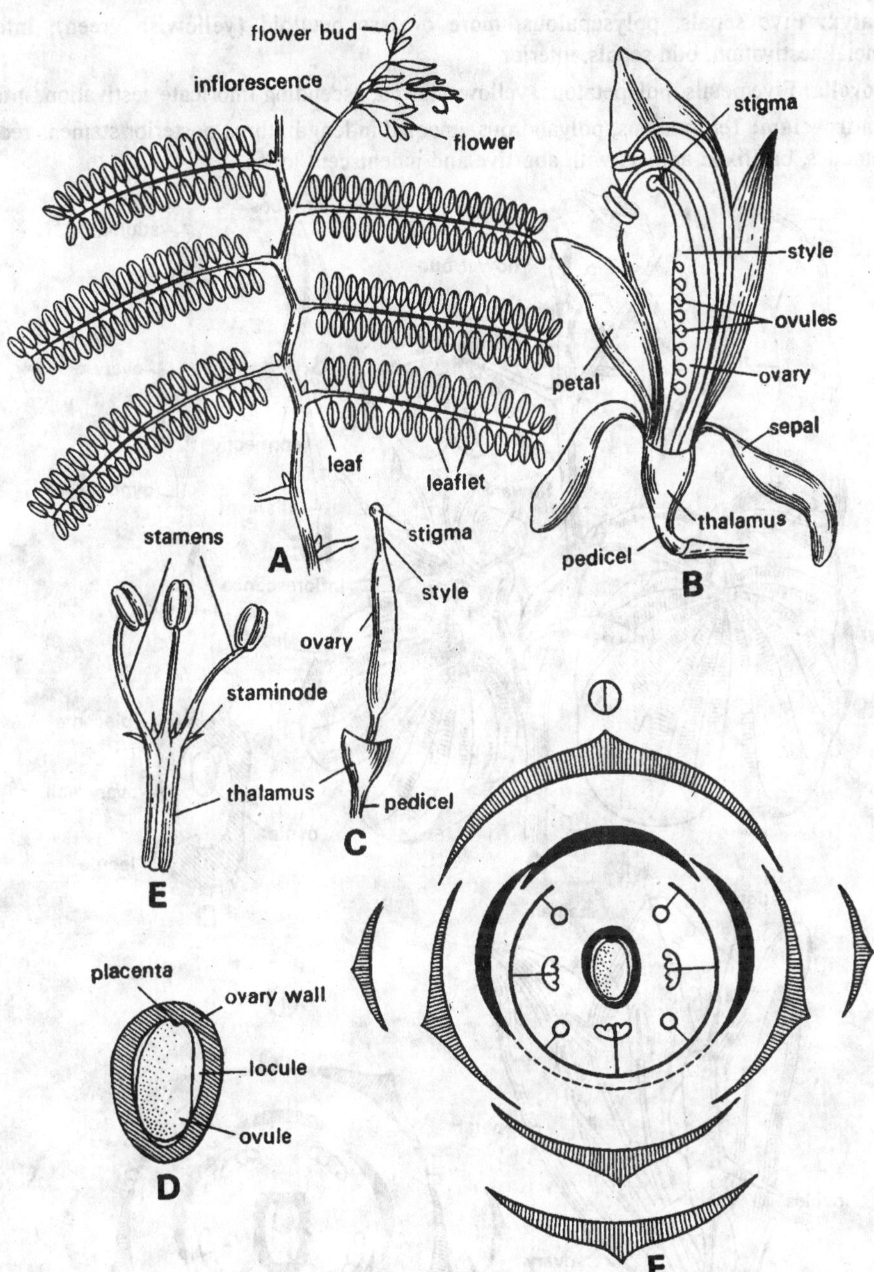

Fig. 10.3. Caesalpiniaceae–*Tamarindus indica* L. (**Imli**). A, flowering twig; B, L.S. of flower, C, gynoecium; D, T.S. of ovary; E, stamens; F, floral diagram.

Leaves: Compound, alternate, paripinnate, pulvinus at the basc, petiolate, stipulate (stipules caducous), leaflets 4-8 pairs, ovate, entire, acute, glabrous, venation unicostate, reticulate.

Inflorescence: Racemose, typical raceme.

Flower: Pedicellate (long pedicels), bracteate (bracts minute and caducous) hermaphrodite, actinomorphic, complete, hypogynous, yellow.

Calyx: Five sepals, polysepalous, more or less petaloid (yellowish green), inferior, quincuncial aestivation, odd sepals anterior.

Corolla: Five petals, polypetalous, yellow, clawed, ascending imbricate aestivation, inferior.

Androecium: Ten stamens, polyandrous, unequal in length, three posterior stamens reduced to staminodes, basifixed anthers with abortive and indehiscent lobes.

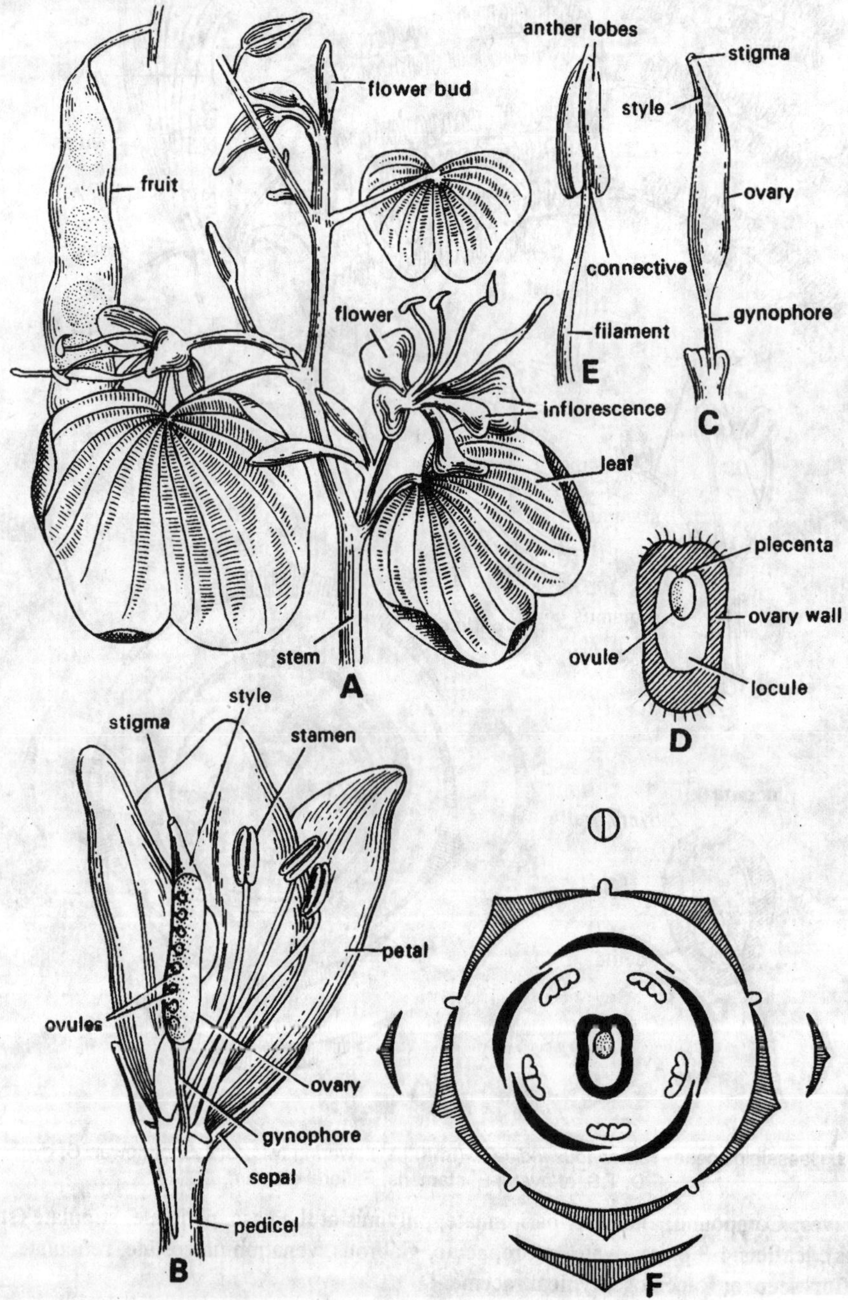

Fig. 10.4. Caesalpiniaceae—*Bauhinia variegata* (**Kachnar**). A, flowering twig and fruit; B, L.S. of flower, C, gynoecium; D, T.S. of ovary; E, stamen, F, floral diagram

Gynoecium: One carpel (monocarpellary), ovary superior, unilocular, marginal placentation, style short, stigma terminal, hairy.

Fruit: A legume, 1-2 feet long, cylindrical, black when ripe.

Floral Formula: Br ·|· ⚥ K 5, C 5, A 7 + 3, G $\underline{1}$.

3. ***Tamarindus indica*** L., Eng. Tamarind; Verna. **Imli. Stem**–woody, aerial, erect, cylindrical, branched, solid. **Leaf**–cauline and ramal, exstipulate, alternate, compound, uni–and paripinnate, petiolate, leaf base pulvinus, leaflet elliptical, entire, obtuse, unicostate reticulate, glabrous. **Inflorescence.** Racemose, axillary raceme. **Flower**–bracteate, bracteolate, pedicellate, complete. zygomorphic, hermaphrodite, perigynous, cyclic. **Calyx–5**, polysepalous, posterior sepal large, imbricate. **Corolla–5**, polysepalous, posterior sepal large, imbricate. **Corolla–5**, polypetalous, anterior two petals reduced, imbricate. **Androecium**–3 stamens, 4 staminodes, all 7 form a staminal column, monadelphous, dithecous, versatile, introrse. **Gynoecium**–1, monocarpellary, half-inferior, marginal, ovules many, style long, stigma globular. **Fruit**–legume, pod.

Floral formula– Br Brl ·|· ⚥ K_4, C_{3+2}, $A_{(3+4)}$ G1—

2. ***Bauhinia variegata*** L., Eng. Mountain ebony; Verna. **Kachnar. Stem**–woody, aerial, erect, branched, cylindrical, solid. **Leaf**–cauline and ramal, alternate, stipulate, simple, petiolate, pulvinus, entire, emarginate, multicostate reticulate. **Inflorescence**–cymose, panicle cyme. **Flower**–bracteate, bracteolate, pedicellate, zygomorphic, hermaphrodite, complete, pentamerous, perigynous, cyclic. **Calyx**–5, gamosepalous, valvate. **Corolla**–5, polypetalous, ascending imbricate, purple. **Androecium**–5, polyandrous, filament, long, dithecous, basifixed, introrse. **Gynoecium**–1, monocarpellary, superior, unilocular, ovules many, marginal, style short, stigma simple, **Fruit.** legume.

Floral formula– Br Brl ·|· ⚥ $K_{(5)}$, $C_{(5)}$, A_5, G_1—.

Economic Importance of Family

Economically the family is fairly important. A list of few important plants is given here.

1. *Saraca indica;* Eng.–Ashok tree; Verna–**Ashok**–This is a tree; native of India. It is grown as an ornamental in the gardens for its scarlet orange flowers. The plant is of great medicinal value.

2. *Saraca declinata*–This is a small tree. It is native of Sumatra. Grown as an ornamental for the bright orange-yellow flowers.

3. *Bauhinia variegata;* Syn. *B candida;* Eng.–Mountain ebony; Verna–**Kachnar**–This is a tree. The white flowers possess red and yellow stripes. The leaves and flower buds are eaten as vegetable. The bark is used for dyeing and tanning. This is found in Punjab, the Western Penisula and Assam.

4. *Bauhinia malabarica;* Eng–Malabar mountain ebony; Verna–**Amli**–This is a common tree of South India, Assam and Bengal. The leaves used for flavouring food stuff. The bark is used for tanning. The tender seeds and leafy tips are eaten as vegetable.

5. *Bauhinia purpurea;* Verna–**Kaliar, Lal Kachnar**–This is found in Northern India. Assam, the Khasia hills and the Western Peninsula. The bark is used for tanning. A dye and a fibre are obtained from its bark.

6. *Bauhinia tomentosa;* Verna–**Kachnar**–This is small tree. The flowers are large yellow and possess a purple dye. The bark is employed as cordage. This is commonly found in South India, Assam and Bihar.

7. *Bauhinia corymbosa;* Verna–**Safed Kachnar**–This is a large ornamental tree.

8. *Bauhinia acuminata*; Verna–**Safed Kachnar**–This is a small ornamental tree.

9. *Bauhinia racemosa;* Verna–**Kachnar, Gurial**–This is a small bushy tree. The flower buds and fruits are eaten as vegetable. The bark yields a fibre.

10. *Bauhinia galpinii*–This is a straggling or prostrate shrub. This is native of South and tropical Africa. This is grown in the gardens as an ornamental for its bright scarlet flowers.

11. *Bauhinia retusa;* Verna–**Kandla**–This is found in Bihar, Madhya Pradesh and Punjab. It yields a gum which is used in paper and textile industry.

12. *Bauhinia vahlii;* Verna–**Malijhan, Jallur**–This is a large climber. It is found in Assam, Madhya Pradesh, Bihar and the Punjab. The bark yields a fibre which is used for making ropes. The bark is also employed for tanning.

13. *Bauhinia monandra;* Eng.–Butterfly flowers-This is a small tree grown as a hedge plant.

14. *Haematoxylon campechianum;* Eng–Logwood; Verna–**Patang**–This is an ornamental thorny tree. It is native of Central America and West Indies. The heartwood contains haeamotoxylin, which is used in dyeing and as nuclear stain.

15. *Cassia fistula;* Verna–**Amaltas**–This is a tree, grown as an ornamental. The pulp of the fruits is used as a purgative and laxative.

16. *Cassia alata;* Verna–**Dadmurdan**– This is a small shrub. It is found in Tamil Nadu, Bengal. The leaves are used for ringworms and skin diseases. Also grown as an ornamental.

17. *Cassia grandis*–This is an ornamental and grown as a hedge plant in the gardens.

18. *Cassia angustifolia;* Verna–**Sena**–This is native of Arabia but now cultivated in Tamil Nadu and Karnataka. The leaves and fruits are used as a laxative and purgative.

19. *Cassia artemisioides*–This is a shrub grown as an ornamental.

20. *Cassia javanica*; Verna–**Java-ki-Rani**–This is small tree, grown as an ornamental for its pink flowers. It is native of Malaya.

21. *Cassia auriculata;* Verna–**Tarvar**–The bark is employed for tanning. It is found in Madhya Pradesh, the Western Peninsula, Tamil Nadu and Rajasthan.

22. *Cassia nodosa*–This is grown as an ornamental. The flowers are pink.

23. *Cassia siamea*–This is an ornamental tree. The flowers are used as vegetable.

24. *Cassia tora*; Verna–**Pamaar, Chukunda**–This is a herb or undershrub, found throughout our country. The seeds are employed in the preparation of sweets. The young leaves are eaten as vegetable.

26. *Cassia sophera;* Verna.–**Kasaundi**–This is herb grown as an ornamental in the gardens.

26. *Cassia renigera;* Eng–pink cassia–this is an ornamental grown as a hedge plant.

27. *Caesalpinia bonducalla;* Verna–**Katkaranj, Karanju**–This is a prickly shrub. The seeds are mixed with black pepper, and used as a tonic. The bark is also used as a tonic. The seeds are used as beads of necklaces, bracelets and rosaries.

28. *Caesalpinia digyna;* Verna–**Vakeri-mul**–This is a climbing shrub. The roots and fruits are employed for tanning. It is found in Bengal, Assam and Andaman islands.

29. *Caesalpinia bonduc;* Eng.–Fever nut–This is a scandent prickly shrub, cultivated as a hedge plant. The seeds are used medicinally to cure diarrhoea and rheumatism.

30. *Caesalpinia crista;* Syn.–*C. nuga*; Verna,– **Karanju, Naktamala**–This is a woody climbing shrub, commonly found in South India and Bengal. A kind of fat is obtained from the seeds which is used for making cosmetics.

31. *Caesalpinia sappan;* Verna–**Patang**–This is native of India and Malaya. It is now cultivated in Bengal and South India. A red dye is obtained from its heart wood. This dye is used for colouring cotton and wooleen cloth. The red ink is also prepared from it.

32. *Caesalpinia coriaria;* Verna–**Libi dibi**–This is a small tree. It is native of South America and West Indies. It is found abundantly in South India. It is also cultivated in North-Western India and Bengal. The dried fruits are used for tanning.

33. *Caesalpinia giliiesii;* Eng–Bird of paradise–This is a shrub. It is native of South America. It is grown as an ornamental.

34. *Poinciana pulcherrima;* Syn.–*Caesalpinia pulcherrima;* Eng–Peacock flower; Verna.–**Gulmohur**–This is a shrub or small tree, cultivated as an ornamental.

35. *Delonix regia;* Syn–*Poinciana regia*; Eng.–Peacock flower; Verna.–**Gulmohur**–This is a tree. It is native of Madagascar but now grown in our country as an ornamental for the large red flowers.

36. *Delonix elata*; Syn–*Poinciana elata*; Eng–white gulmohur; Verna.–**Sankesar**–This is a small tree; native of Abyssinia. It is grown as an ornamental for its yellow-white flowers.

37. *Parkinsonia aculeata;* Verna–**Vilayati Kikar**–This is a small tree. It is native of Panama. It is grown as a hedge plant. The flowers are yellow.

38. *Tamarindus indica;* Eng–Tamarind; Verna–**Imli**–This is a tree found throughout our country. It is grown as an avenue. The fruits are edible and are also used as carminative, and laxative. The bark and leaves are employed in tanning. The seeds yield jellose which is used for sizing jute and cotton. The polyose obtained from the seeds, is good substitute for food pectins. The unripe fruits are rich source of tartaric acid. Tartaric acid and its salts are extensively used in various foods, chemical and pharmaceutical industries.

Family—Mimosaceae

The family Mimosaceae includes about 40 genera, and at the same time has been considered to be the smallest one of all the three families of the order Leguminales.

Distribution: The plants of this family are almost exclusively tropical or sub–tropical in distribution. In our country this family is represented by many important genera, *e.g.*, *Acacia, Mimosa, Neptunia, Prosopis, Senegel*, etc.

Habit: They are either shrubs or trees very rarely herbs, sometimes climbers. Many of them are thorny and xeromorphic, *e.g.*, *Acacia, Prosopis;* hydrophytes are also found, *e.g.*, *Neptunia oleracea.* Mostly wild, but some are grown as sand binders to check soil erosion and desert extension.

Root: Tap and branched, deep rooted in the soil.

Stem: Erect, branched, terete, woody, solid. The stem tissue is often rich in tannin–sacs and gum passages.

Leaves: Alternate, petiolate, usually the base of petiole is provided with pulvinus; stipulate, usually stipules are modified into thorns; compound, pinnate and generally bipinnate. Leaf movements are common, reaching a climax in the ultra–sensitivity of *Mimosa pudica* and *Neptunia oleracea.* In some *Acacias* from Australia the leaves are bipinnate and fall earlier. The petioles are being flattended and developing into phyllodes.

Inflorescence: Racemose, head or spike.

Flower: Usually sessile, actinomorphic, regular, hermaphrodite, hypogynous, complete, small.

Calyx: 5 or 4, gamosepalous, the sepals are more or less connate, green (sepaloid), small, inferior, aestivation generally valvate.

Corolla: 5 or 4. free, polypetalous, valvate, inferior, slightly united towards the base, usually pentamerous.

Androecium: Usually indefinite stamens, sometimes the number is reduced to ten or even four (*e.g.*, in *Mimosa pudica*). Stamens are conspicuous, attractive, bright coloured somewhat scented. Anthers 2 celled, dorsifixed, dehiscence by longitudinal slit. The filaments are long and slender.

Gynoecium: Carpel one (monocarpellary); ovary superior, unilocular; marginal placentation; style long filiform; stigma terminal and simple.

Fruit: Legume or lomentum.

Seeds: Exalbuminous.

Pollination: Usually through the agency of insects (entomophily).

Floral formula: ⊕ ⚥ K (5), C(5), A ∝, <u>G1</u>.

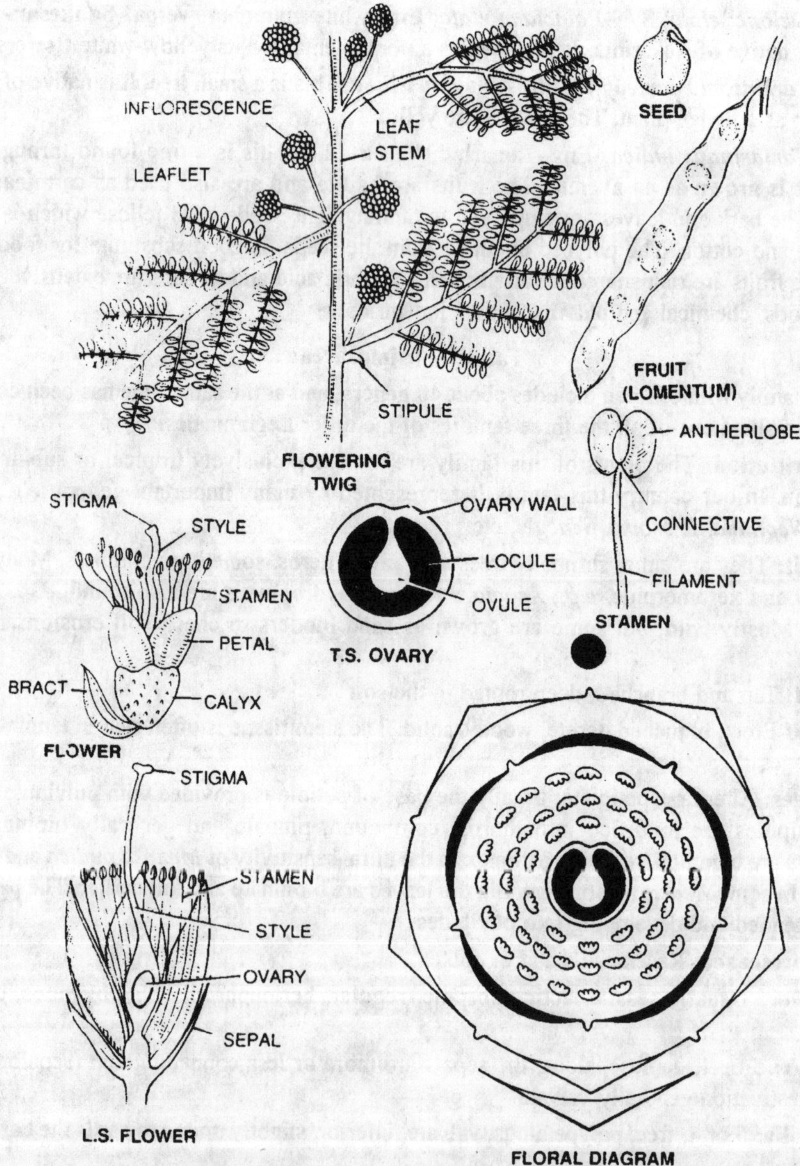

Fig. 10.5. Mimosaceae (Leguminosae). *Acacia nilotica* (Linn) Del

Bentham and Hooker (1862)	**Engler and Prantl (1931)**	**Hutchinson (1959)**
Phanerogams	Phanerogams	Angiospermae
Dicotyledones	Dicotyledones	Dicotyledones
Polypetalae	Archichlamydeae	Lignosae
Calyciflorae	Rosales	Leguminales
Rosales	Leguminosae	Mimosaceae
Leguminosae	Mimosoideae	
Mimosoideae		

Two important plants naming *Acacia nilotica* and *Mimosa pudica* have been described in this text in semi–technical language:

Acacia nilotica; Verna–**Babul, Kikar.**

Habit: A medium sized spiny tree.

Root: Tap, branched.

Stem: Erect, woody, solid, cylindrical, branched.

Leaf: Compound, bipinnate, stipulate, stipules modified into spines, petiolate, rachis possesses glands, leaflet oval, entire, obtuse, unicostate reticulate venation.

Inflorescence: The flowers are arranged in compound cymose heads.

Flower: Sessile, bracteate, actinomorphic, hermaphrodite, hypogynous, complete, fragrant, small yellow.

Calyx: Sepals 4 or 5, minute, gamosepalous, campanulate, valvate aestivation, inferior.

Corolla: Petals 4 or 5, gamopetalous, valvate aestivation, inferior.

Androecium: Stamens indefinite, polyandrous, long filaments, minute anthers, introrse, yellow.

Gynoecium: Monocarpellary, ovary superior, unilocular, marginal placentation, style long, stigma minute.

Fruit: A lomentaceous pod.

Floral Formula: ⊕ ⚥ K(4 – 5), C(4 – 5), A$\propto$ G$\underline{1}$.

Mimosa pudica, Verna–**Chhui–mui, Lajvanti.**

Habit: Perennial herb or shrub.

Root: Tap, branched.

Stem: Erect or prostrate, woody, solid, cylindrical, branched, prickly, densely clothed with deflexed bristles.

Leaf: Compound bipinnate, very sensitive, rachis beset with ascending bristles stipulate, stipules lanceolate, striate margins bristly, pinnae usually four, $2\frac{1}{2}$ to $3\frac{1}{2}$ inches long, leaflet, 12 to 20 pairs, $\frac{1}{2}''$ long, sessile, obliquely narrow–oblong, acute, subcoriaceous, glabrous above, appressed-bristly beneath.

Inflorescence: Heads usually in pairs.

Flower: Small, hermaphrodite, actinomorphic, complete, tetramerous, hypogynous.

Calyx: 4 sepals, minute, gamosepalous, campanulate, shortly toothed.

Corolla: 4 petals, gamopetalous, connate towards the base, small, inconspicous, valvate aestivation, inferior.

Androecium: 4 stamens, conspicuous, brightly coloured, free, filaments filiform, anthers small, bicelled, not gland crested.

Gynoecium: 1, monocarpellary, ovary superior, unilocular, marginal placentation, style filiform, stigma minute, terminal.

Fruit: Pod flat, membranous, joints 3–5 falling away when ripe from the sutures which are armed with weak spreading yellowish bristles.

Floral Formula: ⊕ ⚥ K(4), C(4), A4, G$\underline{1}$.

Economic Importance of the Family

The family is fairly important from the view point of its economic value. A list of important plants is given below:

1. *Acacia catechu;* Eng–Catechu; Verna–**Khair, Kattha, Khadira**–It is found in Himachal Pradesh, Punjab, Madhya Pradesh, Uttar Pradesh, Andhra Pradesh and Bihar. The Kattha industry is mostly located in Bareilly, Gwalior and Bombay. The Kattha tree also serves as a host for the lac insect. The Kattha, obtained from the heartwood is used as a masticatory and as a dyeing and preserving agent. Catechu is also used as a raw material for plywood adhesives. It is extensively used for dyeing canvas, fishing nets and ropes.

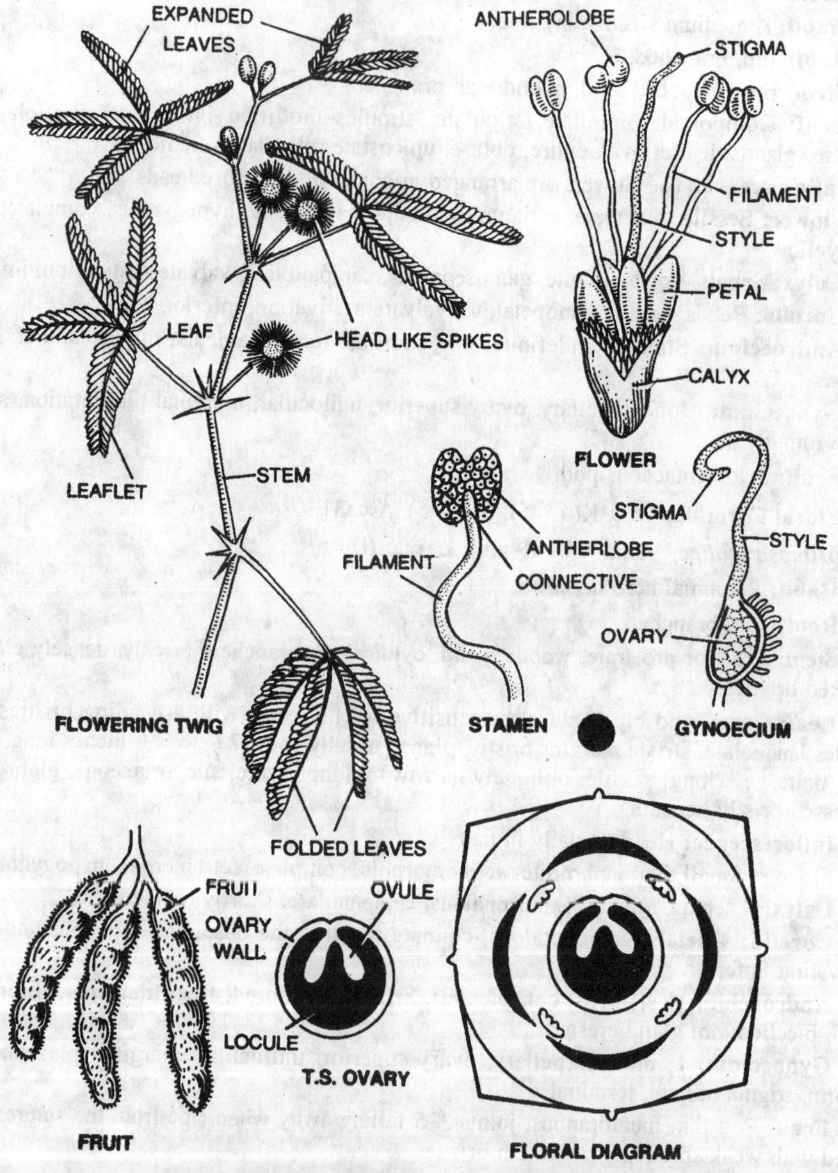

Fig. 10.6. Mimosaceae (Leguminosae), *Mimosa pudica* Linn.

2. *Acacia concinna*; Eng–Soap pod; Verna–**Shikakai, Banritha**–The fruits (pods) are used as a detergent for washing hairs, silk and woollen fabrics. It is found along the West Coast, Andhra Pradesh, Assam and Bihar.

3. *Acacia nilotica;* Syn. *A. arabica*; Eng.–Acacia; Verna–**Babul, Kikar**–This is a tree. The pods (fruits) are used as fodder. The bark is used as a tan. The gum obtained from the bark is of good quality, which is used in confectionery. Baskets are made from young, green twigs. The twigs are used as tooth greenbrushes(Datun). The bark is employed in tanning.

4. *Acacia senegel;* Syn. *A. verek*; Verna.–**Kumta**–This is native of North Africa but now found in the Punjab and Rajasthan. The stem bark, yields a gum which is used in the textile, mucilage, polish, paste and confectionery industries. This is also used as a glaze in painting.

5. *Acacia pycnantha*; Eng.–Goldern wattle–This is native of Australia, but now cultivated in the Nilgiris. The extract of bark is used as **Kattha.**

6. *Acacia decurrens* var. *dealbata;* Syn. *A. dealbata*; Eng.–Silver wattle–A perfume of good quality is obtained from the flowers. The bark is used for tanning and dyeing. It is native of Australia but now cultivated in the Nilgiris.

7. *Acacia decurrens* var. *mollis*; Syn. *A. mollissima.* Eng.–Black wattle–The bark is used for tanning and dyeing. The writing and printing papers are made from its wood pulp.

8. *Acacia farnesiana;* Eng.–Sweet acacia; Verna–**Gandh babul**–This is native of the West Indies. The fragrant flowers yield a perfume. It is found throughout our country.

9. *Acacia ferruginea;* Verna.–**Son Khair**–This is a tree found in Andhra Pradesh, Gujarat and the Western Ghats. The tree yields a gum of good quality.

10. *Acacia aneura*–This is native of Australia. The leaves are used as fodder.

11. *Acacia leucophloea;* Syn. *Mimosa leucophloea;* Verna–**Safed Kikar, Ronj Arinj**–The leaves are used as fodder. The bark yields a tan and a fibre. The bark is also used in the preparation of spirits from sugar and palm juice, acting a clarifying and flavouring agent. It is found in the Punjab, Rajasthan and Madhya Pradesh.

12. *Acacia auriculiformis*; Eng.–Australian Acacia–This is an evergreen tree. It is cultivated as an avenue for its beautiful phyllodes and flowers.

13. *Acacia jacquemontii*; Verna.–**Kinkar, Bouli**–The bark is used as a tan. The branches are used as fodder. It is also grown as a hedge plant.

14. *Albizia lebbeck;* Syn. *Mimosa lebbeck*; *Acacia lebbeck*; Verna.–**Siris.**–This is a common roadside tree. The leaves and twigs are used as fodder. They are grown for shade in tea and coffee plantations. The wood is used for furniture, railway carriages, structural work and interior fittings. The fallen leaves make good manure.

15. *Albizia chinensis;* Syn. *A. stipulata*; Vern.–**Siran**–It is a tree, found in the sub-Himalayan tract, the Western Peninsula and the Nicobar islands. The leaves and twigs are used as fodder. The writing and printing papers are made from wood pulp.

16. *Albizia procera;* Verna.–**Safed siris**–The wood is stronger than that of *A. lebbeck*. The bark yields a poison. The wood is used in making furniture, etc.

17. *Albizia odoratissima*; Syn. *Mimosa odoratissima*; Eng.–black siris; Verna.–**Kala siris.**–The leaves and twigs are used as fodder. The wood is used for making furniture and burnt as fuel.

18. *Albizia amara*; Syn. *A. wightii; Mimosa amara*; Verna.–**Lallei.**–The wood is used for fuel and making tool handles. The dried leaves are used as substitute of soap.

19. *Albizia lucida*; Verna.–**Galwang**–It is found in Assam and Bihar. It makes an important host of lac insect.

20. *Mimosa pudica*, Eng.–Sensitive plant; Verna.–**Lajwanti, Chhui–mui.**–A spreading undershrub, grown as an ornamental.

21. *Mimosa rubicaulis;* Verna.–**Shiah–Kanta, Aila**–This is a shrub, grown as a hedge plant.

22. *Neptunia oleracea*; Syn. *N.prostrata*; Verna.–**Lajalu**–A common aquatic herb. It is used as a vegetable.

23. *Prosopis juliflora*; Syn. *Mimosa juliflora*; Verna. –**Kabuli kikar**–This is a tree. It is native of Mexico and Central America. It is grown as a hedge plant. It is also being planted to check the extension of the desert. The fruits are used as fodder. This is also planted as windbreak.

24. *Prosopis spicigera*; Verna–**Chhonkar Jand**–This is grown as a hedge plant.

25. *Entada phaseoloides*; Syn. *E. scandens; Lens phaseoloides*; Verna.–**Barabi, Chian, Gilla**–This is a large climbing shrub. The seeds are edible. They are also used as a substitute of soap.

26. *Xylia xylocarpa*; Syn. *Xylia dolabriformis* ; Verna–**Jambu**--This is a tree found in Madhya Pradesh, Andhra Pradesh and Orissa. The wood is used for construction work, railway–sleepers, ship–building and bridges.

27. *Pithecolobium dulce*; Syn. *Mimosa dulcis*; *Inga dulcis*; Eng.–Manila tamarind; Verna.–**Jangal jalebi**–This is a tree cultivated throughout India for its edible fruits. They are also grown as hedge plants.

Family–Papilionaceae

The family Papilionaceae is the largest of the three families of the order Leguminales. This includes about 375 genera.

Distribution: The members of the family are being distributed in the temperate regions of both northern and southern hemispheres. In our country the Papilionaceae is represented by many important genera commonly found in hills and plains. Usually the herbs belonging to this family appear and flower in winter season in the Indian plains.

Habit: Generally the plants are herbs or shrubs (often climbing), very rarely trees. Sometimes they are hydrophytes and xerophytes. Most of the members are cultivated in the fields which yield pulses.

Root: The root is tap, branched, bearing nodules containing nitrogen fixing bacteria (*Rhizobia*), and therefore, the plants are being used for green manuring.

Stem: The stem is erect or climbing (it climbs by means of tendrils), branched, angular or cylindrical herbaceous or woody.

Leaves: The leaves are alternate, opposite or whorled, usually compound (digitate or pinnate), rarely even pinnate, sometimes simple, stipulate, stipules occur at the base of the petiole (foliaceous and large), sometimes secondary stipules arise at the base of individual leaflets.

Inflorescence: The inflorescence is racemose, the raceme, the spike, and the contracted raceme or head.

Flower: The flowers are usually pedicellate, zygomorphic, irregular, hermaphrodite (bisexual), complete, perigynous and papilionaceous.

Calyx: It consists of sepals, gamosepalous (united sepals), equal or unequal, below the disc united in a tubular calyx, 5 toothed or five lobed or bi–labiate (the two upper and three lower may unite). Aestivation ascending imbricate.

Corolla: The corolla consists of 5 petals (unequal), the uppermost and the largest petal is known as **standard** or **vexillum**; the two free lateral petals are known as **wings or alae**; the anterior pair of united petals is termed **keel** or **carnia**, this encloses stamens and pistil. They are of various colours. Venation is conspicuous. Aestivation descending imbricate.

Androecium: The stamens are usually 10, inserted on a disc below the calyx, they may be in two bundles (diadelphous) of 9 + 1 or 5 + 5, or in one bundle (monadelphous), rarely free. Anthers bi–celled, dorsifixed, dehiscence by longitudinal slits.

Gynoecium: Carpel one, free, ovary superior, stalked or sessile, unilocular, marginal placentation, style, bent and short, hairy, stigma simple.

Fruit: legume or pod, splitting along both dorsal and ventral sutures. In some species it is indehiscent lomentum.

Seeds: Many, exalbuminous, usually flattened.

Pollination: Entomophilous(insect pollinated)

Floral formula: ⊹ ⚥ K(5), C 1+2+ (2), A(9)+1, G1.

Bentham and Hooker(1862)	**Engler and Prantl (1931)**	**Hutchinson(1959)**
Phanerogams	Phanerogams	Angiospermae
Dicotyledones	Dicotyledones	Dicotyledones
Polypetalae	Archichlamydeae	Lignosae
Calyciflorae	Rosales	Leguminales
Leguminosae	Papilionaceae	
Papilionatae		

Some important plants have been described here in semi–technical language.

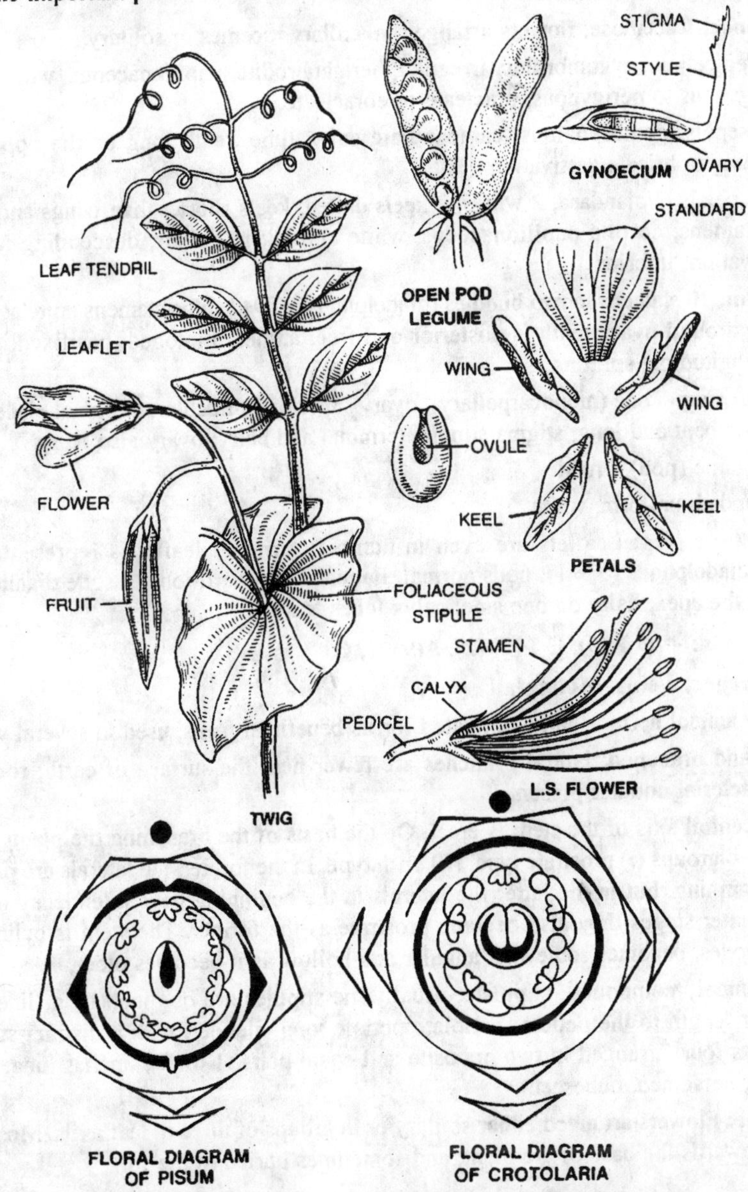

Fig. 10.7. Papilionaceae (Leguminosae)–*Pisum sativum* Linn.

Pisum sativum; Verna.–**Matar**.

Habit: Annual herb; cultivated.

Root: Tap, branched, with nodules containing nitrogen fixing bacteria (*Rhizobium radicicola*).

Stem: Herbaceous, weak, climbing with the help of leaf tendrils, cylindrical, branched, smooth, glaucous.

Leaves: Alternate, compound, imparipinnate, stipulate (stipules large foliaceous, ovate, semicordate, irregularly toothed at the base), leaflets 4 or 6, the common rachis ends in a branched tendril; the leaflets entire, smooth, net veined, oval to oblong, mucronate tips, green and glaucous, the terminal leaflet is always a tendril.

Inflorescence: Racemose, flowers arranged in axillary racemes or solitary.

Flower: Pedicellate, zygomorphic, irregular, hermaphrodite, papilionaceous, white or pink, complete, hypogynous to perigynous, bracteate or ebracteate.

Calyx: 5 sepals, gamosepalous, campanulate calyx tube, teeth long or the upper short; sepaloid, ascending imbricate aestivation.

Corolla: 5 petals, 1 standard, 2 wings, 2 keels united, keels shorter than wings and enclose the pistil and stamens; corolla papilionaceous, white or pink in colour; descending imbricate (vexillary) aestivation; inferior.

Androecium: 10 stamens in two bundles (diadelphous) of 9+1, nine stamens unite at the base and form a tube around ovary tenth is posterior and free; anthers bi--lobed, basifixed, introrse, dehiscence by longitudinal splitting.

Gynoecium: Carpel one (monocarpellary); ovary superior, unilocular; marginal placentation; ovules many; style bent and long, stigma simple; terminal and hairy; ovary also hairy.

Fruit: A legume (pod), broad.

Seeds: Rounded, uniform, white.

Generic Characters: Leaflets are even in number; terminal leaflet is represented by a tendril; stamens diadelphous (9 + 1); pods normal, flattened, not partitioned ; style dialated from the base towards the apex, hairy on one side; calyx lobes leafy.

Floral Formula: ·|· ⚥ K(5), C 1+2+(2), A(9)+1, $\underline{G}1$

Arachis hypogea; Verna–**Moongfali**.

Habit: Low annual herb; usually cultivated for its beneficial fruits, used in several ways.

Root: Tap and branched; lateral branches are fewer near the surface of earth; root hairs usually absent; bacterial nodules present.

Stem: The central axis of the stem is erect. On the basis of the branching the plants can be subdivided into two groups (*i*) prostrate type, (*ii*) bush type. In the former, the laterals are prostrate from the very beginning, but in the latter the laterals in the beginning have a tendency towards erectness, but in later stages they also become prostrate as the former. The stem is cylindrical, hairy, short internodes, becomes somewhat angular and hollow as it becomes older.

Leaves: Pinnately compound, stipulate, usually the stipules are one inch long, linear and fused half of their length to the petiole; petiolate petiole long, slender, with a distinct swelling (pulvinus.); leaflets four, arranged in two opposite and equal pairs. 1 to $2\frac{1}{2}$ inches long, entire, obovate to obtuse, netveined, pubescent.

Inflorescence: Flowers arranged either solitary or in groups of three in leaf axils. Mostly the flowers develop towards the base of the plant, and sometimes buried in the soil.

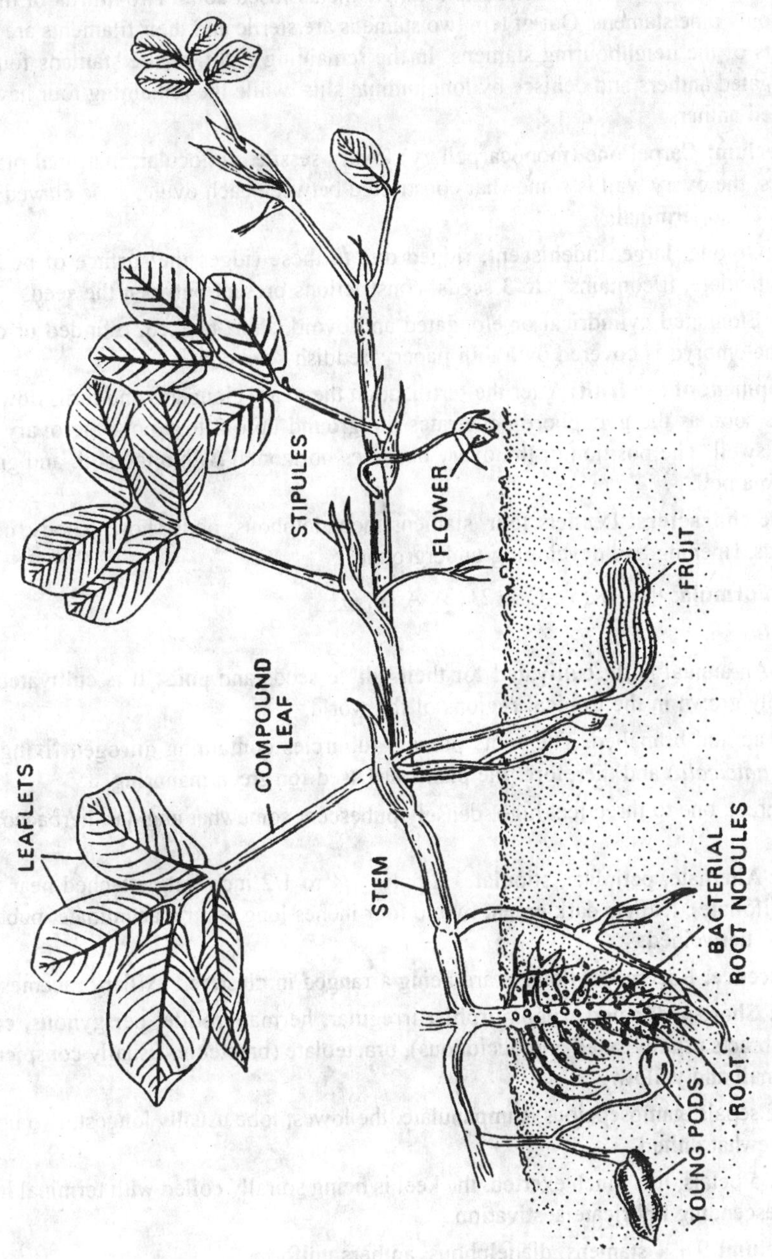

Fig. 10.8. Papilionaceae–*Arachis hypogea* (Verna–Moongfali)–A, a complete plant; B, floral diagram.

Flower: Pedicellate (small pedicels), pedicel usually hairy, zygomorphic, irregular, hypogynous, bisexual, complete, papilionaceous.

Calyx: 5 petals, gamosepalous, the sepals unite to form a calyx tube approximately one inch long; the free lobes are short and irregular.

Corolla: 5 petals, free, one standard, two wings and two keels united, united keels form a beak like structure; the corolla is inserted on the rim of the calyx tube, yellow, papilionaceous

Androecium: 10 stamens, monadelphous, filaments fused about two–thirds of their length. Sometimes only nine stamens. Out of ten, two stamens are sterile and their filaments are fused with the filaments of the neighbouring stamens. In the remaining eight, fertile stamens four have bi-celled, elongated anthers and dehisce by longitudinal slits, while the remaining four have rounded and elongated anthers.

Gynoecium: Carpel one (monocarpellary), ovary sessile, unilocular, marginal placentation, 1 to 5 ovules, the ovary wall is somewhat constricted between each ovule; style curved and hairy, stigma simple and terminal.

Fruit: Oblong, large, indehiscent, ridged due to these ridges the surface of pod becomes somewhat reticulate. It contains 1 to 3 seeds, constrictions present between the seeds.

Seeds: Elongated cylindrical or elongated and ovoid, short beaked, rounded or diagonally flattened. The embryo is covered by a thin papery, reddish seed coat.

Development of the fruit: After the fertilization the ovary elongates and turns down towards the earth. As soon as the gynophore penetrates the ground upto 2 to 5 cm., the ovary begins to develop and swell. The position of the ovary becomes horizontal within the soil, and gradually it develops into a pod.

Generic characters: Leaflets four, stamens monadelphous; pods short. woody, constricted between seeds, ripening and developing underground.

Floral Formula: ·|· ⚥ K(5), C 1+2+(2), A 10, $\underline{G1}$.

***Phaseolus* sp.**

Habit: An annual herb; cultivated for their edible seeds and pods. It is cultivated in rainy season. Mainly grown in the tropical regions of the world.

Root: Tap and branched. The roots possess tuburcles containing nitrogen fixing bacteria (*Rhizobium radicicola*) and therefore, the plants are used for green manuring.

Stem: Erect, one to three feet high, densely pubescent somewhat angular, herbaceous, green branched.

Leaves: Alternate, petiolate, stipulate, stipules 1/4 to 1/2 inch long attached near the base, compound (trifoliate), leaflets dark green, two to four inches long, entire, acuminate, pubescent on both the sides, net veined.

Inflorescence: Racemose, flowers are being arranged in clustered axillary racemes.

Flower: Short peduncled, zygomorphic, irregular, hermaphrodite, perigynous, complete, bracteate(the bracts usually small and deciduous), bracteolate (bracteoles usually conspicuous and persistent); small and yellow.

Calyx: 5 sepals, gamosepalous, campanulate; the lowest lobe usually longest, two uppermost lobes are somewhat united.

Corolla: 5 petals, it is much exerted, the keel is being spirally coiled with terminal horn–like appendage; descending imbricate aestivation.

Androecium: 9 + 1 stamens, diadelphous, anthers uniform.

Gynoecium: One carpel (monocarpellary), ovary superior, unilocular, sessile many ovuled, marginal placentation; style hairy, filiform, twisted round with the keel; stigma oblique and terminal.

Fruit: Legume or pod; linear rarely oblong, green, very often separate between the seeds, usually $1\frac{1}{2}$ to 2 inches long.

Seeds: Exalbuminous, with thick cotyledons, green (*Phaseolus aureus*–**Moong**) or black (*P. mungo*–**Urd**).

Generic Characters: Leaflets three in number; stamens diadelphous; style hairy; inflorescence racemose in clusters;calyx lobes equal to the tube; keel of corolla coiled.

Floral Formula: ·|· ⚥ K(5), C 1+2+(2), A 9+1, G1.

Cajanus cajan; Verna.–**Arhar**.

Habit: It is an erect branched shrub 4 to 10 feet high; cultivated for edible seeds, in tropical and sub–tropical regions.

Root: Tap, branched with bacterial nodules.

Stem: Erect, much-branched, solid, woody, green pubescent.

Leaves: Alternate, petiolate, compound, trifoliate (unipinnate, unijugate), gland dotted below; leaflets are lanceolate to narrow elliptical, two to four inches long, acuminate, softly pubescent on both surfaces, net veined, stipulate, the stipules are minute and caducous.

Inflorescence: Racemose, corymbose raceme or a terminal panicle.

Flower: Pedicellate, zygomorphic, irregular, hermaphrodite, complete, perigynous, yellow or orange in colour but brownish at the back.

Calyx: Two upper lobes united, calyx tube campanulate, toothed or lobed, teeth short.

Corolla: 5 petals, polypetalous; the standard is large and broad, auricled; the keel is obtuse and incurved; descending imbricate aestivation; yellow or pink in colour.

Androecium: Stamens 9+1, diadelphous; anthers uniform, bicelled.

Gynoecium: Carpel one (monocarpellary), ovary superior, pubescent, unilocular, few ovuled, sessile, placentation marginal; style long; thickened, much incurved, beardless; capitate, terminal stigma.

Fruit: Pod or legume (lomentaceous pod), many, compressed, with diagonal depressions, two or three inches long by half an inch broad, pubescent and beaked.

Seeds: Three to five in a single pod; small sized, orbicular and compressed one edge flattened; brown coloured with a white small hilum.

Generic Characters: Leaves gland dotted beneath, leaflets lanceolate oblong, leaflets two inches or more; pods with depressed lines between the seeds; ovules 3 to 5; seeds not strophiolate.

Floral Formula: ·|· ⚥ K(5), C1 + 2+ (2), A9 + 1, G1

***Crotalaria* sp.**

Habit: The plants are either herbs or shrubs; cultivated for fibre (*Crotalaria juncea*–Indian sunn hemp–Verna–**San**) and seeds; some are ornamentals.

Root: Tap and branched; nodulated, nodules contain nitrogen fixing bacteria.

Stem: Erect, branched, cylindrical, pubescent, green, sometimes grooved.

Leaves: Alternate, petiolate, simple or compound, if compound it is trifoliate rarely 5 to 7 foliate; digitate (usually all leaflets born at the apex of the petiole); stipulate (minute stipules); leaflets are linear or oblong, entire, shining silky hairs on both the sides; net veined.

Inflorescence: Flowers arranged in solitary or racemose inflorescences; the flowers are being arranged in either terminal, lateral or leaf opposed racemes.

Flower: Pedicellate (short peduncle), zygomorphic, irregular, hermaphrodite, complete, perigynous, bracteate, usually large and showy.

Calyx: 5 sepals, gamosepalous, sub–bilabiate, the upper lip bifid, the lower cordate trifid; the teeth being linear or lanceolate.

Corolla: 5 petals, polypetalous; standard usually broad and shortly clawed; the wings obovate oblong; the keel is broad and nearly equal to wing in size, petals of keel are united, incurved and beaked; papilionaceous; descending imbricate aestivation; usually yellow coloured, sometimes pink and purplish; big and showy.

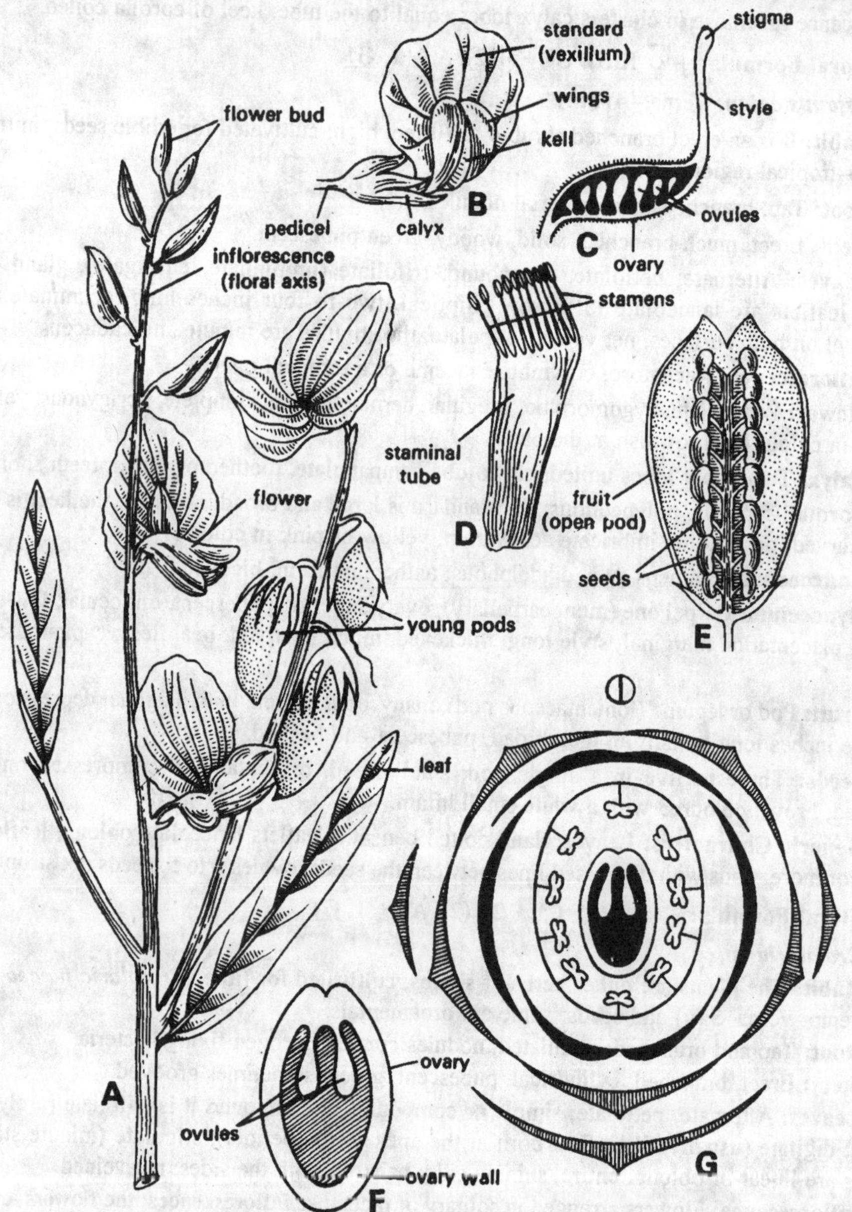

Fig. 10.9. Papilionaceae–*Crotalaria juncea*, A, flowering twig with young pods; B, flower; C, L.S. gynoecium; D, androecium; E, dehisced fruit; F, T.S. of ovary; G, floral diagram.

Androecium: 10 stamens, monadelphous, connate, the anthers are 2–celled, dimorphous, dehiscence by longitudinal slits.

Gynoecium: One carpel (monocarpellary), ovary superior, unilocular, many ovuled sessile; style long, strongly incurved or reflexed, somewhat bearded (hairy); the stigma is small and oblique; placentation marginal.

Fruit: Legume or pod, globose or oblong sessile, turgid or inflated, with many loose seeds.

Seeds: Exalbuminous, small, usually flat, different coloured.

Generic Characters: Leaflets 3, 5 or 7; leaf simple or digitate (leaflets arranged at the apex of the petiole); inflorescence–solitary or racemose; stamens monadelphous; dimorphous; pod swollen or inflated.

Floral Formula: ·|· ⚥ K(2) + (3), C1 + 2 + (2), A 10, $\underline{G1}$.

Cicer arietinum; Verna–**Chana.**

Habit: Annual herb, one to two feet high, cultivated for its seeds which are used as pulses and forage.

Root: Tap root, fairly stout, deeply penetrating the soil and carries many laterals on the upper part. The roots possess nodules containing nitrogen fixing bacteria (*Rhizobium radicicola*).

Stem: Erect, much branched viscous hairy, terete herbaceous, green, solid.

Leaves: Petiolate; compound, bipinnate, stipulate (with large ovate strongly serrate stipules); rachis one to two inches long with small toothed strongly nerved leaflets; leaflets are 9 to 16 in number and each about half an inch long, opposite or alternate with a terminal leaflet, oblong or obovate in outline.

Inflorescence: Solitary axillary, cymose.

Flower: Small; pedicellate (long pedicels in the leaf axils); white or pinkish in colour; zygomorphic, irregular, hermaphrodite, perigynous, complete, papilionaceous.

Calyx: 5 sepals, gamosepalous, calyx-tube oblique, teeth sub-equal lanceolate, sepaloid.

Corolla: 5 petals, polypetalous; standard broad; clawed; wings free; petals of keel united.

Androecium: 9+1 stamens, diadelphous; anthers bi–celled bursting longitudinally.

Gynoecium: Carpel one (monocarpellary), ovary superior, sessile, 2 or more ovuled, unilocular, marginal placentation, style incurved or reflexed, glabrous, stigma capitate.

Fruit: Pod, sessile, oblong, turgid, hairy, pointed, about three fourth an inch long, narrowed into the persistent style.

Seed: Obovate or subglobose, reddish brown, beaked; hilum is visible.

Generic Characters: Herbs, leaf pinnate, the terminal leaflet present, margins of leaflets serrate all–round; stamens diadelphous, style glabrous; pod turgid, conical.

Floral formula: ·|· ⚥ K(5), C1 + 2 + (2), A(9) + 1, $\underline{G1}$.

***Melilotus* sp.**

Habit: An annual herb.

Root: Tap, branched with bacterial nodules.

Stem: Erect, herbaceous, branched, cylindrical, solid green, glabrous.

Leaf: Compound trifoliate, alternate, cauline, petiolate, stipulate (stipules linear), obovate, acuminate, margins serrate, mucronate apex, unicostate reticulate venation.

Inflorescence: Racemose, typical raceme.

Flower: Pedicellate, bracteate, minute, hermaphrodite, zygomorphic (irregular), complete, yellow, hypo or perigynous.

Calyx: Five sepals, gamosepalous, nearly equal lobes, lanceolate, valvate aestivation, odd sepal posterior.

Corolla: Five petals, polypetalous, yellow, one standard, two wings, two innermost used forming keel, aestivation descending imbricate.

Androecium: Ten stamens, diadelphous, filaments of 9 stamens fused forming one bundle, tenth stamen free (9+1), anthers dorsified.

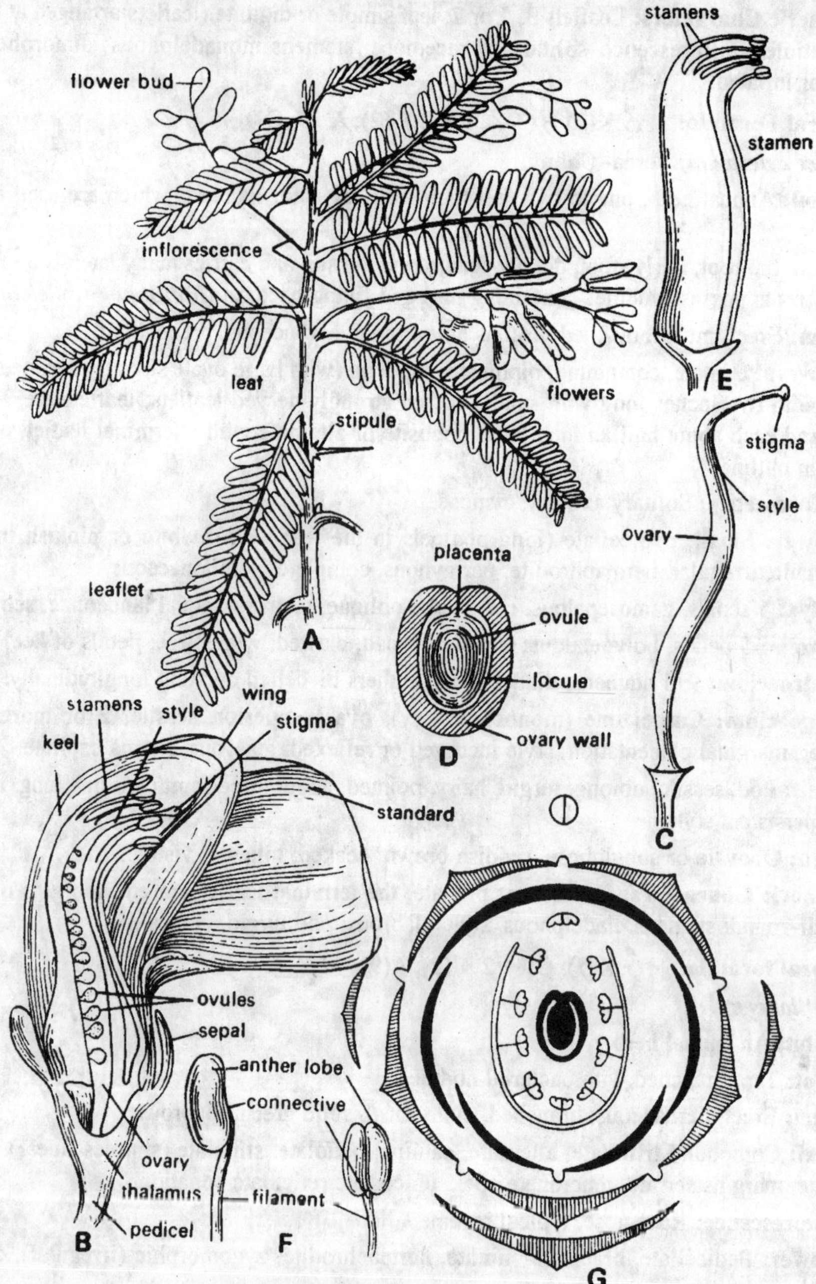

Fig. 10.10. Papilionaceae. *Sesbania sesban* (L.) Merr. (**Jayanti**). A, flowering twig; B, L.S., of flower; C, gynoecium; D, T.S. of ovary; E, androecium; F, stamens; G, floral diagram.

Gynoecium: One carpel (monocarpellary), ovary superior, unilocular, marginal, placentation.

Fruit: A pod.

Floral Formula: ·|· ⚥ K(5), C1 + 2 + (2), A(9) + 1, <u>G1</u>.

Sesbania sesban (L) Merr.; Verna **Jayanti, Jait. Stem**–herbaceous and woody, aerial, erect, cylindrical, branched, solid, smooth, green. **Leaf**–cauline, ramal, alternate, stipulate, leaf base pulvinus, compound, unipinnate; pinna oblong, entire, mucronate, unicostate reticulate. **Inflorescence**–racemose, axillary raceme. **Flower**–bracteolate, pedicellate, complete, zygomorphic, hermaphrodite, pentamerous, perigynous, cyclic, **Calyx**–5, gamosepalous, persistent, green, valvate. **Corolla**–5, polypetalous, vexillary, papilionaceous, yellow. **Androecium** (9)+1, diadelphous, dithecous, dorsifixed, introrse. **Gynoecium**–monocarpellary, half-inferior, unilocular, marginal, style long, stigma capitate. **Fruit**–legume pod. **Floral formula** Br Brl ⚬|⚬ ⚥ K5, C5, A(9) + 1, G1.

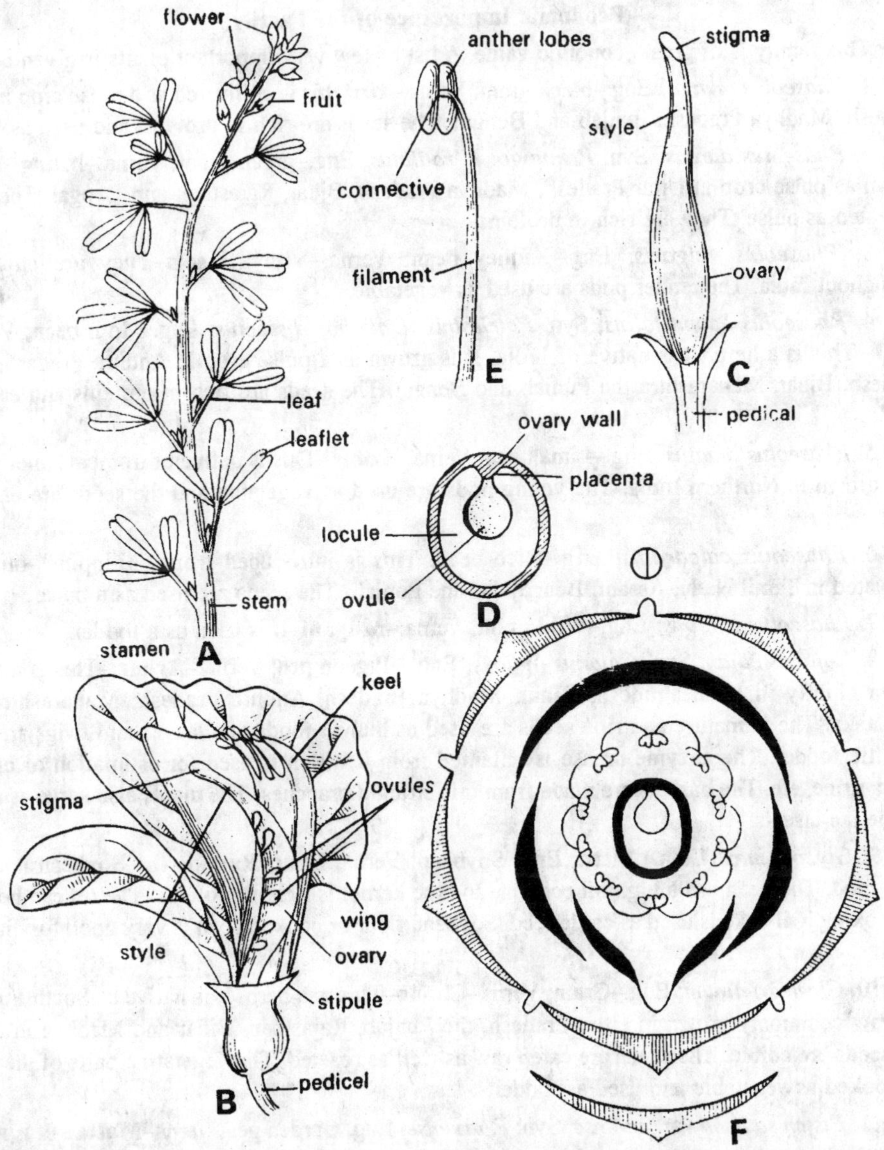

Fig. 10.11. Papilionaceae. *Crotalaria medicaginea* Lamk. (**Gulabi**). A, flowering twig; B, L.S, of flower; C, gynoecium; D, T.S of ovary; E, stamen; F, floral diagram

Crotalaria medicaginea Lamk; Verna **Gulabi. Stem**–herbaceous, aerial erect, branched, cylindrical, solid, pubescent, green. **Leaf**-cauline, ramal, alternate, stipulate, palmately compound trifoliate, petiolate, leaf base pulvinus, leaflet obovate, entire, hairy. **Inflorescence**–racemose, raceme. **Flower**–pedicellate, complete, zygomorphic, hermaphrodite, pentamerous, perigynous, cyclic. **Calyx**–5, gamosepalous, green, valvate. **Corolla**–5, polypetalous, vexillary, papilionaceous, yellow.

Androecium–(10), monadelphous, dithecous, dorsifixed, introrse. **Gynoecium**–1, monocarpellary, half inferior, unilocular, many ovules, marginal, style long, stigma capitate. **Fruit**–legume, pod, **Floral formula**– Br Brl ·|· ⚥ $K_{(5)}$, C_5, $A_{(10)}$, G1–

Economic Importance of the Family

This family is of great economic value. A list of few very important plants is given below :

1. *Phaseolus mungo;* Eng.–black gram; Verna.–**Urd** this is cultivated as a pulse crop in Uttar Pradesh, Madhya Pradesh, Punjab and Bengal. The seeds are rich in proteins and used as **'dal'**.

2. *Phaseolus aureus;* Syn. *P. mungo*; *P. radiatus*; Eng.–green gram; Verna.–**Mung.** This is grown as pulse crop in Uttar Pradesh, Madhya Pradesh, Bihar, Rajasthan and Bengal. The seeds are eaten as pulse. They are rich in proteins.

3. *Phaseolus vulgaris;* Eng.–Kidney bean; Verna.–**Vilayati sem**–They are cultivated throughout India. The tender pods are used as vegetable.

4. *Phaseolus aconitifolius*; Syn. *P. trilobus; Dolichos dissectus*; Eng. Moth bean; Verna.–**Moth**–This is a **herb**. It is native of India. It is grown as a pulse crop in Andhra Pradesh, Uttar Pradesh, Bihar, Maharashtra the Punjab and Bengal. The seeds are rich in proteins and eaten as **'dal'**.

5. *Phaseolus lunatus*; Eng.–Lima bean; Verna.–**Lobia**–This is native of tropical America but now grown in Northern India. The young pods are used as vegetable and the seeds are eaten as pulse.

6. *Phaseolus calcaratus*; Eng.–Rice bean. This is introduced from Philippines but now cultivated in Tamil Nadu, Assam, Bengal and the Punjab. The seeds are used as a pulse.

7. *Phaseolus trilobus*; Eng.–Wild gram; Verna.–**Mugani**–It is used as a fodder.

8. *Cajanus cajan*; Syn. *Cajanus indicus*; Eng.—Pigeon pea; Verna.–**Arhar**.–This is a shrub, grown chiefly in Uttar Pradesh, Bihar, Madhya Pradesh, Andhra pradesh, Maharashtra and Karnataka. The immature and ripe seeds are used as human food. The leaves and twigs are used as cattle fodder. The enzyme urease is obtained from it which is used for estimation of urea in blood, urine, etc. The baskets are made from thin straight branches. The plants also serve as a host for the lac insects.

9. *Glycine max* (Linn.) Merr.; Eng. Soybean; Verna. **Bhat, Ramkurthi**–Soybean makes a good food. The bean has a high oil content, and the kernel is rich in proteins. The oil can be used for cooking. Oil-cake is used as cattle feed. Soybean flour or cooked bean is very good for diabetic patients.

10. *Cicer arietinum*; Eng.–Gram; Verna–**Chana**–This is a herb. It is native of South Europe, but now commonly grown in Uttar Pradesh, the Punjab, Rajasthan, Bihar and Madhya Pradesh. The seeds are edible. The seeds are eaten raw as well as roasted. The vegetative parts of the plant are cooked as vegetable and used as fodder.

11. *Pisum sativum* var. *arvense*; Syn. *P. arvense*; Eng. Garden pea; Verna–**Mattar**–It is native of South Europe but now cultivated in our country chiefly in the Punjab, Uttar Pradesh, Delhi and Himachal Pradesh. The seeds and buds are eaten as vegetable. They are quite rich in proteins.

12. *Lens culinaris*; Syn. *L. esculenta*; *Ervum lens*; *Eng.* Lentil; Verna–**Masur**–This is native of South West Asia but now cultivated in our country chiefly in Andhra Pradesh, Uttar Pradesh, Rajasthan and Bengal for its edible seeds. The seeds are consumed as pulse. The leaves and stems are used as fodder.

13. *Dolichos lablab*; Syn. *Lablab niger*; Eng. Lablab; Verna–**Sem**–This is a twining herb. It is cultivated throughout our country. The green pods and seeds are consumed as vegetable.

14. *Dolichos biflorus*; Eng. Horse gram; Verna–**Kulthi**–This is a trailing herb. It is cultivated throughout our country as fodder crop.

15. *Canavalia gladiata*; Syn. *Dolichos gladiatus*; Eng.–Jack bean; Verna–**Bara sem**–It is cultivated throughout our country. The pods are used as vegetable. The plant is also used as green manure.

16. *Lupinus albus*; Eng–white lupine; Verna–**Turmas**–This is a herb, grown for fodder and green manure.

17. *Cyamopsis tetragonoloba*; Eng.–Cluster bean; Verna–**Gwar**–This is a herb or small shrub. It is cultivated throughout our country. The pods are eaten as vegetable and used as fodder. The seeds are the source of a gum which is used in food, paper and textile industries.

18. *Melilotus alba*; Eng–White sweet clover. This is a herb. It is native of Eurasia, but now grown in Northern India. It is used as fodder.

19. *Melilotus indica*; Syn. *M. parviflora*; Eng.–Sweet clover; Verna–**Banmethi, Senji**–This is native of Europe and Asia but now grown in the Punjab and Uttar Pradesh as a winter fodder crop.

20. *Medicago sativa*; Eng.–**Alfalfa**; Verna–**Vilayati Gawuth**–This herb is used as fodder.

21. *Medicago denticulata*; Eng.–Toothed, bur clover; Verna–**Maina**–This is a common herb and used as fodder.

22. *Trifolium alexandrium*; Eng.–Egyptian clover : Verna–**Berseem**–This is a herb grown as green manure and fodder crop.

23. *Trifolium resupinatum*; Eng.–Persian clover; Verna–**Shaftal**–It is used as green fodder.

24. *Trifolium pratense*; Eng–Red clover; Verna–**Trepatra.** This is used as fodder. A yellow dye is also extracted from the flowers.

25. *Trifolium repens*; Eng.–White clover; Verna–**Shaftal** It is used as fodder. This is grown in the temperate Himalayas and the Nilgiris.

26. *Trigonella foenum–graecum*; Verna–**Methi**–This is a herb. It is native of South Europe, but now grown chiefly in Northern India. The leaves are used as vegetable and fodder. The seeds are used as spice and condiment. The seeds are also used medicinally as carmintive and tonic.

27. *Vicia faba*; *Faba vulgaris*; Eng.–Broad bean, Verna–**Bakla**–This is a herb commonly grown in North Western India. It is used as fodder. The seeds are eaten as vegetable.

28. *Vigna sinensis*; Eng–Cowpea. Verna–**Lobia, Chowli**–It is cultivated throughout our country. The young pods and seeds are used as a vegetable and pulse.

29. *Abrus precatorius*; Verna–**Geumchi, Rati**–This is a climbing shrub. The seeds are used as weights by goldsmiths. They are also used as beads for necklaces.

30. *Butea monosperma;* Syn. *B. frondosa*; Eng.–Flame of the forest; Verna–**Dhak, Palas, Tesu**–This is a small tree. This serves as a host for the lac insect. The flowers yield a yellow dye. The leaves are used as substitute for plates chiefly on festive occasions. The ripe fruits are edible.

31. *Indigofera tinctoria*; Eng.–Indigo; Verna–**Nil**–This is a common small shrub, grown as green manure crop. The plant is of great medicinal value.

32. *Arachis hypogea*; Eng.–Groundnut; Verna–**Mungphali.** This is a native of Brazil but now widely grown in South India, Maharashtra and Uttar Pradesh. The nuts (seeds) are used for the preparation of peanut butter. They are also eaten after roasting. The filtered refined oil is used for cooking and in making margarine. The oil cake is used as fodder. The protein obtained from groundnuts is used in the manufacture of ardil, a synthetic fibre.

33. *Aeschynomene indica*; Verna–**Phulan**–The soft wood is used for making sola hats. It is found in Kashmir, Assam, Bengal and South India.

34. *Aeschynomene aspera*; Eng. Sola pith; Verna–**Sola**–This is a tall herb, found in wet places in Bengal and South India. The soft wood is used for making sun-hats, toys, models, artificial flowers, etc. It is used in the manufacture of swimming jackets and life-belts. The leaves are edible.

35. *Alhagi pseudalhagi*; Syn. *A. maurorum A. camelorum* ; Eng.–Camel's thorn; Verna–**Jawasa**–This is a spiny shrub of the deserts. It is used as camel fodder. The twigs are used in making screens.

36. *Dalbergia sissoo*; Verna–**Shisham**–it is found in the Punjab, Uttar Pradesh, Bengal and Assam. The wood is used for making furniture and carving. The pulp is used for making writing and printing papers.

37. *Dalbergia latifolia*; Verna–**Kala Shisham**- This is a tree found in Bengal, Bihar, Madhya Pradesh and Western Peninsula. The wood is used for cabinet work and furniture.

38. *Dalbergia melanoxylon*; Eng.–African black wood. It is cultivated in Tamil Nadu, Maharashtra and Bengal. The wood is used for musical instruments, walking sticks and fancy articles.

39. *Desmodium triflorum*; Verna–**Kubaliya**–It is grown as a soil, binder and for fodder.

40. *Crotalaria juncea*; Eng.–Sunn hemp; Verna–**San**–This is a tall shrub, cultivated throughout our country. The stem fibre is used for making ropes, cordage, mats and paper. The flowers and fruits are eaten as vegetable. The green stems and leaves make good manure.

41. *Crotalaria mucronata*; Syn. *C. striata*; Verna–**San**–This is a herb used as a cover crop and green manure.

42. *Crotalaria retusa*; Verna–**Ghungbunian**–This is cultivated all over India. A fibre is obtained from its stem bark, which is used for making canvas and cordage.

43. *Crotalaria burhia*; Verna–**Khip, Sis, Booi**–This is a xerophytic under shrub, found in the Punjab, Uttar Pradesh, Rajasthan and Gujarat. The stem fibre is used for making ropes and cordage. The branches and leaves are used as fodder.

44. *Lathyrus odoratus*; Eng.–Sweet pea; Verna **Phul mattar**–It is a native of Italy, cultivated as an ornamental.

45. *Lathyrus aphaca*;Verna.–**Jangli matar**–This is a climbing herb. It is grown in the Punjab, Uttar Pradesh, Bengal and Madhya Pradesh. It is used as a fodder.

46. *Lathyrus sativus*; Verna.–**Khesari**–It is cultivated in the Punjab, Bihar, Baroda, Assam and Orissa. It is used as a fodder. The seeds are eaten as vegetable. The ripe seeds are source of adulteration in pulses.

47. *Erythrina indica* var. *parcellii*; Eng.–Indian coral tree. This is a small tree, grown for its beautiful flowers.

48. *Erythrina blake.* This is a small tree, grown for its beautiful scarlet flowers.

49. *Erythrina crista–galli*; Eng.–Cockspur coral tree. It is grown as an ornamental.

50. *Erythrina variegata* var. *orientalis*; Eng.–Coral tree; Verna–**Pangra, Nasut**–This is a tree found in the Sundarbans, the Andaman and Nicobar Islands, Assam, Tamil Nadu and Orissa. The

tender pods and seeds are eaten as vegetable. The bark fibre is used for making cordage. The flowers yield a red dye. It is cultivated as a hedge plant and also employed to support the black pepper vine.

51. *Erythrina suberosa*; Verna.–**Panga**–This is a small tree found in the Punjab and Rajasthan. A fibre is obtained from its bark which is used for cordage.

52. *Clitoria ternatea*; Eng.–Butterfly pea. A twiner with beautiful blue flowers; grown in the gardens.

53. *Sesbania bispinosa*; Syn. *S. aculeata*; Eng–Dhencha; Verna.–**Jayanti**–This is a tree. Its stem fibre is used for making ropes.

54. *Sesbania cannabina*; Syn. *S. aculeata* var. *cannabina*; Verna.–**Dhaincha**–This is a tall herb, grown for its fibre. The fibre is used in making fishing-nets. The plant is also used as a green manure crop.

55. *Sesbania grandiflora*; Syn. *Robinia grandiflora; Agati grandiflora*; Verna.–**Basna**–This is a tree. It is native of tropical Asia but now grown in Assam, Bengal, Andhra Pradesh, Baroda and Tamil Nadu. The flowers, fruits and leaves are eaten as vegetable. The plant is also used as a support for the betel vine.

56. *Sesbania sesban*; Syn. *S. aegyptiaca*; Verna–**Jayanti**–This is a large shrub, cultivated throughout our country. The stem fibre is used for making ropes and cordage. The leaves are used as fodder.

57. *Pongamia pinnata*; Syn. *P. glabra*; Eng.–Pongam; Verna.–**Karanja**–This is a tree, commonly found near banks of streams in both the Peninsulas, West Bengal and Travancore. An oil is extracted from the seeds which is used for soap making and as an illuminant. It is also used in the treatment of skin diseases and rheumatism. The leaves are used as manure. The fruits are edible.

58. *Alysicarpus longifolius*; Verna–**Jangli gailia**–This is a herb, used as fodder.

59. *Astragalus heratensis.* This is a large shrub. The stem is the source of gum Katira gabina, which is used for glazing.

60. *Astragalus prolixus.*This is an undershrub, found in the Punjab. The stem is the source of gum tragacanth, which is used in cosmetics, calico–printing and confectionery.

61. *Astragalus strobiliferus*; Verna.–**Garmezu**–The gum obtained from the stem is used in confectionery. It is found in Kashmir.

62. *Calopogonium mucunoides*. This is a herb, introduced in the Western Ghats. It is used as fodder and also prevents soil erosion.

63. *Castanospermum australe*; Eng.–Chestnut. This is an introduced tree from Australia, the seeds are edible.

Systematic Relationships : According to Bentham and Hooker and Engler the family Leguminosae consists of three sub–families, *i.e.*, Mimosoideae, Caesalpinioideae and Papilionatae (Lotoideae). Many botanists have treated these sub–families as three distinct families, *i.e.*, the Mimosaceae, the Caesalpiniaceae and the Papilionaceae. Authors who accept them as separate families treat the trio as a single order. Hutchinson has placed these three families within the order Leguminales. The characters that distinguish one sub–family from the other are more apparent than the characters which bond them together. Many botanists bond them together only because of conservatism and tradition. Sen (1943) has pointed out, that as regards their wood anatomy there are no clear cut lines which separate the three sub–families. He also advocated that there was no cytological data in the support or denial of the elevation of each sub–family to the rank of family. The Mimosoideae of the three sub–families are accepted to be most primitive and the Papilionatae to be most evolved.

11

Order-Malvales Family-Malvaceae

According to Hutchinson this is the thirty third order of the phylum Angiospermae, subphylum Dicotyledones and division Lignosae. This order consists of a single family–the Malvaceae, which has been described in this text in detail. Bentham and Hooker have included the family Malvaceae along with other two families, *i.e.*, Sterculiaceae and Tiliaceae in the sixth order–the Malvales of class Dicotyledones, division Polypetalae and series Thalamiflorae. Engler and Prantl have placed the family Malvaceae along with seven families including Tiliaceae and Sterculiaceae in their twenty sixth order–the Malvales of class Dicotyledoneae and subclass Archichlamydeae.

Mainly herbaceous or softly woody and fibrous, calyx always valvate, anthers split in two and therefore 1-locular.

Family–Malvaceae

There are about 82 genera and 1,500 species in this family.

Distribution: The members of this family are cosmopolitan in their distribution. They are specially found in tropical regions of the world. Several species of this family have been reported from our country, *e.g.*, *Hibiscus, Sida, Malvastrum, Malva, Abutilon, Gossypium*, etc. Several of them are wild and some are commonly cultivated in gardens, *e.g.*, *Althaea rosea*. However, the most important member of the family is the genus *Gossypium*, from which various cottons are obtained.

Habit: The plants may be herbs, shrubs or trees. A few plants are annual or sometimes perennial herbs, *e.g.*, *Abutilon, Malva, Malvastrum, Hibiscus* sp., etc., they may be under shrubs, *e.g.*, *Sida cordinifolia, S. carpinifolia*, etc. The species of *Hibiscus*, may be herbs, shrubs or even small trees. *H. rosa sinensis* is a large ornamental shrub with beautiful red flowers. *H. mutabilis* is a small ornamental tree.

Root: Tap and branched.

Stem: The stem is erect, herbaceous or woody, branched, cylindrical, solid usually with stellate hairs. Mucilage sacs occur abundantly in the tissues.

Leaves: The leaves are simple, alternate, round or palmately divided, stipulate, usually palmately veined. The leaves bear stellate hairs. Mucilage sacs are also found.

Inflorescence: The inflorescence is of cymose type, very rarely racemose. This may be solitary axillary terminal, or compound cyme.

Flower: The flowers are usually pedicellate, bracteolate (the bracteoles are arranged in a whorl known as epicalyx), actinomorphic regular, hermaphrodite, complete, hypogynous and pentamerous.

Calyx: It consists of five sepals, gamosepalous, *i.e.*, the sepals are united to each other, sepaloid, inferior, and the aestivation is valvate.

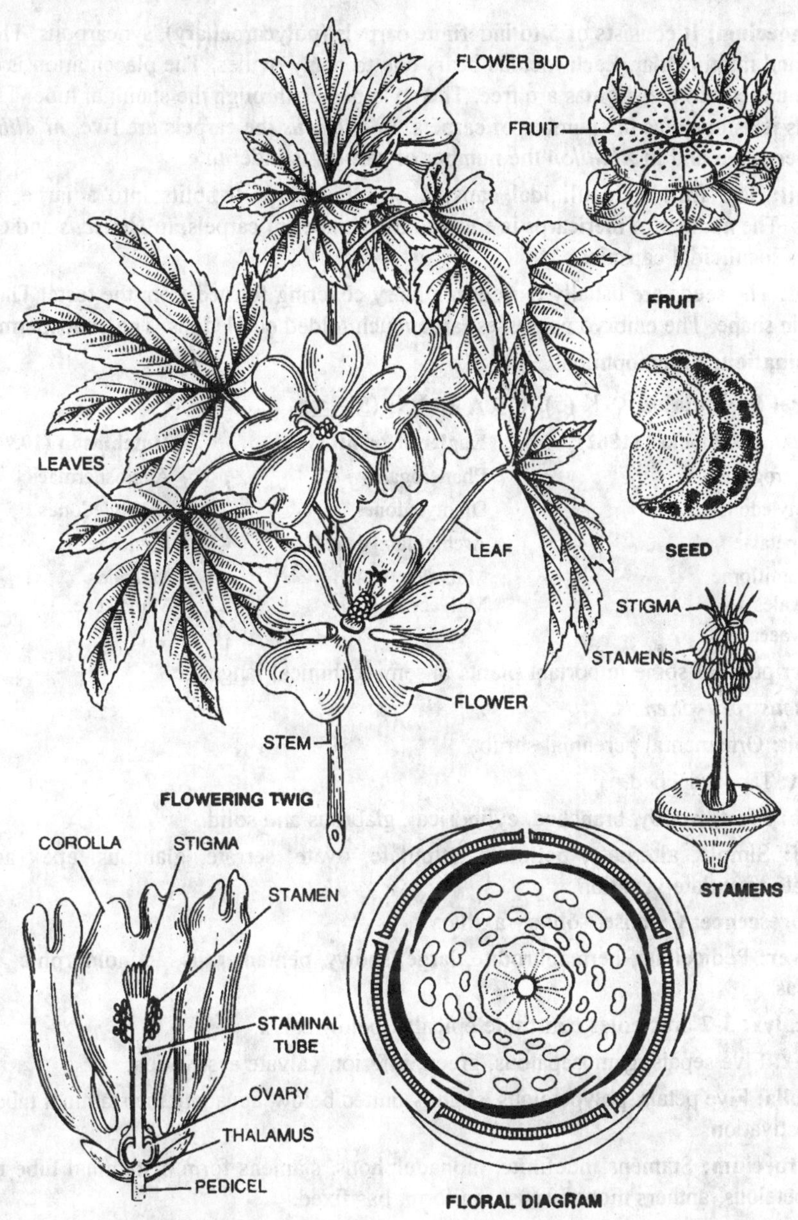

Fig. 11.1. Malvaceae. *Malva sylvestris* Linn., Eng., Common mallow; Verna. **Vilayati Kangi.**

Corolla: It consists of five petals, polypetalous but the petals are inferior and the aestivation is twisted. The petals are variously coloured.

Androecium: It consists of indefinite stamens, the stamens are found arranged on a staminal tube. The stamens are monadelphous. The staminal tube remains united at the base to the petals, and therefore, the condition is known to be epipetalous. The filaments are short and the anthers are monothecous, *i.e.*, one celled and dorsifixed. The anthers dehisce transversely.

The stamens are derived from profuse branching of five antipetalous stamens. The outer whorl has been lost, but however, represented in *Hibiscus* by five staminodes.

Gynoecium: It consists of 5 to indefinite carpels (polycarpellary), syncarpous. The ovary is superior and multilocular. Each loculus bears one to many ovules. The placentation is axile. The styles are united but the stigmas are free. The style passes through the staminal tube. The number of stigmas is as much as the number of carpels. In *Hibiscus* the carpels are five, in *Althaea rosea* the number is ten and in *Abutilon* the number of carpels is indefinite.

Fruit: The fruit is loculicidal capsule or schizocarp. It splits into a large number of mericarps. The number of mericarps is equal to the number of carpels. In *Hibiscus* and *Gossypium* the fruit is loculicidal capsule.

Seed: The seeds are usually buried in a hairy covering formed from the testa. The reniform or ovoid in shape. The embryo possesses large much folded cotyledons. The endosperm is scanty.

Pollination: Entomophilous.

Floral Formula: ⊕ ⚥ K (5), $\widehat{C5,A}$ (∝), G ($\underline{5-\infty}$)

Bentham & Hooker (1862)	**Engler& Prantl (1931)**	**Hutchinson (1959)**
Phanerogams	Phanerogams	Angiospermae
Dicotyledones	Dicotyledones	Dicotyledones
Polypetalae	Archichlamydeae	Lignosae
Thalamiflorae	Malvales	Malvales
Malvales	Malvaceae	Malvaceae.
Malvaceae		

Description of some important plants in semi-techinical language.

1. ***Hibiscus rosa-sinensis.***

Habit: Ornamental perennial shrub.

Root: Tap, branched.

Stem: Erect, woody, branched, cylindrical, glabrous and solid.

Leaf: Simple, alternate, petiolate, stipulate, ovate, serrate, glabrous, apex acuminate, multicostate reticulate venation.

Inflorescence: Cymose, solitary axillary.

Flower: Pedicellate, hermaphrodite, large, showy, pentamerous, actinomorphic, complete, hypogynous.

Epicalyx: 5-7 bracteoles constitute epicalyx below calyx.

Calyx: Five sepals, gamosepalous, green, inferior, valvate aestivation:

Corolla: Five petals, polypetalous, slightly united below, adnate to the staminal tube, inferior twisted aestivation.

Androecium: Stamens indefinite, monadelphous, stamens form a staminal tube round the style, epipetalous, anthers monothecus, reniform, basifixed.

Gynoecium: Five carpels (pentacarpellary), syncarpous ovary superior, pentalocular, axile placentation, style long and passes through the staminal tube ending in five distinct rounded stigmas.

Fruit: In other species of *Hibiscus*, loculicidal capsule.

Floral Formula: ⊕ ⚥ K(5), $\widehat{C5,A}$ (∝), G($\underline{5}$)

2. *Abutilon indicum*; Verna–**Kangi.**

Habit: Annual or perennial herb or undershrub.

Root: Tap, branched.

Stem: Erect, branched, herbaceous, solid, hairy, green.

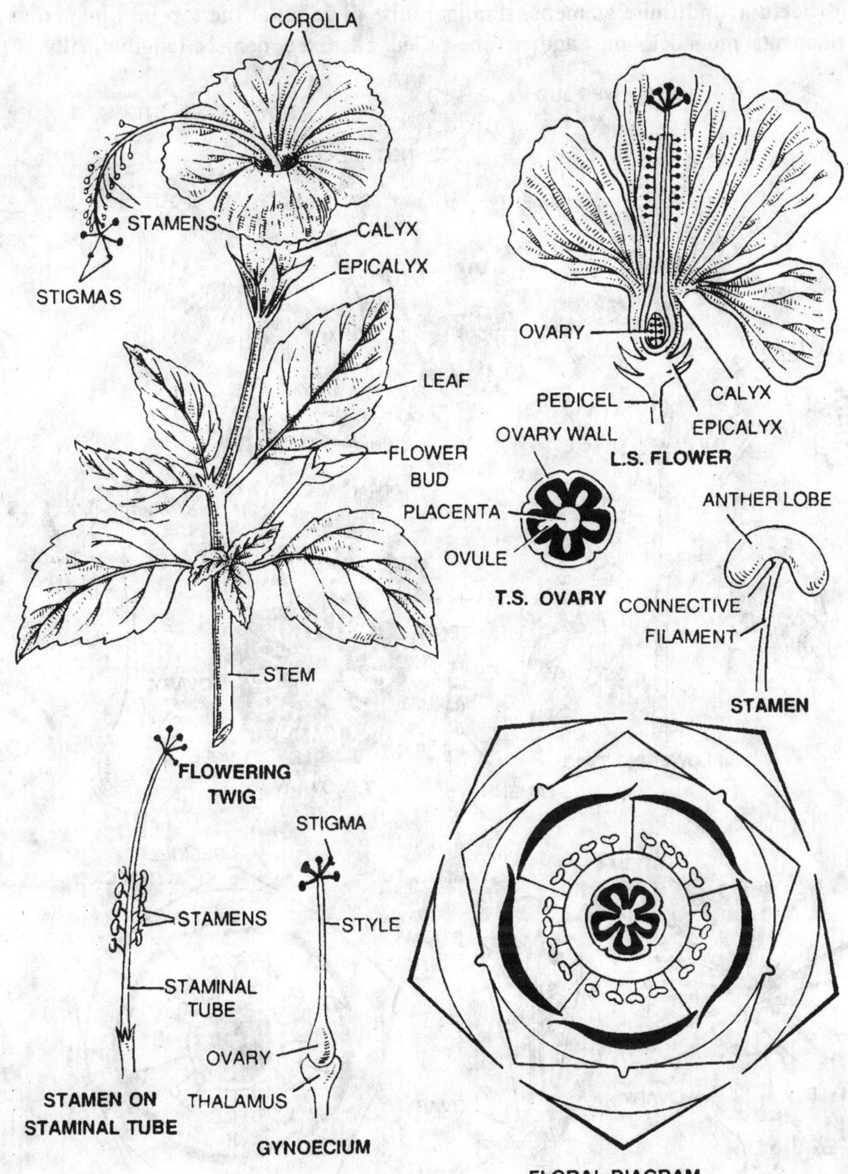

Fig. 11.2. Malvaceae. *Hibiscus rosa sinensis* Linn. Eng., China shoeflower, Verna. **Gurhal.**

Leaves: Simple, alternate, petiolate (petiole usually longer than blade), broadly cordate, acute or acuminate apex, irregularly toothed, pale, minutely pubescent on both surfaces, entire, multicostate reticulate venation.

Inflorescence: Cymose, solitary axillary.

Flower: Pedicellate (pedicels longer than the petioles and jointed near the top), one inch diameter, hermaphrodite, actinomorphic, complete, hypogynous.

Calyx: Five sepals, gamosepalous, shallow, ovate acute, persistent valvate aestivation:

Corolla: Five petals, polypetalous, free above, connate below and adnate to the staminal tube, orange, yellow, petals spreading imbricate aestivation.

Androecium: Indifinite stamens, staminal tube dividing at the top into numerous anther bearing filaments, monadelphous, anthers one-celled, basifixed, dehisce longitudinally.

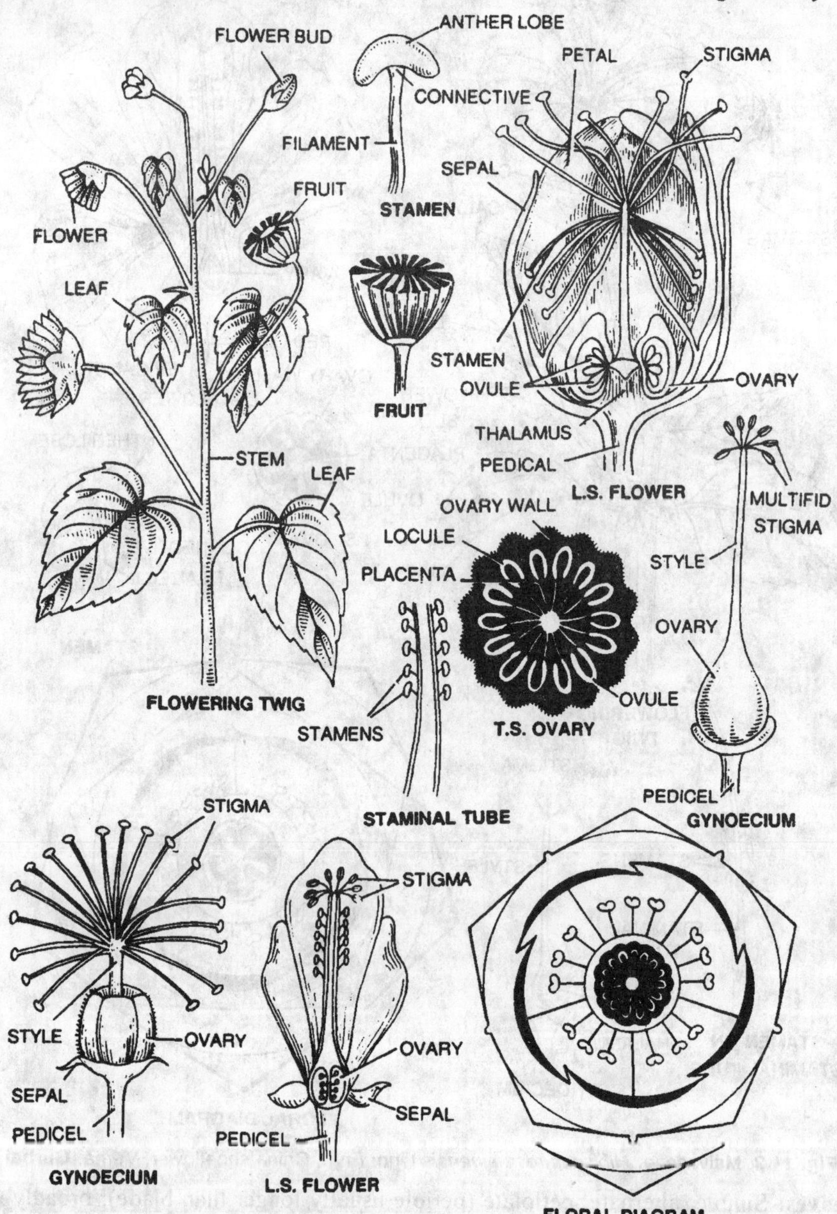

Fig. 11.3. Malvaceae—*Abutilon indicum* (Linn) Sweet. Eng. Country mallow, Verna. **Kangi**

Gynoecium: 15-20 carpels (polycarpellary), carpels longer than the calyx, with short spreading awns, hairy, syncarpous, ovary superior, multilocular, axile placentation, many ovules in each locule.

Floral Formula: ⊕ ⚥ K(5), C 5, A (∝), G $(\underline{\propto})$

Note. The genus *Sida* is closely allied to *Abutilon.*

Malvastrum tricuspidatum. A. Gray.

Habit: An annual or perennial herb.

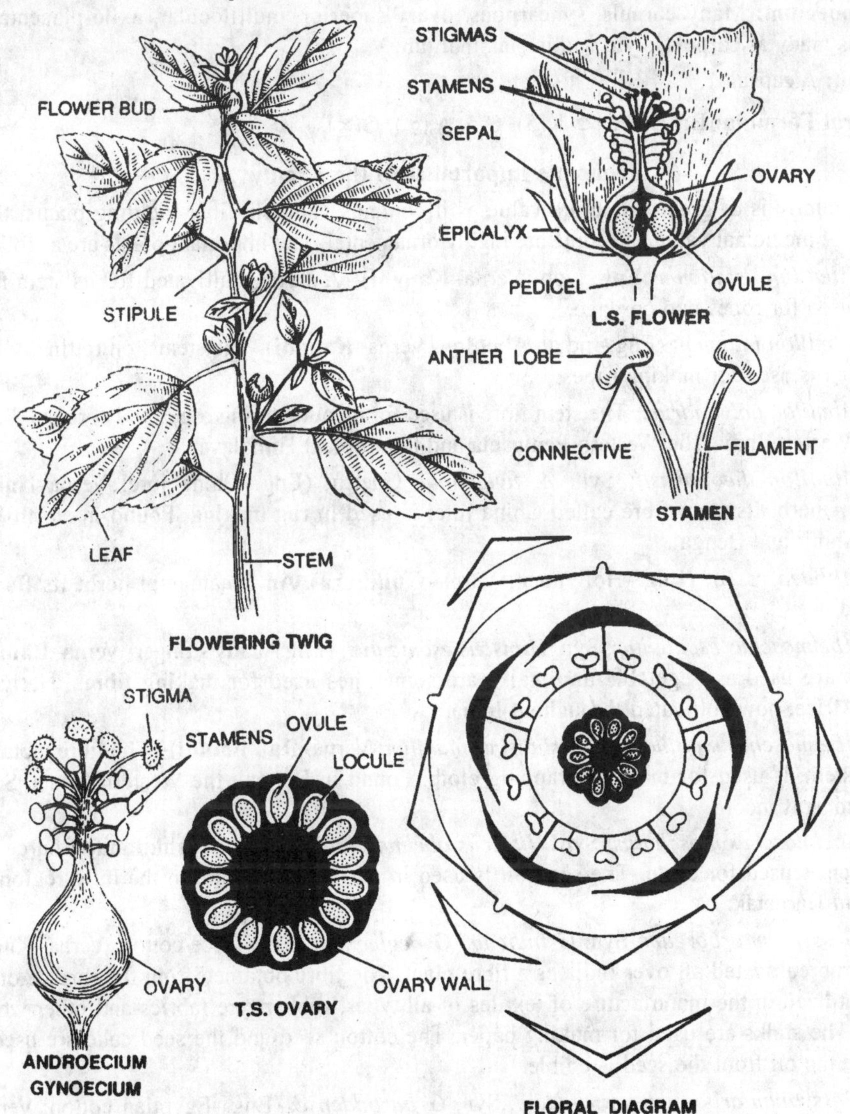

Fig. 11.4. Malvaceae. *Malvastrum tricuspidatum* A. Gray.

Root: Tap, branched

Stem: Erect, aerial, herbaceous, cylinderical, branched, solid; hairy, green.

Leaf: Cauline and ramal, petiolate, stipulate stipules small and free-lateral, alternate simple, ovate-lanceolate, serrate, acute, glabrous, unicostae reticulate.

Inflorescence: Cymose, solitary axillary.

Flower: Bracteate. bracteolate, 3 small bracteoles, pedicellate, actinomorphic, hermaphrodite, hypogynous, pentamerous, cyclic.

Calyx: 5 sepals, gamosepalous, valvate aestivation, persistent.

Corolla: 5 petals, polypetalous, twisted, yellow.

Androecium: Indefinite, monadelphous, a staminal tube is formed around the style, epipetalous, monothecous, reniform, basifixed, extrorse.

Gynoecium: Many carpels, syncarpous, ovary superior, multilocular, axile placentation, stigmas as many as carpels, style within staminal tube.

Fruit: A capsule.

Floral Formula: Br. Brl. ⊕ ⚥ K(5), C 5, A ($\propto$), G($\underline{\propto}$).

Economic Importance of the Family

The family is of great economic value as it contains valuable fibre yielding plants. Some plants are of medicinal value and some are highly ornamental. The important plants are as follows:

1. *Abutilon asiaticum;* Eng. Don (Verna–**Kanghi**)– A shrub, cultivated for its stem fibre, which is used for ropes and cordage.

2. *Abutilon indicum*; (Eng.-Indian Abutilon; Verna–**Kanghi**). The stems, on retting, yield a fibre, which is used for making ropes.

3. *Abutilon polyandrum*; The stem fibre is used for manufacturing ropes and cordage. Found in North Western India, the Western Peninsula and the tropical Himalayas.

4. *Abutilon theophrasti*; Syn. *A. avicennae* Gaertn. (Eng.–China jute; Verna–**Nahani khapat**)- A herb. Its stem fibre called China jute; is used in rug making. Found in North-West India, Kashmir and Bengal.

5. *Althaea rosea*; (Eng.–Hollyhock; Verna–**Gulkhera**). An ornamental herb. Its flowers yield a red dye.

6. *Abelmoschus esculentus*; Syn. *Hibiscus esculentus*; (Eng.–Lady's finger; Verna–**Bhindi**)– The fruits are used as vegetable and stalks are sometimes used for making fibres. Native of Tropical Africa, now cultivated throughout India.

7. *Abelmoschus manihot*; (Syn. *Hibiscus manihot*: Verna–**Ran bhindi**)– The fibre, obtained from the stem, is used for making wrapping cloth. Found in Bengal, the Western Ghats, South Kanara and Travancore.

8. *Abelmoschus moschatus*; Syn. *Hibiscus abelmoschus*; (Verna–**Mushkdana**)–Source of a fibre, which is used for cloth. The seed-oil is used in perfumery. Found in the hilly regions of Deccan and Karnatak.

9. *Gossypium arboreum*; Syn. *G. indicum, G. neglectum*; (Eng.–Tree cotton; Verna-**Kapas, Rui**)-A shrub cultivated all over India as a fibre plant. The fibre obtained from the surface of the seeds, is utilized in the manufacture of textiles of all types, rubber-tyre fabrics and other woven products. The stalks are used for making paper. The cotton seed and the seed cake are used as fodder and the oil from the seed is edible.

10. *Gossypium arboreum* var. *nadam*; Syn. *G. barbadense*; (Eng.–Egyptian cotton; Verna–**Kapas, Rui**).-It is native of tropical America, grown mainly in Malabar and South Kanara. The fibre obtained from the surface of the seeds, is utilized in the manufacture of textiles and the seed oil is edible.

11. *Gossypium herbaceum*; (Eng–Asiatic cotton; Verna–**Kapas**)–Cultivated mainly in Maharashtra, Tamil Nadu, Andhra Pradesh and Karnataka. The fibre obtained from the seeds is used in making cloth and other woven products. The oil–cake is used as fertilizer and fodder.

12. *Gossypium hirsutum*; Syn. *G. mexicanum* (Eng.–Upland cotton; Verna–**Kapas**)– Grown chiefly in the Punjab, Uttar Pradesh, Rajasthan, Madhya Pradesh, Tamil Nadu and Maharashtra. The fibre obtained from the seeds, is utilized in the manufacture of textiles, rubber–tyre fabrics and other woven products. The cotton seed and cake are used as fodder and the seed oil is edible.

13. *Hibiscus cannabinus*; (Eng.–Kenaf, Verna–**Patsan**)–A herb grown in West Bengal, Andhra Pradesh, Assam, Bihar and the Punjab. The fibre obtained from the stem, is used for making ropes cordage and paper pulp and the seed oil is edible.

14. *Hibiscus collinus*; A small tree grown as an ornamental.

15. *Hibiscus ficulneus*; Syn. *Abelmoschus ficulneus*; Verna–**Ran Bhindi**–A herb, found in the Punjab, Bengal and South India. The mucilaginous extract of the green stem is used for clarifying sugar juice and the seeds are used for flavouring purposes.

16. *Hibiscus urcatus* Roxb.–Found in Assam, Madhya Pradesh, Orissa, Andhra Pradesh, Bengal and Karnataka. The leaves are edible and the stem fibre is used for flavouring purposes.

17. *Hibiscus macrophyllus*; Roxb. (Verna–**Baiza**)–A small shrub, found in Bengal and the Eastern Peninsula. The stem fibre is used for ropes and cordage.

18. *Hibiscus mutabilis*; (Eng.–cotton rose; Verna–**Guliajaib**)–A bushy shrub; native of China. The bark yields a fibre, which is used for ropes and cordage.

19. *Hibiscus rosa-sinensis*; (Eng.–Rose of China; Verna–**Gurhal**)–A common ornamental shrub. the flowers yield a dark purplish dye, which is used for making shoe-polishes.

20. *Hibiscus sabdariffa*; (Verna–**Patwa**)– A shrub native of the West Indies, now grown in Andhra Pradesh, Bihar, the Punjab, Assam and Tamil Nadu. The stem fibre is utilized for making ropes and cordage and the red fleshy epicalyx is edible.

21. *Hibiscus schizopetalous*; (Eng–coral Hibiscus)–A shrub, native of East Africa, grown as an ornamental for its orange-red flowers.

22. *Hibiscus surattensis*; (Verna–**Ranbhindi**)–A herb cultivated mainly in Bengal. The leaves are edible and the stem fibre is used for rough cloth and cordage.

23. *Hibiscus syriacus*; (Verna–**Swet jaha**)–A shrub native of China and Japan, now cultivated all over India for its edible flowers.

24. *Hibiscus tiliaceus*; Syn. *Paritium tiliaceum*; (Eng.–Cuban bast; Verna–**Chelwa**)– Commonly found along the coasts of both the Peninsulas and in the Sunderbans. The leaves are edible and the bark fibre is used for ropes and mats.

25. *Hibiscus vitifolius*; (Verna–**Ban kapas**)–Found in Northern India and Madhya Pradesh. The bark is the source of strong fibre.

26. *Malachra capitata*; (Eng.–cork wood: Verna–**Ranbhindi**)–A herb, found throughout the hotter parts of the country. The stem fibre is used for cloth, twines and ropes.

27. *Urena lobata*; Syn. *U. sinuata*; (Verna–**Vilayati san**)–A small shrub, commonly found in Bengal. The stem fibre is a substitute for jute, and is used for cordage ropes and coarse fabrics. The mucilaginous seeds are used for making soap.

Systematic Relationships: According to different taxonomists the arrangement of the families within the order Malvales varies. Engler, Wettstein and Rendle include the common families, Tiliaceae, Bombacaceae, Sterculiaceae and Malvaceae in this order. Hutchinson, however, separated the Malvales to include only the Malvaceae and included the Tiliaceae, Bombacaceae, and Sterculiaceae in his new order the Tiliales. According to Rendle the Malvales show the affinities with the Guttiferae in having hypogynous, regular and pentamerous flowers. He also thinks that on the other hand the Malvales are related to Euphorbiales in some respects. Bessey expanded Eichler's concept of Malvales and included the Balanopsidaceae, Ulmaceae, Moraceae and Urticaceae within this order.

12

Order-Euphorbiales Family-Euphorbiaceae

According to Hutchinson this is the thirty fifth order of the phylum Angiospermae, sub phylum Dicotyledones and division Lignosae. The order consists of a single family, *i.e.*, the Euphorbiaceae which has been described in the present text in detail. Bentham and Hooker have included the Euphorbiaceae along with other eight families including Urticaceae and Casuarinaceae in their seventh series—the Unisexuales of class Dicotyledones, and division Monochlamydeae or Incompletae. Engler and Prantl have included the family Euphorbiaceae along with other nineteen families including Linaceae, Rutaceae and Meliaceae in their twenty third order—the Geraniales of class Dicotyledoneae and sub-class Archichlamydeae.

Hypogynous; actinomorphic; unisexual; petals usually absent, if present sometimes quite sympetalous (*Jatropha*); syncarpous with 1-2 ovules pendulous from the inner angle; stamens various; seeds often with a conspicuous caruncle; endosperms mostly copious; stipules mostly present.

Family—Euphorbiaceae

There are about 283 genera and 7,300 species in this family.

Distribution: The plants of this family are found throughout the world. However, they are not found in arctic regions. In our country the family is represented by several genera such as, *Euphorbia*, *Ricinus*, *Phyllanthus*, *Crotòn*, *Pedilanthus*, etc. In the desert regions of Africa and elsewhere the family is represented by cactus-like plants of different species of *Euphorbia*. Heath like *Euphorbias* are quite common in Australia. In Britain only two genera, *i.e.*, *Euphorbia* and *Mercurialis* are found, which are represented by sixteen and two species respectively.

Habit: The plants exhibit great variation in their habit. The plants may be herbs, shrubs or trees. *Euphorbia hirta, E. thymifolia, E. helioscopica, E. peplus; E. heterophylla, E. cristata, E. elegans; Phyllanthus niruri, Croton* sp., *Acalypha indica*, etc., are annual or prennial herbs. *Euphorbia pulcherrima, E. splendens*, are beautiful shrubs. *Pedilanthus* sp., and *Jatropha* sp., are shrubby plants. *Euphorbia royleana, E. tirucalli* are cactus like shrubs. *Ricinus communis* **(Arand)** is a tall annual and becomes small tree-like in habit. The tree habit of the family is represented by *Phyllanthus emblica* **(Amla)**, *Bischofia javanica, Putranjiva roxburghii*, etc. *Havea brasiliensist* (rubber tree) is a large tree 60 to 100 feet in height and 8-12 feet in girth. Species of the genus *Tragia* are tropical climbers. Majority of the members of the family possess large laticiferous vessels which contain latex.

Root: Tap and branched. The roots of *Manihot utilissima* and *M.palmata* are tuberous and rich in starch.

Stem: Herbaceous or woody, erect, very rarely climbing as in a tropical genus *Tragia*. The species of *Xylophylla* possess flat phylloclades. The stem is branched. It may be cylindrical, angular or flat. Usually solid but sometimes hollow as in *Ricinus communis*. Many stems possess spines. In many *Euphorbia* sp., the stems become fleshy, green and cactus like in appearance.

Leaves: The form and position of leaves are variable. The arrangement is usually alternate but sometimes they are opposite, *e.g., Euphorbia hirta*. In *Pedilanthus* the leaves are arranged alternately in the lower region of the plant whereas opposite in the floral region. Usually the leaves are simple but in some they are deeply incised, *e.g., Ricinus, Manihot*, etc. In many *Euphorbias* the leaves are scaly and caducous. In many cases the leaves are reduced to spines. In few cases the leaves are replaced by cladodes. Usually the leaves are stipulate. In *Jatropha* sp., the stipules become branched and hair-like. In many *Euphorbia* sp., they are represented by glands or spines.

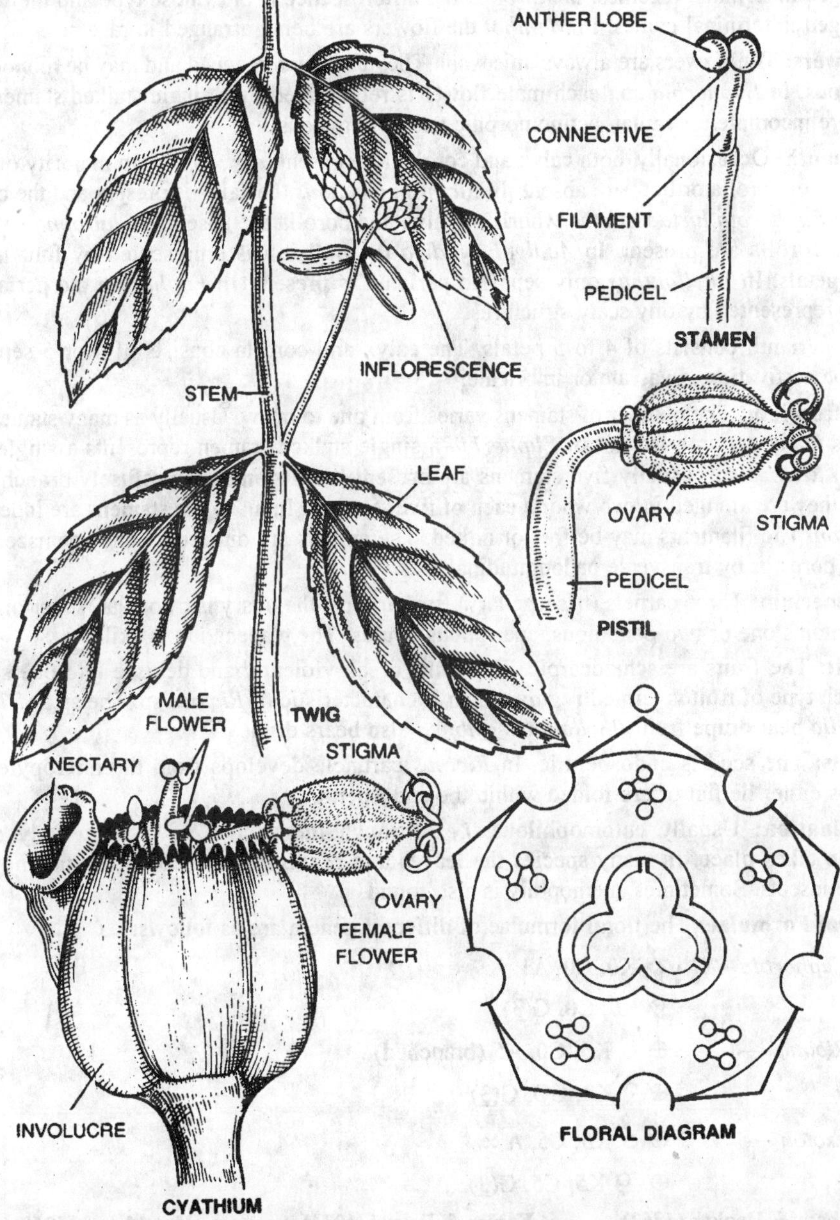

Fig. 12.1. Euphorbiaceae. *Euphorbia* spp., plant with latex and cyathium inflorescence.

Inflorescence: The inflorescence varies greatly. It may be racemose or cymose or sometimes complex. In *Euphorbia,* the inflorescence is peculiar but very characteristic and known as **cyathium**. This is the modification of a cyme. In cyathium inflorescence a large number of male flowers each represented by a stalked stamen are found arranged around a central stalked female flower. The female flower consists of gynoecium only. The complete inflorescence looks like a single flower. The bracts are being arranged like a perianth. The bracts are so united that they form a cup-like structure. In *Acalypha* the inflorescence is catkin type. In *Croton* and *Ricinus* the flowers are arranged in terminal racemes. In *Jatropha* the inflorescence is of cymose type and the flowers are arranged in terminal cymes. In *Manihot* the flowers are being arranged in racemes.

Flowers: The flowers are always unisexual. They are much reduced and may be monoecious or dioecious. In *Euphorbia* sp., each male flower is represented by a single stalked stamen. The flowers are incomplete, regular, actinomorphic and hypogynous.

Perianth: Occasionally, both calyx and corolla are present, *e.g., Croton.*In majority of cases either calyx or corolla or both are absent. In *Ricinus communis* the calyx is present and the corolla absent. In *Euphorbia hirta* both the whorls of calyx and corolla are absent. In *Jatropha* sp., both calyx and corolla are present. In *Acalypha indica* the perianth is represented by four minute sepaloid petals. In *Phyllanthus* only sepaloid perianth is present. In *Euphorbia* the perianth is absent or represented by tiny scaly structures.

The perianth consists of 4 to 5 petals. The calyx and corolla consists of 4 or 5 sepals or petals. The aestivation is valvate or imbricate.

Androecium: The number of stamens varies from one to many. Usually as many stamens are present as many perianth leaves. In *Euphorbia* a single stalked stamen represents a single male flower. In *Ricinus* sp., usually five stamens are present, each stamen is profusely branched. In *Jatropha* they are aranged in two whorls each of five stamens. In many the stamens are indefinite, *e.g., Croton.* The filaments may be free or united. The anthers are dithecous. They dehisce either by apical pores or by transverse or longitudinal slits.

Gynoecium: Three carpels (tricarpellary), syncarpous; the ovary is trilocular, superior. Each locule contains one or two pendulous, anatropous ovules. The placenation is axile.

Fruit: The fruits are schizocarpic. The fruits break violently and dehisce into one seeded cocci. Such type of fruit is termed **regma** which is characteristic of *Ricinus* sp. The sp., of *Trewia* and *Bridelia* bear drupe fruit. *Phyllanthus emblica* also bears drupe.

Seeds: The seed is endospermic. In *Ricinus* caruncle develops from the micropyle. The cotyledons either lie flat or are folded within the endosperm.

Pollination: Usually entomophilous, *i.e.,* through the agency of insects. Only cross-pollination takes place. In many species the leaves and bracts become coloured and showy to attract the insects. Sometimes anemophily is also found.

Floral Formulae: The floral formulae of different genera are as follows:

Euphorbia ♂ K0, C0, A1

♀ K0, C0, G($\underline{3}$).

Ricinus — ⊕ ♂ K5, C0, A5 (branched)

⊕ ♀ K3, CO, G($\underline{3}$).

Croton — ⊕ ♂ K5, C5, A ∞

⊕ ♀ K5, C5, G($\underline{3}$).

Bentham & Hooker (1862)	**Engler & Prantl (1931)**	**Hutchinson (1959)**
Phanerogams	Phanerogams	Angiospermae
Dicotyledones	Dicotyledones	Dicotyledones

Monochlamydeae Incompletae or Unisexuales Euphorbiaceae	Archichlamydeae Geraniales Euphorbiaceae	Lignosae Euphorbiales Euphorbiaceae

A few important plants have been described here in semi-technical language:

1. ***Ricinus communis;*** Eng.–Castor oil plant; Verna.–**Arand**.

   Habit: Tall annual or perennial shrub or small tree.

   Root: Tap and branched.

   Stem: Erect, cylindrical, branched, solid, glabrous.

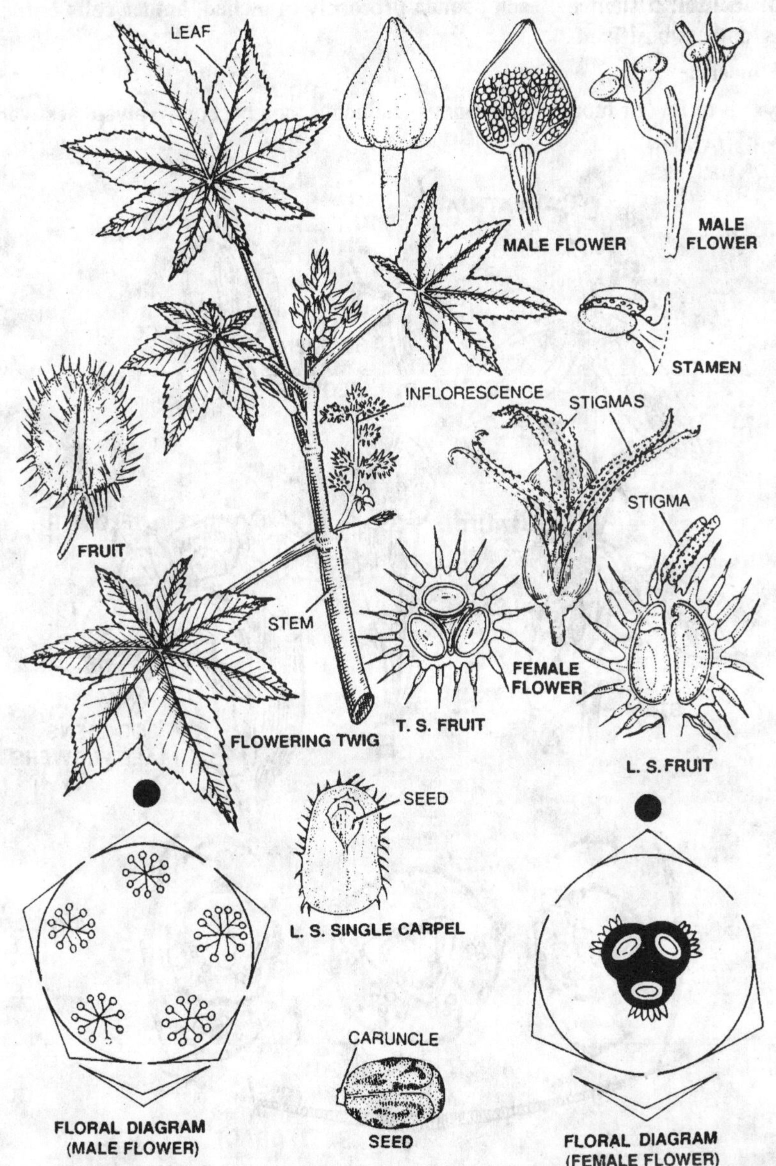

Fig. 12.2. Euphorbiaceae—*Ricinus communis* Linn., Eng., Castor oil plant; Verna. **Arand.**

Leaf: Alternate, simple, petiolate, broad, palmately lobed, lobes usually 7-9, serrate, multicostate reticulate venation.

Inflorescence: Terminal panicled cyme.

Flower: Pedicellate, ebracteate, unisexual, monoecious, male flowers towards lower portion, female flowers towards apical portion of inflorescence, actinomorphic, incomplete, hypogynous.

Male Flowers:

Calyx: 5 or 4, membranous, polysepalous, green, inferior, valvate aestivation.

Corolla: Absent.

Androecium: 5 stamens, each stamen profusely branched, anther cells borne on ultimate branchlets, anthers basifixed.

Female Flowers:

Calyx: 3 to few or more, membranous, caducous, spathaceous, valvate aestivation.

Corolla: Absent.

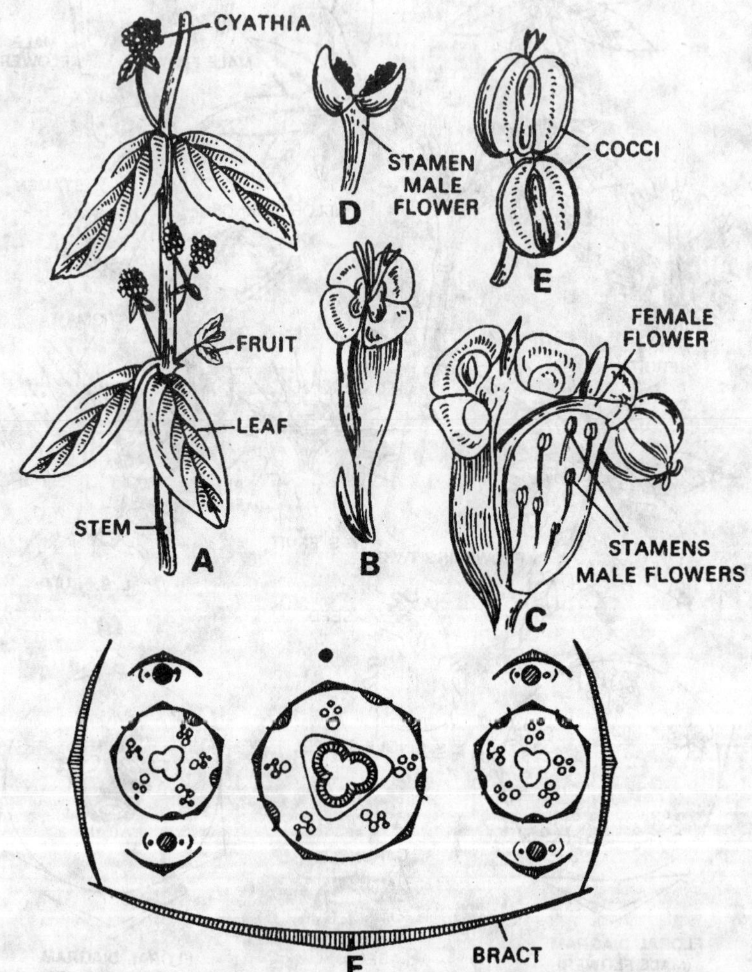

Fig. 12.3. Euphorbiaceae, *Euphorbia* sp. A, a flowering twig; B, cyathium; C, cyathium cut open; D, stamen; E, fruit-capsule showing cocci; F, floral diagram showing three cyathia.

Gynoecium: 3 carpels (tricarpellary), syncarpous, ovary superior, trilocular, covered with spiny outgrowths, axile placentation, single large ovlue in each locule, three styles, three bifid stigmas.

Fruit: A schizocarpic capsule or regma, splitting into three cocci, spiny outgrowths on fruit.

Floral Formula :

Male: ⊕ ♂ K 5, C 0, A 5 (branched)

Female: ⊕ ♀ K 3, C 0, G (3).

2. *Euphorbia hirta.*

Habit: Annual wild herb.

Stem: Erect, usually unbranched (sometimes branched), herbaceous, cylindrical, solid, covered with yellow crisped hairs.

Leaf: Simple, opposite, superposed, subsessile (short petiole), acute, oblong, lanceolate, serrulate, stipulate, stipules caducous, unicostate reticulate venation.

Inflorescence: Large number of cyathia densely crowded and arranged in peduncled axillary cymes.

Flowers: Pedicellate, unisexual, monoecious, enclosed within minute involucre of bracts forming cupular structure.

Male Flower:

Perianth: Absent, naked flower.

Androecium: Single stalked stamen representing male flower, bracteate, anthers 2-celled dehiscing longitudinally.

Female Flower:

Perianth: Absent, naked flower.

Gynoecium: A single stalked, bracteate, tricarpellary pistil represents female flower, it remains surrounded by male flowers (stalked stamens), ovary superior, three-chambered, single ovule in each loculus, axile placentation, three styles, three bifid stigmas.

Fruit: A capsule.

Floral Formulae:

Male flower— ⊕ ♂ K 0, C 0, A I

Female flower— ⊕ ♀ K 0, C 0, G (3).

3. *Croton sparciflorus*

Habit: Annual wild herb.

Stem: Erect, herbaceous, branched, rough, latex present, solid, cylindrical.

Leaf: Alternate, simple, petiolate, exstilpulate, serrate margin, acute, rough, unicostate reticulate venation.

Inflorescence: Racemose, raceme, male flowers on the upper portion, female flowers on the lower portion of inflorescence.

Flower: Pedicellate, bracteate, unisexual, monoecious plant, actinomorphic, incomplete, hypogynous, glands present.

Male Flower:

Calyx: 5 sepals, polysepalous, persistent, imbricate.

Corolla: 5 petals, polypetalous, 5 glands opposite sepals, valvate aestivation.

Androecium: Stamens many, inserted on receptacle, filaments free, incurved in bud, anthers adnate.

Female Flower:

Calyx: As in male flowers.

Corolla: Absent.

Gynoecium: 3 carpels (tricarpellary), syncarpous, 3 styles ending in 3 bifid stigmas, ovary superior, three chambered, single ovule in each loculus, axile placentation.

Fruit: Indehiscent capsule,

Floral Formulae:

Male flower— ⊕ ♂ K 5, C 5, A ∝

Female flower— ⊕ ♀ K 5, C 0, G (3).

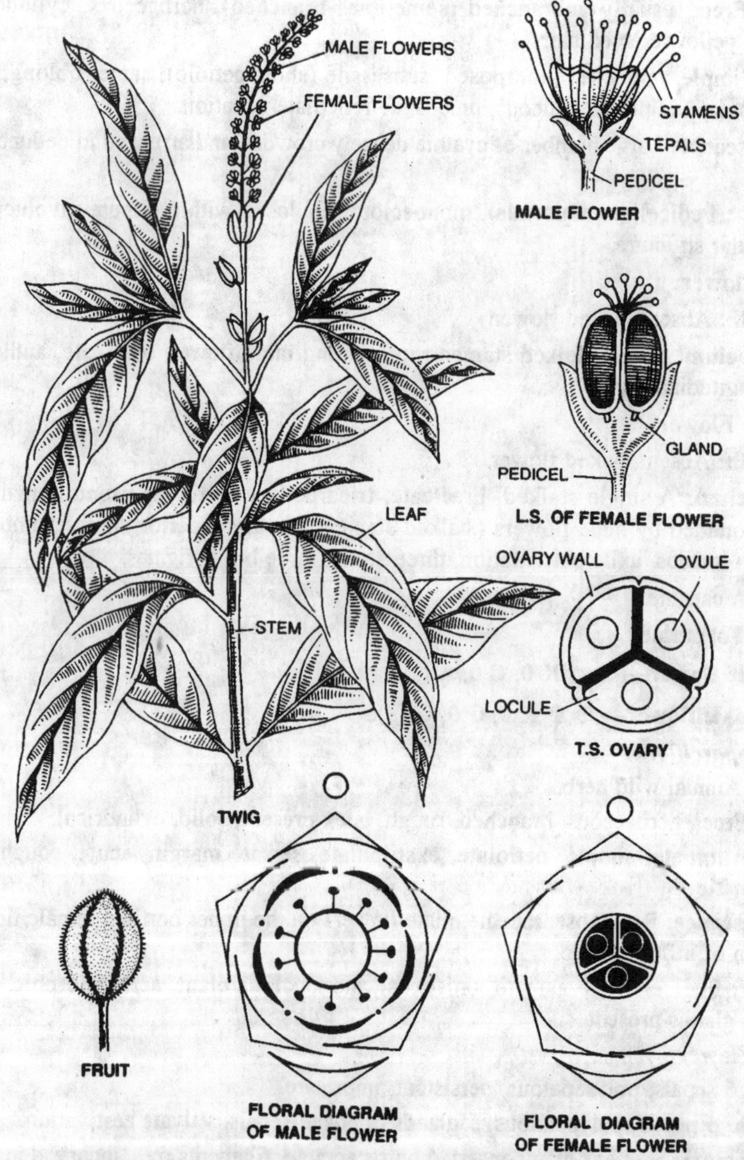

Fig. 12.4. Euphorbiaceae. *Croton sparciflorus.*

4. ***Phyllanthus reticulata;*** Verna. **Panjoli.**

Habit: An annual herb.

Stem: Erect, branched, herbaceous, green, solid, cylindrical 1/2 to 2 feet in height.

Leaf: Simple, alternate, subsessile, elliptic-oblong, rounded obtuse, glaucous beneath, stipulate, distichous often resembling pinnate compound leaves, 1/4 – 1/2 inch long.

Inflorescence: Axillary flowers.

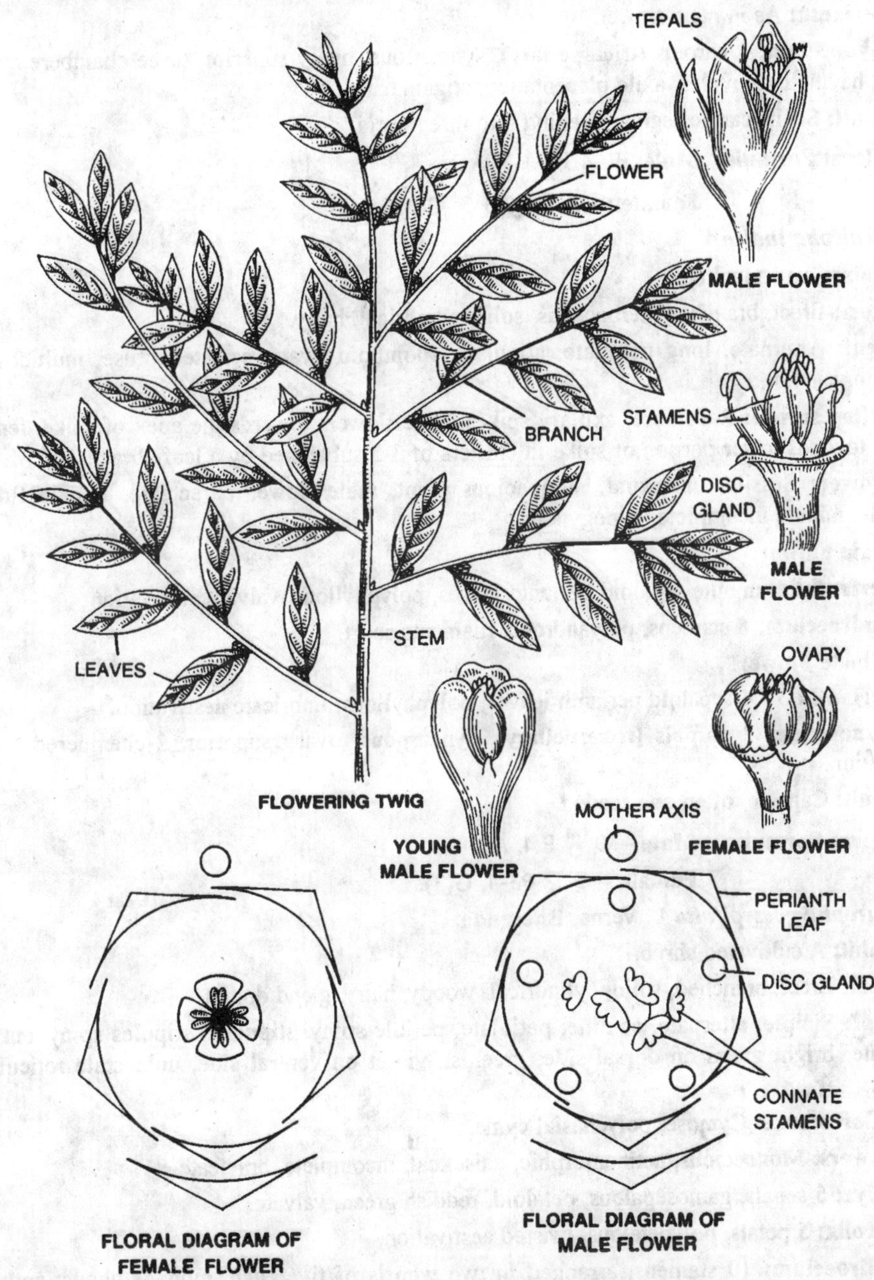

Fig. 12.5. Euphorbiaceae. *Phyllanthus reticulata* Poir.

Flower: Minute, unisexual, trimerous, actinomorphic, bracteate, shortly pedicellate.

Male flower:

Perianth: 6 petals (perianth leaves), arranged in two whorls of three each, polyphyllous, sepaloid, imbricate aestivation, disc of minute glands inside the perianth.

Androecium: 3 stamens, polyandrous, filaments short, borne on a short column.

Female Flower:

Perianth: As in male flower.

Gynoecium: 3 carpels (tricarpellary), syncarpous, ovary superior, three chambered, each loculus having two ovules, axile placentation, stigma trifid.

Fruit: Schizocarpic regma, opening into three cocci.

Floral Formula: **Male:** ⊕ ♂ P3 + 3, A 3

Female :– ⊕ ♀ P3 + 3, G($\underline{3}$).

5. ***Acalypha indica***

Habit: An annual herb.

Stem: Erect, branched, herbaceous, solid cylindrical, hairy.

Leaf: Alternate, long petiolate, stipulate, rhomboid ovate, serrate obtuse, multicostate reticulate venation.

Inflorescence: Racemose, axillary spikes, male flowers towards the apex of spike, female flowers towards lower portion of spike in clusters of 3-5 subtended by a leafy bract.

Flower: Sessile, unisexual, monoecious plant, male flower ebracteate, female flower bracteate, minute incomplete, hypogynous.

Male Flower:

Perianth: 4, minute sepaloid perianth leaves, polyphyllous, valvate aestivation.

Androecium: 8 stamens, polyandrous, filaments short.

Female Flower:

Perianth: 3 - 4 sepaloid perianth leaves, polyphyllous, imbricate aestivation.

Gynoecium: 3 carpels (tricarpellary), syncarpous, ovary superior, 3-chambered, axile placentation.

Fruit: Capsule, often one seeded.

Floral Formula: **Male –** ⊕ ♂ P 4, A8.

Female – ⊕ ♀ P3-4, G($\underline{3}$).

6. ***Jatropha gossypifolia*** L. Verna. **Bherenda.**

Habit: A cultivated shrub.

Stem: Erect, branched, solid, cylindrical, woody, hairy, gland dotted.

Leaf: Simple, alternate, cauline, petiolate, petiole spiny, stipulate, stipules spiny, entire, acuminate, bright green on dorsal side, greenish violet on ventral side, unicostate reticulate venation.

Inflorescence: Cymose, polychasial cyme.

Flowers: Monoecious, actinomorphic, unisexual, incomplete, bracteate.

Calyx: 5 sepals, gamosepalous, petaloid, reddish green, valvate.

Corolla: 5 petals, polypetalous, twisted aestivation.

Androecium: 10 stamens, arranged in two whorls of five each, monadelphous, anthers dorsifixed.

Female Flowers: Pedicellate, bracteate, unisexual, incomplete, hypogynous.

Calyx: As in male flower.

Corolla: As in male flower.

Gynoecium: 3 carpels (tricarpellary), syncarpous, ovary superior, three-chambered, axile placentation, three styles, three bifid stigmas.

Fruit: Three valved capsule.

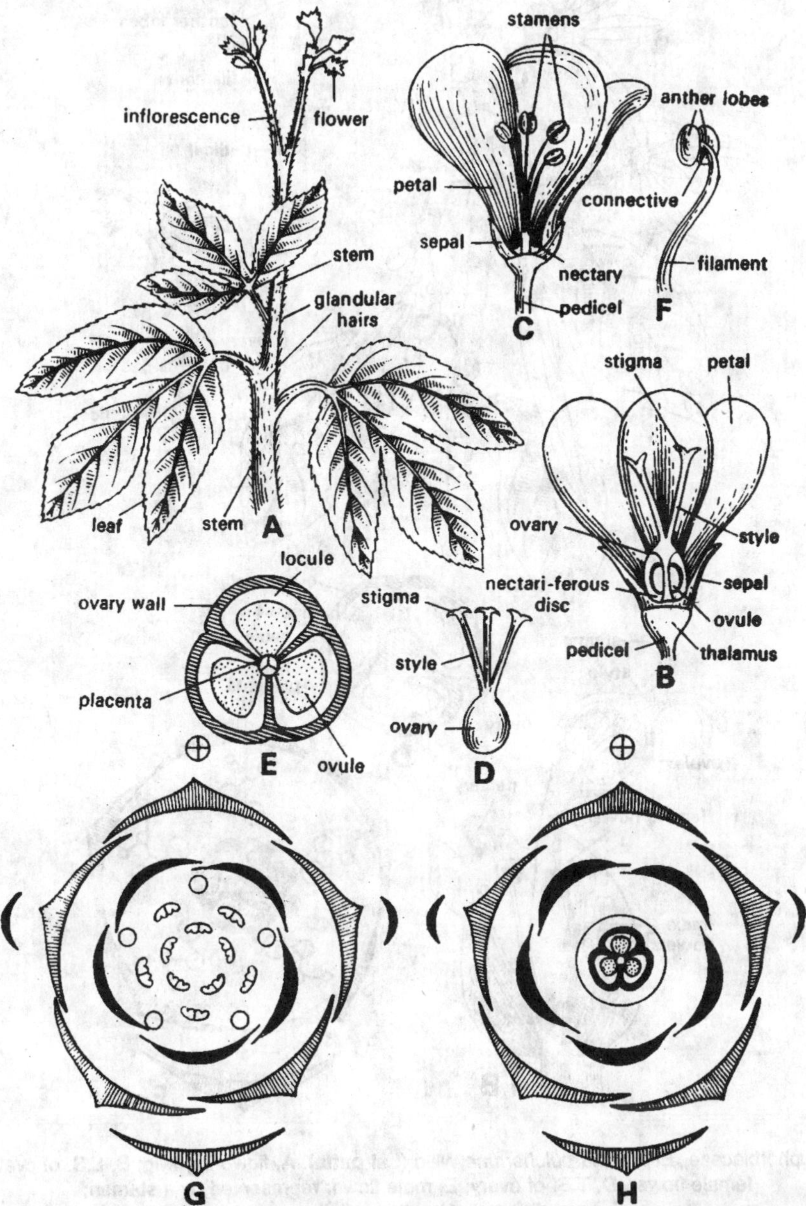

Fig. 12.6. Euphorbiaceae. *Jatropha gossypifolia* L. (**Bherenda**). A, flowering twig; B, L.S. of female flower; C, L.S. of male flower; D, gynoecium; E, T.S. of ovary; F, stamen; G, floral diagram of male flower, H, floral diagram of female flower.

Floral Formulae: **Male:** ⊕ ♂ K (5), C5, A(5 + 5)

Female: ⊕ ♀ K (5), C5, G(3̲).

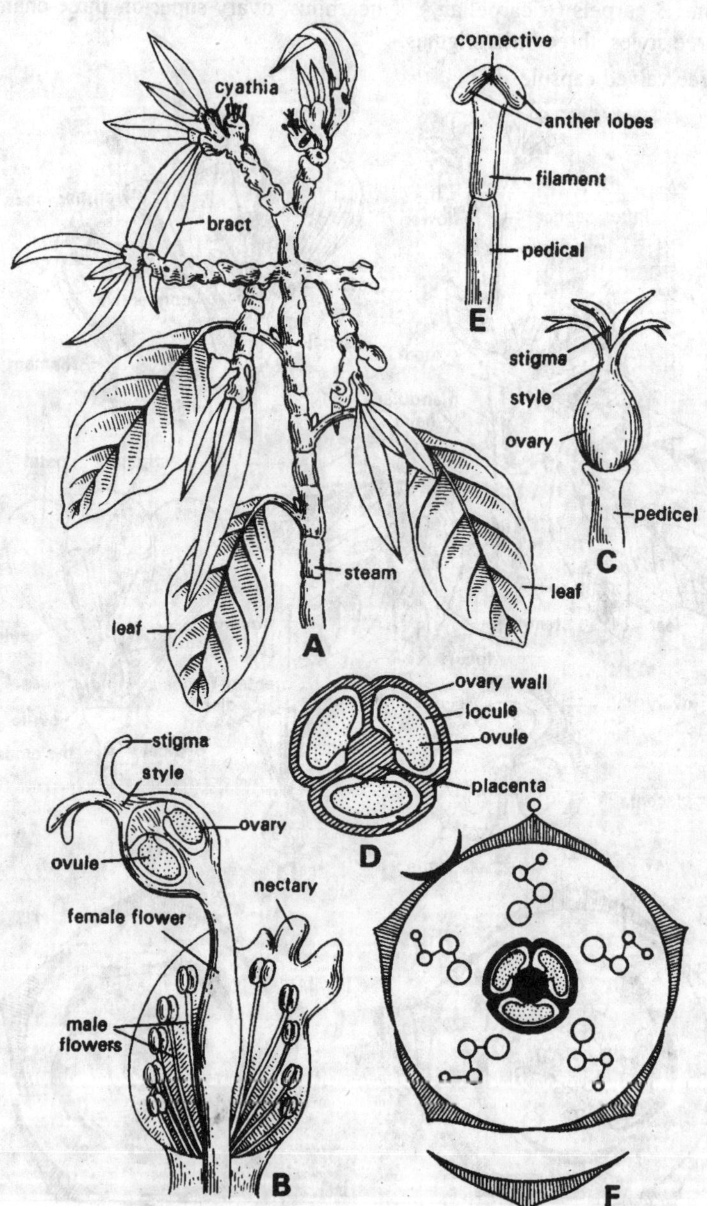

Fig.12.7. Euphorbiaceae, *Euphorbia pulcherrima* Willd (**Lal patta**). A, flowering twig; B, L.S. of cyathium; C, female flower, D, T.S. of ovary; E, male flower represented by a stamen; F, floral diagram of cyathium (inflorescence).

7. ***Euphorbia pulcherrima*** willd., Eng. Poinsettia; Verna. **Lal patta. Stem**– herbaceous or woody, aerial, erect, branched, solid cylindrical, latex present. **Leaf**–cauline and ramal, simple, aternate, extipulate, petiolate, ovate, entire, acute, glabrous, latex present, unicostate reticulate. **Inflorescence**–special, cyathium, bracts red and leaf-like, male flowers surround the female flower. **Male flower**–represented by a single jointed stamen, anther dithecous, basifixed introrse. **Floral formula**–Br ♂ K_0, C_0, A_1, G_0, **Female flower** – represented by a tricarpellary stalked gynoecium, syncarpous, superior, trilocular, one ovule in each locule, axile, style short, stigmas 3 bifid. **Fruit**–capsule. **Floral formula**–Br, ♀ K_0, C_0, A_0, $G_{(3)}$.

Economic Importance of the Family

The family consists of the plants of great economic value. Some of the important ones are given below:

1. *Acalypha hispida;* An ornamental herb.

2. *Acalypha wilkesiana*; (Eng. –Garden Acalypha). Grown in the gardens as ornamental plant.

3. *Bischofia javanica*; (Eng.–Bishop wood; Verna–**Bhillar**)– A tree. A red dye, obtained from the bark is used to stain rattan baskets. The bark is also used as a tan. The wood is exteremely resistant to water effect and therefore, largely utilized in the construction of bridges and boats.

4. *Bridelia retusa*; (Verna.–**Ekdania, Khaja**)–A tree, found in Rajasthān, Madhya Pradesh, the Western Peninsula and Bihar. The bark is used for tanning. The fruits are edible and the leaves are used as fodder.

5. *Cleistanthus collinus*; (Verna–**Garari**)–A small tree, found in Tamil Nadu, Malabar, Bihar, Orissa and Madhya Pradesh. The bark, leaves and green fruits are used as tan.

6. *Croton aromaticus*; An aromatic shrub or small tree found in Andhra Pradesh produces a gum-resin, which is used in varnishes.

7. *Croton oblongifolius*; (Verna–**Chuka**) – The seeds yield an oil, which is used as a purgative and also as an insecticide.

8. *Croton tigllium*; (Verna–**Jamalgota**)–A shrub or small tree, native of South East Asia but cultivated in Assam, Bengal and South India. The seeds are the source of croton oil, which is used as a purgative.

9. *Baccaurea courtallensis*: Found in the Western Ghats. The fruits are edible.

10. *Baccaurea sapida*; (Verna–**Lathua**)–Found in Assam, Bengal and Andaman Islands. The fruits are edible.

11. *Chrozophora prostrata* (Verna–**Subali**)–The root ashes are given to the children for cough treatment. The seeds are used as a purgative.

12. *Baliospermum montanum*; (Verna–**Danti**)–The seeds are used as a strong purgative. The decoction of leaves is used in asthma.

13. *Euphorbia milii*; Syn. *E. splendens*; A small climbing shrub; native of Madagascar. Grown in gardens as ornamental.

14. *Euphorbia pulcherrima*; Syn. *Poinsettia pulcherrima*; Eng–Poinsettia; (Verna.–**Lal Patta**)–A shrub, native of Central America. Grown as an ornamental.

15. *Euphorbia tirucalli*; (Verna.–**Tohar**)–Succulent spineless small tree, native of Africa. The roots are used for poisoning fish and birds.

16. *Euphorbia hirta*; (Verna.–**Dudhi**)–The plant is of medicinal value and used in many diseases of children and adults.

17. *Euphorbia thymifolia*; (Verna.–**Chhoti dudhi**)–The juice of leaves and seeds is used as a strong purgative. The juice is also used for remedy of ringworms and other skin diseases. It is an antidote for snake bite.

18. *Euphorbia royleana*; (Verna.–**Thar**)–The plants are used in hedges. The latex is used medicinally in several ways.

19. *Euphorbia antiquorum*; (Verna.–**Tridhara, sehund**)–The decoction of stem is used as a remedy of gout. The juice of the plant is also used as a strong purgative. The root bark is also used as a purgative.

20. *Glochidion zeylanicum*; (Verna.–**Kumbalm**)–The leaves are used in itches.

21. *Securinega leucopyrus*; Syn. *Phyllanthus leucopyrus*; *Flueggea leucopyrus* (Verna.–**Hartho**)–A large shrub or small tree, found in Uttar Pradesh, the Punjab, Maharashtra and Tamil Nadu. The fruits are edible.

22. *Jatropha curcas*; (Verna.–**Safed arand**)– A shrub or small tree, native of tropical America, now cultivated in Travancore. The seed oil is used for manufacturing candles, soaps and as a lubricant and for illumination. The seed oil is also used as a purgative. The tender shoots are edible. Oil also used as biodiesel.

23. *Jatropha gossypifolia*; (Verna.–**Bherenda**)–A shrub, native of Brazil. Cultivated as an ornamental.

24. *Jatropha hastata*; Grown as an ornamental for bright crimson flowers.

25. *Jatropha padagrica*; (Eng.–Gouty stemmed Jatropha)–Native of Panama, grown for its bright scarlet flowers.

26. *Manihot esculenta*; Syn. *M. utilissima*; (Eng.–Tapioca; Verna.–**Sakarkand**)– A small shrub; native of Brazil, now grown in Kerala, Tamil Nadu and Karnataka. The tapioca tubers are exploited commercially to obtain starch, sago, semolina and flour.

27. *Hevea brasiliensis*; (Eng.–Para rubber; Verna.–**Rabar**) – A tall tree; native of Brazil;, now grown in Kerala, Tamil Nadu, Karnataka and North-Eastern Assam. The latex, obtained from the bark of the tree, is used for preparing rubber, which is used for tyres and inner tubes, waterproof clothing and various electrical goods.

28. *Mallotus philippinensis*; (Verna.–**Kamala, Rauni**)–A small tree found commonly in Bengal, Madhya Pradesh, Maharashtra and Orissa. The red dye, obtained from the surface of the fruits, is used for dyeing silk. Also used medicinally to remove thread worms and Ascaris.

29. *Cicca acida*; Syn. *Phyllanthus acidus*; *Averrhoa acida*; (Eng.–Stargoose-berry; Verna.–**Hariphul**)- A small tree cultivated in Bengal and South India for the edible fruits. The leaves are edible. The bark is used as a tan.

30. *Emblica officinalis*; Syn. *Phyllanthus emblica*; (Verna.–**Amla**)– A common tree with edible fruits. The fruits are also used in diarrhoea and dysentery. The bark, leaves and fruits are used in dyeing and tanning. The wood yields excellent charcoal. The pickle or jam is prepared from the fruits. The fruit is very rich in vitamin C.

31. *Emblica fischeri*; A small tree, found in South India. The fruits are edible.

32. *Kiganelia reticulata*; Syn. *Phyllanthus reticulatus*; (Verna.–**Panjoli**)– A climbing shrub, commonly found in Northern India. The roots are the source of a red dye.

33. *Putranjiva roxburgii*; (Eng.–Child-life tree; Verna.–**Putranjiva**)–A tall tree, grown as a hedge plant. The nuts are made in rosaries. The Hindus believe that if the hard stones of the fruits are made into rosaries and placed around the neck of the children, they keep them in good health. The seeds also yield an oil which is used for burning purposes. The leaves are used as fodder. The leaves, fruits and stone of fruits are used medicinally in colds and fevers.

34. *Ricinus communis*; (Eng.–castor-oil plant; Verna–**Arand**)–A small tree, cultivated chiefly in Andhra Pradesh, Maharashtra, Karnataka and Orissa. The seeds are the source of castor-oil, which is mainly used as a lubricant and as a purgative. It is also used for transparent soap, textile soap, typewriter-inks, perfume, aromatics, varnishes and paints. The seed cake is used as a fertilizer. The writing and printing-papers are made of the wood-pulp. Castor stems are used for strawboards and cheap wrappings.

35. *Tragia involucrata*; (Verna–**Barhaita**)–The roots and fruits are used medicinally.

Systematic Relationships: Bentham and Hooker have included nine families in the series Unisexuales. They are Euphorbiaceae, Balanopseae, Urticaceae, Plantanaceae, Leitnerieae, Juglandeae, Myricaceae, Casuarineae and Cupuliferae.

Hallier included the family Euphorbiaceae in his Passionales, Rendle and Hutchinson placed the family in the order Euphorbiales.

13

Order-Rutales Family-Rutaceae

According to Hutchinson this is the forty seventh order of the phylum Angiospermae, subphylum Dicotyledones and division Lignosae. This order comprises of four families, *i.e.*, Rutaceae, Simaroubaceae, Burseraceae and Averrhoaceae, of which the Rutaceae has been described in the present text in detail. Bentham and Hooker placed the family Rutaceae along with other ten families including Lineae and Meliaceae in their seventh order–the Geraniales of class Dicotyledones, division Polypetalae and series Disciflorae. Engler and Prantl have also placed the Rutaceae along with nineteen other families including Linaceae and Meliaceae in the twenty third order–the Geraniales of class Dicotyledoneae and Subclass Archichlamydeae.

Hypogynous to slightly perigynous; bisexual or unisexual; apocarpous to syncarpous with axile, basal or apical placentation; ovules mostly few; stamens definite in number, free; petals contorted to valvate; disc mostly conspicuous; leaves often gland-dotted; endosperm present.

Family—Rutaceae

There are about 140 genera and 1300 species in this family.

Distribution: They are widely distributed in warm temperate regions. They are also found in tropical regions. In our country the family is represented by several important genera such as, *Citrus, Aegle, Feronia, Murraya, Ruta*, etc.

Habit: Usually the plants are trees or shrubs. Herbs are rarely found. Some of the shrubs are climbing and xerophytic. *Aegle marmelos* (**Bel**) and *Feronia elephantum* (**Kaith**) are large sized trees. *Murraya exotica* is an ornamental shrub with scented white flowers. Sometimes the shrub becomes tree-like in habit. Species of *Citrus* are small trees. They are grown in the gardens for their fruits rich in vitamins. *Ruta graveolens* is a strong-smelling herb which is a common garden shrub. The plant of *Paraminya grifithii* is a woody climber. This possesses strong curved spines.

It is believed that *Citrus aurantium* (**Narangi**) probably originated in Northern India and Assam. It has been cultivated since 1500 B.C.

Root: Tap and branched.

Stem: Usually erect, sometimes climbing, branched, woody or rarely herbaceous, cylindrical, solid, green when young, grey when becomes old.

Leaves: The leaves may be either simple or compound. They are usually alternate and rarely opposite. They are exstipulate. The leaves are dotted with glands. These glands contain volatile oils, which give typical smell to the leaves. In most of the *Citrus* species the leaves possess winged petioles. According to many authors the leaves with winged petioles are regarded as unifoliate palmate type of compound leaves. In early stages actually the leaf possesses three leaflets but on its maturity two of them fall off. In certain Australian genera such as *Citrus, Feronia*, etc., the leaves are reduced to spines. Such plants are extreme xerophytes.

Inflorescence: The inflorescence consists of several variations in different species. Generally it is cymose, but sometimes it is racemose. In some cases the flowers are solitary axillary. In *Murraya exotica* the flowers are arranged in either axillary or terminal corymbs. In *Citrus* sp., the flowers are either solitary axillary or arranged in racemes.

Flower: The flowers are pedicellate, complete, hermaphrodite, regular or occasionally slightly zygomorphic, white or yellow, hypogynous. A disc is present beneath the ovary. Usually pentamerous and rarely tetramerous, or trimerous. Sometimes the flowers are unisexual, *e.g.*, *Evodia, Zanthoxylum*, etc.

Calyx: There are usually five sepals. Sometimes only four sepals are found. Usually polysepalous but in the genera with zygomorphic flowers the calyx becomes either tubular or united. The aestivation is either imbricate or quincuncial.

Corolla: There are usually five petals. Sometimes only four petals are present. The corolla is polypetalous. However, in an Australian species *Correa speciosa*, the petals become united attaining campanulate shape. The petals are coloured–white, yellow or red. The petals are inferior. The aestivation is imbricate.

Androecium: Usually ten or eight stamens are present in obdiplostemonous condition. Sometimes the number of stamens decreases or increases to indefinite. In *Citrus* sp., and *Aegle marmelos* the stamens are indefinite. In such cases they are arranged in irregular bundles, *i.e.*, polyadelphous condition. In *Murraya exotica* there are ten stamens arranged in two whorls of five each.

One whorl of stamens is alternating and the other antipetalous. The filaments are free. The anthers are bilobed, basifixed and introrse.

Gynoecium: The gynoecium consists of 3, 4, or five carpels. Rarely it consists of a fewer number of carpels or indefinite carpels. The carpels are fused together, *i.e.*, syncarpous. In most of cases the placentaion is axile. The ovary is tri to multilocular. Each locule contains one or more anatropous ovules. In *Feronia*, however, the placentation is parietal. In this a large number of ovules are arranged in several rows on the parietal placentas.

Fruit: The fruits are of various kinds in different genera. They may be berry, drupe or capsule. In *Murraya exotica* the fruit is berry. The *Citrus* fruits are known as hespiridium (a berry). In *Acronychia*, the fruit may be capsule or drupe.

Seeds: The seeds are endospermic or non-endospermic. Each seed contains a large straight very often curved embryo.

Pollination: Usually entomophily, *i.e.*, by means of the agency of insects. The nectar secreting disc found beneath the ovary is main source of attraction for the insects. The coloured and scented flowers also attract the insects.

Floral Formula: ⊕ ⚥ K 5, or (5), C 5, A 10, G ($\underline{3-5}$).

Bentham & Hooker (1862)	**Engler & Prantl (1931)**	**Hutchinson (1959)**
Phanerogams	Phanerogams	Angiospermae
Dicotyledones	Dicotyledones	Dicotyledones
Polypetalae	Archichlamydeae	Lignosae
Disciflorae	Geraniales	Rutales
Geraniales	Rutaceae	Rutaceae
Rutaceae		

Description of some important plants in semi-technical language:

1. ***Citrus aurantium.***

Habit: A small tree.

Root: Tap and branched.

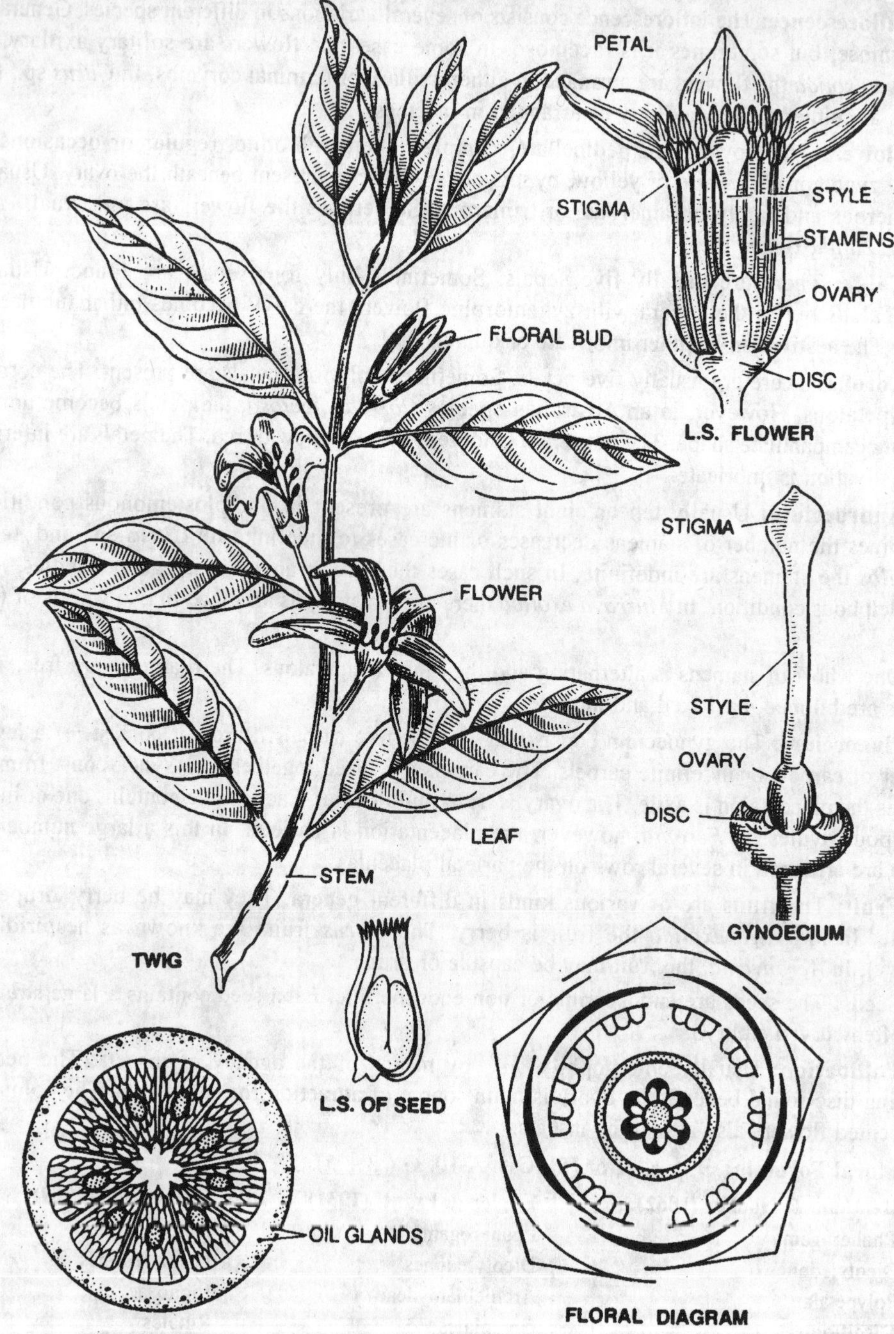

Fig. 13.1. Rutaceae, *Citrus aurantium* Linn., Orange; Verna, **Narangi.**

Stem: Erect, woody, branched, cylindrical, glabrous, solid.

Leaf: Cauline, alternate, petiolate (petiole winged and flattened), stipulate, coriacious, persistent, more or less serrate, acute or obtuse, glabrous, unicostate reticulate venation.

Inflorescence: Cymose, solitary axillary or in small cymes.

Flower: Pedicellate, bracteate, hermaphrodite, white or pinkish, scented, actinomorphic, complete, hypogynous.

Calyx: Five sepals, gamosepalous, inferior, quincuncial aestivation.

Corolla: Five petals, polypetalous, white, scented, inferior, imbricate aestivation, linear oblong.

Androecium: Indefinite, stamens (15 – 60), inserted around, a large cupular or annular disc, polyadelphous, filaments variously connate, compressed at the base, anthers oblong, basifixed, introrse.

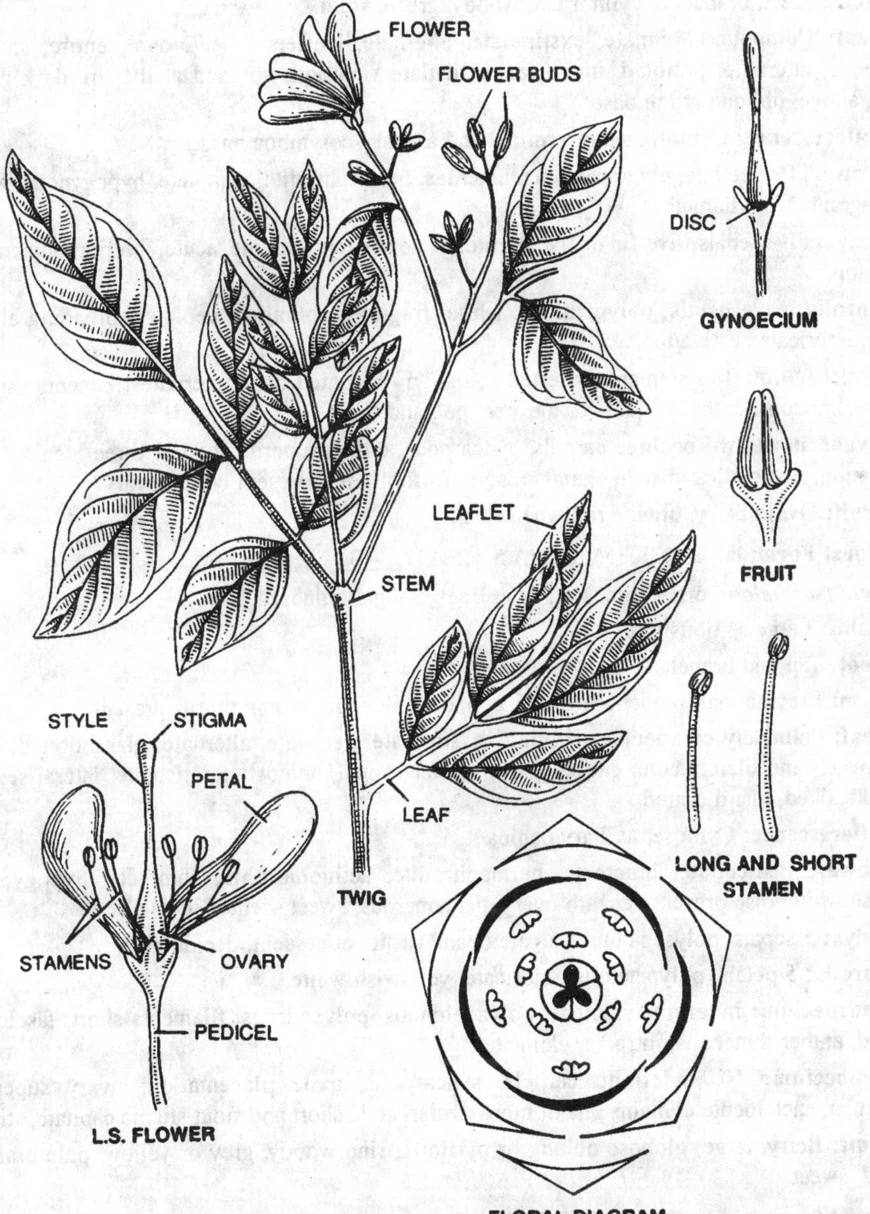

Fig. 13.2. Rutaceae. *Murraya exotica* Spreng.

Gynoecium: Polycarpellary, syncarpous, ovary superior, multilocular, nectar secreting disc present below ovary, axile, placentation, style stout, deciduous, stigma capitate.

Fruit: Berry (hesperidium).

Floral Formula: ⊕ ⚥ K (5), C 5, A ∝, G ($\underline{\infty}$)

2. ***Murraya exotica***; Verna–**Kamini**

Habit: Evergreen large shrub.

Root: Tap, branched.

Stem: Erect, branched, cylindrical, woody, green, solid.

Leaf: Compound, pinnate, exstipulate, alternate, leaflets 3 – 7, oval, entire, smooth, coriaceous, alternate, petioled, unicostate reticulate venation, notched at the tip, dark green, shining above, oblique at the base.

Inflorescence: Cymose, short terminal and axillary corymbocymes.

Flower: Pedicellate; ebracteate, pentamerous, hermaphrodite, complete, hypogynous, white, very fragrant, 1/2″ diameter.

Calyx: Five sepals, five-fid or five partite, gamosepalous, minute, acute, inferior, quincuncial aestivation.

Corolla: Five petals, polypetalous, white, fragrant, oblong lanceolate, spreading above, inferior, imbricate aestivation.

Androecium: Ten stamens, inserted round an elongated disc, alternate filaments shorter, obdiplostemonous, anthers bi-celled, introrse, polyandrous.

Gynoecium: Two or three carpels, syncarpous, ovary, superior, two or three locular, axile placentation, two ovules in each chamber, style filiform, disk present below ovary.

Fruit: Ovoid berry, orange red when ripe.

Floral Formula: ⊕ ⚥ K (5), C 5, A 5 + 5, G ($\underline{2-3}$).

3. ***Aegle marmelos*** Corr. Verna. **Bel**; Eng. Bael, Bengal quince.

Habit: Large spinous tree.

Root: Tap and branched.

Stem: Erect, aerial, branched woody, cylinderical, solid, strong thorns present.

Leaf: Palmately compound, trifoliate, exstipulate, petiolate, alternate, gland dotted, each leaflet ovate-lanceolate, acute, entire, glabrous, net-veined, unicostate reticulate, lateral sessile, terminal stalked, gland dotted.

Inflorescence: Cymose, axillary panicles.

Flower: Pedicellate, bracteate, hermaphrodite, actinomorphic, complete, hypogynous, yellowish white, disc present beneath ovary, pentamerous, sweet scented, cyclic.

Calyx: 5 sepals, polysepalous, valvate or imbricate, pubescent, deciduous.

Corolla: 5 petals, polypetalous, imbricate, yellowish white.

Androecium: Indefinite stamens, polyadelphous, polyandrous, filaments short, fascicled, basifixed, anther dithecous, introrse, elongate.

Gynoecium: 10, indefinite carpels, syncarpous, axile placentation, ovary superior, multilocular, each locule contains one or more ovules, style short and stout stigma capitate, sticky.

Fruit: Berry, large, globose oblong or pyriform, rind woody, grey or yellow, pulp orange-coloured, sweet.

Floral formula: Br. ⊕ ⚥ K 5, C 5, A ∝, G ($\underline{\infty}$)

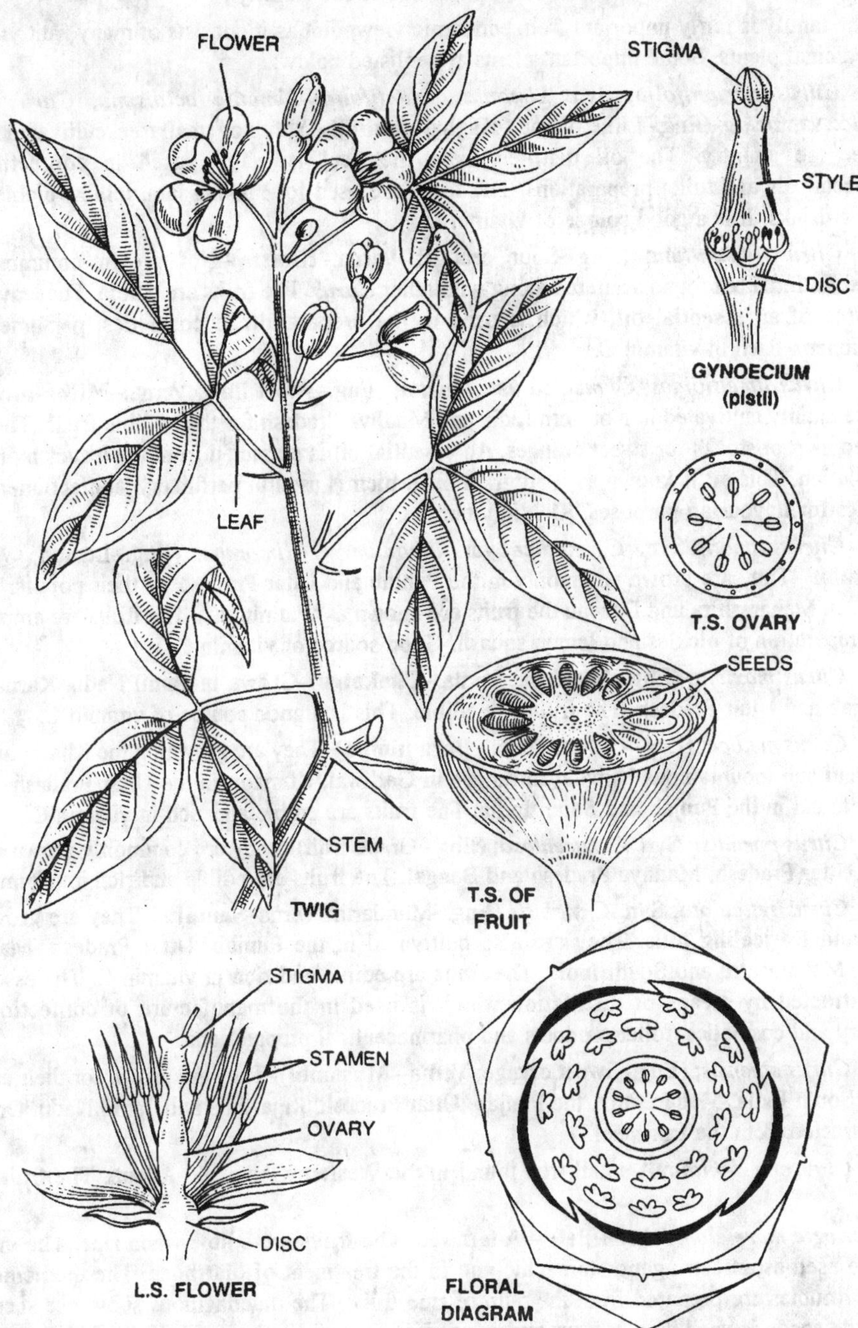

Fig. 13.3. Rutaceae. *Aegle marmelos* Corr.; Eng. Bengal quince; Verna, **bel.**

Economic value. The ripe fruits are edible. The unripe fruits are used as an astringent, stomachic, and also in the treatment of diarrhoea. The mucilaginous substance, secreted round the seeds, is used as a cement and is also employed as a varnish.

Economic Importance of the Family

The family is fairly important from economic viewpoint as it consists of many fruit yieldıng and medicinal plants. Some important plants are enlisted below:

1. *Citrus aurantifolia*; Syn. *Limonia aurantifolia*; *Limonia acidisima; Citrus lima; C. medica* var. *acida*; (Eng.–Lime; Verna.–**Kaghzi-nimbu**). A shrub or small tree, cultivated nearly all over the country. The oil distilled from the peel is mainly used in confectionery, pharmaceuticals and toilet preparations. The fruits are used for making jam, jellies, pickles and alcoholic drinks. It is a good source of vitamin C.

2. *Citrus aurantium*; (Eng.–Sour orange; Verna.–**Khatta**)–Not grown commercially anywhere in India but found scattered along with other *Citrus*. The fruits are edible. The leaves are the source of an essential oil, which is used in the preparation of cosmetics, perfumes and confectionery. Rich in vitamin C.

3. *Citrus limettioides*; *C. medica* var. *limetta*; (Eng.–sweet lime; Verna.–**Mitha-nimbu**)–They are mainly cultivated in Noıthern India and Madhya Pradesh for their edible frutis. They are also used as root stocks for sweet oranges. An essential oil is extracted from the leaves by means of distillation. This oil is known as 'Petitgrain oil' which is used in perfumery, confectionery and cosmetics for flavouring purposes. Rich in vitamin C.

4. *Citrus limon*; Syn. *C. medica* var. *limonum, C. limonum.* (Eng.–Lemon, Verna–**Baranimbu**)–They are grown commonly in the Punjab and Uttar Pradesh by their popular name 'galgal'. In Maharashtra and Deccan the fruits are known as 'Pat nimbu'. The fruits are employed in the preparation of pickles and lemon squash. Good source of vitamin C.

5. *Citrus maxima*; (Eng.–Shadlock, Verna.–**Chakotra**)–Grown in Tamil Nadu, Karnataka, the Punjab and Uttar Pradesh. The fruits are edible. This is a good source of vitamin C.

6. *Citrus medica*; (Eng.–Citron, Verna.–**Bara nimbu**)–They are grown on the Khasia hills of Assam and sub-mountainous Himalayan ranges in Garhwal, Kumaon in the Uttar Pradesh. They are also found in the Punjab and South India. The fruits are edible and rich in vitamin C.

7. *Citrus paradisi*; Syn. *C. racemosa*; (Eng.–Grape fruit)–They are commonly grown in the Punjab, Uttar Pradesh, Madhya Pradesh and Bengal. The fruits are edible and rich in Vitamin C.

8. *Citrus reticulata*; Syn. *C. nobilis* (Eng.–Mandarin; Verna–**Santara**)–They are grown in Khasia and Darjeeling hills. They are also cultivated in the Punjab, Uttar Pradesh, Madhya Pradesh, Maharashtra and South India. The fruits are edible and rich in vitamin C. The essential oil is extracted by means of distillation which is used in the manufacture of confectionery, perfumery and cosmetics, toilet products and pharmaceutical preparations.

9. *Citrus sinensis*; (Eng.–Sweet orange; Verna– **Musambi**)–They are grown for their edible fruits in South India, Maharashtra, the Punjab, Uttar Pradesh, Rajasthan and Tamil Nadu. Orange oil is extracted from the fruit peel.

10. *Clausena dentata*; A small tree found in the Western Ghats and Assam. The fruits are edible.

11. *Aegle marmelos*; (Verna-**Bel**) – A tall tree. The fruits are edible when ripe. The unripe fruits are used as an astringent, stomachic and in the treatment of diarrhoea. The medicines of stomach troubles are prepared from the pulp of ripe fruits. The mucilaginous substance secreted around the seeds is used as a cement and is used as a varnish. The pulp of the fruit is a good laxative. The pulp is often used to strengthen mortar. The seeds yield a dye. The leaves are used as fodder. The leaves are employed in the worship of Lord Shiva by Hindu ladies.

12. *Feronia limonia*; Syn. *F. elephantum*; (Eng.–Elephant apple; Verna–**Kaith**) – A tree found all over India. The fruits are with hard rind, large, globose and edible. The pulp of the fruits is

eaten in many ways. It is also used in the treatment of diarrhoea and dysentery. The leaves are used as fodder. A gum is obtained from the trunk and branches of the tree.

13. *Fortunella japonica*; Syn. *Citrus japonica*; (Eng.–Kumquat)–This is a shrub or a small tree. It is native of East Asia, but now grown mainly in Northern India. The fruits are edible. The shrub is ornamental.

14. *Fortunella margarita*; Syn., *Citrus margarita*–This is an ornamental shrub possessing edible fruits.

15. *Acronychia pedunculata*; A small tree. The leaves yield an essential oil. The tender leaves are used as condiment. It is found in Dehradun, West Bengal and Khasia hills of Assam.

16. *Glycosmis pentaphylla*; (Verna–**Ban-nimbu**)–The ripe fruits are edible. The branches are used as tooth-brushes.

17. *Luvunga scandens*; (Verna–**Long-phul**)–A large, thorny climbing ornamental shrub, with beautiful scented white flowers. They are found in Bengal and Assam. The fruits yield an essential oil which is used in perfumery and pharmaceuticals.

18. *Murraya paniculata*; Syn. *Murraya exotica*; (Verna–**Kamini**)–This is an ornamental shrub possessing white scented flowers. The fruits are edible.

19. *Murraya koenigii;* (Verna–**Mitha neem**)– A large shrub or small tree. Cultivated in both Northern and South India. The leaves are used for flavouring curries.

20. *Ruta graveolens*; (Verna.–**Sadab**)–A herb, cultivated throughout country. The leaves are used as condiment and for flavouring curries. It also possesses medicinal properties.

21. *Toddalia asiatica*; (Verna.–**Kanj**)–A shrub. It is found in sub-tropical Himalayas. The root yields a yellow dye. The leaves and fruits are edible. The pickle is prepared from the fruits.

22. *Zanthoxylum alatum*; (Verna.–**Tejbal**)– A small tree found in sub-tropical Himalayas. The seeds are used as tonic. The walking sticks are made from its wood. The branches are used as tooth brushes.

23. *Fagara budrunga*; Syn. *Zanthoxylum rhesta*; (Verna–**Badrang**)–A large tree. It is found in Western Ghats and West Bengal. An aromatic oil is extracted from its seeds. The fruits are used as spice. The wood is used in making shuttles.

24. *Fagara oxyphylla*; (Verna–**Timar**)–A shrub found in the hills of Kumaon and Assam. The tender shoots are used as vegetable.

Systematic Relationships: Hutchinson has placed the family Rutaceae along with three other families in the order–Rutales. Earlier this view was put forth by Rendle and accepted by Wettstein in principle.

14

Order–Apocynales Family–Apocynaceae & Family–Asclepiadaceae

According to Hutchinson this is the fifty first order of the phylum Angiospermae, sub-phylum Dicotyledones and division Lignosae. This order comprises of only four families, *i.e.*, Plocospermaceae, Apocynaceae, Periplocaceae and Asclepiadaceae. The Apocynaceae and the Asclepiadaceae have been described in the present text in detail. Bentham and Hooker have placed the Apocynaceae and the Asclepiadaceae along with other four families in the seventh order– the Gentianales of class Dicotyledones, division Gamopetalae and series Bicarpellatae. Engler and Prantl have put these families along with three others in the fifth order– the Contortae of class Dicotyledones and sub-class Sympetalae.

Corolla sympetalous; actinomorphic; stamens epipetalous, alternate with the corolla lobes; pollen becoming waxy and collecting into masses as in orchids; ovary superior, carpels-2, free or becoming free in fruit; styles united.; ovules mostly numerous; a parietal or on the septa; seeds usually with endosperm and straight embryo. Leaves opposite, simple; no stipules.

Family—Apocynaceae

There are about 300 genera and 1300 species in this family.

Distribution: The members of this family are found throughout the world, but they are more commonly met with in the tropical regions.

Habit: There is a great variation in the habit of the plants of this family. They may be herbs, erect or twining shrubs or trees. *Vinca rosea* (Verna–**Sadabahar**) is a perennial herb; *Vallaris solanacea* (Verna–**Ramsar**) is a large twining shrub; *Nerium indicum* (*N. odorum*) is a large shrub with beautiful red or white flowers; *Thevetia peruviana* or *Thevetia nerifolia* (Verna–**Pili kaner**) is a large shrub or a small tree; *Plumeria acutifolia* is a small sized tree and *Alstonia scholaris* is a medium sized tree. In some genera, the stem becomes tuber like, *e.g.*, *Adenium*. The species of *Landolphia* and *Clitandra* are climbing shrubs.

The latex is present in most of the genera.

Root: Tap and branched.

Stem: Usually erect, branched, solid, glabrous rarely tuberlike and thick.

Leaves: The leaves are simple, petiolate, usually opposite decussate. In rare cases the leaves are alternate or even whorled (*e.g., in Nerium odorum, Alstonia*, etc.) Usually the leaves are exstipulate and very rarely they may be stipulate.

Inflorescence: Usually the inflorescence is of cymose type. It is very rarely solitary as in *Vinca*.In *Carissa*, the flowers are found to be arranged in corymbose cymes. In *Plumeria*, the flowers are arranged in terminal cymes. In *Alstonia*, the flowers are found to be arranged in umbellate branched panicled cymes. In *Rauvolfia*, the flowers are arranged in umbellate or corymbose cymes.

Flowers: The flowers are pedicellate, bracteate, bracteolate, hermaphrodite, actinomorphic, regular, sometimes slightly, zygomorphic, complete, hypogynous and pentamerous. In rare cases the flowers are tetramerous with reduction to two in the pistil.

Calyx: Usually it consists of five sepals, gamosepalous. The calyx is generally divided almost to the base. The aestivation is quincuncial.

Corolla: Usually the corolla consists of five petals, gamopetalous. It is generally salver or funnel shaped. The corolla tube usually possesses hairy appendge or scales. The aestivation is contorted.

Androecium: It consists of five stamens alternating with the petals. The stamens are situated on the tube or the throat of the corolla (*i.e.*, epipetalous). The filaments are short; anthers introrse, polyandrous or connate and often adhere to the stigma. The antherlobes are sometimes empty at their base and prolonged into spines.

Gynoecium: It consists of two carpels. The carpels may be free (apocarpous) or connate (syncarpous); superior, sometimes partly inferior as in *Plumeria.* The style is simple and the stigma is thick and often bilobed. Rarely the number of carpels exceeds, *i.e.*, 3 to 5. Usually a nectar secreting disc is situated beneath the gynoecium. In syncarpous gynoecium, the ovary may be unilocular with parietal plcentation or it may be bilocular with axile placentation. In the case of separate ovaries the placentation is marginal. The ovary is superior or half–inferior. Numerous ovules are found to be situated on parietal placentas or in two chambered ovaries on marginal walls.

Fruit: In the case of free ovaries, the fruit is a pair of follicles. Sometimes the fruits of separate ovaries are fleshy and indehiscent, or may be one seeded, *e.g.*, *Cameraria.* In the case of syncarpous ovary, usually the fruit is indehiscent, fleshy and berry-like, *e.g.*, in *Landolphia.* In *Cerbera,* it may be a drupe. This fruit is coconut like and distributed by means of water currents. In certain genera, possessing syncarpous ovaries a two-valved capsule is found, *e.g.*, in *Aspidosperma* and *Allamanda.*

Seed: In dry fruits the seeds are generally winged, *e.g.*, in *Plumeria.*Sometimes the seed bears a tuft of hairs at the base, *e.g.*, in *Kickxia*, and sometimes at both ends, *e.g.*, In *Strophanthus.* The embryo is straight, with or without endosperm.

Pollination: Generally it takes place through the agency of insects.

Floral Formula : $\oplus$ ⚥ K (5), $\overline{C(5)\,A(5)}$, G($\underline{2}$)

Bentham & Hooker (1862)	**Engler & Prantl (1931)**	**Hutchinson (1959)**
Phanerogams	Phanerogams	Angiospermae
Dicotyledones	Dicotyledones	Dicotyledones
Gamopetalae	Sympetalae	Lignosae
Bicarpellatae	Contortae	Apocynales
Gentianales	Apocynaceac	Apocynaceae
Apocynaceae		

Description of important plants in semi-technical language:

1. *Vinca rosea*; Verna.-**Sadabahar.**

Habit: A perennial herb.

Stem: Erect, cylindrical, branched, solid, reddish green, glabrous.

Leaf: Cauline, simple, opposite, decussate, petiolate, exstipulate, obovate, entire, glabrous, mucronate apex, unicostate reticulate venation.

Inflorescence: Cymose, flowers arranged in axillary pairs.

Flower: Pedicellate, bracteate, hermaphrodite, actinomorphic, complete, pink, hypogynous.

Calyx: 5, polysepalous, glandular, green, inferior, quincuncial aestivation.

Corolla: 5, gamopetalous forming corolla tube, throat of corolla tube hairy forming a corona, contorted aestivation.

Androecium: 5, free, epipetalous, alternate to petals, almost sessile, anthers dorsifixed, connivent round the stigma, yellowish.

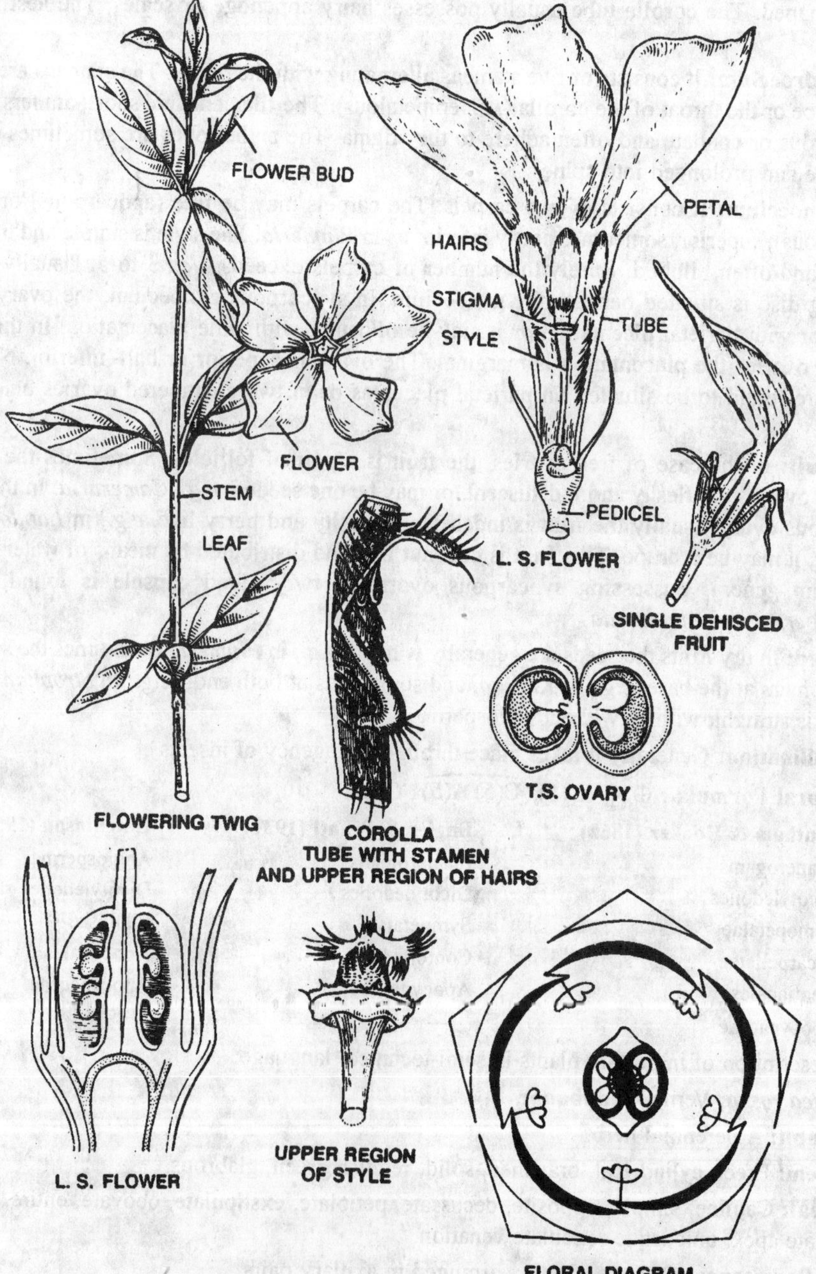

Fig. 14.1. Apocynaceae. *Lochnera rosea;* Syn. *Vinca rosea* Linn.

Gynoecium: 2 carpels (bicarpellary), syncarpous, carpels united above in the region of style and stigma, ovaries free (apocarpous below), with single style and stigma, ovaries superior, nectar secreting disc present beneath ovaries, unilocular, marginal placentation, glands present alternating with the carpels, style filiform, stigma thickened, dumb-bell shaped.

Fruit: A pair of elongated follicles.

Floral Formula: ⊕ ⚥ K 5, $\overline{C(5), A(5)}$, G $(\underline{2})$.

2. ***Plumeria acutifolia***; Verna.–**Goburchampa.**

Habit: A large shrub or small tree.

Stem: Erect, solid, branched, woody, latex present, glabrous, green, cylindrical.

Leaf: Simple, whorled, petiolate, margin smooth, apex acute, unicostate reticulate venation large.

Inflorescence: Cymose.

Flowers: Pedicellate, bracteate, hermaphrodite, actinomorphic, complete, hypogynous.

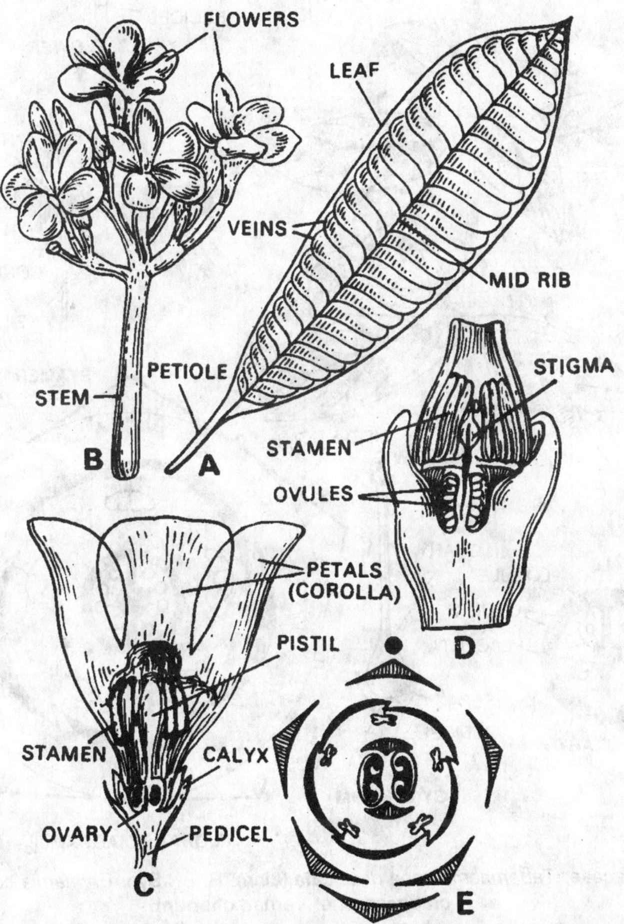

Fig. 14.2. Apocynaceae. *Plumeria* sp. A, a leaf; B, inflorescence; C, flower cut lengthwise; D, lower portion of flower in longitudinal section, E, floral diagram.

Calyx: 5, polysepalous, imbricate aestivation.

Corolla: 5, gamopetalous forming a corolla tube, twisted aestivation.

Androecium: 5, free, epipetalous, included in the corolla tube, introrse, two-celled, basifixed.

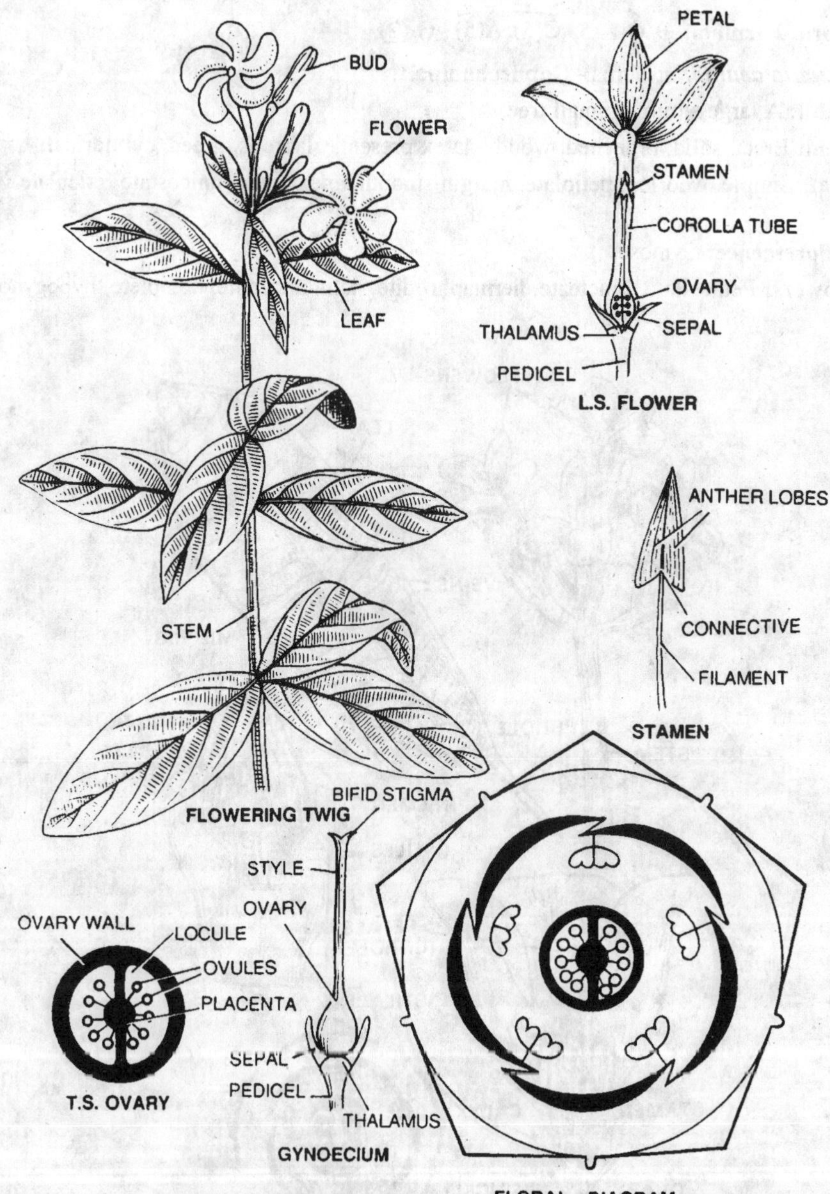

Fig. 14.3. Apocynaceae, *Tabernaemontana divaricata* (Linn). R. Br. Syn. *Ervatamia coronaria* Stapf.; Eng., crape jasmine; Verna., **chandni.**

Economic value. Grown as an ornamental. The red pulp around seed is used as a dye. Wood is refrigerant. Milky juice is used for diseases of eye. Root is acrid, bitter, used as local anodyne and chewed for relief of toothache. (Fig. 14.3)

Gynoecium: 2(bicarpellary), syncarpous, ovary superior or partly inferior, bilocular, axile placentation, style one, stigma simple.

Floral Formula: ⊕ ⚥ K 5, $\overline{C(5), A5}$, G ($\underline{2}$).

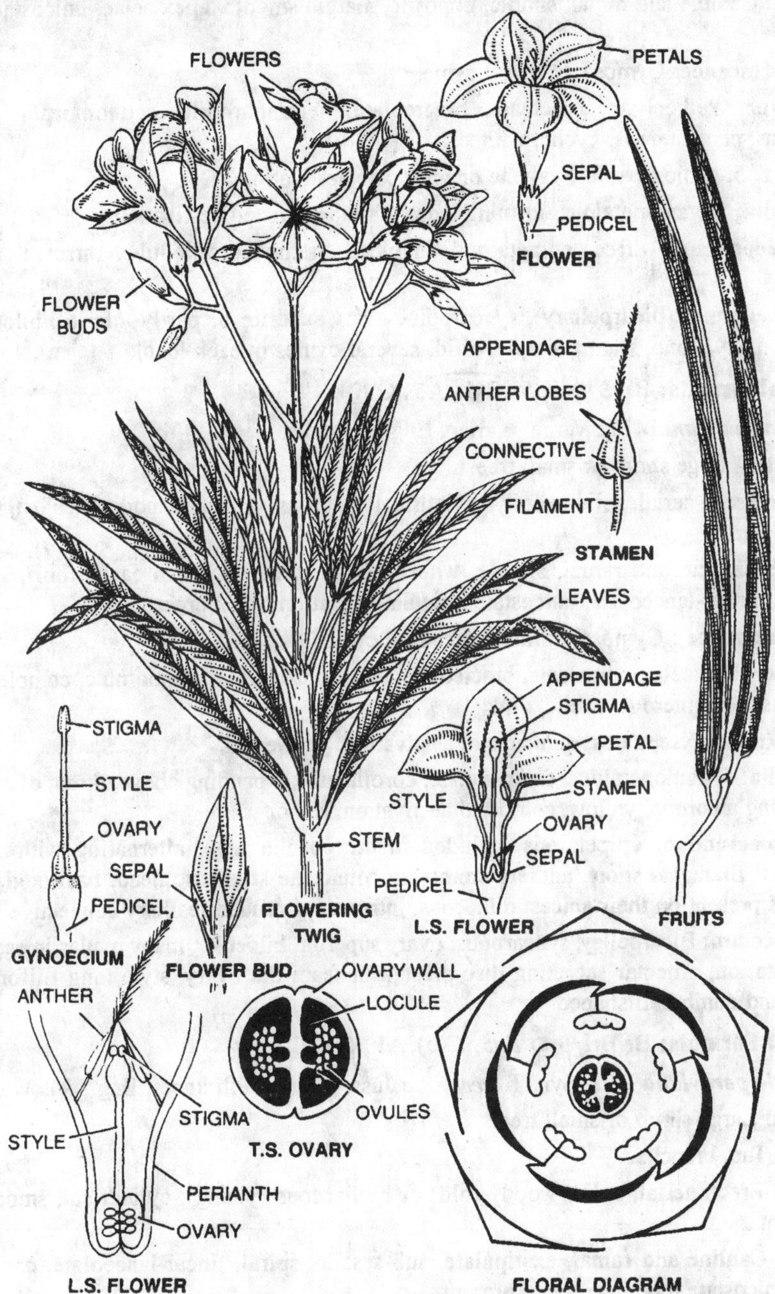

Fig.14.4. Apocynaceae. *Nerium indicum* Mill.; Eng., oleander; Verna, **kaner.**

Economic value. A beautiful shrub; grown as a hedge plant. Plant is poisonous. Root is powerful resolvent and attenuant; used externally. Oil prepared from root bark is used in skin diseases and leprosy.

1. ***Tabernaemontana divaricata*** (Linn); R. Br.; Verna. **Chandni;** Eng. Crape jasmine.

Habit: A large shrub.

Stem: Erect, solid, branched, woody, latex present, glabrous, green, cylindrical.

Leaf: Cauline and ramal, simple, opposite, margin smooth, apex acute, unicostate reticulate venation.

Inflorescence: Cymose, terminal cyme.

Flower: Pedicellate, bracteate, bracteolate, hermaphrodite, actinomorphic, complete, hypogynous, pentamerous, cyclic, white.

Calyx: 5, gamosepalous, valvate or imbricate aestivation.

Corolla: 5, gamopetalous, forming a corolla tube, twisted aestivation.

Androecium: 5, free, epipetalous, included in the corolla tube, introrse, two-celled, basifixed.

Gynoecium: 2(bicarpellary), syncarpous, ovary superior or partly inferior, bilocular, axile placentation, style one, stigma simple, bifid, several ovules in each locule.

Floral formula: Br $\oplus$ ⚥ K 5, $\overline{C(5), A5}$, G ($\underline{2}$).

2. ***Nerium indicum*** **Mill.**; Verna. **Kaner;** Eng. Oleander.

Habit: A large shrub or small tree.

Stem: Erect, aerial, solid, woody, cylindrical, glabrous, branched, nodes swollen, milky latex present.

Leaf: Cauline and ramal, simple, whorled with three leaves in each whorl, sub-sessile, exstipulate, linear-lanceolate; unicostate reticulate venation, latex present.

Inflorescence : Cymose, terminal dichasial cyme.

Flower: Pedicellate, braceate, bracteolate, hermaphrodite, actinomorphic, complete, yellow, hypogynous, pentamerous, cyclic, pink.

Calyx: 5, polysepalous, quincuncial aestivation, purple red.

Corolla: 5, gamopetalous, campanulate, corolla tube expanding above, throat of corolla tube hairy forming a corona, yellow, contorted aestivation, red or white.

Androecium: 5, epipetalous included in the corolla tube, alternating with the petals, polyardrous, filaments short, anthers connivent round the stigmatic head, basifixed, long hair appendages present on their apices, dithecous, introrse, protruded feathery connective present.

Gynoecium: Bicarpellay, syncarpous, ovary superior, bilocular, many ovules in each loculus, axile placetation, a nectar secreting disc present beneath the ovary, style long filiform, stigma thickened and dumb-bell shaped.

Floral Formula: Br Brl $\oplus$ ⚥ K 5, $\overline{C(5), A5}$, G ($\underline{2}$).

3. ***Thevetia peruviana*** Pers. Syn. T. *neriifolia* Juss. : Verna. Pili kaner; Eng. Yellow oleander.

Habit: Large shrub or small tree.

Root: Tap, branched.

Stem: Erect, aerial, solid, woody (old) or herbaceous (young), cylindrical, smooth, milky latex present.

Leaf: Cauline and ramal, exstipulate, sub-sessile, spiral, linear-lanceolate, entire, acute, glabrous, unicostate reticulate, latex present.

Inflorescence : Cymose, axillary dichasial cyme.

Flower: Bracteate, bracteolate, pedicellate, complete, actinomorphic, regular, hermaphrodite, pentamerous, hypogynous, cyclic.

Calyx: 5 sepals, polysepalous, quincuncial aestivation, persistent, green.

Corolla: 5 petals gamopetalous, twisted aestivation, bell-shaped (infundibuliform), yellow fringed coronary outgrowths present.

Androecium: 5 stamens, polyandrous, epipetalous, stamens inserted at the throat of corolla, filament short anther sagittate, dithecous, basifixed and introrse.

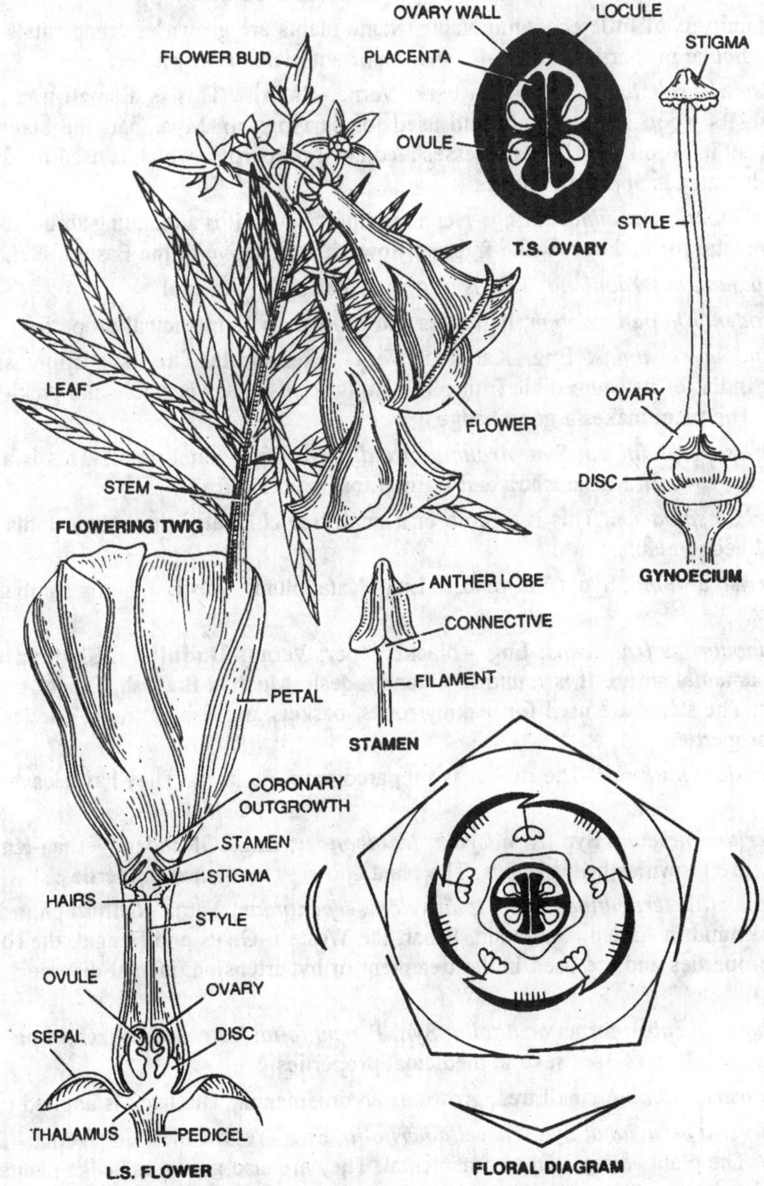

Fig.14.5. Apocynaceae. *Thevetia peruviana* Pers., Syn. *T. neriifolia*. Juss; Eng., yellow oleander; Verna., **pili kaner.**

Economic value. The bark is bitter, cathartic, febrifuge and useful in different kinds of intermittent fevers. Milky juicy and seeds are poisonous. (Fig. 14.5)

Gynoecium: 2 capels (bicarpellary), syncarpous, ovary superior, bilocular, two ovules in each locule, axile placentation, style long, stigma umbrella-shaped, 5-lobed nectar secreting disc present.

Fruit: Drupe.

Floral Formual: Br Brl ⊕ ⚥ K 5, $\overline{C(5), A5}$, G $(\underline{2})$.

Economic Importance of the Family

The family is of little economic value. Some plants are grown as ornamentals, while some possess medicinal properties. A list of some important plants is given here:

1. *Alstonia scholaris*; Eng.–Dita bark; Verna.–**Satwin**. This is a small tree grown as an ornamental. Its wood is quite light and used for carvings. In Myanmar, the black boards are prepared from its wood. The bark possesses medicinal properties, which is used for diarrhoea and dysentery. Its latex is applied to ulcers.

2. *Beaumontia grandiflora*; Eng.–Nepal trumpet flower–It is a climbing shrub, usually grown as an ornamental for its large, white fragrant flowers. It is native of the Eastern Himalayas.

3. *Beaumontia jerdoniana*; This is also grown as an ornamental.

4. *Anodendron paniculatum*; Its leaves and roots possess medicinal properties.

5. *Carissa carandas*; Eng.–Karanda; Verna.–**Karaunda**–This is a spiny shrub grown throughout India for its sour edible fruits. The fruits are used as vegetable and pickle is prepared from them. The plant makes a good hedge.

6. *Carissa grandiflora*; Syn. *Arduina grandiflora*; Eng.–Natal plum.–This is a large spiny shrub usually grown in Maharashtra and Baroda for its edible fruits.

7. *Carissa spinarum*. This is a shrub or a small tree cultivated throughout India for fragrant flowers and hedge plants.

8. *Carrisa arduina*; Syn. *C. bispinosa*; Eng.–Natal plum.–This is a thorny shrub grown for its edible fruits.

9. *Ichnocarpus frutescens*; Eng. –blackcreeper; Verna.–**Dudhilata, Siamalata.** This is a twining ornamental shrub. It is found in Uttar Pradesh, Madhya Pradesh, Bihar, Assam and the Sundarbans. The stems are used for making ropes, baskets and fishing traps. The leaves possess medicinal properties.

10. *Landolphia kirkii*.-The rubber is prepared from its latex. They have leaves with hook tendrils.

11. *Nerium indicum*; Syn. *N. odorum*; *N. oleander*; Eng.–Oleander; Verna.–**Kaner.**–It is a shrub. They are grown as hedge plants. The plants possess medicinal properties.

12. *Rauvolfia serpentina*; Syn. *Ophioxylon serpentinum*; Verna.–**Chhotachand.**–This is a small shrub found in Assam, Dehradun, Bihar, the Western Ghats and Bengal, the roots possess medicinal properties and are used in the treatment of hypertension, mental disorders and related ailments.

13. *Plumeria rubra* forma *acutifolia*; Syn. *P. acutifolia*; Verna.–**Goburchampa**–It is grown as an ornamental. It possesses several medicinal properties.

14. *Plumeria alba*.–A small tree, grown as an ornamental. The latex is applied to ulcers.

15. *Thevetia peruviana*; Syn. *Thevetia nerifolia*; Eng.–Yellow oleander; Verna.– **Pilikaner**.– It is a shrub. The plants are grown as ornamental. They are also grown as hedge plants. The latex is highly poisonous.

16. *Wrightia tinctoria*; Syn. *W. rothii*; *Nerium tinctorum*; Verna.–**Dudhi**.–A tree, found in Rajasthan, Madhya Pradesh and Tamil Nadu. A blue dye is obtained from its flowers and fruits. The fruits are edible. The bark and seeds possess medicinal properties.

17. *Wrightia tomentosa*; Syn. *W. mollissima; Nerium tomentosum*; Verna.–**Dharauli**–The seeds and roots yield a yellow dye. The leaves and fruits are edible. Its soft wood is used for carvings. The bark and roots are used as an antidote for snake bite.

18. *Aganosma dichotoma*; Syn. *Echites dichotoma*; Verna.–**Malati**.–This is a climbing shrub, grown as an ornamental in the gardens.

19. *Allamanda cathartica*; Eng.–Allamanda–This is a beautiful climbing shrub, grown as an ornamental in the gardens. It is native of Central America and Brazil.

Family–Asclepiadaceae

According to Rendle there are about 280 genera and over 1800 species in this family.

Distribution: The members of this family are mainly found in the tropical regions of the world. They occur specially in the drier parts of Central and South America. In our country the family is represented by several genera, such as, *Asclepias, Calotropis, Ceropegia*, etc. *Asclepias curassavica* is commonly found in sub-Himalayan tracts; *Calotropis gigantea* is commonly found in Northern India; *Calotropis procera* is quite common in the Punjab, Uttar Pradesh and Andhra Pradesh. The species of *Ceropegia* are commonly found in South India.

Habit: The plants of this family are mostly erect herbs or woody climbers but some are succulent. The perennial herbs are quite common in the dry regions of South and Central Africa. Some plants perennate by means of tuber-like rhizomes. Large shrubs and trees are rarely found. *Hoya* is a plant of succulent habit, it possesses fleshy stem and leaves. *Stapelia* is a native of South Africa. It is cactus like in habit possessing thick and fleshy stems; the leaves are reduced to thorns or scales. *Dischidia* is an epiphytic plant; it climbs upon its host by means of adventitious roots, the leaves are succulent and covered with a coating of wax. *Cryptostegia grandiflora* is a climbing shrub. The species of *Hemidesmus, Gymnema, Pergularia*, etc., are also climbing shrubs. *Leptadenia spartium* is a shrub. *Calotropis procera*, (Verna.–**Ak**) is a xerophytic shrub and *Calotropis gigantea* (Verna.–**Safed Ak**) is a large shrub. *Asclepias curassavica* is a perennial herb. The species of *Ceropegia, Pentatropis,* etc., possess underground tuber like parts. *Tylophora indica* is a herb. Most of the plants possess xerophytic characters and certain are extreme xerophytes.

Anatomically the plants of this family resemble with that of Apocynaceae. They also possess bicollateral bundles and the latex. According to Busich (1913), the endophytic mycorrhiza is frequently found in the roots of succulent plants of this family.

The species of *Pentatropis* possess tuberous roots; whereas *Ceropegia tuberosa* possesses tubers, by means of which, they perennate.

Leaves: The leaves are simple, sub-sessile and exstipulate. In many cases the leaves are fleshy and covered with a coating of wax, *e.g.*, *Calotropis*. In *Stapelia*, the leaves are very much reduced and represented by spines and scales.

Inflorescence: The inflorescence is of two types, *i.e.*, (1) cymose and (2) racemose. Generally the cymose type of inflorescence is found. In such cases the axis ends in a terminal flower, while branching occurs in the axils of the two branches forming a dichasial cyme, but one of the two branches of the dichasium grows faster than the other and ultimately one of the branches is completely suppressed and the inflorescence becomes monochasial cyme. In certain cases the inflorescence is of racemose type. This is either a raceme or a umbel. Sometimes the flowers are found to be arranged in umbelliform cymes. When the inflorescence is axillary only one flower is found at each node; the other leaf either remains completely sterile or possesses a vegetative shoot. In *Calotropis, Asclepias* the flowers are being arranged in umbellate cymes. In *Hemidesmus*, the flowers are found to be arranged in axillary cymes.

Flowers: The flowers are pedicellate, bracteate, hermaphrodite, actinomorphic, rarely zygomorphic, *e.g.*, in *Ceropegia*, complete. The general plan of the flower is pentamerous with three regularly alternating pentamerous whorls of calyx, corolla and androecium, however, the number of carpels is reduced to two in the gynoecium. Usually the flowers are small in size, but the flowers of *Ceropegia*, *Stapelia* and *Stephanotis* are quite large in size.

Calyx: It consists of five sepals, which are either free (polysepalous) or somewhat connate at the base; with the odd sepal posterior. The aestivation is quincuncial or rarely valvate.

Corolla: It consists of five united petals (*i.e.*, it is gamopetalous). The petals are spreading (*i.e.*, rotate), but in *Stephanotis* the corolla tube is long, forming a salver-shaped corolla. In *Ceropegia* the corolla is pitcher-like in appearance (zygomorphic). The aestivation is contorted and rarely valvate. Sometimes the petaloid appendages arise either from the corolla or from the back of the stamens. Very often these hairs or appendages are found inside or at the mouth of the corolla forming the corolline corona.

Androecium: It consists of five stamens. The anthers are united laterally, giving rise to a five sided blunt cone, which is usually attached on the inside to the stigma head, except in the sub-family Periplocoideae. With the result of this union of anthers and pistils, the **gynostegium** is formed. The pollen grains of each half anther associated in tetrads are found in a sac like structure known as **pollinium.** The anthers are almost quite sessile. In the typical members of the family the **translator** consists of two parts– (1) **corpusculum** and (2) a **pair of arms.** The corpusculum is attached to the margin of stigma head between the anthers and the pair of arms is attached to the pollinia of the adjacent anther halves.

In the members of the sub family Periplocoideae the translator consists of a spoon or funnel-shaped structure in between each anther, terminating below in a sticky disc. The granular pollen grains form one half of each of the two adjacent anthers, on maturity, pour into the concave receptacle of the translator.

Certain appendages or horn-like projections are given out from the backs of the anthers, are known as **cuculi**, which secrete and store honey. These appendages form a corona known as **staminal corona.**

Gynoecium: It consists of two carpels. The ovaries remain free, but styles unite to form a common swollen stigma-head. This stigma-head may be flat or somewhat conical and sometimes even beaked. The receptive surfaces of the stigma are situated on the edge or on the underside of it. The ovary is superior. The placentation is marginal with many ovules.

Fruit: Generally it is a pair of follicles. Sometimes there is only one follicle, because of the suppression of the other.

Seed: Usually the seeds are flat, ovate-oblong and are crowned by a fruit of hairs. These hairs facilitate the dispersal of the seeds by wind. The embryo is large.

Floral Formula: ⊕ ⚥ K5, $\overparen{C(5), A5}$, G($\underline{2}$).

There are two sub-families, which may be distinguished as follows:

1. Periplocoideae–The pollen grains are granulur and arranged in tetrads. The translators are spoon or funnel shaped.

2. Cynanchoideae–Pollen grains are in twos, found in pollinia. The translators are differentiated into a corpusculum and a pair of arms.

Pollination: The pollination takes place strictly by means of the agency of insects. As an insect visits the flower, the sticky basal disc of translator turns outwards and adheres to the head of the insect visitor. The whole translator with its pollinia or pollen contents is carried over by the insect from one flower to another flower where the pollen grains are deposited on the stigmatic

surfaces on the underside of the stigma-head. In the sub family Cynanchoideae, where the pollinia are found the insect visiting the flower in the search of nectar pushes its leg in the slit left between the adjacent anthers, and when it pulls out the leg, it comes in contact with the notched base of the corpusculum, and this way the complete translator with two pollinia is dragged out.

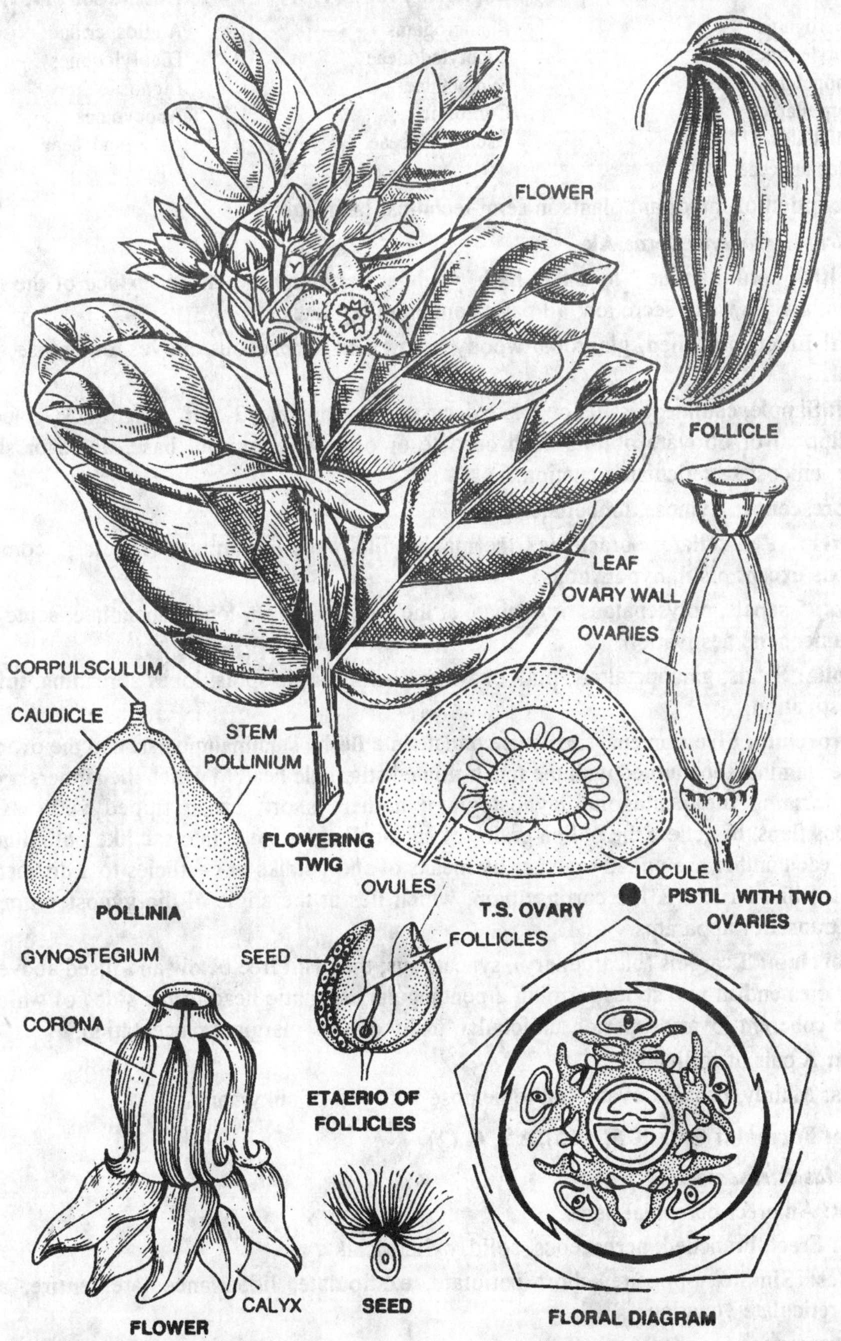

Fig. 14.6. Asclepiadaceae. *Calotropis procera* (Ak.)

At first the pollinia are quite apart from each other, but as the hygroscopic arms of the translator are dried up they become close to each other. If this insect visits another flower, the pollinia are readily pushed into one of the slits, and will slip into the stigmatic chamber, thus effecting cross-pollination. Chiefly the pollination is by two-winged flies.

Bentham & Hooker (1862)	**Engler & Prantl (1931)**	**Hutchinson (1959)**
Phanerogams	Phanerogams	Angiospermae
Dicotyledones	Dicotyledoneae	Dicotyledones
Gamopetalae	Sympetalae	Lignosae
Bicarpellatae	Contortae	Apocynales
Gentianales	Asclepiadaceae	Asclepiadaceae
Asclepiadaceae		

Description of important plants in semi-technical language:

1. ***Calotropis procera***; Verna.**Ak**.

Habit: A shrub, about 3 to 6 feet high, younger parts and the under surface of the leaves covered with white waxy secretion, all parts contain milky latex.

Stem: Erect, branched, glabrous, woody below and herbaceous above, tomentose, solid, cylindrical.

Leaf: Simple, cauline, sessile, opposite decussate, exstipulate, 2-5 in., long, thick, glaucous–green, elliptical or obovate oblong with cordate or often amplexicaul base, acute or shortly acuminate, unicostate reticulate venation.

Inflorescence: Cymose, umbellate cyme.

Flower: Pedicellate, bracteate, hermaphrodite, actinomorphic, regular, complete, pentamerous except pistil, hypogynous.

Calyx: 5 sepals, polysepalous or connate at the base, greenish, lobes lanceolate, acute apex. interior quincuncial aestivation.

Corolla: Petals, gamopetalous, pink or whitish with purple spots, lobes spreading, inferior, twisted aestivation.

Androecium: Five stamens, filaments connate in a fleshy staminal tube around the ovary, the apex of the staminal tube united with the much-dilated stigmatic head to which the anthers are also coherent, forming the pentagonal gynostegium; anthers short, broad tipped with inflexed membranous flaps, bi-celled, the pollen grains of each cell aglutinated into sac like pollinium; the pollinia of each anther are united together by means of short stalks or caudicles to a distinct dark coloured dot-like structure, the corpusculum, which lies at the angle of the gynostegium, thus forming a translator apparatus.

Gynoecium: 2 carpels (bicarpellary), syncarpous; the pistil free below and fused above; two distinct ovaries end in two styles forming a pentangular stigmatic head to the sides of which the anthers are coherent; ovary superior, unilocular, many ovules, marginal placentation.

Fruit: A pair of follicles.

Seeds: Mainly, broadly ovate, flat tomentose with tuft of silky hairs.

Floral Formula: ⊕ ⚥ K 5, $\overline{C(5), A5}$, G $\underline{(2)}$.

2. ***Asclepias curassavica.***

Habit: An erect perennial herb.

Stem: Erect, branched, herbaceous, solid, cylindrical.

Leaves: Simple, opposite, short petiolate, exstipulate, linearlanceolate, entire, acute, unicostate reticulate venation.

Inflorescence: Cymose, umbellate cyme.

Flowers: Pedicellate, purple-yellow, small bisexual, actinomorphic, hypogynous.

Calyx: 5 sepals, polysepalous, imbricate aestivation.

Corolla: 5 petals, gamopetalous, twisted aestivation.

Androecium: Similar to that of *Calotropis procera*. The filaments attached to the base of the corolla are united to form a column around the pistil, the anthers being pressed close to the styles; each stamen bears on the side away from the pistil an appendage which enfolds the anther like a hood; these appendages collectively form a corona.

Gynoecium: 2 carpels (bicarpellary), syncarpous, ovary superior.

Fruits: Follicle.

Floral formula: ⊕ ⚥ K 5, $\overline{C(5), A5}$, G ($\underline{2}$).

3. ***Cryptostegia grandiflora*** R. Br.; Verna. **Vilayati vakhandi**; Eng. Rubber vine.

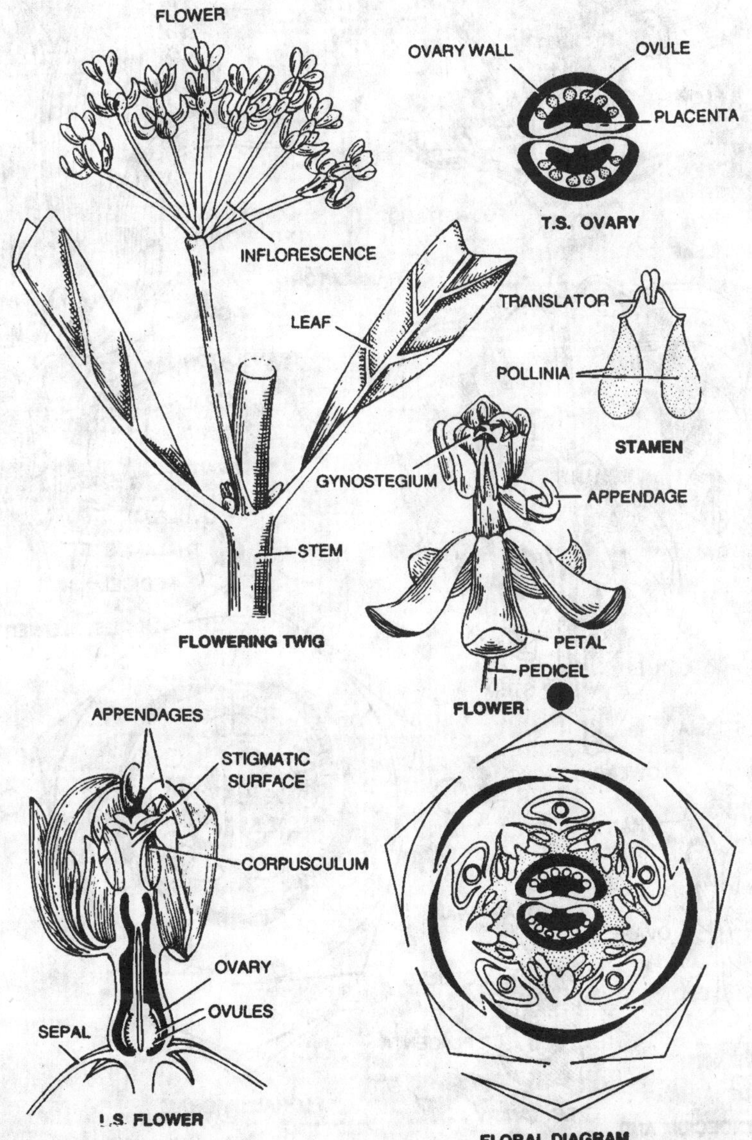

Fig. 14.7. Asclepiadaceae. *Asclepias curassavica* Linn. Eng. Blood flower; Verna. **Kukatundi.**

Habit: A perennial twiner.

Root: Tap and branched.

Stem: Herbaceous, lower part woody, aerial, weak, climbing, branched, twiner, solid, cylindrical, rough, green, latex present, nodes and internodes prominent.

Leaf: Ramal and cauline, opposite decussate, exstipulate, simple, petiolate elliptic ovate, entire, acute, coriaceous, glaucous, unicostate reticulate.

Inflorescence: Cymose dichasial cyme.

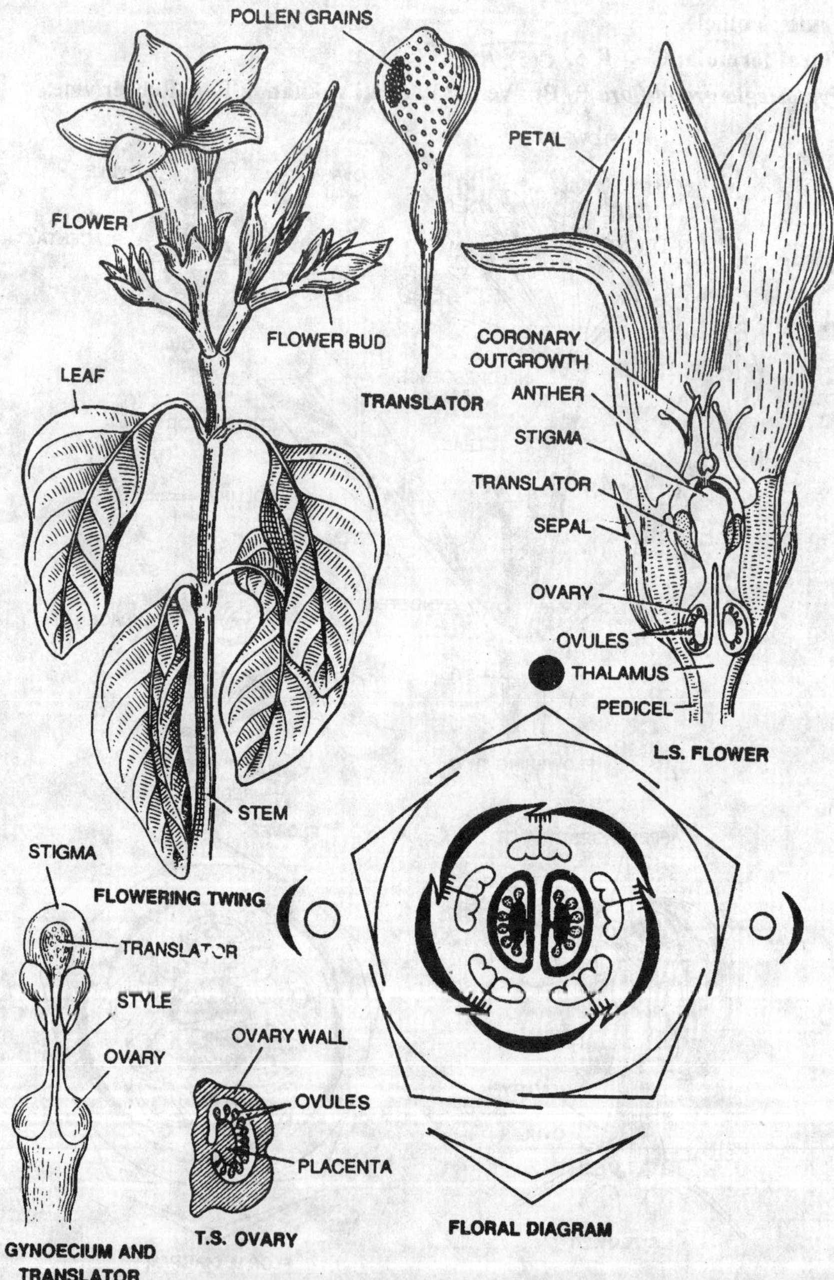

Fig. 14.8. Asclepiadaceae. *Cryptostegia grandiflora* R. Br., rubber vine; Verna, **vilayati vakhandi.**

Flower: Bracteate, bracteolate, pedicellate, actinomorphic, regular, hermaphrodite, pentamerous, hypogynous, cyclic.

Calyx: 5 sepals, polysepalous, quincuncial aestivation, green, margins membranous.

Corolla: 5 petals, gamopetalous, twisted, coronary outgrowths present, violet.

Androecium: 5 stamens, polyandrous, epipetalous, dithecous, introrse; pollen grains are being shed from each anther lobe and are deposited in the translator. Each spoon-shaped translator is found to be situated in between two stamens.

Gynoecium: 2 carpels (bicarpellary), ovary superior, ovaries free but style and stigma fused, many ovules in each locule, marginal placentation, style short, stigma globular.

Floral formula: Br, Brl ⊕ ⚥ K 5, $\overline{C(5), A5}$, G $(\underline{2})$.

Economic Importance of the Family

The family is of little economic value. Some plants are ornamental while certain others are of medicinal value. A list of few important plants is given below:

1. *Asclepias curassavica*; Eng.–Blood flower; Verna–**Kakatundi**–This is a perennial herb or a small shrub. It is cultivated as an ornamental. The plant is of medicinal value. The root is used as a purgative and as a medicine for piles.

2. *Calotropis gigantea*; Eng.–madar; Verna–**Ak**–A perennial undershrub. The floss, obtained from the seed, is used for stuffing purpose. The stem yields a fibre. All parts of the plant are used medicinally. The fresh leaves are used in the fomentation for swellings. The gun powder charcoal is prepared from the wood. It is commonly found in Northern India.

3. *Calotropis procera*; Syn. *Asclepias gigantea*; Eng. Akund; Verna–**Safed Ak.**–The stem fibre is made into cordage and the floss from the seed is used as a stuffing material. The stem and leaves are used medicinally several ways. It is commonly found in Andhra Pradesh and the Punjab.

4. *Ceropegia hirsuta*; Verna–**Khantodi**–This is a climbing shrub. The tubers are edible. It is commonly found in Tamil Nadu, Travancore and Andhra Pradesh.

5. *Ceropegia tuberosa:* Verna.–**Patala tumbi**—The tubers are edible. It is commonly found in Tamil Nadu and Andhra Pradesh.

6. *Ceropegia oculata*; Verna.–**Khantodi**–The tubers are edible. It is a herb, found in South India.

7. *Ceropegia lawii*; Verna–**Kharpundi**–The tubers are edible. It is a succulent herb. Found in South India.

8. *Cryptolepis buchanani*; The plant produces a floss which is used as a stuffing material.

9. *Cryptostegia grandiflora*: Verna.–**Vilayati-vaknandi**–This is a beautiful climbing shrub with red-rose flowers, grown as ornamental.

10. *Daemia extensa*; Syn. *Pergularia extensa*; Verna–**Utsan**–A fibre is obtained from its stem. The juice of leaves is used medicinally in several ways, *e.g.*, as the remedy of diarrhoea, asthama, etc.

11. *Dischidia rafflesiana*; Verna–**Bandikuri**–This is an epiphytic undershrub. The roots are chewed as a remedy for coughs.

Economic value. Grown as an ornamental plant for its beautiful violet-coloured flowers. The leaves are toxic.

12. *Gymnema sylvestre*; Verna.–**Gurmar**–This is a large climbing shrub. The flowers are found to be arranged in umbellate cymes. The roots are used as antidote for snake-bite. The leaves are used as remedy for diabetes.

13. *Hemidesmus indicus*; Eng. –Indian sarsaparilla; Verna.–**Magrabu**–This is a twining shrub. The roots are used as a tonic, which is useful in loss of appetite, fever, skin diseases, as blood purifier, rheumatism and snake-bite. The tonic (sarsaparilla) is diuretic diaphoretic and demulcent.

14. *Oxyrstelma esculentum*; Verna–**Dudhialata**–The fruits are edible; the plant possesses medicinal properties. The latex is used to wash ulcers. The medicine for scabies is prepared from it.

15. *Pentatropis cynanchoides*; Verna–**Kanathodi**–The tuberous roots are edible.

16. *Periploca aphylla*; Verna–**Barri**–The stem yields a fibre for cordage; the latex is applied to swellings and tumours, it is used as a purgative; sometimes the flowers are eaten.

17. *Leptadenia pyrotechnica*; Syn. *L. spartium*–This is a shrub found in the Punjab, Uttar Pradesh, Rajasthan and Gujarat. The slimy fruits and young twigs are eaten as vegetable.

18. *Tylophora indica*; Verna–**Antamul**–This is a herb. Its leaves and roots are used as substitute for ipecacuanha. This is also used as emetic, diaphoretic and expectorant. The plants are commonly found in Bengal, Assam and South India.

19. *Sarcostemma acidum*; Verna–**Somlata**–This is a leafless shrub, commonly found in dry areas of Andhra Pradesh, Tamil Nadu, and Karnataka. It is used to destroy white ants from sugarcane fields.

20. *Caralluma fimbriata*; Verna–**Makedshingi**–This is a fleshy herb, eaten as a vegetable. It is mainly found in Andhra Pradesh.

21. *Cynanchum arnottianum*; This is a herb whose dried leaves are used to destroy maggots. It is found in Kashmir.

22. *Cosmostigma racemosum*–This is a climbing shrub, found in the Western Ghats. Karnataka and Andhra Pradesh. The Leaves are used to cure ulcerous sores.

Systematic Relationships: The various systems of classification differ in their interpretation of the limits of the order Gentianales. Bentham and Hooker included six families in this order. They are–Oleaceae, Salvadoraceae, Apocynaceae, Asclepiadaceae, Loganiaceae and Gentianaceae. Engler and Diels called the Gentianales the Contortae having six families, *i.e.*, Oleaceae, Desfontaineaceae, Loganiaceae, Gentianaceae, Apocynaceae and Asclepiadaceae. Wettstein and Rendle restricted the Contortae to contain five families but Oleaceae. They placed the Oleaceae within Oleales. Hutchinson has placed the families of Contortae into three separate orders. The Loganiales consists of Loganiaceae and Oleceae; Apocynales of Apocynaceae and Asclepiadaceae and Gentianales of Gentianaceae. Originally Engler included the Salvadoraceae in the order Contortae, but later on (Engler and Gilg, 1924) this was transferred to the Sapindales.

The Asclepiadaceae have close affinities with Apocynaceae, Hutchinson considered that the Apocynaceae have been derived from stocks ancestral to the Loganiaceae.

15

Order–Rubiales Family–Rubiaceae

According to Hutchinson this is the fifty second order of the phylum Angiospermae, subphylum Dicotyledones and division Lignosae. This order consists of only two families, *i.e.*, Dialypetalanthaceae and Rubiaceae. The Rubiaceae has been described in the present text in detail. Bentham and Hooker placed the family Rubiaceae along with one more, *i.e.*, Caprifoliaceae, in the first order, the Rubiales of class Dicotyledones, division Gamopetalae and series Inferae. Engler and Prantl have included the Rubiaceae along with four other families in the 8th order– the Rubiales of class Dicotyledones and sub-class Sympetalae.

Corolla sympetalous, more or less zygomorphic; stamens 5, rarely more, epipetalous, alternate with the corolla-lobes; anthers free, opening lengthwise; ovary inferior; axile placentation, rarely parietal; style 1; ovules numerous to solitary; seeds with endosperm. Leaves opposite, simple, entire, stipules inter- or intrapetiolar.

Family—Rubiaceae

There are about 400 genera and 5,300 species in this family.

Distribution: The family is cosmopolitan in distribution but the great majority of the species are found in tropical regions. The species of *Galium*, *Oldenlandia*, etc., are frequently found in the temperate regions. Certain species of *Galium* are found even in the Arctic zone or high elevations on tropical mountain ranges. The species of *Nertera* are found to be spread along the Andes. The genus *Comprosma* is commonly found in the south temperate and Antarctic regions, *Guettarda speciosa* is quite common round the Indian ocean from Africa to Malacca. The family is represented by several genera in our country, *e.g.*, *Cinchona, Coffea, Adina, Hamelia, Ixora, Galium, Gardenia, Mussaenda, Rubia, Morinda* and many others.

Habit: The plants show great variations in habit. Majority of plants are either trees or shrubs. The herbaceous plants are also common and mainly they are confined to certain tribes. The members of the tribe **Galieae** are exclusively herbaceous. The tendency towards the formation of herbaceous forms is widespread, as certain herbaceous genera, such as, *Bouvardia* sp., is found in the typically woody tribe **Cinchoneae.** The climbing habit is also found in the family. Some climbers are herbaceous, whereas certain others are shrubby twiners, *e.g.*, *Manettia. Uncaria* is a typical example of hook climbers. However, *Galium aparine* possesses stiff recurving hairs, which help in climbing on the supporting vegetation. The epiphytes also occur, *e.g.*, *Myrmecodia. Myrmecodia* possesses a tuber-like stem developed from the swollen hypocotyl. This tuberlike stem contains a number of communicating galleries which harbour the ants. Such type of plants are known as **myrmecophilous** but the exact relation of ants to plant is not clear.

In the species of *Nauclea, Duroia*, etc., the ants live just beneath the inflorescence in the swollen hollow internodes. It is thought that the ants protect the flowers from other harmful insects. In the species of *Remijia* and *Duroia*, the ants live in the flask-shaped swellings situated on the leaf base. Some important trees of the family are, *Anthocephalus cadamba* (Verna–**Kadam**), *Adina cordifolia* (Verna–**Haldu**), the species of *Cinchona, Mitragyna parviflora*, etc., *Coffea arabica*,

Hamelia patens, many species of *Ixora, Gardenia, Mussaenda*, etc., are medium sized or large shrubs. Certain species of *Galium, Rubia, Oldenlandia*, etc, are beautiful herbs.

Leaves: They are simple, stipulate, opposite decussate, entire or rarely toothed.

The stipules may be inter or intrapetiolar. In the former case, the stipules are situated at the side of leaf-base and thus lie in between the leaves; whereas in the latter case the stipules are situated in front of the leaf base in between the petiole and the stem. Sometimes the stipules unite to each other and then to the petiole, forming a sheath like structure around the stem. In certain species of the genus *Gardenia*, the stipules form a conical cap like structure which protects the apex of the stem, and is thrown off as the bud develops. Sometimes the stipules split into bristle-like structures, and at the top of each bristle a resin secreting gland develops, *e.g.*, in *Pentas*. In the case of the stipules of *Coffea, Gardenia*, etc., such glands are situated at the base of the stipules. In the case of *Galium* and allied genera, the leaf like (foliaceous) stipules are found. In such cases the stem apparently bears the whorls of leaves at the nodes, and the leaves may only be distinguished from the stipules by the presence of axillary buds.

Inflorescence: Usually the inflorescence is of cymose type, *i.e.*, a dichasial or a panicled cyme. Sometimes the small-flowered dichasia are aggregated into dense globose heads. In these heads the inferior ovaries of the flowers unite together, *e.g.*, in *Sarcocephalus*, *Morinda*, etc. In *Gardenia, Randia*, and some other plants the inflorescence consists of solitary terminal flowers. In *Coffea arabica* the flowers are found to be arranged in axillary cymes, whereas in *Ixora subsessilis*, they are arranged in subsessile corymbose cymes. The flowers of *Oldenlandia* are being arranged in panicled cymes. In *Anthocephaius indicus* and *Adina cordifolia* the flowers are found to be arranged in globose heads. In the species of Australian *Pamax*, the two flowers of a pair are so united, that the two possess a common calyx.

Flowers: The flowers are hermaphrodite, usually actinom-orphic and either tetramerous or pentamerous, with five or four stamens and two carpels. They are complete and epigynous. Sometimes the flowers become unisexual by means of reduction, *e.g.*, *Coprosma*. The flowers of *Henriquezia*, are however zygomorphic.

Calyx: The calyx consists of five or four sepals, gamosepalous. In the case of *Mussaenda, Warscewiezia*, etc., one or rarely more of the sepals become leaf like and white, yellow or variously coloured. The development of such leaf-like sepal increases the attractiveness of the inflorescence and attraction for the insects. In certain cases the sepals increase in size only after fertilization *e.g.*, *Nematostylis, Alberta*, etc. The aestivation is valvate.

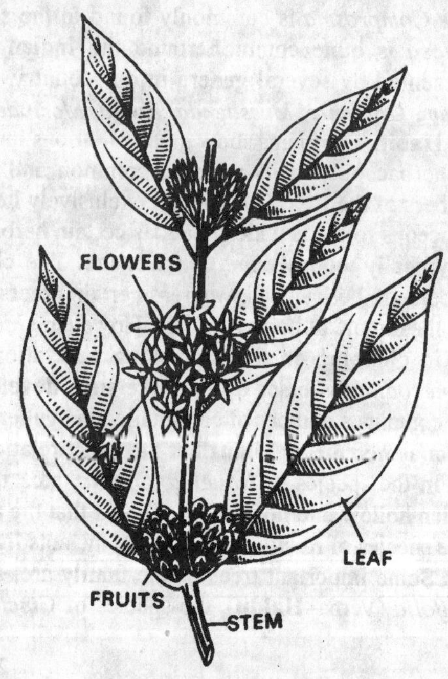

Fig. 15.1. Rubiaceae–a twig of *Coffea arabica* with leaves, flowers and fruits.

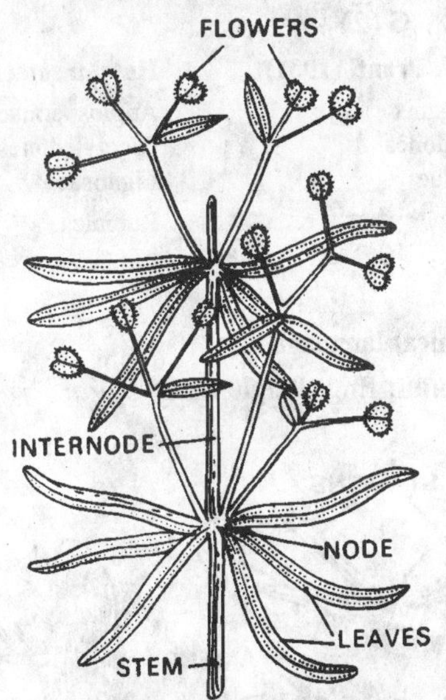

Fig. 15.2. Rubiaceae—A flowering twig of *Galium* sp.

Corolla: The corolla consists of five or four petals, gamopetalous. The corolla is tubular, campanulate or rotate. In *Henriquezia*, the corolla is more or less zygomorphic and bilabiate. The aestivation is either contorted or valvate.

Androecium: It consists of five or four stamens. The stamens are epipetalous, *i.e.*, they are inserted at or near the throat of the corolla tube. The stamens alternate with the petals. The anthers are bicelled, introrse and dehisce by longitudinal slits and sometimes by apical pores. In irregular corollas, the stamens are of unequal length and bent. In *Oldenlandia*, heteroandry occurs (*i.e.*, the stamens are of different length).

Gynoecium: Usually the gynoecium consists of two carpels, syncarpous, Generally the ovary is inferior, but in the Australian genus *Synaptantha*, it is half inferior. However, in *Gaertnera* and *Pogamea*, the ovary is superior. *Generally* the ovary is bilocular but in *Gardenia*, it is unilocular, with parietal placentation. In general, the placentation is axile. One to many ovules are found in each locule of the ovary. The style is simple with globose stigma.

Fruit: The fruit is a drupe, capsule or berry. They may be dry or fleshy. It is a fleshy berry *e.g.*, in the case of *Randia, Mussaenda*, etc., it is a drupe, *e.g.*, in *Coffea* and others; it is a capsule, *e.g.*, in *Silvianthus*, *Ophiorrhiza*, etc. The fruits of the species of *Galium* are sticky, because of the presence of small stiff recurving hairs.

Seeds: Generally the seeds are small and more or less winged, endospermic; the flat cotyledons of embryo lie face to face.

Pollination: The pollination is usually effected by means of insects (*i.e.*, entomophily). The nectar, the flag-like sepals attract the insects.

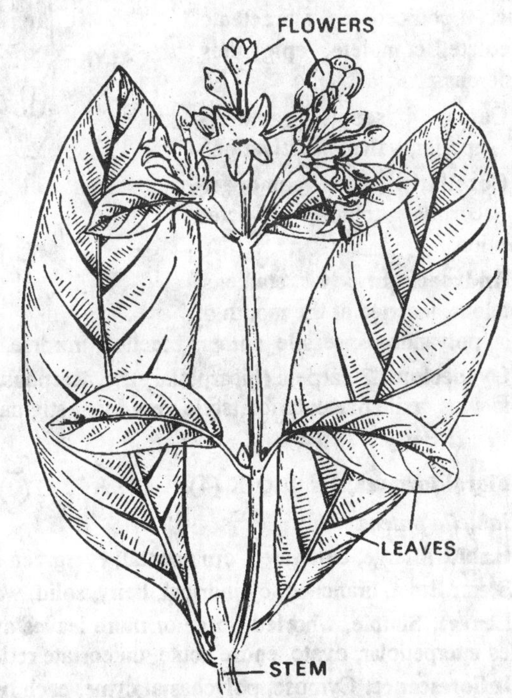

Fig. 15.3. Rubiaceae—A flowering twig of *Cinchona* sp.

Floral formula: $\oplus$ ⚥ K (4–5), $\overline{C\ (4-5), A5}$, G $(\bar{2})$.

Bentham & Hooker (1862)	**Engler & Prantl (1931)**	**Hutchinson (1959)**
Phanerogams	Phanerogams	Angiospermae
Dicotyledones	Dicotyledones	Dicotyledones
Gamopetalae	Sympetalae	Lignosae
Inferae	Rubiales	Rubiales
Rubiales	Rubiaceae	Rubiaceae
Rubiaceae		

Description of important plants in semi-technical language.

1. ***Ixora coccinea*** **Linn**; Verna. **Rangan, Rookmini**; Eng. Jungle flame *Ixora.*

Habit: A large cultivated ornamental shrub.

Stem: Erect, branched, solid, cylindrical, green, herbaeous or somewhat woody.

Leaves: Cauline and ramal, simple, opposite, sessile, stipulate, ovate, smooth entire, acute, unicostate reticulate venation.

Inflorescence: Cymose, corymbose cyme.

Flower: Bisexual, actinomorphic, pedicellate, bracteate, bracteolate, complete, epigynous, tetramerous.

Calyx: 4 sepals, gamosepalous, superior, valvate aestivation.

Corolla: 4 petals, gamopetalous, corolla tubular, twisted aestivation.

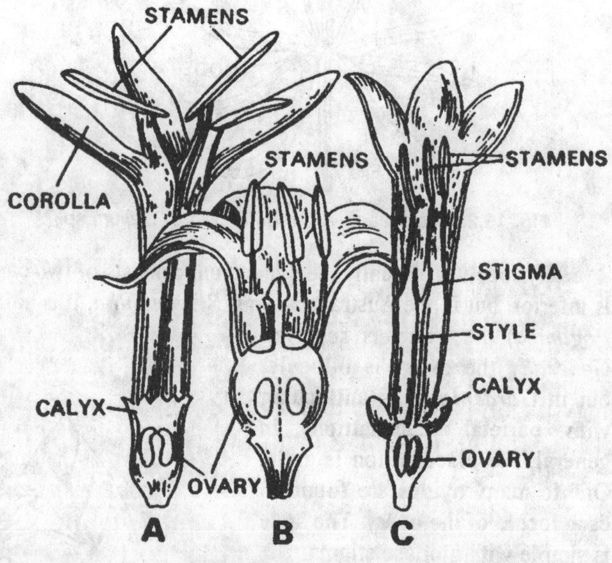

Fig. 15.4. Rubiaceae—flowers in vertical sections : A, *Coffea arabica;* B, *Rubia tinctorum;* C, *Cinchona* sp.

Androecium: 4 stamens, epipetalous, inserted at the mouth of corolla, polyandrous, sessile, anthers bicelled, introrse, dorsifixed.

Gynoecium: 2 carpels (bicarpellary), syncarpous, ovary inferior, bilocular, single ovule in each loculus, axile placentation, style very long, stigma single, bifid.

Fruit: Berry.

Floral formula: Br $\oplus$ ⚥ K (4), $\overline{C\ (4), A4}$, G $(\bar{2})$.

2. ***Hamelia patens.***

Habit: A large, cultivated, ornamental, evergreen shrub.

Stem: Erect, branched, cylindrical, hairy, solid, woody.

Leaves: Simple, whorled (three or more leaves at each node), cauline, petiolate, stipulate, stipules interpetiolar, ovate, entire acute, unicostate reticulate venation.

Inflorescence: Cymose, polychasial cyme, each branch helicoid.

Flowers: Subsessile, ebracteate, bisexual, actinomorphic, complete, orange coloured, epigynous.

Economic value. Grown as a hedge plant.

Calyx: 5 sepals, gamosepalous, petaloid, reddish, adnate to the ovary wall, superior, valvate aestivation.

Corolla: 5 petals, gamopetalous, tubular, orange coloured, superior, twisted aestivation.

Androecium: 5 stamens, polyandrous, alternate to the petals, filaments short, epipetalous, anthers basifixed, long, bicelled, introrse.

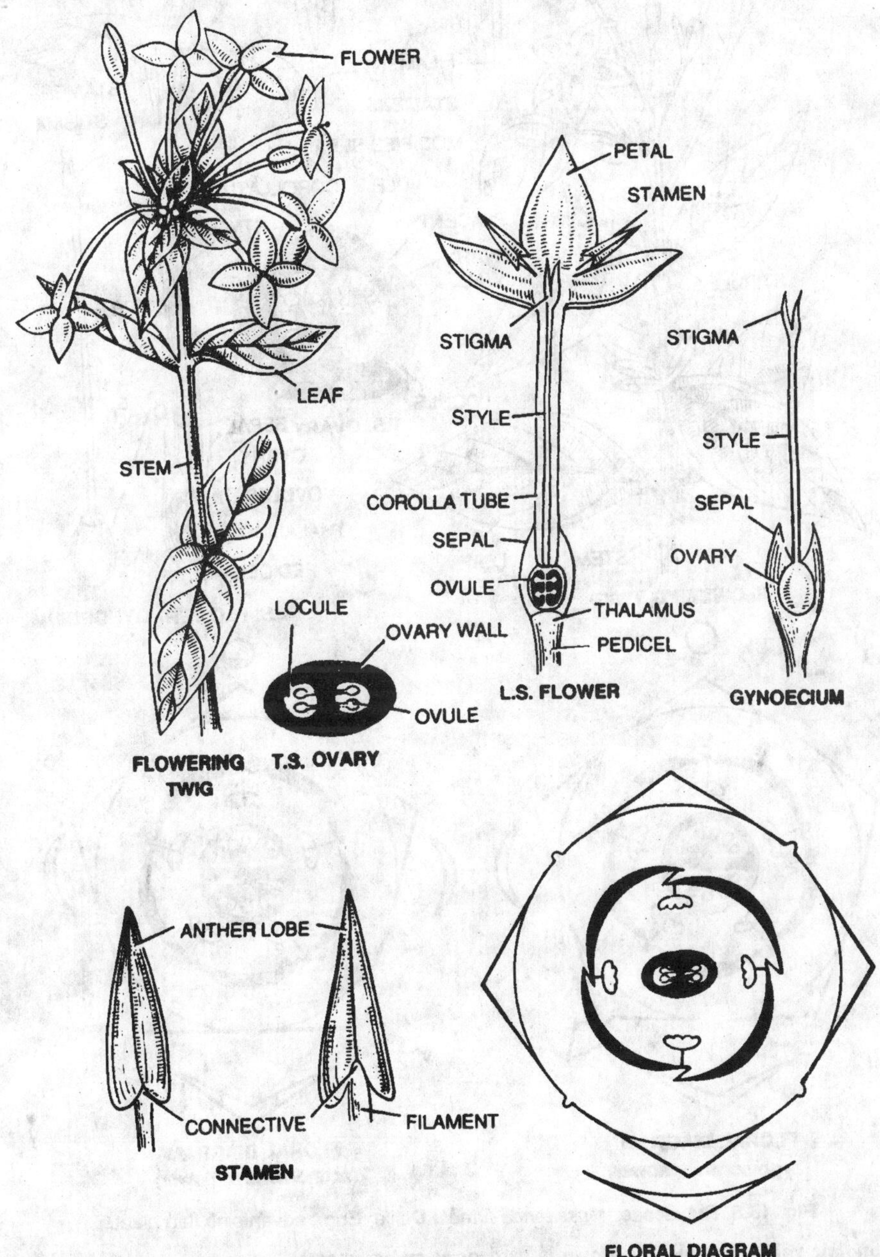

Fig. 15.5. Rubiaceae. *Ixora coccinea* Linn.; Eng. Jungle flame Ixora; Verna. Rangan, rookmini.

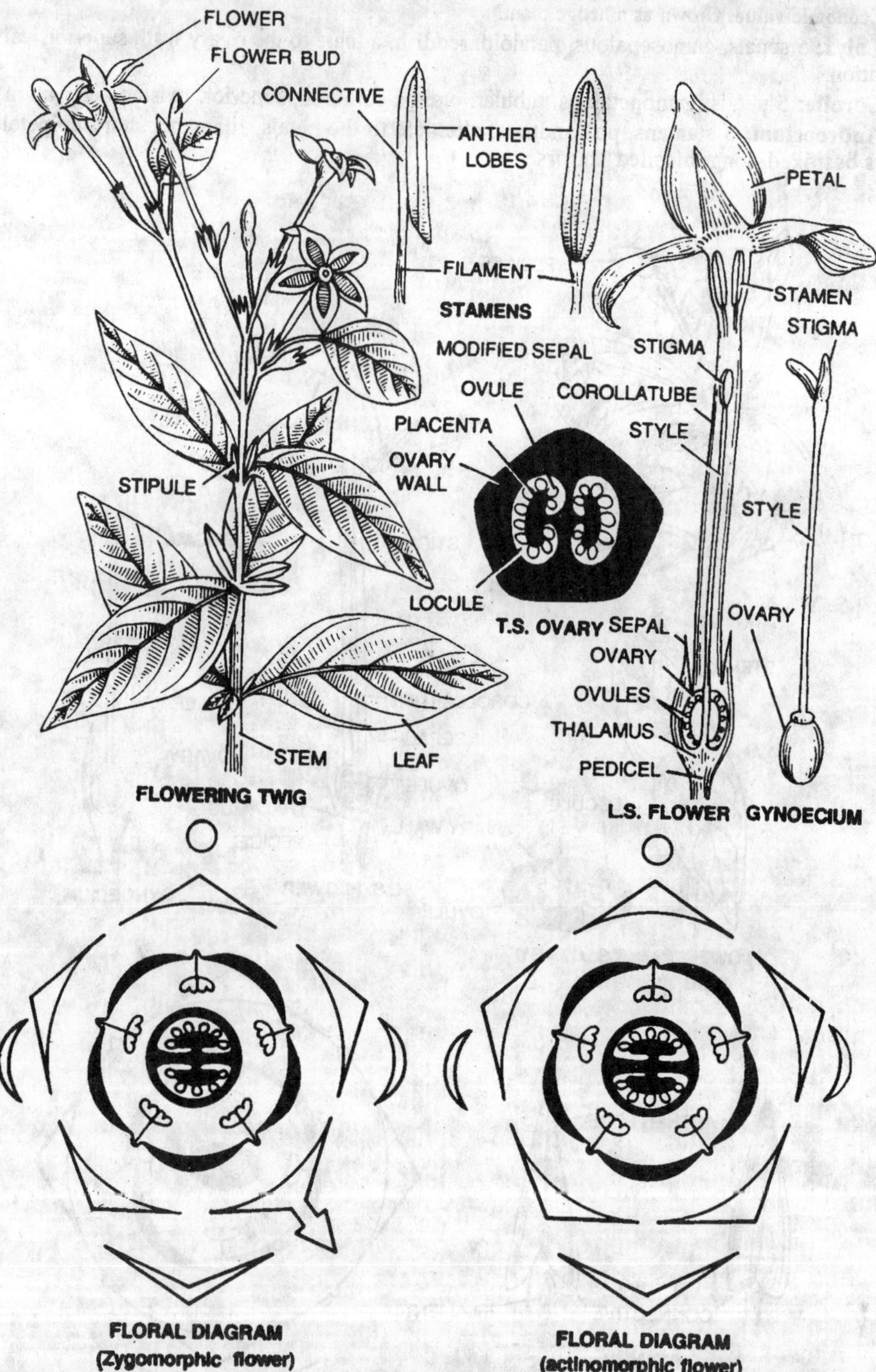

Fig. 15.6. Rubiaceae. *Mussaenda luteola* Delile; Eng., advertising flag plant.

Economic value. The plant is grown in gardens as an ornamental.

Gynoecium: 5 carpels (pentacarpellary), syncarpous, ovary inferior, pentalocular, several ovules in each loculus, axile placentation.

Fruit: Berry.

Floral Formula: ⊕ ⚥ K(5), C(5), A (5), G($\overline{5}$).

3. ***Oldenlandia corymbosa*** L., Eng. Daman–paper. **Root**–simple, tap. **Stem**–herbaceous, aerial, prostrate, branched, solid, angular, green. **Leaf**–cauline and ramal, opposite decussate, stipulate, interpetiolar stipules, simple, sessile, lanceolate, entire, acute, pubescent, unicostate reticulate. **Inflorescence**–cymose, axillary dichasial cyme. **Flower**–Bracteate, bracteolate, pedicellate, actinomorphic, hermaphrodite, complete, tetramerous, epigynous, cyclic. **Calyx**–4, gamosepalous, valvate, persistent. **Corolla**–4, gamopetalous, valvate, violet. **Androecium**–4, polyandrous, epipetalous, filaments short, dithecous, basifixed introrse. **Gynoecium**–2, bicarpellary, syncarpous, inferior, bilocular, many ovules in each locule, axile, style short, stigma simple and globular.

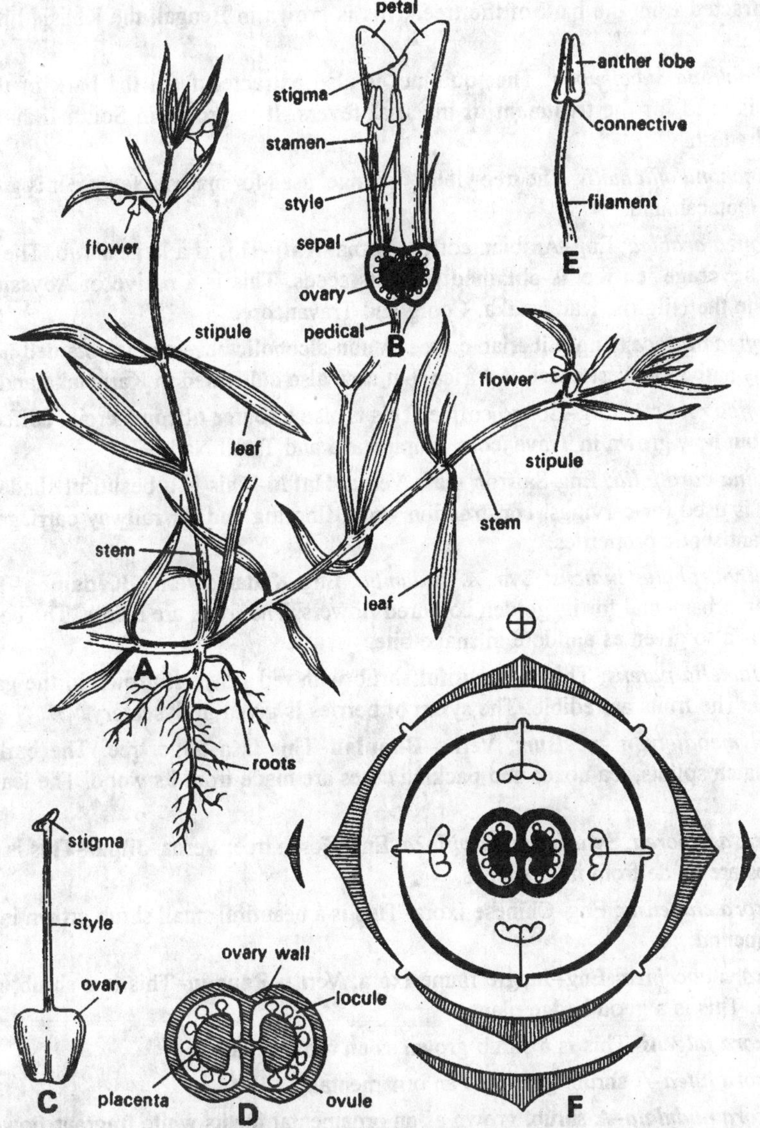

Fig. 15.7. Rubiaceae—*Oldenlandia corymbosa* L.A., flowering plant; B, L.S. of flowers; C, gynoecium, D, T.S. of Ovary; E, Stames; F, Floral diagram.

Fruit–berry.

Floral formula–Br Brl $\oplus$ ⚥ $K_{(4)}, \overline{C_4, A_4}\, G_{(\bar{2})}$.

Economic Importance of the Family

From the point of view of its economic importance, the family is of great value. A list of few important plants is given below:

1. *Cinchona calisaya*; Eng.–Quinine; Verna.-**Cinchona**–This is a tree. The famous medicine of malaria known as quinine is obtained from the bark of this tree. The tree is native of South America, but found in the Nilgiris and Sikkim.

2. *Cinchona ledgeriana*; The medicine known as 'quinine' which is used for malarial fevers is also extracted from the bark of the tree. This is grown in Bengal, the Khasia hills and South India.

3. *Cinchona succirubra*; The 'quinine' is also extracted from the bark of this tree. The medicine is used for the treatment of malarial fevers. It is grown in South India, Sikkim and Madhya Pradesh.

4. *Cinchona officinalis*; The tree yields 'quinine' used for malarial fevers. It is grown in South India and Ootacamund.

5. *Coffea arabica*; Eng–Arabian coffee; Verna–**Kafi**–This is a large shrub. The famous non-alcoholic beverage `coffee' is obtained from its seeds. This is a native of Abyssinia, but now cultivated in the Nilgiris, Karanataka, Coorg and Travancore.

6. *Coffea liberica;* Eng–Liberian coffee–A non-alcoholic beverage is obtained from its seeds. The bush is native of West coast of Africa, but now also cultivated in Karnataka and Travancore.

7. *Coffea robusta:* Eng–Congo coffee–This is also a source of commercial coffee. It is native of Congo but now grown in Travancore, Karanataka and Tamil Nadu.

8. *Adina cordifolia*; Eng-Saffron teak; Verna–**Haldu**–This is a beautiful shade giving tree. The wood is used for carvings, construction work, flooring and for railway carriages. The plant possesses antiseptic properties.

9. *Anthocephalus indicus*; Syn. *A. cadamba*; Eng–Kadam; Verna–**Kadamba**–This is a tree grown as an ornamental for its golden coloured flowers. The fruits are edible. The bark is used as a tonic. It is also given as antidote in snake-bite.

10. *Hamelia patens*; This is beautiful shrub with red flowers, grown in the gardens as an ornamental. The fruits are edible. The syrup of berries is given in dysentery.

11. *Hymeodictyon excelsum*; Verna–**Bhaulan**–This is a large tree. The bark is used in tanning. Match splints, tea boxes and packing cases are made from its wood. The leaves are used as fodder.

12. *Ixora arborea*; Syn. *Ixora parviflora*; Eng–Torch tree; Verna–**Jilpai**–This is a small tree. The torches are made from its branches.

13. *Ixora chinensis*; Eng–Chinese Ixora–This is a beautiful small shrub grown in the gardens as an ornamental.

14. *Ixora coccinea*; Eng–Jungle flame Ixora; Verna–**Rangan**–This is a shrub, grown as an ornamental. This is a good hedge plant.

15. *Ixora fulgens*–This is a shrub grown as an ornamental.

16. *Ixora lutea*–A shrub, grown as an ornamental.

17. *Ixora nudulata*–A shrub, grown as an ornamental for its white fragrant flowers.

18. *Galium* sp. –The plants are very ornamental, and grown in the gardens.

19. *Gardenia campanulata*; Verna–**Bitmara**–A large shrub. The leaves and fruits are edible. The plants are found in Assam and Bihar.

20. *Gardenia gummifera*; Eng–Gummy cape jasmine; Verna–**Dikamli**–The plant yields a gum-resin. It is used as carminative and stimulant, and also given in dyspepsia. The fruits are edible.

21. *Gardenia jasminoides*; Verna–**Gandharaj**–This is an ornamental shrub grown in the gardens. A yellow dye is obtained from the pulp of its fruits. An essential oil is extracted from its flowers, which is used in making of perfumes.

22. *Gardenia latifolia*; Verna–**Papra**–This is a small tree, usually grown as hedge plant.

23. *Gardenia resinifera*; Verna–**Dikamli**–This is a large ornamental shrub with fragrant flowers.

24. *Mussaenda glabrata*; Verna–**Bedina**–This is a small tree, grown as an ornamental in the gardens. The plant is used for asthma, fevers and dropsy.

25. *Mussaenda luteola*; This is a shrub of ornamental value.

26. *Mitragyna parviflora*; Verna–**Kadam**–This is a tree found throughout our country. A fibre is obtained from the bark which is used for cordage. The furniture, agriculture implements and carvings are prepared from its wood. The leaves are used as a fodder.

27. *Morinda angustifolia*; Verna–**Banhardi**–This is a large climbing shrub. The roots yield a yellow dye. The plants are found in the Khasia hills, Assam and Sikkim.

28. *Oldenlandia umbellata*; Eng–Indian madder; Verna–**Saya**–This is a herb, commonly found in Bengal, Orissa and South India. A red dye is obtained from the bark of its root.

29. *Oldenlandia corymbosa;* Verna–**Daman paper**–The decoction is used as a remedy for dysentery and cholera.

30. *Randia dumetorum*; Syn. *Gardenia dumetorum*; Verna–**Mainphal**–This is a large shrub or a small tree. The fruits are edible. The pulp of the fruits is used to cure dysentery. The plants are commonly found in Tamil Nadu, Maharashtra, Gujarat and Uttar Pradesh.

31. *Randia uliginosa*; Syn. *Gardenia uliginosa*; Verna–**Pindalu**–The fruits are edible and used for dyeing purposes. The leaves are used as fodder. The unripe fruits are used as a remedy for dysentery and diarrhoea. The plants are commonly found throughout the country except in the Northern India.

32. *Rubia cordifolia;* Verna–**Manjit**–This is a herb. A dye is extracted from its roots.

33. *Rubia khasiana*; A red dye is obtained from its roots.

34. *Rubia sikkimensis*–A red dye is obtained from its roots. It is used for dyeing both cotton and hair.

35. *Rubia tinctorum*; Verna–**Bacho**–The roots of this plant also yield a red dye which is used for dyeing cloth. The plant is also used as fodder plant.

Systematic Relationships: Bentham and Hooker included two families, *i.e.*, the Caprifoliaceae and the Rubiaceae within the order Rubiales, Engler and Diels included five families in the order. They are–Rubiaceae, Adoxaceae, Dipsacaceae, Caprifoliaceae and Valerianaceae. Hutchinson (1926) however, has placed the Rubiaceae and the Caprifoliaceae of Engler and Diels to his Asterales, but later on (1948) he modified his own view and placed these three families in a new order, the Valerianales. According to him the Rubiales and the Valerianales are not at all related phylogenetically. According to him the Rubiales are derived from Loganiaceae and the Valerianales from Saxifragaceae.

16

Order–Ranales Family–Ranunculaceae

According to Hutchinson this is the fifty-fifth order of the phylum Angiospermae, subphylum Dicotyledones and division Herbaceae. This order comprises of seven families, *i.e.*, Paeoniaceae, Helleboraceae, Ranunculaceae, Nymphaeaceae, Podophyllaceae, Ceratophyllaceae, Cabombaceae, of which Ranunculaceae has been described in the present text in detail. Bentham and Hooker have considered the order–Ranales as the primitive most. According to them this is the first order of class Dicotyledones, division Polypetalaae and series Thalamiflorae. They have placed eight families in this order including Ranunculaceae, Magnoliaceae, Annonaceae, Nymphaeaceae. The Ranunculaceae has been supposed to be most primitive within the order. Engler and Prantl have placed the family Ranunculaceae along with other seventeen families including Nymphaeaceae, Magnoliaceae, Annonaceae, etc., in their eighteenth order– the Ranales of class Dicotyledones and subclass Archichlamydeae.

Hypogynous to rarely perigynous; bisexual; hemicyclic to rarely cyclic; apocarpous; petals usually present; stamens free, often numerous seeds rich in endosperm; embryo minute. Herbs to softly woody; some aquatics.

Family–Ranunculaceae

There are about 35 genera and 1500 species in this family.

Distribution: They are chiefly found in the temperate and other colder parts of the world. In our country, the plants are chiefly found in the hilly regions. The species of *Clematis* are found in the Western Himalayas up to 5,000 feet. *Thalictrum javanicum* is found to be distributed from Shimla to Sikkim hills and also in Khasia hills of Assam. *Thalictrum foliolosum* grows in the Himalayas upto 9000 feet. The species of *Ranunculus* are found in the Indian plains. In Britain the family is represented by several genera, such as, *Ranunculus, Clematis, Anemone, Thalictrum, Paeonia, Aquilegia, Delphinium, Aconitum*, etc.

Habit: Generally the plants are annual or perennial herbs. They perennate either by means of rhizomes or by tuberous roots, *e.g.*, *Ranunculus, Aconitum, Paeonia*, etc. In many plants the primary root perishes and is replaced by adventitious roots. These roots become tuberous and contain reserve food material in them, *e.g.*, *Ranunculus, Aconitum, Paeonia*, etc. A few species are annuals. Undershrubs, and shrubs are also found in this family. Usually the species of *Clematis* are climbing undershrubs. The species of *Naravelia* are also climbing shrubs.

Leaves: The leaves are alternate, exstipulate and more or less divided. In some cases the leaves may be opposite or even radical. In *Clematis*, the leaves are opposite and compound. They possess twining petioles. In *Clematis aphylla* the whole leaf becomes modified into a tendril. In the case of *Naravelia*, the leaves are compound and trifoliate, here the terminal leaflet becomes modified into a tendril. In *Anemone* the leaves are radical. The leaf base usually broadens into a sheath which sometimes elongates into a pair of lateral stipules, *e.g.*, in *Thalictrum*. The leaves of *Thalictrum foliolosum* are pinnately decompound with auricled sheaths. The leaf lamina is sometimes entire. It is narrow in some species of *Ranunculus*, however, in *Ranunculus ficaria* and *Caltha* sp., it is cordate. Heterophylly is found in the aquatic species of *Ranunculus*. In such cases the submerged leaves are very much dissected whereas the floating ones are simply lobed. In *Delphinium* the leaves are palmately lobed and much dissected. The leaves of *Clematis* are climbing.

Inflorescence: In most cases the inflorescence is of cymose type, *e.g.*, *Ranunculus*. However, in some species of *Ranunculus* the floweres are found to be arranged in panicles. The flowers are often solitary and terminal, *e.g.*, *Anemone* sp. In *Anemone vitifolia* the flowers are arranged in decompound cymes. In *Aconitum* and *Delphinium*, the inflorescence is of racemose type. In *Clematis cadmia* the inflorescence is solitary axillary. In the species of *Nigella*, in the beginning a terminal flower is formed but later on the raceme develops.

Flower: The flowers are ebracteate rarely bracteate, hermaphrodite and mostly actinomorphic (regular), rarely they are zygomorphic (irregular) such as *Delphinium* and *Aconitum*, hypogynous and sometimes a disc is found below the gynoecium, *e.g.*, *Paeonia*. The flowers are usually developed on an elongated receptable. The floral parts being arranged wholly or partly spirally.

Perianth: In most of the flowers of this family, the perianth is not distinguished into calyx and corolla. It is simple, petaloid and variously coloured. However, in *Ranunculus* there are two distinct whorls of calyx and corolla, each consisting of five sepals and five petals respectively. The perianth is often associated with nectar secreting structures of various forms. In *Ranunculus*, the petals possess pocket like nectaries at their bases. The nectaries are supposed to be the modification of the petals. In certain genera the nectaries develop in the association of sepals, stamens or even carpels. The number of perianth leaves varies from 4 to 20.

Androecium: Usually the number of stamens is indefinite. The stamens are free (polyandrous). They are hypogynous and arranged spirally. Mostly the anthers are extrorse basifixed and dithecous. They dehisce laterally.

Gynoecium: It consists of indefinite carpels (polycarpellary), the carpels are free, *i.e.*, apocarpous. In *Delphinium* the number of carpels is reduced to one. In *Aconitum* sp., there are three to five carpels. In *Nigella*, there are 5 to 8 carpels, which are more or less united (*i.e.*, syncarpous). In each ovary the number of ovules ranges from one to many. The ovules are anatropous. The placentation is either marginal (*e.g.*, *Delphinium*), or basal (*e.g.*, *Ranunculus*).

Fruit: The fruit is usually either an etaerio of achenes, *e.g.*, *Ranunculus* or an etaerio of follicles, *e.g.*, *Aconitum*. In *Actaea*, it is a berry. In *Nigella* it becomes capsule, because of the fusion of carpels. In some of the cases the achenes possess long persistent feathery styles which help in the distribution of the fruits by means of wind, *e.g.*, *Clematis* and *Anemone*. The seeds of *Helleborus* are disseminated by ants. The ants are attracted by the oil-containing swellings on the raphe. The fruits of certain species of *Ranunculus* possess hooks and are distributed by animals and birds.

Seeds: The seeds are endospermic. Each seed contains only endosperm with a very small embryo.

Pollination: The pollination takes place either by means of insects (entomophily) or by wind (anemophily). In *Ranunculus* and others the flowers are generally protandrous. In *Thalictrum* and *Helleborus* the flowers are protogynous, whereas, in *Anemone* and *Trollius* the flowers are homogamous. The flowers of some species of *Thalictrum* are wind pollinated but most of the other genera are insect pollinated. The flowers of *Delphinium* and *Aconitum* are pollinated by long tongued bees.

Floral Formulae:

Ranunculus: $\oplus$ ⚥ K 5, C 5, A $\propto$, G $\underline{\propto}$

Delphinium: ·|· ⚥ K 5, C (4), A $\propto$, G $\underline{1}$

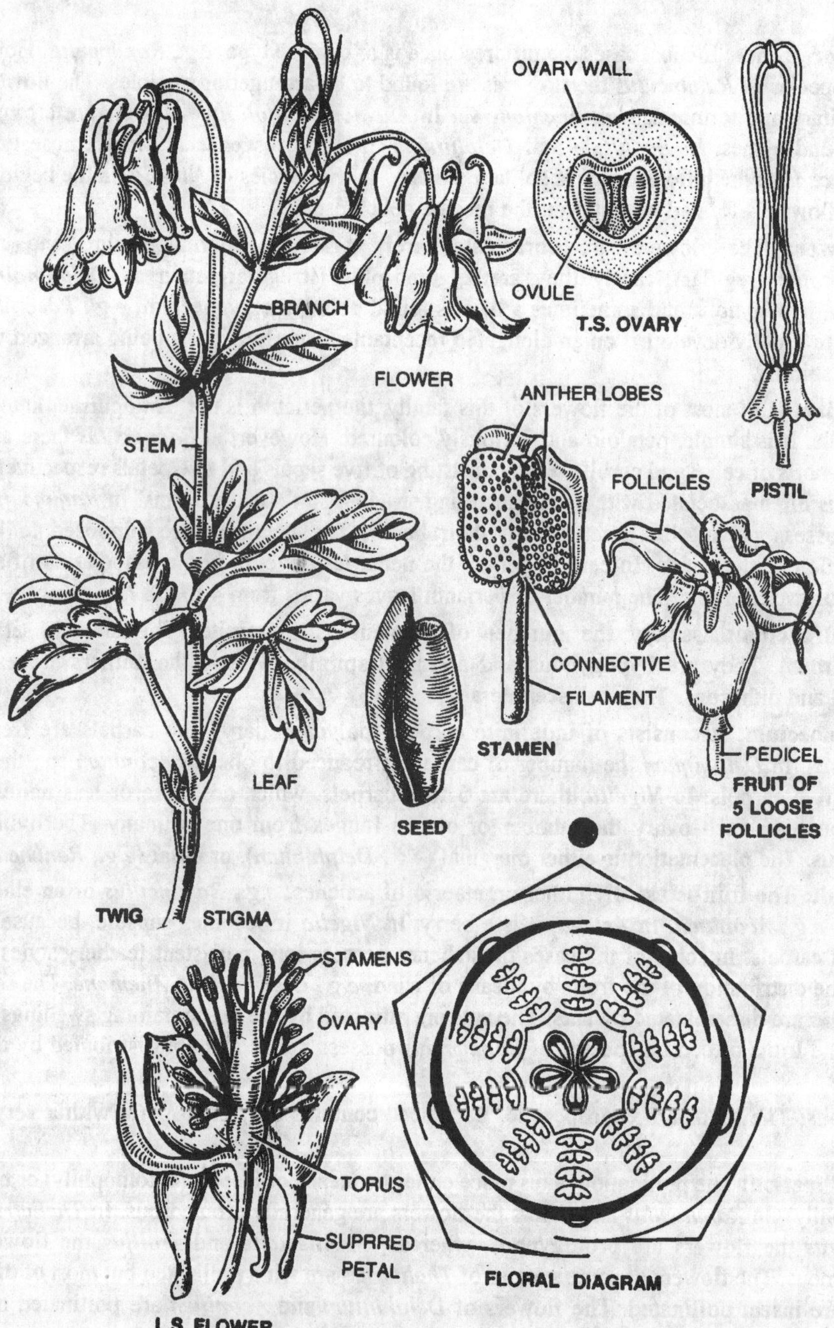

Fig. 16.1. Ranunculaceae. *Aquilegia vulgaris* (Linn.) Thunb.

Bentham & Hooker (1862)	Engler & Prantl (1931)	Hutchinson (1959)
Phanerogams	Phanerogams	Angiospermae
Dicotyledones	Dicotyledones	Dicotyledones
Polypetalae	Archichlamydeae	Herbaceae
Thalamiflorae	Ranales	Ranales
Ranales	Ranunculaceae	Ranunculaceae
Ranunculaceae		

Fig. 16.1(*a*). Ranuculaceae *Clematis paniculata*

Economic value. Grown as on ornamental. It is poisonous.

Description of some important plants in semi-technical language:

1. ***Aquilegia vulgaris*** Linn.

Habit: Annual or perennial herbs.

Root : Tap and branched.

Stem: Erect, herbaceous, aerial green, branched, cylinderical, glabrous.

Leaf: Cauline and ramal, simple, petiolate, much dissected, entire, lobes with acute apices, stipulate, membranous stipules fused with leaf base, multicostate reticulate venation.

Inflorescence: Cymose.

Flower: Pedicellate, bracteate, hermaphrodite, complete, zygomorphic, hypogynous, nectar secreted in spurs of petals, pentamerous.

Calyx: 5 petals, polysepalous, petaloid, imbricate aestivation.

Corolla: 5 petals, polypetalous imbricate aestivation, nectar secreted in spurs of petals.

Androecium: Indefinite, polyandrous, filament long, anthers dithecous, extrorse, basifixed.

Gynoecium: 5 carpels, apocarpous, carpels are connate in a five-chambered ovary, ovary superior.

Fruit: Follicle, made up of 5 loose follicles.

Floral formula: Br. Brl. ·|· ⚥ K 5, C 5, A $\propto$ G $\underline{5}$.

2. ***Clematis paniculatea*** Thunb.

Habit: A climber, annual or perennial.

Root: Tap and branched.

Stem: Herbaceous, aerial weak, climbing with the help of tendrilar petioles, branched, solid, smooth and green.

Leaf: Cauline and ramal, simple, petiolate, petiole tendrilar, exstipulate, opposite decussate, ovate, acute, entire, glabrous, unicostate reticulate, coriaceous.

Inflorescence: Cymose, dichasial cyme.

Flower: Pedicellate, bracteate, bracteolate, complete, actinomorphic, hermaphrodite, tetramerous, hypogynous, hemicyclic and scented.

Calyx: 4 sepals, polysepalous, valvate, petaloid, white.

Corolla: Absent.

Androecium: Indefinite, polyandrous, filamens long, monothecous, extrorse, basifixed.

Gynoecium: 4-6 carpels, apocarpous, ovary superior, unilocular, basal placentation, style short, stigma simple, hairy outgrowths arise from ovary.

Fruit: Achene.

Floral formula: Br. Brl ⊕ ⚥ K 4, C 0, A $\propto$, G $\underline{4\text{–}6}$.

3. ***Ranunculus scleratus:*** Verna–**Shim.**

Habit: Annual herb.

Root: Tap, branched.

Stem: Erect, reduced in early stages, herbaceous, green, glabrous, solid, branched, aerial with nodes and internodes.

Leaf: Simple, petiolate, radical with long petioles, tripartite or trilobed, each lobe further divided with obovate, cuneate segments, membranous stipules fused with the leaf base, multicostate reticulate venation.

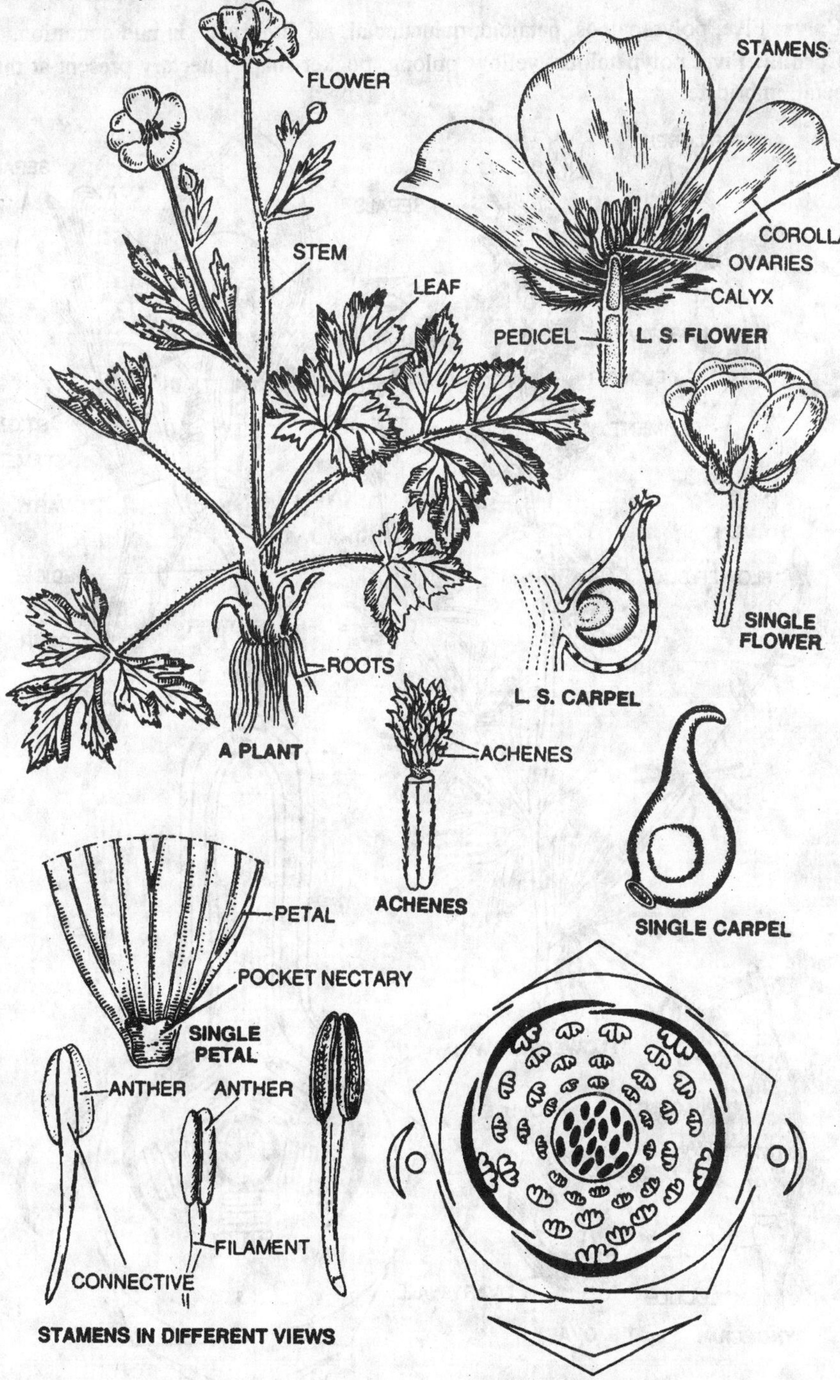

Fig. 16.2. Ranunculaceae–*Ranunçulus sceleratus* Linn.

Inflorescence: Cymose.

Flower: Pedicellate, bracteate, hermaphrodite, complete, actinomorphic, yellow and hypogynous.

Calyx: Five, polysepalous, petaloid, quincuncial, *i.e.*, imbricate in bud condition.

Corolla: Five, polypetalous, yellow, oblong, pocket-shaped nectary present at the base of each petal, imbricate.

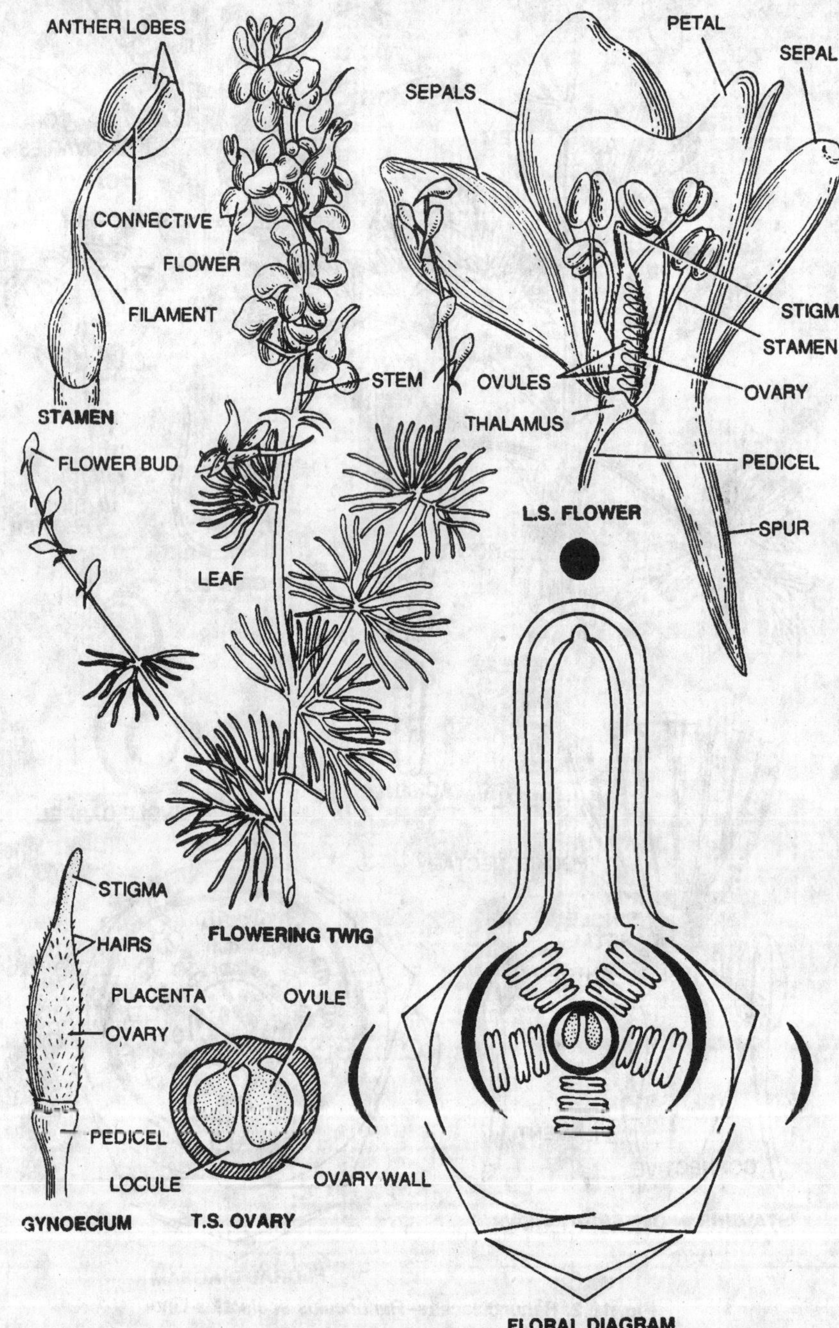

Fig. 16.3. Ranunculaceae. *Delphinium ajacis* Linn.

Androecium: Indefinite, the stamens arranged spirally, polyandrous, filaments long, anthers extrorse, elongated, yellow and basifixed.

Gynoecium: Indefinite (polycarpellary), apocarpous, ovary superior, unilocular, basal placentation.

Fruit: An etaerio of single seeded achenes.

Floral formula; ⊕ ⚥ K 5, C 5, A ∝, G $\underline{\propto}$.

4. ***Delphinium ajacis***; Eng.– Rocket larkspur.

Habit: Annual herb.

Root: Tap, branched and fibrous.

Stem: Erect, herbaceous, green, branched, cylindrical, glabrous and fistular.

Leaf: Simple, cauline, much dissected, lobes linear with acute apex, sessile, exstipulate, multicostate reticulate venation.

Inflorescence: Racemose, typical raceme.

Flower: Pedicellate, bracteate, bracteolate (two bracteoles, hermaphrodite, complete, zygomorphic, hypogynous and violet coloured.

Calyx: Five, polysepalous, petaloid, violet coloured, posterior sepal forms a long spur, quincuncial aestivation.

Corolla: Four, gamopetalous, the two posterior petals are smaller and produced into a long spur which enters into the spur of posterior sepal and the remaining two lobes which are large in size enclose the essential organs, imbricate.

Androecium: Indefinite, stamens arranged spirally in five groups of three stamens each alternating with the petals, polyandrous, filaments flattened at base, basifixed anthers, dehisce by lateral slits.

Gynoecium: One (monocarpellary), ovary superior, unilocular, pubescent, marginal placentation.

Fruit: A follicle.

Floral Formula : Br, Brl ·|· ⚥ K 5, C (4), A∝, G$\underline{1}$.

Economic Importance of the Family

The family is of little economic value. Many plants are grown in the gardens as ornamentals. Some are of medicinal importance. A list of few important plants is given below:

1. *Aconitum chasmanthum*; Verna.–**Banbalnag**–A herb. Its tuberous roots are the source of a drug known as **aconite**. This drug is used externally for rheumatism and internally to relieve pain and fever; usually found in Western Himalayas.

2. *Aconitum heterophyllum*; Verna.–**Atis**—It is found in the Himalayas. The roots of this plant are used medicinally they are astringent, tonic and used in cough and diarrhoea.

3. *Anemone pulsatilla*–They are found in Europe and Russia. This is a herb, which yields a medicine known as **pulsatilla**. This medicine is given for the treatment of nervous and menstrual trouble of women.

4. *Clematis paniculata*; A shrub, grown as an ornamental for its beautiful flowers.

5. *Clematis triloba*; Verna–**Murhari**–It yields a medicine which is used for the treatment of leprosy and other blood diseases.

6. *Delphinium ajacis*; Eng.–Rocket Larkspur–This is native of South Europe. The plants are grown in the gardens as ornamentals. The seeds are insecticidal.

7. *Delphinium brunonianum*; Eng.–Musk larkspur; Verna.–**Laskar.** The juice of leaves is used as an insecticide. This is found in the Western Himalayas.

8. *Delphinium caeruleum*; Verna.–**Dhakangu**–The root is used as an insecticide to kill maggots in the wounds of goats.

9. *Delphinium denundatum*; Verna–**Nirbisi**–The roots are used as a tonic.

10. *Nigella damascena*; Eng.–Love-in-a mist.– A herb, grown as an ornamental in the gardens. It is native of South Europe.

11. *Nigella sativa*; Eng.–Black cumin; Verna.–**Kala jira, Kalonji**–The seeds are used as condiment and spice. The seeds mixed with sesamum oil are used externally in eruption of skin and for scorpion sting. The seeds are stimulant, carminative and diuretic. It is a herb. The plants are cultivated in the Punjab, Assam, Bihar and Bengal.

12. *Ranunculus aquatilis*; Verna.–**Jal dhania**–It is used medicinally in the treatment of rheumatic pain and asthma.

13. *Ranunculus ficaria*; Eng.–Pilewort–The plant is native of Europe. It makes a good medicine of piles. The patent medicine of piles, known as `pilewort ointment' is prepared from this plant.

14. *Thalictrum foliolosum*; Verna.–**Mamira**–The roots are used as a tonic. The plants are found in the Himalayas.

15. *Thalictrum javanicum*; This plant is the alternate host of brown rust of wheat, caused by ***Puccinia recondita.***

17

Order–Brassicales (Cruciales)
Family–Brassicaceae (Cruciferae)

According to Hutchinson this is the sixtieth order of the phylum Angiospermae, sub-phylum Dicotyledones, and division Herbaceae. This order consists of a single family the Cruciferae. This family has been described here in detail. Bentham and Hooker have placed the family Cruciferae along with eight other families including Papaveraceae, Capparidaceae, Violaceae, etc., among the second order–the Parietales of class Dicotyledones, division Polypetalae and series Thalamiflorae. Engler and Prantl have placed the Cruciferae along with other five families including Papaveraceae, Capparidaceae, etc., in the nineteenth order the Rhoeadales of class Dicotyledones and sub-class Archichlamydeae.

Stamens reduced to 6, 4 long and 2 short, petals 4; ovary usually divided by a false septum; no endosperm.

Family–Cruciferae (Brassicaceae)

There are about 350 genera and 2500 species in the family.

Distribution: The members of this family are cosmopolitan in their distribution but majority of them are common in north temperate regions. Some species are sub-arctic. A large number of important vegetable crops as well as many garden and wild flowers are included in this family. The family is represented by several important species in our country, such as *Brassica campestris, Brassica nigra, Raphanus sativus, Iberis amara, Senebiera didyma* and several others. Usually the plants grow in winter season in our country.

Habit: Mostly the plants are annual, bienninal or perennial herbs. Sometimes they are small shrubs. Majority of the species, of *Brassica* are annual, biennial or sometimes perennial herbs. The species of *Raphanus* may be annual or biennial herbs. However, majority of the plants are annuals of a life cycle of only a few weeks. Usually the biennials, *e.g.*, Turnip (**Shalgam**) develop swollen tap roots which store enough of nourishment. *Farsetia* is a small branched undershrub.

Root: Usually tap and branched. The tap root modifications–fusiform root, *e.g.*, *Raphanus sativus*, napiform root, *e.g.*, *Brassica rapa*, are also found. Such roots become swollen and store nourishment in them.

Stem: Usually tap stem is herbaceous. However, in *B. oleracea* var. *caulorapa* (knol khol, **Ganth-Gobhi**). The stem becomes corm like and very much thickened, which is eaten as vegetable. In *Raphanus sativus* (Radish, **Muli**), the stem becomes very much condensed.

Leaves: The leaves are simple, alternate, exstipulate and possess simple or branched hairs. They are usually radial or cauline, sub-sessile or sessile, lyrate. When radical they are found in rosettes, *e.g.*, *Raphanus sativus*.

Sometimes the bulbils develop in the axils of the upper leaves which act as vegetative structures for propagation.

Inflorescence: The inflorescence is generally of racemose type and very often may be a raceme, a corymb or corymboraceme. Bracts and bracteoles are absent. They may be present only in rare cases.

Flowers: Mostly the flowers are ebracteate, pedicellate, regular (actinomorphic), hermaphrodite, complete, cruciform and hypogynous.

Calyx: The calyx is polysepalous and consists of four sepals. The sepals are arranged in two whorls of two each. The lateral inner sepals are sometimes pouched at the base, which serve as nectar containers. The aestivation is imbricate.

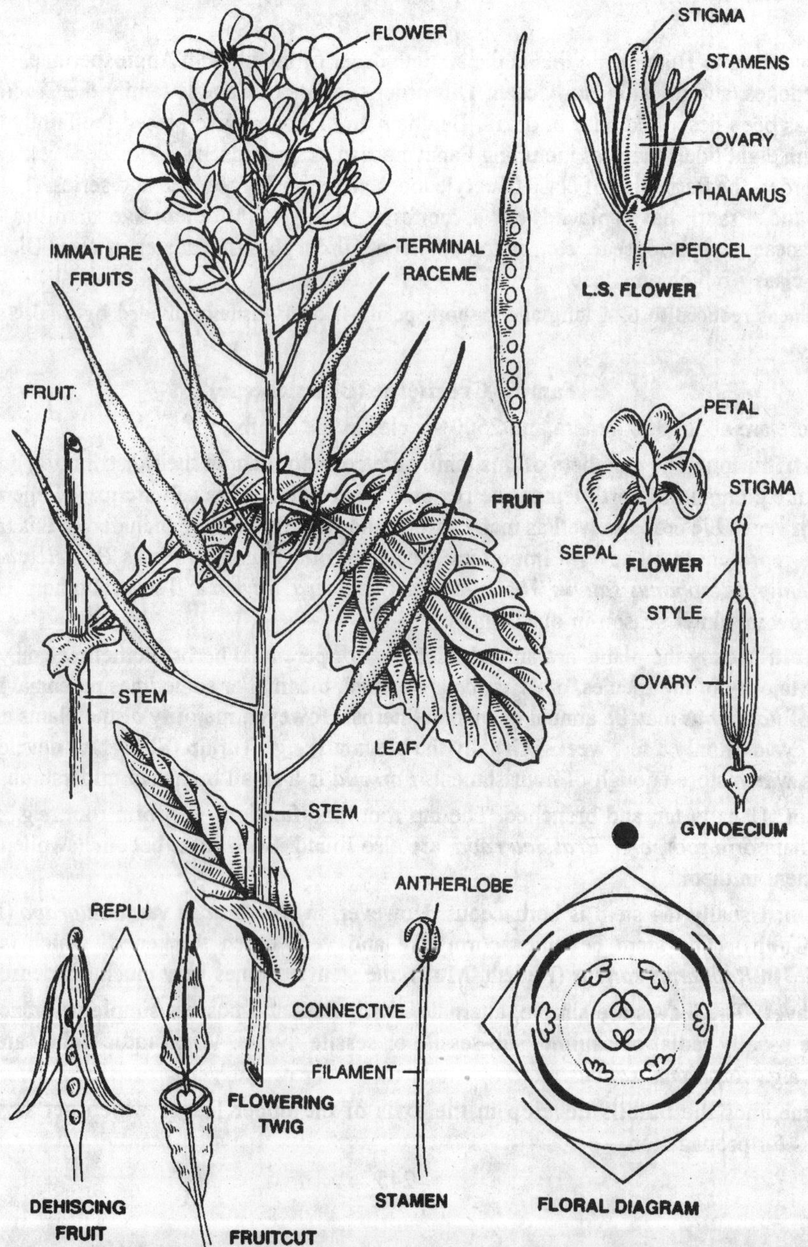

Fig. 17.1. Cruciferae (Brassicaceae). *Brassica campestris* Linn.; Eng. Yellow mustard; Verna. **Sarson.**

Corolla: The corolla is polypetalous and cruciform. It consists of four petals. The petals are found in one whorl. They are alternate to the sepals. Each petal is usually differentiated into a narrow claw and a broad expanded limb. In *Iberis amara* (candytuft) the flowers become zygomorphic by the enlargement of the two outer petals. In *Senebiera* sp., the petals are represented by four minute lobes.

Androecium: Usually the androecium consists of six stamens arranged in two whorls. The outer two stamens are short whereas the inner four are long. This condition of the stamens is known as **tetradynamous** condition and considered to be characteristic of the family. Nectaries are developed as small green glands at the base of the two short stamens. In *Senebiera*, there are only two stamens, the anthers are bilobed; basifixed and introrse.

Gynoecium: It consists of two carpels (bicarpellary), syncarpous. The ovary is superior, unilocular but becomes bilocular because of the development of the **false septum** or **replum** from the ingrowths of the parietal placentas. The placentation is parietal. Many anatropous or campylotropous ovules develop from the parietal placentas. The style is short with two lobed stigmas.

Fruit: The fruit is either siliqua or silicula. Sometimes the fruit is lomentum (lomentaceous siliqua) as in *Raphanus sativus*.

Pollination: The pollination is usually entomophilous, *i.e.*, through the agency of insects. The honey secreting nectaries found at the base of short stamens and anterior sepals are the main centres of the attraction of the insects. The flowers arranged in corymbs also attract the insects and cross pollination is effected. In such cases several flowers are pollinated by one insect at one time.

Floral Formula: ⊕ ⚥ K 2 + 2, C 4, A 2 + 4, G $\underline{(2)}$.

Bentham & Hooker (1862)	**Engler & Prantl (1931)**	**Hutchinson (1959)**
Phanerogams	Phanerogams	Angiospermae
Dicotyledones	Dicotyledoneae	Dicotyledones
Polypetalae	Archichlamydeae	Herbaceae
Thalamiflorae	Rhoeadales	Cruciales (Brassicales)
Parietales	Cruciferae	Cruciferae or Brassicaceae
Cruciferae		

Description of some important plants in semi-technical language:

1. ***Brassica campestris*** Linn.; Verna. **Sarson**; Eng. **Yellow mustard.**

Brassica campestris Linn., very largely grown during the cold season, especially in Uttar Pradesh. The oil **(Karwa tel)** is yielded from its seeds.

Habit: Annual herb, cultivated

Root: Tap, branched.

Stem: Erect, herbaceous, cylindrical, slightly hairy, branched, smooth and green.

Leaf: Cauline and ramal, simple, alternate, sessile, glabrous, upper leaves entire, lower leaves lyrate, exstipulate, unicostate reticulate venation.

Inflorescence: Racemose, typical raceme.

Flower: Pedicellate, ebracteate, hermaphrodite, actinomorphic, yellow, hypogynous, complete, tetramerous, cyclic.

Calyx: Four sepals, polysepalous, arranged in two whorls of two each, the sepals of inner whorl are longer, imbricate aestivation.

Corolla: Four petals, polypetalous, each petal consists of a limb and claw, cruciform, alternating the sepals, imbricate or valvate.

Androecium: Six stamens, polyandrous arranged in two whorls, the two stamens of outer whorl are smaller than the four stamens of inner whorl (*i.e.*, tetradynamous condition), anthers bicelled, basifixed, introrse.

Gynoecium: The carpels (bicarpellary), syncarpous, ovary superior, unilocular but becomes bilocular because of the development of false septum (replum), parietal placentation, style short and stigma bilobed.

Fruit: Siliqua.

Floral Formula: ⊕ ⚥ K 2 + 2, C 4, A 2 + 4, G (2).

Economic value. The tender leaves, and shoots are used as vegetable. Ground mustard seeds are used as a condiment. The seeds yield a fatty oil, which is generally used for cooking purposes.

2. *Senebiera didyma* Pers.

Senebiera didyma Pers. (Lesser Wart-cress). An established weed of cultivation during the cold season. Distribution–plains of N.India.

Habit: Annual herb, wild, weed.

Root: Tap, branched.

Stem: Prostrate, aerial, weak, trailing, cylindrical, branched, branches arranged in rosettes, green, glabrous, herbaceous, solid.

Leaf: Radical, cauline and ramal, compound, exstipulate, unipinnate and imparipinnate, pinnae with cut margins, petiolate, petiole base hairy, alternate, leaf surface glabrous, reticulate venation.

Inflorescence: Racemose, leaf opposed typical raceme.

Flower: Minute, greenish, pedicellate, ebracteate, hermaphrodite, actinomorphic, hypogynous, complete, tetramerous, cyclic.

Calyx: Four sepals, polysepalous, arranged in two whorls of two each, sepals spreading, short, larger than petals, inferior, valvate aestivation.

Corolla: Very much reduced (rudimentary), represented by four minute whitish scaly structures alternating the sepals, valvate.

Androecium: Two stamens, polyandrous, median, anterior-posterior, anthers basifixed, filaments long tapering at apex and broad at base, dithecous, introrse.

Gynoecium: Two carpels (bicarpellary), syncarpous, ovary superior, unilocular but becomes bilocular due to formation of replum, parietal placentation, one ovule on each locule, style reduced, stigma discoid.

Fruit: Compressed silicula.

Floral formula: ⊕ ⚥ K 2 + 2, C 4, A 2 , G (2).

3. *Iberis amara* Linn.; Eng. Candytuft.

Habit: Annual ornamental herb, cultivated.

Stem: Herbaceous, erect, aerial, branched, green, solid, rough surface.

Leaf: Cauline and ramal, alternate, sometimes opposite, sessile, exstipulate, simple, margin somewhat dissected, acute, glabrous, unicostate reticulate.

Inflorescence: Racemose, corymb.

Flower: Ebracteate, pedicellate, complete, irregular (zygomorphic), hermaphrodite, tetramerous, hypogynous, white, cyclic.

Calyx: 4, polysepalous, in two whorls of 2 each, imbricate, petaloid, boat shaped.

Corolla: 4, polypetalous, valvate, 2 anterior petals large, 2 posterior petals small, each petal consists of a claw and limb, cruciform.

Androecium: 6, free stamens (polyandrous), tetradynamous –2 outer lateral short, remaining 4 antero-posterior long, dithecous, dorsifixed, introrse.

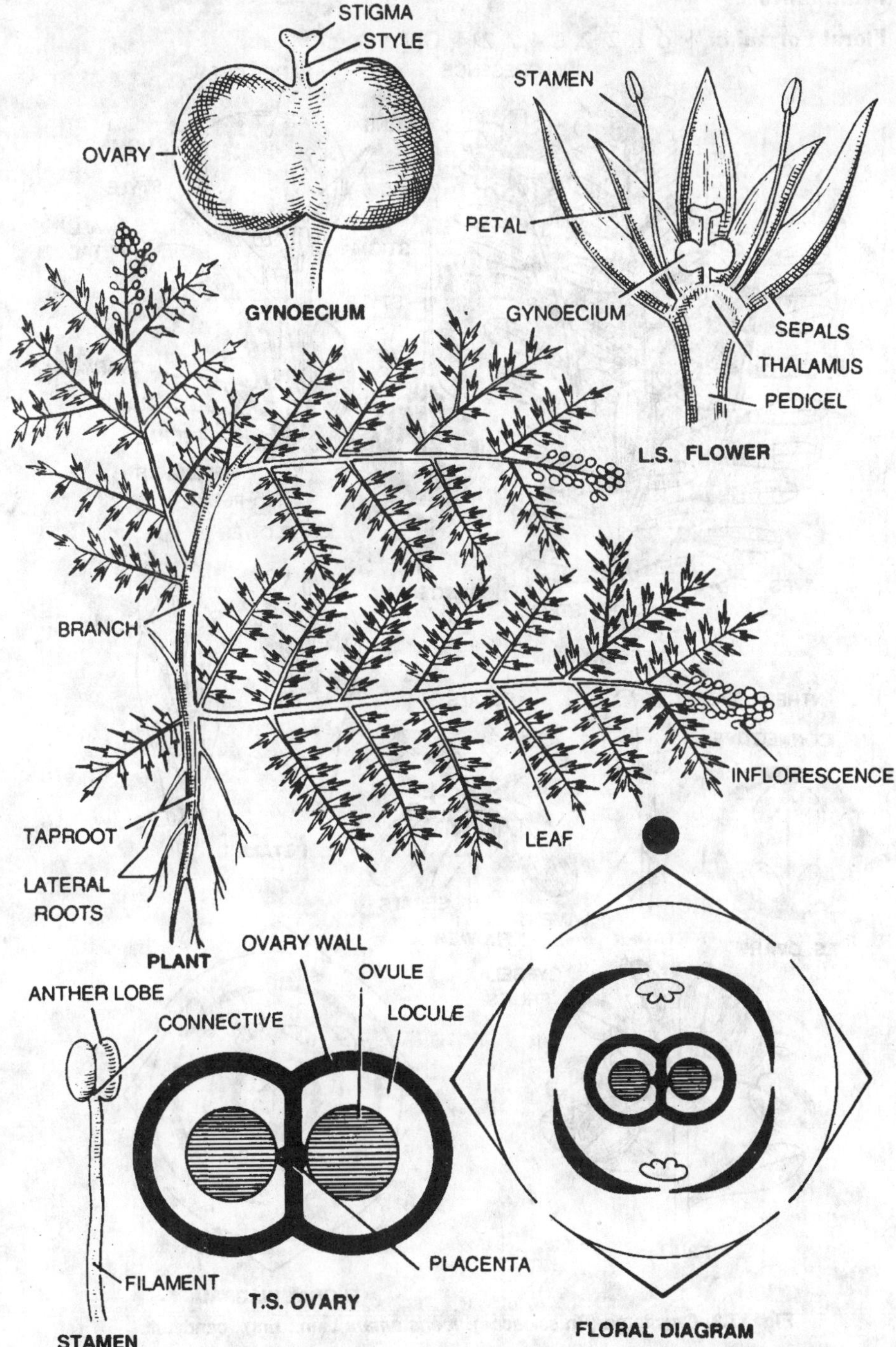

Fig. 17.2. Cruciferae (Brassicaceae). *Senebiera didyma* Pers.; Eng. lesser wart cress

Gynoecium: Bicarpellary, syncarpous, ovary superior, unilocular when young, at maturity becomes bilocular because of the development of false septum, parietal placentation, style long. stigma globular.

Fruit: Silicula.

Floral Formula: ⊕ ⚥ K 2+2, C 4, A 2+4, G(2).

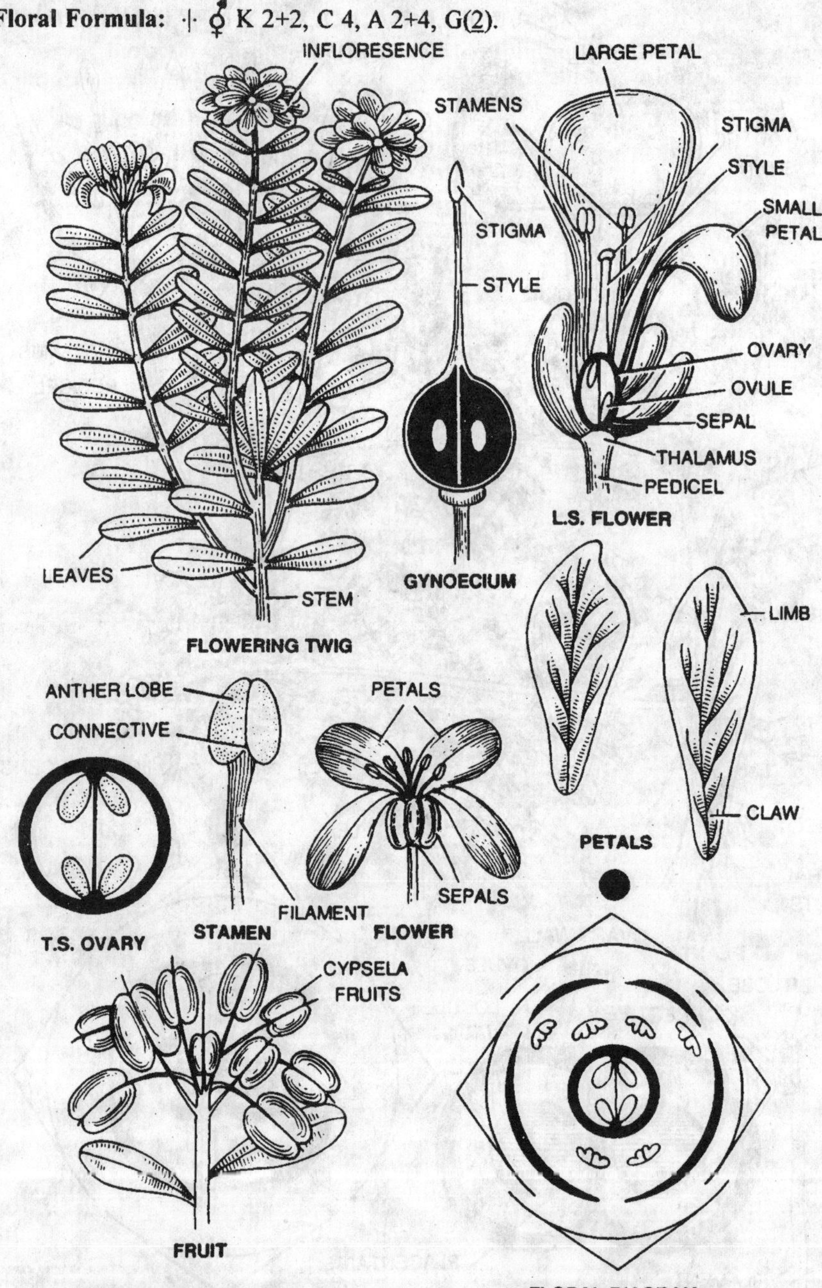

Fig. 17.3. Cruciferae (Brassicacae). *Iberis amara* Linn.; Eng., candytuft.

Economic value. A common ornamental herb. The plants are used in rheumatism and gouts. Seeds are used in asthma and bronchitis.

Economic Importance of the Family

The family is fairly important from the economic point of view. Most of the plants contain sulphur compounds. The seeds of many plants yield vegetable oil of multipurpose use.

1. *Brassica campestris* var. *dichotoma* (Verna.–**Kali Sarson**) cultivated as an oil yielding crop, mostly in the Punjab.

2. *Brassica campestris* var. *sarson*; (Verna.–**Sarson**)–An oil-seed crop grown mainly in Uttar Pradesh, the Punjab, Bihar and Assam. The oil is used for cooking and burning purposes, and the oil cake as a cattle feed. The tender leaves and shoots are used as vegetable.

3. *Brassica campestris* var. *toria* (Verna.–**Toria**)–An oil-yielding crop grown in Uttar Pradesh, West Bengal and the Punjab. The oil is edible and the oil-cake is used as cattle and manure.

4. *Brassica hirta*; Syn. *B. alba* (Eng.–White Mustard; Verna.–**Safed Rai**)–The young leaves and tender shoots are used as vegetable. The seeds yield fatty oil.

5. *Brassica juncea* (Verna.–**Rai**)–The seed oil is used for cooking purposes. Cultivated in the Punjab, West Bengal and Uttar Pradesh.

6. *Brassica juncea* var. *cuneifolia*; Syn. *B. rugosa* var. *cuneifolia* (Verna.–**Rai**)–Commonly grown in the terai areas of Nainital, North Bengal and Assam. The young leaves and tender shoots are used as vegetable.

7. *Brassica napus* (Eng.–Rape; Verna.–**Toriya, Kali sarson**)–The seeds are used as vegetable. Cultivated in the Punjab, Bengal and Bihar.

8. *Brassica nigra* (Eng.–Black mustard; Verna.–**Kali Rai**)–The seeds are used as spice and condiment. Cultivated in the Punjab, Uttar Pradesh and Tamil Nadu.

9. *Brassica oleracea* var. *acephala* (Eng.–Kale; Verna.–**Karam-Sag**)–The young shoots and leaves are eaten as vegetable. Cultivated in Assam, Kashmir and Maharashtra.

10. *Brassica oleracea* var. *botrytis* (Eng.–Cauliflower; Verna.–**Phulgobhi**)–Grown all over Northern India for its edible inflorescence.

11. *Brassica oleracea* var. *capitata* (Eng.–Cabbage; Verna.–**Bandgobhi**)–A common vegetable of Northern India.

12. *Brassica oleracea* var. *gemmifera* (Eng.–Brussels sprouts; Verna.–**Buttamgobhi**)–Grown mainly in Maharashtra. The young shoots, buds and leaves are edible.

13. *Brassica oleracea* var. *gongylodes*; (Eng.–knol–khol: Verna.–**Ganthgobhi**)–The short swollen stem is edible. Grown in Maharashtra, Uttar Pradesh and the Punjab.

14. *Brassica pekinensis* (Eng.–Chinese cabbage)–Grown in Northern India. The leaves are used as vegetable.

15. *Capsella bursa-pastoris*; (Eng.–Shepherd's purse)–A common weed. Found throughout temperate India particularly in Northern Western Himalayas. The plant is of medicinal value.

16. *Cheiranthus cheiri*; (Eng.–Wall flower; Verna.–**Todrisurkh**)–A herb, native of South Europe Grown as an ornamental. The seeds are useful in dry bronchitis, fevers and injuries to the eyes. The flowers are used in paralysis and impotency.

17. *Eruca sativa* (Verna.–**Tara**)–A herb, native of South Europe but grown mainly in Northern India. The oil, obtained from the seeds, is used for burning purposes and the plant is used as fodder.

18. *Iberis amara* (Eng.–Rocket candytuft)–A common ornamental herb. The plants are used medicinally in rheumatism and gouts. Seeds are used medicinally in asthma and bronchites.

19. *Mathiola incana* (Eng.–Stock; Verna.–**Todri safed**)–A herb, native of Europe. Grown as an ornamental. The seeds are used for extraction of an oil, which yields methyl. The seeds mixed with wine are given as an antidote to poisonous bites. They are also used as tonic in stomach complaints.

20. *Lepidium sativum*; (Eng.–Garden cress; Verna.–**Halim**)–A common cultivated herb. The young as well as the ripened seeds are used as vegetable and pulse. The leaves and tender shoots are largely eaten for liver complaints. The plant is also used in the treatment of asthma, cough and bleeding piles.

21. *Nasturtium officinale*; Syn. *N. fontanum* (Eng.–Water-cress; Verna.–**Brahmi Sag**)–A small herb cultivated in Bengal, Orissa and the Punjab. It is used as a vegetable.

22. *Raphanus sativus*; (Eng.–Radish; Verna.–**Muli**)—A herb, cultivated in Uttar Pradesh, the Punjab, Maharashtra, etc. The roots, young leaves and the fruits are used as vegetable.

23. *Raphaņus sativus* var. *candatus* (Eng.–Rat-tail radish; Verna.–**Sengri**)–The fruits are used as vegetable. The fruits are also used for making pickle.

Systematic Relationships: Bentham and Hooker placed the Cruciferae in the order–Parietales. Engler and Diels included the family within the order–Rhoeadales. Hutchinson placed the single family Cruciferae in the order–Cruciales. On the basis of morphology and anatomy of androecium and gynoecium the family is much allied to the Capparidaceae, but Hutchinson does not agree with the statement.

18

Order–Caryophyllales
Family–Caryophyllaceae

According to Hutchinson this is the sixty-second order of the phylum Angiospermae, subphylum Dicotyledones and division Herbaceae. The order comprises of five families, *i.e.*, Elatinaceae, Molluginaceae, Caryophyllaceae, Ficoidaceae and Portulacaceae, of which the Caryophyllaceae has been described in the present text in detail. Bentham and Hooker have placed the Caryophyllaceae along with three others in the fourth order–the Caryophyllinae of class Dicotyledones, division Polypetalae and series Thalamiflorae. Engler and Prantl have placed the Caryophyllaceae along with other eight families including Chenopodiaceae, Amarantaceae, Nyctaginaceae, etc., in the seventeenth order–the Centrospermae of class Dicotyledoneae and sub-class Archichlamydeae.

Hypogynous to perigynous; mostly bisexual; cyclic; syncarpous; axile to free central placentation; stamens mostly definite in number; endosperm copious; embryo curved.

Family—Caryophyllaceae

There are about 80 genera and 2100 species in this family.

Distribution. The members of this family are commonly found in the temperate regions of the Northern hemisphere. Certain genera are found in the Southern hemisphere and few are found in the mountains of tropical regions. In our country the plants of this family are either found in the hilly tracts or they grow in the plains during winter season, *e.g.*, *Stellaria*, *Spergula*, *Dianthus*, etc. The family is well represented in the British Flora and many species are cultivated as ornamental garden flowers.

There are about 500 species in the genus *Silene* of this family, which is supposed to be the largest genus; *Dianthus* contains 350 species and *Stellaria* about 100 species.

Habit: Most of them are annual, while some of them are perennial herbs. Certain small shrubs, *e.g.*, some species of *Acanthophyllum* are also found in the warmer parts of the world.

Stem: The stem is erect, branched, green, herbaceous, solid and mostly swollen at the nodes.

Leaves: The leaves are simple, opposite, entire and exstipulate. The leaves sometimes possess shortly connate perfoliate base, *e.g.*, in *Dianthus*. They are linear to lanceolate in shape. The development of the leaves is unusual. At each node one leaf develops earlier than the other. This leaf bears in its axil a more vigorous bud than on the other side and frequently it is only this bud which devlops later.

The chief anatomical feature is the presence of caryophyllaceous type of stomata. Each such stoma possesses two subsidiary cells placed at the right angles to the guard cells.

Inflorescence: The inflorescence is cymose. Usually it is a dichasium which later on becomes a dichasial cyme ending in a monochasial cyme. This is known as **cincinnus** inflorescence which is characteristic of the family. In certain cases the flowers are arranged in racemes. In some of genera, *e.g.*, *Arenaria, Githago*, etc., the flowers are solitary.

Flower: The flowers are pedicellate, actinomorphic, usually hermaphrodite and pentamerous, but rarely unisexual or tetramerous. They are regular, complete and hypogynous. Each typical flower bears five whorls, each consisting of five members, such as, five sepals, five petals, two pentamerous whorls of stamens, five carpels, five styles and five double rows of ovules and sometimes five partition walls in the basal part of the ovary.

Calyx: It is composed of five and very rarely of four sepals. The sepals may be free or united together into a tube. They are usually persistent with membranous margins. The aestivation is imbricate (quincuncial).

Corolla: It is composed of five and rarely of four petals. The petals are always free (polypetalous). The petals are mostly notched, sometimes bifid, *e.g., Stellaria media.* Usually the aestivation is imbricate.

Androecium: Usually it consists of ten stamens. Sometimes the number of stamens reduces to eight, five, four, three or even one. They are polyandrous, obdiplostemonous, *i.e.*, the stamens are arranged in two whorls of five each, the stamens of the outer whorl are seen to be opposite the petals and of inner whorl alternating the petals. The stamens are hypogynous and occasionally perigynous. The anthers are two-lobed and dehisce longitudinally. Sometimes the petaloid staminodes are present. In *Spergula*, the five or ten stamens are found to be arranged on a perigynous disc.

Gynoecium: It consists of two or five carpels; syncarpous. The styles are free. The ovary is superior and unilocular; the ovules are many, campylotropous and arranged on a central column. The placentation is free-central which is most characteristic of the family. The number of carpels corresponds to the number of styles and stigmas. In *Stellaria*, the number of carpels is reduced to three.

Fruit: Generally the fruit is an unilocular capsule, *e.g., Stellaria, Arenaria, Spergula*, etc. In some cases the fruit, may be an achene or a nut, *e.g., Herniaria, Dysphania, Scleranthus*, etc.

Seeds: The seeds are small and endospermic. The embryo is curved in the endosperm. Sometimes the funicle is conspicuous. They are dispersed by censor mechanism.

Pollination: It usually takes place by means of insects (*i.e.*, entomophily).

Floral Formula: ⊕ ⚥ K 5, C 5, A 5 + 5 or 5, G $\underline{(2-5)}$.

Bentham & Hooker (1862)	**Engler & Prantl (1931)**	**Hutchinson (1959)**
Phanerogams	Phanerogams	Angiospermae
Dicotyledones	Dicotyledoneae	Dicotyledones
Polypetalae	Archichlamydeae	Herbaceae
Thalamiflorae	Centrospermae	Caryophyllales
Caryophyllinae	Caryophyllaceae	Caryophyllaceae
Caryphyllaceae		

Description of some important plants in semi-technical language:

***Saponaria vaccaria* L.** Eng. Cow–herb; Verna. **Musna. Stem**–aerial, erect, herbaceous, cylindrical branched, solid, green glabrous. **Leaf**–cauline, ramal, opposite decussate, exstipulate, simple, sessile, entire, acute, glabrous,lanceolate, coriaceous, unicostate reticulate. **Inflorescence**–cymose, dichasial cyme. **Flower**–bracteate, bracteolate, pedicellate, complete, actinomorphic, hermaphrodite, pentamerous, hypogynous, cyclic. **Calyx**–5, gamosepalous, valvate, tubular, persistent. **Corolla**–5, polypetalous, twisted, caryophyllaceous, petal distinguished into claw and limb. **Androecium**–5+5, obdiplostemonous, polyandrous, filaments, long dithecous, versatile, introrse. **Gynoecium**–bicarpellory, syncarpous, superior, unilocular free-central, ovules many, styles 2, stigmas 2. **Fruit**–capsule.

Floral Formula– Br, Brl ⊕ ⚥ $K_{(5)}\ C_5\ A_{5+5}\ G_{\underline{(2)}}$.

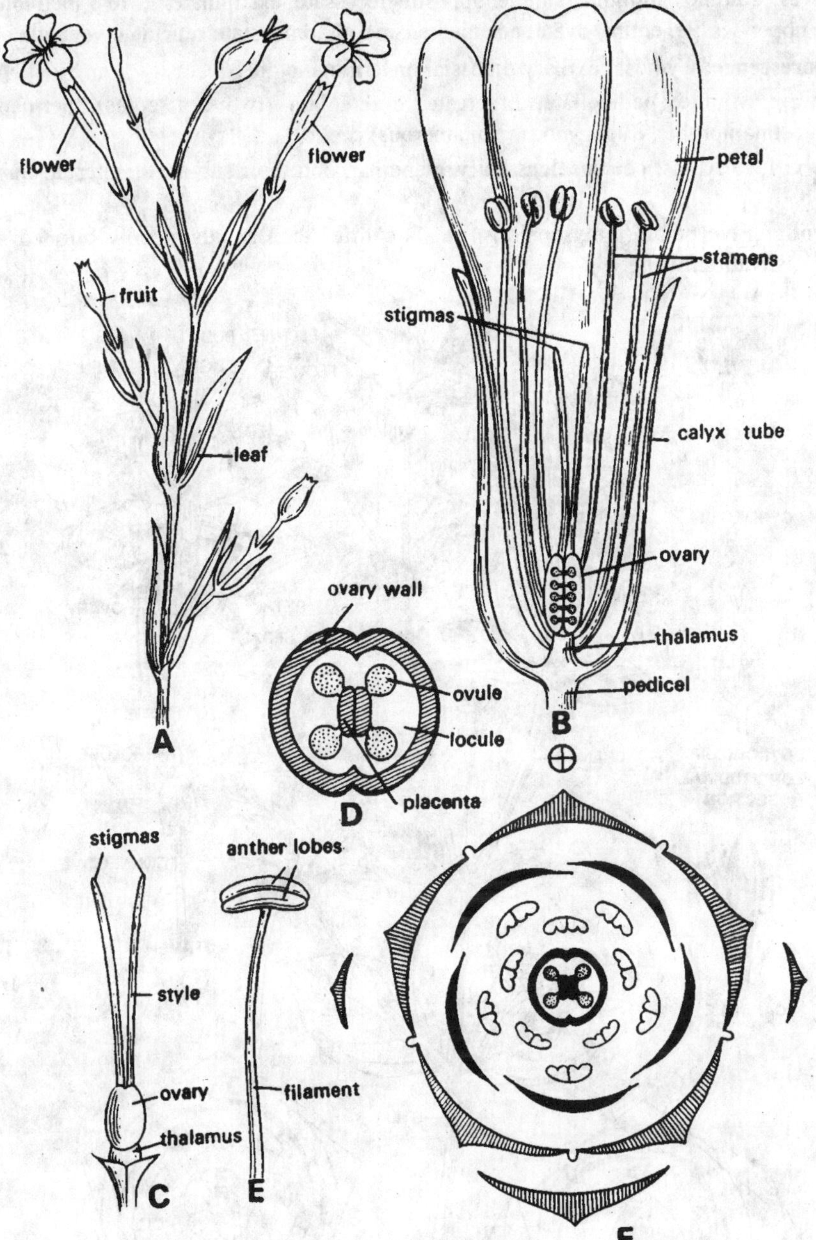

Fig. 18.1. Caryophyllaceae—*Saponaria vaccaria* L. (Musna). A, Flowering twig; B, L.S. of flower; C, gynoecium; D, T.S. of ovary; E, stamen; F, floral diagram.

Stellaria media Linn.; Verna. **Morolia**; Eng. Chickweed.

Stellaria media Linn.(Eng. Chickweed). A common weed. Distribution: throughout N. India.

Habit: An annual herb, weed, six inches to two feet high.

Root: Tap, branched.

Stem: Erect, aerial, herbaceous, cylinderical, branched, green, glabrous, solid, nodes swollen

Leaves: Cauline and ramal, simple, opposite decussate, exstipulate, $\frac{1}{2}$ to 1 inch long, lower petiolate, upper sessile, entire, ovate, acuminate, smooth, unicostate reticulate venation.

Inflorescence: Cymose, axillary dichasial or terminal cyme.

Flower: Minute, pedicellate, bracteate, bracteolate (two bracteoles), hermaphrodite, complete, actinomorphic, hypogynous, pentamerous, cyclic.

Calyx: Five sepals, polysepalous hairy, glandular, obtuse or sub-acute, inferior, quincuncial aestivation.

Corolla: Five petals, polysepalous, petals shorter than sepals, deeply bilobed, inferior, imbricate aestivation, white.

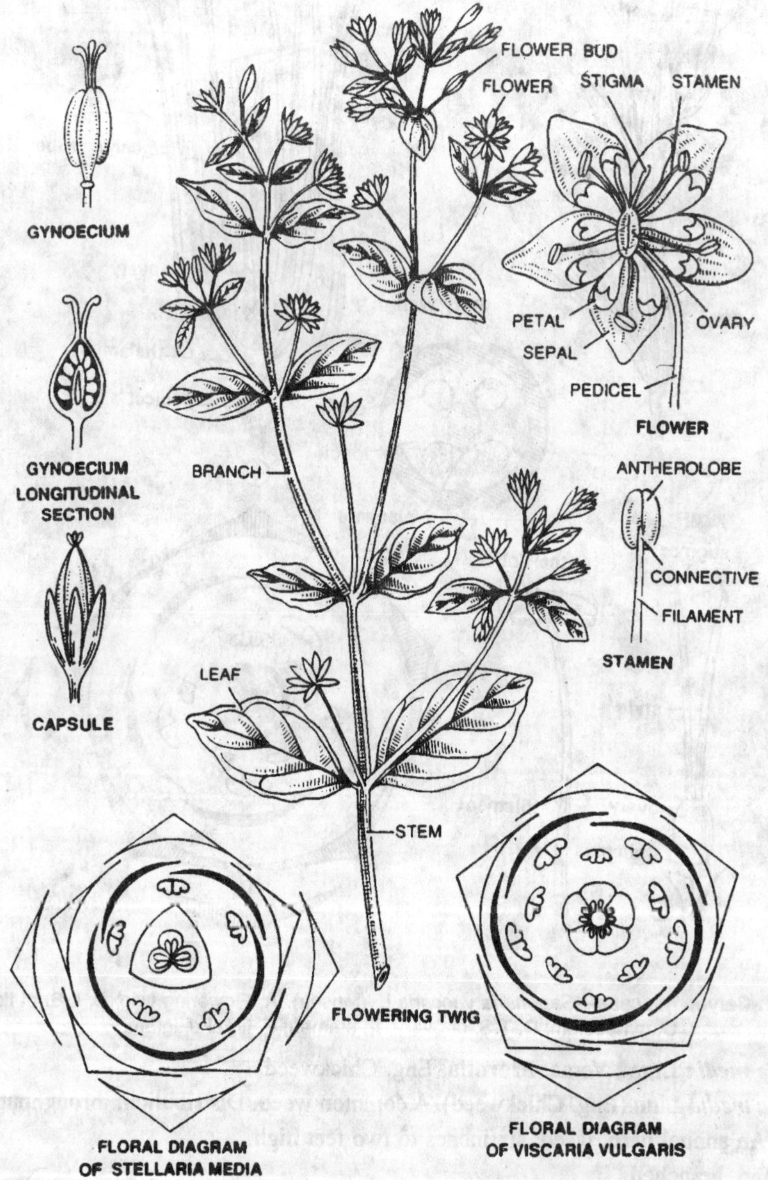

Fig. 18.2. Caryophyllaceae. *Stellaria media* Linn. Eng. Chickweed; Verna. **Morolla.**

Economic value. The natives of Assam eat the leaves and tender stalks, boiled in *Khar* water, either by themselves or with fish. It is used also as vegetable by the natives of the Nilgiris. The plant is employed in plaster for broken bones and swellings, as it is supposed to be cooling and binding.

Androecium: Ten stamens, arranged in two whorls of five each, polyandrous, obdiplostemonous (two alternating whorls of stamens, the stamens of outer whorl being arranged opposite the petals), generally outer three or all five stamens reduced to staminodes, filaments long and slender, dithecous, basifixed, introrse.

Gynoecium: Three carpels, syncarpous, ovary superior, unilocular, free-central placentation, styles three, each with its own stigma.

Fruit: An unilocular capsule.

Floral formula: Br. Brl. ⊕ ⚥ K 5, C 5, A 5+5, G $(\underline{3})$.

Spergula arvensis Linn.; Eng. Corn spurrey.

Spergula arvensis Linn. (Eng. Corn spurrey), plentiful in cultivated ground, flowering during the cold season. Distribution : throughout N. India.

Habit: Annual herb.

Root: Tap, branched.

Stem: Erect, cylindrical, branched, solid or fistular, green, herbaceous, thickened at the joints, glabrous.

Leaf: Cauline and ramal, simple, opposite, linear, sessile, fleshy, leafy buds in their axils and looking apparently whorled, lanceolate, acute, stipulate (stipules small).

Inflorescence: Cymose, dichasial cyme with suppression of branches on one side, *i.e.* cincinnus.

Flower: Pedicellate, bracteate, bracteolate (two bracteoles), hermaphrodite, complete, actinomorphic, hypogynous, greenish white, pentamerous, cyclic.

Calyx: Five sepals, inferior, quincuncial aestivation, persistent.

Corolla: Five petals, polypetalous, white, inferior, entire, imbricate aestivation.

Androecium: Ten stamens, arranged in two whorls of five each obdiplostemonous, anthers bicelled, introrse, dorsifixed.

Gynoecium: 3 or 5 carpels, syncarpous, ovary superior, unilocular, free-central placentation, styles three or five, stigmas 3 or 5.

Fruit: An ovoid capsule.

Floral formula: Br. Brl ⊕ ⚥ K 5, C 5, A 5, G ($\underline{3}$ or $\underline{5}$).

Dianthus caryophyllatus Linn.; Eng. Carnation plant.

Habit: An annual herb, ornamental, cultivated.

Root: Tap, branched.

Stem: Erect, aerial, herbaceous, branched, green smooth, solid.

Leaf: Cauline and ramal, simple, opposite decussate, sessile, exstipulate, sessile, lanceolate, entire, acute, glabrous, unicostate reticulate venation, leaf-base sheathing.

Inflorescence: Cymose, axillary or terminal dichasial cyme.

Flower: Pedicellate, ebracteate, hermaphrodite, actinomorphic, complete, hypogynuous, pentamerous, cyclic.

Epicalyx: 4, in two whorls of 2 each, outer whorl anterio-posterior.

Calyx: Five sepals, polysepalous, inferior, quincuncial aestivation.

Corolla: Five petals, polypetalous, petals fimbriate, inferior, imbricate aestivation, variously coloured, caryophyllaceous, each petal possesses a claw and limb.

Androecium: Ten stamens arranged in two whorls of five each, obdiplostemonous, polyandrous, anthers bicelled, dorsifixed, introrse.

Gynoecium: Two-five carpels, syncarpous, ovary superior, unilocular, free-central placentation, ovules many, styles two, stigmas two, coiled and feathery.

Fruit: A capsule.

Floral formula: ⊕ ⚥ Epi K 2+2, K (5), C 5, A 5+5, G ($\underline{2}$ – $\underline{5}$)

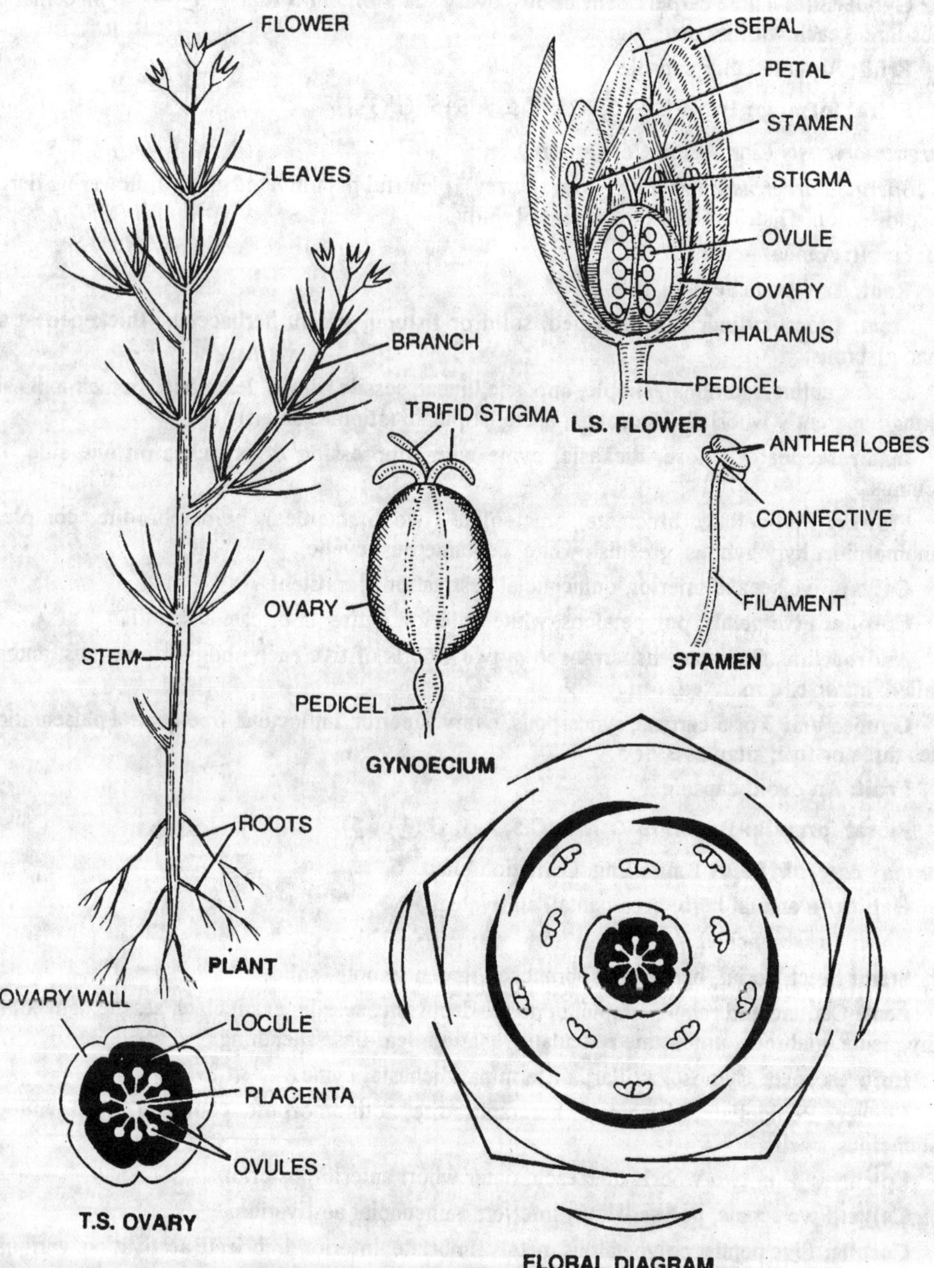

Fig. 18.3. Caryophyllaceae. *Spergula arvensis* Linn.; Eng., corn spurrey.

Economic value. Cultivated in Europe as a fodder plant. Used as a diuretic.

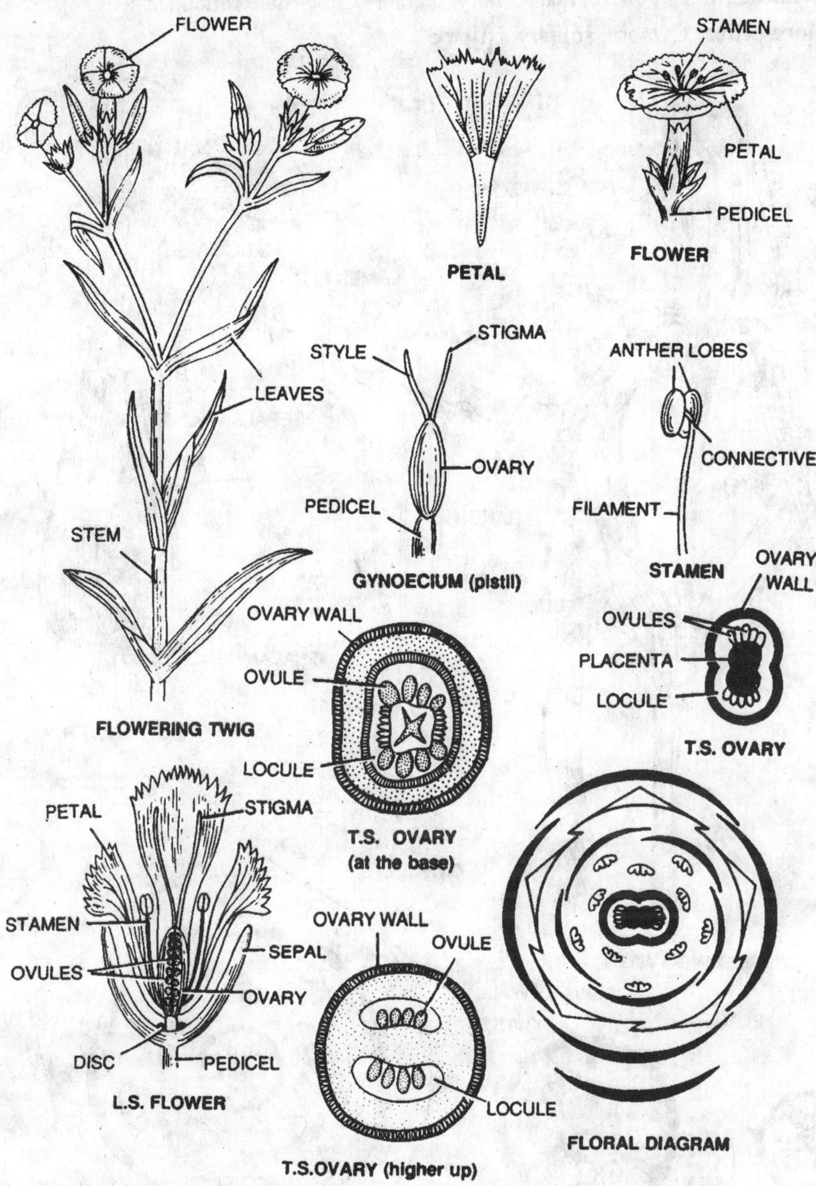

Fig. 18.4. Caryophyllaceae. *Dianthus caryophyllatus* Linn.; Eng., carnation plant.

Economic value. Grown as an ornamental in the gardens. The flowers are considered cardiotonic, diaphoretic, nervine and antiseptic.

Silence conoida **Linn**

Habit: An annual herb.

Root: Tap, branched.

Stem: Erect, herbaceous, aerial, branched, solid, cylindric, flat at the nodes, hairy and green.

Leaf: Cauline and ramal, exstipulate, simple, sessile, opposite decussate, leaf base semiamplexicaul, linear, entire, acute, hairy, leathery, unicostate reticulate.

Inflorescence: Cymose, solitary axillary.

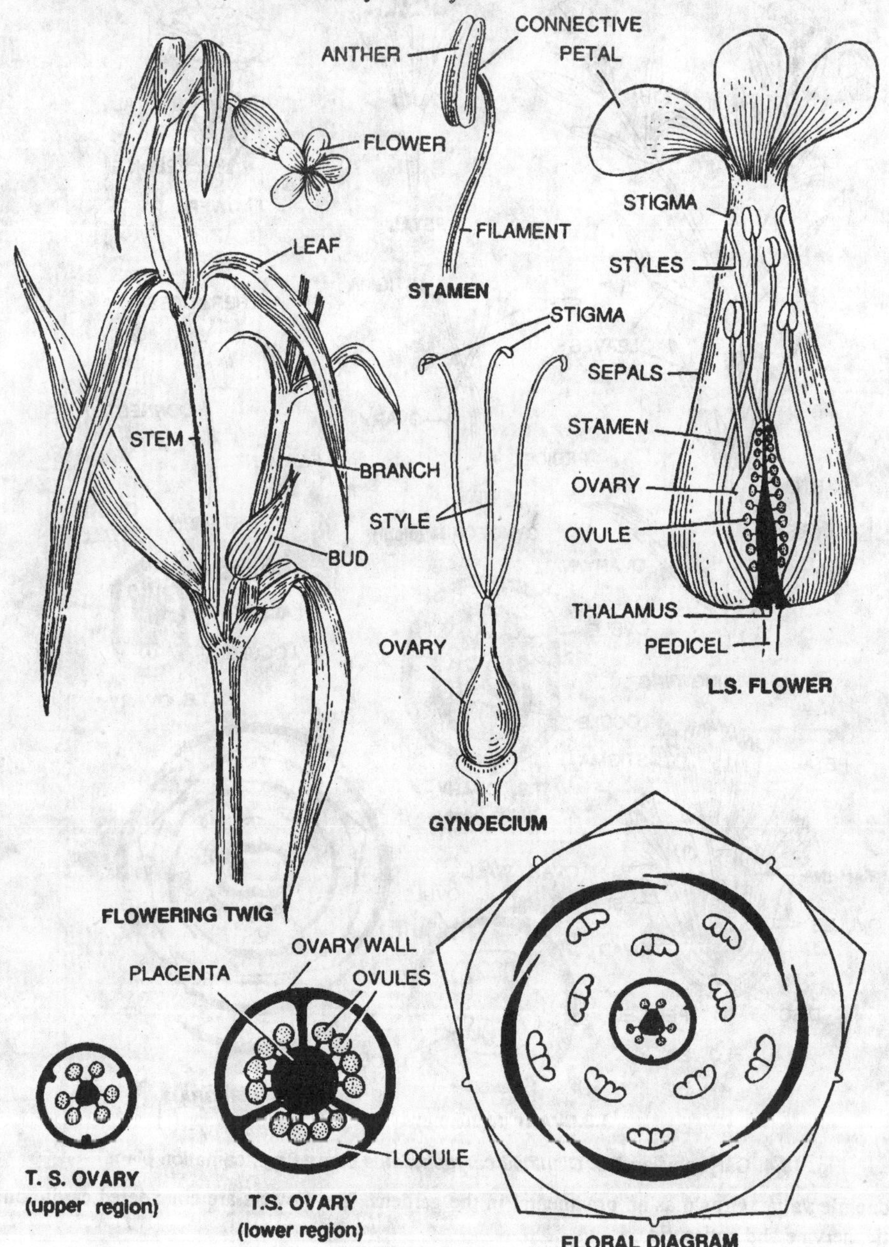

Fig. 18.5. Caryophyllaceae. *Silene conoidea* Linn.

Economic value. The plant is considered emollient and used in baths or as a fumigant. The plant juice is used in opthalmia.

Flower: Pedicellate, ebracteate, hermaphrodite, actinomorphic, complete, cyclic, pentamerous, hypogynous.

Calyx: 5 sepals, gamosepalous, forming a flask-like structure, acute, inferior, green.

Corolla: 5 petals, polypetalous, twisted, caryophyllaceous, pink, inferior.

Androecium: 10 stamens in two whorls, 5 in each whorl, polyandrous, obdiplostemonous, dithecous, basifixed, introrse.

Gynoecium: 3 carpels, syncarpous, ovary superior, upper region of the ovary possesses free central placentation while the lower region possesses axile placentation, many ovules, styles three, stigmas three and globular.

Fruit: A capsule.

Floral formula: ⊕ ⚥ K (5), C5, A 5+5, G $(\underline{3})$.

Economic Importance of the Family

1. *Dianthus barbatus*; Eng.–Sweet William–An annual herb, grown in the gardens as an ornamental.

2. *Dianthus caryophyllatus*; Eng.–Carnation–It is grown as an ornamental.

3. *Dianthus chinensis;* Syn. *D. sinensis*; Eng.–Rainbow pink–This is grown as an ornamental. An annual herb.

4. *Arenaria* sp.–It is grown as an ornamental.

5. *Saponaria vaccaria*; Eng.–Cow herb: Verna.–**Sabuni**–This is an ornamental herb, found throughout India. The plant juice is used as a substitute of soap.

6. *Spergula arvensis*-It is a common winter season weed. In Europe it is used as fodder. It is diuretic.

7. *Stellaria aquatica*; Eng.-Chick weed; Verna.–**Badeola**–The shoots are eaten as vegetable.

8. *Gypsophila* sp.–An extremely ornamental plant.

9. *Lychnis coronaria*–In Spain the decoction of roots is used medicinally for the treatment of lung and liver diseases.

10. *Silene armeria*; Eng.–Sweet William catchfly. This is an annual herb, grown as an ornamental. It is native of South Europe.

11. *Cerastium* sp.–This is an ornamental plant.

Systematic Relationships: The taxonomists have presented many views for the evolution of Caryophyllaceae. Acccording to Eichler (1878) the family has been originated from the Phytolaccaceae by conversion of the outer whorl of stamens to petals and outer whorl of carpels to stamens. Pax, Wettstein and Rendle have also accepted this view in principle. Wernham (1911) proposed another view. According to him the family has been evolved from ranalian ancestors and to have been the source of origin for the Primulaceae and other such reduced families, *e.g.*, Amarantaceae and Chenopodiaceae. This view has been accepted by Bessey, Hutchinson and others. According to Dickson (1936) the family has been derived from the Geraniaceae. Thompson did not agree with this view. The veiw proposed by Wernham, seems more correct in tracing the probable origin and the systematic relationships of the family.

19

Order–Chenopodiales Family–Chenopodiaceae & Family–Amaranthaceae

According to Hutchinson this is the sixty fourth order of the phylum Angiospermae, sub-phylum Dicotyledones, and division Herbaceae. The order comprises of ten families, of which only two (*i.e.*, Chenopodiaceae and Amaranthaceae) have been described here in detail. Bentham and Hooker have placed these families along with five others including Nyctagineae, Polygonaceae, etc., in the first series–the Curvembryae of class Dicotyledones and division Monochlamydeae or Incompletae. Engler and Prantl have placed these families along with seven others including Nyctaginaceae, Caryophyllaceae, etc., among the seventeenth order–the Centrospermae of class Dicotyledoneae and sub-class Archichlamydeae.

Stipules absent or very small; carpels numerous to solitary, free or connate; embryo curved. Flowers actinomorphic; usually pentamerous, hermaphrodite, monoecious or dioecious, perianth either single or double; stamens arranged in one or two whorls, five or ten in number when five opposite the petals; ovary superior, unilocular, 1 to 5 carpels, syncarpous; fruit nut-let, achene or capsule; embryo coiled or curved. perisperm present.

Family—Chenopodiaceae

There are 102 genera and about 1400 species in this family.

Distribution: The plants of this family are cosmopolitan in their distribution, *i.e.*, they are found throughout the world. The plants are particularly found in xerophytic and halophytic areas. They are mostly found in the Prairies, North America, Mediterranean coasts, Red sea coasts, Australia and central Asia. In our country the family is represented by several important genera, such as, *Chenopodium, Beta, Spinacia, Kochia, Suaeda, Salsola*, etc.

Habit: Usually the plants are annual or biennial herbs but sometimes the shrubs are also found. The small trees are very rare. *Haloxylon* is a small tree which reaches a height of 20 feet. *Salsola kali* is known as salt wort and grows near the sea shores. The species of *Salicornia* grow in marshy places.

Root: Tap and branched.

Stem: The stem is usually erect, herbaceous, branched, cylindrical solid hairy and green.

Leaves: The leaves are simple and alternate. In *Salicornia, Nitrophila*, etc., the leaves are opposite. In many cases the leaves are minute, fleshy and covered with hairs, *e.g.*, *Salicornia, Suaeda*, etc.

Inflorescence: Cymose, dichasial cyme.

Flowers: The flowers are minute, greenish in colour and arranged in dense cymose inflorescence. Hermaphrodite or unisexual. The plants may be monoecious (*e.g.*, *Sarcobatus*) or dioecious (*e.g.*, *Grayia*). They are bracteate, actinomorphic and hypogynous.

Perianth: Usually consists of 5 perianth leaves (tepals), sepaloid. Sometimes there are 3 or 4 perianth leaves, *e.g.*, *Salicornia* or only 2, *e.g.*, *Atriplex*. The perianth leaves may be free or united. Usually the perianth leaves remain persistent in the fruit. Sometimes the male flowers are devoid of perianth leaves. The aestivation is imbricate.

Androecium: Five stamens, usually the number of stamens corresponds to the number of perianth leaves. The stamens are usually situated opposite the tepals. The number of stamens varies from species to species and sometimes even from plant to plant of the same species, *e.g.*, *Chenopodium*. They are hypogynous or situated on a disc. The anthers are bicelled and dehisce longitudinally.

Gynoecium: 2 or 3 carpels, syncarpous. The ovary is superior. However, in *Beta* the ovary is inferior or half inferior. It is unilocular with a single basal campylotropous ovule.

Fruit: A nut or achene enclosed in the perianth.

Seeds: The seeds are albuminous with a curved or twisted embryo.

Pollination: The pollination takes place through the agency of insects.

Floral formula: ⊕ ⚥ or ♂ or ⚥ P 5 or (5), A 5, G $(\underline{2-3})$ or $\overline{(2-2)}$

Bentham & Hooker (1862)	Engler & Prantl (1931)	Hutchinson (1959)
Phanerogams	Phanerogams	Angiospermae
Dicotyledones	Dicotyledoneae	Dicotyledones
Monochlamydeae or Incompletae	Archichlamydeae	Herbaceae
Curvembryae	Centrospermae	Chenopodiales
Chenopodiaceae	Chenopodiaceae	Chenopodiaceae

1. ***Beta vulgaris* Linn.**; Verna. **Chukandar;** Eng. Beet root.

Beta vulgaris is cosmopolitan in distribution. The roots and leaves are used as vegetable. In European countries the roots are used as main source for the manufacture of sugar.

Habit: Annual or biennial herbs.

Root: Tap and branched; swollen, possessing dark-red juice.

Stem: Herbaceous, erect, branched, solid, cylindrical, pubescent, purplish green.

Leaves: Cauline and ramal, alternate, exstipulate, simple, petiolate, entire, elliptic lanceolate, unicostate, reticulate, coriaceous.

Inflorescence: Cymose, condensed cyme.

Flower: Bracteate, sessile, complete, actinomorphic, hermaphrodite, pentamerous, peri or epigynous.

Perianth: Tepals 5, polyphyllous, sepaloid; imbricate aestivation.

Androecium: 5 stamens; antetepalous, situated on a disc; anthers dithecous, introrse, dehiscing longitudinally.

Gynoecium: 3 carpels (tricarpellary), syncarpous; ovary half-inferior or inferior; unilocular with a single basal campylotropous ovule.

Fruit: A nut or achene enclosed in the perianth.

Seeds: Albuminous with a curved embryo.

Floral Formula: Br ⊕ ⚥ P5, A5, $G(\bar{3})$

2. ***Chenopodium album*** Linn.; Eng. Pigweed, lambs-quarters; Verna. **Bathua.**

Habit: An annual herb.

Root: Tap and branched.

Stem: Erect, aerial, herbaceous, branched, angular, solid, hairy, reddish green or green.

Leaf: Cauline and ramal, exstipulate, petiolate, alternate, simple, elliptic-lanceolate, coriaceous, unicostate reticulate.

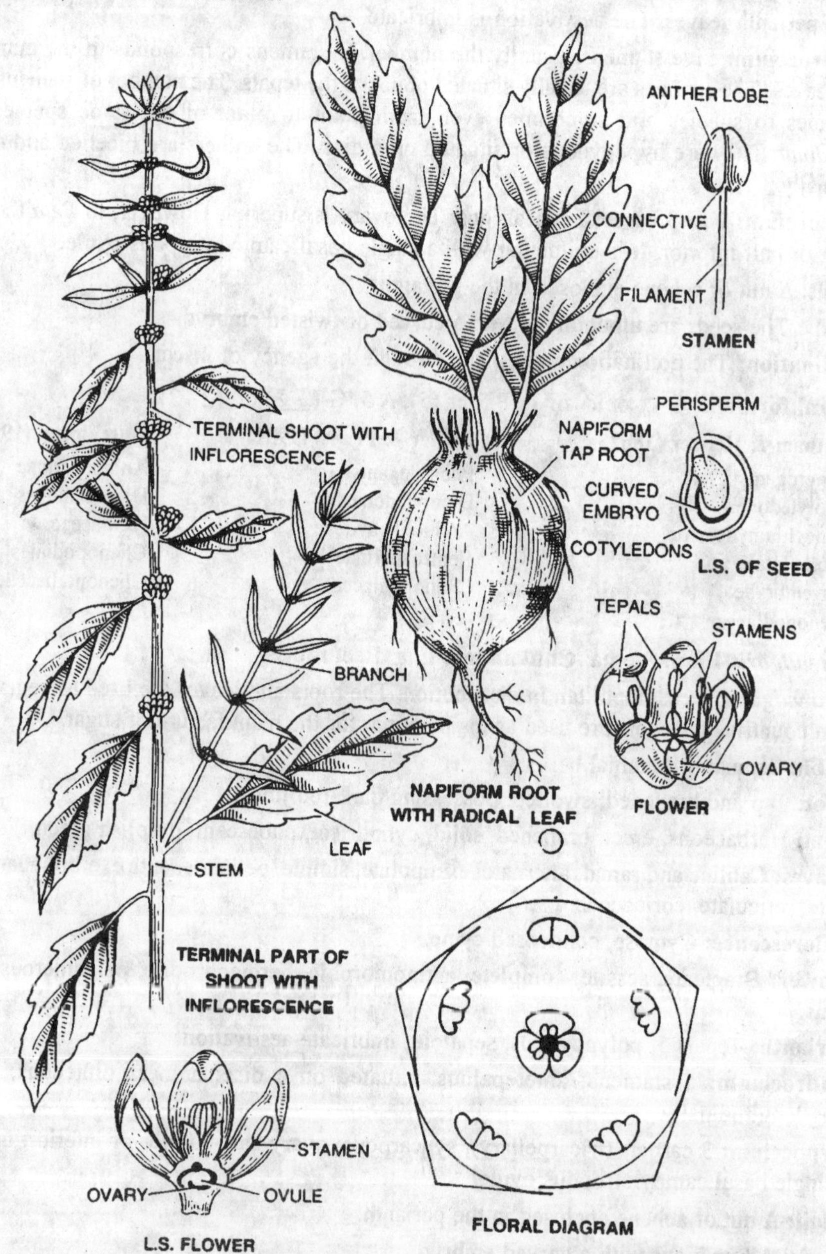

Fig. 19.1. Chenopodiaceae. *Beta vulgaris* Linn.; Eng., beet root; Verna. **chukandar.**

Economic value. Beet root is used in vegetable salad. Refreshing juice is extracted. Sugar is manufactured from juice in European countries. The seeds are cooling and diaphoretic. Leaves are applied to burns and bruises. Beet greens contain more iron and are richer in vitamins, particularly vitamin A.

Inflorescence: Cymose, condensed cyme.

Flower: Bracteate, sessile, actinomorphic, hermaphrodite, complete, pentamerous, hypogynous, cyclic.

Perianth: 5 tepals, polytepalous, quincuncial aestivation, sepaloid.

Androecium: 5 stemens, polyandrous, anteposed, filament long, dithecous, basifixed, introrse.

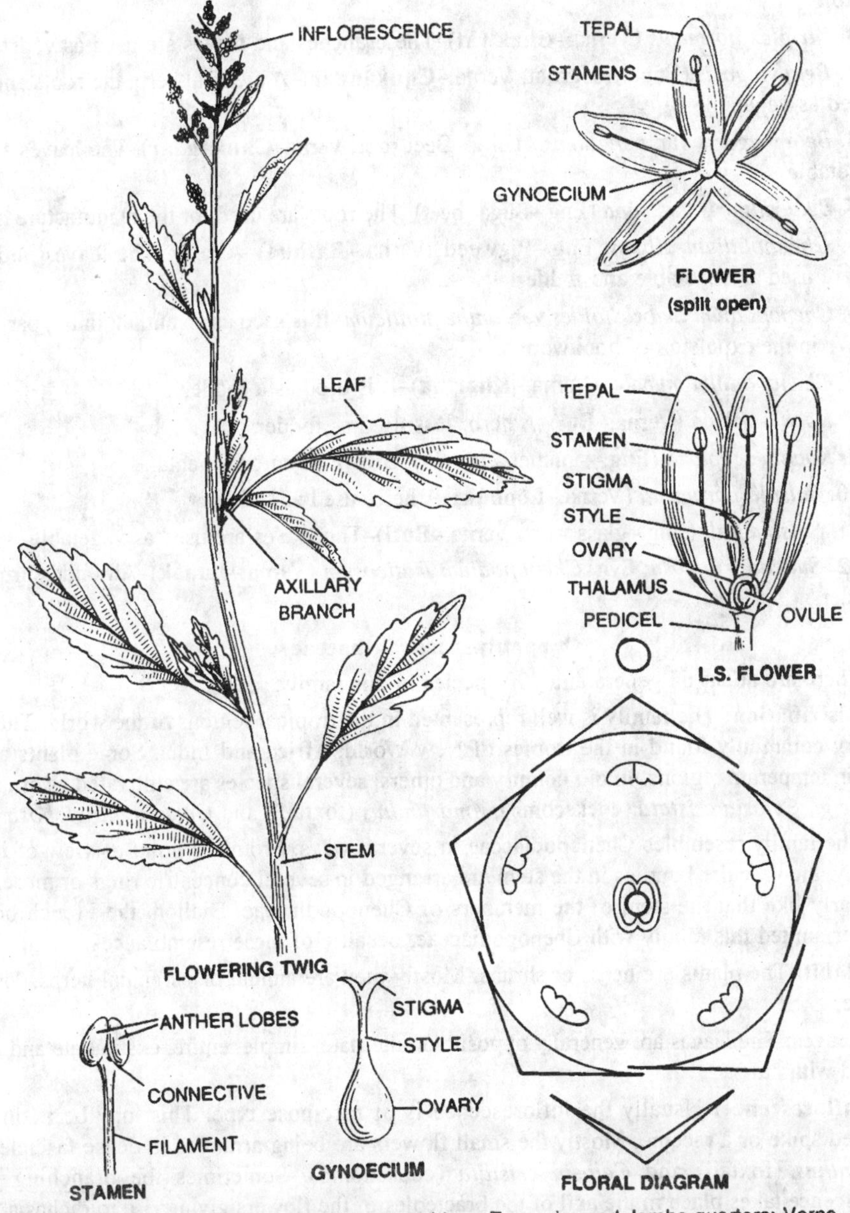

Fig. 19.2. Chenopodiaceae. *Chenopodium album* Linn.; Eng., pigweed, lambs-quarters; Verna., **bathua.**

Economic value. The tender twigs are used as vegetable and fodder. The plant is laxative and anthelmintic.

Gynoecium: 2 carpels (bicarpellary), syncarpous, ovary superior, unilocular, single basal ovule, style short, stigma bifid.

Fruit: Utricle.

Floral Formula: Br ⊕ ⚥ P5, A5, G(2)

Economic Importance of the Family

The family consists of a few plants of economic importance. Most of them are used as vegetable.

1. *Atriplex hortensis* (Verna.–**Chakwat**)–The branches and leaves are used as vegetable.
2. *Beta vulgaris* (Eng.–Beet root; Verna.–**Chukandar**)–A biennial herb, the roots and leaves are used as vegetable.
3. *Beta vulgaris* var. *bengalensis* (Eng.–Beet root; Verna–**Chukandar**)–The leaves are eaten as vegetable.
4. *Beta vulgaris* var. *rapa* (Eng.–Sugar beet)–The roots are used for the manufacture of sugar.
5. *Chenopodium album* (Eng.–Pigweed; Verna.–**Bathua**)–A herb. The leaves and tender twigs are used as vegetable and fodder.
6. *Chenopodium ambrosioides* var. *anthelminticum*–It is used as an anthelmintic, particularly effective in the expulsion of hookworms.
7. *Chenopodium murale*. (Verna.–**Khartua**)–It is used as a fodder.
8. *Kochia indica* (Verna.–**Büi**)–A herb. It is used as fodder.
9. *Kochia scoparia* (Eng.–summer–cypress)–Grown as an ornamental.
10. *Salsola baryosma* (Verna.–**Loonuk**)–A herb, used as fodder.
11. *Salsola kali* (Eng.–glass wort; Verna.–**Buti**)–The leaves are used as vegetable.
12. *Suaeda fruticosa*; Syn. *Chenopodium fruticosum* (Verna.–**Lunak**)–An undershrub, used as fodder.

Family–Amaranthaceae

There are about 64 genera and 800 species in this family.

Distribution: The family is well represented in the tropical regions of the world. The plants are very commonly found in the tropics of New World, Africa and India. Some plants are also found in temperate regions. In our country and others, several species are cultivated as ornamental plants *e.g., Celosia cristata* (cockscomb), *Amaranthus* (foxtail), and *Gomphrena globosa.*

The family resembles Chenopodiaceae in several ways. From the point of view of its stem anatomy, the vascular bundles in the stem are arranged in several concentric rings or more or less irregularly like that the stem of the members of Chenopodiaceae. Biallon, the French botanist, however, united this family with Chenopodiaceae, because of these resemblances.

Habit: The plants are herbs or shrubs. Mostly they are annual or perennial herbs. The trees are rare.

Leaves: The leaves are generally opposite or alternate, simple, entire, exstipulate and usually covered with hairs.

Inflorescence: Usually the inflorescence is of racemose type. This may be a simple or branched spike or a raceme. Mostly the small flowers are being arranged in dense fascicles, *e.g., Amaranthus* (foxtail) and *Celosia cristata* (cockscomb). Sometimes the branching of the inflorescence takes place in the axil of the bracteoles of the flowers giving rise to dichasia, which may be borne on a simple or racemosely or cymosely branched main axis. In certain cases the lateral flowers are being aborted and on their place tufts of hair or prickles are developed which persist in the ripe fruit and help in dissemination.

Flower: Usually the flowers are very minute and arranged in dense facicles, *e.g.*, *Celosia cristata* (cockscomb) and *Amaranthus* (foxtail), etc. Each flower bears a pair of large membranous persistent bracteoles; which are sterile. The flower in its structure quite resembles to that of Chenopodiaceae. Usually the flowers are bracteate and bracteolate, hermaphrodite, rarely unisexual by abortion, actinomorphic (regular) and hypogynous.

Perianth: The perianth is dry, membranous and very often white or coloured. It consists of 4 or 5 perianth leaves. The perianth leaves are green and herbaceous. The membranous perianth leaves are free or more or less united. Sometimes they become hairy.

Androecium: It consists of 4 or 5 stamens. They are situated opposite the perianth leaves. Usually the stamens are united at the base forming a membranous tube. Sometimes these stamens bear fringed outgrowths in between them. Mostly the anthers are two-celled but in certain cases they may become one celled. They dehisce longitudinally.

Gynoecium: It consists of 2 or 3 carpels, syncarpous. The ovary is unilocular and superior. In *Celosia* and certain other allied genera there is a single basal ovule. Generally, the ovules are campylotropous. The styles are one, two or three.

Fruit: The fruit is generally dry. This may be a nut, drupe or berry.

Seeds: The seeds are endospermic usually having a rough or polished testa. The embryo is curved and lies close to the seed–coat.

Pollination: Entomophily.

Floral Formula: Br Brl ⊕ ⚥ P 4–5, A 4–5, C$(\underline{2-3})$.

Bentham & Hooker (1862)	Engler & Prantl (1931)	Hutchinson (1959)
Phanerogams	Phanerogams	Angiospermae
Dicotyledones	Dicotyledoneae	Dicotyledones
Monochlamydeae or Incompletae	Archichlamydeae	Lignosae
Curvembryae	Centrospermae	Chenopodiales
Amaranthaceae	Amaranthaceae	Amaranthaceae

1. ***Achyranthes aspera*** Linn.; Verna. **Chirchita, Latzira, Apmarg.**

Habit: Perennial herbs.

Root: Tap and branched.

Stem: Erect, herbaceous, quadrangular, branched, solid, green, pubescent.

Leaf: Cauline and ramal, opposite, exstipulate, simple, sub-sessile, ovate, entire, acute, unicostate reticulate, rough, coriaceous, hairy.

Inflorescence: Racemose, spike.

Flower: Bracteate, bracteolate, sessile, bracts and bracteoles, spinous and persistent, hermaphrodite, actinomorphic, regular, complete, pentamerous, hypogynous, cyclic.

Perianth: 5 tepals, polyphyllous, dry, membranous, white or coloured, hairy.

Androecium: 5 stamens, antetepalous, stamens alternated by fringed outgrowths, anthers dithecous, introrse, dehiscing longitudinally.

Gynoecium: 2, bicarpellary, syncarpous, ovary unilocular, superior, single basal ovule, style short, stigma bifid.

Fruit: Dry, utricle.

Floral Formula: Br Brl ⊕ ⚥ P5, A5, G(2)

2. ***Amaranthus spinosus*** Linn.; Verna. **Jangli chaulai.**

Habit: A spiny herb.

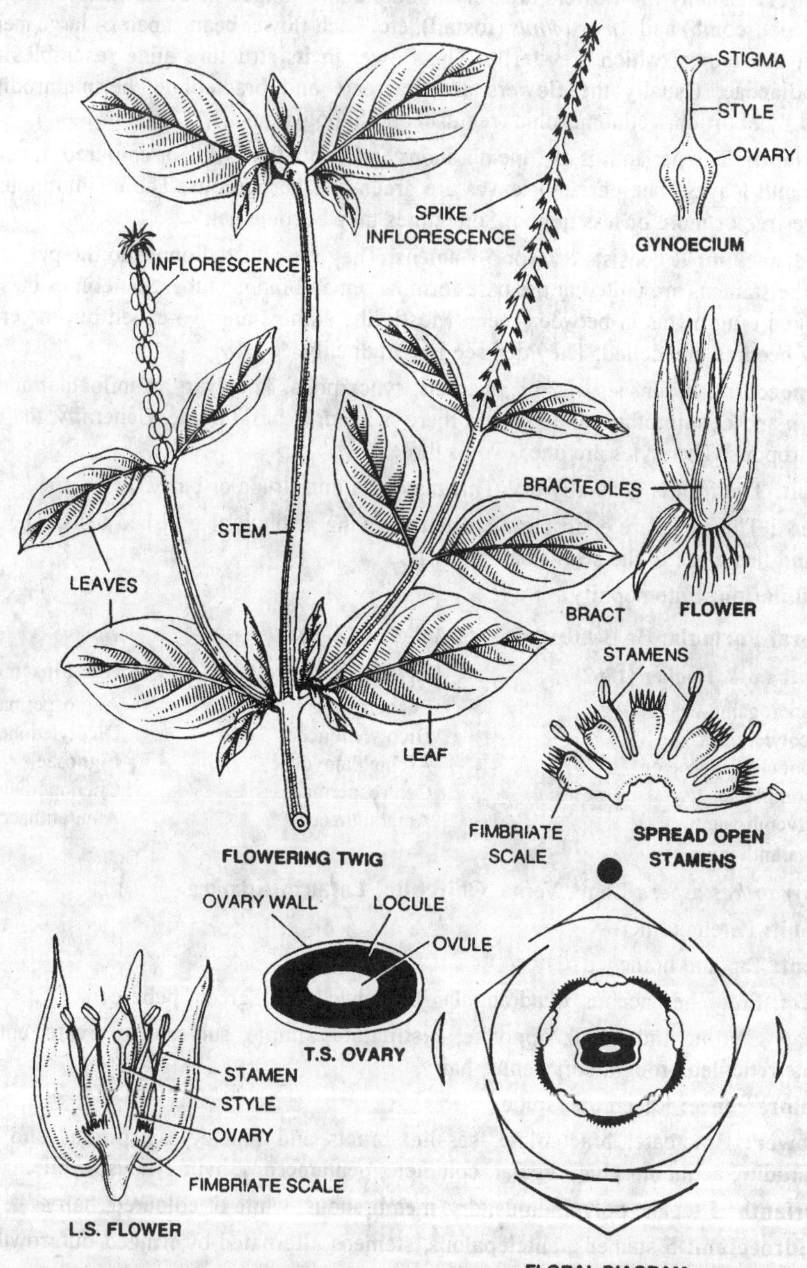

Fig. 19.3. Amaranthaceae. *Achyranthes aspera* Linn.; Verna. **Chirchita, latzira, apmarg.**

Economic value. The plants are used medicinally for piles, colic, boils, etc. It is pungent, purgative, diuretic and astringent. Roots are used for pyrrhoea. Also used in cough and fevers.

Root: Tap and branched.

Stem: Erect, aerial, branched, herbaceous, terete, solid, smooth, green.

Leaf: Cauline and ramal, exstipulate, alternate, simple, petiolate, ovate, entire, acute, unicostate reticulate; 2 spines present in the axil which represent the modified axillary branch.

Inflorescence: Spike.

Male Flower: Bracteate, bracteolate, sessile, actinomorphic, unisexual, incomplete, staminate, cyclic.

Perianth: 5 tepals, polytepalous, quincuncial aestivation, membranous.

Androecium: 4-5 stamens, polyandrous, anteposed, filaments long, dithecous, versatile, introrse.

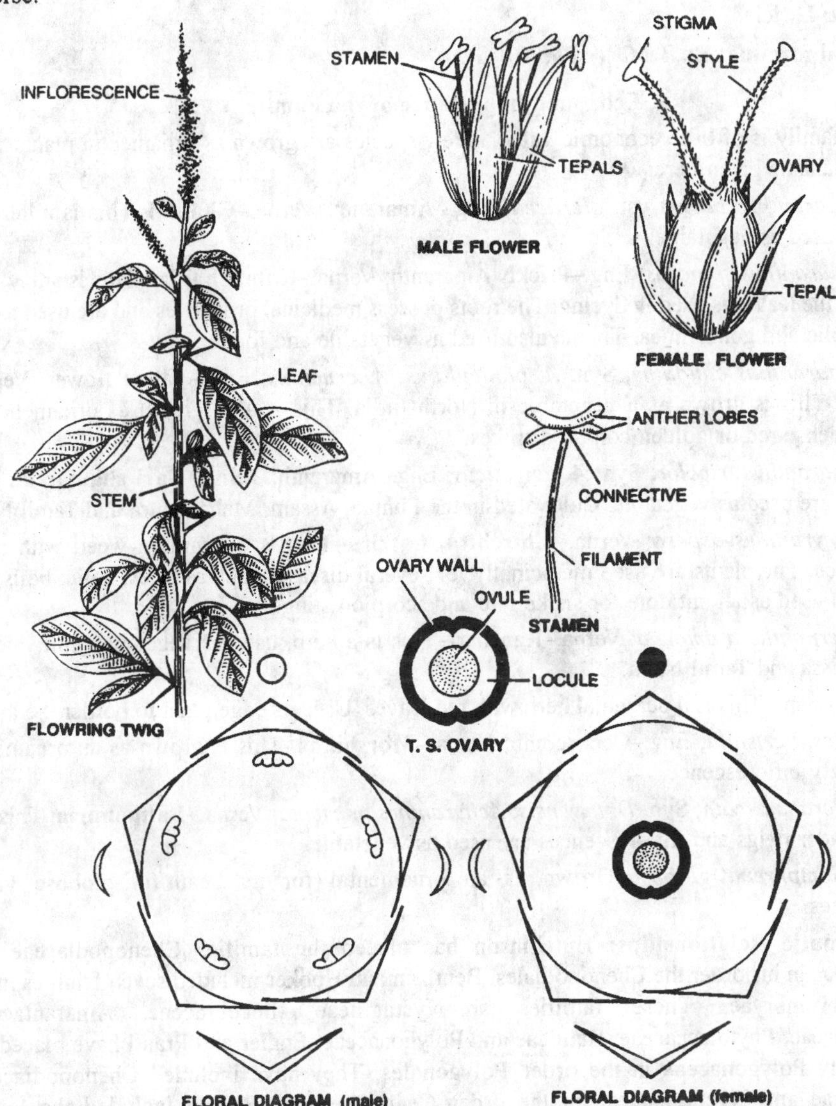

Fig. 19.4. Amaranthaceae. *Amaranthus spinosus* Linn.; Verna. jangli chaulai.

Economic value. The root is used in eczema and colic. The boiled leaves and roots are given to children as laxative and applied as emollient poultice to abscesses, boils and burns. Plant is also used in snakebite.

Gynoecium: Absent.

Floral Formula: Br, Brl ⊕ ♂ P5, A4 or 5, G0

Female flower: Bracteate, bracteolate, sessile, unisexual, incomplete, pistillate, actinomorphic, cyclic.

Perianth: 5 tepals, polytepalous, quincuncial, membranous.

Androecium: Absent.

Gynoecium: 2 carpels (bicarpellary), syncarpous, ovary superior, unilocular, single basal ovule, style short, stigma bifid and hairy.

Fruit: Utricle.

Floral formula: Br Brl ⊕ ♀ P5, A0, G$(\underline{2})$

Economic Importance of the Family

The family is of little economic value. A few species are grown as ornamental plants. A list of some plants is given here:

1. *Amaranthus blitum* var. *oleracea*; Eng.–Amaranth; Verna.–**Chaulai**–This is a herb, its leaves are used as vegetable.

2. *Amaranthus spinosus*; Eng.–Prickly Amaranth; Verna.–**Katailchaulai**–This is spiny herb. The ash of the leaves is used in dyeing. The roots possess medicinal properties and are used to cure eczema, colic and gonorrhoea. This is also used as vegetable and fodder.

3. *Amaranthus caudatus*; Syn. *A. paniculatus*; *A. cruentus*; Eng.–Valvet flower; Verna.–**Ramdana**–This is grown as a vegetable in North India. This is also grown as ornamental. It possesses dense red or golden coloured spikes.

4. *Amaranthus tricolor*; Syn. *A gangeticus*; Eng.–Amaranth; Verna.–**Bari chaulai, lal sag**–The leaves are used as vegetable. Cultivated in the Punjab, Assam, Maharashtra and Tamil Nadu.

5. *Achyranthes aspera*; Verna.–**Chirchita, Latzira**–This is a common weed with spike inflorescence. The plants are used medicinally for several diseases such as piles, colic, boils, etc. This is also used as an antidote for snake bite and scorpion sting.

6. *Alternanthera amoena*; Verna.–Kanchari–This is a herb, used as vegetable, cultivated in Bengal, Orissa and Tamil Nadu.

7. *Aerva* sp.–This is a perennial herb with red leaves. Used as edge plant to border the lawns.

8. *Celosia cristata*; Eng.–Cockscomb; Verna.–**Morshikha**–This is grown as an ornamental for its velvety inflorescence.

9. *Digera muricata*; Syn. *D. arvensis*; *Achyranthes muricata*; Verna.–**Latmohuria**–This is a herb. Its tender twigs and inflorescences are used as vegetable.

10. *Gomphrena globosa*–Grown as an ornamental for its beautiful globose violet inflorescences.

Systematic Relationships: Hutchinson has placed the families Chenopodiaceae and Amarantaceae in his order the Chenopodiales. Bentham and Hooker included seven families in the series Curvembryae. These families are–Nyctagineae Illecebraceae, Amarantaceae, Chenopodiaceae, Phytolaccaceae, Batideae and Polygonaceae. Engler and Prantl have placed the single family Polygonaceae in the order–Polygonales. They have included Chenopodiaceae, Amarantaceae and Nyctaginaceae in the order–Centrospermae. Bessey included the family Polygonaceae as an advanced taxon of his Caryophyllales. Hallier included the family in his Centrospermae. Hutchinson considered his order–Chenopodiales as reduced types of Caryophyllales. This is certain that Polygonales have been derived from Caryophyllaceous ancestors.

20

Order–Umbellales
Family–Apiaceae (Umbelliferae)

According to Hutchinson this is the seventy second order of the phylum Angiospermae, sub-phylum Dicotyledones and division Herbaceae. This order consists of a single family–the Umbelliferae. This family has been described here in detail. Bentham and Hooker have placed the family along with two others in the fifteenth order–the Umbellales of class Dicotyledones, division Polypetalae and series Calyciflorae. Engler and Prantl have put this family along with two others in the 30th order–the Umbelliflorae of class Dicotyledoneae and sub-class Archichlamydeae.

Flowers in umbels; ovary inferior, of two carpels. Herbs, often with hollow stems; leaves without stipules.

Family–Umbelliferae (Apiaceae)

There are about 200 genera and 2900 species in this family.

Distribution: The plants of this family are cosmopolitan in their distribution, however, they are not found in arctic regions. They are very commonly found in northern temperate regions. In the tropical countries they are either found in the hilly tracts or cultivated in the winter season. Majority of the plants are of economic value and many grow as weeds. In our country the family is represented by several important species such as, *Coriandrum sativum, Foeniculum vuigare, Cuminum cyminum, Daucus carota, Carum copticum, Hydrocotyle asiatica*, etc.

Habit: Majority of the plants are annual, biennial or perennial herbs. Sometimes shrubs and undershrubs are also found. *Bupleurum falcatum* is an undershrub and *Pseudocarum* is a climbing plant.

Root: Usually the root is tap and branched.

Stem: Usually, erect, rarely climbing, herbaceous. It becomes fistular at maturity on account of the shrinkage of the pith. The nodes and internodes are quite distinct. The stem is usually green and becomes pinkish as the plants mature. The stems are very often ribbed and angled.

Leaves: The leaves are alternate, exstipulate amplexicaul and much dissected, rarely the leaves are opposite, *e.g., Apiastrum*. Usually the leaves are pinnately compound. However, in some cases the leaves are simple, *e.g., Hydrocotyle, Bupleurum*, etc. The species *Eryngium* and *Aciphylla* possess the leaves with parallel venation and sheathing bases like many monocotyledonous plants. Oil glands are present in all the aerial parts of the plant.

Inflorescence: The inflorescence is umbel. This may be simple or compound. In majority of cases compound umbel is found, *e.g., Coriandrum, Foeniculum*, etc., however, in some cases it is simple umbel, *e.g., Hydrocotyle*. The compound umbels are sometimes cymose in character and a terminal flower may occur, *e.g., Eryngium*. In compound umbels an **involucre** of bracts is present at the top of the main floral axis from where the umbels arise. At the base of each umbel **involucel** of the bracts is found.

Flower: The flowers are usually hermaphrodite but sometimes unisexual flowers are also found. In several genera the outer flowers of the inflorescence are male only. In *Astransia* sp., majority of the flowers of the inflorescence are bisexual and some are only male flowers. In *Echinophora* sp., the monoecious plants bear the unisexual flowers. The plants of *Arctopus* sp. are dioecious and bear unisexual flowers. In majority of the inflorescences the flowers are regular (actinomorphic) but sometimes the outer flowers of the umbles are zygomorphic and irregular, *e.g.,Coriandrum sativum*. The flowers are usually pedicellate, bracteate, complete and epigynous.

Calyx: It consists of five sepals. These sepals are very minute in structure. The odd sepal being posterior. In many the calyx is absent, *e.g., Foeniculum vulgare*.

Corolla: It consists of five petals, polypetalous, *i.e.*, the petals are free. They are usually white or yellow in colour. The petals vary in size. In many cases two of the petals being larger than the other three. The tips of the petals are usually reflexed. Sometimes the petals are bifid, *e.g., Coriandrum sativum*. The aestivation is either valvate or imbricate.

Androecium: It consists of five stamens alternate to the petals. The stamens arise from an epigynous disc. The anthers are bilobed, introrse, basi or dorsifixed. The filaments are free. The anthers split by longitudinal slits.

Gynoecium: It consists of two carpels (bicarpellary), syncarpous. The ovary is inferior, bilocular, each loculus contains a single pendulous, anatropous ovule. The placentation is axile. On the top of the ovary a nectar disc is found which surrounds the two capitate stigmas.

Fruit: The fruit is **cremocarp**. The ovary splits into two **mericarps**. The two mericarps remain temporarily attached to a very thin axial prolongation, between them, known as **carpophore.** Each mericarp is usually covered by five longitudinal ridges known as **costae**. Between costae there are furrows known as **valleculae.** Under these furrows there lie special oil ducts known as **vittae.**

Seed: The seed is endospermic containing a minute embryo.

Pollination: The pollination is usually entomophilous. The insects are attracted towards the beautiful condesed umbels having outer zygomorphic flowers. The nectar secreting disc situated at the top of the ovary is the main attraction of the insects. The flowers are protandrous and thus cross pollination is effected.

Floral Formula: ⊕ or ·|· ⚥ K 5, C5, A 5, G (2)

Bentham and Hooker (1862)	Engler and Prantl (1931)	Hutchinson (1959)
Phanerogams	Phanerogams	Angiospermae
Dicotyledones	Dicotyledoneae	Dicotyledones
Polypetalae	Archichlamydeae	Herbaceae
Calyciflorae	Umbelliflorae	Umbellales
Umbellales	Umbelliferae	Umbelliferae or
Umbelliferae		Apiaceae

Description of some important plants in semi-technical language:

1. ***Trachyspermum ammi*** (Linn.) Sprague. Verna. **Ajwain**; Eng. Lovage, Ammi.

Habit: Annual, aromatic, cultivated herb.

Root: Tap and branched.

Stem: Erect, branched, glabrous, fistular, ribbed, nodes slightly swollen.

Leaf: Cauline and ramal, compound 2, 3-4 pinnate, ultimate segments linear, alternate, petiolate, petiole sheatihing at the base, exstipulate, aromatic.

Inflorescence: Racemose, compound umbel, primary umbels with an involucre of bracts at the base and the secondary umbel with an involucel of bracteoles.

Flower: Pedicellate, ebracteate, hermaphrodite, actinomorphic, complete, yellow, epigynous

Calyx: 5 sepals, gamosepalous, calyx tube adnate to the ovary wall, calyx teeth 0.

Corolla: 5 petals, polypetalous, emarginate, yellow, valvate aestivation, superior, yellow.

Androecium: 5 stamens, polyandrous, alternating with petals, filaments long, anthers introrse, bicelled, basifixed.

Gynoecium: 2 carpels, bicarpellary, syncarpous, ovary inferior, bilocular, one ovule in each loculus, axile placentation, style extremely short, stigma bilobed, vittae present in ovary wall.

Fruit: Cremocarp splitting into two mericarps.

Floral formula: ⊕ ⚥ K (5), C5, A 5, G $\overline{(2)}$.

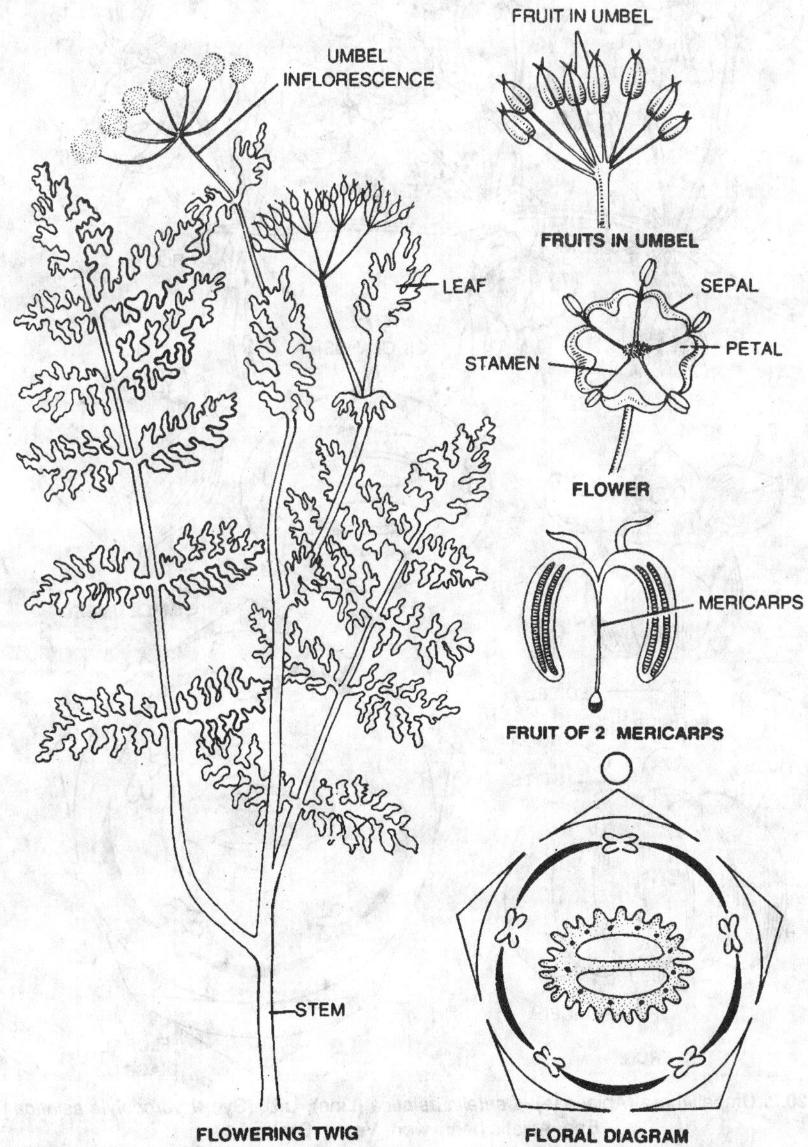

Fig. 20.1. Umbelliferae (Apiaceae). *Trachyspermum ammi* (Linn.) Sprague. (Syn. *Carum copticum*); Eng., lovage ammi, Verna, **Ajwain.**

Economic value: The fruits are used as spice. They are also used medicinally as carminative, stimulant, tonic and in indigestion.

2. ***Centella asiatica*** (Linn.) Urb. (Syn. *Hydrocotyle asiatica* Linn.) Verna. **Brahmi**; Eng. Asiatic pennywort.

Habit: Prostrate herbs, rooting at nodes.

Root: Adventitious roots.

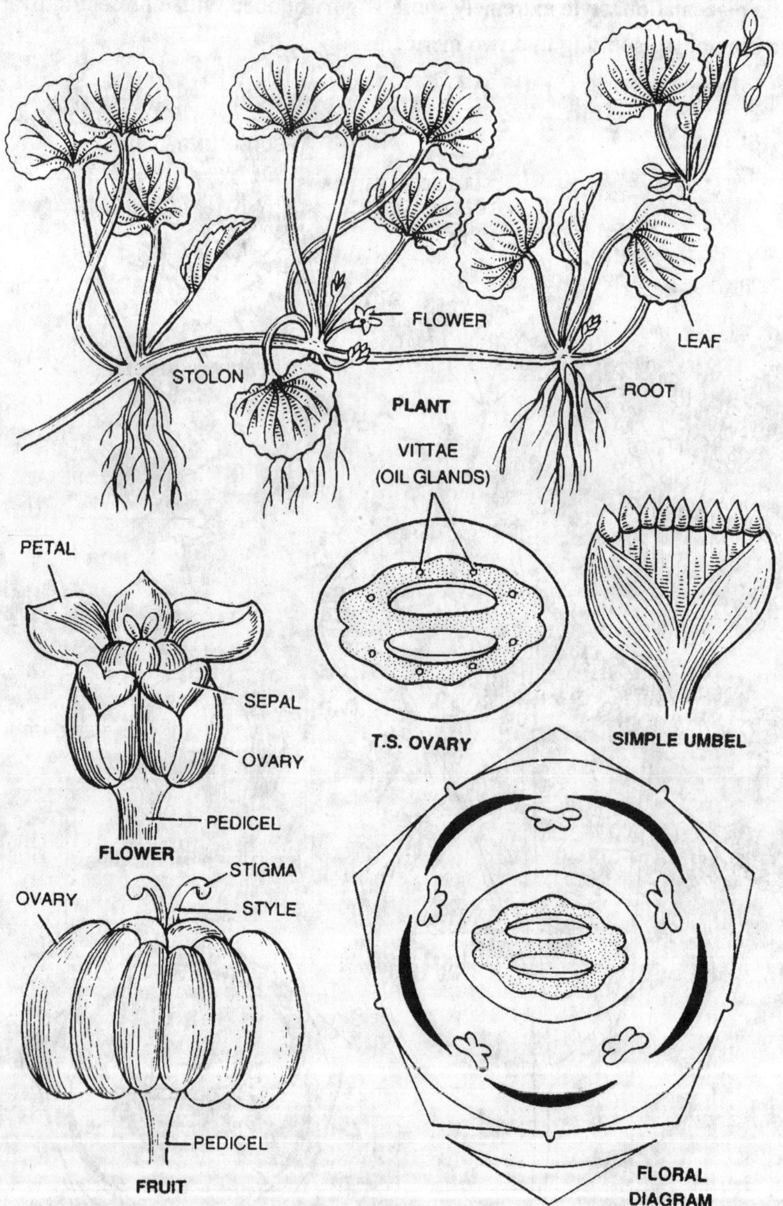

Fig. 20.2. Umbelliferae (Apiaceae). *Centella asiatica* (Linn). Urb. (Syn. *Hydrocotyle asiatica* L); Eng. Asiatic pennywort; Verna. **Brahmi.**

Economic value. The weed is alterative, tonic, diuretic, antiphlogistic, blood purifier and local stimulant. It is a remedy for skin diseases, like chronic eczema, chronic ulcers, etc., enlargement of glands, chronic rheumatism, chronic nervous diseases, madness, cholera, amenorrhoea and piles. An ointmant made of leaf juice and lanoline is of great value in elephantiasis.

Stem: Prostrate, herbaceous, weak, cylindrical, stolon, leaves and roots at nodes.

Leaves: Simple, arising in groups from nodes, orbicular or reniform, crenate and often lobed; glabrous, shining, petiolate, long petioled; stipulate, stipule, adnate to petioles.

Inflorescence: Racemose, simple umbels, 3-6 flowers in each group.

Flower: Pedicellate, small pedicels, bracteate, bracts small, ovate, embracing the flowers, hermaphrodite, actinomorphic, regular, complete, epigynous.

Calyx: 5 sepals, minute, connate, valvate.

Corolla: 5 petals, polypetalous, valvate.

Androecium: 5 stamens, polyandrous, alternate to petals, filament short, anther dithecous, introrse.

Gynoecium: 2 carpels, syncarpous, ovary inferior, 2-celled, disk 2-lobed, vittae present in ovary wall, axile placentation, one ovule in each locule.

Fruit: Cremocarp, laterally compressed.

Floral formula: Br ⊕ ⚥ K (5), C5, A 5, $G\overline{(2)}$

3. ***Coriandrum sativum***; Verna. –Dhaniya.

Habit: An annual cultivated herb.

Root: Tap and branched.

Stem: Erect, herbaceous, green, solid, branched, cylindrical, aromatic, ribbed, glabrous, nodes, slightly swollen.

Leaf: Decompound, much branched when old, cauline, petiolate, base slightly sheathing, unicostate reticulate venation.

Inflorescence: Compound umbel with involucre of bracts and involucel.

Flower: Pedicellate, bracteate, hermaphrodite, complete, epigynous, the outer flowers of the inflorescence are zygomorphic, *i.e.*,with large and unequal petals, whereas the central flowers are actinomorphic with equal and small petals.

Calyx: 5 sepals, gamosepalous, green, acute, valvate aestivation, superior, calyx tube adnate to the ovary wall.

Corolla: 5 petals, polypetalous, in peripheral flowers two posteior petals bilobed and smallest, one anterior deeply bilobed and largest two lateral large, bilobed, one lobe very small and the other large, zygomorphic, imbricate aestivation, purplish white; the petals of central flowers are small and equal in size, each petal consists of two equal lobes.

Androecium: 5 stamens polyandrous, alternating with petals, filaments long, anthers dorsifixed, introrse and yellow.

Gynoecium: 2 carpels, bicarpellary, syncarpous, ovary inferior, bilocular, one ovule ın each loculus, axile placentation, epigynous disc present at the top of ovary prolonged in two short styles, vittae present in ovary wall.

Fruit: Cremocarp.

Floral Formulae:

·|· ⚥ K (5), C 5, A 5, $G_{\overline{(2)}}$ (zygomorphic flowers)

⊕ ⚥ K (5), C5, A 5, $G_{\overline{(2)}}$ (actinomorphic flowers)

4. ***Foeniculum vulgare***; Verna.–**Sonf.**

Habit: Annual aromatic cultivated herb.

Root: Tap and branched.

Stem: Erect, branched, glabrous, fistular, ribbed, nodes slightly swollen.

Leaf: Compound 2, 3-4 pinnate, ultimate segments linear, cauline, alternate, petiolate, petiole sheathing at the base, exstipulate.

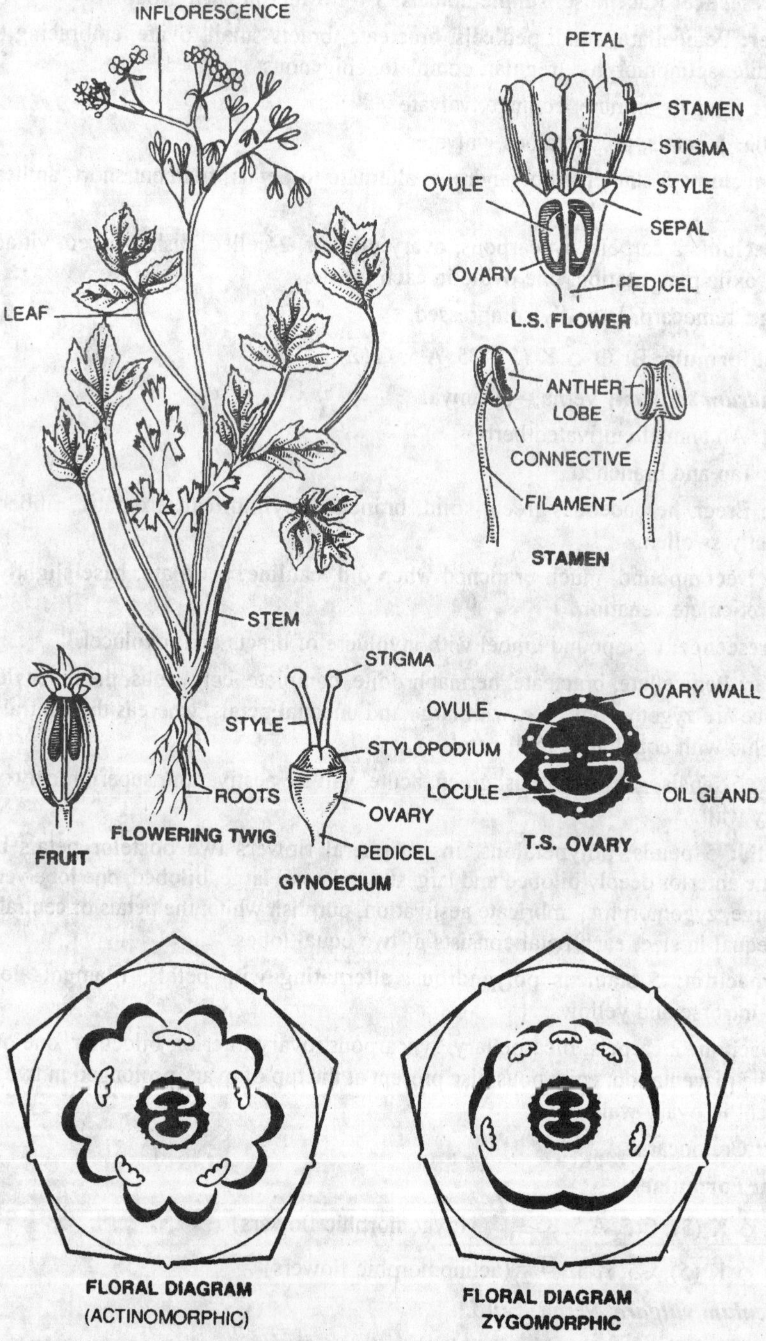

Fig. 20.3. Umbelliferae. *Coriandrum sativum* Linn., Eng. Coriander; Verna. **Dhaniya.**

Inflorescence: Racemose, compound umbel, primary umbels with an involucre of bracts at the base and the secondry umbel with an involucel of bracteoles.

Flower: Pedicellate ebracteate, hermaphrodite, actinomorphic, complete, yellow, epigynous.

Calyx: 5 sepals, gamosepalous, calyx tube adnate to the ovary wall, calyx wall, calyx teeth 0.

Corolla: 5 petals, polypetalous, emarginate, yellow, valvate aestivation, superior, yellow.

Androecium: 5 stamens, polyandrous, alternating with petals filaments long, anthers introrse, bicelled, basifixed.

Gynoecium: 2 carpels, bicarpellary, syncarpous, ovary inferior, bilocular, one ovule in each loculus, axile placentation, style extremely short, stylopodium present, stigma bilobed, vittae present in ovary wall.

Fruit: Cremocarp.

Floral formula: $\oplus$ ⚥ K (5), C 5, A 5, $G(\overline{2})$.

Economic Importance of the Family

The family is very important from the economic point of view. The fruits of some plants are used as condiments whereas the other plants are of medicinal value.

1. *Apium graveolens* var.*dulce* (Eng.–Garden celery; Verna.–**Ajmud,Karas**)–Native of Europe, now cultivated in the North-Western Himalayas and in the hills of Uttaranchal, H.P. and South India. Used as a vegetable. The roots and seeds are used medicinally.

2. *Apium graveolens* var. *rapaceum*; Syn. *A. rapaceum* (Eng.–Celeriac; Verna.–**Ajmud, Salari**)–The roots are eaten as vegetable. The fruits are used for extracting an oil, which is much valued, and the seeds as spice. Cultivated in the hills of Uttaranchal, H.P. and South India.

3. *Coriandrum sativum* (Eng.–Coriander; Verna.–**Dhania**)–An aromatic herb, native of the Mediterranean region, now cultivated chiefly in Madhya Pradesh, Maharashtra, Karnataka, Bihar and Uttar Pradesh for its fruits and leaves, which are used as condiment and spice. The fruits are also used as stimulant, carminative, stomachic and tonic.

4. *Cuminum cyminum*; (Verna.–**Jira**)–A herb, native of the Mediterranean region but commonly grown in the Punjab and Uttar Pradesh, for the aromatic fruits, which are used for flavouring purposes.

5. *Carum carvi*; (Eng.–Caraway; Verna.–**Shiajira**)–A herb, native of Europe and West Asia, now cultivated in Bihar, Orissa, the Punjab, Bengal and Andhra Pradesh. The fruits are used as condiment, and are also used medicinally as stomachic and carminative.

6. *Bunium persicum*; Syn. *Carum bulbocastanum*; (Verna.–**Kala zira**)–A perennial herb, found in Kashmir. The starchy tubers are eaten as vegetable and the seeds are used as spice.

7. *Trachyspermum ammi*; Syn. *Carum copticum*; (Verna.–**Ajwain**)–A herb, cultivated throughout India. The fruits are used as spice, and also used medicinally as carminative, stimulant, tonic and in indigestion.

8. *Trachyspermum roxburghianum*; Syn. *Carum roxburghianum*; (Eng.–**Ajmud**; Verna.–**Ajmuda, Radhuni**); used as spice and condiment.

9. *Daucus carota* var. *sativa* (Eng.–Carrot, Verna.–**Gajar**)– An annual or biennial herb, grown chiefly in the Punjab, Uttar Pradesh and Madhya Pradesh. The roots are edible.

10. *Centella asiatica*; Syn. *Hydrocotyle asiatica* (Eng.–Asiatic pennywort; Verna.–**Brahmi**)–A creeping herb, used as an antidote against cholera and also to cure madness. The plants are used as remedy of certain forms of leprosy.

11. *Ferula assafoetida*; (Verna.–**Hing**)–A perennial herb, commonly grown in the Punjab and Kashmir. The gum-resin obtained from the roots, is used in perfumery and for flavouring food products. It is also used in medicines in the treatment of asthma, cough and indigestion.

12. *Ferula galbaniflua*; (Eng.–Galbanum; Verna.–**Gandhabiroza**)–Found in North Western India. The oleogum resin, obtained from the stem, used in perfumery, and is also used in the treatment of chronic bronchites and ashthma.

13. *Ferula jaeschkena*–Grows in Kashmir. The roots and seeds produce an essential oil.

14. *Ferula narthex*; (Verna.–**Hing**)–Commonly found in Kashmir. It produces an oleo-gum resin, which is used for flavouring food products.

15. *Anethum graveolens*; Syn. *Peucedanum graveolens*; (Verna.–**Soya**)–Native of Eurasia, now cultivated on a small scale in Jammu and Kashmir. The fruits are used as spice and condiment and are also used as carminative.

16. *Anethum sowa*; (Verna.–**Sowa**)–Cultivated all over India. The leaves are used for flavouring and the fruits are used as carminative, stomachic and stimulant.

17. *Pastinaca sativa*; Syn. *Peucedanum sativum*; (Verna–**Gujur**)–A herb, native of Europe. The roots are eaten as vegetable.

18. *Seseli indicum*; (Verna–**Banajowan**)–A herb, commonly found in Uttar Pradesh, Bengal and Assam. The fruits are used as stimulant, stomachic, and are also used in expelling roundworms.

19. *Foeniculum vulgare*; (Verna–**Saunf**)–An aromatic herb, native of the Mediterranean region, now cultivated mainly in the Punjab, Assam, Uttar Pradesh and Maharashtra. The fruits are used as spice and condiment. Also used medicinally as carminative and stimulant. The residue, left after the distillation of an essential oil, makes a valuable fodder.

Systematic Relationships: Engler has put three families, *i.e.*, Araliaceae, Umbelliferae and Cornaceae in the order Umbelliflorae. Bentham and Hooker also placed these three families in the Umbellales. Rendle has also adopted the same arrangement. Hallier recognized only two families, *i.e.*, Cornaceae and the Umbelliferae. He merged the Araliaceae with Umbelliferae. Hutchinson (1948) has included the single family Umbelliferae in the order Umbellales. He has treated the Araliaceae as a separate order the Araliales, derived from the Cornaceae. According to Rickett (1945) there are two orders the Umbellales and the Cornales. The Umbellales consists of the Araliaceae and Umbelliferae and the Cornales consists of the Cornaceae and the Nyssaceae.

21

Order–Solanales Family–Solanaceae

According to Hutchinson this is the seventy-seventh order of the phylum Angiospermae, subphylum Dicotyledones and divison Herbaceae. This order comprises of three families, *i.e*, Solanaceae, Convolvulaceae, and Nolanaceae, of which first two (*i.e.*,Solanaceae and Convolvulaceae) have been described here in detail. Bentham and Hooker have placed these families along with three others in their 8th order–the Polemoniales of class Dicotyledones, division Gamopetalae and series Bicarpellatae. Engler and Prantl have included these families along with eighteen others including Verbenaceae, Labiatae, Scrophulariaceae, Pedaliaceae, Acanthaceae, etc., in the sixth order the Tubiflorae of class Dicotyledoneae and subclass Sympetalae.

Corolla actinomorphic to very slightly zygomorphic; stamens the same number and alternate with the corolla-lobes, epipetalous; ovary superior, 4-1-(mostly 2)-locular with numerous ovules on axile placentas, or rarely basal. Herbaceous to rarely woody often climbing.

Family–Solanaceae

There are about 85 genera and 2200 species in this family.

Distribution: The members of this family are primarily distributed in tropical America and South America, but they are also well represented in temperate regions. The largest genus of the family is *Solanum* which includes about 1500 species. The genera *Lycium*, *Physalis* and *Nicotiana* each containing 100 species.

Habit: The species belonging to this family are either erect or climbing herbs, shrubs or small trees. *Solanum nigrum* is a small annual herb; species of *Datura* are large annual herbs. *Cestrum nocturnum* is a shrub.

Root: Tap or adventitious.

Leaves: Alternate or becoming opposite at or near the inflorescences, rarely whorled, exstipulate, entire, simple, rarely dissected or pinnate.

Inflorescence: Cymose usually a typical axillary cyme or combination of cymes, rarely helicoid.

Flowers: Usually pedicellate, hermaphrodite, actinomorphic, regular (sometimes slightly zygomorphic), hypogynous complete, bracts and bracteoles absent.

Calyx: 5 sepals, gamosepalous (limb usually 5-lobed or-toothed), persistent and often much enlarged in fruit.

Corolla: 5 petals, gamopetalous, funnel shaped, rotate to tubular, corolla of various shapes but rarely 2-lipped, usually plicate or convolute (rarely valvate).

Androecium: 5 stamens, sometimes 4 and didynamous or not so, or only 2, *e.g.*, *Schizanthus*; when the stamens are less than five the lost stamen is usually represented by a reduced (staminode). Stamens epipetalous and alternate, the petals, usually not equal in size. Anthers are 2-celled (dithecous), introrse, dehiscence by longitudinal slit or by apical pore, *e.g.*, in *Solanum*, the connective sometimes enlarged, usually hypogynous disc present which is quite apparent.

Gynoecium: 2 carpels (bicarpellary), syncarpous, ovary superior, bilocular (sometimes becomes tetralocular due to false septation), rarely 3-5 celled ovary, placentation axile, ovules usually many on prominently peltate placentas, anatropous or somewhat amphitropous. The carpels placed obliquely in the flowers. Single style, stigma bi-lobed and capitate.

Fruit: A berry or septicidal capsule.

Seed: Albuminous (endospermic) smooth or pitted, curved embryo embedded in the endosperm.

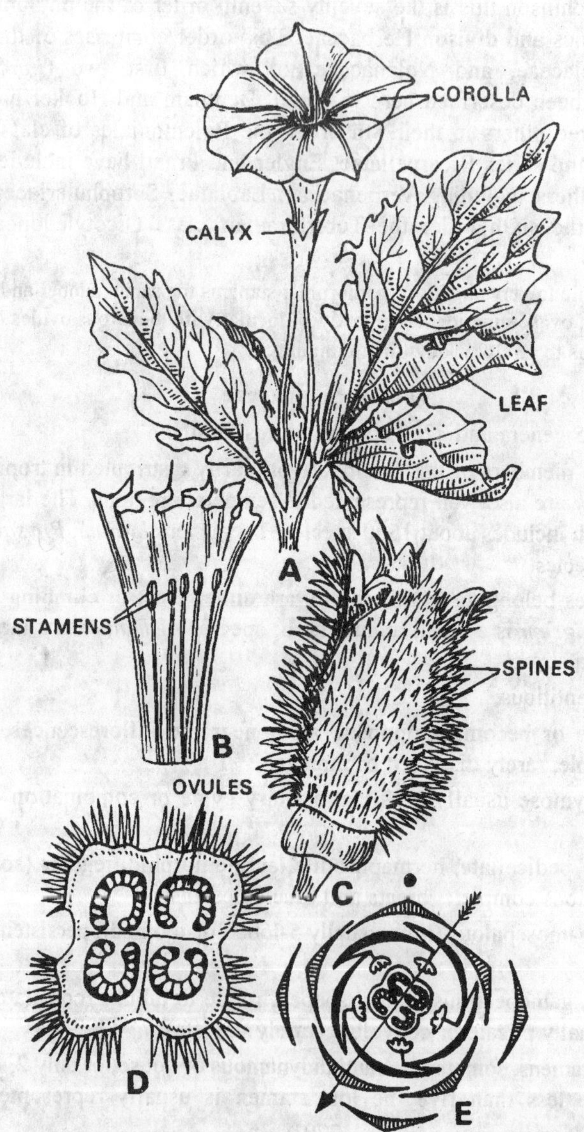

Fig. 21.1. Solanaceae. *Datura stramonium* (Verna–**Dhatura**); A, a flowering twig; B, corolla spread out with epipetalous stamens; C, opening fruit (capsule); D, cross-section of capsule; E, floral diagram.

Pollination: (Entomophilous) through the agency of insects.

Floral Formula: $\oplus$ ⚥ K 5, $\overline{C(5), A5}, G_{(2)}$

Bentham & Hooker (1862)	Engler & Prantl (1931)	Hutchinson (1959)
Phanerogams	Phanerogams	Angiospermae
Dicotyledones	Dicotyledoneae	Dicotyledones
Gamopetalae	Sympetalae	Herbaceae
Bicarpellatae	Tubiflorae	Solanales
Polemoniales	Solanaceae	Solanaceae
Solanaceae		

Some important plants have been described here in semi-technical language.

Solanum sp.

Habit: Erect or climbing herbs, shrubs or small trees. Most of them are pubescent, some are spiny, and some unarmed.

Leaf: The leaves are alternate, or subopposed in unequal pairs, exstipulate and either lobed or pinnatifid.

Inflorescence: The inflorescence may be cymose, racemose, umbellate or rarely panicle, in most cases extra axillary or opposite the leaves.

Calyx: The calyx is bell-shaped or rotate, 5-lobed or 5-parted, sometimes 10-lobed, rarely 4-lobed or subentire accrescent or not in fruit.

Corolla: The corolla is rotate, rarely campanulate, tube short, limb usually 5-lobed, plicate in bud.

Androecium: The stamens are 5, rarely 4 or 6, inserted on the corolla throat, filaments short, anthers oblong, often narrowed upwards, connivent round the style, dehiscence by apical pores or short slits, introrse.

Gynoecium: The ovary is superior bi-celled and many ovuled (rarely 3-4 celled), style simple, linear, stigma small and terminal.

Fruit: The fruit is globose or elongated berry with persistent calyx.

Seeds: Many, smooth, discoid.

Generic Characters: Corolla rotate or campanulate, lobes plicate in bud; anthers connivent in a cone, dehiscence by apical pores; calyx small, persistent sometimes fleshy.

Nicotiana sp.

Habit: Usually herbaceous plants, some species are shrubby or of arborescent type; annual or perennial in habit.

Stem: Usually tall growing, often viscid pubescent, strongly scented with narcotic poisonous properties.

Leaf: Alternate, simple, entire or rarely sinuate, shortly petiolate or sessile, decurrent.

Inflorescence: Either panicle or one sided raceme; sometimes lower flowers solitary axillary.

Flower: Flowers usually large and showy and of different colours (white, yellow, greenish, purple or rosy).

Calyx: Tubular–campanulate and 5–lobed

Corolla: Funnel or salver-shaped, 5–lobed with a long tube.

Androecium: 5 stamens in number, attached near the base of corolla tube, epipetalous; filaments long.

Gynoecium: Two carpels, syncarpous, superior ovary, bilocular, slender style with capitate stigma.

Fruit: 2-celled capsule splitting at maturity into 2 to 4 valves.

Seeds: Many and small.

Generic Characters: Fruit, a capsule; capsule splitting lengthwise; calyx nearly or quite covering the fruit; fruit not spiny.

Solanum nigrum; Verna.–**Makoi**.

Habit: An annual small wild herb.

Root: Tap and branched.

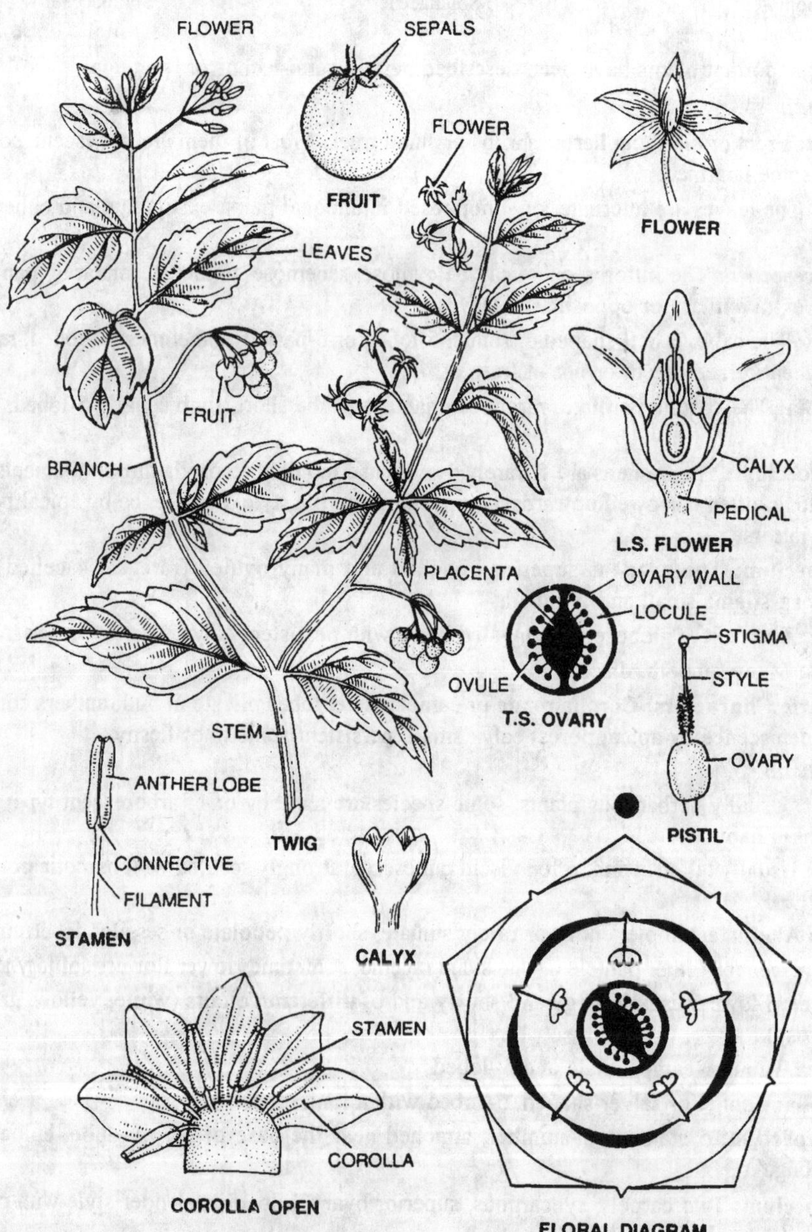

Fig. 21.2. Solanaceae. *Solanum nigrum* Linn. Eng. black nightshade; (Verna. **Makol**)

Stem: Erect, branched, herbaceous, cylindrical, solid, glabrous or somewhat hairy.

Leaf: Simple, alternate, cauline, petiolate or sub-sessile, exstipulate, ovate, dentate, acute, unicostate reticulate venation.

Inflorescence: Cymose, extra-axillary subumbellate cyme.

Flower: Pedicellate, ebracteate, hermaphrodite, actinomorphic, pentamerous except pistil, complete, small, white, hypogynous.

Calyx: 5 sepals, gamosepalous, lobes small, oblong, acute, green, hairy, valvate or imbricate aestivation.

Corolla: 5 petals, gamopetalous, rotate, lobes oblong, acute, white, twisted or valvate aestivation.

Androecium: 5 stamens, alternate to petals, polyandrous, epipetalous, filaments short, anthers oblong, connivent in a cone like structure, bicelled, introrse, yellow, dehiscence by apical pores.

Gynoecium: 2 carpels (bicarpellary), syncarpous, ovary obliquely placed in the flower, ovary superior, globose, bilocular, axile placentation, many shining ovules present on placenta, style simple, hairy at the base, stigma one.

Fruit: Berry with persistent calyx.

Floral Formula: $\oplus$ ⚥ K 5, $\overline{C(5), A5}$, G($\underline{2}$).

Petunia nyctaginifolia Juss.

Habit: An annual ornamental herb.

Root: Tap and branched.

Stem: Erect,branched, herbaceous, solid, cylindrical, green and hairy.

Leaf: Simple, cauline, opposite, decussate, sessile, exstipulate, ovate, entire, acute, multicostate reticulate venation.

Inflorescence: Cymose, solitary axillary.

Flower: Pedicellate, pedicel hairy, bracteate, hermaphrodite, actinomorphic, regular, complete, hypogynous.

Calyx: 5 sepals, gamosepalous, free above and connate below, green, hairy, quincuncial or imbricate aestivation.

Corolla: 5 petals, gamopetalous, tubular, twisted or imbricate aestivation.

Androecium: 5 stamens, polyandrous, alternate to petals, epipetalous, filaments long, anthers two celled, basifixed.

Gynoecium: 2 carpels (bicarpellary), syncarpous, ovary superior, obliquely placed in the flower, bilocular, axile placentation, numerous ovules in each loculus, style simple, long, stigma bifid.

Fruit: Capsule

Floral formula: $\oplus$ ⚥ K 5, $\overline{C(5), A5}$, G($\underline{2}$).

Withania somnifera Dunal; Verna. **Ashwagandha**.

Habit: A perennial shrub.

Root: Tap and branched.

Stem: Erect, aerial basal part woody, terminal part herbaceous, branched, cylindrical, solid, hairy, green.

Leaf: Cauline and ramal, exstipulate, petiolate, alternate, simple, ovate, entire, acute, glabrous, coriaceous, unicostate reticulate.

Inflorescence: Cymose, axillary umbellate cyme.

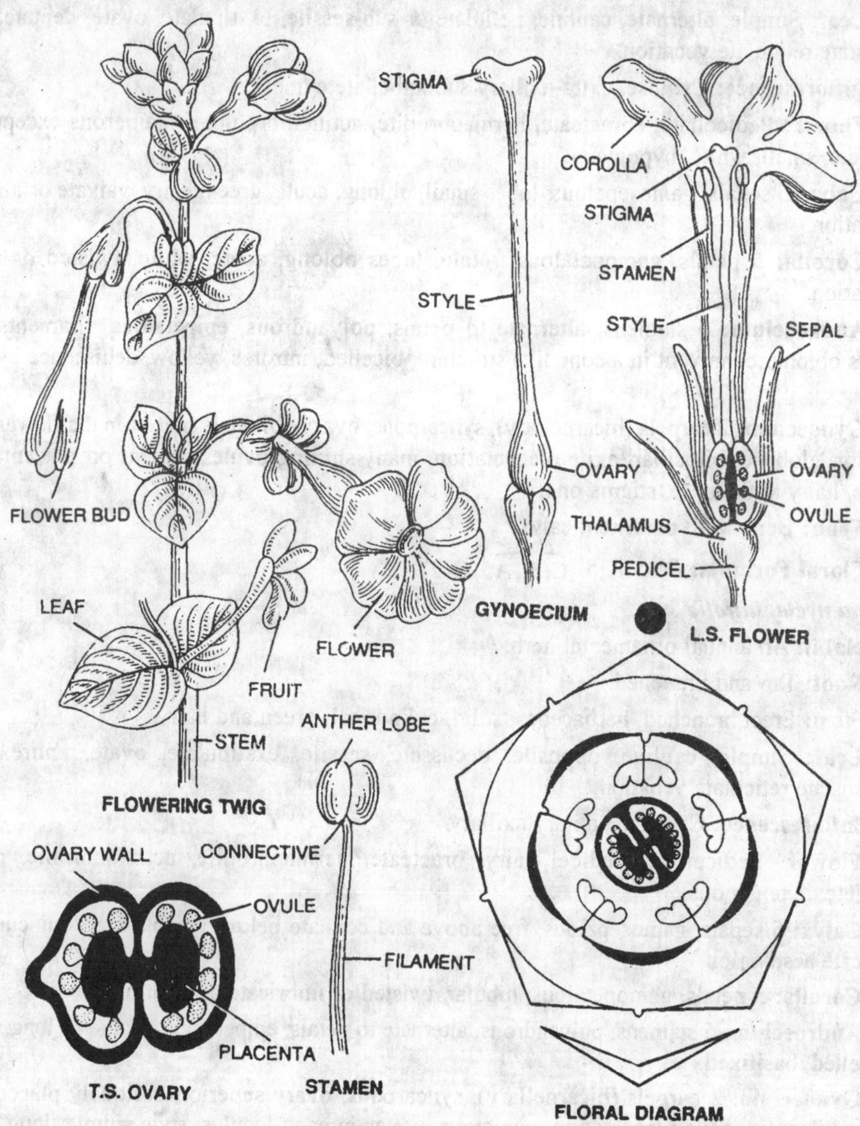

Fig. 21.3. Solanaceae. *Petunia nyctaginifolia* Juss.

Flower: Ebracteate, pedicellate, actinomorphic, regular, hermaphrodite, pentamerous, hypogynous, complete, cyclic.

Calyx: 5 sepals, gamosepalous, valvate, persistent, hairy, green, inferior.

Corolla: 5 petals, gamopetalous, valvate, campanulate, whitish-yellow, inferior.

Androecium: 5 stamens, polyandrous, epipetalous, anthers connivent, dithecous, basifixed, introrse.

Gynoecium: 2 carpels (bicarpellary), syncarpous, ovary superior, bilocular, many ovules in each locule, axile placentation, ovary obliquely placed in the flower, placenta swollen, style long, stigma capitate.

Fruit: Berry

Floral Formula: ⊕ ⚥ K 5, $\overline{C(5), A5}$, G(2).

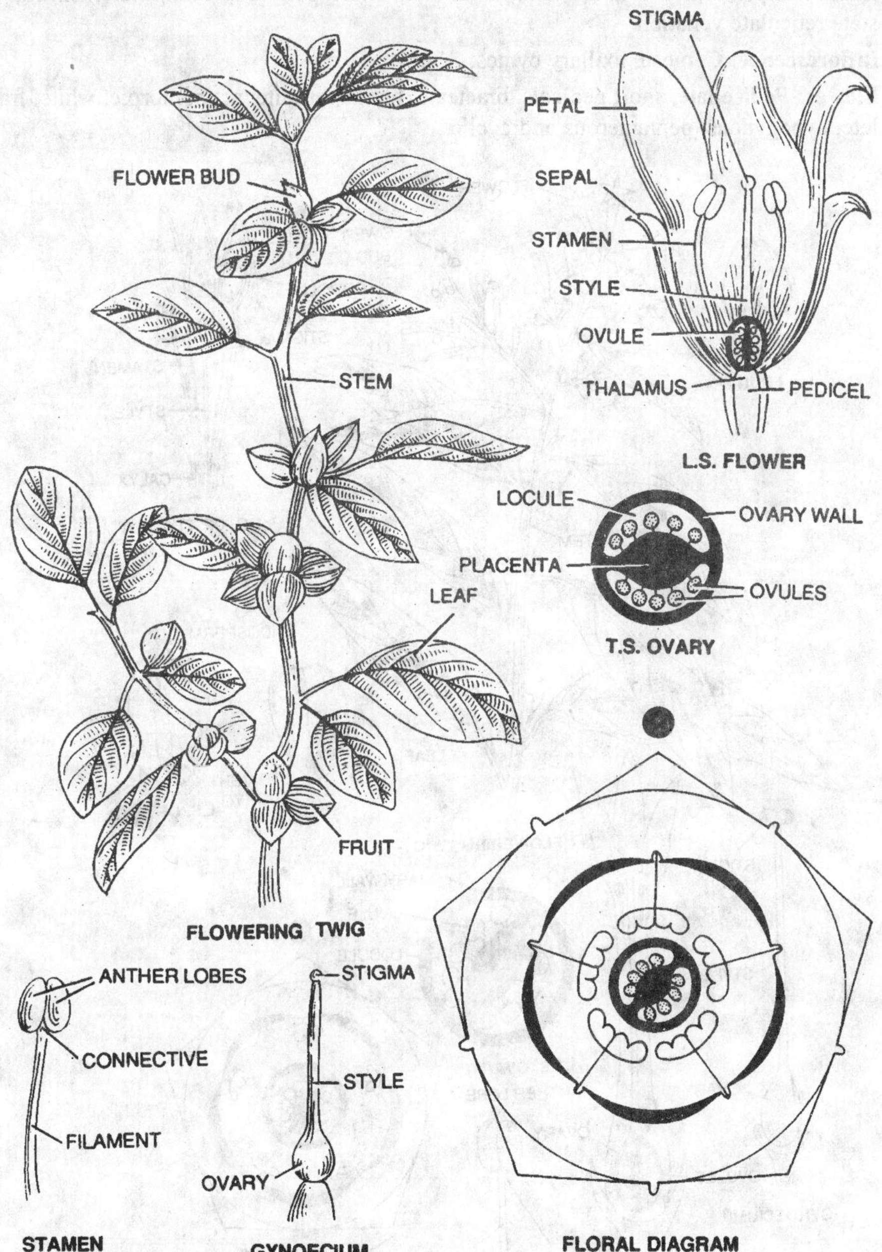

Fig. 21.4. Solanaceae. *Withania somnifera* Dunal; Verna. **Ashwagandha.**

Economic value. The root is considered alterative, aphrodisiac, tonic, deobstruent, diuretic, narcotic, abortifacient; used in rheumatism, consumption, debility from old age, emaciation of children, etc.

Cestrum nocturnum Linn.;Verna.**Rat-ki-Rani;** Eng. Night jessamine.

Habit: An ornamental perennial shrub.

Stem: Erect, branched, woody, solid, glabrous and green.

Leaf: Simple, cauline and ramal, alternate, short petioled, exstipulate, entire, acute, unicostate reticulate venation.

Inflorescence: Cymose, axillary cymes.

Flower: Pedicellate, short pedicels, bracteate, hermaphrodite, actinomorpic, white, fragrant, complete, hypogynous, pentamerous and cyclic.

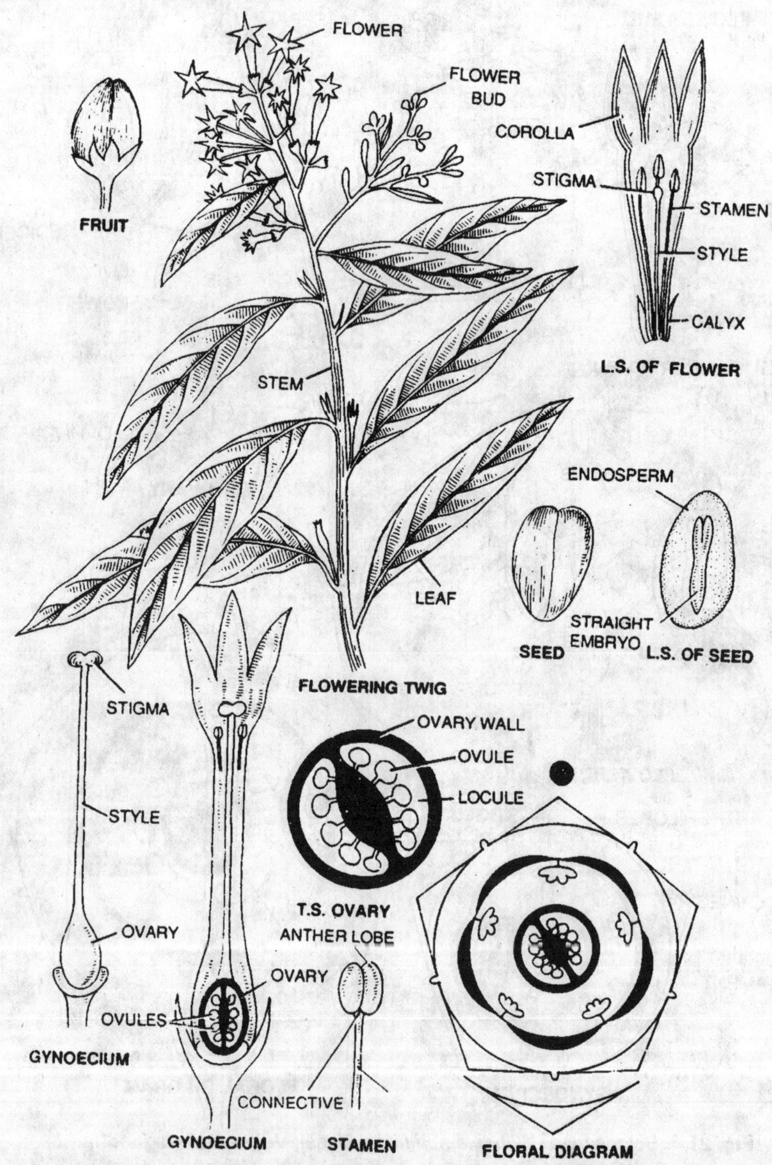

Fig. 21.5. Solanaceae. *Cestrum nocturnum* Linn.; Eng., night jessamine; Verna, **Raat-ki-Rani.**

Economic value. Grown for its flowers, which are seented at night.

Calyx: 5 sepals, gamosepalous, green, valvate aestivation, inferior.

Corolla: 5 petals, gamopetalous, tubular, white, fragrant, valvate aestivation, inferior.

Androecium: 5 stamens, polyandrous, epipetalous, alternate to the petals, filaments long, anthers bicelled, introrse.

Gynoecium: 2 carpels (bicarpellary), syncarpous, ovary superior, obliquely placed in the flower, bilocular, axile placentation, many ovules in each loculus.

Floral formula: Br $\oplus$ ⚥ K 5, $\overline{C(5), A5}$, G($\underline{2}$).

Economic Importance of the Family

Economically the family is fairly important. It comprises of several plants of food value, medicinal, vegetables and ornamentals. A list of few important plants is given here:

1. *Solanum tuberosum*; Eng.–Potato; Verna–**Alu**–The potato is chiefly grown in Assam, Uttar Pradesh, West Bengal, Bihar and the Punjab. The tubers are used as vegetable, and are also the source of starch. It is naive of South America.

2. *Solanum melongena*; Eng.–Brinjal: Verna–**Baigun**–The fruits are eaten as vegetable. It is native of India, and cultivated throughout the country. The seeds are often used as stimulant, and the leaves as a narcotic.

3. *Solanum ferox*; Verna.–**Rambegun**–This is small prickly herb. The fruits are eaten as vegetable. It is found in Assam and South India.

4. *Solanum indicum*; Eng.–Indian night shade; Verna.–**Bari katai**–This is a herb. The fruits are eaten as vegetable. It is found in the sub-Himalayan tract, Assam and the Khasia hills.The roots and the fruits possess several medicinal properties. The fruits are given in colics. The leaves are used in toothache, cough and fevers.

5. *Solanum grandiflorum;* Eng.–Large flowered night shade, potato tree. This is a shrub or small tree. It is native of South America. Grown in gardens as an ornamental for its large bluish. violet flowers.

6. *Solanum jasminoides*; Eng.–Potato vine. This is a climbing shrub. It is native of Brazil. This is a most useful deciduous climber for cool house. It has variegated leaves.

7. *Solanum pseudocapsicum*–This is a shrub grown as an ornamental in the gardens.

8. *Solanum rantonnetii*-This is grown as an ornamental in the gardens for its violet-coloured flowers. It is native of Argentina.

9. *Solanum xanthocarpum*; Verna.–**Kateli**–This is a common prickly herb. The roots are used in cough, asthma and leaves for rheumatism.

10. *Solanum nigrum*; Eng.–Black night shade; Verna.–**Makoi**–It possesses medicinal properties. The fruits are given in fevers; diarrhoea, eye diseases and hydrophobia. The juice of plants is given in chronic enlargements of the liver, in bleeding piles, dysentery, etc. The fruits are edible and very muck liked by the children.

11. *Solanum torvum*; Verna–**Titbaigun**–The fruits are used as vegetable. It is found in Orissa, Bengal, Bihar and Manipur.

12. *Solanum seaforthianum*; Eng.–Potato creeper. It is grown as an ornamental for its blush purple flowers. Native of Brazil.

13. *Cyphomandra betacea*; Syn. *Solanum betaceum*; Eng.–tree tomato–This is a small tree. The fruits are edible. It is native of Peru, but now cultivated in Bengal Assam, Maharashtra, Tamil Nadu, Karnataka and Travancore.

14. *Capsicum frutescens*; Eng.–Red pepper, chillies; Verna.–**Lal Mirch**–This is a herb. The fruits are extremely pungent and used as a spice and condiment. According to some authors this

species is the synonym of *Capsicum annuum*; while others consider the two as separate species. It is native of the West Indies and tropical America, now cultivated throughout our country.

15. *Nicotiana tabacum*; Eng.– Tobacco; Verna–**Tambaku**–The leaves are used for smoking and also contain alkoloids, which are used as insecticides. The oil obtained from the seeds is used for burning purposes, and is also used in the manufacture of paints and varnishes. The oil cake makes a good manure. The snuffs are prepared from tobacco wastes. The tobacco dust is used for making khamira tobacco for smoking and manufacture of nicotine sulphate, an insecticide. It is native of tropical America now cultivated mainly in Andhra Pradesh, Maharashtra, Karnataka, Uttar Pradesh and West Bengal.

16. *Nicotiana rustica*; Eng.– Turkish or East Indian tobacco; Verna.–**Vilaiti–Tambaku**-The leaves are used for smoking and for manufacturing insecticides. This is a herb, native of central America, now cultivated in the Punjab, Bihar and Bengal.

17. *Datura stramonium*; Eng.– Thorn apple; Verna.–**Dhatura**–This is a common weed. The leaves and seeds are narcotic and sometimes used for criminal poisoning. It is used in medicine in the treatment of asthma. It is native of Asia.

18. *Datura innoxia*; Verna.– **Sadahdhatura**–This is a herb. The dried leaves are used in the treatment of asthma. It is native of Mexico, but commonly found in the Western Himalayas and South India.

19. *Datura metel*; Verna.–**Sadahdhatura**–The leaves are used in the treatment of asthma.

20. *Datura suaveolens*; Eng.–Tree datura. This is a small tree. It is grown in gardens for its fragrant creamy-white flowers. This is native of Mexico.

21. *Physalis peruviana*; Syn. *P.edulis*; Eng.–Cape gooseberry; Verna.–**Rasbhari, Tipari, Tankari**–It is cultivated throughout our country for its edible fruits.

22. *Physalis minima*; Eng.–Sunberry; Verna.–**Papotan**, **Tulatipati**–This is a common herb yielding edible fruits.

23. *Physalis ixocarpa*; Eng.–Tomatillo. The fruits are used as vegetable. They are also used in making chilli sauce and sometimes as a dressing for meat. It is native of Mexico.

24. *Atropa belladona*; Eng.–Belladona; Verna.–**Sag-angur**–The roots possess medicinal properties. They are used as sedative stimulant and antispasmodic. It is native of Europe, but now grown in Kashmir.

25. *Cestrum nocturnum*; Eng.–Night Jessamine; Verna.–**Raat-ki-Rani**–It is grown in the gardens for its fragrant flowers, which are scented at night.

26. *Cestrum diurnum*; Eng.–Day Jessamine–Verna.–**Din-ka-Raja**–This is grown in the garden as an ornamental for its fragrant flowers. It is native of the West-Indies.

27. *Cestrum aurantiacum*-This is a bushy shrub, grown as an ornamental.

28. *Cestrum elegans*-It is grown in the gardens as an ornamental. This is native of Mexico.

29. *Cestrum parqui*; Eng.–Willow-leaved jessamine; Verna.–**Rat-ki-Rani**–This is grown as an ornamental. It is native of Chile.

30. *Petunia axillaris;* Syn. *P. nyctaginflora*; Eng.–Large white Petunia–This is an ornamental herb grown for its beautiful flowers.

31. *Lycopersicon esculentum*; Syn. *Lycopersicum esculentum*; *Solanum lycopersicum*; Eng.–Tomato; Verna.–**Tamatar**–This is cultivated for its edible fruits. This is a common herb. It is native of South America.

32. *Lycopersicon glandulosum*; Eng.–Tomato; The fruits are edible.

33. *Lycopersicon peruvianum*; Eng.–Tomato; Verna–**Tamatar**–It is introduced from Great Britain and grown for its edible fruits.

34. *Lycopersicon pimpinellifolium*; Eng.–Current tomato; Verna.–**Tamatar**–It has been introduced from United Kingdom and grown for the edible fruits.

35. *Lycopersicon pissisi*; Eng.–Tomato–The fruits are edible.

36. *Withania somnifera*; Verna.–**Asgand, punir**–It contains substances with narcotic and sporific properties. This is a undershrub, found in the drier parts of the country.

37. *Withania coagulans*; Eng.–Vegetable rennet; Verna.–**Akri**–The fruits used for coagulating milk. It is found in the Punjab.

Systematic Relationships: According to Bentham and Hooker the Polemoniales consists of five families, they are–Polemoniaceae. Hydrophyllaceae, Boragineae, Convolvulaceae and Solanaceae; Engler and Prantl have placed all these families in his Tubiflorae along with fifteen other families. Hutchinson has splitted the order Tubiflorae into a number of smaller groups (orders). He has placed the families Solanaceae, Convolvulaceae and Nolanaceae in the order Solanales. Rendle separated the Convolvulaceae into a distinct order the Convolvulales, which he also separated from the Tubiflorae. According to the Wettstein the Convolvulaceae may have been originated from the other groups of the Tubiflorae and considers them closely related to the Malvales or Geraniales. Hutchinson has also shown the close systematic relationship of the Geraniales with the Convolvulaceae.

Bentham and Hooker and Bessey included the family Solanaceae among the Polemoniales. Hallier considered the Solanaceae to be primitive member of the Tubiflorae and supposed it to have been derived from Linaceae, Hutchinson considers the family to be primitive most of the order Solanales.

22

Order–Personales
Family–Acanthaceae

According to Hutchinson this is the 78th order of the phylum Angiospermae, subphylum Dictoyledones and division Herbaceae. This order comprises of six families, *i.e.*, Scrophulariaceae, Acanthaceae, Gesneriaceae, Orobanchaceae, Lentibulariaceae, Columelliaceae, of which the Acanthaceae has been described in the present text in detail. Bentham and Hooker have placed the families Scrophulariaceae and Acanthaceae along with six others including the Pedalineae and the Acanthaceae in their ninth order the Personales of class Dicotyledones division Gamopetalae, and series Bicarpellatae. Engler and Prantl have included the Scrophulariaceae and the Acanthaceae along with other eighteen families such as Convolvulaceae, Verbenaceae, Labiatae, Solanaceae, Pedaliaceae, etc.,in their sixth order the Tubiflorae of class Dicotyledoneae and subclass Sympetalae.

Corolla nearly always zygomorphic with stamens fewer than the corolla lobes. Mostly herbaceous; leaves alternate to opposite.

Family–Acanthaceae

There are about 240 genera and 2,200 species in this family.

Distribution: The members of this family are mostly found in tropical to subtropical forests especially in the damp and marshy places.

Habit: Usually they are herbs or shrubs. Trees are rarely found. Some are however xerophytic, such as, *Barleria, Blepharis, Acanthus*, etc. Some plants are climbing in habit, *e.g.*, *Mendoncia, Thunbergia* and other allied plants. *Thunbergia alata* is a herbaceous twiner; *Peristrophe bicalyculata, Andrographis paniculata, Runzia parviflora, Ruellia prostrata, Justicia simplex, Hemigraphis* sp; *Haplanthus* sp., etc., are typical examples of herbs. Several plants are undershrubs, *e.g.*, *Barleria cristata, Ruellia tuberosa, Justicia betonica, Strobilanthes coloratus, Eranthemum pulchellum,* etc. Some good examples of shrubs are, *Barleria alba, Barleria prionitis, Adhatoda vasica*, etc. *Acanthus ilicifolius* is a halophyte and represents mangrove vegetation.

Leaves: The leaves of usual mesophytes are simple, usually opposite decussate, entire, exstipulate, thin and delicate. In xerophytes the leaf blade is more or less spiny.

The characteristic anatomical feature of the family is the presence of cystoliths in the epidermal cells of leaves and stems. Sometimes double cystoliths are found. The cystoliths are calcium carbonate crystals.

Inflorescence: The inflorescence is of cymose type. The dichasial cyme passes into monochasial one in the higher branching. Sometimes the flowers are found to be arranged in the short axillary clusters. Spikes and racemes are also frequently found. Bracts and bracteoles generally well developed and often brightly coloured. These bracts, etc., add to the beauty of the inflorescence. In *Mendoncia* sp., and *Thunbergia* sp., the bracteoles are large and form an involucre around the corolla tube. The flowers of *Dicliptera* sp., are found to be arranged in axillary peduncled cymes. The flowers of *Phialacanthus griffithi* are arranged in typical cymes. The flowers of *Strobilanthes discolor* are found to be arranged in panicled cymose heads whereas in *Strobilanthes anisophyllous* these are arranged in cymose heads. In the species of *Barleria*, the flowers are either solitary or arranged in short spikes, or pairs. In the species of *Thunbergia* the flowers are found to be arranged either in cymose or in racemose types of inflorescence. This way, there is great variation in the types of inflorescence in various members of the family.

Flower: The structure of flower is very uniform. The flowers are bracteate; bracteolate, hermaphrodite, complete, zygomorphic (irregular) and hypogynous. Usually the flowers are tetramerous or pentamerous, but generally there is a reduction to 4 or 2 in the androecium, and there are two carpels in the gynoecium.

Calyx: It consists of five, sometimes four or rarely three sepals. Usually they are gamosepalous. In *Thunbergia* the sepals are reduced to narrow beam. In this case large bracteoles serve the purpose of protection. The aestivation is either contorted or imbricate.

Corolla: Usually it consists of five or sometimes four gamopetalous petals. The corolla is often bilabiate, *e.g.*, in *Ruellia* in such cases the upper lip of the corolla is erect and bifid at the apex and the lower lip is horizontal and three–lobed forming the landing–platform for the insects. Usually the corolla possesses a large or short slender corolla tube which passes above into an equally five lobed limb, *e.g.*, in *Thunbergia*. The inner side of the lip of the corolla possesses dense hairs. These hairs often extend to the mouth of the corolla. In *Acanthus* and other allied genera the upper lip of corolla is altogether absent. Here, the corolla is found to be cut upto the base of the corolla tube. The aestivation is either contorted or imbricate.

Androecium: Usually there are 4 didynamous stamens. In certain cases only two stamens are found. Very rarely the number of fertile stamens reaches to five, *e.g.*, *Pentstemonacanthus*. In the cases where there are only four stamens, the fifth posterior stamen reduces to a staminode or disappears completely. In the **Imbricatae** group of the family, there are only two stamens. In such cases, firstly the posterior stamens reduce to staminodes or disappear completely. However, in *Brillantaisia* the anterior pair of stamens reduces to staminodes.

The filaments are generally quite free and project out from the mouth of the corolla tube. The stamens of *Thunbergia*, are however short and remain inside the corolla tube. The stamens are inserted in the corolla, *i.e.*, epipetalous. The anthers may be two or one celled. In the cases where one celled anthers are found, a rudiment of the second cell may be seen, whereas, in the case of two–celled anthers the cells are equal and more or less separated by a connective. Sometimes the lower cell of the anther becomes spurred. The shape and size of pollen grains are quite different from genus to genus.

Gynoecium: The gynoecium consists of two carpels, syncarpous. The ovary superior, bilocular, two or more ovules are found in each loculus. The Placentation is axile. The style is long and slender and projects out from the mouth of the corolla tube. There are two small stigmas; the posterior stigma is usually reduced.

Fruits: Usually the fruit is bilocular capsule, which dehisces loculicidally. In *Mendoncia* the fruit is drupe.

Seed: In *Nelsonia*, the seeds are small and many. In *Nelsonia* and *Thunbergia*, the funicle of the seeds forms a papilla, the seeds are rounded. In the *Acanthus* and allied genera, *e.g.*, *Ruellia*, the funicle of the seeds forms a hook–like projection, known as **jaculator**, in which the seed rests. These hook–like jaculators make the fruits burst and the seeds become dispersed in different directions. The seeds are exalbuminous.

Pollination: The pollination takes place through the agency of insects.

Floral Formula: $\cdot|\cdot$ ⚥ K (4–5), $C\overline{(4-5)}$, A 4 or 2, G($\underline{2}$).

Bentham and Hooker(1862)	Engler and Prantl(1931)	Hutchinson(1959)
Phanerogams	Phanerogams	Angiospermae
Dicotyledones	Dicotyledoneae	Dicotyledones
Gamopetalae	Sympetalae	Herbaceae
Bicarpellatae	Tubiflorae	Personales
Personales	Acanthaceae	Acanthaceae
Acanthaceae		

Description of some important plants in semi–technical language:

1. ***Barleria prionitis.***

Habit: A prickly bushy shrub.

Root: Tap and branched.

Stem: Erect, branched, woody, slightly angular, solid, glabrous, green.

Leaf: Simple, opposite, decussate, upper leaves short petioled, lower leaves long petioled, interpetiolar spines, lanceolate, entire acuté, unicostate reticulate venation.

Inflorescence: A terminal spike, lower flowers arranged in cymes.

Flower: Sessile, bracteate (leaf bracts), bracteolate (six bracteoles–two large and four small narrow and spiny at the tips), hermaphrodite, zygomorphic, hypogynous.

Calyx: 4 sepals, polysepalous, slightly connate at the base, sepaloid, arranged in two whorls of two each, two outer bigger and inner smaller, inferior, imbricate aestivation.

Corolla: 5 petals, gamopetalous, slightly bilabiate, upper lip consisting of four lobes and the lower of single lobe, inferior, imbricate aestivation, yellowish in colour.

Androecium: 4 stamens, two stamens large and two small, epipetalous, sometimes fifth stamen is represented by staminode, anthers bicelled dorsifixed.

Gynoecium: 2 carpels (bicarpellary), syncarpous, ovary superior, bilocular, axile placentation, one ovule in each loculus style terminal, stigma slightly elongated, ovary situated on a honey secreting disc.

Fruit: An elongated capsule.

Floral Formula: $\cdot|\cdot$ ⚥ K 2+2, $C\overline{(4+1)}$, A 4, G($\underline{2}$).

2. ***Ruellia*** **sp.**

Habit: An erect wild annual herb, common in the gardens near moist places.

Stem: Erect, branched, herbaceous, cylindrical, solid green.

Leaf: Simple, opposite, decussate, subsessile, ovate, entire, acute, glabrous, unicostate reticulate venation.

Inflorescence: Cymose, dichasial cyme.

Flower: Pedicellate, bracteate, bracteolate, hermaphrodite, zygomorphic, funnel shaped, bluish pink, complete, hypogynous.

Calyx: 5 sepals, gamosepalous (connate at the base), acute, sepaloid, valvate asetivation, inferior.

Corolla: 5 lobes, funnel shaped, gamopetalous, bluish–pink, twisted aestivation, inferior.

Androecium: 4 stamens, polyandrous, epipetalous, anthers bicelled, basifixed, introrse.

Gynoecium: 2 carpels (bicarpellary), syncarpous, ovary superior, bilocular, axile placentation, one ovule in each loculus, style slender, stigma single.

Fruit: An elongated capsule with tapering end, jaculators present.

Floral formula: $\cdot|\cdot$ ⚥ K (5), $\overline{C(5), A4}$, G($\underline{2}$)

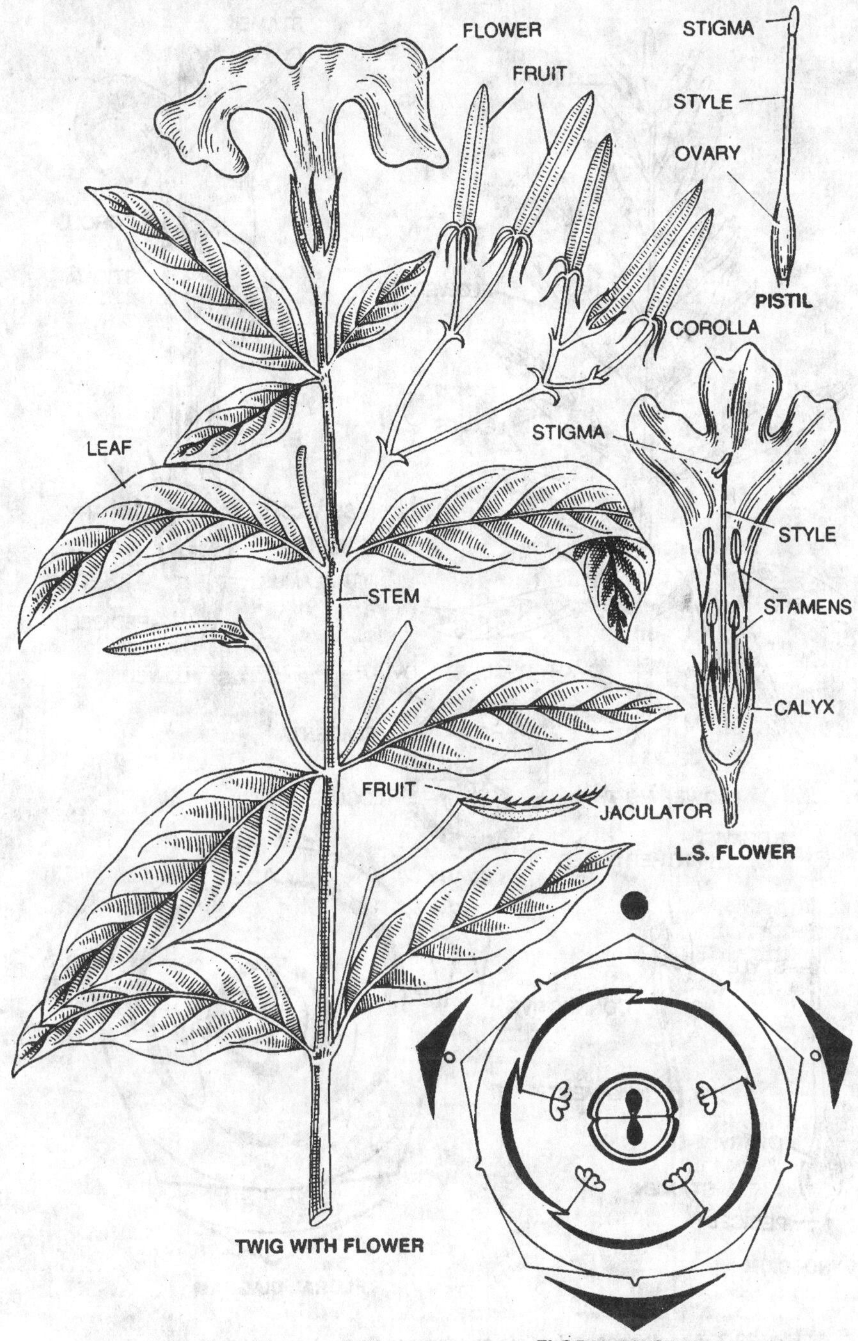

Fig. 22.1. Acanthaceae. *Ruellia prostrata* Lamk.; Verna. **Kalighawani.**

3. ***Justicia*** sp.

Habit: An undershrub, commonly grown as hedge plant.

Stem: Erect, branched, cylindrical, solid, herbaceous or somewhat woody, glabrous.

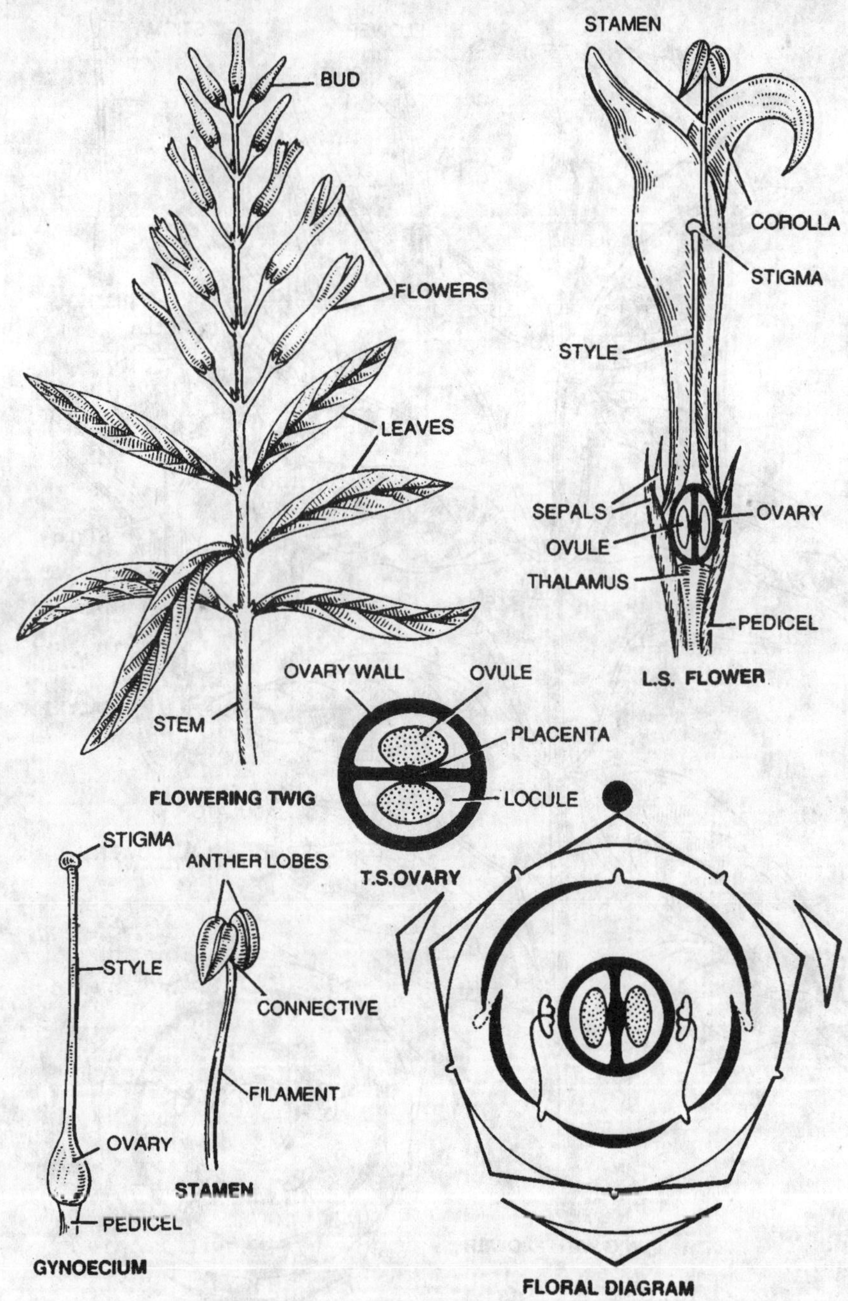

Fig. 22.2. Acanthaceae. *Justicia gendarussa* Brum; **Verna.Nili-Margandi.**

Leaf: Simple, opposite decussate, cauline, petiolate, lanceolate, entire, acute, glabrous, unicostate reticulate venation.

Inflorescence: Racemose and cymose mixed.

Flower: Sessile, bracteate (bracts lanceolate), hermaphrodite, zygomorphic, complete, hypogynous.

Calyx: 5 sepals, gamosepalous, connate at the base and free above, green, inferior.

Corolla: 5 petals, gamopetalous, bilabiate, upper lip consists of two lobes, lower of three lobes, yellowish white with pink or violet dots, imbricate aestivation, inferior.

Androecium: 2 stamens, free, epipetalous, inserted on the upper part of the corolla, filaments long, anthers bicelled, basifixed.

Gynoecium: 2 carpels (bicarpellary), syncarpous, ovary superior, bilocular, axile placentation, two ovules in each loculus, style filiform, stigma bifid.

Fruit: A capsule.

Floral Formula: $\cdot|\cdot$ ⚥ $K\,(5), \overline{C(2+3)\,A\,2}\,G\underline{(2)}$.

4. ***Peristrophe bicalyculata* Nees. Stem**--aerial, erect, angular, branched, solid, pubescent, green, herbaceous. **Leaf**–cauline and ramal, opposite decussate, simple, exstipulate, petiolate, ovate, acute, hairy, unicostate reticulate. **Inflorescence**–cymose, branched or panicled cyme. **Flower**–bracteate, two bracts (anterior and posterior), bracteolate, 4 bracteoles, bracts and bracteoles persistent, pedicellate, bisexual, complete, zygomorphic, pentamerous, hypogynous, cyclic. **Calyx**–5, polysepalous, valvate. **Corolla**–5, gamopetalous, bilabiate 2/3, valvate, purple. **Androecium**–2, polyandrous, epipetalous, filaments long, dithecous, basifixed, introrse. **Gynoecium**–2, bicarpellary, syncarpous, superior, bilocular, 2 ovules in each locule, axile, style filiform stigma bifid. **Fruit**–capsule, **Floral formula**– Br Brl $\cdot|\cdot$ ⚥ $K_5, C_{(2+3)}, A_2, G_{(\underline{2})}$.

Economic Importance of the Family

The family is of little economic value. Some plants are used as hedge plants and some are ornamental. Some plants possess medicinal properties. A list of few important plants is given below.

1. *Adhatoda vasica*; Syn. *Justicia adhatoda*; Eng.–Malabar nut; Verna.–**Arusa, Adulasa**–An Ayurvedic drug is obtained from its leaves, which is used in cough, chronic bronchitis and rheumatism. The twigs and leaves are used as green manure. The plant is also used as an antiseptic and insecticide. This is an evergreen shrub commonly found in Northern India.

2. *Barleria alba*; This is a shrub grown as an ornamental.

3. *Barleria cristata;* This is an ornamental small shrub. The juice of the leaves is given in cough.

4. *Barleria prionitis*; Verna.–**Katsareya**–A plant with medicinal properties. The dried bark is given in the cough.

5. *Barleria trigosa*; Verna.–**Nila kusum**–The root is used as a remedy of severe cough.

6. *Andrographis paniculata*; Verna.–**Kiryat.** The leaves possess medicinal properties. The plants are grown as ornamental herbs.

7. *Asteracantha longifolia*; Verna.–**Tal makhana**–This is a medicinal plant. Its leaves, roots and seeds are given in jaundice, rheumatism and dropsy.

8. *Ruellia prostrata*; the plants possess medicinal properties. They are used as a remedy in ear troubles.

9. *Ruellia tuberosa*; This is herb or a small shrub, grown as an ornamental.

10. *Justicia betonica*; An undershrub, grown as hedge plant.

11. *Thunbergia alata*; It is a herbaceous twiner; grown as an ornamental, as it flowers throughout the year.

12. *Thunbergia coccinea*; This is grown as an ornamental.

13. *Thunbergia fragrans*; Grown in the gardens as ornamental.

14. *Thunbergia grandiflora*; Verna.–**Kukua lata**–This is an ornamental; the leaves are edible.

15. *Aechnanthera wallichii*; The lower surface of the leaves is used in making a kind of cloth by natives.

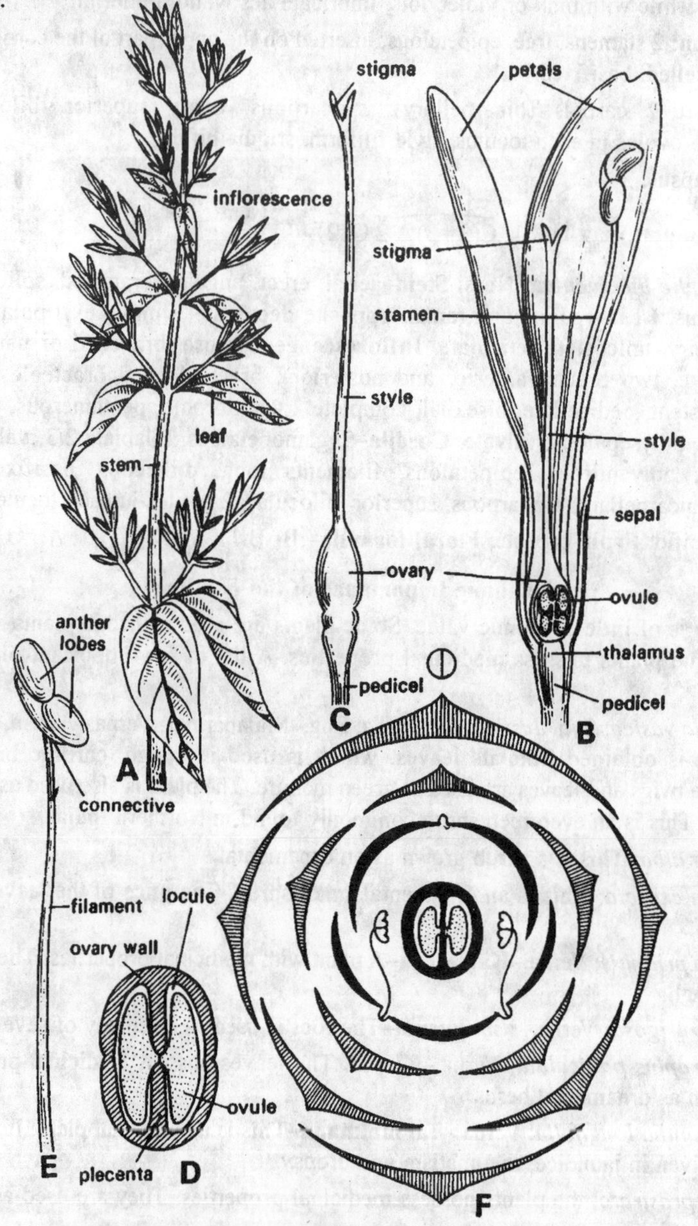

Fig. 22.3. Acanthacea. *Peristrophe bicalyculata* Nees. A, flowering twig; B, L.S. of flower; C, gynoecium; D, T.S. of ovary; E, stamen; F, floral diagram.

16. *Runzia parviflora*; Verna.–**Pindi**–The juice of its leaves is given to children in small-pox.

17. *Peristrophe bicalyculata*; Verna.–**Atrilal.**–Used as an antidote for snake bite.

18. *Peristrophe bivalvis*; Syn. *P. tinctoria*; A shrub; a yellow orange dye is obtained from its leaves and young stems.

19. *Ecbolum viride*; Syn. *E.linneanum*; Verna.–**Udajat**–An ornamental shrub, grown in the gardens. The roots are given as a remedy for jaundice and rheumatism.

20. *Phlogacanthus thyrsiflorus*; Verna.–**Lal bahuk.**–The flowers are used as vegetable. The fruits and leaves are given in fevers. The leaves of *Tubiflorus* are used as a washing soap.

21. *Rhinacanthus communis*; Verna.–**Palakjuhi.**–The leaves and roots are used as a remedy for skin diseases, *e.g.*, ring–worm.

22. *Jacobinia tinctoria*; This is a diffused shrub and grown as an ornamental in the gardens.

23. *Crossandra infundibuliformis*; Verna.–**Priya darsa**– This is a shrub grown as an ornamental.

24. *Asytasia gangetica*; Syn. *Justicia gangetica*; A herb, grown as an ornamental.

Systematic Relationships: Bentham and Hooker have included eight families in the order Personales.-These families are Scrophulariaceae, Orobanchaceae, Lentibulariaceae, Columelliaceae, Gesneraceae, Bignoniaceae; Pedalineae, and Acanthaceae. Engler and Prantl have included all these families in the Tubiflorae along with other twelve families. Hutchinson has included these families in his Personales. He however, puts the family Pedaliaceae in the order Bignoniales which is not related to the Personales.

Hallier considered the Scrophulariaceae to be one of the primitive families of the Tubiflorae. Bessey considered the family as an advanced member of his Scrophulariales, derived perhaps from the Bignoniaceae.

Many botanists have thought that the Acanthaceae has been derived from the Scrophulariaceae. Hutchinson considered the Acanthaceae to be most advanced among his Personales.

23

Order–Lamiales
Family–Lamiaceae (Labiatae)

According to Hutchinson this is the eighty-second order of the phylum Angiospermae, subphylum Dicotyledones and division Herbaceae. This order comprises of four families, *e.g.*, Myoporaceae, Selaginaceae, Globulariaceae and Labiatae, of which the Labiatae has been described in the text in detail, Bentham and Hooker have placed the Labiatae along with three other families including Verbenaceae, in the tenth order the Lamiales of class Dicotyledones, division Gamopetalae and series Bicarpellatae. Engler and Prantl have placed the Labiatae along with nineteen other families including Convolvulaceae, Verbenaceae, Solanaceae, Scrophulariaceae, Pedaliaceae, Acanthaceae, etc., in the sixth order the Tubiflorae of class Dicotyledoneae and sub class Sympetalae.

Leaves mostly opposite or whorled; corolla zygomorphic; stamens 4 or 2; ovary becoming lobed with a gynobasic style; ovules mostly paired. Largely herbaceous, with inflorescence tending to capitula or verticillasters.

Family–Labiatae (Lamiaceae)

There are about 200 general and 3200 species in this family.

Distribution: The members of this family are found throughout the world. They are being distributed both in temperate and tropical regions. The chief centre of distribution of the family is the mediterranean region. In our country the family is represented by several important genera, such as, *Mentha, Ocimum, Leucas, Coleus, Pogostemon*, etc.

Habit: Majority of the plants are annual or perennial herbs, inhabiting the temperate regions. In warmer climates, the plants become shrubby in nature. Trees are rare. Certain plants are xerophytes with extremely reduced leaves. A few species are climbers, but the climbing habit is very rare.

Some important herbs are, *Coleus aromaticus, C. forskohlii, Leucus cephalotes, Mentha arvensis, Ocimum sanctum, O. basilicum*, etc., *Salvia aegyptiaca, Coleus rotundifolius* are undershrubs, *Ocimum gratissimum, Meriandra bengalensis* are shrubs. Certain species of Brazilian genus *Hyptis* and Indian genus *Leucosceptrum* are trees. The certain species of *Mentha* and *Lycopus* are found in marshy places. A few American species of *Scutellaria* are climbers. Certain plants are extreme xerophytes, *e.g., Rosmarinus.* In this plant the leaves are rolled back and stomata are found among the hairs in the grooves on the underside of the leaf.

The propagation of many plants of this family takes place by means of new shoots which remain attached to the plant, *e.g.*, in *Salvia, Ballota.* In *Ajuga reptans* the new shoots form aerial runners with somewhat reduced leaves; in *Lycopus*, the subterranean stolons (suckers) are found with short internodes and scaly colourless leaves. The suckers are found in the species of *Mentha*

Stem: The young shoots are usually four-sided or quadrangular.

Leaves: Usually the leaves are simple, opposite, decussate and exstipulate. They show many variations from an entire blade to toothed, lobed, cut or finely dissected as in many species of

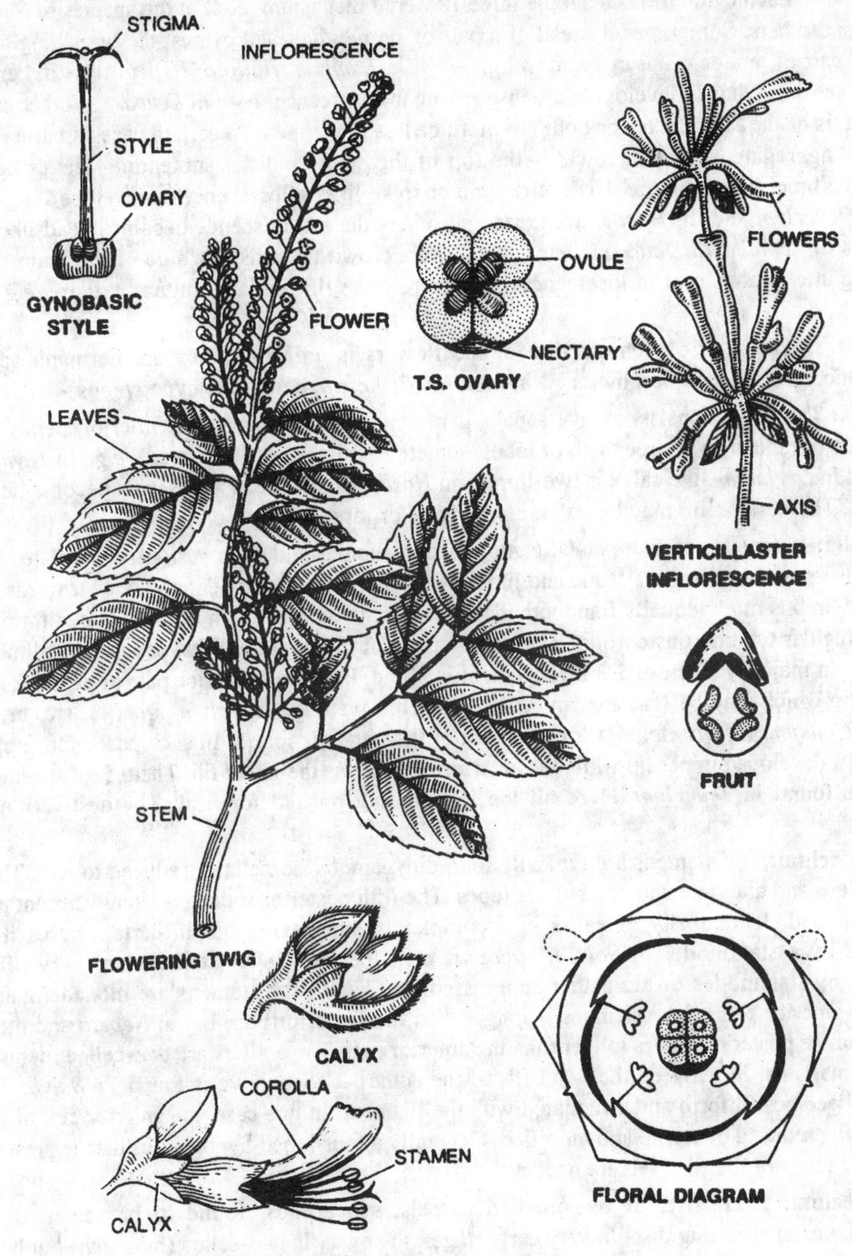

Fig. 23.1. Labiatae–*Ocimum sanctum* Linn. (Verna. tulsi).

Salvia. A whorled leaf arrangement of 3 to 8 leaves is found in some genera, *e.g., Ocimum.* All parts of the plants, such as stem, leaves and inflorescence are more or less hairy and possess glandular hairs, which secrete characteristic scent of the genus or species. Sessile scented oil secreting glands are also found frequently on the epidermis.

Inflorescence: The characteristic inflorescence of the family is **verticillaster.** In this type of inflorescence the whorls of flowers develop at the nodes. This consists of a pair of cymose inflorescences. Each cyme forms a simple three-flowered dichasium, *e.g.*, in the species of *Salvia, Prunella* and others. Sometimes, instead of a pair of simple dichasial cymes, the branching takes place and pair of monochasial cyme, develops, *e.g.*, in *Lamium, Ballota, Nepeta*, etc. In certain cases all axes are reduced developing a dense sessile inflorescence, *e.g.*, in *Leucas*. In other cases the main axis or the lateral axes or both are more or less developed. In certain cases a number of whorls are aggregated together towards the top of the stem, and the subtending leaves being reduced to a bract-like structure. Here, a raceme or spike like inflorescence is developed, *e.g.*, in *Prunella, Teucrium*, etc. In *Hyptis, Monarda* and others the inflorescence becomes head-like. In the species of *Scutellaria, Teucrium,* etc., the flowers crowd towards one side of the main axis, developing into dorsiventral inflorescences. Very rarely the flowers are solitary axillary, *e.g.*, in *Scutellaria.*

Flowers: The plan of construction of the flowers is uniform. They are hermaphrodite, zygomorphic, rarely actinomorphic, *e.g.*, in *Mentha, Elshotzia*, complete, hypogynous.

Calyx: The calyx consists of five sepals, gamosepalous. The sepals are inferior, persistent, campanulate or tubular, with free teeth or lobes, sometimes bilabiate (two-lipped), *e.g.*, in *Thymus, Salvia*, etc. In *Orgianum*, the calyx is two-lipped, in *Hoslundia*, the persistent sepals become fleshy in the fruit. The aestivation may be valvate, imbricate or rarely quincuncial.

Corolla: It consists of five petals, gamopetalous, tubular and limb variously bilabiate. The corolla consists of two pairs, *viz.*, tube and limb. The tube is straight or bent and wide towards the mouth. The limb is rarely equally fine toothed. In *Mentha*, two upper teeth become united forming an almost regular tetramerous corolla. The arrangement of the upper and lower lip of the limb is as follows– In majority of cases 2/3 arrangement is found, here the posterior pair of petals forms the upper lip which may be flat, *e.g.*, in *Thymus*, or concave, *e.g.*, in *Salvia, Stachys*, etc. In the species of *Ocimum, Hyptis* etc., 4/1 arrangement of the petals is found. In such cases four upper petals form a developed upper lip while the anterior petal forms the lower lip. The 0/5 arrangement of petals is found in *Teucrium*. Here all the lobes being pushed forward. The aestivation is imbricate.

Androecium: The stamens are typically four, didynamous, sometimes reduced to two. They are epipetalous and alternate with the corolla lobes. The fifth posterior stamen is usually absent and rarely developed. In majority of cases it is completely suppressed but in certain cases it is represented by a staminode. In *Salvia, Lycopus*, etc., the two upper stamens are very often represented by staminodes or altogether suppressed. In *Coleus*, the stamens are monadelphous. Usually the anterior pair of the stamens is longer than the posterior pair, but in *Nepeta* and other allied genera the posterior pair is longer than the anterior pair. The anthers are two-celled; dehisce by longitudinal slits. In between the two cells of the anther, a connective is found. In *Salvia* this connective becomes filiform and articulated with the filament. In this case, the anterior cell of the anther is being reduced or sterile and modified. Generally a four-lobed hypogynous disc is present. The anterior lobes of the disc secrete nectar.

Gynoecium: It consists of two median carpels, syncarpous, found to be seated on a hypogynous, nectar secreting disc. In very early stages, the ovary is two-celled (bilocular) but later on a constriction develops which divides the two carpels into four one-ovuled segments. The ovules are anatropous. The ovules are attached at the inner corner of each ovary segment. The style is very characteristic of the family. It arises from the base of the ovary in between the four loculi, and is known as **gynobasic style.** The stigmatic papillae of bilobed stigma are situated at the tip of the style arms. The placentation is axile. The ovary is superior.

Fruit: The fruit consists of four one-seeded nutlets, included within the persistent calyx.

Seeds: Exalbuminous or with scanty endosperm. The testa is thin; the embryo possesses flat cotyledons.

Pollination: The pollination takes place through the agency of insects (*i.e.*, entomophilous).

Floral Formula: $\cdot|\cdot$ ⚥ K (5), $\overline{C(5), A}$ 4 or 2, G$\underline{(2)}$.

Bentham & Hooker (1862)	**Engler & Prantl (1931)**	**Hutchinson (1959)**
Phanerogams	Phanerogams	Angiospermae
Dicotyledones	Dicotyledones	Dicotyledones
Gamopetalae	Sympetalae	Herbaceae
Bicarpellatae	Tubiflorae	Lamiales
Lamiales	Labiatae	Labiatae or
Labiatae		Lamiaceae

Description of some important plants in semitechnical language:

1. ***Ocimum sanctum***; (Verna.–**Tulsi**).

Habit: A perennial herb with typical aromatic smell.

Stem: Erect, branched, quadrangular, somewhat woody, solid, branches covered with soft hairs.

Leaf: Simple, opposite, short petioled, exstipulate, ovate, serrate, acute, gland dotted, unicostate reticulate venation.

Inflorescence: Verticillaster, 6-10 flowered whorls present, emitting fine smell.

Flower: Pedicellate, bracteate, bracts small and caducous, hermaphrodite, zygomorphic, complete, purple, hypogynous.

Calyx: 5 sepals, gamosepalous, bilabiate, petaloid (purple coloured), posterior lip broad and boat shaped, anterior lip with 4 small lobes possessing mucronate teeth, gland dotted, imbricate aestivation, inferior.

Corolla: 5 petals, gamopetalous, bilabiate, corolla tube short, upper lip four lobed, lower lip large, imbricate aestivation, inferior.

Androecium: 4 stamens, polyandrous, didynamous, epipetalous, fifth posterior stamen completely suppressed, anthers bicelled, introrse, dorsifixed.

Gynoecium: 2 carpels (bicarpellary), syncarpous, ovary bilocular in early stage but becomes tetralocular in later stage, ovary superior, four chambered, axile placentation, single ovule in each loculus, gynobasic style (*i.e.*, it arises from the base of the ovary), stigma bifid.

Fruit: Schizocarpic, carcerulus, 4 nutlets developed.

Floral Formula: $\cdot|\cdot$ ⚥ K (1 + 4), $\overline{C(4+1), A}$ 2 + 2, G$\underline{(2)}$.

2. ***Salvia officinale***.

Habit: An annual ornamental herb with typical aromatic smell.

Stem: Erect, branched, herbaceous, quadrangular, solid, hairy and green.

Leaf: Simple, opposite, decussate, petiolate, exstipulate, ovate, serrate, acute, unicostate reticulate venation.

Inflorescence: Verticillaster with usually six-flowered whorls.

Flower: Pedicellate, bracteate, hermaphrodite, zygomorphic, complete, coloured, hypogynous.

Calyx: 5 sepals, gamosepalous, bilabiate, 2 lobes consisting of one lip and three the other lip, petaloid, imbricate aestivation, inferior.

Corolla: 5 petals, gamopetalous bilabiate, 2 lobes in one whorl and 3 in the other, imbricate aestivation, inferior.

Androecium: 2 stamens (anterior), posterior, 3 stamens absent or staminodes present, epipetalous, short filament connected to a long connective, lower end of connective is flat, short and sterile, upper end of it with a large, fertile anther lobe, anthers bicelled, introrse.

Gynoecium: 2 carpels (bicarpellary), syncarpous pistil present on nectar secreting disc, ovary superior, bilocular when young, but becomes tetralocular in the later stage, single ovule in each loculus, axile placentation, gynobasic style, bifid stigma.

Fruit: Four nutlets in a group.

Floral Formula: $+ \, ⚥ \, K\,(2+3),\ \overline{C(4+3), A\,2}, G(\underline{2})$.

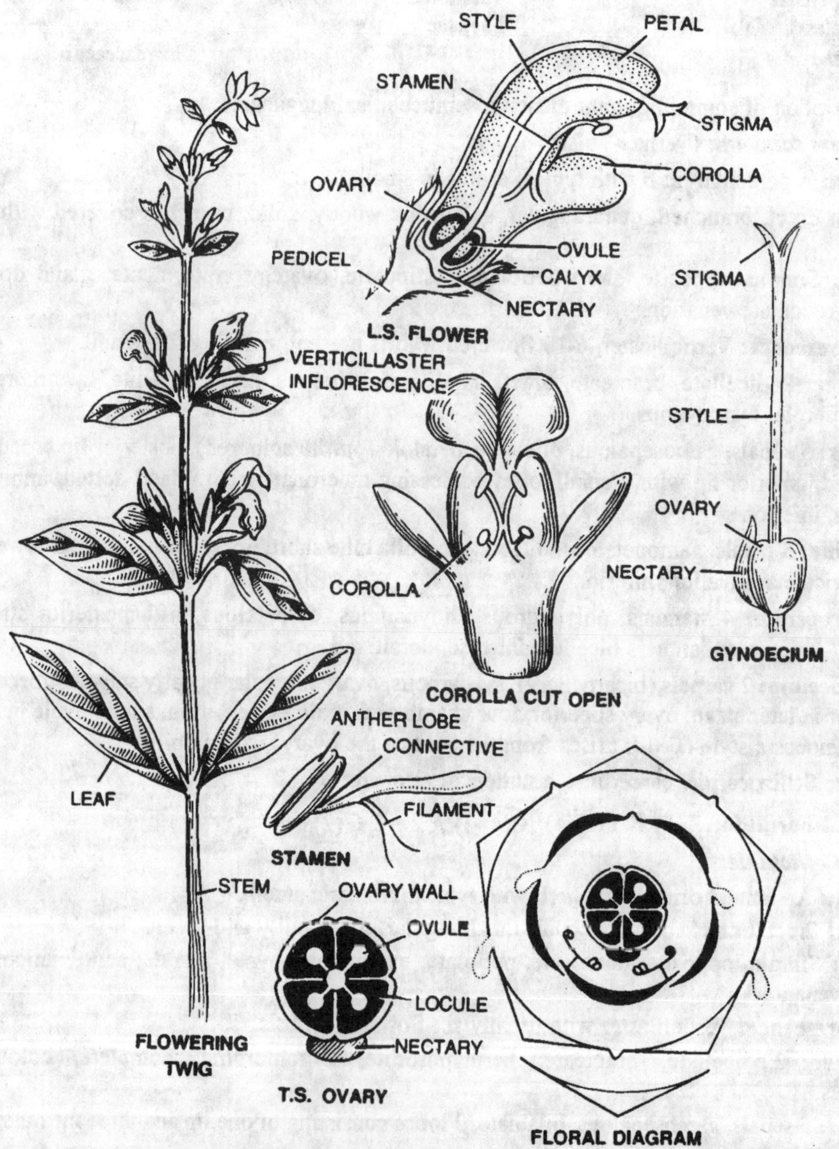

Fig. 23.2. Labiatae (Lamiaceae). *Salvia officinalis* Linn; Eng. sage; Verna. **salbia sefakuss.**

Economic Importance of the Family

The family is of little economic value. Some plants are grown in the gardens as ornamentals, some plants yield essential oil while certain possess medicinal properties. A list of few important plants is given below.

1. *Coleus* sp.; Several species of *Coleus* are grown as ornamentals in the gardens for their variously coloured beautiful foliage.

2. *Coleus amboinicus*; Syn. *C. aromaticus*; Eng.-Indian borage; Verna.–**Pathorchur**–This is a common herb. Its leaves are used for flavouring food products.

3. *Coleus forskohlii*; Syn. *C. barbatus*; This is a herb, cultivated in South India. The roots are used as spice and condiment.

4. *Coleus rotundifolius*; Eng.–Madagascar potato; Verna–**Koorkan**–This is a herb or shrub commonly grown in South India. The underground tuberous stems are eaten as vegetable.

5. *Colebrookea oppositifolia*; Verna.–**Binda**–The leaves are applied to wounds. The wood is used for making gun powder.

6. *Leucas cephalotes*; Verna.–**Goma**–This is a common herb. The flowers are used medicinally for cough and cold.

7. *Leucas zeylanica*; Verna.–**Gattatumba**–The fresh juice of leaves is used as a remedy for headache and cold. It is also applied to skin diseases. The flowers are used for cough and cold.

8. *Meriandra bengalensis*; Verna.–**Kafur ka pat**– This is a shrub grown as an ornamental. The leaves possess medicinal properties. They emit a camphor-like scent. The leaves are used to preserve clothes against insects.

9. *Mentha arvensis*; Eng.–Field mint; Verna.–**Pudina**–This is a perennial herb. The leaves yield an essential oil which is used in the manufacture of certain kinds of cigarettes and pharmaceuticals. It is also used as a carminative, stimulant and refrigerant. Usually grown in the Punjab, Western Himalayas, Kashmir and Kumaon hills.

10. *Mentha arvensis* var *piperascens*; Eng.–Japanese peppermint. The leaves are the source of an essential oil, from which menthol is prepared. It is used in the treatment of colds. The plant is a herb and commonly cultivated in Jammu and Kashmir.

11. *Mentha longifolia*; Syn. *M. sylvestris*; Eng.–Mint; Verna.–**Pudina**–This is an aromatic herb. The leaves are used for flavouring purposes. The dried leaves are used as stimulant and carminative. Usually cultivated in Uttar Pradesh, Kashmir, Maharashtra and the Punjab.

12. *Mentha piperita*; Eng.–Peppermint; Verna.–**Vilayati pudina**–The leaves are used for flavouring purposes. An essential oil is obtained from the leaves, which is used in perfumery and soap making industries. The oil is also used medicinally as stimulant, carminative and for sickness and vomitting. This is a herb, cultivated in the Punjab, Kashmir and Maharashtra.

13. *Mentha pulegium*; Eng.–European pennyroyal–An essential oil is extracted from the leaves and tops of the plants, which is used in perfumery and cosmetics. It is herb, cultivated in Jammu and Kashmir.

14. *Mentha spicata*; Syn. *M. viridis*; Eng.–Spearmint; Verna.–**Pahari pudina**–The leaves yield an essential oil, known as spearmint oil, which is used for flavouring food products. This herb is cultivated in the Punjab, Uttar Pradesh and Maharashtra.

15. *Lavandula officinalis*; Syn. *L.vera*; *L. spica*; Eng.–Lavender–This is a small shrub. It is native of Europe but now grown in Jammu and Kashmir. An essential oil is extracted from its flowers which is used in perfumery.

16. *Ocimum sanctum*; Eng.–Holy basil; Verna.–**Tulsi**–This is a small shrub and treated as holy plant by Hindus. Usually grown in the courtyards of Hindus and by the side of temples. This

plant possesses several medicinal properties. The juice of the leaves is used as a stimulant and as a remedy for bronchitis. The seeds are used as a remedy for urino-genital troubles. The plant is used as an antidote for snake-bite.

17. *Ocimum kilimandscharicum*; Eng.–Camphor basil. The camphor is obtained from the leaves. It is cultivated on a very small scale in West Bengal, Assam, Tamil Nadu and Dehradun.

18. *Ocimum basilicum*; Eng.–Basil; Verna.–**Bantulsi**–This is an aromatic herb cultivated throughout our country. The leaves are used for flavouring purposes. The seeds are used medicinally as a remedy for dysentery and diarrhoea.

19. *Ocimum gratissimum*; Eng.–Shruby basil; Verna.–**Ranitulsi**–This is a common ornamental, aromatic shrub. The plant possesses medicinal properties. The seeds are given in headache. This is also used as a mosquito repellant.

20. *Pogostemon perilloides*; Verna.–**Pacholi**–This is a herb. The leaves yield an essential oil, which is used for making soaps and perfumes. The dried leaves are used for scenting clothes to keep off insects. It is commonly cultivated in the Western Ghats and Nilgiris.

21. *Rosmarinus officinalis*; Verna.–**Rusmori**–This is a herb grown as an ornamental. It yields an essential oil which is used in perfumery.

22. *Roylea calycina*; Verna.–**Patkarru**–The plant possesses medicinal properties and used as a bitter tonic.

23. *Salvia officinalis*; Eng.–Sage; Verna.–**Salbia, Sefakuss, Seesti**–This is a small shrub, grown as an ornamental. This is native of Mediterranean region.

24. *Salvia splendens*; Syn. *S.colorans*; Eng.–Scarlet sage–This is grown as an ornamental.

25. *Satureja hortensis*; This is a herb found in Kashmir. The aromatic stem and leaves are used for flavouring food products.

26. *Anisomeles indica*; Syn. *A. ovata*; Verna.–**Gopali, Kalabhangra**–This is a woody herb, found throughout our country used as fodder and for flavouring purposes.

Systematic Relationships: Bentham and Hooker have included five families in the order Lamiales. They are Myoporineae, Selagineae, Verbenaceae, Labiatae and Plantagineae. Engler and Prantl have included (excluding Selagineae) in the order Tubiflorae. Hutchinson has included the families Myoporaceae, Selagineae, Globulariaceae and Labiatae in his order the Lamiales and the Verbenaceae in the Verbenales.

The Family Verbenaceae is related phylogenetically to the Labiatae. Bessey has also included these families within the Lamiales. Originally Hutchinson (1926) included the Verbenaceae in the Lamiales but later on (1948) he transferred the family Verbenaceae in the Verbenales. He considered that the Verbenaceae and Labiatae are quite unrelated and the Verbenaceae have been derived from Rubiaceae.

24

THE MONOCOTYLEDONES

Order–Liliales

Family–Liliaceae and Allies

According to Hutchinson this is the ninety-fifth order of the phylum Angiospermae, subphylum Monocotyledones and division Corolliferae. The order consists of six families, *i.e.*, Liliaceae, Tecophilaeaceae, Trilliaceae, Pontederiaceae, Smilacaceae and Ruscaceae, of which Liliaceae, Smilacaceae and Ruscaceae have been described in this text in detail. Bentham and Hooker have placed the Liliaceae along with seven other families in the third series –Coronarieae of class Monocotyledones. They have merged the Smilacaceae and Ruscaceae (of Hutchinson) within Liliaceae. Engler and Prantl have placed the Liliaceae along with eight others in the ninth order– the Lilliflorae of class Monocotyledoneae. They have also not recognized the Hutchinson's Smilacaceae and Ruscaceae.

Herbs with rhizomes, corms, as bulbs, rarely climbing; corolla-like perianth a dominant feature, the two series similar and often fusing together into one tube; stamens often 6; ovary superior or rarely semi-inferior; capsule or berry; seeds with copious endosperm.

Family—Liliaceae

There are about 240 genera and 4000 species in this family.

Distribution: The plants belonging to this family are cosmopolitan in distribution. They are abundantly found in temperate and sub-tropical regions. The plants are usually annual and perennial herbs. Sometimes, the soft, wooded shrubs are also found. In our country the family is represented by several genera such as, *Lilium, Aloe, Asparagus, Asphodelus, Gloriosa*, etc.

Habit: The plants are annual, biennial and perennial herbs and shrubs. The plants survive under the most varied conditions. Majority of the plants are xerophytic. The species of *Aloe* are xerophytic and may be either shrubs or tree-like with perennial aerial stems. One of the species of *Lilium* bears the bulbils in the axis of the leaves to furnish the vegetative reproduction. The species of *Aloe* are xerophytic bearing succulent and fleshy leaves. The plants of *Gloriosa* and *Asparagus* are climbers. The plants of *Asparagus* bear cladodes.

Roots: The roots are adventitious and fibrous.

Stem: In some species, the stems are under-ground, bulbs or rhizomes. The inflorescences may develop on a special aerial stem known as **scape** (*e.g.*, in *Asphodelus*).

Leaf: Usually the leaves are simple, radical or cauline. They may be alternate or whorled. The leaves of *Aloe* are fleshy and succulent. The plants of *Asparagus* bear scale-like leaves. However, they possess phylloclades for the purpose of carbon assimilation. The plants possess parallel veined leaves.

Inflorescence: It is usually of racemose type, but the flowers are never arranged in umbels. In *Rohdea*, the native of China and Japan the flowers are found to be arranged spirally on a spadix.

In the species of *Gloriosa*, the flowers are solitary and terminal. In *Tulipa*, *Lilium* and *Fritillaria* flowers are solitary and terminal. In *Gonioscypha*, the flowers are arranged in spikes.

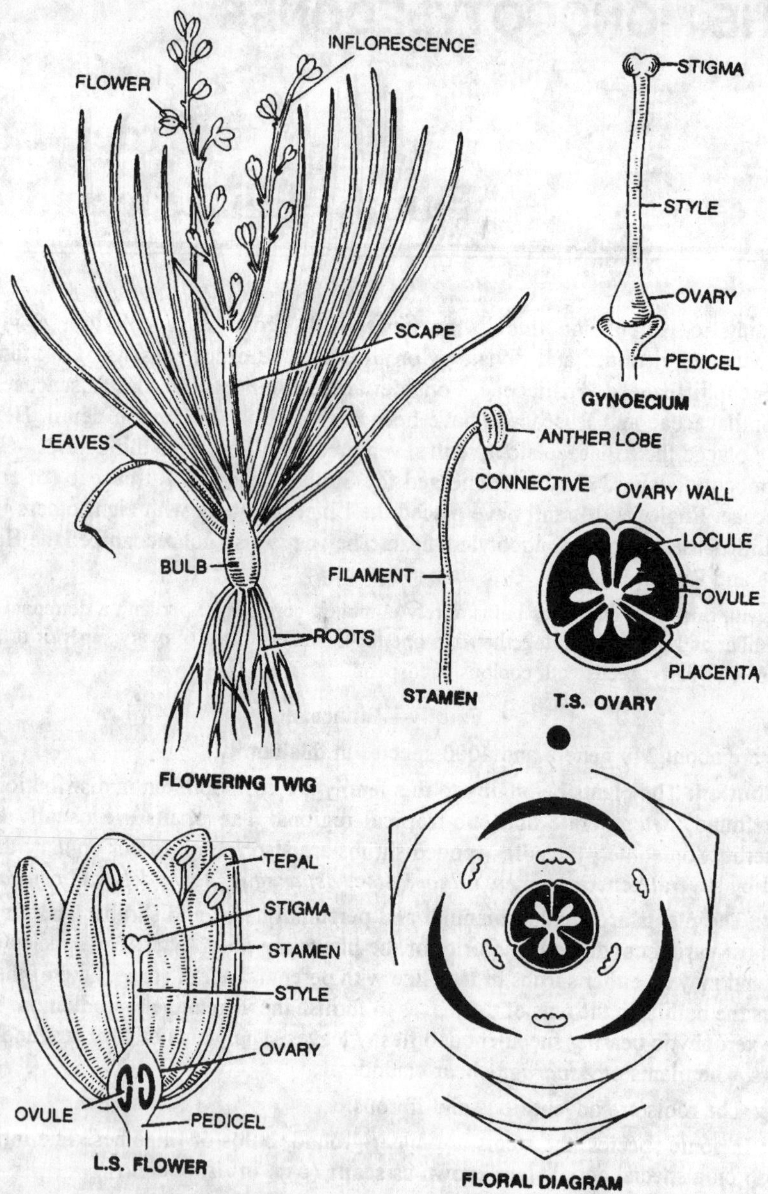

Fig. 24.1. Liliaceae. *Asphodelus tenuifolius* Cav. Vema.–Piazl.

Flower: The flowers are pedicellate, bracteate (sometimes ebracteate), hermaphrodite, complete, regular, hypogynous and usually trimerous. In *Maianthemum bifolium* (British species) the flowers are tetramerous and in *Paris polyphylla* pentamerous.

Perianth: Six perianth leaves arranged in two whorls each whorl consisting of three segments. The segments may be free or united, valvate or imbricate, petaloid or sepaloid and regular. In *Haworthia*, a native of South Africa, the perianth is bilipped.

Androecium: It consists of six stamens arranged in two whorls; the stamens are opposite to the perianth leaves; the stamens may be free or adnate to the perianth leaves. The filaments may be free or connate; the anthers are dithecous, introrse and basifixed, they open usually by a longitudinal slit and rarely by a terminal pore.

Gynoecium: It is tricarpellary, syncarpous. The ovary is superior and trilocular. Placentation axile; style single; stigma trilobed; usually the ovules are numerous and arranged in two rows in each loculus.

Fruit: Fleshy, berry or a capsule. In *Asparagus* it is berry. In *Gloriosa* and *Asphodelus* they are septicidal and loculicidal capsules respectively.

Seeds: Endospermic or albuminous with straight embryo.

Pollination: It takes place through the agency of insects (*i.e.* entomophilous).

Floral Formula: ⊕ ⚥ P_{3+3}, A_{3+3}, $G_{(\underline{3})}$.

Bentham & Hooker (1862)	**Engler & Prantl (1931)**	**Hutchinson (1959)**
Phanerogams	Phanerogams	Angiospermae
Monocotyledones	Monocotyledoneae	Monocotyledones
Coronarieae	Lilliflorae	Corolliferae
Liliaceae	Liliaceae	Liliales
		Liliaceae

Description of *Asphodelus tenuifolius* in semi-technical language:

Asphodelus tenuifolius; Verna.–**Piazi**

Habit: Annual weed.

Root: Adventitious and fibrous.

Stem: Very small bulb.

Leaves: Radical, slender, erect, semiterete, fistular, glabrous and pointed.

Scape: One or more than one; one to two feet long having racemes at their terminal ends.

Inflorescence: Racemose, raceme.

Flower: Pedicellate, bracteate, small, actinomorphic, regular, hermaphrodite and hypogynous.

Perianth: Six perianth leaves in two whorls of three each; white; free or united at the base.

Androecium: Six stamens in two whorls of three each; epiphyllous; filaments broad at base and surround the ovary; anthers dithecous, basifixed or versatile, introrse, brown.

Gynoecium: Three, tricarpellary, syncarpous; ovary superior, trilocular, each loculus bearing two ovules; placentation axile; style short, stigma three-lobed.

Fruit: Small capsule.

Seeds: Albuminous with straight embryo.

Floral Formula: ⊕ ⚥ $P_{(3+3)}$, A_{3+3}, $G_{(\underline{3})}$

Economic Importance of the Family

The family is of little economic value. Some are grown as ornamentals while others possess medicinal properties. A list of few important plants is given below:

1. *Lilium candidum*; Eng.–Madonna lily.–This is a perennial herb, grown as an ornamental in the gardens. It is native of South Europe and South West Asia.

2. *Gloriosa superba*; Eng.–Glory lily; Verna.–**Kalihari**–This is an ornamental climbing herb with beautiful flowers, grown in the gardens.

3. *Asparagus adscendens*; Verna.–**Safed Musli**–This is a small herb. The root is used as a demulcent and also in diarrhoea and dysentery. It is found in Maharashtra and Madhya Pradesh.

4. *Asparagus racemosus*; Verna.–**Satavar**-This is a common herb. It is consumed as vegetable.

5. *Asparagus officinalis*; Eng.–Garden Asparagus; Verna.–**Seetmuli, Merchuba, Halyum**.–The shoots are consumed as vegetable. It is native of Europe and West Asia.

6. *Asparagus plumosus*; Eng.–Asparagus fern.–This is a beautiful climbing herb, grown as an ornamental in the gardens. It is native of South Africa.

7. *Asparagus sarmentosus*; The roots are eaten as vegetable. The roots are also used as demulcent. It is found in Northern India.

8. *Aloe barbadensis*; Syn. *A.vera:* Verna.–**Ghikanvar**–This is a succulent shrub. The leaves yield a resinous drug, which is used mainly as purgative. It is native of West Indies but grows wild in South India.

9. *Colchicum luteum*; Verna.–**Hirantutiya**–This is a perennial herb. The roots are used medicinally in the treatment of rheumatism and gout. It is found in the temperate Himalayas from Kashmir to Chamba valley and the Punjab.

10. *Chlorophytum tuberosum*; Verna.–**Kulai**–This is an undershrub. The leaves are consumed as vegetable. It is found in Madhya Pradesh, Andhra Pradesh, Tamil Nadu and Travancore.

11. *Fritillaria* sp.; Many species are grown as ornamentals in the gardens. Some possess medicinal properties.

12. *Tulipa* sp. The plant is grown as ornamental in the garden.

13. *Convallaria majalis*; They are highly ornamental plants and grown in the gardens for their beautiful flowers.

14. *Urginea indica*; Syn. *Scilla indica*; Eng.–White squill; Verna.–**Jangli piyaz**–This is a herb. The bulbs yield a raticide, and a drug, which is used as an expectorant and stimulant. It is found in the Western Himalayas, Bihar and along the Coromandal coast.

15. *Asphodelus tenuifolius*; Verna.–**Piazi**–This is a main weed of Rabi crop. The seeds are used medicinally.

Family—Smilacaceae

Hutchinson has recognised four genera (*i.e., Rhipogonum; Smilax; Pseudosmilax* and *Heterosmilax*) in this family. There are about 200 species in the genus *Smilax*.

Distribution: The members of the family are widely distributed in the temperate and tropical regions of the World.

Habit: The plants are climbing or straggling shrubs. Usually they possess tendril-like petioles and spiny, branched stems. The spines of the stem are very often found in the form of recurved hooks which help in climbing.

Root: The adventitious roots arise from the rhizome.

Leaves: The leaves are small petioled, alternate or opposite, leathery and net veined, *i.e.*, an unusual character in Monocotyledons. In many species the leaves possess two tendril-like stipules at their base.

Inflorescence: Racemose; generally the flowers are being arranged in axillary umbels. Sometimes they are found to be arranged in racemes of spikes.

Flower: The flowers are small, dioecious, unisexual and rarely hermaphrodite, trimerous and hypogynous.

Perianth: The perianth consists of six perianth leaves arranged in two whorls of three each. Perianth leaves are generally similar and petaloid. They are either free or rarely united to form a dentate tube.

Androecium: In the male flowers it consists of six stamens arranged in two whorls of three each. The filaments are either free or united. The anthers are introse and basifixed. Rudimentary ovary is not found.

Gynoecium: In female flowers it consists of three carpels which are syncarpous. The ovary is superior, trilocular with axile placentation. One to two ovules are found in each loculus. Styles separate or united. Staminodes are present.

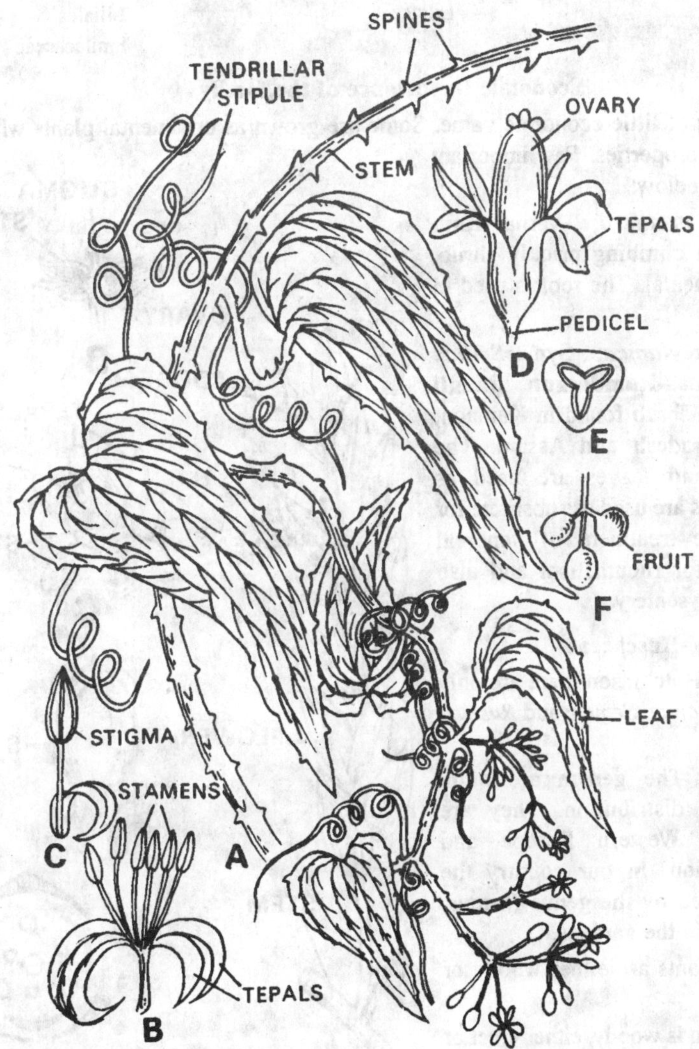

Fig. 24.2. Smilacaceae-*Smilax aspera*. A, a flowering twig with tendril-like stipules; B, male flower; C, perianth leaf and stamen; D, female flower, E, cross-section of ovary; F, fruits.

Fruit: A berry.

Seeds: The seeds contain a copious fleshy endosperm (*i.e.*, endospermic), and a small embryo.

Pollination: It takes place through the agency of insects (*i.e.*, entomophilous).

Floral Formulae:

Male: ⊕ ♂ P_{3+3}, A_{3+3}.

Female: ⊕ ♀ P_{3+3}, $G_{\underline{(3)}}$

Bentham & Hooker (1862)	Engler & Prantl (1931)	Hutchinson (1959)
Phanerogams	Phanerogams	Angiospermae
Monocotyledones	Monocotyledoneae	Monocotyledones
Coronarieae	Lilliflorae	Corolliferae
Liliaceae	Liliaceae	Liliales
		Smilacaceae

Economic Importance of the family

The family is of little economic value. Some are grown as ornamental plants while others possess medicinal properties. Few important plants are enlisted below:

1. *Smilax prolifera*; Verna.–**Ram dataun.**–This is a climbing prickly shrub, grown as an ornamental. The root is used in dysentery.

2. *Smilax zeylanica*; Syn. *Smilax macrophylla*; Verna.–**Ramdataun, Jangli aushbah**–This is a shrub found in Kumaon, Bihar, Madhya Pradesh and Assam. The young branches and leaves are used as vegetable. The roots are used as substitute for sarsaparilla in the treatment of venereal diseases, applied for rheumatism and also used in bloodless dysentery.

Family–Ruscaceae

According to Hutchinson there are only three genera (*i.e., Danae, Semele* and *Ruscus*) in this family.

Distribution: The genera are quite restricted in their distribution. They are chiefly found in Western Europe and Mediterranean region. In our country the family is represented by the genus *Ruscus*, which is cultivated in the gardens.

Habit: The plants are either woody or climbing shrubs.

Stem: The stem is woody, either erect or climbing. The stems are also modified into leaf like cladodes (*e.g.*, in *Ruscus*), which are acute and sharply pointed.

Leaves: The leaves are reduced to small scale-like structures developing in the axils of the leaf-cladodes.

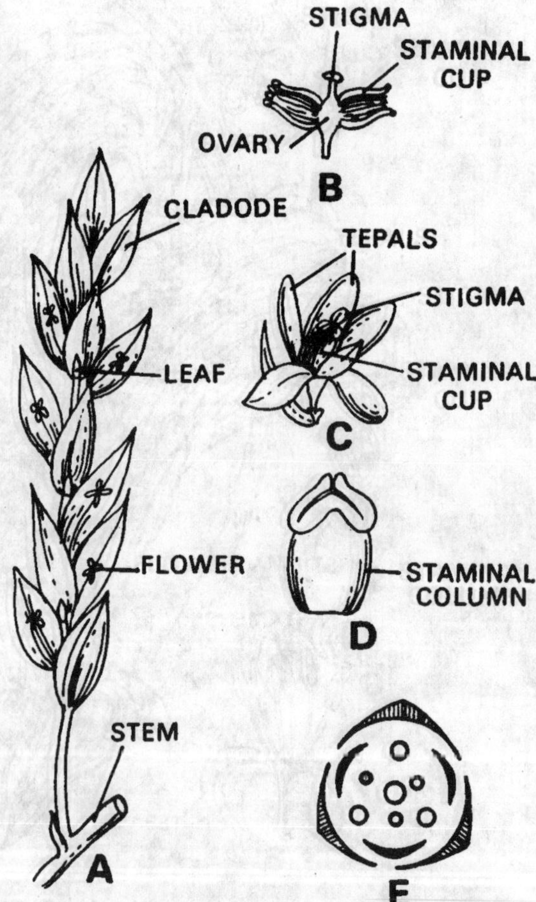

Fig. 24.3. Ruscaceae—*Ruscus* sp.: A, flowering twig; B, pistil with cut away staminal cup; C, female flower with pistil covered by fleshy staminal cup bearing aborted anthers; D, staminal column, formed by the union of three stamens; E, floral diagram of male flower.

Inflorescence: Cymose, the flowers are mostly solitary axillary. Sometimes terminal racemes are developed which remain free from cladodes.

Flowers: The flowers are small, hermaphrodite or unisexual, when unisexual the plants are dioecious.

Perianth: It consists of six perianth leaves arranged in two whorls of three each. The perianth leaves are either free or somewhat connate. When connate, there is a fleshy corona.

Androecium: In male flowers, the androecium consists of six or three stamens. The filaments form a column. The anthers are extrorse. Sometimes rudimentary ovary is found in the male flower.

Gynoecium: In female flowers, the gynoecium consists of three carpels, which are syncarpous. The ovary is superior uni or trilocular, two ovules in each loculus when three chambered. The staminodes are present.

Fruit: A roundish berry.

Seeds: The seeds are endospermic with a small cylindrical embryo.

Pollination: The pollination takes place through the agency of insects.

Floral Formulae:

Male: ⊕ ♂ P_{3+3}, A_{3+3}, or (3)

Female: ⊕ ♀ P_{3+3}, $G_{(\underline{3})}$

Bentham & Hooker (1862)	Engler & Prantl (1931)	Hutchinson (1959)
Phanerogams	Phanerogams	Angiospermae
Monocotyledones	Monocotyledoneae	Monocotyledones
Coronarieae	Lilliflorae	Corolliferae
Liliaceae	Liliaceae	Liliales
		Ruscaeae

Economic Importance of the Family

The family is of little economic value, having a few ornamental plants grown in the gardens and homes, *e.g.*, *Ruscus*.

Systematic Relationships. Engler and Prantl have placed the Liliaceae in the order–the Lilliflorae. Bentham and Hooker have placed the Liliaceae in the series—Coronarieae, Hutchinson has placed the Liliaceae together with other five families (*i.e.*, Trilliaceae, Smilacaceae, Tecophilaeaceae, Pontederiaceae and Ruscaceae) in his order—the Liliales. He included the genera, *Lilium*, *Aloe*, *Asparagus*, *Asphodelus*, *Gloriosa*, etc., in the Liliaceae; *Smilax* in the Smilacaceae and *Ruscus* in the Ruscaceae. He included the genus *Allium* in Amaryllidaceae and *Dracaena*, *Yucca*, *Agave*, *Cordyline*, etc., in the Agavaceae of his order—Agavales. Hutchinson, however has distinguished his Liliaceae and Amaryllidaceae on the basis of the inflorescence character which is probably supposed to be more stable and fundamental. Hutchinson, however, transferred the umbellate spathaceous genera to Amaryllidaceae. The genera *Agapanthus* and *Allium* have umbellate inflorescences and are closely related anatomically and, therefore, Hutchinson has placed them in the Amaryllidaceae. According to him the flowers in Liliaceae are never arranged in umbellate inflorescences. The morphologists, such as Cheadle (1942) the Maia (1941), supported the transfer of Agapantheae, Allieae and Gilliesieae to the Amaryllidaceae on the basis of their pollen studies. Lawrence stated, that from the stand-point of its phylogeny, the Liliaceae is now believed to represent basic monocotyledonous stock from which ancestral liliaceous stocks have evolved the majority of present day petaloid monocot families, such as Amaryllidaceae, Iridaceae, Palmae, Araceae and Orchidaceae. It is also thought that its ancestral stocks have been possibly derived from those of Commelinales and Butomales. There is a general belief that within the Liliaceae the groups that possess rhizomes are more primitive than the groups which possess bulbs.

25

Order–Graminales
Family–Poaceae (Gramineae)

According to Hutchinson this is the 111th order (the last one) of the phylum Angiospermae, subphylum Monocotyledones and division Glumiflorae. He has considered this order to be highest evolved among the phylum Angiospermae. This order comprises of a single family the Gramineae. The family has been described here in detail. Bentham and Hooker have also considered the Gramineae to be highest evolved among Phanerogams. They have placed the Gramineae along with four other families including Cyperaceae in their seventh series – the Glumaceae of class Monocotyledones. Engler and Prantl have placed the Gramineae along with the Cyperaceae in the fourth order– the Glumiflorae of Monocotyledoneae.

Flower subtended by 2 bracts (lemma and palea); perianth usually represented by lodicules; fruit a caryopsis, rarely a nut, berry, or utricle; stems mostly hollow between the nodes.

Characteristic Features: World-wide in distribution; annual or perennial herbs, rarely shrubs or trees, stems erect, ascending or prostrate creeping, usually branched at the base in perennials forming sterile shoots (innovations) and flowering stems (culms), in annuals only the culms present; culms cylindrical rarely flat jointed. Usually hollow in the internodes, solid at the nodes; leaves solitary at nodes, sometimes crowded at the base of the stems, alternate, two rowed, consisting of sheath, ligule and blade; sheath encircles the culm, margins of sheath free and overlapping and more or less connate, swollen at the base; ligule situated at the junction of sheath and blade, membranous or reduced to fringe of hairs, rarely absent; blades long, narrow, rarely broad, passing gradually into the sheath, rarely with a petiole like base, flat, convolute or terete, parallel venation; flowers generally hermaphrodite, rarely unisexual, small and inconspicuous, perianth represented by 2 or 3 minute hyaline or fleshy scales known as lodicules, subsessile between two bracts, *i.e.*, lemma and palea, the whole structure known as floret or a false flower; florets one to many, distichous, sessile on a short slender axis, the rachilla and bearing at the base two empty bracts, the upper and lower glumes, the florets and glumes forming a spikelet; spikelets pedicelled on open or contracted panicles or racemes, or sessile in spikes; stamens 1 to 6, rarely more, usually three, hypogynous, filaments delicate, anthers bicelled, versatile, dehiscing by longitudinal slit; ovary unilocular, with one anatropous ovule, often adnate to the adaxial side of the carpel; styles usually two, rarely 1 or 3; stigmas 2, feathery (plumose); fruit caryopsis, with a thin pericarp adnate to the seed, rarely a nut or berry; seeds endospermic, embryo small.

Family—Gramineae (Poaceae)

There are about 500 genera and 4,500 species in this family (Willis).

Distribution: The members (grasses, cereals and millets, etc.) belonging to this family are cosmopolitan in distribution. The plants are commonly distributed in temperate regions, but they are also not uncommon in tropical and sub-tropical areas. In our country the family is represented by many food crops, *e.g.*, *Triticum aestivum* (Wheat), *Oryza sativa* (Rice), *Hordeum vulgare*

(Barley), *Zea mays* (Maize), *Sorghum vulgare* (Millet **Jowar**), *Avena sativa* (Oat), *Saccharum officinarum* (Sugarcane), etc. The grasses, *e.g., Poa, Festuca, Andropogon, Cynodon, Cymbopogon,* etc., and bamboos are very common.

The great adaptability of the different species has enabled them to survive under the most varied conditions, some are aquatic while some are xerophytic and found on sea-shores and in arid regions.

Habit: Mostly the plants are annual, biennial or perennial herbs or shrubs. The largest woody species are the bamboos, which may grow to more than 100 feet in height and several inches in diameter. Most of the plants, *e.g.*, grasses, are wild while some are cultivated for valuable food.

Root: Generally adventitious, fibrous, fasciculed, stilt (*e.g., Zea mays, Sorghum vulgare*, etc.).

Stem: The stem may be erect, prostrate or even creeping. Usually fistular (hollow), rarely solid (*e.g., Zea, Saccharum*, etc.). The internodes are long and nodes are conspicuous. In many grasses the runners and suckers are also developed. Among the perennial grasses, usually the rhizomes and root stocks are formed, which help in vegetative propagation. Small tubers and corms are also found in so many species.

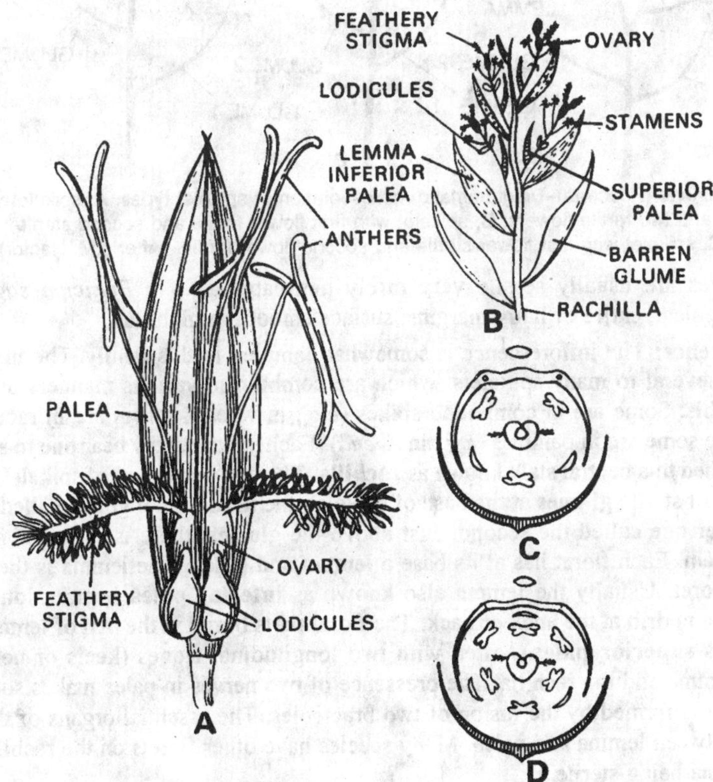

Fig. 25.1. Gramineae (Poaceae)–A, a spikelet of *Festuca* sp., after the removal of the glumes showing lodicules, ovary, anthers, stigma, etc; B, spikelet of a grass; C, floral diagrm of a grass; D, floral diagram of *Bambusa*.

In this family the aerial stem is commonly known as **'culm'** which is usually terete (cylindrical) but in few cases flattened.

In most grasses the main axis only develops lateral branches from the basal buds. These branches are **'tillers'**.

Leaves: In most of the members of this family, the foliage leaf can easily be divided into two parts, the sheath and the blade (lamina).

The leaf sheath usually covers the internode completely or partially. The inner surface of leaf sheath is usually glabrous (smooth) while the upper surface is variously grooved and either glabrous or hairy.

At the junction of the leaf sheath and leaf blade there occurs a thin membranous outgrowth, the **ligule**. In some species the ligule is represented by only few hairs.

In many species there are two lateral pointed outgrowths above the ligule are known as **auricles.** They are well developed in *Hordeum* and *Oryza.*

The leaf blade is long, narrow, acuminate and with parallel venation.

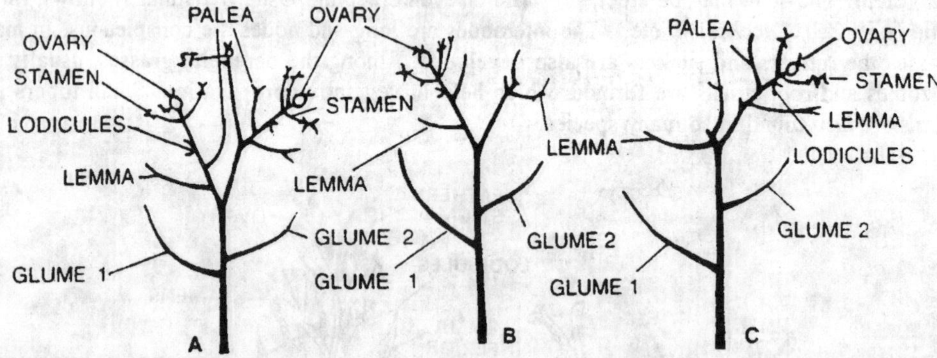

Fig. 25.2. Gramineae (Poaceae)–Diagrammatic representation of spikelet types. A, spikelet with one sterile and two fertile flowers; B, spikelet with first flower fertile and second sterile; C, spikelet with first flower sterile and second flower fertile. (After J.M. Hector)

The leaves are usually sessile very rarely petiolate (*e.g.*, in *Panicum sagittaefolium*), alternate, lanceolate, entire or hairy margins, surface smooth or glabrous.

Inflorescence: The inflorescence is somewhat complex in this family. The inflorescence is composed of several to many spikelets, which are combind in various manners on a main axis called the **rachis.** Some are in compound spikes (*e.g.*, in wheat), others are in racemes (*e.g.*, in *Festuca*), while some are in panicles (*e.g.*, in *Avena*). Each spikelet may bear one to several florets (flowers) attached to a central stalk known as **rachilla.** The usual structure of spikelet is as follows: There is a pair of sterile **glumes** at the base of spikelet, the lower, outer glume called the first, and the upper, inner one called the second. Just above the glumes, there is series of florets, partly enclosed by them. Each floret has at its base a **lemma** and **palea**. The lemma is the lower, outer bract of the floret. Usually the lemma also known as **inferior palea** bears a long **awn** as an extension of the midrib at the apex or back. The floral parts borne in the axil of lemma. The palea (also known as **superior palea**), often with two longitudinal ridges (keels or nerves), stands between the lemma and the rachilla. The pressence of two nerves in palea makes some people to conclude that it is formed by the fusion of two bracteoles. The essential organs of the flower are protected in between lemma and palea. Many species have other florets on the rachilla, the upper or lower of some being sterile.

Flower: Sessile, bracteate (the bracts are represented by lemma and palea), hermaphrodite or unisexual, zygomorphic, hypogynous, irregular.

Perianth: Sometimes absent or represented by two or rarely three (*Bambusa*) minute, membranous scale like **lodicules** (here the perianth leaves are known as lodicules).

Androecium: Usually three stamens, sometimes six, *e.g.*, in *Oryza, Bambusa*, etc., and sometimes reduced to two or one, the stamens are with free, long slender filaments and versatile anthers. Anthers two-celled, dehisce longitudinally.

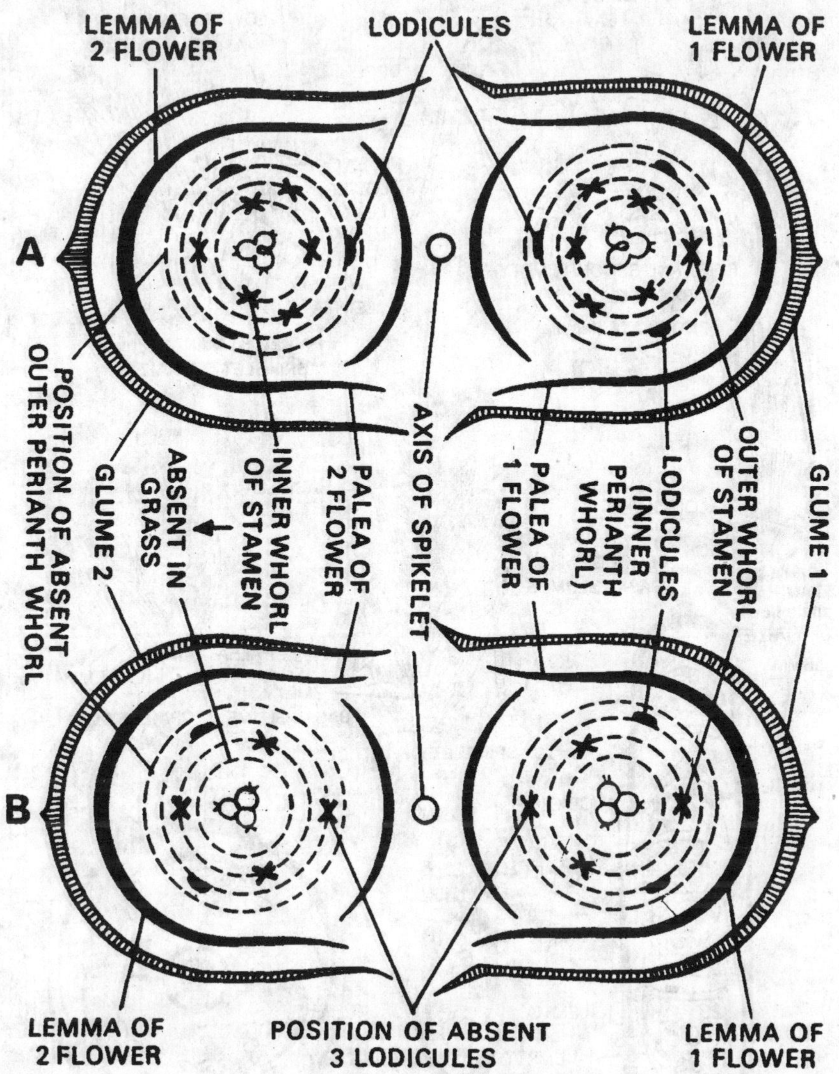

Fig. 25.3. Gramineae (Poaceae)–Diagrammatic representation of the Gramineous spikelets; A, the Bamboo type: (3 lodicules; 2 whorls of 6 stamens; 6 fused carpels each with two styles); B, the grass type: (2 lodicules; 1 whorl of 3 stamens inner whorl absent); 3 fused carpels each with two styles, front style absent. (After J.M. Hector)

Gynoecium: One carpel (monocarpellary) (though the pistil is tricarpellary but only one carpel is functional, syncarpous); unilocular ovary, superior, single ovuled (anatropous ovule); style short, usually two, stigma usually two arise from the carpellary wall, feathery. In maize the style is long and silky; placentation basal.

Fruit: Usually caryopsis (pericarp completely united with the seed coat), rarely a nut or berry (*e.g.*, in *Bambusa*). In *Dendrocalamus* it is a nut.

Seeds: Endospermic (albuminous).

Pollination: Usually anemophilous (wind pollination), cross pollination, may also take place.

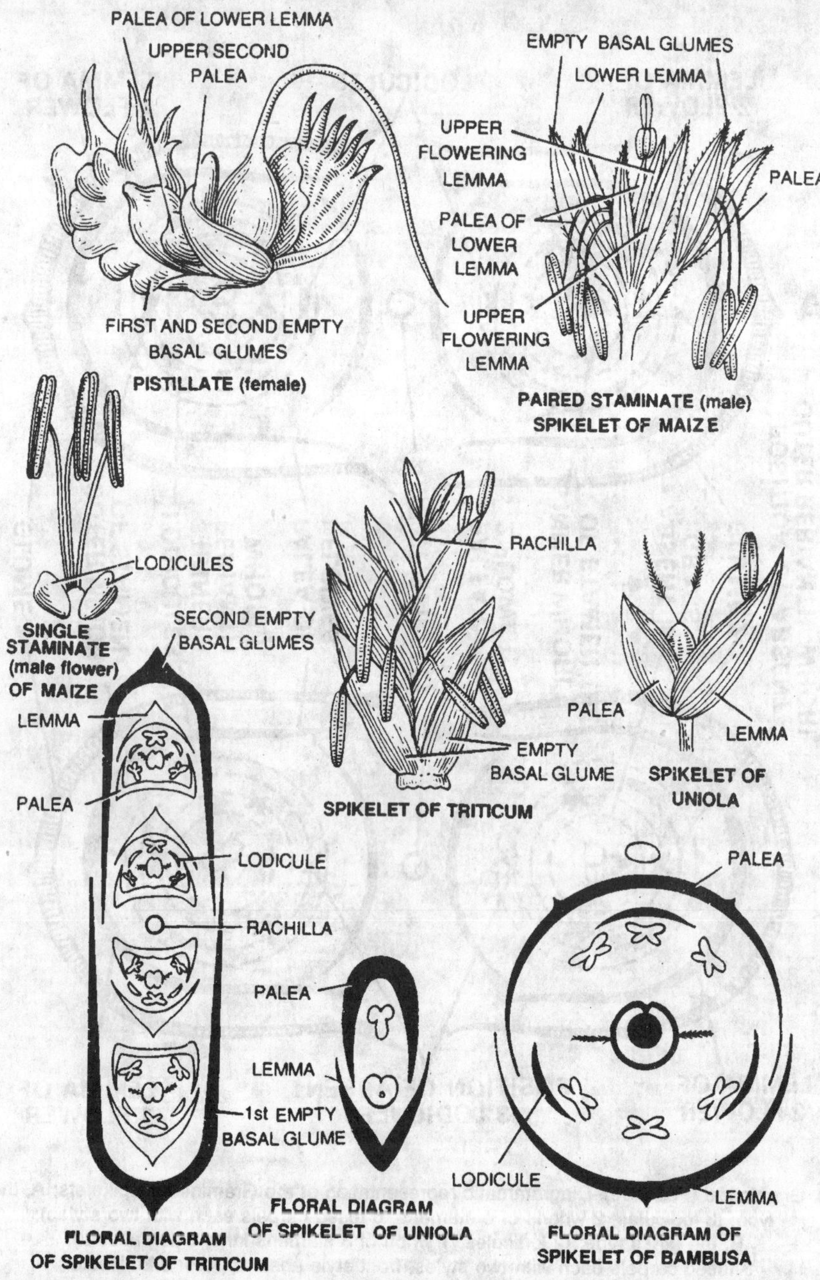

Fig. 25.4. Gramineae (Poaceae). *Zea mays; Triticum; Uniola; Bambusa*, etc.

Floral formula: ·|· ⚥, ♂ or ♀, Po or 2 lodicules, A3 or 6, G$(\underline{1})$.

Bentham & Hooker (1862)	**Engler & Prantl (1931)**	**Hutchinson (1959)**
Phanerogams	Phanerogams	Angiospermae
Monocotyledones	Monocotyledoneae	Dicotyledones
Glumaceae	Glumiflorae	Glumiflorae
Gramineae	Gramineae	Graminales
		Gramineae/Poaceae

Triticum aestivum (Verna.–**Gehun**):

Habit: The wheat plant is an annual cereal. In India all wheats are winter grown, except in Nilgiri and Palini hills where two crops (winter and summer) are taken.

Root: Adventitious, fibrous.

Stem: The stem is 3 to 4 feet in height or more. The stem (culm) is erect, cylindrical sometimes furrowed, and either glabrous or scabrous. The nodes are 5 to 7 in number, the average number is six. Each node is swollen and solid. The internodes are generally hollow. The lower and upper internodes are covered by the leaf-sheaths. Branching by tillers.

Leaves: The normal leaf of wheat is divisible into two parts, the leaf-sheath and the leaf **blade (lamina)**. The leaf also possesses two accessory organs, the ligule and the auricle.

The sheath is inserted on the node and envelops the stem. The leaf sheath is usually thicker than the leaf blade with thin transparent margins. The surface is either glabrous or hairy.

The leaf blade is long, narrow, lanceolate, acuminate with parallel venation, sometimes hair also present along the veins.

At the junction of leaf blade and leaf sheath thin membranous **ligule** encircles the stem. It is colourless and hairy.

Two claw-like appendages near the ligule region are prominent, these are known as **auricles.** They are usually hairy and pale green.

Inflorescence: Spike of spikelets. This is a compound spike bearing two rows of lateral spikelets on its axis and a single terminal spikelet. There are many short internodes on the main axis. Each internode is narrow at the base and broader at the apex. One side of internode is convex while the other side is concave or flattened. The position of concave side alternates with convex side and the whole axis has a *zig zag* appearance. The spikelets are sessile and arranged alternately, the spikelet is inserted on the apex of the internode.

Spikelets: The solitary and sessile flowers (florets) are arranged on a short, jointed axis, the rachilla. The flowers are alternately placed on the spikelet. At the base of each spikelet, two glumes occur. They appear to be opposite of each other. Actually one glume overlaps the other. The normal glume is somewhat boat shaped, with a thick main nerve, dividing it into two unequal halves. The base of the glume is rounded while sometimes the nerve extends in an awn 2 to 5 cms. long. The surface is either glabrous or hairy. In each spikelet the number of flowers varies. Usually two grains, sometimes three, and very rarely four mature in a single spikelet.

Flower: Sessile, bracteate two bracts -a lemma (inferior palea) and a palea (superior palea). They are situated opposite to each other. The **lemma** is somewhat boat-shaped with many nerves. The upper margin is notched and ends in a pointed awn. The colour of lemma is greenish white, sometimes pink. The **palea** is a thin membranous bract just opposite the lemma. It is slipper-shaped with two prominent nerves. The flower is small, zygomorphic, hermaphrodite, hypogynuous, incomplete, irregular and not showy.

Perianth: The perianth is represented by two thin membranous structure, the lodicules. They are colourless, narrow and scale-like, and sometimes hairy.

Androecium: Three stamens. The filaments are long, slender and free; the anthers are bi-celled and versatile.

Gynoecium: Single median carpel (monocarpellary). The ovary is superior, unilocular, hairy and triangular. From the tip of the ovary two styles arise, stigma feathery. Single ovule, basal placentation.

Fruit: A caryopsis, the seed coat firmly united to the ovary wall.

Seed: Albuminous.

Generic characters: Ligule and auricles present, inflorescence spike of spikelets, two rows of lateral spikelets on the axis and a single terminal spikelet, lemma boat shaped, palea slipper shaped.

Floral Formula: ·|· ⚥ P 2 (lodicules), A 3, G $\underline{(1)}$.

Hordeum vulgare; Verna.–**Jau.**

Habit: Annual herb, annual winter cereal crop (in India).

Root: Adventitious, fibrous, branched. The whole root system may be divided into two types of roots, the seminal roots and the adventitious roots. The seminal roots are branched throughout their length and penetrate to a depth of 5 to 6 feet. The true adventitious roots arise from nodes near ground level. They are unbranched and comparatively thicker. They also penetrate to a depth of 3 to 5 feet. Their colour is white and covered throughout their length by root hairs.

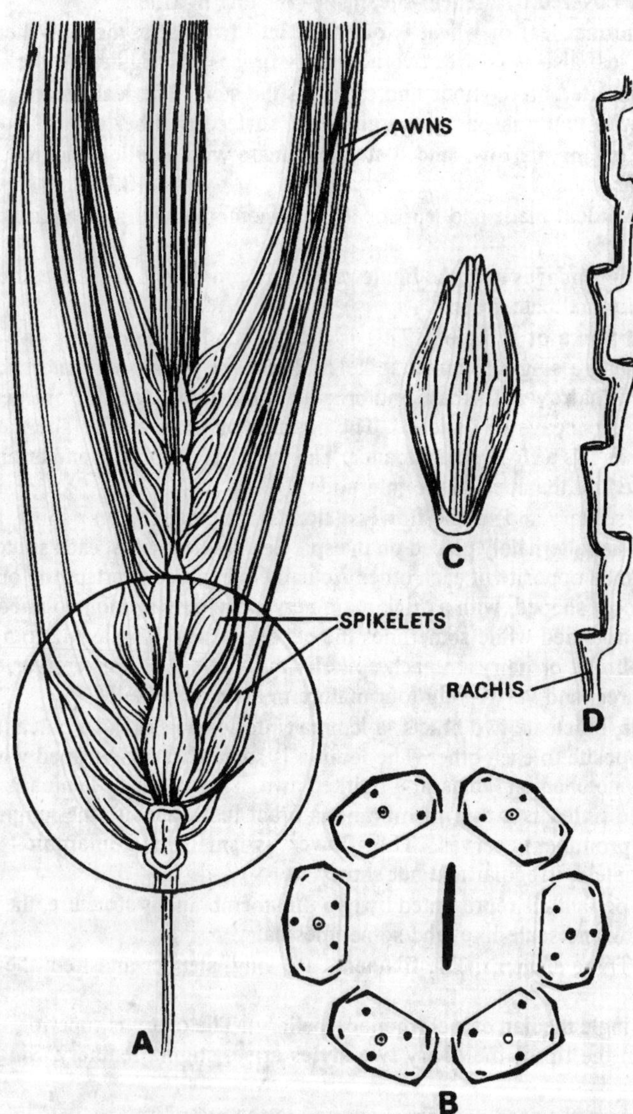

Fig. 25.5. Gramineae (Poaceae)–*Hordeum vulgare*–A, the ear; B, floral diagram; C, husked grain; D, a part of barren rachis.

Stem: Erect, cylindrical, fistular, solid at nodes, 5 to 8 conspicuous nodes. Branching by tillers (tillers develop fewer than wheat). Usually glabrous and covered by leaf sheaths.

Leaves: The leaf can easily be divided into two parts, the leaf sheath and the leaf blade. The leaf sheath covers the stem and inserted at the node, glabrous and parallel veined. The leaf blade is long, linear lanceolate, acuminate, somewhat hairy, margins also hairy, multicostate parallel venation. At the junction of the leaf sheath and the leaf blade a prominent ligule and well developed auricles are present.

Inflorescence: Spike of spikelets. The inflorescence is a cylindrical spike carrying an alternating series of spikelets. The main rachis is compressed and truncate. The rachis consists of a series of straight joints whose ends are more or less flat and upon these flat cushions the spikelets are inserted. The rachis at the base of the ear is surrounded by a ridge of tissue which has been termed the '**collar**'.

Spikelet: There are two long narrow bracts, almost parallel to each other at the base of the spikelet. These bracts may be considered as glumes. Behind these glumes the solitary flower is inserted, enveloped in lemma (inferior palea) and palea (superior palea). The lemma is broad, rounded on the back, 5-nerved and with a barbed awn. The innermost bract known as palea (superior palea) is less obvious, narrow and mostly covered by enveloping lemma. It is awnless and two ridged.

Flower: Sessile, zygomorphic, irregular, hermaphrodite, incomplete, hypogynous.

Perianth: The perianth is represented by two small membranous lodicules, arranged opposite the palea.

Androecium: Three stamens; filaments long, slender, free; anthers 2-celled, versatile.

Gynoecium: Carpel one (monocarpellary); ovary superior, unilocular, single basal ovule, more or less ovate; two feathery stigmas.

Fruit: A caryopsis with pointed ends, usually invested by lemma and palea.

Seed: Albuminous.

Generic characters: Prominent ligules and well developed auricles present; inflorescence spike of spikelets, cylindrical spike carrying alternate series of spikelets; the 'collar' present at the base of rachis, each spikelet represented by a single flower, two long narrow bracts (glumes) with each spikelet; caryopsis usually invested by lemma and palea.

Floral Formula: ·|· ⚥ , P2 (lodicules), A3, $\underline{G(1)}$.

Oryza sativa **(Rice–Chawal)**

Habit: Annual herb, cultivated as annual cereal in both tropical and temperate regions. Also sown upto 6,000 feet in Himalayan region (but in extremely humid conditions). This cereal crop is always sown under water and in the rain faded areas. In India, Bengal and Bihar are under rice cultivation. Usually the rice is cultivated in lowland areas.

Root: The whole root system can be divided into two parts, the seminal or primary roots and the adventitious or secondary roots.

The adventitious roots arise from the lower nodes of the plant and its tillers. They are fibrous and branched.

The rice grown in the deep waters, possess adventitious roots at the nodes above ground level.

Stem: Erect, cylindrical, hollow except at nodes, 6 to 8 mm. thick, nodes and internodes conspicuous, the number of internodes may vary from 10 to 20, branching by tillers.

Leaf: They always emerge from the nodes, so the number of leaves on an axis is equal to the number of nodes. A normal leaf possesses the leaf sheath, ligule, auricles and the leaf blade (lamina). The leaf sheath always encircles the internode from the node (pulvinus) upwards. The leaf sheath is split at the base, finely ribbed, glabrous. The ligule is well developed, membranous, colourless or pink coloured. The auricles are well developed, sickle shaped 3 mm. in length, present at the junction of the sheath and the blade (lamina). The leaf blade is long, narrow, lanceolate, acuminate, usually pubescent, parallel venation with a distinct midrib. The uppermost leaf, the flag is always shorter and broader than the lower leaves.

Inflorescence: A terminal panicle. The inflorescence borne on the peduncle or on the uppermost internode. Usually the inflorescence is partly covered by leaf sheath, in extreme cases it is fully covered. The main axis or the rachis of the panicle is slightly angular, more or less glabrous, somewhat hairy at the nodes. It may be erect or curved. One to many branches are given out from each node of the rachis. The panicle may be compact or loose. On the end of each branch the spikelets are borne.

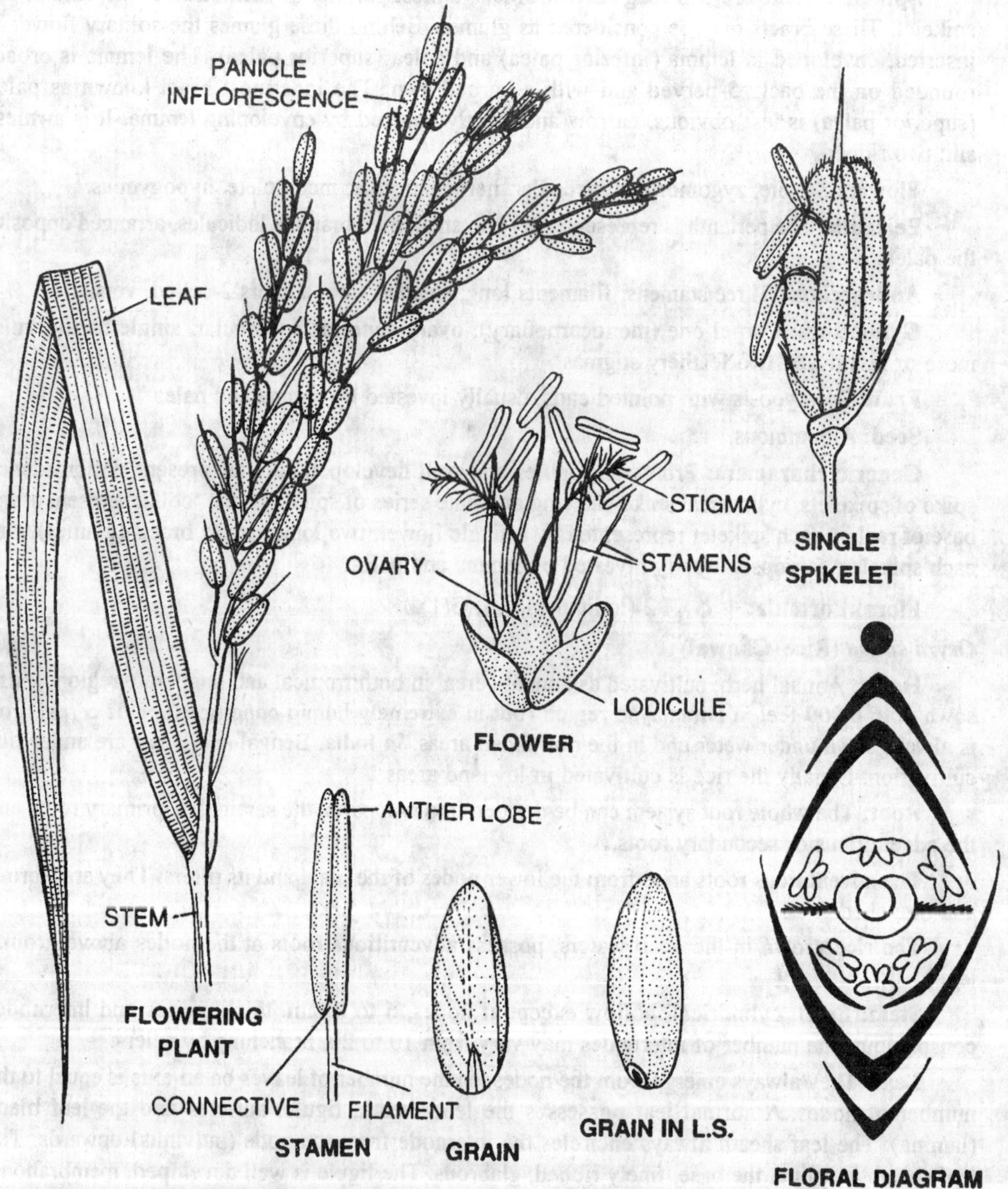

Fig. 25.6. Gramineae (Poaceae)–*Oryza sativa* Linn. Verna. Dhan; Eng. Rice.

Spikelets: The spikelets are borne either singly on the ends of the branches or in clusters of 2 to 7. The number of spikelets on a panicle varies from 50 to 500. Each spikelet is short pedicelled, generally single flowered, very rarely two flowerd. The apex of the pedicel is enlarged and oblique and is termed as **facet.** On the facet two glumes occur one outer and the other inner. The glumes are small, lanceolate, shiny, coriaceous approximately equal in size, coloured or colourless (white, yellow or pink).

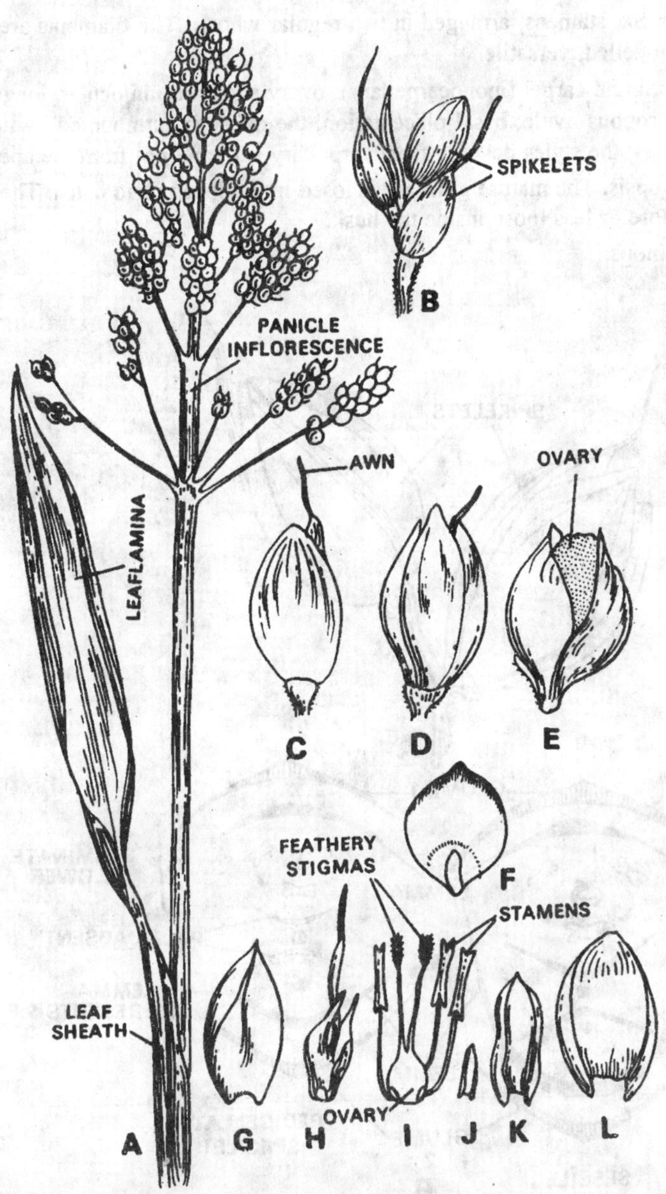

Fig. 25.7. Gramineae (Poaceae)–*Sorghum vulgare* (Jowar): A, a panicle; B, spikelets; C, dorsal view of a spikelet; D, ventral view; E, opened caryopsis; F, a caryopsis; G, glume; H, lemma; I, ovary and stamens; J, pale; K, pale of fertile spikelet; L, glume.

Flower: Bracteate, a large conspicuous bract, the **lemma.** The lemma is boat shaped, 5-nerved and hairy on the nerves. The apex is pointed and usually prolonged into an awn. On each side of the apex or awn two tooth like projections. Lemma is usually pigmented (coloured). The **palea** is situated just opposite the lemma, it is also boat shaped and 3-nerved. The end point of the palea develops into **apiculus.** The apiculus is usually pigmented (purple coloured).

The flower is zygomorphic, hypogynous, irregular, incomplete and sessile.

Perianth: The perianth is represented by two broad, thick and fleshy lodicules.

Androecium: Six stamens, arranged in two regular whorls. The filaments are long, slender and free; anthers bi-celled, versatile.

Gynoecium: Single carpel (monocarpellary); ovary superior, unilocular, longer than broad, single ovuled, anatropous ovule; basal placentation; the ovary is surmounted by two long styles with feathery stigmas; the styles develop from carpellary wall and not from the apex.

Fruit: A caryopsis. The mature grain is enclosed in the lemma and palea. The grain may be firmly placed or more or less loose inside the husk.

Seed: Albuminous.

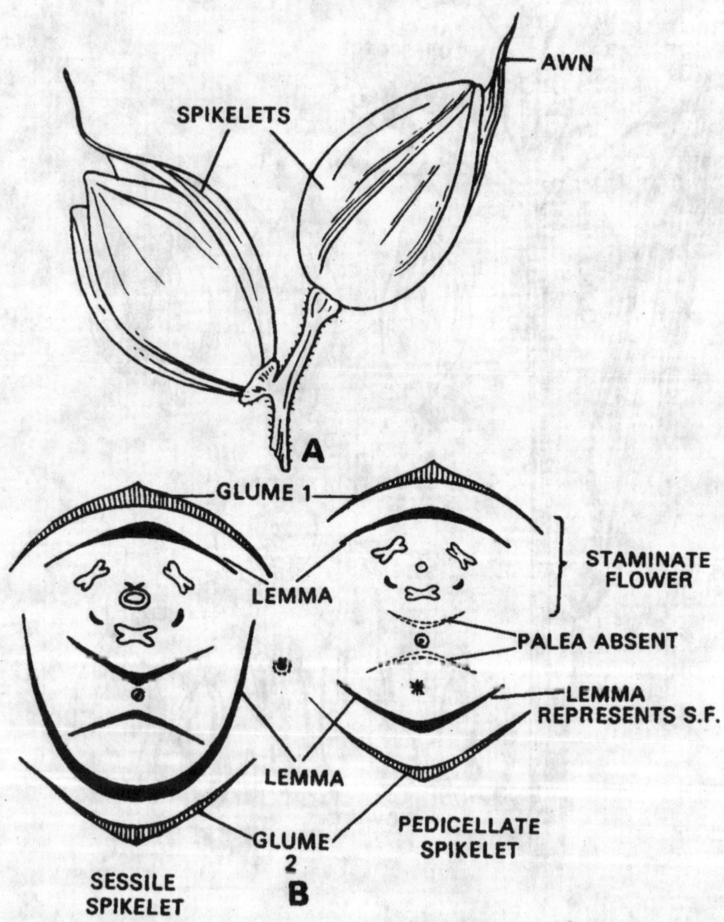

Fig. 25.8. Gramineae (Poaceae)–*Sorghum vulgare;* Verna.–**Jowar**–A, paired spikelets, one sessile another pedicelled; B, floral diagrams of respective spikelets.

Generic characters: The ligule is well developed, membranous and sometimes pink coloured, the auricles are well developed and sickle shaped; inflorescence a terminal pancile, rachis of panicle is angular, the flat apex of the pedicel is termed as `facet'. Six stamens. Caryopsis enclosed in lemma and palea (husk).

Floral formula: ·|· ⚥ P 2 (lodicules), A3+3, G1.

Sorghum vulgare; Verna.–**Jowar.**

Habit: All grain producing Sorghums are annual. They are herbaceous and sown in rainy season.

Root: The root system may be divided into two parts, the seminal roots (primary roots) and the adventitious roots (secondary roots). The seminal roots develop first and penetrate vertically downwards in the soil. The adventitious roots develop from the basal nodes of the stem and extend both vertically downwards and horizontally. The adventitious roots are fibrous and very well developed. They are profusely branched. Stilt roots are also present usually above the ground level.

Stem: Erect, solid, several feet high, slightly furrowed on alternate sides. The nodes and internodes are short at the base and longer above. The terminal internode which bears head is longest of all, those of middle region are equal in length. Branching by **suckers**, each sucker may develop into a head. At each node there is bud except the terminal node.

Leaves: Simple, alternate, sessile; long, lanceolate; acuminate, surface glabrous, flat, distinct sheathing base (leaf sheath), the margins of leaf sheath are membranous and overlapping, amplexicaul, multicostate parallel venation with a very well developed mid-rib. Ligule, the ligule is short, membranous and fringed, present at the junction of the leaf sheath and the leaf blade. The surface of leaf blade is glabrous.

Inflorescence: Usually compact panicle, sometimes loose and spreading panicle. The main axis of rachis narrow towards the apex and it is not identified in the terminal and sub-terminal branches. The rachis is somewhat angular and furrowed, hairy at the nodes. The internodes may be hairy or may not. At the nodes, lateral branches may develop into whorls or in spirals. Each lateral branch may rebranch and on the ultimate branchlets one to many paired spikelets are produced. The lateral branches are somewhat angular and hairy at the base and at the axils, sometimes the heads (panicles) become inverted.

Spikelets: The spikelets usually occur in pairs on the lateral branches of panicles, but sometimes they occur in three towards the tip. When the spikelets are in pairs, one is sessile and perfect while the other one is pedicelled and staminate (male). When they are in threes, one is sessile and perfect while the rest two are pedicellate and staminate (male).

Fertile (perfect) spikelet: The fertile or perfect spikelets are comparatively larger than the staminate spikelets. They may be ovate, oval, elliptical or obovate. Each fertile spikelet possesses two glumes approximately equal in size. Both glumes are nerved. The nerves are more conspicuous on the outer glume than those of the inner one. The outer glume partially encloses the inner one which is narrow and pointed towards the apex. They are thick, hairy and with membranous margins.

Flower: On each spikelet there are two florets (flowers) found within the two glumes. The lower floret is sterile while the upper one is perfect. There is a lemma (inferior palea) of lower floret which also partially envelops the perfect floret. The lemma is usually broad, hairy and membranous. The second and perfect floret has a lemma which is membranous and hairy and two cleft at the apex. In the protruding cleft an awn arises. The palea is usually small and thin.

The flower (perfect) is sessile, bracteate, zygomorphic, irregular and hypogynous.

Perianth: The perianth is represented by two short and broad, truncate and fleshy, hairy lodicules.

Androecium: Three stamens; the filaments are free; anthers bi-celled, versatile.

Gynoecium: Single carpel (monocarpellary); ovary superior, unilocular, with single basal anatropous ovule; basal placentation; two long styles ending in short bushy stigma.

Staminate and Pedicellate Spikelet: The spikelet is narrower than the sessile one; the pedicel is hairy on the margins; two glumes enclose the two florets (flowers). The lower floret is represented by the lemma only; the upper floret is staminate, with short awned lemma; palea absent; two lodicules; three stamens; pistil absent.

Fruit: A caryopsis, vary in shape and colour usually white or pale.

Seed: Albuminous.

Generic characters: Fringed membranous ligule, no auricles; inflorescence loose or compact panicle, rachis angular and furrowed and hairy at nodes, spikelets in pairs on lateral branches, one spikelet is sessile and perfect while the other is pedicelled and staminate, two florets in each spikelet, one fertile other sterile, usually a white oval caryopsis.

Floral Formula: ·|· ⚥ P2 (lodicules), A3, G(1).

Zea mays; Eng.–Maize; Verna.–**Makka.**

Habit: An annual cereal, usually cultivated for its foodgrains.

Root: The roots are fibrous and may be divided into three types, (1) the seminal roots, (2) the adventitious roots, and (3) the prop or stilt roots. The seminal roots penetrate deep in the

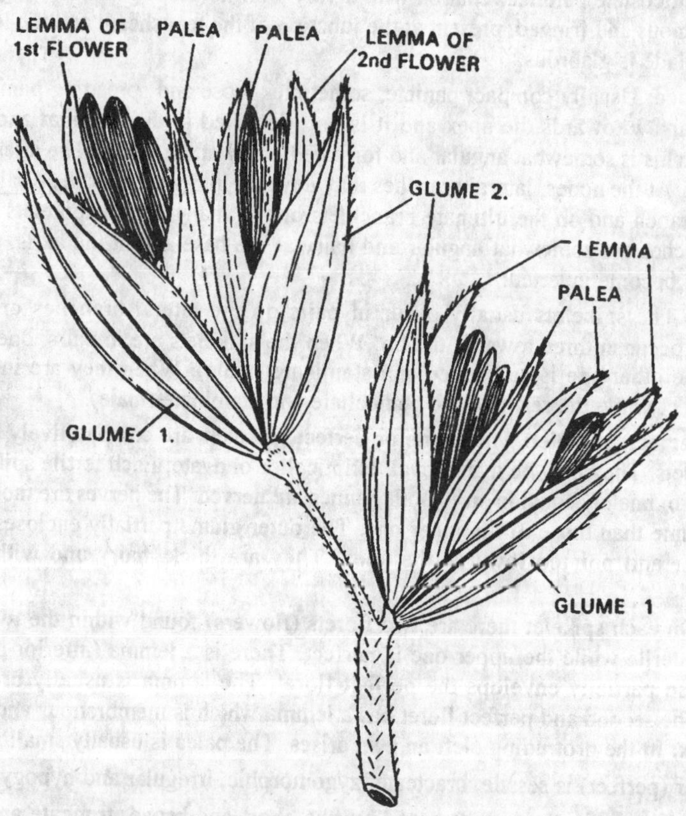

Fig. 25.9. Gramineae (Poaceae)–*Zea mays* (Maize: Verna.–**Makka**) paired male spikelets, one sessile other pedicelled.

soil and are branched profusely. The adventitious roots develop from the nodes near ground level, they run horizontally in the ground and are also profusely branched. The colour is either brown or white. The stilt or prop roots arise from the nodes sufficiently above the ground level. The nodes may give out the whorls of prop roots. These roots are thicker and stronger than the normal roots, and somewhat pigmented.

Stem: Erect, several feet long, well defined nodes and internodes, the stem is thickened at the base of nodes. Both nodes and internodes are solid. The stem is furrowed on the other side of the leaf, position of furrow alternates at each node. The last node ends in the **tassel.** Branching sometimes by tillers. A bud is formed in the axil of each leaf, the buds near ground level develop into tillers while at higher levels they form the **'pistillate branches'.**

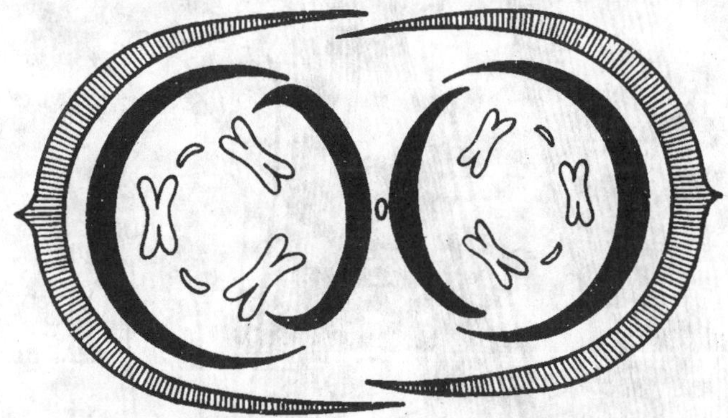

Fig. 25.10. Gramineae (Poaceae)–Floraldiagram–**Maize**(male)—one spikelet.

Leaves: Sessile, simple alternate. The leaf may be divided into four parts, the leaf sheath, the leaf blade, the ligule and the auricles. The leaf sheath encloses the internode (amplexicaul). It is entire below and split above. The surface may be glabrous or hairy. The margins glabrous below and hairy above. The leaf blade is thin, flat, ribbon like, acuminate, lanceolate, parallel venation with a thick and prominent mid-rib. At the junction of the leaf sheath and the leaf blade is membranous, colourless, quarter of an inch long **ligule** is present. At the junction there are two lateral outgrgowths, the **auricles.**

Inflorescence: Maize is monoecious, diclinous, *i.e.*, the stamens and pistils are borne on separate inflorescences, but on the same individual plant. The male flowers are carried on terminal panicles known as "**tassel**", while the female flowers are on the "**ear**".

Staminate or Male Inflorescence: The male or staminate inflorescence is a panicle at the top of the plant; the panicle consists of a central axis and lateral branches of it; the central axis bears several rows (4 to 11) of paired spikelets, while the lateral branches have only two rows of paired spikelets, one sessile and the other pedicelled.

Pistillate or Female Inflorescence: The bud found in the axil of foliage leaf develops in the pistillate inflorescence. Sometimes two and very rarely three buds may develop. In structure the pistillate inflorescence is somewhat similar to the main shoot. All its internodes became very short. The uppermost internode is shortest while the lower ones are progressively bigger in comparison. The leaf at the basal node is the **prophyllum.** The outer leaves constitute the **'husk'** and protect the 'ear'. These protective leaves may be termed as bracts or spathes. Each spathe or bract is a modified leaf possessing leaf sheath, blade and ligule. The ear is a modified spike with thickened central axis, the 'cob'. The 'cob' carries a series of a paired spikelets in longitudinal rows. Each row of paired spikelets develops two grains, and so the rows of grains are always in even number.

Staminate or Male Spikelets: The paired male (staminate) spikelets are quite similar to each other except that one is sessile and the other is pedicelled. At the base of each spikelet there is a pair of glumes. The glumes are somewhat longer than the spikelets and nearabout equal in size. They are glabrous but hairy at magins, nerved (7 to 12 nerves). The outer or the lower glume

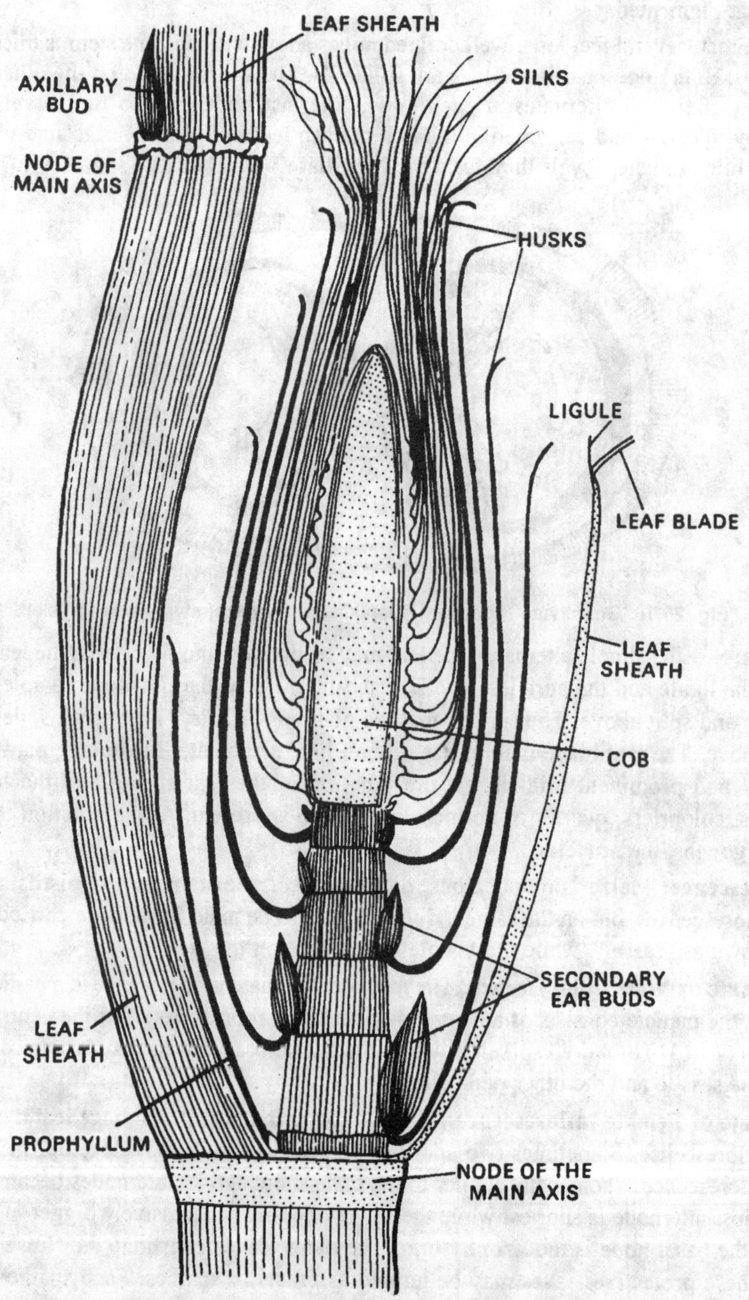

Fig. 25.11. Gramineae (Poaceae)–*Zea mays;* (Maize; Verna–**Makka**) L.S. ear bearing branch; A, leaf sheath; B, axillary bud; C, silks; D, node of main axis; E, husks; F, ligule; G, leaf blade; H, leaf sheath; I, cob; J, secondary ear buds; K, node of the main axis; L, prophyllum; M, leaf sheath.

overlaps the inner or upper one. The apex is acuminate. Within each two glumes there are two staminate florets (flowers) borne on a short common rachilla.

Staminate or Male Flower: Bracteate, the outer bract is the lemma, while the inner one is the palea. The lemma is oval, concave 3-nerved, glabrous, hairy at margins with somewhat rounded apex. The palea is inserted on the opposite side of the lemma. It is thin, membranous, 2-nerved flat and with incurved margins. In the upper flower the palea is larger than lemma while in the lower it is vice versa. The flower is sessile, zygomorphic, unisexual, male.

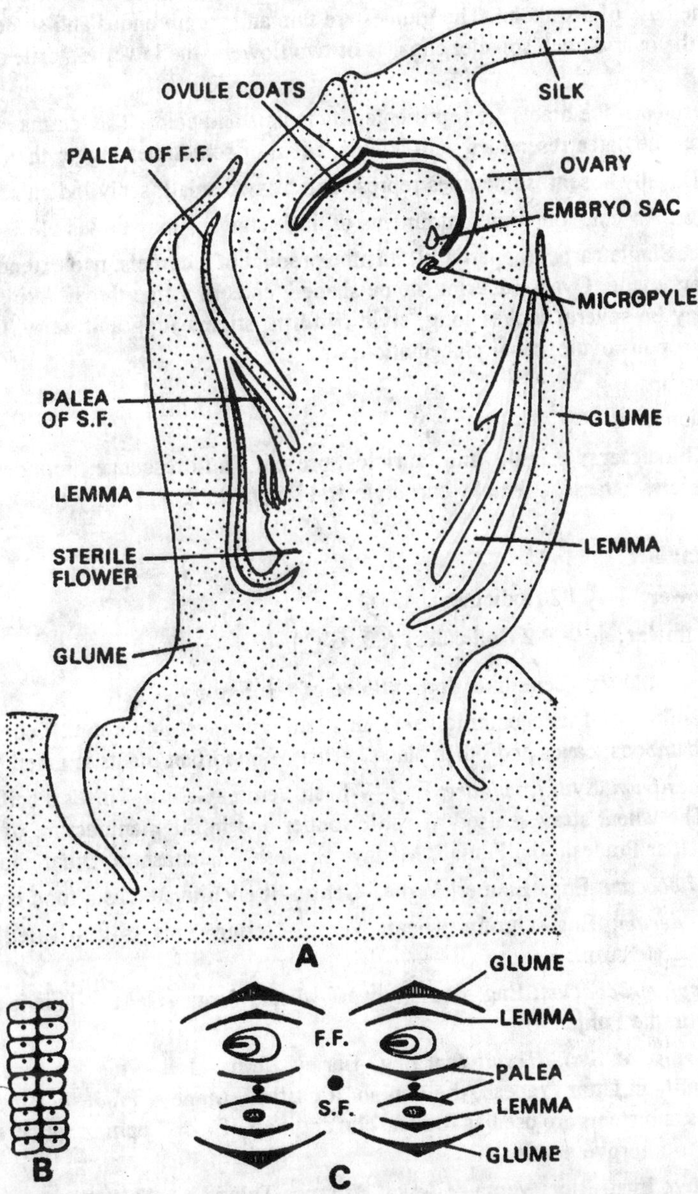

Fig. 25.12. Gramineae (Poaceae)–*Zea mays;* (Maize; Verna.–**Makka**): A, a female spikelet with one sterile and other fertile flower; B, paired female spikelets; C, floral diagrams of a female spikelet.

Perianth: The perianth is represented by two fleshy lodicules. They are inserted opposite the lemma and alternate with the stamens.

Androecium: Three stamens (one opposite the palea, remaining two opposite the lemma); the filaments free and short, anthers bilobed, versatile, purple in colour, dehiscence, by a slit at the terminal end.

Gynoecium: Absent, represented by a rudiment.

Pistillate or Female Spikelets: The pistillate paired spikelets are arranged completely in rows on the axis or cob. Both the spikelets are sessile and quite similar to each other. There are two glumes at the base of a spikelet. The glumes are thin and membranous and somewhat thicker near the base of the ovary. Each spikelet consists of two flowers, the lower is sterile and the upper one is fertile.

Flower: Bracteate, the bracts are represented by lemma and palea. The lemma is short, broad and membranous. The palea resembles with lemma, equal or sometimes bigger than lemma.

Perianth: Usually absent, sometimes represented by two small, scaly lodicules.

Androecium: Absent, sometimes rudiments of three stamens may be visible.

Gynoecium: Single carpel (apparently) it is the product of 3 carpels, two extend to form silk and the third bears ovule. Ovary superior, dome shaped, ends in a fine thread like structure, the silk. The silk may be several inches long, style filiform, stigma long and hairy. Ovary single ovuled, campylotropous ovule, basal placentation.

Fruit: A caryopsis.

Grain: Albuminous.

Generic Characters: Ligule and auricles present; inflorescence, monoecious, male inflorescence panicles– 'tassel', female–ear; style and stigma long and hairy (silk); ovary dome shaped.

Floral Formulae:

Male flower: ·|· ♂ P2 (lodicules), A3, G 0

Female flower: ·|· ♀ P 2 (lodicules) A 0, G $\underline{(1)}$.

Economic Importance of the Family

From the standpoint of its economic value, this family is most important. It contains all the cereals, grasses, bamboos, canes and other plants. A list of important plants is given below:

1. *Triticum aestivum*; Syn. *T. vulgare*; Eng.–Wheat; Verna.–**Gehun**–This is a herb, cultivated as a food crop. The wheat straw is used as cattle fodder and in the manufacture of paper. It is mainly grown in Uttar Pradesh, the Punjab, Madhya Pradesh, Maharashtra, Bihar and Rajasthan.

2. *Triticum dicoccum*; Eng.–Emmer; Verna.–**Gehun**–It is cultivated as a food crop.

3. *Triticum durum*; Eng.–Durum wheat; Verna.–**Gehun**–This is cultivated mainly in Maharashtra as a cereal crop.

4. *Triticum sphaerococcum*; Eng.–Indian dwarf wheat; Verna.–**Gehun**–This is grown as a grain crop mainly in the Punjab.

5. *Hordeum vulgare*; Syn.–*H. sativum*; Eng.–Barley; Verna.–**Jau.**–This is a herb, cultivated as a food crop mainly in Uttar Pradesh, the Punjab, Rajasthan, Madhya Pradesh, Bihar and West Bengal. The leaves and stems are used as fodder. The addition of 1 or 2 ppm of gibberellic acid to barley is claimed to improve maltin.

6. *Oryza sativa*; Eng.–Rice; Verna.–**Chaval, Dhan**:– This is a herb, grown as a food crop. The rice straw is used for making strawboards, paper, mats. Rice-bran oil is used for soaps and cosmetics, and as an anticorrosion oil. It is grown all over India.

7. *Avena sativa*; Eng.–Oats; Verna.–**Jai.**–It is grown to a limited extent in North-Western Himalayas. The grains are used by poor people as food.

8. *Avena sterilis*; Eng.–Indian oat, Red oat; Verna.–**Jai.**–This is a herb used as a fooder. A hybrid has been raised between this and *A. sativa.* It is grown in Uttar Pradesh, the Punjab, Bihar, Bengal, Orissa and in the hills of Tamil Nadu.

9. *Sorghum vulgare*; Syn. *Andropogon sorghum*; *Holcus sorghum;* Eng.–Sorghum; Verna.–**Jowar.**–The grains are used as food and the stem and leaves are consumed as a cattle fodder. A spirit is distilled from the grain. It is cultivated mainly in Uttar Pradesh, Punjab, Madhya Pradesh, Andhra Pradesh, Maharashtra and Rajasthan.

10. *Sorghum sudanense*; Eng.–Sudan grass.–This is an annual grass found in Maharashtra, Assam and the Punjab. It is used as a fodder.

11. *Sorghum halepense*; Syn. *Holcus halepensis*: Eng.–Johnson grass; Verna.–**Baru Kala mucha.**–This is a perennial grass, used as fodder.

12. *Pennisetum typhoides*; Syn. *P. typhoideum*; Eng.–Pearl millet; Verna.–**Bajra.**– It is grown for its edible grains. The straw is used as a fodder. It is cultivated in Uttar Pradesh, the Punjab, Andhra Pradesh, Rajasthan and Maharashtra.

13. *Pennisetum purpureum*; Eng.–Elephant grass.–It is grown in Assam, Bengal, Bihar, Uttar Pradesh and the Punjab. It is used as a fodder.

14. *Panicum miliaceum*; Verna.–**Chin, Morha, Anu.**–The grains are edible and the straw is used as a fodder. It is grown mainly in Uttar Pradesh.

15. *Panicum miliare*; Eng.–Little millet; Verna.–**Sawa, Kutki.**–The grains are edible. The stems and leaves are used as fodder. It is cultivated in Uttar Pradesh and Madhya Pradesh.

16. *Panicum antidotale*; Eng.–Blue panicum; Verna.–**Bansi, Gunara**–This is a tall grass, used as a fodder and also for the fixation and reclamation of sand dunes. It is found in the Punjab, Uttar Pradesh and the Western Peninsula.

17. *Zea mays*; Eng.–Maize, Corn, Indian corn; Verna.–**Makai, Makka, Bhutta.**–This is a tall annual herb. It is grown as a food crop mainly in Uttar Pradesh, the Punjab, Madhya Pradesh, Bihar, Andhra Pradesh, Jammu and Kashmir. The immature cobs are largely eaten after roasting. The grains are also used in making corn starch and industrial alcohol. It is native of South America.

18. *Setaria italica*; Syn. *Panicum italicum*; *Chaetochloa italica*; Eng.–Italian millet; Verna.–**Kangni, Kangu, Kakun.**–It is chiefly cultivated in Andhra Pradesh and Tamil Nadu for fodder and food. It is used as diuretic and astringent also used externally for rheumatism.

19. *Setaria glauca*; Syn. *Panicum glaucum*; Eng.–Cat tailmillet; Verna.–**Bandra.**–This is a herb, used as a fodder. The grains are also eaten as food.

20. *Eleusine coracana*; Eng.–Finger millet, Ragi; Verna.–**Mandal, Mandua.**–This is a herb, cultivated as a food crop. This is the main food of the farmers in South India; Maharashtra and of hill-men in Northern India. This is made into cakes, porridge and sweetmeats. A beer is brewed from the grain by the hill tribes. The enzymic extract of the germinated grain can be used for unhairing hides and skins. It is chiefly cultivated in Andhra Pradesh, Tamil Nadu, Karnataka, Orissa, Bihar, Uttaranchal and Maharashtra.

21. *Saccharum officinarum*; Eng.–Sugarcane; Verna.–**Ganna, Ikh**–The stems yield the cane syrup. The by-products of sugar industry are commercially used for various purposes, *e.g.*, molasses is used in cooking, also for rum, industrial alcohol, synthetic rubber. A mixture of refuge canes and molasses called molas-cuit is used as fodder. Cane refuge is also utilized for manufacturing bagpaper, wrapping-writing and printing papers, card-board and as fuel. It is chiefly grown in Uttar Pradesh, Bihar and the Punjab.

22. *Saccharum spontaneum*; Eng.–Thatch grass; Verna.–**Kans.**–It is used as a sand binder and also for paper pulp. It is found throughout India. The stems and leaves are used for making thatches.

23. *Saccharum fuscum*-This grass is used for making cheaper grades of papers, wrapping-boards, and other cellulose products.

24. *Agropyron repens*; Syn. *Triticum repens*. The decoction of the rhizome is a demulcent and diuretic and used in the treatment of catarrhal diseases of the urino-genital tract.

25. *Aristida depressa*–The panicles are made into brooms.

26. *Aristida setacea*–The panicles are used for making brooms and brushes.

27. *Arundinaria falcata*; Syn. *Bambusa falcata*; Eng.–Himalayan bamboo–The stems are used for making baskets, arrows, fishing rods and the lining of roofs of houses. It is found in Western Himalayas and Sikkim.

28. *Arundinaria spathiflora*–This is a small bamboo growing the undergrowth of the fir, oak and deodar forests of the Western Himalayas at an elevation of 6,500 feet to 9,500 feet. The culms are used for pipe stems and baskets.

29. *Arundinaria racemosa*–It is found in Bhutan and Western Himalayas. The culms are used for making mats and roofs of houses.

30. *Arundo donax*; Eng.–Giant reed; Verna.–**Baranal, Narhal.**–This is a tall perennial grass. The stems are used in the manufacture of baskets, mats, pipes, writing and printing papers. It is chiefly found in Kashmir, Assam and the Nilgiris.

31. *Bambusa arundinacea*; Syn. *B. spinosa*; Eng.–Thorny bamboo; Verna.–**Bans.**–This is a tall woody grass found throughout our country, particularly along river valleys and in moist situations. The culms are used for making paper, thatches, etc. The young buds and grains are consumed as food.

32. *Bambusa vulgaris*; Eng.–Feathery bamboo; Verna.–**Basini bans**–The split culms are made into mats and baskets and the young buds are consumed as vegetable. It is cultivated in the hotter parts of our country.

33. *Bambusa tulda*; Eng.–Bamboo; Verna.–**Peka**–The mature culms are used for manufacturing paper, baskets and fans. The young culms are eaten as vegetable. It is found in Assam and Bengal.

34. *Bambusa baleooa*; Eng.–Plains bamboo; Verna.–**Bhaluka**–It is used for building purposes. It is cultivated in Assam, Bengal, Bihar and Uttar Pradesh.

35. *Bambusa polymorpha*- It is used for making paper, in Bengal and Assam.

36. *Cephalostachyum pergracile*.–This is a tufted bamboo and frequently found in association with teak. The culms are used for building and mat-making, and cooked rice is often carried in the internodes on a journey.

37. *Cymbopogon citratus*; Syn. *Andropogon citratus*; Eng.–West Indian lemon grass; Verna.–**Sugundh rohisha**–The leaves yield an aromatic oil which is used in perfumery, cosmetics and as a flavouring substance. It is chiefly grown in Travancore and Malabar.

38. *Cymbopogon martini*; Syn. *Andropogon martini*; Eng.–Ginger grass; Verna.–**Gandhejghas**–The leaves are the source of an aromatic oil, which is used in perfumery and cosmetics. This occurs in two varieties known as the **Motia** and **Sofia.** The essential oil, obtained from the Motia variety, is of a superior quality and is referred to as palmarosa oil, while the oil, from the Sofia variety is inferior in quality and is known as Ginger grass oil. Next to Sandalwood and lemon-grass oils, palmrosa oil is the important of essential oils produced in our country.

39. *Cymbopogon nardus*; Syn. *Andropogon nardus*; Eng.–Citronella grass; Verna.–**Ganjni**–The oil known as Ceylon type, obtained from the leaves; is used in perfumery and cosmetics. It is also used in mosquito-repellent preparations. Also used in an infusion, as a stomachic and carminative.

40. *Cymbopogon schoenanthus*; Syn. *Andropogon schoenanthus* Eng.–Camel hay; Verna.–**Rousa ghas.**–The leaves yield an essential oil which is used in perfumery.

41. *Cymbopogon winterianus*; Eng.–Citronella grass. The oil commercially known as Java type contains high proportions of geraniol, citronella and citronollol and hence used in the synthesis of organic compounds such as synthetic menthol.

42. *Cymbopogon flexuosus*; Syn. *Andropogon flexuosus*; Eng.–East Indian lemon grass.–The leaves yield an oil which is used in perfumery and cosmetics.

43. *Cymbopogon jwarancusa*–The oil smells like peppermint but the yield is low. The roots are used in the treatment of fevers.

44. *Cymbopogon caesius*; Syn. *Andropogon caesius;* Eng. Ginger grass–This is a perennial grass. The leaves yield an essential oil, which is used in soap manufacture. The grass is grown in Tamil Nadu and Travancore.

45. *Cynodon dactylon;* Eng.–Bermuda-grass; Verna.–**Doob.** This is a perennial grass used as fodder. This is the commonest used for tennis lawns.

46. *Dendrocalamus giganteus*–This is the largest of the Indian bamboos. It produces culms upto twenty five cm. in diameter. They are used for water-buckets and boxes. It is cultivated in Assam, Bengal and Malabar.

47. *Dendrocalamus hamiltonii*; Verna.–**Kagzi bans**–This is a tall tufted grass, used in the manufacture of paper, baskets and mats. It is mainly cultivated in Dehradun.

48. *Dendrocalamus strictus*; Eng.–Solid bamboo; Verna.–**Bans kaban**–It is used in the manufacture of paper, baskets and brushes. It is found in Northern India, Madhya Pradesh and South India.

49. *Echinochloa colonum*; Syn. *Panicum colonum*; Eng.–Shama millet; Verna.–**Samak**–The grains are eaten in times of scarcity.

50. *Echinochloa crus-galli*; Syn. *Panicum crus-galli*; Eng.–Baryard millet; Verna.–**Samak**–It is used as a fodder and food crop. It is chiefly cultivated in Maharashtra and Andhra Pradesh.

51. *Echinochloa frumentacea*; Syn. *Panicum crus galli* var. *frumentacea*; Eng.–Japanese millet; Verna.–**Sanwa, Sawa**–It is used as a fodder and food crop. It is used as a basis of a very potent beer. It is mainly grown in Madhya Pradesh and Uttar Pradesh.

52. *Echinochloa stagnina*; Verna.–**Banti**–It is used as a fodder. The grains are edible. It is found throughout India.

53. *Eragrostis curvula*–This is an introduced grass from United States. It is grown for erosion control and fodder.

54. *Erianthus arundinaceus*; Syn. *Saccharam arundinaceum*; Eng.–Pin reed grass; Verna.–**Ramsar, Sarkanda**–This is a perennial grass. The leaf sheaths yield a fibre which is used for making ropes, twines and paper, and the stems are used for making chairs, stools, tables, baskets and screen.

55. *Erianthus munja*; Syn. *Saccharum munja*; Verna.–**Munj sentha, Sarkanda, Munja**–The stem fibre is used for making baskets, mats and cordage. The leaves are used for thatching, and are also the source of paper. It is also used as soil-binder. It is commonly found in the Punjab and Uttar Pradesh.

56. *Erianthus ravennae*; Syn. *Andropogon ravennae; Saccharum ravennae*; Verna.–**Moonj**–The stem fibre is used for making chairs, muddas, chhappars and ropes.

57. *Heteropogon contortus*, Syn. *Andropogon contortus*; Verna.–**Kumeria**–This is a common perennial grass used as a fodder and the fibre is made into mats. It is also used to check soil erosion and in the manufacture of paper and board.

58. *Imperata cylindrica*; Syn. *I. arundinacea*; Verna.–**Ulu.**–This is a perennial grass, the stem is used for thatching and a raw material for paper making. The pillows and cushions are stuffed with its fruits. It is also used in soil conservation.

59. *Indocalamus wightianus*; Syn. *Arundinaria wightiana*–It is used for making baskets, mats and walking sticks.

60. *Phragmites karka*; Syn. *P. maxima*; Verna.–**Narkul.**–This is a perennial grass, the stems are made into pipes, and are also used for making writing and printing paper. They are also used for making chairs, baskets and mats.

61. *Pseudostachyum polymorphum*–This is a shrubby bamboo found in Sikkim, Assam. It is used by tea planters for making baskets. It is also used for making umbrella handles and walking sticks.

62. *Urochloa panicoides*–A herb used as fodder. The grains are also edible.

63. *Urochloa reptans*–A herb used as a fodder. The grains are also eaten.

Systematic Relationships: Bentham and Hooker have included five families in the series Glumaceae, *i.e.*, the Eriocauleae, Centrolepideae, Restiaceae, Cyperaceae and Gramineae. Engler and Prantl have placed the only two families, *i.e.*, Gramineae and Cyperaceae in the order–Glumiflorae.

Hutchinson has placed the family Cyperaceae in the order–Cyperales; and Gramineae in the order– Graminales. Hutchinson has treated the Graminales to be more highly advanced than the Cyperales. According to him both the Cyperales and Graminales have been derived from liliaceous ancestral stocks in a Juncaceae complex. According to Wettstein the Cyperaceae and Gramineae have been derived on parallel lines from Juncaceae. It is generally agreed that the grasses and sedges are very advanced groups whose simple inflorescence and floral structures represent drastic reductions from unknown ancestral types.

The Gramineae (grasses) and the Cyperaceae (sedges) are not as closely related as was formerly believed to be the case. It has been shown that the Gramineae differ in having terminal flowers whose ovaries evolved from ancestral types having parietal placentation, from that of the Cyperaceae which possess axillary flowers whose ovaries evolved from types having free-central placentation. The spikelet, present in both the families is now not considered to be homologous and it does not indicate any phylogenetic relationship.

Question Bank

Systematics of Flowering Plants

Long Answer Questions

1. Write a short essay on the 'Origin of Angiosperms' (15 minutes:15 marks).

2. Enumerate the important criteria used for classifying plants. (10 mts.; 10 mks.).

3. Give a broad outline of Hutchinson's system of classification of angiosperms and mention its merits and demerits. (15 mts.; 15 mks.).

4. What is a natural system of classification? Is Bentham and Hooker's system a natural system? Explain. (10 mts.; 10 mks.).

5. Explain the principles on which Bentham and Hooker based system of classification. Why is it considered to be a natural system of classification? (10 mts.; 10 mks.).

6. Summarise the staminal variations observed in the different families of sub-class Gamopetalae according to Bentham and Hooker. (30 mts. ; 10 mks.).

7. What is Disciflorae? Give the distinguishing characters of any *three* families included in the group and mention two examples of economic to morphological interest from each of the family you select. (15 mts. ; 15 mks.).

8. Name any *three* similarities you find between the families Papilionaceae, Caesalpiniaceae and Mimosaceae to include them in one main family. (5 mts. ; 3 mks.).

9. Compare the sub-families under Leguminosae with reference to the floral characters. (20 mts. ; 15 mks.).

10. Give the salient features of family Solanaceae and discuss its economic importance. (15 mts. ; 5+2 mks.).

11. Compare the floral structures of Apiaceae (Umbelliferae) with that of Lamiaceae (Labiatae). Which of the two families is highly evolved and why? Give suitable reasons. (15 mts.; 15 mks.).

12. Show how the floral structures are integrated with special mechanism of pollination in Asclepiadaceae. (8 mts. ; 5 mks.).

13. Give a comparative account of Acanthaceae and Solanaceae with reference to their floral characters. (20 mts. ; 8 mks.).

14. Describe the range of floral structure in Liliaceae. Give floral diagram and floral formula. Mention botanical name of any plant belonging to this family. (15 mts. ; 15 mks.).

15. (*a*) What are the different types of roots of Poaceae ?

(*b*) Relate their morphological structure in the function

(*c*) Give an example of each type. (15 mts. ; 8 mks.

16. What is meant by classification of plants? State the difference between 'artificial'. 'natural' and 'phylogenetic' systems of classification of plants. Discuss the principles of any one phylogenetic system of classification which you have studied. (15 mts. ; 15 mks.).

Short Answer Questions

1. Give scientific name of the following plant and assign it to its respective family giving reasons : Castor. (5 minutes : 5 marks).

2. To the generic name Aster, if you add the endings-aceae and **-ales**, what will the two derived names signify? What will be the importance of the generic name itself ? (3 mts. ; 3 mks.).

3. In plant classification features of the reproductive structures are considered more reliable than those of the vegetative structures. Why? (3 mts.: 3 mks.).

4. Give the botanical names of any three plants that are now in flower on your college campus. (2 mts. ; 2 mks.).

5. Write critical notes on the following : Main concepts in Engler and Prantl's system of classification of angiosperms. (5 mts. ; 5 mks.).

6. What is the relationship between phylogeny and classification of flowering plants. (5 mts.; 5 mks.).

7. Write short note on : A.P. de Candolle. (5 mts.; 5 mks.).

8. Write short note on : Linnaeus. (5 mts. ; 5 mks.).

9. State any *two* demerits of Bentham and Hooker's classification. (5 mts. ; 2 mks.).

10. State the basis of Linnaeus's classification of plants. (3 mts.; 3 mks.).

11. Mention any five differences between the classification of Bentham and Hooker and Engler and Prantl. (8 mts. ; 5 mks.).

12. What is chemotaxonomy? (3 mts. ; 2 mks.).

13. Write critical notes on the following : Rules of nomenclature. (5 mts. ; 5 mks.).

14. Write critical notes on the following : Merits of Engler's system of classification. (6 mts. ; 6 mks.).

15. Why is it not advisable to separate two species on the basis of gross morphology alone? What other criteria should be taken into consideration? (5 mts. ; 5 mks.).

16. What do you understand by a type specimen? (3 mts. ; 3 mks.).

17. Give the important floral characters which distinguish the two families in any *one* of the pairs in the following :

(*a*) Rutaceae and Apiaceae (Umbelliferae).

(*b*) Solanaceae and Acanthaceae. (6 mts. ; 6 mks.).

18. Mention the significant features you come across in any *one* of the following genera and assign it to its respective family : (*a*) *Calotropis*, (*b*) *Salvia*. (5 mts. ; 5 mks.).

19. Comment upon any *one* of the following :

(*a*) Corolla in Leguminosae.

(*b*) Flower of Cruciferae.

20. Assign any *one* of the following plants to the respective family:

(*a*) *Leucas aspera*.

(*b*) *Asclepias curassavica* (1mt. ; 1 mk.).

21. Given only the androecium and gynoecium, how would you distinguish any *one* pair of the following families from each other:

(*a*) Rosaceae from Malvaceae.

(*b*) Apocynaceae from Asclepiadaceae. (5 mts. ; 5 mks.).

22. Give *four* important floral characters of Malvaceae. Mention the name of *two* economically important plants belonging to the family. (4 mts. ; 4 mks.).

23. Refer the following genus to its respective family giving reasons : *Triticum* (5 mts.; 5 mks.).

24. How will you differentiate the following from each other : Cruciferae (Brassicaceae) and Caryophyllaceae (5 mts. ; 5 mks.).

25. Give floral diagrams and floral formulae with examples to distinguish between following pair of families.

Acanthaceae and Lamiaceae (5 mts. ; 5 mks.).

26. Mention one plant each under Cruciferae (Brassicaceae) where :

(*a*) There is no underground storage organ.

(*b*) There are less number of stamens than the usual number. (2 mts. ; 2mks.).

27. Describe the structure of fruit in Umbelliferae (Apiaceae). (3 mts. ; 2 mks.).

28. Comment upon the following :

Thalamus of Rosaceae. (5 mts. ; 5 mks.).

29. Give the distinguishing features of any *one* of the following families. Give botanical names of any *three* economically important plants from each family.

(*a*) Umbelliferae (Apiaceae)

(*b*) Rutaceae. (5 mts. ; 5 mks.).

30. Explain how Compositae (Asteraceae) plants are well adapted for rapid spread. (5 mts.; 3 mks.).

31. Mention any *three* distinguishing features of Rubiaceae. (5 mts. ; 3 mks.).

32. When only the fruits of Compositae and Rubiaceae are given, how will you differentiate between the two? (3 mts. ; 2 mks.).

33. List out *five* economically important plants of Solanaceae. (3 mts. ; 2 1/2 mks.).

34. Construct the floral diagram and give the formula of a typical member of Asclepiadaceae. (5 mts. ; 5 mks.).

35. Give floral diagrams and floral formulae with examples to distinguish between the following pair of families :

Asclepiadaceae and Apocynaceae. (5 mts. ; 5 mks.).

36. Give the distinguishing features of the following family. Give botanical names of any *three* economically important plants from : Solanaceae. (5 mts. ; 5 mks.).

37. How will you differentiate the following from each other :

Solanaceae and Apocynaceae (5 mts. ; 5 mks.).

38. Taking into consideration only the vegetative characters how would you differentiate Acanthaceae from Labiatae (Lamiaceae). (5 mts. ; 5 mks.).

39. Write notes on the following :

Inflorescence in Euphorbiaceae. (5 mts. ; 5 mks.).

40. Give a general account and economic importance of the following family Euphorbiaceae. (6 mts. ; 6 mks.).

41. Mention any *three* types of perennating organs in Liliaceae family. Give an example for each type. (8 mts. ; 6 mks.).

42. Comment upon the following :

Roots of *Sorghum* (5 mts. ; 5 mks.).

43. Write critical notes on the following :

Flowers of *Oryza* (6 mts. ; 6 mks.).

44. Comment upon the following :
Economic importance of Poaceae (5 mts. ; 5 mks.).

45. List *three* diagnostic features of family Gramineae (Poaceae). (2 mts. ; 4 mks.).

46. Write notes on the following : spikelet in Gramineae (Poaceae). (5 mts. ; 5 mks.).

47. Give a general account and economic importance of the following family : Gramineae (Poaceae). (6 mts. : 6 mks.).

Simple Questions

1. Which is the most famous botanical garden in the world? (1 minute ; 1 mark.).

2. What is the technical term for the following description : Stamens are in two whorls wherein the outer is opposite to the petals? (5 mts. ; 5 mks.).

3. Who is known as father of Botany? What is his contribution to Botany ? (2 mts. ; 1 mk.).

4. Name the systematic classification mainly based on evolutionary trends. (2 mts. ; 1 mk.).

5. Assign any *one* of the following plants to the respective family:
(*a*) *Solanum tuberosum* ; (*b*) *Calotropis procera* (1 mt. ; 1 mk.).

6. Describe the nature of fruits in the family Papilionaceae. (5 mts. ; 3 mks.).

7. Give the binomials of *two* economically important plants each from :
(*a*) Papilionaceae ; (*b*) Caesalpiniaceae. (5 mts. ; 3 mks.)

8. Name any one of the plants in Papilionaceae having monadelphous androecium. (1 mt. ; 1 mk.).

9. Name an essential oil-yielding plant in Poaceae (1 mt. ; 1 mk.).

10. Describe the nature of fruits and seeds of family Brassicaceae (5 mts. ; 2+1 mks.).

11. Give the binomials of two economically important plants each from : (*a*) Umbelliferae (Apiaceae) ; (*b*) Cruciferae and say how it is useful in everyday life. (2 mts. ; 2 mks.).

12. What is the special modification seen in *Mussaenda* flowers? (2 mts. ; 1 mk.).

13. Name a Compositae (Asteraceae) plant which is used as an insecticide? (1 mt. ; 1 mk.).

14. Describe the nature of fruits and seed of family Umbelliferae (Apiaceae). (5 mts., 2 mks.).

15. Name *four* herbs (give botanical names) of your locality belonging to the family Rubiaceae and mention approximate flowering season for them. (3 mts ; 3 mks.).

16. Describe gynosteigium with a suitable example. (3 mts. ; 3 mks.).

17. Name the plant and the family which yields Quinine. (3 mts. ; 2 mks.).

18. How are the stamens modified in Asclepiadaceae. (2 mts. ; 2 mks.).

19. Draw a sketch of a translator and label its parts. (3 mts. ; 2 mks.).

20. Name the economically important plants of Asclepiadaceae. (3 mts. 3 mks.).

21. Distinguish the families of the following pair : Acanthaceae and Labiatae (Lamiaceae). (5 mts. ; 5 mks.).

22. What is the distinguishing feature of Labiatae (Lamiaceae) gynoecium? (2 mts. ; 1 mk.).

23. Relate the structure of stamens and type of pollination mechanism seen in Labiatae (Lamiaceae). (3 mts. ; 2 mks.).

24. Describe the type of fruit in the Acanthaceae and dispersal mechanism of seeds. (5 mts. 3 mks.).

25. Name the type of pollination mechanism seen in Labiatae (Lamiaceae). (1 mt. ; 1 mk.).

26. Name a grass that yields aromatic essence. (1 mt. ; 1 mk.).

27. What is the botanical name of Lady's finger? (1 mt. ; 1 mk.).
28. Mention two climbers from Liliaceae family. (1 mt. ; 2 mks.).
29. Write botanical names of Ragi, rye, paddy and wheat. (1 mt. ; 2 mks.).

Multiple Choice Questions

(1 minute ; 1 mark, each question)

1. Which characters among the Angiosperms of the following are most important taxonomically to differentiate it from the Gymnosperms:
   (*a*) Triploid endosperm ❑
   (*b*) Seeds enclosed in fruit ❑
   (*c*) Attractive petals ❑
   (*d*) Presence of vessels in xylem ❑
2. *Genera Plantarum* is written by :
   (*a*) de Candolle ❑
   (*b*) A. L. de Jussieu ❑
   (*c*) Engler and Prantl ❑
   (*d*) Bessey ❑
3. Which of the following is a phylogenetic system of classification :
   (*a*) Lindlay ❑
   (*b*) Bentham and Hooker ❑
   (*c*) Hutchinson ❑
   (*d*) Linnaeus ❑
4. The classification formed by the Swedish Botanist Linnaeus is mainly based on :
   (*a*) Morphology ❑
   (*b*) Anatomy ❑
   (*c*) Embryology ❑
   (*d*) Serum diagnosis ❑
5. Botanical name of carrot weed is :
   (*a*) *Argemone mexicana* ❑
   (*b*) *Daucus carota* ❑
   (*c*) *Parthenium* ❑
   (*d*) None of the above ❑
6. 'Families of the Flowering Plants' book is written by:
   (*a*) Bentham and Hooker ❑
   (*b*) John Hutchinson ❑
   (*c*) Engler and Prantl ❑
   (*d*) None of the above ❑
7. Botanical name of Larkspur is:
   (*a*) *Ranunculus sceleratus* ❑
   (*b*) *Abelmoschus esculentus* ❑
   (*c*) *Delphinium ajacis* ❑
   (*d*) None of the above ❑

8. Out of the four widely known systems of classification one remains less phylogenetic and more natural. That is of:

   (*a*) Engler and Prantl ❑
   (*b*) Bentham and Hooker ❑
   (c) Bessey ❑
   (*d*) Hutchinson ❑

9. The condition of androecium in Cucurbitaceae is :

   (*a*) Syngenesious ❑
   (*b*) Synandrous ❑
   (*c*) Diadelphous ❑
   (*d*) Free stamens ❑

10. Obliquely zygomorphic nature of the flower in Solanaceae is due to one of the following whorls:

   (*a*) calyx ❑
   (*b*) corolla ❑
   (*c*) gynoecium ❑
   (*d*) androecium ❑

11. Find out the false statement with regards to Compositae (Asteraceae):

   (*a*) Head inflorescence ❑
   (*b*) Hypogynous flowers ❑
   (*c*) Syngenesious anthers ❑
   (*d*) Fruit a cypsela ❑

12. The largest family among the dicotyledons given below is:

   (*a*) Rubiaceae ❑
   (*b*) Rosaceae ❑
   (*c*) Compositae (Asteraceae) ❑
   (*d*) Cucurbitaceae ❑

13. Epigynous, gamopetalous family is :

   (*a*) Umbelliferae (Apiaceae) ❑
   (*b*) Rosaceae ❑
   (*c*) Compositae (Asteraceae) ❑
   (*d*) Myrtaceae ❑

14. Tetramerous flowers are seen in the gamopetalous family of :

   (*a*) Cruciferae ❑
   (*b*) Mimosaceae ❑
   (*c*) Rubiaceae ❑
   (*d*) Rutaceae ❑

15. Obdiplostamenous condition of androecium is common in the family of :

   (*a*) Anacardiaceae ❑
   (*b*) Rubiaceae ❑
   (*c*) Rosaceae ❑
   (*d*) Rutaceae ❑

16. *Sida*, a member of the family Malvaceae does not show this prominent trait :

(*a*) axile placentation ❑
(*b*) monadelphous stamens ❑
(*c*) alternate stipulate leaves ❑
(*d*) presence of epicalyx ❑

17. Usual fruit in the family of Cucurbitaceae is :

(*a*) Pome ❑
(*b*) Pepo ❑
(*c*) Samara ❑
(*d*) Hesperidium ❑

18. Synandrous and sinuous anthers are common in the family of:

(*a*) Umbelliferae (Apiaceae) ❑
(*b*) Rosaceae ❑
(*c*) Malvaceae ❑
(*d*) Cucurbitaceae ❑

19. Spot out the statement applicable to the family Myrtaceae:

(*a*) zygomorphic flowers ❑
(*b*) hypogynous flowers ❑
(*c*) stamens limited ❑
(*d*) inferior ovary ❑

20. *Rubus* a member of Rosaceae produces:

(*a*) an etaerio of achenes ❑
(*b*) an etaerio of drupes ❑
(*c*) an etaerio of follicles ❑
(*d*) an etaerio of berries ❑

21. Polypetalous, irregular corolla is seen in :

(*a*) Rosaceae ❑
(*b*) Malvaceae ❑
(*c*) Papilionaceae ❑
(*d*) Cruciferae (Brassicaceae) ❑

22. Out of the four statements one doesn't fit into the characters of the family of Anacardiaceae:

(*a*) Resin passages in the parenchyma ❑
(*b*) Well developed intrastaminal disc ❑
(*c*) Presence of lateral style ❑
(*d*) Hesperidium type of fruit ❑

23. Gynobasic style arises from the oase of the ovary. This is characteristic of the family:

(*a*) Labiatae (lamiaceae) ❑
(*b*) Solanaceae ❑
(*c*) Compositae (Asteraceae) ❑
(*d*) Ranunculaceae ❑

24. In *Leucas aspera* the type of inflorescence is:
   (*a*) Panicle ❑
   (*b*) Verticillaster ❑
   (*c*) Thyrsus ❑
   (*d*) Raceme ❑

25. Gymnosperm has been placed in between Dicots and Monocots in the classification of plants by:
   (*a*) Charles E. Bessey ❑
   (*b*) Bentham and Hooker ❑
   (*c*) Alfred Barton Rendle ❑
   (*d*) Hutchinson ❑

26. Cleistogamous flower is seen in:
   (*a*) Annonaceae ❑
   (*b*) Tiliaceae ❑
   (*c*) Rutaceae ❑
   (*d*) Cucurbitaceae ❑
   (*e*) None of the above ❑

27. Apocynaceae differs from Asclepiadaceae in having :
   (*a*) latex ❑
   (*b*) translators ❑
   (*c*) corona ❑
   (*d*) bicarpellary ovary ❑

28. Bicarpellary ovary with parietal placentation is found in :
   (*a*) Cruciferae (Brassicaceae) ❑
   (*b*) Convolvulaceae ❑
   (*c*) Labiatae (Lamiaceae) ❑
   (*d*) Solanaceae ❑

29. With reference to the habit the members of Nymphyaceae are :
   (*a*) Terrestrial plants ❑
   (*b*) Aquatic plants ❑
   (*c*) Epiphytic plants ❑
   (*d*) Xerophytic plants ❑

30. Numerous stamens are generally present in:
   (*a*) Nymphyaceae ❑
   (*b*) Papilionaceae ❑
   (*c*) Rutaceae ❑
   (*d*) Cucurbitaceae ❑

31. In *Brassica*, the placentation is :
   (*a*) axile ❑
   (*b*) basal ❑
   (*c*) parietal ❑

32. Tetradynamous condition of androecium is one of the striking features of:
    (*a*) Cruciferae (Brassicaceae) ❑
    (*b*) Caesalpiniaceae ❑
    (*c*) Papilionaceae ❑
    (*d*) Rubiaceae ❑
33. The condition of stamens in the cotton flower is:
    (*a*) free ❑
    (*b*) monadelphous ❑
    (*c*) diadelphous ❑
    (*d*) polyadelphous ❑
34. In *Mangifera indica* the fruit is a :
    (*a*) Berry ❑
    (*b*) Drupe ❑
    (*c*) Capsule ❑
    (*d*) Pepo ❑
    (*e*) Hesperidium ❑
35. Tetradynamous androecium characteristic of Cruciferae (Brassicaceae) consists of:
    (*a*) Outer 2 long and inner 2 short stamens ❑
    (*b*) Outer 2 short and inner 4 long stamens ❑
    (*c*) Outer 4 long and inner 2 short stamens ❑
    (*d*) Outer 4 short and inner 2 long stamens ❑
36. Monocarpellary condition is one of the striking features of:
    (*a*) Cucurbitaceae ❑
    (*b*) Rubiaceae ❑
    (*c*) Annonaceae ❑
    (*d*) Papilionaceae ❑
37. Pulses are group of plants belonging to the family :
    (*a*) Leguminosae (Papilionaceae) ❑
    (*b*) Cruciferae (Brassicaceae) ❑
    (*c*) Compositae (Asteraceae) ❑
    (*d*) Solanaceae ❑
38. In Papilionaceae the placentation is:
    (*a*) Parietal ❑
    (*b*) Axile ❑
    (*c*) Marginal ❑
    (*d*) Free central ❑
39. In *Daucus carota* the edible part is a :
    (*a*) corm ❑
    (*b*) rhizome ❑
    (*c*) tap root ❑
    (*d*) lateral root ❑
    (*e*) stem tuber ❑

40. In Mustard flowers are:

(*a*) perigynous ❑
(*b*) pentamerous ❑
(*c*) ebracteate ❑
(*d*) slightly zygomorphic ❑

41. In *Launea* the flowers are arranged in the form of head or capitulum. These florets are:

(*a*) Bisexual, tubular and homogamous ❑
(*b*) Bisexual, ligulate and homogamous ❑
(*c*) Outer ligulate and pistillate and inner tubular and bisexual ❑
(*d*) Unisexual on the same head ❑

42. Seeds of sunflower plant contain :

(*a*) Alkaloids ❑
(*b*) Oils ❑
(*c*) Dyes ❑
(*d*) Crystals ❑

43. Which of the following represents the condition seen in the family Compositae (Asteraceae)?

(*a*) superior ovary, axile placentation ❑
(*b*) inferior ovary, axile placentation ❑
(*c*) superior ovary, single basal ovule ❑
(*d*) inferior ovary, single basal ovule ❑

44. In Asclepiadaceae the stamens are modified and known as :

(*a*) Syngenesious ❑
(*b*) Sinuous ❑
(*c*) Translators ❑
(*d*) Androphore ❑

45. The end of the translator that sticks to the body of the insect is:

(*a*) caudicle ❑
(*b*) pollinium ❑
(*c*) retinaculum ❑
(*d*) corpusculum ❑

46. The botanical name of chilly is:

(*a*) *Solanum tuberosum* ❑
(*b*) *Lycopersicon esculentum* ❑
(*c*) *Capsicum frutescens* ❑
(*d*) *Nicandra physaloides* ❑

47. In *Coleus* the stamens are:

(*a*) Monadelphous ❑
(*b*) Diadelphous ❑
(*c*) Epipetalous ❑
(*d*) Polyadelphous ❑

48. The botanical name of sage plant is:
    (*a*) *Mentha piperita* ❑
    (*b*) *Salvia splendens* ❑
    (*c*) *Lavandula vera* ❑
    (*d*) *Thymus vulgaris* ❑
49. In Labiatae (Lamiaceae) the fruits are usually:
    (*a*) Follicle ❑
    (*b*) Capsule ❑
    (*c*) Nutlets ❑
    (*d*) Achene ❑
50. *Lantana* is an example of a :
    (*a*) huge tree ❑
    (*b*) prickly shrub ❑
    (*c*) twiner ❑
    (*d*) soft herb. ❑
51. *Avicennia* is a :
    (*a*) Mesophyte ❑
    (*b*) Epiphyte ❑
    (*c*) Hydrophyte ❑
    (*d*) Mangrove ❑
52. Explosive fruits are seen in :
    (*a*) *Ruellia tuberosa* ❑
    (*b*) *Solanum nigrum* ❑
    (*c*) *Nicotiana tabacum* ❑
    (*d*) *Asclepias curassavica* ❑
53. A plant that belongs to Liliaceae in which stipules are modified into tendrils is:
    (*a*) *Gloriosa* ❑
    (*b*) *Yucca* ❑
    (*c*) *Lilium* ❑
    (*d*) *Smilax* ❑
54. The botanical name of the 'bird's nest 'orchid' is :
    (*a*) *Dendrobium* ❑
    (*b*) *Neottia* ❑
    (*c*) *Cypripedium* ❑
    (*d*) *Orchis* ❑
55. Sugarcane inflorescence is a :
    (*a*) Spike ❑
    (*b*) Catkin ❑
    (*c*) Panicle ❑
    (*d*) Verticillaster ❑

56. The inflorescence of paddy is:

(*a*) racemose ❑

(*b*) catkin ❑

(*c*) panicle ❑

(*d*) verticillaster ❑

ANSWERS

1. (*b*)	2. (*b*)	3. (*c*)	4. (*a*)	5. (*c*)	6. (*b*)	7. (*c*)
8. (*a*)	9. (*b*)	10. (*c*)	11. (*b*)	12. (*c*)	13. (*c*)	14. (*c*)
15. (*d*)	16. (*d*)	17. (*b*)	18. (*d*)	19. (*d*)	20. (*a*)	21. (*c*)
22. (*d*)	23. (*a*)	24. (*b*)	25. (*b*)	26. (*e*)	27. (*b*)	28. (*a*)
29. (*b*)	30. (*a*)	31. (*c*)	32. (*a*)	33. (*b*)	34. (*b*)	35. (*b*)
36. (*d*)	37. (*a*)	38. (*c*)	39. (*c*)	40. (*c*)	41. (*b*)	42. (*b*)
43. (*d*)	44. (*c*)	45. (*d*)	46. (*c*)	47. (*c*)	48. (*b*)	49. (*c*)
50. (*b*)	51. (*d*)	52. (*a*)	53. (*d*)	54. (*b*)	55. (*c*)	56. (*c*)

U.P.S.C. Civil Services Examination (C.S.E.) and Indian Forest Service (I.F.S.) Examination, Question Papers

Long Answer Questions

1. Describe any *two* systems of classification of Angiosperms. Which of them is considered better by you? Give reasons in support of your answer.

2. Discuss the value of cytological characters in improving the existing systems of classification.

3. Write an essay on : classification of flowering plants.

4. Outline briefly the main characters of Amentiferae and Scitaminae and add a note on their distribution in India.

5. What are the principal functions of different parts of an angiospermic plant? Mention two angiosperms which do not develop (*a*) stems; (*b*) roots; (*c*) green leaves.

6. Discuss the evolutionary trends in Tubiflorae or Centrospermae.

7. Compare the floral characters exhibited by Compositae and Umbelliferae and explain their evolutionary trends.

8. Discuss 'modern trends in taxonomy' giving suitable examples.

9. What are the main features of the system of classification proposed by Bentham and Hooker? Discuss its merits and demerits.

10. Discuss the differences between the systems of classification of Engler and Hutchinston. Which of the systems do you consider more phylogenetic? Give reasons.

Short Answer Questions

1. Give briefly the main characters and affinities of the following and indicate their distribution in India: (a) Rutaceae; (*b*) Podostemonaceae; (*c*) Umbelliferae.

2. Exclusively on the basis of the gynoecium and androecium how would you distinguish the following families? (*i*) Rosaceae from Malvaceae; (*ii*) Apocynaceae from Asclepiadaceae; (*iii*) Scrophulariaceae from Acanthaceae; (*iv*) Cruciferae from Capparidaceae.

3. Enumerate the salient features of any two families listed below and discuss their systematic position : (*i*) Nymphaeaceae; (*ii*) Scrophulariaceae; (*iii*) Orchidaceae.

4. Write a critical commentary on : Merits and drawbacks of Hutchinson's system of classification.

5. Write what is unique in the following : (*a*) Pollination in *Vallisneria*; (*b*) Flower of an orchid.

6. Answer the following, using not more than two hundred words foı each part :

(*a*) What is the concept of species in plant taxonomy.

(*b*) What is the scientific basis of botanical nomenclature? What are the most important rules in this system?

(*c*) What are the comparative adaptive advantages of woody and herbaceous conditions?

7. Discuss the special features of the following :

(*a*) Reproductive structures in Asclepiadaceae.

(*b*) Flower of Magnoliaceae.

(*c*) Pollination in Hydrocharitaceae.

(*d*) Floral morphology of Orchidaceae.

8. What are the unique features of the following?

(*a*) Inflorescence of Gramineae.

(*b*) Gynoecium of Rosaceae

(*c*) Androecium of Cucurbitaceae

(*d*) Fruit of Umbelliferae

(*e*) Flowers of Orchidaceae

9. Answer the following using not more than two hundred words for each part :

(*a*) Mention the most essential features of the International Code of Botanical Nomenclature and their significance.

(*b*) Discuss the role of Herbaria in modern taxonomical research.

(*c*) Highlight the importance of chemotaxonomy in the aid of systematics of Angiosperms.

10. Answer in not more than 200 words: What is biosystematics? Explain the role of this approach in modern plant taxonomy.

11. Describe the uniqueness of the following :

(*a*) Androecium in Asclepiadaceae.

(*b*) Corolla in Scrophulariaceae.

(*c*) Pollination in Labiatae.

(*d*) Gynoecium in Cyperaceae

(*e*) Fruit and seed in Umbelliferae.

12. Answer the following, using not more than 200 words for each part :

(*a*) Discuss the merits and demerits of the system of classification proposed by Hutchinson.

(*b*) Give the most important features of the family Palmae and discuss its phylogenetic relationships.

(*c*) Bring out the importance of palynology in the phylogenetic systems of classification.

13. Answer the following, using not more than 200 words for each part.

(*a*) What are the salient features of the system of classification proposed by Takhtajan? How does it compare with that of Engler.

(*b*) Explain the unique features of the family Liliaceae. Discuss its phylogeny.

(*c*) Discuss the role of phytochemistry in the classification and the determination of interrelationships.

14. Describe the uniqueness of the following : (*a*) Flower in Asteraceae; (*b*) Inflorescence in Euphorbiaceae; (*c*) Gynoecium in Rosaceae; (*d*) Androecium in Cucurbitaceae; (*e*) Spikelet in Gramineae.

15. Write the salient features of the following:

(*a*) Inflorescence of Euphorbiaceae.

(*b*) Androecium in the Asclepiadaceae.

(*c*) The primitive carpel.

(*d*) Hypanthodium.

16. Describe the characteristics of the following:

(*a*) Flower of Magnoliaceae;

(*b*) Gynoecium of Cruciferae;

(*c*) Androccium of Cucurbitaceae;

(*d*) Fruit of Umbelliferae;

(*e*) Inflorescence of Gramineae.

17. Place the following *genera* to their *respective* families and indicate their economic importance:

(*i*) *Helianthus;*

(*ii*) *Digitalis;*

(*iii*) *Atropa;*

(*iv*) *Vanilla.*

18. What is numerical taxonomy? Briefly state the methods employed in and importance of their taxonomy.

19. Answer any *two* of the following:

(*i*) Outline the salient features of the classification by Bentham and Hooker. What are the merits and demerits of this system?

(*ii*) Briefly comment on the origin of angiosperms.

(*iii*) Importance of the study of phytochemistry in taxonomy.

20. State the distinguishing characters of Rosaceae and the affinities of the taxon with Ranunculaceae and Leguminosae. Mention *five* Indian fruit plants of the family.

21. (*a*) Place the following genera to their respective families and indicate their economic importance:

(*i*) *Elaeis*;

(*ii*) *Manihot*;

(*iii*) *Arachis*;

(*iii*) *Nigella.*

(*b*) Distinguish between :

(*i*) Cereals and millets;

(*ii*) *Utricularia* and *Aldrovanda*;

(*iii*) Sorosis and syconus.

22. Answer the following in not more than 200 words:

Which characteristics of flowers are considered to be evolutionarily advanced and which are considered to be primitive.

23. Explain very briefly the economic importance of

(*i*) Euphorbiaceae;

(*ii*) Compositae;

(*iii*) Coconut palm.

24. Answer the following, in not more than 200 words each : (*a*) How does serology help in taxonomy of plants; (*b*) Explain the significance of palynological characters in systematics; (*c*) Outline the scope, usefulness and limitations of Chemotaxonomy and Numerical Taxonomy.

25. Why International rules of botanical nomenclature are important and when were they propounded? What are the main principles governing these rules.

26. Answer in not more than 200 words : Discuss the role of biosystematics in the understanding of the taxonomy of angiosperms.

27. Why the Compositae are regarded as the most highly evolved taxon?

28. Answer the following in not more than 200 words each : (*a*) "Magnoliales possess evolutionarily primitive flower"— substantiate; (*b*) Indicate the role of anatomy in angiosperm systematics.

29. Give the floral formula and draw the floral diagram of a diandrous orchid.

30. Compare the following:

(*a*) Systems of Classification of plants by Bentham and Hooker, and Engler and Prantl.
(*b*) Inflorescence of *Euphorbia* and *Helianthus*.

31. Answer in not more than 200 words : Describe the floral structure of *Musa* along with floral diagram and floral formula.

32. Enumerae with brief notes the cytogenetic criteria in biosystematics.

33. Describe the characteristic features of the family Cucurbitaceae. Give the floral formula and floral diagram. Write botanical names of *five* plants of the family having economic importance.

34. Write short notes on : (*i*) Orchid flower; (*ii*) Flower of a grass; (*iii*) Flower of Umbelliferae.

Indian Forest Service (I.F.S.) Examination Question Papers

Q.1. (*a*) Which family is the most primitive one among the dicots? Give reasons in support of your answer.

(*b*) Which family is the most highly evolved among the monocots? Give reasons in support of your answer.

2. (*a*) (*i*) Justify the inclusion of the families Papilionaceae, Caesalpiniaceae, and Mimosaceae in the order Leguminales.

(*ii*) Why Cucurbitaceae is placed in Polypetalae though they are having gamopetalous condition?

(*iii*) Distinguish between : Apocynaceae and Asclepiadaceae : Cucurbitaceae and Rutaceae.

(*b*) (*i*) State any *two* merits of Bentham and Hooker's classification.

(*ii*) Draw illustrated diagrams to show the following : Inflorescence of Labiatae; Androecium of Orchidaceae: Gynoecium of *Triticum*: Corolla of Papilionaceae.

3. Enumerate important differences between the symptoms of classification proposed by Bentham and Hooker, Engler and Prantl and Hutchinson. Which of these systems do you consider to be the most satisfactory? Give reasons in support of your answer.

4. (*a*) Adduce evidences to prove that Asteraceae is an advanced family.

(*b*) Compare the flower and inflorescence of Umbelliferae with that of Gramineae.

(*c*) Outline the scope, usefulness and limitations of Chemotaxonomy and Numerical taxonomy.

5. (*a*) Give the distinguishing features of the following families : (*i*) Loranthaceae; (*ii*) Magnoliaceae; (*iii*) Orchidaceae.

(*b*) What do you understand by Principles of classification? Even though Hutchinson's system of classification is natural, it has not been accepted by most of the taxonomists, comment.

6. (*a*) Describe the floral and diagnostic characters of Euphorbiaceae. Mention *three* plants of economic importance. Mention *three* plants of economic importance and name the products obtained from them.

(*b*) Describe the merits of Takhtajan's system of classification.

7. Write briefly : (*a*) International code of Botanical Nomenclature; (b) Hutchinson's contributions to Taxonomy; (*c*) Palynological criteria in systematics; (*d*) α-Taxonomy; (*e*) Flower of orchid.

8. Describe the floral and diagnostic features of Asteraceae and Palmaceae. Mention the botanical names of *two* plants of economic importance and name the products obtained from them.

9. Explain the special features of the following: (*a*) Androecium in Leguminosae; (*b*) Gynoecium in Papaveraceae; (*c*) Mangrove vegetation.

10. Write critical notes on the following:

(*a*) Merits and demerits of Bentham and Hooker's system of classification; (*b*) Importance of cytology in taxonomy; (*c*) Spikelet of Poaceae; (*d*) Phylogenetic system of classification; (*e*) Numerical taxonomy.

11. Describe the distinguishing features of Apocynaceae and Euphorbiaceae. Mention at least *three* important genera of each family and give their economic significance.

12. Give a brief account of the following:

(*a*) Flower in Asteraceae; (*b*) Pollination mechanism in Orchidaceae; (*c*) Floral diagram and floral formula of Cruciferae and Malvaceae.

13. Give brief account of the following:

(*a*) Evolutionary trends in Angiosperm flower;

(*b*) International botanical nomenclature;

(*c*) Natural system of classification of plants;

(*d*) Discuss how chemical parameters help in the classification of plants.

14. (*a*) Describe the distinguishing characters of Leguminosae. Discuss the criteria based on which this family has been divided into *three* sub-families.

(*b*) Describe the distinguishing features of Labiatae. Mention economic importance of the family. Add a note on pollination mechanism.

15. Give a brief description of the following:

(*a*) Inflorescence, flower and floral diagram of Euphorbiaceae. Mention the economic importance of the family.

(*b*) Give the distinguishing features and affinities of Magnoliaceae.

(*c*) State the distinguishing features, affinities and economic importance of Scitamineae.

(*d*) Distinguish between Rutaceae and Meliaceae.

16. (*a*) Describe the range of floral structure in Rosaceae.

(*b*) Make floral diagrams and write floral formulae of representative species (one each) of Labatae and Rutaceae.

17. (*a*) Compare the floral characters of Rubiaceae with those of Myrtaceae.

(*b*) Describe the various rules of botanical nomenclature.

18. Write short notes on the following:

Illustrate your answer with diagrams.

(*a*) Androecium in Cucurbitaceae.

(*b*) Hypanthodium

(*c*) Translator mechanism

(*d*) Pollination in *Salvia.*

19. Compare the floral characters of Rutaceae with that of Meliaceae. Draw floral diagrams and write floral formulae in each case.

20. Write critical notes on any *four* of the following: (*i*) Inflorescence in *Euphorbia*; (*ii*) Botanical Survey of India; (*iii*) Spikelet of *Poaceae*; (*iv*) *Loranthus*; (*v*) Coconut and its importance; (*vi*) Principle of priority in nomenclature.

21. Describe a system of classification of angiosperms used in practical classes. Give reasons for its use.

22. (*a*) Describe the distinguishing characters of the families Gramineae and Cyperaceae.

(*b*) How does family Acanthaceae differ from Labiatae? Give two economic uses of each.

23. Describe the following:

(*a*) Placentation

(*b*) Primitive flower

(*c*) Distinguishing features of sub-families of Leguminosae.

(*d*) Fruits in Rutaceae and Rosaceae.

24. What is a natural system of classification? Describe a natural system of classification. How is it useful in field studies.

25. (*a*) Describe the floral characters of the family Magnoliaceae.

(*b*) How does family Rutaceae differ from Meliaceae?

26. Describe the following:

(*a*) Spike structure in Gramineae.

(*b*) Pollinia and Gynostegium.

(*c*) Botanical gardens.

(*d*) Fruits in Cucurbitaceae.

26

Basic Body Plan of Flowering Plant

THE PLANT BODY AND ITS DEVELOPMENT

The plant body consists of a number of organs, *i.e.*, root, stem, leaf and flower. The flower consists of sepals, petals, stamens, carpels and sometimes also sterile members. Each organ, is made up of a number of tissues. Each tissue consists of many cells of one kind.

The complex multicellular body of the seed plant is a result of evolutionary specialization of long duration. This specialization has given rise to the establishment of morphological and

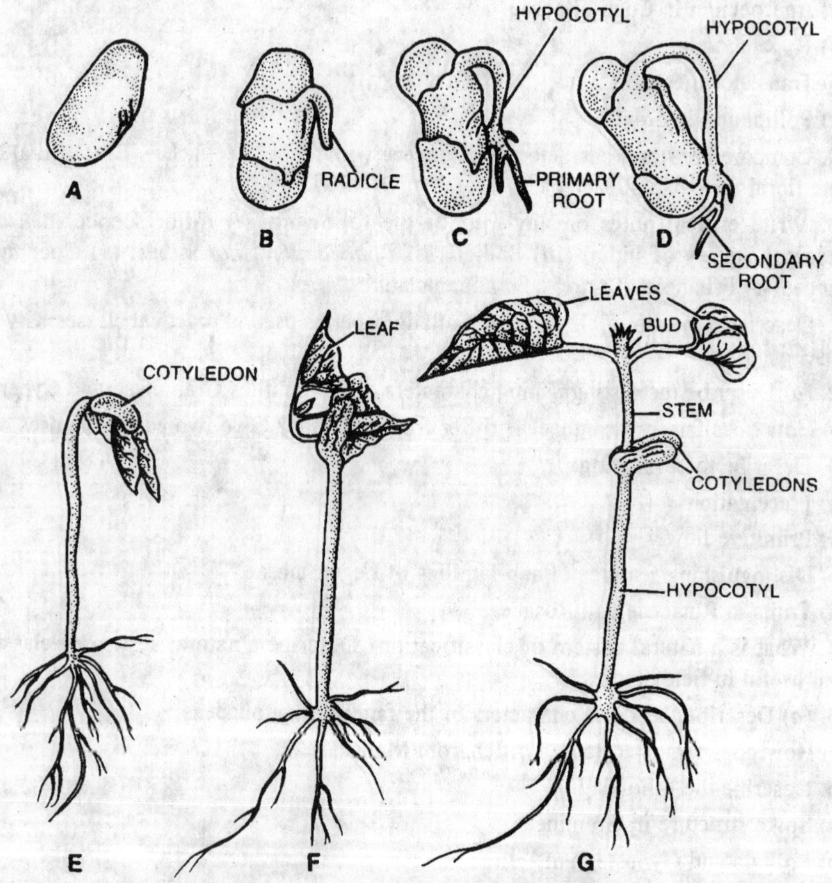

Fig. 26.1. Different stages in the germination of the seed leading to the formation of young seedling (A–G); G, seedling with well developed roots, two young leaves, hypocotyl, cotyledons, stem and young apical bud.

physiological differences between the various parts of the plant body and also caused the development of the concept of **plant organs.**

The organization of the plant body of the oldest known land plants, the Psilophytales, suggests that the differentiation of the vegetative plant into leaf, stem and root is a result of evolutionary development from an originally simple axial structure (Arnold, 1947; Eames, 1936).

As regards the morphologic nature of the flower it is thought that the flower is homologous with a shoot and the floral parts with leaves.

Fundamental parts of the plant body: The axis, consists of two parts—that portion which is normally aerial is know as the **stem,** and the portion which is subterranean is called the **root.** There are three types of appendages arising from the axis. **1. Leaves**—The strands of vascular tissue pass through the leaves. The leaves are characteristic of the stem and do not occur on the root. The leaves are found to be arranged on the stem in a definite manner, and bear an intimate structural relation to the skeleton of the axis. The leaf is looked upon as the lateral expansion of the stem, continuous with it. All fundamental parts of the stem are concerned with the formation of the leaf. **2. Emergences**—In the appendages of the second rank only the outermost layers of stem, the cortex and the epidermis, are usually present which are known as emergences. The prickles of the rose make a good example of it. **3. Hairs**—The appendages of the third rank are hairs. These are projections of the outermost layer of the cells. The emergences and hairs occur on both axis and leaves, usually without definite arrangement.

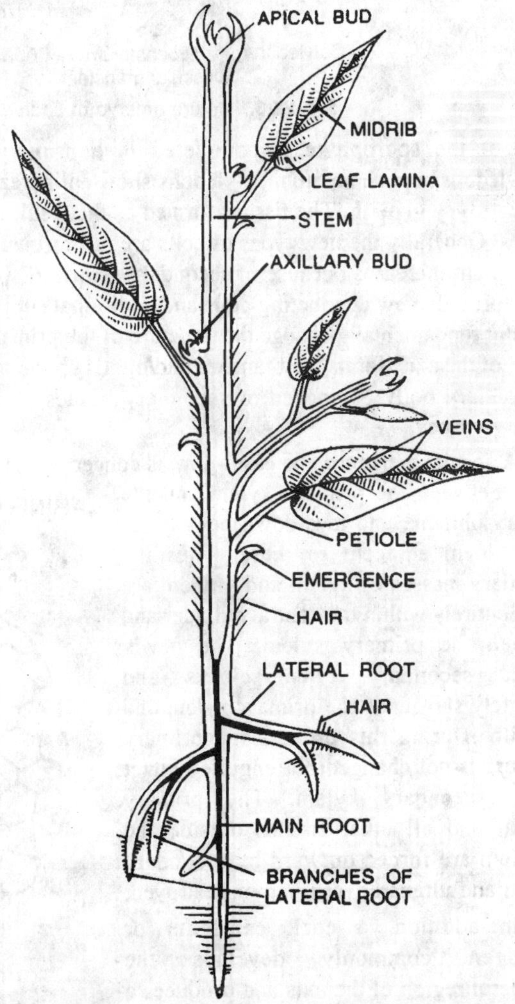

Fig. 26.2. The plant body, showing fundamental parts.

DEVELOPMENT OF THE PLANT BODY

A vascular plant begins its existence as a morphologically simple unicellular zygote (2n). The zygote develops into the embryo and thereafter into the mature sporophyte. The development of the sporophyte involves division and differentiation of cells, and an organization of cells into the tissues and tissue systems. The embryo of the seed plant possesses relatively a simple structure as compared with the mature sporophyte. The embryo bears a limited number of parts—generally only an axis bearing one or more cotyledons. The cells and tissues of this structure are less differentiated. However the embryo grows further, because of the presence of meristems, at two opposite ends of the axis, of future shoot and root. After the germination of the seed, during the development of shoot and root, the new apical meristems appear which cause a repetitive branching of these organs. After a certain period of vegetative growth, the reproductive stage of the plant is attained.

Primary and secondary growth. As mentioned above, this first-formed plant body is known as the **primary plant body**, since it is built up by means of first or **primary growth**. The tissues of this first-formed body are known as primary tissues ; for example the first-formed xylem is called **primary xylem.** In most vascular cryptogams and monocotyledons, the entire life

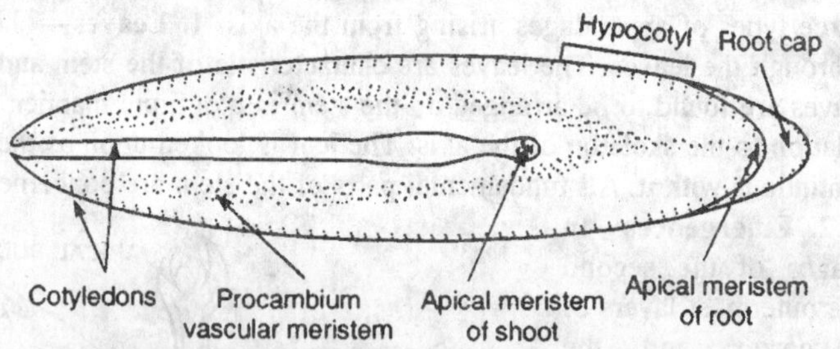

Fig. 26.3. Mature embryo of *Lactuca sativa* (After Esau).

cycle of the sporophyte is completed in a primary plant body. The gymnosperms, most dicotyledons, and some monocotyledons show an increase in thickness of stem and root by means of **secondary growth.** The tissues formed as the result of secondary growth are called **secondary tissues**. Generally the new types of cells are not formed by means of secondary growth. The bulk of the plant increases because of secondary growth. Especially the vascular tissues are developed which provide new conducting cells and additional support and protection. The secondary growth does not fundamentally change the structure of the primary body. The primary growth increases the length of the axis, forms the branches and builds up the new or young parts of the plant body. Thus, a secondary body composed of secondary tissues is added to the primary body composed of primary tissues.

A special meristem, the **cambium**, is concerned with the secondary thickening. The cambium arises between the primary xylem and the primary phloem, and lays down new xylem and phloem adjacent to these. Thus the secondary masses of xylem and phloem are found entirely within the central cylinder and between the primary xylem. The newly formed secondary xylem cloaks and ultimately surrounds the primary xylem and the pith. During this process the primary structure is not changed but engulfed intact within secondary xylem. The primary phloem and all other tissues outside the cambium are forced outward by secondary growth and ultimately crushed or destroyed.

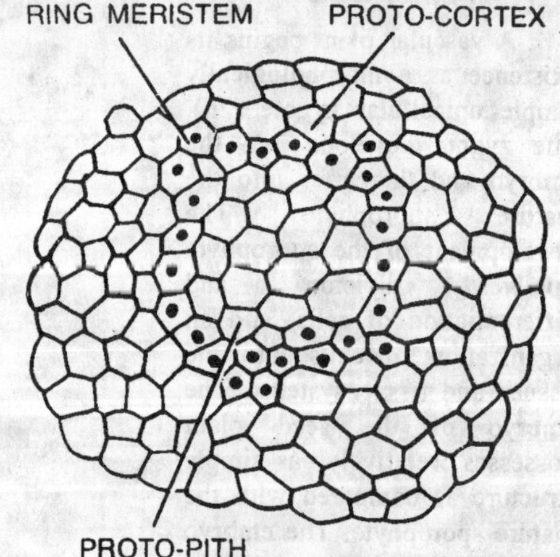

Fig. 26.4. T.S. through the shoot apex of *Ranunculus acris.*

In addition, a **cork cambium** or **phellogen** commonly develops the peripheral region of the axis and produces a periderm, a secondary tissue system assuming a protective function when the primary epidermal layer is disrupted during the secondary growth in thickness.

3. Meristems Based on Position in Plant Body

As regards their position in plant body, the meristems may be classified into three groups **apical meristem**, **intercalary meristem** and **lateral meristem.**

Apical meristem: The apical meristem lies at the apex of the stem and the root of vascular plants. Very often they are also found at the apices of the leaves. Due to the activity of these meristems, the organs increase in length. The initiation of growth takes place by one or more cells situated at the tip of the organ. These cells always maintain their individuality and position and are called **'apical cells'** or **'apical initials'.** Solitary apical cells occur in pteridophytes, whereas in higher vascular plants they occur in groups which may be terminal or terminal and sub-terminal in position.

Intercalary meristems: The intercalary meristems are merely portions of apical meristems that have become separated from the apex during development by layers of more mature or permanent tissues and left behind as the apical meristem moves on in growth. The intercalary meristems are internodal in their position. In early stages, the internode is wholly or partially meristematic, but later on some of its part, becomes mature more rapidly than the rest and in the internode a definite continuous sequence of development is maintained. The intercalary meristems are found lying in between masses of permanent tissues either at the leaf base or at the base of internode. Such meristems are commonly found in the stems of grasses and other monocotyledonous plants and horsetails, where they are basal. Leaves of many monocotyledons (grasses) and some other plants, such as *Pinus,* have basal meristematic regions. These meristematic regions are short living and ultimately disappear, ultimately, they become permanent tissues.

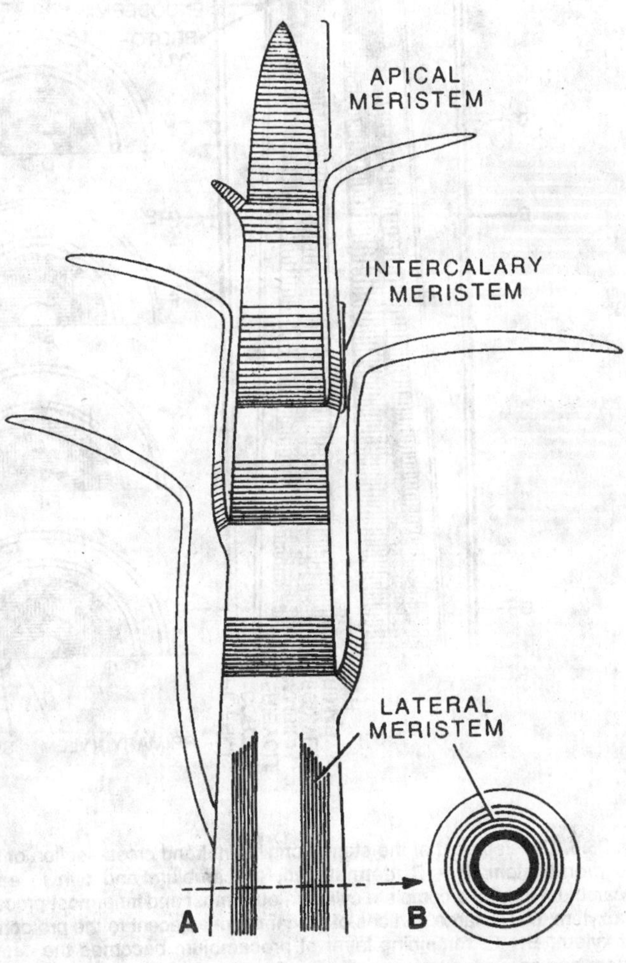

Fig. 26.5. Position of meristems. A, longitudinal view; B, cross-section.

Lateral meristems: The lateral meristems are composed of such initials which divide mainly in one plane (periclinally) and increase the diameter of an organ. They add to the bulk of existing tissues or give rise to new tissues. These tissues are responsible for growth in thickness of plant body. The cambium and the cork cambium are the examples of this type.

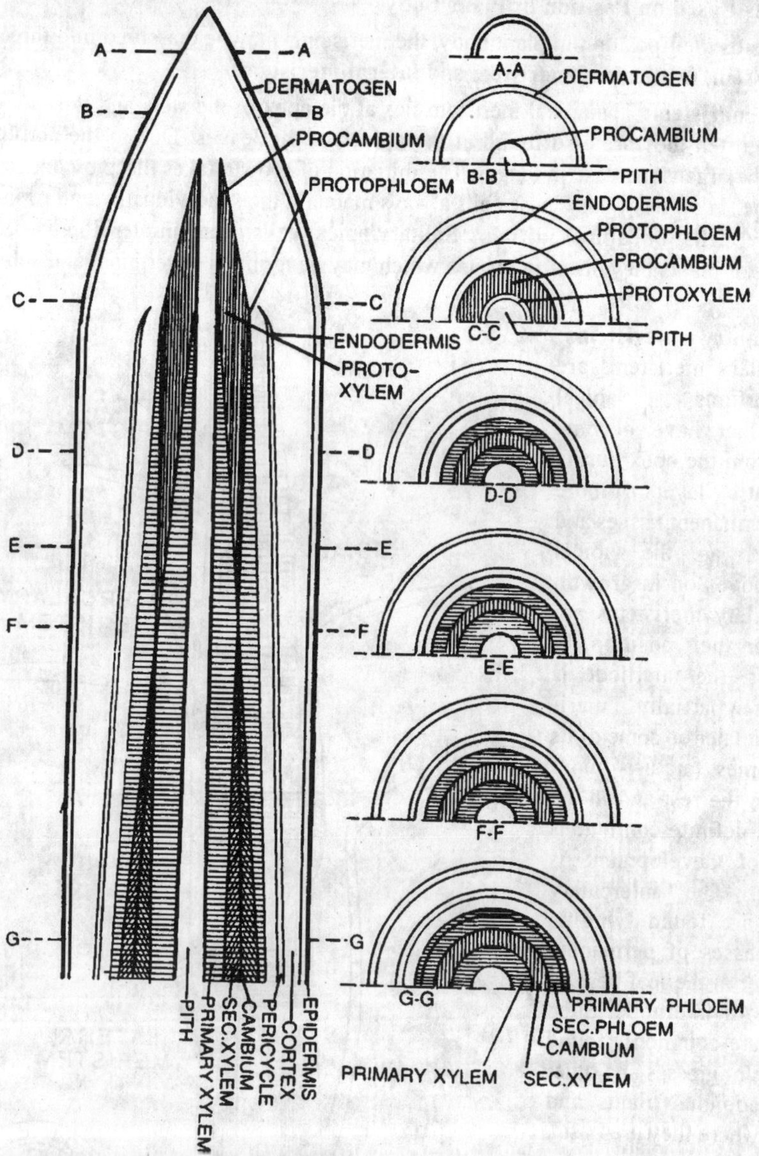

Fig. 26.6. Development of the stem. Longitudinal and cross-section of tip of elongating axis. At A – A, entire axis promeristem; B – B, dermatogen, procambium and pith in early stages of development; C – C, procambium formed a complete cylinder, outermost and innermost procambium has become protophloem and protoxylem; D – D, large portions of procambium adjacent to the protophloem and protoxylem become phloem and xylem; E – E, remaining layer of procambium becomes the cambium and forms the first secondary phloem and secondary xylem cells; F – F, secondary tissues increased in amount; G – G, further secondary growth occurred. (After Eames and MacDaniels).

The cambium does not fall definitely in either group (primary and secondary). It arises from apical meristem of which it is late and specialized stage. However, the accessory cambia are secondary. The tissues formed by the cambium are secondary, whereas the primary meristems form only primary tissues.

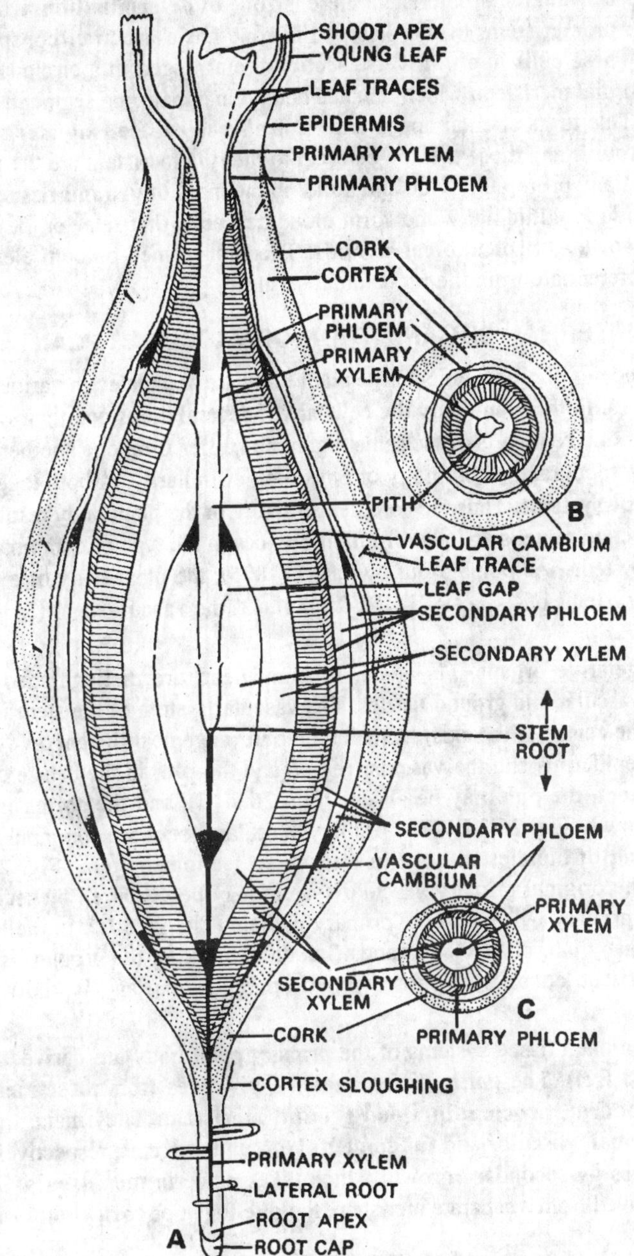

Fig. 26.7. Primary and secondary growth in a dicot plant *A*, longitudinal view of plant; *B*, transverse section of stem; *C*, transverse section of root.

The primary growth of an axis is completed in a relatively short period, whereas the secondary growth persists for a longer period and in a perennial axis the secondary growth continues indefinitely.

The stem apex like the root apex consists of a meristematic zone of cells that remain in a continuous and rapid state of division. This is called **promeristem** having cells with very thin walls. Immediately beneath the promeristem there is **zone of determination** which has no visible boundary with the promeristem. In the dicots, this zone has a group of conspicuous cells with dense cytoplasm. These cells in a transverse section are arranged in a circle (Fig. 31.4.). It is a remnant of a primordial meristem, which remains behind in a maturing segment, and it retained its activity to divide. Due to its circular appearance it is also called the **ring meristem**. The cells in the centre are **protopith** and those that are external to the ring meristem are the **protocortex**. The cells of protocortex and protopith divide and build up the mass of **ground tissue**. The cells of the ring meristem divide longitudinally and form elongated cells that later on develop in vascular bundles and are known as **procambical strands**. The first formed phloem elements and so the xylem elements differentiate from the procambial strands.

INTERNAL ORGANIZATION OF VASCULAR PLANT

The **cells** or the morphologic units of the plant body are associated in various ways with each other forming **tissues.** In the plant body the cells are of several kinds and their combinations into tissues are such that different parts of the same organ may differ from one another. The larger units of tissues may show topographic continuity or physiologic similarity, or both together. Such tissue units are called **tissue systems.** Thus the complex structure of the plant body results from variation in the form and function of cells and also from differences in the type of combination of cells into tissues and tissue systems. As pointed out by Sachs (1875), the plant body of a vascular plant is composed of three systems of tissues—(1) the **dermal,** (2) the **vascular** and (3) the **fundamental** or **ground** system.

The three vegetative organs, *i.e.*, stem, root and leaf, are distinguished by the relative distribution of the vascular and ground tissues. The vascular system of the stem is found between the epidermis and the centre of the axis. In such type of arrangement the **cortex** (ground tissue) is found between the epidermis and the vascular region and the **pith** in the centre of the stem (Fig. 26.8 B,C.). In the root, the pith may be absent (Fig. 26.8 E), and the cortex is generally shed during secondary growth (Fig. 26.8 D). The primary vascular tissues are commonly being arranged in the form of a ring of bundles as seen in transverse section of stem (Fig. 26.8 B). During secondary growth the original primary vascular system may be obscured by secondary vascular tissues between the primary xylem and the primary phloem (Fig. 26.8 C). In the leaf the vascular system consists of many interconnected strands (bundles) found in the ground tissue. In the case of leaf the ground tissue consists of photosynthetic parenchyma, and is known as **mesophyll** (Fig. 26.8 G).

The above mentioned tissue systems of the primary plant body are derived from the apical meristems (Fig. 26.8 F,H). The partly differentiated derivatives from these meristems may be classified as—**protoderm, procambium** and **ground meristem.** They make the meristematic precursors of the **dermal, vascular** and **fundamental** (ground) systems, respectively. The vascular tissue system enlarges by secondary growth which takes place in the **vascular cambium.** The periderm may be derived from a separate meristem, the **phellogen** or **cork cambium** (Fig. 26.8 D).

MODULAR GROWTH

Variation in organisational form that arises from size dependent relationship among parts is a fundamental aspect of development and evolutionary change. **Allometric analysis** is one of the earliest morphometric tools that help in revealing ontogenetic or evolutionary changes in shape as a consequence of changes in organ or body size (Gayon, 2000).

In plant population the individuals change widely in size may be due to asymmetric competition for light (Weimer, *et al*, 2001) or meagre distribution of other resources. The variability is dependent on the way plants are constructed. Plants grow through the repeated

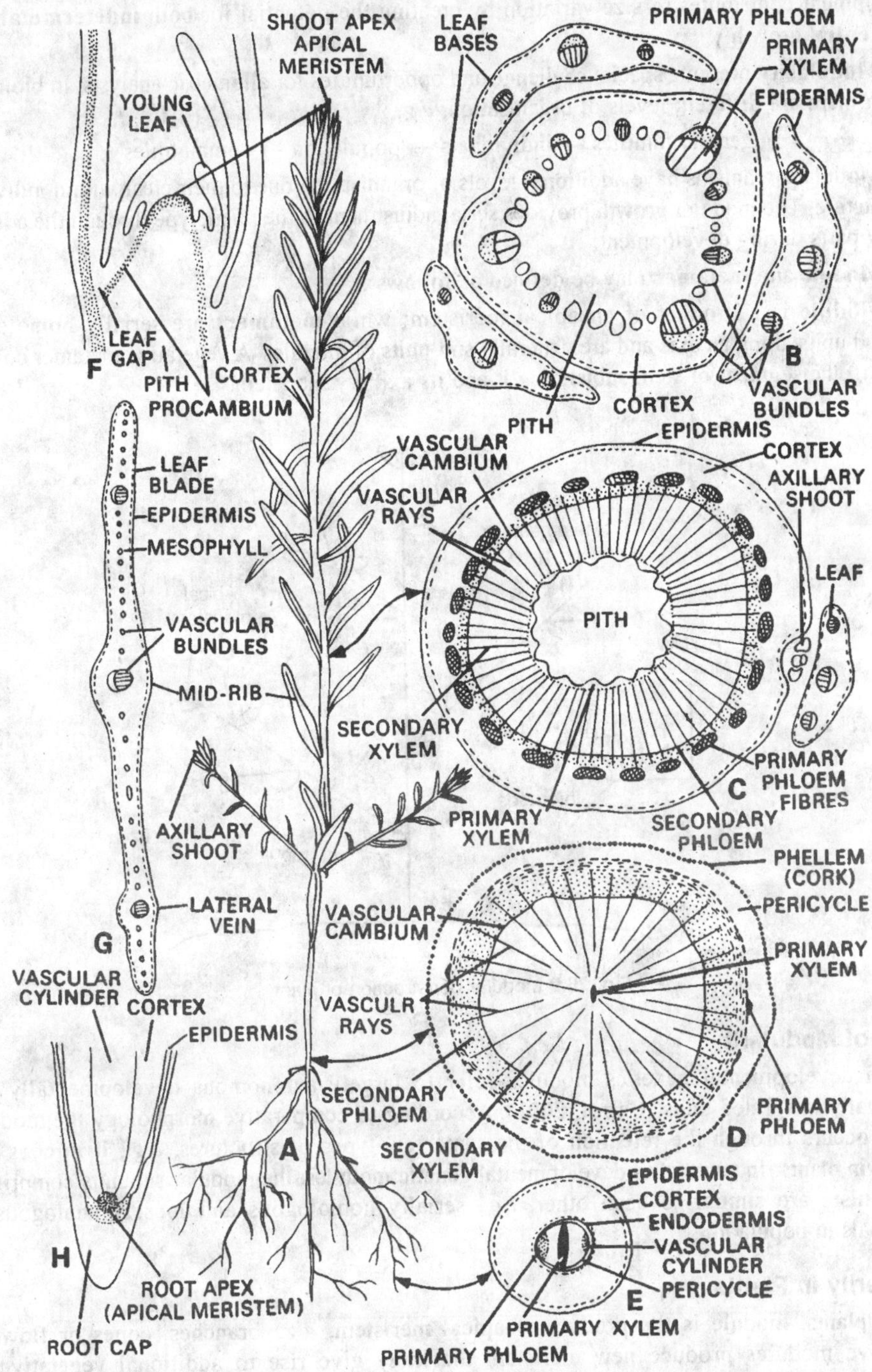

Fig. 26.8. Organization of a vascular plant. A, habit sketch of linseed plant (*Linum usitatissimum*) in vegetative state; B-C, transverse sections of stem; D-E, transverse sections of root; F, L.S. of shoot apex with apical meristem and developing leaves; G, transection of leaf lamina; H, L.S. of root apex with apical meristem and other root regions.

addition of similar morphological sub-units called **metamers** and **modules**. This mode of development contributes to size variation by creating the potential for both **indeterminate** and **exponential** growth.

Modularity presents special challenge and opportunities for allometric analysis. In biological system there are different levels of organisation, *e.g.*,

genetic families → individuals → population → communities

Modular organisms have additional levels of organisation due to their clear within individual substructure. The modular growth provides serial adjustment to the phenotype through the addition of new parts during development.

Module and **metamer** may be defined as follows :

Module is a product of an apical meristem; while **metamers** are serially homologous repeated units along an axis and are generally sub-units of modules. A vegetative metamer consists of a leaf, the segment of stem subtending it and its axillary meristem.

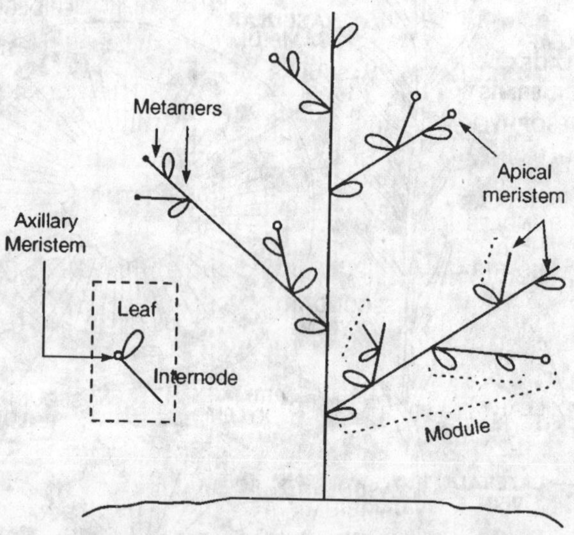

Fig. 26.9 Modular construction of plant

Types of Modularity

On developmental genetics modules refer to largely autonomous developmentally and functionally integrated units (Carrol, 2001). According to comparative morphology the modular growth occurs through the retention of the same few types of structures, *e.g.*, **branches** and **flowers** in plants. In contrast to developmental genetic modules, the modular subunits comprising an organism are similar to each other, *i.e.*, serially homologous and loosely analogous to individuals in population.

Modularity in Plants

In plants, module is the product of apical meristem, *e.g.*, branches, cones or flowers. Vegetative modules produce new meristem that may give rise to additional vegetative or reproductive modules, creating a modular hierarchy.

Usually modules themselves are composed of repeated units or **metamers**. For example, a typical vegetative metamer consists of a node and an internode, a leaf and an axillary meristem (Fig. 26.9).

Modular Development

In seed plants, *e.g.*, gymnosperms or angiosperms, the plant body is essentially **modularly** organised. The smallest elementary unit of construction in the shoot system is called a **metamer**. Metamers make branches and branches construct branching system. Each shoot apex is a potential point of further development. Development is iterative or repetitive process. The finite bit of programme is repeated to add new **metamers**. The number of iterations can be infinite in some species. In annual plants iterations certainly terminate, but their number is not strictly defined and may depend on environment (Schlichting and Smith, 2002). The number of leaves of a tree can vary with age and habitat conditions. It is due to capability of formation of organs continuously. Organs, metamers and branches are disposable without death of organism.

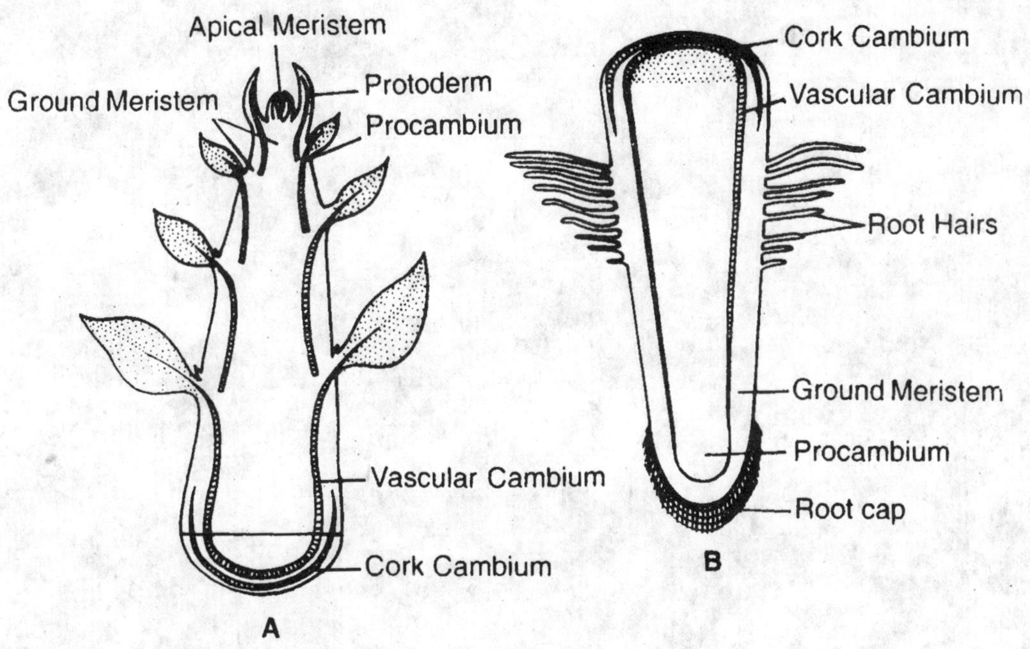

Fig. 26.10 Modular development. Shoot system and root system modules.

Shoot System Modules

For the first time, the term **metamer** was used in animals. **Metamers** unite to form **module**. Modules divide due to outgrowths of axillary buds or adventitious buds. When the axillary bud grows continually without undergoing rest period is called **sylleptic sub-module** while the axillary bud that undergoes some dormancy period is called **proleptic sub-module.**

The activity of module is directly related with that of apical meristem. After getting a stimulus, the axillary bud activates and module branches. New modules may arise some time playing an important role in **regeneration**, *e.g.*, in *Eucalyptus*, after fire new modules grow out rapidly from hidden buds produced by modification of the base of stem.

Any module or sub-module terminates in a flower or inflorescence depending upon the portion or orientation of metamers in shoot apex.

Root System Modules

Primary and lateral roots are modules of root system. Lateral roots are well developed in herbs. Various types of roots, such as respiratory roots, storage roots, etc., are types of modules.

27

Convergence of Evolution of Tree Habit

Convergence is a phenomenon in nature found in all organisms. It is observed tendency of living forms that are quite unrelated phylogenetically, to respond to similar contingencies of life by developing similar structures. In the process of evolution this similarity way arise along two evolutionary lines, *i.e.*, **parallelism** and **convergence** (fig. 27.1.).

Influence of environmental factors may also create similarities by structural modification, *e.g.*, in desert habitat most of the plants show cactus like habit and known as **cacti**. But only on flowering they can be separated in euphorbias, asclepias and cacti. Members of distantly related taxa attained similar structural features which are attained independently by various groups of plants, is called **convergent evolution.**

Parallelism and Convergence

According to Heywood (1967) **parallelism** is the development of similar features separately in two or more genetically similar, fairly closely related lineages. While Dahlgren *et al* suggested parallelism implies that basically corresponding structures in two or more plant groups have evolved in a similar way so that the derived states show more resemblances to one another than to their ancestral stages.

On the other hand **convergence** of development of similar features separately in two or more genetically diverse and not closely related lineages and not due to common ancestry (Heywood, 1967). While Dahlgren *et al* suggested that convergence denotes the condition where more or less different ancestral structures or different, generally distantly related taxa have attained a similar appearance in the course of their evolution.

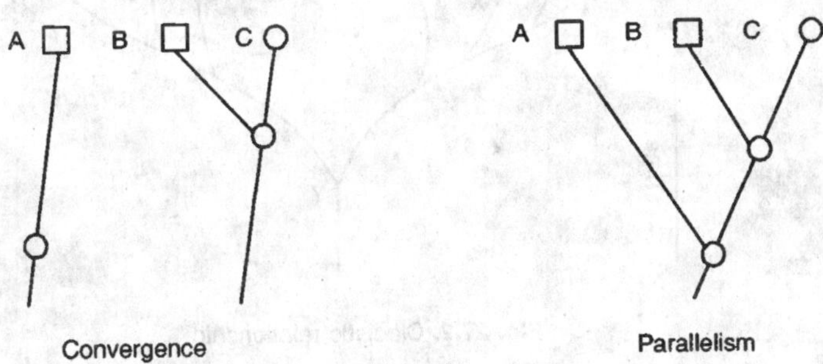

Fig. 27.1. Convergence and parallelism.

Thus, when two or more closely related groups tend to evolve by acquiring similar characters is called **parallelism** and when distantly related groups acquire similar characters is called **convergence.**

Thorne (1973) suggested that germination, nutrition, pollination, fertilisation, dispersal, etc., are the channels causing many diverse unrelated groups to proceed along the same evolutionary channel leading to **convergence**. According to Dahlgren *et al.* (1985), in homologous structures convergence results in superficial similarity of the whole plant, *e.g.*, succulents; and in non homologous structures it leads to superficially similar organs in different taxa, *e.g.*, presence of spines and tendrils in different groups of plants.

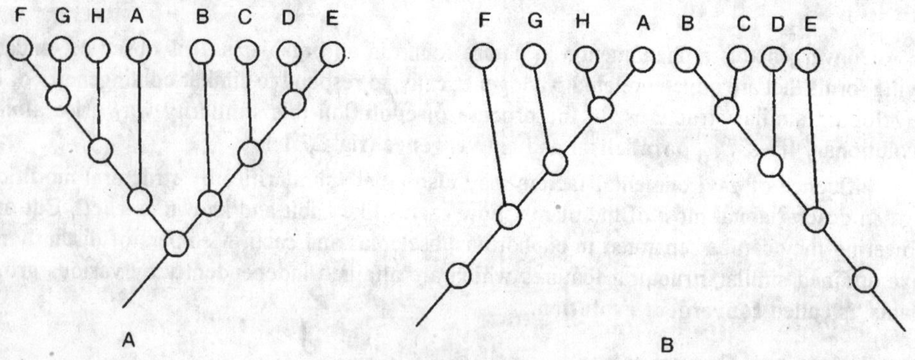

Fig. 27.2. Patristic relationship.

Convergence is generally brought about by adaptation to similar climate or habitats mostly affecting the vegetative parts, or even to similar methods of pollination or dispersal affecting the flower, fruit or seed.

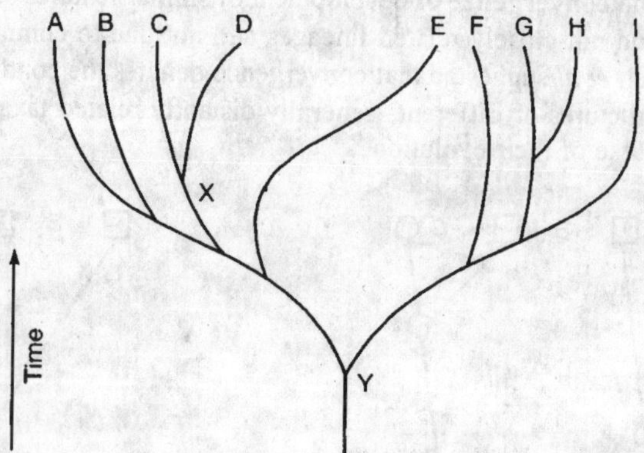

Fig. 27.2. Cladistic relationship.

For example, *Arenaria* and *Minuartia* form natural groups of species which were earlier placed in genus *Arenaria*. However, *Arenaria leptocladus* and *Manuartia hybrida* show much similarities. If similarity is **patristic**, *i.e.*, from common ancestry; the two species are most primitive members and show a single monophyletic group. But on the other hand, if they are most advanced members, they show distinct line of ancestry and similarity between them will be superficial and purely due to convergence. Thus, similarity between two plants due to a known or informed common ancestry is called **patristic** similarity in terms of phyletic lines, the plants are directly related. The pathways by which a given similarity is attained or closeness of relationship in terms of phyletic lines is called **cladistic** similarity.

Parasitic habit in different families, such as Lauraceae (*Cassytha*), Convolvulaceae (*Cuscuta*), Scrophulariaceae (*Striga*), Orobanchaceae (*Orobanche*) and Balanophoraceae (*Balanophora*); Carnivorous accessory nutrition in Droseraceae (*Drosera*), Utriculariaceae (*Utricularia*), Sarraceniaceae (*Sarracenia*), Nepenthaceae (*Nepenthes*) and Lentibulariaceae are some good examples of convergence.

Convergence of Evolution of Tree Habit

Trees are perennials. The groups of trees looking alike are few, *e.g.*, palms (monocots), conifers (gymnosperms), etc. Generally, the convergence in tree habit can be seen in their height. The primitive pteridophytes, gymnosperms and angiosperms, etc., were trees. Progymnosperms were present in upper Devonian and lower Carboniferous period of Palaeozoic era. Many palaeobotamists believe in origin of gymnosperms from pteridosperms.

The link between fern like *Archaeopteris* and gymnospermous *Callixylon* was discovered in 1960. *Archaeopteris* was a 18 m high tree with crown of leaves. The members of Lyginopteridaceae, *e.g.*, *Lyginopteris oldhamia* were large fern like trees with well branched weak stems. *Medullosa*, a member of Medullaceae (fossil gymnosperm) was large fern like but small tree. The fossil gymnosperms had fern like leaves.

However, members of Cycadales are medium sized trees with fern like leaves. *Macrozamia hopei* is tallest cycad 60 feet in height. Mexican cycad *Dioon spinulosum* is 50 feet in height while 1000 year old *D. edule* is only 6 feet high. *Ginkgo biloba*, a member of Ginkgoales is called living fossil. This is oldest living seed plant and there is great similarity between fossil gymnosperm *Ginkgoites* and *Ginkgo biloba* of today.

The members of Coniferales are of great heights, *e.g.*, *Cedrus*, *Cupressus*, *Abies*, *Picea*, *Pinus*, *Taxodium*, *Taxus*, *Podocarpus*, etc.

Gnetales include *Welwitschia*, *Gnetum* and *Ephedra* which make link between gymnosperms and angiosperms.

Angiosperms include some families of high woody trees. Angiosperms first appeared in Cretaceous period; they were divided into dicots and monocots. Trees are found in both dicots and monocots. Palms, Dracaenas, Pandanus and Bamboos are included in monocots. Palms show similarity in appearance with cycads. Some of dicots, such as *Polyalhia* trees show similarity with conifers. Similarities between Gnetales and Angiosperms on the basis of number of cotyledons is due to **convergent evolution**. The flowers of Orchidaceae, Zingiberaceae, Polygalaceae and Papilionaceae are convergently adapted in respone to similar methods of pollination.

Living Fossils

There are several living ancient plants which have undergone little or no change through the ages. Although they may have acquired specialized features later, they have more or less retained their original features. Such survivors are called **living fossils.**

These include the maiden-hair tree, *Ginkgo biloba* which has continued to live for over one hundred million years. *Cycas revoluta*, a cycad, and conifers such as Downredwood (*Metasequoia glyptostroboides*, California Big Tree (*Sequoia sempervirens*) are also considered living fossils.

Perhaps consistency of living habitat and isolation of these forms (living fossils) from their competitors can be two factors that may account for their survival.

28

Trees : Largest and Longest Lived Organisms

It is to be noted that a tree starts its life as a sapling growing from a seed as in most angiosperms. Forests are big assemblage of trees and other plants. In tropical rain forests, typically, the forest canopy is composed of three stories characterized by different types of trees. The trees of the top, or dominant, story form a nearly closed canopy which is frequently 60 metres or more

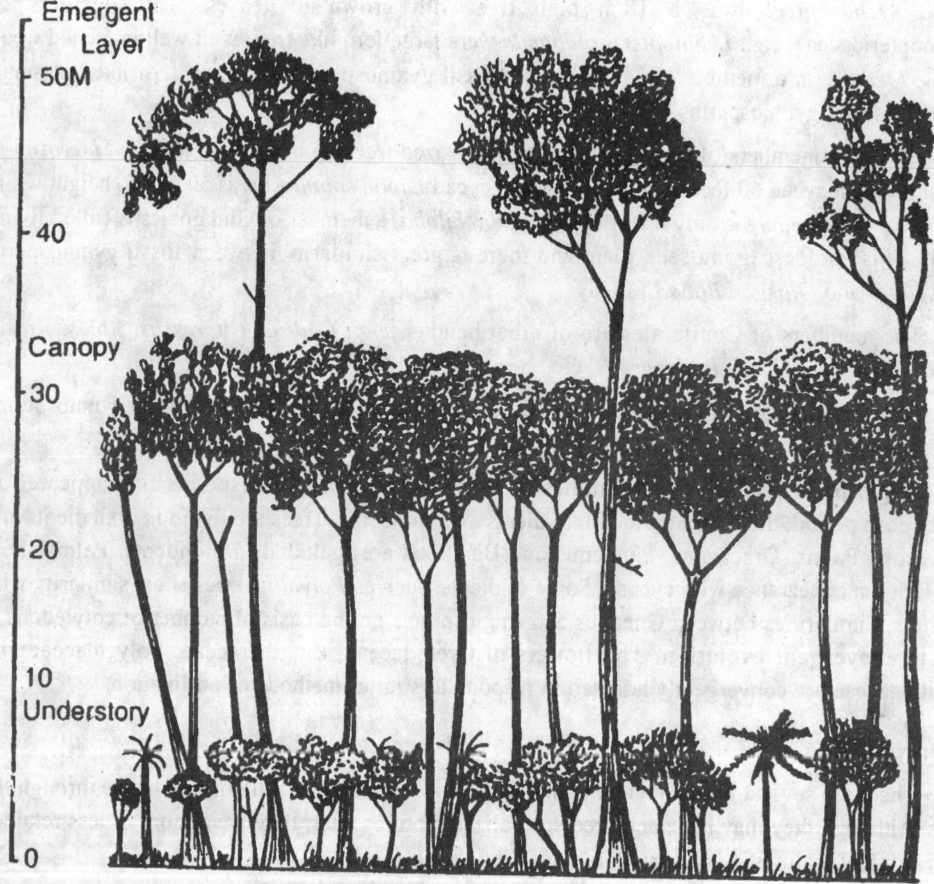

Fig. 28.1. Different strata of canopy

in height. The crowns of the second story are beneath those of the dominant story and like those of the dominant story, frequently form a nearly closed canopy. The trees of the third or lowest story are usually small and slender and have small open crowns. The presence of these three stories of different trees is not usually evident on casual observation for the composition of all the stories is very complex and few of the trees present any striking peculiarities. Moreover, smaller trees of a higher story always occur in a lower story as well as between the different stories, while the different species of a story have different heights. Erect palms are frequently numerous in the lower stories.

Beneath the tree stories there is a ground covering. In dry areas, it may consist largely of woody plants, while in moist situation herbs and ferns are abundant. Some individual tall trees are called **emergents** which protrude above the general level of the **canopy** in tropical rain forests.

There are certain forests which are dominated by single species of trees such as oak (*Quercus*), beech (*Fagus*), pine (*Pinus*) or spruce (*Picea*), etc.

Structural diversity of trees can also be noticed. For example Banyan tree (*Ficus benghalensis* Linn., Family-Moraceae) is an enormous tree, 70 to 100 feet high, sending down roots from the branches, which enter the ground and form trunks, thus extending the growth of the tree indefinitely. Banyan is a common tree planted near temples and shrines, and an open ground near villages. The roots of trees reaching ground act as pillars to support the branches. The trees have the biggest crowns of any in the world and with their aerial prop roots, a single tree looks like a small grove. The Banyan tree of botanical garden, Kolkata, has about 3000 trunks of which more than 2000 are 2 to 3 feet in diameter. About ten thousand people can get shelter under this tree at a time.

Fig. 28.2. The Banyan tree. Botanical garden Kolkata.

There are several other trees, *e.g.*, Pipal tree (*Ficus religiosa*), Shisham (*Dalbergia latifolia*), Indian coral tree (*Erythrina indica*), Mahua tree (*Madhuca indica*), Jamun tree (*Syzygium cumini*), Tamarind tree (*Tamarindus indica*), Silk cotton tree (*Bombax ceiba*), etc., which are commonly found in north Indian plains and peninsular India.

In strict sense, all trees which are not cone bearing are flowering trees (angiosperms) while cone bearing trees are gymnosperms, *e.g.*, *Pinus*, *Cedrus*, *Cupressus*, *Picea*, *Abies*, *Araucaria*, *Cryptomeria*, *Ginkgo* and Cycads.

There are several species of palms (Monocots: family palmaceae), such as oil palm (*Elaeis guineensis*), date palm (*Phoenix dactylifera*), coconut palm (*Cocos nucifera*), palmyra palm (*Borassus flabellifer*), betelnut palm (*Areca catechu*), Toddy palm (*Caryota urens*), etc. All palms have unbranched stems and crown of leaves on their tops.

The Baobab tree (*Adansonia digitata* Linn., Family–Bombacaceae) is known in India for many centuries, particularly along the western coast of this country. This is a strange looking tree. The great size attained by the trunk makes it strange looking. A soft-wooded tree, its trunk is about thickest in the world, suddenly tapering into thick branches.

In those parts of tropical Africa where the tree is indigenous, the trunk reaches 10 metres or more in diameter and this makes the trunk about 30 m in girth. This is an erect tree, seldom going over 20 m high.

Fig. 28.3. The Baobab tree (*Adansonia digitata*)

Oldest habit of seed plants : Tree habit is oldest habit of seed plants. They are perennial in habit living for many years. Due to secondary growth in dicots and gymnosperms they increase in thickness or girth.

California Coastal – redwood (*Sequoia sempervirens*) is considered a living fossil, the tallest tree of world, 112.6 m high. California Big Tree (*Sequoiadendron giganteum*) is also considered a living fossil, populary known as 'general sherman' of 83.82 m height and 24 m girth or circumference.

The tree with greatest girth is sweet chestnut (*Castanea sativa*) of Mount Etna, Sicily. The tree is about 4000 years old. The girth or circumference is about 60 m or more.

Taxodium distichum var. *mucronatum* is also a tree with large girth of about 40 m, situated in St. Maria, Oaxaca.

Abies nobilis (noble fir) is another big tree about 84.73 m high and situated in Gifford Natural Forest, Washington.

Among angiosperms *Eucalyptus regnans* is tallest tree. The tallest tree, 150 m high is situated in Victoria, Australia.

Trees: Longest-lived organisms. Trees are longest-lived organisms (plants). Some of long lived trees are mentioned here : *Shorea laprosula*, a tree of Dipterocarpaceae is more than 250 years old. Average age of *Parashorea* spp., is considered about 300 years. Most trees of tropical rain forests are 100 years old or so. Generally gymnosperms, such as *Sequoiadendron giganteum* are long-lived having average age of 3000 years.

Fig. 28.4. *Dracaena draco* of Sumatra.

Betula spp., (Birch tree) about 100–200 yrs; *Acer rubrum* (red maple) 100 years or more; *Fraxinus* spp., (Ash tree) about 150 years ; *Quercus* spp., (Oak tree), 800 to 1000 years; *Ficus benghalensis* (Banyan tree) 500 years or more ; *Sequoia sempervirens* (California Coastal redwood) more than 2000 years ; *Sequoiadendron giganteum* (California Big Tree) more than 3000 years ; *Pinus aristata* (Bristle cone fern) more than 4500 years; *Taxus baccata* (Yew) more than 5000 years ; Dragon trees (*Dracaena draco*) of Sumatra, more than 10,000 years of age.

Taxus baccata (Yew tree) of Scotoland is supposed to be of 5000 years age with a girth of about 17 metres. English Yew (*Taxus baccata*) lives upto 1000 years. English Oak (*Quercus ruber*) is also a long-lived tree and lives upto 400–500 years. Gymnosperms, such as coast and mountain redwoods of Western United States, some conifers and dragon trees of Sumatra are well establised long-lived trees.

In some trees, during process of ageing in natural life cycle the branches and leaves fall off. For example in Dougluss fir or Xmas tree (*Pseudotsuga mengies*) The leaves and branches fall off as shown in fig. 33.5. A, 8 year old tree; B, 30-40 years old; C, 50-60 years; D, 100 year old tree and E, 200 year mature tree, which may attain a height of 325 feet; F – G, after about 400 years where branches and leaves fall off and H, shows toppled down tree.

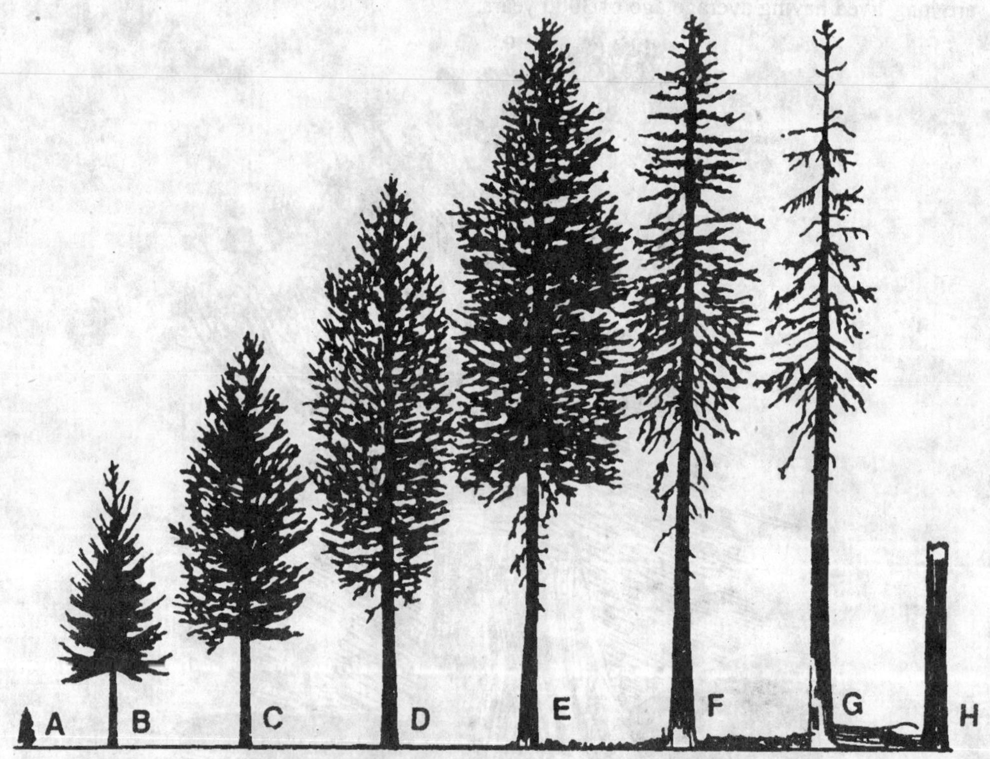

Fig. 28.5. Ageing of Dougllus fir tree (*Pseudotsuga menzies*).

Different Kinds of Trees

On the basis of their leaves, trees may be divided into three important categories. They are as follows :

1. Conifers. They possess needle and scale like leaves, *e.g., Pinus, Cedrus, Cupressus,* etc.

2. Broad-leaved trees. They have flattened simple or compound leaves. Most of Dicotyledons, such as *Bombax, Ficus, Butea,* etc.

3. Monocotyledonous trees. They are palms with crown-leaves, feathery in appearance, *e.g., Borassus, Roystenia, Phoenix,* etc.

Today broad leaved trees are dominant in tropical rain forests. Palms, generally occur in coastal regions. They are supposed to be ancient plants.

Conifers make the gymnospermous trees of today. However, they were dominant 300 million years ago but today they are found only in colder regions. Cycadales and Ginkgoales were dominant in Mesozoic era. Today *Cycas revoluta* and *Ginkgo biloba* are typical examples of living fossils.

Tree ferns are now rare and most of them are found in fossilised forms.

Tree cacti are common plants of deserts.

29

Diversity and Morphology of Flowering Plants

Angiosperms or flowering plants form the largest group of plant kingdom, including about 300 families, 8,000 genera and 300,000 species. They are considered to be highest evolved plants on the surface of the earth. Geologically, they are young. From Cretaceous age, the angiosperms eclipsed all other vegetation and now they are dominant. They are found almost every where in each possible type of habitat and climate. They occur in deep lakes, deserts, in beds of seas and even on high peaks of mountains. The species of *Opuntia* (a desert plant-xerophyte) can survive without water in acute desert conditions, whereas on the other hand the species of *Hydrilla* (an aquatic plant) are extremely sensitive to drought conditions. Some species are found on rocks, some in water falls and some are marine. The species of *Rhizophora*, popularly known as mangrove vegetation, are found near the water of the sea. The epiphytes (*Vanda*- an orchid), parasites (*Cuscuta*), saprophytes (*Monotropa*) and even insectivorous plants (*Utricularia, Drosera*, etc.) are also not uncommon.

They may be annual, biennial or perennial herbs, shrubs, trees, climbers, twiners and lianes. On one hand angiosperms may be as minute in size as a pin head, *e.g., Wolffia microscopica*, on the other extremity like eucleptiles (*Eucalyptus*) may reach upto 300 feet in height. For example, plants like pea and gram live only for a few weeks, whereas the famous **Bodhi** tree (*Ficus religiosa*) at Gaya is about 2,500 years old.

The study of various external features of the plants is known as **plant morphology.** The morphology (*morpho*, form; *logos*, discourse or study) deals with the study of forms and features of different plant organs, such as roots, stems, leaves, flowers, seeds and fruits.

DIVERSITY IN THE PLANT BODY

The plant body of an angiosperm consists of a number of organs, *i.e.*, roots, stem, leaf and flower. The flower consists of sepals, petals, stamens, carpels and sometimes also sterile members. Each organ, is made up of a number of tissues. Each tissue consists of many cells of one kind.

A vascular plant begins its existence as a morphologically simple unicellular zygote (2n). The zygote develops into the embryo and thereafter into the mature sporophyte. The development of the complex multicellular body of the seed plant is a result of evolutionary specialization of long duration. The specialization has given rise to the establishment of morphological and physiological differences between the various parts of the plant body and also caused the development of the concept of **plant organs.**

As regards the morphologic nature of the flower it is thought that the flower is homologous with a shoot and the floral parts with leaves.

Fundamental parts of the plant body. The axis, consists of two parts — that portion which is normally aerial is known as the **stem**, and that portion which is subterranean is called the **root**. There are three types of appendages arising from the axis. **1. Leaves**—The strands of vascular tissue pass through the leaves. The leaves are characteristic of the stem and do not occur on the root. The leaves are found to be arranged on the stem in a definite manner, and bear an intimate structural relation to the skeleton of the axis. The leaf is looked upon as the lateral expansion of the stem, continuous with it. All fundamental parts of the stem are concerned with the formation of the leaf. **2. Emergences**—In the appendages of the second rank only the outermost layers of stem, the cortex and the epidermis, are usually present which are known as emergences. The prickles of the rose make a good example of it. **3. Hairs**.—The appendages of the third rank are hairs. These are projections of the outermost layer of the cells. The emergences and hairs occur on both axis and leaves, usually without definite arrangement.

DIVERSITY IN THE ROOT

The seed plants possess a radicle or simply a root meristem at the root end or root pole of the embryo from which the first root of the plant develops upon germination of the seed. The roots are generally divided into two categories: **(*i*) primary**, normal roots, which originate from the embryo and usually persist throughout life, and **(*ii*) adventitious roots**, which arise secondarily from stem, leaf or other tissues and which may be either permanent or temporary.

Characteristic Features of the Roots

1. The root is the descending portion of the axis of the plant, and grows away from light. Usually, the root is not green in colour. However, in certain cases when the root is exposed to light for a prolonged period it turns green in colour, *e.g.*, in *Tinospora*, some orchids and water chestnut.

2. Root growth is generally directed towards gravity, (*i.e.*, positively geotropic) except for the breathing roots (pneumatophores) of mangrove vegetation.

3. The root does not commonly bear buds. However, in certain cases the roots are seen to bear vegetative buds for vegetative propagation, *e.g.*, in *Aegle*, *Trichosanthes*, sweet potato, etc. Such plants are sometimes propagate by root cuttings.

4. The root ends in and is protected by a cap-like structure, known as the root cap. A distinct multiple root cap is seen in the aerial root of screwpine.

5. The root bears unicellular hairs. Root hairs occur in a cluster in the tender part of the root a little behind the apex. Root hairs possess very thin walls made of cellulose. Root hairs absorb water and minerals from the soil.

6. Lateral roots always develop from an inner layer, and are called endogenous. They are produced endogenously from pericycle.

7. Roots are much variable in their shape and structure. It is related either to their function or environmental conditions.

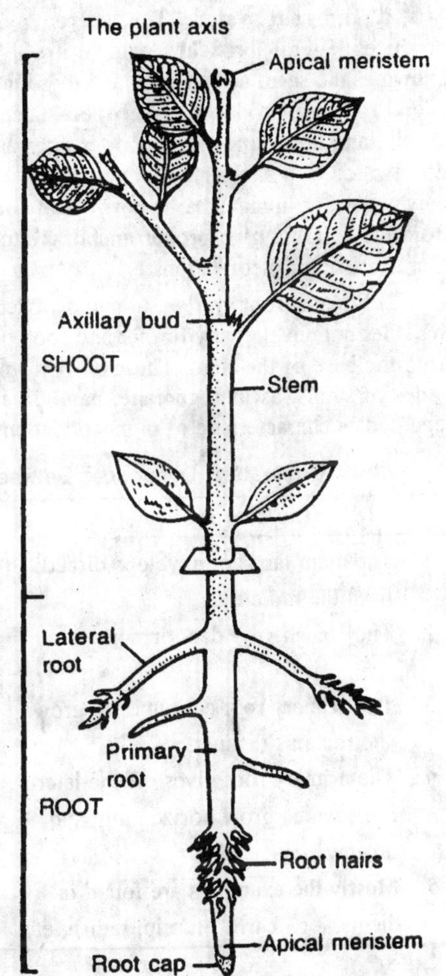

Fig. 29.1. The plant body showing fundamental parts.

ROOT SYSTEMS

The whole extent of the roots of a plant is called the root system. The development of the root system differs fundamentally in vascular plants and may be classified in two categories: **1.** the **tap root system** and **2.** the **fibrous root system.** The tap root system is normally found in dicotyledons and gymnosperms whereas the fibrous root system is commonly met with in monocotyledons. The direct prolongation of the radicle forms the **primary root.** If it persists and continues to grow it is called the **tap root.**

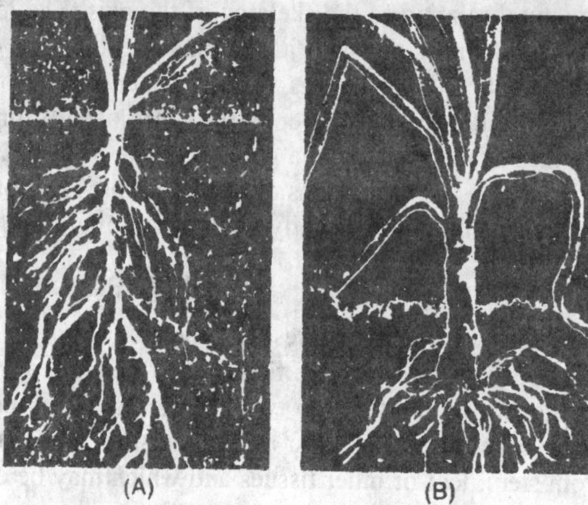

Fig. 29.2. Root systems. A, tap root system; B, fibrous root system of maize.

1. Tap root system. The tap root produces lateral branches which are known as the **secondary roots**, and these in turn produce the **tertiary roots**, and so on. All these lateral roots are produced in **acropetal** succession, *i.e.,* the older and longer roots are away from the tip, and the younger and shorter ones are towards it. The tap root normally grows vertically downwards to a shorter or longer depth while the secondary or tertiary roots grow obliquely downwards, or in many cases horizontally outwards. The tap roots absorb water and mineral salts from the soil and give proper anchorage to the plant. The tap root system may be regarded as characteristic of dicotyledons.

2. Fibrous root system. In monocotyledons the radicle also gives rise to the primary root, but this does not develop any further and soon perishes and is replaced by many thin roots developed from the base of the stem. These are known as the **fibrous roots.** Such roots also develop from nodes of stems, as in sugarcane, bamboo and other grasses. Thus fibrous root system may be regarded as characteristic of monocotyledons.

Table 29.1. Differences Between Tap Root and Fibrous Root Systems

Tap root system	Fibrous root system
1. The main tap root develops directly from the radicle.	1. The radicle does not dominate and its growth stops early during germination.
2. They reach very deep in soil.	2. They are comparatively shallow and spread in a large area than depth.
3. The primary root continues to grow and remains distinct.	3. The primary root stops growing and is not distinct.
4. The primary root gives off the lateral roots which grow horizontally and downwards.	4. Numerous lateral roots arise from the stem base and give rise to a mass of branching adventitious roots.
5. Mostly the examples are found in dicots, *e.g.,* carrot, turnip, radish, etc.	5. The examples are common in monocots, *e.g.,* grasses.

REGIONS OR ZONES OF THE ROOT

The tip of each root is covered by a protective root cap, a thimble-shaped covering of cells which fits over the rapidly growing meristematic region (the calyptrogen). The outer part of the root cap is rough and uneven because its cells are constantly being worn away as the root pushes through the soil. The growing point consists of actively dividing meristematic cells from which all the other tissues of the root are formed. The growing point also gives rise to new root cap cells to replace the ones worn away. Immediately behind the growing point is the **zone of elongation**, here the cells remain undifferentiated but grow rapidly in length by taking in large amounts of water. The growing point is about 1 mm. in length and the zone of elongation is 3 to 5 mm. long; these two are the only parts of the root that account for the continued elongation of the root.

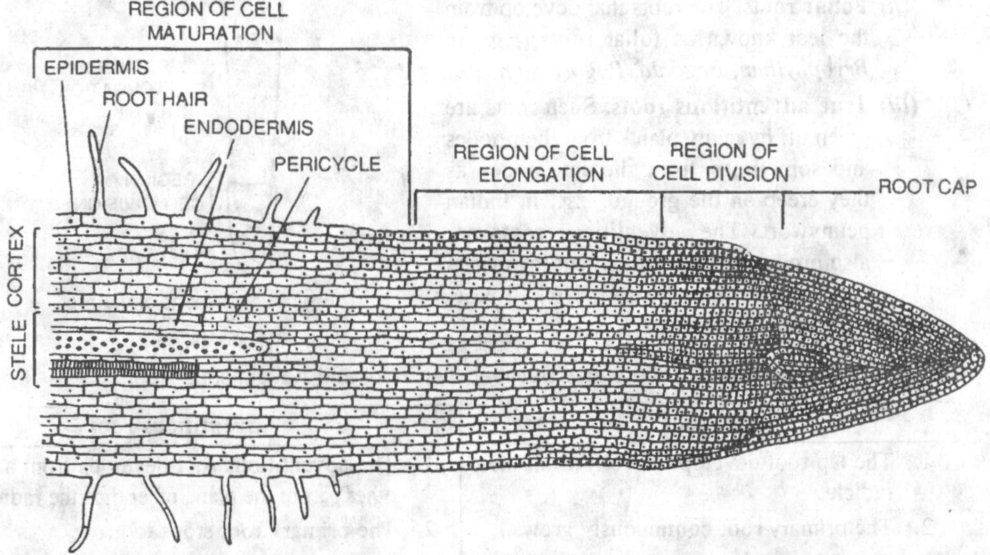

Fig. 29.3. Longitudinal section of root tip showing various regions of growth and root cap.

Above the zone of elongation is the **zone of maturation** characterized externally by a downy covering of whitish root hairs. In this zone the cells differentiate into the permanent tissues of the roots. Each root hair is a slender, elongated, lateral projection from a single epidermal cell, through which most of water and minerals are absorbed. Root hairs are delicate and short-lived; new hairs are constantly formed just behind the zone of elongation and old hairs farther back wither and die as the root elongates. Only a short segment of the root, perhaps one to six cm. long has root hairs.

KINDS OF ROOTS

On the basis of place of origin there are two types of roots; **1.** Tap roots and **2.** Adventitious roots.

1. Tap roots. The root that develops directly from the radicle is known as **primary root.** In most of the cases primary root persists and becomes stronger to form tap root. The tap root normally grows vertically downwards in a shorter or longer depths. Tap root usually produces lateral branches known as **secondary roots.** The secondary roots may further branch to give rise to **tertiary roots.** The primary root may be sparingly or profusely branched according to the need of the plant.

2. Adventitious roots. The roots that develop from any part of the plant other than the radicle are known as **adventitious roots.** They may develop from the base of the stem replacing the primary root or in addition to it, or from any node and internode of the stem or the branch or even from the leaf. On the basis of the nature of their development the adventitious roots may be categorized as follows:

(*i*) **Fibrous roots.** These roots may be given off in clusters from the base of the stem, *e.g.*, in onion, or from the nodes of creeping branches of grasses, or from the lower nodes of the stem, *e.g.*, in maize, sugarcane, etc. Fibrous roots of monocotyledons are all adventitious roots.

(*ii*) **Foliar roots.** The roots that develop from the leaf known as foliar roots, *e.g.*, in *Bryophyllum*, *Begonia*, *Pogostemon*, etc.

(*iii*) **True adventitious roots.** Such roots are given off by many plants from their nodes and sometimes from the internodes as they creep on the ground, *e.g.*, in Indian pennywort. The adventitious roots are also produced from branch cuttings when they are put into the soil, *e.g.*, in rose, sugarcane, tapioca, etc.

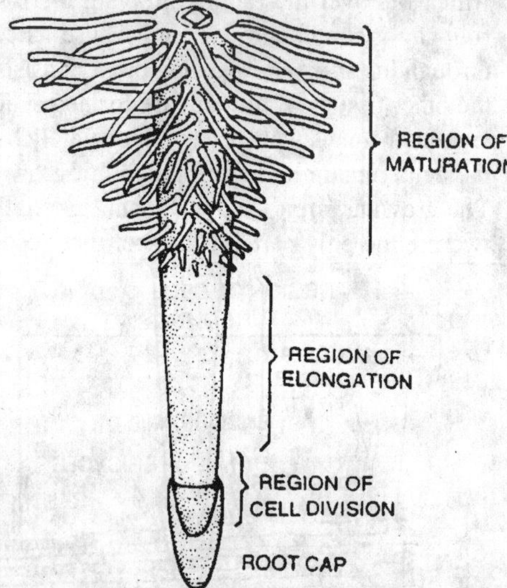

Fig. 29.4. Regions of the root.

Table 29.2. Differences Between Tap Root and Adventitious Root.

Tap root	Adventitious root
1. The tap root develops directly from the radicle.	1. The adventitious root develops from any other part of the plant, other than the radicle.
2. The primary root continuously grows.	2. The primary root stops to grow.
3. They always develops below the soil in the ground.	3. They may be underground or aerial.
4. The single main root gives rise to the fine lateral roots.	4. They form a cluster of roots which may arise from the same point.

MODIFICATIONS OF ROOTS

Many roots of peculiar form and function are known among angiosperms. All these are derived from normal type by the specialized development of one or other of its characteristic functions or structures, in a manner suitable to special circumstances. When the 'modification' of the roots are being considered, it is implied that the simple type of absorbing root in the soil was the earliest type to be evolved and that specialized forms have been derived from it.

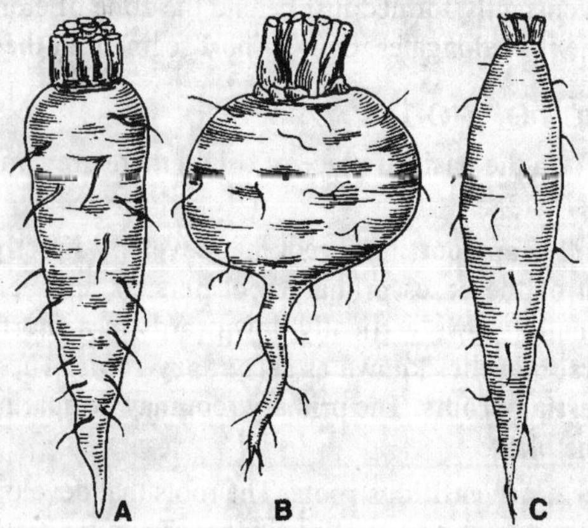

Fig. 29.5. Storage tap roots. A, conical; B, napiform; C, fusiform.

1. MODIFICATIONS OF TAP ROOT

For storage of food

***(i)* Fusiform root.** The primary root is swollen in the middle and gradually tapering towards the apex and the base, (*e.g., Raphanus* = radish).

***(ii)* Napiform root.** The primary root becomes considerably swollen at the upper part (hypocotyl). It becomes almost spherical, and sharply tapering towards lower end, it is called napiform root, *e.g.*, turnip and beet.

***(iii)* Conical root.** In this case the root is broad at the base and it gradually tapers towards the apex like a cone, *e.g.*, carrot.

***(iv)* Tuberous or tubercular root.** Here the root becomes thick and fleshy but does not maintain any specific shape. It is known as tuberous or tubercular root, *e.g.*, in four 'o' clock plant (*Mirabilis*).

***(v)* Branched root modification-Pneumatophores.** Many plants growing in marshy places especially in the tidal swamp forests produce special roots which grow up into the air from beneath the slime are known **pneumatophores** or beathing roots. They arise from long horizontal roots, and as they are negatively geotropic they grow vertically upwards, projecting several inches into the air. They often occur in large numbers around the trunk of the tree. There is a corky layer over each root, which covers even the apex, and the portion in the air bears numerous lenticels. Root hairs are not produced either by the pneuma-tophores themselves or by the horizontal roots from which they arise, but the pneumatophores form short absorbing branches, from their lower portions and hairs are borne on these. They are the centres of active respiration, and their vertical growth may be necessary to place the absorbing roots at the most favourable level. Examples are seen in *Rhizophora, Sonneratia, Heritiera, Avicennia*, etc.

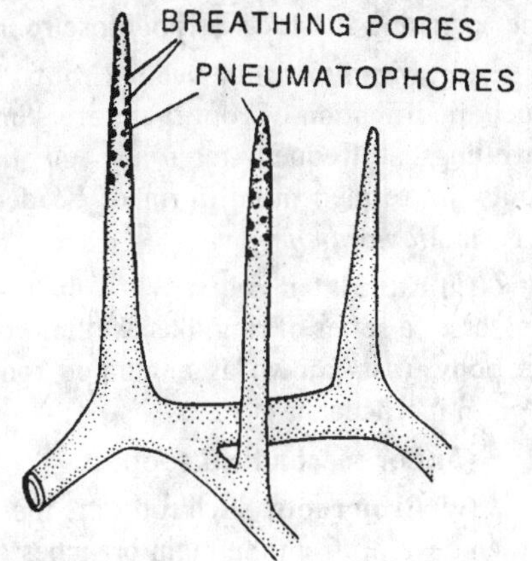

Fig. 29.6. Roots. Pneumatophores arising vertically upwards from an underground root.

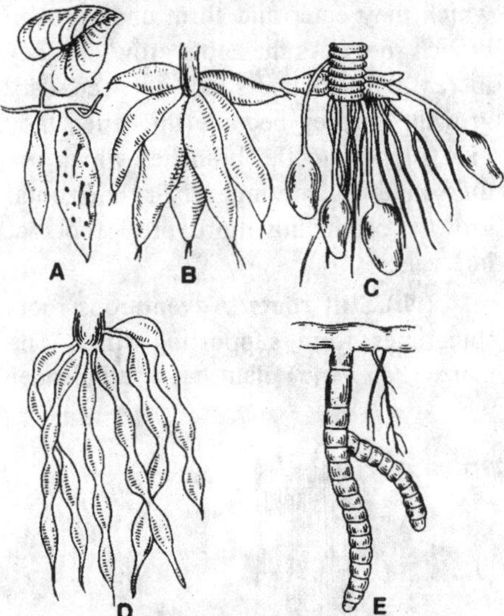

Fig. 29.7. Adventitious modified roots. A, Tuberous roots of sweet potato; B, Fasciculated tuberous roots of *Dahlia*; C, modulated roots of mango ginger; D, Moniliform roots of *Momordica*; E, Annulated roots.

2. MODIFICATIONS OF ADVENTITIOUS ROOTS

(*a*) For storage of food

***(i)* Tuberous or tubercular roots.** Such adventitious roots are swollen without any definite shape, *e.g.*, sweet potato.

***(ii)* Fasciculated roots.** In this type of modification of roots several tubercular roots occur in a cluster at the base of the stem, they are known as fasciculated roots, *e.g.*, in *Asparagus, Dahlia*, etc.

(*iii*) Nodulose roots. In such modification the slender root becomes suddenly swollen near the apex, and is known as nodulose root, *e.g.*, in mango ginger.

(*iv*) Moniliform or beaded roots. In such modification of root there are some swellings at frequent intervals, and the roots are termed moniliform or beaded, *e.g.*, in *Momordica*.

(*v*) Annulated roots. When the root possesses a series of ring like swellings on its body, it is known as annulated root, *e.g.*, in ipecacuanha.

(*b*) For mechanical support

(*vi*) Prop roots. Such roots are those which develop from the main branches of a tree. The best example of this is *Ficus benghalensis*, the Banyan tree. The prop roots grow straight downwards to the soil, which they enter and form underground branch roots. As the roots grow old they increase in thickness due to secondary growth. As they become thickened they give support to the branches which are thus able to continue their horizontal growth, producing more prop roots at intervals.

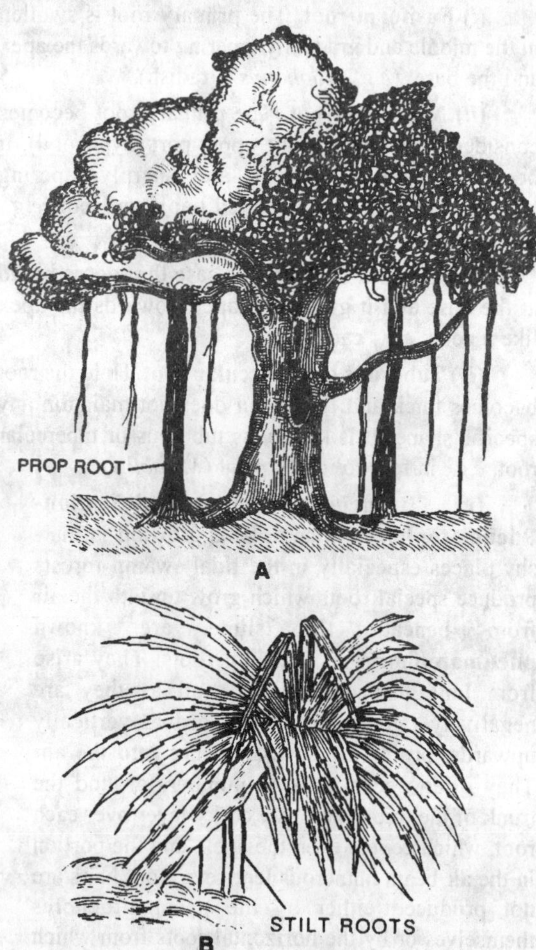

Fig. 29.8. Adventitious modified roots. A, prop root of banyan tree; B, stilt roots of *Pandanus*.

(*vii*) Stilt roots. Adventitious roots sometimes form supporting stilts. The maize (*Zea mays*) plant has a tall slender stem. It develops a cluster of roots from the first one or two nodes above ground level. They grow obliquely downwards into the soil and give added support to the stem. Stilt roots are also well known in *Pandanus*. The stilt roots are less commonly found in Mangroves such as *Rhizophora*, *Sonneratia* and *Avicennia*. Besides the functions of absorption such roots provide additional support to the stem.

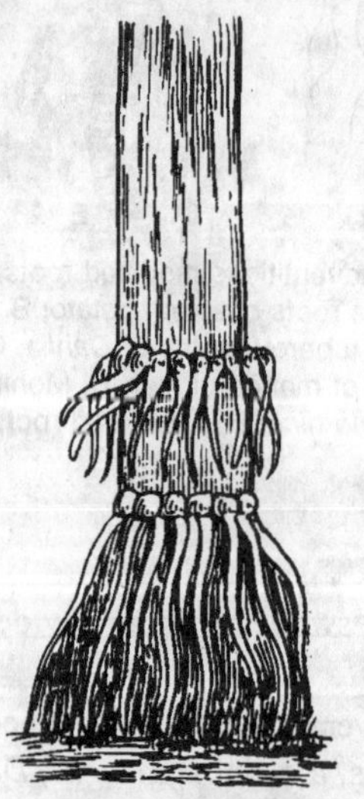

Fig. 29.9. Stilt roots of *Zea mays*.

(*viii*) Climbing roots. Many tropical climbers like *Piper betle*, *P. longum*, *P. nigrum*, *Pothos*, etc., produce roots from their nodes and often from the internodes, by means of which such plants attach themselves to their

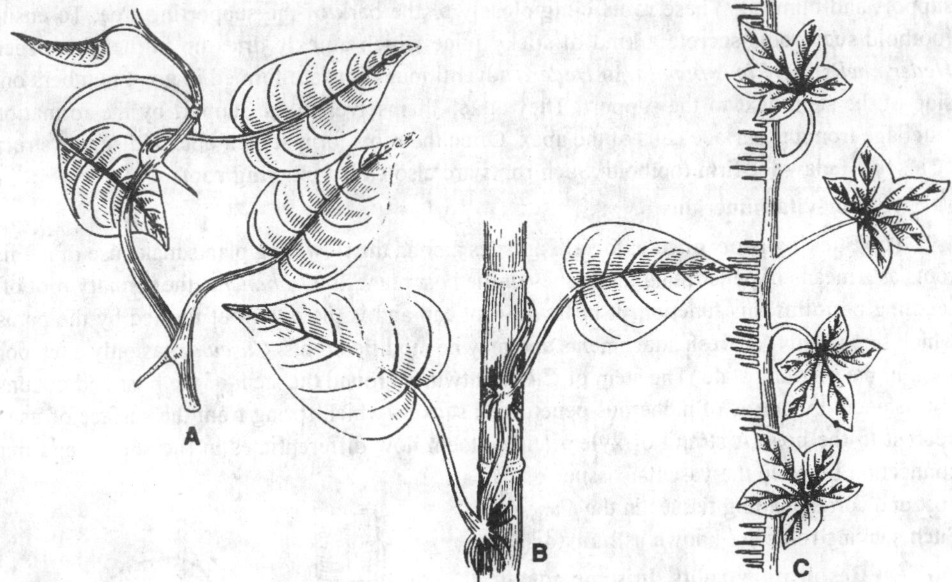

Fig. 29.10. Clinging roots of *Pothos*.

Fig. 29.11. Clinging roots of *Hedera helix*.

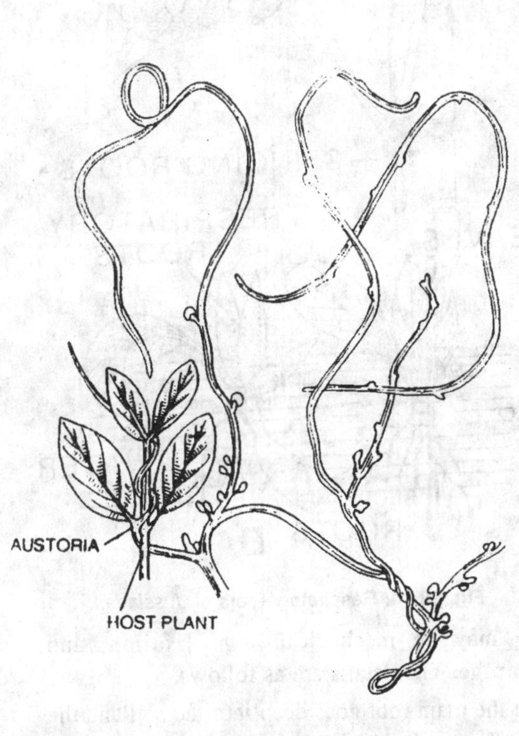

Fig. 29.12. P·rasitic roots of *Cuscuta*.

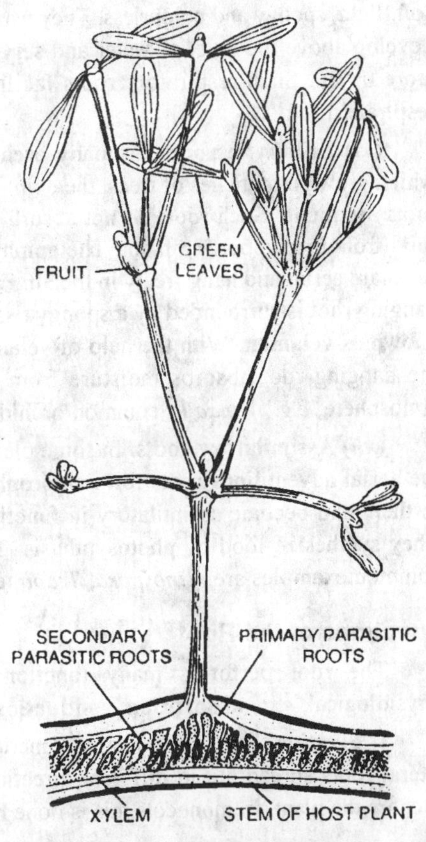

Fig. 29.13. Parasitic roots of *Viscum*.

support and climb it. These roots cling closely to the bark of the supporting tree. To ensure a foothold such roots secrete a kind of sticky juice which quickly dries up in the air as seen in *Hedera helix* and *Ficus pumila*. In *Hedera* adventitious roots are formed in great numbers on the side of the stem next to the support. They attach themselves to the support by the formation of mucilage from the surface cells of the apex. Often the roots form at their apex a disc like structure or a sort of claw for firm foothold. Such roots are also called **clinging roots.**

(c) For vital functions

(ix) Sucking roots or haustoria. Parasites among the flowering plants make use of modified roots as a means of penetrating the tissues of the host plant. In *Orobanche*, the primary root of the seedling performs this function. It bears no root cap and is the only root formed by the parasite, which serves only for fresh attachments and may be modified roots. *Cuscuta*, has only a temporary root, in the seedling state. The stem of *Cuscuta* twines around the stem of the host and obtains its nourishment by means of numerous penetrating suckers which spring from the surface of its stem nearest to the host. A strand of xylem and phloem now differentiates in the sucker and makes connection between the vascular tissues of the host and corresponding tissues in the *Cuscuta*. Such sucking roots are known as **haustoria.**

(x) Respiratory roots. In some aquatic plants such as *Jussiaea* the floating branches develop a kind of adventitious roots which are soft, light, spongy and colourless. They usually develop above the level of water and serve to store up air. Such respiratory roots facilitate respiration.

(xi) Epiphytic roots. In many orchids, which grow on branches of trees, the epiphytic roots are found. Such roots do not absorb any food from the supporting plant. The epiphytic roots are aerial and hang freely in the air. Each hanging root is surrounded by a spongy tissue, known as **velamen.** With the help of velamen the hanging root absorbs moisture from the atmosphere, *e.g., Vanda* (a common orchid).

(xii) Assimilatory roots. In some plants, the aerial adventitious roots form chlorophyll in them and become assimilatory in function. They synthesize food by photosynthesis. The common examples are: *Tinospora, Trapa*, etc.

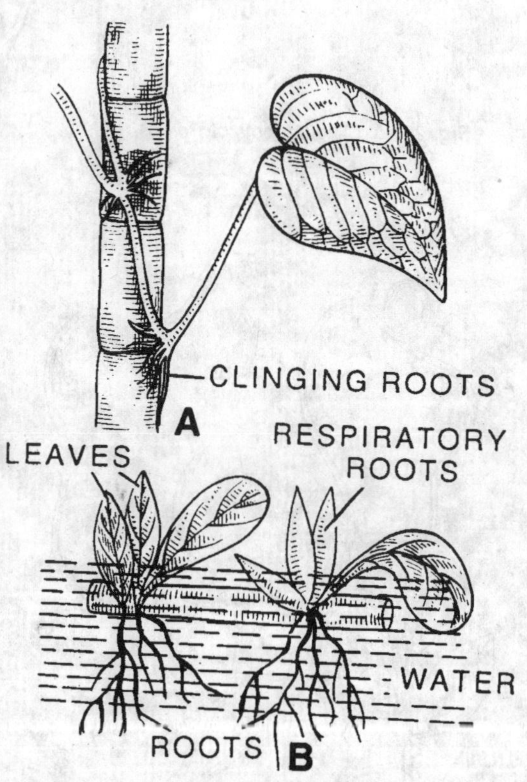

Fig. 29.14. Respiratory roots of *Jussiaea*.

Functions of the Root

The root performs many functions. They may be mechanical, *e.g.,* **fixation**, and physiological, *e.g.,* **absorption**, **conduction** and **storage.** The details are as follows:

1. Fixation. This is mechanical function. Here the main root goes deep into the soil and the lateral roots spread out in different directions, with the result the root system as a whole firmly anchors the plant. In monocots this is done by the fibrous roots.

2. Absorption. The absorption of water and minerals from the soil is done with the help of root hairs which develop in a cluster at a little distance behind the root apex. These root hairs adhere to the soil particles and absorb water and soluble mineral salts from them.

3. Conduction. The root is also concerned in the conduction of water and mineral salts sending them upwards into the stem and the leaves.

4. Storage. There is a certain amount of food stored up in the root, usually in its mature region. As the root grows further the stored food is utilized.

5. Assimilation. On branches of *Tinospora*, long, slender, hanging roots are developed, which form chlorophyll in them and turn green in colour. These roots are assimilatory in function.

Fig. 29.15. Epiphytic roots of *Vanda*.

6. The **hygroscopic roots** are found in epiphytic orchids, *e.g.*, in *Vanda*. They absorb moisture directly from the atmosphere.

7. In most of the genera of Papilionaceae (Fabaceae) (*e.g.*, in gram, pea, many other pulses, etc.), the **symbiotic root nodules** are present. They contain nitrogen fixing bacteria, and help in fixing the atmospheric nitrogen in the soil to increase its fertility.

8. The **assimilatory roots** are green and prepare food by photosynthesis. Such roots are commonly found in *Tinospora*, water chestnut, certain orchids, etc.

9. Some roots help in vegetative propagation, *e.g.*, in sweet potato.

10. In some aquatic plants, *e.g.*, in *Jussiaea*, roots are spongy and store much air in them. They are buoyant and act as **root floats.**

11. The **pneumatophores** of mangrove vegetation (*e.g.*, in *Avecinnia*, *Sonneratia*, etc.) are negatively geotropic and possess lenticels on them for exchange of gases.

12. Roots of several plants develop mutually beneficial partnership with some fungal hyphae to form **mycorrhiza**. *e.g.*, in *Pinus*, *Quercus*, etc.

Table 29.3. Classification of Roots

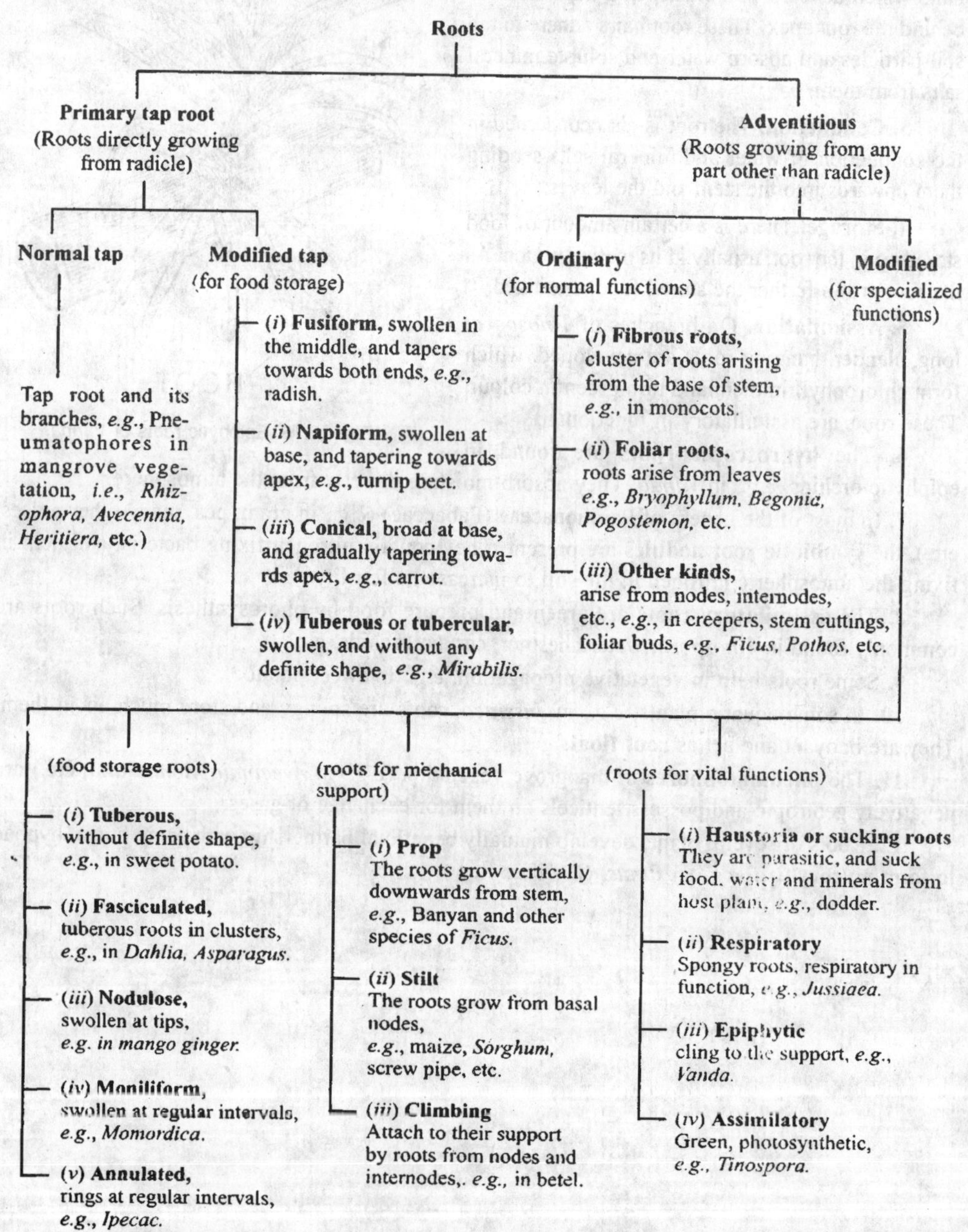

DIVERSITY IN THE STEM

Characteristics of the Stem

1. The stem is the ascending portion of the axis of the plant.
2. It develops directly from the plumule and bears leaves, branches and flowers.
3. When young, it is normally green in colour.
4. The growing apex of the stem remains covered and protected by a number of tiny leaves which arch over it.
5. The stem usually bears multicellular hairs of different kinds.
6. Stem branches and leaves develop exogenously.
7. The stem is provided with nodes and internodes which may not be distinct in all cases.
8. Leaves and branches normally develop from the nodes.
9. When the stem or the branch ends in a vegetative bud it continues to grow upwards or sideways.
10. The stem when ends in a floral bud the growth stops.
11. They are negatively geotropic and positively phototropic.
12. When the plant matures the stem bears branches, flowers and fruits.
13. The stem bears different kinds of buds, *e.g.*, axillary, terminal and floral buds.

Table 29.4. Differences Between Root and Stem

Root	Stem
1. The root is the descending portion of the axis of the plant and grows away from light.	1. The stem is the ascending portion of the axis and it turns towards light.
2. Usually the root is not green in colour.	2. The young stem is normally green in colour.
3. Commonly the root does not bear **buds.**	3. Normally the stem bears buds both floral and vegetative.
4. The root terminates in a cap like structure, known as the **root-cap.**	4. However, the stem ends in a bud-the terminal bud, that consists of growing apex, and surrounded by many young leaves.
5. The root bears unicellular hairs. Root hairs occur in a cluster, a little behind the root apex. As the root grows older root hairs die off, and new root hairs are formed towards the apex.	5. The stem bears mostly multicellular hairs. Shoot hairs are of various kinds and remain scattered all over the surface of the shoot.
6. Root hairs absorb water and minerals from the soil.	6. The shoot hairs prevent evaporation of water from the surface of the stem.
7. The lateral roots are endogenous in origin.	7. The branches are exogenous in origin.
8. The nodes and internodes are absent.	8. The nodes and internodes are present.

Nodes and Internodes. The place on the stem or branch where one or more leaves arise is known as the **node**, and the space between two successive nodes is called the **internode.** In some cases they are very conspicuous, *e.g.*, in bamboos and grasses, while in others they are not clearly marked.

DIVERSITY IN HABIT OF THE PLANT

The plants can be classified into different categories according to their height or according to their life-cycle.

1. According to the height

They may be **(*i*) herbs, (*ii*) shrubs** and **(*iii*) trees.**

(*i*) Herbs. These are small plants with soft stems. They may be **annuals**, *e.g.*, mustard, pea, rice, etc., **biennials**, *e.g.*, beet, carrot, turnip, etc., **perennials**, *e.g.*, canna, ginger, banana, etc.

The **annuals** are those herbs that attain their full growth in one season, living for few months or at the most for one year only. The **biennials** are those herbs that live for two years and they produce flowers and seeds in the second year after which they die off. The **perennials** are those herbs that persist for a number of years. The aerial parts of such plants die every year at the end of the season but new shoots develop again from the underground stem.

(*ii*) Shrubs. They are medium sized plants with hard and woody stems. They branch profusely from near the ground, and thus the plants become bushy in habit without a clear trunk. The examples are *Hibiscus rosa-sinensis*, night jasmine (*Nyctanthes arbortristis*), garden croton, etc.

(*iii*) Trees. They are tall plants with a clear trunk and possess hard and woody stem and branches, *e.g.*, mango, sissoo, nim, teak, jack, etc.

The shrubs and trees are perennials

2. According to the life-cycle

They may be **(*i*) annuals, (*ii*) biennials** and **(*iii*) perennials.**

(*i*) Annuals. Complete their whole life, from seed to fruit in one year or less. In some cases even in a few weeks such as in *Senecio vulgaris*, so that several generations may be passed through in one summer while nothing but the seeds remain through the winter. Biennials and annuals are therefore, typically **monocarpic**, fruiting but once.

(*ii*) Biennials. Last only for two years. In the first season they produce at soil level a very contracted stem bearing a rosette of leaves. During the second season the stem elongates and bears the flowers and fruit, after which the whole plant dies, *e.g.*, carrot, radish, turnip, etc.

(*iii*) Perennials. May be woody, either tree like with one main trunk, or else shrub-like with a cluster of stems. They may also be herbaceous lying down to the ground level each winter and persisting only by under ground organs. The duration of perennials is very variable. Some herbs live only five to six years. Large trees on the other hand may take twenty five to thirty years to attain flowering. The common factor in all perennials is that they are **polycarpic**, *i.e.*, they flower and fruit again and again.

Certain monocotyledonous perennials, on the other hand, are naturally monocarpic. For example, bamboos flower every twenty to thirty years and then die. *Agave americana* flowers once but only after hundred years producing a 50 feet inflorescence and thereafter dies.

DIVERSITY IN BRANCHING OF THE STEM

The mode of arrangement of the branches on the stem is known as **branching.** The branches develop exogenously from the lateral vegetative buds. The lateral branching is of two main types, *i.e.*, **racemose** and **cymose.**

1. Racemose type. In this type of branching the growth of the main stem is indefinite, *i.e.*, it continues to grow indefinitely by its terminal bud and gives off branches in acropetal succession.

Here the lower branches are older and longer than the upper branches, *e.g., Polyalthia, Casuarina, Eucalyptus*, etc.

2. Cymose type. In such type of branching the main axis or the stem does not grow indefinitely due to the limited growth of the terminal or apical bud. Here the growth is definite, and the main stem produces one or more lateral branches which grow more vigorously than the terminal one. The process may be repeated again and again.

The cymose branching may be subdivided as follows: Uniparous cyme and biparous cyme.

A. Uniparous cyme. In such type, only one lateral branch is produced at a time and the branching is known as uniparous or monochasial. It is further subdivided into two types, *i.e.,* helicoid cyme and scorpioid cyme.

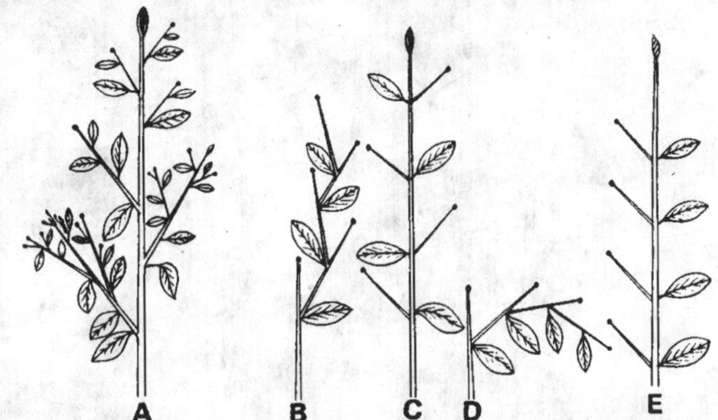

Fig. 29.16. Branching of stem. A, racemose branching; B–C, scorpioid-uniparous cymose branching; D—E, helicoid-uniparous cymose branching.

(*a*) Helicoid cyme. Here the successive branches develop on one side only, *e.g.,* in *Saraca indica.*

(*b*) Scorpioid cyme. Here the successive lateral branches develop on alternate sides, forming a zig-zag, *e.g.,* in *Vitis vinifera.*

(*c*) Biparous cyme. In such type of cymose branching, two lateral axes develop at a time, and it is called biparous or dichasial cyme, *e.g.,* in *Mirabilis, Viscum, Stellaria*, etc.

DIVERSITY IN MODIFICATIONS OF STEMS

In most of the plants the stem is aerial, vertically upward and bears leaves, flowers and fruits. However, some stems are modified into various shapes to carry on special functions, such as perennation, vegetative propagation, and synthesis and storage of food, etc. The various modifications of the stems may be categorized as follows: **1.** Underground modifications of stems, **2.** Sub-aerial modification of stems and **3.** Aerial modifications of stems.

1. Underground Modifications of stems

Some stems develop underground in the soil for the purpose of perennation, vegetative propagation and storage of food. These modifications of stems can be differentiated from the roots as having **(*a*)** nodes and internodes, **(*b*)** scale leaves and **(*c*)** buds of axillary and terminal types. Various types of underground stems are as follows:

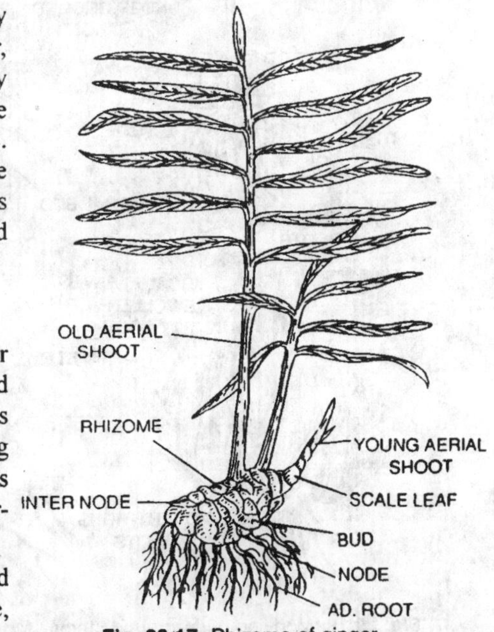

Fig. 29.17. Rhizome of ginger.

(*i*) Rhizomes. They are plagiotropic underground stems of the most varied kind. It is a prostrate, thickened stem, creeping horizontally under the

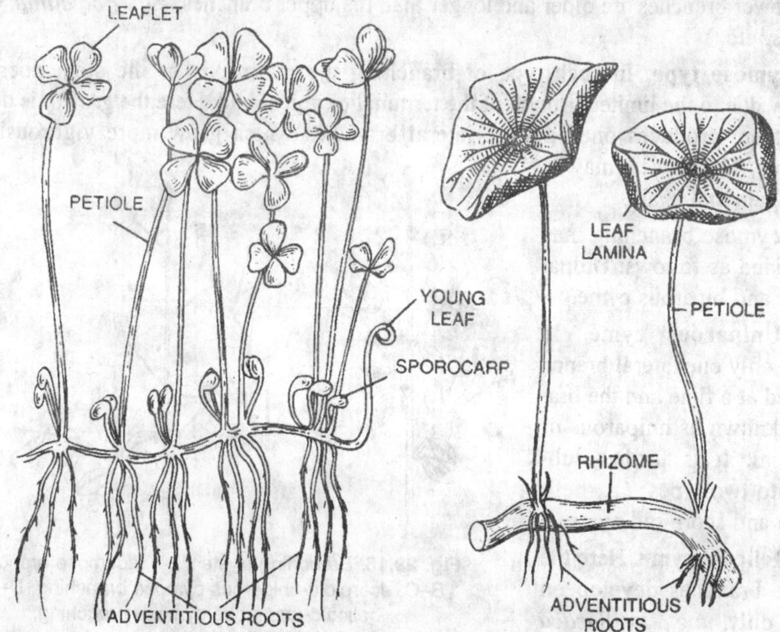

Fig. 29.18. Rhizome of Marsilea and *Nelumbo*.

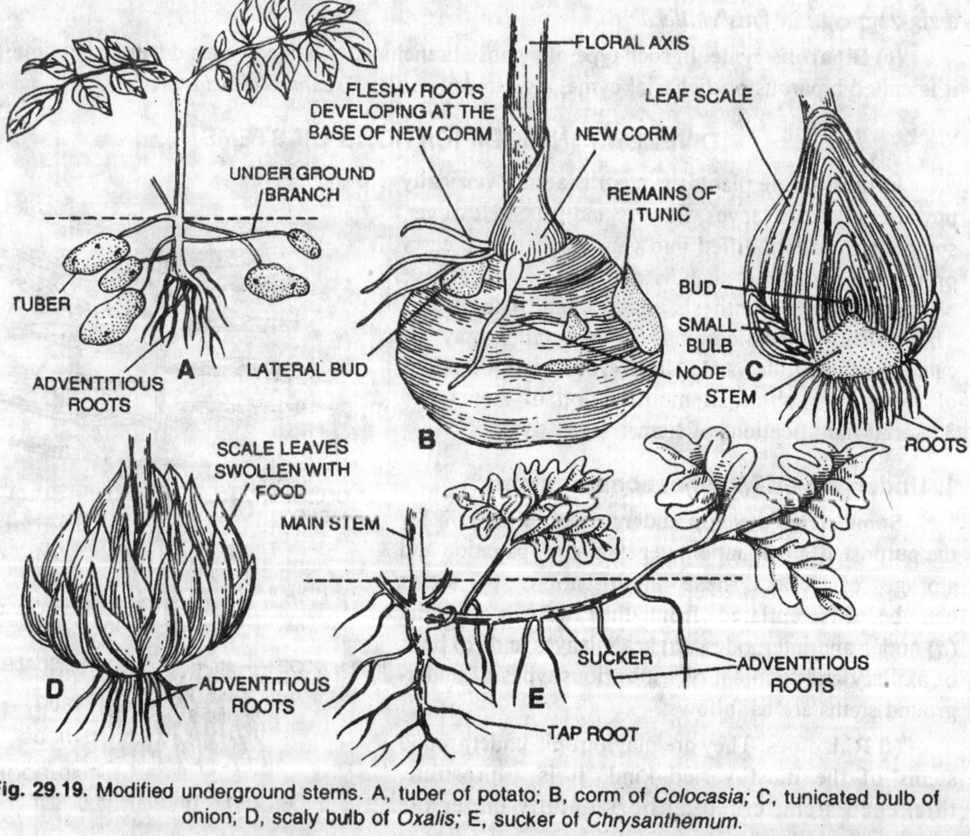

Fig. 29.19. Modified underground stems. A, tuber of potato; B, corm of *Colocasia;* C, tunicated bulb of onion; D, scaly bulb of *Oxalis;* E, sucker of *Chrysanthemum*.

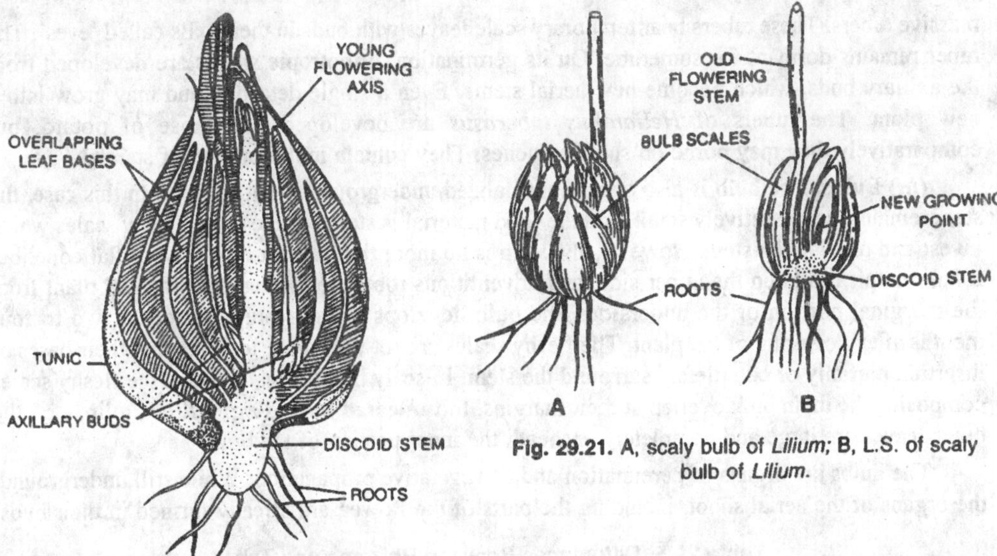

Fig. 29.20. L.S. tunicated bulb of onion (*Allium cepa.*)

Fig. 29.21. A, scaly bulb of *Lilium;* B, L.S. of scaly bulb of *Lilium.*

surface of the soil. It is provided with distinct nodes and long or short internodes. It possesses a bud in the axil of scaly leaf and it ends in a terminal bud. The apex of a rhizome is always eventually trans-formed into an upright shoot, which becomes aerial. As this apex grows through the soil it requires protection, which is afforded by smooth, hard and pointed scales which provide a boring point, as in *Ammophila*. The rhizome may be unbranched or sometimes the axillary buds grow out into short stout branches. It remains dormant underground and on the approach of favourable conditions the terminal bud grows into aerial shoot. In certain cases the branches are separated off, each growing into a new independent plant. In the end of the season or after flowering the aerial parts die down every year and in the following year growth is resumed by one or more lateral buds and this process of growth continues year after year. Normally the rhizomes are horizontal but sometimes they grow vertically upwards, *e.g.,* in *Alocasia*. Examples of rhizomes are found in *Canna*, ginger, *Curcuma*, water lily, *Elattaria*, *Amomum*, lotus, ferns and many aroids.

***(ii)* Tubers.** The tubers are solid, thickened stems or branches serving for storage of food and also vegetative propagation. They are generally formed on rhizomes, on axillary branches or on main stems, either below or above ground. In the case of *Cyperus esculentus* **(kaseru)** the tubers are terminal on rhizome branches. Branch tubers are found in potato. The tuber bearing branches come from the lowest axils of the aerial stem. These branches are weakly geotropic, and when formed above ground they bury themselves and swell at their apices to form the

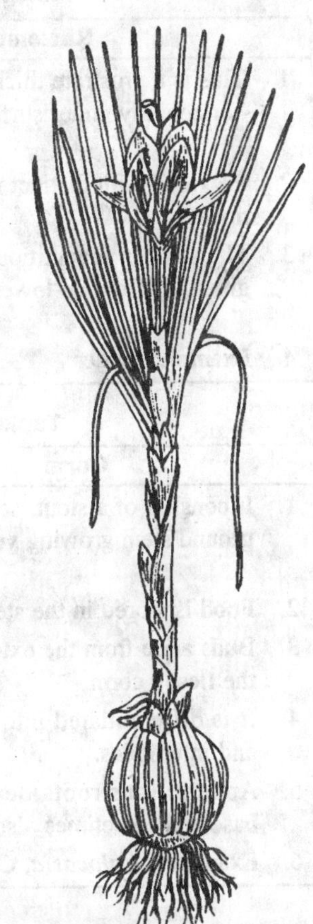

Fig. 29.22. A laminate or tunicated bulb of *Crocus*.

massive tubers. These tubers bear temporary scale leaves with buds in their axils called 'eyes'. The tuber remains dormant for sometime. On its germination, orthotropic shoots are developed from the axiliary buds, which become new aerial stems. Even a single detached bud may grow into a new plant. The tubers of *Helianthus tuberosus* are developed like those of potato, but comparatively they may borne on short branches. They contain inulin instead of starch.

***(iii)* Bulbs.** The bulb is also a short specialized underground shoot or stem. In this case, the stem remains, comparatively small, and the food material is stored in the large, fleshy scales which invest and overlap the stem. However, the stem is no more than a thick disc or very flat cone, but it has an apical bud on the upper side, and adventitious roots are formed in an annual plant from the marginal portion of the underside. The bulb develops in the resting period or two to four months after flowering of the plant. The fleshy scales are set very close together and their bases of insertion partially or completely surround the stem. In scaly bulbs (*e.g., Lilium*), the fleshy scales composing the main bulk overlap at their margins. In **tunicated bulbs** (*e.g., Allium*, tulip, etc.) the outer leaves are large and completely ensheath the inner portion of the bulb.

The bulbs are organs of perennation and of vegetative propagation. While still underground, the organs of the aerial shoot, including the parts of the flower, are already formed in their buds.

Table 29.5. Differences Between Rhizome and Tuber

Rhizome	Tuber
1. This is a prostrate thick stem creeping horizontally under surface of the soil.	1. This is the swollen end of a special underground branch that arises from the axil of a lower leaf, and grows horizontally.
2. Provided with distinct nodes and internodes.	2. A number of eyes are present on its surface which grow in new plants.
3. The slender adventitious roots are given off from its lower side.	3. Here the adventitious roots are absent, and develop only when tuber is planted from the base of aerial shoots.
4. Example: ginger	4. Example: potato

Table 29.6. Differences Between Corm and Bulb

Corm	Bulb
1. It consists of a stout, solid, fleshy underground stem growing vertically.	1. It consists of a short convex disc, from the upper surface of which fleshy scale leaves arise.
2. Food is stored in the stem.	2. Food is stored in the fleshy leaves.
3. Buds arise from the external surface of the fleshy corm.	3. Buds are internal and remain covered by fleshy leaves.
4. It is differentiated into circular nodes and internodes.	4. Being discoid, it is not differentiated into nodes and internodes.
5. Adventitious roots develop from the base and sometimes also from the sides.	5. Adventitious roots develop downward from the bulbous disc.
6. Examples: *Colocasia*, *Crocus*.	6. Examples: onion, garlic.

Table 29.7. Differences Between Rhizome and Corm

Rhizome	Corm
1. This is a prostrate, thickened underground stem.	1. This is stout, solid, fleshy underground stem growing vertically.
2. It is usually branched and rarely unbranched.	2. It is usually unbranched.
3. It is elongated in shape.	3. It is spherical in shape.
4. It contains less amount of food material.	4. It contains a heavy deposit of food material and often grows to a large size.
5. They are generally perennial and live for several years.	5. They can live only for few (1-3) years.
6. Growth is continuous so that succession of rhizome is not produced.	6. New corm is produced every year either above or on the sides of old corm.
7. Examples: ginger, water lily, etc.	7. Examples: *Amorphophallus*, *Colocasia*, *Crocus*, etc.

(*iv*) Corms. The corm is a condensed form of rhizome and consists of a stout, solid, fleshy, underground stem growing in the vertical direction. The swollen stem bears a number of loose, more or less sheathing scale leaves. The size of the stem is due to the large amount of food material stored in it. On top it bears an apical bud, from which summer shoots with leaves and flowers are produced, and it bears the annual adventitious roots, usually from the lower end. The examples are seen in *Crocus* and *Gladiolus*. In these cases the corm is considerably flattened and its length remains shorter than its diameter. The axillary buds are occasionally found at the nodes on the corm, from which side shoots and new corms are given out. After flowering, the leaves of *Crocus* remain active for some time, and the material assimilated by them is stored at the base of the flowering stem, producing a new corm, on top of the old one, from an axillary bud. Other examples are seen in *Colocasia*, *Amorphophallus*, *Colchicum*, etc.

2. Sub-aerial Modifications of stems

In some plants for the purpose of vegetative propagation some of the lower dormant buds of the stem grow out into lateral branches which according to their origin, nature and mode of propagation are known by different names, such as runner, stolon, offset and sucker. They may be subaerial or partly subterranean.

(*i*) Runner. This type of slender, prostrate branch possesses distinct nodes and long internodes. It creeps on the ground and the roots are produced from the nodes. The runner develops

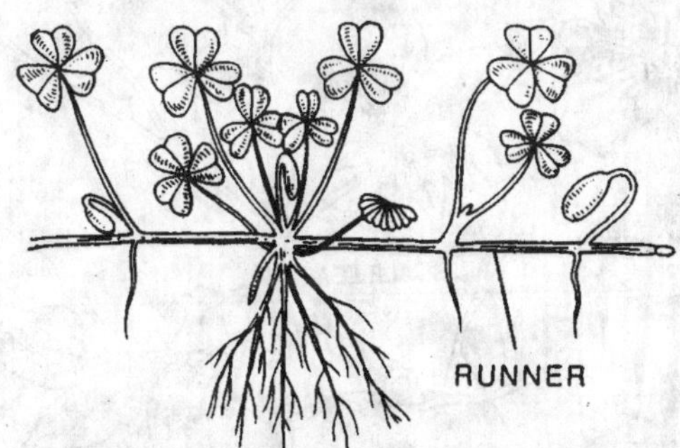

Fig. 29.23. Sub-aerial modification of stem. Runner of *Oxalis*.

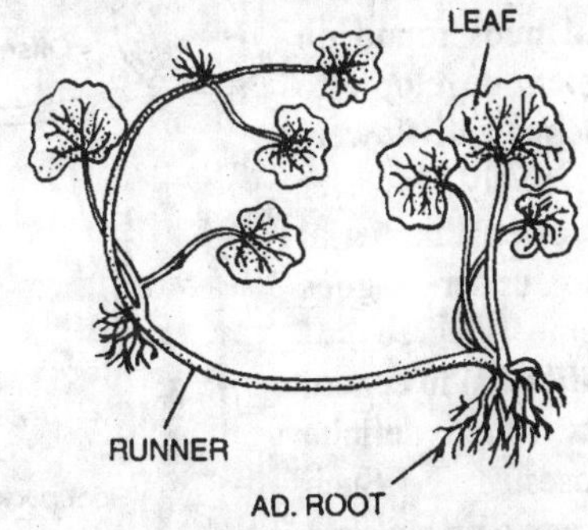

Fig. 29.24. Runner of *Centella asiatica*.

as an axillary bud and creeps some distance away from the mother plant. The roots are given from the nodes and thus new plants develop, *e.g., Oxalis, Cynodon, Fragaria, Centella*, etc.

(ii) Stolon. It is a weak, slender, lateral branch that arises from the base of the shoot buried in the soil. It grows horizontally outwards, and where it touches the ground, the adventitious roots are given out. The branches grow out in different directions, and its terminal bud emerges out of the soil and develops into a new plant. The branch is subterranean. Examples are seen in *Alocasia, Colocasia*, jasmine, etc.

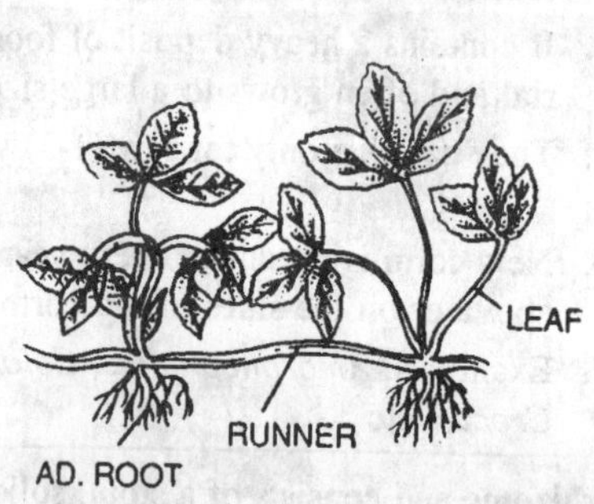

Fig. 29.25. Runner of *Fragaria vesca.*

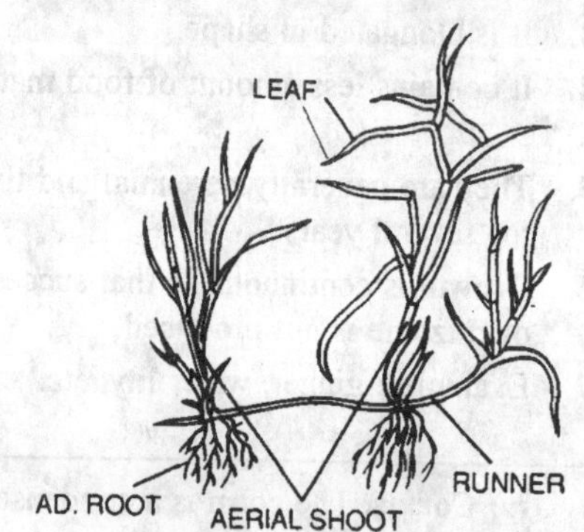

Fig. 29.26. Runner of doob grass.

(iii) Offset. The offset of a plant originates in the axil of a leaf as a short, thick, prostrate branch, and produces at the apex a tuft of leaves above and a cluster of roots below. It often breaks away from the mother plant and an independent daughter plant develops, *e.g., Pistia, Eichornia* (water hyacinth), etc.

(iv) Sucker. Like the stolon the sucker also develops as a lateral branch from the underground part of the stem. It grows obliquely upwards and develops into a new plant. It is always much shorter than a stolon. The sucker develops roots at the base.

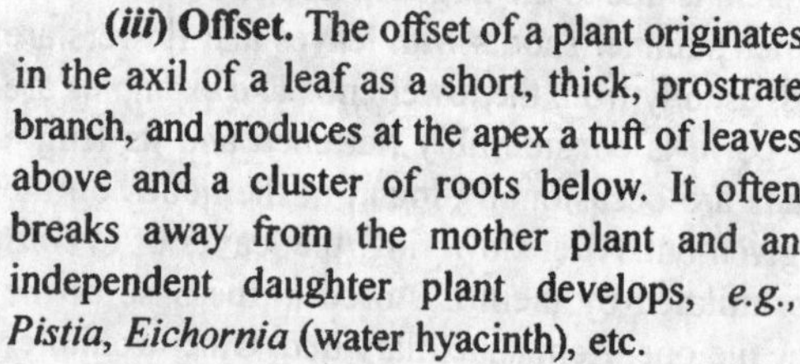

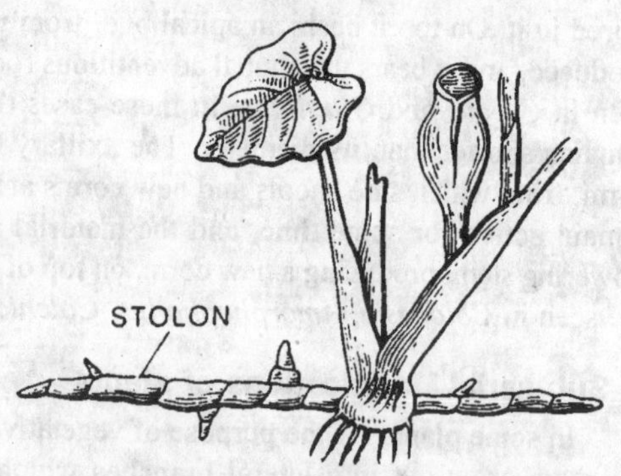

Fig. 29.27. Stolon of *Colocasia.*

This is device for vegetative propagation, as seen in *Chrysanthemum*, mint, etc.

3. Aerial Modifications of stems

Vegetative and floral buds found on the stem develop into branches and flowers respectively. However very often this aerial stem under- goes extreme degree of modification in certain plants for definite purposes. Such metamorphosed organs may be stem

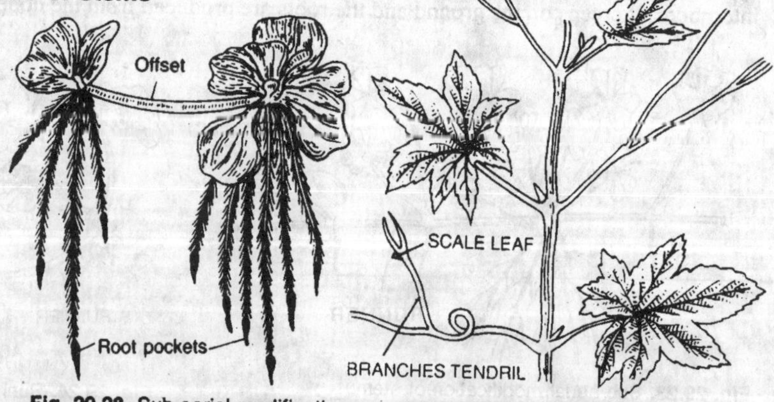

Fig. 29.28. Sub-aerial modification of stem. Offset of *Pistia.*

Fig. 29.29. Leaf opposed tendrils of vine— *Vitis vinifera.*

tendril for climbing, thorn for protection, phylloclade for food manufacture, and bulbil for vegetative propagation.

***(i)* Stem-tendril.** These are highly specialized climbing organs. They are slender often branched and may bear small scale leaves. The climbers by means of these tendrils attach themselves to neighbouring objects and climb them. The stem tendril is the modification of a stem is evident from the fact that it arises in the axil of a leaf or at the apex of a branch. For example in passion flower (*Passiflora*) the axillary bud is modified into the tendril, whereas in vine (*Vitis*) it is the terminal bud which is modified into tendril.

***(ii)* Thorns.** True stem thorns arise from leaf-axils and often bear leaves or flowers as evidence of their branch nature. Examples are seen in *Crataegus* and *Hippophae*. The thorns arise in spring as axillary shoots with normal leaves and with an apical bud. The apex soon stops growth and hardens into a woody point, from which the undeveloped leaves fall away, leaving it naked. In *Ulex*, the main shoots are thickly set with compound thorn branches, but here the leaves are also reduced to spines and the function of photosynthesis develops on the branches including the thorns, which are all green.

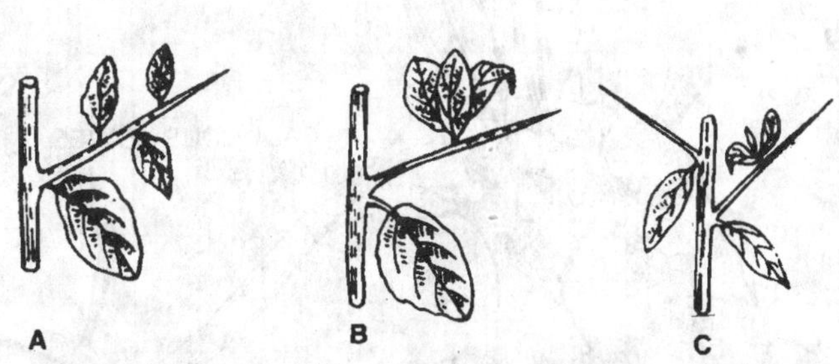

Fig. 29.30. Stem thorns. A, stem thorn of *Pyrus coronaria* bearing leaves; B, the same bearing a bud that has started developing; C, stem thorn of *Alhagi* bearing flowers.

***(iii)* Phylloclade.** This is a green, flattened rounded stem, usually found in the plants of dry and arid habitats. This stem structure has taken on the general appearance and functions of a leaf. Usually the phylloclades represent lateral branches.

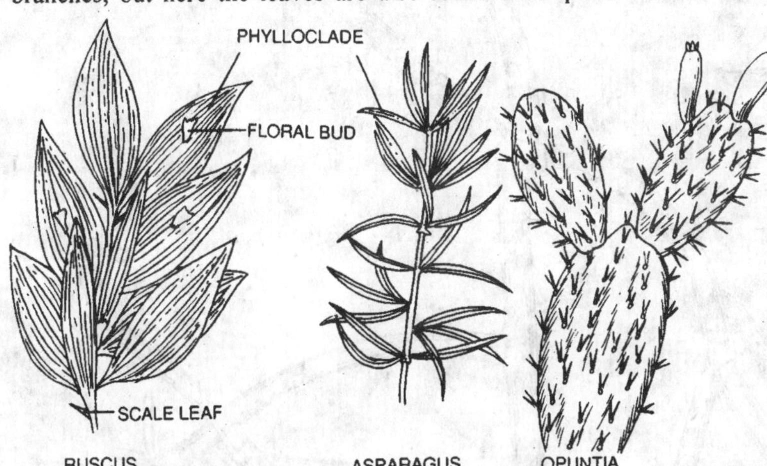

Fig. 29.31. Stem modifications, Cladodes of *Ruscus*, *Asparagus* and phylloclade of *Opuntia*.

They possess distinct nodes and internodes and sometimes they bear modified leaves in the form of scales or spines. Examples are found in *Opuntia*, *Cocoloba*, *Euphorbia*, etc.

The phylloclade of one internode is known as **cladode**, *e.g.*, in *Asparagus*, duckweed (*Lemna*).

In *Ruscus* it consists of two internodes. Here the cladodes are flat, leaf like and perform the function of photosynthesis like foliage leaves.

***(iv)* Bulbils.** They may be described as axillary buds, which become large and fleshy owing to the storage of food material in their leaves. it is a special multicellular body essentially meant for the reproduction of the plant. The bulbils differ from ordinary buds in the fact that they separate from the parent plant, fall to the ground and produce new plants, thus serving for reproduction. The examples are found in *Dioscorea*, *Globba*, *Agave*, onion, etc.

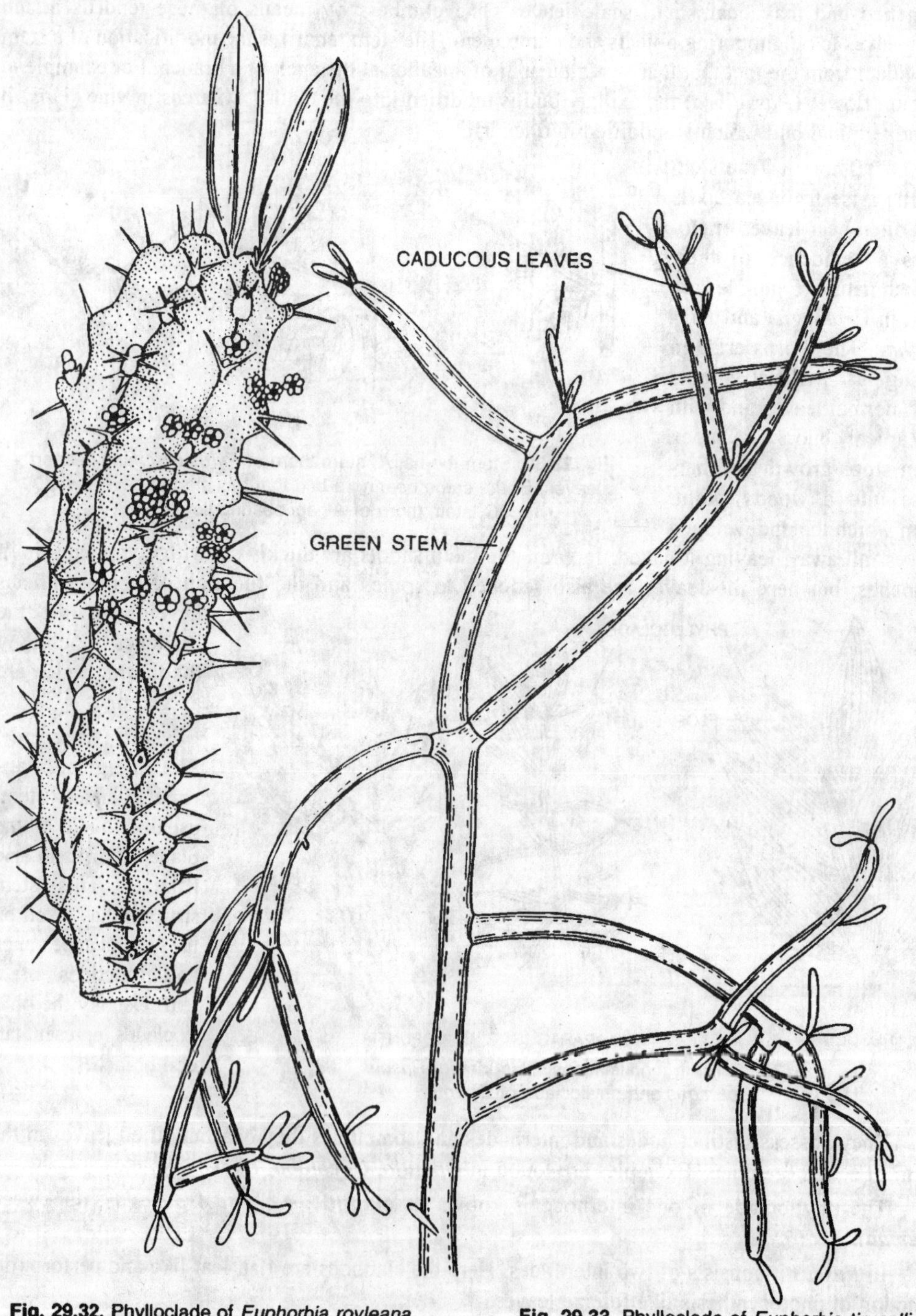

Fig. 29.32. Phylloclade of *Euphorbia royleana*.

Fig. 29.33. Phylloclade of *Euphorbia tirucalli*.

Table 29.8. Differences Between Phylloclade and Cladode

Phylloclade	Cladode
1. This is of unlimited growth.	1. This is of limited growth.
2. It possesses many nodes and internodes.	2. It has only one or two internodes.
3. The main stem and all branches of it are modified into phylloclades.	3. In this case only a branch of the stem that arises from the axil of a leaf becomes modified.
4. **Examples.** *Opuntia*, cocoloba, *Euphorbia*, etc.	4. **Examples:** *Asparagus*, duckweed (*Lemna*).

Functions of the stem

1. Mechanical support. The main stem is usually thick, aerial and stout and keeps the plant in erect position. The main stem and the branches bear leaves and distribute them out on all sides so that all of them may get the maximum sunlight for photosynthesis. They also bear flowers and fruits.

2. Conduction. The stem conducts water and minerals from the root to the leaf. It also transports the prepared food material from the leaf to the different parts of plant body, specifically the storage organs and the growing regions.

3. Food storage. In many cases, particularly in the underground modified stems, it serves as a storehouse of food material (*e.g.,* in potato, ginger, taro, onion, *Amorphophallus*, etc.

4. Water storage. In many fleshy stems of cacti and euphorbias water is stored in sufficient quantity that is used by the plant in extreme dry conditions.

5. Photosynthesis. The young green stems and the modified stems of many cacti and other fleshy plants manufacture food material in the presence of sun light with the help of chloroplasts present in them.

6. Protection. Several modified stems into thorns protect the plants from grazing animals.

7. Perennation. Several modified underground food laden stems also serve as perennating organs, *e.g.,* potato, ginger, onion, taro, etc.

8. Vegetative propagation. Several aerial, sub-aerial and underground modifications of the stem help in vegetative propagation, *e.g.,* runner of doob grass, sucker of mint and setts of sugarcane.

DIVERSITY IN THE LEAF

The leaf is a flattened, lateral outgrowth of the stem in the branch, developing from a node and having a bud in its axil. It is normally green in colour and manufactures food for the whole plant. The leaves take up water and carbon dioxide and convert them into carbohydrates in the presence of sunlight and chlorophyll. Leaves always follow an **acropetal** development and are exogenous in origin.

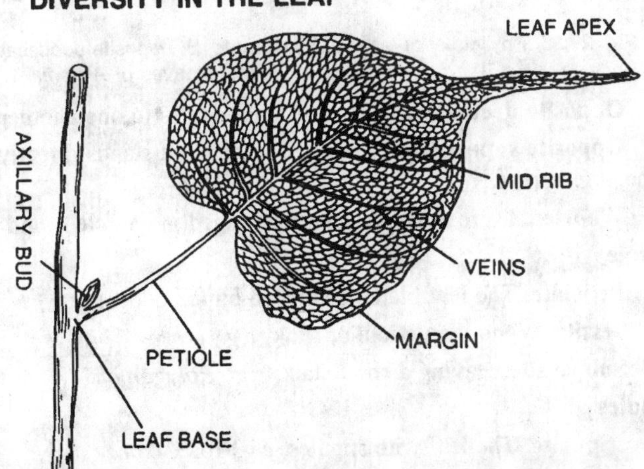

Fig. 29.34. Leaf of *Ficus religiosa* (**pipal**) showing various parts of the leaf.

Parts of a leaf

A typical leaf of *Ficus religiosa* (pipal) has a broad thin, flat structure called the **lamina**. The thin stalk below the lamina is the **petiole.** The lamina possesses a network of **veins.** The veins have both xylem and phloem elements which are continuous with similar tissues of the stem through those of the petiole. A strong vein, known as the **midrib**, runs centrally through the leaf-blade from its base to the apex; this produces thinner lateral veins which in their turn give rise to still thinner veins or veinlets. The lamina is the most important part of the leaf since this is the seat of food manufacture for the whole plant.

Types of a Leaf

Radical. Proceeding from or near the root, *e.g.,* onion, radish, etc.

Cauline. Pertaining to the stem, *e.g.,* palms.

Cauline and ramal. Pertaining to the main stem as well as its branches, *e.g.,* mango.

Phyllotaxy of Leaves

Alternate. A single leaf arising at each node, *e.g.,* *Hibiscus rosa-sinensis.*

Opposite. On different sides of the axis with the bases at the same level.

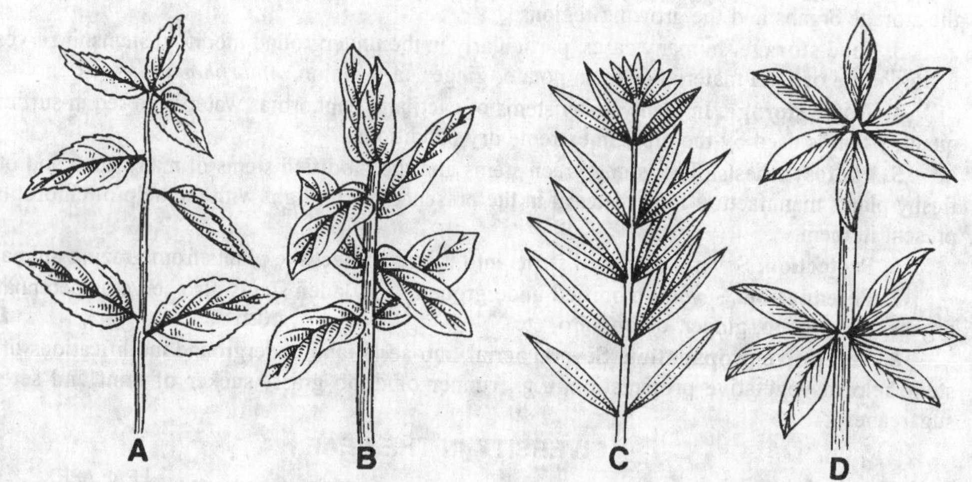

Fig. 29.35. Phyllotaxy of leaves. A, alternate; B, opposite decussate; C, whorled leaves of *Nerium*; D, whorled leaves of *Alstonia*.

Opposite decussate. In pairs at right angles to one another, *e.g., Calotropis.*

Opposite superposed. A pair of leaves that stands directly over the lower pair in the same plane, *e.g.,* guava.

Whorled. More than two leaves arranged in a circle round an axis, *e.g., Spergula, Alstonia.*

Petiole

Petiolate. The leaf blade is situated on the petiole, *e.g., Hibiscus, Ficus,* etc.

Sessile. Without a petiole or stalk, *e.g., Ixora*

Sub-sessile. Having a short stalk, *e.g., Polygonum.*

Stipules

Stipulate. The leaf with stipules, *e.g.,* rose, *Ixora,*

Exstipulate. The leaves having no stipules, *e.g., Ipomoea.*

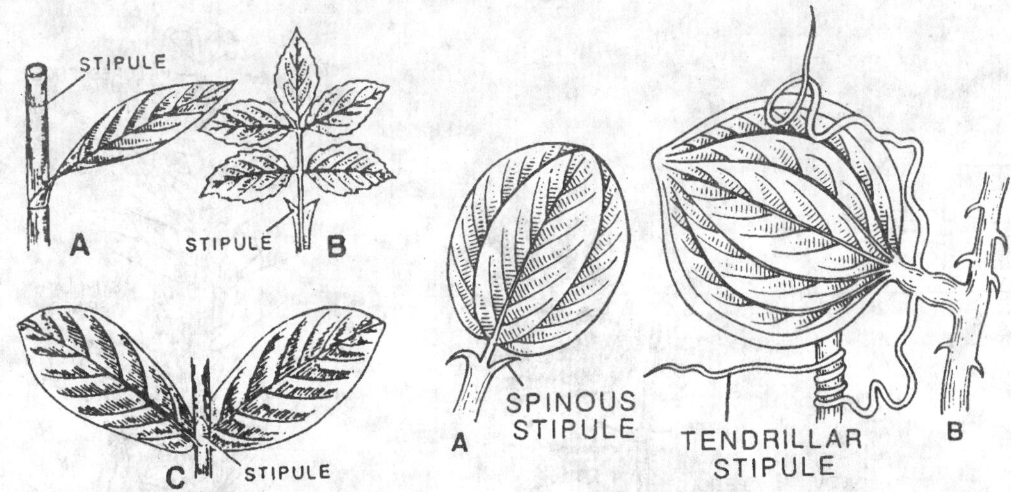

Fig. 29.36. Stipules. A, ochreate stipule of *Polygonum;* B, adnate stipule of rose; C, interpetiolar stipule of *Ixora.*

Fig. 29.37. Stipules. A, spinous stipules of *Zizyphus;* B, tendrillar stipule of *Smilax.*

The stipules may be of several types. They are as follows:

Normally two stipules are developed at the base of a leaf petiole; they may be **foliaceous**, *e.g.*, in *Lathyrus*; **free lateral**, *e.g.*, in China rose; **adnate**, *e.g.*, in rose; **interpetiolar**. *e.g.*, in *Ixora, Spergula*; **spiny**, *e.g.*, in *Acacia, Euphorbia splendens*; **tendrillar**, *e.g.*, in *Smilax*.

Leaf Base of Sessile Leaves

Connate. Two sessile opposite leaves meeting each other across the stem and fusing together, *e.g.*, *Lomicera flava.*

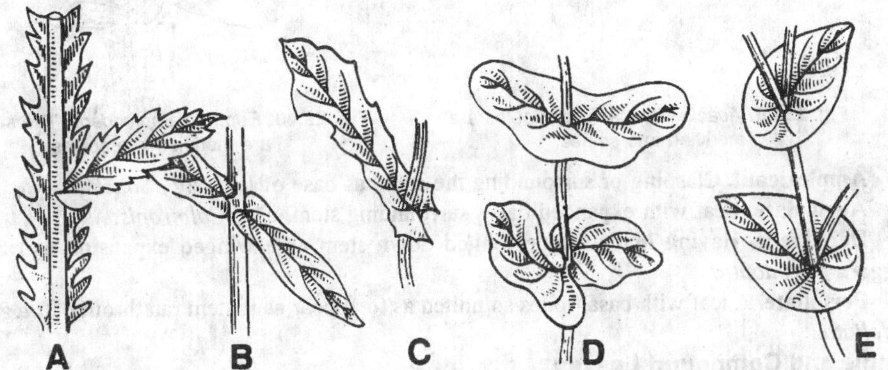

Fig. 29.38. Leaf base of sessile leaves. A, decurrent leaf of *Laggera;* B, auriculate leaf of *Calotropis;* C,'amplexicaul leaf of *Emilia;* D, connate leaf of *Lomicera*; E, perfoliate leaves.

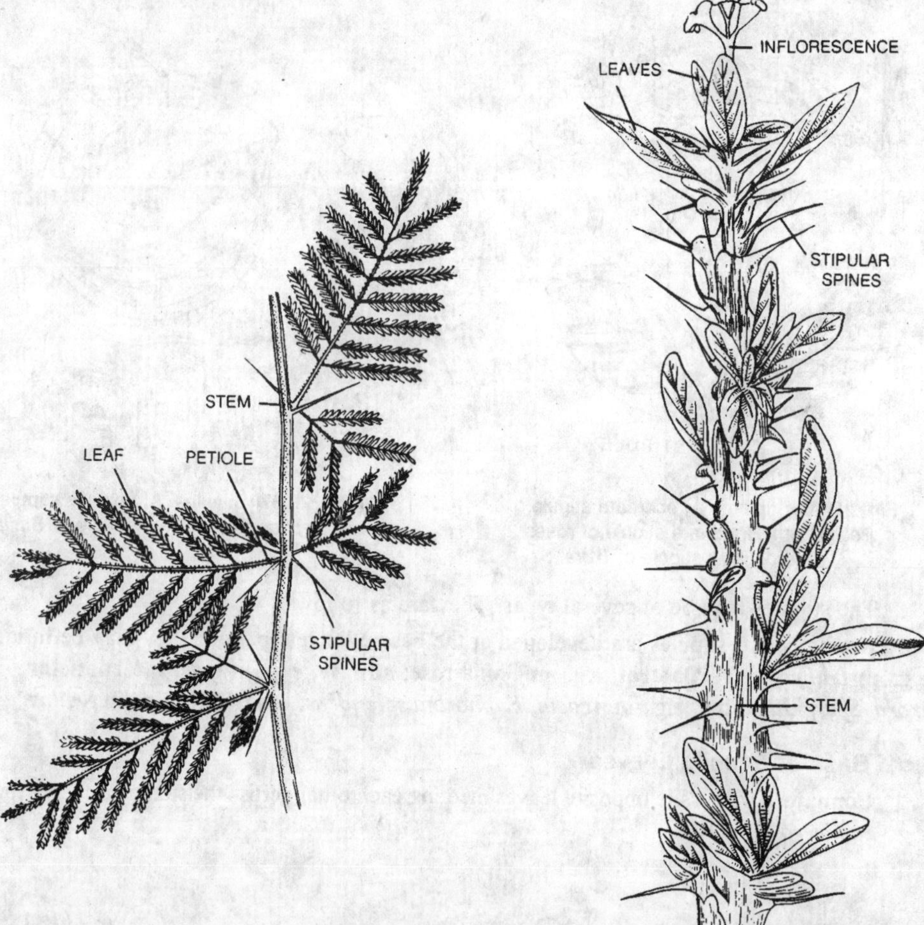

Fig. 29.39. *Acacia nilotica.* The stipules are modified into spines.

Fig. 29.40. *Euphorbia splendens.* The stipules are modified into spines.

Amplexicaul. Clasping or surrounding the stem, as base of leaf, *e.g., Sonchus.*

Auriculate. Leaf with expanded bases surrounding stem, *e.g., Calotropis.*

Decurrent. Having leaf base prolonged down stem as a winged expansion or rib, *e.g., Laggera pterodonta.*

Perfoliate. A leaf with basal lobes so united as to appear as if stem ran through it, *e.g., Aloe perfoliata.*

Simple and Compound Leaves

Simple. A leaf which may be entire or incised to any depth, but not down to the midrib or petiole.

Compound. A leaf made up of two or more leaflets, *e.g.,* pea, and several other members of Leguminosae. The compound leaves may be of several types. They are as follows:

A. Palmately compound Leaf

(*i*) **Unifoliate.** Having one leaflet only, *e.g., Citrus.*

(*ii*) **Bifoliate.** Palmate compound leaf with two leaflets, *e.g., Prinsepia, Balanites.*

(*iii*) **Trifoliate.** Such palmate compound leaf having three leaflets growing from same point, *e.g., Oxalis, Vigna, Trifolium, Melilotus*, etc.

(*iv*) **Quadrifoliate.** Compound palmate leaf with four leaflets arising at a common point, *e.g., Marsilea* (a pteridophyte).

(*v*) **Multifoliate.** Compound palmate leaf with five or more leaflets arising at a common point, *e.g., Gynandropsis pentaphylla, Bombax ceiba.*

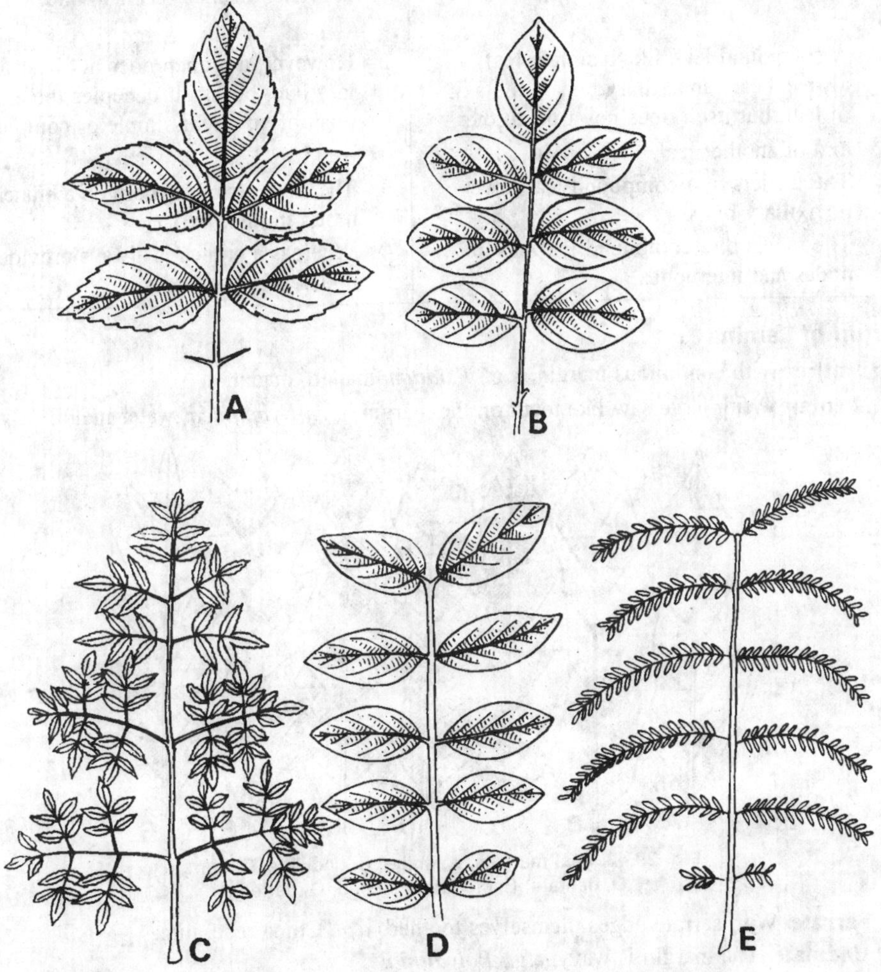

Fig. 29.41. Compound leaves. A, imparipinnate compound leaf of rose; B, imparipinnate compound leaf with alternate leaflets in *Murraya;* D, paripinnate compound leaf of *Cassia fistula;* E, bipinnate compound leaf of *Acacia;* C, tripinnate compound leaf of *Melia azadarach.*

B. Pinnately Compound Leaf

(*a*) **Pinnate.** A compound leaf having leaflets on each side on an axis or midrib.

(*b*) **Unipinnate.** Having leaflets on each side of an axis, *e.g., Cassia.*

(*c*) **Bipinnate.** The central axis produces secondary axis which bears the leaflets, *e.g., Acacia.*

(*d*) **Tripinnate.** The secondary axes produce the tertiary axis which bear the leaflets, *e.g., Moringa.*

(*e*) **Decompound.** More than thrice pinnate, *e.g.*, old leaves of coriander.

(*f*) **Paripinnate.** Pinnately compound without a terminal leaflet, *e.g.*, *Cassia*.

(*g*) **Imparipinnate.** Pinnately compound leaf with an odd terminal leaflet, *e.g.*, pea.

Table 29.9. Differences between compound leaf and leafy branch

Compound leaf	Leafy branch
1. A compound leaf never bears a terminal bud.	1. A branch bears a terminal bud.
2. A compound leaf, like a simple leaf, always bears an axillary bud in the axil of leaf, but itself does not arise in the axil of another leaf.	2. However, a branch does not bear an axillary bud, but itself occupies the axillary position of a leaf-simple or compound.
3. The leaflets of a compound leaf have no axillary buds.	3. The leaves (simple) born on a branch have a bud in their axil.
4. The rachis of a compound leaf has no nodes and internodes.	4. Whereas a branch is always provided with nodes and internodes.

Margin of Lamina

Entire. With continuous margin, *e.g.*, *Psidium*, mango, madar.

Dentate. With large saw like teeth on the margin, *e.g.*, *Nymphaea*, water melon.

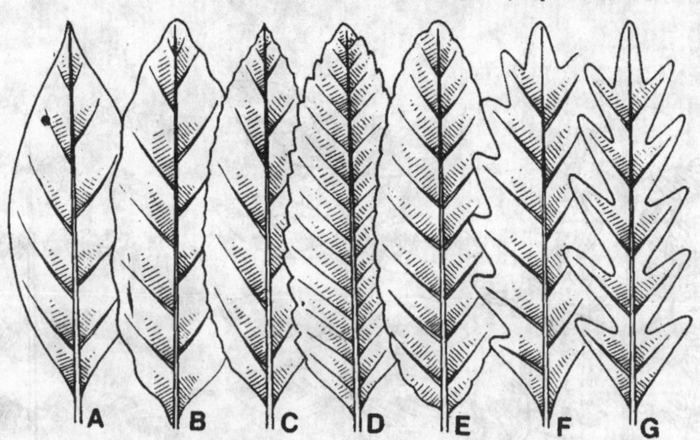

Fig. 29.42. Leaf margins. A, entire; B, undulate; C, serrate; D, dentate; E, crenate; F, lobed; G, parted.

Serrate. With serrate edges themselves toothed, *e.g.*, China rose, nim.

Undulate. The margin is wavy, *e.g.*, *Polyalthia.*

Convolute. Rolled together.

Crenate. With obtusely toothed margin, *e.g.*, *Bryophyllum*, *Centella.*

Lacerate. Having margin or apex deeply cut into irregular lobes, *e.g.*, many members of Ranunculaceae.

Laciniate. Irregularly incised, fringed.

Laciniolate. Minutely incised or fringed.

Ciliate. Bearing fine hairs on the margin, *e.g.*, *Cleome viscosa.*

Crispate. Curled or extremely undulate margin.

Spinous. Bearing many spines, *e.g.*, *Argemone.*

Pectinate. Comb like margin.

Lobed. Leaf margin divided into many lobes, *e.g.*, *Ranunculus.*

Leaf Apex

Acute. Ending in a sharp point forming an acute angle, *e.g.,* mango.

Acuminate. Drawn out into long point; tapering; pointed, *e.g., Ficus religiosa.*

Obtuse. With blunt or rounded end, *e.g.,* Banyan.

Emarginate. Having a notch at apex, *e.g., Bauhinia.*

Truncate. Terminating abruptly, as if tapering end were cut off, *e.g., Caryota urens,*

Mucronate. Abruptly terminated by a sharp spine, *e.g.,* apex of leaflet of *Cassia obtusifolia.*

Cuspidate. Terminating in a point.

Aristate. Provided with awns or with a well developed bristle.

Retuse. Obtuse with a broad shallow notch in middle, *e.g., Oxalis.*

Cirrhose. Leaf with prolongation or mid-rib forming a tendril, *e.g., Gloriosa.*

Apiculate. Forming abruptly to a small tip, *e.g., Dalbergia.*

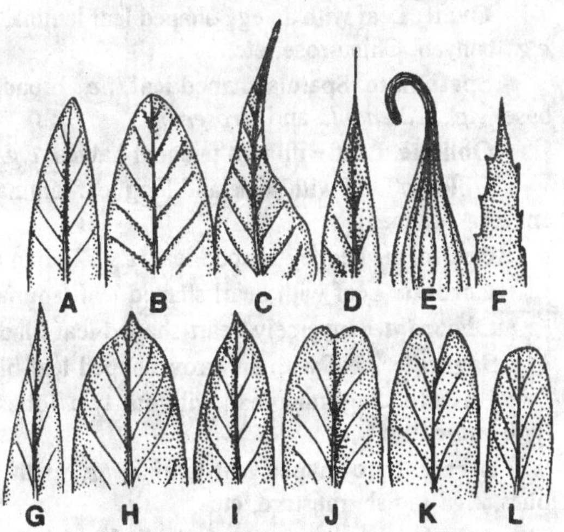

Fig. 29.43. Leaf apices. A, acute; B, obtuse; C, acuminate; D, aristate; E, cirrhose; F, spinous or cuspidate; G, caudate; H, mucronate; I, same as H; J, retuse; K, emarginate; L, truncate.

Leaf Venation

System or disposition of veins in the leaves. They are of several types.

Reticulate. Like net work, *e.g.,* in most of dicots.

Parallel. Parallel veined, *e.g.,* most of monocots.

(*a*) Unicostate reticulate. Having only one principal vein, *e.g.,* mango, banyan, etc.

(*b*) Multicostate reticulate. Having many principal veins, *e.g.,* castor, cucumber, etc.

Shape of Leaf Lamina

Linear. Long and narrow leaf, *e.g.,* many grasses.

Lanceolate. Lance-shaped leaf, *e.g.,* bamboo, *Nerium,* etc.

Round or orbicular. Leaf with a circular leaf blade, *e.g.,* lotus, garden nasturtium, etc.

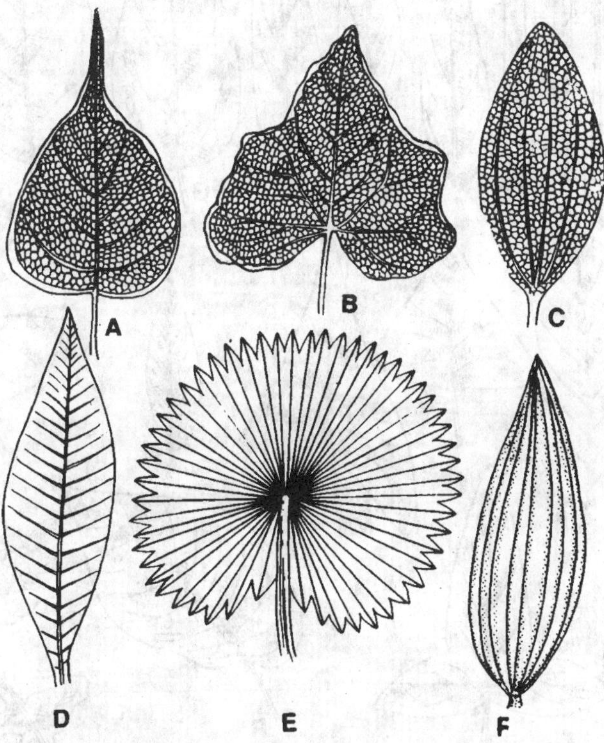

Fig. 29.44. Leaf venation. A, reticulate unicostate venation; B, reticulate multicostate—divergent venation; C, reticulate multicostate—convergent venation; D, parallel unicostate; E, parallel; F, multicostate—convergent.

Elliptical. An ellipse-shaped leaf, *e.g.,* guava, jack, etc.

Ovate. Leaf with an egg-shaped leaf lamina, *i.e.,* slightly broader at the base than at the apex, *e.g.,* banyan, China rose, etc.

Spathulate. Spatula-shaped leaf, *i.e.,* broad and round at the top and narrower towards the base, *e.g., Calendula* and *Drosera.*

Oblique. Leaf with two unequal halves, *e.g., Begonia.*

Oblong. Leaf with wide and long leaf lamina. Here the two margins run more or less straight up, *e.g.,* banana.

Reniform. Kidney-shaped leaf, *e.g.,* Indian pennywort.

Cordate. Leaf with heart shaped leaf lamina, *e.g.,* betel.

Obcordate. Inversely heart-shaped leaf blade, *e.g.,* wood-sorrel.

Sagittate. Leaf with an arrow shaped leaf blade, *e.g.,* arrow-head and some aroids.

Hastate. Sagittate leaf with its two lobes directed outside, *e.g.,* water bindweed and *Typhonium.*

Lyrate. Lyre-shaped leaf lamina, *i.e.,* with a large terminal lobe and some smaller lateral lobes, *e.g.,* radish, mustard, etc.

Acicular. Long, narrow and cylindrical leaf, *i.e.,* needle-shaped, *e.g.,* pine (a gymnosperm).

Cuneate. Wedge shaped leaf, *e.g.,* water lettuce.

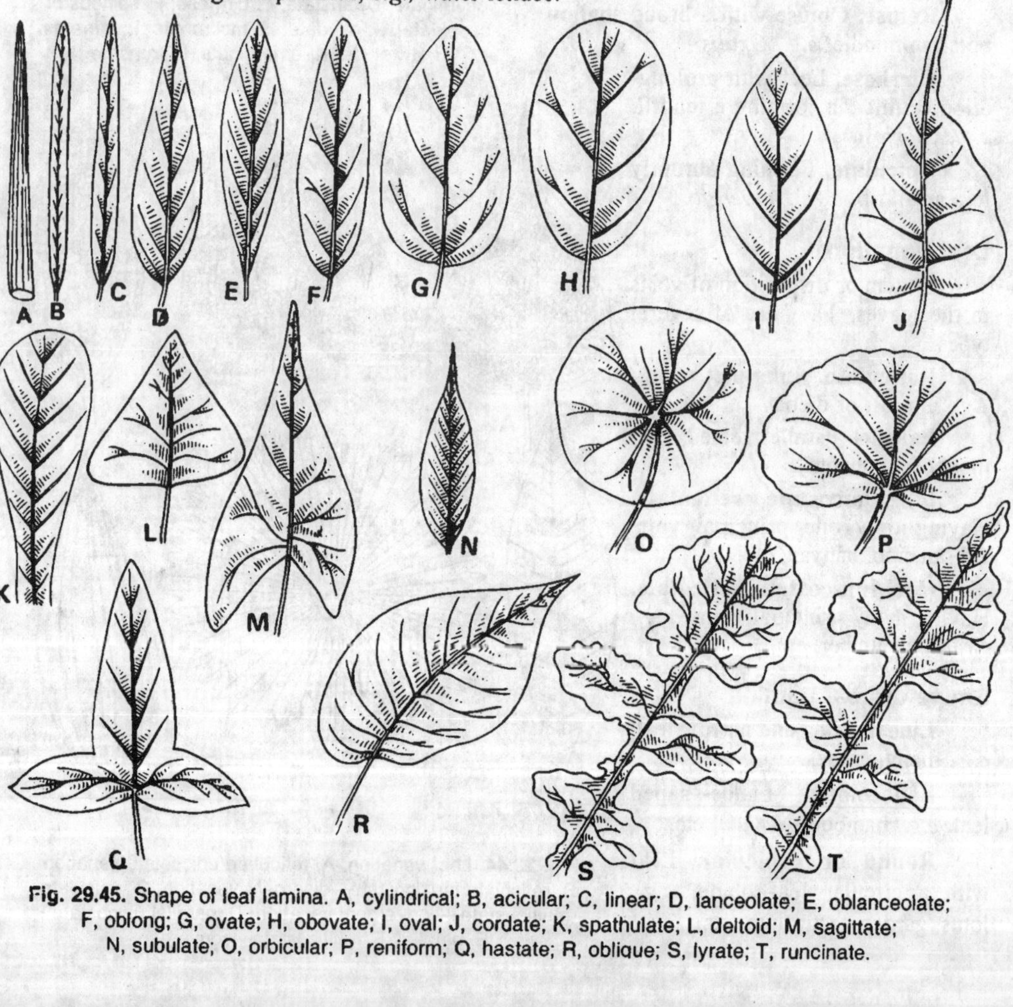

Fig. 29.45. Shape of leaf lamina. A, cylindrical; B, acicular; C, linear; D, lanceolate; E, oblanceolate; F, oblong; G, ovate; H, obovate; I, oval; J, cordate; K, spathulate; L, deltoid; M, sagittate; N, subulate; O, orbicular; P, reniform; Q, hastate; R, oblique; S, lyrate; T, runcinate.

DIVERSITY IN MODIFICATIONS OF LEAVES

The leaves of several plants are metamorphosed into certain special structures which carry on special functions. They are as follows:

1. Bladder. In bladderwort (*Utricularia*) the leaves are very much segmented and they simulate roots excepting that they are green in colour. Some of these segments become modified into **bladders.** Each bladder is about 3 mm in diameter and is provided with a trapdoor entrance. The trap door acts as a short of valve which can be pushed open inwards from outside, but never from inside to outside. This trap-door entrance allows aquatic animalcules to pass in, but never to come out. The inner surface of the bladder is dotted all over with numerous digestive glands. These glands secrete the digestive agent and absorb the digestive products.

2. Pitcher. In the pitcher plant (*Nepenthes*) the leaf becomes modified into a **pitcher.** The morphology of the leaf of pitcher plant is that the pitcher itself is the modification of leaf blade, the tendrillar stalk supporting the pitcher is the modification of the petiole, and the laminated structure that of the leaf base. The inner surface of the pitcher corresponds to the upper surface of the leaf and the lid of the pitcher arises as an outgrowth of leaf apex. The function of the pitcher is to capture and digest insect. When young the mouth of the pitcher remains closed by its lid which later on opens and stands erect. The inner side of the pitcher remains covered with numerous, smooth and sharp hairs, all pointing downwards. Lower down the inner surface numerous digestive glands are found. The digestive agent, secreted by glands, is trypsin which helps in digesting the proteins. In the sundew (*Drosera*) the upper surface of the leaf is covered with glandular hairs which are sensitive to touch and capture insects.

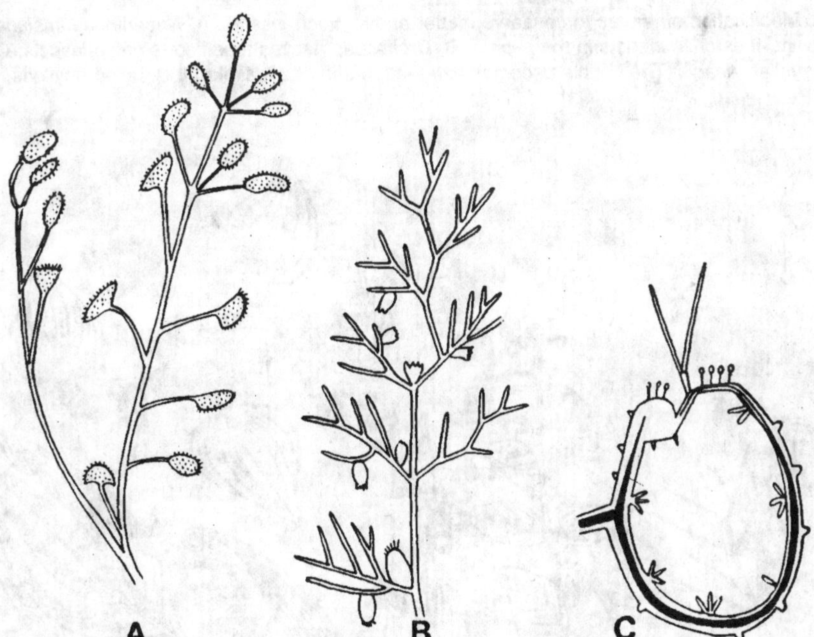

Fig. 29.46. Leaf modification to catch insects. A, *Drosera;* B,—C, *Utricularia.* In this case the leaf segments become modified into bladders that catch insects. In C a bladder is shown in L.S. to illustrate its internal structure.

3. Phyllode. In some species of Australian *Acacia* the lamina of the leaf is absent but the petiole is so flattened as to appear leaf-like. These flattened petioles are known as phyllodes and they are so developed as to place their surfaces in the vertical plane. The normal leaf is pinnately compound and only develops in the seedling stage. The phyllode then carries all the functions of the leaf.

4. Leaf-tendrils. In certain plants the leaves become modified into slender, wire-like-coiled structures known as **tendrils.** The leaf may be partially or wholly modified into tendrils.

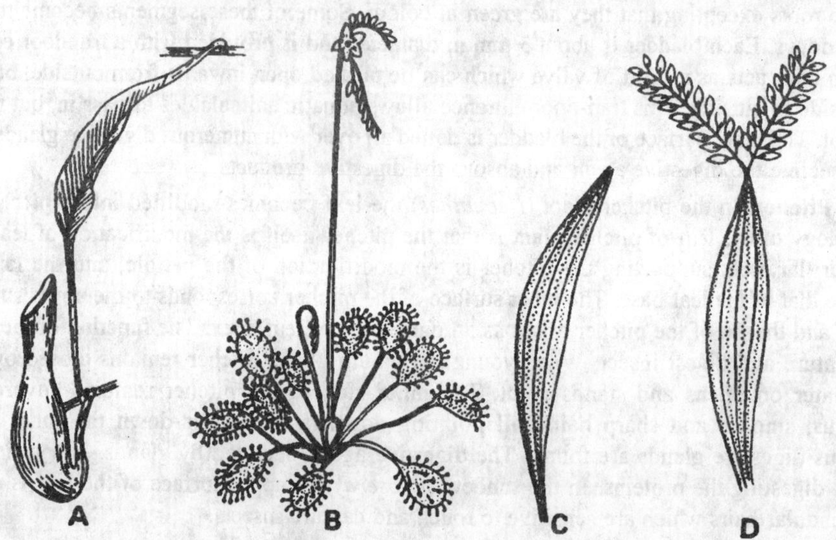

Fig. 29.47. Modification of leaves to conserve matter and to catch insects. A, *Nepenthes khasiana* or the pitcher plant. It catches insects in the pitcher; B, *Drosera* sp. leaves modified to catch insects; C—D, Australian *Acacia*. The petiole becomes flattened, green and leaf like. It is called phyllode.

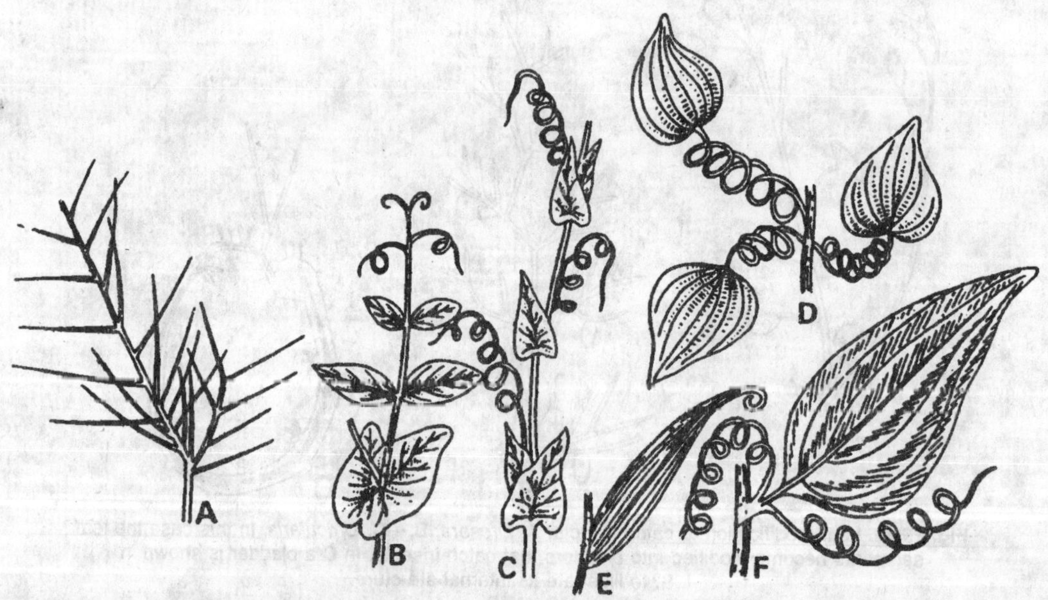

Fig. 29.48. Leaf tendrils and spines. A, leaf spines of *Hakea* ; B, leaflet tendril of *Pisum sativum;* C, leaf tendril or *Lathyrus aphaca.* Note the leaf like stipules; D, petiolar tendrils of *Clematis;* E,tendrillar leaf tip of *Gloriosa superba* ; F, stipular tendrils of *Smilax.*

For example, in pea only the upper leaflets are modified into tendrils. In *Naravelia* and *Bignonia* the terminal leaflet converts into a tendril. In *Gloriosa* the leaf apex becomes modified into a tendril. In *Nepenthes* the petiole acts as a tendrillar structure. In *Lathyrus aphaca* the whole leaf is being converted into a single tendril while the two foliaceous stipules act like the leaves. Tendrils are always climbing organs and are sensitive to contact with any solid body. Whenever a tendril comes in contact with a neighbouring object it coils around it and helps the plant to climb.

5. Leaf-spines. In *Hakea* and *Opuntia* the whole leaves are modified into spines. The morphological nature of such spines can be pointed out by the presence of a bud in their axis. In such cases the stems become green and carry on photosynthesis. In *Acacia nilotica* and *Zizyphus* the stipules are modified into spines. The position of such spines on either side of the leaf base shows their morphological nature. In *Solanum xanthocarpum, Argemone mexicana, Aloe, Acanthus*, etc., the surface and margins of leaf are covered with spines. Morphologically, they are the modified parts of the leaves.

6. Scale-leaves. They are thin, dry, papery, stalkless membranous structures usually brown in colour. Sometimes scale-leaves are thick and fleshy as found in onion. In *Casuarina, Tamarix, Asparagus, Ruscus*, etc., the leaves are reduced to scales. In such cases the stem becomes green, flattened and leaf like to perform functions of leaf. Scale leaves are common on underground stems where they cover and protect the axillary buds under unfavourable conditions. The scale-leaves are also common on angiospermic parasites where they replace the green vegetative leaves.

Functions of the Leaves

The functions of the leaves are as follows:

1. Manufacture of carbohydrates. The main function of the leaf is to manufacture food particularly carbohydrates. Chloroplasts found in the leaf cells, trap the solar energy which is then utilized in the synthesis of carbohydrates from carbon dioxide and water by the process of photosynthesis. The upper side of the leaf contains abundance of the chloroplasts and the sun rays fall directly on the upper surface and normally the manufacture of food takes place in this region of the leaf.

2. Exchange of gases. To facilitate the exchange of gases between the atmosphere and the plant body numerous minute openings called **stomata**, develop, usually on the undersurface of the leaf. The stomata remain open during day light. In the process of respiration of all the living cells the oxygen is taken in and carbon dioxide is given out while in photosynthesis the green cells absorb carbon dioxide and give out oxygen. The respiration of the living cells goes on round the clock, while the photosynthesis takes place only in daytime.

3. Transpiration. Although large quantities of water are absorbed by plants from the soil but only a small amount of it is utilized. The excess of water is lost from the aerial parts of plants in the form of water vapours. This is called transpiration. It occurs mostly through stomata, but sometimes it also takes place through cuticle and lenticels. The transpiration is necessary as it helps in the transport of water within the plant body and also regulates its temperature.

4. Storage of food. Fleshy leaves of succulents, such as Indian aloe, purslane and fleshy scale leaves of onion store up water and food material for the future use of the plants. Fleshy leaves of many desert plants store a large quantity of water, mucilage and food material.

5. Vegetative propagation. The leaves of *Bryophyllum, Begonia* and *Kalanchoe* produce buds on their margins. Each such bud develops into a new plant.

6. Protection. The leaves also give necessary protection to the axillary bud. The leaves modified into thorns and spines (*e.g.*, in *Berberis, Aegle*), give protection to the plants from animals.

DIVERSITY IN THE INFLORESCENCE

In many cases the vegetative axis bears a solitary flower either at its apex (*i.e.,* terminal flower) or in the axil of a leaf (*i.e.,* axillary flower). In so many other cases the floral region remains quite distinct from the vegetative region. Thus, the floral region consisting of a collection of flowers is known as the **inflorescence.** The inflorescence may be terminal or axillary, and may be branched in several ways. This way, depending on the type of branching several kinds of inflorescence have been developed, and, they may be classified into two distinct groups—**(1) Racemose** (*i.e.,* indefinite) and **(2) Cymose** (*i.e.,* definite)

1. Racemose Inflorescence

In this type of inflorescence the main axis does not end in a flower, but it grows continuously and develops flowers on its lateral sides in acropetal succession (*i.e.,* the lower or outer flowers are older than the upper or inner ones). The various forms of racemose inflorescence may be described under three heads. They are as follows:

(*i*) With the main axis elongated, *i.e.,* (*i*) **raceme;** (*ii*) **spike;** (*iii*) **spikelets;** (*iv*) **catkin** and (*v*) **spadix.**

(*ii*) With the main axis shortened, *i.e.,* **(*i*) corymb and (*ii*) umbel.**

(*iii*) With the main axis flattened, *i.e.,* **capitulum or head.**

(*i*) Main Axis Elongated

(*a*) Raceme. In such cases the main axis remains elongated and it bears laterally a number of stalked flowers. The lower or older flowers possess longer stalks than the upper or younger ones, *e.g.,* radish (*Raphanus sativus*), mustard (*Brassica campestris*), etc. When the main axis of raceme is branched and the lateral branches bear the flowers, the inflorescence is known as **compound raceme** or **panicle**, *e.g.,* **neem** (*Azadirachta indica*), **gul-mohar** (*Delonix regia*), etc.

The main axis of the inflorescence together with the latest axes, if present, is termed as the **peduncle.** The stalk of the individual flower of the inflorescence is called the **pedicel.**

(*b*) Spike. In this type of racemose inflorescence the main axis remains elongated and the lower flowers are older, *i.e.,* opening earlier than the upper ones, as found in raceme, but here the flowers are sessile, *i.e.,* without pedicel or stalk, *e.g.,* amaranth (*Amaranthus* spp.), **latjira** (*Achyranthes aspera*), etc.

Table 29.10. Differences Between Racemose and Cymose Inflorescences

Racemose	Cymose
1. The main axis is of unlimited growth.	1. The main axis is of limited growth.
2. It does not terminate into a flower.	2. It terminates into a flower.
3. May be branched or unbranched.	3. It is usually branched.
4. The flowers are arranged in acropetal succession. The lower or outermost flowers are older and open earlier than the upper or inner ones.	4. The flowers develop in basipetal succession. The terminal flower is older and opens earlier than the lateral ones.
5. The opening of flowers is centripetal.	5. The opening of flowers is centrifugal.

(*c*) Spikelets. Each spikelet may bear one to several flowers (florets) attached to a central stalk known as **rachilla.** Spikeletes are arranged in a spike inflorescence which is composed of several to many spikelets which are combined in various manners on a main axis called the **rachis.** Some are in compound spikes (*i.e.,* in wheat—*Triticum aestivum*), others are in racemes (*e.g.,* in *Festuca*), while some are in panicles (*e.g.,* in *Avena*). The usual structure of spikelet is as follows–

–There is a pair of sterile **glumes** at the base of spikelet, the lower, outer glume called the **first,** and the upper, inner one called the **second.** Just above the glumes, there is series of florets, partly enclosed by them. Each floret has at its base a **lemma** and **palea.** The lemma is the lower, outer bract of the floret. Usually the lemma also known as **inferior palea** bears a long **awn** as an extension of the mid-rib at the apex or back. The floral parts borne in the axil of lemma. The palea (also known as **superior palea**) often with two longitudinal ridges (keels or nerves), stands between the lemma and the rachilla. Flowers and glumes are arranged on the spikelet in two opposite rows. Spikeletes are characteristic of Poaceae (Gramineae) or Grass family, *e.g.,* grasses, wheat, barley, oats, sorghum, sugarcane, bamboo, etc.

***(d)* Catkin.** This is a modified spike with a long and drooping axis bearing unisexual flowers, *e.g.,* mulberry (*Morus alba*), birch (*Betula* spp.), oak (*Quercus* spp.), etc.

***(e)* Spadix.** This is also a modification of spike inflorescence having a fleshy axis, which remains enclosed by one or more large, often brightly coloured bracts, the **spathes,** *e.g.,* in members of Araceae, Musaceae and Palmaceae. This inflorescence is found only in monocotyledonous plants.

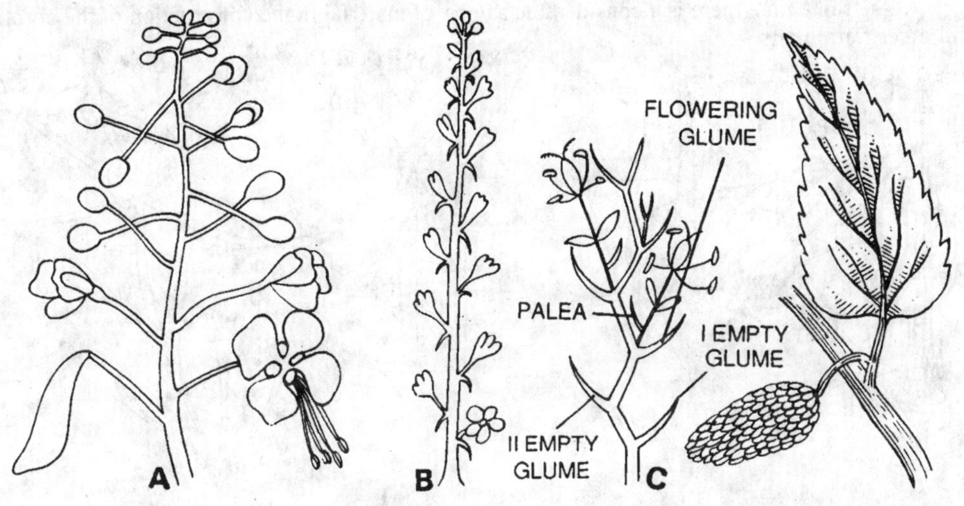

Fig. 29.49. Inflorescence-racemose. A, raceme of **gul-mohar** ; B, spike ; C, spikelet of a grass; D, female catkin of mulberry.

(*ii*) Main axis shortened

***(a)* Corymb.** In this inflorescence the main axis remains comparatively short and the lower flowers possess much longer stalks or pedicels than the upper ones so that all the flowers are brought more or less to the same level, *e.g.,* in candytuft (*Iberis amara*).

***(b)* Umbel.** In this inflorescence the primary axis remains comparatively short, and it bears at its tip a group of flowers which possess pedicels or stalks of more or less equal lengths so that the flowers are seen to spread out from a common point. In this inflorescence a whorl of bracts forming an involucre is always present, and each individual flower develops from the axil of a bract. Generally the umbel is branched and is known as **umbel of umbels (compound umbel),** and the branches bear flowers, *e.g.,* in coriander (*Coriandrum sativum*), fennel, carrot, etc. Sometimes, the umbel is unbranched and known as **simple umbel,** *e.g.,* **Brahmi** (*Centella asiatica*). This inflorescence (umbel) is characteristic of Apiaceae (Umbelliferae) family.

(iii) Main axis flattened

Capitulum or head. In this type of inflorescence the main axis or receptacle becomes suppressed, and almost flat, and the flowers (also known as florets) are sessile (without stalk) so that they become crowded together on the flat surface of the receptacle. The florets are arranged in a centripetal manner on the receptacle, *i.e.,* the outer flowers are older and open earlier than the inner ones.

The individual flowers (florets) are bracteate. In addition the whole inflorescence remains surrounded by a series of bracts arranged in two or three whorls. The flowers (florets) are usually of two kinds—**(i) ray florets** (marginal strap-shaped flowers) and **(ii) disc florets** (central tubular flowers). The capitulum (head) may also consist of only one kind of florets, *e.g.,* only tubular florets in *Ageratum* or only ray or strap-shaped florets in *Sonchus*. A capitulum or head is characteristic of Asteraceae (Compositae) family, *e.g.* sunflower (*Helianthus annuus*), marigold (*Tagetes indica*), safflower (*Carthamus tinctorius*), *Zinnia, Cosmos, Tridax, Vernonia,* etc. Besides, it is also found in *Acacia* and sensitive plant (*Mimosa pudica*) of Mimosaceae family.

The capitulum inflorescence has been considered to be the most perfect. The reasons are as follows:

The individual flowers are quite small, and massed together in heads, and therefore, they add to greater conspicuousness to attract the insects and flies for pollination.

At the same time there is a considerable saving of material in the construction of the corolla and other floral parts.

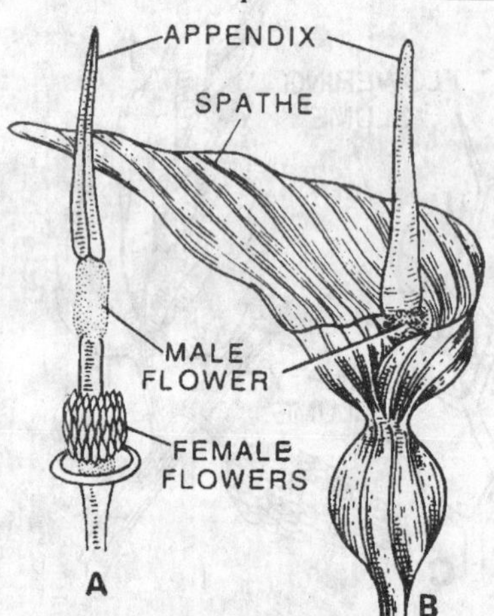

Fig. 29.50. Inflorescence-spadix–A, spadix of an aroid without spathe; B, same with spathe.

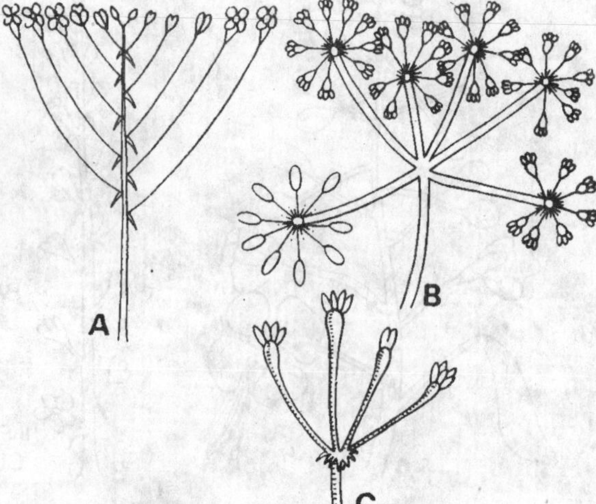

Fig. 29.51. Inflorescence, A, corymb; B, compound umbel (umbel of umbels); C, simple umbel.

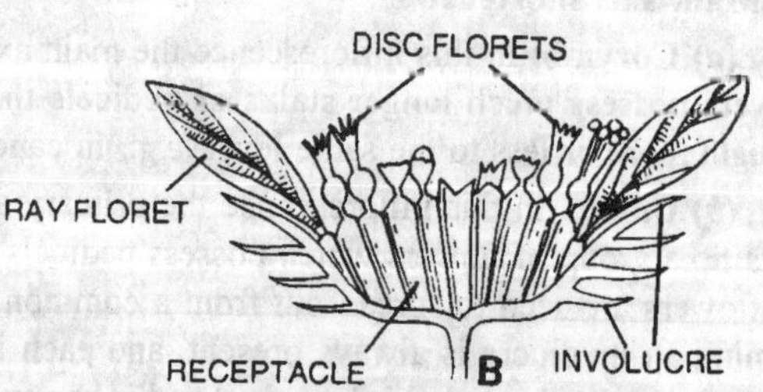

Fig. 29.52. Inflorescence. Head (capitulum)—A, a head; B, a head in L.S.

A single insect may pollinate flowers in a short time without flying from one flower to another.

2. Cymose Inflorescence

In this type of inflorescence the growth of the main axis is ceased by the development of a flower at its apex, and the lateral axis which develops the terminal flower also culminates in a flower and its growth is also ceased. The flowers may be pedicellate (stalked) or sessile (without stalk). Here the flowers develop in basipetal succession, *i.e.,* the terminal flower is the oldest and the lateral ones younger. This type of opening of flowers is known as **centrifugal.** The cymose inflorescence may be of four main types ***(i)*** Uniparous or monochasial cyme; (*ii*) Biparous or dichasial cyme; ***(iii)*** Multiparous or polychasial cyme and ***(iv)*** Cymose capitulum.

(i) Uniparous or Monochasial cyme

Here the main axis ends in a flower and it produces only one lateral branch at a time ending in a flower. The lateral and succeeding branches again produce only one branch at a time like the primary one. There are three forms of uniparous cyme — (*a*) **helicoid**, (*b*) **scorpioid**, and (*c*) **sympodial**

(*a*) Helicoid cyme. When the lateral axes develop successively on the same side, forming a sort of helix, the cymose inflorescence is known as **helicoid** or **one-sided cyme**, *e.g.*, in *Begonia, Juncus, Hemerocallis* and some members of Solanaceae.

(*b*) Scorpioid cyme. When the lateral branches develop on alternate sides, forming a zigzag, the cymose inflorescence is known as **scorpioid** or **alternate-sided cyme**, *e.g.*, in *Gossypium* (cotton), *Drosera* (sundew), *Heliotropium, Freesia,* etc.

(*c*) Symopodial cyme. Sometimes, in monochasial or uniparous cyme successive axes may be at first curved or zig-zag (as in scorpioid cyme) but later on it becomes straight due to rapid growth, thus forming a central or pseudoaxis. This type of inflorescence is known as **sympodial cyme** as found in some members of Solanaceae (*e.g., Solanum nigrum*).

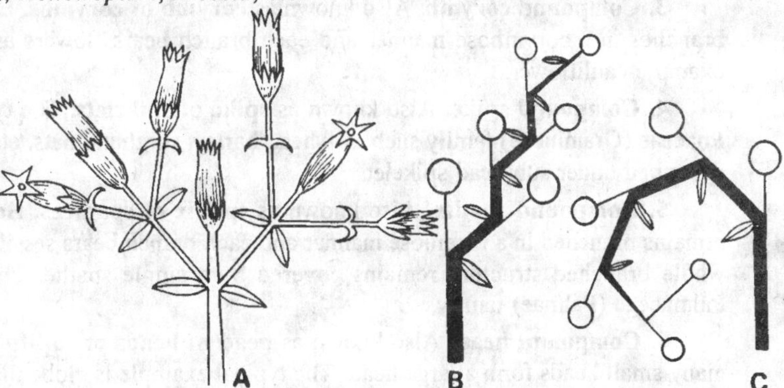

Fig. 29.53. Inflorescence. Cymose—A, biparous (dichasial) cyme; B, scorpioid cyme; C, helicoid cyme.

(ii) Biparous or Dichasial Cyme

In this type of inflorescence the peduncle bears a terminal flower and stops growing. At the same time the peduncle produces two lateral younger flowers or two lateral branches each of which terminates in a flower. There are three flowers; the oldest one is in the centre. The lateral and succeeding branches in their turn behave in the same manner, *e.g.*, jasmine, teak, *Ixora, Saponaria,* etc. This is also known as **true cyme or compound dichasium.**

(iii) Multiparous or Polychasial cyme

In this type of cymose inflorescence the main axis culminates in a flower, and at the same time it again produces a number of lateral flowers around. The oldest flower is in the centre and ends the main floral axis (peduncle). This is a **simple polychasium.** The whole inflorescence looks like an umbel, but is readily distinguished from the latter by the opening of the middle flower first, *e.g.,* **Ak** (*Calotropis procera*), *Hamelia patens*, etc.

(iv) Cymose Capitulum

This type of inflorescence is found in *Acacia, Mimosa* and *Albizzia*. In such cases the peduncle is reduced or condensed to a circular disc. It bears sessile or sub-sessile flowers on it.

The oldest flowers develop in the centre and youngest towards the periphery of the disc, such arrangement is known as **centrifugal.** The flowers make a globose head, which is also called **glomerule.**

Compound Inflorescence

In this type of inflorescence the main axis (peduncle) branches repeatedly once or twice in racemose or cymose manner. In the former case it becomes a compound raceme and in the latter case it becomes a compound cymose inflorescence. The main types of compound inflorescence are as follows.

1. Compound raceme or panicle. In this case the raceme is branched, and the branches bear flowers in a racemose manner, *e.g., Delonix regia, Azadirachta indica, Clematis buchaniana, Cassia fistula*, etc.

2. Compound umbel. Also known as **umbel of umbels.** Here the peduncle (main axis) is short and bears many branches which arise in an umbellate cluster. Each such branch bears a group of flowers in an umbellate manner. Usually a whorl of leafy bracts is found at the base of branches and also at the bases of flowers arranged in umbellate way. The former whorl of bracts is called **involucre** and the latter **involucel.** Typical examples of compound umbel are—*Daucus carota* (carrot), *Foeniculum vulgare* (fennel), *Coriandrum sativum* (coriander), etc.

3. Compound corymb. Also known as **corymb of corymbs.** Here the main axis (peduncle) branches in a corymbose manner and each branch bears flowers arranged in corymbs. Typical example-cauliflower.

4. Compound spike. Also known as **spike of spikelets.** The typical examples are found in Poaceae (Gramineae) family such as-wheat, barley, sorghum, oats, etc. This type has already been described under sub-head spikelets.

5. Compound spadix. Also known as **spadix of spadices.** Here the main axis (peduncle) remains branched in a racemose manner and each branch bears sessile and unisexual flowers. The whole branched structure remains covered by a single spathe. The examples are common in Palmaceae (Palmae) family.

6. Compound head. Also known as **head of heads** or **capitulum of capitula.** In this case many small heads form a large head. The typical example is globe thistle (*Echinops*). In this plant the heads are small and one-flowered and are arranged together forming a big compound head.

Special Types of Inflorescence

1. Cyathium. This type of inflorescence is found in genus *Euphorbia* of family Euphorbiaceae; also found in genus *Pedilanthus* of the family. In this inflorescence there is a cup-shaped involucre, often provided with nectar secreting glands. The involucre encloses a single female flower, represented by a pistil, in the centre, situated on a long stalk. This female flower remains surrounded by a number of male flowers arranged centrifugally. Each male flower is reduced to a solitary stalked stamen. It is evident that each stamen is a single male flower from the facts that it is articulated to a stalk and that it possesses a scaly bract at the base. The examples can be seen in poinsettia (*Euphorbia*), *Pedilanthus*, etc.

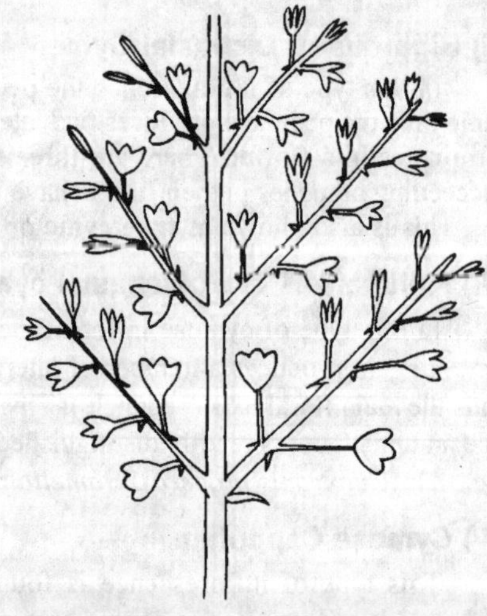

Fig. 29.54. Inflorescence. Panicle.

2. Verticillaster. This type of inflorescence is a condensed form of dichasial (biparous) cyme with a cluster of sessile or sub-sessile flowers in the axil of a leaf, forming a false whorl of

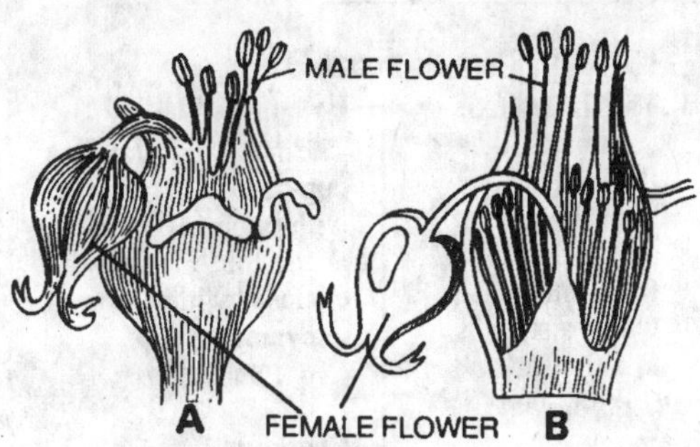

Fig. 29.55. Inflorescence-special. Cyathium—A, cyathium of *Euphorbia;* B, L.S. of same.

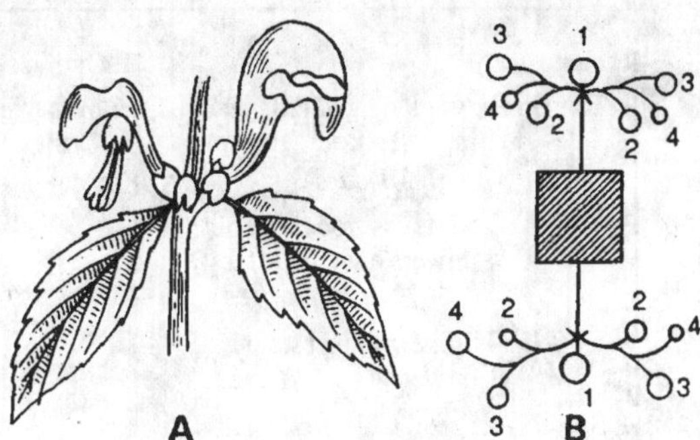

Fig. 29.56. Inflorescence—special verticillaster—A verticillaster of Labiatae; B, diagram of same.

flowers at the node. The first of main floral axis gives rise to two lateral branches and these branches and the succeeding branches bear only one branch each on alternate sides. The type of inflorescence is characteristic of Lamiaceae (Labiatae) family. Typical examples, are—*Ocimum*, *Coleus*, *Mentha*, *Leucas*, etc.

3. Hypanthodium. In this type of inflorescence the receptacle forms a hollow cavity with an apical opening guarded by scales. Here the flowers are borne on the inner wall of the cavity. The flowers are unisexual, the female flowers develop at the base of the cavity and the male flowers towards the apical pore. The examples, are found in genus *Ficus* of Moraceae family, *e.g.*, *Ficus carica*, *F. glomerata*, *F. benghalensis*, *F. religiosa*, etc.

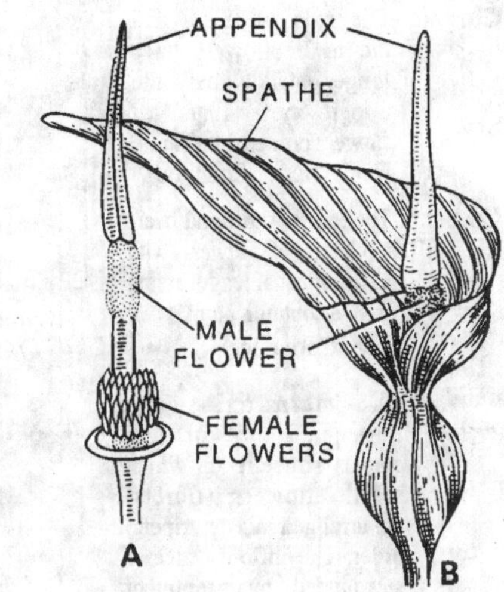

Fig. 29.57. Inflorescence. A, hypanthodium of *Ficus;* B, male flower; C, female flower.

Table 29.11. Differences Between Cyathium, Verticillaster and Hypanthodium Inflorescence.

1. Cyathium	2. Verticillaster	3. Hypanthodium
The cupular structure is developed by an involucre of bracts. There is a single female flower in the centre, surrounded by numerous stalked male flowers. Each flower is represented by a single stamen, e.g., *Euphorbia*.	This is special type of cymose inflorescence. It consists of two clusters of flowers that develop from each of the two opposite axils of the leaves. Each cluster represents a dichasial cyme. Flowers are sessile and appear as a cluster around the node, *e.g.*, *Salvia, Ocimum,* etc.	The main axis develops in a cupular receptacle, with an apical opening at the apex. Both male and female flowers are present on the inner wall of cavity enclosed in a cupular structure, *e.g.*, *Ficus*.

Table 29.12. Main Types of Inflorescences

- **Racemose**
  - **Raceme.** An elongated axis bearing stalked flowers (*e.g., Mustard*).
  - **Spike.** Same as raceme, but flowers have no stalks, (*e.g., Achyranhes*).
  - **Catkin.** Spike with unisexual flowers, (*e.g., Morus*).
  - **Spadix.** Spike with fleshy axis enclosed by one or more large bracts, (*e.g., Colocasia, Musa,* etc.)
  - **Corymb.** The axis is short, and the lower flowers have longer stalks than the upper ones. Thus all flowers come to the same level, (*e.g.,* Candytuft.)
  - **Umbel.** The axis is short and bears a cluster of flowers with stalks of equal length arsing from a common point, (*e.g.,* Coriander)
  - **Head or Capitulum.** The main axis is a flattened, more or less convex structure on which sessile flowers (florets) are arranged in a centripetal order. The inflorescence is surrounded by prominent bracts, (*e.g.,* Sunflower, marigold, etc.)
- **Cymose**
  - **Monochasial Cyme** The main axis terminates in a flower and one lateral branch axis develops from its base which also ends in a flower, (*e.g., Begonia, cotton,* etc.)
  - **Dichasial Cyme.** Two lateral branches develop on either side of the terminal flower of the main axis; the lateral branches also end in a flower and may again branch similarly (*e.g.,* jasmine, *Dianthus,* etc.)
  - **Polychasial Cyme.** More than two lateral branches arise from the base of the terminal flower, (*e.g., Calotropis*).
- **Special Types**
  - **Hypanthodium.** The main axis forms a cup-shaped receptacle with a small opening at the top. Flowers are enclosed within the cup in cymose groups. (*e.g. Ficus*).
  - **Cyathium.** Here the involucre forms a cup. Single female flower without perianth arises in the middle surrounded by a large number of male flowers represented by stalked stamens. *e.g., Euphorbia.*
  - **Verticillastor.** In this type of inflorescence, typical of plants with opposite leaves, a cyme arises in each leaf axil. The first axis ends in a flower. Two branches arise below it bearing branches in an alternating manner. Flowers are sessile and appear as a cluster around the node, *e.g., Ocimum.*

DIVERSITY IN THE FLOWER

The flower, in its basic design, is a shoot in which the internodes (the regions of stem between nodes) do not elongate. As a result, the floral parts that occur successively along this shoot, develop as a closely packed cluster. The enlarged portion of the stalk on which the flower develops is known as the **receptacle,** and the stalk of flower is the **pedicel.**

Flower develops from a bud known as **flower bud.** This bud usually develops in the axil of a leaf called the **bract.** Sometimes the floral bud arises from the apical bud when the bract is absent. The flower buds are homologous with the vegetative buds. It consists of a stem tip, the receptacle, resembling in ontogeny and fundamental structure a vegetative tip. It consists of nodes and internodes, and bears appendages. The nodes are crowded and brought together by the suppression of internodes. The apical growth is determinate in its growth and typically only four nodes are formed bearing floral leaves at the nodes, the calyx, corolla, androecium and gynoecium, from the base towards the apex respectively. The calyx and corolla are more leaf like, and are known as the **non-essential organs** as they are only indirectly concerned in the process of reproduction and the other two, the androecium and gynoecium are more unlike leaves, and are known as the **essential organs**, as they are directly required in the process of reproduction.

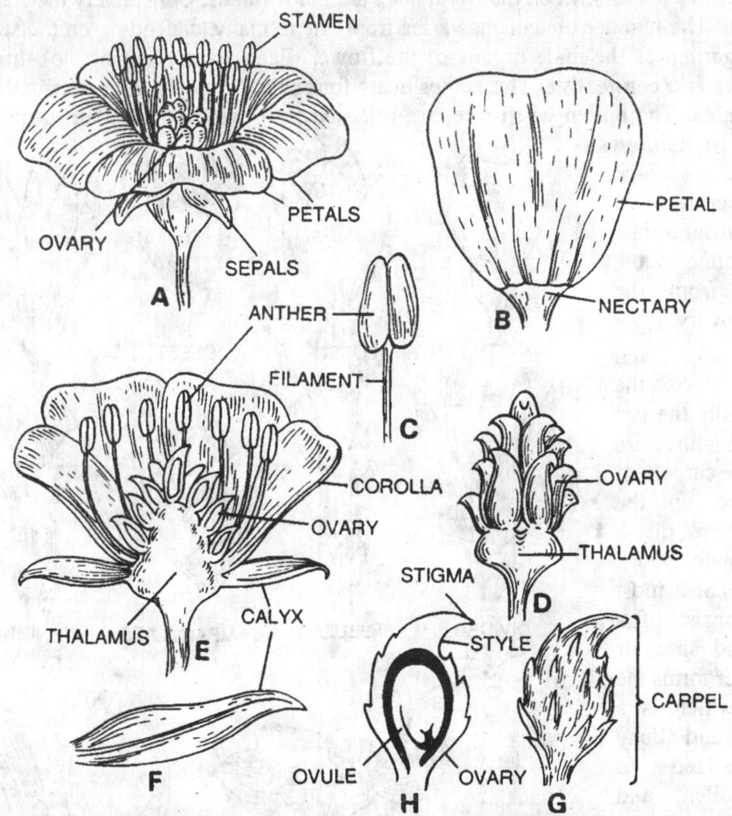

Fig. 29.58. Various parts of a flower of *Ranunculus*. A complete flower; B, a petal; C, stamen; D, thalamus with apocarpous gynoecium; E, a sepal; F, V.S. of flower, G, a carpel; H, V.S. of carpel.

The various groups of floral parts develop one after another from the receptacle. The first to form are **sepals**, which are usually green in colour and leaf-like in structure. Collectively the sepals are known as **calyx.** The calyx encloses the flower bud. Next to form are the **petals**, which are also usually shaped somewhat like leaves but are often brightly coloured. Collectively the petals make up the **corolla.** The special function of the corolla is to advertise the presence of the flower to potential pollinators. In some flowers, such as the lily and tulip, the sepals and petals cannot be distinguished separately, in such cases, they are known as **tepals.** This situation is commonly found in monocotyledons, in which the flower parts are often borne in three's or multiples of three.

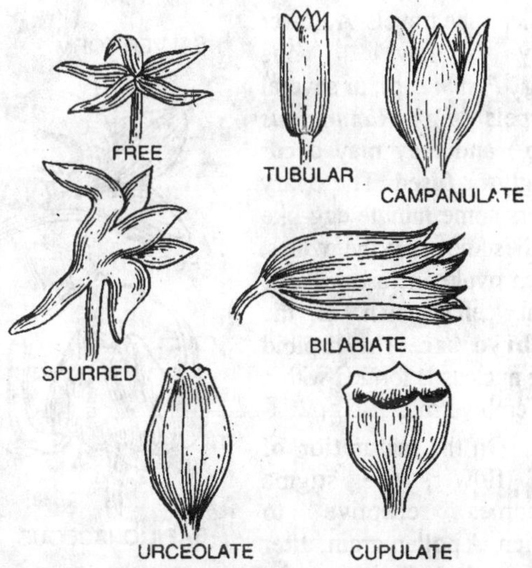

Fig. 29.59. Calyx. Forms of calyx.

Collectively the tepals are known as the **perianth.** The perianth consists of the sterile, or non-reproductive, parts of the flower.

The next primordia to form on the floral apex are the **stamens.** Collectively the stamens form the **androecium.** The number of stamens varies from one to many, depending on the species. The stamens are regarded as the male organs of the flower. Each stamen consists of three parts—**filament**, **anther** and **connective.** The anther bears four chambers or **pollen sacs**, each filled up with **pollen grains.** The pollen when ripe, are released from within the anther, usually through narrow slits. The haploid male nuclei, which function like sperm but do not swim, differentiate within the pollen tube, which grows from the pollen into the ovary.

The upper and innermost parts of the flower, and usually the last floral parts to mature, are the **carpels.** The carpels of a flower make up the **gynoecium** or the **pistil.** This is the female whorl. The carpels of most angiosperms consist of a hollow expanded base, or ovary, which later forms the fruit; **stigma**, which is a flat, often hairy and sticky surface specialized to receive the pollen; and connecting stigma and ovary, a slender stalk, the **style**, down which the pollen tube travels. A flower may have one (*e.g., Pisum* spp), or several carpels (*e.g., Ranunculus* spp.) and they may occur singly or fused. The ovary bears some minute egg-like bodies known as the **ovules.** Each ovule encloses a large oval cell known as the **embryo sac.** The haploid egg nucleus is formed within the embryo sac.

On the maturation of the flower, the stigma becomes receptive to pollen. A pollen grain, after being deposited on the stigma, produces a long

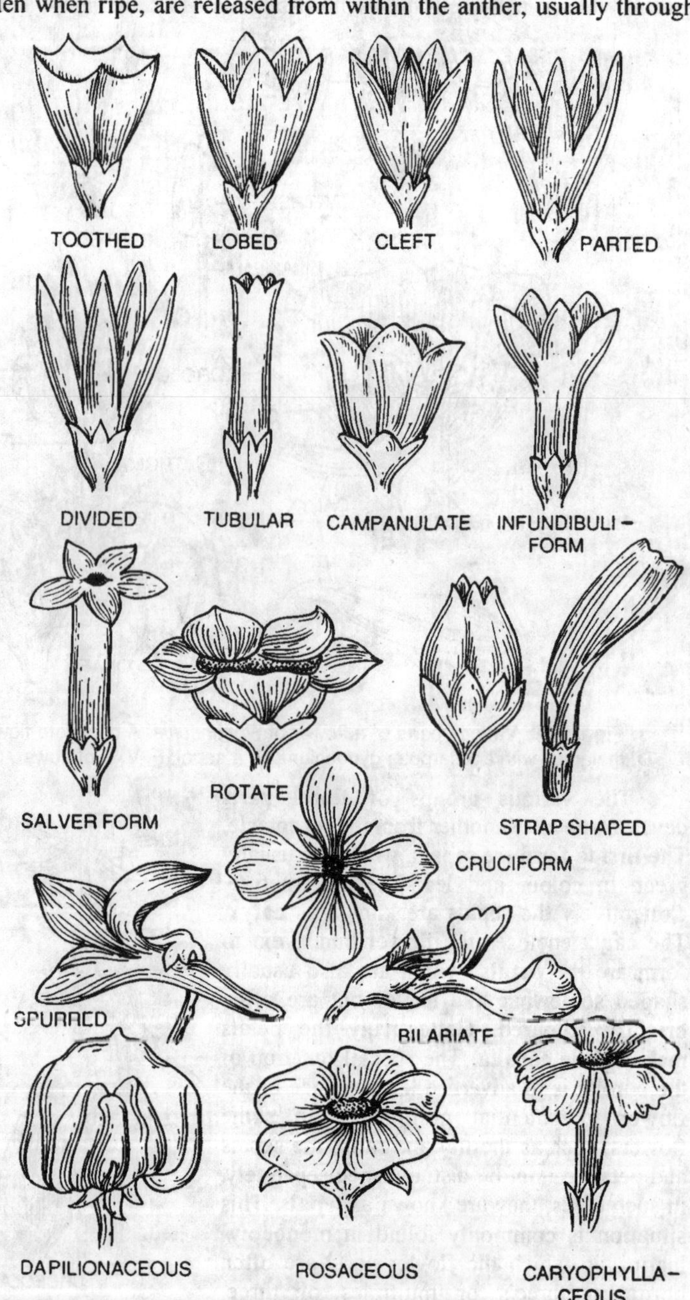

Fig. 29.60. Forms of corolla.

tube which grows down through the stigma and the style and enters one of the ovules in the ovary. Two sperm nuclei are found at the end of this tube. One of these enters to egg of an ovule and fuses with the nucleus of the egg to form the zygote. This fusion of male and female nuclei is called **fertilization**, and the flower is said to be **fertilized** when this has taken place. After fertilization the ovule develops into a seed, while the whole ovary becomes a **fruit.** The mature fruit contains one to many seeds, each with an **embryo** inside it. From these embryos grow the young plants of the next generation, and the life cycle of angiosperm begins again.

STRUCTURE OF FLOWER

The vegetative shoot shows unlimited growth, whereas the flower shows the limited growth. In flower, the apical meristem ceases to be active after the formation of floral parts. Anatomically the flower is a determinate stem with crowded appendages with internodes much shortened or obliterated. The appendages are of leaf rank but differ from those of the vegetative stem, in function and shape. Here the flower has been treated on the basis of the concept of homology between the flower and the shoot in their phylogeny and ontogeny.

As already mentioned, the flower consists of an axis, also known as receptacle and lateral appendages. The appendages are known as **floral parts** or **floral organs.** They are sterile and reproductive. The sepals and petals which constitute the calyx and corolla respectively are the sterile parts. The stamens compose the **androecium,** whereas the free or united carpels compose the **gynoecium.**

Pedicel

The pedicel and receptacle have typical structure, with a normal vascular cylinder. The cylinder may be unbroken or it may contain a ring of vascular bundles. In the region where floral organs are borne, the pedicel expands into the **receptacle.** The vascular cylinder also expands and the vascular bundles increase somewhat in number, and finally traces begin to diverge. In the simplest cases vasular traces for different organs and whorls of organs arise quite independently, (*e.g., Aquilegia*).

Sepals

The sepals resemble leaves in their anatomy. Each sepal consists of ground parenchyma a branched system and epidermis. The chloroplasts are found in the green sepals but usually there is no differentiation into palisade and spongy parenchyma. The epidermis of sepals may possess stomata and trichomes.

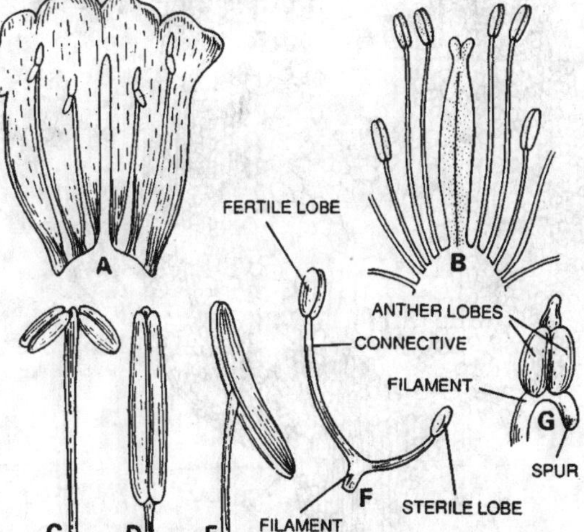

Fig. 29.61. Androecium; A, didynamous; B, tetradynamous; C, innate fixation; D, adnate fixation; E, varsatile; F, stamen of *Salvia* with an elongated connective; G, spurred stamens of *Viola*.

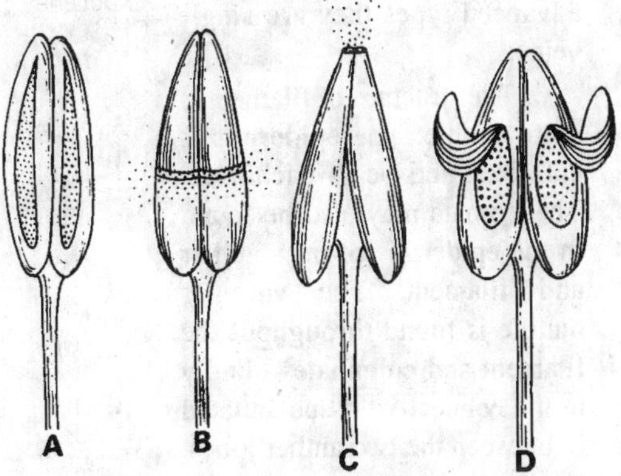

Fig. 29.62. Various types of dehiscence of anther. A, longitudinal slit; B, transverse slit; C, apical pore (porous); D, valvular.

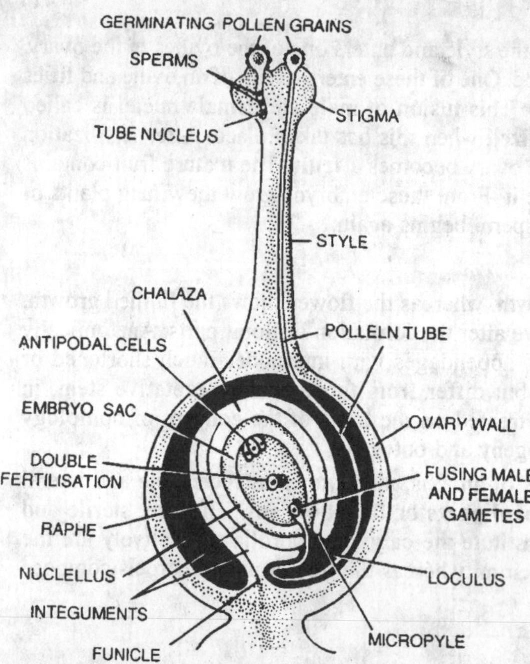

Fig. 29.64. A monocarpellary pistil cut vertically to show its external and internal parts. It also shows pollen grains on stigma.

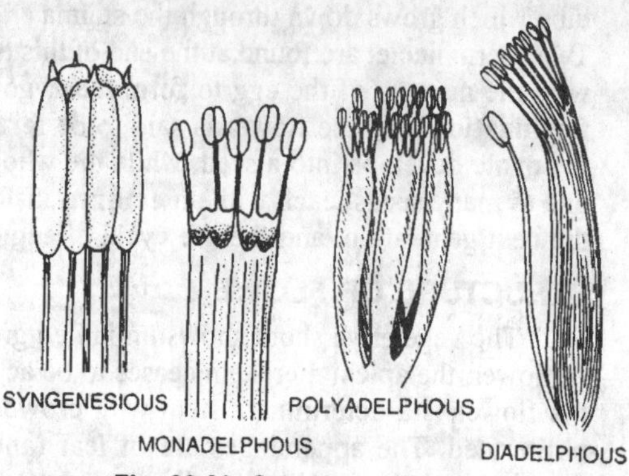

Fig. 29.63. Cohesion of stamens.

Petals

The petals also resemble leaves in their internal structure. They contain ground parenchyma, a more or less branched vascular system, and an epidermis. They contain pigments-containing chromoplasts. Very often, the epidermal cells of the petals contain volatile oils which emit the characteristic fragrance of the flowers.

Stamen

Commonly the stamen consists of a two-lobed four loculed anther. The anther is found to be situated on a slender filament which bears vascular bundle. In certain primitive dicotyledonous families the stamens are leaf-like and possess three veins, whereas in advanced types they are single veined.

The structure of filament is quite simple. The epidermis is cutinized and bears trichomes. The stomata may also be found in the epidermis of both anther and filament. The vascular bundle is found throughout the filament and culminates blindly in the connective tissue situated in between the two anther lobes.

The outermost wall layer of the anther is the epidermis. Just beneath the epidermis there is endothecium which usually

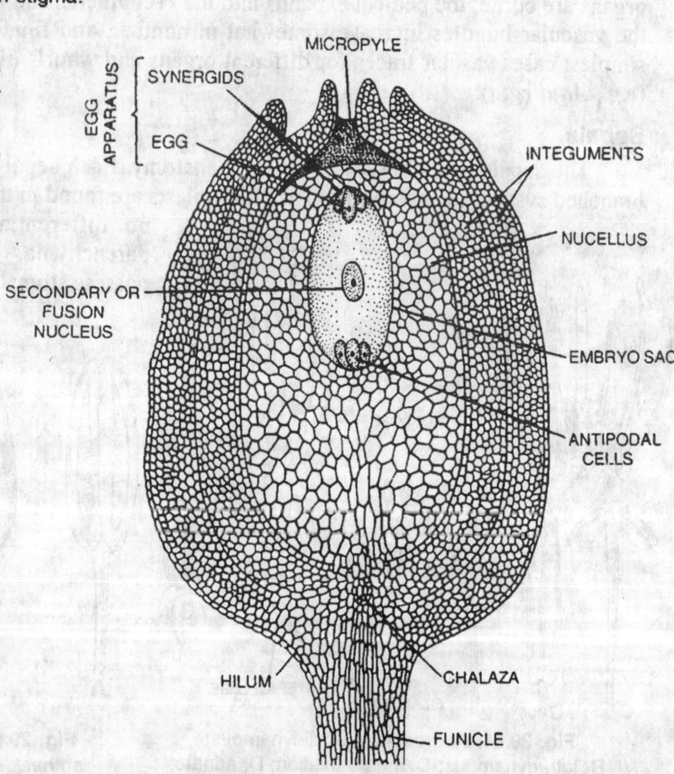

Fig. 29.65. V.S. ovule showing its structure just before fertilization.

possesses strips or ridges of secondary wall material mainly on those walls which do not remain in contact with the epidermis. The innermost layer is composed of multinucleate cells; this is nutritive in function and known as **tapetum.** On the maturation of the pollen the tapetum disintegrates and the outer wall of the pollen sac now consists of only epidermis and endothecium. At the time of dehiscence of the anthers the pollen are released out through stomium.

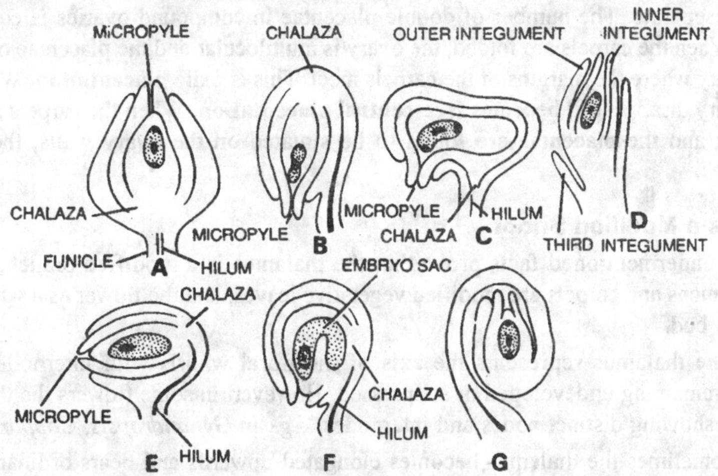

Fig. 29.66. Types of ovules. A, orthotropous; B, anatropous; C, campylotropous; D, ovule with three integuments; E, hemianatropous; F, amphitropous; G, circinotropous.

Gynoecium

The unit of gynoecium is called the **carpel.** A flower may possess one carpel or more than one. If two or more carpels are present they may be united or free from one another. When the carpels are united the gynoecium is known as **syncarpous**; when they are free the gynoecium is said to be **apocarpous.** The carpel is commonly interpreted as foliar structure. The carpel of an apocarpous or syncarpous gynoecium is being differentiated into the ovary and the style. The upper part of the style is differentiated as a stigma. The stigma is sessile.

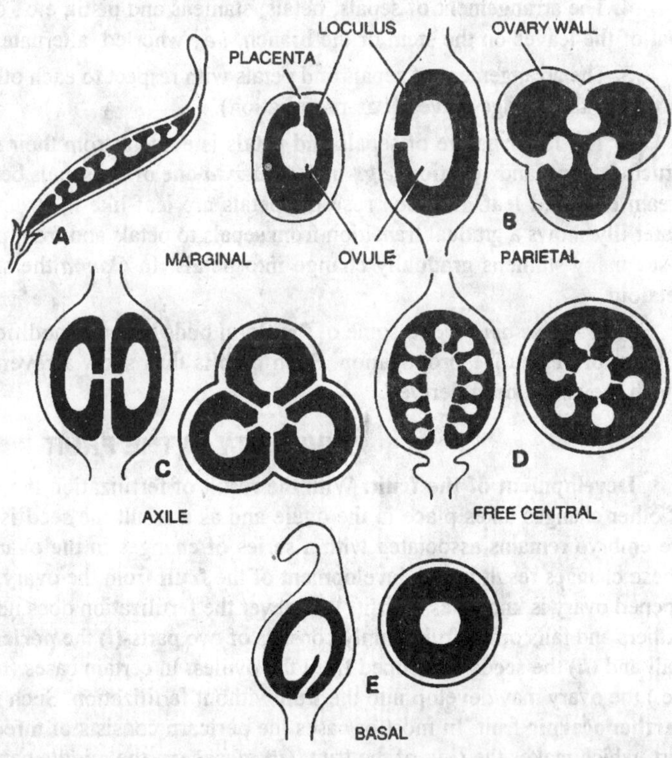

Fig. 29.67. Various types of placentation in angiosperms (A—E).

Placentation

The ovary consists of the ovary wall, the locule or locules, and in a multilocular ovary, the partitions. The ovules are found to be situated on the inner or adaxial (ventral) side of the ovary wall. The ovule-bearing region forms the **placenta.** In a carpel the placenta occurs close to the margin. Since there are two margins, the placenta is double in nature. The two halves may be united or separate. The number of double placentae in compound ovaries is equal to number of carpels. When the carpels are folded, the ovary is multilocular and the placentae occur in the centre of the ovary where the margins of the carpels meet. This is **axile placentation.** When the partitions of the ovary disappear, it becomes **free-central placentation.** When the carpels are joined margin to margin and the placentae are found to be situated on the ovary walls, the placentation is **parietal.**

Flower is a Modified Shoot

The undermentioned facts prove that the thalamus is a modified branch, whereas sepals, petals, stamens and carpels are modified vegetative leaves, and the flower as a whole is a modified vegetative bud.

1. The thalamus represents the axis of the floral whorls with internodes between them normally remaining undeveloped or suppressed. However, in some flowers the thalamus becomes elongated showing distinct nodes and internodes, *e.g.,* in *Gynandropsis*, *Capparis*, etc.

2. Sometimes the thalamus becomes elongated upwards and bears ordinary foliage leaves, and behaves as a branch, *e.g.,* in rose.

3. A floral bud like a vegetative bud is either terminal or axillary in position.

4. The arrangement of sepals, petals, stamens and pistil, etc., on the thalamus is the same as that of the leaves on the stem or the branch, *i.e.,* whorled, alternate, spiral or opposite.

5. The arrangement of sepals and petals with respect to each other (*i.e.,* aestivation) is similar to that of the foliage leaves (*i.e.,* prefoliation).

6. The foliar nature of sepals and petals is evident from their similarity to leaves as regards structure, form and venation, *e.g.,* in *Mussaenda* one of the sepals becomes modified into a distinct cream coloured leaf. In green rose the petals are leaf-like in structure and green in colour. The water lily shows a gradual transition from sepals to petals and from petals to stamens. In cultivated roses many stamens gradually change into petals. In *Canna* the stamens and the style become petaloid.

7. In *Agave americana* some of the floral buds become modified into vegetative buds, called bulbils for vegetative propagation. Such bulbils thus show a reversion to ancestral forms, from which they have been derived.

DIVERSITY IN THE FRUIT

Development of the fruit. With the result of fertilization the embryo develops and a series of other changes takes place in the ovule and as a result the **seed** is formed. The development of the embryo remains associated with a series of changes in the ovary wall and other floral parts. These changes result in the development of the **fruit** from the ovary. In other words the mature or ripened ovary is known as a fruit. Whenever the fertilization does not take place, the ovary simply withers and falls off. A fruit mainly consists of two parts-***(i)*** the **pericarp** developed from the ovary wall and ***(ii)*** the **seeds** developed from the ovules. In certain cases (*e.g.,* oranges, grapes, bananas, etc.) the ovary may develop into the fruit without fertilization. Such fruit is seedless and known as **parthenocarpic** fruit. In most of cases the pericarp consists of three parts ***(i)*** **epicarp** - the outer part, which makes the skin of the fruit; ***(ii)*** **mesocarp** the middle part, which makes the pulpy part of the fruit, as in mango, peach, etc., and ***(iii)*** **endocarp**, the inner part which is very often thin and membranous, as in orange, or it may be hard and stony as in mango, plums, etc. In some cases the pericarp is not differentiated into three parts.

In the cases where only the ovary of the flower develops into the fruit, it is called the **true fruit**, but in other cases where the other floral parts such as the thalamus, receptacle or calyx also take part in the development of the fruit; the fruit is said to be **false fruit**, **spurious fruit** or **pseudocarp.** For example, in apple and pear the thalamus grows round the ovary and becomes fleshy. In *Dillenia*, the calyx is persistent and fleshy and makes the main part of the fruit. In rose the enlarged thalamus bears small true fruits on its inner concave surface. In cashewnut the peduncle and thalamus develop into a false fruit or pseudocarp, whereas the true fruit is a nut, and develops from the ovary seated on the swollen peduncle. The aggregate fruit of custard apple having number of small true fruits fused together, makes a spurious fruit. Lastly, the fruits which develop from an inflorescence as those of mulberry, pine apple, jack fruit, fig, banyan, peepal, etc., also known as spurious fruits.

In the development of a fruit both cell division and cell expansion are involved. The growth of the ovary after fertilization may be rapid. For example, the pumpkin ovary shows a 20 fold increase in about two weeks time.

All the flowers borne on a plant do not always mature into fruit. In mango for example, fruit-set is extremely low as compared to the total number of flowers produced. Shedding of flowers may occur before or after anthesis; even young fruits may drop. Dropping of blossoms and young fruits is of advantage to the cultivator because it results in the production of fruits of large size. In certain orchard trees, such as apples, pears and plums, flowers thinning by hormonal spray is a regular practice.

Fruit ripening. As soon as the growth of the ovary wall due to cell division and cell enlargement stops, the fruit is said to be mature. This is followed by the final phase of fruit development called fruit ripening. During this period the conversion of starch into sugar, the reduction in the concentration of acids, the production of various esters, and the breakdown of chlorophyll lead to changes in colour, texture, taste, and flavour of the fruit. A mature fruit of mango is hard and green and its edible portion white and sour. On ripening, the mesocarp becomes yellow-orange, juicy and sweet. A characteristic feature of ripening of some fruit (like banana) is a sudden increase in respiration which is known as the **climacteric.** After this stage, the fruit starts decaying. Although from the point of view of the fruit, ripening leads to death, for man and other animals it is a stage which is most useful.

Parthenocarpy. Some plants are able to form fruits without fertilization. Such fruits are called parthenocarpic and the phenomenon is known as **parthenocarpy.** Parthenocarpic fruits are either seedless or contain empty or nonviable seeds. In these fruits, the "seed-factor" for fruit growth is provided by the tissue of the ovary wall itself. Seedless varieties of grapes and oranges have been reported to have up to seven times as much auxin in the ovaries of unpollinated flowers as seeded varieties. This may explain the normal development of parthenocarpic fruits even in the absence of healthy seeds. Most commonly cultivated varieties of banana are parthenocarpic. Parthenocarpic oranges and watermelons are also very common. Even in those plants which normally bear seeded fruits, parthenocarpy could be induced by the application of low concentrations of axuin or gibberellin.

Parthenocarpy is of great commercial value. Parthenocarpic fruits are ideal for consumption as such, or in preparation of jams and fruit juices on a commercial scale. You can imagine the annoyance caused by seed while eating a watermelon or guava.

Biological Significance of Fruit Formation

Fruits (of all categories) were used by the ancient man as his main food and even today they form an important part of man's diet. However, we might ask the question: "why does a plant invest so much of its food in the production of fruit?" Certainly not for human consumption. The fruit serves the plant in various ways. It protects the immature seeds against hostile climatic conditions and animals. The seeds remain enclosed in the fruit until they are ready to germinate or, at least withstand the possible unfavourable environmental conditions. The function of seed protection by the fiuit wall is also achieved through its colour. When young, most fruits are green, and remain hidden in the green foliage. As the seeds mature, the fruit acquires bright colours and attract the seed dispersing agents. Immature fruits also offer chemical defence against the animals as they often contain such unpalatable and repelling substances as astringents, tannins, bitter alkaloids and sour acids which disappear upon ripening.

Another important role played by the fruit tissues is in the dispersal of seeds to distant localities which is of great biological significance.

Classification of Fruits (Types of Fruits)

The fruits are classified into three main groups, *viz.,* **(1) simple (2) aggregate** and **(3) multiple or composite.**

1. Simple Fruits

When the ovary of a flower with or without other accessory floral parts converts into a single fruit, the fruit is said to be a **simple fruit.** It may be **dry** or **fleshy.** The dry fruits may further be classified into ***(i)* dehiscent, *(ii)* indehiscent** or ***(iii)* schizocarpic** fruits.

***(i)* Dry dehiscent fruits**

***(a)* Legume or pod.** This type of simple dry fruit is monocarpellary, developing from a superior, one chambered ovary. It dehisces by both the sutures. Typical examples are found in Leguminosae family, *e.g.,* pea, bean, pulses, gram and many others.

***(b)* Follicle.** This type of simple dry fruit is monocarpellary, developing from a superior, one-chambered ovary like the legume fruit, but it dehisces by one suture only as in *Calotropis* **(Ak),** *Asclepias, Vinca rosea* **(Sadabahar),** *Michelia,* etc.

***(c)* Siliqua.** This is a long, narrow, many seeded fruit which develops from a superior bicarpellary ovary with two parietal placentae. It dehisces from below upwards by both the sutures. Here the ovary is one-chambered at first, but later on it becomes two-chambered because of the development of a flase septum, the **replum.** which extends from one placenta to another. This type of fruit is commonly found in Brassicaceae (Cruciferae), *e.g.,* mustard, radish, *Eruca,* etc.

***(d)* Silicula.** When a siliqua fruit becomes much shorter and flattened and as broad as it is long with a few seeds, it is called **silicula,** *e.g., Capsella* and candytuft (*Iberis amara*).

***(e)* Capsule.** This is a many-seeded, uni-or multilocular fruit which develops from a superior bi-or polycarpellary ovary. It dehisces in many ways. The dehiscent fruits which develop from a syncarpous ovary are commonly called capsule fruits. A capsule may dehisce by pores, as in poppy; or transversely in cock's comb; or loculicidally, as in cotton and **Bhindi** septicidally, as in linseed and septifragally as in *Datura.* In *Datura* the dehiscence lines are irregular and expose the central column bearing seeds.

***(ii)* Dry indehiscent or achenial fruits.** These fruits are dry and one-seeded. Here the pericarp does not split or break open to allow the seed to escape. They develop from monocarpellary to polycarpellary syncarpous pistil with one chamber and one ovule. The main kinds of such fruits are as follows:

***(a)* Caryopsis.** This is a very small, dry, one-chambered and one-seeded fruit which develops from a superior monocarpellary ovary. Here the pericarp remains fused with the seed coat. The well known examples are found in Poaceae (Gramineae) family, *e.g.,* maize, wheat, barley, millets, etc.

(*b*) **Achene.** An achene is a small, dry ındehiscent fruit. It develops from a superior monocarpellary, one-chambered and single ovuled ovary. The pericarp is thin and leathery and encloses a single seed. The pericarp remains free from the testa or seed coat. The achenes are commonly developed from an apocarpous pistil, and therefore a single flower produces as many achenes as many free carpels are there, *e.g.*, *Clematis*, *Naravelia*, *Ranunculus*, etc.

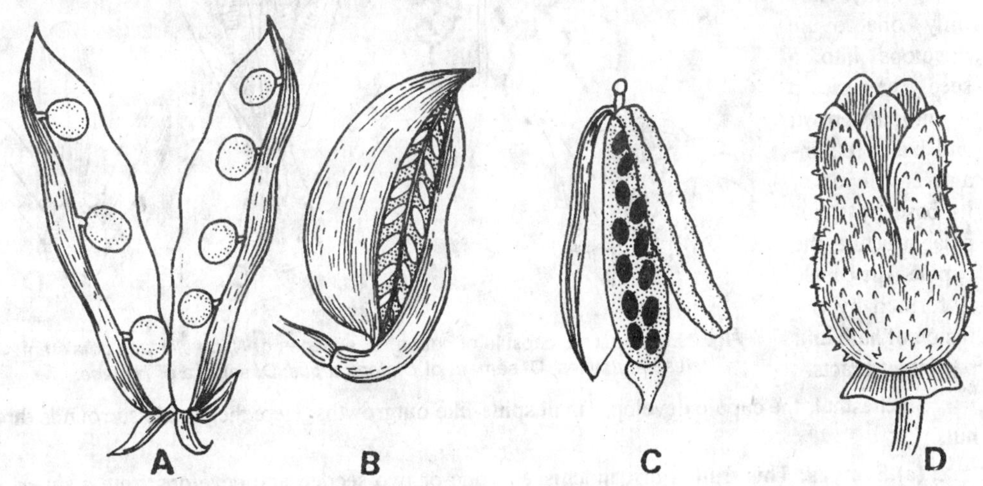

Fig. 29.68. Fruit. A, legume or pod of pea; B, follicle of *Calotropis;* C, siliqua of mustard; D, capsule of *Datura.*

(*c*) **Cypsela.** This is a dry, one-chambered and one-seeded fruit. It develops from an inferior, bicarpellary ovary. The pericarp and the seed coat remain free from each other. The examples are commonly found in Asteraceae (Compositae family, *e.g.*, sunflower, *Cosmos*, *Tagetes*, *Ageratum*, *Sonchus*, etc.

(*d*) **Nut.** This is a dry, one-chambered and one-seeded fruit with hard and woody epicarp. It develops from a superior bi-or polycarpellary ovary, *e.g.*, chestnut, oak, walnut, etc.

In *Litchi* the nut develops from a tricarpellary, syncarpous, superior and trilocular ovary with more than one ovules. On the maturity of the fruit only one ovule develops into a mature seed. The juicy aril which is an outgrowth of the testa from micropylar end surrounds the whole seed. The fleshy and juicy aril is edible. The pericarp is hard and leathery and bears numerous echinulate tubercles on it. It becomes brownish red at maturity.

In *Juglans regia* (English walnut) the fruit is a nut which develops from bicarpellary, syncarpous, inferior and one-chambered ovary with one basal ovule. Here the fruit remains surrounded by an **exocarp** of fused involucre and perianth surrounding the ovary. It makes the husk of the fruit and is removed by burning it before marketing the fruits. The thalamus and pericarp are hard and woody. The seed is large with convoluted cotyledons. which make the edible part of the nut. The stony pericarp is produced into ingrowths.

In cashewnut (*Anacardium occidentale*) the fruit (nut) develops from a monocarpellary superior, single-chambered, single-ovuled ovary. The thalamus becomes fleshy and edible. The kernel of true nut is also edible.

In *Chenopodium* the nut remains surrounded by persistent perianth and is known as **utricle.**

The fruit (nut) develops from a bicarpellary, syncarpous superior and unilocular ovary with single basal ovule.

In oak (*Quercus* spp.) the nut develops from a tri-carpellary syncarpous and inferior ovary. Here only one ovule develops into a seed. The pericarp is smooth, brown and hard. It remains enveloped at its base by a cupular structure, the **cupule**. The cupule is formed by the fusion of hard and persistent bracts.

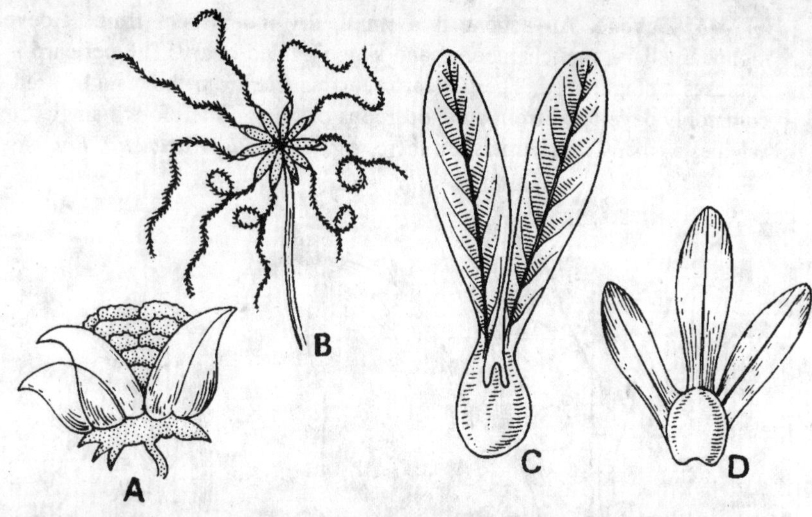

Fig. 29.69. Fruit. A, capsule of cotton; B, achenes of *Naravelia* with persistent feathery styles; C, samara of *Dipterocarpus;* D, samara of *Hiptage.*

In chestnut, the cupule develops small spine-like outgrowths. Here the cupule surrounds three nuts.

***(e)* Samara.** This fruit is dry, indehiscent, one-or two seeded and develops from a superior, bi-or tricarpellary ovary, with flattened wing-like outgrowths, *e.g., Hiptage, Dipterocarpus, Acer*, etc. Here the wings always develop from the pericarp, and the fruit breaks into its component parts, each enclosing a seed.

The fruit of *Shorea* is also winged, but here the wings are dry and persistent sepals. Such winged fruits are called **samaroid** fruits.

***(iii)* Splitting or schizocarpic fruits.** They are dry, many seeded, indehiscent fruits. The fruits generally break into many or few one-seeded portions, the **mericarps.** The mericarps are indehiscent. In certain schizocarpic fruits, *e.g.,* castor the one seeded parts of the fruit are dehiscent, and are called **cocci.** They are of following types.

***(a)* Lomentum.** It is characteristic of Mimosaceae family, *e.g., Acacia, Mimosa*, etc., the fruit is derived from a monocarpellary superior, unilocular ovary with marginal placentation. Here the legume is constricted or partitioned, into a number of one-seeded **mericarps.**

***(b)* Cremocarps.** This type of fruit is dry indehiscent and two-chambered. It develops from an inferior, bicarpellary ovary. On maturity the fruit splits apart into two indehiscent, one seeded portions, the **mericarps.** The mericarps remain attached to the prolonged end of the axis, known as **carpophore.** This type of fruit is characteristic of Apiaceae (Umbelliferae) family, *e.g.*, coriander, fennel, carrot, cumin, etc.

***(c)* Carcerulus.** This type of fruit is derived from a syncarpous and superior ovary with many chambers and axile placentation. Each loculus may possess one or two ovules. The mature fruit splits into many single seeded and indehiscent mericarps, *e.g., Althaea rosea* (hollyhock), *Abutilon indicum* (**Kanghi**), *Salvia*, *Ocimum*, etc.

***(d)* Regma.** This type of fruit is characteristic of castor plant (*Ricinus communis*). It develops from tricarpellary, syncarpous, superior, trilocular ovary with axile placentation. The fruit wall bears spine like tubercles. The locules are separated from each other on maturity by an explosive mechanism. These dehiscent locules are known as **cocci**, and remain attached to a central column, the **carpophore.**

The regma fruit is also found in *Geranium*. In this case it is derived from a pentacarpellary, syncarpous and superior ovary. The carpels remain fused around the base of a central axis the **carpophore.** On maturity, the fruit breaks into free single seeded cocci.

(*e*) Double samara. It is derived from a superior, bi-or-tricarpellary, syncarpous ovary with a single ovule in each loculus. Two or three wings are produced from the pericarp. The mature fruit breaks into two or three, indehiscent, single seeded mericarps. Examples are *Acer*, *Dodonaea*.

Fleshy Fruits

In these fruits the pericarp becomes fleshy and edible. Such fruits may be one-or many chambered, one- or many-seeded or inferior, with axile or parietal placentation. Commonly they are indehiscent fruits, and their seeds may be liberated only on the decay of the pulp. The pericarp of fleshy fruits is generally distinguished into three parts, (*a*) **epicarp,** (*b*) **mesocarp** and (*c*) **endocarp**. Epicarp is the outermost layer of the pericarp. It is the protective layer and may be thin or smooth or thin and papery. In *Aegle marmelos* it becomes hard and stony. Just beneath the epicarp, the **mesocarp** is found. In some (*e.g.*, mango) it is fleshy, juicy and edible, whereas in others it is fibrous (*e.g.*, coconut). In orange and banana it is thread like. The endocarp is the inner most layer of the pericarp. It is hard and stony in mango; fleshy and edible in banana and tomato. In orange the endocarp is thin and papery and bears juicy hair on its inner surface. Here the juicy hairs are edible.

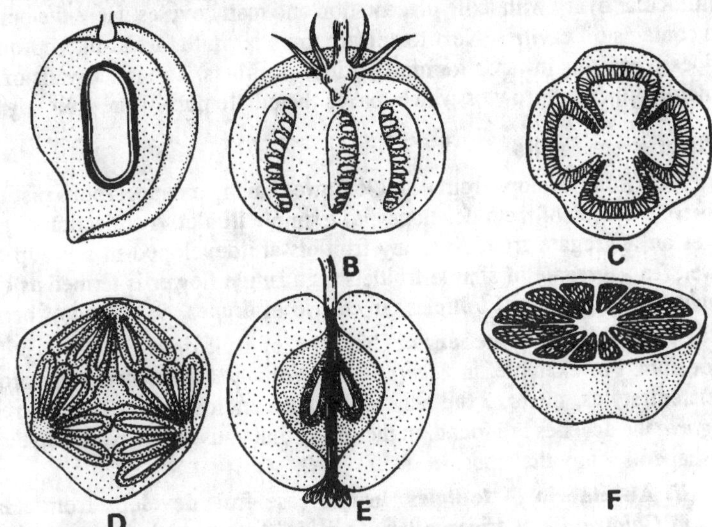

Fig. 29.70. Fruits. A, drupe of mango; B, berry of tomato, C; same in T.S.; D, pepo of cucumber in T.S.; E, pome of apple; F, hesperidium of orange.

1. Drupe. This is a fleshy, one-or more chambered and one-or more seeded fruit. It develops from a monocarpellary or syncarpous pistil. The pericarp of the fruit remains differentiated into thin epicarp, fleshy mesocarp and stony endocarp. Because of the presence of stony endocarp the fruit is also known as **stone fruit**, *e.g.*, mango, peach, plum, coconut, almond, etc.

2. Berry or **Bacca.** This is a superior or inferior, indehiscent, many seeded and fleshy fruit. It develops from a single carpel or from a syncarpous pistil with axile or parietal placentation, *e.g.*, tomato, grapes, brinjal, guava, papaw, etc. In this type of fruit the seeds at first remain attached to the placentae, but later on they are detached from the placentae and lie free in the pulp. The date-palm and *Artabotrys* are the examples of one-seeded berry. The pericarp of berry is also differentiated into epicarp, mesocarp and endocarp but like drupe the endocarp is not stony.

3. Pepo. This type of fruit is characteristic of Cucurbitaceae family, *e.g.*, gourd, cucumber, melon, watermelon, squash, etc., the fruit is fleshy or pulpy and many seeded. It develops from an inferior, syncarpous pistil with parietal placentation. Here the seeds remain embedded in the pulp and attached to the placentae.

4. Pome. This type of fruit is found in Rosaceae family, *e.g.*, apple, pear, etc., the fruit develops from pentacarpellary, syncarpous and inferior ovary with many seeds. The fruits remain surrounded by fleshy thalamus. The fleshy thalamus makes the edible part, whereas the actual fruit lies within it. The outer portion of the pericarp (*i.e.*, epicarp and mesocarp) is fleshy and fused with the thick edible portion (thalamus). The inner layer of pericarp (*i.e.*, endocarp) is cartilaginous and forms the core surrounding the seeds. Each part of the core represents one of the five carpels. The remains of the perianth may be seen at the top of the fruit.

5. Hesperidium. This type of berry is derived from a polycarpellary, syncarpous superior, multilocular ovary with axile placentation and many ovules. Here the epicarp is thick and leathery and contains oil cavities. Next to epicarp there is white thread-like fibrous layer of mesocarp. The endocarp projects inwards forming distinct chambers. The inner walls of the endocarp give rise to numerous juicy outgrowths which makes the edible part of the fruit, *e.g.*, orange, lemon, etc.

2. Aggregate Fruits

Such fruit develops from a single flower having an apocarpous pistil. In such case the carpels are free, and each of them develops into a simple fruitlet. A collection or a group of simple fruitlets makes an aggregate fruit. As many fruitlets are developed in a group as carpels are there in a flower. An aggregate of simple fruitlets on a single flower is termed as etaerio-such as an etaerio of achenes, an etaerio of follicles, an etaerio of drupes, an etaerio of berries, etc.

***(i)* An etaerio of achenes.** This type of fruit develops from polycarpellary apocarpous gynoecium. For example, in *Clematis* and *Naravelia* the achenes are provided with feathery and persistent styles, in *Rosa* the achenes remain enclosed in a hollow, receptacular thalamus, in *Fragaria* the achenes are found on the fleshy thalamus and in *Nelumbo* the fruitlets occur on a flat, top-shaped spongy thalamus.

***(ii)* An etaerio of follicles.** In this case fruit develops from a bicarpellary apocarpous (*e.g.*, in *Calotropis*) or tricarpellary (*e.g.*, in *Aconitum*) or polycarpellary apocarpous (*e.g.*, in *Michelia*) gynoecium. In *Asclepias* and *Calotropis* each etaerio consists of a pair of follicles. In *Aconitum* an aggregate of three follicles is produced on the top of the thalamus, whereas in *Michelia* numerous follicles are developed on an elongated thalamus.

***(iii)* An etaerio of drupes.** The characteristic example of this type is found in *Rubus* (raspberry). It develops from superior, polycarpellary and apocarpous gynoecium. The carpels are borne on a conical thalamus. Here a number of small drupelets develop from separate carpels of a flower and they are grouped together on a fleshy thalamus.

***(iv)* An etaerio of berries.** In this type of fruit each carpel of an apocarpous pistil develops into a berry. During development the margins of the carpels may be fused (*e.g.*, in *Annona squamosa*) or they may remain separate from each other (*e.g.*, in *Artabotrys odoratissimus*). In *Annona squamosa* the mesocarp of berries is edible.

3. Composite or Multiple Fruits

A composite or multiple fruit develops from the entire inflorescence. Here the flowers as well as the peduncles on which they are borne take part in the development of the fruit. Such a fruit is also known as **infructescence.** There are two main types of such fruits.

***(i)* Sorosis.** This type of multiple fruit develops from a spike or spadix. Here the flowers fuse together by their perianth lobes and simultaneously the axis bearing them becomes fleshy and juicy, and as a result the whole inflorescence forms a compact mass, *e.g.*, pine apple, jack fruit, etc. In *Morus* (mulberry), the perianth lobes become thick and juicy and are edible. Here the ovules do not mature into seeds and therefore, the carpels develop into small nutlets that are seedless.

***(ii)* Syconus.** This type of composite fruit develops from a hollow, pear-shaped, fleshy receptacle which encloses numerous minute male and female flowers. The receptacle develops further and converts into the so-called edible fleshy fruit. It really encloses a number of true fruits or achenes which develop from the female flowers of the hypanthodium inflorescence, *e.g.*, fig, banyan and many species of *Ficus*.

Table 29.13. Common Edible Fruits

English name	Botanical name and family	Type of fruit	Edible part
Apple	*Malus sylvestris* Rosaceae	pome	thalamus
Banana	*Musa paradisiaca* Musaceae	berry	mesocarp and endocarp
Cashew-nut	*Anacardium occidentale* Anacardiaceae	nut	peduncle and cotyledons
Coconut	*Cocos nucifera* Palmaceae	fibrous drupe	endosperm
Cucumber	*Cucumis sativus* Cucurbitaceae	pepo	mesocarp, endocarp and placentae
Custard apple	*Annona squamosa* Annonaceae	etaerio of berries	fleshy pericarp of individual berries
Date palm	*Phoenix sylvestris* Palmaceae	one-seeded berry	pericarp
Fig	*Ficus carica* Moraceae	syconus	fleshy receptacle
Jack fruit	*Artocarpus heterophyllus* Moraceae	sorosis	bracts, perianth and seeds
Grape	*Vitis vinifera* Vitaceae	berry	pericarp and placentae
Guava	*Psidium guava* Myrtaceae	berry	thalamus and pericarp
Indian plum	*Zizyphus jujuba* Rhamnaceae	drupe	mesocarp and epicarp
Litchi	*Litchi chinensis* Sapindaceae	one-seeded nut	fleshy aril
Wheat and other cereals	*Triticum aestivum* Poaceae	caryopsis	starchy endosperm
Mango	*Mangifera indica* Anacardiaceae	drupe	mesocarp
Melon	*Cucumis melo* Cucurbitaceae	pepo	mesocarp
Orange	*Citrus aurantium* Rutaceae	hesperidium	juicy placental hair
Papaw	*Carica papaya* Caricaceae	berry	mesocarp
Pea	*Pisum sativum* Papilionaceae	legume	cotyledons
Pear	*Pyrus communis* Rosaceae	pome	fleshy thalamus
Pine-apple	*Ananas comosus* Bromeliaceae	sorosis	outer portion of receptacle bracts and perianth
Pomegranate	*Punica granatum* Punicaceae	special	juicy testa (balusta)
Strawberry	*Fragaria vesca* Rosaceae	etaerio of achenes or nuts	succulent thalamus
Tomato	*Lycopersicon esculentum* Solanaceae	berry	pericarp and placentae
Wood apple	*Aegle marmelos* Rutaceae	berry	mesocarp, endocarp and placentae

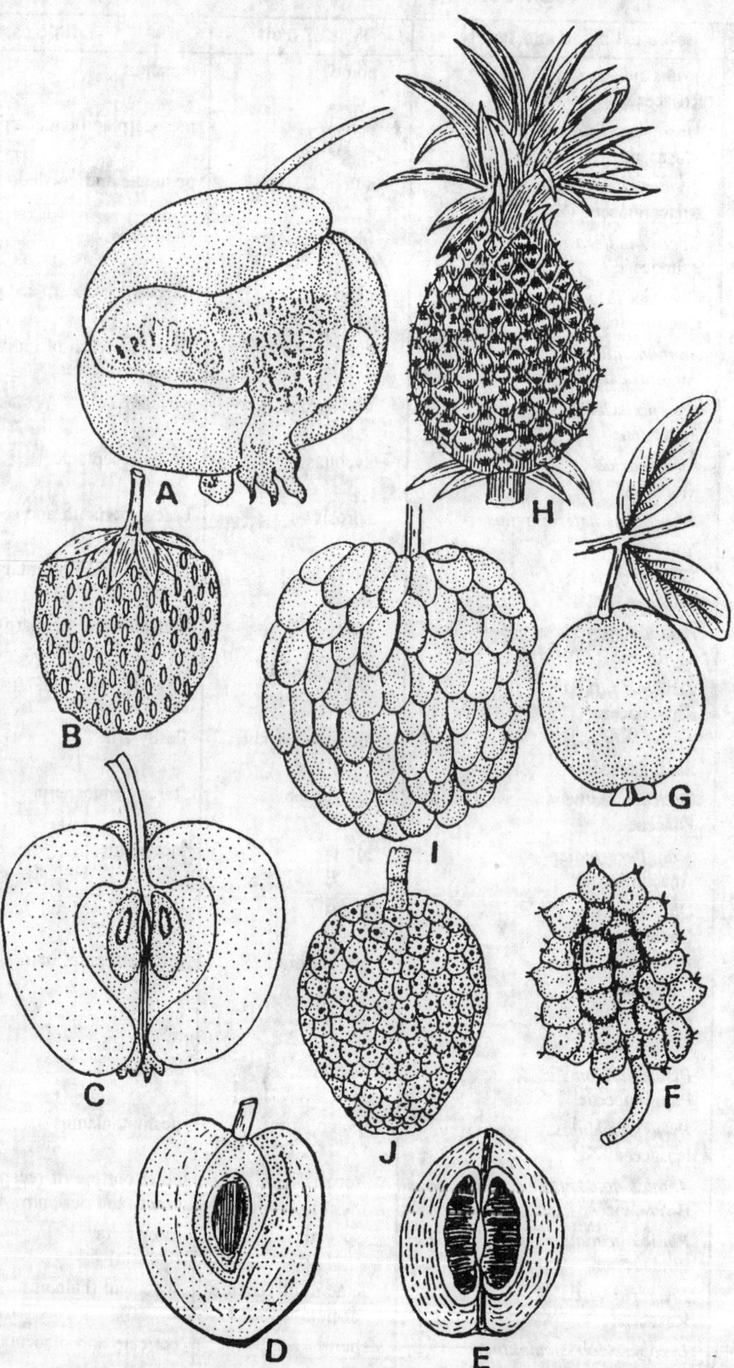

Fig. 29.71. Some edible fruits. A, Pomegranate (*Punica granatum*) ; B, Strawberry (*Fragaria vesca*) ; C, Apple (*Malus sylvestris*); D, Peach (*Prunus persica*) ; E, Sapodilla (*Achras zapota*) ; F, Mulberry (*Morus alba*) ; G, Guava (*Psidium guava*) ; H, Pineapple (*Ananas comosus*). ; I, Custard apple (*Annona squamosa*); J, Litchi (*Litchi chinensis*).

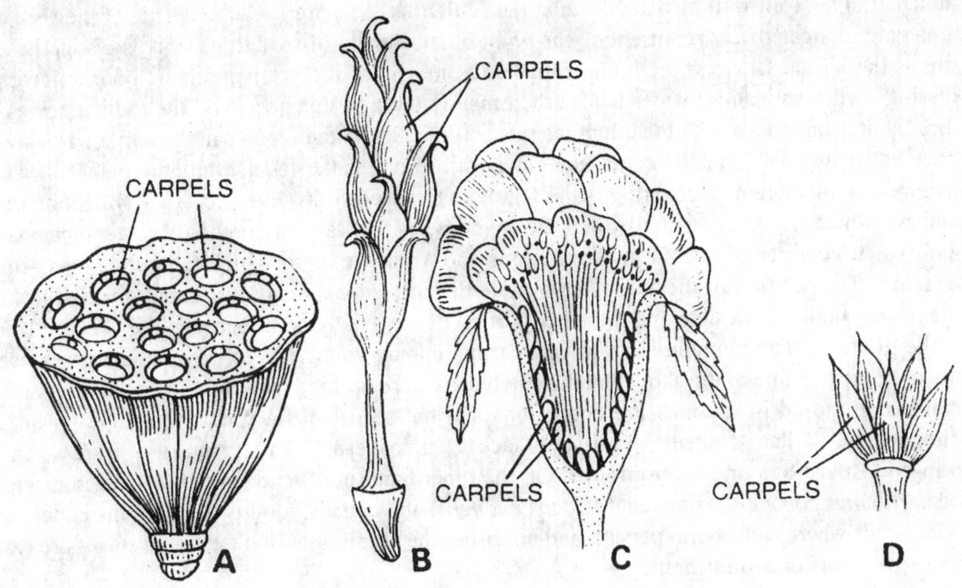

Fig. 29.72. Apocarpous pistil. A, lotus; B, *Michelia;* C, Rose; D, *Sedum.*

Functions of Fruits

The fruits are meant for the protection of the seeds. They are the main instruments of seed dispersal. The fruit wall is variously modified in many fruits for dispersal by wind, animals and water. The fleshy fruits by their decay enrich the soil with inorganic acids and thus provide a good start to the young seedings which develop from the seed. Many fruits are edible and make the main food of human beings.

DIVERSITY IN SEED

A true seed is defined as a fertilized mature ovule that possesses embryonic plant, stored material, and a protective coat or coats. Seed is the reproductive structure characteristic of all phanerogams. The structure of seeds may be studied in such common types of pea, gram, bean, almond or sunflower. They are all built on the same plan although there may be differences in the shape or size of the seed the relative proportion of various parts.

There are hundreds of variations in the seed size, shape, colour and surface. The seeds range in size from tiny dust particles, as found in some orchids, to large double-coconuts. The seed surface may be smooth, wrinkled, striate, ribbed, furrowed, reticulate, tuberculate, alveolate, hairy, pulpy or having patterns like finger prints.

In the seed, life activities are temporarily suspended in order to enable the plant to successfully pass through unfavourable and injurious climatic conditions. On the approach of favourable conditions, the seed resumes active life and grows into full plant. In the form of seeds, a plant can be carried to long distances without special precautions.

Structure of Seed

The various parts of a seed may be easily studied after it has been soaked in water for a day or so varying according to the nature of the seeds. A mature seed contains an embryonic plant (with a radicle and plumule), and is provided with reserve food materials and protective seed coats. A mature pod of pea (*Pisum sativum*) has a number of seeds arranged in two rows. The seeds are attached to the fruit wall by a small stalk, the **funiculus.** At maturity, on one side of the seed coat a narrow, elongated scar representing the point of attachment of seed to its stalk is distinctly seen, this is the **hilum.** Close to the hilum situated at one end of it there is a minute pore, **micropyle.** During seed germination, water is absorbed mainly through this pore, and the radicle comes out through it. Continuous with the hilum there is sort of ridge in the seed coat, the **raphe.** The seed is covered by two distinct **seed coats**; the outer whitish one is the **testa**, while the other inner thin, hyaline and membranous covering is the **tegmen.** The seed coats give necessary protection to the embryo which lies within. The whitish fleshy body, as seen after removing the seed-coats is the embryo. It consists of two fleshy cotyledons and a short **axis** to which the cotyledons remain attached. The position of the axis lying outside the cotyledons, bent inward and directed towards the micropyle is the **radicle** and the other portion of the axis lying in between the two cotyledons is the **plumule.** The plumule is crowned by some minute young leaves. The radicle gives rise to the root, the plumule to the shoot and the cotyledons store up food material. Since the reserve food material is stored in the massive cotyledons and the seed lacks a special nutritive tissue, the **endosperm.** The seeds which lack endosperm at maturity are called **non-endospermous** or **exalbuminous.** On the other hand in several other plants such as castor bean (*Ricinus communis*), coconut (*Cocos nucifera*) and cereals, food is stored in the endosperm. Such seeds where endosperm persists and nourishes the seedling during the initial stages are called **endospermous** or **albuminous.**

On the basis of the number of cotyledons in the embryo the angiosperms have been divided into two large groups;

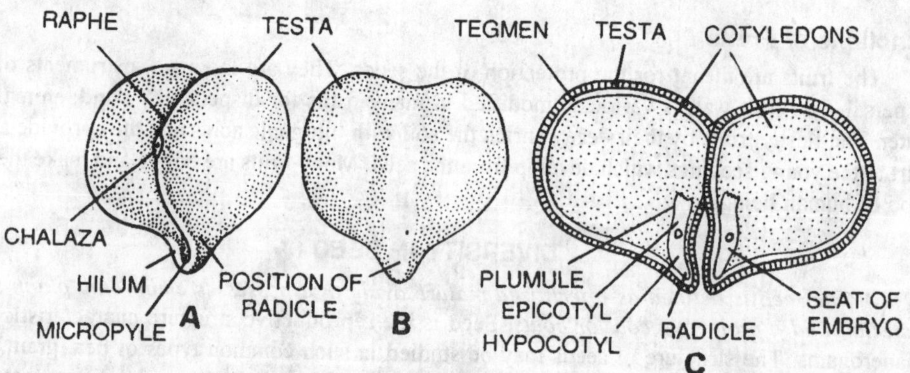

Fig. 29.73. Structure of gram seed. A, complete seed showing various parts; B, seed as seen after removing the testa; C, L.S. of seed.

1. **Dicotyledons**, having embryos with two cotyledons, and
2. **Monocotyledons**, with only one cotyledon

A maize grain is a single-seeded fruit in which the seed coat and the fruit wall are unseparable. On one side of the grain a small, opaque, whitish, deltoid area is seen to be distinctly marked out from the region. The embryo lies embedded in this area. There is only a thin layer surrounding the whole grain. This layer is made up of the seed-coat and the wall of the fruit fused together. The grain remains divided into two unequal portions by a definite layer known as the **epithelium.** The bigger portion is the **endosperm**, and the smaller portion, the **embryo.** The endosperm, is the food storage tissue. The embryo consists of one shield shaped cotyledon, known as the **scutellum** and **axis.** The upper portion of the axis, with minute leaves arching over it, is the **plumule**, and the lower portion provided with the root cap the **radicle.** The plumule is surrounded by a leaf-sheath or coleoptile and the radicle is surrounded by a root sheath or **coleorhiza.** These are the protective sheaths of the plumule and the radicle respectively.

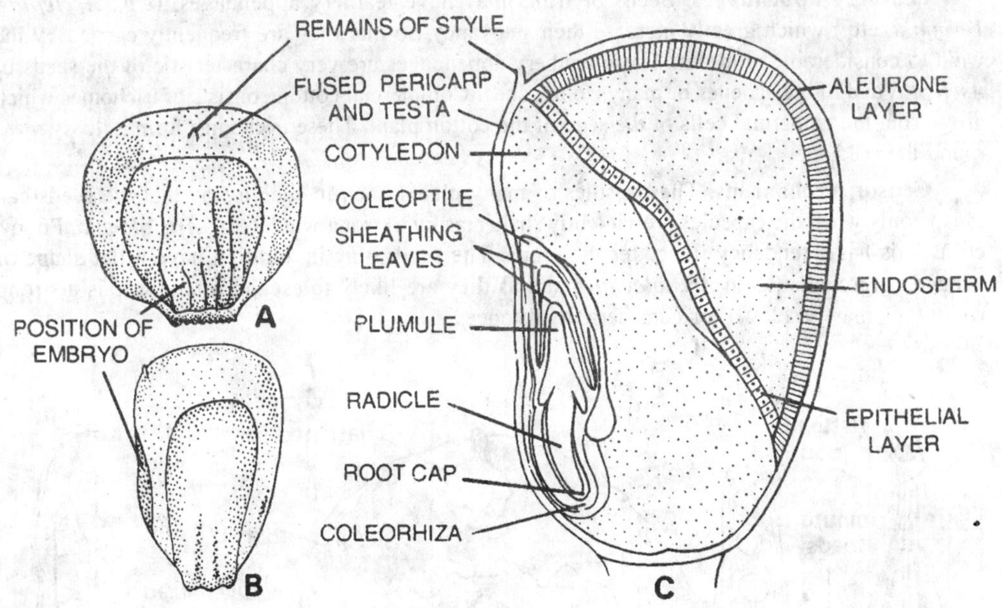

Fig. 29.74. Structure of maize grain. A, grain illustrating of external characters; B, grain as seen from lateral side; C, section of the grain showing its internal structure.

Besides the basic structures (endosperm, embryo and seed-coat) certain special structures may arise during seed development. In castor bean a fleshy whitish tissue, the **caruncle,** develops at one end of the seed. It is derived from the integument. The juicy edible part of the litchi fruit (**aril**) is an outgrowth of the funiculus that develops after fertilization. The cotton fibres are the elongated epidermal cells of the seed-coat. These fibres are single-celled and thin walled. They attain a length of upto 45 mm and have characteristic twists.

DISPERSAL OF FRUITS AND SEEDS

The seed consists of an embryo, stored food material and a protective covering. The embryo is capable of growing into a plant and the stored food material furnishes it with supply of food for its growth in the period before it has become established. Most seeds are also provided with some means of dispersal. Without this the great majority of seeds would fall under the parent plant and not be carried to a location favourable to germination and growth.

Dispersal by Wind

Either the whole fruit or the individual seeds may be suited to dispersal by wind. Seeds that are thus disseminated are characteristically light. The means of adaptation to wind dispersal may be grouped under the headings of **minute seeds, flattened fruits** or **seeds, wing like outgrowths, feathery appendages** and the so-called **censor mechanisms.**

Minute seeds. The seeds of orchids are very small and besides, have a light, inflated outer covering. These dust-like seeds can be blown by the wind for great distance.

Wings. In many cases, seeds and in others, whole fruits are flattened or have wing-like outgrowths or they may be both flattened and winged. This type of structure results in the scattering of the seed by the wind, (*e.g.,* the fruits of *Terminalia*, ash, maple, *Macrozanonia*, *Pterocymbidium*, *Dipterocarpus*, linden, etc.)

Feathery appendages. Seeds or fruits may have feathery appendages, (*e.g., Asclepias, Vernonia*, etc.) which greatly increase their buoyancy, so that they are frequently carried by the wind to considerable altitudes. These feathery appendages are very characteristic of the seeds of *Asclepias* and of the achenes or many Compositae. Commercial cotton consists of trichomes which grow from the epidermal cells of the seed of the cotton plant. These trichomes form a flossy mass round the seed.

Censor mechanisms. The capsules of many plants open in such a way that the seeds can escape only when the capsules are violently shaken as by a strong wind, (*e.g., Aristolochia*, Poppy, etc.). This has a tendency to scatter the seeds. The seeds may in addition have a flat shape or winged outgrowths (as in *Aristolochia*); and as they are likely to escape when there is a strong wind, they may be blown for considerable distances.

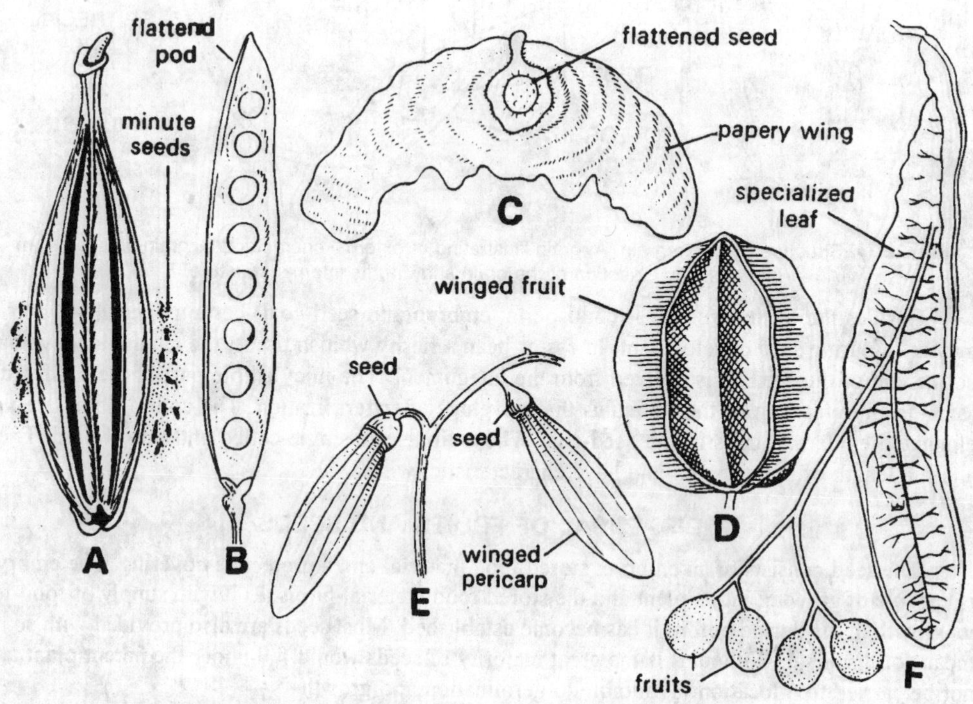

Fig. 29.75. Dispersal by wind. A, capsule of an orchid (*Cymbidium* spp.) releasing minute seeds; B, flattened pod of *Albizzia* spp.; C, winged seed of *Macrozanonia;* D, winged fruit of *Terminalia arjuna;* E. fruits of *Tilia americana* attached to a flattened bract that help in wind dispersal; F, the fruit with seeds attached to a specialized leaf for dispersal.

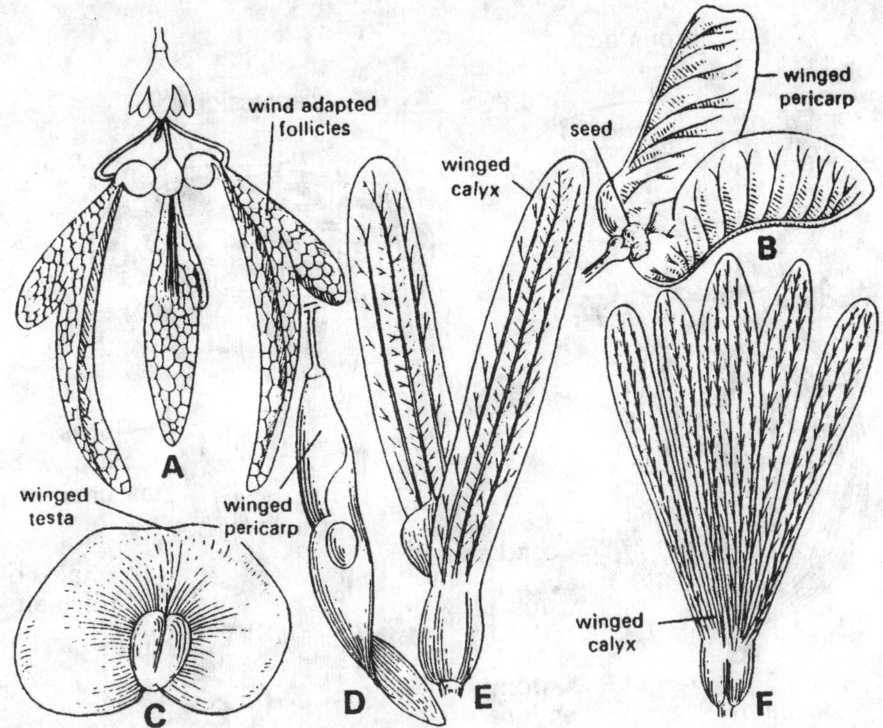

Fig. 29.76. Dispersal by wind. A, follicles of *Pterocymbidium*, with pericarp developed into wing like structures; B, winged fruit of mapple; C, winged seed of *Jacranda;* D, pod of *Alianthus* showing propellar mechanism; E, fruit of *Dipterocarpus* having calyx lobes (*i.e.*, winged calyx) ; F, winged calyx of *Parashorea stellata*.

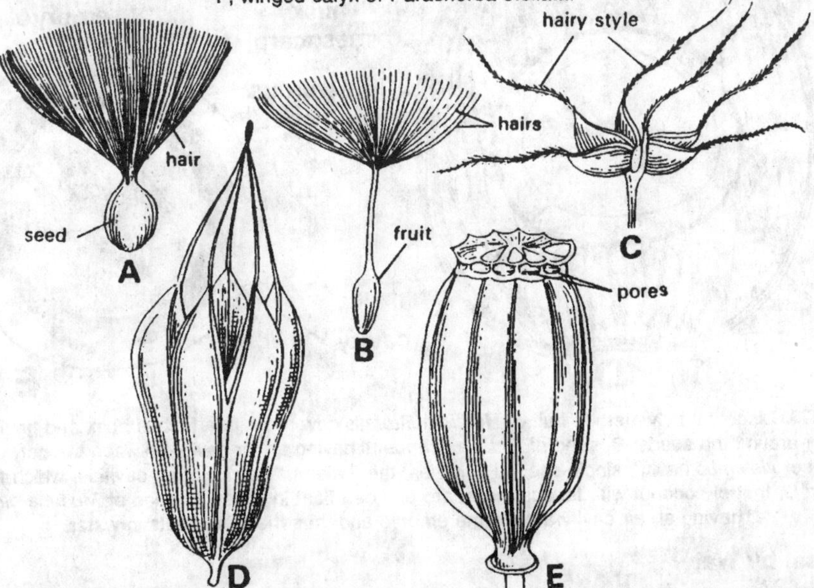

Fig. 29.77. Dispersal by wind. A, plumed seeds of *Calotropis procera;* B, fruit of *Taraxacum* with hairy pappus; C, an eaterio of achenes of *Nigella;* D, capsule of *Aristolochia,* the seeds are being dispersed when the capsule is shaken by wind; E, capsule of poppy, the minute seeds are released through pores in capsule and are dispersed by wind.

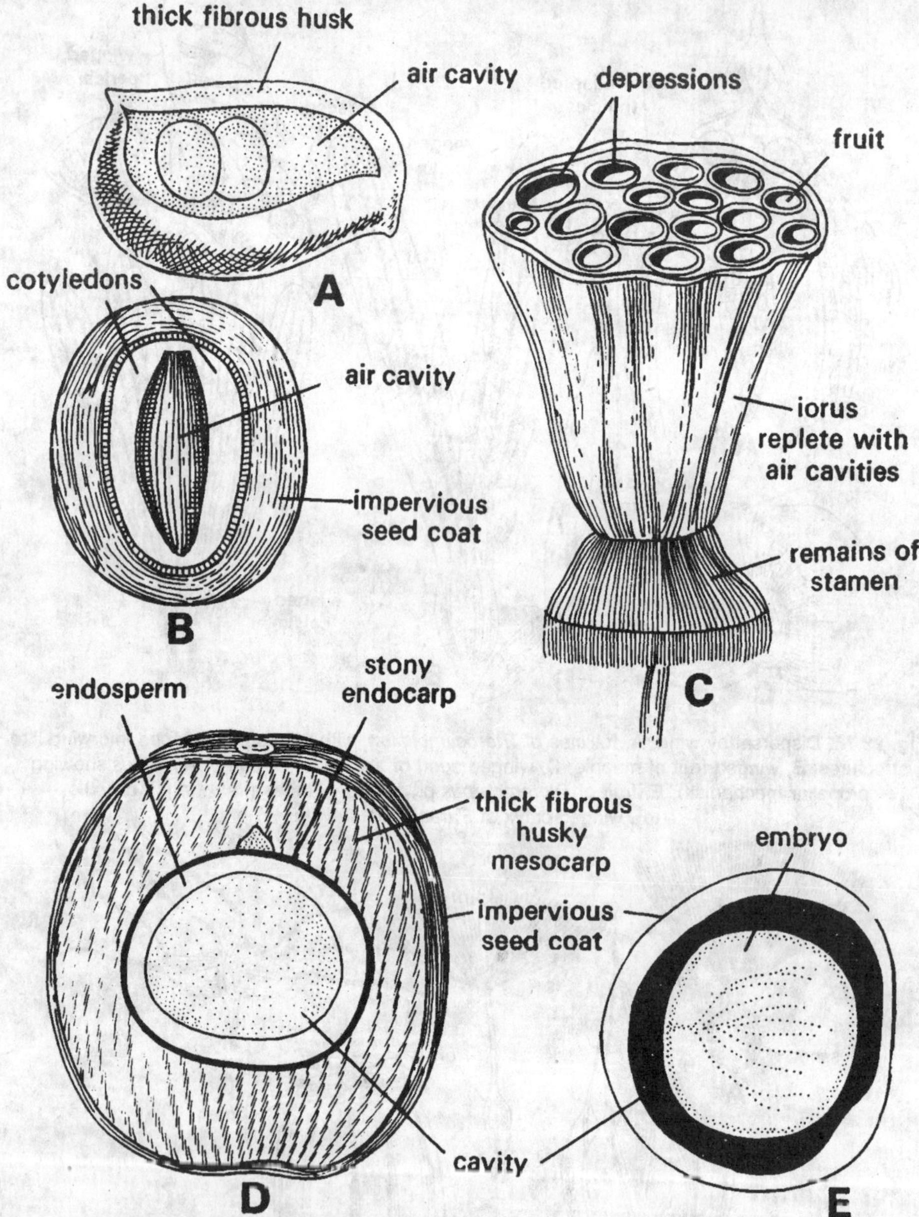

Fig. 29.78. Dispersal by water. A, fruit of *Heritiera littoralis* covered with a fibrous husk and having an air cavity around the seeds; B, seed of *Entada scandens* having an air cavity between the cotyledons; C, fruit of *Nelumbo* having single-seeded achenes, the thalamus has lot of air cavities, which floats in water; D, fruit of coconut with fibrous mesocarp and can float in water; E, seed of *Mucuna gigantea* having an air cavity around the embryo and thus the seed floats in water.

Dispersal by Water

Adaptations for dispersal by water occur in many seashore and aquatic plants. Either the whole fruit or the seed may be adapted for floating. The pericarp of a fruit may be composed of light tissue, (*e.g.*, in *Terminalia catappa*) or the fruit may be inflated, (*e.g.*, in *Heritiera littoralis*).

The coconut is an excellent example of a fruit with a light pericarp. Floating seeds may likewise contain either a mass of light tissue of large air spaces (*e.g., Entada scandens*). In the lotus fruit the torus is a greatly enlarged mass of loose, air-filled tissue which floats readily, while the individual fruits are also adapted for floating. Many seeds that are not especially fitted for floating may occasionally float for short distances, or seeds that by themselves would not float may be carried in floating debris.

Dispersal by Animals

Seeds that are adapted for dispersal by animals are disseminated in two general ways: in the case of fleshy fruits, a portion of the fruit is eaten by the animals, while any dry fruits adhere to animals.

Fleshy fruits. Fleshy fruits are generally adapted to being eaten by animals. Such fruits are usually constructed so that the fleshy part may be eaten without injury to the seed. In many cases the seed coat is very hard.

While in drupes the seed is protected by the stony endocarp. Owing to this protection a seed is protected by the stony endocarp. Owing to this protection a seed may pass without injury through the digestive tract of an animal. Birds are particularly prominent in disseminating the seeds of fleshy fruits. Sometimes they eat the fleshy portion of a fruit and throw the seeds away. Dry fruits are often carried off for food by seed-eating animals which lose them in one way or another and leave them to grow.

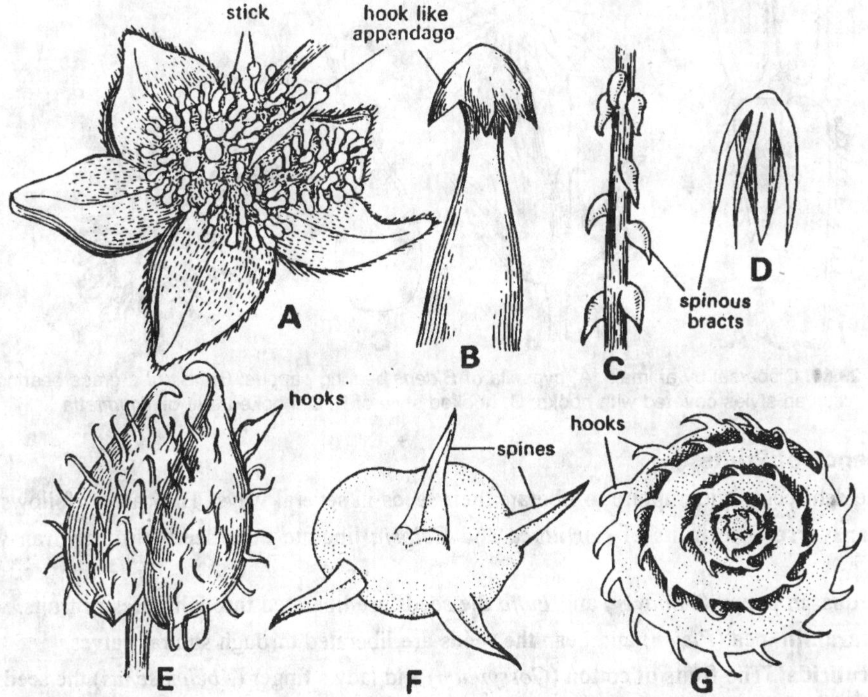

Fig. 29.79. Dispersal by animals. A, fruit of Chinese forget-me-not bearing sticky glands; B, hook-like appendage of A; C, fruits of *Achyranthes aspera* bearing spinous bracts; D, spinous bract of C; E, fruit of *Xanthium* bearing hooks on the outer covering of fruit, made up of thalamus; F, fruit of *Tribulus terrestris* bearing spinous outgrowths; G, spirally coiled pod of *Medicago* bearing hook-like out-growths.

Adhesive fruits. Many dry fruits have hooklike appendages (*e.g., Cosmos, Bidens, Mimosa pudica, Triumfetta,* etc.). which are particularly fitted for grasping the hair of animals. Animals to which the fruits adhere carry them about and thus distribute the seeds. In a similar way fruit may adhere to clothing and thus be disseminated by man. Some seeds and fruits have a sticky covering which will adhere to the hair of an animal. Fruits of Chinese forget-me-not adhere by sticky hooklike appendages. Feathery appendages are usually capable of adhering to fur as well as of flying on the wind.

Minute seeds. Many plants have minute seeds which are disseminated by being caught in mud that adhere to the feet or other parts of birds or other animals.

Dispersal by Explosive Mechanism

Many fruits are so constructed that they explode when ripe and scatter the seeds. This method of dispersal is frequently conspicuous in members of the bean family, where the explosive forces are due to stresses arising from the drying of the valves of the pod. The balsam has somewhat fleshy capsules which are very turgid. When these are disturbed by contact the segments of the pericarps roll up with considerable force and in such a way that they scatter the seeds. An unusual explosive mechanism is found in the squirting cucumber (*Ecaballium elaterium*).

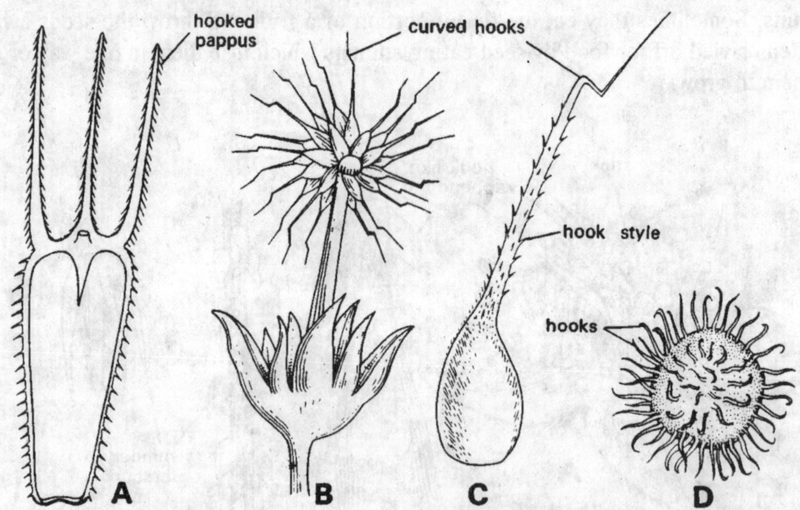

Fig. 29.80. Dispersal by animals. A, cypsella of *Bidens* bearing pappus; B, fruit of a grass bearing curved styles covered with hooks; C, hooked style of B; D, hooked fruit of *Triumfetta*.

Dehiscence of Fruits

The dehiscent fruits rupture to liberate their seeds in several ways. They are as follows:

Transverse. The fruits of *Portulaca* and *Celosia* liberate their seeds through transverse rupture.

Porous. In poppy (*Papaver*) and *Luffa* the seeds are liberated through pores in fruits.

Sutural. In pea (*Pisum*) and bean the seeds are liberated through sutural valves.

Loculicidal. The fruits of cotton (*Gossypium*) and lady's finger (*Abelmoschus*) the seeds are liberated through locules.

Septicidal. In linseed (*Linum*) and mustard (*Brassica*) the seeds are liberated through septa or partition wall.

Septifragal. In *Datura* and other such plants, the fruits rupture loculicidally throwing valves away from fruit and leaving seeds attached to the central axis.

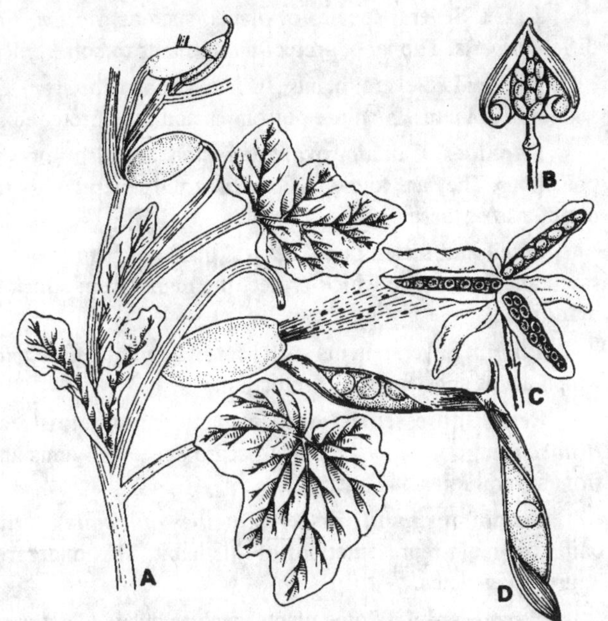

Fig. 29.81. Dispersal by explosive mechanism. A, explosive dehiscence in the fruit of *Ecaballium;* B, explosive dehiscence due to turgor movements in the fruit of *Impatilens balsamina;* C, dehiscence in the capsule of *Viola;* D, dihiscence in the legume of *Lathyrus odoratus.*

Defense Mechanisms in Plants

Many plants develop certain devices or special organs which give protection to them from their enemies. Some devices are mentioned here.

Thorns. Many plants, such as lemon (*Citrus*), pomegranate (*Punica*), bael (*Aegle marmelos*), *Duranta*, and ber (*Zizyphus*) possess thorns to protect them from their enemies.

Pointed spines. The plants like pineapple, datepalm (*Phoenix*), *Agave* and *Yucca* have pointed spines at the leaf ends which protect them to some extent from enemies.

Prickles. The rose plants and silk cotton tree possess prickles which act as defence organs.

Spines. *Opuntia* and several other members of Cactaceae (cacti) have spines on their phylloclades which protect them from attack of animals.

Stinging hairs. The stinging hairs with sharp and siliceous apices occur on all parts of the body in nettles (*Laportea*), *Urtica dioica* and several other members of Urticaceae which protect the plants from their expected enemies.

Glandular hairs. Glandular hair with sticky substances are found on plant body of *Jatropha*, *Boerhaavia* and tobacco (*Nicotiana*) which protect them from outer agencies.

Stiff hair. The dense coating of such stiff hair is repulsive to attacking animals, they are found on cud weed (*Gnaphalium*) and many cucurbits.

Besides, above mentioned special organs certain other defense devices like poisonous and irritating substances are found in many plants. They are as follows.

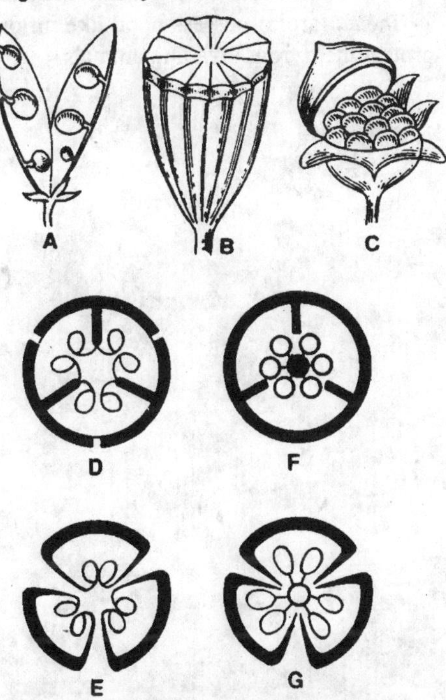

Fig. 29.82. Dehiscence of fruits, A, sutural, B, porous; C, transverse; D, loculicidal; E, septicidal; F and G, septifragal.

Latex. Several species of plants, such as *Nerium*, *Ficus* and *Euphorbia* have latex in their different parts. The latex protect these plants to some extent from his enemies.

Alkaloids. Several plants, such as Poppy (*Papaver*), *Datura* and tobacco (*Nicotiana*) possess alkaloids in them which are poisonous and give protection to the plants.

Raphides. Calcium oxalate crystals are also known as raphides which are irritating and poisonous. They are found in *Colocasia* and several other members of Araceae (aroids). They help in defense of these plants.

Smell and bitter taste. Several plants of Lamiaceae, such as *Ocimum* and mint have typical smell and bitter taste which protect the plants from attack of insects. Neem and Karela have also bitter taste.

Tannin. Several plants such as *Acacia nitotica, Acacia catechu,* etc., possess tannins in them which act as substances of protection.

Resin and essential oils. In many angiosperms resins and essential oils are present. In *Pistacea* and other members of Anacardiaceae the gums and resins are found which protect them from several injurious agencies.

Geophilous habit. Zinger (*Zingiber officinale*), turmeric (*Curcuma longa*), colocasia and onion are main representatives of this habit. This underground habit protects them from several injurious agencies.

Myrmecophily. Some plants, such as guava (*Psidium* spp.), mango (*Mangifera indica*) have a habit of harbouring ants which save the plants from the damage by other organisms.

Mimicry. This is habit of initiating the general appearance, colour and shape of other plants, generally disliked by attackers. For example, the aroids (*Cladium*) and *Sansevieria* resemble spotted snakes and thus by this habit keep away from plant-eating animals. The species of *Arisaema* (aroid) possess hood like inflorescence which resembles cobra in appearance and thus protect them from invading animals.

QUESTION BANK

Very Short Answer Questions (1 mark each)

1. Which part of the embryo develops into a root?
2. What is the function of pneumatophores?
3. What type of roots are found in *Cuscuta*?
4. Which kind of plants harbour nitrogen fixing bacteria in their root nodules?
5. Give the function of root hair.
6. Which plants bear the stilt roots?
7. Is phyllode a stem or a petiole?
8. Name kinds of underground stems.
9. What kind of stem modification is found in cactus?
10. Write the main functions of leaf lamina.
11. What type of venation is found in monocot leaves?
12. Give the function of tendrils.
13. What kind of underground stem is found in potato?
14. Give the main parts of a leaf.
15. What kind of leaf is found in pea?
16. What kind of leaves are found in *Acacia nilotica*?
17. Name two insectivorous plants.
18. Name the plant in which the leaves are modified to capture and digest insects.
19. Which same plant part has transformed into the following different modifications?
    (*i*) tendril of pumpkin
    (*ii*) thorn of *Citrus*.
20. *Cuscuta* develops roots to penetrate the host tissue and obtain nutrition. What are here roots called?
21. What is heterophylly?
22. What are the male and female reproductive parts of a flower?
23. What are the three parts of a carpel?
24. Name the four whorls of a flower.
25. What is the method of arrangement of ovules within the ovary wall known as?
26. What is the arrangement of flowers on the floral axis known as?
27. What is the type of arrangement of flowers in which oldest flowers are nearest the growing tip?
28. What is the main difference between a raceme and spike?
29. What type of fruit is tomato?
30. Give an example each of catkin, capitulum, and spikelet.
31. What are the two main types of inflorescences?
32. What is an indehiscent fruit.
33. Give an example of spadix inflorescence.
34. What is a false fruit? Give an example.
35. Name the outer covering of the seed.
36. Which group of plants shows double fertilization?
37. Mention the kind of arrangement of flowers in a capitulum.
38. What are the main functions of a flower?
39. Which agency helps in the dispersal of fruits and seeds in coconut?

ANSWERS

1. Radicle; **2.** Respiration; **3.** Haustoria; **4.** Leguminous plants; **5.** Absorption of water and dissolved minerals; **6.** Maize, Sorghum, *Pandanus*, etc.; **7.** Petiole; **8.** Rhizome, corn, tuber, bulb; **9.** Phylloclade; **10.** Photosynthesis-carbon assimilation. **11.** Parallel; **12.** Climbing; **13.** Tuber; **14.** Petiole and lamina; **15.** Imparipinnate compound; **16.** Compound bipinnate; **17.** *Drosera* and *Utricularia*; **18.** Sundew plant. 19. Axillary bud; **20.** Haustoria; **21.** Two types of leaves on same plant, *e.g., Trapa*; **22.** Androecium and gynoecium; **24.** Ovary, style and stigma; **25.** Placentation; **26.** Inflorescence; **27.** Basipetal succession; **28.** The flowers in a raceme are pedicellate while the flowers in a spike are sessile or sub-sessile; **29.** Fleshy berry; **30.** *Morus*, *Helianthus*, wheat; **31.** Racemose and cymose; **32.** A fruit that does not split open to release its seeds; **33.** *Colocasia*; **34.** Fruits where other parts, other than gynoecium participate in the formation of fruit, *e.g.*, apple, pear, cashewnut, etc; **35.** Testa; **36.** Angiosperms; **37.** In centripetal order; **38.** Sexual reproduction; **39.** Water.

Short Answer Questions (2 marks each)

1. Give three main functions of stem.
2. Give three primary functions of a root.
3. Give three principal functions of a leaf.
4. What are stilt roots? Give the names of two plants in which they are found.
5. What is venation? Give main types of venation with suitable examples.
6. Most types of potatoes and sugarcanes do not produce any flower in plains of North India. Will you still call the potato and sugarcane flowering plants? If so, why?
7. How would you differentiate an underground stem from a root?
8. How would you differentiate a pinnate compound leaf from a branch having simple leaves?
9. How would you differentiate the dicot plants from monocot ones by seeing their leaves only?
10. What are the main differences between annuals, biennials and perennials?
11. What type of stems are found in: (*a*) *Colocasia* (*b*) *Opuntia* (*c*) *Chrysanthemum* (*d*) Doob grass.
12. Give differences between corm and bulb.
13. What features are found in orchid roots?
14. Of what importance is the study of leaf morphology for a physiologist?
15. Give any two reasons to justify that onion bulb is a modified stem.
16. What do you understand by hypogynous, epigynous and perigynous flowers? Explain the terms giving suitable diagrams.
17. Draw a labelled diagram of the L.S of an anatropous ovule.
18. What is placentation? Define the main types.
19. How does the coriander inflorescence differ from the mustard inflorescence in arrangement? Give the technical term for each.
20. Explain the biological and economic importance of fruits.
21. Describe the nature of three kinds of angiospermic fruits. Give an example of each.
22. Draw a labelled diagram of L.S. of pistil of typical flower.

Short Answer Questions (3 marks each)

1. Define fruit give the structure of a drupe fruit.
2. Write a note on fundamental parts of plant body.
3. Give the main modifications of the roots.
4. Describe the modifications of the aerial stem.
5. How can you distinguish a simple leaf from a compound leaf?
6. Pick out the plant with unipinnate type of compound leaves from the list given below and say why it is categorised so? *Trifolium, Moringa, Neem.*

7. Four 'O clock plant and sweet potato are similar in one aspect and dissimilar in another aspect with reference to the nature and modification of roots. Mention these two aspects.
8. Amongst pea tendrils opuntia spines, lemon thorns and cucurbit tendrils which ones are homologous structures? Why do you call them homologous?
9. What are the main kinds of fruits?
10. Name two plants showing dichasial cyme inflorescence.
11. What are the main functions of a flower?
12. What are the two main types of inflorescences?
13. What is fruit? Differentiate between true and false fruits.
14. Write notes on Cyathium and Hypanthodium.
15. Write short notes on (*a*) Spadix (*b*) Catkin.

Long Answer Questions (5 marks each)

1. Write a note on multicellularity in plants. Give the structure of apical meristem of shoot.
2. Write a detailed note on fundamental parts of plant body.
3. Give the structure and function of the primary and adventitious roots.
4. What are root systems? Describe the types of root systems.
5. Give the main modifications of the roots and stems.
6. Describe the types of branching in stems. Also give modifications of aerial stems.
7. Describe the modifications of the aerial stems and leaves.
8. Describe the modifications of the underground stem.
9. How can you distinguish a simple leaf from a compound leaf? A compound leaf from a branch.
10. Describe some interesting forms of modified leaves.
11. Write short notes on : (*a*) storage roots, (*b*) contractile roots, (*c*) buttress roots, (*d*) haustoria, (*d*) stilt roots, (*f*) pneumatophores.
12. Describe the nature of a vegetative bud. Describe briefly different kinds of buds.
13. Write short notes on. (*a*) cladode, (*b*) phylloclade, (*c*) bulb, (*d*) dropper, (*e*) corm.
14. What are the main functions of a leaf ?
15. What are the main parts of a typical simple leaf? Write a note on venation.
16. Write short notes on. (*a*) heterophylly, (*b*) pitcher, (*c*) phyllode, (*d*) bladder.
17. What is inflorescence? Classify it into two main categories with their definitions and suitable examples.
18. Give the special types of inflorescence with suitable examples. Give a comprehensive account of fruits.
19. Define racemose inflorescence. Give its main types with examples.
20. What is cymose inflorescence? Mention its main types with examples.
21. What is compound inflorescence? Write brief account of its main types with examples.
22. Show the placentations of various types in transverse sections of ovaries of different angiosperms.
23. Give the diagrammatic structures of (*a*) spadix of an aroid (*b*) corymb (*c*) capitulum and (*d*) hypanthodium.
24. Write short notes on; (*a*) legume (*b*) siliqua (*c*) caryopsis (*d*) samara (*e*) lomentum (*f*) cremocarp and (*g*) regma.
25. What are schizocarpic fruits? Give suitable examples.
26. Give the characteristic inflorescence of the following (*a*) *Helianthus* (*b*) coriander (*c*) wheat (*d*) holy basil (*e*) *Euphorbia* (*f*) *Ficus* (*g*) *Musa*.
27. Write short notes on: (*a*) spike, (*b*) spadix, (*c*) umbel of umbels, (*d*) cyathium, (*e*) hypanthodium and (*f*) capitulum.
28. Write an essay on dispersal of seeds.

29. Describe the structure of a monocot and dicot seed with suitable diagrams.
30. What do you understand by seed fruit dispersal? Name the various agencies that aid in dispersal. Mention the characteristics of fruits and seeds dispersed by wind.

True or False

Write whether the following statements are True or False:

1. *Drosera* obtains proteins by digesting the insects.
2. The leaf of *Citrus* represents a simple leaf.
3. *Utricularia* is provided with trap door mechanism to store the food material.
4. *Nepenthes* is a parasitic plant.
5. Orchids are generally found as epiphytes.
6. The flower with superior ovary is called hypogynous.
7. Wall of fruit having epicarp, mesocarp and endocarp is called periderm.
8. A cotyledon of bean seed is called scutellum.
9. A bean seed is albuminous, while a maize seed is exalbuminous.

ANSWERS

1. True 2. False 3. True 4. False 5. True 6. True
7. False 8. False 9. False

Fill in the Blanks

1. A protective sheath of radicle in monocots is called ....
2. The flower with superior ovary is called ....
3. Study of pollen grains is known as ....
4. The fruit that develops from the whole of inflorescence is called ....
5. A stamen consists of .... and ....
6. The flowers in catkin are ....
7. The tip of the pedicel on which different floral parts are inserted is known as ....
8. The one-chambered, one-seeded fruit developed from a bicarpellary inferior ovary, often with persistent pappus is called ....
9. The anther that consists of only one anther lobe is called ....
10. The lateral appendages of the leaf borne at its base are called ....
11. Plants that capture insects and feed upon them absorbing only nitrogenous compounds from their bodies are called .... plants.
12. The plants that grow upon other plants but do not suck them are called ....

ANSWERS

1. Coleorhiza, 2. Hypogynous, 3. Palynology, 4. Sorosis, 5. Filament and anther, 6. Unisexual, 7. Thalamus, 8. Cypsella, 9. Monothecous, 10. Stipules, 11. Insectivorous, 12. Epiphytes.

Multiple Choice Questions (Objective Type Questions)

1. The roots develop in
   (*a*) Bryophytes (*b*) Algae (*c*) Angiosperms (*d*) Fungi
2. The fibrous roots are commonly found in
   (*a*) Dicotyledons (*b*) Monocotyledons (*c*) Gymnosperms (*d*) None of above
3. The tap root system is commonly found in
   (*a*) Dicotyledons (*b*) Bryophytes (*c*) Monocotyledons (*d*) All of them

4. The roots developing from any part of the plant other than radicle are known as
(*a*) Tap roots (*b*) Adventitious roots (*c*) Fleshy tap roots (*d*) None of above
5. The edible roots are found in
(*a*) Wheat (*b*) Rice (*c*) Sweet potato (*d*) Potato
6. The root hairs are
(*a*) Unicellular (*b*) Bicellular (*c*) Multicellular (*d*) None of above
7. Conical root is found in
(*a*) Radish (*b*) Turnip (*c*) Carrot (*d*) *Asparagus*
8. Napiform root is found in
(*a*) Radish (*b*) Turnip (*c*) Carrot (*d*) Sweet potato
9. Fusiform root is found in
(*a*) Radish (*b*) Turnip (*c*) Carrot (*d*) Wheat
10. Moniliform roots are found in
(*a*) *Dahlia* (*b*) *Ipomoea batatas* (*c*) *Asparagus* (*d*) *Momordica*
11. The haustorial roots are found in
(*a*) Orchid (*b*) *Pothos* (*c*) *Monstera* (*d*) *Cuscuta*
12. The bacterial nodules are found in the roots of
(*a*) Wheat (*b*) Rice (*c*) Pea (*d*) Maize
13. Fasciculated roots are found in
(*a*) Sweet potato (*b*) Dahlia (*c*) Radish (*d*) Carrot
14. Tap root develops from
(*a*) Plumule (*b*) Radicle (*c*) Endosperm (*d*) Axillary bud
15. Pneumatophores are found in
(*a*) Leguminous plants (*b*) Desert plants
(*c*) Mangrove vegetation (*d*) None of above
16. The aerial roots are found in
(*a*) Wheat (*b*) Groundnut (*c*) Orchids (*d*) None of above
17. Prop roots are found in
(*a*) Mango tree (*b*) Guava tree (*c*) Banyan tree (*d*) Gulmohar tree
18. The stilt roots are found in
(*a*) Egg plant (*b*) Gram (*c*) Maize (*d*) Potato
19. The respiratory roots are found in
(*a*) Wheat (*b*) Gram (*c*) *Mimosa* (*d*) *Jussiaea*
20. The mycorrhizic roots are the association of
(*a*) Algae and fungi (*b*) Fungi and bryophytes
(*c*) *Nostoc* and *Cycas* (*d*) Fungi and *Pinus*
21. The nodes and internodes are found on
(*a*) Tap root (*b*) Adventitious root (*c*) Stem (*d*) None of above
22. Unbranched, erect, cylindrical stout axis with distinct nodes and internodes with jointed appearance is called
(*a*) Runner (*b*) Sucker (*c*) Stolon (*d*) Culm
23. Phylloclade is a modification of
(*a*) Stem (*b*) Root (*c*) Leaf (*d*) Petiole
24. Select the edible stem from the following
(*a*) Potato (*b*) Sweet potato (*c*) Egg plant (*d*) Bitter gourd

25. Plants which complete their life-cycle in a single season are called
(*a*) Annuals (*b*) Biennials (*c*) Perennials (*d*) None of above
26. Cladode is a modification of
(*a*) Stem (*b*) Root (*c*) Leaf (*d*) None of above
27. Phyllode is a modification of
(*a*) Root (*b*) Stem (*c*) Petiole (*d*) None of above
28. Leaf tendrils are found in
(*a*) *Cucurbita* (*b*) *Vitis* (*c*) *Pisum sativum* (*d*) *Passiflora*
29. Stipular tendrils are found in
(*a*) *Cucurbita* (*b*) *Lathyrus* (*c*) *Smilax* (*d*) *Vitis*
30. The runners are found in
(*a*) *Fragaria vesca* (*b*) *Colocasia* (*c*) *Pistia* (*d*) None of above
31. Rhizome is found in
(*a*) Ginger (*b*) Doob grass (*c*) Potato (*d*) Onion
32. Bulb is found in
(*a*) Onion (*b*) *Colocasia* (*c*) Potato (*d*) Ginger
33. Parallel venation is characteristic of
(*a*) Pea (*b*) Banyan (*c*) Grasses (*d*) Shoe flower
34. The bladders are found in
(*a*) *Drosera* (*b*) *Utricularia* (*c*) *Nepenthes* (*d*) None of the above
35. The pitcher plant is
(*a*) *Utricularia* (*b*) *Vallisneria* (*c*) *Nepenthes* (*d*) *Drosera*
36. Bipinnate compound leaf is found in
(*a*) *Cassia* (*b*) *Acasia* (*c*) *Pisum* (*d*) Rose
37. Phyllode is found in
(*a*) Australian acacia (*b*) *Cassia fistula* (*c*) *Opuntia* (*d*) *Ruscus*
38. When the apex of leaf is round, it is called
(*a*) Acute (*b*) Obtuse (*c*) Acuminate (*d*) Cuspidate
39. The capacity for vegetative reproduction is found in
(*a*) Leaves (*b*) Roots (*c*) Stem (*d*) All the above
40. The vegetative propagation in *Bryophyllum* takes place by
(*a*) Root (*b*) Stem (*c*) Leaves (*d*) None of above
41. The stem cuttings for vegetative propagation are used in
(*a*) *Coleus* (*b*) *Pothos* (*c*) Rose (*d*) All the above
42. The sugarcane plants are propagated by
(*a*) Roots (*b*) Leaves (*c*) Stems (*d*) All the above
43. The methods of vegetative propagation are commonly used in the production of the plants of
(*a*) Wheat (*b*) Rose (*c*) Marigold (*d*) Sunflower
44. The layering is practiced in
(*a*) Jasminum (*b*) Groundnut (*c*) Soybean (*d*) All the above
45. Grafting is done in
(*a*) Monocots (*b*) Dicots (*c*) Pedicel (*d*) Pteridophytes
46. The axis of inflorescence is known as
(*a*) Petiole (*b*) Peduncle (*c*) Bryophytes (*d*) None of above
47. The main axis is elongated in
(*a*) Raceme (*b*) Spike (*c*) Catkin (*d*) All the above

48. The main axis is flattened in
(*a*) Raceme (*b*) Spike (*c*) Catkin (*d*) Capitulum

49. The shortened main axis is found in
(*a*) Raceme (*b*) Corymb (*c*) Umbel (*d*) Both (*b*) and (*c*)

50. Spadix inflorcesence is found in
(*a*) *Alocasia* (*b*) *Morus* (*c*) *Mangifera* (*d*) *Magnolia*

51. Corymb inflorescence is found in
(*a*) Coriander (*b*) Candytuft (*c*) *Brassica* (*d*) *Helianthus*

52. Simple umbel is found in
(*a*) *Centella* (*b*) *Foeniculum* (*c*) *Coriandrum* (*d*) All the above

53. Capitulum is found in
(*a*) *Helianthus* (*b*) *Tagetes* (*c*) *Zinnia* (*d*) All the above

54. Cyathium inflorescence is found in
(*a*) *Euphorbia* (*b*) *Croton* (*c*) *Ricinus* (*d*) *Coleus*

55. The ratio of male: Female flowers in Cyathium inflorescence is
(*a*) One : one (*b*) One : many (*c*) Many : one (*d*) Many : many

56. Zig-zag inflorescence axis is formed in
(*a*) Spadix (*b*) Scorpioid cyme (*c*) Helicoid cyme (*d*) Dichasial cyme

57. Which of the following is actually not a flower
(*a*) Shoe flower (*b*) Rose (*c*) Sun flower (*d*) Pea

58. The essential parts of the flower are
(*a*) The sepals and petals (*b*) The sepals and stamens
(*c*) The petals and pistils (*d*) The stamens and pistils.

59. In racemose inflorescence the flowers are
(*a*) Arranged in acropetal succession (*b*) Arranged in basipetal succession
(*c*) Of the same sex (*d*) Of separate sexes.

60. The gynandrophore is present in
(*a*) *Brassica* (*b*) *Hibiscus* (*c*) *Gynandropsis* (*d*) *Iberis*

61. Hypanthodium inflorescence is found in
(*a*) *Ficus* (*b*) *Morus* (*c*) *Helianthus* (*d*) *Ranunculus*

62. Select out the special type of inflorescence
(*a*) Raceme (*b*) Spadix (*c*) Spike (*d*) Verticillaster

63. When the stamens are adnate to the tepals it is known as
(*a*) Epipetalous (*b*) Episepalous (*c*) Epiphyllous (*d*) Epigynous

64. The pollinia are found in
(*a*) *Calotropis* (*b*) *Gynandropsis* (*c*) *Raphanus* (*d*) *Physalis*

65. Anemophilous flowers possess
(*a*) Small smooth stigma (*b*) Large feathery stigma
(*c*) Sessile stigma (*d*) Attractive coloured bracts.

66. The phenomenon of double fertilization was discovered by
(*a*) Calvin (*b*) Nawaschin (*c*) Strasburger (*d*) Warburg

67. The oil is stored in the endosperm of
(*a*) Groundnut (*b*) Coconut (*c*) Sesame (*d*) Soybean

68. Albuminous seeds are those which possess
(*a*) Endosperm (*b*) Cotyledons (*c*) Perisperm (*d*) Nucellus

69. Spike of spikelets inflorescence is found in
(*a*) Mustard (*b*) Wheat (*c*) Buck wheat (*d*) Gorgan nut

70. Catkin inflorescence is found in
(*a*) Mulberry (*b*) Fig (*c*) Wheat (*d*) Radish

71. Scorpioid inflorescence is found in
(*a*) *Dianthus* (*b*) *Tropaeolum* (*c*) *Heliotropium* (*d*) *Begonia*

72. Select out the special type of inflorescence
(*a*) Compound umbel (*b*) Cyathium (*c*) Capitulum (*d*) Compound spadix

73. The bilabiate corolla is found in
(*a*) *Rose* (*b*) *Salvia* (*c*) *Dianthus* (*d*) *Datura*

74. Syngenesious cohesion of stamens is found in
(*a*) *Helianthus* (*b*) Coriander (*c*) *Centella* (*d*) *Cedrella*

75. The fruit of mango is
(*a*) Drupe (*b*) Nut (*c*) Pome (*d*) Berry

76. Ornithophily refers to pollination by
(*a*) Insects (*b*) Wind (*c*) Water (*d*) Birds

77. The fruit of candytuft is
(*a*) Lomentum (*b*) Nut (*c*) Silicula (*d*) Siliqua

78. The fruit that develops from hypanthodium is called.
(*a*) Drupe (*b*) Berry (*c*) Achene (*d*) Syconus

79. Primary endosperm nucleus is
(*a*) Haploid (*b*) Diploid (*c*) Triploid (*d*) Tetraploid

80. Fragrance, attractive colours and pr sence of necter are the main characteristic of the flowers showing
(*a*) Entomophily (*b*) Ornithopnily (*c*) Anemophily (*d*) Hydrophily

81. It will not make any sense to produce seedless fruits in
(*a*) Grapes (*b*) Papaya (*c*) Guava (*d*) Pomegranate.

82. The study of fruits is called
(*a*) Ornithology (*b*) Anthology (*c*) Pomology (*d*) Mycology

83. Cleistogamy is found in
(*a*) *Calotropis* (*b*) *Vanda* (*c*) *Commelina bengalensis* (*d*) *Ficus bengalensis*

84. In coconut the dispersal is due to fibrous
(*a*) Epicarp (*b*) Mesocarp (*c*) Endocarp (*d*) None of above

85. In poppy dispersal of seeds takes place through
(*a*) Pores (*b*) Longitudinal slits (*c*) Valves (*d*) Septa

86. When the pollens from one flower are deposited on the stigma of another flower borne on the same plant the pollination is known as
(*a*) Self pollination (autogamy) (*b*) Self pollination (geitonogamy)
(*c*) Cross pollination (allogamy) (*d*) Anemophily.

87. The dispersal of seeds of *Cinchona* and orchids take place by
(*a*) Wind (*b*) Water (*c*) Bats (*d*) Birds

88. Marginal placentation is found in
(*a*) Poppy (*b*) Mustard (*c*) Sunflower (*d*) Pea

89. Cypsela fruit develops from
(*a*) Superior monocarpellary ovary (*b*) Superior tricarpellary ovary
(*c*) Inferior bicarpellary ovary (*d*) Inferior pentacarpellary ovary

90. The juicy edible aril is found in
(*a*) Plum (*b*) Litchi (*c*) Peach (*d*) Jujube

91. Pepo fruit characteristic of
(*a*) Cucurbits (*b*) Mango (*c*) Pomegranate (*d*) Orange

92. Select out the composite fruit
(*a*) Pome (*b*) Hespiridium (*c*) Regma (*d*) Sorosis

93. Double fertilization is characteristic of
(*a*) Gymnosperms (*b*) Angiosperms (*c*) Pteridophytes (*d*) Bryophytes

94. How many male cells are there in the pollen tube of angiosperms
(*a*) 1 (*b*) 2 (*c*) 3 (*d*) 4

95. When fruits are developed without fertilization the process is called
(*a*) Parthenogenesis (*b*) Parthenocarpy (*c*) Both (*a*) and (*b*) (*d*) None of the above

ANSWERS

1. (*c*)	**2.** (*b*)	**3.** (*a*)	**4.** (*b*)	**5.** (*c*)	**6.** (*a*)	**7.** (*c*).	**8.** (*b*)
9. (*a*)	**10.** (*d*)	**11.** (*d*)	**12.** (*c*)	**13.** (*b*)	**14.** (*b*)	**15.** (*c*)	**16.** (*c*)
17. (*c*)	**18.** (*c*)	**19.** (*d*)	**20.** (*d*)	**21.** (*c*)	**22.** (*d*)	**23.** (*a*)	**24.** (*a*)
25. (*a*)	**26.** (*a*)	**27.** (*c*)	**28.** (*c*)	**29.** (*c*)	**30.** (*a*)	**31.** (*a*)	**32.** (*a*)
33. (*c*)	**34.** (*b*)	**35.** (*c*)	**36.** (*b*)	**37.** (*a*)	**38.** (*b*)	**39.** (*d*)	**40.** (*c*)
41. (*d*)	**42.** (*c*)	**43.** (*b*)	**44.** (*a*)	**45.** (*b*)	**46.** (*b*)	**47.** (*d*)	**48.** (*d*)
49. (*d*)	**50.** (*a*)	**51.** (*b*)	**52.** (*a*)	**53.** (*d*)	**54.** (*a*)	**55.** (*c*)	**56.** (*b*)
57. (*c*)	**58.** (*d*)	**59.** (*a*)	**60.** (*c*)	**61.** (*a*)	**62.** (*d*)	**63.** (*c*)	**64.** (*a*)
65. (*b*)	**66.** (*b*)	**67.** (*b*)	**68.** (*a*)	**69.** (*b*)	**70.** (*a*)	**71.** (*c*)	**72.** (*b*)
73. (*b*)	**74.** (*a*)	**75.** (*a*)	**76.** (*d*)	**77.** (*c*)	**78.** (*d*)	**79.** (c)	**80.** (*a*)
81. (*d*)	**82.** (*c*)	**83.** (*c*)	**84.** (*b*)	**85.** (*a*)	**86.** (*b*)	**87.** (*a*)	**88.** (*d*)
89. (*c*)	**90.** (*b*)	**91.** (*a*)	**92.** (*d*)	**93.** (*b*)	**94.** (*b*)	**95.** (*b*).	

30

Reproduction in Flowering Plants

The capacity to reproduce is one of the most important characteristics of all living beings, *i.e.*, both plants and animals. Reproduction is a process of producing offspring, and is aimed to preserve individual species. This process is known for self perpetuation. There are several modes of reproduction which vary from species to species, in available conditions. For example, in lower organisms, this may take place by division of cell or by budding, while the higher organisms may possess fully developed sex organs.

However, in higher plants reproduction may be **asexual** or **sexual**. In asexual reproduction, sex organs are not involved, and reproduction takes place by means of asexual or vegetative methods, while in sexual reproduction, fusion of male and female gametes takes place. Generally, the male and female sex organs are called **stamen** and **pistil**, respectively, in angiosperms. In liverworts, mosses and ferns, they are called **antheridia** (male) and **archegonia** (female).

In certain other cases of plants, special modes of reproduction such as parthenogenesis, sporophytic budding and polyembryony, are also found. In subsequent paragraphs, you will study various methods of reproduction involved in flowering plants (angiosperms).

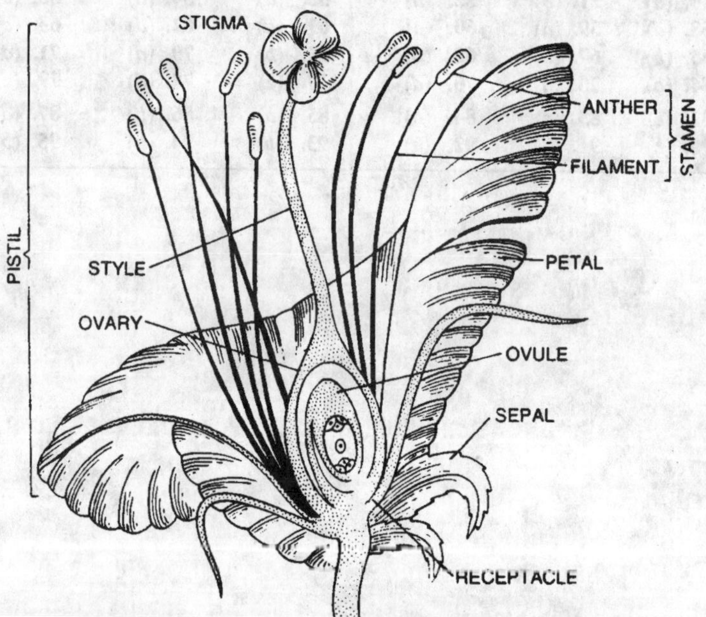

Fig. 30.1. General plan of a typical flower. Flower in sectional view (L.S.).

MODES OF REPRODUCTION IN ANGIOSPERMS

In angiosperms or flowering plants, there are several modes of reproduction. Generally, they are arranged in two large groups of reproduction, *i.e.*, (*i*) asexual or vegetative and (*ii*) sexual types.

In asexual or vegetative reproduction, the offspring are produced from the somatic cells, while in sexual reproduction there is fusion of male and female gametes.

In the case of vegetative reproduction, any part of the plant, *i.e.*, stem, root or leaf, is capable of growing into a new plant, in addition to performing its sexual functions. Sometimes, in certain plants, buds and bulbils are developed, which develop into new plants.

In sexual reproduction, the **gametes** from male and female organs of the flower are fused to produce a **zygote.** In angiosperms, these organs are generally called, **androecium** and **gynoecium**, respectively.

In some plants, certain special modes of reproduction are found, which are commonly known as parthenogenesis, sporophytic budding, polyembryony, apomixis, apospory, and micro-propagation. The production of synthetic or artificial seeds is also possible through tissue culture.

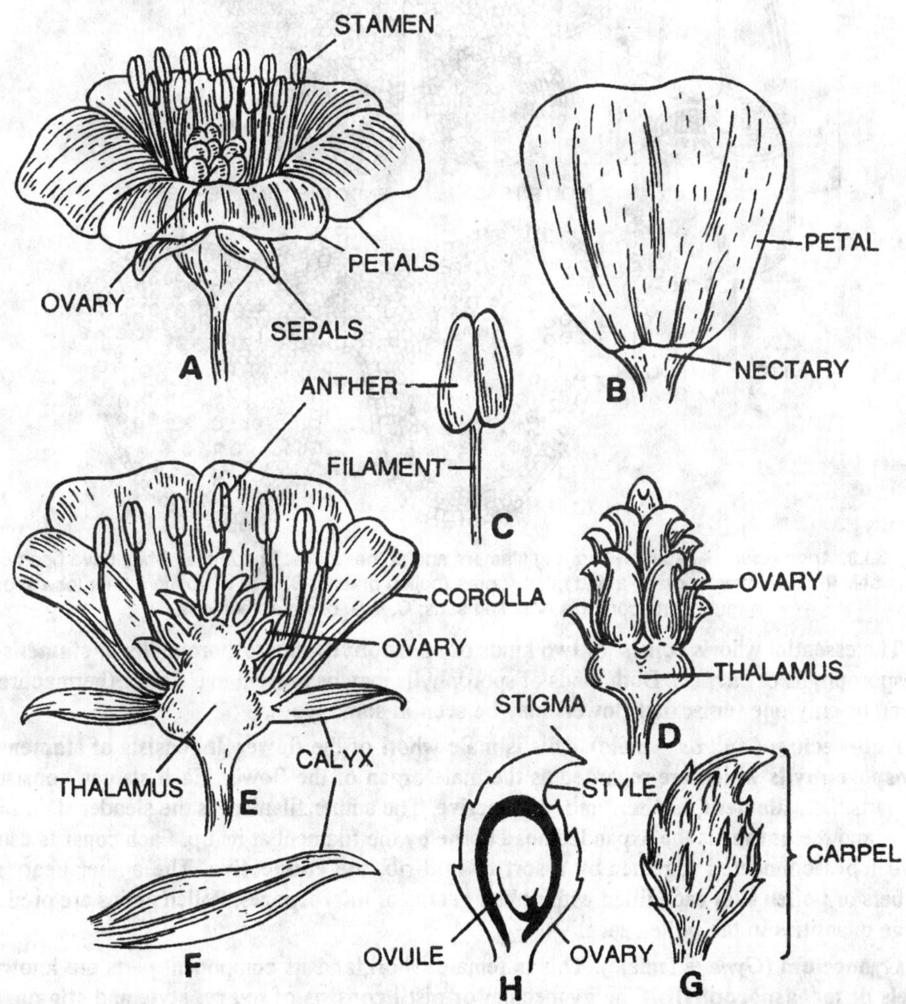

Fig. 30.2. Various parts of a flower of *Ranunculus*. A, complete flower; B, a petal; C, stamen; D, thalamus with apocarpous gynoecium; E, a sepal; F, V.S. of flower; G, a carpel; H, V.S. of carpe

SEXUAL REPRODUCTION

The flower. The flower is a highly specialized reproductive shoot. Each typical flower consists of four distinct types of members arranged in four separate but closely set whorls, one above the other, on the top of a long or short stalk. The lower two whorls are called **accessory** whorls, and the upper two **essential** or **reproductive** whorls because only these two are directly concerned in reproduction.

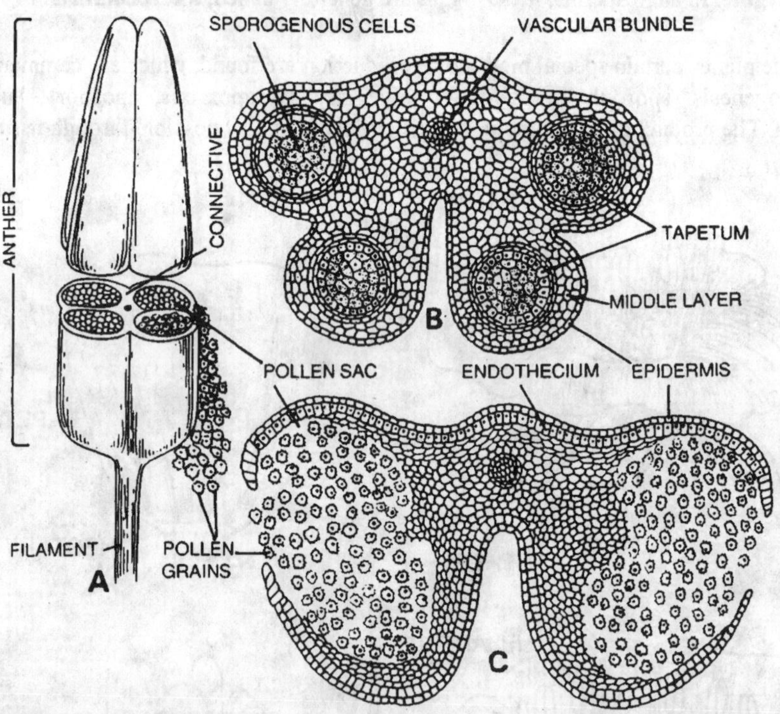

Fig. 30.3. *Androecium.* A, stamen showing filament and anther lobes. The anther lobes have been cut transversely to show microsporangia and microspores (pollen grains); B, T,S. of young anther lobe showing four microsporangia or pollen sacs; C, T.S. of old anther lobe.

The essential whorls consist of two kinds of **sporophylls**: microsporophylls or **stamens** and megasporophylls or **carpels**. Both kinds of sporophylls may be present in a flower (hermaphrodite flower), or only one (unisexual flower) may be seen in some types.

Androecium (*Andros* = male). This is male whorl of the flower. It consists of **stamens** or **microsporophylls** which are regarded as the male organ of the flower. Each stamen consists of three parts, *i.e.*, **filament**, **anther** and **connective**. The anther filament is the slender stalk of the stamen, and the **anther** is the expanded head borne by the filament at its tip. Each consists usually of two lobes connected together by a sort of mid-rib, the **connective**. The anther bears four chambers or **pollen sacs** each filled with **pollen grains** or **microspores**. Pollen grains are produced in large quantities in the pollen sacs.

Gynoecium (*Gyne* = female). This is female whorl, and its component parts are known as **carpels** or **megasporophylls**. The gynoecium or pistil consists of **ovary, style** and **stigma**. The swollen basal part of the gynoecium which forms one or more chambers is known as **ovary**. The small rounded head of the gynoecium is known as the **stigma**. The ovary contains one or more little, roundish or oval, egg-like bodies which are the rudiments of seeds and are known as **ovules**. Each ovule encloses a large oval cell known as the **embryo sac**. On maturation, the ovary gives rise to the **fruit** and the ovules to **seeds**.

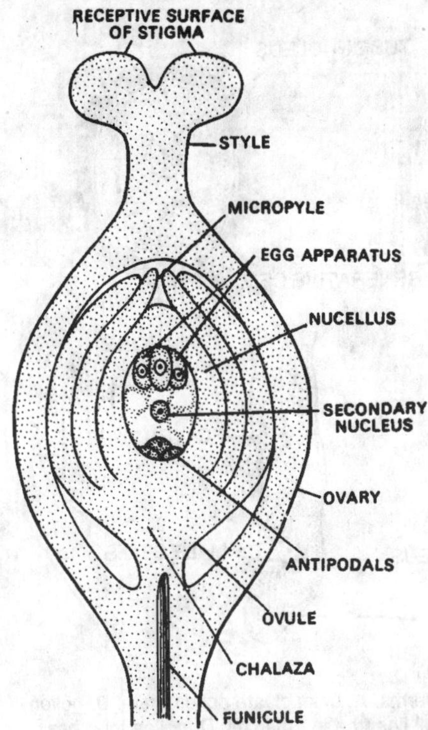

Fig. 30.4. Gynoecium in longitudinal section showing different parts of mature ovule attached to a basal placenta.

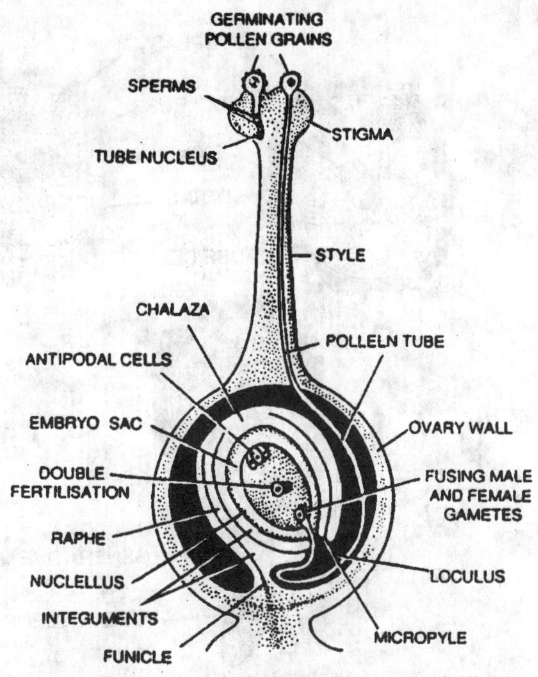

Fig. 30.5. A monocarpellary pistil cut vertically to show its external and internal parts. It also shows pollen grains on stigma.

Homologies.

Carpel = Megasporophyll

Ovule = Megasporangium

Embryo sac = Female gametophyte

Egg = Female gamete

Stamen = Microsporophyll

Anther (pollen sac) = Microsporangium

Pollen grain = Microspore

Germinating pollen grain = Male gametophyte

The reproductive process in angiosperms is as follows:

Development of the Male Gametophyte

Microsporangium (pollen sac). The **epidermis** is the outermost layer of microsporangium. The cells of epidermis are generally stretched and flattened. The layer next to the epidermis is the

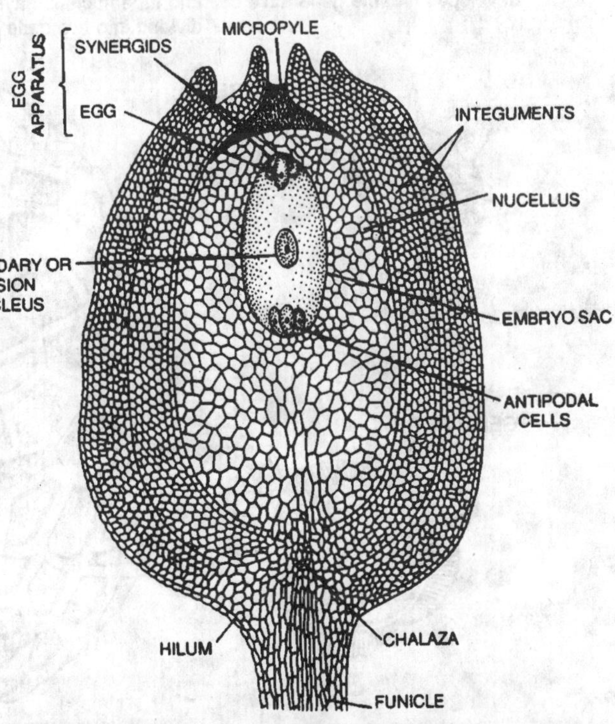

Fig. 30.6. V.S. Ovule showing its structure just before fertilization.

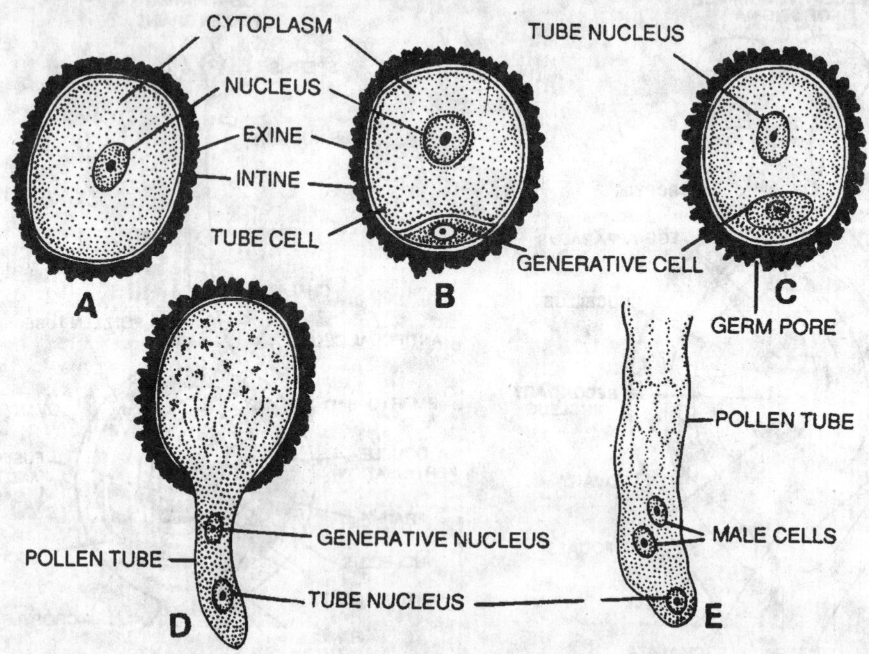

Fig. 30.7. Structure and germination of pollen grain in angiosperms. A, uninucleate pollen grain; B, pollen grain with generative cell and tube nucleus; C, generative cell lies free in tube cell, D, pollen tube has developed and the generative cell and tube nucleus have migrated into it; E, generative nucleus has divided into two male gametes.

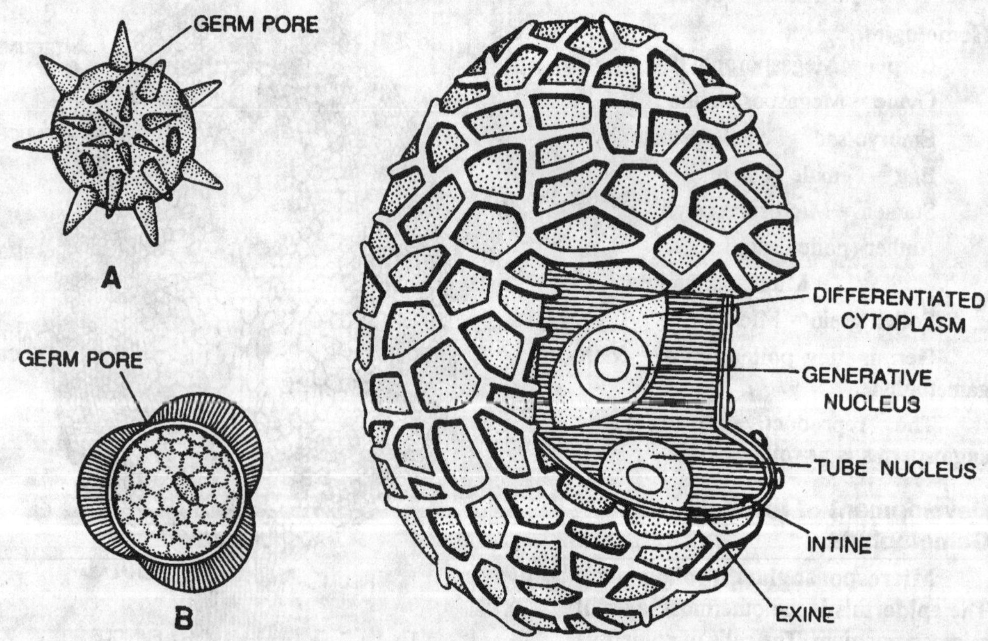

Fig. 30.8. Pollen grains. A sculptured pollen grain; B, sectional view.

Fig. 30.9. Pollen grain of lily. Stereoscopic view.

endothecium or **fibrous layer**. As a rule, by the development of the fibrous bands of thickening the endothecium becomes hygroscopic and is, therefore, mainly responsible for the dehiscence of mature anther.

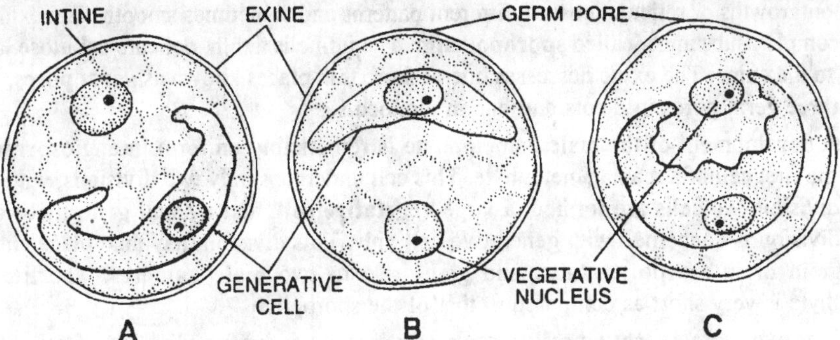

Fig. 30.10. Male gametophyte. Pollen grains showing vegetative nucleus and generative cells. In A and B, the generative cells are much elongated; where as in C the generative cell is of amoeboid nature.

The anther is generally bilobed, containing two longitudinally running chambers or pollen sacs per lobe. Each chamber contains a large number of pollen grains. The anther wall is composed of 4 to 5 layers. The innermost layer of these wall layers develops into a single layered **tapetum**. The tapetal layer is of great physiological significance as all the food material entering into the sporogenous tissue diffuses through, this layer. Ultimately the cells of tapetal layer disorganise. Thus, tapetum makes a nutritive layer for the developing microspores.

The cells of endothecium are thin walled along the line of dehiscence of each anther lobe. The opening through which the pollen grains are discharged from the pollen sac is called **stomium.** On the maturity of the anther, a strain is exerted on the stomium due to the loss of water by the cells of endothecium, with the result the stomium ruptures and the anther dehisces. Generally, the mature anther dehisces by means of slits or apical pores.

The pollen grains. The pollen grains or microspores are the male reproductive bodies of

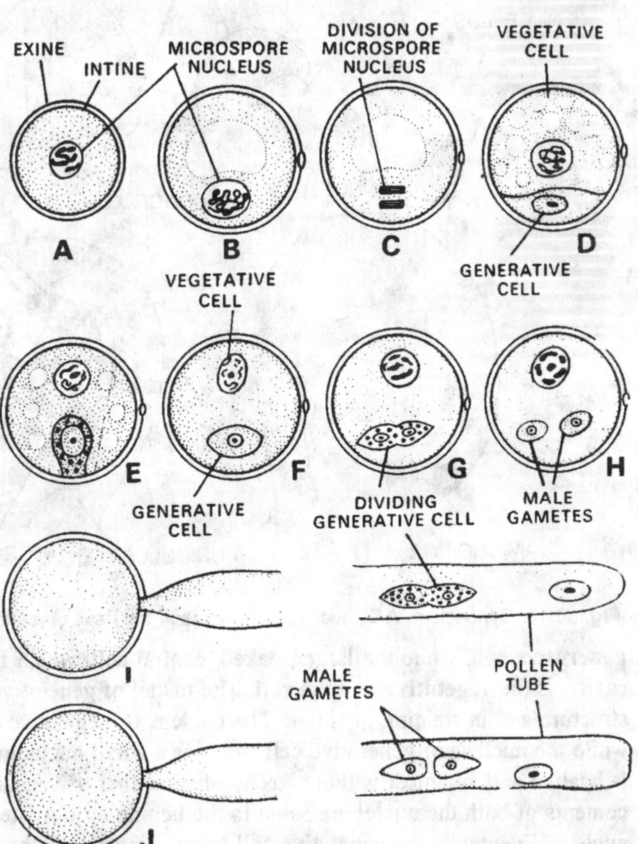

Fig. 30.11. Male gametophyte, A-J, different stages in the development of male gametophyte.

a flower, and are contained in the pollen sac or microsporangia. They are very minute in size and are like particles of dust. Each pollen grain consists of a single microscopic cell, possessing two coats: the **exine** and the **intine**. The exine, tough cutinized layer, which is often provided with spinous outgrowths or reticulations of different patterns and sometimes smooth. The exine is made up of a complex substance, called **sporopollenin**. The intine is a thin, delicate cellulose layer lying internal to the exine. The exine possesses one or more thin places known as **germ pores.** There are usually three germ pores in dicots and one in monocots.

The development of the male gametophyte is remarkably uniform in angiosperms. Pollen grain is the first cell of a male gametophyte. This cell undergoes only two divisions, with the result of first division two cells are formed: a large **vegetative cell** and a small **generative cell.** The second division is concerned with generative cell only. This division may take place either in the pollen grain or in the pollen tube, and give rise to two **male gametes**. The life of male gametophyte is very short as compared to that of the sporophyte.

Vegetative and generative cells. As already described, the first division of the pollen grain gives rise to the **vegetative** and **generative** cells. The first formed walled and peripheral cell is the

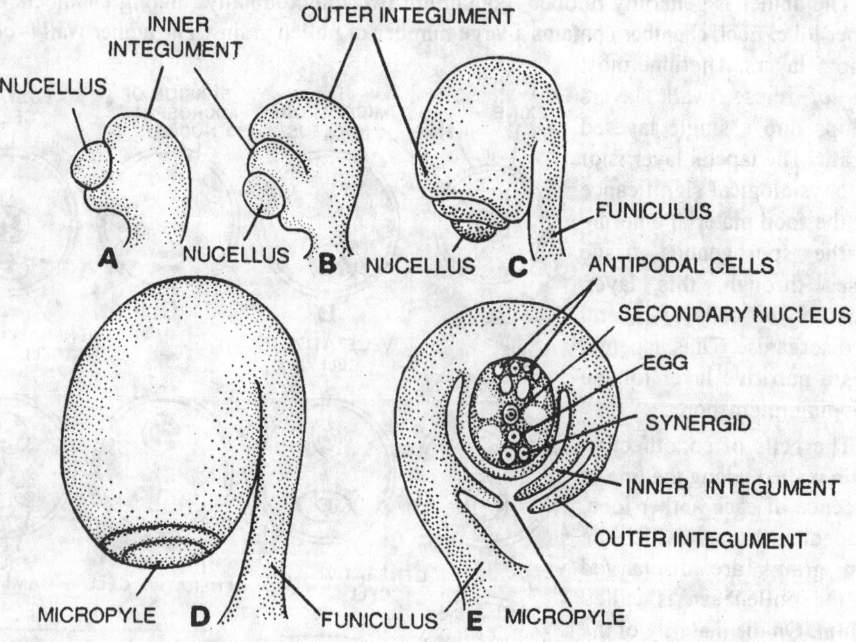

Fig. 30.12. Gynoecium, A-F, successive stages in the development of anatropous ovule, and its structure.

generative cell, while the larger, naked, central cell, which fills the remainder of the spore-wall cavity, is the **vegetative** or **tube cell**. The nuclei of generative and vegetative cells differ in size, structure and in staining qualities. The nucleus of vegetative cell possesses a prominent nucleus, while the nucleus of generative cell contains a small nucleolus. The cytoplasm of generative cell is hyaline and is almost without RNA, whereas that of vegetative cell is rich in RNA. The DNA contents of both the nuclei are same in the beginning but later on they increase in the generative nucleus. Eventually the generative cell loses contact with the microspore (pollen grain) wall, and is being shifted into the vegetative cell, where it may lie in any part of it. Thus, the pollen grain becomes two-celled. Generally, the pollen grains are being shed from the microsporangium (pollen sac) in two-celled stage for pollination.

Development of the Female Gametophyte

The megasporangium or ovule. An ovule or megasporangium develops from the base or the inner surface of the ovary. It is a small generally oval structure and consists chiefly of a central body of tissue, the **nucellus** and one or two **integuments.** Each ovule is attached in the placenta by a small stalk called the **funiculus.** The place of attachment of the stalk with the main body of the ovule is called the **hilum**. In an inverted ovule, the funicle fuses with the main body of the ovule, forming a sort of ridge, known as the **raphe**. The upper end of the raphe which is the junction of the integuments and the nucellus is called the **chalaza**. The nucellus makes the main body of the ovule, which is made up of parenchyma tissue. Nucellus is the megasporangium proper, and is surrounded by two coats, the **integuments**. A small opening is left at the apex of the integuments; this is called the **micropyle**. When there are two integuments then the inner integument is formed first and followed by the formation of the outer integument. A large oval cell lying embedded in the nucellus towards the micropylar end is the **embryo sac**. This makes the most important part of the mature ovule. It is the embryo sac, which bears the **embryo** later on.

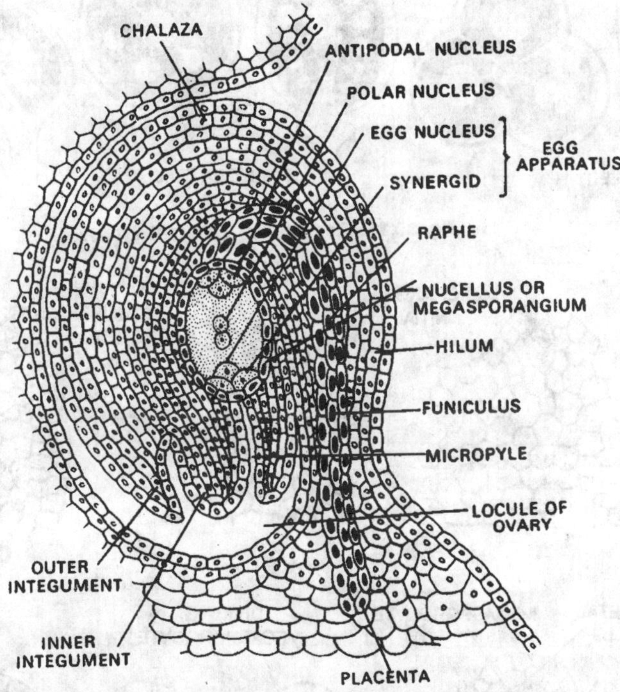

Fig. 30.13. Ovule. L.S. of a typical ovule of angiosperm.

Placentation. The **placenta** is an outgrowth of a parenchymatous tissue in the inner wall of the ovary to which the ovule or ovules (megasporangia) remain attached. The placentae usually develop on the margins of carpels, either along their whole line of union, called the **suture** or at their base or apex. The manner in which the placentae are distributed in the cavity of the ovary is known as **placentation.**

In the simple ovary (*i.e.*, of one carpel), there is one common type of placentation, known as **marginal**, and in the compound ovary (*i.e.*, of two or more carpels united together) placentation may be **axile, central** and **free-central, basal, parietal** and **superficial** as shown in fig. 46.14.

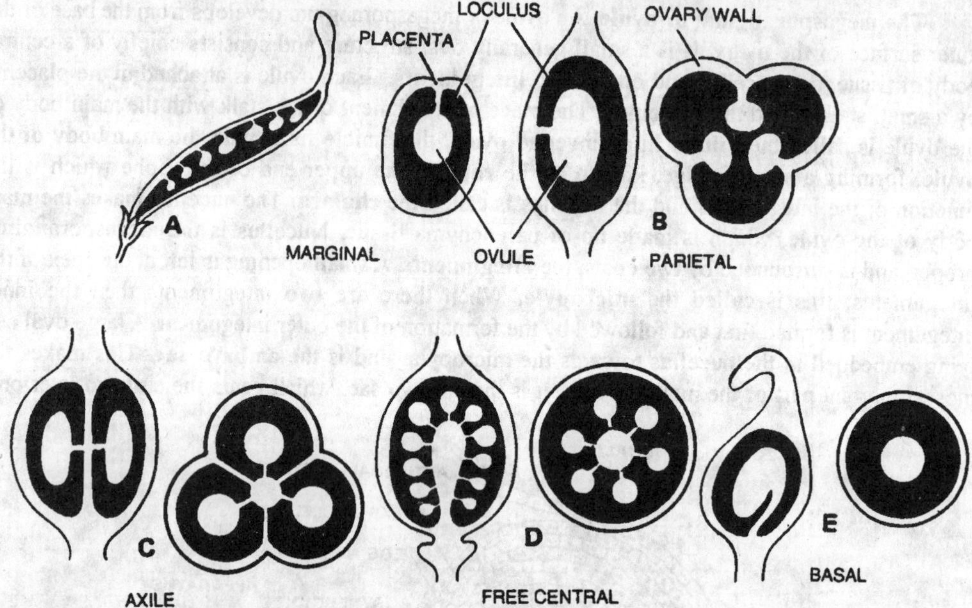

Fig. 30.14. Placentation. Different Types.

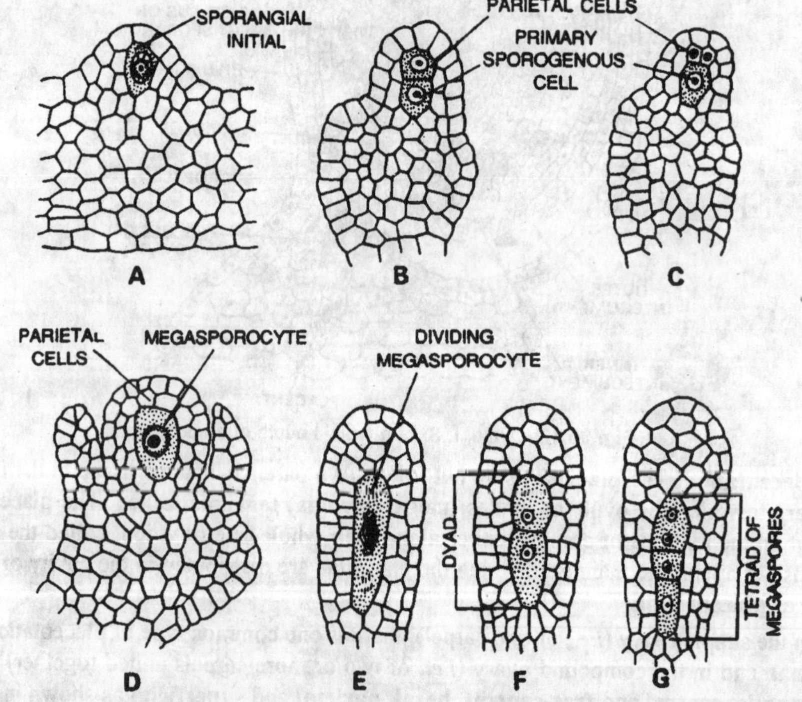

Fig. 30.15. Megasporogenesis. A, hypodermal sporangial initial; B-C, periclinal division of sporangial initial forming parietal cell D, megasporocyte being situated below the layer of wall cells; E-F, formation of dyad cells; G, linear tetrad of megaspores.

P. Maheshwari, F.R.S. (1904–1966). He worked on embryological aspects, especially the embryo sac of several plants belonging to more than 100 angiospermous families. His famous books of international fame are:

1. Introduction to the Embryology of Angiosperms
2. Recent advances in Embryology of Angiosperms (1963, edited by P. Maheswari).

He devoted his life for plant embryology, and very often referred to as 'Father of Indian Plant Embryology.' He was honoured with fellowship of Royal Society.

Megasporogenesis. The **archesporium** is hypodermal in origin. At some early stage in the development of the ovule, usually at the time of the initiation of the integumentary primordia, single hypodermal cell known as **primary archesporial cell**, becomes differentiated at the apex of the nucellus beneath the epidermis. It can be distinguished from other neighbouring cells owing to its large size, conspicuous deeply staining nucleus, and dense cytoplasm. Usually this primary archesporial cell divides periclinically forming an outer **primary parietal cell** and an inner **primary sporogenous cell.** The primary parietal cell may divide further several times both by anticlinal and periclinical divisions forming a variable amount of parietal tissue, or sometimes it remains undivided. The primary sporogenous cell usually does not divide further and functions directly as the **megaspore mother cell**.

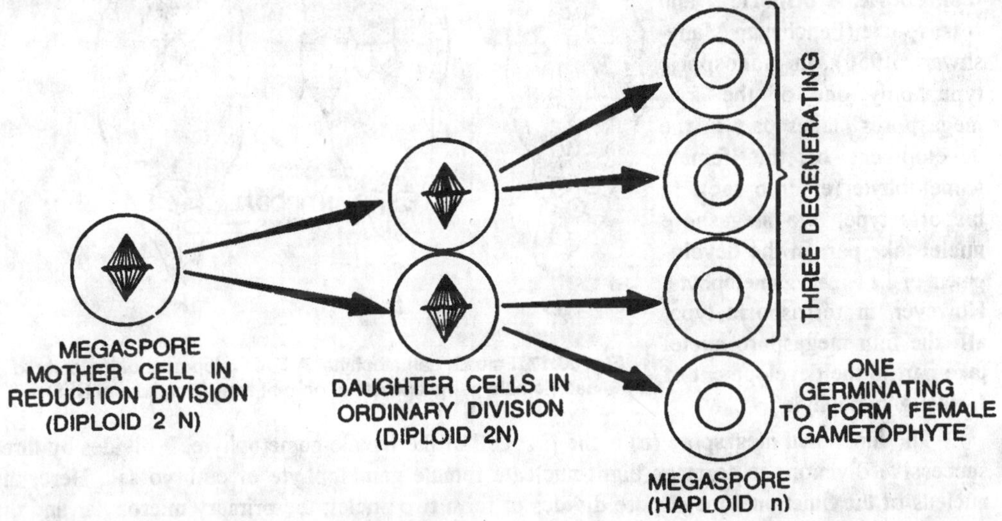

Fig. 30.16. Formation of megaspores from a megaspore mother cell. Only one megaspore is functional.

Usually the megaspore mother cell divides meiotically forming a tetrad of four megaspores. This usual process of meiotic division is termed **megasporogenesis.** Here the first division (*i.e.*, meiosis I) is always transverse and gives rise to two cells. The second is also transverse (*i.e.*, meiosis II), and thus in total four cells are being formed. The four megaspores thus formed in an axial row within the nucellus forming a **linear tetrad.**

Of the four megaspores, so formed, each with half (n) the usual number (2n) of chromosomes, the three upper ones degenerate and appear as dark caps, while the lowest one functions, and gives rise to the embryo sac. The developing megaspore encroaches upon and absorbs the other three degenerating megaspores of tetrad and the neighbouring cells of the nucellus.

Female gametophyte. The megaspore (n) makes the beginning of the female gametophyte. The nucleus of the functional megaspore divides and develops into the female gametophyte or the embryo sac. The female gametophyte of angiosperms is very much reduced and totally dependent for its nutrition upon the tissue of the sporophyte.

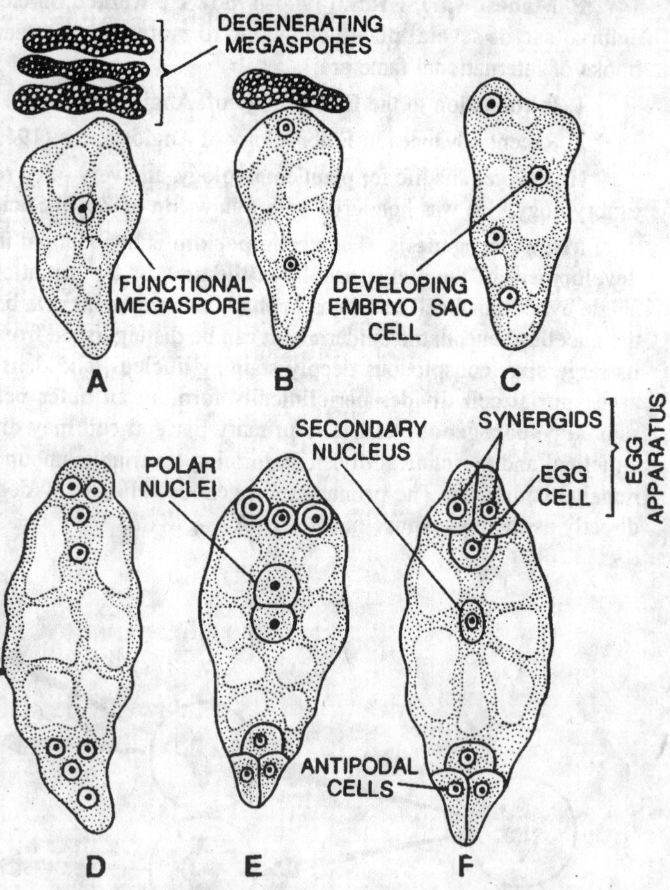

Fig. 30.17. Female gametophyte. A-F, development of the embryo sac (female gametophyte) of normal type (*Polygonum* type).

Depending on the number of megaspores taking part in the development, the embryo sacs (female gametophytes) of angiosperms may be classified into three main categories; **monosporic, bisporic** and **tetrasporic** (Panchanan Maheshwari, 1950). In monosporic type, only one of the four megaspores takes part in the development of the female gametophyte (embryo sac). In bisporic type, two megaspore nuclei take part in the development of the female gametophyte. However, in tetrasporic type, all the four megaspore nuclei take part in the development of female gametophyte.

The functional megaspore (n) is the first cell of the female gametophyte. It divides by three successive divisions to form an eight-nucleate female gametophyte or embryo sac. Here, the nucleus of the functional megaspore divides to form two nuclei; the primary micropylar and the primary chalazal nuclei. These nuclei again divide, so that the number is increased to four. Each of these nuclei divides, so that altogether eight nuclei are formed in the embryo sac, four at each end. The female gametophyte increases in size. Now, one nucleus from each end or pole passes inwards, and the two polar nuclei fuse together somewhere in the middle of the embryo-sac, forming the **secondary nucleus** (2n). The remaining three nuclei at the micropylar end, each surrounded by a very thin wall, form the **egg apparatus.** The egg apparatus consists of two synergids and an egg cell. The other three nuclei at the opposite or chalazal end, lying in a group, often surrounded by very thin walls, form the **antipodal cells**. This type of embryo-sac is the most common and generally known as the **normal type.**

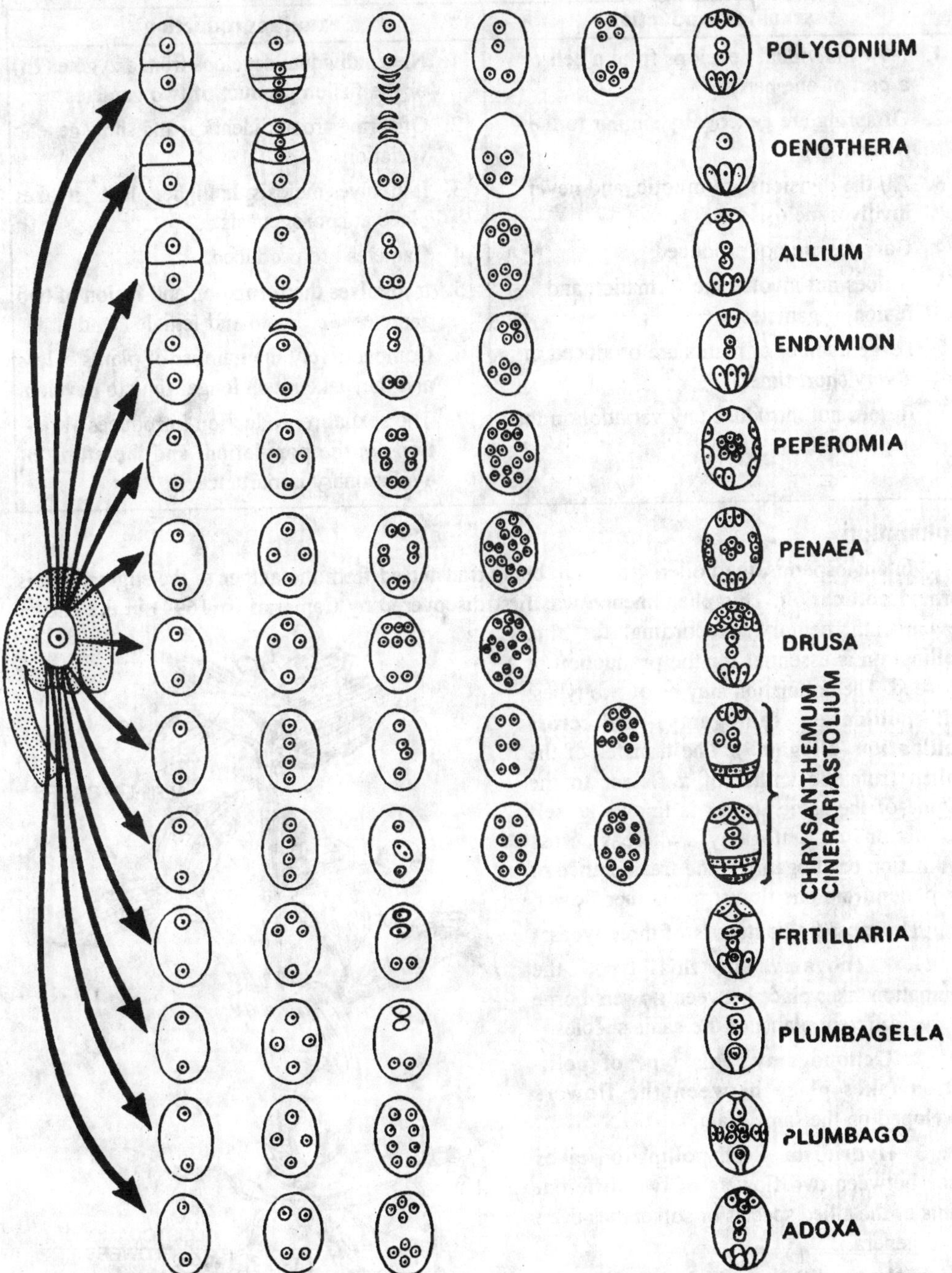

Fig. 30.18. Female gametophyte. Development of different types of embryo sacs in angiosperms.

Table 30.1. Differences Between Asexual and Sexual Reproduction in Angiosperms

Asexual reproduction	Sexual reproduction
1. New individual develops from a cell or a part of one parent.	1. New individual develops from a zygote (2n) or the fusion product of two gametes.
2. Offspring are genetically similar to the parent.	2. Offspring are not identical and show genetic variations.
3. All the divisions are mitotic, and never involves meiosis.	3. It involves meiosis. In higher plants, it takes place at sporogenesis.
4. Gametes are not produced.	4. Gametes are produced.
5. It does not involve the formation and fusion of gametes.	5. It involves the formation and fusion of two gametes, *i.e.*, male and female (and).
6. Large number of plants are produced in a very short time.	6. Comparatively the number of plants is less and they take much longer time to develop.
7. It does not introduce any variation in the population,	7. The sexual reproduction introduces variations in the population, and therefore, of evolutionary importance.

Pollination

In angiosperms, the pollen grains are being transferred from the anther to the stigma, and is termed **pollination**. This phenomenon was first discovered by Camerarius (1694) in the end of seventeenth century. According to him, pollination is essential for the production of the seed. The pollination may be of two types: **self pollination (autogamy)** and **cross pollination (allogamy)**. The transfer of the pollen from the anther of a flower to the stigma of the same flower is known as self pollination or autogamy, whereas cross pollination or allogamy is the transference of the pollen from one flower to another flower.

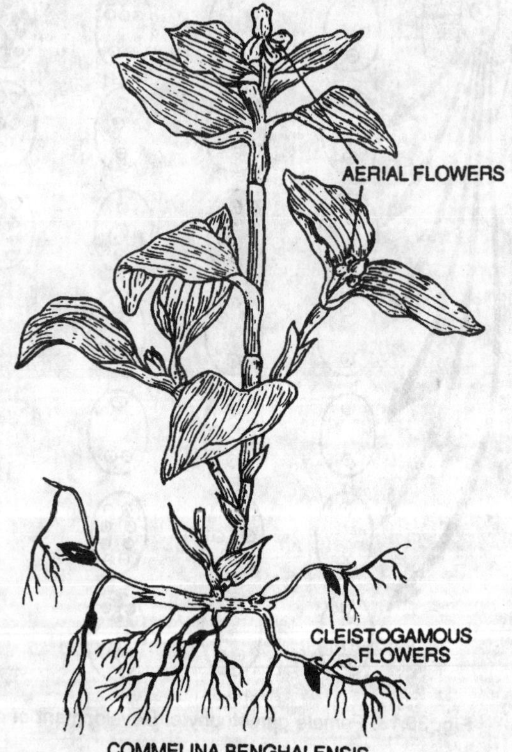

Fig. 30.19. Homogamy-cleistogamy. The cleistogamous flowers are born on underground branches of *Commelina benghalensis*.

The cross pollination is of three types:

1. Xenogamy. In this type, the pollination takes place between flowers borne on two different plants of the same species.

2. Geitonogamy. This type of pollination takes place between the flowers developed on the same plant.

3. Hydridism. Such pollination takes place between two flowers of two different plants of the allied species or sometimes even allied genera.

In the condition in which the pollen are discharged from the anther, they show considerable resistance to environmental changes. Sometimes, they remain viable for several weeks.

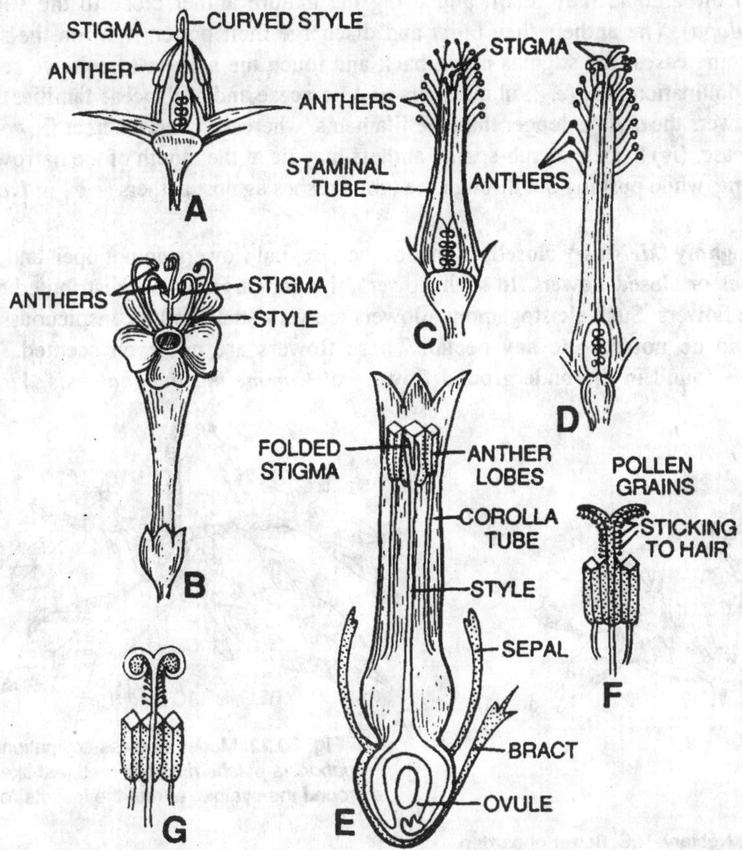

Fig. 30.20. Modes of self pollination. A, in *Solanum tuberosum* the stigma bent over the staments; B, in flower of *Mitrabilis jalapa* the anthers bent over the stigma; C-D, in the flowers of *Hibiscus rosa-sinensis* the stigma brushes against the anther lobes to receive pollen; E-G, in flower of *Helianthus annuus* the stigmatic lobes curl backward to receive the pollen sticking to hair on the outer stigmatic surface and the style.

In certain cases even in hermaphrodite flowers, self-pollination does not take place. This happens because of **heterostyly,** *e.g.*, in *Primula vulgaris;* **dichogamy,** where the maturity of male and female sex organs of the flowers is attained at different times, *e.g.*, in *Impatiens*; **herkogamy**, in which the structure of male and female sex organs in the flowers acts as barrier for self-pollination, and **self-sterility**, as found in *Petunia axillaris.*

As mentioned, pollination is of two kinds: (1) **self pollination** or **autogamy** (auto = self; *gamos* = marriage) and (2) **Cross-pollination** or **allogamy** (*allos* = different).

1. Self -pollination (autogamy). This kind of pollination is the transference of pollen grains from the anther of a flower to the stigma of the same flower or from a flower (male or bisexual) to a flower (female or bisexual), both found on the same individual plant. Here only one parent plant is concerned to give rise to the offspring. The self-pollination is, however, presented in unisexual flowers borne by two separate plants, and also in many bisexual flowers. The undermentioned adaptations are commonly found in flowers to achieve self-pollination.

Homogamy (*homos* = same). In this condition, the anthers and the stigmas of a bisexual flower mature at the same time. (*i*) Here some of the pollen grains may reach the stigma of the same flower through the agency of wind or insects, thus effecting self-pollination. (*ii*) The filaments of the anthers may recoil and bring the mature anther close to the stigma (*e.g.*, in *Mirabilis jalapa*). The anthers then burst and discharge their pollen right on the surface of the stigma. In some cases, the stigmas move back and touch the anthers to achieve self-pollination when cross-pollination fails (*e.g.*, in members of Asteraceae and Malvaceae families). (*iii*) in some drooping flowers the style is longer than the filaments, whereas in certain erect flowers the reverse may be the case. (*iv*) Sessile or sub-sessile anthers may lie at the mouth of the narrow corolla tube and the stigma, while pushing out through the tube brushes against anthers (*e.g.*, in *Ixora, Gardenia* and *Vinca*).

Cleistogamy (*kleistos* = closed). Some of the bisexual flowers do not open and are known as cleistogamous or closed flowers. In such flowers, the pollen grains are distributed on the stigma of the same flowers. Such cleistogamous flowers are very small and inconspicuous. They are not coloured, and do not secrete any nectar. These flowers are not even scented. This type of pollination is found in the underground flowers of *Commelina benghalensis, Viola, Drosera, Oxalis,* etc.

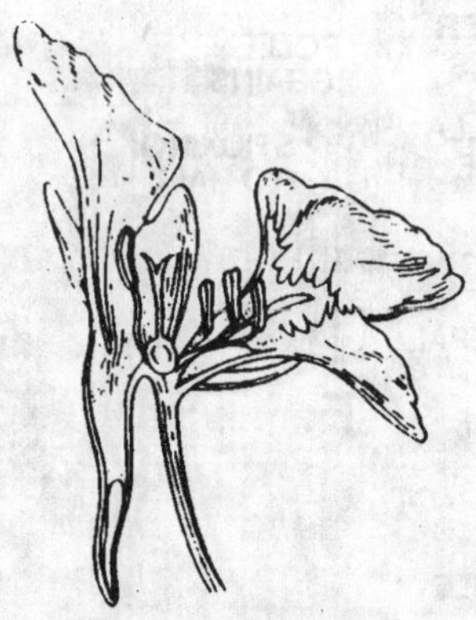

Fig. 30.21. Nectary. V.S. flower of garden **nasturtium** showing the nectary concealed in the spur.

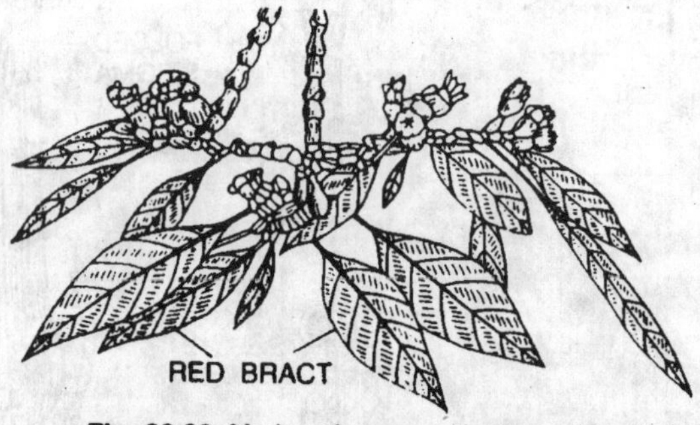

Fig. 30.22. Modes of cross-pollination-colour. In *Euphoribia pulcherrima* coloured leaf-like breacts that surround the cyathia, attract the insects for the purpose.

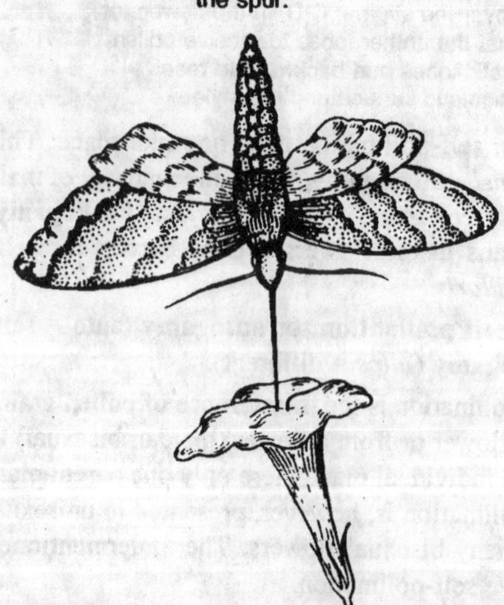

Fig. 30.23. *Entomophily.* Flower of *Petunia* being visited by a moth.

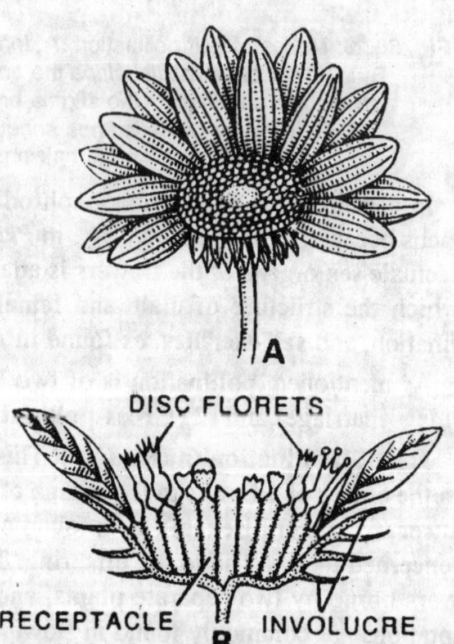

Fig. 30.24. In capitutum inflorescence of *Helianthus* (A–B), the closely arranged flowers have every chance of pollination.

2. Cross-pollination (allogamy). The cross-pollination is induced by external agents which carry the pollen grains of one flower and deposit them on the stigma of another flower, the two being borne by two separate plants of the same or closely allied species. These agencies may be **insects** (*e.g.*, bees, flies, moths, etc.), **animals** (*e.g.*, birds, snails, etc.), **wind** and **water**. The allogamy (cross-pollination) is the rule in unisexual flowers borne by two separate plants, while in bisexual flowers, it also occurs generally. Nature favours cross-pollination and there are so many adaptations in flowers to achieve this type of pollination.

Entomophily (*entomon* = insect, *phileo* = to love). This type of pollination takes place through the agency of insects. It is of general occurrence among plants. The insect-loving flower possesses various adaptations by which they attract insects and use them as carriers of pollen grains for the purpose of cross-pollination. The main such adaptations are **colour, nectar** and **scent.** The flowers of Asteraceae and Lamiaceae families are generally pollinated by the bees and butterflies.

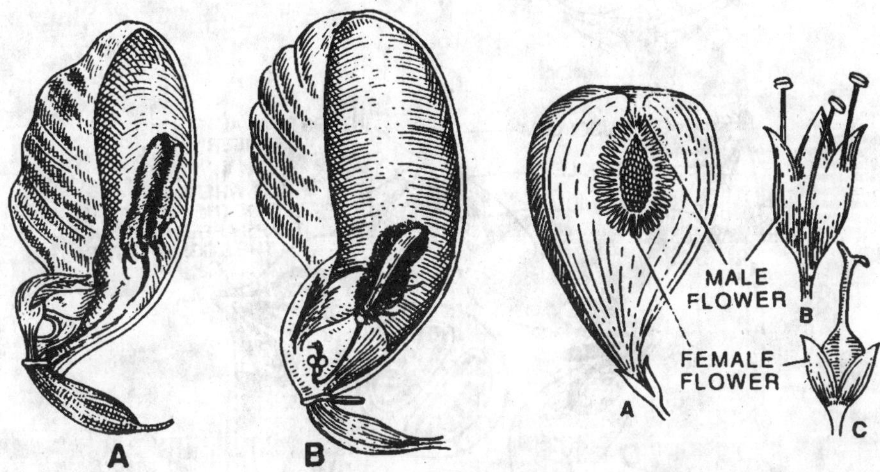

Fig. 30.25. *Entomophily* A-B, V.S. flowers of Cypripedium (can orchid) showing pollination by bee.

Fig. 30.26. The insects enter through the apical pore of hypanthodium in *Ficus*, to ensure pollination (A-C).

Generally, the pollen grains of entomophilous flowers are sticky. The stigma is also sticky. Pollen grains and nectar are very often used as food materials by the insects. Flowers generally attract insects by their colour, nectar or scent, or they visit the flowers in search of food, or shelter from sun and rain. Thus, as the insects visit the flowers, their body gets dusted with pollen grains, and when they fly and visit other flowers, they brush against the stigma which being sticky at once receives the pollen grains from their body. Thus, cross-pollination is achieved.

In several species of *Ficus,* the insects enter the chamber of the inflorescence (hypanthodium) through the apical pore, and as they move over the unisexual flowers inside the chamber, the pollination is achieved. Female flowers lie at the base of the cavity and open earlier, whereas male flowers lie near the apical opening and open later so that pollen grains have to be brought over from another inflorescence.

Flowers are generally adapted for pollination by some specific insects. For example, in snapdragon and other such flowers with saccate corolla, only the insects of particular size and weight can open the mouth of the corolla. On the other hand, long-tongued insects can pollinate the flowers with long corolla tubes.

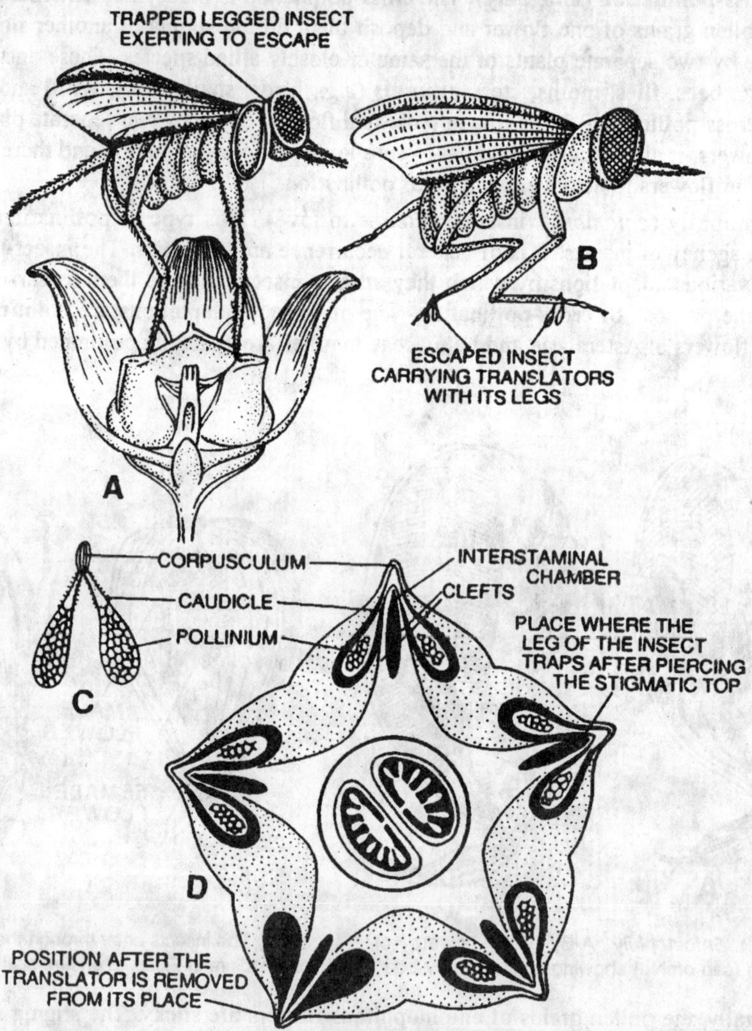

Fig. 30.27. Pollination in *Calotropis*.

Pollination in *Calotropis*. This is member of Asclepiadaceae and is pollinated by bees. In this flower, the filaments of stamens form a tube around gynoecium. The anthers are fused with the stigma to form a 5-angled disc called **gynostegium**. The staminal tube gives out distinct lobes called the **corona**. It is fused with the petals. The anther lobes, that are fused with the stigmatic disc, have straight and parallel sides and are separated only by long narrow clefts. Within these clefts there are **interstaminal chambers**. The pollen grains in the anthers are grouped in the form of mass called **pollinium**. The pollinium develops a rider mechanism or the **translator.** It consists of two arms called the **caudicles**. The caudicles carry the two pollinia on one side and are fused to form a black and sticky dot on the other side. This dot-like structure is called the **corpusculum.** There are five corpuscula at the angles of gynostegium from two adjacent anthers. An insect crawling about over the flowers is sooner or later trapped, through one of its legs becoming caught in one of the clefts between adjacent anthers. The insect can release itself only by drawing the leg upwards through the clefts and this it does, but as the leg becomes free at the top of the cleft, it catches in the notch of the corpusculum so that further movements pull this together with its attached pollinia, away from its anchorage on the gynostegium. The released insect in due course visits another flower and again becomes caught by the leg in the similar way. While drawing the leg, this time, through the anther cleft the pollinia brought from the previous flower are torn away from corpusculum and are deposited in the interstaminal chamber. The new translators are carried away and there is repetition of the whole process.

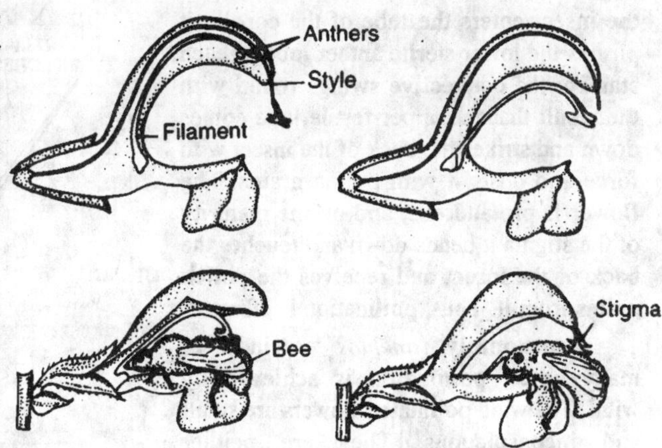

Fig. 30.28. Entomophily. An interesting type of cross-pollination in *Salvia* by insects.

Pollination in *Salvia*. An interesting type of cross-pollination takes place in *Salvia* by insects. In this flower, there are two stamens. The two anther lobes of each stamen are widely separated by the elongated curved connective which plays freely on the filament. The upper lobe is fertile and the lower one sterile. In the natural position, the connective remains upright. When

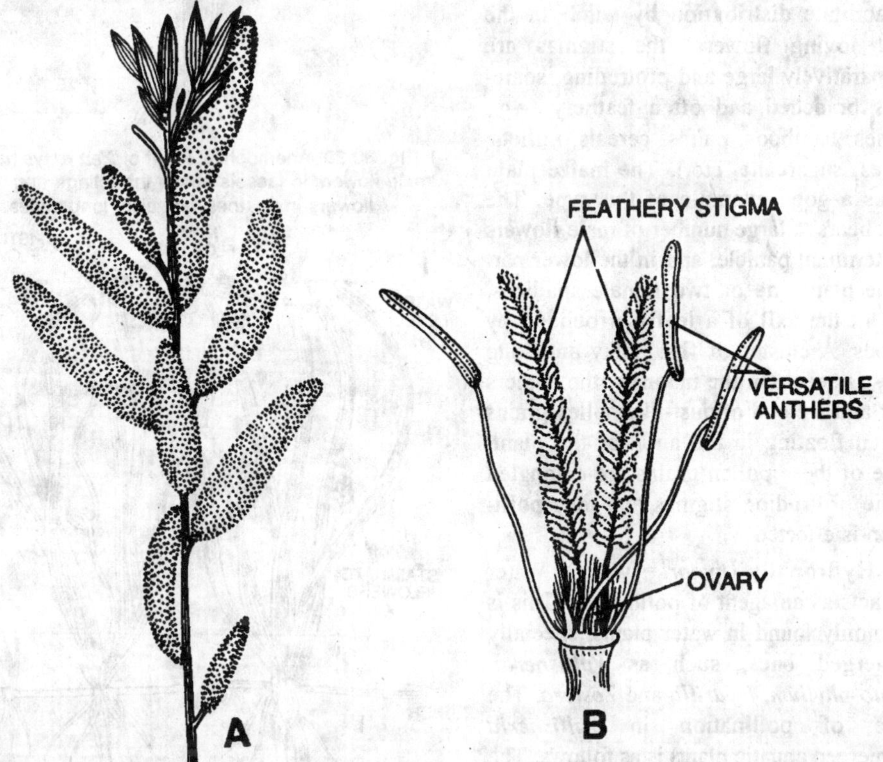

Fig. 30.29. Anemophily. A, the male catkin of *Salix;* B, the flower of *Avena sativa* with feathery stigmas.

the insect enters the tube of the corolla it pushes the lower sterile anther lobe of each stamen; the connective swings round with the result that the upper fertile lobe comes down and strikes the back of the insect with force and dusts it with pollen grains. The flower is protandrous, and on the maturity of the stigma it bends down and touches the back of the insect and receives the pollen grains from it. Thus, pollination is effected.

Anemophily (*anemos* = wind). In many cases, pollination is achieved by wind. The wind pollinated flowers are small and inconspicuous. They are neither coloured nor showy. They do not have any smell and they do not secrete any nectar. The anthers produce an immense quantity of pollen grains. A large quantity of pollen grains is being wasted during transit from one flower to another. The pollen grains are quite light and dry, and sometimes provided with wings (*e.g., in Pinus,* a gymnosperm) to facilitate distribution by wind. In the wind loving flowers, the stigmas are comparatively large and protruding, sometimes branched and often feathery (*e.g.,* grasses, bamboos, palms, cereals, millets, sedges, sugarcane, etc.). The maize plant makes a good example of this type. The plant bears a large number of male flowers in a terminal panicle, and in the lower part of the plant one or two female spadices, each in the axil of a leaf, surrounded by spathes. A cluster of fine, silky and long styles is seen. On the maturity, the anthers burst and a cloud of dust-like pollen grains is seen floating in the air near the plant. Some of these pollen grains are entangled by the protruding stigmas and thus pollination is effected.

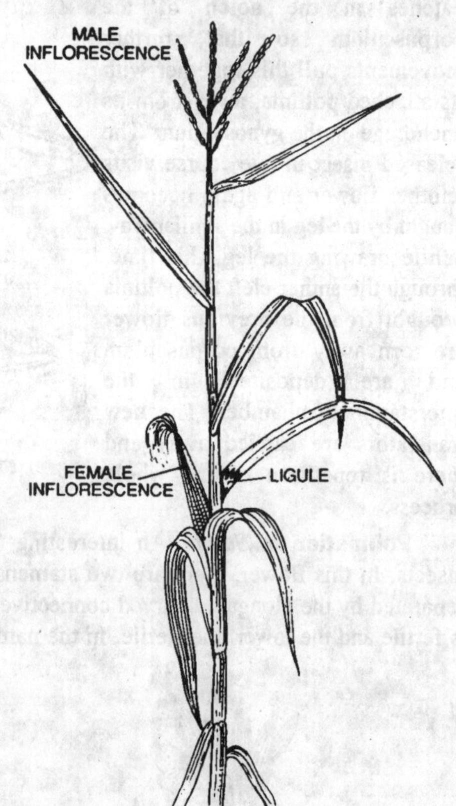

Fig. 30.30. Anemophily. Plant of *Zea mays* having male flowers in tassels above the foliage and female flowers in spathes with silky long styles.

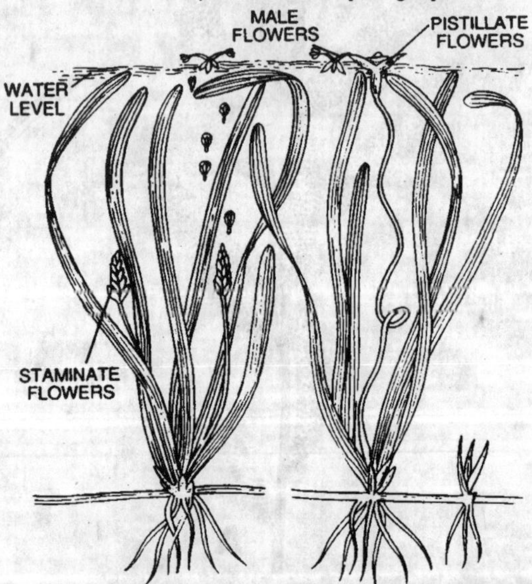

Fig. 30.31. Hydrophily. Pollination by water in *Vallisneria.*

Hydrophily (*hydor* = water). Water also acts as an agent of pollination. This is commonly found in water plants, specially submerged ones, such as *Vallisneria, Cerato-phyllum, Hydrilla* and *Zostera.* The mode of pollination in *Vallisneria* (submerged aquatic plant) is as follows: The plant is dioecious. The male plant bears a large number of minute male flowers in a small spadix surroun-ded by a spathe and borne on a short stalk, whereas the female plant bears solitary female flowers each on a long slender pedicel. This stalk of the flower elongates and takes the female flower to the surface of the water. The spathe bursts releasing the male flowers from the spadix, while still closed, and float on the surface of the water. The anthers burst and the sticky pollen grains adhere to the surface of trifid stigmas which thereafter close. As soon the pollination is over, the stalk of the female flower becomes spirally coiled and pulls the female flower down into the water. The fruit develops and matures under water a little above the bottom.

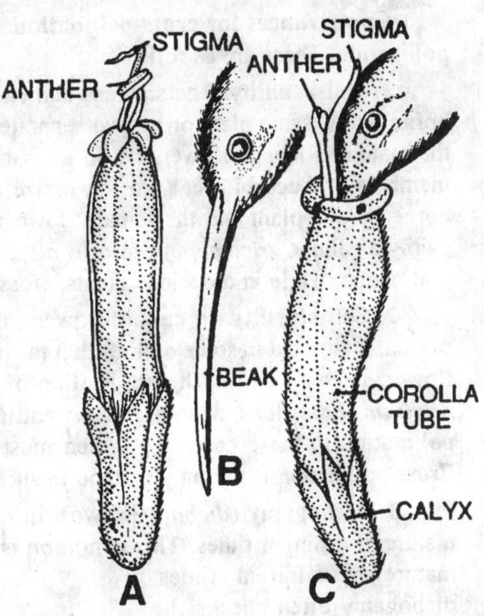

Fig. 30.32. Zoophily. Many flowers are pollinated by long-beaked birds. (ornithophily)

Zoophily (*zoo* = animals). There are so many animals, such as birds, squirrels, bats, snails, etc., which are involved in cross-pollination. The pollination by birds, generally called **ornithophily**, is common in coral tree (*Erythrina*), bottle brush *(Callistemon), Butea monosperma* and silk cotton tree (*Bombax ceiba*). The pollination in *Adansonia, Kigelia* and *Anthocephalus* are carried out by bats. This type is called **chiropterophily.** Snails are involved in pollination of several aroids (members of family Araceae). In aroids, the inflorescence is a spadix; the female flowers remain situated at the base of the spadix and the male flowers towards top. The stigmas mature first and the pollen grains are brought from another spadix.

Table 30.2. Agencies and types of Pollination

Agencies of Pollination		Types of Pollination
Abiotic	wind	Anemophily
	Water	Hydrophily
Biotic	Insects	Entomophily
	Birds	Ornithophily
	Bats	Cheiropteriphily
	Snails	Malacophily

Significance of Pollination

1. Pollination is an important process which leads to fertilisation and production of seeds and fruits, which are responsible for continuity of plant life.

2. The seeds and fruits are also used as food both for animal and humans. They make source of vitamins and minerals.

3. The pollination, especially cross pollination is important for production of plants with a combination of characters from two plants.

4. Pollination is also important in the production of hybrid seeds.

Contrivances for cross-pollination. Certain structural devices in the flowers favour cross-pollination. These are as follows:

1. Unisexuality. The stamens and carpels lie in separate flowers—male and female, either borne by the same plant or by two separate plants. There are two kinds of unisexuality: (*i*) where the male and female flowers lie on the same plant and the plant is said to be **monoecious** (*e.g.*, members of Cucurbitaceae, castor, maize, etc.), (*ii*) where the male and the female flowers are borne by one plant and the female flowers lie on another plant, it is known as **dioecious** (*e.g.*, palmyra palm, *Carica papaya, Morus alba,* etc.). In monoecious plants, there may be self-or cross-pollination, while in dioecious plants, cross-pollination is a basic necessity.

2. Self-sterility. In certain flowers, the pollen grains are unable to germinate on its own stigma. It is noted in some orchids that the pollen has an injurious effect on the stigma of the same flower. In this case on the application of pollen to stigma, the stigma dries up and falls off, *Abutilon, Passiflora, Malva, Prunus* and *Pyrus* are self-sterile. To effect the successful cross-pollination in these cases the pollen must be from two such parents which differ genetically. Cross-pollination is the only method in such cases for the setting of seeds.

3. Dichogamy (*dicha* = in two). In many bisexual flowers, the anthers and stigmas often mature at different times. This condition is known as **dichogamy**. As the anther and the stigma mature at different times, dichogamy often checks the self-pollination. There are two types of dichogamy: (*i*) **protogyny** (*protos* = first; *gyne* = female) where the gynoecium matures earlier than the anthers of the same flower, and in such cases, the stigma receives the pollen grains brought from another flower and thus cross-pollination becomes indispensable (*e.g.*, *Ficus, Mirabilis, Magnolia, Annona,* etc.) and (*ii*) **protandry** (*protos* = first; *andros* = male) where the anthers mature earlier than the stigma of the same flower and hence the pollen grains, are carried over to the stigma of another flower (*e.g.*, *Clerodendron, Hibiscus rosa-sinensis, Abelmoschus esculentus, Helianthus annuus, Coriandrum sativum,* etc.)

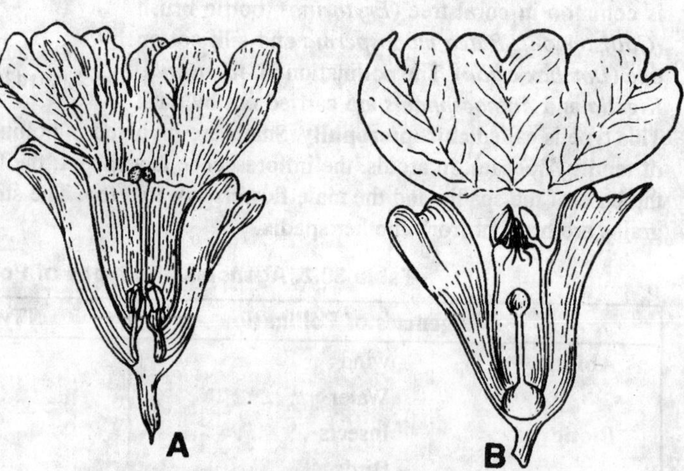

Fig. 30.33. Heterostyly in the flowers of *Primula*.

4. Heterostyly (*heteros* = different). Some plants bear flowers of two different forms. One form possesses long stamens and a short style, while the other form possesses short stamens and a long style. This kind of bearing of stamens and styles is known as **dimorphic heterostyly**. In such cases, the chances of self-pollination decrease whereas chances of cross-pollination increase. In the flowers of this type, cross-pollination readily takes place between stamens and styles of the same length borne by different flowers. Dimorphic heterostyly is seen in *Oxalis, Linum, Polygonum fagopyrum, Woodfordia,* etc.

5. Herkogamy (*hercos* = barrier). In some homogamous flowers, there are certain structural peculiarities of the floral parts which act as a barrier to self-pollination and thus favour cross-pollination by insects. Here are cited some important examples:

For example, in *Calotropis* and orchids, the pollinia are located at places where they can never come in contact with the stigma by themselves and can only be carried away by insects. The lever mechanism in *Salvia* also promotes cross-pollination and avoids self-pollination. In *Viola tricolor* the stigma is protected by a flap or a lid, that prevents contact between the stigma and anther. This flap is pushed aside by the insect and thus cross-pollination is effected.

Advantages and disadvantages of self and cross-pollination. Some of more important features are given here:

Advantages of self-pollination. This type of pollination is almost certain in a bisexual flower, if the stamens and carpels of the flower mature at the same time.

Table 30.3. Differences Between Self-pollination and Cross-pollination

Self-Pollination	Cross-Pollination
1. Pollen grains are transferred from the anther to the stigma of the same flower (autogamy) or another flower borne on the same plant (geitonogamy)	1. Pollen grains are transferred from the anther of one flower to the stigma of another flower borne on a different plant of the same species (allogamy).
2. It can occur even when the flowers are closed (cleistogamy).	2. It occurs only when the flowers are open.
3. Both the anthers and the stigmas mature at the same time.	3. The anthers and the stigmas of a flower generally mature at different times.
4. Self-pollination ultimately results in progenies which are pure lines, *i.e.*, homozygous	4. Cross-pollination produces the heterozygous offspring, and they show variations in characteristics.
5. It cannot eliminate harmful or useful characters.	5. It can eliminate harmful or useless characters.
6. External agent (insects, wind, etc.) is not required for self-pollination.	6. An external agent is required for cross-pollination.
7. Self-pollination does not help in the production of new varieties and species.	7. Cross-pollination is responsible for producing new varieties and species among plants.
8. Here the yield of the plant gradually falls with time.	8. The yield of the plant usually does not fall.
9. This type of pollination is economical for the plant.	9. This type of pollination is not economical.
10. Immunity of the race towards disease falls in the succeeding progenies.	10. Immunity of the race towards disease is generally maintained in the succeeding progenies.
11. It leads to less vigorous offspring.	11. Here the offspring are vigorous and exhibit genetic recombination.
12. Self-pollination does not introduce any variations, and therefore, the offspring are unable to adapt to the changed environment.	12. Cross-pollination introduces variations in the offspring, and therefore, the plants are better adapted in the changed environment.

Disadvantages of self-pollination. Continued self-pollination generation after generation definitely results in weaker progeny.

Advantages of cross-pollination. (*i*) It always gives rise to healthier offspring in subsequent generations which are better adapted in the struggle for existence; (*ii*) more abundant and viable seeds are produced; (*iii*) new varieties can be developed by this method; (*iv*) the adaptability of the plants to their environment is definitely better by this method.

Disadvantages of cross-pollination. (*i*) The plants have to depend upon external agencies for pollination (such as wind, water, insects, animals); (*ii*) various devices are needed to fulfil the needs of outer agencies; (*iii*) there is always a considerable waste of pollen where wind is the pollinating agent in cross-pollination.

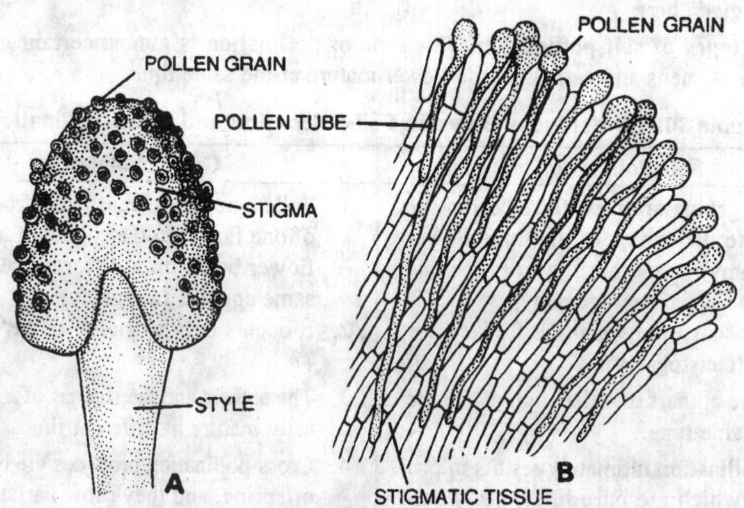

Fig. 30.34. Fertilization. A, pollen grains deposited on stigma; B, pollen grains germinating through stigmatic tissue forming pollen tubes.

Pollen Germination and Fertilisation

The fusion of two dissimilar sexual reproductive units or gametes is termed as **fertilization.** In gymnosperms, the pollen grains usually land directly on the nucellus, while in angiosperms, they fall on the stigma. In angiosperms, the fertilization is being completed as follows:

Germination of pollen grain. After being deposited on the stigma, the pollen grain absorbs liquid from the moist surface of the stigma, expands in size, and the intine protrudes out through a germ pore. The small tubular structure also known as pollen tube continues to elongate, and makes its way down the tissues of the stigma and style. Only the distal part of the pollen tube possesses living cytoplasm.

The stigma plays an important role in the germination of pollen grain. The stigma, secretes fluid containing liquids, gums, sugar and resins. The chief function of the stigmatic secretion is to protect the pollen as well as the stigma from desiccation.

After arriving to the wall of the ovary, the pollen tube enters the ovule either through the micropyle or by some other route. The entrance of the pollen tube through the micropyle is the normal condition and is known as **porogamy.** In some cases the pollen tube enters the ovule through the chalaza. This condition is known as **chalagogamy.** The chalagogamy was first reported by Treub (1891), in *Casuarina*.

The period between pollination and fertilisation varies from plant to plant. Usually this period varies from 12 to 48 hours. According to P. Maheshwari (1949), the temperature is responsible for controlling the growth of pollen tube.

Gametic fusion. In normal case, one male gamete unites with the egg to form the zygote and the second travels a little farther and unites with the secondary nucleus. This process is known as

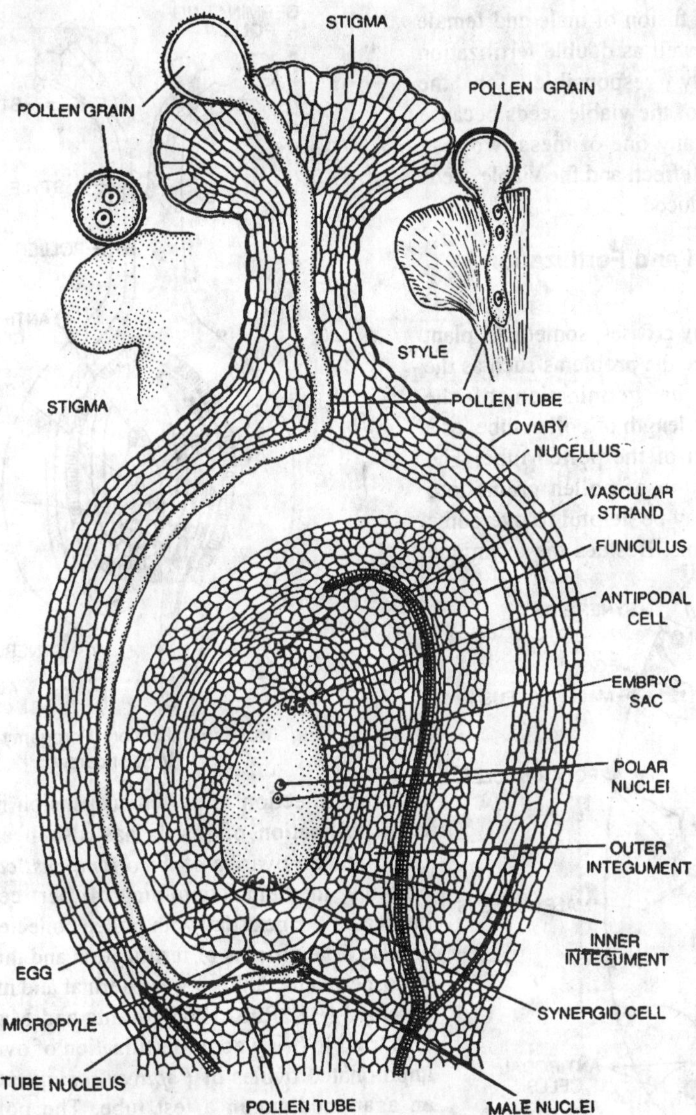

Fig. 30.35. Fertilization. C, porogamous fertilization in an anatropous ovule as seen in V.S. of carpel. A, the pollen grain is on the stigmatic surface; B, formation of pollen tube, that enters through the stigmatic tissue.

double fertilisation (Navaschin, 1898; Guignard, 1899). As the second male gamete fuses with the secondary diploid (2n) nucleus, producing a triploid (3n) primary endosperm nucleus, this is called **triple fusion.** Thus in an embryo sac there occur two sexual fusions; one in **syngamy** (*i.e.*, fusion of male gamete with egg), and the other in triple fusion, and therefore, the phenomenon is called **double fertilisation.**

Significance of double fertilization. The embryo, during its growth and development receives its nourishment from endosperm, which is a product of double fertilization. This process also gives required energy to the polar nuclei, which fail to divide further. Since the endosperm nuclei are the resultants of double feritlization, they are characterized by maternal and paternal chromosomes and thus endosperm represents the physiological aggressiveness due to hybrid vigour. The fusion of male and female gametes as well as double fertilization are equally responsible for the production of the viable seeds because absence of any one of these two may cause lethal effect, and the viable seeds are not produced.

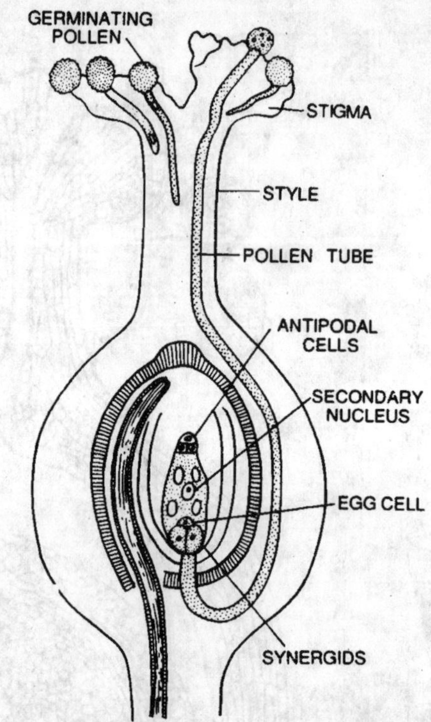

Fig. 30.36. Fertilization. Porogamy as seen in V.S. of carpel.

Pollination and Fertilization in Vitro

In many crosses, sometimes plant breeders face the problems such as the failure of the germination of the pollen, short length of pollen tube, or a slow growth of the pollen tube. The direct injection of pollen grains into the ovary may be helpful to overcome such problems. Besides the technique of direct injection of pollen suspension into the ovary, *in vitro* pollination of pistils has also been accomplished. Unpollinated ovaries of *Nicotiana rustica* were grown on Nitsch's medium containing 4 per cent sucrose and pollinated the next day with pollen collected from dehiscing anthers. The process of fertilization and the development of endosperm and embryo were normal and mature seeds were obtained (P.S. Rao, 1965). Kanta and Maheshwari (1963) also tried to bring about fertilization of ovules *in vitro*. The unpollinated ovules of *Papaver somniferum* were sown on an agar medium in a test tube. The pollen grains were dusted over the implanted ovules. The pollen grains germinated, and the pollen tubes grew rapidly and covered the ovules. Successful fertilization occurred in many ovules.

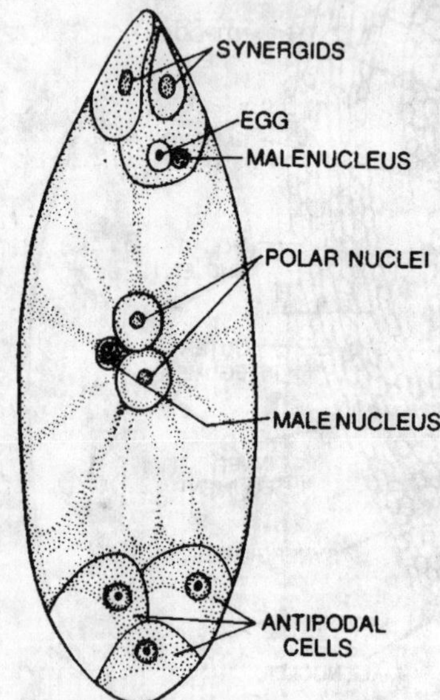

Fig. 30.37. Double fertilization in *Nicotiana*.

DEVELOPMENT OF ENDOSPERM

The Endosperm

The endosperm makes the main source of food for the embryo. In gymnosperms, the endosperm is haploid (n) and forms a continuation of the female gametophyte. On the other hand, in angiosperms it is formed mostly as the result of a fusion of the two polar nuclei and one of the male gametes. Since all the three nuclei taking part in the fusion are haploid, the endosperm becomes triploid (3n). In normal cases, the endosperm is triploid but haploid, tetraploid and polyploid endosperms are also known. Generally the endosperm nucleus divides after the division of the oospore, but in several cases the endosperm is formed to a great extent even before the first

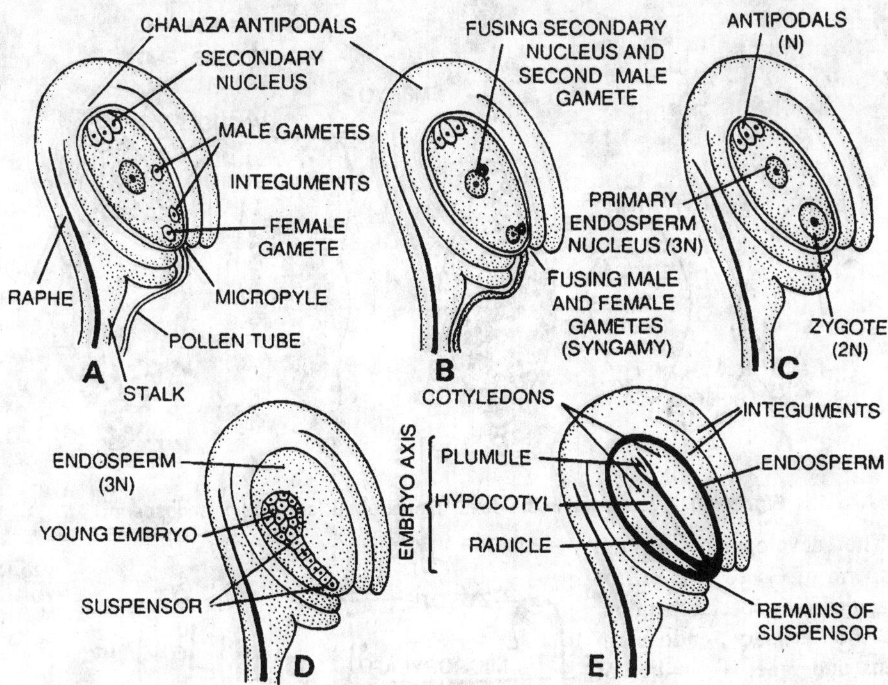

Fig. 30.38. Fertilization and post-fertilization changes. A, two male gametes discharged in the embryo sac; B, syngamy and double fertilization; C, formation of zygote (2n) and primary endospern nucleus (3n); D-E, post fertilization changes.

division of the oospore. However, endosperm formation is suppressed in two angiospermic families, the Orchidaceae and Podostemonaceae.

Types of Endosperm Formation

There are three general types of endosperm formation: (*a*) **nuclear type,** (*b*) **cellular type** and (*c*) **helobial type.**

Nuclear type. In this type, the first division and usually several of the following divisions are unaccompanied by wall formation. The nuclei may either remain free or in later stages, they may become separated by walls.

As divisions progress, the nuclei are being pushed towards the periphery, thus a large central vacuole is formed. Often the nuclei are especially aggregated at the micropylar and chalazal ends of the sac and form only a thin layer at the sides. Generally the endosperm nuclei in the chalazal part of the embryo sac have been observed to be larger than those in the micropylar end. The number of free nuclear divisions varies in different plants.

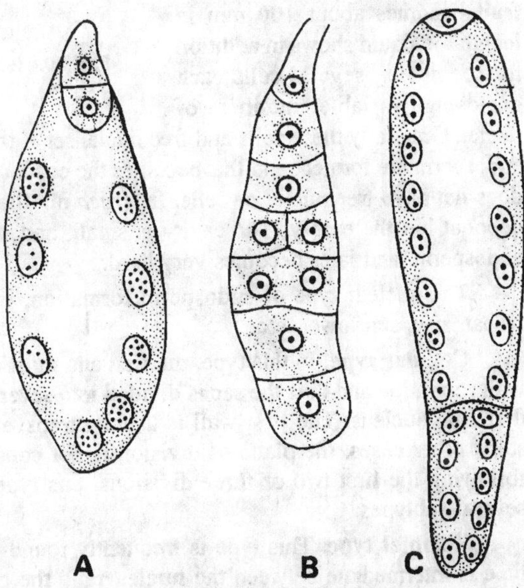

Fig. 30.39. Development of endosperm. A, nuclear type; B, cellular type and C, helobial type.

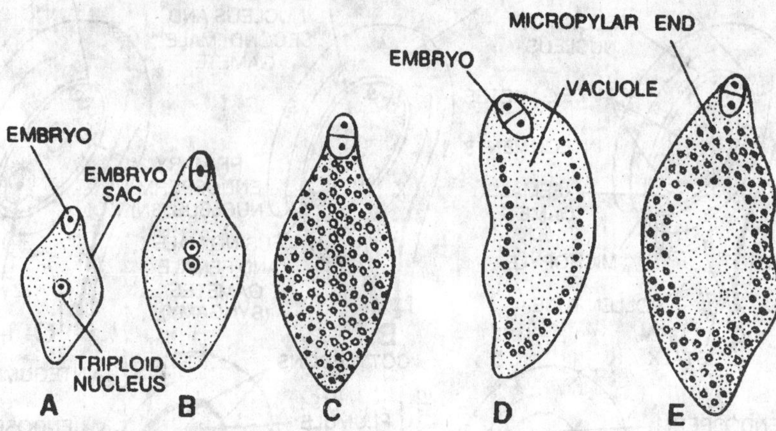

Fig. 30.40. A-E, stages in the development of nuclear type of endosperm.

The development of the endosperm of *Cocos nucifera* of Palmae deserves special mention. Here the primary endosperm nucleus undergoes a number of free nuclear divisions. When the fruit is about 50 mm long the embryo sac remains filled with a watery fluid or milk containing free nuclei and fine cytoplasmic particles. At a later stage when the fruit becomes about 100 mm in length the liquid shows in addition to free nuclei, several cells each enclosing variable number of nuclei. Gradually these cells and free nuclei set at the periphery of the cavity, and layers of cellular endosperm are formed, and this becomes the coconut meat. On maturity of coconuts the endosperm does not have free nuclei or cells. In *Areca* nut the development of the endosperm is like that of coconut but the embryo sac cavity is small, and it is completely filled up by the growth of the endosperm, and later becomes very hard.

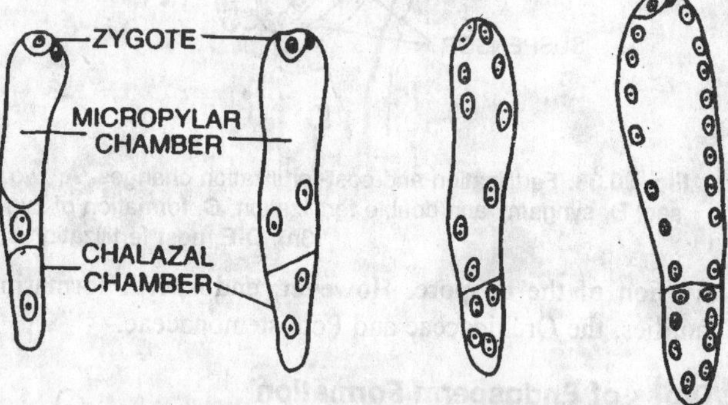

Fig. 30.41. Development of helobial type of endosperm in *Eremurus*.

The nuclear type of endosperm formation, is the most common type and found in maize, wheat, rice, sunflower, etc.

Cellular type. In this type, the first and most of the following divisions are accompanied by wall formation and thus the sac is divided into several chambers, some of which may contain more than one nucleus. The first wall is usually transverse but sometimes vertical or oblique, and in some other cases, the plane of division is not constant. On the basis of the orientation of walls following the first two or three divisions, this type of endosperm has been further divided into several subtypes.

Helobial type. This type is frequently found in the members of the order Helobiales. This type is intermediate between the **nuclear** and the **cellular** types. In this type the first division is followed by a transverse wall resulting in a micropylar and chalazal chamber. Further divisions are generally free nuclear and may be formed by the micropylar chamber only.

Eremurus is an example of a typical Helobial endosperm. Here the primary endosperm nucleus divides tranversely forming two chambers, a large micropylar and a small chalazal. Free nuclear divisions occur in both but are more rapid in micropylar chamber. Thus, when four nuclei are formed in the chalazal chamber, eight nuclei are produced in the micropylar chamber. When the chalazal chamber has eight nuclei, the micropylar chamber contains sixteen nuclei, and when there are 30 to 32 nuclei in chalazal chamber the micropylar chamber has considerably a large number of nuclei. In older ovules, the chalazal chamber begins to degenerate. Finally, when cell formation takes place in the micropylar chamber, the chalazal chamber is almost crushed and shows only a few disorganized nuclei.

DEVELOPMENT OF EMBRYO

After fertilization, the fertilized egg is called zygote or oospore which develops into an **embryo**. The oospore before it actually enters into the process undergoes a period of rest which may vary from few hours to few months. Generally the zygote (oospore) divides immediately after the first division of the primary endosperm nucleus. Unlike gymnosperms where the early stages of the development show free nuclear divisions the first division of zygote is always followed by wall-formation resulting in a two-celled proembryo. Practically there are no fundamental differences in the early stages of the development of the embryos of monocots and dicots. But in late stages, there is a marked difference between the embryos of dicotyledonous and monocotyledonous plants, hence their embryogenesis has been considered here separately.

Development of Embryo in Dicots

According to Soueges, the mode of origin of the four-celled proembryo and the contribution made by each of these cells makes the base for the classification of the embryonal type. However, Schnarf (1929), Johansen (1945) and Maheshwari (1950) have recognized *five* main types of embryos in dicotyledons. They are as follows:

I. The terminal cell of the two-celled proembryo divides by longitudinal wall.

(*i*) **Crucifer type.** Basal cell plays little or no role in the development of the embryo.

(*ii*) **Asterad type.** Basal and terminal cells play an important role in the development of the embryo.

II. The terminal cell of the two-celled proembryo divides by a transverse wall. (*a*) Basal cell plays a little or no role in the development of the embryo.

III. Solanad type. Basal cell usually forms a suspensor of two or more cells.

IV. Caryophyllod type. Basal cell does divide further.

V. Chenopodiad type. Both basal and terminal cells take part in the development of the embryo.

Here citing the example of *Capsella bursa-pastoris* (Shepherd's purse), the detailed study of **Crucifer type** of the development of the embryo has been given.

Development of dicot embryo in *Capsella bursa-pastoris* (Crucifer type). For the first time Hanstein (1870) worked out the details of the development of embryo in *Capsella bursa-pastoris*, a member of Crucifeae.

The oospore divides transversely forming two cells, a **terminal cell** and **basal cell.** The cell towards the micropylar end of the embryo sac is the **suspensor cell** (*i.e.*, basal cell) and the other one makes to the **embryo cell** (*i.e.*, terminal cell). The terminal cell by subsequent divisions gives rise to the embryo while the basal cell contributes the formation of **suspensor**. The terminal cell divides by a vertical division forming a 4-celled 1-shaped embryo. In certain plants the basal cell also forms the hypocotyl (*i.e.*, the root end of the embryo) in addition of suspensor. The terminal

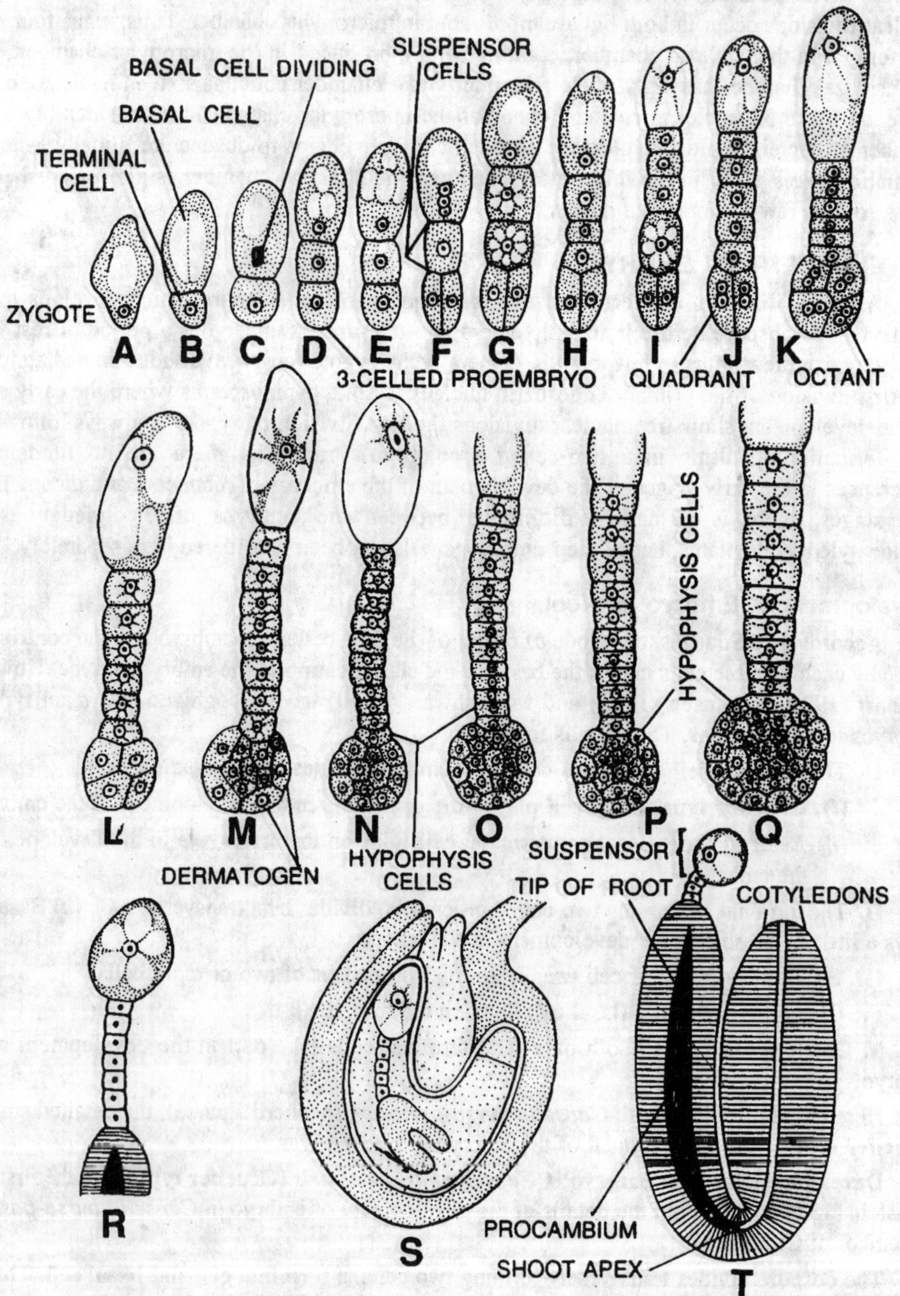

Fig. 30.42. Stages in the development of a typical dicot embryo in *Capsella bursa-pastoris*.

cells of the four-celled proembryo divide vertically at right angle to the first vertical wall forming four cells. Now each of the four cells divides transversely forming the **octant stage** (8-celled) of the embryo.

The four cells next to the suspensor are termed the **hypobasal** or **posterior octants** while the remaining four cells make the **epibasal** or **anterior octants.** The epibasal octants give rise to **plumule** and the **cotyledons,** whereas the hybobasal octants give rise to the hypocotyl with the exception of its tip. Now all the eight cells of the octant divide periclinally forming outer and inner cells. The outer cells divide further by anticlinal division forming a peripheral layer of epidermal cells, the **dermatogen.** The inner cells divide by longitudinal and transverse divisions forming **periblem** beneath the dermatogen and **plerome** in the central region. The cells of periblem give rise to the cortex while that of plerome form the stele.

At the time of the development of the octant stage of embryo the two basal cells divide transversely forming a 6-10 celled filament, the suspensor which attains its maximum development by the time embryo attains globular stage. The suspensor pushes the embryo cells down into the endosperm. The distal cell of the suspensor is much larger than the other cells and acts as a **haustorium.** The lowermost cell of the suspensor is known as **hypophysis.** By further divisions, the hypophysis gives rise to the embryonic root and root cap.

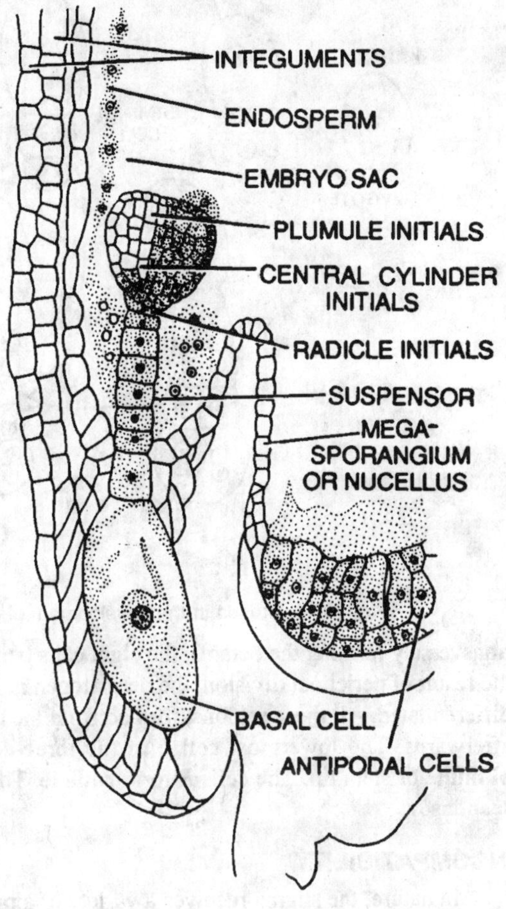

Fig. 30.43. The embryo. L.S. showing differentiation of embryo in *Capsella*.

With the continuous growth, the embryo becomes heart-shaped which is made up of two primordia of cotyledons. The mature embryo consists of a **short axis** and two **cotyledons**. Each cotyledon appears on either side of the hypocotyl. In most of dicotyledons, the general course of embryogenesis is followed as seen in *Capsella bursa-pastoris*.

Development of Embryo in Monocots

There is no essential difference between the monocotyledons and the dicotyledons regarding the early cell divisions of the proembryo, but the mature embryos are quite different in two groups. Here the embryogeny of *Sagittaria sagittifolia* has been given as one of the examples.

The zygote divides transversely forming the terminal cell and the basal cell. The basal cell, which is the larger and lies towards the micropylar end, does not divide again but becomes transformed directly into a large vesicular cell. The terminal cell divides transversely forming the two cells. Of these, the lower cell divides vertically forming a pair of juxtaposed cells, and the middle cell divides transversely into two cells. In the next stage, the two cells once again divide vertically forming quadrants. The cell next to the quadrants also divides vertically and the cell next to the upper vesicular divides several times transversely. The quadrants now divide

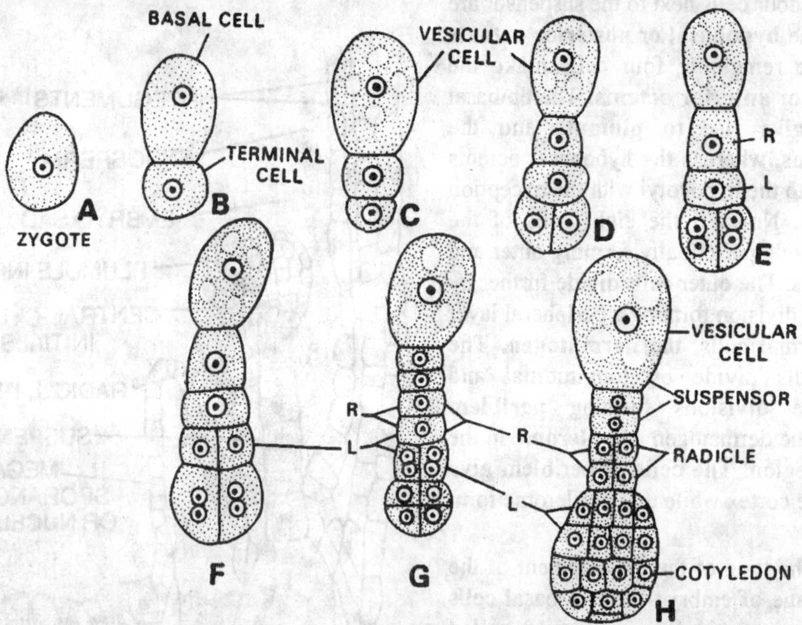

Fig. 30.44. Stages in the development of a typical monocot embryo in *Sagittaria*.

transversely forming the octants, the eight cells being arranged in two tiers of four cells each. With the result of periclinal division, the dermatogen is formed. Later the periblem and plerome are also differentiated. All these regions, formed from the octants develop into a single terminal cotyledon afterwards. The lowermost cell *L* of the three-celled suspensor divides vertically to form the **plumule** or stem tip. The cells *R* form **radicle**. The upper 3-6 cells contribute to the formation of **suspensor.**

INCOMPATIBILITY

In nature, the stigma receives a variety of air-borne or insect-carried pollen but not all pollen that reach the stigma succeed in effecting fertilisation. The pistil allows the pollen of only the right mating type to function normally, others are discarded. Thus, incompatibility is the inability of certain gametes, even from genetically similar plant species, to fuse with each other. If a pistil carrying functional female gamete fails to set seed following pollination with viable and fertile pollen, capable of bringing about fertilisation in another pistil, the two are said to be incompatible.

Incompatibility is also called **intraspecific incompatibility, self-sterility** or **self-incompatibility**. The phenomenon of self-incompatibility is geneally due to the prevention of some physiological or morphological mechanisms. This involves many complex mechanisms associated with interactions of pollen and stigmatic tissues.

Self-Incompatibility

A large number of flowering plants are successfully fertilised only by the pollen of other plants and not by their own. Such flowering plants are called **outbreeders.** In nature different floral adaptations, such as dichogamy, herkogamy and unisexuality have evolved to prevent self-pollination but the most widespread and effective natural device in such plants is self-incompatibility, which refers to the inability of a plant producing functional male and female gametes to set seeds when self-pollinated.

Depending on the origin of factors determining the mating types on the pollen side, two categories of self-incompatibility have been recognised:

(*i*) Sporophytic incompatibility. Where the incompatibility is due to the genotype of the sporophytic stigmatic tissues, *e.g.*, in Asteraceae, Brassicaceae, etc

(*ii*) Gametophytic incompatibility. Where the incompatibility is due to the genotype of pollen, *e.g.,* in Poaceae, Liliaceae, Solanaceae, etc. This may be due to prevention of pollen germination, retardation of growth, deorientation of pollen tube, or even failure of nuclear fusion. It is controlled by genes with multiple alleles (S-allele). Usually, it develops with the maturation of stigma.

Biological significance of incompatibility. In nature a balanced breeding of the plants is regulated by incompatibility. Extensive selfing of the plants leads to highly homozygous individuals which have a very low survival value. To overcome this, nature has imposed self-incompatibility.

Despite the natural value, incompatibility may turn out to be a serious hurdle in a plant improvement programme. For example, before the introduction of anther culture technique selfing was one of the chief methods to obtain homozygous individuals. For several plants where anther culture has failed to yield haploids, self-pollination continues to be an important approach to achieve this. In this context, self-incompatibility is a serious problem.

SPECIAL MODES OF REPRODUCTION

The vegetative and sexual modes of reproduction are generally normal and occur in nature. There are cases of special modes of reproduction where propagation takes place without the act of fertilisation. Some special modes are mentioned here:

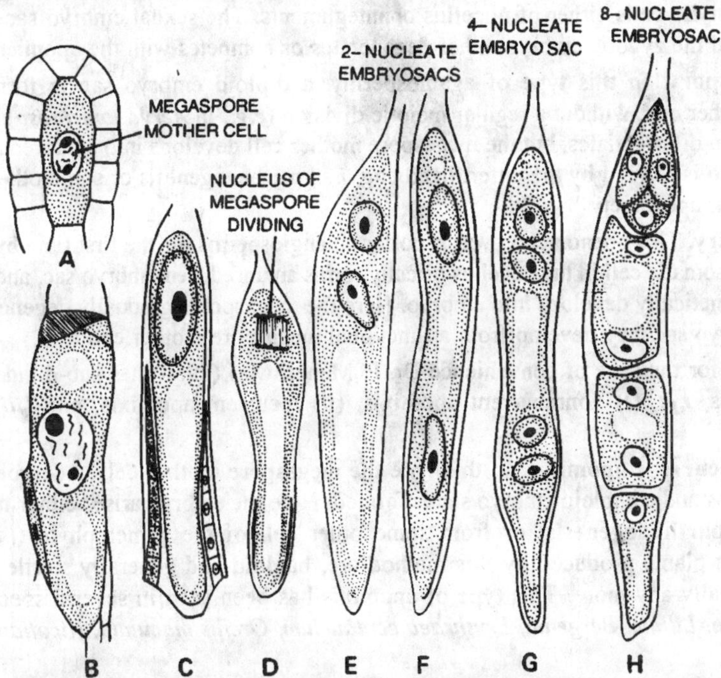

Fig. 30.45. Apomixis. Development of embryo-sac in *Eupatorium glandulosum* (A-H).

Apomixis

According to Winkler (1908, 1934), the substitution for sexual reproduction of an asexual process which does not involve any nuclear fusion. The term **apomixis** (*apo* = without; *mixis* = mingling) is a general one, and covers all types of asexual reproduction which tend to replace, or to act as substitute for the sexual method. This phenomenon covers all the non-sexual methods of reproduction in which the reproducing spores are developed without fusion, whereas the normal sexual cycle (**amphimixis**) involves two important processes : (*a*) **meiosis,** that transforms a diploid (2n) sporophytic cell into four haploid (*n*) gametophytic cells, and (*b*) **fertilization,** where two haploid gametes of opposite sex fuse and re-establish the diploid sporophytic generation. Thus, in a sexual cycle, a diploid (2*n*) or sporophytic generation alternates with the haploid (*n*) or gametophytic generation. In angiosperms, the gametophytic generations are very short and are represented by embryo sac on the female side and microspore or pollen grain on the male side. The remaining part of the life-cycle represents the sporophytic generation.

The plants where the usual sexual reproduction has been completely replaced by a type of asexual reproduction are called **apomictic**, and the phenomenon, **apomixis.**

There are two main categories of apomixis, *i.e.,* **1. Agamospermy** and **2. Vegetative reproduction.**

Agamospermy

In this category of apomixis the plants have retained seed as the agent of propagation but the embryo is formed by some process in which normal meiosis and syngamy have been eliminated. This phenomenon is known as **agamospermy**. There are three different types of agamospermy.

1. Adventive embryony. In this type of agamospermy, embryos arise directly from the diploid sporophytic cells either of nucellus or integuments. The sexual embryo sac develops in normal way and the zygotic embryo either degenerates or competes with the apomictic embryos.

2. Diplospory. In this type of agamospermy, a diploid embryo sac is formed from a megaspore mother cell, without a regular meiotic division (*e.g.*, in *Aerva tomentosa*). In this type an archesporium differentiates, but the megaspore mother cell develops into an unreduced embryo sac. The embryo is formed by the unfertilized egg, *i.e.*, parthenogenesis or some other cell of the embryo sac, *i.e.,* apogamety.

3. Apospory. This phenomenon was reported in angiosperms for the first time by Rosenberg (1907). Here a somatic cell in the nucellus directly forms an unreduced embryo sac, and the diploid egg parthenogenetically develops into embryo. Here the megaspores gradually degenerate and the aposporic embryo sac may develop from a functional megaspore mother cell.

However, for the sake of convenience Dr. P. Maheshwari (1950) has sub-divided apomixis into three types, *i.e.*, (*i*) nonrecurrent apomixis, (*ii*) recurrent apomixis, and (*iii*) adventive embryony.

1. Non-recurrent apomixis. In this type the megaspore mother cell undergoes the usual meiotic divisions and a haploid embryo sac is formed. Here the embryo arises either from the egg (*i.e.*, **haploid parthenogenesis**) or from some other cell of the gametophyte (*i.e.*, **haploid apogamy**). The plants produced by this method are, haploid and generally sterile and do not reproduce sexually any more. This type of apomixis has been seen in several species such as *Solanum nigrum, Lilium, Bergenia, Erythraea centaurium, Orchis maculata, Nicotiana tabacum,* etc.

2. Recurrent apomixis. In this type, the embryo sac generally arises either from an archesporial cell (*i.e.*, **generative apospory**) or from some other part of the nucellus (*i.e.,* **somatic apospory**). Here all the nuclei of the embryo sac are diploid, and there is no meiotic division. The embryo arises either from the egg (**diploid parthenogenesis**) or from some other cell of the gametophyte (**diploid apogamy**). Generative apospory has been observed in *Eupatorium glandulosum, Parthenium argentatum,* etc. Somatic apospory has been reported in *Hieracium excelens, H. flagellare* and *H. aurantiacum.*

3. Adventive embryony. This type of apomixis is also known as **sporophytic budding**. Here, the developed embryo sacs may be haploid or diploid, but the embryos do not arise from the cells of nucellus or the integument. There is no alternation of generations, because the diploid tissues of the present sporophyte directly give rise to the new embryo. Adventive embryony has been frequently reported in *Citrus, Euphorbia dulcis, Capparis frondosa, Mangifera indica* and *Hiptage madablota.*

Significance of apomixis. The most important apomictic crop plants are citrus, mango, mangosteen and black berries. As a reproductive system, it offers the possibility of the indefinite propagation of specially favourable biotypes, which may be highly heterozygous or sexually sterile. In obligate apomixis, this advantage is served at the expense of the long term evolutionary flexibility which is the gift of sexuality. However, in facultative apomixis, where sexual and apomictic members co-exist, the phenomenon is of special significance.

Vegetative Reproduction

In this type of apomixis, the new individual arises from a group of undifferentiated or differentiated cells, where neither embryo nor seed are produced. This type of reproduction takes place by means of bulbs, bulbils, tubers, runners, suckers, etc. Such propagules are formed by the sporophyte only. Gustafsson (1946) has distinguished three types of vegetative reproduction in angiosperms : 1. In this type, the propagules are formed outside the floral regions; though the sex organs are formed, yet no fertilization or seed setting takes place, *e.g.*, in *Agave americana* and *Elodea canadensis*. 2. In this type, the propagules are formed outside the floral regions, and the plants are sexually sterile, as found in *Fritillaria imperialis* and *Lilium bulbiferum*. 3. In this type, the propagules are formed on the floral branches either in addition to the flowers or in place of them. This phenomenon is commonly known as **vivipary**. This term is also used for those plants (*e.g.*, mangrove vegetation), in which sexually formed seeds germinate on the mother plant. Here the term vivipary will be used as **vegetative vivipary.** The vegetative vivipary is commonly found in grasses, *e.g., Festuca, Poa,* and in *Allium*. The vegetative vivipary is actually an adaptation for the multiplication of a genotype under a set of environmental conditions which prevent or check to some extent the opportunity for normal pollination. However, most of the vegetative viviparous races have not lost the capacity for flowering and normal seed setting, *e.g.*, a grass, *Deschampsia caespitosa* is sexually reproducing in Sweden, while growing by vegetative vivipary in California.

Apospory

The apospory (*apo* = without) is the development of the gametophyte from a cell of the sporophyte without the intervention of a spore. In angiosperms, this phenomenon was for the first time reported by Rosenberg (1907) in *Hieracium* spp. Here the megaspore mother cell undergoes the usual meiotic divisions and forms a tetrad. In the angiosperms it is sometimes seen that an embryo may be formed from the diploid cells of the nucellus, as in *Citrus, Mangifera, Opuntia,* etc., or even from those of integument, as in *Allium cepa*. The embryo thus formed is pushed into the embryo sac during the course of its development. Since the nucellus or the integument belongs to the sporophyte, the production of the embryo from the tissue of the sporophyte without the intervention of the spore is a case of apospory.

In the aposporic members of the Asteraceae, only one nucellar cell acts as the mother cell and gives rise to a normal 8- nucleate embryo sac, while in the grasses, more than one aposporic embryo sac may develop in the same nucellus, and the organization of mature embryo sac is 4-nucleate.

Parthenogenesis

The development of the zygote from the egg-cell without the act of fertilization, or in other words, formation of embryo from an unfertilized egg is called **parthenogenesis**. In some species of angiosperms, the embryo develops parthenogenetically. In such cases the embryo may develop from haploid egg-cell or diploid egg-cell. In apomicts of Asteraceae and Rubiaceae, the development of embryo is independent of the pollination stimulus. However, in many other apomicts, the embryo develops only after pollination, and the phenomenon is known as **pseudogamy**, *e.g.*, many apomictic grasses. According to Heslop-Harrison (1972) there are three roles of pollination in pseudogamy: (*i*) to activate the growth of ovary and ovule, (*ii*) to supply the male nucleus for the development of the endosperm, and (*iii*) to stimulate the parthenogenesis.

In most of grass apomicts, the egg may divide parthenogenetically but the proembryo ceases to grow unless and until the endosperm develops. Here, the development of endosperm occurs only after the fusion of the male nucleus with the polars, and therefore, a mature apomictic embryo in these plants is formed only after pollination.

Significance of parthenogenesis. The role of parthenogenesis in nature is severely limited; the zygogenesis and other methods of reproduction are predominant. The reason is thought to be that parthenogenetic species are not adaptable enough. In ameiotic parthenogenesis, genetic variability is practically nil. In meiotic parthenogenesis, individuals tend to become homozygous, and cause the accompanying disadvantage of this condition.

The advantages of this phenomenon are several fold. In meiotic parthenogenesis which applies to homozygous genotypes, advantageous combinations of genes are maintained instead of dispersed by meiosis as in zygogenetic species. Well adapted forms thus spread rapidly, as long as the enviornmental conditions remain unchanged. As Peacock (1961) reports, the existence of hardly triploid organism becomes possible in ameiotic parthenogenesis.

Parthenocarpy

In certain cases of angiosperms the ovary normally develops into a fruit without pollination and fertilization. This type of free development of fruit is known as **parthenocarpy**. Such fruits (parthenocarpic fruits) are always seedless. Sometimes the fruit formation may be induced by artificial pollination by foreign pollen from another species, but without subsequent fertilization. The parthenocarpy may also be induced by the spraying of growth promoting substances, such as naphthalene acetic acid NAA. This is called **induced parthenocarpy.** The examples of parthenocarpy are commonly found in banana, papaya, pineapple, guava, grapes, apple, *Thalictrum, Alchemilla*, etc.

In modern days, a lot of work has been done to produce seedless fruit on seeded varieties by controlling pollination and applying certain chemical subtances to the pistil. Thimann (1934) suggested that many pollen grains possess considerable quantities of growth substances. Gustaffson (1938) used several growth substances, such as IAA, IBA, α-NAA and phenylacetic acid mixed in the lanolin paste of about 0.5 to 1 per cent strength and smeared on the stigma to produce parthenocarpic fruits. Among the species where the fruits can be set by auxins are tomato, tobacco, pepper, figs and blackberry. Here, the fruits which have been set by treating unpollinated flowers are seedless. Nitsch and other, workers used α-NAA in aqueous form to induce parthenocarpy by atomizer. Crane (1964) induced parthenocarpy in tomatoes, apples and pears by gibberellins. Cytokinins have also been used in the induction of parthenocarpy in certain fruits.

In addition to induced parthenocarpy, the natural parthenocarpy may occur in certain species. Many horticultural varieties of bananas, pineapples, cucumber, tomatoes and figs exist, where seedless fruits are normally produced without the need for any exogenous hormone. In some species, the fruits are formed without pollination while in others, pollination is necessary but fertilization does not occur, in still others fertilization occurs but the embryos abort before the fruits mature. It is not understood clearly how the growth of these parthenocarpic fruits is controlled, but it is thought that in some cases the maternal tissues, such as the placentae may produce auxin in the absence of normal embryos. It has been noticed that the ovaries of unopened flowers of parthenocarpic varieties of orange and *Vitis vinifera* (grape) possess a higher auxin content than those of normal seeded varieties.

POLYEMBRYONY

The occurrence of more than one embryo in the seed is known as **polyembryony**. Polyembryony is quite common among conifers (gymnosperms), but many species both of dicots and monocots (angiosperms) exhibit this phenomenon. Usually there are two main types of polyembryony, *i.e.*, 1. **true polyembryony** and 2. **false polyembryony.**

1. True Polyembryony

The true polyembryony may be subdivided into two types: (*i*) **cleavage polyembryony,** where the embryos arise within an embryo sac, either by a cleavage of the egg, or from the synergids, antipodals or endosperm; (*ii*) **adventive polyembryony,** where the embryos arise from the tissues living outside the embryo sac, *i.e.*, the cells of the nucellus or the integuments, but generally they come to lie within the embryo sac.

2. False Polyembryony

Sometimes the polyembryony occurs due to the presence of multiple embryo sacs within the ovule. They may arise from: (*a*) the derivatives of the same megaspore mother cell; (*b*) from two or more megaspore mother cells, or (*c*) from nucellus cells (*i.e.,* apospory).

Importance of Polyembryony

This phenomenon plays an important role in plant breeding and horticulture. Nucellar adventive polyembryony is of great value in horticulture, where the nucellar seedlings of *Citrus* have been proved to be better clones of orchard stock than cuttings. On the other hand, the nucellar embryos are supposed to be free from disease, and the nucellar seedlings rejuvenate the vigour that is generally lost after continued cutting propagation. The adventive polyembryony is much useful in the propagation of the fruit trees, such as *Citrus* and *Mango*. The application of adventive embryos is also important for providing genetically uniform seedlings in fruit trees.

The haploids can be used for the development of homozygous diploid, which are of much value. Due to the practical value of haploids in plant breeding, the methods have been recognized for the artificial production of these embryos from the eggs or synergids.

VEGETATIVE REPRODUCTION

In vegetative propagation, any part of the plant, *i.e.*, stem, root, leaf or even buds are capable to give rise to new plants. The most striking example of such reproductive capacity is seen in the leaves of *Bryophyllum*. Like other leaves they too carry on photosynthesis, their usual function, but in addition they have the capacity to produce one or more plants from the notches in their margins. If one keeps a leaf of *Bryophyllum* on damp soil, within a day or two the buds appear in the notches of the margin of leaf. These buds would later form new plants, complete with roots, leaves and stem.

More often the capacity for vegetative propagation resides in roots, *e.g.*, in sweet potato, asparagus and dahlia, or stems, *e.g.*, in ginger, turmeric, banana, potato, onion, sugarcane and canna. In all these cases the plant parts possess the power of **regeneration,** which means that a part of the plant can either replace its lost portion or bring forth a new adult plant.

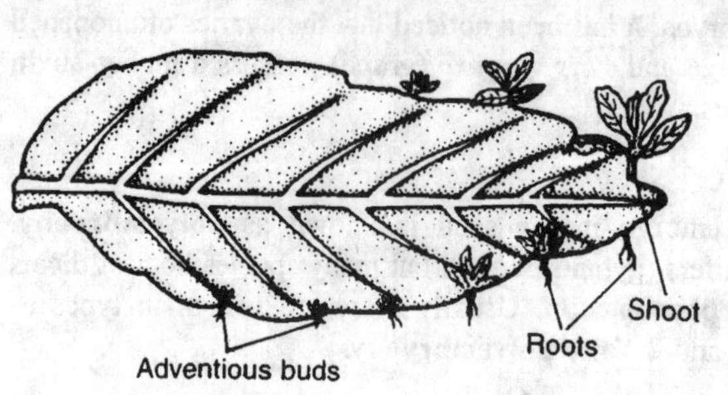

Fig. 30.46. *Bryophyllum.* Buds on margin of leaf.

Fig. 30.47. Tuberous roots of sweet potato.

The capacity for regeneration is quite common in most higher plants and has been put to great use by mankind since time immemorial. Thus, rose, coleus, pothos and glory of garden (*Bougainvillaea*) are generally propagated by placing their stem cuttings in moist soil. The cuttings soon strike roots and grow into new plants.

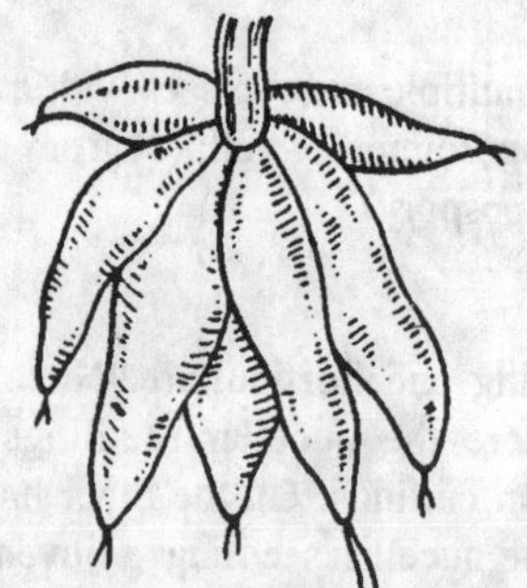

Fig. 30.48. Fasciculated tuberous roots of *Dahlia*.

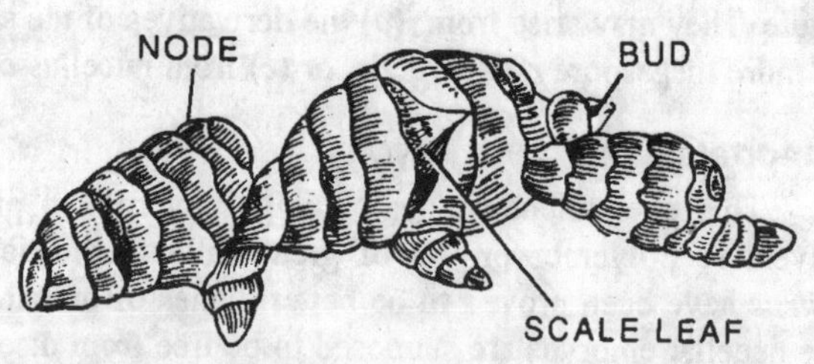

Fig. 30.49. Underground modified stem rhizome of ginger with bud.

Cutting, grafting and layering are generally referred to as artificial methods of vegetative propagation. The resulting offspring are normally identical, and resemble the parent forms in almost all respects. Gardeners generally use these methods for getting plants of same types.

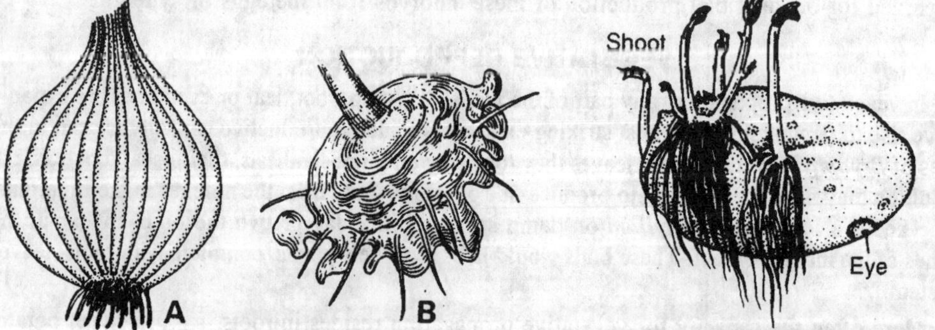

Fig. 30.50. Underground modified stems.

Fig. 30.51. Underground modified stem of potato.

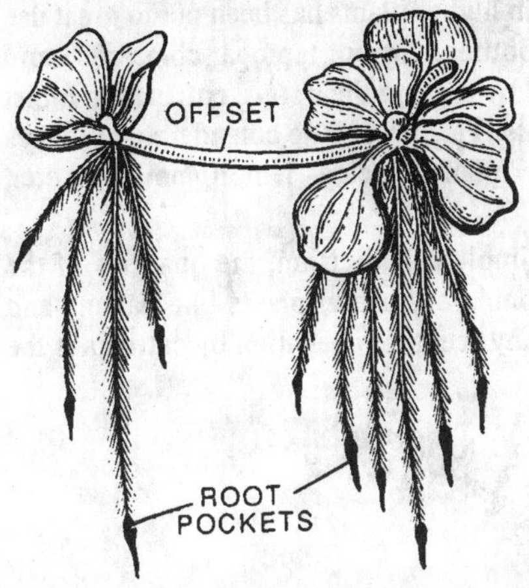

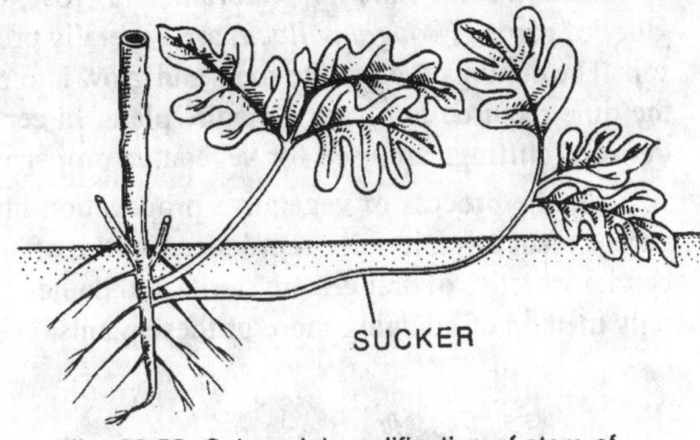

Fig. 30.53. Sub-aerial modification of stem of *Chrysanthemum*.

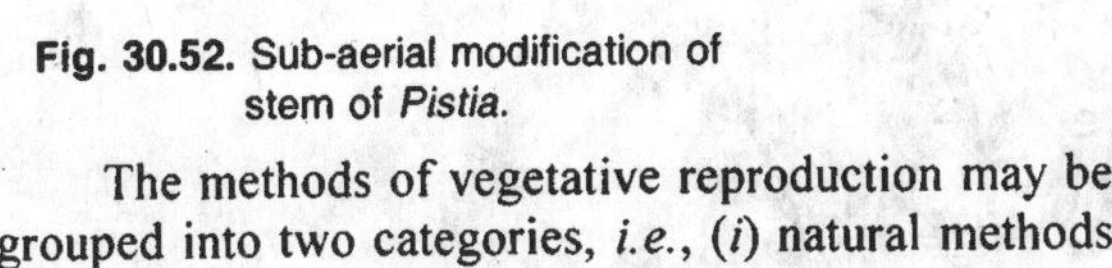

Fig. 30.52. Sub-aerial modification of stem of *Pistia*.

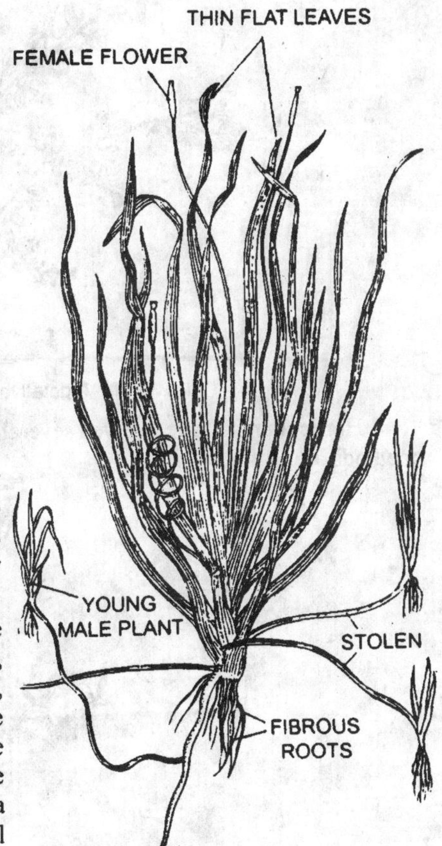

Fig. 30.54. Vegetative propagation in *Vallisneria spiralis* by means of stolons.

The methods of vegetative reproduction may be grouped into two categories, *i.e.*, (*i*) natural methods and (*ii*) artificial methods.

1. Natural methods. In such methods of vegetative propagation, a portion of the plant gets detached from the mother plant and develops into a new independent plant under suitable environmental conditions. The detached parts may be roots, stems, leaves or flowers.

The most striking example of such reproductive capacity is seen in the leaves of *Bryophyllum*. Like other leaves they too carry on photosynthesis, their usual function, but in addition they have the capacity to produce one or more plants from the notches in their margins. If one keeps a leaf of *Bryophyllum* on damp soil, within a day or two the buds appear in the notches of the margin of leaf. These buds would later form new plants, complete with roots, leaves and stem.

More often the capacity for vegetative propagation resides in roots (*e.g.*, in sweet potato, asparagus and dahlia) or stems (*e.g.*, in ginger, turmeric, banana, potato, onion, zamikand, sugarcane and canna). In all these cases the plant parts possess the power of **regeneration**, which means that a part of the plant can either replace its lost portion or bring forth a new adult plant. The stem of plants with subaerial modifications, as in *Pistia*, *Chrysanthemum*, *Eichhornia* and pine apple are also used for propagating plants.

In plants, such as *Agave*, *Dioscorea* and *Oxalis*, small buds develop near the flower, leaf axil and tuberous root, respectively, to give rise to new plants. The capacity of regeneration is quite common in most higher plants. In *Vallisneria* (a submerged water plant), vegetative propagation takes place by means of stolons.

2. Artificial methods. The capacity for regeneration in higher plants has been put to great use by mankind since time immemorial. Thus rose, coleus, pothos, croton, tapioca, china rose and glory of garden (*Bougainvillaea*) are generally propagated by placing their **stem cuttings** in moist soil. The cuttings soon strike roots and grow into new plants. This procedure considerably reduces the time required to obtain a mature plant. In certain other plants, such as lemon, tamarind, etc., the **root cuttings** are used for vegetative propagation.

As the process of vegetative propagation involves simple cell division, the qualities of the parent plant are faithfully reproduced in the offspring. Some very useful plants like banana and certain varieties of oranges are sterile and do not produce any seeds. Propagation by cuttings is the only method of attaining more of these plants.

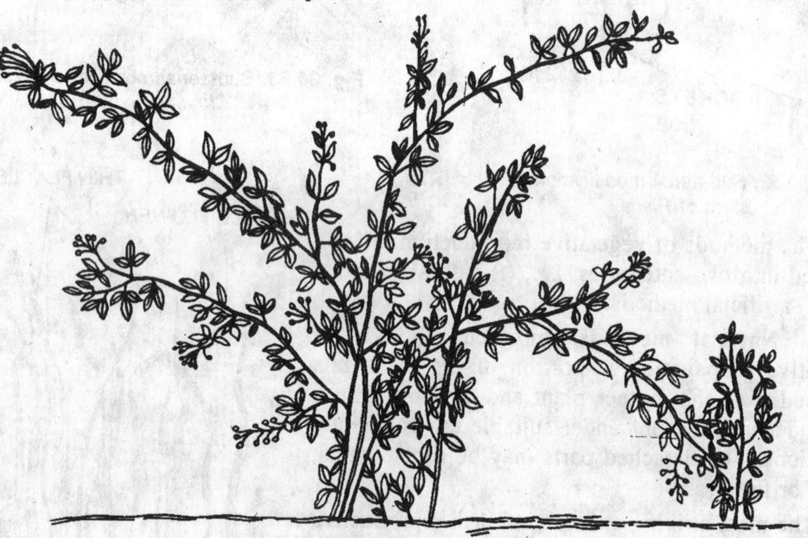

Fig. 30.55. Vegetative propagation. Tip layering in Raspberry.

Horticulturists have evolved several other methods to produce new plants in a short time. The methods are as follows:

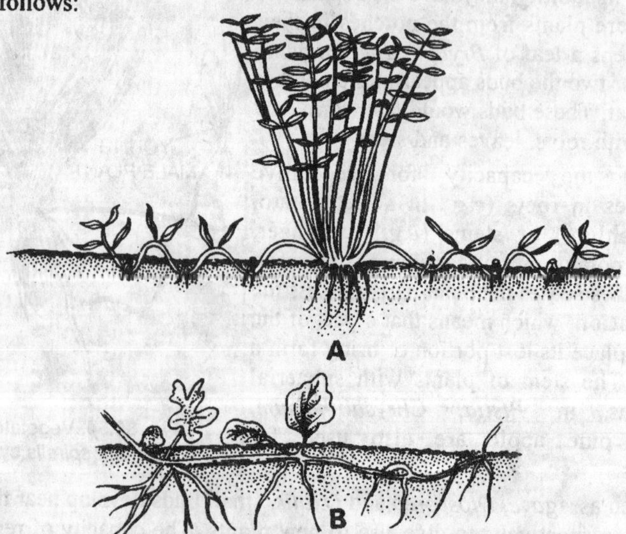

Fig. 30.56. Layering. A, serpentine layering; B, trench layering.

Layering. This is one of the most common methods used for propagating plants such as rose, jasmine, grape vine, lemon, rhododendrons and magnolias.

Layering is the development of roots on a stem while it is still attached to the parent plant. The stem or branch that develops adventitious roots while still attached to the parent plant is called a **layer**. This is a natural means of reproduction in black raspberries and trailing blackberries. Layering can be artificially induced in many plants. It can be induced by bending the branch to the ground and covering a medium portion of it below the soil. The apical portion of the branch is left exposed to the air to form the leaves. The part below the soil develops adventitious roots. Now the connection of the layered branch can be cut from the parent plant and grown independently.

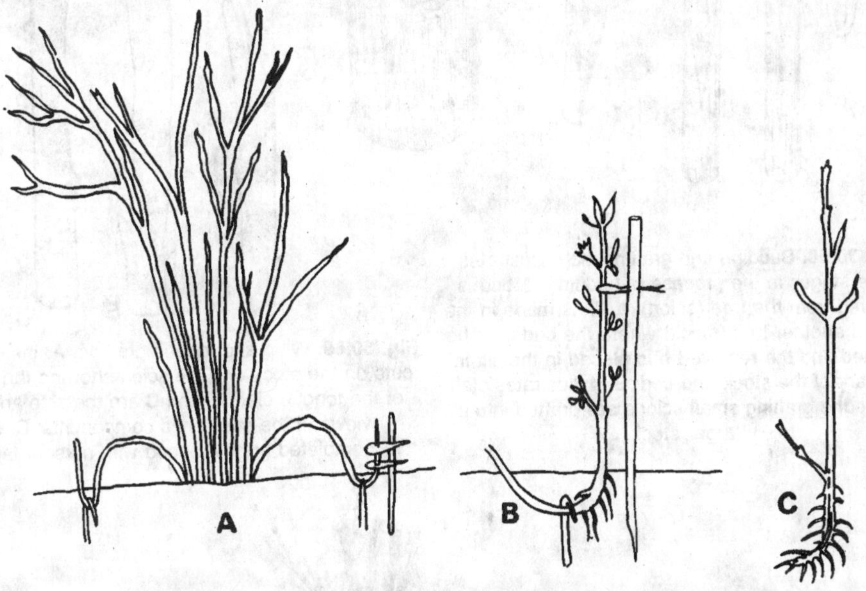

Fig. 30.57. Common method of layering. A, the branches of the plants are bent and buried under the soil; B, buried part, after it develops roots; C, the cut part is separately planted.

Grafting. This is another method in which man has remarkably exploited the regenerative capacity of higher plants. It is generally practiced in plants which do not root easily or have a weak root system. It is an art of joining parts of two plants of the same or allied species in such a way as to bring about an organic union or fusion of the tissues. This method is practiced in mango. apple, pear, citrus, guava, rose and rubber plant.

It is a common experience that from a mango tree grown a good seed will not always bear fruit of the same quality as the parent tree. Other cultivated fruit trees also show such variations. Under such conditions one may resort to grafting. In this process, a detached part of one plant is inserted into the stem or root system of another plant. The former is called **scion** (short piece of detached shoot containing several dormant buds) and the latter **stock** (longer portion of the plant which is fixed to the soil by its root system). The **scion** is generally taken from the plant having superior characters.

In successful grafting, the object is to connect or fit two pieces of living plants in such a way that they will unite and subsequently behave as one plant. This is done in such a way that the cambium tissue of both (scion and stock) comes in contact and forms a cambium layer common to both. Consequently, the two **scion and stock** grow together. The place where they are grafted is covered with grafting wax to avoid infection. The scion becomes a part of the plant into which it is grafted. This plant bears flowers and fruits characteristic of the **scion**. At this junction of graft,

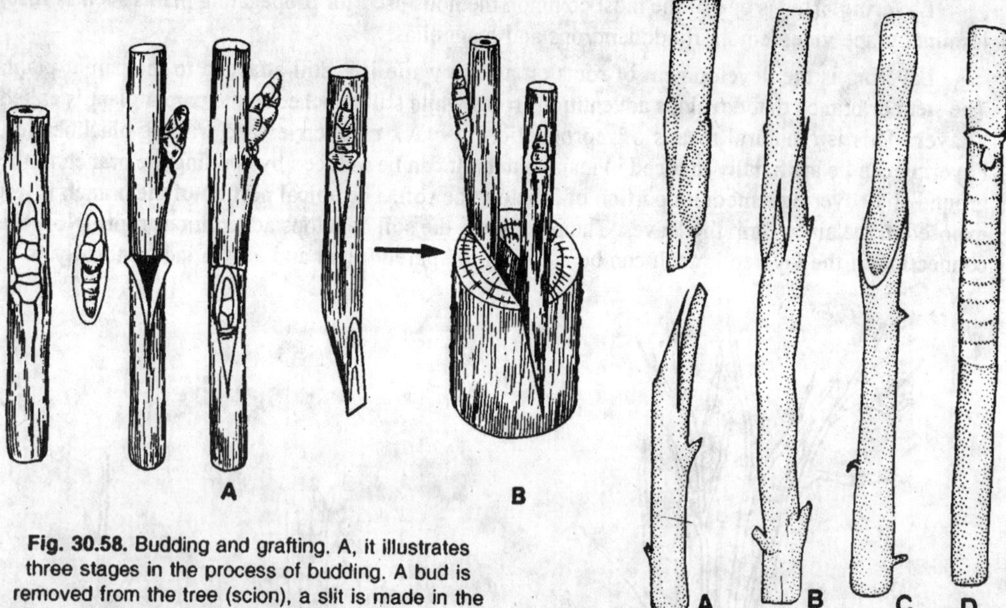

Fig. 30.58. Budding and grafting. A, it illustrates three stages in the process of budding. A bud is removed from the tree (scion), a slit is made in the bark of another tree (stock) where the bud is to be planted and the removed bud placed in the slit in the bark of the stock and tied; B, it illustrates cleft or wedge grafting small scions are grafted into a larger stock.

Fig. 30.59. Whip and tongue grafting. A, the diagonal cuts on the stock and the scion showing the portion of the tongue cleft. B and C are the side and face views of the positioned components; D, the completed graft wrapped with nursery tape.

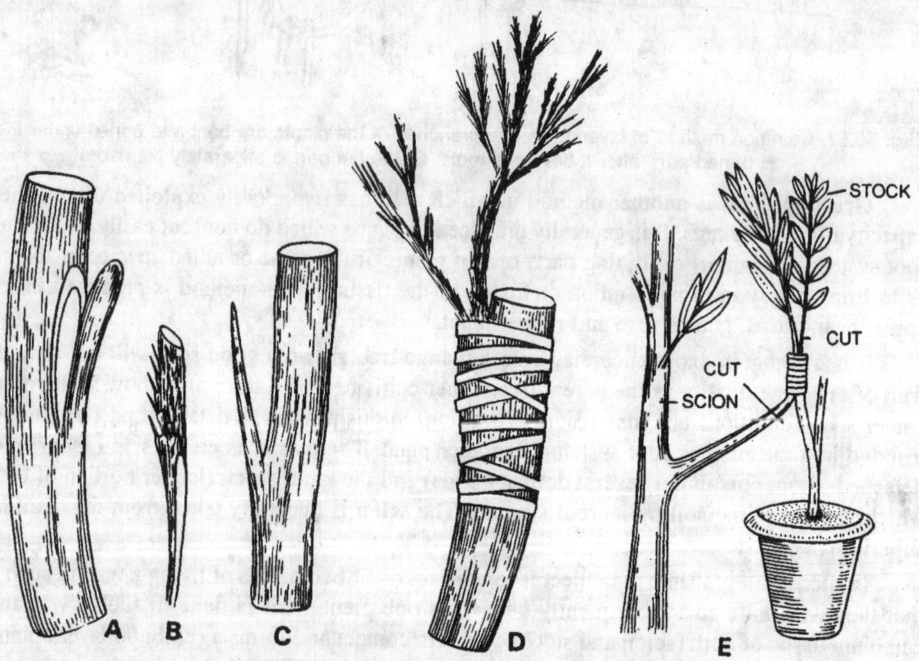

Fig. 30.60. Side grafting, A and C illustrate the position of the cut on the understock; B, scion prepared for insertion; D, complete graft; E, spliced grafting.

union a mass of parenchyma tissue develops from the living cells of both stock and the scion. This tissue heals the wound and is called **callus**. Its appearance marks successful grafting.

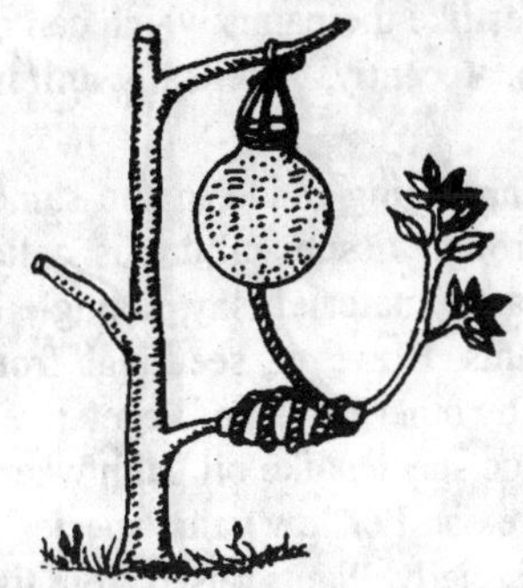

Fig. 30.61. Gootee. Method usually employed for vegetative propagation of lemon, orange, litichi, etc.

Gootee. This method is usually employed for vegetative propagation of lemon, orange, guava and litchi during rainy season. In this process, a healthy and woody branch is selected and the bark is sliced off in a ring form of about 3-5 cm in length. A thick plaster of grafting clay is wrapped up with rag and tied around the debarked portion. The 'grafting clay' is prepared by mixing cow-dung, finely cut hay and water. The tied up portion is kept moist by suitable supply of water. In about 2 or 3 months the roots come out and 'gootee' is ready to be cut below the tied rag for propagation.

Significance of Vegetative Reproduction

1. Plants in which useful characters of parents have to be maintained, are propagated vegetatively.
2. The plants which do not produce viable seeds are propagated by this method.
3. Plants with reduced power of sexual reproduction, long dormant period of seed or poor viability are multiplied easily through this method.
4. The plants which are unable to produce seeds as a result of self-pollination cannot breed true. Such plants bear flowers that are only cross-pollinated and thus there is always mixture of characters. Such plants cannot produce seeds that will germinate to give rise to plants exactly like the parent. In such cases, vegetative propagation is the only method to get true to type plants.
5. Vegetative reproduction also helps in removing common infections from the parent plant.
6. Plants raised through vegetative propagation from a single plant constitute a genetically uniform population, the **clone**. Whereas plants developed from seeds may show variations due to genetic recombination and segregation thus by propagating plants through vegetative means a large stock of selected strains can be built up and maintained without losing the desirable characters.
7. It helps to perpetuate particular form of plants. In some individuals young stage of plants is quite different from the adult form. In order to maintain this young or juvenile form, vegetative propagation is essential.
8. Grafting helps in getting an economically important plant having useful characters of two different individuals in a short time.
9. It is a cheap and rapid method of propagation of the plants with long periods of seed dormancy.

Micropropagation

Tissue culture serves as an alternative to vegetative propagation by conventional methods, such as by rooting of cuttings. This is particularly valuable in the case of difficult-to-root species and those species like the palms which can neither be rooted by cuttings nor grafted because they lack a cambium. Recently, genetically uniform oil palms (*Elaeis guineensis*) have been produced by tissue culture.

Attempts are being made in the same direction with the coconut palm also. Vegetative propagation through tissue culture is called **micropropagation**, because it uses very small quantities of starter material, say, a single cutting or even a few grams of tissue to regenerate hundreds of plants. However, seedlings from tissue cultures are relatively more expensive than those produced by other means of vegetative propagation such as cuttings. Therefore, it is applied only to high value species like oil palm where, moreover, other alternative vegetative propagation methods do not exist. For low value species such as *Eucalyptus* rooted coppice shoot cuttings are still used commercially. Plants raised from tissue cultures require a longer hardening period before they become fit for transplanting, especially in harsh forest environments. Genetic instability of callus cultures is another drawback. Thus chromosome abnormalities are common in tissue cultures of Gymnosperms which constitute important components of temperate and arboreal forests.

In recent years, the application of micropropagation techniques as an alternative means of asexual propagation of important plants has increased the interest of workers in various fields. The micropropagation techniques are preferred over the conventional asexual propagation methods due to following reasons:

(*i*) In this method only a small amount of tissue is needed as the initial explant for regeneration of millions of clonal plants in a year.

(*ii*) This method provides a possible alternative method for developing resistance in many species.

(*iii*) It provides a mean for international exchange of plant materials, hence the problem for introduction of disease can be solved in quarantine.

(*iv*) Stock can be quickly proliferated *in vitro* as it is not season dependent.

(*v*) Valuable germplasm can be stored for a long time.

Application in horticulture and forestry. Regeneration of plantlets in cultured plant cells and tissues has been achieved in many trees of high economic value. Many of the studies are aimed at large scale micropropagation of important trees yielding fuel, pulp, timber, oil or fruits.

Hence, clonal forestry and horticulture are gaining an increasing recognition as an alternative for tree improvement. However, strategies for transferring cultured plants *in vitro* to field conditions are based on relatively higher priced horticultural species rather than agricultural and forestry species.

Regeneration of plantlets in cultured tissue has been described by Murashige (1974) into three undermentioned stages:

(*i*) Establishment of explant aseptically.

(*ii*) Multiplication of propagules by repeated subcultures on a specific nutrient medium.

(*iii*) Rooting and hardening of plantlets and planting into soil.

Later on Fossard (1987) has described the following four stages of micropropagation:

(*i*) Selection of a suitable explant and inoculation into nutrient medium.

(*ii*) Multiplication and growth of culture which takes about two months, followed by repeated subcultures.

(*iii*) Cultures are obtained by changing the medium to planting out.

(*iv*) This is 'planting out' stage where plantlets are aseptically removed from test tube environment to natural and harsh environment. At this stage, roots should be fully functional in potting mix, the soil where plantlets are transplanted.

However, at planting out stage, the plantlets fail to survive due to sudden change in the environment and invasion by soil microbes. Scientists have suggested that the 'regenerates' should be transferred first to green house and then to field. The humidity should be controlled by covering the plants with transparent polyethylene sheets.This acclimatization requires several weeks which should be followed by potting into sterile peat or soil.

In recent years, the interest has created in commercialising the *in vitro* propagation of forest trees. For betterment and improvement of tree plants of high economic value, a breakthrough in forestry research has come with production of artificial seeds in *Eucalyptus*, and genetic transformation and *in vitro* regeneration in conifers. Moreover, micropropagation has been successfully done in many trees.

Advantages of Micropropagation (Tissue Culture)

Micropropagation, an essential component of plant biotechnology, has been proved to be means of producing millions of identical plants by culturing plant tissues or organs under aseptic conditions. The advantages of micropropagation over conventional methods, of a sexual reproduction are as follows:

(*i*) Greater output
(*ii*) Economy of time and space
(*iii*) Freedom from seasonal constraints
(*iv*) Multiplication under disease-free conditions.

Hence, this technique is ideal for the species where:

(*i*) Traditional methods of propagation are inadequate to meet the demand for planting material
(*ii*) Marked variability exists in the natural population and productivity can be significantly increased by cloning of superior individuals.
(*iii*) Virus can be eliminated by culturing the shoot meristem, thus pathogen-free stock can be maintained by routine micropropagation.

In the modern days, tissue culture has been successfully employed for the multiplication of orchids and many other ornamental plants. In the developed countries, tissue culture is a routine method of multiplication while in developing countries, the techniques are largely used for producing plants for export markets.

The Department of Biotechnology (DBT), Government of India, set up two pilot-scale national facilities for large scale production of elite planting material of forest trees through tissue culture, one at the Tata Energy Research Institute (TERI), New Delhi, and the other at the National Chemical Laboratory (NCL), Pune.

QUESTION BANK

Very Short Answer Questions (1 mark each)

1. Define parthenocarpy.
2. Which are the nuclei that fuse to form endosperm?
3. What is double fertilization?
4. What is bulbil?
5. What is coleorhiza?
6. What are parthenocarpic fruits?
7. What does the term monoecious mean?
8. What is the term used for plant bearing female flowers?
9. Name a bat pollinated flower.
10. What is scutellum?
11. What type of fruit tomato is?
12. Give three examples of false fruit.
13. What is funiculus?
14. How does the mustard inflorescence differ from banana inflorescence in arrangement? Give the technical term for each.
15. Why are cucurbits referred to as monoecious?
16. What is geitonogamy?
17. What is allogamy?
18. What is ornithophily?
19. What is chiropterophily?
20. What is agamospermy?
21. Which technique is used for propagating improved varieties of mango and rose.
22. (*i*) Give the term for pollination by bat. (*ii*) Give one example of bat pollinated flowers.
23. Why are date palm plants referred to as dioecious?
24. One of the plants bearing anemophilous flowers is (*a*) *Vallisneria* (*b*) *Salvia* (*c*) Kigelia (*d*) Maize.

ANSWERS

1. The formation of fruits without fertilization; **2.** Two polar nuclei (secondary nucleus) and one male gamete; **3.** Here one male gamete fuses with egg (*i.e.,* syngamy) to form zygote, while second male gamete fuses with two polar nuclei to form primary endosperm nucleus (triple fusion), the process is double fertilization; **4.** Here floral buds become fleshy, which can give rise to new plants, *e.g., Agave*; **5.** Protective cap on radicle of maize; **6.** The fruits formed without fertilization, *e.g.,* seedless fruits of papaya or banana; **7.** When both male and female flowers are borne on the same plant; **8.** Bisexual or hermaphrodite; **9.** Kadamba; **10.** Single cotyledon in caryopsis of maize is called scutellum; **11.** Berry; **12.** Apple, pine apple, jack fruit; **13.** The stalk of ovule; **14.** In mustard, the inflorescence is raceme, flowers are arranged in acropetal order on floral axis, whereas in banana, inflorescence is spadix, where fleshy axis remains covered by spathe; **15.** Because male and female flowers are borne on same plant; **16.** A type of pollination, where pollen is transferred from anther of a flower to stigma of another flower of the same plant; **17.** A type of pollination, where the pollens move from the anther of one flower to the stigma of another flower of a different plant, it is referred to as cross pollination, xenogamy or allogamy; **18.** The pollination by birds; **19.** Pollination by bats; **20.** A process of reproduction, where seeds are formed without fusion of gametes; **21.** Grafting; **22.** (*i*) Chiropterophily (*ii*) *Kigelia pinnata;* **23.** Date palms are unisexual, where male and female flowers are borne on different plants, and therefore, they are referred to as dioecious; **24.** (*d*).

Short Answer Questions (2 marks each)

1. Name the pollinating agents of flowers like maize and wheat. Give any two favourable features of such a flower.
2. Define triple fusion. What is the product of this process? What does the product develop into?
3. Name the pollinating agent of flowers like *Salvia*, Nasturtium and sunflower. Give two favourable features of such a flower for pollination.
4. Some crop plants can be grown from a seed as well as vegetatively from stem cuttings. List any four advantages of vegetative propagation in such cases.
5. Draw a labelled diagram of the L.S. of an anatropous ovule.
6. How does the coriander inflorescence differ from the mustard inflorescence in arrangement?
7. Describe the nature of three kinds of angiospermic fruits. Give an example of each.
8. Give characteristics of insect pollinated flowers.
9. Write the significance of double fertilization.
10. What is parthenogenesis? Give one example.
11. Give four characteristics of wind pollinated flowers.

Short Answer Questions (3 marks each)

1. Trace the development of megasporocyte in the ovary to a mature ovule.
2. What do you understand by ornithophily and chiropterophily.
3. Describe the structure of a typical monosporic embryo sac found in flowering plants.
4. Write a note on the need and significance of pollination in plants.
5. What do you understand by double fertilization.
6. Draw a well labelled diagram of a mature ovule, showing its internal structure.
7. Describe the structure of a pollen grain and the process of its germination.
8. What do you understand by vegetative methods of reproduction? Explain with examples.
9. Describe the importance of artificial methods of vegetative reproduction.
10. Write short notes on the following:
    (*a*) hydrophily (*b*) entomophily
    (*c*) anemophily (*d*) cross-pollination
11. Name the advantages of propagating plants without using seeds.
12. Define
    (*a*) Siphonogamy (*b*) Endosperm (*c*) Double fertilization.
13. (*i*) Give the term for pollination by bats (*ii*) Give an example of bat pollinated plant.
14. Draw labelled diagram of the longitudinal section of a pistil showing pollen germination.
15. Differentiate between geitonogamy and allogamy.
16. Trace the development of a mature ovule from a megaspore mother cell.

Long Answer Questions (5 marks each)

1. Trace the development of a megaspore mother cell of a flower into a mature ovule. Give a labelled diagram of the final stage.
2. Draw a L.S. of a pistil showing pollen germination. Explain the events in the embryo sac during the process of fertilization. Which resulting stages give rise to the embryo and endosperm respectively?
3. (*i*) Draw a neat labelled diagram of the longitudinal section of an anatropous ovule.
   (*ii*) Which cell of the ovule gets transformed into megaspore mother cell?
4. Define double fertilization. Explain the process of double fertilization in an angiospermic plant.
5. Describe the development of male gametophyte in an angiosperm.
6. Describe the development of female gametophyte in angiosperms.

7. What develops into a microspore mother cell in a flower? Trace the development of this cell into a pollen grain which is ready for germination. Draw a labelled figure of a mature pollen grain.
8. Describe in brief the various special modes of reproduction. Mention their economic significance.
9. "Incompatibility is a natural barrier in the fusion of gametes". Justify the statement.
10. Write a note on the development of endosperm. Mention the types with examples.
11. Write an essay on the development of female gametophyte. Illustrate the answer with suitable diagrams.
12. What do you understand by the development of an embryo? Support the answer with suitable diagrams.

True or False

1. Nutritive layer of cells around pollen sacs in anther is tapetum.
2. A carbohydrate layer present on the outer side of grains of maize, wheat, rice, etc. is called aleurone layer.
3. The flower with superior ovary is called hypogynous.
4. Wall of fruit having epicarp, mesocarp and endocarp is called periderm.
5. The part of the plant that is grafted on the stalk is called scion.
6. A cotyledon of bean seed is called scutellum.
7. The maintenance of purity of a given variety of plants by vegetative propagation is called cloning.
8. The transference of the pollen grains from the anther to stigma is called fertilization.
9. Double fertilization or triple fusion is characteristic of angiosperms.
10. A bean seed is albuminous, while a maize seed is exalbuminous.
11. The main causes of seed dormany are hardness of seed coat and presence of inhibitory substances.
12. Two cells present on either side of egg cell in an embryo sac are termed as antipodals.
13. Cross pollination is also known as allogamy.
14. The process of entry of pollen tube into ovule through integuments for fertilization is called porogamy.
15. Three cells present at the chalazal end in the embryo sac are known as antipodals.
16. Within each ovule, a haploid embryo sac usually containing nuclei is formed.
17. Cross-pollination requires abiotic or biotic agent.
18. Only meiotic divisions are involved in asexual reproduction.
19. Microspores are diploid.
20. In *Vallisneria,* the male flowers are released in air.

ANSWERS

1. True ;	2. False ;	3. True ;	4. False ;	5. True ;	6. False ;
7. True ;	8. False ;	9. True ;	10. Flase ;	11. True ;	12. False ;
13. True ;	14. False ;	15. True ;	16. True ;	17. True ;	18. False ;
19. False ;	20. False ;				

Fill in the Blanks

1. A technique to produce genetically identical individuals from a single cell is known as ________.
2. A population of genetically identical plants derived from an individual is called ________.
3. A protective sheath of radicle in monocots is called ________.
4. The flower with superior ovary is called ________.
5. Study of pollen grains is known as ________.
6. Phenomenon of the formation of more than one embryo per ovule is called ________.

7. During grafting, the part that becomes the supporting portion (usually the root) is called ________________.
8. The fruit that develops from the whole of inflorescence is called ____________.
9. A stamen consists of ____________ and ____________.
10. Genetically, the megaspore mother cell is ____________.
11. A small pore in the ovule through which the pollen tube enters is called ____________.
12. The ovary wall develops into a fruit wall is called ____________.
13. Nutritive layer of cells around pollen sacs in anther is called as ____________.
14. The flowers in catkin are ____________.
15. The tip of the pedicel on which different floral parts are inserted is known as ____________.
16. Stem cuttings are frequently used for ____________.
17. *Bryophyllum* is the best example of propagation by ____________.
18. Each microspore mother cell has ________________ of chromosomes and therefore is ________________.
19. The process of nuclear fusion to form zygote is called ____________.
20. Inside ovary, ovules develop from a special tissue called ____________.

ANSWERS

1. Cloning	2. Clone	3. Coleorhiza	4. Hypogynous
5. Palynology	6. Polyembryony	7. Stock	8. Sorosis
9. Filament and anther	10. Diploid	11. Micropyle	12. Pericarp
13. Tapetum	14. Unisexual	15. Thalamus	16. Propagation
17. Leaf	18. 2n, diploid	19. Syngamy	20. Placenta.

Multiple Choice Questions (Objective Type Questions)

1. Layering is used in vegetative propagation of :
(*a*) Jasmine (*b*) Rose
(*c*) Mango (*d*) All the above
2. Egg apparatus consists of:
(*a*) Egg and antipodals (*b*) Polar nuclei
(*c*) Egg and synergids (*d*) Egg
3. Sexual reproduction of flowering plants was discovered by:
(*a*) Camerarius (*b*) Nawaschin
(*c*) Strasburger (*d*) Maheshwari
4. In banana, edible part is:
(*a*) Fleshy epicarp (*b*) Rudimentary mesocarp and fleshy endocarp
(*c*) Pericarp (*d*) Rudimentary endocarp and fleshy mesocarp
5. A spike of unisexual flowers is found in:
(*a*) *Amaranthus* (*b*) Mulberry
(*c*) Wheat (*d*) Bottle brush
6. Cleistogamous condition present in:
(*a*) *Brassica* (*b*) *Solanum tuberosum*
(*c*) *Commelina* (*d*) *Allium cepa*.
7. Fusion of two dissimilar gametes is known as:
(*a*) Fertilisation (*b*) Autogamy
(*c*) Allogamy (*d*) Dichogamy

8. The edible part of ripe mango is morphologically:
(*a*) Epicarp (*b*) Mesocarp
(*c*) Pericarp (*d*) Endocarp

9. In monocots, grafting is almost impossible be cause they lack:
(*a*) Cambium (*b*) Ground tissue
(*c*) Vascular bundle (*d*) Parenchymatous cells

10. The outer whorl of *Bougainvillaea* flower consists of :
(*a*) Bracts (*b*) Epicalyx
(*c*) Sepals (*d*) Petals

11. *Ficus* is an example of which kind of inflorescence?
(*a*) Spike (*b*) Raceme
(*c*) Cyathium (*d*) Hypanthodium

12. Anemophilous flowers have:
(*a*) Sessile stigma (*b*) Small smooth stigma
(*c*) Coloured flower (*d*) Large feathery stigma

13. Pollination by snails and slugs is:
(*a*) Ornithophily (*b*) Cheiropterophily
(*c*) Entomophily (*d*) Malacophily

14. Pollination by ants is:
(*a*) Malacophily (*b*) Myrmecophily
(*c*) Entomophily (*d*) Ornithophily

15. In *Kigelia*, Pollination occurs by:
(*a*) Wind (*b*) Bats
(*c*) Birds (*d*) Humans

16. Feathery stigmas occur in:
(*a*) Pea (*b*) *Datura*
(*c*) *Caesalpinia* (*d*) Wheat

17. Function of tapetum is:
(*a*) Protective (*b*) Nutritive
(*c*) Respiratory (*d*) All the above

18. Formation of embryo directly from nucellus and integument is:
(*a*) Adventive polyembryory (*b*) Apospory
(*c*) Apogamy (*d*) Apomixis

19. Transfer of pollen to the stigma of another flower of the same plant is:
(*a*) Autogamy (*b*) Allogamy
(*c*) Xenogamy (*d*) Geitonogamy

20. Bicarpellary gynoecium and oblique ovary occurs in:
(*a*) Banana (*b*) Brinjal
(*c*) *Pisum* (*d*) Mustard

21. Anemophily occurs in:
(*a*) *Salvia* (*b*) *Vallisneria*
(*c*) Coconut (*d*) Bottle brush

22. The correct match for edible part of fruit is:
(*a*) Guava-pericarp with thalamus fused (*b*) Tomato-thalamus
(*c*) Maize-cotyledon (*d*) Date palm-epicarp

23. Type of placentation in which ovary is syncapous unilocular and ovules on sutures is called:
(*a*) Apical placentation (*b*) Parietal placentation
(*c*) Marginal placentation (*d*) Superficial placentation

24. Aril is the edible part of:
(*a*) Banana (*b*) Litchi
(*c*) Mango (*d*) Apple

25. The hairs present in maize corn cob is :
(*a*) Seed hairs (*b*) Modified hairs of bract
(*c*) Stigma (*d*) Style

26. A piece of potato tuber will form a new plant if it possesses:
(*a*) Roots (*b*) Eyes
(*c*) Stored food (*d*) Branches

27. Winged fruits are commonly known as:
(*a*) Cypsela (*b*) Regma
(*c*) Samara (*d*) Siliqua

28. Maturation of male and female sex organs at different times is known as:
(*a*) Herkogamy (*b*) Dichogamy
(*c*) Polygamy (*d*) Apogamy

29. Which one of the following events in a botanical garden is never directly influenced by light:
(*a*) Flowering (*b*) Photosynthesis
(*c*) Transpiration (*d*) Fertilization

30. Development of egg without fertilization:
(*a*) Parthenocarpy (*b*) Polyembryony
(*c*) Parthenogenesis (*d*) Adventive embryony

31. Food material in exalbuminous seed is mainly stored in;
(*a*) Nucellus (*b*) Testa
(*c*) Endosperm (*d*) Embryo

32. Apomixis is a type of reproduction that results in the development of a/an:
(*a*) New organism without fusion of gametes
(*b*) New organism from fusion products of gametes
(*c*) Embryo from endosperm
(*d*) Embryo from nucellus

33. The arrangement of megaspores in a tetrad in an angiosperm is:
(*a*) Decussate (*b*) Tetrahedral
(*c*) Linear (*d*) Isobilateral

34. Which among the following are pollinated by bat:
(*a*) *Salvia* (*b*) *Ophrys*
(*c*) *Kigelia* (*d*) *Brassica*

35. Fruit of *Abelmoschus esculentus* is:
(*a*) Cypsella (*b*) Capsule
(*c*) Legume (*d*) Pome

36. Father of Indian Angiosperm Embryology is:
(*a*) B.M. Johri (*b*) B.G.L. Swamy
(*c*) R.N. Kapil (*d*) P. Maheshwari

37. Censer mechanism of seed dispersal is found in:
(a) Papaveraceae (b) Liliaceae
(c) Leguminosae (d) Rosaceae
38. Allogamy is best favoured by:
(a) Dichogamy (b) Dicliny
(c) Cleistogamy (d) Homogamy
39. Cleistogamy occurs in:
(a) *Pisum sativum* (b) *Oryza sativa*
(c) *Allium cepa* (d) *Commelina bengalensis*
40. In *Bryophyllum*, vegetative propagation is by:
(a) Roots (b) Leaves
(c) Rhizomes (d) Stem
41. The development of fruit without fertilization is:
(a) Parthenogenesis (b) Parthenocarpy
(c) Apomixis (d) Apogamy
42. Translator apparatus is found in:
(a) Mustard (b) Mango
(c) Pea (d) *Calotropis*
43. Germ pore is the region where the exine is:
(a) Thin (b) Uniform
(c) Thick and uniform (d) Absent
44. Double fertilization in angiosperms was dicovered by:
(a) Strasburger (b) J.C. Bose
(c) Maheshwari (d) Nawaschin
45. In a young anther the four rows of cells, which later produce pollen, are called:
(a) Antheridium (b) Archesporium
(c) Tapetum (d) Zoosporangium
46. Double fertilization was first discovered by Nawaschin (1898) in:
(a) *Lilium* and *Fritillaria* (b) Mango and sugarcane
(c) Papaya and pea (d) *Brassica* and candytuft
47. The point where funicle joins with ovule is known as:
(a) Chalaza (b) Micropyle
(c) Hilum (d) Integument
48. Axillary buds are used to raise crop of:
(a) Wheat (b) Rice
(c) Groundnut (d) Sugarcane
49. The female gametophyte of a typical dicot at the time of fertilization is:
(a) 8-celled (b) 7-celled
(c) 6-celled (d) 4-celled
50. Embryo sac is also known as:
(a) Microgametophyte (b) Megagametophyte
(c) Microsporangium (d) Megasporangium
51. An inflorescence always forms a:
(a) Simple fruit (b) Dry dehiscent fruit
(c) Aggregate fruit (d) Multiple or composite fruit

52. Which plant propagates tnrough leaves:

(*a*) *Agave* (*b*) *Kalonchoe*

(*c*) *Gladiolus* (*d*) Potato

53. Fertilization is depicted by the condition:

(*a*) N → 2N (*b*) 2N → N

(*c*) 2N → 4N (*d*) 4N → 2N

54. Pollen grains are able to withstand extremes of temperature and desiccation because there exine is composed of :

(*a*) Cutin (*b*) Suberin

(*c*) Sporopollenin (*d*) Callose

55. The plant part which consists of two generations one within the other is:

(*a*) Germinated pollen grain (*b*) Seed

(*c*) Unfertilized ovule (*d*) Embryo

56. Synergids are:

(*a*) Haploid (*b*) Diploid

(*c*) Diploid after fertilization (*d*) Triploid after fertilization

57. Pick the odd one out:

(*a*) Chalazogamy (*b*) Porogamy

(*c*) Mesogamy (*d*) Allogamy

58. When the filament runs along the back of anther it is called:

(*a*) Adnate (*b*) Versatile

(*c*) Longitudinal (*d*) Syngenecious

59. Vegetative propagation by bulbils occur in:

(*a*) *Agave* (*b*) *Zingiber*

(*c*) *Vallisneria* (*d*) *Curcuma*

60. If any somatic cell of sporophyte produces gametophyte without reduction division. It is:

(*a*) Parthenogenesis (*b*) Apogamy

(*c*) Apospory (*d*) Amphimixis

ANSWERS

1. (*a*)	**2.** (*c*)	**3.** (*a*)	**4.** (*d*)	**5.** (*b*)	**6.** (*c*)	**7.** (*a*)	**8.** (*b*)
9. (*a*)	**10.** (*a*)	**11.** (*d*)	**12.** (*d*)	**13.** (*d*)	**14.** (*b*)	**15.** (*b*)	**16.** (*d*)
17. (*b*)	**18.** (*a*)	**19.** (*d*)	**20.** (*b*)	**21.** (*c*)	**22.** (*a*)	**23.** (*b*)	**24.** (*b*)
25. (*d*)	**26.** (*b*)	**27.** (*c*)	**28.** (*b*)	**29.** (*d*)	**30.** (*c*)	**31.** (*d*)	**32.** (*a*)
33. (*c*)	**34.** (*c*)	**35.** (*b*)	**36.** (*d*)	**37.** (*a*)	**38.** (*a*)	**39.** (*d*)	**40.** (*b*)
41. (*b*)	**42.** (*d*)	**43.** (*a*)	**44.** (*d*)	**45.** (*b*)	**46.** (*a*)	**47.** (*c*)	**48.** (*d*)
49. (*b*)	**50.** (*b*)	**51.** (*d*)	**52.** (*b*)	**53.** (*a*)	**54.** (*c*)	**55.** (*b*)	**56.** (*a*)
57. (*d*)	**58.** (*a*)	**59.** (*a*)	**60.** (*c*).				